HEAT TRANSFER

The single objective of this book is to provide engineers with the capability, tools, and confidence to solve real-world heat transfer problems. The textbook includes many advanced topics, such as Bessel functions, Laplace transforms, separation of variables, Duhamel's theorem, and complex combination, as well as high-order explicit and implicit numerical integration algorithms. These analytical and numerical solution methods are applied to topics not considered in most textbooks. Examples are heat exchangers involving fluids with varying specific heats or phase changes, regenerators, semi-gray surface radiation exchange, and numerical solutions to internal flow problems. To improve readability, derivations of important results are presented completely, without skipping steps, which reduces student frustration and improves retention. The examples in the book are ubiquitous, not trivial "textbook" exercises. They are rather complex and timely real-world problems that are inherently interesting. This textbook integrates the computational software packages Maple, MATLAB, FEHT, and Engineering Equation Solver (EES) directly with the heat transfer material.

Gregory Nellis is an Associate Professor of Mechanical Engineering at the University of Wisconsin–Madison. He received his M.S. and Ph.D. at the Massachusetts Institute of Technology and is a member of the American Society of Heating, Refrigeration, and Air-Conditioning Engineers (ASHRAE), the American Society of Mechanical Engineers (ASME), the International Institute of Refrigeration (IIR), and the Cryogenic Society of America (CSA). Professor Nellis carries out applied research that is related to energy systems with a focus on refrigeration technology and he has published more than 40 journal papers. Professor Nellis's focus has been on graduate and undergraduate education, and he has received the Polygon, Pi Tau Sigma, and Woodburn awards for excellence in teaching as well as the Boom Award for excellence in cryogenic research.

Sanford Klein is the Bascom Ouweneel Professor of Mechanical Engineering at the University of Wisconsin–Madison. He has been on the faculty at Wisconsin since 1977. He is associated with the Solar Energy Laboratory and has been involved in many studies of solar and other types of energy systems. He is the author or co-author of more than 160 publications relating to the analysis of energy systems. Professor Klein's current research interests are in solar energy systems and applied thermodynamics and heat transfer. In addition, he is also actively involved in the development of engineering computer tools for both instruction and research. He is the primary author of a modular simulation program (TRNSYS), a solar energy system design program (F-CHART), a finite element heat transfer program (FEHT), and a general engineering equation solving program (EES). Professor Klein is a Fellow of ASME, ASHRAE, and the American Solar Energy Society (ASES).

Heat Transfer

GREGORY NELLIS

University of Wisconsin–Madison

SANFORD KLEIN

University of Wisconsin–Madison

CAMBRIDGE
UNIVERSITY PRESS

CAMBRIDGE
UNIVERSITY PRESS

32 Avenue of the Americas, New York NY 10013-2473, USA

Cambridge University Press is part of the University of Cambridge.

It furthers the University's mission by disseminating knowledge in the pursuit of
education, learning and research at the highest international levels of excellence.

www.cambridge.org
Information on this title: www.cambridge.org/9781107671379

First published 2009
Reprinted 2010 (twice), 2012
First paperback edition 2012

A catalogue record for this publication is available from the British Library

Library of Congress Cataloguing in Publication data

Nellis, Gregory.
Heat transfer / Gregory Nellis, Sanford Klein
    p.   cm.
Includes bibliographical references and index.
ISBN 978-0-521-88107-4 (hardback)
1. Heat – Transmission. I. Klein, Sanford A., 1950– II. Title.
TJ260.N45   2008
621.402'2 – dc22   2008021961

ISBN 978-0-521-88107-4 Hardback
ISBN 978-1-107-67137-9 Paperback

Additional resources for this publication at www.cambridge.org/nellisandklein

This book is dedicated to Stephen H. Nellis...thanks Dad.

CONTENTS

* Section can be found on the website that accompanies this book (www.cambridge.org/nellisandklein)

* Section can be found on the website that accompanies this book (www.cambridge.org/nellisandklein)

3   TRANSIENT CONDUCTION • 302

* Section can be found on the website that accompanies this book (www.cambridge.org/nellisandklein)

* Section can be found on the website that accompanies this book (www.cambridge.org/nellisandklein)

* Section can be found on the website that accompanies this book (www.cambridge.org/nellisandklein)

* Section can be found on the website that accompanies this book (www.cambridge.org/nellisandklein)

* Section can be found on the website that accompanies this book (www.cambridge.org/nellisandklein)

6 NATURAL CONVECTION • 735

7 BOILING AND CONDENSATION • 778

* Section can be found on the website that accompanies this book (www.cambridge.org/nellisandklein)

* Section can be found on the website that accompanies this book (www.cambridge.org/nellisandklein)

* Section can be found on the website that accompanies this book (www.cambridge.org/nellisandklein)

* Section can be found on the website that accompanies this book (www.cambridge.org/nellisandklein)

* Section can be found on the website that accompanies this book (www.cambridge.org/nellisandklein)

PREFACE

The single objective of this book is to provide engineers with the capability, tools, and confidence to solve real-world heat transfer problems. This objective has resulted in a textbook that differs from existing heat transfer textbooks in several ways. First, this textbook includes many topics that are typically not covered in undergraduate heat transfer textbooks. Examples are the detailed presentations of mathematical solution methods such as Bessel functions, Laplace transforms, separation of variables, Duhamel's theorem, and Monte Carlo methods as well as high order explicit and implicit numerical integration algorithms. These analytical and numerical solution methods are applied to advanced topics that are ordinarily not considered in a heat transfer textbook.

Judged by its content, this textbook should be considered as a graduate text. There is sufficient material for two-semester courses in heat transfer. However, the presentation does not presume previous knowledge or expertise. This book can be (and has been) successfully used in a single-semester undergraduate heat transfer course by appropriately selecting from the available topics. Our recommendations on what topics can be included in a first heat transfer course are provided in the suggested syllabus. The reason that this book can be used for a first course (despite its expanded content) and the reason it is also an effective graduate-level textbook is that all concepts and methods are presented in detail, starting at the beginning. The derivation of important results is presented completely, without skipping steps, in order to improve readability, reduce student frustration, and improve retention. You will not find many places in this textbook where it states that "it can be shown that ... " The use of examples, solved and explained in detail, is ubiquitous in this textbook. The examples are not trivial, "textbook" exercises, but rather complex and timely real-world problems that are of interest by themselves. As with the presentation, the solutions to these examples are complete and do not skip steps.

Another significant difference between this textbook and most existing heat transfer textbooks is its integration of modern computational tools. The engineering student and practicing engineer of today is expected to be proficient with engineering computer tools. Engineering education must evolve accordingly. Most real engineering problems cannot be solved using a sequential set of calculations that can be accomplished with a pencil or hand calculator. Engineers must have the ability to quickly solve problems using the powerful computational tools that are available and essential for design, parametric study, and optimization of real-world systems. This book integrates the computational software packages Maple, MATLAB, FEHT, and Engineering Equation Solver (EES) directly with the heat transfer material. The specific commands and output associated with these software packages are presented as the theory is developed so that the integration is seamless rather than separated.

The computational software tools used in this book share some important characteristics. They are used in industry and have existed for more than a decade; therefore, while this software will certainly continue to evolve, it is not likely to disappear. Educational versions of these software packages are available, and therefore the use of these tools should not represent an economic hardship to any academic institution or student. Useful versions of EES and FEHT are provided on the website that accompanies this textbook (www.cambridge.org/nellisandklein). With the help provided in the book, these tools are easy to learn and use. Students can become proficient with all of them in a reasonable amount of time. Learning the computer tools will not detract significantly from material coverage. To facilitate this learning process, tutorials for each of the software packages are provided on the companion website. The book itself is structured so that more advanced features of the software are introduced progressively, allowing students to become increasingly proficient using these tools as they progress through the text.

Most (if not all) of the tables and charts that have traditionally been required to solve heat transfer problems (for example, to determine properties, view factors, shape factors, convection relations, etc.) have been made available as functions and procedures in the EES software so that they can be easily accessed and used to solve problems. Indeed, the library of heat transfer functions that has been developed and integrated with EES as part of the preparation of this textbook enables a profound shift in the focus of the educational process. It is trivial to obtain, for example, a shape factor, a view factor, or a convection heat transfer coefficient using the heat transfer library. Therefore, it is possible to assign problems involving design and optimization studies that would be computationally impossible without the computer tools.

Integrating the study of heat transfer with computer tools does not diminish the depth of understanding of the underlying physics. Conversely, our experience indicates that the innate understanding of the subject matter is enhanced by appropriate use of these tools for several reasons. First, the software allows the student to tackle practical and relevant problems as opposed to the comparatively simple problems that must otherwise be assigned. Real-world engineering problems are more satisfying to the student. Therefore, the marriage of computer tools with theory motivates students to understand the governing physics as well as learn how to apply the computer tools. The use of these tools allows for coverage of more advanced material and more interesting and relevant problems. When a solution is obtained, students can carry out a more extensive investigation of its behavior and therefore obtain a more intuitive and complete understanding of the subject of heat transfer.

This book is unusual in its linking of classical theory and modern computing tools. It fills an obvious void that we have encountered in teaching both undergraduate and graduate heat transfer courses. The text was developed over many years from our experiences teaching Introduction to Heat Transfer (an undergraduate course) and Heat Transfer (a first-year graduate course) at the University of Wisconsin. It is our hope that this text will not only be useful during the heat transfer course, but also provide a life-long resource for practicing engineers.

G. F. Nellis
S. A. Klein
May, 2008

Acknowledgments

The development of this book has taken several years and a substantial effort. This has only been possible due to the collegial and supportive atmosphere that makes the Mechanical Engineering Department at the University of Wisconsin such a unique and impressive place. In particular, we would like to acknowledge Tim Shedd, Bill Beckman, Doug Reindl, John Pfotenhauer, Roxann Engelstad, and Glen Myers for their encouragement throughout the process.

Several years of undergraduate and graduate students have used our initial drafts of this manuscript. They have had to endure carrying two heavy volumes of poorly bound paper with no index and many typographical errors. Their feedback has been invaluable to the development of this book.

We have had the extreme good fortune to have had dedicated and insightful teachers. These include Glen Myers, John Mitchell, Bill Beckman, Joseph Smith Jr., John Brisson, Borivoje Mikic, and John Lienhard V. These individuals, among others, have provided us with an indication of the importance of teaching and provided an inspiration to us for writing this book.

Preparing this book has necessarily reduced the "quality time" available to spend with our families. We are most grateful to them for this indulgence. In particular, we wish to thank Jill, Jacob, and Spencer and Sharon Nellis and Jan Klein. We could have not completed this book without their continuous support.

Finally, we are indebted to Cambridge University Press and in particular Peter Gordon for giving us this opportunity and for helping us with the endless details needed to bring our original idea to this final state.

STUDY GUIDE

This book has been developed for use in either a graduate or undergraduate level course in heat transfer. A sample program of study is laid out below for a one-semester graduate course (consisting of 45 class sessions).

Graduate heat transfer class

Day	Sections in Book	Topic
1	1.1	Conduction heat transfer
2	1.2	1-D steady conduction and resistance concepts
3	2.8	Resistance approximations
4	1.3	1-D steady conduction with generation
5	1.4, 1.5	Numerical solutions with EES and MATLAB
6	1.6	Fin solution, fin efficiency, and finned surfaces
7	1.7	Other constant cross-section extended surface problems
8	1.8	Bessel function solutions
9	2.2	2-D conduction, separation of variables
10	2.2	2-D conduction, separation of variables
11	2.4	Superposition
12	3.1	Transient, lumped capacitance problems – analytical solutions
13	3.2	Transient, lumped capacitance problems – numerical solutions
14	3.3	Semi-infinite bodies, diffusive time constant
15	3.3	Semi-infinite bodies, self-similar solution
16	3.4	Laplace transform solutions to lumped capacitance problems
17	3.4	Laplace transform solutions to 1-D transient problems
18	3.5	Separation of variables for 1-D transient problems
19	3.8	Numerical solutions to 1-D transient problems
20	4.1	Laminar boundary layer concepts
21	4.2, 4.3	The boundary layer equations & dimensionless parameters
22	4.4	Blasius solution for flow over a flat plate
23	4.5, 4.6	Turbulent boundary layer concepts, Reynolds averaged equations
24	4.7	Mixing length models and the laws of the wall
25	4.8	Integral solutions
26	4.8, 4.9	Integral solutions, external flow correlations
27	5.1, 5.2	Internal flow concepts and correlations
28	5.3	The energy balance
29	5.4	Analytical solutions to internal flow problems
30	5.5	Numerical solutions to internal flow problems
31	6.1, 6.2	Natural convection concepts and correlations

A sample program of study is laid out below for a one-semester undergraduate course (consisting of 45 class sessions).

Undergraduate heat transfer class

NOMENCLATURE

a_i	i^{th} coefficient of a series solution
A_c	cross-sectional area (m^2)
A_{min}	minimum flow area (m^2)
A_p	projected area (m^2)
A_s	surface area (m^2)
$A_{s,fin}$	surface area of a fin (m^2)
A_{tot}	prime (total) surface area of a finned surface (m^2)
AR	aspect ratio of a rectangular duct
AR_{tip}	area ratio of fin tip to fin surface area
Att	attenuation (-)
B	parameter in the blowing factor (-)
BF	blowing factor (-)
Bi	Biot number (-)
Bo	boiling number (-)
Br	Brinkman number
c	specific heat capacity (J/kg-K)
	concentration (-)
	speed of light (m/s)
c_a''	specific heat capacity of an air-water mixture on a unit mass of air basis (J/kg$_a$-K)
$c_{a,sat}''$	specific heat capacity of an air-water mixture along the saturation line on a unit mass of air basis (J/kg$_a$-K)
c_{eff}	effective specific heat capacity of a composite (J/kg-K)
c_{ms}	ratio of the energy carried by a micro-scale energy carrier to its temperature (J/K)
c_v	specific heat capacity at constant volume (J/kg-K)
C	total heat capacity (J/K)
$\dot{C}$	capacitance rate of a flow (W/K)
$C_1, C_2, \ldots$	undetermined constants
C_{crit}	dimensionless coefficient for critical heat flux correlation (-)
C_D	drag coefficient (-)
C_f	friction coefficient (-)
$\overline{C_f}$	average friction coefficient (-)
C_{lam}	coefficient for laminar plate natural convection correlation (-)
C_{nb}	dimensionless coefficient for nucleate boiling correlation (-)
C_R	capacity ratio (-)
$C_{turb,U}$	coefficient for turbulent, horizontal upward plate natural conv. correlation (-)
$C_{turb,V}$	coefficient for turbulent, vertical plate natural convection correlation (-)

Co	convection number (-)
CTE	coefficient of thermal expansion (1/K)
D	diameter (m)
	diffusion coefficient (m²/s)
D_h	hydraulic diameter (m)
dx	differential in the x-direction (m)
dy	differential in the y-direction (m)
e	size of surface roughness (m)
err	convergence or numerical error
$\dot{E}$	rate of thermal energy carried by a mass flow (W)
E	total emissive power (W/m²)
E_b	total blackbody emissive power (W/m²)
E_λ	spectral emissive power (W/m²-μm)
$E_{b,\lambda}$	blackbody spectral emissive power (W/m²-μm)
Ec	Eckert number (-)
f	frequency (Hz)
	dimensionless stream function, for Blasius solution (-)
	friction factor (-)
$\bar{f}$	average friction factor (-)
f_l	friction factor for liquid-only flow in flow boiling (-)
F	force (N)
	correction-factor for log-mean temperature difference (-)
$F_{0-\lambda_1}$	external fractional function (-)
$F_{i,j}$	view factor from surface i to surface j (-)
$\hat{F}_{i,j}$	the "F-hat" parameter characterizing radiation from surface i to surface j (-)
fd	fractional duty for a pinch-point analysis (-)
Fo	Fourier number (-)
Fr	Froude number (-)
Fr_{mod}	modified Froude number (-)
g	acceleration of gravity (m/s²)
$\dot{g}$	rate of thermal energy generation (W)
$\dot{g}'''$	rate of thermal energy generation per unit volume (W/m³)
$\dot{g}'''_{eff}$	effective rate of generation per unit volume of a composite (W/m³)
$\dot{g}'''_{v}$	rate of thermal energy generation per unit volume due to viscous dissipation (W/m³)
G	mass flux or mass velocity (kg/m²-s)
	total irradiation (W/m²)
G_λ	spectral irradiation (W/m²-μm)
Ga	Galileo number (-)
Gr	Grashof number (-)
Gz	Graetz number (-)
h	local heat transfer coefficient (W/m²-K)
$\bar{h}$	average heat transfer coefficient (W/m²-K)
$\tilde{h}$	dimensionless heat transfer coefficient for flow boiling correlation (-)
h_D	mass transfer coefficient (m/s)
$\bar{h}_D$	average mass transfer coefficient (m/s)
h_l	superficial heat transfer coefficient for the liquid phase (W/m²-K)

$\bar{h}_{rad}$	the equivalent heat transfer coefficient associated with radiation (W/m^2-K)
i	index of node (-)
	index of eigenvalue (-)
	index of term in a series solution (-)
	specific enthalpy (J/kg-K)
	square root of negative one, $\sqrt{-1}$
i''_a	specific enthalpy of an air-water mixture on a per unit mass of air basis (J/kg$_a$)
I	current (ampere)
Ie	intensity of emitted radiation (W/m^2-μm-steradian)
Ii	intensity of incident radiation (W/m^2-μm-steradian)
j	index of node (-)
	index of eigenvalue (-)
J	radiosity (W/m^2)
j_H	Colburn j_H factor (-)
k	thermal conductivity (W/m-K)
k_B	Bolzmann's constant (J/K)
k_c	contraction loss coefficient (-)
k_e	expansion loss coefficient (-)
k_{eff}	effective thermal conductivity of a composite (W/m-K)
Kn	Knudsen number (-)
l_1	Lennard-Jones 12-6 potential characteristic length for species 1 (m)
$l_{1,2}$	characteristic length of a mixture of species 1 and species 2 (m)
L	length (m)
L^+	dimensionless length for a hydrodynamically developing internal flow (-)
L^*	dimensionless length for a thermally developing internal flow (-)
L_{char}	characteristic length of the problem (m)
$L_{char,vs}$	the characteristic size of the viscous sublayer (m)
L_{cond}	length for conduction (m)
L_{flow}	length in the flow direction (m)
L_{ml}	mixing length (m)
L_{ms}	distance between interactions of micro-scale energy or momentum carriers (m)
Le	Lewis number (-)
M	number of nodes (-)
	mass (kg)
m	fin parameter (1/m)
$\dot{m}$	mass flow rate (kg/s)
$\dot{m}''$	mass flow rate per unit area (kg/m^2-s)
m_{ms}	mass of microscale momentum carrier (kg/carrier)
mf	mass fraction (-)
MW	molar mass (kg/kgmol)
n	number density (#/m^3)
n_{ms}	number density of the micro-scale energy carriers (#/m^3)
$\dot{n}''$	molar transfer rate per unit area (kgmol/m^2-s)
N	number of nodes (-)
	number of moles (kgmol)
N_s	number of species in a mixture (-)
Nu	Nusselt number (-)

$\overline{Nu}$	average Nusselt number (-)
NTU	number of transfer units (-)
p	pressure (Pa)
	pitch (m)
P	$LMTD$ effectiveness (-)
	probability distribution (-)
p_∞	free-stream pressure (Pa)
$\tilde{p}$	dimensionless pressure (-)
Pe	Peclet number (-)
per	perimeter (m)
Pr	Prandtl number (-)
Pr_{turb}	turbulent Prandtl number (-)
$\dot{q}$	rate of heat transfer (W)
$\dot{q}_{i\ to\ j}$	rate of radiation heat transfer from surface i to surface j (W)
$\dot{q}_{max}$	maximum possible rate of heat transfer, for an effectiveness solution (W)
$\dot{q}''$	heat flux, rate of heat transfer per unit area (W/m^2)
$\dot{q}''_s$	surface heat flux (W/m^2)
$\dot{q}''_{s,crit}$	critical heat flux for boiling (W/m^2)
Q	total energy transfer by heat (J)
$\tilde{Q}$	dimensionless total energy transfer by heat (-)
r	radial coordinate (m)
	radius (m)
$\tilde{r}$	dimensionless radial coordinate (-)
R	thermal resistance (K/W)
	ideal gas constant (J/kg-K)
	$LMTD$ capacitance ratio (-)
$R_{\overline{A}}$	thermal resistance approximation based on average area limit (K/W)
R_{ac}	thermal resistance to axial conduction in a heat exchanger (K/W)
R_{ad}	thermal resistance approximation based on adiabatic limit (K/W)
R_{bl}	thermal resistance of the boundary layer (K/W)
R_c	thermal resistance due to solid-to-solid contact (K/W)
R_{conv}	thermal resistance to convection from a surface (K/W)
R_{cyl}	thermal resistance to radial conduction through a cylindrical shell (K/W)
R_e	electrical resistance (ohm)
R_f	thermal resistance due to fouling (K/W)
R_{fin}	thermal resistance of a fin (K/W)
$R_{i,j}$	the radiation space resistance between surfaces i and j (1/m^2)
R_{iso}	thermal resistance approximation based on isothermal limit (K/W)
$R_{\overline{L}}$	thermal resistance approximation based on average length limit (K/W)
R_{pw}	thermal resistance to radial conduction through a plane wall (K/W)
R_{rad}	thermal resistance to radiation (K/W)
$R_{s,i}$	the radiation surface resistance for surface i (1/m^2)
$R_{semi-\infty}$	thermal resistance approximation for a semi-infinite body (K/W)
R_{sph}	thermal resistance to radial conduction through a spherical shell (K/W)
R_{tot}	thermal resistance of a finned surface (K/W)
R_{univ}	universal gas constant (8314 J/kgmol-K)
R''_c	area-specific contact resistance (K-m^2/W)
R''_f	area-specific fouling resistance (K-m^2/W)
Ra	Rayleigh number (-)

Re	Reynolds number (-)
Re_{crit}	critical Reynold number for transition to turbulence (-)
RH	relative humidity (-)
RR	radius ratio of an annular duct (-)
s	Laplace transformation variable (1/s)
	generic coordinate (m)
S	shape factor (m)
	channel spacing (m)
Sc	Schmidt number (-)
Sh	Sherwood number (-)
$\overline{Sh}$	average Sherwood number (-)
St	Stanton number (-)
t	time (s)
t_{sim}	simulated time (s)
th	thickness (m)
tol	convergence tolerance
T	temperature (K)
T_b	base temperature of fin (K)
T_{film}	film temperature (K)
T_m	mean or bulk temperature (K)
T_s	surface temperature (K)
T_{sat}	saturation temperature (K)
T_∞	free-stream or fluid temperature (K)
T^*	eddy temperature fluctuation (K)
T'	fluctuating component of temperature (K)
$\overline{T}$	average temperature (K)
TR	temperature solution that is a function of r, for separation of variables
Tt	temperature solution that is a function of t, for separation of variables
TX	temperature solution that is a function of x, for separation of variables
TY	temperature solution that is a function of y, for separation of variables
th	thickness (m)
U	internal energy (J)
	utilization (-)
u	specific internal energy (J/kg)
	velocity in the x-direction (m/s)
u_{char}	characteristic velocity (m/s)
u_f	frontal or upstream velocity (m/s)
u_m	mean or bulk velocity (m/s)
u_∞	free-stream velocity (m/s)
u^*	eddy velocity (m/s)
u^+	inner velocity (-)
$\tilde{u}$	dimensionless x-velocity (-)
u'	fluctuating component of x-velocity (m/s)
$\overline{u}$	average x-velocity (m/s)
UA	conductance (W/K)
v	velocity in the y- or r-directions (m/s)
v_δ	y-velocity at the outer edge of the boundary layer, approximate scale of y-velocity in a boundary layer (m/s)
v_{ms}	mean velocity of micro-scale energy or momentum carriers (m/s)

$\tilde{v}$	dimensionless y-velocity (-)
v'	fluctuating component of y-velocity (m/s)
$\bar{v}$	average y-velocity (m/s)
V	volume (m³)
	voltage (V)
$\dot{V}$	volume flow rate (m³/s)
vf	void fraction (-)
w	velocity in the z-direction (m/s)
$\dot{w}$	rate of work transfer (W)
W	width (m)
	total amount of work transferred (J)
x	x-coordinate (m)
	quality (-)
$\tilde{x}$	dimensionless x-coordinate (-)
X	particular solution that is only a function of x
$x_{fd,h}$	hydrodynamic entry length (m)
$x_{fd,t}$	thermal entry length (m)
X_{tt}	Lockhart Martinelli parameter (-)
y	y-coordinate (m)
	mole fraction (-)
y^+	inner position (-)
$\tilde{y}$	dimensionless y-coordinate (-)
Y	particular solution that is only a function of y
z	z-coordinate (m)

Greek Symbols

α	thermal diffusivity (m²/s)
	absorption coefficient (1/m)
	absorptivity or absorptance (-), total hemispherical absorptivity (-)
	surface area per unit volume (1/m)
α_{eff}	effective thermal diffusivity of a composite (m²/s)
α_λ	hemispherical absorptivity (-)
$\alpha_{\lambda,\theta,\phi}$	spectral directional absorptivity (-)
β	volumetric thermal expansion coefficient (1/K)
δ	film thickness for condensation (m)
	boundary layer thickness (m)
δ_d	mass transfer diffusion penetration depth (m)
	concentration boundary layer thickness (m)
δ_m	momentum diffusion penetration depth (m)
	momentum boundary layer thickness (m)
δ_{vs}	viscous sublayer thickness (m)
δ_t	energy diffusion penetration depth (m)
	thermal boundary layer thickness (m)
Δi_{fus}	latent heat of fusion (J/kg)
Δi_{vap}	latent heat of vaporization (J/kg)
Δp	pressure drop (N/m²)
Δr	distance in r-direction between adjacent nodes (m)

ΔT	temperature difference (K)
ΔT_e	excess temperature (K)
ΔT_{lm}	log-mean temperature difference (K)
Δt	time step (s)
	time period (s)
Δt_{crit}	critical time step (s)
Δx	distance in x-direction between adjacent nodes (m)
Δy	distance in y-direction between adjacent nodes (m)
ε	heat exchanger effectiveness (-)
	emissivity or emittance (-), total hemispherical emissivity (-)
ε_{fin}	fin effectiveness (-)
ε_H	eddy diffusivity for heat transfer (m^2/s)
ε_λ	hemispherical emissivity (-)
$\varepsilon_{\lambda,\theta,\phi}$	spectral, directional emissivity (-)
ε_M	eddy diffusivity of momentum (m^2/s)
ε_1	Lennard-Jones 12-6 potential characteristic energy for species 1 (J)
$\varepsilon_{1,2}$	characteristic energy parameter for a mixture of species 1 and species 2 (J)
ϕ	porosity (-)
	phase angle (rad)
	spherical coordinate (rad)
η	similarity parameter (-)
	efficiency (-)
η_{fin}	fin efficiency (-)
η_o	overall efficiency of a finned surface (-)
κ	von Kármán constant
λ	dimensionless axial conduction parameter (-)
	wavelength of radiation (μm)
λ_i	i^{th} eigenvalue of a solution (1/m)
μ	viscosity (N-s/m^2)
υ	frequency of radiation (1/s)
θ	temperature difference (K)
	angle (rad)
	spherical coordinate (rad)
$\tilde{\theta}$	dimensionless temperature difference (-)
θ^+	inner temperature difference (-)
θR	temperature difference solution that is only a function of r, for separation of variables
θt	temperature difference solution that is only a function of t, for separation of variables
θX	temperature difference solution that is only a function of x, for separation of variables
θXt	temperature difference solution that is only a function of x and t, for reduction of multi-dimensional transient problems
θY	temperature difference solution that is only a function of y, for separation of variables
θYt	temperature difference solution that is only a function of y and t, for reduction of multi-dimensional transient problems
θZt	temperature difference solution that is only a function of z and t, for reduction of multi-dimensional transient problems

ρ	density (kg/m^3)
	reflectivity or reflectance (-), total hemispherical reflectivity (-)
ρ_e	electrical resistivity (ohm-m)
ρ_{eff}	effective density of a composite (kg/m^3)
ρ_λ	hemispherical reflectivity (-)
$\rho_{\lambda,\theta,\varphi}$	spectral, directional reflectivity (-)
σ	surface tension (N/m),
	molecular radius (m)
	ratio of free-flow to frontal area (-)
	Stefan-Boltzmann constant (5.67×10^{-8} W/m^2-K^4)
τ	time constant (s)
	shear stress (Pa)
	transmittivity or transmittance (-), total hemispherical transmittivity (-)
τ_{diff}	diffusive time constant (s)
τ_{lumped}	lumped capacitance time constant (s)
τ_λ	hemispherical transmittivity (-)
$\tau_{\lambda,\theta,\varphi}$	spectral, directional transmittivity (-)
τ_s	shear stress at surface (N/m^2)
υ	kinematic viscosity (m^2/s)
ω	angular velocity (rad/s)
	humidity ratio (kg$_v$/kg$_a$)
	solid angle (steradian)
Ω_D	dimensionless collision integral for diffusion (-)
Ψ	stream function (m^2/s)
ζ	tilt angle (rad)
	curvature parameter for vertical cylinder, natural convection correlation (-)
ζ_i	the ith dimensionless eigenvalue (-)

Superscripts

o	at infinite dilution

Subscripts

a	air
abs	absorbed
ac	axial conduction (in heat exchangers)
an	analytical
app	apparent
	approximate
b	blackbody
bl	boundary layer
$bottom$	bottom
c	condensate film
	corrected
C	cold
	cold-side of a heat exchanger
cc	complex conjugate, for complex combination problems
$char$	characteristic
cf	counter-flow heat exchanger

cond	conduction, conductive
conv	convection, convective
crit	critical
CTHB	cold-to-hot blow process
dc	dry coil
df	downward facing
diff	diffusive transfer
eff	effective
emit	emitted
evap	evaporative
ext	external
f	fluid
fc	forced convection
fd,h	hydrodynamically fully developed
fd,t	thermally fully developed
fin	fin, finned
h	homogeneous solution
H	hot
	hot-side of a heat exchanger
	constant heat flux boundary condition
hs	on a hemisphere
HTCB	hot-to-cold blow process
i	node *i*
	surface *i*
	species *i*
in	inner
	inlet
ini	initial
int	internal
	interface
	integration period
j	node *j*
	surface *j*
l	liquid
lam	laminar
LHS	left-hand side
lumped	lumped-capacitance
m	mean or bulk
	melting
max	maximum or maximum possible
min	minimum or minimum possible
mod	modified
ms	micro-scale carrier
n	normal
nac	without axial conduction (in heat exchangers)
nb	nucleate boiling
nc	natural convection
no-fin	without a fin
out	outer
	outlet
p	particular (or non-homogeneous) solution

pf	parallel-flow heat exchanger
pp	pinch-point
r	regenerator matrix
	at position *r*
rad	radiation, radiative
ref	reference
RHS	right-hand side
s	at the surface
sat	saturated
	saturated section of a heat exchanger
sat,l	saturated liquid
sat,v	saturated vapor
sc	sub-cooled section of a heat exchanger
semi-∞	semi-infinite
sh	super-heated section of a heat exchanger
sph	sphere
sur	surroundings
sus	sustained solution
T	constant temperature boundary condition
	at temperature *T*
top	top
tot	total
turb	turbulent
uf	upward-facing
unfin	not finned
v	vapor
	vertical
	viscous dissipation
w	water
wb	wet-bulb
wc	wet coil
x	at position *x*
	in the *x*-direction
x^-	in the negative *x*-direction
x^+	in the positive *x*-direction
y	at position *y*
	in the *y*-direction
∞	free-stream, fluid
90°	solution that is 90° out of phase, for complex combination problems

Other notes

A	arbitrary variable
A'	fluctuating component of variable *A*
	value of variable *A* on a unit length basis
A''	value of variable *A* on a unit area basis
A'''	value of variable *A* on a unit volume basis
$\tilde{A}$	dimensionless form of variable *A*
$\hat{A}$	a guess value or approximate value for variable *A*
$\hat{A}$	Laplace transform of the function *A*

$\overline{A}$	average of variable A
$\underline{A}$	denotes that variable A is a vector
$\underline{\underline{A}}$	denotes that variable A is a matrix
dA	differential change in the variable A
δA	uncertainty in the variable A
ΔA	finite change in the variable A
$O(A)$	order of magnitude of the variable A

HEAT TRANSFER

1 One-Dimensional, Steady-State Conduction

1.1 Conduction Heat Transfer

1.1.1 Introduction

Thermodynamics defines heat as a transfer of energy across the boundary of a system as a result of a temperature difference. According to this definition, heat by itself is an energy transfer process and it is therefore redundant to use the expression 'heat transfer'. Heat has no option but to transfer and the expression 'heat transfer' reinforces the incorrect concept that heat is a property of a system that can be 'transferred' to another system. This concept was originally proposed in the 1800's as the caloric theory (Keenan, 1958); heat was believed to be an invisible substance (having mass) that transferred from one system to another as a result of a temperature difference. Although the caloric theory has been disproved, it is still common to refer to 'heat transfer'.

Heat is the transfer of energy due to a temperature gradient. This transfer process can occur by two very different mechanisms, referred to as conduction and radiation. Conduction heat transfer occurs due to the interactions of molecular (or smaller) scale energy carriers within a material. Radiation heat transfer is energy transferred as electromagnetic waves. In a flowing fluid, conduction heat transfer occurs in the presence of energy transfer due to bulk motion (which is not a heat transfer) and this leads to a substantially more complex situation that is referred to as convection.

1.1.2 Thermal Conductivity

Conduction heat transfer occurs due to the interactions of micro-scale energy carriers within a material; the type of energy carriers depends upon the structure of the material. For example, in a gas or a liquid, the energy carriers are individual molecules whereas the energy carriers in a solid may be electrons or phonons (i.e., vibrations in the structure of the solid). The transfer of energy by conduction is fundamentally related to the interactions of these energy carriers; more energetic (i.e., higher temperature) energy carriers transfer energy to less energetic (i.e., lower temperature) ones, resulting in a net flow of energy from hot to cold (i.e., heat transfer). Regardless of the type of energy carriers involved, conduction heat transfer can be characterized by Fourier's law, provided that the length and time scales of the problem are large relative to the distance and time between energy carrier interactions. Fourier's law relates the heat flux in any direction to the temperature gradient in that direction. For example:

$$\dot{q}'' = -k \frac{\partial T}{\partial x} \tag{1-1}$$

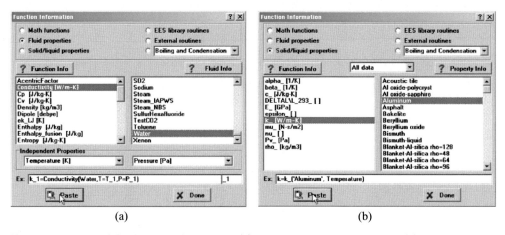

Figure 1-1: Conductivity functions in EES for (a) compressible substances and (b) incompressible substances.

where $\dot{q}''$ is the heat flux in the x-direction and k is the thermal conductivity of the material. Fourier's law actually provides the definition of thermal conductivity:

$$k = \frac{-\dot{q}''}{\frac{\partial T}{\partial x}} \tag{1-2}$$

Thermal conductivity is a material property that varies widely depending on the type of material and its state. Thermal conductivity has been extensively measured and values have been tabulated in various references (e.g., NIST (2005)). The thermal conductivity of many substances is available within the Engineering Equation Solver (EES) program. It is suggested that the reader go through the tutorial that is provided in Appendix A.1 in order to become familiar with EES. Appendix A.1 can be found on the web site associated with this book (www.cambridge.org/nellisandklein). To access the thermal conductivity functions in EES, select Function Info from the Options menu and select the Fluid Properties button; this action displays the properties that are available for compressible fluids. Navigate to Conductivity in the left hand window and select the fluid of interest in the right hand window (e.g., Water), as shown in Figure 1-1(a). Select Paste in order to place the call to the Conductivity function into the Equations Window. Select the Solid/liquid properties button in order to access the properties for incompressible fluids and solids, as shown in Figure 1-1(b).

The EES code below specifies the unit system to be used (SI) and then computes the conductivity of water, air, and aluminum ($k_w, k_a,$ and k_{al}) at $T = 20°C$ and $p = 1.0$ atm,

```
$UnitSystem SI MASS RAD PA K J
$Tabstops 0.2 0.4 0.6 3.5 in

T=converttemp(C,K,20 [C])              "temperature"
P=1.0 [atm]*convert(atm,Pa)            "pressure"
k_w=Conductivity(Water,T=T,P=P)        "conductivity of water at T and P"
k_a=Conductivity(Air,T=T)              "conductivity of air at T and P"
k_al=k_('Aluminum', T)                 "conductivity of aluminum at T"
```

which leads to $k_w = 0.59$ W/m-K, $k_a = 0.025$ W/m-K, and $k_{al} = 236$ W/m-K. The conductivity of aluminum (an electrically conductive metal) is approximately 10,000x that

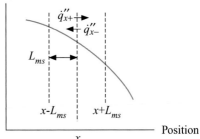

Figure 1-2: Energy flow through a plane at position x.

of air (a dilute gas, at these conditions), with water (a liquid) falling somewhere between these values.

It is possible to understand the thermal conductivity of various materials based on the underlying characteristics of their energy carriers, the microscopic physical entities that are responsible for conduction. For example, the kinetic theory of gases may be used to provide an estimate of the thermal conductivity of a gas and the thermal conductivity of electrically conductive metals can be understood based on a careful study of electron behavior.

Consider conduction through a material in which a temperature gradient has been established in the x-direction, as shown in Figure 1-2. We can evaluate (approximately) the net rate of energy transferred through a plane that is located at position x. The flux of energy carriers passing through the plane from left-to-right (i.e., in the positive x-direction) is proportional to the number density of the energy carriers (n_{ms}) and their mean velocity (v_{ms}). The energy carriers that are moving in the positive x-direction experienced their last interaction at approximately $x–L_{ms}$ (on average), where L_{ms} is the distance between energy carrier interactions. (Actually, the last interaction would not occur exactly at this position since the energy carriers are moving relative to each other and also in the y- and z-directions.) The energy associated with these left-to-right moving carriers is proportional to the temperature at position x-L_{ms} ($T_{x-L_{ms}}$). The energy per unit area passing through the plane from left-to-right ($\dot{q}''_{x+}$) is given approximately by:

$$\dot{q}''_{x+} \approx \underbrace{n_{ms}\, v_{ms}}_{\substack{\text{\#carriers}\\\text{area-time}}}\ \underbrace{c_{ms}\, T_{x-L_{ms}}}_{\substack{\text{energy}\\\text{carrier}}} \tag{1-3}$$

where c_{ms} is the ratio of the energy of the carrier to its temperature. Similarly, the energy per unit area carried through the plane in the negative x-direction by the energy carriers that are moving from right-to-left ($\dot{q}''_{x-}$) is given approximately by:

$$\dot{q}''_{x-} \approx n_{ms}\, v_{ms}\, c_{ms}\, T_{x+L_{ms}} \tag{1-4}$$

The net conduction heat flux passing through the plane ($\dot{q}''$) is the difference between $\dot{q}''_{x+}$ and $\dot{q}''_{x-}$,

$$\dot{q}'' \approx n_{ms}\, v_{ms}\, c_{ms}\, (T_{x-L_{ms}} - T_{x+L_{ms}}) \tag{1-5}$$

which can be rearranged to yield:

$$\dot{q}'' \approx -n_{ms}\, v_{ms}\, c_{ms}\, L_{ms}\, \frac{(T_{x+L_{ms}} - T_{x-L_{ms}})}{L_{ms}} \tag{1-6}$$

Recall from calculus that the definition of the temperature gradient is:

$$\frac{\partial T}{\partial x} = \lim_{dx \to 0} \frac{T_{x+dx} - T_{x-dx}}{2\, dx} \tag{1-7}$$

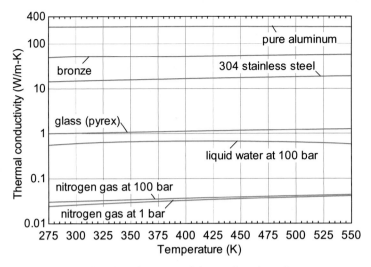

Figure 1-3: Thermal conductivity of various materials as a function of temperature.

In the limit that the length between energy carrier interactions (L_{ms}) is much less than the length scale that characterizes the problem (L_{char}):

$$\frac{(T_{x+L_{ms}} - T_{x-L_{ms}})}{2\,L_{ms}} \approx \lim_{dx \to 0} \frac{T_{x+dx} - T_{x-dx}}{2\,dx} = \frac{\partial T}{\partial x} \tag{1-8}$$

Equation (1-8) can be substituted into Eq. (1-6) to yield:

$$\dot{q}'' \approx -2\ \underbrace{n_{ms}\,v_{ms}\,c_{ms}\,L_{ms}}_{\propto k}\ \frac{\partial T}{\partial x} \tag{1-9}$$

The ratio of the length between energy carrier interactions to the length scale that characterizes the problem is referred to as the Knudsen number. The Knudsen number (Kn) should be calculated in order to ensure that continuum concepts (like Fourier's law) are applicable:

$$Kn = \frac{L_{ms}}{L_{char}} \tag{1-10}$$

If the Knudsen number is not small then continuum theory breaks down. This limit may be reached in micro- and nano-scale systems where L_{char} becomes small as well as in problems involving rarefied gas where L_{ms} becomes large. Specialized theory for heat transfer is required in these limits and the interested reader is referred to books such as Tien et al. (1998), Chen (2005), and Cercignani (2000).

Comparing Eq. (1-9) with Fourier's law, Eq. (1-1), shows that the thermal conductivity is proportional to the product of the number of energy carriers per unit volume, their average velocity, the mean distance between their interactions, and the ratio of the amount of energy carried by each energy carrier to its temperature:

$$k \propto n_{ms}\,v_{ms}\,c_{ms}\,L_{ms} \tag{1-11}$$

The scaling relation expressed by Eq. (1-11) is informative. Figure 1-3 illustrates the thermal conductivity of several common materials as a function of temperature.

Notice that metals have the largest thermal conductivity, followed by other solids and liquids, while gases have the lowest conductivity. Gases are diffuse and thus the number density of the energy carriers (gas molecules) is substantially less than for other forms of matter. Pure metals have the highest thermal conductivity because energy is carried primarily by electrons which are numerous and fast moving. The thermal conductivity and electrical resistivity of pure metals are related (by the Wiedemann-Franz law) because both electricity and thermal energy are transported by the same mechanism, electron flow. Alloys have lower thermal conductivity because the electron motion is substantially impeded by the impurities within the structure of the material; this effect is analogous to reducing the parameter L_{ms} in Eq. (1-11). In non-metals, the energy is carried by phonons (or lattice vibrations), while in liquids the energy is carried by molecules.

Thermal Conductivity of a Gas

This extended section of the book can be found on the website (www.cambridge.org/nellisandklein) and discusses the application of Eq. (1-11) to the particular case of an ideal gas where the energy carriers are gas molecules.

1.2 Steady-State 1-D Conduction without Generation

1.2.1 Introduction

Chapters 1 through 3 examine conduction problems using a variety of conceptual, analytical, and numerical techniques. We will begin with simple problems and move eventually to complex problems, starting with truly one-dimensional (1-D), steady-state problems and working finally to two-dimensional and transient problems. Throughout this book, problems will be solved both analytically and numerically. The development of an analytical or a numerical solution is accomplished using essentially the same steps regardless of the complexity of the problem; therefore, each class of problem will be solved in a uniform and rigorous fashion. The use of computer software tools facilitates the development of both analytical and numerical solutions; therefore, these tools are introduced and used side-by-side with the theory.

1.2.2 The Plane Wall

In general, the temperature in a material will be a function of position (x, y, and z, in Cartesian coordinates) and time (t). The definition of steady-state is that the temperature is unchanging with time. There are certain idealized problems in which the temperature varies in only one direction (e.g., the x-direction). These are one-dimensional (1-D), steady-state problems. The classic example is a plane wall (i.e., a wall with a constant cross-sectional area, A_c, in the x-direction) that is insulated around its edges. In order for the temperature distribution to be 1-D, each face of the wall must be subjected to a uniform boundary condition. For example, Figure 1-4 illustrates a plane wall in which the left face ($x = 0$) is maintained at T_H while the right face ($x = L$) is held at T_C.

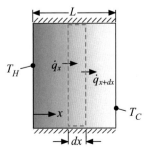

Figure 1-4: A plane wall with fixed temperature boundary conditions.

The first step toward developing an analytical solution for this, or any, problem involves the definition of a differential control volume. The control volume must encompass material at a uniform temperature; therefore, in this case it must be differentially small in the x-direction (i.e., it has width dx, see Figure 1-4) but can extend across the entire cross-sectional area of the wall as there are no temperature gradients in the y- or z-directions. Next, the energy transfers across the control surfaces must be defined as well as any thermal energy generation or storage terms. For the steady-state, 1-D case considered here, there are only two energy transfers, corresponding to the rate of conduction heat transfer into the left side (i.e., at position x, $\dot{q}_x$) and out of the right side (i.e., at position $x + dx$, $\dot{q}_{x+dx}$) of the control volume. A steady-state energy balance for the differential control volume is therefore:

$$\dot{q}_x = \dot{q}_{x+dx} \tag{1-19}$$

A Taylor series expansion of the term at $x + dx$ leads to:

$$\dot{q}_{x+dx} = \dot{q}_x + \frac{d\dot{q}}{dx} dx + \frac{d^2\dot{q}}{dx^2} \frac{dx^2}{2!} + \frac{d^3\dot{q}}{dx^3} \frac{dx^3}{3!} + \cdots \tag{1-20}$$

The analytical solution proceeds by taking the limit as dx goes to zero so that the higher order terms in Eq. (1-20) can be neglected:

$$\dot{q}_{x+dx} = \dot{q}_x + \frac{d\dot{q}}{dx} dx \tag{1-21}$$

Substituting Eq. (1-21) into Eq. (1-19) leads to:

$$\dot{q}_x = \dot{q}_x + \frac{d\dot{q}}{dx} dx \tag{1-22}$$

or

$$\frac{d\dot{q}}{dx} = 0 \tag{1-23}$$

Equation (1-23) is typical of the initial result that is obtained by considering a differential energy balance: a differential equation that is expressed in terms of energy rather than temperature. This form of the differential equation should be checked against your intuition. Equation (1-23) indicates that the rate of conduction heat transfer is not a function of x. For the problem in Figure 1-4, there are no sources or sinks of energy and no energy storage within the wall; therefore, there is no reason for the rate of heat transfer to vary with position.

The final step in the derivation of the governing equation is to substitute appropriate rate equations that relate energy transfer rates to temperatures. The result of this substitution will be a differential equation expressed in terms of temperature. The rate equation for conduction is Fourier's law:

$$\dot{q} = -k A_c \frac{\partial T}{\partial x} \tag{1-24}$$

For our problem, the temperature is only a function of position, x, and therefore the partial differential in Eq. (1-24) can be replaced with an ordinary differential:

$$\dot{q} = -k A_c \frac{dT}{dx} \tag{1-25}$$

Substituting Eq. (1-25) into Eq. (1-23) leads to:

$$\frac{d}{dx}\left[-k A_c \frac{dT}{dx} \right] = 0 \tag{1-26}$$

If the thermal conductivity is constant then Eq. (1-26) may be simplified to:

$$\frac{d^2T}{dx^2} = 0 \tag{1-27}$$

The derivation of Eq. (1-27) is trivial and yet the steps are common to the derivation of the governing equation for more complex problems. These steps include: (1) the definition of an appropriate control volume, (2) the development of an energy balance, (3) the expansion of terms, and (4) the substitution of rate equations.

In order to completely specify a problem, it is necessary to provide boundary conditions. Boundary conditions are information about the solution at the extents of the computational domain (i.e., the limits of the range of position and/or time over which your solution is valid). A second order differential equation requires two boundary conditions. For the problem shown in Figure 1-4, the boundary conditions are:

$$T_{x=0} = T_H \tag{1-28}$$

$$T_{x=L} = T_C \tag{1-29}$$

Equations (1-27) through (1-29) represent a well-posed mathematical problem: a second order differential equation with boundary conditions. Equation (1-27) is very simple and can be solved by separation and direct integration:

$$d\left[\frac{dT}{dx}\right] = 0 \tag{1-30}$$

Equation (1-30) is integrated according to:

$$\int d\left[\frac{dT}{dx}\right] = \int 0 \tag{1-31}$$

Because Eq. (1-31) is an indefinite integral (i.e., there are no limits on the integrals), an undetermined constant (C_1) results from the integration:

$$\frac{dT}{dx} = C_1 \tag{1-32}$$

Equation (1-32) is separated and integrated again:

$$\int dT = \int C_1\, dx \tag{1-33}$$

to yield

$$T = C_1 x + C_2 \tag{1-34}$$

Equation (1-34) shows that the temperature distribution must be linear; any linear function (i.e., any values of the constants C_1 and C_2) will satisfy the differential equation, Eq. (1-27). The constants of integration are obtained by forcing Eq. (1-34) to also satisfy the two boundary conditions, Eqs. (1-28) and (1-29):

$$T_H = C_1\, 0 + C_2 \tag{1-35}$$

$$T_C = C_1\, L + C_2 \tag{1-36}$$

Equations (1-35) and (1-36) are solved for C_1 and C_2 and substituted into Eq. (1-34) to provide the solution:

$$T = \frac{(T_C - T_H)}{L} x + T_H \tag{1-37}$$

The heat transfer at any location within the wall is obtained by substituting the temperature distribution, Eq. (1-37), into Fourier's law, Eq. (1-25):

$$\dot{q} = -kA_c \frac{dT}{dx} = \frac{kA_c}{L}(T_H - T_C) \tag{1-38}$$

Equation (1-38) shows that the heat transfer does not change with the position within the wall; this behavior is consistent with Eq. (1-23).

The development of analytical solutions is facilitated using a symbolic software package such as Maple. It is suggested that the reader stop and go through the tutorial provided in Appendix A.2 which can be found on the web site associated with the book (www.cambridge.org/nellisandklein) in order to become familiar with Maple. Note that the Maple Command Applet that is discussed in Appendix A.2 is available on the internet and can be used even if you do not have access to the Maple software. The mathematical solution to the 1-D, steady-state conduction problem associated with a plane wall is easy enough that there is no reason to use Maple. However, it is worthwhile to use the problem in order to illustrate some of the basic steps associated with using Maple in anticipation of more difficult problems. Start a new problem in Maple (select New from the File menu). Enter the governing differential equation, Eq. (1-27), and assign it to the function ODE; note that the second derivative of T with respect to x is obtained by applying the diff command twice.

```
> restart;
> ODE:=diff(diff(T(x),x),x)=0;
```
$$ODE := \frac{d^2}{dx^2}T(x) = 0$$

The solution to the ordinary differential equation is obtained using the dsolve command and assigned to the function Ts.

```
> Ts:=dsolve(ODE);
```
$$Ts := T(x) = \_C1 x + \_C2$$

The solution identified by Maple is consistent with Eq. (1-34), except that Maple uses the variables _C1 and _C2 rather than C_1 and C_2 to represent the constants of integration. The two boundary conditions, Eqs. (1-35) and (1-36), are obtained symbolically using the eval command to evaluate the solution at a particular position and assigned to the functions BC1 and BC2.

```
> BC1:=eval(Ts,x=0)=T_H;
```
$$BC1 := (T(0) = \_C2) = T\_H$$
```
> BC2:=eval(Ts,x=L)=T_C;
```
$$BC2 := (T(L) = \_C1 L + \_C2) = T\_C$$

The result of the eval command is almost, but not quite what is needed to solve for the constants. The expressions include the extraneous statements T(0) and T(L); use the rhs function in order to return just the expression on the right hand side.

```
> BC1:=rhs(eval(Ts,x=0))=T_H;
```
$$BC1 := \_C2 = T\_H$$

```
> BC2:=rhs(eval(Ts,x=L))=T_C;
```
$$BC2 := \_C1L + \_C2 = T\_C$$

The constants are explicitly determined using the solve command. Note that the solve command requires two arguments; the first is the equation or, in this case, set of equations to be solved (the boundary conditions, BC1 and BC2) and the second is the variable or set of variables to solve for (the constants _C1 and _C2).

```
> constants:=solve({BC1,BC2},{_C1,_C2});
```
$$constants := \{\_C2 = T\_H, \_C1 = -\frac{T\_H - T\_C}{L}\}$$

The constants are substituted into the general solution using the subs command. The subs command requires two arguments; the first is the set of definitions to be substituted and the second is the set of equations to substitute them into.

```
> Ts:=subs(constants,Ts);
```
$$Ts := T(x) = -\frac{(T\_H - T\_C)x}{L} + T\_H$$

This result is the same as Eq. (1-37).

1.2.3 The Resistance Concept

Equation (1-38) is the solution for the rate of heat transfer through a plane wall. The equation suggests that, under some limiting conditions, conduction of heat through a solid can be thought of as a flow that is driven by a temperature difference and resisted by a thermal resistance, in the same way that electrical current is driven by a voltage difference and resisted by an electrical resistance. Inspection of Eq. (1-38) suggests that the thermal resistance to conduction through a plane wall (R_{pw}) is given by:

$$R_{pw} = \frac{L}{k A_c} \tag{1-39}$$

allowing Eq. (1-38) to be rewritten:

$$\dot{q} = \frac{(T_H - T_C)}{R_{pw}} \tag{1-40}$$

The concept of a thermal resistance is broadly useful and we will often return to this idea of a thermal resistance in order to help develop a conceptual understanding of various heat transfer processes. The usefulness of Eqs. (1-39) and (1-40) go beyond the simple situation illustrated in Figure 1-4. It is possible to approximately understand conduction heat transfer in most any situation provided that you can identify the distance that heat must be conducted and the cross-sectional area through which the conduction occurs.

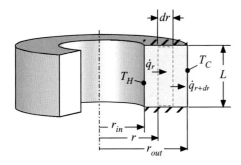

Figure 1-5: A cylinder with fixed temperature boundary conditions.

Resistance equations provide a method for succinctly summarizing a particular solution and we will derive resistance solutions for a variety of physical situations. By cataloging these resistance equations, it is possible to quickly use the solution in the context of a particular problem without having to go through all of the steps that were required in the original derivation. For example, if we are confronted with a problem involving steady-state heat transfer through a plane wall then it is not necessary to rederive Eqs. (1-19) through (1-38); instead, Eqs. (1-39) and (1-40) conveniently represent all of this underlying math.

1.2.4 Resistance to Radial Conduction through a Cylinder

Figure 1-5 illustrates steady-state, radial conduction through an infinitely long cylinder (or one with insulated ends) without thermal energy generation. The analytical solution to this problem is derived using the steps described in Section 1.2.2. The differential energy balance (see Figure 1-5) leads to:

$$\dot{q}_r = \dot{q}_{r+dr} \tag{1-41}$$

The $r + dr$ term in Eq. (1-41) is expanded and Fourier's law is substituted in order to reach:

$$\frac{d}{dr}\left[-k\,A_c\,\frac{dT}{dr} \right] = 0 \tag{1-42}$$

The difference between the plane wall geometry considered in Section 1.2.2 and the cylindrical geometry considered here is that the cross-sectional area for heat transfer, A_c in Eq. (1-42), is not constant but rather varies with radius:

$$\frac{d}{dr}\left[-k\,\underbrace{2\,\pi\,r\,L}_{A_c}\,\frac{dT}{dr} \right] = 0 \tag{1-43}$$

where L is the length of the cylinder. Assuming that the thermal conductivity is constant, Eq. (1-43) is simplified to:

$$\frac{d}{dr}\left[r\,\frac{dT}{dr} \right] = 0 \tag{1-44}$$

and integrated twice according to the following steps:

$$\int d\left[r\,\frac{dT}{dr} \right] = \int 0 \tag{1-45}$$

$$r\,\frac{dT}{dr} = C_1 \tag{1-46}$$

$$\int dT = \int \frac{C_1}{r}\, dr \tag{1-47}$$

$$T = C_1 \ln(r) + C_2 \tag{1-48}$$

where C_1 and C_2 are constants of integration, evaluated by applying the boundary conditions:

$$T_H = C_1 \ln(r_{in}) + C_2 \tag{1-49}$$

$$T_C = C_1 \ln(r_{out}) + C_2 \tag{1-50}$$

After some algebra, the temperature distribution in the cylinder is obtained:

$$T = T_C + (T_H - T_C)\frac{\ln\left(\dfrac{r_{out}}{r}\right)}{\ln\left(\dfrac{r_{out}}{r_{in}}\right)} \tag{1-51}$$

The rate of heat transfer is given by:

$$\dot{q} = -k\,2\pi r L\,\frac{dT}{dr} = \underbrace{\frac{2\pi L k}{\ln\left(\dfrac{r_{out}}{r_{in}}\right)}}_{1/R_{cyl}}(T_H - T_C) \tag{1-52}$$

Therefore, the thermal resistance to radial conduction through a cylinder (R_{cyl}) is:

$$R_{cyl} = \frac{\ln\left(\dfrac{r_{out}}{r_{in}}\right)}{2\pi L k} \tag{1-53}$$

It is worth noting that the thermal resistance to radial conduction through a cylinder must be computed using the ratio of the outer to the inner radii in the numerator, regardless of the direction of the heat transfer.

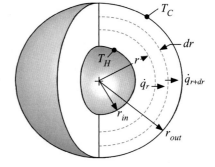

Figure 1-6: A sphere with fixed temperature boundary conditions.

1.2.5 Resistance to Radial Conduction through a Sphere

Figure 1-6 illustrates steady-state, radial conduction through a sphere without thermal energy generation. The differential energy balance (see Figure 1-6) leads to:

$$\dot{q}_r = \dot{q}_{r+dr} \tag{1-54}$$

which is expanded and used with Fourier's law to reach:

$$\frac{d}{dr}\left[-k\,A_c\,\frac{dT}{dr}\right] = 0 \qquad (1\text{-}55)$$

The cross-sectional area for heat transfer is the surface area of a sphere:

$$\frac{d}{dr}\left[-k\,\underbrace{4\,\pi\,r^2}_{A_c}\,\frac{dT}{dr}\right] = 0 \qquad (1\text{-}56)$$

Assuming that k is constant allows Eq. (1-56) to be simplified:

$$\frac{d}{dr}\left[r^2\,\frac{dT}{dr}\right] = 0 \qquad (1\text{-}57)$$

Equation (1-57) is entered in Maple:

```
> restart;
> ODE:=diff(r^2*diff(T(r),r),r)=0;
```

$$ODE := 2r\left(\frac{d}{dr}T(r)\right) + r^2\left(\frac{d^2}{dr^2}T(r)\right) = 0$$

and solved:

```
> Ts:=dsolve(ODE);
```

$$Ts := T(r) = \_C1 + \frac{\_C2}{r}$$

The boundary conditions are:

$$T_{r=r_{in}} = T_H \qquad (1\text{-}58)$$

$$T_{r=r_{out}} = T_C \qquad (1\text{-}59)$$

These equations are entered in Maple:

```
> BC1:=rhs(eval(Ts,r=r_in))=T_H;
```

$$BC1 := \_C1 + \frac{\_C2}{r\_in} = T\_H$$

```
> BC2:=rhs(eval(Ts,r=r_out))=T_C;
```

$$BC2 := \_C1 + \frac{\_C2}{r\_out} = T\_C$$

The constants are obtained by solving this system of two equations and two unknowns:

```
> constants:=solve({BC1,BC2},{_C1,_C2});
```

$$constants := \{ \_C1 = \frac{T\_Hr\_in - T\_Cr\_out}{-r\_out + r\_in}, \_C2 = -\frac{r\_in\,r\_out(-T\_C + T\_H)}{-r\_out + r\_in} \}$$

and substituted into the general solution:

```
> Ts:=subs(constants,Ts);
```

$$Ts := T(r) = \frac{T\_Hr\_in - T\_Cr\_out}{-r\_out + r\_in} - \frac{r\_in\,r\_out(-T\_C + T\_H)}{(-r\_out + r\_in)r}$$

The heat transfer at any radial location is given by Fourier's law:

$$\dot{q} = -k\,4\,\pi\,r^2\,\frac{dT}{dr} \tag{1-60}$$

```
> q_dot:=-k*4*pi*r^2*diff(Ts,r);
```

$$q\_dot := -4\,k\,\pi\,r^2\left(\frac{d}{dr}\mathrm{T}(r)\right) = -\frac{4\,k\,\pi\,r\_in\,r\_out(-T\_C + T\_H)}{-r\_out + r\_in}$$

and used to compute the thermal resistance for steady-state, radial conduction through a sphere:

$$R_{sph} = \frac{T_H - T_C}{\dot{q}} \tag{1-61}$$

```
> R_sph:=(T_H-T_C)/rhs(q_dot);
```

$$R\_sph := -\frac{-r\_out + r\_in}{4\,k\,\pi\,r\_in\,r\_out}$$

which can be simplified to:

$$R_{sph} = \frac{\left(\dfrac{1}{r_{in}} - \dfrac{1}{r_{out}}\right)}{4\,\pi\,k} \tag{1-62}$$

1.2.6 Other Resistance Formulae

Many heat transfer processes may be cast in the form of a resistance formula, allowing problems involving various types of heat transfer to be represented conveniently using thermal resistance networks. Resistance networks can be solved using techniques borrowed from electrical engineering. Also, it is often possible to obtain a physical feel for the problem by inspection of the thermal resistance network. For example, small resistances in series with large ones will tend to be unimportant. Large resistances in parallel with small ones can also be neglected. This type of understanding is important and can be obtained quickly using thermal resistance networks.

Convection Resistance

Convection is discussed in Chapters 4 through 7 and refers to heat transfer between a surface and a moving fluid. The rate equation that characterizes the rate of convection heat transfer ($\dot{q}_{conv}$) is Newton's law of cooling:

$$\dot{q}_{conv} = \underbrace{\overline{h} A_s}_{1/R_{conv}} (T_s - T_\infty) \tag{1-63}$$

where $\overline{h}$ is the average heat transfer coefficient, A_s is the surface area at temperature T_s that is exposed to fluid at temperature T_∞. Note that the heat transfer coefficient is not a material property like thermal conductivity, but rather a complex function of the geometry, fluid properties, and flow conditions. By inspection of Eq. (1-63), the thermal resistance associated with convection (R_{conv}) is:

$$R_{conv} = \frac{1}{\overline{h} A_s} \tag{1-64}$$

Contact Resistance

Contact resistance refers to the complex phenomenon that occurs when two solid surfaces are brought together. Regardless of how well prepared the surfaces are, they are not flat at the micro-scale and therefore energy carriers in either solid cannot pass through the interface unimpeded. The energy carriers in two dissimilar materials may not be the same in any case. The result is a temperature change at the interface that, at the macro-scale, appears to occur over an infinitesimally small spatial extent and grows in proportion to the rate of heat transfer across the interface. In reality, the temperature does not drop discontinuously but rather over some micro-scale distance that depends on the details of the interface. This phenomenon is usually modeled by characterizing the interface as having an area-specific contact resistance (R_c'', often provided in K-m^2/W). The resistance to heat transfer across an interface (R_c) is:

$$R_c = \frac{R_c''}{A_s} \tag{1-65}$$

where A_s is the contact area (the projected area of the surfaces, ignoring their microstructure). Contact resistance is not a material property but rather a complex function of the micro-structure, the properties of the two materials involved, the contact pressure, the interstitial material, etc. The area-specific contact resistance for interface conditions that are commonly encountered has been measured and tabulated in various references, for example Schneider (1985). Some representative values are listed in Table 1-1.

The area-specific contact resistance tends to be reduced with increasing clamping pressure and smaller surface roughness. One method for reducing contact resistance is to insert a soft metal (e.g., indium) or grease into the interface in order to improve the heat transfer across the interstitial gap. The values listed in Table 1-1 can be used to determine whether contact resistance is likely to play an important role in a specific application. However, if contact resistance is important, then more precise data for the interface of interest should be obtained or measurements should be carried out.

Table 1-1: Area-specific contact resistance for some interfaces, from Schneider (1985) and Fried (1969).

Materials	Clamping pressure	Surface roughness	Interstitial material	Temperature	Area-specific contact resistance
copper-to-copper	100 kPa	0.2 μm	vacuum	46°C	1.5×10^{-4} K-m^2/W
copper-to-copper	1000 kPa	0.2 μm	vacuum	46°C	1.3×10^{-4} K-m^2/W
aluminum-to-aluminum	100 kPa	0.3 μm	vacuum	46°C	2.5×10^{-3} K-m^2/W
aluminum-to-aluminum	100 kPa	1.5 μm	vacuum	46°C	3.3×10^{-3} K-m^2/W
stainless-to-stainless	100 kPa	1.3 μm	vacuum	30°C	4.5×10^{-3} K-m^2/W
stainless-to-stainless	1000 kPa	1.3 μm	vacuum	30°C	2.4×10^{-3} K-m^2/W
stainless-to-stainless	100 kPa	0.3 μm	vacuum	30°C	2.9×10^{-3} K-m^2/W
stainless-to-stainless	1000 kPa	0.3 μm	vacuum	30°C	7.7×10^{-4} K-m^2/W
stainless-to-aluminum	100 kPa	1.2 μm	air	93°C	3.3×10^{-4} K-m^2/W
aluminum-to-aluminum	1000 kPa	0.3 μm	air	93°C	6.7×10^{-5} K-m^2/W
aluminum-to-aluminum	100 kPa	10 μm	air	20°C	2.8×10^{-4} K-m^2/W
aluminum-to-aluminum	100 kPa	10 μm	helium	20°C	1.1×10^{-4} K-m^2/W
aluminum-to-aluminum	100 kPa	10 μm	hydrogen	20°C	0.72×10^{-4} K-m^2/W
aluminum-to-aluminum	100 kPa	10 μm	silicone oil	20°C	0.53×10^{-4} K-m^2/W

Radiation Resistance

Radiation heat transfer occurs between surfaces due to the emission and absorption of electromagnetic waves, as described in Chapter 10. Radiation heat transfer is complex when many surfaces at different temperatures are involved; however, in the limit that a single surface at temperature T_s interacts with surroundings at temperature T_{sur} then the radiation heat transfer from the surface can be calculated according to:

$$\dot{q}_{rad} = A_s\,\sigma\,\varepsilon\left(T_s^4 - T_{sur}^4\right) \tag{1-66}$$

where A_s is the area of the surface, σ is the Stefan-Boltzmann constant (5.67×10^{-8} W/m^2-K^4), and ε is the emissivity of the surface. Emissivity is a parameter that ranges between near 0 (for highly reflective surfaces) to near 1 (for highly absorptive surfaces). Note that both T_s and T_{sur} must be expressed as absolute temperature (i.e., in units K rather than °C) in Eq. (1-66).

Equation (1-66) does not seem to resemble a resistance equation because the heat transfer is not driven by a difference in temperatures but rather by a difference in temperatures to the fourth power. However, Eq. (1-66) may be expanded to yield:

$$\dot{q}_{rad} = A_s\,\underbrace{\underbrace{\sigma\,\varepsilon\left(T_s^2 + T_{sur}^2\right)\left(T_s + T_{sur}\right)}_{\bar{h}_{rad}}\left(T_s - T_{sur}\right)}_{1/R_{rad}} \tag{1-67}$$

Comparing Eq. (1-67) for radiation to Eq. (1-63) for convection shows that a 'radiation heat transfer coefficient', $\bar{h}_{rad}$, can be defined as:

$$\bar{h}_{rad} = \sigma\,\varepsilon\left(T_s^2 + T_{sur}^2\right)\left(T_s + T_{sur}\right) \tag{1-68}$$

The radiation heat transfer coefficient is a useful quantity for many problems because it allows convection and radiation to be compared directly, as discussed in Section 10.6.2.

Equation (1-67) suggests that an appropriate thermal resistance for radiation heat transfer (R_{rad}) is:

$$R_{rad} = \frac{1}{A_s\,\sigma\,\varepsilon\left(T_s^2 + T_{sur}^2\right)\left(T_s + T_{sur}\right)} \tag{1-69}$$

Because the absolute surface and surrounding temperatures are both typically large and not too different from each other, Eq. (1-69) can often be approximated by:

$$R_{rad} \approx \frac{1}{A_s\,\sigma\,\varepsilon\,4\,\overline{T}^3} \tag{1-70}$$

where $\overline{T}$ is the average of the surface and surrounding temperatures:

$$\overline{T} = \frac{T_s + T_{sur}}{2} \tag{1-71}$$

With this approximation, the radiation heat transfer coefficient is:

$$\bar{h}_{rad} \approx \sigma\,\varepsilon\,4\,\overline{T}^3 \tag{1-72}$$

The resistance associated with radiation and the radiation heat transfer coefficient are both clearly temperature-dependent. However, the conductivity, contact resistance, and average heat transfer coefficient that are required to compute other types of resistances are also temperature-dependent and therefore the resistance concept can only be approximate in any case. A summary of the thermal resistance associated with several common situations is presented in Table 1-2.

Table 1-2: A summary of common resistance formulae.

Situation	Resistance formula	Nomenclature
Plane wall	$$R_{pw} = \frac{L}{k\,A_c}$$	L = wall thickness (‖ to heat flow) k = conductivity A_c = cross-sectional area (⊥ to heat flow)
Cylinder (radial heat transfer)	$$R_{cyl} = \frac{\ln\left(\dfrac{r_{out}}{r_{in}}\right)}{2\,\pi\,L\,k}$$	L = cylinder length k = conductivity r_{in} and r_{out} = inner and outer radii
Sphere (radial heat transfer)	$$R_{sph} = \frac{1}{4\,\pi\,k}\left[\frac{1}{r_{in}} - \frac{1}{r_{out}}\right]$$	k = conductivity r_{in} and r_{out} = inner and outer radii
Convection	$$R_{conv} = \frac{1}{\overline{h}\,A_s}$$	$\overline{h}$ = average heat transfer coefficient A_s = surface area exposed to convection
Contact between surfaces	$$R_c = \frac{R''_c}{A_s}$$	R''_c = area specific contact resistance A_s = surface area in contact
Radiation (exact)	$$R_{rad} = \frac{1}{A_s\,\sigma\,\varepsilon\,(T_s^2 + T_{sur}^2)(T_s + T_{sur})}$$	A_s = radiating surface area σ = Stefan-Boltzmann constant ε = emissivity T_s = absolute surface temperature T_{sur} = absolute surroundings temperature
Radiation (approximate)	$$R_{rad} \approx \frac{1}{A_s\,\sigma\,\varepsilon\,4\,\overline{T}^3}$$	A_s = radiating surface area σ = Stefan-Boltzmann constant ε = emissivity $\overline{T}$ = average absolute temperature

Note that useful reference information, such as Table 1-2, is included in the Heat Transfer Reference Section of EES in order to facilitate solving heat transfer problems without requiring that you locate a written reference book. To access this section, select the Reference Material from the Heat Transfer menu. This will open an online document that contains material from this book. Notice that the Heat Transfer menu also includes all of the examples that are associated with the book.

EXAMPLE 1.2-1: LIQUID OXYGEN DEWAR

Figure 1 illustrates a spherical dewar containing saturated liquid oxygen that is kept at pressure $p_{LOx} = 25$ psia; the saturation temperature of oxygen at this pressure is $T_{LOx} = 95.6$ K.

The dewar consists of an inner and outer metal liner separated by polystyrene foam insulation. The inner metal liner has inner radius $r_{mli,in} = 10.0$ cm and thickness $th_m = 2.5$ mm. The outer metal liner also has thickness $th_m = 2.5$ mm. The conductivity of both metal liners is $k_m = 15$ W/m-K. The heat transfer coefficient between the oxygen within the dewar and the inner surface of the dewar is $\overline{h}_{in} = 150$ W/m²-K. The outer surface of the dewar is surrounded by air at $T_\infty = 20°C$ and radiates to surroundings that are also at $T_\infty = 20°C$. The emissivity of the outer surface of the dewar is $\varepsilon = 0.7$. The heat transfer coefficient between the outer surface of the dewar and the surrounding air is $\overline{h}_{out} = 6$ W/m²-K. The area-specific

EXAMPLE 1.2-1: LIQUID OXYGEN DEWAR

contact resistance that characterizes the interfaces between the insulation and the adjacent metal liners is $R_c'' = 3.0 \times 10^{-3}$ K-m^2/W.

The thickness of the insulation between the two metal liners is $th_{ins} = 1.0$ cm. You are trying to evaluate the impact of using polystyrene foam insulation in place of the more expensive insulation that is currently used. Flynn (2005) suggests that the conductivity of polystyrene foam at cryogenic temperatures is approximately $k_{ins} = 330$ μW/cm-K.

a) Draw a network that represents this situation using 1-D resistances.

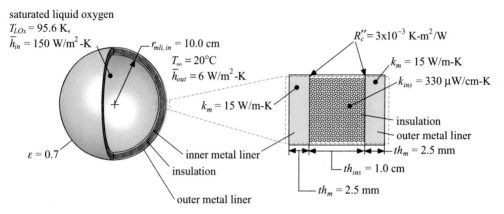

Figure 1: Spherical dewar containing saturated liquid oxygen.

The resistance network is illustrated in Figure 2.

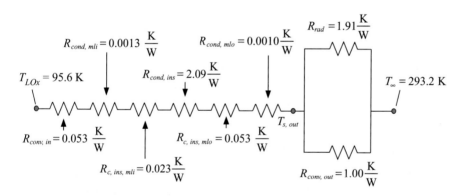

The resistances include:

$R_{conv, in}$ = convection from inside surface

$R_{cond, mli}$ = conduction through inner metal liner

$R_{c, ins, mli}$ = contact between inner metal liner & insulation

$R_{cond, ins}$ = conduction through insulation

$R_{c, ins, mlo}$ = contact between outer metal liner & insulation

$R_{cond, mlo}$ = conduction through outer metal liner

R_{rad} = radiation

$R_{conv, out}$ = convection from outer surface

Figure 2: Resistance network representing the dewar.

The resistance network interacts with the surrounding air and surroundings (at T_∞) and the saturated liquid oxygen (at T_{LOx}).

b) Estimate the rate of heat transfer to the liquid oxygen.

The solution will be carried out using EES. It is assumed that you have already been exposed to the EES software by completing the self-guided tutorial contained in Appendix A.1. The first step in preparing a successful solution to any problem with EES is to enter the inputs to the problem and set their units. Experience has shown that it is generally best to work exclusively in SI units (m, J, K, kg, Pa, etc.) because this unit system is entirely self-consistent. If the problem statement includes parameters in other units, they can be converted to SI units within the "Inputs" section of the code. The upper section of your EES code should look something like:

```
"EXAMPLE 1.2-1: Liquid Oxygen Dewar"

$UnitSystem SI MASS RAD PA K J
$Tabstops 0.2 0.4 0.6 3.5 in

"Inputs"
p_LOx=25 [psia]* convert(psia,Pa)              "pressure of liquid oxygen"
T_LOx=95.6 [K]                                 "temperature of liquid oxygen"
h_bar_in =150 [W/m^2-K]
   "heat transfer coefficient between the liquid oxygen and the inner wall"
r_mli_in=10 [cm]* convert(cm,m)                "inner radius of the inner metal liner"
th_m=2.5 [mm]* convert(mm,m)                   "thickness of inner metal liner"
th_ins_cm=1.0 [cm]                             "thickness of insulation, in cm"
th_ins=th_ins_cm* convert(cm,m)                "thickness of insulation"
e=0.7 [-]                                      "emissivity of outside surface"
T_infinity=converttemp(C,K,20 [C])     "temperature of surroundings and surrounding air"
R"_c=3.0e-3 [K-m^2/W]                         "area-specific contact resistance"
k_ins=330 [microW/cm-K]*convert(microW/cm-K,W/m-K)
                                               "mean conductivity of insulation"
k_m=15 [W/m-K]                                 "conductivity of metal"
h_bar_out=6 [W/m^2-K]       "heat transfer coefficient between outer wall and surrounding air"
```

The resistance to convection between the inner surface of the dewar and the oxygen is:

$$R_{conv,in} = \frac{1}{\overline{h}_{in}\, 4\, \pi\, r_{mli,in}^2}$$

```
R_conv_in=1/(4*pi*r_mli_in^2*h_bar_in)         "convection resistance to liquid oxygen"
```

The inner radius of the insulation is:

$$r_{ins,in} = r_{mli,in} + th_m$$

The resistance to conduction through the inner metal liner is:

$$R_{cond,mli} = \frac{1}{4\,\pi\,k_m}\left(\frac{1}{r_{mli,in}} - \frac{1}{r_{ins,in}}\right)$$

EXAMPLE 1.2-1: LIQUID OXYGEN DEWAR

and the contact resistance resistance between the inner metal liner and the insulation is:

$$R_{c,ins,mli} = \frac{R_c''}{4\,\pi\,r_{ins,in}^2}$$

r_ins_in=r_mli_in+th_m   "inner radius of insulation"
R_cond_mli=(1/r_mli_in-1/r_ins_in)/(4*pi*k_m)   "conduction resistance of inner metal liner"
R_c_ins_mli=R"_c/(4*pi*r_ins_in^2)   "contact resistance between inner metal liner and insulation"

The outer radius of the insulation is:

$$r_{ins,out} = r_{ins,in} + th_{ins}$$

The resistance to conduction through the insulation is:

$$R_{cond,ins} = \frac{1}{4\,\pi\,k_{ins}}\left(\frac{1}{r_{ins,in}} - \frac{1}{r_{ins,out}}\right)$$

and the contact resistance resistance between the insulation and the outer metal liner is:

$$R_{c,ins,mlo} = \frac{R_c''}{4\,\pi\,r_{ins,out}^2}$$

r_ins_out=r_ins_in+th_ins   "outer radius of insulation"
R_cond_ins=(1/r_ins_in-1/r_ins_out)/(4*pi*k_ins)   "conduction resistance of insulation"
R_c_ins_mlo=R"_c/(4*pi*r_ins_out^2)
    "contact resistance between insulation and outer metal liner"

The outer radius of the outer metal liner is:

$$r_{mlo,out} = r_{ins,out} + th_m$$

The resistance to conduction through the outer metal liner is:

$$R_{cond,mlo} = \frac{1}{4\,\pi\,k_m}\left(\frac{1}{r_{ins,out}} - \frac{1}{r_{mlo,out}}\right)$$

and the convection resistance between the outer surface of the dewar and the air is:

$$R_{conv,out} = \frac{1}{\bar{h}_{out}\,4\,\pi\,r_{mli,out}^2}$$

r_mlo_out=r_ins_out+th_m   "outer radius of outer metal liner"
R_cond_mlo=(1/r_ins_out-1/r_mlo_out)/(4*pi*k_m)   "conduction resistance of outer metal liner"
R_conv_out=1/ (4*pi*r_mlo_out^2*h_bar_out)   "convection resistance to surrounding air"

The surface temperature on the outside of the dewar ($T_{s,out}$ in Figure 2) cannot be known until the problem is solved and yet it must be used to calculate the resistance to radiation, R_{rad}. One of the nice things about using the EES software to solve this problem is that the software can deal with this type of nonlinearity and provide the solution to the implicit equations. It is this capability that simultaneously makes

EXAMPLE 1.2-1: LIQUID OXYGEN DEWAR

EES so powerful and yet sometimes, ironically, difficult to use. EES should be able to solve equations regardless of the order in which they are entered. However, you should enter equations in a sequence that allows you to solve them as you enter them; this is exactly what you would be forced to do if you were to solve the problem using a typical programming language (e.g., MATLAB, FORTRAN, etc.). This technique of entering your equations in a systematic order provides you with the opportunity to debug each subset of equations as you move along, rather than waiting until all of the equations have been entered before you try to solve them. Another benefit of approaching a problem in this sequential manner is that you can consistently update the guess values associated with the variables in your problem. EES solves your equations using a nonlinear relaxation technique and therefore the closer the guess values of the variables are to "reasonable" values, the more likely it is that EES will find the correct solution.

To proceed with the solution to this problem using EES, it is helpful to initially assume a reasonable surface temperature (e.g., a reasonable guess might be the average of the surrounding and the liquid oxygen temperatures) so that it is possible to calculate a value of the radiation resistance:

$$R_{rad} = \frac{1}{4 \pi r_{mli,out}^2 \sigma \varepsilon \left(T_{s,out}^2 + T_\infty^2\right)\left(T_{s,out} + T_\infty\right)}$$

and continue with the solution.

```
T_s_out = (T_LOx+T_infinity)/2
                          "guess for the surface temperature (removed to complete problem)"
R_rad=1/(4*pi*r_mlo_out^2*sigma#*e* (T_s_out^2+T_infinity^2)* (T_s_out+T_infinity))
                          "radiation resistance"
```

Solve the equations that have been entered (select Calculate from the Solve menu) and check that your answers make sense. Verify that the variables and equations have a consistent set of units by setting the units for each of the variables. The best way to do this is to go to the Variable Information window (select Variable Info from the Options menu) and enter the units for each variable in the Units column. Once this is done, check the units for your problem (select Check Units from the Calculate menu) in order to make sure that all of the units are consistent with the equations. Alternatively, units can be set by right-clicking on the variables in the Solution Window.

The total resistance separating the liquid oxygen from the surroundings is:

$$R_{total} = R_{conv,in} + R_{cond,mli} + R_{c,ins,mli}$$

$$+ R_{cond,ins} + R_{c,ins,mlo} + R_{cond,mlo} + \left(\frac{1}{R_{conv,out}} + \frac{1}{R_{rad}}\right)^{-1}$$

and the heat transfer rate from the surroundings to the liquid oxygen is:

$$\dot{q} = \frac{\left(T_\infty - T_{LOx}\right)}{R_{total}}$$

```
R_total=R_conv_in+R_cond_mli+R_c_ins_mli+R_cond_ins+R_c_ins_mlo&
    +R_cond_mlo+(1/R_conv_out+1/R_rad)^(-1)          "total resistance"
q_dot=(T_infinity-T_LOx)/R_total                     "heat flow"
```

EXAMPLE 1.2-1: LIQUID OXYGEN DEWAR

At this point, we can use the heat transfer rate to recalculate the surface temperature (as opposed to assuming it).

$$T_{s,out} = T_{LOx} + \dot{q}\left(R_{conv,in} + R_{cond,mli} + R_{c,ins,mli} + R_{cond,ins} + R_{c,ins,mlo} + R_{cond,mlo}\right)$$

It is necessary to comment out or delete the equation that provided the assumed surface temperature and instead calculate the surface temperature correctly. This step creates an implicit set of nonlinear equations. Before you ask EES to solve the set of equations, it is a good idea to update the guess values for each variable (select Update Guesses from the Calculate menu).

```
{T_s_out=(T_LOx+T_infinity)/2}                    "guess for the surface temperature"
T_s_out=T_LOx+q_dot*(R_conv_in+ R_cond_mli+R_c_ins_mli+R_cond_ins+R_c_ins_mlo+R_cond_mlo)
                                                  "surface temperature"
```

The rate of heat transfer to the liquid oxygen is $\dot{q} = 69.4$ W.

Resistance networks provide substantial insight into the problem. Figure 2 shows the magnitude of each of the resistances in the network. The resistances associated with conduction through the insulation, radiation from the surface of the dewar, and convection from the surface of the dewar are of the same order of magnitude and large relative to the others in the circuit. Conduction through the insulation is much more important than conduction through the metal liners, convection to the liquid oxygen or the contact resistance; these resistances can probably be neglected in a rough analysis and certainly very little effort should be expended to better understand these aspects of the problem.

Both radiation and convection from the outer surface are important, as they are of similar magnitude. The convection resistance is smaller and therefore more heat will be transferred by convection from the surface than is radiated from the surface. If the radiation resistance had been much larger than the convection resistance (as is often the case in forced convection problems where the convection heat transfer coefficient is much larger) then radiation could be neglected. The smallest resistance in a parallel network will dominate because most of the energy will tend to flow through that resistance.

It is almost always a good idea to estimate the size of the resistances in a heat transfer problem prior to solving the problem. Often it is possible to simplify the analysis considerably, and the size of the resistances can certainly be used to guide your efforts. For this problem, a detailed analysis of conduction through the metal liner or a lengthy search for the most accurate value of the thermal conductivity of the metal would be a misguided use of time whereas a more accurate measurement of the conductivity of the insulation might be important.

c) Plot the rate of heat transfer to the liquid oxygen as a function of the insulation thickness.

In order to generate the requested plot, it is necessary to parametrically vary the insulation thickness. The specified value of the insulation thickness is commented out

```
{th_ins_cm=1.0 [cm]}                              "thickness of insulation, in cm"
```

EXAMPLE 1.2-1: LIQUID OXYGEN DEWAR

and a parametric table is generated (select New Parametric Table from the Tables menu) that includes the variables th_ins_cm and q_dot (Figure 3).

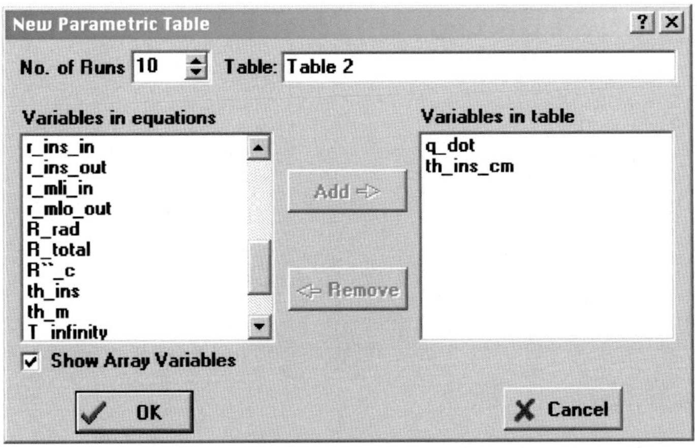

Figure 3: New Parametric Table Window.

Right-click on the th_ins_cm column and select Alter Values; vary the thickness from 0 cm to 10 cm and solve the table (select Solve Table from the Calculate menu). Prepare a plot of the results (select New Plot Window from the Plots menu and then select X-Y Plot) by selecting the variable th_ins_cm for the X-Axis and q_dot for the Y-Axis (Figure 4).

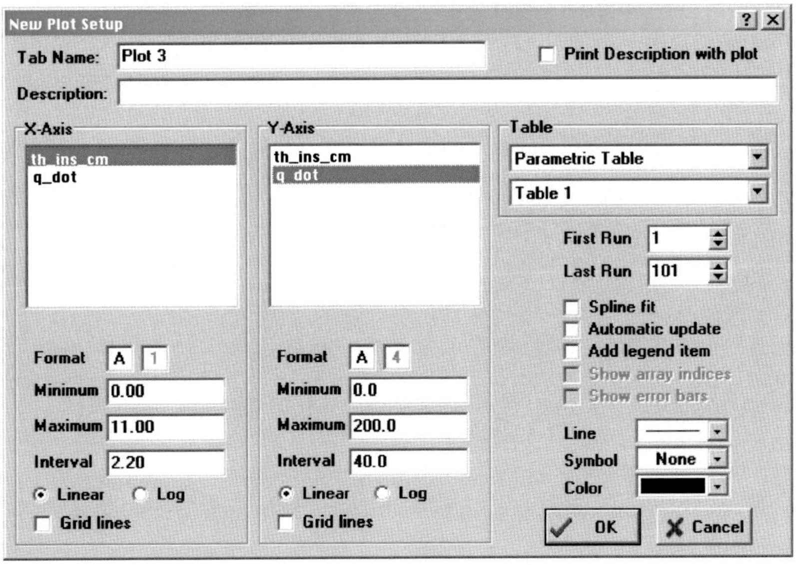

Figure 4: New Plot Setup Window.

EXAMPLE 1.2-1: LIQUID OXYGEN DEWAR

Figure 5 illustrates the rate of heat transfer as a function of the insulation thickness.

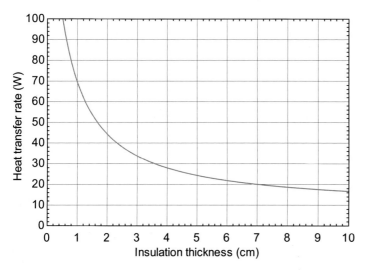

Figure 5: Heat transfer rate as a function of insulation thickness.

1.3 Steady-State 1-D Conduction with Generation

1.3.1 Introduction

The generation of thermal energy within a conductive medium may occur through ohmic dissipation, chemical or nuclear reactions, or absorption of radiation. According to the first law of thermodynamics, energy cannot be generated (excluding nuclear reactions); however, it can be converted from other forms (e.g., electrical energy) to thermal energy. The energy balance that we use to solve these problems is then strictly a thermal energy conservation equation. The addition of thermal energy generation to the 1-D steady-state solutions considered in Section 1.2 is straightforward and the steps required to obtain an analytical solution are essentially the same.

1.3.2 Uniform Thermal Energy Generation in a Plane Wall

Consider a plane wall with temperatures fixed at either edge that experiences a volumetric generation of thermal energy, as shown in Figure 1-7.

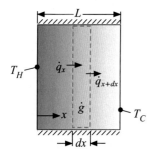

Figure 1-7: Plane wall with thermal energy generation and fixed temperature boundary conditions.

The problem is assumed to be 1-D in the x-direction and therefore an appropriate differential control volume has width dx (see Figure 1-7). Notice the additional energy term in the control volume that is related to the generation of thermal energy. A steady-state energy balance includes conduction into the left-side of the control volume ($\dot{q}_x$, at position x), generation within the control volume ($\dot{g}$), and conduction out of the right-side of the control voume ($\dot{q}_{x+dx}$, at position $x + dx$):

$$\dot{q}_x + \dot{g} = \dot{q}_{x+dx} \tag{1-73}$$

or, after expanding the right side:

$$\dot{q}_x + \dot{g} = \dot{q}_x + \frac{d\dot{q}}{dx}dx \tag{1-74}$$

The rate of thermal energy generation within the control volume can be expressed as the product of the volume enclosed by the control volume and the rate of thermal energy generation per unit volume, $\dot{g}'''$ (which may itself be a function of position or temperature):

$$\dot{g} = \dot{g}''' A_c \, dx \tag{1-75}$$

where A_c is the cross-sectional area of the wall. The conduction term is expressed using Fourier's law:

$$\dot{q} = -k A_c \frac{dT}{dx} \tag{1-76}$$

Substituting Eqs. (1-76) and (1-75) into Eq. (1-74) results in

$$\dot{g}''' A_c \, dx = \frac{d}{dx}\left(-k A_c \frac{dT}{dx}\right) dx \tag{1-77}$$

which can be simplified (assuming that conductivity is constant):

$$\frac{d}{dx}\left(\frac{dT}{dx}\right) = -\frac{\dot{g}'''}{k} \tag{1-78}$$

Equation (1-78) is separated and integrated:

$$\int d\left(\frac{dT}{dx}\right) = \int -\frac{\dot{g}'''}{k}dx \tag{1-79}$$

If the volumetric rate of thermal energy generation is spatially uniform, then the integration leads to:

$$\frac{dT}{dx} = -\frac{\dot{g}'''}{k}x + C_1 \tag{1-80}$$

where C_1 is a constant of integration. Equation (1-80) is integrated again:

$$\int dT = \int \left(-\frac{\dot{g}'''}{k}x + C_1\right) dx \tag{1-81}$$

which leads to:

$$T = -\frac{\dot{g}'''}{2k}x^2 + C_1 x + C_2 \tag{1-82}$$

where C_2 is another constant of integration. Note that the same solution can be obtained using Maple. The governing differential equation, Eq. (1-78), is entered in Maple:

```
> restart;
> GDE:=diff(diff(T(x),x),x)=-gv/k;
```

$$GDE := \frac{d^2}{dx^2} T(x) = -\frac{gv}{k}$$

and solved:

```
> Ts:=dsolve(GDE);
```

$$Ts := T(x) = -\frac{gvx^2}{2k} + \_C1x + \_C2$$

Equation (1-82) satisfies the governing differential equation, Eq. (1-78), throughout the computational domain (i.e., from $x = 0$ to $x = L$) for arbitrary values of C_1 and C_2. It is easy to use Maple to check that this is true:

```
> rhs(diff(diff(Ts,x),x))+gv/k;
```

$$0$$

Note that it is often a good idea to use Maple to quickly doublecheck that an analytical solution does in fact satisfy the original governing differential equation.

All that remains is to force the general solution, Eq. (1-82), to satisfy the boundary conditions by adjusting the constants C_1 and C_2. The fixed temperature boundary conditions shown in Figure 1-7 correspond to:

$$T_{x=0} = T_H \tag{1-83}$$

$$T_{x=L} = T_C \tag{1-84}$$

Substituting Eq. (1-82) into Eqs. (1-83) and (1-84) leads to:

$$C_2 = T_H \tag{1-85}$$

$$-\frac{\dot{g}'''}{2k}L^2 + C_1 L + T_H = T_C \tag{1-86}$$

Solving Eq. (1-86) for C_1 leads to:

$$C_1 = \frac{\dot{g}'''}{2k}L - \frac{(T_H - T_C)}{L} \tag{1-87}$$

Substituting Eqs. (1-85) and (1-87) into Eq. (1-82) leads to:

$$T = \frac{\dot{g}''' L^2}{2k} \left[\frac{x}{L} - \left(\frac{x}{L}\right)^2 \right] - \frac{(T_H - T_C)}{L} x + T_H \qquad (1\text{-}88)$$

Again, Maple can be used to achieve the same result. The boundary condition equations are defined in Maple:

> BC1:=rhs(eval(Ts,x=0))=T_H;

$$BC1 := \_C2 = T\_H$$

> BC2:=rhs(eval(Ts,x=L))=T_C;

$$BC2 := -\frac{gvL^2}{2k} + \_C1 L + \_C2 = T\_C$$

and solved for the two constants:

> constants:=solve({BC1,BC2},{_C1,_C2});

$$constants := \{\_C2 = T\_H, \_C1 = \frac{gvL^2 - 2T\_Hk + 2T\_Ck}{2Lk}\}$$

The constants are substituted into the general equation:

> Ts:=subs(constants,Ts);

$$Ts := T(x) = -\frac{gvx^2}{2k} + \frac{(gvL^2 - 2T\_Hk + 2T\_Ck)x}{2Lk} + T\_H$$

It is good practice to examine any solution and verify that it makes sense. By inspection, it is clear that Eq. (1-88) limits to T_H at $x = 0$ and T_C at $x = L$; therefore, the boundary conditions were implemented correctly. Also, in the absence of any generation Eq. (1-88) limits to Eq. (1-37), the solution that was derived in Section 1.2.2 for steady-state conduction through a plane wall without generation.

Figure 1-8 illustrates the temperature distribution for $T_H = 80°C$ and $T_C = 20°C$ with $L = 1.0$ cm and $k = 1$ W/m-K for various values of $\dot{g}'''$. The temperature profile becomes more parabolic as the rate of thermal energy generation increases because energy must be transferred toward the edges of the wall at an increasing rate. The temperature gradient is proportional to the local rate of conduction heat transfer; as $\dot{g}'''$ increases, the heat transfer rate at $x = L$ increases while the heat transfer rate entering at $x = 0$ decreases and eventually becomes negative (i.e., heat actually begins to leave from both edges of the wall). This effect is evident in Figure 1-8 by observing that the temperature gradient at $x = 0$ changes from a negative to a positive value as $\dot{g}'''$ is increased.

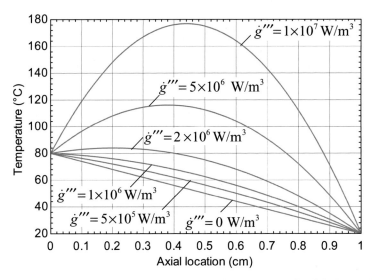

Figure 1-8: Temperature distribution within a plane wall with thermal energy generation ($k =$ 1 W/m-K, $T_H = 80°C$, $T_C = 20°C$, $L = 1.0$ cm).

The rate of heat transfer by conduction in the wall is obtained by applying Fourier's law to the solution for the temperature distribution:

$$\dot{q} = -k A_c \frac{dT}{dx} \tag{1-89}$$

Substituting Eq. (1-88) into Eq. (1-89) leads to:

$$\dot{q} = A_c \dot{g}''' L \left(\frac{x}{L} - \frac{1}{2} \right) + \frac{k A_c}{L} (T_H - T_C) \tag{1-90}$$

The heat transfer rate is not constant with position. Therefore, the plane wall with generation cannot be represented as a thermal resistance in the manner discussed in Section 1.2. However, it is always a good idea to carry out a number of sanity checks on your solution and the resistance concepts discussed in Section 1.2 provide an excellent mechanism for doing this.

Equation (1-88) and Figure 1-8 both show that there is a maximum temperature elevation (relative to the zero generation case) that occurs at the center of the wall. Substituting $x = L/2$ into the 1st term in Eq. (1-88) shows that the magnitude of the maximum temperature elevation is:

$$\Delta T_{max} = \frac{\dot{g}''' L^2}{8 k} \tag{1-91}$$

Maple can provide this result as well. The zero-generation solution is obtained by substituting $\dot{g}''' = 0$ into the original solution:

```
> Tsng:=subs(gv=0,Ts);
```

$$Tsng := T(x) = \frac{(-2T\_Hk + 2T\_Ck)x}{2 Lk} + T\_H$$

Then the temperature elevation is the difference between the original solution and the zero-generation solution:

> DeltaT:=rhs(Ts-Tsng);

$$DeltaT := -\frac{gvx^2}{2\,k} + \frac{(gvL^2 - 2T\_Hk + 2T\_Ck)x}{2\,Lk} - \frac{(-2T\_Hk + 2T\_Ck)x}{2\,Lk}$$

and the maximum value of the temperature elevation is evaluated at $x = L/2$:

> DeltaTmax=eval(DeltaT,x=L/2);

$$DeltaTmax = -\frac{gvL^2}{8\,k} + \frac{gvL^2 - 2T\_Hk + 2T\_Ck}{4\,k} - \frac{-2T\_Hk + 2T\_Ck}{4\,k}$$

This result can be simplified using the **simplify** command:

> simplify(%);

$$DeltaTmax = \frac{gvL^2}{8\,k}$$

Note that the % character refers to the result of the previous command (i.e., the output of the last calculation).

The temperature elevation occurs because the energy that is generated within the wall must be conducted to one of the external surfaces. Therefore, the temperature elevation must be consistent, in terms of its order of magnitude if not its exact value, with the temperature rise that is associated with the rate of thermal energy generation passing through an appropriately defined resistance. This is an approximate analysis and is only meant to illustrate the process of providing a "back of the envelope" estimate or a sanity check on a solution.

The total energy that is generated in one-half of the wall material must pass to the adjacent edge. A very crude estimate of the temperature elevation is:

$$\Delta T_{max} \sim \underbrace{\left(\dot{g}''' A_c \frac{L}{2} \right)}_{\substack{\text{rate of thermal energy} \\ \text{generated in half the wall}}} \underbrace{\left(\frac{L}{2\,k\,A_c} \right)}_{\substack{\text{thermal resistance} \\ \text{of half the wall}}} = \frac{\dot{g}''' L^2}{4\,k} \tag{1-92}$$

which, in this case, is within a factor of two of the exact analytical solution (the "back of the envelope" calculation is too large because the energy generated near the surfaces does not need to pass through half of the wall material). Again, the intent of this analysis was not to obtain exact agreement, but rather to provide a quick check that the solution makes sense.

1.3.3 Uniform Thermal Energy Generation in Radial Geometries

The area for conduction through the plane wall discussed in Section 1.3.2 is constant in the coordinate direction (x). If the conduction area is a function of position, then it cannot be canceled from each side of Eq. (1-77) and therefore the differential equation becomes more complicated. Figure 1-9 illustrates the differential control volumes that should be defined in order to analyze radial heat transfer in (a) a cylinder and (b) a sphere with thermal energy generation.

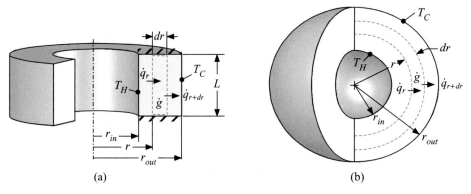

Figure 1-9: Differential control volume for (a) a cylinder and (b) a sphere with volumetric thermal energy generation.

The differential control volume suggested by either Figure 1-9(a) or (b) is:

$$\dot{q}_r + \dot{g} = \dot{q}_{r+dr} \tag{1-93}$$

which is expanded and simplified:

$$\dot{g} = \frac{d\dot{q}}{dr}dr \tag{1-94}$$

The rate equations for $\dot{q}$ and $\dot{g}$ in a cylindrical geometry, Figure 1-9(a), are:

$$\dot{q} = -k\,2\,\pi\,r\,L\,\frac{dT}{dr} \tag{1-95}$$

$$\dot{g} = 2\,\pi\,r\,L\,dr\,\dot{g}''' \tag{1-96}$$

where L is the length of the cylinder and $\dot{g}'''$ is the rate of thermal energy generation per unit volume. Substituting Eqs. (1-95) and (1-96) into Eq. (1-94) leads to:

$$r\,\dot{g}''' = -\frac{d}{dr}\left(k\,r\,\frac{dT}{dr}\right) \tag{1-97}$$

which is integrated twice (assuming that k and $\dot{g}'''$ are constant) to achieve:

$$T = -\frac{\dot{g}'''\,r^2}{4\,k} + C_1\,\ln(r) + C_2 \tag{1-98}$$

where C_1 and C_2 are constants of integration that depend on the boundary conditions.

The rate equations for $\dot{q}$ and $\dot{g}$ in a spherical geometry, Figure 1-9(b), are:

$$\dot{q} = -k\,4\,\pi\,r^2\,\frac{dT}{dr} \tag{1-99}$$

$$\dot{g} = 4\,\pi\,r^2\,dr\,\dot{g}''' \tag{1-100}$$

Substituting Eqs. (1-99) and (1-100) into Eq. (1-94) leads to:

$$\dot{g}'''\,r^2 = \frac{d}{dr}\left(-k\,r^2\,\frac{dT}{dr}\right) \tag{1-101}$$

which is integrated twice to achieve:

$$T = -\frac{\dot{g}'''}{6\,k}r^2 + \frac{C_1}{r} + C_2 \tag{1-102}$$

Table 1-3: Summary of formulae for 1-D uniform thermal energy generation cases.

	Plane wall	Cylinder	Sphere
Governing differential equation	$\dfrac{d^2 T}{dx^2} = -\dfrac{\dot{g}'''}{k}$	$\dot{g}''' \, r = \dfrac{d}{dr}\left(-k\,r\,\dfrac{dT}{dr}\right)$	$\dot{g}''' \, r^2 = \dfrac{d}{dr}\left(-k\,r^2\,\dfrac{dT}{dr}\right)$
Temperature gradient	$\dfrac{dT}{dx} = -\dfrac{\dot{g}'''}{k}x + C_1$	$\dfrac{dT}{dr} = -\dfrac{\dot{g}''' \, r}{2\,k} + \dfrac{C_1}{r}$	$\dfrac{dT}{dr} = -\dfrac{\dot{g}'''}{3\,k}r - \dfrac{C_1}{r^2}$
General solution	$T = -\dfrac{\dot{g}'''}{2\,k}x^2 + C_1 x + C_2$	$T = -\dfrac{\dot{g}''' \, r^2}{4\,k} + C_1 \ln(r) + C_2$	$T = -\dfrac{\dot{g}'''}{6\,k}r^2 + \dfrac{C_1}{r} + C_2$

The governing differential equation and general solutions for these 1-D geometries with a uniform rate of thermal energy generation are summarized in Table 1-3.

EXAMPLE 1.3-1: MAGNETIC ABLATION

Thermal ablation is a technique for treating cancerous tissue by heating it to a lethal temperature. A number of thermal ablation techniques have been suggested in order to apply heat locally to the cancerous tissue and therefore spare surrounding healthy tissue. One interesting technique utilizes ferromagnetic thermoseeds, as discussed by Tompkins (1992). Small metallic spheres (thermoseeds) are embedded at precise locations within the cancer tumor and then the region is exposed to an oscillating magnetic field. The magnetic field does not generate thermal energy in the tissue. However, the spheres experience a volumetric generation of thermal energy that causes their temperature to increase and results in the conduction of heat to the surrounding tissue. Precise placement of the thermoseed can be used to control the application of thermal energy. The concept is shown in Figure 1.

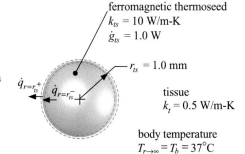

ferromagnetic thermoseed
$k_{ts} = 10$ W/m-K
$\dot{g}_{ts} = 1.0$ W

$r_{ts} = 1.0$ mm

tissue
$k_t = 0.5$ W/m-K

body temperature
$T_{r\to\infty} = T_b = 37°C$

$\dot{q}_{r=r_{ts}^+}$ $\dot{q}_{r=r_{ts}^-}$

Figure 1: A thermoseed used for ablation of a tumor.

It is necessary to determine the temperature field associated with a single thermoseed placed in an infinite medium of tissue. The thermoseed has radius $r_{ts} = 1.0$ mm and conductivity $k_{ts} = 10$ W/m-K. A total of $\dot{g}_{ts} = 1.0$ W of generation is uniformly distributed throughout the sphere. The temperature far from the thermoseed is the body temperature, $T_b = 37°C$. The tissue has thermal conductivity $k_t = 0.5$ W/m-K and is assumed to be in perfect thermal contact with the thermoseed. The effects of metabolic heat generation (i.e., volumetric generation in the tissue) and blood perfusion (i.e., the heat removed by blood flow in the tissue) are not considered in this problem.

EXAMPLE 1.3-1: MAGNETIC ABLATION

a) Prepare a plot showing the temperature in the thermoseed and in the tissue (i.e., the temperature from $r = 0$ to $r \gg r_{ts}$).

This problem is 1-D because the temperature varies only in the radial direction. There are no circumferential non-uniformities that would lead to temperature gradients in any dimension except r. However, the problem includes two, separate computational domains that share a common boundary (i.e., the thermoseed and the tissue). Therefore, there will be two different governing equations that must be solved and additional boundary conditions that must be considered.

It is always good to start your problem with an input section in which all of the given information is entered and, if necessary, converted to SI units.

```
"EXAMPLE 1.3-1: Magnetic Ablation"

$UnitSystem SI MASS RAD PA K J
$Tabstops 0.2 0.4 0.6 3.5 in

"Inputs"
r_ts=1 [mm]*convert(mm,m)              "radius of the thermoseed"
k_ts=10 [W/m-K]                        "thermal conductivity of thermoseed"
g_dot_ts=1.0 [W]                       "total generation of thermal energy in thermoseed"
T_b=converttemp(C,K,37 [C])            "body temperature"
k_t=0.5 [W/m-K]                        "tissue thermal conductivity"
```

Notice a few things about the EES code. First, comments are provided to define the nomenclature and make the code understandable; this type of annotation is important for clarity and organization. Also, units are not ignored but rather explicitly specified and dealt with as the problem is set up, rather than as an afterthought at the end. The unit system that EES will use can be specified in the Properties Dialog (select Preferences from the Options menu) from the Unit System tab or by using the $UnitSystem directive, as shown in the EES code above. The units of numerical constants can be set directly in square brackets following the value. For example, the statement

```
r_ts=1 [mm]*convert(mm,m)                         "radius of the thermoseed"
```

tells EES that the constant 1 has units of mm and these should be converted to units of m. Therefore the variable r_ts will have units of m. The units of r_ts are not automatically set, as the equation involving r_ts is not a simple assignment. If you check units at this point (select Check Units from the Calculate menu) then EES will indicate that there is a unit conversion error. This unit error occurs because the variable r_ts has not been assigned any units but the equations are consistent with r_ts having units of m. It is possible to have EES set units automatically with an option in the Options tab in the Preferences Dialog. However, this is not recommended because the engineer doing the problem should know and set the units for each variable.

The Formatted Equations window (select Formatted Equations from the Windows menu) shows the equations and their units more clearly (Figure 2).

EXAMPLE 1.3-1: MAGNETIC ABLATION

$\mathbf{E_{ES}}$ Formatted Equations

EXAMPLE 1.3-1: Magnetic Ablation

Inputs

r_{ts} = 1 [mm] · $\left| 0.001 \cdot \dfrac{m}{mm} \right|$ radius of the thermoseed

k_{ts} = 10 [W/m-K] thermal conductivity of thermoseed

$\dot{g}_{ts}$ = 1 [W] total generation of thermal energy in thermoseed

T_b = ConvertTemp (C , K , 37 [C]) body temperature

k_t = 0.5 [W/m-K] tissue thermal conductivity

Figure 2: Formatted Equations window.

It is good practice to set the units of all variables. One method of accomplishing this is to right-click on a variable in the Solutions window, which brings up the Format Selected Variables dialog. The units can be typed directly into the Units input box. Note that right-clicking in the Units input box and selecting the Unit List menu item provides a partial list of the SI units that EES recognizes. All of the units that are recognized by EES can be examined by selecting Unit Conversion Info from the Options menu. Once the units of r_ts are set to m, then a unit check (select Check Units from the Calculate menu) should reveal no errors.

It is necessary to work with a different governing equation in each of the two computational domains. An appropriate differential control volume for the spherical geometry with uniform thermal energy generation was presented in Section 1.3.3 and leads to the general solution listed in Table 1-3. The general solution that is valid within the thermoseed (i.e., from $0 < r < r_{ts}$) is:

$$T_{ts} = -\frac{\dot{g}_{ts}'''}{6\,k_{ts}}r^2 + \frac{C_1}{r} + C_2 \tag{1}$$

where C_1 and C_2 are undetermined constants of integration and $\dot{g}_{ts}'''$ is the volumetric rate of generation of thermal energy in the thermoseed, which is the ratio of the total rate of thermal energy generation to the volume of the thermoseed:

$$\dot{g}_{ts}''' = \frac{3\,\dot{g}_{ts}}{4\,\pi\,r_{ts}^3}$$

g‴_dot_ts=3*g_dot_ts/(4*pi*r_ts^3)   "volumetric rate of generation in the thermoseed"

The general solution that is valid within the tissue (i.e., for $r > r_{ts}$) is:

$$T_t = \frac{C_3}{r} + C_4 \tag{2}$$

because the volumetric rate of thermal energy generation in the tissue is zero; note that C_3 and C_4 are undetermined constants of integration that are different from C_1 and C_2 in Eq. (1).

The next step is to define the boundary conditions. There are four undetermined constants and therefore there must be four boundary conditions. At the center of the thermoseed ($r = 0$) the temperature must remain finite. Substituting $r = 0$ into Eq. (1) leads to:

$$T_{ts,r=0} = -\frac{0^2}{6\,k_{ts}}\,\dot{g}'''_{ts} + \frac{C_1}{k_{ts}\,0} + C_2$$

which indicates that C_1 must be zero:

$$C_1 = 0 \tag{3}$$

Alternatively, specifying that the temperature gradient at the center of the sphere is equal to zero leads to the same conclusion, $C_1 = 0$. As the radius approaches infinity, the tissue temperature must approach the body temperature. Substituting $r \to \infty$ into Eq. (2) leads to:

$$T_b = -\frac{C_3}{\infty} + C_4$$

which indicates that:

$$C_4 = T_b \tag{4}$$

The remaining boundary conditions are defined at the interface between the thermoseed and the tissue. It is assumed that the sphere and tissue are in perfect thermal contact (i.e., there is no contact resistance) so that the temperature must be continuous at the interface:

$$T_{ts,r=r_{ts}} = T_{t,r=r_{ts}}$$

or, substituting $r = r_{ts}$ into Eqs. (1) and (2):

$$-\frac{r_{ts}^2}{6\,k_{ts}}\,\dot{g}'''_{ts} + \frac{C_1}{r_{ts}} + C_2 = \frac{C_3}{r_{ts}} + C_4 \tag{5}$$

An energy balance on the interface (see Figure 1) requires that the heat transfer rate at the outer edge of the thermoseed ($\dot{q}_{r=r_{ts}^-}$ in Figure 1) must equal the heat transfer rate at the inner edge of the tissue ($\dot{q}_{r=r_{ts}^+}$ in Figure 1).

$$\dot{q}_{r=r_{ts}^-} = \dot{q}_{r=r_{ts}^+} \tag{6}$$

According to Fourier's law, Eq. (6) can be written as:

$$-4\,\pi\,r_{ts}^2\,k_{ts}\,\frac{dT_{ts}}{dr}\bigg|_{r=r_{ts}} = -4\,\pi\,r_{ts}^2\,k_t\,\frac{dT_t}{dr}\bigg|_{r=r_{ts}}$$

or

$$k_{ts}\,\frac{dT_{ts}}{dr}\bigg|_{r=r_{ts}} = k_t\,\frac{dT_t}{dr}\bigg|_{r=r_{ts}} \tag{7}$$

Substituting Eqs. (1) and (2) into Eq. (7) leads to:

$$-k_{ts}\left(-\frac{r_{ts}}{3k_{ts}}\,\dot{g}'''_{ts} - \frac{C_1}{r_{ts}^2}\right) = -k_t\left(-\frac{C_3}{r_{ts}^2}\right) \tag{8}$$

Entering Eqs. (3), (4), (5), and (8) into EES will lead to the solution for the four constants of integration without the algebra and the associated opportunities for error.

```
"Determine constants of integration"
C_1=0                                      "temperature at center must be finite"
C_4=T_b                                    "temperature far from the thermoseed"
-r_ts^2*g'''_dot_ts/(6*k_ts)+C_1/r_ts+C_2=C_3/r_ts+C_4
                                           "continuity of temperature at the interface"
-k_ts*(-r_ts*g'''_dot_ts/(3*k_ts)-C_1/r_ts^2)=-k_t*(-C_3/r_ts^2)   "equal heat flux at the interface"
```

Finally, we can generate a plot using the solution. Displaying the radius in millimeters in the plot will make a much more reasonable scale than in meters and the temperature should be displayed in °C. New variables, r_mm, T_ts_C, and T_t_C, are defined for this purpose.

```
"Prepare a plot"
r=r_mm*convert(mm,m)                       "radius"
T_ts=-r^2*g'''_dot_ts/(6*k_ts)+C_1/r+C_2   "thermoseed temperature"
T_ts_C=converttemp(K,C,T_ts)               "in C"
T_t=C_3/r+C_4                              "tissue temperature"
T_t_C=converttemp(K,C,T_t)                 "in C"
```

The units of the variables C_1, C_2, etc. should be set in the Variable Information window and the set of equations subsequently checked for unit consistency. The relationship between temperature and radial position will be determined using two parametric tables. The first table will include variables r_mm and T_ts_C and the second will include the variables r_mm and T_t_C. In the first table, r_mm is varied from 0 to 1.0 mm (i.e., within the thermoseed) and in the second it is varied from 1.0 mm to 10.0 mm (i.e., within the tissue). A plot in which the information contained in the two tables is overlaid leads to the temperature distribution shown in Figure 3.

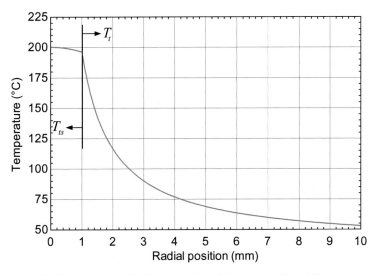

Figure 3: Temperature distribution through thermoseed and tissue.

Figure 3 agrees with physical intuition. The temperature decays toward the body temperature with increasing distance from the thermoseed. The rate of energy being transferred through the tissue is constant, but the area for conduction is growing as r^2 and therefore the gradient in the temperature is dropping. The conduction heat transfer rates at the outer edge of the sphere (i.e., $r = r_{sp}^-$) and at the inner edge of the tissue (i.e., $r = r_{sp}^+$) are identical; the discontinuity in slope is related to the fact that the thermoseed is more conductive than the tissue.

b) Determine the maximum temperature in the tissue and the extent of the lesion as a function of the rate of thermal energy generation in the thermoseed. The extent of the lesion (r_{lesion}) is defined as the radial location where the tissue temperature reaches the lethal temperature for tissue, approximately $T_{lethal} = 50°C$ according to Izzo (2003).

The maximum tissue temperature ($T_{t,max}$) is the temperature at the interface between the thermoseed and the tissue and is obtained by substituting $r = r_{ts}$ into Eq. (2):

$$T_{t,max} = \frac{C_3}{r_{ts}} + C_4$$

```
T_t_max=C_3/r_ts+C_4              "maximum tissue temperature"
T_t_max_C=converttemp(K,C,T_t_max)   "in C"
```

The extent of the lesion can be obtained by determining the radial location where $T_t = T_{lethal}$:

$$T_{lethal} = \frac{C_3}{r_{lesion}} + C_4$$

```
T_lethal=converttemp(C,K, 50 [C])    "lethal temperature for cell death"
T_lethal=C_3/r_lesion+C_4            "determine the extent of the lesion"
r_lesion_mm=r_lesion*convert(m,mm)   "in mm"
```

In order to investigate the maximum tissue temperature and the extent of the lesion as a function of the thermal energy generation rate, it is necessary to prepare a parametric table that includes the variables T_t_max_C, r_lesion, and g_dot_sp and then vary the value of g_dot_sp within the table. Figure 4 illustrates the maximum temperature and the lesion extent as a function of the rate of thermal energy generation in the thermoseed. Note that $T_{t,max}$ and r_{lesion} have very different magnitudes and therefore it is necessary to plot the variable r_lesion on a secondary y-axis.

EXAMPLE 1.3-1: MAGNETIC ABLATION

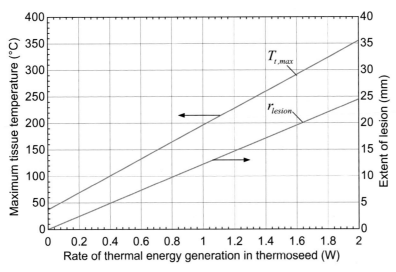

Figure 4: Maximum tissue temperature and the extent of the lesion as a function of the rate of thermal energy generation in the thermoseed.

1.3.4 Spatially Non-Uniform Generation

The first few steps in solving conduction heat transfer problems include setting up an energy balance on a differential control volume and substituting in the appropriate rate equations. This process results in one or more differential equations that must be solved in order to determine the temperature distribution and heat transfer rates. The governing equations resulting from 1-D conduction with constant properties and constant internal generation are provided in Table 1-3 for the Cartesian, cylindrical, and spherical geometries. These equations are relatively simple and we have demonstrated how to solve them analytically by hand and, in some cases, using Maple.

The complexity of the governing equation can increase significantly if the thermal conductivity or internal generation depends on position or temperature. In these cases, an analytical solution to the governing equation may not be possible and the numerical solution techniques presented in Sections 1.4 and 1.5 must be used. In many of these cases, however, an analytical solution is possible but may require more mathematical expertise than you have or more effort than you'd like to expend. In these cases, the combined use of a symbolic software tool to identify the solution and an equation solver to manipulate the solution provides a powerful combination of tools. Analytical solutions are concise and elegant as well as being accurate and therefore preferable in many ways to numerical solutions. It is often best to have both an analytical and a numerical solution to a problem; their agreement constitutes the best possible double-check of a solution.

EXAMPLE 1.3-2 demonstrates the combined use of the symbolic solver Maple (to derive the symbolic solution and boundary condition equations) and the equation solver EES (to carry out the algebra required to obtain the constants of integration and implement the solution).

EXAMPLE 1.3-1: MAGNETIC ABLATION

EXAMPLE 1.3-2: ABSORPTION IN A LENS

EXAMPLE 1.3-2: ABSORPTION IN A LENS

A lens is used to focus the illumination energy (i.e., radiation) that is required to develop the resist in a lithographic manufacturing process, as shown in Figure 1. The lens can be modeled as a plane wall with thickness $L = 1.0$ cm and thermal conductivity $k = 1.5$ W/m-K. The lens is not perfectly transparent but rather absorbs some of the illumination energy that is passed through it. The absorption coefficient of the lens is $\alpha = 0.1$ mm^{-1}. The flux of radiant energy that is incident at the lens surface ($x = 0$) is $\dot{q}''_{rad} = 0.1$ W/cm^2. The top and bottom surfaces of the lens are exposed to air at $T_\infty = 20°C$ and the average heat transfer coefficient on these surfaces is $\bar{h} = 20$ W/m^2-K.

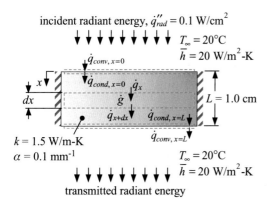

Figure 1: Lens absorbing radiant energy.

The volumetric rate at which absorbed radiation is converted to thermal energy in the lens ($\dot{g}'''$) is proportional to the local intensity of the radiant energy flux, which is reduced in the x-direction by absorption. The result is an exponentially distributed volumetric generation that can be expressed as:

$$\dot{g}''' = \dot{q}''_{rad} \, \alpha \, \exp\left(-\alpha \, x\right) \tag{1}$$

a) Determine and plot the temperature distribution within the lens.

The inputs are entered into EES:

```
"EXAMPLE 1.3-2: Absorption in a lens"

$UnitSystem SI MASS RAD PA K J
$Tabstops 0.2 0.4 0.6 3.5 in

k=1.5 [W/m-K]                                      "conductivity"
L=1 [cm]*convert(cm,m)                             "lens thickness"
alpha=0.1 [1/mm]*convert(1/mm,1/m)                 "absorption coefficient"
qf_dot_rad=0.1 [W/cm^2]*convert(W/cm^2,W/m^2)      "incident energy flux"
h_bar=20 [W/m^2-K]                                 "average heat transfer coefficient"
T_infinity=converttemp(C,K,20 [C])                 "ambient air temperature"
A_c=1 [m^2]             "carry out the problem on a per unit area basis"
```

EXAMPLE 1.3-2: ABSORPTION IN A LENS

An energy balance on an appropriate, differential control volume (see Figure 1) provides:

$$\dot{q}_x + \dot{g} = \dot{q}_{x+dx}$$

which is expanded and simplified:

$$\dot{g} = \frac{d\dot{q}}{dx}dx$$

Substituting the rate equations for $\dot{q}$ and $\dot{g}$ leads to:

$$\dot{g}''' A_c = \frac{d}{dx}\left(-k A_c \frac{dT}{dx}\right) \tag{2}$$

where A_c is the cross-sectional area of the lens. Substituting Eq. (1) into Eq. (2) and simplifying leads to the governing differential equation for this problem.

$$\frac{d}{dx}\left(\frac{dT}{dx}\right) = -\frac{\dot{q}''_{rad}\,\alpha}{k}\,\exp\left(-\alpha\,x\right) \tag{3}$$

The governing differential equation is entered in Maple.

```
> restart;
> GDE:=diff(diff(T(x),x),x)=-qf_dot_rad*alpha*exp(-alpha*x)/k;
```

$$GDE := \frac{d^2}{dx^2}T(x) = -\frac{qf\_dot\_rad\,\alpha\,e^{(-\alpha\,x)}}{k}$$

The general solution to the equation is obtained using the **dsolve** command:

```
> Ts:=dsolve(GDE);
```

$$Ts := T(x) = -\frac{qf\_dot\_rad\,e^{(-\alpha\,x)}}{\alpha\,k} + \_C1\,x + \_C2$$

We can check that this solution is correct by integrating Eq. (3) by hand:

$$\int d\left(\frac{dT}{dx}\right) = -\int \frac{\dot{q}''_{rad}\,\alpha}{k}\,\exp\left(-\alpha\,x\right)dx$$

which leads to:

$$\frac{dT}{dx} = \frac{\dot{q}''_{rad}}{k}\,\exp\left(-\alpha\,x\right) + C_1 \tag{4}$$

Equation (4) is integrated again:

$$\int dT = \int \left[\frac{\dot{q}''_{rad}}{k}\,\exp\left(-\alpha\,x\right) + C_1\right]dx$$

which leads to:

$$T = -\frac{\dot{q}''_{rad}}{k\,\alpha}\,\exp\left(-\alpha\,x\right) + C_1\,x + C_2 \tag{5}$$

EXAMPLE 1.3-2: ABSORPTION IN A LENS

Note that Eq. (5) is consistent with the result from Maple. The constants of integration, C_1 and C_2, are obtained by enforcing the boundary conditions. The boundary conditions for this problem are derived from "interface" energy balances at the two edges of the computational domain ($x = 0$ and $x = L$, as shown in Figure 1). It would be correct to include the radiant energy flux in the interface balances. However, the interface thickness is zero and therefore no radiant energy is absorbed at the interface. The amount of radiant energy entering and leaving the interface is the same and these terms would immediately cancel.

$$\dot{q}_{conv,x=0} = \dot{q}_{cond,x=0}$$

$$\dot{q}_{cond,x=L} = \dot{q}_{conv,x=L}$$

Substituting the rate equations for convection and conduction into the interface energy balances leads to the boundary conditions:

$$\bar{h} A_c (T_\infty - T_{x=0}) = -k A_c \left. \frac{dT}{dx} \right|_{x=0} \tag{6}$$

$$-k A_c \left. \frac{dT}{dx} \right|_{x=L} = \bar{h} A_c (T_{x=L} - T_\infty) \tag{7}$$

Note that it is important to consider the direction of the energy transfers during the substitution of the rate equations. For example, $\dot{q}_{conv,x=0}$ is defined in Figure 1 as being *into* the top surface of the lens and therefore it is driven by $(T_\infty - T_{x=0})$ while $\dot{q}_{conv,x=L}$ is defined as being *out of* the bottom surface of the lens and therefore it is driven by $(T_{x=L} - T_\infty)$. The general solution for the temperature distribution, Eq. (5), must be substituted into the boundary conditions, Eqs. (6) and (7), and solved algebraically to determine the constants C_1 and C_2. Maple and EES can be used together in order to solve the differential equation, derive the symbolic expressions for the boundary conditions, carry out the required algebra to obtain C_1 and C_2, and manipulate the solution.

The temperature gradient is obtained symbolically from Maple using the diff command.

```
> dTdx:=diff(Ts,x);
```

$$dTdx := \frac{d}{dx} T(x) = -\frac{qf\_dot\_rad\, e^{(-\alpha x)}}{k} + \_C1$$

The first boundary condition, Eq. (6), requires both the temperature and the temperature gradient evaluated at $x = 0$. The eval function in Maple is used to symbolically determine these quantities (T0 and dTdx0):

```
> T0:=eval(Ts,x=0);
```

$$T0 := T(0) = -\frac{qf\_dot\_rad}{\alpha k} + \_C2$$

EXAMPLE 1.3-2: ABSORPTION IN A LENS

Use the **rhs** function to redefine T0 to be just the expression on the right-hand side of the above result.

```
> T0:=rhs(T0);
```

$$T0 := -\frac{qf\_dot\_rad}{\alpha k} + \_C2$$

The **rhs** function can also be applied directly to the **eval** function. The statement below determines the symbolic expression for the temperature gradient evaluated at $x = 0$.

```
> dTdx0:=rhs(eval(dTdx,x=0));
```

$$dTdx0 := -\frac{qf\_dot\_rad}{k} + \_C1$$

These expressions for T0 and dTdx0 are copied and pasted into EES in order to specify the first boundary condition. The equation format used by Maple is similar to that used in EES and therefore only minor modifications are required. Select the desired symbolic expressions in Maple (note that the selected text will appear to be highlighted), use the Copy command from the Edit menu to place the selection on the Clipboard. When pasted into EES, the equations will appear as:

```
"boundary condition at x=0"
T0 := -1/alpha*qf_dot_rad/k+_C2        "temperature at x=0, copied from Maple"
dTdx0 := qf_dot_rad/k+_C1              "temperature gradient at x=0, copied from Maple"
```

All that is necessary to use this equation in EES is to change the := to = and change the constants _C1 and _C2 to C_1 and C_2, respectively. For lengthier expressions, the search and replace feature in EES (select Replace from the Search manu) facilitates this process. After modification, the expressions should be:

```
T0= -1/alpha*qf_dot_rad/k+C_2          "temperature at x=0"
dTdx0= qf_dot_rad/k+C_1                "temperature gradient at x=0"
```

The boundary condition at $x = 0$, Eq. (6), is specified in EES:

```
h_bar*A_c*(T_infinity-T0)=-k*A_c*dTdx0      "boundary condition at x=0"
```

EXAMPLE 1.3-2: ABSORPTION IN A LENS

The same process is used for the boundary condition at $x = L$, Eq. (7). The temperature and temperature gradient at $x = L$ are determined using Maple:

```
> TL:=rhs(eval(Ts,x=L));
```

$$TL := -\frac{qf\_dot\_rad\, e^{(-\alpha\, L)}}{\alpha\, k} + \_C1\, L + \_C2$$

```
> dTdxL:=rhs(eval(dTdx,x=L));
```

$$dTdxL := -\frac{qf\_dot\_rad\, e^{(-\alpha\, L)}}{k} + \_C1$$

These expressions are copied into EES:

```
"boundary condition at x=L"
TL := -1/alpha*qf_dot_rad*exp(-alpha*L)/k+_C1*L+_C2   "temperature at x=L, copied from Maple"
dTdxL := qf_dot_rad*exp(-alpha*L)/k+_C1               "temperature gradient at x=L, copied from Maple"
```

and modified for consistency with EES:

```
TL= -1/alpha*qf_dot_rad*exp(-alpha*L)/k+C_1*L+C_2   "temperature at x=L, copied from Maple"
dTdxL= qf_dot_rad*exp(-alpha*L)/k+C_1               "temperature gradient at x=L, copied from Maple"
```

The boundary condition at $x = L$, Eq. (7), is specified in EES:

```
-k*A_c*dTdxL=h_bar*A_c*(TL-T_infinity)          "boundary condition at x=L"
```

Solving the EES program will provide numerical values for both of the constants. The units should be set for each of the variables, including the constants, in order to ensure that the expressions are dimensionally consistent.

The general solution is copied from Maple to EES:

```
T(x) = -1/alpha*qf_dot_rad*exp(-alpha*x)/k+_C1*x+_C2   "general solution, copied from Maple"
```

and modified for consistency with EES:

```
x_mm=0 [mm]                                        "x-position, in mm"
x=x_mm*convert(mm,m)                               "x_position"
T=-1/alpha*qf_dot_rad*exp(-alpha*x)/k+C_1*x+C_2    "general solution for temperature"
T_C=converttemp(K,C,T)                             "in C"
```

The temperature as a function of position is shown in Figure 2. Notice the asymmetry that is produced by the non-uniform volumetric generation (i.e.,

EXAMPLE 1.3-2: ABSORPTION IN A LENS

more thermal energy is generated towards the top of the lens than the bottom and so the temperature is higher near the top surface of the lens).

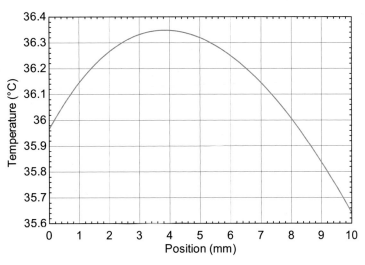

Figure 2: Temperature distribution in the lens.

b) Determine the location of the maximum temperature (x_{max}) and the value of the maximum temperature (T_{max}) in the lens.

The location of the maximum temperature in the lens can be determined by setting the temperature gradient, Eq. (4), to zero. The value of the maximum temperature (T_{max}) is obtained by substituting the resulting value of x_{max} into the equation for temperature, Eq. (5). The expression for the temperature gradient is copied from Maple, pasted into EES and modified for consistency.

```
qf_dot_rad*exp(-alpha*x_max)/k+C_1=0        "temperature gradient is zero at position x_max"
x_max_mm=x_max*convert(m,mm)                "in mm"
```

The temperature at x_{max} is determined using the general solution.

```
T_max=-1/alpha*qf_dot_rad*exp(-alpha*x_max)/k+C_1*x_max+C_2
                                            "maximum temperature in the lens"
T_max_C=converttemp(K,C,T_max)              "in C"
```

It is not likely that solving the code above will immediately result in the correct value of either x_{max} or T_{max}. These are non-linear equations and EES must iterate to find a solution. The success of this process depends in large part on the initial guesses and bounds used for the unknown variables. In most cases, any reasonable values of the guess and bounds will work. For this problem, set the lower and upper bounds on x_{max} to 0 (the top of the lens) and 0.01 m (1.0 cm, the bottom of the lens), respectively, and the guess for x_{max} to 0.005 m (the middle of the lens) using the

EXAMPLE 1.3-2: ABSORPTION IN A LENS

Variable Information dialog (Figure 3). Upon solving, EES will correctly predict that $x_{max} = 0.38$ cm and $T_{max} = 36.35°C$.

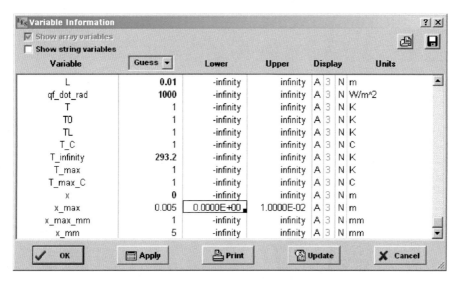

Figure 3: Variable Information window showing the limits and guess value for the variable x_max.

It is also possible to determine the maximum temperature within the lens using EES' built-in optimization routines. There are several sophisticated single- and multi-variable optimization algorithms included with EES that can be accessed with the Min/Max command in the Calculate menu.

1.4 Numerical Solutions to Steady-State 1-D Conduction Problems (EES)

1.4.1 Introduction

The analytical solutions examined in the previous sections are convenient since they provide accurate results for arbitrary inputs with minimal computational effort. However, many problems of practical interest are too complicated to allow an analytical solution. In such cases, numerical solutions are required. Analytical solutions remain useful as a way to test the validity of numerical solutions under limiting conditions.

Numerical solutions are generally more computationally complex and are not unconditionally accurate. Numerical solutions are only approximations to a real solution, albeit approximations that are extremely accurate when done correctly. It is relatively straightforward to solve even complicated problems using numerical techniques.

The steps required to set up a numerical solution using the finite difference approach remain the same even as the problems become more complex. The result of a numerical model is not a functional relationship between temperature and position but rather a prediction of the temperatures at many discrete positions. The first step is to define small control volumes that are distributed through the computational domain and to specify the locations where the numerical model will compute the temperatures (i.e., the nodes). The control volumes used in the numerical model are small but finite, as opposed to the infinitesimally small (differential) control volume that is defined in order to derive an analytical solution. It is necessary to perform an energy balance on each control volume. This requirement may seem daunting given that many control volumes will be required to provide an accurate solution. However, computers are very good at repetitive calculations. If your numerical code is designed in a systematic manner, then these operations can be done quickly for 1000's of control volumes.

Once the energy balance equations for each control volume have been set up, it is necessary to include rate equations that approximate each term in the energy balance based upon the nodal temperatures or other input parameters. The result of this step will be a set of algebraic equations (one for each control volume) in an equal number of unknown temperatures (one for each node). This set of equations can be solved in order to provide the numerical prediction of the temperature at each node.

It is tempting to declare victory after successfully solving the finite difference equations and obtaining a set of results that looks reasonable. In reality, your work is only half done and there are several important steps remaining. First, it is necessary to verify that you have chosen a sufficiently large number of control volumes (i.e., an adequately refined mesh) so that your numerical solution has converged to a solution that no longer depends on the number of nodes. This verification can be accomplished by examining some aspect of your solution (e.g., a temperature or an energy transfer rate that is particularly important) as the number of control volumes increases. You should observe that your solution stops changing (to within engineering relevance) as the number of control volumes (and therefore the computational effort) increases. Some engineering judgment is required for this step. You have to decide what aspect of the solution is most important and how accurately it must be predicted in order to determine the level of grid refinement that is required.

Next, you should make sure that the solution makes sense. There are a number of 'sanity checks' that can be applied to verify that the numerical model is behaving according to physical expectations. For example, if you change the input parameters, does the solution respond as you would expect?

Finally, it is important that you verify the numerical solution against an analytical solution in some appropriate limit. This step may be the most difficult one, but it provides the strongest possible verification. If the numerical model is to be used to make decisions that are important (e.g., to your company's bottom line or to your career) then you should strive to find a limit where an analytical solution can be derived and show that your numerical model matches the analytical solution to within numerical error.

The numerical solutions considered in this section are for steady-state, 1-D problems. More complex problems (e.g., multi-dimensional and transient) are discussed in subsequent sections and chapters. This section also focuses on solving these problems using EES whereas subsequent sections discuss how these solutions may be implemented using a more formal programming language, specifically MATLAB.

1.4.2 Numerical Solutions in EES

The development of a numerical model is best discussed in the context of a problem. Figure 1-10 illustrates an aluminum oxide cylinder that is exposed to fluid at its internal and external surfaces. The temperature of the fluid exposed to the internal surface is $T_{\infty,in} = 20°C$ and the average heat transfer coefficient on the internal surface is $\bar{h}_{in} = 100 \text{ W/m}^2\text{-K}$. The temperature of the fluid exposed to the outer surface is $T_{\infty,out} = 100°C$ and the average heat transfer coefficient on the outer surface is $\bar{h}_{out} = 200 \text{ W/m}^2\text{-K}$. The thermal conductivity of the aluminum oxide is $k = 9.0 \text{ W/m-K}$ in the temperature range of interest. The rate of thermal energy generation per unit

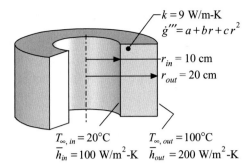

Figure 1-10: Cylinder with volumetric gene-ration.

volume within the cylinder varies with radius according to:

$$\dot{g}''' = a + br + cr^2 \qquad (1\text{-}103)$$

where $a = 1 \times 10^4$ W/m^3, $b = 2 \times 10^5$ W/m^4, and $c = 5 \times 10^7$ W/m^5. The inner and outer radii of the cylinder are $r_{in} = 10$ cm and $r_{out} = 20$ cm, respectively.

The inputs are entered in EES:

```
"Section 1.4.2"
$UnitSystem SI MASS RAD PA K J
$Tabstops 0.2 0.4 0.6 3.5 in

"Inputs"
r_in=10 [cm]*convert(cm,m)                    "inner radius of cylinder"
r_out=20 [cm]*convert(cm,m)                   "outer radius of cylinder"
L=1 [m]                                       "unit length of cylinder"
k=9.0 [W/m-K]                                 "thermal conductivity"
T_infinity_in=converttemp(C,K,20 [C])         "temperature of fluid inside cylinder"
h_bar_in= 100 [W/m^2-K]                       "average heat transfer coefficient at inner surface"
T_infinity_out=converttemp(C,K,100 [C])       "temperature of fluid outside cylinder"
h_bar_out=200 [W/m^2-K]                        "average heat transfer coefficient at outer surface"
```

It is convenient to define a function that provides the volumetric rate of thermal energy generation specified by Eq. (1-103). An EES function is a self-contained code segment that is provided with one or more input parameters and returns a single value based on these parameters. Functions in EES must be placed at the top of the Equations Window, before the main body of equations. The EES code required to specify a function for the rate of thermal energy generation per unit volume is shown below.

```
function gen(r)
   "This function defines the volumetric heat generation in the cylinder
   Inputs: r, radius (m)
   Output: volumetric heat generation at r (W/m^3)"
   a=1e4 [W/m^3]                              "coefficients for generation function"
   b=2e5 [W/m^4]
   c=5e7 [W/m^5]
   gen=a+b*r+c*r^2                            "generation is a quadratic"
end
```

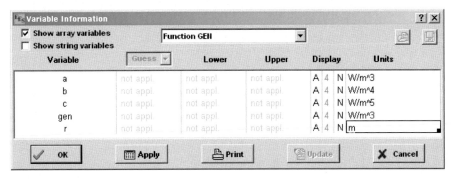

Figure 1-11: Variable Information Window showing the variables for the function GEN.

The function begins with a header that defines the name of the function (gen) and the input arguments (r) and is terminated by the statement end. None of the variables in the main EES program are accessible within the function other than those that are explicitly passed to the function as an input. Unlike equations in the main program, the equations within a function are executed in the order that they are entered and all variables on the right hand side of each expression must be defined (i.e., the equations are assignments rather than relationships). The statements within the function are used to define the value of the function (gen, in this case). Units for the variables in the function should be set using the Variable Information dialog in the same way that units are set for variables in the main program. The most direct way to enter the units for the function is to select Variable Info from the Options menu. Then select Function GEN from the pull down menu at the top of the Variable Information dialog and enter the units for the variables (Figure 1-11).

The finite difference technique divides the continuous medium into a large (but not infinite) number of small control volumes that are treated using simple approximations. The computational domain in this problem lies between the edges of the cylinder ($r_{in} < r < r_{out}$). The first step in the solution process is to locate nodes (i.e., the positions where the temperature will be predicted) throughout the computational domain. The easiest way to distribute the nodes is uniformly, as shown in Figure 1-12 (only nodes $1, 2, i - 1, i,$ $i + 1, N - 1$, and N are shown). The extreme nodes (i.e., nodes 1 and N) are placed on the surfaces of the cylinder.

In some problems, it may not be computationally efficient to distribute the nodes uniformly. For example, if there are large temperature gradients at some location then it may be necessary to concentrate nodes in that region. Placing closely-spaced nodes throughout the entire computational domain may be prohibitive from a

Figure 1-12: Nodes and control volumes for the numerical model.

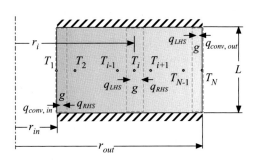

computational viewpoint. For the uniform distribution shown in Figure 1-12, the radial location of each node (r_i) is:

$$r_i = r_{in} + \frac{(i-1)}{(N-1)} (r_{out} - r_{in}) \quad \text{for } i = 1..N \quad (1\text{-}104)$$

where N is the number of nodes. The radial distance between adjacent nodes (Δr) is:

$$\Delta r = \frac{(r_{out} - r_{in})}{(N-1)} \quad (1\text{-}105)$$

It is necessary to specify the number of nodes used in the numerical solution. We will start with a small number of nodes, $N = 6$, and increase the number of nodes when the solution is complete.

```
N=6 [-]                                    "number of nodes"
DELTAr=(r_out-r_in)/(N-1)                  "distance between adjacent nodes (m)"
```

The location of each node will be placed in an array, i.e., a variable that contains more than one element (rather than a scalar, as we've used previously). EES recognizes a variable name to be an element of an array if it ends with square brackets surrounding an array index, e.g., r[4]. Array variables are just like any other variable in EES and they can be assigned to values, e.g., r[4]=0.16. Therefore, one way of setting up the position array would be to individually assign each value; r[1]=0.1, r[2]=0.12, r[3]=0.14, etc. This process is tedious, particularly if there are a large number of nodes. It is more convenient to use the duplicate command which literally duplicates the equations in its domain, allowing for varying array index values. The duplicate command must be followed by an integer index, in this case i, that passes through a range of values, in this case 1 to 6. The EES code shown below will copy (duplicate) the statement(s) that are located between duplicate and end N times; each time, the value of i is incremented by 1. It is exactly like writing:

```
r[1]=r_in+(r_out-r_in)*0/(N-1)
r[2]=r_in+(r_out-r_in)*1/(N-1)
r[3]=r_in+(r_out-r_in)*2/(N-1)
etc.
```

```
"Set up nodes"
duplicate i=1,N                            "this loop assigns the radial location to each node"
   r[i]=r_in+(r_out-r_in)*(i-1)/(N-1)
end
```

Be careful not to put statements that you do not want to be duplicated between the duplicate and end statements. For example, if you accidentally placed the statement N=6 inside the duplicate loop it would be like writing N=6 six times, which corresponds to six equations in a single unknown and is not solvable.

Units should be assigned to arrays in the same manner as for other variables. The units of all variables in array r can be set within the Variable Information dialog. Note that each element of the array r appears in the dialog and that the units of each element can be set one by one. In most arrays, each element will have the same units and therefore it is not convenient to set the units an element at a time. Instead, deselect the Show array variables box in the upper left corner of the window, as shown in

De-select the Show array variables button
in order to collapse the arrays

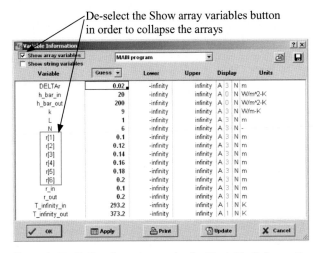

Figure 1-13: Collapsing an array in the Variable Information dialog.

Figure 1-13. The array is collapsed onto a single entry r[] and you can assign units to all of the elements in the array r[].

A control volume is defined around each node. You have some freedom relative to the definition of a control volume, but the best control volume for this problem is defined by bisecting the distance between the nodes, as shown in Figure 1-12.

The second step in the numerical solution is to write an energy balance for the control volume associated with every node. The internal nodes (i.e., nodes 2 through $N-1$) must be considered separately from the nodes at the edge of the computational domain (i.e., nodes 1 and N). The control volume for an arbitrary, internal node (node i) is shown in Figure 1-12. There are three energy terms associated with each control volume: conduction heat transfer passing through the surface on the left-hand side ($\dot{q}_{LHS}$), conduction heat transfer passing through the surface on the right-hand side ($\dot{q}_{RHS}$), and generation of thermal energy within the control volume ($\dot{g}$). A steady-state energy balance for the internal control volume is:

$$\dot{q}_{LHS} + \dot{q}_{RHS} + \dot{g} = 0 \qquad (1\text{-}106)$$

Note that Eq. (1-106) is rigorously correct since no approximations have been used in its development. In the next step, however, each of the terms in the energy balance are modeled using a rate equation that is only approximately valid; it is this step that makes the numerical solution only an approximation of the actual solution. Conduction through the left-hand surface is driven by the temperature difference between nodes $i-1$ and i through the material that lies between these nodes. If there are a large number of nodes, then Δr is small and the effect of curvature within any control volume is small. In this case, $\dot{q}_{LHS}$ can be modeled using the resistance of a plane wall, given in Table 1-2:

$$\dot{q}_{LHS} = \frac{k\,L\,2\pi}{\Delta r}\left(r_i - \frac{\Delta r}{2}\right)(T_{i-1} - T_i) \qquad (1\text{-}107)$$

where L is the length of the cylinder and k is the thermal conductivity of the material, which is assumed to be constant here. It is relatively easy to consider a material with temperature- or spatially dependent thermal conductivity, as discussed in Section 1.4.3. Note that it does not matter which direction the heat flow arrow associated with $\dot{q}_{LHS}$ is drawn in Figure 1-12; that is, the heat transfer could have been defined either as an input or an output to the control volume. However, once the direction is defined, it is absolutely necessary that the energy balance on the control volume and the model of the heat transfer rate be consistent with this selection. For the energy balance shown in Figure 1-12, the heat transfer rate was written on the inflow side of the energy balance in Eq. (1-106) and the heat transfer was written as being driven by $(T_{i-1} - T_i)$ in Eq. (1-107).

The rate of conduction into the right-hand surface can be approximated in the same manner:

$$\dot{q}_{RHS} = \frac{kL2\pi}{\Delta r}\left(r_i + \frac{\Delta r}{2}\right)(T_{i+1} - T_i) \tag{1-108}$$

The rate of generation of thermal energy is the product of the volume of the control volume and the rate of thermal energy generation per unit volume, which is approximately:

$$\dot{g} = \dot{g}'''_{r=r_i}\, 2\pi r_i L\, \Delta r \tag{1-109}$$

The generation term is spatially dependent in this problem, but it is approximated by assuming that the volumetric generation evaluated at the position of the node (r_i) can be applied throughout the entire control volume. This approximation improves as Δr is reduced. Equations (1-106) through (1-109) can be conveniently written for each internal node using a duplicate loop:

```
"Internal control volume energy balances"
duplicate i=2,(N-1)
   q_dot_LHS[i]=2*pi*L*k*(r[i]-DELTAr/2)*(T[i-1]-T[i])/DELTAr   "conduction in from inner radius"
   q_dot_RHS[i]=2*pi*L*k*(r[i]+DELTAr/2)*(T[i+1]-T[i])/DELTAr   "conduction in from outer radius"
   g_dot[i]=gen(r[i])*2*pi*r[i]*L*DELTAr                         "generation"
   q_dot_LHS[i]+g_dot[i]+q_dot_RHS[i]=0                          "energy balance"
end
```

Attempting to solve the EES program at this point will result in a message indicating that there are two more variables than equations and so the problem is under-specified. We have not yet written energy balance equations for the two nodes that lie on the boundaries.

Figure 1-12 illustrates the control volume associated with the node that is placed on the inner surface of the cylinder (i.e., node 1). The energy balance for the control volume associated with node 1 includes a conduction term $(\dot{q}_{RHS})$, a generation term $(\dot{g})$, and a convection term $(\dot{q}_{conv,in})$. For steady-state conditions and energy terms having the signs indicated in Figure 1-12:

$$\dot{q}_{conv,in} + \dot{q}_{RHS} + \dot{g} = 0 \tag{1-110}$$

The conduction term model is the same as it was for internal nodes:

$$\dot{q}_{RHS} = \frac{kL2\pi}{\Delta r}\left(r_1 + \frac{\Delta r}{2}\right)(T_2 - T_1) \tag{1-111}$$

Even though the control volume for node 1 is half as wide as the others, the distance between nodes 1 and 2 is still Δr and therefore the resistance to conduction between nodes 1 and 2 does not change.

The generation term is slightly different because the control volume is half as large as the internal control volumes.

$$\dot{g} = \dot{g}'''_{r=r_1}\, 2\pi r_1 L\, \frac{\Delta r}{2} \tag{1-112}$$

Convection from the fluid is given by:

$$\dot{q}_{conv,in} = \overline{h}_{in} \, 2\pi \, r_1 \, L \, (T_{\infty,in} - T_1) \tag{1-113}$$

Equations (1-110) through (1-113) are programmed in EES:

```
"Energy balance for node on internal surface"
q_dot_RHS[1]=2*pi*L*(r[1]+DELTAr/2)*k*(T[2]-T[1])/DELTAr   "conduction in from outer radius"
q_dot_conv_in=2*pi*L*r_in*h_bar_in*(T_infinity_in-T[1])    "convection from internal fluid"
g_dot[1]=gen(r[1])*2*pi*r[1]*L*DELTAr/2                    "generation"
q_dot_RHS[1]+q_dot_conv_in+g_dot[1]=0                      "energy balance for node 1"
```

A similar procedure applied to the control volume associated with node N (see Figure 1-12) leads to:

$$\dot{q}_{conv,out} + \dot{q}_{LHS} + \dot{g} = 0 \tag{1-114}$$

where

$$\dot{q}_{conv,out} = \overline{h}_{out} \, 2\pi \, r_N \, L \, (T_{\infty,out} - T_N) \tag{1-115}$$

$$\dot{q}_{LHS} = \frac{k \, L \, 2\pi}{\Delta r} \left(r_N - \frac{\Delta r}{2} \right) (T_{N-1} - T_N) \tag{1-116}$$

$$\dot{g} = \dot{g}'''_{r=r_N} \, 2\pi \, r_N \, L \, \frac{\Delta r}{2} \tag{1-117}$$

```
"Energy balance for node on external surface"
q_dot_LHS[N]=2*pi*L*(r[N]-DELTAr/2)*k*(T[N-1]-T[N])/DELTAr
                                                          "conduction in from from inner radius"
q_dot_conv_out=2*pi*L*r_out*h_bar_out*(T_infinity_out-T[N])  "convection from external fluid"
g_dot[N]=gen(r[N])*2*pi*r[N]*L*DELTAr/2                   "generation"
q_dot_LHS[N]+q_dot_conv_out+g_dot[N]=0                    "energy balance for node N"
```

There are now an equal number of equations as unknowns and therefore solving the problem will provide the temperature at each node. The calculated temperatures are converted to °C:

```
duplicate i=1,N
   T_C[i]=converttemp(K,C,T[i])      "temperature in C"
end
```

Computers are very good at solving large systems of equations, particularly with linear equations such as these. There are a number of programs other than EES that can be used to solve these equations (e.g., MATLAB, FORTRAN, and C++). With most of these software packages, you must take the system of equations, carefully put them into a matrix format, and then decompose the matrix in order to obtain the solution. This additional step is not necessary with EES, which saves considerable effort for the user (although EES must do this step internally). In Section 1.5, we will look at how these equations have to be set up in a formal programming environment, specifically MAT-LAB, in order to obtain a solution.

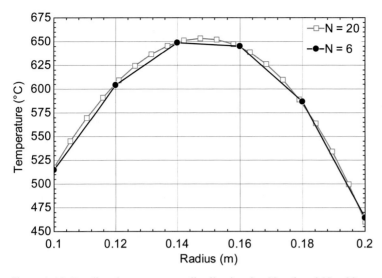

Figure 1-14: Predicted temperature distribution for $N = 6$ and $N = 20$.

It is good practice to assign the units for all variables including the arrays before attempting to solve the equations. The unit consistency of each equation is checked when the equations are solved. The solution is provided in the Solution Window for the scalar quantities, and the Arrays Window for the array of predicted temperatures. A plot showing the predicted temperature as a function of radius is shown in Figure 1-14 for $N = 6$ and $N = 20$.

Note that EES calculates the individual energy transfer rates for each of the control volumes. To see these energy transfer rates, select Arrays from the Windows menu in order to view the Arrays table (Figure 1-15). It is useful to examine these energy transfer rates and make sure that they agree with your intuition. For example, energy should not be created or destroyed at the interfaces between the control volumes; that is, $\dot{q}_{LHS}$ for node i should be equal and opposite $\dot{q}_{RHS}$ for node $i - 1$. It is easy to inadvertently use incorrect rate equations for the conduction terms where this is not true, and you can quickly identify this type of problem using the Arrays window. Figure 1-15 shows that the rate of energy transfer by conduction in the positive radial direction (i.e., $\dot{q}_{LHS}$) becomes more positive with increasing radius, as it should due to the volumetric generation.

Before the numerical solution can be used with confidence, it is necessary to verify its accuracy. The first step in this process is to ensure that the mesh is adequately refined. Figure 1-14 shows that the solution becomes smoother and represents the actual

Sort	r_i [m]	T_i [K]	$T_{C,i}$ [C]	$\dot{g}_i$ [W]	$\dot{q}_{LHSi}$ [W]	$\dot{q}_{RHSi}$ [W]
[1]	0.1	366.4	93.27	62.83		857.9
[2]	0.12	369.2	96.03	150.8	-857.9	707.1
[3]	0.14	371.1	97.95	175.9	-707.1	531.2
[4]	0.16	372.4	99.21	201.1	-531.2	330.1
[5]	0.18	373	99.89	226.2	-330.1	103.9
[6]	0.2	373.2	100.1	125.7	-103.9	

Figure 1-15: Arrays table.

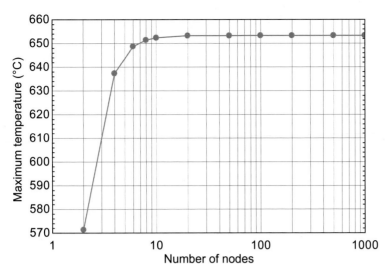

Figure 1-16: Predicted maximum temperature as a function of the number of nodes.

temperature distribution better as the number of nodes is increased. The general approach to choosing a mesh is to pick an important characteristic of the solution and examine how this characteristic changes as the number of nodes in the computational domain is increased. In this case, an appropriate characteristic is the maximum predicted temperature within the cylinder. The following EES code extracts this value from the solution using the max function, which returns the maximum of the arguments provided to it.

```
T_max_C=max(T_C[1..N])          "max temperature in the cylinder, in C"
```

The maximum temperature as a function of the number of nodes is shown in Figure 1-16 (Figure 1-16 was created by making a parametric table that includes the variables N and T_max). Notice that the solution has converged after approximately 20 nodes and further refinement is not likely to be necessary.

The next step is to check that the solution agrees with physical intuition. For example, if the heat transfer coefficient on the internal surface is reduced, then the temperatures within the cylinder should increase. Figure 1-17 illustrates the temperature as a function of radius for various values of $\bar{h}_{in}$ (with $N = 20$) and shows that reducing the heat transfer coefficient does tend to increase the temperature in the cylinder.

There are many additional 'sanity checks' that could be tested. For example, decrease the thermal conductivity or increase the generation rate and make sure that the temperatures in the computational domain increase as they should.

Finally, it is important that the numerical solution be verified against an analytical solution in an appropriate limit. In this case, it is not easy (although it is possible) to develop an analytical solution to the problem with the spatially varying volumetric generation. However, it is relatively straightforward to develop an analytical solution for the problem in the limit of a constant volumetric generation rate. It is also very easy to adapt the numerical model to this limiting case so that the analytical and numerical solutions can be compared. The variables b and c in the generation function, Eq. (1-103),

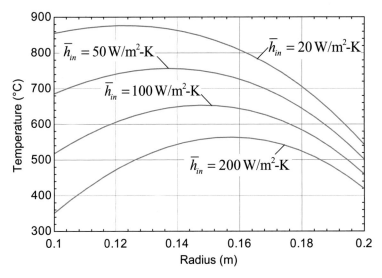

Figure 1-17: Temperature as a function of radius for various values of the heat transfer coefficient.

are temporarily set to 0 in order to implement the numerical solution using a spatially uniform volumetric generation:

```
function gen(r)
    "This function defines the volumetric heat generation in the cylinder
    Inputs: r: radius
    "Output: volumetric heat generation at r (W/m^3)"
    a=1e4 [W/m^3]                        "coefficients for generation function"
    b=0{2e5} [W/m^4]
    c=0{5e7} [W/m^5]
    gen=a+b*r+c*r^2                      "generation is a quadratic"
end
```

The analytical solution is solved using the technique presented in Section 1.3. Table 1-3 provides the temperature distribution and temperature gradient in a cylinder with constant generation, to within the constants of integration:

$$T = -\frac{\dot{g}''' r^2}{4k} + C_1 \ln(r) + C_2 \tag{1-118}$$

$$\frac{dT}{dr} = -\frac{\dot{g}''' r}{2k} + \frac{C_1}{r} \tag{1-119}$$

The constants of integration may be determined using the boundary conditions at the inner and outer surfaces, which are obtained from energy balances at these interfaces. At the inner surface, convection and conduction must balance:

$$\bar{h}_{in} \left(T_{\infty,in} - T_{r=r_{in}} \right) = -k \left. \frac{dT}{dr} \right|_{r=r_{in}} \tag{1-120}$$

Substituting Eqs. (1-118) and (1-119) into Eq. (1-120) leads to:

$$\bar{h}_{in} \left[T_{\infty,in} - \left(-\frac{\dot{g}''' r_{in}^2}{4k} + C_1 \ln(r_{in}) + C_2 \right) \right] = -k \left(-\frac{\dot{g}''' r_{in}}{2k} + \frac{C_1}{r_{in}} \right) \tag{1-121}$$

At the outer surface, the interface energy balance leads to:

$$-k \left. \frac{dT}{dr} \right|_{r=r_{out}} = \overline{h}_{out} \, (T_{r=r_{out}} - T_{\infty,out}) \tag{1-122}$$

Substituting Eqs. (1-118) and (1-119) into Eq. (1-122) leads to:

$$-k \left(-\frac{\dot{g}''' \, r_{out}}{2 \, k} + \frac{C_1}{r_{out}} \right) = \overline{h}_{out} \left[-\frac{\dot{g}''' \, r_{out}^2}{4 \, k} + C_1 \ln (r_{out}) + C_2 - T_{\infty,out} \right] \tag{1-123}$$

The constants of integration are determined by using EES to solve Eqs. (1-121) and (1-123):

```
"Analytical Solution"
g_dot_c=gen(r_in)                          "constant volumetric generation rate for verification"
h_bar_in*(T_infinity_in-(-g_dot_c*r_in^2/(4*k)+C_1*ln(r_in)C_2))=-k*(-g_dot_c*r_in/(2*k)+C_1/r_in)
                                           "boundary condition at inner surface"
-k*(-g_dot_c*r_out/(2*k)+C_1/r_out)=h_bar_out*(-g_dot_c*r_out^2/(4*k)+&
    C_1*ln(r_out)+C_2-T_infinity_out)      "boundary condition at outer surface"
```

The ampersand character appearing in the equation above is a line break character that is only needed for formatting (the equation that is terminated with the ampersand continues on the following line). The analytical solution is evaluated at the locations of the nodes in the numerical solution. The absolute error between the analytical and numerical solution is calculated at each position and the maximum error over the computational domain is computed using the max command.

```
duplicate i=1,N
    T_an[i]=-g_dot_c*r[i]^2/(4*k)+C_1*ln(r[i])+C_2       "analytical temperature at node i"
    T_an_C[i]=converttemp(K,C,T_an[i])                   "in C"
    err[i]=abs(T_an[i]-T[i])
    "absolute value of the discrepancy between numerical and analytical temperature"
end
err_max=max(err[1..N])                                   "maximum error"
```

Figure 1-18 illustrates the temperature distribution predicted by the analytical and numerical models in the limit where the volumetric generation is constant (i.e., the coefficients b and c in the volumetric generation function are set equal to zero). The agreement is nearly exact for 20 nodes, confirming that the numerical model is valid.

Figure 1-19 illustrates the maximum value of the error between the analytical and numerical models as a function of the number of nodes in the solution. Note that the constants b and c in the generation function were set to zero in order to provide a uniform generation. Figure 1-19 provides a more precise method of selecting the number of nodes. A required level of accuracy (i.e., agreement with the analytical model) can be used to specify a required number of nodes. For example, if the problem requires temperature estimates that are accurate to within 0.01 K, then you should use at least 7 nodes. However, predictions accurate to within 0.1 mK will require more than 60 nodes.

1.4.3 Temperature-Dependent Thermal Conductivity

The thermal conductivity of most materials is a function of temperature, although it is often approximated to be constant. This approximation is appropriate for situations in

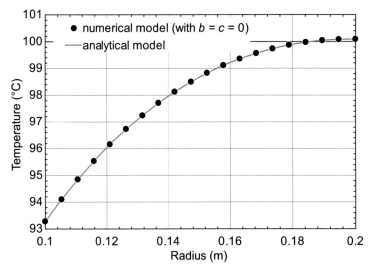

Figure 1-18: Temperature as a function of position predicted by the analytical and numerical models in the limit of constant volumetric generation.

which the temperature range is small provided that the thermal conductivity is evaluated at an average temperature. A constant value of thermal conductivity is usually assumed when solving a problem analytically; otherwise the differential equation is intractable. A major advantage of a numerical solution is that the temperature dependence of physical properties can be considered with little additional effort.

The consideration of temperature-dependent thermal conductivity in a numerical model is demonstrated in the context of the problem that was considered previously in Section 1.4.2. The thermal conductivity of the aluminum oxide cylinder was assumed to be 9 W/m-K, independent of temperature. However, the temperature of the aluminum

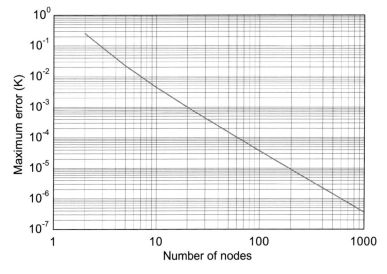

Figure 1-19: Maximum discrepancy between the analytical and numerical solutions (in the limit that $b = c = 0$) as a function of N.

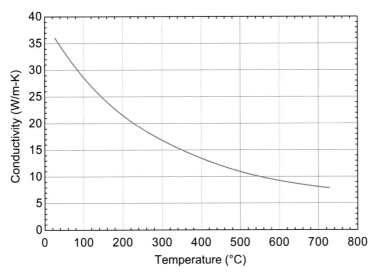

Figure 1-20: Thermal conductivity of polycrystalline aluminum oxide as a function of temperature.

oxide varies by 200°C within the cylinder (see Figure 1-14) and the thermal conductivity of aluminum oxide varies substantially over this range of temperatures, as shown in Figure 1-20.

It is possible to alter the EES code that was developed in Section 1.4.2 so that the temperature-dependent thermal conductivity of polycrystalline aluminum oxide is considered. A function k is written in EES that returns the conductivity. The function must be placed at the top of the Equations window, either above or below the previously defined function gen, and has a single input (the temperature). EES' internal function for the thermal conductivity of solids is used in function k to evaluate the thermal conductivity of the polycrystalline aluminum oxide.

```
function k(T)
    "This function provides the thermal conductivity of the aluminum oxide
    Inputs: T: temperature (K)
    Output: thermal conductivity (W/m-K)"
    k=k_('Al oxide-polycryst',T)
end
```

The temperature-dependent conductivity can cause a problem. It is tempting to evaluate the conduction terms for each control volume using the thermal conductivity evaluated at the temperature of the node. However, doing so will result in an error in the energy balance. Each conduction term, for example $\dot{q}_{LHS}$, represents an energy exchange between node i and its adjacent node to the left, node $i - 1$. The value of $\dot{q}_{LHS}$ evaluated at node i must be equal and opposite to $\dot{q}_{RHS}$ evaluated at node $i - 1$; if the thermal conductivity is evaluated using the nodal temperature, then this will not be true and energy will be artificially generated or destroyed at the boundaries between the nodal control volumes.

To avoid this problem, the thermal conductivity should be evaluated at the average temperature of the two nodes that are involved in the conduction heat transfer (i.e., the temperature at the boundary). The energy balance on the internal nodes (see Figure 1-12) remains:

$$\dot{q}_{LHS} + \dot{q}_{RHS} + \dot{g} = 0 \qquad (1\text{-}124)$$

However, the conduction terms must be calculated according to:

$$\dot{q}_{LHS} = \frac{k_{T=(T_i+T_{i-1})/2}\, L\, 2\, \pi}{\Delta r} \left(r_i - \frac{\Delta r}{2}\right)(T_{i-1} - T_i) \qquad (1\text{-}125)$$

$$\dot{q}_{RHS} = \frac{k_{T=(T_i+T_{i+1})/2}\, L\, 2\, \pi}{\Delta r} \left(r_i + \frac{\Delta r}{2}\right)(T_{i+1} - T_i) \qquad (1\text{-}126)$$

where $k_{T=(T_i+T_{i-1})/2}$ is the thermal conductivity evaluated at the average of T_i and T_{i-1} and $k_{T=(T_i+T_{i+1})/2}$ is the thermal conductivity evaluated at the average of T_i and T_{i+1}. A similar process is used for nodes 1 and N. The modified energy balances in EES are shown below, with the modified code indicated in bold:

```
"Internal control volume energy balances"
duplicate i=2,(N-1)
  k_LHS[i]=k((T[i-1]+T[i])/2)              "thermal conductivity at LHS boundary"
  k_RHS[i]=k((T[i+1]+T[i])/2)              "thermal conductivity at RHS boundary"
  q_dot_LHS[i]=2*pi*L*k_LHS[i]*(r[i]-DELTAr/2)*(T[i-1]-T[i])/DELTAr
                                           "conduction in from inner radius"
  q_dot_RHS[i]=2*pi*L*k_RHS[i]*(r[i]+DELTAr/2)*(T[i+1]-T[i])/DELTAr
                                           "conduction in from outer radius"
  g_dot[i]=gen(r[i])*2*pi*r[i]*L*DELTAr    "generation"
  q_dot_LHS[i]+g_dot[i]+q_dot_RHS[i]=0     "energy balance"
end

"Energy balance for node on internal surface"
k_RHS[1]=k((T[2]+T[1])/2)                  "thermal conductivity at RHS boundary"
q_dot_RHS[1]=2*pi*L*(r[1]+DELTAr/2)*k_RHS[1]*(T[2]-T[1])/DELTAr
                                           "conduction in from outer radius"
q_dot_conv_in=2*pi*L*r_in*h_bar_in*(T_infinity_in-T[1]) "convection from internal fluid"
g_dot[1]=gen(r[1])*2*pi*r[1]*L*DELTAr/2    "generation"
q_dot_RHS[1]+q_dot_conv_in+g_dot[1]=0      "energy balance"

"Energy balance for node on external surface"
k_LHS[N]=k((T[N-1]+T[N])/2)                "thermal conductivity at LHS boundary"
q_dot_LHS[N]=2*pi*L*(r[N]-DELTAr/2)*k_LHS[N]*(T[N-1]-T[N])/DELTAr
                                           "conduction in from from inner radius"
q_dot_conv_out=2*pi*L*r_out*h_bar_out*(T_infinity_out-T[N])
g_dot[N]=gen(r[N])*2*pi*r[N]*L*DELTAr/2    "generation in node N"
q_dot_LHS[N]+q_dot_conv_out+g_dot[N]=0     "energy balance for node N"
```

Select Solve from the Calculate menu and you are likely to find that the problem either fails to converge or converges to some ridiculously high temperatures. This is not surprising, since the use of a temperature-dependent conductivity has transformed the algebraic equations from a set of equations that are linear in the unknown temperatures to a set of equations that are nonlinear. Therefore, EES must start from some guess value for the unknown temperatures and attempt to iterate until a solution is obtained.

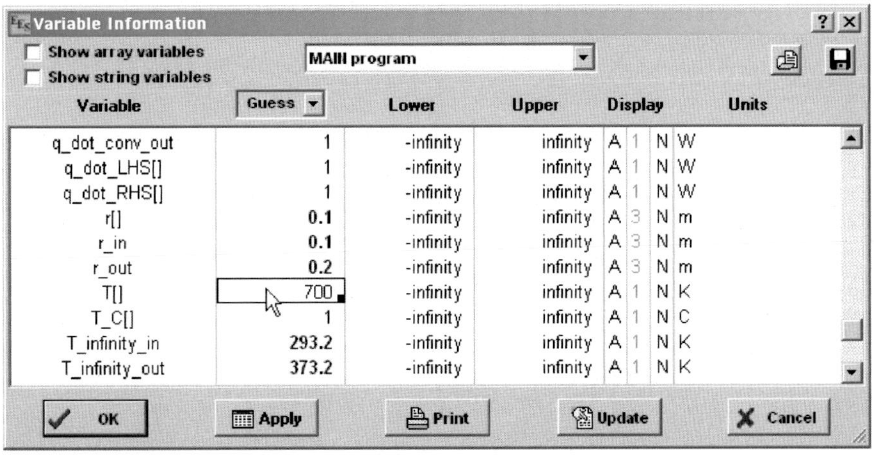

Figure 1-21: Setting the guess values for the array T[] in the Variable Information window.

The success of this process is highly dependent on the guess values that are used. It may be possible to simply set better guess values for the unknown temperatures. Select Variable Info from the Options menu and deselect the Show array variables button. Set the guess value for the array T[] to something more reasonable than its default value of 1 K (e.g., 700 K), as shown in Figure 1-21.

This strategy of manually setting reasonable guess values will not always work. A more reliable strategy uses the solution with constant conductivity, from Section 1.4.2, in order to provide guess values for the non-linear problem associated with temperature-dependent conductivity. This is an easy process; modify the conductivity function (k) as shown below, so that it returns a constant value rather than the temperature-dependent value:

```
function k(T)
  "This function provides the thermal conductivity of the aluminum oxide
  Inputs: T: temperature (K)
  Output: thermal conductivity (W/m-K)"
  {k=k_('Al oxide-polycryst',T)}
  k=9 [W/m-K]
end
```

Solve the problem and EES will converge to a solution. To use this constant conductiviy solution as the guess value for the non-linear problem, select Update Guesses from the Calculate menu and then return the conductivity function to its original form. Solve the problem and your EES code will converge to the actual solution. Examine the Arrays Table (Figure 1-22) and notice that energy is conserved at each boundary (i.e., $\dot{q}_{RHS}$ for node i is equal and opposite to $\dot{q}_{LHS}$ for node $i + 1$ for every node); this simple check shows that your rate equations have been set up appropriately.

Figure 1-23 illustrates the temperature of the aluminum oxide as a function of radius for the case in which the thermal conductivity is evaluated as a function of temperature. The temperature distribution calculated in Section 1.4.2 using a constant thermal conductivity of 9 W/m-K is also shown. This example illustrates that the temperature dependence of thermal conductivity may be an important factor in some problems.

energy is conserved
at each boundary

Sort	$\dot{g}_i$ [W]	$k_{LHS,i}$ [W/m-k]	$k_{RHS,i}$ [W/m-k]	$\dot{q}_{LHS,i}$ [W]	$\dot{q}_{RHS,i}$ [W]	r_i [m]	T_i [K]	$T_{C,i}$ [C]
[1]	3330		10.12		27426	0.1	782.7	509.5
[2]	11370	10.12	9.353	-27426	16056	0.12	861.1	587.9
[3]	17910	9.353	9.113	-16056	-1853	0.14	903.1	629.9
[4]	26580	9.113	9.498	1853	-28434	0.16	898.8	625.6
[5]	37684	9.498	10.65	28434	-66118	0.18	842.7	569.6
[6]	25761	10.65		66118		0.2	738.7	465.6

Figure 1-22: Arrays Window.

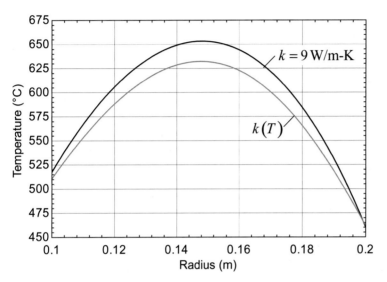

Figure 1-23: Temperature as a function of radius for the case where conductivity is a function of temperature, $k(T)$, and conductivity is constant, $k = 9.0$ W/m-K.

1.4.4 Alternative Rate Models

The rate equations used to model the conduction between adjacent nodes in Section 1.4.2 were based on separating these nodes with a thin cylindrical shell of material that is modeled as a plane wall. An even better approximation for these conduction terms uses the resistance to conduction through a cylinder. According to the resistance formula listed in Table 1-2, the rate equations become:

$$\dot{q}_{LHS} = \frac{k L 2 \pi \, (T_{i-1} - T_i)}{\ln\left(\dfrac{r_i}{r_{i-1}}\right)} \tag{1-127}$$

$$\dot{q}_{RHS} = \frac{k L 2 \pi \, (T_{i+1} - T_i)}{\ln\left(\dfrac{r_{i+1}}{r_i}\right)} \tag{1-128}$$

Also, the volume of the control volume may be computed more exactly in order to provide a better estimate of the thermal energy generation term:

$$\dot{g} = \dot{g}'''_{r=r_i} \pi \left[\left(r_i + \frac{\Delta r}{2}\right)^2 - \left(r_i - \frac{\Delta r}{2}\right)^2 \right] L \tag{1-129}$$

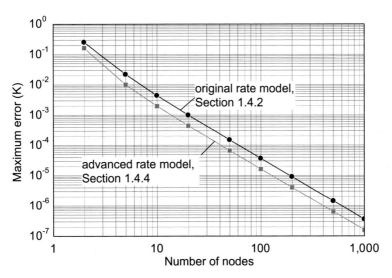

Figure 1-24: Maximum error between the analytical and numerical models (original and advanced) as a function of the number of nodes.

The portion of the EES code from Section 1.4.2 that is modified to use these more advanced rate models (with modifications indicated in bold), is shown below.

```
"Internal control volume energy balances"
duplicate i=2,(N-1)
   q_dot_LHS[i]=2*pi*L*k*(T[i-1]-T[i])/ln(r[i]/r[i-1])          "conduction in from inner radius"
   q_dot_RHS[i]=2*pi*L*k*(T[i+1]-T[i])/ln(r[i+1]/r[i])          "conduction in from outer radius"
   g_dot[i]=gen(r[i])*pi*((r[i]+DELTAr/2)^2-(r[i]-DELTAr/2)^2)*L   "generation"
   q_dot_LHS[i]+g_dot[i]+q_dot_RHS[i]=0                         "energy balance"
end


"Energy balance for node on internal surface"
q_dot_RHS[1]=2*pi*L*k*(T[2]-T[1])/ln(r[2]/r[1])                 "conduction in from outer radius"
q_dot_conv_in=2*pi*L*r_in*h_bar_in*(T_infinity_in-T[1])         "convection from internal fluid"
g_dot[1]=gen(r[1])*pi*((r[1]+DELTAr/2)^2-r[1]^2)*L             "generation"
q_dot_RHS[1]+q_dot_conv_in+g_dot[1]=0                           "energy balance"


"Energy balance for node on external surface"
q_dot_LHS[N]=2*pi*L*k*(T[N-1]-T[N])/ln(r[N]/r[N-1])             "conduction in from from inner radius"
q_dot_conv_out=2*pi*L*r_out*h_bar_out*(T_infinity_out-T[N])     "convection from external fluid"
g_dot[N]=gen(r[N])*pi*(r[N]^2-(r[N]-DELTAr/2)^2)*L            "generation in node N"
q_dot_LHS[N]+q_dot_conv_out+g_dot[N]=0                          "energy balance for node N"
```

The more advanced numerical solution will approach the actual solution somewhat more quickly (i.e., with fewer nodes) than the original numerical solution. Figure 1-24 illustrates the difference between the advanced numerical solution (in the limit of a constant generation rate, $b = c = 0$) and the analytical solution as well as the error associated with the original solution derived in Section 1.4.2. Notice that the use of the advanced rate models provides a substantial improvement in accuracy for any given number of nodes, N.

The use of advanced rate models for a spherical problem is illustrated in EXAMPLE 1.4-1.

EXAMPLE 1.4-1: FUEL ELEMENT

EXAMPLE 1.4-1: FUEL ELEMENT

A nuclear fuel element consists of a sphere of fissionable material (fuel) with radius $r_{fuel} = 5.0$ cm and conductivity $k_{fuel} = 1.0$ W/m-K that is surrounded by a shell of metal cladding with outer radius $r_{clad} = 7.0$ cm and conductivity $k_{clad} = 300$ W/m-K. The outer surface of the cladding is exposed to helium gas that is being heated by the reactor. The average convection coefficient between the gas and the cladding surface is $\overline{h} = 100$ W/m^2-K and the temperature of the gas is $T_\infty = 500°$C.

Inside the fuel element, fission fragments are produced that have high velocities. The products collide with the atoms of the material and provide the thermal energy for the reactor. This process can be modeled as a non-uniform volumetric thermal energy generation ($\dot{g}'''$) that can be approximated by:

$$\dot{g}''' = \dot{g}_0''' \exp\left(-b\frac{r}{r_{fuel}}\right) \tag{1}$$

where $\dot{g}_0''' = 5 \times 10^5$ W/m^3 is the volumetric rate of heat generation at the center of the sphere and $b = 1.0$ is a dimensionless constant that characterizes how quickly the generation rate decays in the radial direction.

a) Develop a numerical model for the spherical fuel element using EES.

A function is defined that returns the volumetric generation given the radial position and the radius of the fuel element.

```
function gen(r, r_fuel)
   "This function defines the volumetric heat generation in the fuel element
   Inputs: r: radius (m)
           r_fuel: radius of fuel sphere (m)
   Output: volumetric heat generation at r (W/m^3)"
   g'''_dot_0=5e5 [W/m^3]                        "volumetric generation rate at the center"
   b=1.0 [-]                                      "constant that describes rate of decay"
   gen=g'''_dot_0*exp(-b*r/r_fuel)               "volumetric rate of generation"
end
```

The next section of the EES code provides the problem inputs.

```
"EXAMPLE 1.4-1: Fuel Element"

$UnitSystem SI MASS RAD PA K J
$Tabstops 0.2 0.4 0.6 3.5 in

"Inputs"
r_fuel=5.0 [cm]*convert(cm,m)                    "fuel radius"
r_clad=7.0 [cm]*convert(cm,m)                    "cladding radius"
k_fuel=1.0 [W/m-K]                               "fuel conductivity"
k_clad=300 [W/m-K]                               "cladding conductivity"
h_bar=100 [W/m^2-K]                              "average convection coefficient"
T_infinity=converttemp(C,K,500 [C])             "helium temperature"
```

The numerical solution proceeds by distributing nodes throughout the computational domain that stretches from $r = 0$ to $r = r_{fuel}$. There is no reason to include the metal cladding in the numerical model. The cladding increases the thermal

EXAMPLE 1.4-1: FUEL ELEMENT

resistance that is already present due to convection with the gas; however, this effect can be included using a conduction thermal resistance.

The positions of a uniformly distributed set of nodes are obtained from:

$$r_i = \frac{(i-1)}{(N-1)} r_{fuel} \quad \text{for } i = 1..N$$

and the distance between adjacent nodes is:

$$\Delta r = \frac{r_{fuel}}{(N-1)}$$

```
"Setup nodes"
N=50 [-]                                    "number of nodes"
duplicate i=1,N
     r[i]=r_fuel*(i-1)/(N-1)               "radial position of each node"
end
DELTAr=r_fuel/(N-1)                         "distance between adjacent nodes"
```

A control volume for an arbitrary internal node is shown in Figure 1.

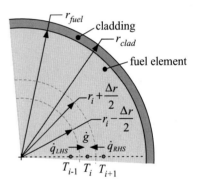

Figure 1: Control volume for an internal node.

The energy balances for the internal control volumes are:

$$\dot{q}_{LHS} + \dot{q}_{RHS} + \dot{g} = 0$$

The control volumes are spherical shells. Therefore, it is appropriate to use a conduction model that is consistent with conduction through a spherical shell (see Table 1-2). Note that this problem could also be solved by approximating the spherical shells as plane walls with different surface areas, as was done in Section 1.4.2. However, building the proper geometry into the control volume energy balances will allow the problem to be solved to a specified accuracy with fewer nodes.

$$\dot{q}_{LHS} = \frac{(T_{i-1} - T_i)}{\dfrac{1}{4\pi\, k_{fuel}} \left[\dfrac{1}{r_{i-1}} - \dfrac{1}{r_i} \right]}$$

and

$$\dot{q}_{RHS} = \frac{(T_{i+1} - T_i)}{\dfrac{1}{4\pi\, k_{fuel}} \left[\dfrac{1}{r_i} - \dfrac{1}{r_{i+1}} \right]}$$

The temperature differences used to evaluate $\dot{q}_{LHS}$ and $\dot{q}_{RHS}$ are consistent with the direction of the conduction heat transfer terms shown in Figure 1 whereas the resistance values in the denominators are written in the form of $1/r_{in} - 1/r_{out}$ (e.g., $1/r_{i-1} - 1/r_i$ and $1/r_i - 1/r_{i+1}$) so that the resistances are positive. The generation in each control volume is given by:

$$\dot{g}_i = \frac{4}{3}\pi \left[\left(r_i + \frac{\Delta r}{2}\right)^3 - \left(r_i - \frac{\Delta r}{2}\right)^3\right]\dot{g}'''_{r_i}$$

Combining these equations allows the control volume energy balances for the internal nodes to be written as:

$$\frac{4\pi\, k_{fuel}\,(T_{i-1} - T_i)}{\left[\dfrac{1}{r_{i-1}} - \dfrac{1}{r_i}\right]} + \frac{4\pi\, k_{fuel}\,(T_{i+1} - T_i)}{\left[\dfrac{1}{r_i} - \dfrac{1}{r_{i+1}}\right]} + \frac{4}{3}\pi\left[\left(r_i + \frac{\Delta r}{2}\right)^3 - \left(r_i - \frac{\Delta r}{2}\right)^3\right]\dot{g}'''_{r_i} = 0$$

$$\text{for } i = 2..(N-1) \quad (2)$$

```
"Internal control volume energy balance"
duplicate i=2,(N-1)
    4*pi*k_fuel*(T[i-1]-T[i])/(1/r[i-1]-1/r[i])+4*pi*k_fuel*(T[i+1]-T[i])/(1/r[i]-1/r[i+1])+&
    4*pi*((r[i]+DELTAr/2)^3-(r[i]-DELTAr/2)^3)*gen(r[i],r_fuel)/3=0
end
```

The energy balance for the node that is placed at the outer edge of the fuel (i.e., node N) is:

$$\underbrace{\frac{4\pi\, k_{fuel}\,(T_{N-1} - T_N)}{\left[\dfrac{1}{r_{N-1}} - \dfrac{1}{r_N}\right]}}_{\substack{\text{conduction between the outermost}\\\text{and adjoining nodes}}} + \underbrace{\frac{(T_\infty - T_N)}{R_{cond,clad} + R_{conv}}}_{\substack{\text{combined thermal}\\\text{resistance of cladding}\\\text{and convection}}} + \underbrace{\frac{4}{3}\pi\left[r_N^3 - \left(r_N - \frac{\Delta r}{2}\right)^3\right]\dot{g}'''_{r_N}}_{\text{generation in outer shell}} = 0$$

$$(3)$$

where R_{clad} is the resistance to conduction through the cladding:

$$R_{cond,clad} = \frac{1}{4\pi\, k_{clad}}\left[\frac{1}{r_{fuel}} - \frac{1}{r_{clad}}\right]$$

and R_{conv} is the resistance to convection from the surface of the cladding to the gas:

$$R_{conv} = \frac{1}{4\pi\, r_{clad}^2\, \bar{h}}$$

```
R_cond_clad=(1/r_fuel-1/r_clad)/(4*pi*k_clad)    "conduction resistance of cladding"
R_conv=1/(4*pi*r_clad^2*h_bar)                   "convection resistance from surface of cladding"
4*pi*k_fuel*(T[N-1]-T[N])/(1/r[N-1]-1/r[N])+(T_infinity-T[N])/(R_cond_clad+R_conv)+&
    4*pi*(r[N]^3-(r[N]-DELTAr/2)^3)*gen(r[N],r_fuel)/3=0    "node N energy balance"
```

If the same conduction model is used, then the energy balance for the node placed at the center of the fuel (i.e., node 1) is:

$$\frac{4\pi\, k_{fuel}\,(T_2 - T_1)}{\left[\dfrac{1}{r_1} - \dfrac{1}{r_2}\right]} + \frac{4}{3}\pi\left[\left(r_1 + \frac{\Delta r}{2}\right)^3 - r_1^3\right]\dot{g}'''_{r_1} = 0 \quad (4)$$

EXAMPLE 1.4-1: FUEL ELEMENT

```
4*pi*k_fuel*(T[2]-T[1])/(1/r[1]-1/r[2])+4*pi*((r[1]+DELTAr/2)^3-r[1]^3)*gen(r[1],r_fuel)/3=0
    "node 1 energy balance"
```

Executing the EES code will lead to a division by zero error message. The radial location of node 1, r_1, is equal to 0 and therefore the $1/r_1$ term in the denominator of Eq. (4) is infinite. (The actual resistance associated with conducting energy to a point is infinite.) A similar error will be encountered when computing $\dot{q}_{LHS}$ for node 2 in Eq. (2). This problem can be dealt with by calculating the conduction between nodes 1 and 2 using a plane wall approximation. The energy balance for node 1 becomes:

$$4 \pi \, k_{fuel} \left(r_1 + \frac{\Delta r}{2} \right)^2 \frac{(T_2 - T_1)}{\Delta r} + \frac{4}{3} \pi \left[\left(r_1 + \frac{\Delta r}{2} \right)^3 - r_1^3 \right] \dot{g}_{r_1}''' = 0$$

```
{4*pi*k_fuel*(T[2]-T[1])/(1/r[1]-1/r[2])+4*pi*((r[1]+DELTAr/2)^3-r[1]^3)*gen(r[1],r_fuel)/3=0}
4*pi*k_fuel*(r[1]+DELTAr/2)^2*(T[2]-T[1])/DELTAr+4*pi*((r[1]+DELTAr/2)^3-r[1]^3)*gen(r[1],r_fuel)/3=0
    "node 1 energy balance"
```

The energy balance for node 2 has to be rewritten in the same way:

$$4 \pi \, k_{fuel} \left(r_1 + \frac{\Delta r}{2} \right)^2 \frac{(T_1 - T_2)}{\Delta r} + \frac{4 \pi \, k_{fuel} \, (T_3 - T_2)}{\left[\dfrac{1}{r_2} - \dfrac{1}{r_3} \right]}$$

$$+ \frac{4}{3} \pi \left[\left(r_2 + \frac{\Delta r}{2} \right)^3 - \left(r_2 - \frac{\Delta r}{2} \right)^3 \right] \dot{g}_{r_2}''' = 0$$

```
"Internal control volume energy balance"
{duplicate i=2,(N-1)
    4*pi*k_fuel*(T[i-1]-T[i])/(1/r[i-1]-1/r[i])+4*pi*k_fuel*(T[i+1]-T[i])/(1/r[i]-1/r[i+1])+&
        4*pi*((r[i]+DELTAr/2)^3-(r[i]-DELTAr/2)^3)*gen(r[i],r_fuel)/3=0
end}
duplicate i=3,(N-1)
    4*pi*k_fuel*(T[i-1]-T[i])/(1/r[i-1]-1/r[i])+4*pi*k_fuel*(T[i+1]-T[i])/(1/r[i]-1/r[i+1])+&
        4*pi*((r[i]+DELTAr/2)^3-(r[i]-DELTAr/2)^3)*gen(r[i],r_fuel)/3=0
end
4*pi*k_fuel*(r[1]+DELTAr/2)^2*(T[1]-T[2])/DELTAr+4*pi*k_fuel*(T[3]-T[2])/(1/r[2]-1/r[3])+&
    4*pi*((r[2]+DELTAr/2)^3-(r[2]-DELTAr/2)^3)*gen(r[2],r_fuel)/3=0    "node 2 energy balance"
```

With these changes, the program can be solved. The solution is converted to °C:

```
duplicate i=1,N
    T_C[i]=converttemp(K,C,T[i])            "temperature in C"
end
```

EXAMPLE 1.4-1: FUEL ELEMENT

Figure 2 illustrates the temperature in the fuel element as a function of radius.

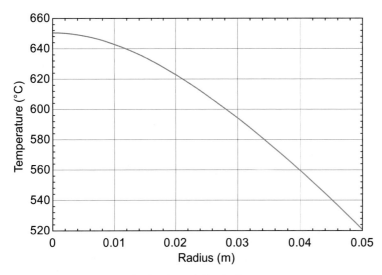

Figure 2: Temperature distribution within fuel.

The maximum temperature in the fuel element is obtained with the **max** function.

T_max_C=max(T_C[1..N])   "maximum temperature of the fuel, in C"

Figure 3 shows the maximum temperature in the fuel as a function of the number of nodes and indicates that the solution converges if at least 100 nodes are used.

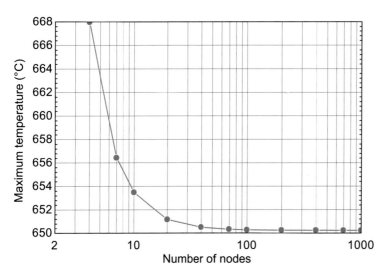

Figure 3: Maximum temperature within fuel as a function of the number of nodes.

Several sanity checks can be carried out in order to verify that the solution is physically correct. Figure 4 shows the maximum temperature in the fuel as a function of the fuel conductivity for various values of the volumetric generation at the center of

EXAMPLE 1.4-1: FUEL ELEMENT

the fuel ($\dot{g}_0'''$). The maximum temperature increases as either the fuel conductivity decreases or the volumetric rate of generation increases.

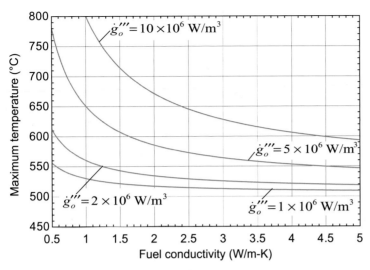

Figure 4: Maximum temperature in the fuel as a function of k_{fuel} for various values of $\dot{g}_0'''$.

Finally, we can compare the numerical model with the analytical solution for the limiting case where $b = 0$ in Eq. (1) (i.e., the fuel experiences a uniform rate of volumetric generation). The general solution for the temperature distribution and temperature gradient within a sphere exposed to a uniform generation rate is given in Table 1-3:

$$T = -\frac{\dot{g}'''}{6 \, k_{fuel}} r^2 + \frac{C_1}{r} + C_2 \tag{5}$$

and

$$\frac{dT}{dr} = -\frac{\dot{g}'''}{3 \, k_{fuel}} r - \frac{C_1}{r^2} \tag{6}$$

where C_1 and C_2 are constants of integration. The temperature at the center of the sphere must be bounded and therefore C_1 must be equal to 0 by inspection of Eq. (5); alternatively, the temperature gradient at the center must be zero, which would also require that $C_1 = 0$ according to Eq. (6). The second boundary condition is related to an energy balance at the interface between the cladding and the fuel:

$$-k_{fuel} \, 4 \, \pi \, r_{fuel}^2 \left. \frac{dT}{dr} \right|_{r=r_{fuel}} = \frac{(T_{r=r_{fuel}} - T_\infty)}{R_{cond,clad} + R_{conv}} \tag{7}$$

Combining Equations (5) through (7) leads to:

$$-k_{fuel} \, 4 \, \pi \, r_{fuel}^2 \left(-\frac{\dot{g}'''}{3 \, k_{fuel}} r_{fuel} \right) = \frac{\left(-\dfrac{\dot{g}'''}{6 \, k_{fuel}} r_{fuel}^2 + C_2 - T_\infty \right)}{R_{cond,clad} + R_{conv}}$$

EXAMPLE 1.4-1: FUEL ELEMENT

which can be solved for C_2.

```
"Analytical solution"
g'''_dot=gen(0 [m],r_fuel)                "rate of volumetric generation to use in analytical solution"
-k_fuel*4*pi*r_fuel^2*(-g'''_dot*r_fuel/(3*k_fuel))= &
     (-g'''_dot*r_fuel^2/(6*k_fuel)+C_2-T_infinity)/(R_cond_clad+R_conv)
                                          "boundary condition at r=r_fuel"
```

The analytical solution is obtained at the same radial locations as the numerical solution.

```
duplicate i=1,N
  T_an[i]=-g'''_dot*r[i]^2/(6*k_fuel)+C_2                "analytical solution"
  T_an_C[i]=converttemp(K,C,T_an[i])                     "in C"
end
```

Figure 5 shows the analytical and numerical solutions in the limit that $b = 0$ for 50 nodes; the agreement is nearly exact, indicating that the numerical solution is adequate.

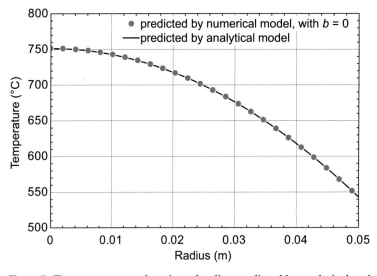

Figure 5: Temperature as a function of radius predicted by analytical and numerical models in the limit that $b = 0$.

1.5 Numerical Solutions to Steady-State 1-D Conduction Problems using MATLAB

1.5.1 Introduction

Numerical models of 1-D steady-state conduction problems are introduced and implemented using EES in Section 1.4. EES internally provides all of the numerical manipulations that are needed to solve the system of algebraic equations that constitutes a numerical model. This capability reduces the complexity of the problem. However, there are disadvantages to using EES. For example, EES will generally require significantly

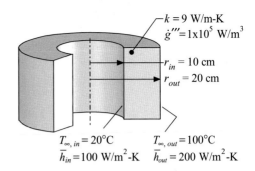

Figure 1-25: Cylinder with volumetric generation.

more time to solve the equations than is required by a compiled computer language. There is an upper limit to the number of variables that EES can handle, which places an upper bound on the number of nodes that can be used in the model. The structure of EES requires that every variable be retained in the final solution. Therefore, you cannot define and then erase intermediate variables in the course of obtaining a solution and, as a result, numerical solutions in EES often require a lot of memory. Finally, some models require logic statements (e.g., if-then-else statements) that are difficult to include in EES. For these reasons, it is useful to learn how to implement numerical models in a formal programming language, e.g., FORTRAN, C++, or MATLAB. The steps required to solve the algebraic equations associated with a numerical model are demonstrated in this section using the MATLAB software. It is suggested that the reader stop and go through the tutorial provided in Appendix A.3 in order to become familiar with MATLAB. Appendix A.3 can be found on the web site associated with this book (www.cambridge.org/nellisandklein).

1.5.2 Numerical Solutions in Matrix Format

The cylinder problem that is considered in Section 1.4 in order to illustrate numerical methods using EES is shown again in Figure 1-25. An aluminum oxide cylinder is exposed to fluid on its internal and external surfaces. The temperature of the fluid that is exposed to the internal surface is $T_{\infty,in} = 20°C$ and the average heat transfer coefficient on this surface is $\bar{h}_{in} = 100$ W/m²-K. The temperature of the fluid exposed to the external surface is $T_{\infty,out} = 100°C$ and the average heat transfer coefficient on this surface is $\bar{h}_{out} = 200$ W/m²-K. The thermal conductivity is assumed to be constant and equal to $k = 9.0$ W/m-K. We will begin by solving the problem for the case where the rate of volumetric generation of thermal energy within the cylinder is uniform and equal to $\dot{g}''' = 1 \times 10^5$ W/m³. The inner and outer radii of the cylinder are $r_{in} = 10$ cm and $r_{out} = 20$ cm, respectively.

The development of the system of equations proceeds as discussed in Section 1.4.2. A uniform distribution of nodes is used and therefore the radial location of each node (r_i) is:

$$r_i = r_{in} + \frac{(i-1)}{(N-1)}(r_{out} - r_{in}) \quad i = 1..N \tag{1-130}$$

where N is the number of nodes used for the simulation. The radial distance between adjacent nodes (Δr) is:

$$\Delta r = \frac{(r_{out} - r_{in})}{(N-1)} \tag{1-131}$$

The energy balances for the internal nodes are:

$$\dot{q}_{LHS} + \dot{q}_{RHS} + \dot{g} = 0 \qquad (1\text{-}132)$$

where

$$\dot{q}_{LHS} = \frac{k \, L \, 2 \pi \left(r_i - \dfrac{\Delta r}{2} \right) (T_{i-1} - T_i)}{\Delta r} \qquad (1\text{-}133)$$

$$\dot{q}_{RHS} = \frac{k \, L \, 2 \pi \left(r_i + \dfrac{\Delta r}{2} \right) (T_{i+1} - T_i)}{\Delta r} \qquad (1\text{-}134)$$

$$\dot{g} = \dot{g}''' \, 2 \pi \, r_i \, L \, \Delta r \qquad (1\text{-}135)$$

Equations (1-132) through (1-135) are combined:

$$\frac{k \, L \, 2 \pi \left(r_i - \dfrac{\Delta r}{2} \right)}{\Delta r} (T_{i-1} - T_i) + \frac{k \, L \, 2 \pi \left(r_i + \dfrac{\Delta r}{2} \right)}{\Delta r} (T_{i+1} - T_i) + \dot{g}''' \, 2 \pi \, r_i \, L \, \Delta r = 0$$
$$\text{for } i = 2..\,(N-1) \qquad (1\text{-}136)$$

The energy balance for the control volume associated with node 1 is:

$$\overline{h}_{in} \, 2 \pi \, r_1 \, L \, (T_{\infty,in} - T_1) + \frac{k \, L \, 2 \pi \left(r_1 + \dfrac{\Delta r}{2} \right)}{\Delta r} (T_2 - T_1) + \dot{g}''' \, 2 \pi \, r_1 \, L \, \frac{\Delta r}{2} = 0$$
$$(1\text{-}137)$$

The energy balance for the control volume associated with node N is:

$$\overline{h}_{out} \, 2 \pi \, r_N \, L \, (T_{\infty,out} - T_N) + \frac{k \, L \, 2 \pi \left(r_N - \dfrac{\Delta r}{2} \right)}{\Delta r} (T_{N-1} - T_N) + \dot{g}''' \, 2 \pi \, r_N \, L \, \frac{\Delta r}{2} = 0$$
$$(1\text{-}138)$$

Equations (1-136) through (1-138) represent N linear algebraic equations in an equal number of unknown temperatures. In order to solve these equations using a formal programming language, it is necessary to represent this set of equations as a matrix equation. Recall from linear algebra that a linear system of equations, such as:

$$\begin{aligned} 2x_1 + 3x_2 + 1x_3 &= 1 \\ 1x_1 + 5x_2 + 1x_3 &= 2 \\ 7x_1 + 1x_2 + 2x_3 &= 5 \end{aligned} \qquad (1\text{-}139)$$

can be written as a matrix equation:

$$\underbrace{\begin{bmatrix} 2 & 3 & 1 \\ 1 & 5 & 1 \\ 7 & 1 & 2 \end{bmatrix}}_{\underline{\underline{A}}} \underbrace{\begin{bmatrix} x_1 \\ x_2 \\ x_3 \end{bmatrix}}_{\underline{X}} = \underbrace{\begin{bmatrix} 1 \\ 2 \\ 5 \end{bmatrix}}_{\underline{b}} \qquad (1\text{-}140)$$

or

$$\underline{\underline{A}} \, \underline{X} = \underline{b} \qquad (1\text{-}141)$$

where $\underline{\underline{A}}$ is a matrix and $\underline{X}$ and $\underline{b}$ are vectors:

$$\underline{\underline{A}} = \begin{bmatrix} 2 & 3 & 1 \\ 1 & 5 & 1 \\ 7 & 1 & 2 \end{bmatrix}, \quad \underline{X} = \begin{bmatrix} x_1 \\ x_2 \\ x_3 \end{bmatrix}, \quad \text{and} \quad \underline{b} = \begin{bmatrix} 1 \\ 2 \\ 5 \end{bmatrix} \tag{1-142}$$

Most programming languages, including MATLAB, have built-in or library routines for decomposing the system of equations and solving for the vector of unknowns, $\underline{X}$. This is a mature area of research and advanced methods exist for quickly solving matrix equations, particularly when most of the entries in $\underline{\underline{A}}$ are 0 (i.e., $\underline{\underline{A}}$ is a sparse matrix).

MATLAB is specifically designed to handle large matrix equations. In this section, we will use MATLAB to solve heat transfer problems. This requires an understanding of how to place large systems of equations, corresponding to the energy balances, into a matrix format. Each *row* of the $\underline{\underline{A}}$ matrix and $\underline{b}$ vector correspond to an equation whereas each *column* of the $\underline{\underline{A}}$ matrix is the coefficient that multiplies the corresponding unknown (typically a nodal temperature) in that equation. To set up a system of equations in matrix format, it is necessary to carefully define how the rows and energy balances are related and how the columns and unknown temperatures are related.

The first step is to define the vector of unknowns, the vector $\underline{X}$ in Eq. (1-140). It does not really matter what order the unknowns are placed in $\underline{X}$, but the implementation of the solution is much easier if a logical order is used. In this problem, the unknowns are the nodal temperatures. Therefore, the most logical technique for ordering the unknown temperatures in the vector $\underline{X}$ is:

$$\underline{X} = \begin{bmatrix} X_1 = T_1 \\ X_2 = T_2 \\ \cdots \\ X_N = T_N \end{bmatrix} \tag{1-143}$$

Equation (1-143) shows that the unknown temperature at node i (i.e., T_i) corresponds to element i of vector $\underline{X}$ (i.e., X_i).

The next step is to define how the rows in the matrix $\underline{\underline{A}}$ and the vector $\underline{b}$ correspond to the N control volume energy balances that must be solved. Again, it does not matter what order the equations are placed into the $\underline{\underline{A}}$ matrix, but the solution is easiest if a logical order is used:

$$\underline{\underline{A}} = \begin{bmatrix} \text{row 1} = \text{control volume 1 equation} \\ \text{row 2} = \text{control volume 2 equation} \\ \cdots \\ \text{row } N = \text{ control volume } N \text{ equation} \end{bmatrix} \tag{1-144}$$

Equation (1-144) shows that the equation for control volume i is placed into row i of matrix $\underline{\underline{A}}$.

1.5.3 Implementing a Numerical Solution in MATLAB

We can return to the numerical problem that was discussed in Section 1.5.2. Open a new M-file (select New M-File from the File menu) which will bring up the M-file editor. Save the script as cylinder (select Save As from the File menu) in a directory that is in your search path (you can specify the directories in your search path by typing pathtool in the Command window).

Enter the inputs to the problem at the top of the script and save it. Note that the % symbol indicates that anything that follows on that line will be a comment.

MATLAB will not assign units to any of the variables; they are all dimensionless as far as the software is concerned. This limitation puts the burden squarely on the user to clearly understand the units of each variable and ensure that they are consistent. The use of a semicolon after each assignment statement prevents the variables from being echoed in the working environment. The clear command at the top of the script clears all variables from the workspace.

```
clear;    %clear all variables from the workspace

% Inputs
r_in=0.1;                        %inner radius of cylinder (m)
r_out=0.2;                       %outer radius of cylinder (m)
g_dot_tp=1e5;                    %constant volumetric generation (W/m^3)
L=1;                             %unit length of cylinder (m)
k=9;                             %thermal conductivity of cylinder material (W/m-K)
T_infinity_in=20+273.2;          %average temperature of fluid inside cylinder (K)
h_bar_in=100;                    %heat transfer coefficient inside cylinder (W/m^2-K)
T_infinity_out=100+273.2;        %temperature of fluid outside cylinder (K)
h_bar_out=200;                   %heat transfer coefficient at outer surface (W/m^2-K)
```

In order to run your script from the MATLAB working environment, type cylinder at the command prompt:

```
>> cylinder
```

Nothing appears to have happened. However, all of the variables that are defined in the script are now available in the work space. For example, if the name of any variable is entered at the command prompt then its value is displayed.

```
>> h_bar_out
h_bar_out =
   200
```

For a complete list of variables in the workspace, use the command who.

```
>> who
Your variables are:
L       T_infinity_out h_bar_in      k       r_out
T_infinity_in   g_dot_tp       h_bar_out  r_in
```

The number and location of the nodes for the solution must be specified. A vector of radial locations (r) is setup using a for loop. Each of the statements between the for and end statements is executed each time through the loop. Enter the following lines into the cylinder M-file:

```
% Setup nodes
N=10;                            %number of nodes (-)
Dr=(r_out-r_in)/(N-1);           %distance between nodes (m)
for i=1:N
    r(i)=r_in+(i-1)*(r_out-r_in)/(N-1);   %radial position of each node (m)
end
```

The energy balances must be setup in an appropriately sized matrix, the variable A, and vector, the variable b. Recall that our matrix $\underline{A}$ needs to have as many rows as there are equations (the N control volume energy balances) and as many columns as there are unknowns (the N unknown temperatures) and $\underline{b}$ is a vector with as many elements as there are equations (N). We'll start with $\underline{A}$ and $\underline{b}$ composed entirely of zeros and subsequently add non-zero elements according to the equations. The $\underline{A}$ matrix ends up being composed almost entirely of zeros and thus it is referred to as a sparse matrix. Initially we will not take advantage of this sparse characteristic of $\underline{A}$. However, in Section 1.5.5, it will be shown that the solution can be accelerated considerably by using specialized matrix solution techniques that are designed for sparse matrices. MATLAB makes it extremely easy to use these sparse matrix solution techniques.

The zeros function used below returns a matrix filled with zeros with its size determined by the input arguments; the variable A will be an $N \times N$ matrix filled with zeros and the variable b will be an $N \times 1$ vector filled with zeros where the variable N is the number of nodes.

```
%Setup A and b
A=zeros(N,N);
b=zeros(N,1);
```

The most difficult step in the process is to fill in the non-zero elements of $\underline{A}$ and $\underline{b}$ so that the solution of the system of equations can be obtained through a matrix decomposition process. According to Eq. (1-144), the 1st row in $\underline{A}$ must correspond to the energy balance for control volume 1, which is given by Eq. (1-137), repeated below:

$$\bar{h}_{in}\, 2\,\pi\, r_1\, L\, (T_{\infty,in} - T_1) + \frac{k\, L\, 2\, \pi}{\Delta r}\left(r_1 + \frac{\Delta r}{2}\right)(T_2 - T_1) + \dot{g}'''\, 2\,\pi\, r_1\, L\, \frac{\Delta r}{2} = 0$$

$$(1\text{-}137)$$

It is necessary to algebraically manipulate Eq. (1-137) so that the coefficients that multiply each of the unknowns in this equation (i.e., T_1 and T_2) and the constant term in the equation (i.e., terms that are known) can be identified.

$$T_1 \underbrace{\left[-\frac{k\, L\, 2\, \pi}{\Delta r}\left(r_1 + \frac{\Delta r}{2}\right) - \bar{h}_{in}\, 2\,\pi\, r_1\, L\right]}_{A_{1,1}} + T_2 \underbrace{\left[\frac{k\, L\, 2\, \pi}{\Delta r}\left(r_1 + \frac{\Delta r}{2}\right)\right]}_{A_{1,2}}$$

$$= \underbrace{-\dot{g}'''\, 2\,\pi\, r_1\, L\, \frac{\Delta r}{2} - \bar{h}_{in}\, 2\,\pi\, r_1\, L\, T_{\infty,in}}_{b_1}$$

$$(1\text{-}145)$$

Equation (1-145) corresponds to the 1st row of $\underline{A}$ and $\underline{b}$. The coefficient in the first equation that multiplies the first unknown in $\underline{X}$, T_1 according to Eq. (1-143), must be $A_{1,1}$:

$$A_{1,1} = -\frac{k\, L\, 2\, \pi}{\Delta r}\left(r_1 + \frac{\Delta r}{2}\right) - \bar{h}_{in}\, 2\,\pi\, r_1\, L \qquad (1\text{-}146)$$

The coefficient in the first equation that multiplies the second unknown in $\underline{X}$, T_2 according to Eq. (1-143), must be $A_{1,2}$:

$$A_{1,2} = \frac{k\, L\, 2\, \pi}{\Delta r}\left(r_1 + \frac{\Delta r}{2}\right) \qquad (1\text{-}147)$$

Finally, the constant terms associated with the first equation must be b_1:

$$b_1 = -\dot{g}''' \, 2 \pi \, r_1 \, L \, \frac{\Delta r}{2} - \bar{h}_{in} \, 2 \pi \, r_1 \, L \, T_{\infty,in} \qquad (1\text{-}148)$$

These assignments are accomplished in MATLAB:

```
%Energy balance for control volume 1
A(1,1)=-k*L*2*pi*(r(1)+Dr/2)/Dr-h_bar_in*2*pi*r(1)*L;
A(1,2)=k*L*2*pi*(r(1)+Dr/2)/Dr;
b(1)=-h_bar_in*2*pi*r(1)*L*T_infinity_in-g_dot_tp*2*pi*r(1)*L*Dr/2;
```

According to Eq. (1-144), rows 2 through $N-1$ of matrix $\underline{A}$ correspond to the energy balances for the corresponding internal control volumes; these equations are given by Eq. (1-136), which is repeated below:

$$\frac{k\,L\,2\pi}{\Delta r}\left(r_i - \frac{\Delta r}{2}\right)(T_{i-1} - T_i) + \frac{k\,L\,2\pi}{\Delta r}\left(r_i + \frac{\Delta r}{2}\right)(T_{i+1} - T_i) + \dot{g}''' \, 2 \pi \, r_i \, L \, \Delta r = 0$$

$$\qquad (1\text{-}136)$$

$$\text{for } i = 2..(N-1)$$

Again, Eq. (1-136) must be rearranged to identify coefficients and constants.

$$T_i \underbrace{\left[-\frac{k\,L\,2\pi}{\Delta r}\left(r_i - \frac{\Delta r}{2}\right) - \frac{k\,L\,2\pi}{\Delta r}\left(r_i + \frac{\Delta r}{2}\right)\right]}_{A_{i,i}} + T_{i-1}\underbrace{\left[\frac{k\,L\,2\pi}{\Delta r}\left(r_i - \frac{\Delta r}{2}\right)\right]}_{A_{i,i-1}}$$

$$+ T_{i+1}\underbrace{\left[\frac{k\,L\,2\pi}{\Delta r}\left(r_i + \frac{\Delta r}{2}\right)\right]}_{A_{i,i+1}} = \underbrace{-\dot{g}''' \, 2 \pi \, r_i \, L \, \Delta r}_{b_i} \quad \text{for } i = 2..(N-1) \qquad (1\text{-}149)$$

All of the coefficients for control volume i must go into row i of $\underline{A}$, the column depends on which unknown they multiply. Therefore:

$$A_{i,i} = -\frac{k\,L\,2\pi}{\Delta r}\left(r_i - \frac{\Delta r}{2}\right) - \frac{k\,L\,2\pi}{\Delta r}\left(r_i + \frac{\Delta r}{2}\right) \quad \text{for } i = 2\ldots(N-1) \qquad (1\text{-}150)$$

$$A_{i,i-1} = \frac{k\,L\,2\pi}{\Delta r}\left(r_i - \frac{\Delta r}{2}\right) \quad \text{for } i = 2\ldots(N-1) \qquad (1\text{-}151)$$

$$A_{i,i+1} = \frac{k\,L\,2\pi}{\Delta r}\left(r_i + \frac{\Delta r}{2}\right) \quad \text{for } i = 2\ldots(N-1) \qquad (1\text{-}152)$$

The constant for control volume i must go into row i of $\underline{b}$.

$$b_i = -\dot{g}''' \, 2 \pi \, r_i \, L \, \Delta r \quad \text{for } i = 2\ldots(N-1) \qquad (1\text{-}153)$$

These equations are programmed most conveniently in MATLAB using a for loop:

```
%Energy balances for internal control volumes
for i=2:(N-1)
  A(i,i)=-k*L*2*pi*(r(i)-Dr/2)/Dr-k*L*2*pi*(r(i)+Dr/2)/Dr;
  A(i,i-1)=k*L*2*pi*(r(i)-Dr/2)/Dr;
  A(i,i+1)=k*L*2*pi*(r(i)+Dr/2)/Dr;
  b(i)=-g_dot_tp*2*pi*r(i)*L*Dr;
end
```

Finally, the last row of $\underline{\underline{A}}$ (i.e., row N) corresponds to the energy balance for the last control volume (node N), which is given by Eq. (1-138), repeated below:

$$\bar{h}_{out}\, 2\,\pi\, r_N\, L\, (T_{\infty,out} - T_N) + \frac{k\, L\, 2\,\pi \left(r_N - \dfrac{\Delta r}{2}\right)}{\Delta r}\, (T_{N-1} - T_N) + \dot{g}'''\, 2\,\pi\, r_N\, L\, \frac{\Delta r}{2} = 0$$

$$(1\text{-}138)$$

Equation (1-138) is rearranged:

$$T_N \underbrace{\left[-\bar{h}_{out}\, 2\,\pi\, r_N\, L - \frac{k\, L\, 2\,\pi}{\Delta r}\left(r_N - \frac{\Delta r}{2}\right)\right]}_{A_{N,N}} + T_{N-1} \underbrace{\left[\frac{k\, L\, 2\,\pi}{\Delta r}\left(r_N - \frac{\Delta r}{2}\right)\right]}_{A_{N,N-1}}$$

$$= \underbrace{-\bar{h}_{out}\, 2\,\pi\, r_N\, L\, T_{\infty,out} - \dot{g}'''\, 2\,\pi\, r_N\, L\, \frac{\Delta r}{2}}_{b_N}$$

$$(1\text{-}154)$$

The coefficients in the last row of $\underline{\underline{A}}$ and $\underline{b}$ are:

$$A_{N,N} = -\frac{k\, L\, 2\,\pi}{\Delta r}\left(r_N - \frac{\Delta r}{2}\right) - \bar{h}_{out}\, 2\,\pi\, r_N\, L$$

$$(1\text{-}155)$$

$$A_{N,N-1} = \frac{k\, L\, 2\,\pi}{\Delta r}\left(r_N - \frac{\Delta r}{2}\right)$$

$$(1\text{-}156)$$

and

$$b_N = -\bar{h}_{out}\, 2\,\pi\, r_N\, L\, T_{\infty,out} - \dot{g}'''\, 2\,\pi\, r_N\, \frac{\Delta r}{2}\, L$$

$$(1\text{-}157)$$

```
%Energy balance for control volume N
A(N,N)=-k*L*2*pi*(r(N)-Dr/2)/Dr-h_bar_out*2*pi*r(N)*L;
A(N,N-1)=k*L*2*pi*(r(N)-Dr/2)/Dr;
b(N)=-h_bar_out*2*pi*r(N)*L*T_infinity_out-g_dot_tp*2*pi*r(N)*Dr*L/2;
```

At this point, the matrix $\underline{\underline{A}}$ and vector $\underline{b}$ are completely set up and can be used to determine the unknown temperatures. The solution to the matrix equation, Eq. (1-141), is:

$$\underline{X} = \underline{\underline{A}}^{-1}\, \underline{b}$$

$$(1\text{-}158)$$

where $\underline{\underline{A}}^{-1}$ is the inverse of matrix $\underline{\underline{A}}$. The solution is obtained using the **backslash** operator in MATLAB (note that this is much more efficient than explicitly solving for the inverse of $\underline{\underline{A}}$ using the **inv** command):

```
%Solve for unknowns
X=A\b;                          %solve for unknown vector, X
```

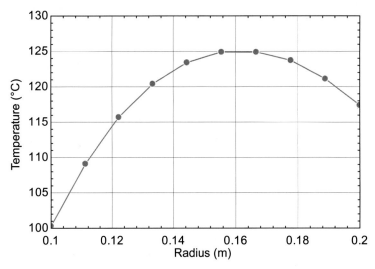

Figure 1-26: Temperature as a function of radius predicted by the numerical model.

The vector of unknowns, $\underline{X}$, is identical to the temperatures for this problem.

```
T=X;                          %assign temperatures from X
T_C=T-273.2;                  %in C
```

The script cylinder can be executed from the working environment by typing cylinder at the command prompt. After execution, the variables that were defined in the M-file and the solution will reside in the workspace; for example, you can view the solution vector (T_C) by typing T_C and you can plot temperature as a function of radius using the plot command.

```
>> cylinder
>> T_C

T_C =

   100.3213
   109.0658
   115.6711
   120.3899
   123.4143
   124.8938
   124.9468
   123.6689
   121.1383
   117.4197

>> plot(r,T_C)
```

Figure 1-26 illustrates the temperature predicted by the MATLAB numerical model as a function of radius. The solution is exactly equal to the solution that would be obtained using the EES code in Section 1.4.2 if the same rate of volumetric generation of thermal energy is used.

1.5.4 Functions

The problem in Section 1.5.3 was solved using a script, which is a set of MATLAB instructions that can be stored and edited, but otherwise operates on the main workspace just as if you typed the instructions in one at a time. It is often more convenient to use a function. A function will solve the problem in a separate workspace that communicates with the main workspace (or any workspace from which the function is called) only through input and output variables.

There are a few advantages to using a function rather than a script. The function may embody a sequence of operations that must be repeated several times within a larger program. For example, suppose that there are multiple sections of a cylinder, but each section has different material properties or boundary conditions. It would be possible to cut and paste the script that was written in Section 1.5.3 over and over again in order to solve this problem. A more attractive option (and the one least likely to result in an error) is to turn the script cylinder.m into a function that can be called whenever you need to consider conduction through a cylinder with generation. The function can be debugged and tested until you are sure that it works and then applied with confidence at any later time. The use of functions provides modularity and elegance to a program and facilitates parametric studies and optimizations.

Any computer code that is even moderately complicated should be broken down into smaller, well-defined sub-programs (functions) that can be written and tested separately before they are integrated through well-defined input/output protocols. MATLAB (or EES) programs are no different. It is convenient to develop your code as a script, but you will likely need to turn your script into a function at some point.

Let's turn the script cylinder.m into a function. Save the file cylinder as cylinderf. The first line of the function must declare that it is a function and define the input/output protocol. For the cylinder problem, the inputs might include the number of nodes (the variable N), the cylinder radii (the variables r_in and r_out), and the material conductivity (the variable k). Any other parameter that you are interested in varying could also be provided as an input. The logical outputs include the vector of radial positions that define the nodes (the vector r) and the predicted temperature at these positions (the vector T_C).

```
function[r,T_C]=cylinderf(N,r_in,r_out,k)
```

The keyword function declares the M-file to be a function and the variables in square brackets are outputs; these variables should be assigned in the body of the function and are passed to the calling workspace. The function name follows the equal sign and the variables in parentheses are the inputs. Note that nothing you do within the function can affect any variable that is external to the function other than those that are explicitly defined as output variables. You should also comment out the clear command that was used to develop the script. The clear statement is not needed since the function operates with its own variable space. The function is terminated with an end statement. The cylinderf function is shown below; the modifications to the original script cylinder are indicated in bold.

```
function[r,T_C]=cylinderf(N,r_in,r_out,k)

    %clear;                         %clear all variables from the workspace

    % Inputs
    %r_in=0.1;                      %inner radius of cylinder (m)
    %r_out=0.2;                     %outer radius of cylinder (m)
    g_dot_tp=1e5;                   %constant volumetric generation (W/m^3)
    L=1;                            %unit length of cylinder (m)
    %k=9;                           %thermal conductivity of cylinder material (W/m-K)
    T_infinity_in=20+273.2;         %average temperature of fluid inside cylinder (K)
    h_bar_in=100;                   %heat transfer coefficient inside cylinder (W/m^2-K)
    T_infinity_out=100+273.2;       %temperature of fluid outside cylinder (K)
    h_bar_out=200;                  %average heat transfer coefficient at outer surface (W/m^2-K)

    % Setup nodes
    %N=10;                          %number of nodes (-)
    Dr=(r_out-r_in)/(N-1);          %distance between nodes (m)
    for i=1:N
        r(i)=r_in+(i-1)*(r_out-r_in)/(N-1);    %radial position of each node (m)
    end

    %Setup A and b
    A=zeros(N,N);
    b=zeros(N,1);

    %Energy balance for control volume 1
    A(1,1)=-k*L*2*pi*(r(1)+Dr/2)/Dr-h_bar_in*2*pi*r(1)*L;
    A(1,2)=k*L*2*pi*(r(1)+Dr/2)/Dr;
    b(1)=-h_bar_in*2*pi*r(1)*L*T_infinity_in-g_dot_tp*2*pi*r(1)*L*Dr/2;

    %Energy balances for internal control volumes
    for i=2:(N-1)
        A(i,i)=-k*L*2*pi*(r(i)-Dr/2)/Dr-k*L*2*pi*(r(i)+Dr/2)/Dr;
        A(i,i-1)=k*L*2*pi*(r(i)-Dr/2)/Dr;
        A(i,i+1)=k*L*2*pi*(r(i)+Dr/2)/Dr;
        b(i)=-g_dot_tp*2*pi*r(i)*L*Dr;
    end

    %Energy balance for control volume N
    A(N,N)=-k*L*2*pi*(r(N)-Dr/2)/Dr-h_bar_out*2*pi*r(N)*L;
    A(N,N-1)=k*L*2*pi*(r(N)-Dr/2)/Dr;
    b(N)=-h_bar_out*2*pi*r(N)*L*T_infinity_out-g_dot_tp*2*pi*r(N)*Dr*L/2;

    % Solve for unknowns
    X=A\b;                          %solve for unknown vector, X
    T=X;                            %assign temperatures from X
    T_C=T-273.2;                    %in C
end
```

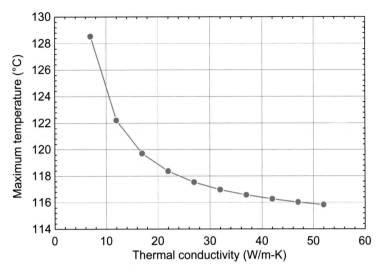

Figure 1-27: Maximum temperature as a function of thermal conductivity.

The following code, typed in the main workspace, will call the function cylinderf for a specific set of values of the input parameters: $N = 10$, $r_{in} = 0.1$ m, $r_{out} = 0.2$ m, and $k = 9$ W/m-K.

```
>> N=10;
>> r_in=0.1;
>> r_out=0.2;
>> k=9;
>> [r,T_C]=cylinderf(N,r_in,r_out,k);
```

The vectors r and T_C are the same as those determined in Section 1.5.3.

It is easy to carry out a parametric study using functions. For example, create a new script called varyk that calls the function cylinderf for a range of conductivity and keeps track of the maximum temperature in the wall that is predicted for each value of conductivity.

```
N=50;                                    %number of nodes
r_in=0.1;                                %inner radius (m)
r_out=0.2;                               %outer radius (m)
Nk=10;                                   %number of values of k to investigate
for i=1:Nk
    k(i,1)=2+i*50/Nk;        %a vector consisting of the conductivities to be considered (W/m-K)
    [r,T_C]=cylinderf(N,r_in,r_out,k(i,1));      %call the cylinderf function
    T_max_C(i,1)=max(T_C);            %determine the maximum temperature
end
```

Call the script varyk from the main workspace and plot the maximum temperature as a function of conductivity (Figure 1-27).

```
>> varyk
>> plot(k,T_max_C)
```

It is possible to call a function from within a function. The volumetric generation in the cylinder was specified in Section 1.4.2 as a function of radius according to:

$$\dot{g}''' = a + br + cr^2 \qquad (1\text{-}159)$$

where $a = 1 \times 10^4$ W/m^3, $b = 2 \times 10^5$ W/m^4, and $c = 5 \times 10^7$ W/m^5. Generate a sub-function (a function that is only visible to other functions in the same M-file) that has one input (r, the radial position) and one output (g, the volumetric rate of thermal energy generation). The function below, generation, placed at the bottom of the cylinderf file will be callable from within the function cylinderf.

```
function[g]=generation(r)
    %the generation function returns the volumetric
    %generation (W/m^3) as a function of radius (m)

    %constants for generation function
    a=1e4;          %W/m^3
    b=2e5;          %W/m^4
    c=5e7;          %W/m^5
    g=a+b*r+c*r^2;  %volumetric generation
end
```

Replacing the constant generation within the cylinderf code with calls to the function generation will implement the solution for non-uniform generation; the altered portion of the code is shown in bold.

```
%Energy balance for control volume 1
A(1,1)=-k*L*2*pi*(r(1)+Dr/2)/Dr-h_bar_in*2*pi*r(1)*L;
A(1,2)=k*L*2*pi*(r(1)+Dr/2)/Dr;
b(1)=-h_bar_in*2*pi*r(1)*L*T_infinity_in-generation(r(1))*2*pi*r(1)*L*Dr/2;

%Energy balances for internal control volumes
for i=2:(N-1)
    A(i,i)=-k*L*2*pi*(r(i)-Dr/2)/Dr-k*L*2*pi*(r(i)+Dr/2)/Dr;
    A(i,i-1)=k*L*2*pi*(r(i)-Dr/2)/Dr;
    A(i,i+1)=k*L*2*pi*(r(i)+Dr/2)/Dr;
    b(i)=-generation(r(i))*2*pi*r(i)*L*Dr;
end

%Energy balance for control volume N
A(N,N)=-k*L*2*pi*(r(N)-Dr/2)/Dr-h_bar_out*2*pi*r(N)*L;
A(N,N-1)=k*L*2*pi*(r(N)-Dr/2)/Dr;
b(N)=-h_bar_out*2*pi*r(N)*L*T_infinity_out-generation(r(N))*2*pi*r(N)*Dr*L/2;
```

Figure 1-28 illustrates the temperature as a function of radius predicted by the model, modified to account for the non-uniform volumetric generation. Figure 1-28 is identical to Figure 1-14, the solution obtained using EES in Section 1.4.2.

1.5.5 Sparse Matrices

The problem considered in Section 1.5.3 can be solved more efficiently (from a computational time standpoint) using sparse matrix solution techniques. Sparse matrices are matrices with mostly zero elements. The matrix $\underline{A}$ that was setup in Section 1.5.3 has

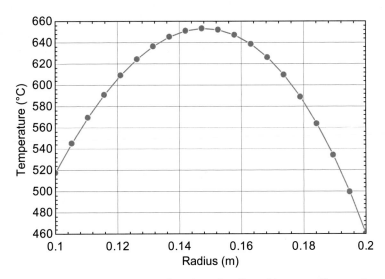

Figure 1-28: Temperature as a function of radius with non-uniform volumetric generation.

many zero elements and the fraction of the entries in the matrix that are zero increases with increasing N.

Matrix $\underline{\underline{A}}$ is tridiagonal, which means that it only has non-zero values on the diagonal and on the super- and sub-diagonal positions. This type of banded matrix will occur frequently in numerical solutions of conduction heat transfer problems because the non-zero elements are related to the coefficients in the energy equations and therefore represent thermal interactions between different nodes. Typically, only a few nodes can directly interact and therefore there will be only a few non-zero coefficients in any row.

The script below (varyN) keeps track of the time required to run the cylinderf function as the number of nodes in the solution increases. The MATLAB function toc returns the elapsed time relative to the time when the function tic was executed. These functions provide a convenient way to keep track of how much time various parts of a MATLAB program are consuming. A more complete delineation of the execution time within a function can be obtained using the profile function in MATLAB.

```
clear;
r_in=0.1;                              %inner radius (m)
r_out=0.2;                             %outer radius (m)
k=9;                                   %thermal conductivity (W/m-K)
for i=1:9
    N(i,1)=2^(i+1);                    %number of nodes (-)
    tic;                               %start time
    [r,T_C]=cylinderf(N(i,1),r_in,r_out,k);   %call cylinder function
    time(i,1)=toc;                     %end timer and record time
end
```

The elapsed time as a function of the number of nodes is shown in Figure 1-29. The computational time grows approximately with the number of nodes to the second power. There is an upper limit to the number of nodes that can be considered that depends on the amount of memory installed in your personal computer. It is likely that you cannot set N to be greater than a few thousand nodes.

The problem can be solved more efficiently if sparse matrices are used. Rather than initializing $\underline{\underline{A}}$ as a full matrix of zeros, it can be initialized as a sparse matrix using the

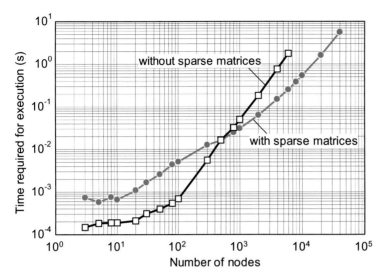

Figure 1-29: Time required to run the cylinderf.m function as a function of the number of nodes with and without sparse matrices.

spalloc (sparse matrix allocation) command. The spalloc command requires three arguments, the first two are the dimensions of the matrix and the last is the number of non-zero entries. Equations (1-145), (1-149), and (1-154) show that each equation will include at most three unknowns and therefore each row of $\underline{\underline{A}}$ will have at most three non-zero entries. Therefore, the matrix $\underline{\underline{A}}$ can have no more than $3\,N$ non-zero entries.

```
%A=zeros(N,N);
A=spalloc(N,N,3*N);
```

When the variable A is defined by spalloc, only the non-zero entries of the matrix are tracked. MATLAB operates on sparse matrices just as it does on full matrices. The remainder of the function cylinderf does not need to be modified, however the function is now much more efficient for large numbers of nodes. If the script varyN is run again, you will find that you can use much larger values of N before running out of memory and also that the code executes much faster at large values of N. The execution time as a function of the number of nodes for the cylinderf function using sparse matrices is also shown in Figure 1-29. Notice that the sparse matrix code is actually somewhat less efficient for small values of N due to the overhead required to set up the sparse matrices; however, for large values of N, the code is much more efficient. The computation time grows approximately with N to the first power when sparse matrices are used.

The use of sparse matrices may not be particularly important for the steady-state, 1-D problem investigated in this section. However, the 2-D and transient problems investigated in subsequent chapters require many more nodes in order to obtain accurate solutions and therefore the use of sparse matrices becomes important for these problems.

1.5.6 Temperature-Dependent Properties

The cylinder problem considered in Section 1.5.3 is an example of a linear problem. The set of equations in a linear problem can be represented in the form $\underline{\underline{A}}\,\underline{X} = \underline{b}$, where $\underline{\underline{A}}$ is a matrix and $\underline{b}$ is a vector. Linear problems can be solved in MATLAB without iteration. The inclusion of temperature-dependent conductivity (or generation, or any aspect of the problem) causes the problem to become non-linear. The introduction of temperature-dependent properties did not cause any apparent problem for the EES model discussed in Section 1.4.3, although it became important to identify a good set of guess values. EES automatically detected the non-linearity and iterated as necessary to solve the non-linear system of equations. However, non-linearity complicates the solution using MATLAB because the equations can no longer be put directly into matrix format. The coefficients multiplying the unknown temperatures themselves depend on the unknown temperatures. It is necessary to use some type of a relaxation process in order to use MATLAB to solve the problem. There are a few options for solving this kind of nonlinear problem; in this section, a technique that is sometimes referred to as successive substitution is discussed.

The successive substitution process begins by assuming a temperature distribution throughout the computational domain (i.e., assume a value of temperature for each node, $\hat{T}_i$ for $i = 1..N$). The assumed values of temperature are used to compute the coefficients that are required to set up the matrix equation (e.g., the temperature-dependent conductivity). The matrix equation is subsequently solved, as discussed in Section 1.5.3, which results in a prediction for the temperature distribution throughout the computational domain (i.e., a predicted value of the temperature for each node, T_i for $i = 1..N$). The assumed and predicted temperatures at each node are compared and used to compute an error; for example, the sum of the square of the difference between the value of $\hat{T}_i$ and T_i at every node. If the error is greater than some threshold value, then the process is repeated, this time using the solution T_i as the assumed temperature distribution $\hat{T}_i$ in order to calculate the coefficients of the matrix equation. The implementation of the successive substitution process carries out the solution that was developed in Sections 1.5.2 through 1.5.4 within a while loop that terminates when the error becomes sufficiently small. This process is illustrated schematically in Figure 1-30 and demonstrated in EXAMPLE 1.5-1.

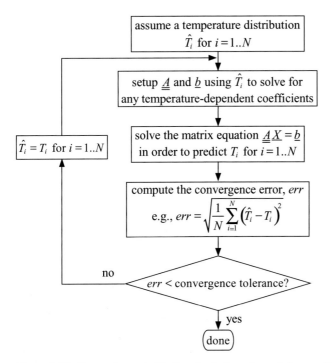

Figure 1-30: Successive substitution technique for solving problems with temperature-dependent properties.

EXAMPLE 1.5-1: THERMAL PROTECTION SYSTEM

EXAMPLE 1.5-1: THERMAL PROTECTION SYSTEM

The kinetic energy associated with the atmospheric entry of a space vehicle results in extremely large heat fluxes, large enough to completely vaporize the vehicle if it were not adequately protected. The outer structure of the vehicle is called its aeroshell and the outer layer of the material on the aeroshell is called the Thermal Protection System (or TPS). The heat flux experienced by the aeroshell can reach $100 \, \text{W/cm}^2$, albeit for only a short period of time.

Consider a TPS consisting of a non-metallic ablative layer with thickness, $th_{ab} = 5$ cm that is bonded to a layer of steel with thickness, $th_s = 1$ cm, as shown in Figure 1. The outer edge of the ablative heat shield ($x = 0$) reaches the material's melting temperature ($T_m = 755$ K) under the influence of the heat flux. The melting limits the temperature that is reached at the outer surface of the shield and protects the internal air until the shield is consumed. In this problem, we will assume that the shield is consumed very slowly so that a quasi-steady temperature distribution is set up in the ablative shield. The latent heat of fusion of the ablative shield is $\Delta i_{fus,ab} = 200 \, \text{kJ/kg}$ and its density is $\rho_{ab} = 1200 \, \text{kg/m}^3$.

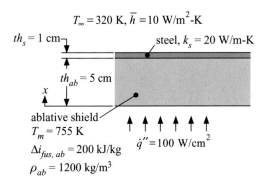

Figure 1: A Thermal Protection System.

The thermal conductivity of the ablative shield is highly temperature dependent; thermal conductivity values at several temperatures are provided in Table 1.

Table 1: Thermal conductivity of ablative shield material in the solid phase

Temperature	Thermal conductivity
300 K	0.10 W/m-K
350 K	0.15 W/m-K
400 K	0.19 W/m-K
450 K	0.21 W/m-K
500 K	0.22 W/m-K
550 K	0.24 W/m-K
600 K	0.28 W/m-K
650 K	0.33 W/m-K
700 K	0.38 W/m-K
755 K	0.45 W/m-K

EXAMPLE 1.5-1: THERMAL PROTECTION SYSTEM

EXAMPLE 1.5-1: THERMAL PROTECTION SYSTEM

The thermal conductivity of the steel may be assumed to be constant at $k_s = 20$ W/m-K. The internal surface of the steel is exposed to air at $T_\infty = 320$ K with average heat transfer coefficient, $\bar{h} = 10$ W/m²-K.

Assume that the TPS reaches a quasi-steady-state under the influence of a heat flux $\dot{q}'' = 100$ W/cm² and that the surface temperature of the ablation shield reaches its melting point.

a) Develop a numerical model using MATLAB that can determine the heat flux that is transferred to the air and the rate that the ablative shield is being consumed.

The solution is developed as a MATLAB function called Ablative_shield; the input to the function is the number of nodes to use in the solution while the outputs include the position of the nodes and the predicted temperature at each node as well as the two quantities specifically requested, the heat flux incident on the air and the rate of shield ablation. Select New and M-File from the File menu and save the M-file as Ablative_shield (the .m extension is added automatically). The first line of the function establishes the input/output protocol:

```
function[x,T,q_flux_in_Wcm2,dthabdt_cms]=Ablative_shield(N)

%EXAMPLE 1.5-1: Thermal Protection System for Atmospheric Entry
%
% Inputs:
% N: number of nodes in solution (-)
%
% Outputs:
% x: position of nodes (m)
% T: temperatures at nodes (K)
% q_flux_in_Wcm2: heat flux to air (W/cm^2)
% dthabdt_cms: rate of shield consumption (cm/s)
```

The next section of the code establishes the remaining input parameters (i.e., those not provided as arguments to the function); note that each input is converted immediately to SI units.

```
th_ab=0.05;                %ablation shield thickness (m)
th_s=0.01;                 %steel thickness (m)
k_s=20;                    %steel conductivity (W/m-K)
q_flux=100*100^2;          %heat flux (W/m^2)
T_m=755;                   %melting temperature (K)
DELTAi_fus_ab=200e3;       %latent heat of fusion (J/kg)
h_bar=10;                  %heat transfer coefficient (W/m^2-K)
T_infinity=320;            %internal air temperature (K)
rho_ab=1200;               %density (kg/m^3)
A_c=1;                     %per unit area of wall (m^2)
```

In order to solve this problem, it is necessary to create a function that returns the conductivity of the ablative shield material. The easiest way to do this is to enter the data from Table 1 into a sub-function and interpolate between the data points.

EXAMPLE 1.5-1: THERMAL PROTECTION SYSTEM

```
function[k]=k_ab(T)

    %data
    Td=[300,350,400,450,500,550,600,650,700,755];
    kd=[0.1,0.15,0.19,0.21,0.22,0.24,0.28,0.33,0.38,0.45];
    k=interp1(Td,kd,T,'spline'); %interpolate data to obtain conductivity
end
```

The interp1 function in MATLAB is used for the interpolation. The interp1 function requires three arguments; the first two are the vectors of the independent and dependent data, respectively, and the third is the value of the independent variable at which you want to find the dependent variable. An optional fourth argument specifies the type of interpolation to use. To obtain more detailed help for this (or any) MATLAB function, use the help command:

```
>> help interp1
INTERP1 1-D interpolation (table lookup)
    YI=INTERP1(X,Y,XI) interpolates to find YI, the values of the
    underlying function Y at the points in the array XI. X must be a
    vector of length N.
    If Y is a vector, then . . .
```

The numerical model of the TPS will consider the ablative material; the steel will be considered as part of the thermal resistance between the inner surface of the shield and the air and therefore will affect the boundary condition at $x = th_{ab}$. There is no reason to treat the steel with the numerical model since the steel has, by assumption, constant properties and is at steady state. Therefore, the analytical solution derived in Section 1.2.3 for the resistance of a plane wall holds exactly.

The nodes are distributed uniformly from $x = 0$ to $x = th_{ab}$, where $x = 0$ corresponds to the outer surface of the shield, as shown in Figure 2.

$$x_i = (i - 1)\frac{th_{ab}}{(N - 1)} \quad \text{for } i = 1 \ldots N$$

The distance between adjacent nodes (Δx) is:

$$\Delta x = \frac{th_{ab}}{(N - 1)}$$

The nodes are setup in the MATLAB code according to:

```
%setup nodes
DELTAx=th_ab/(N-1);            %distance between nodes
for i=1:N
    x(i,1)=th_ab*(i-1)/(N-1);  %position of each node
end
```

EXAMPLE 1.5-1: THERMAL PROTECTION SYSTEM

An internal control volume, shown in Figure 2, experiences only conduction; therefore, a steady-state energy balance on the control volume is:

$$\dot{q}_{top} + \dot{q}_{bottom} = 0 \tag{1}$$

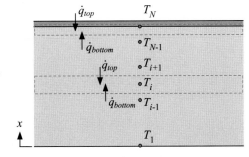

Figure 2: Distribution of nodes and control volumes.

The conductivity used to approximate the conduction heat transfer rates in Eq. (1) must be evaluated at the temperature of the boundaries, i.e., the average of the temperatures of the nodes involved in the conduction process, as discussed previously in Section 1.4.3. With this understanding, these rate equations become:

$$\dot{q}_{top} = k_{ab,T=(T_{i+1}+T_i)/2} \frac{A_c}{\Delta x} (T_{i+1} - T_i) \tag{2}$$

$$\dot{q}_{bottom} = k_{ab,T=(T_{i-1}+T_i)/2} \frac{A_c}{\Delta x} (T_{i-1} - T_i) \tag{3}$$

where A_c is the cross-sectional area. Substituting Eqs. (2) and (3) into Eq. (1) leads to:

$$k_{ab,T=(T_{i+1}+T_i)/2} \frac{A_c}{\Delta x} (T_{i+1} - T_i) + k_{ab,T=(T_{i-1}+T_i)/2} \frac{A_c}{\Delta x} (T_{i-1} - T_i) = 0 \quad \text{for } i = 2..(N-1) \tag{4}$$

The node on the outer surface (i.e., node 1) has a specified temperature, the melting temperature of the ablative material:

$$T_1 = T_m \tag{5}$$

The energy balance for the node on the inner surface (i.e., node N) is also shown in Figure 2:

$$\dot{q}_{top} + \dot{q}_{bottom} = 0 \tag{6}$$

where

$$\dot{q}_{bottom} = k_{ab,T=(T_{N-1}+T_N)/2} \frac{A_c}{\Delta x} (T_{N-1} - T_N) \tag{7}$$

$$\dot{q}_{top} = \frac{(T_\infty - T_N)}{R_{cond,s} + R_{conv}} \tag{8}$$

EXAMPLE 1.5-1: THERMAL PROTECTION SYSTEM

where $R_{cond,s}$ and R_{conv} are the thermal resistances associated with conduction through the steel and convection from the internal surface of the steel to the air.

$$R_{cond,s} = \frac{th_s}{k_s \, A_c}$$

$$R_{conv} = \frac{1}{\bar{h} \, A_c}$$

```
R_cond_s=th_s/(A_c*k_s);              %conduction resistance of steel (K/W)
R_conv=1/(A_c*h_bar);                 %convection resistance (K/W)
```

Substituting Eqs. (7) and (8) into Eq. (6) leads to:

$$\frac{(T_\infty - T_N)}{R_{cond,s} + R_{conv}} + k_{ab,T=(T_{N-1}+T_N)/2} \frac{A_c}{\Delta x} (T_{N-1} - T_N) = 0 \qquad (9)$$

Note that Eqs. (4), (5), and (9) are a complete set of equations in the unknown temperatures T_i for $i = 1..N$; however, these equations cannot be written as a linear combination of the unknown temperatures because the conductivity of the ablative shield depends on temperature. In order to apply successive substitution, the conductivity will be evaluated using guess values for these temperatures ($\hat{T}$). A linear variation in temperature from T_m to T_∞ is used as the guess values to start the process:

$$\hat{T}_i = T_m + (T_\infty - T_m) \frac{(i-1)}{(N-1)} \quad \text{for } i = 1..N$$

```
%initial guess for temperature distribution
for i=1:N
    Tg(i,1)=T_m+(T_infinity-T_m)*(i-1)/(N-1); %linear from melting to air (K)
end
```

The matrix A and vector b are initialized according to:

```
%setup matrices
A=spalloc(N,N,3*N);
b=zeros(N,1);
```

Equation (5) is rewritten to make it clear what the coefficient and the constants are:

$$T_1 \underbrace{[1]}_{A_{1,1}} = \underbrace{T_m}_{b_1}$$

```
%node 1
A(1,1)=1;
b(1,1)=T_m;
```

EXAMPLE 1.5-1: THERMAL PROTECTION SYSTEM

Equation (4) is rewritten, using the guess temperatures to compute the conductivity of the ablative shield and also to clearly identify the coefficients and constants:

$$T_i \underbrace{\left[-k_{ab,T=(\hat{T}_{i+1}+\hat{T}_i)/2} \frac{A_c}{\Delta x} - k_{ab,T=(\hat{T}_{i-1}+\hat{T}_i)/2} \frac{A_c}{\Delta x} \right]}_{A_{i,i}} + T_{i+1} \underbrace{\left[k_{ab,T=(\hat{T}_{i+1}+\hat{T}_i)/2} \frac{A_c}{\Delta x} \right]}_{A_{i,i+1}}$$

$$+ T_{i-1} \underbrace{\left[k_{ab,T=(\hat{T}_{i-1}+\hat{T}_i)/2} \frac{A_c}{\Delta x} \right]}_{A_{i,i-1}} = 0 \quad \text{for } i = 2..(N-1)$$

```
%internal nodes
for i=2:(N-1)
    A(i,i)=-k_ab((Tg(i)+Tg(i+1))/2)*A_c/DELTAx-k_ab((Tg(i)+Tg(i-1))/2)*A_c/DELTAx;
    A(i,i+1)=k_ab((Tg(i)+Tg(i+1))/2)*A_c/DELTAx;
    A(i,i-1)=k_ab((Tg(i)+Tg(i-1))/2)*A_c/DELTAx;
end
```

Equation (9) is rewritten:

$$T_N \underbrace{\left[-\frac{1}{R_{cond,s} + R_{conv}} - k_{ab,T=(\hat{T}_{N-1}+\hat{T}_N)/2} \frac{A_c}{\Delta x} \right]}_{A_{N,N}}$$

$$+ T_{N-1} \underbrace{\left[k_{ab,T=(\hat{T}_{N-1}+\hat{T}_N)/2} \frac{A_c}{\Delta x} \right]}_{A_{N,N-1}} = \underbrace{-\frac{T_\infty}{R_{cond,s} + R_{conv}}}_{b_N}$$

```
%node N
A(N,N)=-k_ab((Tg(N)+Tg(N-1))/2)*A_c/DELTAx-1/(R_cond_s+R_conv);
A(N,N-1)=k_ab((Tg(N)+Tg(N-1))/2)*A_c/DELTAx;
b(N,1)=-T_infinity/(R_cond_s+R_conv);
```

The matrix equation is solved:

```
X=A\b;                                    %solve matrix equation
T=X;
```

The solution is not complete, it is necessary to iterate until the solution (T) matches the assumed temperature ($\hat{T}$). This is accomplished by placing the commands that setup and solve the matrix equation within a while loop that terminates when the rms error (*err*) is below some tolerance (*tol*). The rms error is computed according to:

$$err = \sqrt{\frac{1}{N} \sum_{i=1}^{N} (T_i - \hat{T}_i)^2}$$

EXAMPLE 1.5-1: THERMAL PROTECTION SYSTEM

```
err=sqrt(sum((T-Tg).^2)/N)                    %compute rms error
```

Note that the sum command computes the sum of all of the elements in the vector provided to it and the use of .^2 indicates that each element in the vector should be squared (as opposed to ^2 which would multiply the vector by itself). Also notice that the error computation is not terminated with a semicolon so that the value of the rms error will be reported after each iteration.

To start the iteration process, the value of the error is set to a large number (larger than *tol*) in order to ensure that the while loop executes at least once. After the solution has been obtained, the rms error is computed and the vector Tg is reset to the vector T. The result is shown below, with the new lines highlighted in bold.

```
err=999;                           %initial value of error (K), must be larger than tol
tol=0.01;                          %tolerance for convergence (K)
while (err>tol)

    %node 1
    A(1,1)=1;
    b(1,1)=T_m;

    %internal nodes
    for i=2:(N-1)
        A(i,i)=-k_ab((Tg(i)+Tg(i+1))/2)*A_c/DELTAx-k_ab((Tg(i)+Tg(i-1))/2)*A_c/DELTAx;
        A(i,i+1)=k_ab((Tg(i)+Tg(i+1))/2)*A_c/DELTAx;
        A(i,i-1)=k_ab((Tg(i)+Tg(i-1))/2)*A_c/DELTAx;
    end

    %node N
    A(N,N)=-k_ab((Tg(N)+Tg(N-1))/2)*A_c/DELTAx-1/(R_cond_s+R_conv);
    A(N,N-1)=k_ab((Tg(N)+Tg(N-1))/2)*A_c/DELTAx;
    b(N,1)=-T_infinity/(R_cond_s+R_conv);

    X=A\b;           %solve matrix equation
    T=X;

    err=sqrt(sum((T-Tg).^2)/N)         %compute rms error
    Tg=T;                              %reset guess values used to setup A and b
end
```

The heat flux to the air ($\dot{q}''_{in}$) is computed.

$$\dot{q}''_{in} = \frac{(T_N - T_\infty)}{A_c\,(R_s + R_{conv})}$$

The rate at which the ablative shield is consumed can be determined using an energy balance at the outer surface; the heat flux related to re-entry either consumes the shield or is transferred to the air. Note that this is actually a simplification of the problem; this is a moving boundary problem and this solution is valid only in the limit that the energy carried by the motion of interface is small relative to the energy removed by its vaporization.

$$\dot{q}'' = \rho_{ab}\,\Delta i_{fus,ab}\,\frac{dth_{ab}}{dt} + \dot{q}''_{in}$$

EXAMPLE 1.5-1: THERMAL PROTECTION SYSTEM

or

$$\frac{dth_{ab}}{dt} = \frac{\dot{q}'' - \dot{q}''_{in}}{\rho_{ab}\,\Delta i_{fus,ab}}$$

These calculations are provided by adding the following lines to the Ablative_shield function:

```
q_flux_in=(T(N)-T_infinity)/(R_cond_s+R_conv)/A_c;    %heat flux to air (W/m^2)
q_flux_in_Wcm2=q_flux_in/100^2;                        %heat flux to air (W/cm^2)
dthabdt=(q_flux-q_flux_in)/(DELTAi_fus_ab*rho_ab);     %rate of shield consumption (m/s)
dthabdt_cms=dthabdt*100;                               %rate of shield consumption (cm/s)
end
```

Calling the function Ablative_shield from the workspace leads to:

```
>> [x,T,q_flux_in_Wcm2,dthabdt_cms]=Ablative_shield(100);
err =
    105.7742
err =
    9.0059
err =
    0.8979
err =
    0.1798
err =
    0.0238
err =
    0.0035
```

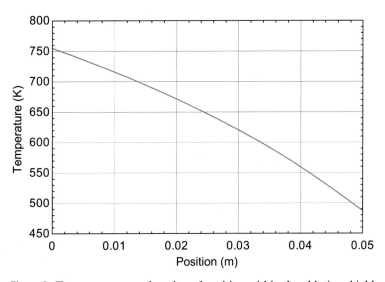

Figure 3: Temperature as a function of position within the ablative shield.

Figure 3 shows the temperature as a function of position within the shield. The temperature gradient agrees with intuition; the temperature gradient is smaller where the conductivity is largest (i.e., at higher temperatures), which is consistent

EXAMPLE 1.5-1: THERMAL PROTECTION SYSTEM

with a constant heat flow. It is important to verify that the number of nodes used in the solution is adequate. Figure 4 shows the heat flux to the air as a function of the number of nodes in the solution and indicates that at least 20 nodes are required. The heat flux to the air at the inner surface of the TPS is 0.166 W/cm², nearly three orders of magnitude less than the heat flux at the outer surface. The TPS is being consumed at a rate of 0.416 cm/s suggesting that the atmospheric entry process cannot last more than ten seconds without consuming the entire shield. The thermal analysis of this problem does not consider the loss of ablative material with time and it is therefore a very simplified model of the TPS.

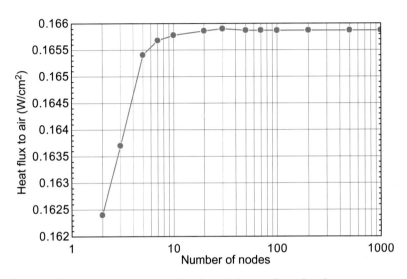

Figure 4: Heat flux to the air as a function of the number of nodes.

1.6 Analytical Solutions for Constant Cross-Section Extended Surfaces

1.6.1 Introduction

The situations that were examined in Sections 1.2 through 1.5 were truly one-dimensional; that is, the geometry and boundary conditions dictated that the temperature could only vary in one direction. In this section, problems that are only approximately 1-D, referred to as extended surfaces, are considered. Extended surfaces are thin pieces of conductive material that can be approximated as being isothermal in two dimensions with temperature variations in only one direction. Extended surfaces are particularly relevant to a large number of thermal engineering applications because the fins that are used to enhance heat transfer in heat exchangers can often be treated as extended surfaces.

1.6.2 The Extended Surface Approximation

An extended surface is not truly 1-D; however, it is often approximated as being such in order to simplify the analysis. Figure 1-31 shows a simple extended surface, sometimes called a fin. The fin length (in the x-direction) is L and its thickness (in the y-direction) is th. The width of the fin in the z-direction, W, is assumed to be much larger than its thickness in the y-direction, th. The conductivity of the fin material is k. The base of

Figure 1-31: A constant cross-sectional fin.

the fin (at $x = 0$) is maintained at temperature T_b and the fin is surrounded by fluid at temperature T_∞ with heat transfer coefficient $\overline{h}$.

Energy is conducted axially from the base of the fin. As energy moves along the fin in the x-direction, it is also conducted laterally to the fin surface where it is finally transferred by convection to the surrounding fluid. Temperature gradients always accompany the transfer of energy by conduction through a material; thus, the temperature must vary in both the x- and y-directions and the temperature distribution in the extended surface must be 2-D. However, there are many situations where the temperature gradient in the y-direction is small and therefore can be neglected in the solution without significant loss of accuracy.

Figure 1-32 illustrates the temperature as a function of lateral position (y) at various axial locations (x) for an arbitrary set of conditions. (The 2-D solution for this fin is derived in EXAMPLE 2.2-1.) Notice that the temperature decreases in both the x- and y-directions as conduction occurs in both of these directions.

At every value of axial location, x, there is a temperature drop through the material in the y-direction due to conduction. (For $x/L = 0.25$, this temperature drop is labeled $\Delta T_{cond,y}$ in Figure 1-32.) There is another temperature drop from the surface of the material to the surrounding fluid due to convection (labeled ΔT_{conv} in Figure 1-32 for $x/L = 0.25$). The extended surface approximation refers to the assumption that the temperature in the material is a function only of x and not of y; this approximation turns a 2-D problem into a 1-D problem, which is easier to solve. The extended surface approximation is valid when the temperature drop due to conduction in the y-direction is much less than the temperature drop due to convection (i.e, $\Delta T_{cond,y} \ll \Delta T_{conv}$). The

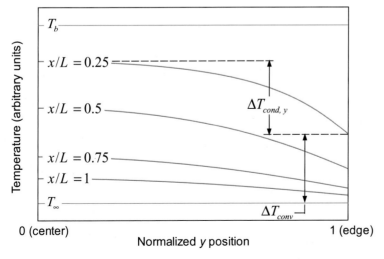

Figure 1-32: Temperature as a function of $y/(th/2)$ for various values of x/L.

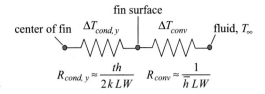

Figure 1-33: Heat transfer in the y-direction within the fin may be approximately represented by two thermal resistances that are related to conduction and convection.

extended surface approximation would not be appropriate for the situation illustrated in Figure 1-32.

The best way to compare the magnitude of these two temperature drops is to think in terms of a resistance network; there are two thermal resistances that oppose heat transfer in the y-direction, conduction and convection, as shown in Figure 1-33. The resistance network shown in Figure 1-33 is clearly only approximate, but it is a useful conceptual tool for understanding the problem. The resistance due to conduction in the y-direction ($R_{cond,y}$) is:

$$R_{cond,y} = \frac{th}{2\,k\,W\,L} \qquad (1\text{-}160)$$

and the resistance due to convection (R_{conv}) is:

$$R_{conv} = \frac{1}{\bar{h}\,W\,L} \qquad (1\text{-}161)$$

The temperature drop across a thermal resistance is proportional to the magnitude of the resistance; therefore, the ratio of the temperature drops is approximately equal to the ratio of the resistances:

$$\frac{\Delta T_{cond,y}}{\Delta T_{conv}} \approx \frac{R_{cond,y}}{R_{conv}} \qquad (1\text{-}162)$$

The validity of the extended surface approximation increases as the ratio of the two resistances becomes small relative to unity; a ratio of resistances used for this purpose is referred to as the Biot number (Bi):

$$Bi = \frac{R_{cond,y}}{R_{conv}} \qquad (1\text{-}163)$$

Substituting Eqs. (1-160) and (1-161) into Eq. (1-163) leads to:

$$Bi = \frac{th}{2\,k\,W\,L}\,\frac{\bar{h}\,W\,L}{1} = \frac{th\,\bar{h}}{2\,k} \qquad (1\text{-}164)$$

As the Biot number becomes smaller, there is less error introduced by the extended surface approximation. In many textbooks it is stated that the Biot number should be less than 0.1 in order to use the extended surface approximation; however, this is clearly a matter of engineering judgment and the threshold for an allowable Biot number cannot be stated without some knowledge of the application and the required accuracy of the solution.

The Biot number will show up often in heat transfer in different contexts. The Biot number is really a concept; it represents the ratio of two resistances, one resistance

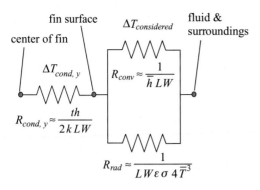

Figure 1-34: Conceptual resistance network for heat transfer in the *y*-direction within the fin when radiation and convection from the fin surface are both considered.

captures a phenomenon that you'd like to neglect and the other resistance captures a phenomenon that you are considering:

$$Bi = \frac{\text{resistance you'd like to neglect in your model}}{\text{resistance that you are considering in your model}} \qquad (1\text{-}165)$$

If the Biot number is much less than unity, then the simpler model can be justified because the resistance you are neglecting is suitably small; however, 'small' is a relative term and it can only be judged in relation to other quantities. The resistance you are neglecting must be small in relation to those that you are considering. In the case of the extended surface problem, the resistance that we'd like to neglect is conduction in the *y*-direction and the resistance that we are going to consider is convection.

There will be situations where Eq. (1-163) is not the correct Biot number to calculate. The resistances that are involved in a problem are not limited to conduction in the *y*-direction and convection. It is important, therefore, that you do not attempt to memorize Eq. (1-164) and apply it to every situation but rather understand the underlying concept of a Biot number. For example, an extended surface model might consider both convection and radiation from the surface of the fin. In this case, the conceptual resistance diagram shown in Figure 1-33 should be modified to include radiation, as shown in Figure 1-34.

The radiation resistance (R_{rad}) is calculated using the approximate formula provided in Table 1-2:

$$R_{rad} = \frac{1}{L\,W\,\varepsilon\,4\,\overline{T}^3} \qquad (1\text{-}166)$$

where ε is the emissivity of the surface, σ is the Stefan-Boltzmann constant, and $\overline{T}$ is the average of the absolute temperature of the fin and the surroundings. The appropriate Biot number that should be calculated in order to evaluate the extended surface approximation in this situation is:

$$Bi = \frac{R_{cond,y}}{\left[\dfrac{1}{R_{conv}} + \dfrac{1}{R_{rad}}\right]^{-1}} = \frac{th\left(\overline{h} + \varepsilon\sigma 4\,\overline{T}^3\right)}{2\,k} \qquad (1\text{-}167)$$

1.6.3 Analytical Solution

The analytical solution to the extended surface problem begins with the derivation of the governing differential equation; this is accomplished using a differential control volume.

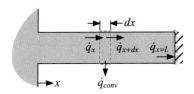

Figure 1-35: Differential control volume used to derive the governing differential equation for an extended surface.

Note that the differential control volume should include material that is at a uniform temperature and therefore it must be differential in x but not in y or z, as shown in Figure 1-35.

The energy balance suggested by Figure 1-35 is:

$$\dot{q}_x = \dot{q}_{conv} + \dot{q}_{x+dx} \tag{1-168}$$

Expanding the higher order term and simplifying leads to:

$$0 = \dot{q}_{conv} + \frac{d\dot{q}}{dx}dx \tag{1-169}$$

The convection term is given by:

$$\dot{q}_{conv} = per\,dx\,\overline{h}\,(T - T_\infty) \tag{1-170}$$

where per is the perimeter of the fin; for the rectangular cross-section shown in Figure 1-31, $per = 2(W + th)$. The conduction term is given by Fourier's law:

$$\dot{q} = -k\,A_c\,\frac{dT}{dx} \tag{1-171}$$

where A_c is the cross-sectional area of the fin; for the fin in Figure 1-31, $A_c = Wth$. Substituting Eqs. (1-170) and (1-171) into Eq. (1-169) leads to:

$$0 = per\,dx\,\overline{h}\,(T - T_\infty) + \frac{d}{dx}\left(-k\,A_c\,\frac{dT}{dx}\right)dx \tag{1-172}$$

The cross-sectional area and conductivity are assumed to be constant, allowing Eq. (1-172) to be simplified:

$$\frac{d^2T}{dx^2} - \frac{per\,\overline{h}}{k\,A_c}\,T = -\frac{per\,\overline{h}}{k\,A_c}\,T_\infty \tag{1-173}$$

Equation (1-173) is a second order, non-homogeneous, linear ordinary differential equation (ODE). It is worth understanding what each of these terms mean before proceeding. The order of the equation refers to order of the highest order derivative; in Eq. (1-173), the highest order derivative is second order. A homogeneous equation is one where any multiple of a solution ($C\,T$ where C is some arbitrary constant and T is a solution) is itself a solution. Substituting $C\,T$ into Eq. (1-173) for T leads to:

$$C\left(\frac{d^2T}{dx^2} - \frac{per\,\overline{h}}{k\,A_c}\,T\right) = -\frac{per\,\overline{h}}{k\,A_c}\,T_\infty \tag{1-174}$$

Substituting Eq. (1-173) into Eq. (1-174) leads to:

$$C\left(-\frac{per\,\overline{h}}{k\,A_c}\,T_\infty\right) = -\frac{per\,\overline{h}}{k\,A_c}\,T_\infty \tag{1-175}$$

which is only true for arbitrary C if $T_\infty = 0$; therefore, Eq. (1-173) is non-homogeneous. A linear equation does not contain any products of the dependent variable or its derivative; therefore, Eq. (1-173) is linear.

Equation (1-173) cannot be solved by direct integration (as was possible for the problems encountered in Sections 1.2 and 1.3) because it is not possible to separate the x and T portions of the differential equation. A differential equation like Eq. (1-173) is typically solved by "separating" it into homogeneous and non-homogeneous (or particular) differential equations. We do this because mathematicians have defined functions that solve many homogeneous differential equations; therefore, we can be confident that it will be possible to deal with the homogeneous differential equation. We are then left with the non-homogeneous part, which is often trivial to solve.

To separate the differential equation, assume that the solution (T) can be expressed as the sum of a homogeneous solution (T_h) and a particular (non-homogeneous) solution (T_p):

$$T = T_h + T_p \tag{1-176}$$

Substituting Eq. (1-176) into Eq. (1-173) leads to:

$$\frac{d^2\,(T_h + T_p)}{dx^2} - \frac{per\,\overline{h}}{k\,A_c}\,(T_h + T_p) = -\frac{per\,\overline{h}}{k\,A_c}\,T_\infty \tag{1-177}$$

or

$$\underbrace{\frac{d^2 T_h}{dx^2} - \frac{per\,\overline{h}}{k\,A_c}\,T_h}_{\substack{=0 \\ \text{for homogeneous} \\ \text{differential equation}}} + \underbrace{\frac{d^2 T_p}{dx^2} - \frac{per\,\overline{h}}{k\,A_c}\,T_p = -\frac{per\,\overline{h}}{k\,A_c}\,T_\infty}_{\substack{\text{whatever is left over must be the} \\ \text{particular differential equation}}} \tag{1-178}$$

Extract from Eq. (1-178) the homogeneous differential equation for T_h:

$$\frac{d^2 T_h}{dx^2} - \frac{per\,\overline{h}}{k\,A_c}\,T_h = 0 \tag{1-179}$$

and whatever is left over must be the particular differential equation:

$$\frac{d^2 T_p}{dx^2} - \frac{per\,\overline{h}}{k\,A_c}\,T_p = -\frac{per\,\overline{h}}{k\,A_c}\,T_\infty \tag{1-180}$$

Let's start with the homogeneous differential equation, Eq. (1-179). How do we solve this equation? Actually, functions have been defined specifically to solve various types of homogeneous equations. The function that solves Eq. (1-179) is the exponential. To see that this is true, assume a solution with an exponential form:

$$T_h = C \exp(m\,x) \tag{1-181}$$

where m and C are both arbitrary constants. Substitute Eq. (1-181) into Eq. (1-179):

$$C\,m^2 \exp(m\,x) - \frac{per\,\overline{h}}{k\,A_c}\,C \exp(m\,x) = 0 \tag{1-182}$$

Equation (1-182) is satisfied if:

$$m^2 = \frac{per\,\overline{h}}{k\,A_c} \tag{1-183}$$

There are actually two exponential equations ($T_{h,1}$ and $T_{h,2}$) that solve Eq. (1-179), corresponding to the positive and negative roots of Eq. (1-183):

$$T_{h,1} = C_1 \exp(m\,x) \tag{1-184}$$

and

$$T_{h,2} = C_2 \exp(-mx) \tag{1-185}$$

where

$$m = \sqrt{\frac{per\,\overline{h}}{k\,A_c}} \tag{1-186}$$

Because Eq. (1-179) is a linear, homogeneous ODE, the sum of the two solutions is also a solution:

$$T_h = C_1 \exp(mx) + C_2 \exp(-mx) \tag{1-187}$$

Equation (1-187) is the homogeneous solution and it will solve the homogeneous differential equation regardless of the choice of C_1 and C_2.

Next, the non-homogeneous (particular) differential equation must be solved. Any solution to Eq. (1-180) will do and it is usually a good idea to start with the simplest possibility. By inspection of Eq. (1-180), it seems likely that a constant will solve the differential equation:

$$T_p = C_3 \tag{1-188}$$

where C_3 is a constant. Substituting Eq. (1-188) into Eq. (1-180) leads to:

$$-\frac{per\,\overline{h}}{k\,A_c}C_3 = -\frac{per\,\overline{h}}{k\,A_c}T_\infty \tag{1-189}$$

or

$$C_3 = T_\infty \tag{1-190}$$

Substituting Eq. (1-190) into Eq. (1-188) leads to the particular solution:

$$T_p = T_\infty \tag{1-191}$$

Substituting the homogeneous and particular solutions, Eqs. (1-187) and (1-191), into Eq. (1-176) leads to:

$$T = C_1 \exp(mx) + C_2 \exp(-mx) + T_\infty \tag{1-192}$$

Equation (1-192) represents the solution to Eq. (1-173) to within two undetermined constants (C_1 and C_2) in the same way that the equation:

$$T = -\frac{\dot{g}'''}{2k}x^2 + C_1 x + C_2 \tag{1-193}$$

from Table 1-3 represents the solution for the temperature in a plane wall with thermal energy generation to within the two constants of integration.

Maple is very good at recognizing the solution to differential equations like Eq. (1-173). Enter the governing differential equation into Maple:

```
> restart;
> ODE:=diff(diff(T(x),x),x)-per*h_bar*T(x)/(k*A_c)=-per*h_bar*T_infinity/(k*A_c);
```

$$ODE := \left(\frac{d^2}{dx^2}T(x)\right) - \frac{per\,h\_bar\,T(x)}{k\,A\_c} = -\frac{per\,h\_bar\,T\_infinity}{k\,A\_c}$$

and obtain the solution using the **dsolve** command:

> Ts:=dsolve(ODE);

$$Ts := T(x) = e^{\left(\frac{\sqrt{per}\sqrt{h\_bar}\,x}{\sqrt{k}\sqrt{A\_c}}\right)}\_C2 + e^{\left(\frac{-\sqrt{per}\sqrt{h\_bar}\,x}{\sqrt{k}\sqrt{A\_c}}\right)}\_C1 + T\_infinity$$

The solution identified by Maple is identical to Eq. (1-192).

The constants C_1 and C_2 are obtained by enforcing the boundary conditions. One boundary condition is clear; the base temperature is specified and therefore:

$$T_{x=0} = T_b \tag{1-194}$$

Substituting Eq. (1-192) into Eq. (1-194) leads to:

$$C_1 + C_2 + T_\infty = T_b \tag{1-195}$$

The boundary condition at the tip of the fin is less clear and there are several possibilities. The most common model assumes that the tip is adiabatic. In this case, an interface balance at the tip (see Figure 1-35) leads to:

$$\dot{q}_{x=L} = 0 \tag{1-196}$$

Substituting Fourier's law into Eq. (1-196) leads to:

$$\left.\frac{dT}{dx}\right|_{x=L} = 0 \tag{1-197}$$

Substituting Eq. (1-192) into Eq. (1-197) leads to:

$$C_1\, m\, \exp\,(m\, L) - C_2\, m\, \exp\,(-m\, L) = 0 \tag{1-198}$$

Note that Eqs. (1-195) and (1-198) are together sufficient to determine C_1 and C_2. If the solution is implemented in EES then no further algebra is required. However, it is worthwhile to obtain the explicit form of the solution for this common problem. Equation (1-195) is multiplied by $m \exp(m\, L)$ and rearranged:

$$C_1\, m \exp\,(m\, L) + C_2\, m \exp\,(m\, L) = (T_b - T_\infty)\, m \exp\,(m\, L) \tag{1-199}$$

Equation (1-198) is added to Eq. (1-199):

$$C_1\, m\, \exp\,(m\, L) - C_2\, m\, \exp\,(-m\, L) = 0$$

$$\frac{+\,[C_1\, m\, \exp(m\, L) + C_2\, m\, \exp(m\, L) = (T_b - T_\infty)\, m\, \exp(m\, L)]}{-C_2\, m\, \exp(-m\, L) - C_2\, m\, \exp(m\, L) = -(T_b - T_\infty)\, m\, \exp(m\, L)} \tag{1-200}$$

Equation (1-200) can be solved for C_2:

$$C_2 = \frac{(T_b - T_\infty)\, \exp\,(m\, L)}{\exp\,(-m\, L) + \exp\,(m\, L)} \tag{1-201}$$

A similar sequence of operations leads to:

$$C_1 = \frac{(T_b - T_\infty)\, \exp\,(-m\, L)}{\exp\,(-m\, L) + \exp\,(m\, L)} \tag{1-202}$$

These constants can also be obtained from Maple; the governing differential equation, Eq. (1-173) is entered and solved, this time in terms of m:

```
> restart;
> ODE:=diff(diff(T(x),x),x)-m^2*T(x)=-m^2*T_infinity;
```

$$ODE := \left(\frac{d^2}{dx^2} T(x) \right) - m^2 T(x) = -m^2 \, T\_infinity$$

```
> Ts:=dsolve(ODE);
```

$$Ts := T(x) = e^{(-mx)} \_C2 + e^{(mx)} \_C1 + T\_infinity$$

The boundary conditions, Eqs. (1-194) and (1-197), are defined:

```
> BC1:=rhs(eval(Ts,x=0))=T_b;
```

$$BC1 := \_C2 + \_C1 + T\_infinity = T\_b$$

```
> BC2:=rhs(eval(diff(Ts,x),x=L))=0;
```

$$BC2 := -m \, e^{(-mL)} \_C2 + m \, e^{(mL)} \_C1 = 0$$

and solved symbolically:

```
> constants:=solve({BC1,BC2},{_C1,_C2});
```

$$constants := \left\{ \_C2 = -\frac{e^{(mL)}(T\_infinity - T\_b)}{e^{(-mL)} + e^{(mL)}}, \ \_C1 = -\frac{e^{(-mL)}(T\_infinity - T\_b)}{e^{(-mL)} + e^{(mL)}} \right\}$$

The constants identified by Maple are identical to Eqs. (1-201) and (1-202). Substituting the constants of integration into the general solution, Eq. (1-192), leads to:

$$T = \frac{(T_b - T_\infty) \exp(-mL)}{\exp(-mL) + \exp(mL)} \exp(mx) + \frac{(T_b - T_\infty) \exp(mL)}{\exp(-mL) + \exp(mL)} \exp(-mx) + T_\infty$$

$$(1\text{-}203)$$

or, using Maple:

```
> Ts:=subs(constants,Ts);
```

$$Ts := T(x) = -\frac{e^{(-mx)}e^{(mL)}(T\_infinity - T\_b)}{e^{(-mL)} + e^{(mL)}} - \frac{e^{(mx)}e^{(-mL)}(T\_infinity - T\_b)}{e^{(-mL)} + e^{(mL)}} + T\_infinity$$

Equation (1-203) can be simplified to:

$$T = (T_b - T_\infty) \frac{[\exp(-m(L-x)) + \exp(m(L-x))]}{[\exp(-mL) + \exp(mL)]} + T_\infty \qquad (1\text{-}204)$$

Equation (1-204) can be stated more concisely using hyperbolic functions as opposed to exponentials; hyperbolic functions are functions that have been defined in terms of exponentials. The combinations:

$$\frac{1}{2}\left[\exp\left(A\right)+\exp\left(-A\right)\right] \tag{1-205}$$

$$\frac{1}{2}\left[\exp\left(A\right)-\exp\left(-A\right)\right] \tag{1-206}$$

occur so frequently in math and science that they are given special names, the hyperbolic cosine (cosh, pronounced "kosh") and the hyperbolic sine (sinh, pronounced "cinch"), respectively.

$$\cosh\left(A\right)=\frac{1}{2}\left[\exp\left(A\right)+\exp\left(-A\right)\right] \tag{1-207}$$

$$\sinh\left(A\right)=\frac{1}{2}\left[\exp\left(A\right)-\exp\left(-A\right)\right] \tag{1-208}$$

These hyperbolic functions behave in much the same way that the cosine and sine functions do. For example:

$$\cosh^2\left(A\right)-\sinh^2\left(A\right)$$

$$=\frac{1}{4}\left[\exp\left(A\right)+\exp\left(-A\right)\right]^2-\frac{1}{4}\left[\exp\left(A\right)-\exp\left(-A\right)\right]^2$$

$$=\frac{1}{4}\left[\exp^2\left(A\right)+2\exp\left(A\right)\exp\left(-A\right)+\exp^2\left(-A\right)\right] \tag{1-209}$$

$$-\frac{1}{4}\left[\exp^2\left(A\right)-2\exp\left(A\right)\exp\left(-A\right)+\exp^2\left(-A\right)\right]$$

$$=\exp\left(A\right)\exp\left(-A\right)=1$$

or

$$\cosh^2\left(A\right)-\sinh^2\left(A\right)=1 \tag{1-210}$$

which is analogous to the trigonometric identity:

$$\cos^2\left(A\right)+\sin^2\left(A\right)=1 \tag{1-211}$$

Furthermore, the derivative of cosh is sinh and vice versa, which is analogous to derivatives of sine and cosine (albeit, without the sign change):

$$\frac{d}{dx}\left[\sinh\left(A\right)\right]=\frac{d}{dx}\left\{\frac{1}{2}\left[\exp\left(A\right)-\exp\left(-A\right)\right]\right\}$$

$$=\frac{1}{2}\left[\exp\left(A\right)+\exp\left(-A\right)\right]\frac{dA}{dx}=\cosh\left(A\right)\frac{dA}{dx} \tag{1-212}$$

or

$$\frac{d}{dx}\left[\sinh\left(A\right)\right]=\cosh\left(A\right)\frac{dA}{dx} \tag{1-213}$$

A similar set of operations leads to:

$$\frac{d}{dx}\left[\cosh\left(A\right)\right]=\sinh\left(A\right)\frac{dA}{dx} \tag{1-214}$$

Equation (1-204) is rearranged so that it can be expressed in terms of hyperbolic cosines:

$$T = (T_b - T_\infty) \underbrace{\frac{[\exp(-m\,(L-x)) + \exp(m\,(L-x))]}{2}}_{\cosh(m(L-x))} \underbrace{\frac{2}{[\exp(-m\,L) + \exp(m\,L)]}}_{1/\cosh(m\,L)} + T_\infty$$

$$(1\text{-}215)$$

$$T = (T_b - T_\infty)\frac{\cosh(m\,(L-x))}{\cosh(m\,L)} + T_\infty \qquad (1\text{-}216)$$

Equation (1-216) is much more concise but functionally identical to Eq. (1-204). Note that Maple can convert from exponential to hyperbolic form as well, using the convert command with the 'trigh' identifier:

```
> T_s:=convert(Ts,'trigh');
```

$$T\_s := T(x) = (-T\_infinity + T\_b)\,\cosh(m\,x) + T\_infinity$$
$$+ \frac{\sinh(m\,x)\,\sinh(m\,L)\,(T\_infinity - T\_b)}{\cosh(m\,L)}$$

The rate of heat transfer to the base of the fin ($\dot{q}_{fin}$) is obtained from Fourier's law evaluated at $x = 0$:

$$\dot{q}_{fin} = -k\,A_c\,\frac{dT}{dx}\bigg|_{x=0} \qquad (1\text{-}217)$$

Substituting Eq. (1-216) into Eq. (1-217) leads to:

$$\dot{q}_{fin} = -k\,A_c\,\frac{d}{dx}\left[(T_b - T_\infty)\frac{\cosh(m\,(L-x))}{\cosh(m\,L)} + T_\infty\right]_{x=0}$$
$$= -\frac{k\,A_c\,(T_b - T_\infty)}{\cosh(m\,L)}\frac{d}{dx}[\cosh(m\,(L-x))]_{x=0} \qquad (1\text{-}218)$$

Recalling that the derivative of cosh is sinh, according to Eq. (1-214):

$$\dot{q}_{fin} = \frac{k\,A_c\,(T_b - T_\infty)}{\cosh(m\,L)}m\,[\sinh(m\,(L-x))]_{x=0} \qquad (1\text{-}219)$$

or

$$\dot{q}_{fin} = (T_b - T_\infty)\,k\,A_c\,m\,\frac{\sinh(m\,L)}{\cosh(m\,L)} \qquad (1\text{-}220)$$

The same result may be obtained from Maple:

```
> q_dot_fin:=-k*A_c*eval(diff(T_s,x),x=0);
```

$$q\_dot\_fin := -k\,A\_c\left(\frac{d}{dx}T(x)\right)\bigg|_{x=0} = -\frac{k\,A\_c\,m\,\sinh(m\,L)(T\_infinity - T\_b)}{\cosh(m\,L)}$$

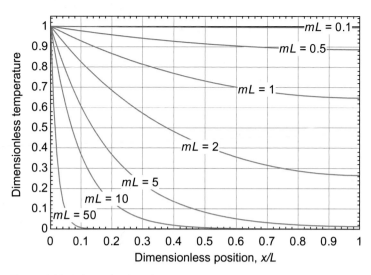

Figure 1-36: Dimensionless fin temperature as a function of dimensionless position for various values of the parameter $m\,L$.

The ratio of sinh to cosh is the hyperbolic tangent (just as the ratio of sine to cosine is tangent); therefore, Eq. (1-220) may be written as:

$$\dot{q}_{fin} = (T_b - T_\infty)\,\sqrt{\overline{h}\;per\;k\,A_c}\;\tanh{(m\,L)} \tag{1-221}$$

The temperature distribution and heat transfer rate provided by Eqs. (1-216) and (1-221) represent the most important aspects of the solution. The solutions for fins with other types of boundary conditions at the tip are summarized in Table 1-4.

1.6.4 Fin Behavior

The temperature distribution within a fin with an adiabatic tip, Eq. (1-216), can be expressed as a ratio of the temperature elevation with respect to the fluid temperature to the base-to-fluid temperature difference:

$$\frac{T - T_\infty}{T_b - T_\infty} = \frac{\cosh\left(m\,L\left(1 - \dfrac{x}{L}\right)\right)}{\cosh{(m\,L)}} \tag{1-222}$$

The dimensionless temperature as a function of dimensionless position (x/L) is shown in Figure 1-36 for various values of $m\,L$.

Regardless of the value of $m\,L$, the solutions satisfy the boundary conditions; the curves intersect at $(T - T_\infty)/(T_b - T_\infty) = 1.0$ at $x/L = 0$ and the slope of each curve is zero at $x/L = 1.0$. However, the shape of the curves changes with $m\,L$. Smaller values of $m\,L$ result in a smaller temperature drop due to conduction along the fin (and therefore more due to convection from the fin surface) whereas large values of $m\,L$ have a corresponding large temperature drop due to conduction and little for convection.

The functionality of an extended surface is governed by two processes; conduction along the fin (in the x-direction) and convection from its surface. (Conduction in the y-direction was neglected in the derivation of the solution.) The parameter $m\,L$ represents the balance of these two effects. The resistance to conduction along the fin ($R_{cond,x}$)

Table 1-4: Solutions for constant cross-section extended surfaces with different end conditions.

	Tip condition	Solution
Adiabatic tip	$\bar{h}, T_\infty$ ... T_b ... x	$\dfrac{T - T_\infty}{T_b - T_\infty} = \dfrac{\cosh\left(m\,(L-x)\right)}{\cosh\left(m\,L\right)}$ $\dot{q}_{fin} = (T_b - T_\infty)\sqrt{\bar{h}\ per\ k\,A_c}\ \tanh\left(m\,L\right)$ $\eta_{fin} = \tanh\left(m\,L\right)/\left(m\,L\right)$
Convection from tip	$\bar{h}, T_\infty$ ... T_b ... x	$\dfrac{T - T_\infty}{T_b - T_\infty} = \dfrac{\cosh\left(m\,(L-x)\right) + \dfrac{\bar{h}}{m\,k}\sinh\left(m\,(L-x)\right)}{\cosh\left(m\,L\right) + \dfrac{\bar{h}}{m\,k}\sinh\left(m\,L\right)}$ $\dot{q}_{fin} = (T_b - T_\infty)\sqrt{\bar{h}\ per\ k\,A_c}\ \dfrac{\sinh\left(m\,L\right) + \dfrac{\bar{h}}{m\,k}\cosh\left(m\,L\right)}{\cosh\left(m\,L\right) + \dfrac{\bar{h}}{m\,k}\sinh\left(m\,L\right)}$ $\eta_{fin} = \dfrac{\left[\tanh\left(m\,L\right) + m\,L\,AR_{tip}\right]}{m\,L\left[1 + m\,L\,AR_{tip}\tanh\left(m\,L\right)\right]\left(1 + AR_{tip}\right)}$
Specified tip temperature	$\bar{h}, T_\infty$ ... T_L ... T_b ... x	$\dfrac{T - T_\infty}{T_b - T_\infty} = \dfrac{\left[\dfrac{T_L - T_\infty}{T_b - T_\infty}\right]\sinh\left(m\,x\right) + \sinh\left(m\,(L-x)\right)}{\sinh\left(m\,L\right)}$ $\dot{q}_{fin} = (T_b - T_\infty)\sqrt{\bar{h}\ per\ k\,A_c}\ \dfrac{\left(\cosh\left(m\,L\right) - \left[\dfrac{T_L - T_\infty}{T_b - T_\infty}\right]\right)}{\sinh\left(m\,L\right)}$
Infinitely long	$\bar{h}, T_\infty$ ... to ∞ ... T_b ... x	$\dfrac{T - T_\infty}{T_b - T_\infty} = \exp\left(-m\,x\right)$ $\dot{q}_{fin} = (T_b - T_\infty)\sqrt{\bar{h}\ per\ k\,A_c}$

where:

T_b = base temperature
T_∞ = fluid temperature
per = perimeter
L = length
T = temperature

$\bar{h}$ = heat transfer coefficient
A_c = cross-sectional area
k = thermal conductivity
$\dot{q}_{fin}$ = fin heat transfer rate
x = position (relative to base of fin)

$m\,L = \sqrt{\dfrac{per\,\bar{h}}{k\,A_c}}\ L$ = fin constant $AR_{tip} = \dfrac{A_c}{per\,L}$ = tip area ratio

is given by:

$$R_{cond,x} = \frac{L}{k\,A_c} \qquad (1\text{-}223)$$

and the resistance to convection from the surface (R_{conv}) is:

$$R_{conv} = \frac{1}{\bar{h}\ per\ L} \qquad (1\text{-}224)$$

Note that the resistance in Eq. (1-223) is related to conduction in the x-direction and should not be confused with $R_{cond,y}$ in Eq. (1-160), which was used to define the Biot number. Clearly the behavior of the fin cannot be exactly represented using these thermal resistances; the analysis in Section 1.6.3 was complex and showed that the conduction along the fin is gradually reduced by convection. Nevertheless, the relative value of these resistances provides substantial insight into the qualitative characteristics of the fin:

$$\frac{R_{cond,x}}{R_{conv}} = \frac{per\,\overline{h}}{k\,A_c}L^2 \qquad (1\text{-}225)$$

The resistance ratio in Eq. (1-225) is related to the parameter $m\,L$:

$$(m\,L)^2 = \left[\sqrt{\frac{per\,\overline{h}}{k\,A_c}}L\right]^2 = \frac{per\,\overline{h}}{k\,A_c}L^2 = \frac{R_{cond,x}}{R_{conv}} \qquad (1\text{-}226)$$

In the light of Eq. (1-226), Figure 1-36 begins to make sense. A small value of $m\,L$ represents a fin with a small resistance to conduction in the x-direction relative to the resistance to convection. The temperature drop due to the conduction heat transfer along the fin must therefore be small. At the other extreme, a large value of $m\,L$ indicates that the resistance to conduction in the x-direction is much larger than the resistance to convection and therefore most of the temperature drop is related to the conduction heat transfer.

Before starting an analysis of an extended surface, it is helpful to calculate the two dimensionless parameters discussed thus far. The value of the Biot number will indicate whether it is possible to treat the situation as a 1-D problem and the value of $m\,L$ will determine whether it is even worth the time. If $m\,L$ is either very small or very large, then the behavior can be understood with no analysis: the fin temperature will be very close to the base temperature or the fluid temperature, respectively.

1.6.5 Fin Efficiency and Resistance

The fin efficiency is defined as the ratio of the heat transfer to the fin ($\dot{q}_{fin}$) to the heat transfer to an ideal fin. An ideal fin is made of an infinitely conductive material and therefore this limit corresponds to a fin that is everywhere at a temperature of T_b. Note that an ideal fin with infinite conductivity corresponds to the limit of $m\,L = 0$ in Figure 1-36.

$$\eta_{fin} = \frac{\text{heat transfer to fin}}{\text{heat transfer to fin as } k \to \infty} \qquad (1\text{-}227)$$

or

$$\eta_{fin} = \frac{\dot{q}_{fin}}{\overline{h}\,A_{s,fin}\,(T_b - T_\infty)} \qquad (1\text{-}228)$$

where the denominator of Eq. (1-228) is the product of the average heat transfer coefficient, the surface area of the fin that is exposed to fluid, and the base-to-fluid temperature difference. The fin efficiency represents the degree to which the temperature drop along the fin due to conduction has reduced the average temperature difference driving convection from the fin surface.

The fin efficiency depends on the boundary condition at the tip and the geometry of the fin. For a constant cross-sectional area fin with an adiabatic tip, Eq. (1-221) can be

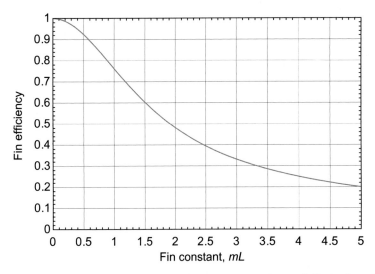

Figure 1-37: Fin efficiency for a constant cross-section, adiabatic tipped fin as a function of the parameter $m\,L$.

substituted into Eq. (1-228):

$$\eta_{fin} = \frac{(T_b - T_\infty)\,\sqrt{\overline{h}\ per\ k\,A_c}\ \tanh{(m\,L)}}{\overline{h}\ per\ L\ (T_b - T_\infty)} \tag{1-229}$$

or

$$\eta_{fin} = \frac{\tanh{(m\,L)}}{\sqrt{\dfrac{\overline{h}\,per}{k\,A_c}}\,L} \tag{1-230}$$

which can be simplified by substituting in the definition of m:

$$\eta_{fin} = \frac{\tanh{(m\,L)}}{m\,L} \tag{1-231}$$

Figure 1-37 illustrates the fin efficiency for a fin having a constant cross-sectional area and an adiabatic tip as a function of $m\,L$. Notice that the fin efficiency drops as $m\,L$ increases. This is consistent with the discussion in Section 1.6.4; a large value of $m\,L$ corresponds to a large temperature drop due to conduction along the fin, as seen in Figure 1-36.

The fin efficiency is the most useful format for presenting the results of a fin solution because it allows the calculation of a fin resistance (R_{fin}). The fin resistance is the thermal resistance that opposes heat transfer from the base of the fin to the surrounding fluid. Equation (1-228) can be rearranged:

$$\dot{q}_{fin} = \underbrace{\eta_{fin}\,\overline{h}\,A_{s,fin}}_{1/R_{fin}}\ (T_b - T_\infty) \tag{1-232}$$

where $A_{s,fin}$ is the surface area of the fin exposed to the fluid. Note that without the fin efficiency, Eq. (1-232) is equivalent to Newton's law of cooling and therefore the thermal resistance is defined in basically the same way:

$$R_{fin} = \frac{1}{\eta_{fin}\,\overline{h}\,A_{s,fin}} \tag{1-233}$$

The fin efficiency is always less than one and therefore the fin resistance will be larger than the corresponding convection resistance; this increase in resistance is related to the conduction resistance within the fin. The concept of a fin resistance is convenient since it allows the effect of fins to be incorporated into a more complex problem (for example, one in which fins are attached to other structures) as additional resistances in a network.

For most extended surfaces, the surface area that is available for convection at the tip is insignificant relative to the total area for convection and therefore the adiabatic tip solution for fin efficiency is sufficient. However, the solution for the heat transfer from a fin that experiences convection from its tip (see Table 1-4) can also be used to provide an expression for fin efficiency:

$$\dot{q}_{fin} = (T_b - T_\infty)\sqrt{\bar{h}\, per\, k\, A_c}\; \frac{\sinh{(m\,L)} + \dfrac{\bar{h}}{m\,k}\cosh{(m\,L)}}{\cosh{(m\,L)} + \dfrac{\bar{h}}{m\,k}\sinh{(m\,L)}} \tag{1-234}$$

So the fin efficiency is:

$$\eta_{fin} = \frac{\dot{q}_{fin}}{\bar{h}\,(per\,L + A_c)} = \frac{\sqrt{\bar{h}\, per\, k\, A_c}}{\bar{h}\,(per\,L + A_c)}\; \frac{\sinh{(m\,L)} + \dfrac{\bar{h}}{m\,k}\cosh{(m\,L)}}{\cosh{(m\,L)} + \dfrac{\bar{h}}{m\,k}\sinh{(m\,L)}} \tag{1-235}$$

which can be simplified somewhat to:

$$\eta_{fin} = \frac{\left[\tanh{(m\,L)} + m\,L\,AR_{tip}\right]}{m\,L\left[1 + m\,L\,AR_{tip}\tanh{(m\,L)}\right](1 + AR_{tip})} \tag{1-236}$$

where AR_{tip} is the ratio of the area for convection from the tip to the surface area along the length of the fin:

$$AR_{tip} = \frac{A_c}{per\,L} \tag{1-237}$$

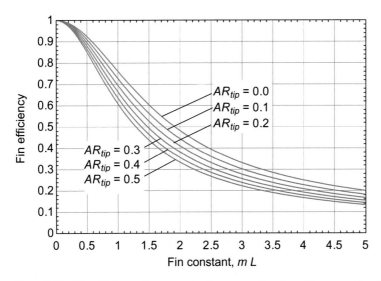

Figure 1-38: Fin efficiency for a constant cross-section fin with convection from the tip as a function of the parameter $m\,L$ and various values of the tip area ratio.

One-Dimensional, Steady-State Conduction

Figure 1-38 illustrates the fin efficiency associated with a fin with a convective tip as a function of the fin parameter ($m\,L$) for various values of the tip area ratio (AR_{tip}). Note that the fin efficiency is reduced as the tip area is larger. This counterintuitive result is related to the fact that the tip area is included in the surface area that is available for convection from an ideal fin in the definition of the fin efficiency. The heat transfer rate from the fin will increase as the tip area is increased; however, the rate that heat could be transferred from an ideal (i.e., isothermal) fin would increase by a larger amount.

It is possible to approximately correct for convection from the tip and use the simpler adiabatic tip fin efficiency equation by modifying the length of the fin slightly (e.g., adding the half-thickness of a fin with a rectangular cross-section). In most cases the correction associated with the tip convection is so small that it is not worth considering. In any case, neglecting convection from the tip is slightly conservative and other uncertainties in the problem (e.g., the value of the heat transfer coefficient) are likely to be more important.

The fin efficiency solutions for many common fin geometries have been determined. For fins without a constant cross-section, the solution requires the use of more advanced techniques, such as Bessel functions, which are covered in Section 1.8. Several common fin solutions are listed in Table 1-5.

A more comprehensive set of fin efficiency solutions has been programmed in EES. To access these solutions, select Function Info from the Options menu and then select the radio button in the lower right side of the top box and scroll to the Fin Efficiency category (Figure 1-39).

It is possible to scroll through the various functions that are available or see more detailed information about any of these functions by pressing the Info button. Note that the fin efficiency can be accessed either in dimensional form (in which case the geometric parameters, conductivity, and heat transfer coefficient must be supplied) or nondimensional form (in which case the nondimensional parameters, such as $m\,L$, must be supplied).

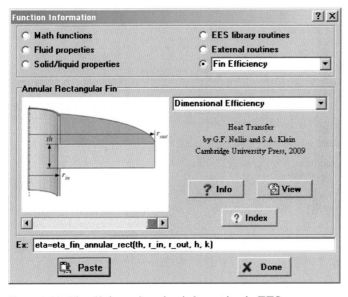

Figure 1-39: Fin efficiency function information in EES.

Table 1-5: Solutions for extended surfaces.

Shape		Solution
Straight rectangular	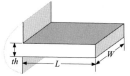	$\eta_{fin} = \dfrac{\tanh(mL)}{mL}$ $A_{s,fin} = 2WL$ $mL = \sqrt{\dfrac{2\bar{h}}{k\,th}}\,L$
Straight triangular	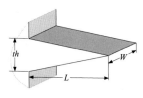	$\eta_{fin} = \dfrac{\text{BesselI}(1, 2mL)}{mL\,\text{BesselI}(0, 2mL)}$ $A_{s,fin} = 2W\sqrt{L^2 + \left(\dfrac{th}{2}\right)^2}$ $mL = \sqrt{\dfrac{2\bar{h}}{k\,th}}\,L$
Straight parabolic	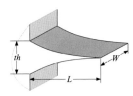	$\eta_{fin} = \dfrac{2}{\left[\sqrt{4(mL)^2 + 1} + 1\right]}$ $A_{s,fin} = W\left[C_1 L + \dfrac{L^2}{th}\ln\left(\dfrac{th}{L} + C_1\right)\right]$ $mL = \sqrt{\dfrac{2\bar{h}}{k\,th}}\,L,\quad C_1 = \sqrt{1 + \left(\dfrac{th}{L}\right)^2}$
Spine rectangular	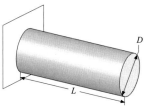	$\eta_{fin} = \dfrac{\tanh(mL)}{mL}$ $A_{s,fin} = \pi D L$ $mL = \sqrt{\dfrac{4\bar{h}}{kD}}\,L$
Spine triangular	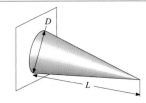	$\eta_{fin} = \dfrac{2\,\text{BesselI}(2, 2mL)}{mL\,\text{BesselI}(1, 2mL)}$ $A_{s,fin} = \dfrac{\pi D}{2}\sqrt{L^2 + \left(\dfrac{D}{2}\right)^2}$ $mL = \sqrt{\dfrac{4\bar{h}}{kD}}\,L$
Rectangular annular	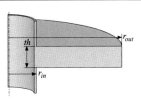	$A_{s,fin} = 2\pi\left(r_{out}^2 - r_{in}^2\right)$ $mr_{out} = \sqrt{\dfrac{2\bar{h}}{k\,th}}\,r_{out}$ $mr_{in} = \sqrt{\dfrac{2\bar{h}}{k\,th}}\,r_{in}$

$$\eta_{fin} = \frac{2\,mr_{in}}{\left[(mr_{out})^2 - (mr_{in})^2\right]}\,\frac{\left[\text{BesselK}(1, mr_{in})\,\text{BesselI}(1, mr_{out}) - \text{BesselI}(1, mr_{in})\,\text{BesselK}(1, mr_{out})\right]}{\left[\text{BesselI}(0, mr_{in})\,\text{BesselK}(1, mr_{out}) + \text{BesselK}(0, mr_{in})\,\text{BesselI}(1, mr_{out})\right]}$$

where $\bar{h}$ = heat transfer coefficient k = thermal conductivity

EXAMPLE 1.6-1: SOLDERING TUBES

Two large pipes must be soldered together using a propane torch, as shown in Figure 1.

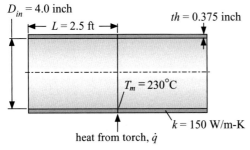

$T_\infty = 20°C, \bar{h} = 20$ W/m^2-K

Figure 1: Two bare pipes being soldered together.

Each of the two pipes is $L = 2.5$ ft long with inner diameter $D_{in} = 4.0$ inches and a thickness $th = 0.375$ inch. The pipe material has conductivity $k = 150$ W/m-K. The surrounding air is at $T_\infty = 20°C$ and the heat transfer coefficient between the external surface of the pipe and the air is $\bar{h} = 20$ W/m^2-K. Assume that convection from the internal surface of the pipe can be neglected.

a) **The temperature of the interface between the two pipes must be elevated to $T_m = 230°C$ in order to melt the solder; estimate the heat transfer rate, $\dot{q}$, that must be applied by the propane torch in order to accomplish this process.**

This problem is solved using EES. The initial section of the code provides the stated inputs (converted to SI units).

```
"EXAMPLE 1.6-1: Soldering Tubes"

$UnitSystem SI MASS DEG PA C J
$Tabstops 0.2 0.4 0.6 0.8 3.5

"Inputs"
D_in=4.0 [inch]*convert(inch,m)            "Inner diameter"
th=0.375 [inch]*convert(inch,m)            "Pipe thickness"
k=150 [W/m-K]                              "Conductivity"
h_bar=20 [W/m^2-K]                         "Heat transfer coefficient"
T_infinity=converttemp(C,K,20 [C])         "Air temperature"
L=2.5 [ft]*convert(ft,m)                   "Pipe length"
T_m=converttemp(C,K,230 [C])               "Melt temperature"
```

The two pipes can be treated as constant cross-sectional area fins; the solutions obtained in Section 1.6 are valid provided that the Biot number characterizing the temperature gradient within the pipe in the radial direction is sufficiently small:

$$Bi = \frac{\bar{h}\, th}{k}$$

EXAMPLE 1.6-1: SOLDERING TUBES

```
Bi=h_bar*th/k                                    "Biot number"
```

The Biot number is 0.0013, which is much less than unity. The cross-sectional area for conduction (A_c) is:

$$A_c = \frac{\pi}{4}\left[(D_{in} + 2\,th)^2 - D_{in}^2\right]$$

and the perimeter exposed to air (*per*) is:

$$per = \pi(D_{in} + 2\,th)$$

Notice that the internal surface of the pipe (which is assumed to be adiabatic) is not included in the perimeter. The ratio of the area of the exposed ends of the pipe to the external surface area (AR_{tip}) is calculated according to:

$$AR_{tip} = \frac{A_c}{per\,L}$$

```
A_c=pi*((D_in+2*th)^2-D_in^2)/4                  "area for conduction"
per=pi*(D_in+2*th)                               "perimeter"
AR_tip=A_c/(per*L)                               "area ratio"
```

The tip area ratio is $AR_{tip} = 0.012$ and therefore, according to Figure 1-38, the adiabatic tip fin solution can be used with no loss of accuracy. The fin constant ($m\,L$) is:

$$m\,L = \sqrt{\frac{per\,\overline{h}}{k\,A_c}}\,L$$

and the fin efficiency (η_{fin}) is:

$$\eta_{fin} = \frac{\tanh(m\,L)}{m\,L}$$

Therefore, the resistance of each fin (R_{fin}) is:

$$R_{fin} = \frac{1}{\eta_{fin}\,\overline{h}\,per\,L}$$

```
mL=sqrt(h_bar*per/(k*A_c))*L                     "fin parameter"
eta_fin=tanh(mL)/mL                              "fin efficiency"
R_fin=1/(eta_fin*h_bar*per*L)                    "fin resistance"
```

The problem can be represented by the resistance network shown in Figure 2; the two pipes correspond to the two resistors connecting the interface to the air and the

EXAMPLE 1.6–1: SOLDERING TUBES

heat input from the propane torch enters at the interface. In order for the solder to melt, the interface temperature must reach T_m.

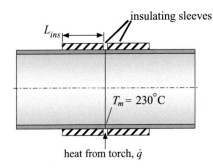

Figure 2: Resistance network associated with soldering two bare pipes.

The heat transfer required from the torch is therefore:

$$\dot{q} = 2\frac{(T_m - T_\infty)}{R_{fin}}$$

q_dot=2*(T_m-T_infinity)/R_fin   "required torch heat transfer rate"

The factor 2 appears because there are two pipes, each of which acts as a fin. The EES solution indicates that the propane torch must provide at least 812 W to accomplish this job.

b) Unfortunately, the propane torch cannot provide 812 W and it is not possible to melt the solder. Therefore, you decide to place insulating sleeves over the pipes adjacent to the soldering torch, as shown in Figure 3. If the insulation is perfect (i.e., convection is eliminated from the section of the pipe covered by the insulating sleeves), then how long must the sleeves be (L_{ins}) in order to reduce the heat required to $\dot{q} = 500$ W?

insulating sleeves

L_{ins}

$T_m = 230°C$

Figure 3: Pipes with insulating sleeves placed over them to reduce the heat transfer required.

heat from torch, $\dot{q}$

The pipes with insulating sleeves can be represented by a resistance network similar to the one shown in Figure 2, but with additional resistances inserted between the interface and the base of the fins. These additional resistances correspond to the insulated sections of pipes, as shown in Figure 4.

$\dot{q}$

T_∞ —\/\/\/—\/\/\/— T_m —\/\/\/—\/\/\/— T_∞
R_{fin} $R_{cond,ins}$ $R_{cond,ins}$ R_{fin}

Figure 4: Resistance network with additional resistors associated with the insulated sections of the pipe.

EXAMPLE 1.6-1: SOLDERING TUBES

The resistance of the insulated sections of the pipe is:

$$R_{cond,ins} = \frac{L_{ins}}{k\,A_c}$$

R_cond_ins=L_ins/(k*A_c)   "resistance of insulated portion of pipe"

The length of the un-insulated section of pipe is reduced and therefore the fin efficiency and fin resistance must be recalculated. The fin efficiency (η_{fin}) becomes:

$$\eta_{fin} = \frac{\tanh\left[m\,(L - L_{ins})\right]}{m\,(L - L_{ins})}$$

and the resistance of each fin (R_{fin}) is:

$$R_{fin} = \frac{1}{\eta_{fin}\,\overline{h}\,per\,(L - L_{ins})}$$

mL=sqrt(h_bar*per/(k*A_c))*(L-L_ins)   "fin parameter"
eta_fin=tanh(mL)/mL   "fin efficiency"
R_fin=1/(eta_fin*h_bar*per*(L-L_ins))   "fin resistance"

Using the resistance network shown in Figure 4, the required heat transfer rate can be expressed as:

$$\dot{q} = 2\,\frac{(T_m - T_\infty)}{R_{cond,ins} + R_{fin}}$$

EES can solve for the length of insulation that is required by setting the heat transfer rate to the available heat transfer rate,

q_dot=2*(T_m-T_infinity)/(R_fin+R_cond_ins)   "required torch heat transfer"
q_dot=500 [W]   "available torch heat transfer"
L_ins_ft=L_ins*convert(m,ft)   "length of insulation in ft"

The solution indicates that the length of the insulating sleeves must be at least 0.16 m (0.52 ft) in order to reduce the required heat transfer rate to the point where 500 W will suffice.

1.6.6 Finned Surfaces

Fins are often placed on surfaces in order to improve their heat transfer capability. Examples of finned surfaces can be found within nearly every appliance in your house, from the evaporator and condenser on your refrigerator and air conditioner to the processor in your personal computer. Fins are essential to the design of economical but high-performance thermal devices.

Figure 1-40 illustrates a single fin installed on a surface; the temperature of the surface is the base temperature of the fin, T_b. The fin and surface are surrounded by fluid at T_∞ with heat transfer coefficient $\overline{h}$. The fin has perimeter per, length L, conductivity k, and cross-sectional area A_c.

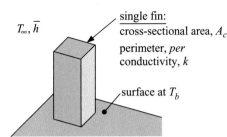

$T_\infty, \overline{h}$

single fin:
cross-sectional area, A_c
perimeter, per
conductivity, k

surface at T_b

Figure 1-40: Single fin placed on a surface.

It is of interest to determine the heat transfer rate from an area of the surface that is equal to the base area of the fin, both with and without the fin installed. If there were no fin, then the heat transfer rate from area A_c is:

$$\dot{q}_{no-fin} = \overline{h} A_c (T_b - T_\infty) \qquad (1\text{-}238)$$

while the heat transfer rate from the fin, assuming an adiabatic tip and the same heat transfer coefficient, is given by Eq. (1-221):

$$\dot{q}_{fin} = (T_b - T_\infty) \sqrt{\overline{h} \, per \, k \, A_c} \, \tanh \left(\sqrt{\frac{\overline{h} \, per}{k \, A_c}} L \right) \qquad (1\text{-}239)$$

The fin effectiveness (ε_{fin}) is defined as the ratio of the heat transfer rate from the fin ($\dot{q}_{fin}$) to the heat transfer rate that would have occurred from the surface area occupied by the fin without the fin attached ($\dot{q}_{no-fin}$):

$$\varepsilon_{fin} = \frac{\dot{q}_{fin}}{\dot{q}_{no-fin}} = \frac{(T_b - T_\infty) \sqrt{\overline{h} \, per \, k \, A_c} \, \tanh \left(\sqrt{\dfrac{\overline{h} \, per}{k \, A_c}} L \right)}{\overline{h} A_c (T_b - T_\infty)} \qquad (1\text{-}240)$$

which can be simplified to:

$$\varepsilon_{fin} = \sqrt{\frac{k \, per}{\overline{h} \, A_c}} \, \tanh \left(\sqrt{\frac{\overline{h} \, A_c}{k \, per}} \frac{1}{AR_{tip}} \right) \qquad (1\text{-}241)$$

where AR_{tip} is the ratio of the area of the tip to the exposed surface area of the fin. The fin effectiveness predicted by Eq. (1-241) is illustrated in Figure 1-41 as a function of the dimensionless group $(k \, per)/(\overline{h} \, A_c)$ for various values of the area ratio AR_{tip}.

The effectiveness of the fin provides a measure of the improvement in thermal performance that is achieved by placing fins onto the surface. Equation (1-241) and Figure 1-41 are useful in that they clarify the characteristics of an application that would benefit substantially from the use of fins. If the heat transfer coefficient is low, then the group $(k \, per)/(\overline{h} \, A_c)$ is large and there is a substantial benefit associated with the addition of fins. This explains why many devices that transfer thermal energy to air or other low conductivity gases with correspondingly low heat transfer coefficients are finned. For example, the air-side of a domestic refrigerator condenser will certainly be finned while the refrigerant side is typically not finned, since the heat transfer coefficient associated with the refrigerant condensation process is very high (as discussed in Chapter 7). Fins are typically thin structures (with a large perimeter to cross-sectional area ratio, per/A_c) made of high conductivity material; these features enhance the fin effectiveness by increasing the parameter $(k \, per)/(\overline{h} \, A_c)$.

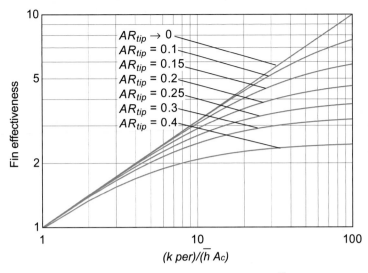

Figure 1-41: Fin effectiveness as a function of $(k \, per)/(\overline{h} \, A_c)$ for various values of AR_{tip}.

The concept of a fin resistance makes it possible to approximately consider the thermal performance of an array of fins that are placed on a surface. For example, Figure 1-42 illustrates an array of square fins placed on a base with area $A_{s,b}$ at temperature T_b. The surface is partially covered with fins; in Figure 1-42, the number of fins is $N_{fin} = 16$. Each fin has a cross-sectional area at its base of $A_{c,b}$ and surface area $A_{s,fin}$. The surfaces are exposed to a surrounding fluid with temperature T_∞ and average heat transfer coefficient $\overline{h}$.

The heat transferred from the base can either pass through one of the N_{fin} fins (each with resistance, R_{fin}) or from the un-finned surface area on the base (with resistance, $R_{un\text{-}finned}$). The resistance of a single fin is given by Eq. (1-233):

$$R_{fin} = \frac{1}{\eta_{fin} \, \overline{h} \, A_{s,fin}} \tag{1-242}$$

where η_{fin} is the fin efficiency, computed using the formulae or function specific to the geometry of the fin. The resistance of the un-finned surface of the base is:

$$R_{un-finned} = \frac{1}{\overline{h} \, (A_{s,b} - N_{fin} \, A_{c,b})} \tag{1-243}$$

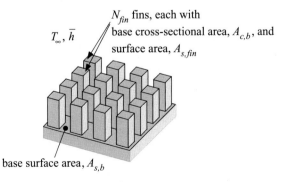

N_{fin} fins, each with base cross-sectional area, $A_{c,b}$, and surface area, $A_{s,fin}$

$T_\infty, \overline{h}$

base surface area, $A_{s,b}$

Figure 1-42: An array of fins on a base.

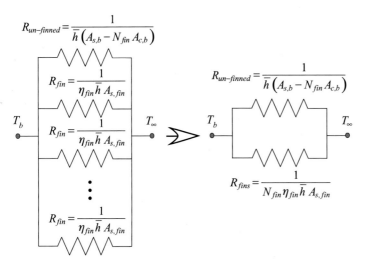

Figure 1-43: Resistance network associated with a finned surface.

where the area in the denominator of Eq. (1-243) is the area of the exposed portion of the base, i.e., the area not covered by fins. These heat transfer paths are in parallel and therefore the thermal resistance network that represents the situation is shown in Figure 1-43.

The total resistance of the finned surface is therefore:

$$R_{tot} = \left[\frac{1}{R_{un-finned}} + \frac{N_{fin}}{R_{fin}} \right]^{-1} \tag{1-244}$$

or, substituting Eqs. (1-242) and (1-243) into Eq. (1-244):

$$R_{tot} = [\overline{h}\,(A_{s,b} - N_{fin}\,A_{c,b}) + N_{fin}\,\eta_{fin}\,\overline{h}\,A_{s,fin}]^{-1} \tag{1-245}$$

The total rate of heat transfer is:

$$\dot{q}_{tot} = \frac{(T_b - T_\infty)}{R_{tot}} = [\overline{h}\,(A_{s,b} - N_{fin}\,A_{c,b}) + N_{fin}\,\eta_{fin}\,\overline{h}\,A_{s,fin}](T_b - T_\infty) \tag{1-246}$$

The overall surface efficiency (η_o) is defined as the ratio of the total heat transfer rate from the surface to the heat transfer rate that would result if the entire surface (the exposed base and the fins) were at the base temperature; as with the fin efficiency, this limit corresponds to using a material with an infinite conductivity.

$$\eta_o = \frac{\dot{q}_{tot}}{\overline{h}\,\underbrace{[(A_{s,b} - N_{fin}\,A_{c,b}) + N_{fin}\,A_{s,fin}]}_{\text{prime surface area, } A_{tot}}(T_b - T_\infty)} \tag{1-247}$$

The area in the denominator of Eq. (1-247) is often referred to as the prime surface area (A_{tot}):

$$A_{tot} = A_{s,b} - N_{fin}\,A_{c,b} + N_{fin}\,A_{s,fin} \tag{1-248}$$

Substituting Eq. (1-246) into Eq. (1-247) leads to:

$$\eta_o = \frac{[(A_{s,b} - N_{fin}\,A_{c,b}) + N_{fin}\,\eta_{fin}\,A_{s,fin}]}{[(A_{s,b} - N_{fin}\,A_{c,b}) + N_{fin}\,A_{s,fin}]} \tag{1-249}$$

Equation (1-249) can be rearranged:

$$\eta_o = \frac{[(A_{s,b} - N_{fin} A_{c,b}) + N_{fin} A_{s,fin} + N_{fin} \eta_{fin} A_{s,fin} - N_{fin} A_{s,fin}]}{[(A_{s,b} - N_{fin} A_{c,b}) + N_{fin} A_{s,fin}]} \tag{1-250}$$

or

$$\eta_o = 1 - \frac{N_{fin} A_{s,fin}(1 - \eta_{fin})}{[(A_{s,b} - N_{fin} A_{c,b}) + N_{fin} A_{s,fin}]} \tag{1-251}$$

which can be expressed in terms of the prime surface area:

$$\boxed{\eta_o = 1 - \frac{N_{fin} A_{s,fin}}{A_{tot}}(1 - \eta_{fin})} \tag{1-252}$$

Rearranging Eq. (1-247), the total resistance to heat transfer from a finned surface can be expressed in terms of the overall surface efficiency and the prime surface area:

$$\boxed{R_{tot} = \frac{1}{\eta_o \bar{h} A_{tot}}} \tag{1-253}$$

EXAMPLE 1.6-2: THERMOELECTRIC HEAT SINK

Heat rejection from a thermoelectric cooling device is accomplished using a 10×10 array of $D_{fin} = 1.5$ mm diameter pin fins that are $L_{fin} = 15$ mm long. The fins are attached to a square base plate that is $W_b = 3$ cm on a side and $th_b = 2$ mm thick, as shown in Figure 1. The conductivity of the fin material is $k_{fin} = 70$ W/m-K and the thermal conductivity of the base material is $k_b = 25$ W/m-K. There is a contact resistance of $R_c'' = 1 \times 10^{-4}$ m²-K/W at the interface between the base of the fins and the base plate. The hot end of the thermoelectric cooler is at $T_{hot} = 30°$C and the surrounding air temperature is $T_\infty = 20°$C. The average heat transfer coefficient between the air and the surface of the heat sink is $\bar{h} = 50$ W/m²-K.

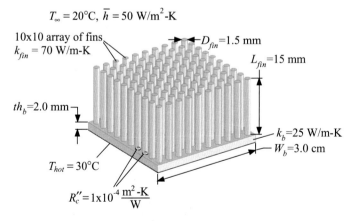

$T_\infty = 20°$C, $\bar{h} = 50$ W/m²-K

10x10 array of fins, $k_{fin} = 70$ W/m-K

$D_{fin} = 1.5$ mm

$L_{fin} = 15$ mm

$th_b = 2.0$ mm

$T_{hot} = 30°$C

$k_b = 25$ W/m-K

$W_b = 3.0$ cm

$R_c'' = 1 \times 10^{-4} \frac{m^2\text{-}K}{W}$

Figure 1: Heat sink mounted on a thermoelectric cooler.

a) What is the total thermal resistance between the hot end of the thermoelectric cooler and the air? What is the rate of heat rejection that can be accomplished under these conditions?

EXAMPLE 1.6-2: THERMOELECTRIC HEAT SINK

The first section of the EES code provides the inputs for the problem.

EXAMPLE 1.6-2: THERMOELECTRIC HEAT SINK

```
"EXAMPLE 1.6-2: Thermoelectric Heat Sink"

$UnitSystem SI MASS RAD PA K J
$Tabstops 0.2 0.4 0.6 3.5 in

"Inputs"
T_infinity=converttemp(C,K,20 [C])              "Air temperature"
T_hot=converttemp(C,K,30 [C])                   "Hot end of thermoelectric cooler"
D_fin=1.5[mm]*convert(mm,m)                     "Fin diameter"
L_fin=15[mm]*convert(mm,m)                      "Fin length"
N_fin=100                                       "Number of fins"
W_b=3 [cm] *convert(cm,m)                       "Width of base (square)"
th_b=2[mm]*convert(mm,m)                        "Thickness of base"
k_fin=70 [W/m-K]                                "Conductivity of fin"
k_b=25 [W/m-K]                                  "Conductivity of base"
h_bar=50 [W/m^2-K]                              "Heat transfer coefficient"
R"_c=1e-4 [m^2-K/W]                             "Contact resistance"
```

The constant cross-sectional area fins can be treated using the solutions presented in Section 1.6. The perimeter (*per*), cross-sectional area (A_c), and surface area for convection ($A_{s,fin}$, assuming adiabatic ends) associated with each fin are calculated according to:

$$per = \pi\, D_{fin}$$

$$A_c = \frac{\pi}{4}\, D_{fin}^2$$

$$A_{s,fin} = \pi\, L\, D_{fin}$$

```
per=pi*D_fin                      "Perimeter of fin"
A_c=pi*D_fin^2/4                  "Cross-sectional area for conduction"
A_s_fin=pi*L_fin*D_fin            "Surface area of fin for convection"
```

The fin constant and fin efficiency for an adiabatic tip, constant cross-sectional area fin are computed according to:

$$m = \sqrt{\frac{per\,\overline{h}}{k_{fin}\, A_c}}$$

$$\eta_{fin} = \frac{\tanh(m\, L_{fin})}{m\, L_{fin}}$$

The resistance of any type of fin (R_{fin}) can be obtained from its efficiency:

$$R_{fin} = \frac{1}{\eta_{fin}\,\overline{h}\, A_{s,fin}}$$

EXAMPLE 1.6-2: THERMOELECTRIC HEAT SINK

```
mL=sqrt(h_bar*per/(k_fin*A_c))*L_fin        "Fin parameter"
eta_fin=tanh(mL)/mL                         "Fin efficiency"
R_fin=1/(h_bar*A_s_fin*eta_fin)             "Fin resistance"
```

The resistance network that represents the entire heat sink (Figure 2) extends from the hot end of the cooler to the air and includes conduction through the base ($R_{cond,b}$) followed by two paths in parallel, corresponding to the heat that is transferred by convection from the unfinned upper surface of the base ($R_{un-finned}$) and the heat that is transferred through the contact resistance at the base of the fins (R_c) and then through the resistance associated with the fin itself (R_{fin}). Note that R_c and R_{fin} are in parallel N_{fin} times and therefore the value these resistances in the circuit is reduced by $1/N_{fin}$.

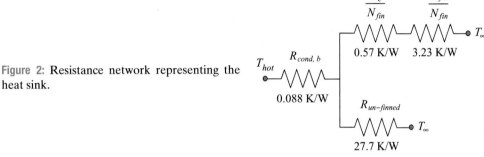

Figure 2: Resistance network representing the heat sink.

The resistance to conduction through the base is:

$$R_{cond,b} = \frac{th_b}{k_b W_b^2}$$

The contact resistance associated with each fin-to-base interface is:

$$R_c = \frac{R_c''}{A_c}$$

The resistance of the unfinned upper surface of the base is:

$$R_{un-finned} = \frac{1}{\overline{h}\left(W_b^2 - N_{fin} A_c\right)}$$

These resistances are calculated in EES:

```
R_b=th_b/(k_b*W_b^2)                        "Resistance due to conduction through the base"
R_unfinned=1/((W_b^2-N_fin*A_c)*h_bar)      "Resistance of unfinned base"
R_c=R''_c/A_c                               "Fin-to-base contact resistance"
```

The total resistance (R_{tot}) and heat transfer ($\dot{q}_{tot}$) from the heat sink are obtained according to:

$$R_{tot} = R_b + \left(\frac{1}{\left(\dfrac{R_c}{N_{fin}} + \dfrac{R_{fin}}{N_{fin}}\right)} + \frac{1}{R_{un-finned}}\right)^{-1}$$

$$\dot{q}_{tot} = \frac{(T_{hot} - T_\infty)}{R_{tot}}$$

EXAMPLE 1.6-2: THERMOELECTRIC HEAT SINK

and calculated in EES.

```
R_tot=R_b+(1/(R_c/N_fin+R_fin/N_fin)+1/R_unfinned)^(-1)        "Total resistance"
q_dot_tot=(T_hot-T_infinity)/R_tot                             "Total rate of heat transfer"
```

The total resistance is 3.42 K/W and the rate of heat transfer is 2.92 W. The numerical values of each resistance are included in Figure 2 in order to understand the mechanisms that are governing the behavior of the heat sink. Notice that the resistance of the base is not very important, as it is a small resistor in series with larger ones. The resistance of the unfinned portion of the base is also not critical, since it is a large resistor in parallel with smaller ones. On the other hand, both the contact resistance and the fin resistance are important as these two resistors dominate the problem and are of the same order of magnitude. The fin resistance is the most critical parameter in the problem and any attempt to improve performance should focus on this element of the heat sink.

b) **Through material selection and manipulation of the air flow across the heat sink, it is possible to affect design changes to k_{fin} and $\bar{h}$. Generate a contour plot that illustrates contours of constant heat rejection in the parameter space of k_{fin} (ranging from 5 W/m-K to 150 W/m-K) and $\bar{h}$ (ranging from 10 W/m²-K to 200 W/m²-K).**

One of the nice things about solving problems using a computer program as opposed to pencil and paper is that parametric studies and optimization are relatively straightforward. In order to prepare a contour plot with EES, it is necessary to setup a parametric table in which both of the parameters of interest vary over a specified range. Open a new parametric table and include the two independent variables (the variables k_fin and h_bar) as well as the dependent variable of interest (the variable q_dot_tot). In order to run the simulation for 20 values of k_{fin} and 20 values of $\bar{h}$, 20 × 20 = 400 runs must be included in the table. (Add runs using the Insert/Delete Runs option from the Tables menu.)

It is necessary to set the values of k_fin and h_bar in the table. It is possible to vary k_fin from 5 to 150 W/m-K, 20 times by using the "Repeat pattern every" option in the Alter Values dialog that appears when you right-click on the k_fin column, as shown in Figure 3.

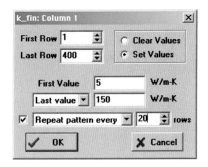

Figure 3: Vary k_{fin} from 5 to 150 W/m-K 20 times.

EXAMPLE 1.6-2: THERMOELECTRIC HEAT SINK

In order to completely cover the parameter space, it is necessary to evaluate the solution over a range of h_bar at each unique value of k_fin; this can be accomplished using the "Apply pattern every" option in the Alter Values dialog for the h_bar column of the table, see Figure 4.

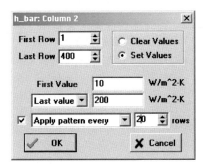

Figure 4: Vary $\bar{h}$ from 10 to 200 W/m-K with 20 runs for each of 20 values.

When the specified values of the variables k_fin and h_bar are commented out in the Equations window, it is possible to run the parametric table using the Solve Table command in the Calculate menu (F3); 400 values of $\dot{q}$ are determined, one for each combination of k_{fin} and $\bar{h}$ set in the parametric table. To generate a contour plot, select X-Y-Z plot from the New Plot Window option in the Plots menu. Select k_fin as the variable on the x-axis, h_bar as the y-axis variable and q_dot_tot as the contour variable. The appearance of the resulting contour plot can be adjusted by altering the resolution, smoothing, color options, and the type of function used for interpolation. A contour plot generated using isometric lines is shown in Figure 5.

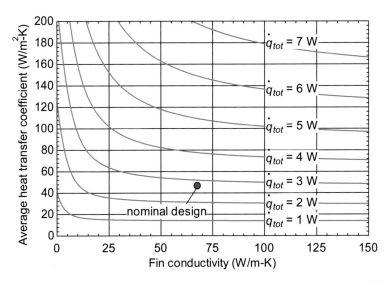

Figure 5: Contours of constant heat transfer rate in the parameter space of fin material conductivity and heat transfer coefficient.

EXAMPLE 1.6-2

The nominal design point shown in Figure 1 is also indicated in Figure 5. Contour plots are useful in that they can clarify the impact of design changes. For example, Figure 5 shows that it would be more beneficial to explore methods to increase the heat transfer coefficient than the fin conductivity at the nominal design conditions (i.e., moving from the nominal design point towards higher heat transfer will result in much larger performance gains than moving toward higher fin conductivity).

1.6.7 Fin Optimization

This extended section of the book, which can be found on the website (www. cambridge.org/nellisandklein), presents an optimization of a constant cross-sectional area fin in order to maximize the rate of heat transfer per unit volume of fin material. The process illustrates the use of EES' single-variable optimization capability and shows that a well-optimized fin is characterized by $m\,L$ that is approximately equal to 1.4. Fins with $m\,L$ much less than 1.4 are shorter than optimal and therefore have very small temperature gradients due to conduction; additional length will provide a substantial benefit and therefore the available volume of fin material should be stretched, providing additional length at the expense of cross-sectional area. Fins with $m\,L$ much greater than 1.4 are longer than optimal and therefore have large temperature gradients due to conduction; additional length will not provide much benefit as the tip temperature is approaching the ambient temperature. Therefore, the available volume should be compressed, reducing the length but providing more cross-sectional area for conduction.

1.7 Analytical Solutions for Advanced Constant Cross-Section Extended Surfaces

1.7.1 Introduction

The constant cross-section fins that were investigated in Section 1.6 are certainly the most common type of extended surface used in practice. However, other extended surface problems (with alternative boundary conditions, more complex thermal loadings, multiple computational domains, etc.) are also encountered. Extended surfaces represent 2-D heat transfer situations that can be approximated as being 1-D and these problems can be solved analytically using the techniques that were introduced in Section 1.6.

1.7.2 Additional Thermal Loads

An extended surface can be subjected to additional thermal loads such as thermal energy generation (due to ohmic heating, for example) or an external heat flux. These additional effects show up in the governing differential equation but do not affect the character of the solution. Figure 1-46 illustrates an extended surface with cross-sectional area A_c and perimeter *per* that has a uniform volumetric generation ($\dot{g}'''$) and is exposed to a uniform heat flux ($\dot{q}''_{ext}$, for example from solar radiation). The extended surface is surrounded by fluid at T_∞ with average heat transfer coefficient $\overline{h}$.

A differential control volume is used to derive the governing differential equation (see Figure 1-46) and provides the energy balance:

$$\dot{q}_x + \dot{g} + \dot{q}_{ext} = \dot{q}_{conv} + \dot{q}_{x+dx} \qquad (1\text{-}267)$$

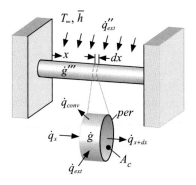

Figure 1-46: Extended surface with additional thermal loads related to generation and an external heat flux.

The final term can be expanded:

$$\dot{g} + \dot{q}_{ext} = \dot{q}_{conv} + \frac{d\dot{q}}{dx}dx \qquad (1\text{-}268)$$

Substituting the appropriate rate equation for each term results in:

$$\dot{g}''' A_c\, dx + \dot{q}''_{ext}\, per\, dx = \overline{h}\, per\, dx\, (T - T_\infty) + \frac{d}{dx}\left[-k\, A_c \frac{dT}{dx}\right] dx \qquad (1\text{-}269)$$

where k is the conductivity of the material and T is the temperature at any axial position. Note that temperature is assumed to be only a function of x, which is consistent with the extended surface approximation. This assumption should be verified using an appropriately defined Biot number, as discussed in Section 1.6.2. After some simplification, the governing differential equation for the extended surface becomes:

$$\frac{d^2 T}{dx^2} - \frac{\overline{h}\, per}{k\, A_c}T = -\frac{\overline{h}\, per}{k\, A_c}T_\infty - \frac{\dot{g}'''}{k} - \frac{\dot{q}''_{ext}\, per}{k\, A_c} \qquad (1\text{-}270)$$

Equation (1-270) is a nonhomogeneous, linear, second order ODE. The solution is assumed to be the sum of a homogeneous and particular solution:

$$T = T_h + T_p \qquad (1\text{-}271)$$

Equation (1-271) is substituted into Eq. (1-270):

$$\underbrace{\frac{d^2 T_h}{dx^2} - \frac{\overline{h}\, per}{k\, A_c}T_h}_{\substack{=0 \\ \text{for homogeneous} \\ \text{differential equation}}} + \underbrace{\frac{d^2 T_p}{dx^2} - \frac{\overline{h}\, per}{k\, A_c}T_p = -\frac{\overline{h}\, per}{k\, A_c}T_\infty - \frac{\dot{g}'''}{k} - \frac{\dot{q}''_{ext}\, per}{k\, A_c}}_{\text{whatever is left over must be the particular differential equation}} \qquad (1\text{-}272)$$

The homogeneous differential equation is:

$$\frac{d^2 T_h}{dx^2} - \frac{\overline{h}\, per}{k\, A_c}T_h = 0 \qquad (1\text{-}273)$$

which is solved by:

$$T_h = C_1 \exp(m x) + C_2 \exp(-m x) \qquad (1\text{-}274)$$

where

$$m = \sqrt{\frac{per\, \overline{h}}{k\, A_c}} \qquad (1\text{-}275)$$

The particular differential equation is:

$$\frac{d^2 T_p}{dx^2} - \frac{\overline{h}\,per}{k\,A_c} T_p = -\frac{\overline{h}\,per}{k\,A_c} T_\infty - \frac{\dot{g}'''}{k} - \frac{\dot{q}''_{ext}\,per}{k\,A_c} \tag{1-276}$$

Notice that the right side of Eq. (1-276) is a constant; therefore, the particular solution is a constant:

$$T_p = C_3 \tag{1-277}$$

Substituting Eq. (1-277) into Eq. (1-276) leads to:

$$-\frac{\overline{h}\,per}{k\,A_c} C_3 = -\frac{\overline{h}\,per}{k\,A_c} T_\infty - \frac{\dot{g}'''}{k} - \frac{\dot{q}''_{ext}\,per}{k\,A_c} \tag{1-278}$$

Solving for C_3:

$$C_3 = T_\infty + \frac{\dot{g}'''\,A_c}{\overline{h}\,per} + \frac{\dot{q}''_{ext}}{\overline{h}} \tag{1-279}$$

Substituting Eq. (1-279) into Eq. (1-277) leads to:

$$T_p = T_\infty + \frac{\dot{g}'''\,A_c}{\overline{h}\,per} + \frac{\dot{q}''_{ext}}{\overline{h}} \tag{1-280}$$

Substituting Eqs. (1-280) and (1-274) into Eq. (1-271) leads to:

$$T = C_1 \exp(mx) + C_2 \exp(-mx) + T_\infty + \frac{\dot{g}'''\,A_c}{\overline{h}\,per} + \frac{\dot{q}''_{ext}}{\overline{h}} \tag{1-281}$$

The boundary conditions at either edge of the extended surface should be used to evaluate C_1 and C_2 for a specific situation.

It is possible to use Maple to solve this problem (and therefore avoid the mathematical steps discussed above). Enter the governing differential equation:

```
> restart;
> ODE:=diff(diff(T(x),x),x)-h_bar*per*T(x)/(k*A_c)=-h_bar*per*T_infinity/
  (k*A_c)-gv/k-qf_ext*per/(k*A_c);
```

$$ODE := \left(\frac{d^2}{dx^2} T(x)\right) - \frac{h\_bar\,per\,T(x)}{k\,A\_c} = -\frac{h\_bar\,per\,T\_infinity}{k\,A\_c} - \frac{gv}{k} - \frac{qf\_ext\,per}{k\,A\_c}$$

and solve it:

```
> Ts:=dsolve(ODE);
```

$$Ts := T(x) =$$
$$\mathrm{e}^{\left(\frac{\sqrt{h\_bar}\sqrt{per}\,x}{\sqrt{k}\sqrt{A\_c}}\right)}\_C2 + \mathrm{e}^{\left(-\frac{\sqrt{h\_bar}\sqrt{per}\,x}{\sqrt{k}\sqrt{A\_c}}\right)}\_C1 + \frac{(h\_bar\,T\_infinity + qf\_ext)\,per + gv\,A\_c}{h\_bar\,per}$$

The solution identified by Maple is functionally identical to Eq. (1-281).

For situations where the volumetric generation or external heat flux is not spatially uniform, it will not be as easy to identify the particular solution. For example, suppose that the volumetric generation varies sinusoidally from $x = 0$ to $x = L$, where L is the length of the extended surface.

$$\dot{g}''' = \dot{g}'''_{max} \sin\left(\pi \frac{x}{L}\right) \tag{1-282}$$

where $\dot{g}'''_{max}$ is the volumetric generation at the center of the extended surface. The resulting governing differential equation is:

$$\frac{d^2 T}{dx^2} - \frac{\overline{h}\, per}{k\, A_c} T = -\frac{\overline{h}\, per}{k\, A_c} T_\infty - \frac{\dot{g}'''_{max}}{k} \sin\left(\pi \frac{x}{L}\right) - \frac{\dot{q}''_{ext}\, per}{k\, A_c} \tag{1-283}$$

The solution is assumed to be the sum of a homogeneous and particular solution. Substituting Eq. (1-271) into Eq. (1-283) leads to:

$$\underbrace{\frac{d^2 T_h}{dx^2} - \frac{\overline{h}\, per}{k\, A_c} T_h}_{\substack{=0 \\ \text{for homogeneous} \\ \text{differential equation}}} + \underbrace{\frac{d^2 T_p}{dx^2} - \frac{\overline{h}\, per}{k\, A_c} T_p = -\frac{\overline{h}\, per}{k\, A_c} T_\infty - \frac{\dot{g}'''_{max}}{k} \sin\left(\pi \frac{x}{L}\right) - \frac{\dot{q}''_{ext}\, per}{k\, A_c}}_{\text{the particular differential equation}}$$

$$\tag{1-284}$$

The homogeneous differential equation has not changed and therefore the homogeneous solution remains Eq. (1-274). However, the particular differential equation has become more complex:

$$\frac{d^2 T_p}{dx^2} - \frac{\overline{h}\, per}{k\, A_c} T_p = -\frac{\overline{h}\, per}{k\, A_c} T_\infty - \frac{\dot{g}'''_{max}}{k} \sin\left(\pi \frac{x}{L}\right) - \frac{\dot{q}''_{ext}\, per}{k\, A_c} \tag{1-285}$$

Identifying the particular solution can take some skill. Equation (1-285) involves both a constant and a sinusoidal term on the right hand side and the governing differential equation involves both the solution and its derivatives. Therefore, it seems likely that a particular solution that includes sines, cosines, and constants as well as their derivatives (cosines, sines, and 0) might work. One method of obtaining the particular solution is to assume such a solution with appropriate, undetermined constants (C_3, C_4, and C_5):

$$T_p = C_3 \sin\left(\pi \frac{x}{L}\right) + C_4 \cos\left(\pi \frac{x}{L}\right) + C_5 \tag{1-286}$$

and substitute it into the particular differential equation. The first and second derivatives of the particular solution, Eq. (1-286), are:

$$\frac{dT_p}{dx} = \frac{C_3\, \pi}{L} \cos\left(\pi \frac{x}{L}\right) - \frac{C_4\, \pi}{L} \sin\left(\pi \frac{x}{L}\right) \tag{1-287}$$

$$\frac{d^2 T_p}{dx^2} = -\frac{C_3\, \pi^2}{L^2} \sin\left(\pi \frac{x}{L}\right) - \frac{C_4\, \pi^2}{L^2} \cos\left(\pi \frac{x}{L}\right) \tag{1-288}$$

Substituting Eqs. (1-288) and (1-286) into Eq. (1-285) leads to:

$$-\frac{C_3\, \pi^2}{L^2} \sin\left(\pi \frac{x}{L}\right) - \frac{C_4\, \pi^2}{L^2} \cos\left(\pi \frac{x}{L}\right) - \frac{\overline{h}\, per}{k\, A_c}\left[C_3 \sin\left(\pi \frac{x}{L}\right) + C_4 \cos\left(\pi \frac{x}{L}\right) + C_5\right]$$

$$= -\frac{\overline{h}\, per}{k\, A_c} T_\infty - \frac{\dot{g}'''_{max}}{k} \sin\left(\pi \frac{x}{L}\right) - \frac{\dot{q}''_{ext}\, per}{k\, A_c} \tag{1-289}$$

In order for the particular solution to work, the sine, cosine and constant terms in Eq. (1-289) must add up correctly. By considering the coefficients of the sine terms, it is possible to obtain the equation:

$$-\frac{C_3\,\pi^2}{L^2} - \frac{\overline{h}\,per}{k\,A_c}C_3 = -\frac{\dot{g}'''_{max}}{k} \tag{1-290}$$

which can be solved for C_3:

$$C_3 = \frac{\dfrac{\dot{g}'''_{max}}{k}}{\left(\dfrac{\pi^2}{L^2} + \dfrac{\overline{h}\,per}{k\,A_c}\right)} \tag{1-291}$$

The sum of the coefficients of the cosine terms provides an additional equation:

$$-\frac{C_4\,\pi^2}{L^2} - \frac{\overline{h}\,per}{k\,A_c}C_4 = 0 \tag{1-292}$$

which indicates that

$$C_4 = 0 \tag{1-293}$$

Finally, the sum of the coefficients for the constant terms leads to:

$$-\frac{\overline{h}\,per}{k\,A_c}C_5 = -\frac{\overline{h}\,per}{k\,A_c}T_\infty - \frac{\dot{q}''_{ext}\,per}{k\,A_c} \tag{1-294}$$

so

$$C_5 = T_\infty + \frac{\dot{q}''_{ext}}{\overline{h}} \tag{1-295}$$

Substituting Eqs. (1-291), (1-293), and (1-295) into Eq. (1-286) leads to:

$$T_p = \frac{\dfrac{\dot{g}'''_{max}}{k}}{\left(\dfrac{\pi^2}{L^2} + \dfrac{\overline{h}\,per}{k\,A_c}\right)} \sin\left(\pi\,\frac{x}{L}\right) + T_\infty + \frac{\dot{q}''_{ext}}{\overline{h}} \tag{1-296}$$

The solution to the differential equation is the sum of the homogeneous and the particular solutions.

$$T = C_1\exp(m\,x) + C_2\exp(-m\,x) + \frac{\dfrac{\dot{g}'''_{max}}{k}}{\left(\dfrac{\pi^2}{L^2} + \dfrac{\overline{h}\,per}{k\,A_c}\right)} \sin\left(\pi\,\frac{x}{L}\right) + T_\infty + \frac{\dot{q}''_{ext}}{\overline{h}} \tag{1-297}$$

where the boundary conditions determine the values of C_1 and C_2.

It is somewhat easier to obtain the solution using Maple. Enter the governing differential equation, Eq. (1-283):

```
> restart;
> ODE:=diff(diff(T(x),x),x)-h_bar*per*T(x)/(k*A_c)=-h_bar*per*T_infinity/
(k*A_c)-gv_max*sin(pi*x/L)/k-qf_ext*per/(k*A_c);
```

$$ODE := \left(\frac{d^2}{dx^2}T(x)\right) - \frac{h\_bar\,per\,T(x)}{k\,A\_c}$$
$$= -\frac{h\_bar\,per\,T\_infinity}{k\,A\_c} - \frac{gv\_max\,\sin\left(\dfrac{\pi x}{L}\right)}{k} - \frac{qf\_ext\,per}{k\,A\_c}$$

and then solve the differential equation:

> Ts:=dsolve(ODE);

$$Ts := T(x) = e^{\left(\frac{\sqrt{h\_bar}\sqrt{per}\,x}{\sqrt{k}\sqrt{A\_c}}\right)} \_C2 + e^{\left(\frac{\sqrt{h\_bar}\sqrt{per}\,x}{\sqrt{k}\sqrt{A\_c}}\right)} \_C1$$
$$+ \frac{gv\_max\, A\_c\, L^2\, h\_bar\, \sin\left(\frac{\pi x}{L}\right) + (h\_bar\, T\_infinity + qf\_ext)(h\_bar\, per\, L^2 + \pi^2\, k\, A\_c)}{h\_bar\,(h\_bar\, per\, L^2 + \pi^2\, k\, A\_c)}$$

The solution identified by Maple is functionally equivalent to Eq. (1-297). This result from Maple can be copied and pasted directly into EES for evaluation and manipulation.

EXAMPLE 1.7-1: BENT-BEAM ACTUATOR

One design of a micro-scale, lithographically fabricated (i.e., MEMS) device that can produce in-plane motion is called a bent-beam actuator (Que (2000)). A V-shaped structure (the bent-beam in Figure 1) is suspended between two anchors. The anchors are thermally staked to the underlying substrate and therefore keep the ends of the bent-beam at room temperature ($T_a = 20°C$). An elevated voltage is applied to one pillar and the other is grounded. The voltage difference causes current (I) to flow through the bent-beam structure. The temperature of the bent-beam rises as a result of ohmic heating and the thermally induced expansion causes the apex of the bent-beam to move outwards. The result is a voltage-controlled actuator capable of producing in-plane motion.

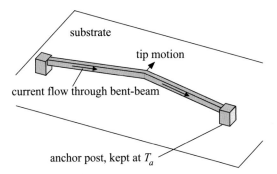

Figure 1: Bent-beam actuator.

The anchors of the bent-beam actuator are placed $L_a = 1$ mm apart and the beam structure has a cross-section of $W = 10\,\mu$m by $th = 5\,\mu$m. The slope of the beams (with respect to a line connecting the two pillars) is $\theta = 0.5$ rad, as shown in Figure 2. The bent-beam material has conductivity $k = 80$ W/m-K, electrical resistivity $\rho_e = 1 \times 10^{-5}$ ohm-m and coefficient of thermal expansion $CTE = 3.5 \times 10^{-6}\,$K^{-1}. You may neglect radiation from the beam and assume all of the heat that is generated is convected to the surrounding air at temperature $T_\infty = 20°C$ with average heat transfer coefficient $\overline{h} = 100$ W/m^2-K or transferred conductively to the pillars (which remain at $T_a = 20°C$). The actuator is activated with $I = 10$ mA of current.

EXAMPLE 1.7-1: BENT-BEAM ACTUATOR

EXAMPLE 1.7-1: BENT-BEAM ACTUATOR

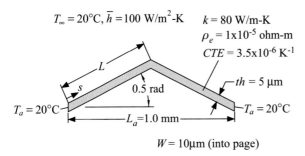

$T_\infty = 20°C, \bar{h} = 100$ W/m^2-K $k = 80$ W/m-K
$\rho_e = 1 \times 10^{-5}$ ohm-m
$CTE = 3.5 \times 10^{-6}$ K^{-1}

L

s 0.5 rad $th = 5$ μm

$T_a = 20°C$ $T_a = 20°C$

$\leftarrow L_a = 1.0$ mm $\rightarrow$

$W = 10$μm (into page)

Figure 2: Dimensions and conditions associated with bent-beam actuator.

a) Is it appropriate to treat the bent-beam as an extended surface?

The input parameters for the problem are entered into EES:

```
"EXAMPLE 1.7-1: Bent-beam Actuator"

$UnitSystem SI MASS RAD PA C J
$Tabstops 0.2 0.4 0.6 0.8 3.5

"Inputs"
L_a=1 [mm]*convert(mm,m)                    "distance between anchors"
w=10 [micron]*convert(micron,m)            "width of beam"
th=5 [micron]*convert(micron,m)            "thickness of beam"
I=0.010 [Amp]                               "current"
theta=0.5 [rad]                             "slope of beam"
T_a=converttemp(C,K,20 [C])                 "temperature of pillars"
T_infinity=converttemp(C,K,20 [C])         "temperature of air"
h_bar=100 [W/m^2-K]                         "heat transfer coefficient"
k=80 [W/m-K]                                "conductivity"
rho_e=1e-5 [ohm-m]                          "electrical resistivity"
CTE=3.5e-6 [1/K]                            "coefficient of thermal expansion"
```

The extended surface approximation requires that the 3-D temperature distribution within the bent-beam be approximated as 1-D; that is, temperature gradients within the beam that are perpendicular to the surface will be ignored so that the temperature may be approximated as a function only of s, the coordinate that follows the beam (see Figure 2). The resistance that must be neglected in order to use the extended surface approximation is conduction in the lateral direction ($R_{cond,lat}$). The extended surface approximation is justified provided that the lateral conduction resistance is small relative to the resistance that is being considered, convection from the outer surface (R_{conv}). The Biot number is therefore:

$$Bi = \frac{R_{cond,lat}}{R_{conv}}$$

The heat transfer will take the shortest path to the surface and therefore it is appropriate to use the smallest lateral dimension ($th/2$) to compute the lateral conduction resistance.

$$Bi = \left(\frac{th}{2 \, k \, W \, L}\right)\left(\frac{\bar{h} \, W \, L}{1}\right) = \frac{th \, \bar{h}}{2 \, k}$$

EXAMPLE 1.7-1: BENT-BEAM ACTUATOR

where L is the length of the beam from pillar to apex (see Figure 2).

```
Bi=th*h_bar/(2*k)                              "Biot number"
```

The Biot number is small (3×10^{-6}) and therefore the extended surface approximation is justified.

b) Develop an analytical solution that can predict the temperature of one leg of the bent-beam as a function of position along the beam, s.

The general solution for an extended surface with a constant cross-sectional area and spatially uniform generation is derived in Section 1.7.2:

$$T = C_1 \exp(ms) + C_2 \exp(-ms) + T_\infty + \frac{\dot{g}''' A_c}{\overline{h}\, per} \tag{1}$$

For the bent-beam actuator, the perimeter (per), cross-sectional area (A_c), and fin parameter (m) are

$$per = 2\,(W + th)$$

$$A_c = W\,th$$

$$m = \sqrt{\frac{\overline{h}\, per}{k\, A_c}}$$

```
per=2*(W+th)                                   "perimeter"
A_c=W*th                                        "area"
m=sqrt(h_bar*per/(k*A_c))                       "fin parameter"
```

The volumetric generation, $\dot{g}'''$, is related to ohmic heating. The electrical resistance of the bent-beam structure (R_e) is:

$$R_e = \frac{\rho_e\, 2\, L}{A_c}$$

where

$$L = \frac{L_a}{2\,\cos(\theta)}$$

The volumetric rate of electrical dissipation is the ratio of ohmic dissipation to the volume of the structure:

$$\dot{g}''' = \frac{I^2\, R_e}{2\, L\, A_c}$$

```
L=L_a/(2*cos(theta))                           "length of half-beam"
R_e=rho_e*L*2/A_c                               "resistance of beam structure"
g'''_dot=I^2*R_e/(2*L*A_c)                      "volumetric generation"
```

EXAMPLE 1.7-1: BENT-BEAM ACTUATOR

The constants C_1 and C_2 in Eq. (1) are determined using the boundary conditions. The temperature of the beam where it meets the pillar is specified:

$$T_{s=0} = T_a \qquad (2)$$

Substituting Eq. (1) into Eq. (2) leads to:

$$C_1 + C_2 + T_\infty + \frac{\dot{g}''' A_c}{\bar{h}\,per} = T_a \qquad (3)$$

A half-symmetry model of the bent-beam actuator considers only one leg. Because both legs of the bent-beam see identical conditions, there is nothing to drive heat from one leg to the other and therefore there will be no conduction through the end of the leg (at $s = L$):

$$\dot{q}_{s=L} = -k \left.\frac{dT}{ds}\right|_{s=L} = 0$$

or

$$\left.\frac{dT}{ds}\right|_{s=L} = 0 \qquad (4)$$

Substituting Eq. (1) into Eq. (4) leads to:

$$C_1\, m \exp(m\,L) - C_2\, m \exp(-m\,L) = 0 \qquad (5)$$

Equations (3) and (5) can be entered in EES and used to determine C_1 and C_2.

```
T_infinity+C_1+C_2+g'''_dot*A_c/(h_bar*per)=T_a        "from boundary condition at s=0"
C_1*m*exp(m*L)-C_2*m*exp(-m*L)=0                       "from boundary condition at s=L"
```

A variable s_bar is defined as s/L so that s_bar $= 0$ corresponds to the pillar and s_bar $= 1$ to the apex. The variable s_bar is defined for convenience, so that it is easy to generate a parametric table in which s is varied from 0 to L even if parameters such as θ and L_a change.

```
s_bar=s/L                      "non-dimensional position"
```

The temperature is evaluated using Eq. (1).

```
T=T_infinity+C_1*exp(m*s)+C_2*exp(-m*s)+g'''_dot*A_c/(h_bar*per)     "temperature"
T_C=converttemp(K,C,T)                                              "in C"
```

A parametric table is generated that includes the variables s_bar and T_C. The temperature distribution through one leg of the beam is shown in Figure 3.

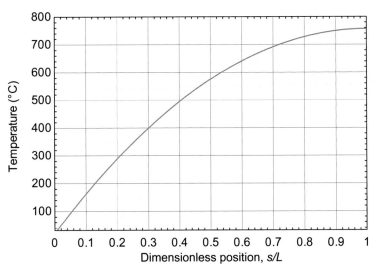

Figure 3: Temperature as a function of dimensionless position along one leg of beam.

c) **The thermally induced elongation of a differential segment of the beam (of length *ds*) is given by:**

$$dL = CTE \, (T - T_a) \, ds$$

Estimate the displacement of the apex of the beam. Plot the displacement as a function of voltage.

The total elongation of the beam (ΔL) is obtained by integrating the differential elongation along the beam:

$$\Delta L = \int_0^L CTE \, (T - T_a) \, ds \qquad (6)$$

Substituting the solution for the temperature distribution, Eq. (1) into Eq. (6) leads to:

$$\Delta L = \int_0^L CTE \left(C_1 \exp\,(ms) + C_2 \exp\,(-ms) + T_\infty + \frac{\dot{g}''' A_c}{\bar{h} \, per} - T_a \right) ds$$

Evaluating the integral:

$$\Delta L = CTE \left[\left(T_\infty - T_a + \frac{\dot{g}''' A_c}{\bar{h} \, per} \right) s + \frac{C_1}{m} \exp\,(ms) - \frac{C_2}{m} \exp\,(-ms) \right]_0^L$$

EXAMPLE 1.7-1: BENT-BEAM ACTUATOR

Substituting the integration limits:

$$\Delta L = CTE \left\{ \left(T_\infty - T_a + \frac{\dot{g}''' A_c}{h \, per} \right) L + \frac{C_1}{m} [\exp(mL) - 1] - \frac{C_2}{m} [\exp(-mL) - 1] \right\}$$

DELTAL=CTE*((T_infinity-T_a+g'''_dot*A_c/(h_bar*per))*L+C_1*(exp(m*L)-1)/m-C_2*(exp(-m*L)-1)/m)
"displacement of beam"

Assuming that the joint associated with the apex does not provide a torque on either leg of the beam, the displacement of the apex can be estimated using trigonometry (Figure 4).

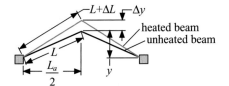

Figure 4: Trigonometry associated with apex motion.

The original position of the apex (y) is given by:

$$y = \sqrt{L^2 - \left(\frac{L_a}{2} \right)^2}$$

therefore, the motion of the apex (Δy) is:

$$\Delta y = \sqrt{(L + \Delta L)^2 - \left(\frac{L_a}{2} \right)^2} - \sqrt{L^2 - \left(\frac{L_a}{2} \right)^2}$$

DELTAy=sqrt((L+DELTAL)^2-(L_a/2)^2)-sqrt(L^2-(L_a/2)^2)   "displacement of apex"
DELTAy_micron=DELTAy*convert(m,micron)   "in μm"

The voltage across the beam (V) is:

$$V = I \, R_e$$

V=I*R_e   "voltage"

Figure 5 illustrates the actuator displacement as a function of voltage. This plot was generated using a parametric table including the variables **DELTAy_micron** and **V**; the variable **I** was commented out in order to make the table.

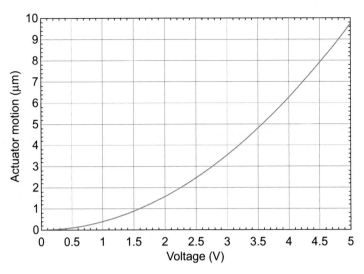

Figure 5: Actuator displacement as a function of the applied voltage.

1.7.3 Moving Extended Surfaces

An interesting class of problems arises in situations where an extended surface is moving with respect to the frame of reference of the problem. Problems of this type occur in rotating systems, (such as in drum and disk brakes), extrusions, and in manufacturing systems. The energy carried by the moving material represents an additional energy transfer into and out of the differential control volume and provides an additional term in the governing differential equation. Figure 1-47 illustrates an extended surface (i.e., a material for which temperature is only a function of x) that is moving with velocity u through fluid with temperature T_∞ and $\overline{h}$.

An energy balance on the differential control volume (Figure 1-47) includes conduction and energy transport due to material motion ($\dot{E}$) at either edge, as well as convection to the surrounding fluid.

$$\dot{q}_x + \dot{E}_x = \dot{q}_{conv} + \dot{q}_{x+dx} + \dot{E}_{x+dx} \qquad (1\text{-}298)$$

or

$$0 = \dot{q}_{conv} + \frac{d\dot{q}}{dx}dx + \frac{d\dot{E}}{dx}dx \qquad (1\text{-}299)$$

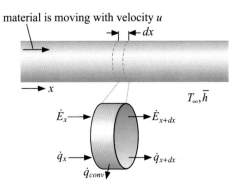

Figure 1-47: An extended surface moving with velocity u.

The conduction and convection terms are represented by the familiar rate equations:

$$\dot{q}_{cond} = -k\,A_c\,\frac{dT}{dx} \tag{1-300}$$

$$\dot{q}_{conv} = per\,dx\,\overline{h}\,(T - T_\infty) \tag{1-301}$$

where k is the conductivity of the material and per and A_c are the perimeter and cross-sectional area of the extended surface, respectively. The rate of energy transfer due to the motion of the material, $\dot{E}$, is the product of the enthalpy of the material (i) and its mass flow rate. The mass flow rate is the product of the velocity, density (ρ), and cross-sectional area.

$$\dot{E} = u\,A_c\,\rho\,i \tag{1-302}$$

Substituting Eqs. (1-300) through (1-302) into Eq. (1-299) leads to:

$$0 = per\,dx\,\overline{h}\,(T - T_\infty) + \frac{d}{dx}\left[-k\,A_c\,\frac{dT}{dx}\right]dx + \frac{d}{dx}[u\,A_c\,\rho\,i]\,dx \tag{1-303}$$

Assuming constant properties:

$$0 = per\,\overline{h}\,(T - T_\infty) - k\,A_c\,\frac{d^2T}{dx^2} + u\,A_c\,\rho\,\frac{di}{dx} \tag{1-304}$$

The enthalpy gradient is expanded:

$$0 = per\,\overline{h}\,(T - T_\infty) - k\,A_c\,\frac{d^2T}{dx^2} + u\,A_c\,\rho\,\frac{di}{dT}\,\frac{dT}{dx} \tag{1-305}$$

Assuming that the material is incompressible, the derivative of enthalpy with respect to temperature is the specific heat capacity (c):

$$0 = per\,\overline{h}\,(T - T_\infty) - k\,A_c\,\frac{d^2T}{dx^2} + u\,A_c\,\rho\,c\,\frac{dT}{dx} \tag{1-306}$$

Equation (1-306) is rearranged in order to obtain the governing differential equation:

$$\frac{d^2T}{dx^2} - \frac{u\,\rho\,c}{k}\,\frac{dT}{dx} - \frac{per\,\overline{h}}{k\,A_c}\,T = -\frac{per\,\overline{h}}{k\,A_c}\,T_\infty \tag{1-307}$$

The solution is again divided into a homogeneous and particular solution:

$$T = T_h + T_p \tag{1-308}$$

Substituting Eq. (1-308) into Eq. (1-307) leads to:

$$\underbrace{\frac{d^2T_h}{dx^2} - \frac{u\,\rho\,c}{k}\,\frac{dT_h}{dx} - \frac{per\,\overline{h}}{k\,A_c}\,T_h}_{=0\text{ for homogeneous differential equation}} + \underbrace{\frac{d^2T_p}{dx^2} - \frac{u\,\rho\,c}{k}\,\frac{dT_p}{dx} - \frac{per\,\overline{h}}{k\,A_c}\,T_p = -\frac{per\,\overline{h}}{k\,A_c}\,T_\infty}_{\text{whatever is left is the particular differential equation}} \tag{1-309}$$

The particular differential equation:

$$\frac{d^2 T_p}{dx^2} - \frac{u\,\rho\,c}{k}\frac{dT_p}{dx} - \frac{per\,\overline{h}}{k\,A_c}T_p = -\frac{per\,\overline{h}}{k\,A_c}T_\infty \tag{1-310}$$

is solved by a constant:

$$T_p = T_\infty \tag{1-311}$$

The homogeneous differential equation is:

$$\frac{d^2 T_h}{dx^2} - \frac{u\,\rho\,c}{k}\frac{dT_h}{dx} - \frac{per\,\overline{h}}{k\,A_c}T_h = 0 \tag{1-312}$$

The fin parameter (m) is defined as in Section 1.6:

$$m = \sqrt{\frac{\overline{h}\,per}{k\,A_c}} \tag{1-313}$$

The group of properties, $k/\rho c$, appearing in Eq. (1-312) is encountered often in heat transfer and is defined as the thermal diffusivity (α):

$$\alpha = \frac{k}{\rho\,c} \tag{1-314}$$

With these definitions, Eq. (1-312) can be written as:

$$\frac{d^2 T_h}{dx^2} - \frac{u}{\alpha}\frac{dT_h}{dx} - m^2\,T_h = 0 \tag{1-315}$$

Equation (1-315) is solved by an exponential:

$$T_h = C\,\exp(\lambda\,x) \tag{1-316}$$

where C and λ are both arbitrary constants. Equation (1-316) is substituted into Eq. (1-315):

$$C\lambda^2\,\exp(\lambda\,x) - \frac{u}{\alpha}C\lambda\,\exp(\lambda\,x) - m^2\,C\,\exp(\lambda\,x) = 0 \tag{1-317}$$

which can be simplified:

$$\lambda^2 - \frac{u}{\alpha}\lambda - m^2 = 0 \tag{1-318}$$

Equation (1-318) is quadratic and therefore has two solutions (λ_1 and λ_2):

$$\lambda_1 = \frac{u}{2\,\alpha} + \sqrt{\frac{1}{4}\left(\frac{u}{\alpha}\right)^2 + m^2} \tag{1-319}$$

$$\lambda_2 = \frac{u}{2\,\alpha} - \sqrt{\frac{1}{4}\left(\frac{u}{\alpha}\right)^2 + m^2} \tag{1-320}$$

Because the governing equation is linear, the sum of the two solutions ($T_{h,1}$ and $T_{h,2}$):

$$T_{h,1} = C_1\,\exp(\lambda_1\,x) \tag{1-321}$$

$$T_{h,2} = C_2\,\exp(\lambda_2\,x) \tag{1-322}$$

is also a solution and therefore the general solution to the homogeneous governing differential equation is:

$$T_h = C_1 \exp(\lambda_1 x) + C_2 \exp(\lambda_2 x) \tag{1-323}$$

The solution to the differential equation is the sum of the homogeneous and particular solutions:

$$T = C_1 \exp(\lambda_1 x) + C_2 \exp(\lambda_2 x) + T_\infty \tag{1-324}$$

where the constants C_1 and C_2 are determined according to the boundary conditions.

The solution may also be obtained using Maple by entering and solving the governing differential equation:

```
> restart;
> ODE:=diff(diff(T(x),x),x)-u*diff(T(x),x)/alpha-m^2*T(x)=-m^2*T_infinity;
```

$$ODE := \left(\frac{d^2}{dx^2} T(x)\right) - \frac{u\left(\frac{d}{dx} T(x)\right)}{\alpha} - m^2 T(x) = -m^2 T\_infinity$$

```
> T_s:=dsolve(ODE);
```

$$T\_s := T(x) = e^{\left(\frac{(u+\sqrt{u^2+4m^2\alpha^2})x}{2\alpha}\right)} \_C2 + e^{\left(\frac{(u-\sqrt{u^2+4m^2\alpha^2})x}{2\alpha}\right)} \_C1 + T\_infinity$$

which is equivalent to Eq. (1-324).

EXAMPLE 1.7-2: DRAWING A WIRE

Figure 1 illustrates a wire drawn from a die. The wire diameter is $D = 0.5$ mm. The temperature of the material at the exit of the die is $T_{draw} = 600°C$ and it has a draw velocity of $u = 10$ mm/s. The properties of the wire are $\rho = 2700\,kg/m^3$, $k = 230\,W/m\text{-}K$, and c = 1000 J/kg-K. The wire is surrounded by air at $T_\infty = 20°C$ with an average heat transfer coefficient of $\bar{h} = 25\,W/m^2\text{-}K$. The wire travels for $L = 25$ cm before entering a pool of water that is kept at $T_w = 20°C$; you may assume that the water-to-wire heat transfer coefficient is very high so that the wire equilibrates essentially instantaneously with the water as it enters the pool.

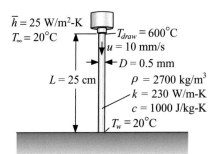

Figure 1: Wire drawn from a die.

a) **Develop an analytical model that can predict the temperature distribution in the wire.**

EXAMPLE 1.7-2: DRAWING A WIRE

The input parameters are entered in EES:

```
"EXAMPLE 1.7-2: Drawing a Wire"

$UnitSystem SI MASS DEG PA C J
$Tabstops 0.2 0.4 0.6 0.8 3.5

"Inputs"
D=0.5 [mm]*convert(mm,m)                    "diameter"
u=10 [mm/s]*convert(mm/s,m/s)              "draw velocity"
c=1000 [J/kg-K]                            "specific heat capacity"
k=230 [W/m-K]                              "conductivity"
rho=2700 [kg/m^3]                          "density"
h_bar=25 [W/m^2-K]                         "heat transfer coefficient"
T_infinity=converttemp(C,K,20 [C])        "air temperature"
T_draw=converttemp(C,K,600 [C])           "draw temperature"
T_w=converttemp(C,K,20 [C])               "water temperature"
L=25 [cm]*convert(cm,m)                    "length of wire"
```

The governing differential equation for a moving extended surface was derived in Section 1.7.3:

$$\frac{d^2 T}{dx^2} - \frac{u}{\alpha}\frac{dT}{dx} - m^2 T = -m^2 T_\infty$$

where α is the thermal diffusivity:

$$\alpha = \frac{k}{\rho c}$$

m is the fin constant:

$$m = \sqrt{\frac{h\, per}{k\, A_c}}$$

and per and A_c are the perimeter and cross-sectional area, respectively, of the moving surface:

$$per = \pi D$$

$$A_c = \pi \frac{D^2}{4}$$

```
A_c=pi*D^2/4                              "cross-sectional area"
per=pi*D                                  "perimeter"
alpha=k/(rho*c)                           "thermal diffusivity"
m=sqrt(h_bar*per/(k*A_c))                 "fin parameter"
```

The general solution derived in Section 1.7.3 is:

$$T = C_1 \exp(\lambda_1 x) + C_2 \exp(\lambda_2 x) + T_\infty \tag{1}$$

EXAMPLE 1.7-2: DRAWING A WIRE

where C_1 and C_2 are undetermined constants and:

$$\lambda_1 = \frac{u}{2\,\alpha} + \sqrt{\frac{1}{4}\left(\frac{u}{\alpha}\right)^2 + m^2}$$

$$\lambda_2 = \frac{u}{2\,\alpha} - \sqrt{\frac{1}{4}\left(\frac{u}{\alpha}\right)^2 + m^2}$$

```
lambda_1=u/(2*alpha)+sqrt((u/alpha)^2/4+m^2)        "solution parameter 1"
lambda_2=u/(2*alpha)-sqrt((u/alpha)^2/4+m^2)        "solution parameter 2"
```

The constants are evaluated using the boundary conditions. The temperatures at $x = 0$ and $x = L$ are specified:

$$T_{x=L} = T_w \qquad\qquad (2)$$

$$T_{x=0} = T_{draw} \qquad\qquad (3)$$

Substituting Eq. (1) into Eqs. (2) and (3) leads to two algebraic equations for C_1 and C_2:

$$C_1 \exp\left(\lambda_1 L\right) + C_2 \exp\left(\lambda_2 L\right) + T_\infty = T_w$$

$$C_1 + C_2 + T_\infty = T_{draw}$$

which are entered in EES:

```
C_1*exp(lambda_1*L)+C_2*exp(lambda_2*L)+T_infinity=T_w    "boundary condition at x=L"
C_1+C_2+T_infinity=T_draw                                 "boundary condition at x=0"
```

The solution is evaluated in EES and converted to Celsius.

```
x=x_bar*L                                               "position"
T=C_1*exp(lambda_1*x)+C_2*exp(lambda_2*x)+T_infinity    "solution"
T_C=converttemp(K,C,T)                                  "in C"
```

A parametric table in EES can be used to provide the temperature as a function of position. It is convenient to define the variable x_bar, the axial position normalized by the length of the wire. Including x_bar in the table and varying it from 0 to 1 is equivalent to varying the position from 0 to L. One advantage of using the variable x_bar is that as the length of the wire is changed, it is not necessary to adjust the parametric table, only to re-run it. Figure 2 illustrates the temperature as a function of dimensionless position for various values of the length.

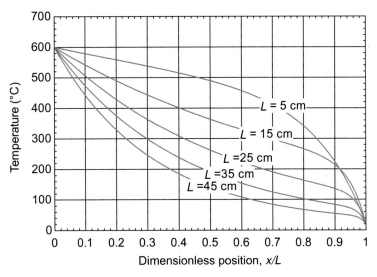

EXAMPLE 1.7-2: DRAWING A WIRE

Figure 2: Temperature as a function of dimensionless position for various values of length.

1.8 Analytical Solutions for Non-Constant Cross-Section Extended Surfaces

1.8.1 Introduction

Sections 1.6 and 1.7 showed how the differential equation describing constant cross-section fins and other extended surfaces is derived. Analytical solutions for these differential equations take the form of an exponential function. In this section, extended surface problems are considered for which the cross-sectional area for conduction and the wetted perimeter for convection are not constant. The resulting differential equation is solved by Bessel functions.

1.8.2 Series Solutions

It is worthwhile asking what the "exponential function" really is; we take it for granted in terms of its properties (i.e., how it can be integrated and differentiated). With some experience, it is possible to see that it solves a certain type of differential equation. In fact, that is its purpose: the exponential is really a polynomial series that has been defined so that it solves a commonly encountered differential equation. There are other types of differential equations that appear in engineering problems; series solutions to these differential equations have been defined and given formal names like "Bessel function" and "Kelvin function".

The homogeneous differential equation that results from the analysis of a constant cross-sectional area fin is derived in Section 1.6.3

$$\frac{d^2 T_h}{dx^2} - m^2 T_h = 0 \qquad (1\text{-}325)$$

Provided that the solution to Eq. (1-325) is continuous, it can be represented by a series of the form:

$$T_h = a_0 + a_1 x + a_2 x^2 + a_3 x^3 + a_4 x^4 + \cdots = \sum_{i=0}^{\infty} a_i x^i \qquad (1\text{-}326)$$

By substituting Eq. (1-326) into Eq. (1-325), it is possible to identify the characteristics of the series that solves this class of differential equation. The second derivative of the solution is required:

$$\frac{dT_h}{dx} = a_1 + 2\,a_2\,x + 3\,a_3\,x^2 + 4\,a_4\,x^3 + 5\,a_5\,x^4 + \cdots = \sum_{i=1}^{\infty} a_i\,i\,x^{i-1} \qquad (1\text{-}327)$$

$$\frac{d^2T_h}{dx^2} = 2\,(1)\,a_2 + 3\,(2)\,a_3\,x + 4\,(3)\,a_4\,x^2 + 5\,(4)\,a_5\,x^3 + 6\,(5)\,a_6\,x^4 + \cdots$$

$$= \sum_{i=2}^{\infty} a_i\,i\,(i-1)\,x^{i-2} \qquad (1\text{-}328)$$

Substituting Eqs. (1-328) and (1-326) into Eq. (1-325) leads to:

$$2\,(1)\,a_2 + 3\,(2)\,a_3\,x + 4\,(3)\,a_4\,x^2 + 5\,(4)\,a_5\,x^3 + 6\,(5)\,a_6\,x^4 + \cdots$$
$$-m^2[a_0 + a_1\,x + a_2\,x^2 + a_3\,x^3 + a_4\,x^4 + \cdots] = 0 \qquad (1\text{-}329)$$

or

$$\sum_{i=2}^{\infty} a_i\,i\,(i-1)\,x^{i-2} - m^2 \sum_{i=0}^{\infty} a_i\,x^i = 0 \qquad (1\text{-}330)$$

Since x is an independent variable that can assume any value, Eqs. (1-329) or (1-330) can only be generally satisfied if the coefficients that multiply each term of the series (i.e., each power of x) each sum to zero. Examining Eq. (1-329), this requirement leads to:

$$2\,(1)\,a_2 - m^2\,a_0 = 0$$
$$3\,(2)\,a_3 - m^2 a_1 = 0$$
$$4\,(3)\,a_4 - m^2 a_2 = 0 \qquad (1\text{-}331)$$
$$5\,(4)\,a_5 - m^2 a_3 = 0$$
$$6\,(5)\,a_6 - m^2 a_4 = 0$$
$$\cdots$$

The even coefficients are therefore related according to:

$$a_2 = \frac{m^2\,a_0}{2\,(1)}$$

$$a_4 = \frac{m^2 a_2}{4\,(3)} = \frac{m^4 a_0}{4\,(3)\,(2)\,(1)} \qquad (1\text{-}332)$$

$$a_6 = \frac{m^2 a_4}{6\,(5)} = \frac{m^6 a_0}{6\,(5)\,(4)\,(3)\,(2)\,(1)}$$
$$\cdots$$

or, more generally

$$a_{2i} = \frac{m^{2i} a_0}{(2i)!} \quad \text{where } i = 0 \cdots \infty \qquad (1\text{-}333)$$

The odd coefficients are also related:

$$a_3 = \frac{m^2 a_1}{3\,(2)}$$

$$a_5 = \frac{m^2 a_3}{5\,(4)} = \frac{m^4 a_1}{5\,(4)\,(3)\,(2)} \tag{1-334}$$

$$a_7 = \frac{m^2 a_5}{7\,(6)} = \frac{m^6 a_1}{7\,(6)\,(5)\,(4)\,(3)\,(2)}$$

$$\cdots$$

or, more generally

$$a_{2i+1} = \frac{m^{2i} a_1}{(2i+1)!} \quad \text{where } i = 0 \cdots \infty \tag{1-335}$$

Therefore, we have determined two functions that both solve Eq. (1-325), related to the even and odd terms of the series; let's call them F_{even} and F_{odd}:

$$F_{even} = a_0 \sum_{i=0}^{\infty} \frac{(mx)^{2i}}{(2i)!} \tag{1-336}$$

$$F_{odd} = a_1 \sum_{i=0}^{\infty} \frac{m^{2i} x^{2i+1}}{(2i+1)!} \tag{1-337}$$

F_{even} and F_{odd} are two solutions to the governing equation regardless of the particular values of the constants a_0 and a_1 in the same way that the functions $C_1 \exp(mx)$ and $C_2 \exp(-mx)$ (or, equivalently, $C_1 \sinh(mx)$ and $C_2 \cosh(mx)$) were identified in Section 1.6 as solutions to Eq. (1-325) regardless of the values of C_1 and C_2. The constants are determined in order to make the solution match the boundary conditions. In fact, if Eqs. (1-336) and (1-337) are rearranged slightly we see that they are identical to the series expansion of the functions $\sinh(mx)$ and $\cosh(mx)$:

$$C_1 \cosh(mx) = C_1 \sum_{i=0}^{\infty} \frac{(mx)^{2i}}{(2i)!} = \frac{C_1}{a_0} F_{even} \tag{1-338}$$

$$C_2 \sinh(mx) = C_2 \sum_{i=0}^{\infty} \frac{(mx)^{2i+1}}{(2i+1)!} = \frac{m\,C_2}{a_1} F_{odd} \tag{1-339}$$

The functions cosh and sinh are useful because they solve a particular differential equation, Eq. (1-325), which appears in many engineering problems; they are in fact nothing more than shorthand for the series given by Eqs. (1-336) and (1-337). To see this clearly, compute each of the terms in Eqs. (1-336) and (1-337) using EES:

```
mx=1                                              "argument of function"
Nterm=10                                          "number of terms in series"
duplicate i=0,Nterm
      F_even[i]=(mx)^(2*i)/Factorial(2*i)         "term in F_even"
      F_odd[i]=(mx)^(2*i+1)/Factorial(2*i+1)      "term in F_odd"
end
```

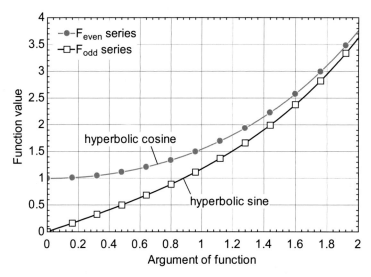

Figure 1-48: Comparison of the functions F_{even} and F_{odd} to the functions sinh and cosh.

and sum these terms using the sum command:

```
F_even=sum(F_even[i],i=0,Nterm)              "sum of all terms in F_even"
F_odd=sum(F_odd[i],i=0,Nterm)                "sum of all terms in F_odd"
```

The results can be compared to the functions sinh and cosh:

```
sinh=sinh(mx)                                "sinh function"
cosh=cosh(mx)                                "cosh function"
```

A parametric table is created that includes the variables mx, F_even, F_odd, sinh and cosh; the variable mx is varied from 0 to 2.0 and the results are shown in Figure 1-48.

This exercise is meant to show that "solving" the homogeneous ordinary differential equation, Eq. (1-325), was really a matter of recognizing that it is solved by the series solutions that we call sinh and cosh (or equivalently exponentials with positive and negative arguments). Maple is very good at recognizing the solutions to differential equations.

In this section, extended surfaces that do not have a constant cross-sectional area are considered. The governing differential equations that apply to these problems are more complex than Eq. (1-325). However, these differential equations are also solved by correctly defined series that are given different names: Bessel functions.

1.8.3 Bessel Functions

Extended surfaces with non-uniform cross-section may arise in many engineering applications. For example, tapered fins are of interest since they may provide heat transfer rates comparable to constant cross-section fins but require less material. Figure 1-49 illustrates a wedge fin, an extended surface that has a thickness that varies linearly from its value at the base (th) to zero at the tip. The fin is surrounded by fluid at T_∞ with average heat transfer coefficient $\bar{h}$. The fin material has conductivity k and the base of the fin is kept at T_b.

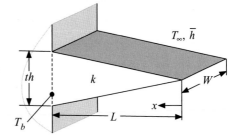

Figure 1-49: Wedge fin (not drawn to scale, $W \gg th$).

The width of the fin is W and its length is L. We will assume that $W \gg th$ so that convection from the edges of the fin may be ignored. Also, we will assume that the criteria for the extended approximation is satisfied (in this case, the Biot number $\bar{h}\,th/(2\,k) \ll 1$) so that the temperature can be assumed to be spatially uniform at any axial location x. It is convenient to define the origin of the axial coordinate at the tip of the fin (see Figure 1-49) so that the cross-sectional area for conduction (A_c) can be expressed as:

$$A_c = th\,W\,\frac{x}{L} \tag{1-340}$$

The differential control volume used to derive the governing differential equation is shown in Figure 1-50 and suggests the energy balance:

$$\dot{q}_x = \dot{q}_{x+dx} + \dot{q}_{conv} \tag{1-341}$$

or, after expanding the $x + dx$ term and simplifying:

$$0 = \frac{d\dot{q}}{dx}dx + \dot{q}_{conv} \tag{1-342}$$

The convection heat transfer rate is, approximately

$$\dot{q}_{conv} = 2\,W\,dx\,\bar{h}\,(T - T_\infty) \tag{1-343}$$

Note that Eq. (1-343) is only valid if $th/L \ll 1$ so that the surface area within the control volume that is exposed to fluid is approximately $2\,W\,dx$. The conduction heat transfer rate is:

$$\dot{q} = -k\,A_c\,\frac{dT}{dx} = -k\,th\,W\,\frac{x}{L}\frac{dT}{dx} \tag{1-344}$$

Substituting Eqs. (1-343) and (1-344) into Eq. (1-342) leads to:

$$0 = \frac{d}{dx}\left[-k\,th\,W\,\frac{x}{L}\frac{dT}{dx}\right]dx + 2\,W\,dx\,\bar{h}\,(T - T_\infty) \tag{1-345}$$

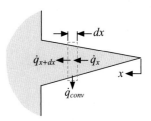

Figure 1-50: Differential control volume.

which can be simplified to:

$$\frac{d}{dx}\left[x\frac{dT}{dx}\right] - \frac{2\bar{h}L}{k\,th}T = -\frac{2\bar{h}L}{k\,th}T_\infty \qquad (1\text{-}346)$$

The solution is divided into a homogeneous and particular component:

$$T = T_h + T_p \qquad (1\text{-}347)$$

Substituting Eq. (1-347) into Eq. (1-346) leads to:

$$\underbrace{\frac{d}{dx}\left[x\frac{dT_h}{dx}\right] - \frac{2\bar{h}L}{k\,th}T_h}_{\substack{= 0 \text{ for homogeneous} \\ \text{differential equation}}} + \underbrace{\frac{d}{dx}\left[x\frac{dT_p}{dx}\right] - \frac{2\bar{h}L}{k\,th}T_p = -\frac{2\bar{h}L}{k\,th}T_\infty}_{\text{whatever is left is the particular differential equation}} \qquad (1\text{-}348)$$

The solution to the particular differential equation is:

$$T_p = T_\infty \qquad (1\text{-}349)$$

The homogeneous differential equation is:

$$\frac{d}{dx}\left[x\frac{dT_h}{dx}\right] - \beta\,T_h = 0 \qquad (1\text{-}350)$$

where the parameter β is defined for convenience to be:

$$\beta = \frac{2\bar{h}L}{k\,th} \qquad (1\text{-}351)$$

Note that the homogeneous differential equation, Eq. (1-350), is fundamentally different from the homogeneous differential equation that was obtained for a constant cross-section fin, Eq. (1-179). Equation (1-350) is not solved by exponentials; to make this clear, assume an exponential solution:

$$T_h = C\exp(mx) \qquad (1\text{-}352)$$

and substitute it into the homogeneous differential equation:

$$\frac{d}{dx}[xCm\exp(mx)] - \beta C\exp(mx) = 0 \qquad (1\text{-}353)$$

or, using the chain rule:

$$Cm\exp(mx) + xCm^2\exp(mx) - \beta C\exp(mx) = 0 \qquad (1\text{-}354)$$

which can be simplified to:

$$m + xm^2 - \beta = 0 \qquad (1\text{-}355)$$

Unfortunately, there is no value of m that will satisfy Eq. (1-355) for all values of x.

There must be some other function that solves Eq. (1-350). A series solution is again assumed:

$$T_h = a_0 + a_1 x + a_2 x^2 + a_3 x^3 + a_4 x^4 + \cdots = \sum_{i=0}^{\infty} a_i x^i \qquad (1\text{-}356)$$

The series is substituted into the governing differential equation in order to identify the characteristics of the series that solves this new class of differential equation. Expanding

Eq. (1-350) using the chain rule shows that both the first and second derivatives are required:

$$x\frac{d^2 T_h}{dx^2} + \frac{dT_h}{dx} - \beta T_h = 0 \qquad (1\text{-}357)$$

These derivatives were derived in Section 1.8.2:

$$\frac{dT_h}{dx} = a_1 + 2 a_2 x + 3 a_3 x^2 + 4 a_4 x^3 + 5 a_5 x^4 + \cdots = \sum_{i=1}^{\infty} a_i i x^{i-1} \qquad (1\text{-}358)$$

$$\frac{d^2 T_h}{dx^2} = 2 (1) a_2 + 3 (2) a_3 x + 4 (3) a_4 x^2 + 5 (4) a_5 x^3 + 6 (5) a_6 x^4 + \cdots$$

$$= \sum_{i=2}^{\infty} a_i i (i-1) x^{i-2} \qquad (1\text{-}359)$$

Substituting Eqs. (1-356), (1-358), and (1-359) into Eq. (1-357) leads to:

$$x\frac{d^2 T_h}{dx^2} \rightarrow 2 (1) a_2 x + 3 (2) a_3 x^2 + 4 (3) a_4 x^3 + 5 (4) a_5 x^4 + 6 (5) a_6 x^5 + \cdots$$

$$+\frac{dT_h}{dx} \rightarrow a_1 + 2 a_2 x + 3 a_3 x^2 + 4 a_4 x^3 + 5 a_5 x^4 + \cdots \qquad (1\text{-}360)$$

$$\underline{-\beta T_h \rightarrow -\beta a_0 - \beta a_1 x - \beta a_2 x^2 - \beta a_3 x^3 - \beta a_4 x^4 + \cdots}$$

$$= 0$$

or, collecting like terms:

$$[a_1 - \beta a_0] + [2 (1) a_2 + 2 a_2 - \beta a_1] x + [3 (2) a_3 + 3 a_3 - \beta a_2] x^2$$

$$+ [4 (3) a_4 + 4 a_4 - \beta a_3] x^3 + [5 (4) a_5 + 5 a_5 - \beta a_4] x^4 \cdots = 0 \qquad (1\text{-}361)$$

For the series to solve the differential equation, each of the coefficients must be zero; again, considering the coefficients one at a time leads to a recursive formula that defines the series.

$$a_1 = \beta a_0$$

$$\underbrace{(2 (1) + 2)}_{2^2} a_2 = \beta a_1$$

$$\underbrace{(3 (2) + 3)}_{3^2} a_3 = \beta a_2$$

$$\underbrace{(4 (3) + 4)}_{4^2} a_4 = \beta a_3 \qquad (1\text{-}362)$$

$$\underbrace{(5 (4) + 5)}_{5^2} a_5 = \beta a_4$$

$$\cdots$$

or

$$a_1 = \beta\, a_0$$

$$a_2 = \frac{\beta}{2^2}\, a_1 = \frac{\beta^2}{2^2}\, a_0$$

$$a_3 = \frac{\beta}{3^2}\, a_2 = \frac{\beta^3}{[3\,(2)]^2}\, a_0$$

$$a_4 = \frac{\beta}{4^2}\, a_3 = \frac{\beta^4}{[4\,(3)\,(2)]^2}\, a_0 \qquad (1\text{-}363)$$

$$a_5 = \frac{\beta}{5^2}\, a_4 = \frac{\beta^5}{[5\,(4)\,(3)\,(2)]^2}\, a_0$$

$$\cdots$$

More generally, the coefficients are defined by the equation:

$$a_i = \frac{\beta^i}{(i!)^2}\, a_0 \quad \text{where } i = 1 \cdots \infty \qquad (1\text{-}364)$$

Equation (1-364) defines a function (let's call it F) that provides a general solution to the homogeneous differential equation, Eq. (1-350).

$$F = a_0 \sum_{i=0}^{\infty} \frac{(\beta x)^i}{(i!)^2} \qquad (1\text{-}365)$$

The function F is actually a combination of Bessel functions; this is simply the name given to the series solutions of a particular class of differential equations (just as hyperbolic sine and hyperbolic cosine are names given to series solutions of a different class of differential equations). The Bessel functions behave according to a set of formalized rules (just as hyperbolic sines and cosines do) that must be carefully obeyed when using them to solve problems. Bessel functions and the rules for manipulating them are completely recognized by Maple. Therefore, the combination of Maple and EES together allow you to avoid much of tedium associated with recognizing the correct Bessel function and then manipulating it to satisfy the boundary conditions of the problem.

For the wedge fin problem considered here, it is possible to enter the governing differential equation, Eq. (1-346), in Maple:

```
> restart;
> ODE:=diff(x*diff(T(x),x),x)-beta*T(x)=-beta*T_infinity;
```

$$ODE := \left(\frac{d}{dx}\mathrm{T}(x)\right) + x\left(\frac{d^2}{dx^2}\mathrm{T}(x)\right) - \beta\,\mathrm{T}(x) = -\beta\,T\_\mathit{infinity}$$

and solve it:

```
> Ts:=dsolve(ODE);
```

$$Ts := \mathrm{T}(x) = \mathrm{BesselJ}(0,\, 2,\, \sqrt{-\beta}\,\sqrt{x})\_C2 + \mathrm{BesselY}(0,\, 2\sqrt{-\beta}\,\sqrt{x})\_C1 + T\_\mathit{infinity}$$

Maple has recognized that the solution to the governing differential equation includes two Bessel functions: the functions BesselJ and BesselY corresponding to Bessel functions of the first and second kind, respectively. The first argument indicates the order of the Bessel function and the second indicates the argument. The parameter β was defined in Eq. (1-351) and involves the product of only positive quantities. Therefore, β must be positive and the arguments of both of the Bessel functions are complex (i.e., they involve the square root of a negative number); Bessel functions evaluated with a complex argument result in what are called modified Bessel functions (BesselI and BesselK are the modified Bessel functions of the first and second kind, respectively). Maple will identify this fact for you, provided that you specify that the variable beta must be positive (using the assume command) and solve the equation again:

```
> assume(beta>0);
> Ts:=dsolve(ODE);
```

$$Ts := \mathrm{T}(x) = \mathrm{BesselI}(0,\ 2\sqrt{\beta}\sim\sqrt{x})\_C2 + \mathrm{BesselK}(0,\ 2\sqrt{\beta}\sim\sqrt{x})\_C1 + T\_infinity$$

The trailing tilde ($\sim$) notation is used in Maple to indicate that the variable is associated with an assumption regarding its value; this convention can be changed in the preferences dialog in Maple. The solution is expressed in terms of modified Bessel functions with real arguments rather than Bessel functions with complex arguments. It is worth noting that if the argument of the function cosh is complex, then the result is cosine; thus the cosine function can be thought of as the modified hyperbolic cosine. The same behavior occurs for the sine and hyperbolic sine functions. All of the Bessel functions are built into EES and therefore the solution from Maple can be copied and pasted into EES for evaluation and manipulation.

Maple has identified the general solution to the wedge fin problem:

$$T = C_2\,\mathrm{BesselI}(0, 2\sqrt{\beta x}) + C_1\,\mathrm{BesselK}(0, 2\sqrt{\beta x}) + T_\infty \tag{1-366}$$

All that remains is to determine the constants C_1 and C_2 so that Eq. (1-366) also satisfies the boundary conditions. It is tempting to assume that one boundary condition must force the rate of heat transfer at the tip to be zero; however, the fact that the cross-sectional area at the tip is zero guarantees this fact, provided that the temperature gradient and therefore the temperature at the tip (i.e., at $x = 0$) is finite.

$$T_{x=0} < \infty \tag{1-367}$$

Substituting Eq. (1-366) into Eq. (1-367) leads to:

$$C_2\,\mathrm{BesselI}(0, 0) + C_1\,\mathrm{BesselK}(0, 0) + T_\infty < \infty \tag{1-368}$$

Figure 1-51 illustrates the behavior of the zeroth order modified Bessel function of the first (BesselI) and second (BesselK) kind.

Notice that the zeroth order modified Bessel function of the second kind is unbounded at zero and therefore the solution cannot include BesselK; the constant C_1 must be zero. The remaining boundary condition corresponds to the specified base temperature of the fin:

$$T_{x=L} = T_b \tag{1-369}$$

Substituting Eq. (1-366) into Eq. (1-369) leads to:

$$C_2\,\mathrm{BesselI}(0, 2\sqrt{\beta L}) + T_\infty = T_b \tag{1-370}$$

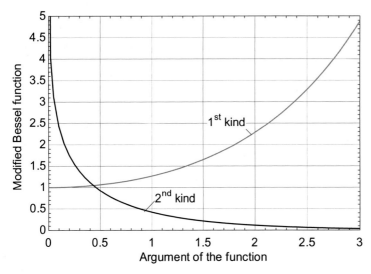

Figure 1-51: Modified zeroth order Bessel functions of the first and second kind.

Solving for the constant C_2 leads to:

$$C_2 = \frac{(T_b - T_\infty)}{\text{BesselI}\left(0, 2\sqrt{\beta L}\right)} \tag{1-371}$$

Substituting Eq. (1-371) into Eq. (1-366) (with $C_1 = 0$) leads to the solution for the temperature distribution for a wedge fin:

$$T = (T_b - T_\infty)\frac{\text{BesselI}(0, 2\sqrt{\beta x})}{\text{BesselI}(0, 2\sqrt{\beta L})} + T_\infty \tag{1-372}$$

which can be expressed as:

$$\frac{(T - T_\infty)}{(T_b - T_\infty)} = \frac{\text{BesselI}\left(0, 2\sqrt{\beta L\frac{x}{L}}\right)}{\text{BesselI}(0, 2\sqrt{\beta L})} \tag{1-373}$$

Figure 1-52 illustrates the dimensionless temperature, $(T - T_\infty)/(T_b - T_\infty)$, as a function of the dimensionless position, x/L, for various values of βL.

Note that according to Eq. (1-351), the dimensionless parameter βL is:

$$\beta L = \frac{2\bar{h}L^2}{k\,th} \tag{1-374}$$

and resembles the fin parameter, mL, for a constant cross-sectional area fin. The parameter βL plays a similar role in the solution. The resistance to conduction in the x-direction is approximately:

$$R_{cond,x} \approx \frac{2L}{kW\,th} \tag{1-375}$$

and the resistance to convection is approximately:

$$R_{conv} \approx \frac{1}{\bar{h}\,2LW} \tag{1-376}$$

The ratio of $R_{cond,x}$ to R_{conv} is related to βL:

$$\frac{R_{cond,x}}{R_{conv}} \approx \frac{2L}{kW\,th}\frac{\bar{h}\,2LW}{1} = 4\frac{\bar{h}L^2}{k\,th} = 2\beta L \tag{1-377}$$

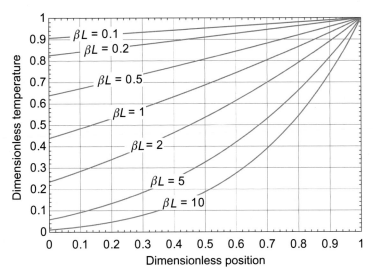

Figure 1-52: Dimensionless temperature distribution as a function of dimensionless position for various values of βL.

As βL is reduced, the resistance to conduction in the x-direction becomes small relative to the resistance to convection and so the fin becomes nearly isothermal at the base temperature. If βL is large, then the convection resistance is large relative to the conduction resistance and so the fin temperature approaches the fluid temperature.

The rate of heat transfer to the fin is given by:

$$\dot{q}_{fin} = k \, W \, th \, \left. \frac{dT}{dx} \right|_{x=L} \tag{1-378}$$

Substituting Eq. (1-372) into Eq. (1-378) leads to:

$$\dot{q}_{fin} = k \, W \, th \, \frac{d}{dx} \left[(T_b - T_\infty) \frac{\text{BesselI}(0, 2\sqrt{\beta x})}{\text{BesselI}(0, 2\sqrt{\beta L})} + T_\infty \right]_{x=L} \tag{1-379}$$

or

$$\dot{q}_{fin} = \frac{k \, W \, th \, (T_b - T_\infty)}{\text{BesselI}\,(0, 2\sqrt{\beta L})} \frac{d}{dx} [\text{BesselI}(0, 2\sqrt{\beta x})]_{x=L} \tag{1-380}$$

Rules for manipulating Bessel functions are presented in Section 1.8.4; however, Maple can be used to work with Bessel functions. The derivative required by Eq. (1-380) is computed easily using Maple:

```
> restart;
> diff(BesselI(0,2*sqrt(beta*x)),x);
```

$$\frac{\text{BesselI}(1, 2\sqrt{\beta x})\,\beta}{\sqrt{\beta x}}$$

so that Eq. (1-380) can be written as:

$$\dot{q}_{fin} = k \, W \, th \, (T_b - T_\infty) \frac{\text{BesselI}(1, 2\sqrt{\beta L})}{\text{BesselI}(0, 2\sqrt{\beta L})} \sqrt{\frac{\beta}{L}} \tag{1-381}$$

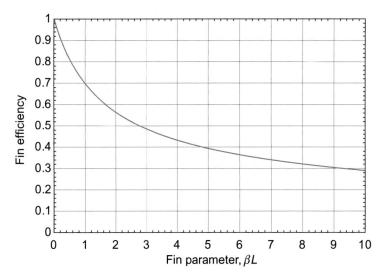

Figure 1-53: Fin efficiency of a wedge fin as a function of βL.

The efficiency of the wedge fin is defined as discussed in Section 1.6.5:

$$\eta_{fin} = \frac{\dot{q}_{fin}}{\bar{h}\,2\,W\,L\,(T_b - T_\infty)} \tag{1-382}$$

Substituting Eq. (1-381) into Eq. (1-382) leads to:

$$\eta_{fin} = \frac{k\,th}{\bar{h}\,2\,L}\frac{\text{BesselI}(1, 2\sqrt{\beta L})}{\text{BesselI}(0, 2\sqrt{\beta L})}\sqrt{\frac{\beta}{L}} \tag{1-383}$$

or

$$\eta_{fin} = \frac{\text{BesselI}(1, 2\sqrt{\beta L})}{\text{BesselI}(0, 2\sqrt{\beta L})\sqrt{\beta L}} \tag{1-384}$$

Figure 1-53 illustrates the fin efficiency of a wedge fin as a function of the parameter βL.
    Notice that the fin efficiency approaches unity when βL approaches zero because the fin is nearly isothermal and the efficiency decreases as βL increases.

1.8.4 Rules for using Bessel Functions

Bessel functions are well-defined functions with specific rules for integration and differentiation. These rules are summarized in this section; however, the use of Maple will greatly reduce the need to know these rules.
    The differential equation:

$$\frac{d}{dx}\left(x^p\frac{d\theta}{dx}\right) \pm c^2\,x^s\,\theta = 0 \tag{1-385}$$

or, equivalently

$$x^p\frac{d^2\theta}{dx^2} + p\,x^{p-1}\frac{d\theta}{dx} \pm c^2\,x^s\,\theta = 0 \tag{1-386}$$

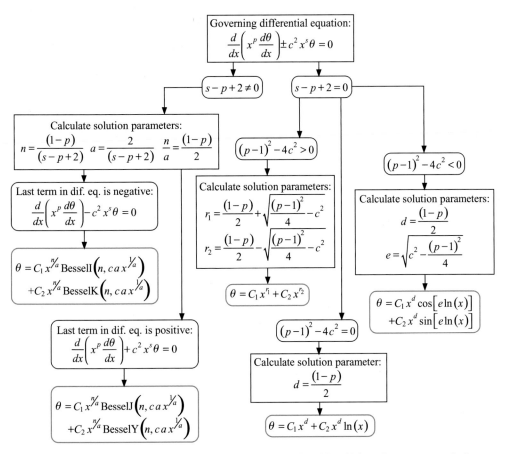

Figure 1-54: Flowchart illustrating the steps involved with identifying the correct solution to Bessel's equation.

where θ is a function of x and p, c, and s are constants is a form of Bessel's equation that has been solved using power series. The rules for identifying the appropriate solution given the form of the equation are laid out in flowchart form in Figure 1-54.

Following the path outlined in Figure 1-54, the first step is to evaluate the quantity $s - p + 2$; if $s - p + 2$ is not equal to zero, then the intermediate solution parameters n and a should be calculated.

$$n = \frac{1 - p}{s - p + 2} \tag{1-387}$$

$$a = \frac{2}{s - p + 2} \tag{1-388}$$

$$\frac{n}{a} = \frac{1 - p}{2} \tag{1-389}$$

The solution depends on the sign of the last term in Eq. (1-385); if the sign of the last term is negative, then the solution is expressed as:

$$\theta = C_1 x^{n/a} \, \mathrm{BesselI}\left(n, c\, a x^{1/a}\right) + C_2 x^{n/a} \, \mathrm{BesselK}\left(n, c\, a x^{1/a}\right) \tag{1-390}$$

where C_1 and C_2 are the undetermined constants that depend on the boundary conditions. The functions BesselI and BesselK are modified Bessel functions of the first and second kind, respectively. The first parameter in the function is the order of the modified Bessel function and the second parameter is the argument of the function. The EES code below provides the zeroth order modified Bessel function of the second kind evaluated at 2.5 (0.06235).

```
y=BesselK(0,2.5)
```

The order of the Bessel function can either be integer (e.g., 0, 1, 2, ...) or fractional (e.g. 0.5).

If the sign of the last term in Eq. (1-385) is positive, then the solution is:

$$\theta = C_1 x^{n/a} \, \text{BesselJ}\left(n, c\,ax^{1/a}\right) + C_2 x^{n/a} \, \text{BesselY}\left(n, c\,ax^{1/a}\right) \tag{1-391}$$

where the functions BesselJ and BesselY are Bessel functions of the first and second kind, respectively.

If $s - p + 2$ is equal to zero, then the solution depends on the sign of the parameter $(p-1)^2 - 4c^2$. If $(p-1)^2 - 4c^2$ is positive, then the solution is:

$$\theta = C_1 x^{r_1} + C_2 x^{r_2} \tag{1-392}$$

where

$$r_1 = \frac{(1-p)}{2} + \sqrt{\frac{(p-1)^2}{4} - c^2} \tag{1-393}$$

and

$$r_2 = \frac{(1-p)}{2} - \sqrt{\frac{(p-1)^2}{4} - c^2} \tag{1-394}$$

If $(p-1)^2 - 4c^2$ is zero, then the solution is:

$$\theta = C_1 x^d + C_2 x^d \, \ln(x) \tag{1-395}$$

where

$$d = \frac{(1-p)}{2} \tag{1-396}$$

Finally, if $(p-1)^2 - 4c^2$ is negative, then the solution is:

$$\theta = C_1 x^d \cos(e\,\ln(x)) + C_2 x^d \sin(e\,\ln(x)) \tag{1-397}$$

where

$$e = \sqrt{c^2 - \frac{(p-1)^2}{4}} \tag{1-398}$$

The zeroth and first order modified Bessel functions of the first and second kind are shown in Figure 1-55. Notice that the modified Bessel functions of the second kind are unbounded at zero while the modified Bessel functions of the first kind are unbounded as the argument tends towards infinity; this characteristic can be helpful to determine the undetermined constants.

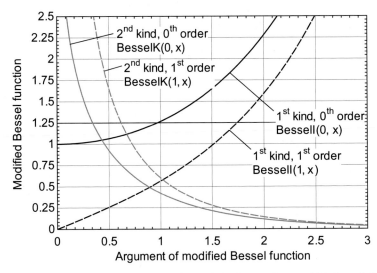

Figure 1-55: Modified Bessel functions of the first and second kinds and the zeroth and first orders.

The zeroth and first order Bessel functions of the first and second kind are shown in Figure 1-56. Notice that the Bessel functions of the second kind, like the modified Bessel functions of the second kind, are unbounded at zero.

The rules for differentiating zeroth order Bessel and zeroth order modified Bessel functions are:

$$\frac{d}{dx}[\text{BesselI}(0, u)] = \text{BesselI}(1, u)\frac{du}{dx} \qquad (1\text{-}399)$$

$$\frac{d}{dx}[\text{BesselK}(0, u)] = -\text{BesselK}(1, u)\frac{du}{dx} \qquad (1\text{-}400)$$

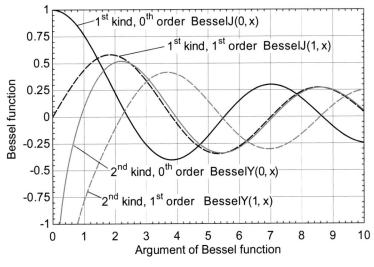

Figure 1-56: Bessel functions of the first and second kinds and the zeroth and first orders.

$$\frac{d}{dx}\left[\text{BesselJ}(0, u)\right] = -\text{BesselJ}(1, u)\frac{du}{dx} \tag{1-401}$$

$$\frac{d}{dx}\left[\text{BesselY}(0, u)\right] = -\text{BesselY}(1, u)\frac{du}{dx} \tag{1-402}$$

For arbitrary order Bessel and modified Bessel functions with positive integer order n, the rules for differentiation are:

$$\frac{d}{dx}\text{BesselI}(n, m\,x) = m\,\text{BesselI}(n-1, m\,x) - \frac{n}{x}\text{BesselI}(n, m\,x) \tag{1-403}$$

$$\frac{d}{dx}\text{BesselK}(n, m\,x) = -m\,\text{BesselK}(n-1, m\,x) - \frac{n}{x}\text{BesselK}(n, m\,x) \tag{1-404}$$

$$\frac{d}{dx}\text{BesselJ}(n, m\,x) = m\,\text{BesselJ}(n-1, m\,x) - \frac{n}{x}\text{BesselJ}(n, m\,x) \tag{1-405}$$

$$\frac{d}{dx}\text{BesselY}(n, m\,x) = m\,\text{BesselY}(n-1, m\,x) - \frac{n}{x}\text{BesselY}(n, m\,x) \tag{1-406}$$

Finally, the following differentials are also sometimes useful:

$$\frac{d}{dx}\left[x^n\,\text{BesselI}(n, m\,x)\right] = m\,x^n\,\text{BesselI}(n-1, m\,x) \tag{1-407}$$

$$\frac{d}{dx}\left[x^n\,\text{BesselK}(n, m\,x)\right] = -m\,x^n\,\text{BesselK}(n-1, m\,x) \tag{1-408}$$

$$\frac{d}{dx}\left[x^n\,\text{BesselJ}(n, m\,x)\right] = m\,x^n\,\text{BesselJ}(n-1, m\,x) \tag{1-409}$$

$$\frac{d}{dx}\left[x^n\,\text{BesselY}(n, m\,x)\right] = -m\,x^n\,\text{BesselY}(n-1, m\,x) \tag{1-410}$$

$$\frac{d}{dx}\left[x^{-n}\,\text{BesselI}(n, m\,x)\right] = m\,x^{-n}\,\text{BesselI}(n+1, m\,x) \tag{1-411}$$

$$\frac{d}{dx}\left[x^{-n}\,\text{BesselK}(n, m\,x)\right] = -m\,x^{-n}\,\text{BesselK}(n+1, m\,x) \tag{1-412}$$

$$\frac{d}{dx}\left[x^{-n}\,\text{BesselJ}(n, m\,x)\right] = -m\,x^{-n}\,\text{BesselJ}(n+1, m\,x) \tag{1-413}$$

$$\frac{d}{dx}\left[x^{-n}\,\text{BesselY}(n, m\,x)\right] = -m\,x^{-n}\,\text{BesselY}(n+1, m\,x) \tag{1-414}$$

EXAMPLE 1.8-1: PIPE IN A ROOF

EXAMPLE 1.8-1: PIPE IN A ROOF

A pipe with outer radius $r_p = 5.0$ cm emerges from a metal roof carrying hot gas at $T_{hot} = 90°C$. The pipe is welded to the roof, as shown in Figure 1. Assume that the temperature at the interface between the pipe and the roof is equal to the gas temperature, T_{hot}. The inside of the roof is well-insulated, but the outside of the roof is exposed to ambient air at $T_\infty = 20°C$. The average heat transfer coefficient between the outside of the roof and the ambient air is $\bar{h} = 50$ W/m²-K. The outside of the roof is also exposed to a uniform heat flux due to the incident solar radiation, $\dot{q}''_s = 800$ W/m². The spatial extent of the roof is large with respect to the outer radius of the pipe. The metal roof has thickness $th = 2.0$ cm and thermal conductivity $k = 50$ W/m-K.

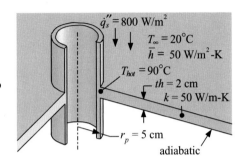

Figure 1: Pipe passing through a roof exposed to solar radiation.

a) Can the roof be modeled using an extended surface approximation?

The input parameters are entered in EES:

```
"EXAMPLE 1.8-1: Pipe in a Roof"

$UnitSystem SI MASS RAD PA K J
$Tabstops 0.2 0.4 0.6 3.5 in

"Input Parameters"
r_p=5.0 [cm]*convert(cm,m)          "Pipe radius"
T_hot=converttemp(C,K,90[C])        "Hot gas temperature"
T_infinity=converttemp(C,K,20[C])   "Air temperature"
h_bar=50 [W/m^2-K]                  "Heat transfer coefficient"
qf_s=800 [W/m^2]                    "Solar flux"
th=2.0 [cm]*convert(cm,m)           "Roof thickness"
k=50 [W/m-K]                        "Roof conductivity"
```

The extended surface approximation ignores any temperature gradients across the thickness of the roof. This is equivalent to ignoring the resistance to conduction across the thickness of the roof while considering the resistance associated with convection from the top surface of roof. The ratio of these resistances is calculated using an appropriately defined Biot number:

$$Bi = \frac{th\,\bar{h}}{k}$$

EXAMPLE 1.8-1: PIPE IN A ROOF

which is calculated in EES:

```
Bi=h_bar*th/k                          "Biot number to check extended surface approximation"
```

The Biot number is 0.02, which is sufficiently less than 1 to justify the extended surface approximation.

b) Develop an analytical model for the roof that can be used to predict the temperature distribution in the roof and also determine the rate of heat loss from the pipe by conduction to the roof.

Because the roof is large relative to the spatial extent of our problem, the edge of the roof will have no effect on the temperature distribution in the metal around the pipe and the temperature distribution will be axisymmetric; the problem can be solved in radial coordinates.

An energy balance on a differential segment of the roof is shown in Figure 2.

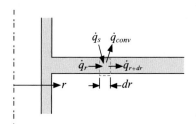

Figure 2: Differential energy balance.

The energy balance includes conduction, convection and solar irradiation:

$$\dot{q}_r + \dot{q}_s = \dot{q}_{r+dr} + \dot{q}_{conv}$$

or

$$\dot{q}_s = \frac{d\dot{q}_r}{dr} dr + \dot{q}_{conv}$$

Substituting the rate equations:

$$\dot{q}_r = -k\, 2\,\pi\, r\, th \frac{dT}{dr}$$

$$\dot{q}_s = \dot{q}_s''\, 2\,\pi\, r\, dr$$

$$\dot{q}_{conv} = 2\,\pi\, r\, dr\, \bar{h}\, (T - T_\infty)$$

into the energy balance leads to:

$$\dot{q}_s''\, 2\,\pi\, r\, dr = \frac{d}{dr}\left[-k\, 2\,\pi\, r\, th \frac{dT}{dr} \right] dr + 2\,\pi\, r\, dr\, \bar{h}\, (T - T_\infty)$$

Simplifying leads to:

$$\frac{d}{dr}\left[r \frac{dT}{dr} \right] - \frac{\bar{h}}{k\, th} r\, T = -\frac{\bar{h}}{k\, th} r\, T_\infty - \frac{\dot{q}_s''}{k\, th} r$$

EXAMPLE 1.8-1: PIPE IN A ROOF

The solution is split into its homogeneous and particular components:

$$T = T_h + T_p$$

which leads to:

$$\underbrace{\frac{d}{dr}\left[r\frac{dT_h}{dr}\right] - \frac{\overline{h}}{k\,th}\,r\,T_h}_{= 0 \text{ for homogeneous differential equation}} + \underbrace{\frac{d}{dr}\left[r\frac{dT_p}{dr}\right] - \frac{\overline{h}}{k\,th}\,r\,T_p = -\frac{\overline{h}}{k\,th}\,r\,T_\infty - \frac{\dot{q}_s''}{k\,th}\,r}_{\text{whatever is left is the particular differential equation}}$$

The solution to the particular differential equation:

$$\frac{d}{dr}\left[r\frac{dT_p}{dr}\right] - \frac{\overline{h}}{k\,th}\,r\,T_p = -\frac{\overline{h}}{k\,th}\,r\,T_\infty - \frac{\dot{q}_s''}{k\,th}\,r$$

is a constant:

$$T_p = T_\infty + \frac{\dot{q}_s''}{\overline{h}}$$

The homogeneous differential equation is:

$$\frac{d}{dr}\left[r\frac{dT_h}{dr}\right] - m^2\,r\,T_h = 0 \tag{1}$$

where

$$m = \sqrt{\frac{\overline{h}}{k\,th}}$$

Equation (1) is a form of Bessel's equation:

$$\frac{d}{dx}\left(x^p\frac{d\theta}{dx}\right) \pm c^2\,x^s\,\theta = 0 \tag{2}$$

where $p = 1$, $c = m$, and $s = 1$. Referring to the flow chart presented in Figure 1-54, the value of $s - p + 2$ is equal to 2 and therefore the solution parameters n and a must be computed:

$$n = \frac{1-1}{1-1+2} = 0$$

$$a = \frac{2}{1-1+2} = 1$$

The last term in Eq. (1) is negative and therefore the solution to Eq. (2), as indicated by Figure 1-54, is given by:

$$\theta = C_1\,x^{n/a}\,\text{BesselI}\left(n, c\,a\,x^{1/a}\right) + C_2\,x^{n/a}\,\text{BesselK}\left(n, c\,a\,x^{1/a}\right)$$

where $x = r$ and $c = m$ for this problem. The homogeneous solution is:

$$T_h = C_1\,\text{BesselI}(0, mr) + C_2\,\text{BesselK}(0, mr)$$

The temperature distribution is the sum of the homogeneous and particular solutions:

$$T = C_1\,\text{BesselI}(0, mr) + C_2\,\text{BesselK}(0, mr) + T_\infty + \frac{\dot{q}_s''}{\overline{h}} \tag{3}$$

EXAMPLE 1.8-1: PIPE IN A ROOF

Maple can be used to obtain the same result:

```
> restart;
> ODE:=diff(r*diff(T(r),r),r)-m^2*r*T(r)=-m^2*r*T_infinity-qf_s*r/(k*th);
```

$$ODE := \left(\frac{d}{dr}T(r)\right) + r\left(\frac{d^2}{dr^2}T(r)\right) - m^2\, r\, T(r) = -m^2\, r\, T\_infinity - \frac{qf\_s\, r}{k\, th}$$

```
> Ts:=dsolve(ODE);
```

$$Ts := T(r) = \text{BesselI}(0, mr)\_C2 + \text{BesselK}(0, mr)\_C1 + \frac{m^2\, T\_infinity\, k\, th + qf\_s}{m^2\, k\, th}$$

Note that the constants C_1 and C_2 are interchanged in the Maple solution but it is otherwise the same as Eq. (3).

The boundary conditions must be used to obtain C_1 and C_2. As r approaches ∞, the effect of the pipe disappears. In this limit, the heat gain from the sun exactly balances convection, therefore:

$$\dot{q}_s'' = \bar{h}\,(T_{r\to\infty} - T_\infty) \tag{4}$$

Substituting Eq. (3) into Eq. (4) leads to:

$$\dot{q}_s'' = \bar{h}\left[C_1\,\text{BesselI}(0, \infty) + C_2\,\text{BesselK}(0, \infty) + T_\infty + \frac{\dot{q}_s''}{\bar{h}} - T_\infty\right]$$

or

$$C_1\,\text{BesselI}(0, \infty) + C_2\,\text{BesselK}(0, \infty) = 0$$

Figure 1-55 shows that the zeroth order modified Bessel function of the first kind (i.e., BesselI(0,x)) limits to ∞ as x approaches ∞ while the zeroth order modified Bessel function of the second kind (i.e., BesselK(0,x)) approaches 0 as x approaches ∞. This information can also be obtained using Maple and the limit command:

```
> limit(BesselI(0,x),x=infinity);
```

$$\infty$$

```
> limit(BesselK(0,x),x=infinity);
```

$$0$$

Therefore, C_1 must be zero while C_2 can be any finite value.

$$T = C_2\,\text{BesselK}(0, mr) + T_\infty + \frac{\dot{q}_s''}{\bar{h}} \tag{3}$$

The temperature where the roof meets the pipe is specified:

$$T_{r=r_p} = T_{hot}$$

or

$$C_2\,\text{BesselK}(0, mr_p) + T_\infty + \frac{\dot{q}_s''}{\bar{h}} = T_{hot}$$

EXAMPLE 1.8-1: PIPE IN A ROOF

The solution is programmed in EES:

```
m=sqrt(h_bar/(k*th))                                    "fin parameter"
C_2*BesselK(0,m*r_p)=T_hot-T_infinity-qf_s/h_bar        "boundary condition"
T=C_2*BesselK(0,m*r)+T_infinity+qf_s/h_bar              "solution"
T_C=converttemp(K,C,T)                                  "in C"
```

The temperature in the roof as a function of position is shown in Figure 3 for $\overline{h} = 50\,\text{W/m}^2\text{-K}$ (as specified in the problem statement) and also for $\overline{h} = 5\,\text{W/m}^2\text{-K}$.

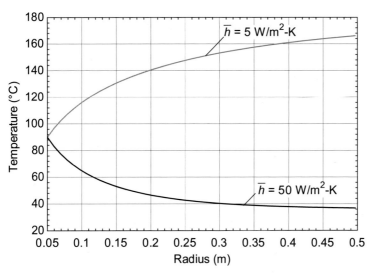

Figure 3: Temperature as a function of radius for $\overline{h} = 50\,\text{W/m}^2\text{-K}$ and $\overline{h} = 5\,\text{W/m}^2\text{-K}$ with $\dot{q}_s'' = 800\,\text{W/m}^2$.

The heat transfer between the pipe and the roof ($\dot{q}_p$) is evaluated using Fourier's law at $r = r_p$:

$$\dot{q}_p = -k\,th\,2\,\pi\,r_p\,\frac{dT}{dr}\bigg|_{r=r_p} \tag{4}$$

Substituting Eq. (3) into Eq. (4) leads to:

$$\dot{q}_p = -k\,th\,2\,\pi\,r_p\,C_2\,\frac{d}{dr}[\text{BesselK}\,(0,\,mr)]_{r=r_p}$$

which can be evaluated using the differentiation rule provided by Eq. (1-400):

$$\dot{q}_p = k\,th\,2\,\pi\,r_p\,C_2\,m\,\text{BesselK}(1,\,mr_p)$$

or using Maple:

```
> q_dot_p:=-k*th*2*pi*r_p*C_2*eval(diff(BesselK(0,m*r),r),r=r_p);
```

$$q\_dot\_p := 2\,k\,th\pi\,r\_p\,C\_2\,\text{BesselK}\,(1,\,mr\_p)\,m$$

EXAMPLE 1.8-1: PIPE IN A ROOF

The solution is programmed in EES:

```
q_dot_p=k*th*2*pi*r_p*C_2*m*BesselK(1,m*r_p)              "heat transfer into pipe"
```

Figure 4 illustrates the rate of heat transfer into the pipe as a function of the heat transfer coefficient and for various values of the solar flux.

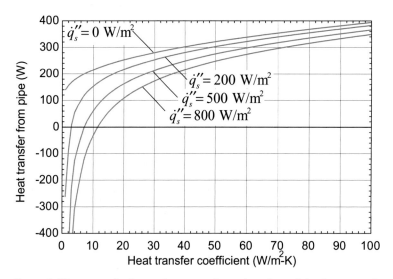

Figure 4: Heat transfer from pipe to roof as a function of the heat transfer coefficient for various values of the solar flux.

It is always important to understand your solution after it has been obtained. Notice in Figure 4 that the rate of heat transfer to the roof tends to increase with increasing heat transfer coefficient. This makes sense, as the temperature gradient at the interface between the roof and the pipe will increase as the heat transfer coefficient increases. However, when there is a non-zero solar flux, the heat transfer rate will change direction (i.e., become negative) at low values of the heat transfer coefficient indicating that the heat flow is into the pipe under these conditions. This effect occurs when the solar flux elevates the temperature of the roof to the point that it is above the hot gas temperature. Figure 4 shows that we can expect this behavior for $\bar{h} = 5$ W/m^2-K and $\dot{q}_s'' = 800$ W/m^2 and Figure 3 illustrates the temperature distribution under these conditions.

EXAMPLE 1.8-2: MAGNETIC ABLATION WITH BLOOD PERFUSION

EXAMPLE 1.8-2: MAGNETIC ABLATION WITH BLOOD PERFUSION

EXAMPLE 1.3-1 examined an ablative technique for locally heating cancerous tissue using small, conducting spheres (thermoseeds) that are embedded at precise locations and exposed to a magnetic field. Each thermoseed experiences a volumetric generation of thermal energy that causes its temperature and the temperature of the adjacent tissue to rise. In EXAMPLE 1.3-1, blood perfusion in the tissue was neglected; blood perfusion refers to the volumetric removal of energy in the tissue by the blood flowing in the microvascular structure.

The blood perfusion may be modeled as a volumetric heat sink that is proportional to the difference between the local temperature and the normal body temperature ($T_b = 37°C$); the constant of proportionality, β, is nominally 20,000 W/m^3-K. The thermoseed has a radius $r_{ts} = 1.0$ mm and it experiences a total rate of thermal energy generation of $\dot{g}_{ts} = 1.0$ W. The temperature far from the thermoseed is the body temperature, T_b. The tissue has thermal conductivity $k_t = 0.5$ W/m-K.

a) **Determine the steady-state temperature distribution in the tissue associated with a single sphere placed in an infinite medium of tissue considering blood perfusion.**

The input parameters are entered in EES:

```
"EXAMPLE 1.8-2: Magnetic Ablation with Blood Perfusion"

$UnitSystem SI MASS DEG PA C J
$Tabstops 0.2 0.4 0.6 0.8 3.5

"Inputs"
r_ts=1.0 [mm]*convert(mm,m)        "radius of thermoseed"
T_b=converttemp(C,K,37 [C])        "blood and body temperature"
g_dot_ts=1.0 [W]                   "generation in the thermoseed"
beta=20000 [W/m^3-K]               "blood perfusion constant"
k_t=0.5 [W/m-K]                    "tissue conductivity"
```

Figure 1 illustrates a differential control volume in the tissue that balances conduction with blood perfusion. The energy balance on the control volume is:

$$\dot{q}_r = \dot{q}_{r+dr} + \dot{g}$$

where $\dot{q}$ is conduction and $\dot{g}$ is the rate of energy *removed* by blood perfusion.

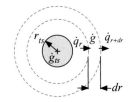

Figure 1: Differential control volume in the tissue.

The conduction through the tissue is given by:

$$\dot{q}_r = -k_t \, 4 \, \pi \, r^2 \, \frac{dT}{dr}$$

and the rate of energy removal by blood perfusion is:

$$\dot{g} = 4\,\pi\,r^2\,dr\,\beta\,(T - T_b)$$

Combining these equations leads to:

$$0 = \frac{d}{dr}\left[-k_t\,4\,\pi\,r^2\,\frac{dT}{dr}\right]dr + 4\,\pi\,r^2\,dr\,\beta\,(T - T_b)$$

which can be simplified:

$$\frac{d}{dr}\left[r^2\,\frac{dT}{dr}\right] - \frac{\beta}{k_t}r^2\,T = -\frac{\beta}{k_t}r^2\,T_b$$

The solution is divided into its homogeneous and particular components:

$$T = T_h + T_p$$

so that:

$$\underbrace{\frac{d}{dr}\left[r^2\,\frac{dT_h}{dr}\right] - \frac{\beta}{k_t}r^2\,T_h}_{=\,0\text{ for homogeneous differential equation}} + \underbrace{\frac{d}{dr}\left[r^2\,\frac{dT_p}{dr}\right] - \frac{\beta}{k_t}r^2\,T_p = -\frac{\beta}{k_t}r^2\,T_b}_{\text{whatever is left is the particular differential equation}}$$

The particular solution is:

$$T_p = T_b$$

The homogeneous differential equation is:

$$\frac{d}{dr}\left[r^2\,\frac{dT_h}{dr}\right] - m^2\,r^2\,T_h = 0 \qquad (1)$$

where

$$m = \sqrt{\frac{\beta}{k_t}}$$

Equation (1) is a form of Bessel's equation:

$$\frac{d}{dx}\left(x^p\frac{d\theta}{dx}\right) \pm c^2\,x^s\,\theta = 0$$

where $p = 2$, $c = m$, and $s = 2$. Referring to the flow chart presented in Figure 1-54, the value of $s - p + 2$ is equal to 2 and therefore the solution parameters n and a must be computed:

$$n = \frac{1 - 2}{2 - 2 + 2} = -\frac{1}{2}$$

$$a = \frac{2}{2 - 2 + 2} = 1$$

The last term in Eq. (1) is negative and therefore the solution is given by:

$$\theta = C_1\,x^{n/a}\,\text{BesselI}\left(n, c\,a\,x^{1/a}\right) + C_2\,x^{n/a}\,\text{BesselK}\left(n, c\,a\,x^{1/a}\right)$$

or

$$T_h = C_1\,r^{-1/2}\,\text{BesselI}\left(-\frac{1}{2}, mr\right) + C_2\,r^{-1/2}\,\text{BesselK}\left(-\frac{1}{2}, mr\right)$$

EXAMPLE 1.8-2: MAGNETIC ABLATION WITH BLOOD PERFUSION

The solution is the sum of the homogeneous and particular solutions:

$$T = C_1 \, r^{-1/2} \, \text{BesselI}\left(-\frac{1}{2}, mr\right) + C_2 \, r^{-1/2} \, \text{BesselK}\left(-\frac{1}{2}, mr\right) + T_b \qquad (2)$$

The constants are obtained by applying the boundary conditions. As r approaches ∞, the temperature must approach the body temperature:

$$T_{r\to\infty} = T_b \qquad (3)$$

Substituting Eq. (2) into Eq. (3) leads to:

$$C_1 \frac{\text{BesselI}\left(-\frac{1}{2}, \infty\right)}{\sqrt{\infty}} + C_2 \frac{\text{BesselK}\left(-\frac{1}{2}, \infty\right)}{\sqrt{\infty}} = 0 \qquad (4)$$

At first glance it is unclear how Eq. (4) helps to establish the constants; however, Maple can be used to show that C_1 must be zero because the first term limits to ∞ while the second term limits to 0:

```
> limit(BesselI(-1/2,r)/sqrt(r),r=infinity);
```
$$\infty$$

```
> limit(BesselK(-1/2,r)/sqrt(r),r=infinity);
```
$$0$$

The second boundary condition is obtained from an interface energy balance at $r = r_{ts}$; the rate of conduction heat transfer into the tissue must equal the rate of generation within the thermoseed:

$$-4\pi \, r_{ts}^2 \, k_t \left.\frac{dT}{dr}\right|_{r=r_{ts}} = \dot{g}_{ts} \qquad (5)$$

Substituting Eq. (2) with $C_1 = 0$ into Eq. (5) leads to:

$$-4\pi \, r_{ts}^2 \, k_t C_2 \frac{d}{dr}\left[r^{-1/2} \, \text{BesselK}\left(-\frac{1}{2}, mr\right)\right]_{r=r_{ts}} = \dot{g}_{ts}$$

Using Eq. (1-408) leads to:

$$-C_2 \, r_{ts}^{-1/2} \, m \, \text{BesselK}\left(-\frac{3}{2}, mr_{ts}\right) = -\frac{\dot{g}_{ts}}{4\pi \, r_{ts}^2 \, k_t}$$

The constant C_2 is evaluated in EES:

```
"Determine constant"
m=sqrt(beta/k_t)                                           "solution parameter"
-C_2*m*BesselK(-3/2,m*r_ts)/sqrt(r_ts)=-g_dot_ts/(4*pi*r_ts^2*k_t)    "determine constant"
```

EXAMPLE 1.8-2: MAGNETIC ABLATION WITH BLOOD PERFUSION

The solution is programmed in EES and converted to Celsius:

```
"Solution"
T=C_2*BesselK(-0.5,m*r)/sqrt(r)+T_b          "temperature"
T_C=converttemp(K,C,T)                       "in C"
r_mm=r*convert(m,mm)                         "radius in mm"
```

Figure 2 illustrates the temperature in the tissue as a function of radial position for various values of blood perfusion. Note that the temperature distribution as $\beta \to 0$ (i.e., in the absence of blood perfusion) agrees exactly with the solution for the tissue temperature obtained in EXAMPLE 1.3-1 (which is overlaid onto Figure 2) although the mathematical form of the solution looks very different. Figure 2 shows that the effect of blood perfusion is to reduce the extent of the elevated temperature region and therefore diminish the amount of tissue killed by the thermoseed.

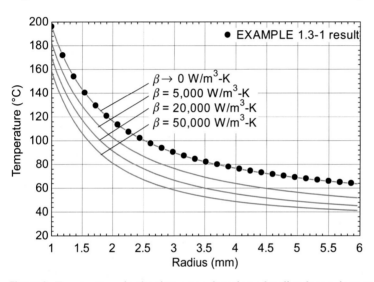

Figure 2: Temperature in the tissue as a function of radius for various values of blood perfusion; also shown is the result from EXAMPLE 1.3-1 which was derived for the same problem in the absence of blood perfusion ($\beta = 0$).

1.9 Numerical Solution to Extended Surface Problems

1.9.1 Introduction

Sections 1.6 through 1.8 present analytical solutions to extended surface problems. Only simple problems with constant properties can be considered analytically. There will be situations where these simplifications are not justified and it will be necessary to use a numerical model. Numerical modeling of extended surface problems is a straightforward extension of the numerical modeling techniques that are described in Sections 1.4 and 1.5.

If the extended surface approximation discussed in Section 1.6.2 is valid, then it is possible to obtain a numerical solution by dividing the computational domain into many small (but finite) one-dimensional control volumes. Energy balances are written for each control volume; the energy balances can include convective and/or radiative terms in addition to the conductive and generation terms that are considered in Sections 1.4 and

1.5. Each term in the energy balance is represented by a rate equation that reflects the governing heat transfer mechanism; the result is a system of algebraic equations that can be solved using EES or MATLAB. The solution should be checked for convergence, checked against your physical intuition, and compared with an analytical solution in the limit where one is valid.

EXAMPLE 1.9-1: TEMPERATURE SENSOR ERROR DUE TO MOUNTING & SELF HEATING

A resistance temperature detector (RTD) utilizes a material that has an electrical resistivity that is a strong function of temperature. The temperature of the RTD is inferred by measuring its electrical resistance. Figure 1 shows an RTD that is mounted at the end of a metal rod and inserted into a pipe in order to measure the temperature of a flowing liquid. The RTD is monitored by passing a known current through it and measuring the voltage drop across it. This process results in a constant amount of ohmic heating that will cause the RTD temperature to rise relative to the temperature of the surrounding liquid; this effect is referred to as a self-heating measurement error. Also, conduction from the wall of the pipe to the temperature sensor through the metal rod can result in a temperature difference between the RTD and the liquid; this effect is referred to as a mounting measurement error.

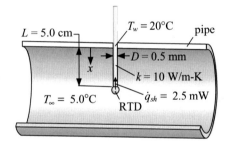

Figure 1: Temperature sensor mounted in a flowing liquid.

The thermal energy generation associated with ohmic heating is $\dot{q}_{sh} = 2.5$ mW. All of this ohmic heating is assumed to be transferred from the RTD into the end of the rod at $x = L$. The rod has a thermal conductivity $k = 10$ W/m-K, diameter $D = 0.5$ mm, and length $L = 5.0$ cm. The end of the rod that is connected to the pipe wall (at $x = 0$) is maintained at a temperature of $T_w = 20°C$.

The liquid is at a uniform temperature, $T_\infty = 5°C$. However, the local heat transfer coefficient between the liquid and the rod (h) varies with x due to the variation of the liquid velocity in the pipe. This problem resembles external flow over a cylinder, which will be discussed in Chapter 4; however, you may assume that the heat transfer coefficient between the rod surface and the fluid varies according to:

$$h = 2000 \left[\frac{W}{m^{2.8}\,K} \right] x^{0.8} \tag{1}$$

where h is the heat transfer coefficient in W/m²-K and x is position along the rod in m.

EXAMPLE 1.9-1: TEMPERATURE SENSOR ERROR DUE TO MOUNTING & SELF HEATING

EXAMPLE 1.9-1: TEMPERATURE SENSOR ERROR DUE TO MOUNTING & SELF HEATING

a) Can the rod be treated as an extended surface?

The input parameters are entered in EES; note that the heat transfer coefficient is computed using a function defined at the top of the EES code.

```
"EXAMPLE 1.9-1: Temperature Sensor Error due to Mounting and Self Heating"

$UnitSystem SI MASS DEG PA C J
$Tabstops 0.2 0.4 0.6 0.8 3.5


"Function for heat transfer coefficient"
function h(x)
   h=2000 [W/m^2.8-K]*x^0.8
end

"Inputs"
q_dot_sh=2.5 [milliW]*convert(milliW,W)          "self-heating power"
k=10 [W/m-K]                                      "conductivity of mounting rod"
D=0.5 [mm]*convert(mm,m)                           "diameter of mounting rod"
L=5.0 [cm]*convert(cm,m)                           "length of mounting rod"
T_w=converttemp(C,K,20 [C])                        "temperature of wall"
T_infinity=converttemp(C,K,5 [C])                  "temperature of liquid"
```

The appropriate Biot number for this case is:

$$Bi = \frac{hD}{2k}$$

The Biot number will be largest (and therefore the extended surface approximation least valid) when the heat transfer coefficient is largest. According to Eq. (1), the highest heat transfer coefficient occurs at the tip of the rod; therefore, the Biot number is calculated according to:

```
Bi=h(L)*D/(2*k)                                    "Biot number"
```

The Biot number calculated by EES is 0.0046, which is much less than 1.0 and therefore the extended surface approximation is justified.

b) Develop a numerical model of the rod that will predict the temperature distribution in the rod and therefore the error in the temperature measurement; this error is the difference between the temperature at the tip of the rod (i..e, the temperature of the RTD) and the liquid.

The development of the numerical model follows the same steps that are discussed in Section 1.4. Nodes (i.e., locations where the temperature will be determined) are positioned uniformly along the length of the rod, as shown in Figure 2. The location of each node (x_i) is:

$$x_i = \frac{(i-1)}{(N-1)}L \quad i = 1..N$$

EXAMPLE 1.9-1: TEMPERATURE SENSOR ERROR DUE TO MOUNTING & SELF HEATING

where N is the number of nodes used for the simulation. The distance between adjacent nodes (Δx) is:

$$\Delta x = \frac{L}{(N-1)}$$

This distribution is entered in EES:

```
N=100                              "number of nodes"
duplicate i=1,N
   x[i]=(i-1)*L/(N-1)              "position of each node"
end
DELTAx=L/(N-1)                     "distance between adjacent nodes"
```

A control volume is defined around each node; the control surface bisects the distance between the nodes, as shown in Figure 2.

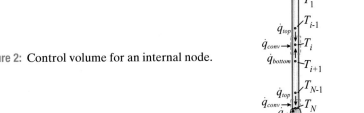

Figure 2: Control volume for an internal node.

The control volume for internal node i shown in Figure 2 is subject to conduction heat transfer at each edge ($\dot{q}_{top}$ and $\dot{q}_{bottom}$) and convection ($\dot{q}_{conv}$). The energy balance is:

$$\dot{q}_{top} + \dot{q}_{bottom} + \dot{q}_{conv} = 0$$

The conduction terms are approximated according to:

$$\dot{q}_{top} = \frac{k\,\pi\,D^2}{4\,\Delta x}\,(T_{i-1} - T_i)$$

$$\dot{q}_{bottom} = \frac{k\,\pi\,D^2}{4\,\Delta x}\,(T_{i+1} - T_i)$$

The convection term is modeled using the convection coefficient evaluated at the position of the node:

$$\dot{q}_{conv} = h_{x_i}\,\pi\,D\,\Delta x\,(T_\infty - T_i)$$

EXAMPLE 1.9-1: TEMPERATURE SENSOR ERROR DUE TO MOUNTING & SELF HEATING

Combining these equations leads to:

$$\frac{k \pi D^2}{4 \Delta x}(T_{i-1} - T_i) + \frac{k \pi D^2}{4 \Delta x}(T_{i+1} - T_i) + h_{x_i} \pi D \Delta x (T_\infty - T_i) = 0 \quad \text{for } i = 2..(N-1)$$
(2)

```
"internal control volume energy balances"
duplicate i=2,(N-1)
    k*pi*D^2*(T[i-1]-T[i])/(4*DELTAx)+k*pi*D^2*(T[i+1]-T[i])/(4*DELTAx)+ &
        pi*D*DELTAx*h(x[i])*(T_infinity-T[i])=0
end
```

The nodes at the edges of the domain must be treated separately. At the pipe wall, the temperature is specified:

$$T_1 = T_w$$
(3)

```
T[1]=T_w                                    "boundary condition at wall"
```

The ohmic dissipation, $\dot{q}_{sh}$ is assumed to enter the half-node at the tip (i.e., node N) and therefore is included in the energy balance for this node (see Figure 2):

$$\frac{k \pi D^2}{4 \Delta x}(T_{N-1} - T_N) + \frac{h_{x_N} \pi D \Delta x}{2}(T_\infty - T_N) + \dot{q}_{sh} = 0$$
(4)

Note the factor of 2 in the denominator of the convection term that arises because the half-node has half the surface area of the internal nodes.

```
k*pi*D^2*(T[N-1]-T[N])/(4*DELTAx)+pi*D*DELTAx*h(x[N])*(T_infinity-T[N])/2+q_dot_sh=0
    "boundary condition at tip"
```

Equations (2) through (4) are a system of N equations in an equal number of unknown temperatures that are entered in EES. The solution is converted to Celsius:

```
duplicate i=1,N
    T_C[i]=converttemp(K,C,T[i])            "solution in Celsius"
end
```

Figure 3 illustrates the temperature distribution in the rod for $N = 100$ nodes. The temperature elevation of the tip relative to the fluid is about 3.4 K and represents the measurement error. For the conditions in the problem statement, it is clear that the measurement error is primarily due to self-heating because the effect of the wall (the temperature elevation at the base) has died off after about 2.0 cm.

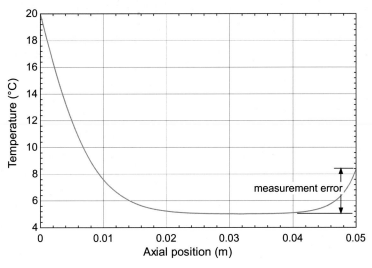

Figure 3: Temperature distribution in the mounting rod.

As with any numerical solution, it is important to verify that a sufficient number of nodes have been used so that the numerical solution has converged. The key result of the solution is the tip-to-fluid temperature difference, which is the measurement error for the sensor (δT):

$$\delta T = T_N - T_\infty$$

deltaT=T[N]-T_infinity   "measurement error"

Figure 4 illustrates the tip-to-fluid temperature difference as a function of the number of nodes and shows that the solution has converged for N greater than 100 nodes.

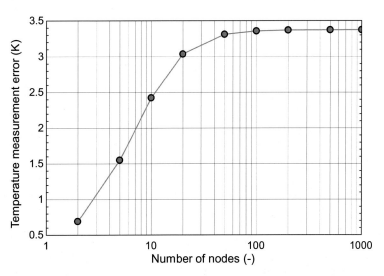

Figure 4: Tip-to-fluid temperature difference as a function of the number of nodes.

EXAMPLE 1.9-1: TEMPERATURE SENSOR ERROR DUE TO MOUNTING & SELF HEATING

EXAMPLE 1.9-1: TEMPERATURE SENSOR ERROR DUE TO MOUNTING & SELF HEATING

The analytical solution for this problem in the limit of a constant heat transfer coefficient and an adiabatic tip was derived in Section 1.6.3 and is included in Table 1-4:

$$\frac{T - T_\infty}{T_w - T_\infty} = \frac{\cosh\left(m\left(L - x\right)\right)}{\cosh\left(m\,L\right)}$$

where

$$m = \sqrt{\frac{4\,\overline{h}}{k\,D}}$$

The analytical solution is programmed in EES:

```
"Analytical solution for verification in the limit q_dot_sh=0 and h=constant"
m=sqrt(4*h(L)/(k*D))                                              "fin parameter"
duplicate i=1,N
   T_an[i]=T_infinity+(T_w-T_infinity)*cosh(m*(L-x[i]))/cosh(m*L)   "analytical solution"
   T_an_C[i]=converttemp(K,C,T_an[i])                              "in C"
end
```

The numerical solution is obtained in this limit by setting the variable q_dot_sh equal to zero and modifying the function h so that it returns 100 W/m²-K regardless of position.

```
"Function for heat transfer coefficient"
function h(x)
   {h=2000 [W/m^2.8-K]*x^0.8}
   h=100 [W/m^2-K]
end

"Inputs"
q_dot_sh=0 [W] {2.5 [milliW]*convert(milliW,W)}              "self-heating power"
```

The temperature distribution predicted by the numerical model is compared with the analytical solution in Figure 5.

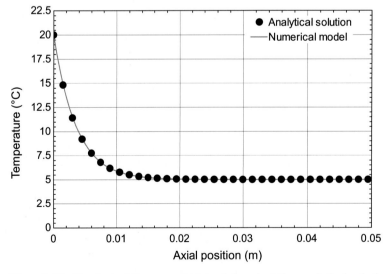

Figure 5: Verification of the numerical model against the analytical solution in the limit that the heat transfer coefficient is constant at $h = 100$ W/m²-K and there is no self-heating, $\dot{q}_{sh} = 0$ W.

EXAMPLE 1.9-1: TEMPERATURE SENSOR ERROR DUE TO MOUNTING & SELF HEATING

c) Investigate the effect of thermal conductivity on the temperature measurement error. Identify the optimal thermal conductivity and explain why an optimal thermal conductivity exists.

Figure 6 illustrates the temperature measurement error as a function of the thermal conductivity of the rod material; note that the function h has been set back to its original form and the variable q_dot_sh restored to 2.5 mW. Figure 6 shows that the optimal thermal conductivity, corresponding to the minimum measurement error, is around 100 W/m-K. Below the optimal value, the self-heating error dominates as the local temperature rise at the tip of the rod is large. Above the optimal value, the conduction from the wall dominates.

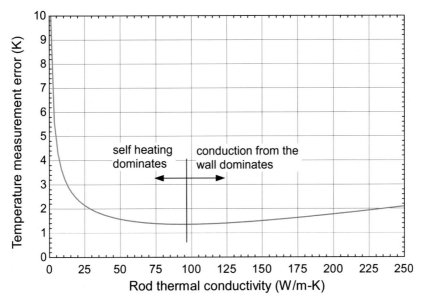

Figure 6: Temperature measurement error as a function of rod thermal conductivity.

EXAMPLE 1.9-2: CRYOGENIC CURRENT LEADS

It is often necessary to supply a cryogenic experiment or apparatus with electrical current. Some examples include current for superconducting electronics and magnets, to energize a resistance-based temperature sensor, and to energize a heater used for temperature control. In any of these cases, careful design of the wires that are used to supply and return the current to the facility is important. The heat transfer to the cryogenic device from these wires should be minimized as this energy must be removed either by a refrigeration system (i.e., a cryocooler) or by consumption of a relatively expensive cryogen (e.g., by the boil off of liquid helium or liquid nitrogen). There is an optimal wire diameter for any given application that minimizes this parasitic heat transfer to the device.

Figure 1 illustrates two current leads, each carrying $I = 100$ ampere (one supply and the other return). These current leads extend from the room temperature wall of the vacuum vessel, where the wire material is at $T_H = 20°C$, to the experiment, where the wire material is at $T_C = 50\,K$. The length of both current leads is $L = 1.0\,m$ and their diameter, D, should be optimized. The vacuum in the vessel prevents any convection heat transfer from the surface of the wires. However, the surface of the

EXAMPLE 1.9-2: CRYOGENIC CURRENT LEADS

EXAMPLE 1.9-2: CRYOGENIC CURRENT LEADS

wires radiate to their surroundings, which may be assumed to be at $T_H = 20°C$. The external surface of the wires has emissivity, $\varepsilon = 0.5$.

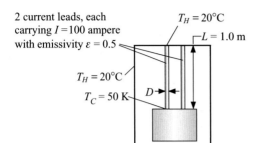

2 current leads, each carrying $I = 100$ ampere with emissivity $\varepsilon = 0.5$

$T_H = 20°C$

$T_C = 50$ K

$T_H = 20°C$

$L = 1.0$ m

D

Figure 1: Cryogenic current leads.

The leads are made of oxygen free, high-conductivity copper; the thermal conductivity and resistivity of copper can vary substantially at cryogenic temperatures depending on the purity and history of the material (e.g., whether it has been annealed or not). The purity of the metal is often expressed as the Residual Resistivity Ratio (RRR), which is defined as the ratio of the metal's electrical resistivity at 273 K to that at 4.2 K. Oxygen free, high conductivity copper (OFHC) has an RRR of approximately 200. The thermal conductivity and electrical resistivity of RRR 200 copper as a function of temperature is provided in Table 1 (Iwasa (1994)).

Table 1: Thermal conductivity and electrical resistivity of OFHC copper.

Temperature	Thermal conductivity	Electrical resistivity
500 K	4.31 W/cm-K	3.19 μohm-cm
400 K	4.15 W/cm-K	2.49 μohm-cm
300 K	3.99 W/cm-K	1.73 μohm-cm
250 K	4.04 W/cm-K	1.39 μohm-cm
200 K	4.11 W/cm-K	1.06 μohm-cm
150 K	4.24 W/cm-K	0.72 μohm-cm
125 K	4.34 W/cm-K	0.54 μohm-cm
100 K	4.71 W/cm-K	0.36 μohm-cm
90 K	4.98 W/cm-K	0.29 μohm-cm
80 K	5.43 W/cm-K	0.22 μohm-cm
70 K	6.25 W/cm-K	0.15 μohm-cm
60 K	7.83 W/cm-K	0.098 μohm-cm
55 K	9.11 W/cm-K	0.076 μohm-cm
50 K	11.0 W/cm-K	0.057 μohm-cm

a) **Develop a numerical model in MATLAB that can predict the rate of heat transfer to the cryogenic experiment from the pair of current leads.**

The input conditions are entered in a MATLAB function **EXAMPLE1p9_2.m**; the two arguments to the function are diameter (the variable **D**) and number of nodes (the variable **N**) as we know that these parameters will be varied during the verification

EXAMPLE 1.9-2: CRYOGENIC CURRENT LEADS

and optimization process. Any of the other parameters can be added in order to facilitate additional parametric studies or optimization.

```
function [ ]=EXAMPLE1p9_2(D,N)

    I=100;                        %current (amp)
    T_H=20+273.2;                 %hot temperature (K)
    T_C=50;                       %cold temperature (K)
    L=1;                          %length of lead (m)
    eps=0.5;                      %emissivity of lead surface (-)
    sigma=5.67e-8;                %Stefan-Boltzmann constant (W/m^2-K^4)
```

Notice that we have not, to this point, specified what parameters are returned when the function executes (i.e., there are no variables listed between the square brackets in the function header).

Functions are defined (at the bottom of the M-file) that return the conductivity and electrical resistivity of the OFHC copper; the interp1 function is used to carry out interpolation on the data provided in Table 1 using a cubic spline technique.

```
%———Property functions———
function[k]=k_cu(T)
    %returns the thermal conductivity (W/m-K) given temperature (K)
    Td=[500,400,300,250,200,150,125,100,90,80,70,60,55,50];    %temperature data (K)
    kd=[4.31,4.15,3.99,4.04,4.11,4.24,4.34,4.71,4.98,5.43,6.25,7.83,9.11,11.0]*100;
                                  %conductivity data (W/m-K)
    k=interp1(Td,kd,T);
end

function[rho_e]=rho_e_cu(T)
    %returns the electrical resistivity (ohm-m) given temperature (K)
    Td=[500,400,300,250,200,150,125,100,90,80,70,60,55,50];       %temperature data (K)
    rho_ed=[3.19,2.49,1.73,1.39,1.06,0.72,0.54,0.36,0.29,0.22,0.15,0.098,0.076,0.057]/(1e6*100);
        %electrical resistivity data (ohm-m)
    rho_e=interp1(Td,rho_ed,T);
end
```

The first step is to position the nodes throughout the computational domain. For this problem, the nodes will be distributed uniformly, as shown in Figure 2:

$$x_i = \frac{(i-1)}{(N-1)}L \text{ for } i = 1..N$$

The distance between adjacent nodes (Δx) is:

$$\Delta x = \frac{L}{(N-1)}$$

EXAMPLE 1.9-2: CRYOGENIC CURRENT LEADS

The MATLAB code that accomplishes these assignments is:

```
%Position nodes
for i=1:N
    x(i,1)=(i-1)*L/(N-1);                    %position of each node (m)
end
DELTAx=L/(N-1);                             %distance between adjacent nodes (m)
```

A control volume for an internal node is shown in Figure 2; the control volume experiences conduction heat transfer from the adjacent nodes above and below ($\dot{q}_{top}$ and $\dot{q}_{bottom}$, respectively), as well as radiation ($\dot{q}_{rad}$) and generation due to the ohmic dissipation associated with the current ($\dot{g}$). An energy balance for the control volume is:

$$\dot{q}_{top} + \dot{q}_{bottom} + \dot{g} = \dot{q}_{rad} \tag{1}$$

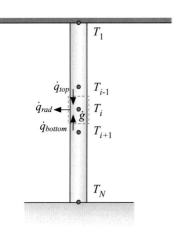

Figure 2: Control volume for an internal node and associated energy terms.

The conductivity used to approximate the conduction heat transfer rates must be evaluated at the temperature of the boundaries in order to avoid energy balance violations, as discussed in Section 1.4.3. With this understanding, these rate equations become:

$$\dot{q}_{top} = k_{T=(T_i+T_{i-1})/2} \frac{\pi D^2}{4 \Delta x} (T_{i-1} - T_i) \tag{2}$$

$$\dot{q}_{bottom} = k_{T=(T_i+T_{i+1})/2} \frac{\pi D^2}{4 \Delta x} (T_{i+1} - T_i) \tag{3}$$

The rate of thermal energy generation is calculated using the resistivity evaluated at the temperature of each node:

$$\dot{g} = \rho_{e,T=T_i} \frac{4 \Delta x}{\pi D^2} I^2 \tag{4}$$

The rate of radiation heat transfer is approximately given by:

$$\dot{q}_{rad} = \varepsilon \sigma \pi D \Delta x \left(T_i^4 - T_H^4\right) \tag{5}$$

EXAMPLE 1.9-2: CRYOGENIC CURRENT LEADS

where ε is the emissivity of the surface of the leads and σ is the Stefan-Boltzmann constant. Substituting Eqs. (2) through (5) into Eq. (1) leads to:

$$k_{T=(T_i+T_{i-1})/2}\frac{\pi\,D^2}{4\,\Delta x}\,(T_{i-1}-T_i)+k_{T=(T_i+T_{i+1})/2}\frac{\pi\,D^2}{4\,\Delta x}\,(T_{i+1}-T_i)+\rho_{e,T=T_i}\frac{4\,\Delta x}{\pi\,D^2}\,I^2$$

$$=\varepsilon\,\sigma\,\pi\,D\,\Delta x\,\left(T_i^4-T_H^4\right)\quad\text{for }i=2..\,(N-1)\tag{6}$$

The remaining equations specify the boundary temperatures:

$$T_1=T_H\tag{7}$$

$$T_N=T_C\tag{8}$$

Equations (6) through (8) are a set of N equations in the N unknown temperatures; however, the temperature dependence of the material properties (k and ρ_e) as well as the non-linear rate equation associated with radiation heat transfer cause the system of equations to be non-linear. Therefore, a relaxation technique will be employed in order to obtain the solution; the relaxation process is discussed in Section 1.5.6. The assumed solution ($\hat{T}$) will be successively substituted with the predicted solution. This process will continue until the assumed and predicted solutions agree to within an acceptable tolerance. The solution proceeds by assuming a temperature distribution that can be used to evaluate the coefficients in the linearized equations. A linear temperature distribution provides a reasonable start for the iteration:

$$\hat{T}_i=T_H+\frac{(i-1)}{(N-1)}\,(T_H-T_C)\quad\text{for }i=1..N$$

```
%Start relaxation with a linear temperature distribution
for i=1:N
  Tg(i,1)=T_H-(T_H-T_C)*(i-1)/(N-1);
end
```

In order to solve this problem using MATLAB, we will need a set of linear equations; linear equations cannot contain products of the unknown temperatures with other unknown temperatures or functions of the unknown temperatures. Therefore, the temperature-dependent material properties must be evaluated at the assumed temperatures ($\hat{T}$):

$$\dot{q}_{top}=k_{T=(\hat{T}_i+\hat{T}_{i-1})/2}\frac{\pi\,D^2}{4\,\Delta x}\,(T_{i-1}-T_i)\tag{9}$$

$$\dot{q}_{bottom}=k_{T=(\hat{T}_i+\hat{T}_{i+1})/2}\frac{\pi\,D^2}{4\,\Delta x}\,(T_{i+1}-T_i)\tag{10}$$

$$\dot{g}=\rho_{e,T=\hat{T}_i}\frac{4\,\Delta x}{\pi\,D^2}\,I^2\tag{11}$$

The fourth power temperature terms cause the radiation equation, Eq. (5), to be non-linear. Therefore, it is necessary to linearize the radiation equation so that it can be placed in matrix format, as was done for the material properties. The radiation

EXAMPLE 1.9-2: CRYOGENIC CURRENT LEADS

terms can be linearized most conveniently using the same factorization that was previously used to define a radiation resistance in Section 1.2.6:

$$\dot{q}_{rad} = \sigma \, \varepsilon \, \pi \, D \, \Delta x \big(\hat{T}_i^2 + T_H^2\big)\big(\hat{T}_i + T_H\big)\big(T_i - T_H\big) \tag{12}$$

Substituting the linearized rate equations, Eqs. (9) through (12), into the energy balance for an internal node, Eq. (1), leads to:

$$k_{T=(\hat{T}_i+\hat{T}_{i-1})/2}\frac{\pi \, D^2}{4\,\Delta x}(T_{i-1}-T_i) + k_{T=(\hat{T}_i+\hat{T}_{i+1})/2}\frac{\pi \, D^2}{4\,\Delta x}(T_{i+1}-T_i) + \rho_{e,T=\hat{T}_i}\frac{4\,\Delta x}{\pi \, D^2}I^2$$
$$= \sigma \, \varepsilon \, \pi \, D \, \Delta x \, \big(\hat{T}_i^2 + T_H^2\big)\big(\hat{T}_i + T_H\big)\big(T_i - T_H\big) \quad \text{for } i = 2..(N-1) \tag{13}$$

Equations (7), (8), and (13) must be placed in matrix format:

$$\underline{\underline{A}}\,\underline{X} = \underline{b}$$

where $\underline{X}$ is a vector of unknown temperatures, $\underline{\underline{A}}$ is a matrix containing the coefficients of each equation, and $\underline{b}$ is a vector containing the constant terms for each equation. The matrix $\underline{\underline{A}}$ is declared as sparse in MATLAB; note that Eq. (13) indicates that there are at most three nonzero entries in each row of $\underline{\underline{A}}$:

```
%Setup A and b
A=spalloc(N,N,3*N);
b=zeros(N,1);
```

Equation (7) can be placed in row 1 of the matrix equation:

$$T_1 \underbrace{[1]}_{A_{1,1}} = \underbrace{T_H}_{b_1} \tag{14}$$

and Eq. (8) can be placed in row N of the matrix equation:

$$T_N \underbrace{[1]}_{A_{N,N}} = \underbrace{T_C}_{b_N} \tag{15}$$

Equation (13) must be rearranged to make it clear which row and column each coefficient should be entered into the matrix:

$$T_i \underbrace{\left[-k_{T=(\hat{T}_i+\hat{T}_{i-1})/2}\frac{\pi \, D^2}{4\,\Delta x} - k_{T=(\hat{T}_i+\hat{T}_{i+1})/2}\frac{\pi \, D^2}{4\,\Delta x} - \sigma \, \varepsilon \, \pi \, D \, \Delta x \, \big(\hat{T}_i^2 + T_H^2\big)\big(\hat{T}_i + T_H\big)\right]}_{A_{i,i}}$$
$$+ T_{i-1}\underbrace{\left(k_{T=(\hat{T}_i+\hat{T}_{i-1})/2}\frac{\pi \, D^2}{4\,\Delta x}\right)}_{A_{i,i-1}} + T_{i+1}\underbrace{\left(k_{T=(\hat{T}_i+\hat{T}_{i+1})/2}\frac{\pi \, D^2}{4\,\Delta x}\right)}_{A_{i,i+1}} \tag{16}$$
$$= \underbrace{-\sigma \, \varepsilon \, \pi \, D \, \Delta x \, \big(\hat{T}_i^2 + T_H^2\big)\big(\hat{T}_i + T_H\big) T_H - \rho_{e,T=\hat{T}_i}\frac{4\,\Delta x}{\pi \, D^2}I^2}_{b_i} \quad \text{for } i = 2..(N-1)$$

The numerical solution is placed within a while loop that checks for convergence of the relaxation scheme. The variable err is used to terminate the while loop and represents the average, absolute error between the assumed and predicted temperature distribution. (There are other criteria that could be used, but this is sufficient for most problems.) The while loop is terminated when the variable err decreases to less than the input parameter tol, which represents the convergence

EXAMPLE 1.9-2: CRYOGENIC CURRENT LEADS

tolerance for the relaxation process. Initially, the value of **err** is set to a value greater than tol to ensure that the while loop executes at least one time.

```
err=999;                        %error that terminates the while loop (K)
tol=0.1;                        %criteria for terminating the while loop (K)
while(err>tol)

    end
end
```

Within the while loop, the matrix is filled in using the coefficients suggested by Eq. (14),

```
%specify the hot end temperature
A(1,1)=1;
b(1,1)=T_h;
```

Eq. (15),

```
%specify the cold end temperature
A(N,N)=1;
b(N,1)=T_C;
```

and Eq. (16).

```
%internal nodes
for i=2:(N-1)
    A(i,i)=-k_cu((Tg(i+1,1)+Tg(i,1))/2)*pi*D^2/(4*DELTAx)- ...
        k_cu((Tg(i-1,1)+Tg(i,1))/2)*pi*D^2/(4*DELTAx)-...
        sigma*eps*pi*D*DELTAx*(Tg(i,1)^2+T_H^2)*(Tg(i,1)+T_H);
    A(i,i-1)=k_cu((Tg(i-1,1)+Tg(i,1))/2)*pi*D^2/(4*DELTAx);
    A(i,i+1)=k_cu((Tg(i+1,1)+Tg(i,1))/2)*pi*D^2/(4*DELTAx);
    b(i,1)=-rho_e_cu(Tg(i,1))*4*DELTAx*I^2/(pi*D^2)- ...
        sigma*eps*pi*D*DELTAx*(Tg(i,1)^2+T_H^2)*(Tg(i,1)+T_H)*T_H;
end
```

Note that the three periods in the above code is a line break; it indicates that the code is continued on the subsequent line. The matrix equation is solved and the error between the assumed and predicted temperature is computed.

$$err = \frac{1}{N} \sum_{i=1}^{N} |T_i - \hat{T}_i|$$

The final step in the while loop is to update the assumed temperature distribution with the predicted temperature distribution.

```
T=full(A/b);
err=sum(abs(T-Tg))/N                %compute the error
Tg=T;                               %update the guess temperature array
```

EXAMPLE 1.9-2: CRYOGENIC CURRENT LEADS

Note that the full command in the above code converts the sparse matrix, T, that results from the operation on the sparse matrix A into a full matrix.

The header of the function is modified to specify the output arguments, x and T:

```
function[x,T]=EXAMPLE1p9_2(D,N)
```

Because the statement that calculates the variable err is not terminated with a semicolon, the result of the calculation will be echoed in the workspace allowing you to keep track of the progress. If you call this program with a diameter of 5.0 mm you should see:

```
>> [x,T]=EXAMPLE1p9_2(0.005,100);
err =
   41.2789
err =
   16.1165
err =
   6.2792
err =
   2.4864
err =
   1.0023
err =
   0.4110
err =
   0.1685
err =
   0.0693
```

The relaxation process had to iterate several times in order to converge due to the nonlinearity of the problem. Figure 3 illustrates the temperature distribution in the current lead for several different values of the diameter.

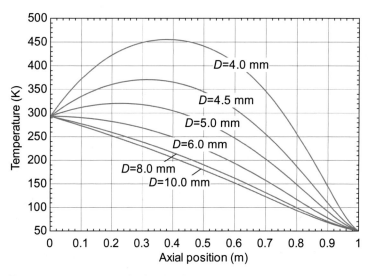

Figure 3: Temperature distribution in the current lead for various values of the diameter.

Notice that the smaller diameter leads result in large amounts of ohmic dissipation and therefore the wire tends to become hotter and the temperature gradient at the cold end increases. For a given temperature gradient, larger diameter leads will result in a higher rate of heat transfer to the cold end due to the larger area for conduction. There is a balance between these effects that results in an optimal diameter.

The heat transferred to the cold end ($\dot{q}_c$) is calculated using an energy balance on node N:

$$\dot{q}_c = k_{T=(T_N+T_{N-1})/2}\frac{\pi D^2}{4\,\Delta x}(T_{N-1}-T_N) + \rho_{e,T=T_N}\frac{2\,\Delta x}{\pi D^2}I^2 - \sigma\,\varepsilon\,\pi\,D\,\frac{\Delta x}{2}\left(T_H^4 - T_N^4\right)$$

or, in MATLAB:

```
q_dot_c=k_cu((T(N)+T(N-1))/2)*pi*D^2*(T(N-1)-T(N))/(4*DELTAx)+...
    rho_e_cu(T(N))*2*DELTAx*I^2/(pi*D^2)-sigma*eps*pi*D*DELTAx*(T_H^4-T(N)^4)/2;
%heat transfer to cold end
```

The function header is modified so that $\dot{q}_c$ is also returned:

```
function[q_dot_c,x,T]=EXAMPLE1p9_2(D,N)
```

It is necessary to verify that the solution has a sufficient number of nodes and, if possible, verify the result against an analytical solution. The critical parameter for the solution is the rate of heat transfer to the experiment per current lead; therefore Figure 4 illustrates $\dot{q}_c$ as a function of the number of nodes in the solution, N. The information shown in Figure 4 was generated quickly using the script varyN (below), which calls the function EXAMPLE1p9_2 multiple times with varying values of N:

```
%Script varyN.m
clear all;
D=0.005;
N=[2,5,10,20,50,100,200,500,1000,2000]';
for i=1:10
    i
    [q_dot_c(i,1),x,T]=EXAMPLE1p9_2(D,N(i));
end
```

The clear all statement at the beginning of the script clears all variables from memory. Using the clear all statement is often a good idea as it prevents previous elements (e.g., from previous runs) of the variables q_dot_c or N from being retained. If you had previously run the script varyN with more than 10 runs, then N and q_dot_c would exist in memory with more than 10 elements. Running the script varyN as shown above would overwrite the first 10 elements of these variables, but leave all subsequent elements which could lead to confusion.

EXAMPLE 1.9-2: CRYOGENIC CURRENT LEADS

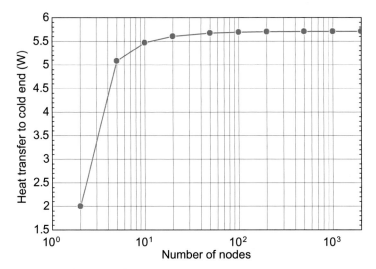

Figure 4: Heat transferred to the cold end per lead as a function of the number of nodes for a 5.0 mm lead.

Figure 4 suggests that at least 100 nodes should be used for sufficient accuracy. In the absence of any radiation heat transfer ($\varepsilon = 0$) and with constant resistivity and conductivity, it is possible to compare the numerical solution to the analytical solution for the temperature in a generating wall with fixed end conditions. This result was derived in Section 1.3.2 and is repeated below:

$$T = \frac{\dot{g}''' L^2}{2\,k}\left[\frac{x}{L} - \left(\frac{x}{L}\right)^2\right] - \frac{(T_H - T_C)}{L}\,x + T_H$$

where the volumetric generation is given by:

$$\dot{g}''' = \frac{16\,I^2\,\rho_e}{\pi^2\,D^4}$$

```
%constant property analytical solution
g_dot_vol=16*I^2*rho_e_cu(T_H)/(pi^2*D^4);
for i=1:N
    T_an(i,1)=g_dot_vol*L^2*((x(i)/L)-(x(i)/L)^2)/(2*k_cu(T_H))- ...
        (T_H-T_C)*x(i)/L+T_H;
end
```

The property functions in MATLAB are modified to return, temporarily, constant values of $k = 200$ W/m-K and $\rho_e = 1 \times 10^{-8}$ ohm-m.

```
%----Property functions---------
function[k]=k_cu(T)
    %returns the thermal conductivity (W/m-K) given temperature (K)
    Td=[500,400,300,250,200,150,125,100,90,80,70,60,55,50];      %temperature data (K)
    kd=[4.31,4.15,3.99,4.04,4.11,4.24,4.34,4.71,4.98,5.43,6.25,7.83,9.11,11.0]*100;
        %conductivity data (W/m-K)
    %k=interp1(Td,kd,T);
    k=200;
end
```

EXAMPLE 1.9-2: CRYOGENIC CURRENT LEADS

```
function[rho_e]=rho_e_cu(T)
  %returns the electrical resistivity (ohm-m) given temperature (K)
  Td=[500,400,300,250,200,150,125,100,90,80,70,60,55,50];    %temperature data (K)
  rho_ed=[3.19,2.49,1.73,1.39,1.06,0.72,0.54,0.36,0.29,0.22,0.15,0.098,0.076,0.057]/(1e6*100);
    %electrical resistivity data (ohm-m)
  %rho_e=interp1(Td,rho_ed,T);
  rho_e=1e-8;
end
```

The emissivity is set to 0 and the MATLAB code is run for a 5.0 mm diameter lead.
The temperature distribution predicted by the MATLAB code is compared with the
analytical solution in Figure 5. Note that with these modifications (i.e., constant k
and ρ_e and $\varepsilon = 0$), the problem becomes linear and therefore a single iteration is
required in order to reduce the relaxation error to 0.

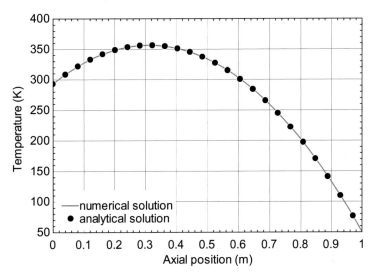

Figure 5: Comparison of the analytical and numerical solutions in the limit that $k = 200$ W/m-K
(constant), $\rho_e = $ 1e-8 ohm-m (constant) and $\varepsilon = 0$ for a 5.0 mm diameter wire.

Finally, it is possible to parametrically vary the wire diameter, D, in order to minimize the heat flow to the cold end of the current lead. The code is returned to its
original, non-linear form. A script (varyd) is used to call the function multiple times
with various diameters in order to carry out a parametric study of this parameter:

```
%Script varyd.m
clear all;
N=100;
D=linspace(0.0038,0.01,100)';    %generate 100 values of D between 0.0038 and 0.01 [m]
for i=1:100
    [q_dot_c(i,1),x,T]=EXAMPLE1p9_2(D(i),N);
end
```

EXAMPLE 1.9–2: CRYOGENIC CURRENT LEADS

Figure 6 illustrates the heat leak to the cold end as a function of wire diameter (note that the functions for k and ρ_e were reset and the value of ε was reset to 0.50) and shows that there is a clear optimal diameter around 5.1 mm for this application. Smaller values of D lead to excessive self-heating whereas larger values provide a large path for conduction heat transfer.

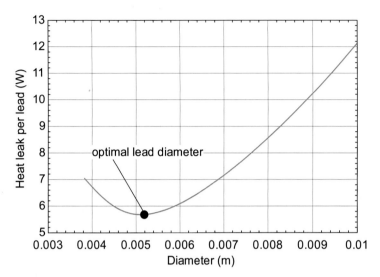

Figure 6: Heat leak to the cold end of each current lead as a function of diameter.

MATLAB has powerful, built-in optimization algorithms that allow you to automate the process of determining the optimal diameter. The MATLAB function fminbnd is the simplest available and carries out a bounded, 1-D minimization. The function fminbnd is called according to:

```
x_opt=fminbnd(function,x1,x2)
```

where function is the name of a function that requires a single argument and provides a single output (that should be minimized) and x1 and x2 are the lower and upper bounds of the argument to use for the minimization. First, it is necessary to modify the function EXAMPLE1p9_2 so that it takes a single argument (D) and returns a single output (q_dot_c):

```
function[q_dot_c]=EXAMPLE1p9_2(D)

N=100;                              %number of nodes (-)
```

Then the fminbnd function can be called directly from the workspace:

```
>> D_opt=fminbnd('EXAMPLE1p9_2',0.004,0.01);
```

in order to identify the optimal diameter.

EXAMPLE 1.9-2: CRYOGENIC CURRENT LEADS

```
>> D_opt
D_opt =
   0.0051
```

Note that the calculation of the variable err in the function EXAMPLE1p9_2 is terminated with a semicolon so that the error is not echoed to the workspace during each iteration. The function fminbnd will return the optimized value of the heat leak as well by adding an additional output argument to the fminbnd call:

```
>>[D_opt,q_dot_c_min]=fminbnd('EXAMPLE1p9_2',0.004,0.01);
```

which indicates that the optimal value of the heat leak is 5.57 W.

```
>> q_dot_c_min
q_dot_c_min =
   5.6815
```

It is possible to control the details of the optimization using an optional fourth input argument to the function fminbnd that sets the optimization parameters; the easiest way to set this last argument is using the optimset command. If you enter

```
>> help optimset
```

into the workspace then a complete list of the parameters that can be controlled is returned. It is possible, for example, to display the progress of the optimization using:

```
>> [D_opt,q_dot_c_min]=fminbnd('EXAMPLE1p9_2',0.004,0.01,optimset('Display','iter'))
```

Func-count	x	f(x)	Procedure
1	0.0062918	6.35363	initial
2	0.0077082	8.12533	golden
3	0.00541641	5.73753	golden
4	0.0043798	6.14038	parabolic
5	0.00523819	5.68985	parabolic
6	0.00515231	5.6824	parabolic
7	0.00511897	5.68148	parabolic
8	0.00508564	5.68212	parabolic

Optimization terminated:
the current x satisfies the termination criteria using OPTIONS.TolX of 1.000000e-004

```
D_opt =
   0.0051
q_dot_c_min

   =
   5.6815
```

EXAMPLE 1.9-2: CRYOGENIC CURRENT LEADS

The first argument to **optimset** specifies the parameter to be controlled ('Display', which controls the level of display) and the second indicates its new value ('iter', which indicates that the progress should be displayed after each iteration).

It would be inconvenient to use the **fminbnd** function to carry out a parametric variation of how the optimal value of the variables D and q_dot_c are affected by current or some other parameter. In its current format, it is not possible to pass the value of the current (I) to the **fminbnd** function and therefore the study would have to be carried out manually by running the **fminbnd** function and then changing the value of the variable I within the function **EXAMPLE1p9_2**. This process would become tedious and can be avoided by parameterizing the function. For example, suppose you want to determine how the optimal value of diameter changes with current. First, include current as an additional argument to the function:

```
function[q_dot_c]=EXAMPLE1p9_2(D,I)

N=100;              %number of nodes (-)
% I=100;            %current (amp)
```

If you try to repeat the optimization using the previous protocol you will receive an error:

```
>> [D_opt,q_dot_c_min]=fminbnd('EXAMPLE1p9_2',0.004,0.01)
??? Input argument "I" is undefined.

Error in ==> EXAMPLE9_2 at 42
   b(i,1)=-rho_e_cu(Tg(i,1))*4*DELTAx*I^2/(pi*D^2)-...

Error in ==> fminbnd at 182
x=xf; fx=funfcn(x,varargin{:});
```

However, you can parameterize the function using a one-argument anonymous function that captures the value of I (set in the workspace) and calls **EXAMPLE1p9_2** with two arguments:

```
>> I=100;
>> [D_opt,q_dot_c_min]=fminbnd(@(D) EXAMPLE1p9_2(D,I),0.004,0.01)
D_opt =
   0.0051
q_dot_c_min =
   5.6815
```

Now it is possible to generate a script, **varyI**, that evaluates the optimized diameter and heat flow as a function of current.

```
%Script varyI.m
clear all;
I=linspace(1,100,10)';
for i=1:10
   [d_opt(i,1),q_dot_min(i,1)]=fminbnd(@(d) EXAMPLE1p9_2(d,I(i,1)),0.0025,0.01)
end
```

EXAMPLE 1.9-2: CRYOGENIC CURRENT LEADS

Figure 7 illustrates the optimal diameter and the associated heat leak as a function of current. Note that the optimal diameter and minimized heat leak are both approximately linear functions of the current.

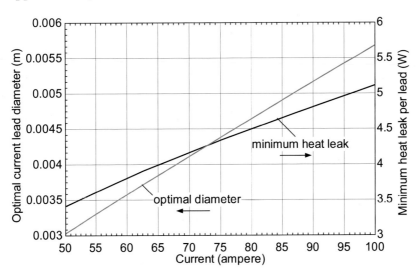

Figure 7: Optimal diameter and minimized heat leak to cold end of each current lead as a function of current.

Chapter 1: One-Dimensional, Steady-State Conduction

The website associated with this book (www.cambridge.org/nellisandklein) provides many more problems than are included here.

Conduction Heat Transfer

1–1 Section 1.1.2 provides an approximation for the thermal conductivity of a monatomic gas at ideal gas conditions. Test the validity of this approximation by comparing the conductivity estimated using Eq. (1-18) to the value of thermal conductivity for a monotonic ideal gas (e.g., low pressure argon) provided by the internal function in EES. Note that the molecular radius, σ, is provided in EES by the Lennard-Jones potential using the function sigma_LJ.

   a.) What are the value and units of the proportionality constant required to make Eq. (1-18) an equality?

   b.) Plot the value of the proportionality constant for 300 K argon at pressures between 0.01 and 100 MPa on a semi-log plot with pressure on the log scale. At what pressure does the approximation given in Eq. (1-18) begin to fail?

Steady-State 1-D Conduction without Generation

1–2 Figure P1-2 illustrates a plane wall made of a thin ($th_w = 0.001$ m) and conductive ($k = 100$ W/m-K) material that separates two fluids. Fluid A is at $T_A = 100°C$ and the heat transfer coefficient between the fluid and the wall is $\bar{h}_A = 10$ W/m²-K while fluid B is at $T_B = 0°C$ with $\bar{h}_B = 100$ W/m²-K.

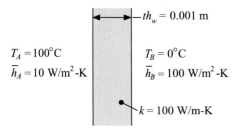

Figure P1-2: Plane wall separating two fluids.

a.) Draw a resistance network that represents this situation and calculate the value of each resistor (assuming a unit area for the wall, $A = 1\,\text{m}^2$).
b.) If you wanted to predict the heat transfer rate from fluid A to fluid B very accurately then which parameters (e.g., th_w, k, etc.) would you try to understand/measure very carefully and which parameters are not very important? Justify your answer.

1–3 You have a problem with your house. Every spring at some point the snow immediately adjacent to your roof melts and runs along the roof line until it reaches the gutter. The water in the gutter is exposed to air at temperature less than $0°C$ and therefore freezes, blocking the gutter and causing water to run into your attic. The situation is shown in Figure P1-3.

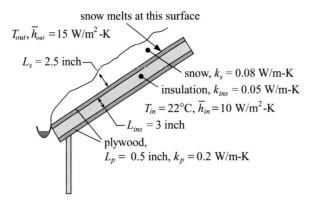

Figure P1-3: Roof of your house.

The air in the attic is at $T_{in} = 22°C$ and the heat transfer coefficient between the inside air and the inner surface of the roof is $\bar{h}_{in} = 10\,\text{W/m}^2\text{-K}$. The roof is composed of a $L_{ins} = 3.0$ inch thick piece of insulation with conductivity $k_{ins} = 0.05\,\text{W/m-K}$ that is sandwiched between two $L_p = 0.5$ inch thick pieces of plywood with conductivity $k_p = 0.2\,\text{W/m-K}$. There is an $L_s = 2.5$ inch thick layer of snow on the roof with conductivity $k_s = 0.08\,\text{W/m-K}$. The heat transfer coefficient between the outside air at temperature T_{out} and the surface of the snow is $\bar{h}_{out} = 15\,\text{W/m}^2\text{-K}$. Neglect radiation and contact resistances for part (a) of this problem.
a.) What is the range of outdoor air temperatures where you should be concerned that your gutters will become blocked by ice?
b.) Would your answer change much if you considered radiation from the outside surface of the snow to surroundings at T_{out}? Assume that the emissivity of snow is $\varepsilon_s = 0.82$.

1–4 Figure P1-4(a) illustrates a composite wall. The wall is composed of two materials (A with $k_A = 1\,\text{W/m-K}$ and B with $k_B = 5\,\text{W/m-K}$), each has thickness $L =$

1.0 cm. The surface of the wall at $x = 0$ is perfectly insulated. A very thin heater is placed between the insulation and material A; the heating element provides $\dot{q}'' = 5000\ \text{W/m}^2$ of heat. The surface of the wall at $x = 2L$ is exposed to fluid at $T_{f,in} = 300\ \text{K}$ with heat transfer coefficient $\bar{h}_{in} = 100\ \text{W/m}^2\text{-K}$.

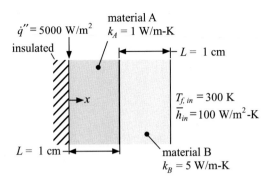

Figure P1-4 (a): Composite wall with a heater.

You may neglect radiation and contact resistance for parts (a) through (c) of this problem.

a.) Draw a resistance network to represent this problem; clearly indicate what each resistance represents and calculate the value of each resistance.

b.) Use your resistance network from (a) to determine the temperature of the heating element.

c.) Sketch the temperature distribution through the wall. Make sure that the sketch is consistent with your solution from (b).

Figure P1-4(b) illustrates the same composite wall shown in Figure P1-4(a), but there is an additional layer added to the wall, material C with $k_C = 2.0\ \text{W/m-K}$ and $L = 1.0\ \text{cm}$.

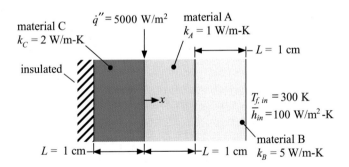

Figure P1-4 (b): Composite wall with material C.

Neglect radiation and contact resistance for parts (d) through (f) of this problem.

d.) Draw a resistance network to represent the problem shown in Figure P1-4(b); clearly indicate what each resistance represents and calculate the value of each resistance.

e.) Use your resistance network from (d) to determine the temperature of the heating element.

f.) Sketch the temperature distribution through the wall. Make sure that the sketch is consistent with your solution from (e).

Figure P1-4(c) illustrates the same composite wall shown in Figure P1-4(b), but there is a contact resistance between materials A and B, $R_c'' = 0.01$ K-m^2/W, and the surface of the wall at $x = -L$ is exposed to fluid at $T_{f,out} = 400$ K with a heat transfer coefficient $\bar{h}_{out} = 10$ W/m^2-K.

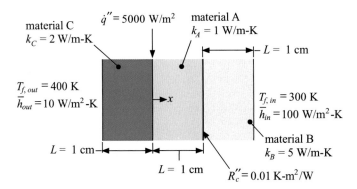

material C
$k_C = 2$ W/m-K

$\dot{q}'' = 5000$ W/m^2 material A
$k_A = 1$ W/m-K

$\leftarrow\!\!-\!\!|\; L = 1$ cm

$T_{f,out} = 400$ K
$\bar{h}_{out} = 10$ W/m^2-K

$\rightarrow x$

$T_{f,in} = 300$ K
$\bar{h}_{in} = 100$ W/m^2-K

material B
$k_B = 5$ W/m-K

$L = 1$ cm

$L = 1$ cm

$R_c'' = 0.01$ K-m^2/W

Figure P1-4 (c): Composite wall with convection at the outer surface and contact resistance.

Neglect radiation for parts (g) through (i) of this problem.

g.) Draw a resistance network to represent the problem shown in Figure P1-4(c); clearly indicate what each resistance represents and calculate the value of each resistance.

h.) Use your resistance network from (g) to determine the temperature of the heating element.

i.) Sketch the temperature distribution through the wall.

1–5 You have decided to install a strip heater under the linoleum in your bathroom in order to keep your feet warm on cold winter mornings. Figure P1-5 illustrates a cross-section of the bathroom floor. The bathroom is located on the first story of your house and is $W = 2.5$ m wide $\times L = 2.5$ m long. The linoleum thickness is $th_L = 5.0$ mm and has conductivity $k_L = 0.05$ W/m-K. The strip heater under the linoleum is negligibly thin. Beneath the heater is a piece of plywood with thickness $th_P = 5$ mm and conductivity $k_P = 0.4$ W/m-K. The plywood is supported by $th_s = 6.0$ cm thick studs that are $W_s = 4.0$ cm wide with thermal conductivity $k_s = 0.4$ W/m-K. The center-to-center distance between studs is $p_s = 25.0$ cm. Between each stud are pockets of air that can be considered to be stagnant with conductivity $k_a = 0.025$ W/m-K. A sheet of drywall is nailed to the bottom of the studs. The thickness of the drywall is $th_d = 9.0$ mm and the conductivity of drywall is $k_d = 0.1$ W/m-K. The air above in the bathroom is at $T_{air,1} = 15°$C while the air in the basement is at $T_{air,2} = 5°$C. The heat transfer coefficient on both sides of the floor is $\bar{h} = 15$ W/m^2-K. You may neglect radiation and contact resistance for this problem.

a.) Draw a thermal resistance network that can be used to represent this situation. Be sure to label the temperatures of the air above and below the floor ($T_{air,1}$ and $T_{air,2}$), the temperature at the surface of the linoleum (T_L), the temperature of the strip heater (T_h), and the heat input to the strip heater ($\dot{q}_h$) on your diagram.

b.) Compute the value of each of the resistances from part (a).

c.) How much heat must be added by the heater to raise the temperature of the floor to a comfortable 20°C?

d.) What physical quantities are most important to your analysis? What physical quantities are unimportant to your analysis?

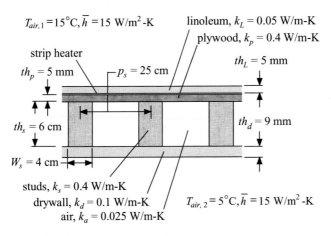

Figure P1-5: Bathroom floor with heater.

e.) Discuss at least one technique that could be used to substantially reduce the amount of heater power required while still maintaining the floor at 20°C. Note that you have no control over $T_{air,1}$ or $\bar{h}$.

1–6 You are a fan of ice fishing but don't enjoy the process of augering out your fishing hole in the ice. Therefore, you want to build a device, the super ice-auger, that melts a hole in the ice. The device is shown in Figure P1-6.

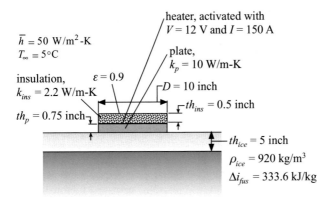

Figure P1-6: The super ice-auger.

A heater is attached to the back of a $D = 10$ inch plate and electrically activated by your truck battery, which is capable of providing $V = 12$ V and $I = 150$ A. The plate is $th_p = 0.75$ inch thick and has conductivity $k_p = 10$ W/m-K. The back of the heater is insulated; the thickness of the insulation is $th_{ins} = 0.5$ inch and the insulation has conductivity $k_{ins} = 2.2$ W/m-K. The surface of the insulation experiences convection with surrounding air at $T_\infty = 5°C$ and radiation with surroundings also at $T_\infty = 5°C$. The emissivity of the surface of the insulation is $\varepsilon = 0.9$ and the heat transfer coefficient between the surface and the air is $\bar{h} = 50$ W/m²-K. The super ice-auger is placed on the ice and activated, causing a heat transfer to the plate-ice interface that melts the ice. Assume that the water under the ice is at $T_{ice} = 0°C$ so that no heat is conducted away from the plate-ice interface; all of the energy transferred to the plate-ice interface goes into melting the ice. The thickness of the ice is

$th_{ice} = 5$ inch and the ice has density $\rho_{ice} = 920$ kg/m^3. The latent heat of fusion for the ice is $\Delta i_{fus} = 333.6$ kJ/kg.

a.) Determine the heat transfer rate to the plate-ice interface.

b.) How long will it take to melt a hole in the ice?

c.) What is the efficiency of the melting process?

d.) If your battery is rated at 100 amp-hr at 12 V then what fraction of the battery's charge is depleted by running the super ice-auger.

Steady-State 1-D Conduction with Generation

1–7 One of the engineers that you supervise has been asked to simulate the heat transfer problem shown in Figure P1-7(a). This is a 1-D, plane wall problem (i.e., the temperature varies only in the x-direction and the area for conduction is constant with x). Material A (from $0 < x < L$) has conductivity k_A and experiences a uniform rate of volumetric thermal energy generation, $\dot{g}'''$. The left side of material A (at $x = 0$) is completely insulated. Material B (from $L < x < 2L$) has *lower* conductivity, $k_B < k_A$. The right side of material B (at $x = 2L$) experiences convection with fluid at room temperature (20°C). Based on the facts above, critically examine the solution that has been provided to you by the engineer and is shown in Figure P1-7(b). There should be a few characteristics of the solution that do not agree with your knowledge of heat transfer; list as many of these characteristics as you can identify and provide a clear reason why you think the engineer's solution must be wrong.

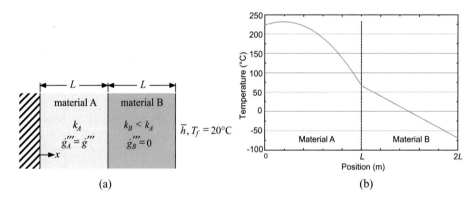

(a)   (b)

Figure P1-7 (a): Heat transfer problem and (b) "solution" provided by the engineer.

1–8 Freshly cut hay is not really dead; chemical reactions continue in the plant cells and therefore a small amount of heat is released within the hay bale. This is an example of the conversion of chemical to thermal energy. The amount of thermal energy generation within a hay bale depends on the moisture content of the hay when it is baled. Baled hay can become a fire hazard if the rate of volumetric generation is sufficiently high and the hay bale sufficiently large so that the interior temperature of the bale reaches 170°F, the temperature at which self-ignition can occur. Here, we will model a round hay bale that is wrapped in plastic to protect it from the rain. You may assume that the bale is at steady state and is sufficiently long that it can be treated as a one-dimensional, radial conduction problem. The radius of the hay bale is $R_{bale} = 5$ ft and the bale is wrapped in plastic that is $t_p = 0.045$ inch thick with conductivity $k_p = 0.15$ W/m-K. The bale is surrounded by air at $T_\infty = 20$°C with $\overline{h} = 10$ W/m^2-K. You may neglect radiation. The conductivity of the hay is $k = 0.04$ W/m-K.

a.) If the volumetric rate of thermal energy generation is constant and equal to $\dot{g}''' = 2$ W/m^3 then determine the maximum temperature in the hay bale.

b.) Prepare a plot showing the maximum temperature in the hay bale as a function of the hay bale radius. How large can the hay bale be before there is a problem with self-ignition?

Prepare a model that can consider temperature-dependent volumetric generation. Increasing temperature tends to increase the rate of chemical reaction and therefore increases the rate of generation of thermal energy according to: $\dot{g}''' = a + bT$ where $a = -1$ W/m^3 and $b = 0.01$ W/m^3-K and T is in K.

c.) Enter the governing equation into Maple and obtain the general solution (i.e., a solution that includes two constants).

d.) Use the boundary conditions to obtain values for the two constants in your general solution. (hint: one of the two constants must be zero in order to keep the temperature at the center of the hay bale finite). You should obtain a symbolic expression for the boundary condition in Maple that can be evaluated in EES.

e.) Overlay on your plot from part (b) a plot of the maximum temperature in the hay bale as a function of bale radius when the volumetric generation is a function of temperature.

1–9 Figure P1-9 illustrates a simple mass flow meter for use in an industrial refinery.

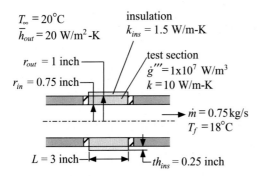

Figure P1-9: A simple mass flow meter.

A flow of liquid passes through a test section consisting of an $L = 3$ inch section of pipe with inner and outer radii, $r_{in} = 0.75$ inch and $r_{out} = 1.0$ inch, respectively. The test section is uniformly heated by electrical dissipation at a rate $\dot{g}''' = 1 \times 10^7$ W/m^3 and has conductivity $k = 10$ W/m-K. The pipe is surrounded with insulation that is $th_{ins} = 0.25$ inch thick and has conductivity $k_{ins} = 1.5$ W/m-K. The external surface of the insulation experiences convection with air at $T_\infty = 20°$C. The heat transfer coefficient on the external surface is $\bar{h}_{out} = 20$ W/m^2-K. A thermocouple is embedded at the center of the pipe wall. By measuring the temperature of the thermocouple, it is possible to infer the mass flow rate of fluid because the heat transfer coefficient on the inner surface of the pipe ($\bar{h}_{in}$) is strongly related to mass flow rate ($\dot{m}$). Testing has shown that the heat transfer coefficient and mass flow rate are related according to:

$$\bar{h}_{in} = C \left(\frac{\dot{m}}{1 \ [kg/s]} \right)^{0.8}$$

where $C = 2500$ W/m^2-K. Under nominal conditions, the mass flow rate through the meter is $\dot{m} = 0.75$ kg/s and the fluid temperature is $T_f = 18°$C. Assume that the ends of the test section are insulated so that the problem is 1-D. Neglect radiation and assume that the problem is steady state.

a.) Develop an analytical model in EES that can predict the temperature distribution in the test section. Plot the temperature as a function of radial position for the nominal conditions.

b.) Using your model, develop a calibration curve for the meter; that is, prepare a plot of the mass flow rate as a function of the measured temperature at the mid-point of the pipe. The range of the instrument is 0.2 kg/s to 2.0 kg/s.

The meter must be robust to changes in the fluid temperature. That is, the calibration curve developed in (b) must continue to be valid even as the fluid temperature changes by as much as $10°C$.

c.) Overlay on your plot from (b) the mass flow rate as a function of the measured temperature for $T_f = 8°C$ and $T_f = 28°C$. Is your meter robust to changes in T_f?

In order to improve the meters ability to operate over a range of fluid temperature, a temperature sensor is installed in the fluid in order to measure T_f during operation.

d.) Using your model, develop a calibration curve for the meter in terms of the mass flow rate as a function of ΔT, the difference between the measured temperatures at the mid-point of the pipe wall and the fluid.

e.) Overlay on your plot from (d) the mass flow rate as a function of the difference between the measured temperatures at the mid-point of the pipe wall and the fluid if the fluid temperature is $T_f = 8°C$ and $T_f = 28°C$. Is the meter robust to changes in T_f?

f.) If you can measure the temperature difference to within $\delta \Delta T = 1$ K then what is the uncertainty in the mass flow rate measurement? (Use your plot from part (d) to answer this question.)

g.) Set the temperature difference to the value you calculated at the nominal conditions and allow EES to calculate the associated mass flow rate. Now, select Uncertainty Propagation from the Calculate menu and specify that the mass flow rate as the calculated variable while the temperature difference is the measured variable. Set the uncertainty in the temperature difference to 1 K and verify that EES obtains an answer that is approximately consistent with part (f).

h.) The nice thing about using EES to determine the uncertainty is that it becomes easy to assess the impact of multiple sources of uncertainty. In addition to the uncertainty $\delta \Delta T$, the constant C has relative uncertainty of $\delta C = 5\%$ and the conductivity of the material is only known to within $\delta k = 3\%$. Use EES' built-in uncertainty propagation to assess the resulting uncertainty in the mass flow rate measurement. Which source of uncertainty is the most important?

i.) The meter must be used in areas where the ambient temperature and heat transfer coefficient may vary substantially. Prepare a plot showing the mass flow rate predicted by your model for $\Delta T = 50$ K as a function of T_∞ for various values of $\overline{h}_{out}$. If the operating range of your meter must include $-5°C < T_\infty < 35°C$ then use your plot to determine the range of $\overline{h}_{out}$ that can be tolerated without substantial loss of accuracy.

Numerical Solutions to Steady-State 1-D Conduction Problems using EES

1–10 Reconsider the mass flow meter that was investigated in Problem 1-9. The conductivity of the material that is used to make the test section is not actually constant, as was assumed in Problem 1-9, but rather depends on temperature according to:

$$k = 10\,\frac{\text{W}}{\text{m-K}} + 0.035\left[\frac{\text{W}}{\text{m-K}^2}\right](T - 300\,[\text{K}])$$

a.) Develop a numerical model of the mass flow meter using EES. Plot the temperature as a function of radial position for the conditions shown in Figure P1-9 with the temperature-dependent conductivity.

b.) Verify that your numerical solution limits to the analytical solution from Problem 1-9 in the limit that the conductivity is constant.

c.) What effect does the temperature dependent conductivity have on the calibration curve that you generated in part (d) of Problem 1-9.

Numerical Solutions to Steady-State 1-D Conduction Problems using MATLAB

1–11 Reconsider Problem 1-8, but obtain a solution numerically using MATLAB. The description of the hay bale is provided in Problem 1-8. Prepare a model that can consider the effect of temperature on the volumetric generation. Increasing temperature tends to increase the rate of reaction and therefore increase the rate of generation of thermal energy; the volumetric rate of generation can be approximated by: $\dot{g}''' = a + bT$ where $a = -1$ W/m^3 and $b = 0.01$ W/m^3-K and T is in K.

a.) Prepare a numerical model of the hay bale. Plot the temperature as a function of position within the hay bale.

b.) Show that your model has numerically converged; that is, show some aspect of your solution as a function of the number of nodes and discuss an appropriate number of nodes to use.

c.) Verify your numerical model by comparing your answer to an analytical solution in some, appropriate limit. The result of this step should be a plot that shows the temperature as a function of radius predicted by both your numerical solution and the analytical solution and demonstrates that they agree.

1–12 Reconsider the mass flow meter that was investigated in Problem 1-9. Assume that the conductivity of the material that is used to make the test section is not actually constant, as was assumed in Problem 1-9, but rather depends on temperature according to:

$$k = 10\,\frac{\text{W}}{\text{m-K}} + 0.035\left[\frac{\text{W}}{\text{m-K}^2}\right](T - 300\,[\text{K}])$$

a.) Develop a numerical model of the mass flow meter using MATLAB. Plot the temperature as a function of radial position for the conditions shown in Figure P1-9 with the temperature-dependent conductivity.

b.) Verify that your numerical solution limits to the analytical solution from Problem 1-9 in the limit that the conductivity is constant.

Analytical Solutions for Constant Cross-Section Extended Surfaces

1–13 A resistance temperature detector (RTD) utilizes a material that has a resistivity that is a strong function of temperature. The temperature of the RTD is inferred by measuring its electrical resistance. Figure P1-13 shows an RTD that is mounted at the end of a metal rod and inserted into a pipe in order to measure the temperature of a flowing liquid. The RTD is monitored by passing a known current through it and measuring the voltage across it. This process results in a constant amount of ohmic heating that may tend to cause the RTD temperature to rise relative to the temperature of the surrounding liquid; this effect is referred to as a self-heating error. Also, conduction from the wall of the pipe to the temperature sensor through the metal rod can result in a temperature difference between the RTD and the liquid; this effect is referred to as a mounting error.

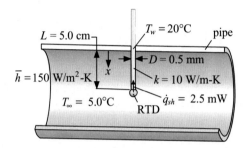

Figure P1-13: Temperature sensor mounted in a flowing liquid.

The thermal energy generation associated with ohmic heating is $\dot{q}_{sh} = 2.5$ mW. All of this ohmic heating is assumed to be transferred from the RTD into the end of the rod at $x = L$. The rod has a thermal conductivity $k = 10$ W/m-K, diameter $D = 0.5$ mm, and length $L = 5.0$ cm. The end of the rod that is connected to the pipe wall (at $x = 0$) is maintained at a temperature of $T_w = 20°$C. The liquid is at a uniform temperature, $T_\infty = 50°$C and the heat transfer coefficient between the liquid and the rod is $\bar{h} = 150$ W/m²-K.

a.) Is it appropriate to treat the rod as an extended surface (i.e., can we assume that the temperature in the rod is a function only of x)? Justify your answer.

b.) Develop an analytical model of the rod that will predict the temperature distribution in the rod and therefore the error in the temperature measurement; this error is the difference between the temperature at the tip of the rod and the liquid.

c.) Prepare a plot of the temperature as a function of position and compute the temperature error.

d.) Investigate the effect of thermal conductivity on the temperature measurement error. Identify the optimal thermal conductivity and explain why an optimal thermal conductivity exists.

1–14 Your company has developed a micro-end milling process that allows you to easily fabricate an array of very small fins in order to make heat sinks for various types of electrical equipment. The end milling process removes material in order to generate the array of fins. Your initial design is the array of pin fins shown in Figure P1-14. You have been asked to optimize the design of the fin array for a particular application where the base temperature is $T_{base} = 120°$C and the air temperature is $T_{air} = 20°$C. The heat sink is square; the size of the heat sink is $W = 10$ cm.

The conductivity of the material is $k = 70$ W/m-K. The distance between the edges of two adjacent fins is a, the diameter of a fin is D, and the length of each fin is L.

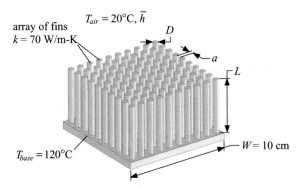

Figure P1-14: Pin fin array.

Air is forced to flow through the heat sink by a fan. The heat transfer coefficient between the air and the surface of the fins as well as the unfinned region of the base, $\bar{h}$, has been measured for the particular fan that you plan to use and can be calculated according to:

$$\bar{h} = 40 \left[\frac{\text{W}}{\text{m}^2\text{K}}\right] \left(\frac{a}{0.005\,[\text{m}]}\right)^{0.4} \left(\frac{D}{0.01\,[\text{m}]}\right)^{-0.3}$$

Mass is not a concern for this heat sink; you are only interested in maximizing the heat transfer rate from the heat sink to the air given the operating temperatures. Therefore, you will want to make the fins as long as possible. However, in order to use the micro-end milling process you cannot allow the fins to be longer than 10x the distance between two adjacent fins. That is, the length of the fins may be computed according to: $L = 10\,a$. You must choose the optimal values of a and D for this application.

a.) Prepare a model using EES that can predict the heat transfer coefficient for a given value of a and D. Use this model to predict the heat transfer rate from the heat sink for $a = 0.5$ cm and $D = 0.75$ cm.

b.) Prepare a plot that shows the heat transfer rate from the heat sink as a function of the distance between adjacent fins, a, for a fixed value of $D = 0.75$ cm. Be sure that the fin length is calculated using $L = 10\,a$. Your plot should exhibit a maximum value, indicating that there is an optimal value of a.

c.) Prepare a plot that shows the heat transfer rate from the heat sink as a function of the diameter of the fins, D, for a fixed value of $a = 0.5$ cm. Be sure that the fin length is calculated using $L = 10\,a$. Your plot should exhibit a maximum value, indicating that there is an optimal value of D.

d.) Determine the optimal values of a and D using EES' built-in optimization capability.

Analytical Solutions for Advanced Constant Cross-Section Extended Surfaces

1–15 Figure P1-15 illustrates a material processing system.

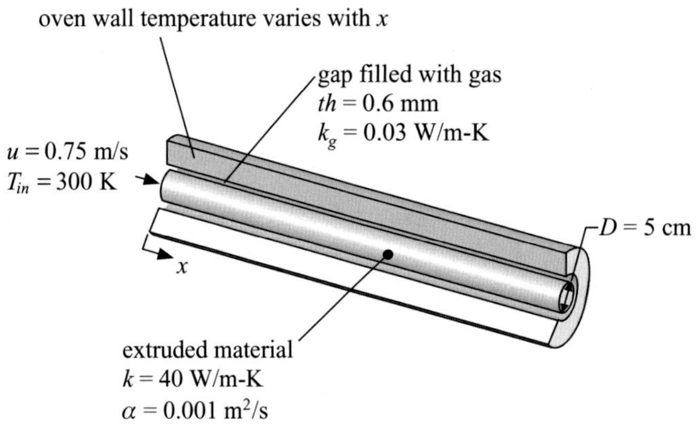

oven wall temperature varies with x

gap filled with gas
$th = 0.6$ mm
$k_g = 0.03$ W/m-K

$u = 0.75$ m/s
$T_{in} = 300$ K

$D = 5$ cm

x

extruded material
$k = 40$ W/m-K
$\alpha = 0.001$ m^2/s

Figure P1-15: Material processing system.

Material is extruded and enters the oven at $T_{in} = 300$ K with velocity $u = 0.75$ m/s. The material has diameter $D = 5$ cm. The conductivity of the material is $k = 40$ W/m-K and the thermal diffusivity is $\alpha = 0.001$ m^2/s.

In order to precisely control the temperature of the material, the oven wall is placed very close to the outer diameter of the extruded material and the oven wall temperature distribution is carefully controlled. The gap between the oven wall and the material is $th = 0.6$ mm and the oven-to-material gap is filled with gas that has conductivity $k_g = 0.03$ W/m-K. Radiation can be neglected in favor of convection through the gas from the oven wall to the material. For this situation, the heat flux experienced by the material surface can be approximately modeled according to:

$$\dot{q}''_{conv} \approx \frac{k_g}{th} (T_w - T)$$

where T_w and T are the oven wall and material temperatures at that position, respectively. The oven wall temperature varies with position x according to:

$$T_w = T_f - (T_f - T_{w,0}) \exp\left(-\frac{x}{L_c}\right)$$

where $T_{w,0}$ is the temperature of the wall at the inlet (at $x = 0$), $T_f = 1000$ K is the temperature of the wall far from the inlet, and L_c is a characteristic length that dictates how quickly the oven wall temperature approaches T_f. Initially, assume that $T_{w,0} = 500$ K, $T_f = 1000$ K, and $L_c = 1$ m. Assume that the oven can be approximated as being infinitely long.

a.) Is an extended surface model appropriate for this problem?

b.) Assume that your answer to (a) was yes. Develop an analytical solution that can be used to predict the temperature of the material as a function of x.

c.) Plot the temperature of the material and the temperature of the wall as a function of position for $0 < x < 20$ m. Plot the temperature gradient experienced by the material as a function of position for $0 < x < 20$ m.

The parameter L_c can be controlled in order to control the maximum temperature gradient and therefore the thermal stress experienced by the material as it moves through the oven.

d.) Prepare a plot showing the maximum temperature gradient as a function of L_c. Overlay on your plot the distance required to heat the material to $T_p = 800$ K (L_p). If the maximum temperature gradient that is allowed is 60 K/m, then what is the appropriate value of L_c and the corresponding value of L_p?

1–16 The receiver tube of a concentrating solar collector is shown in Figure P1-16.

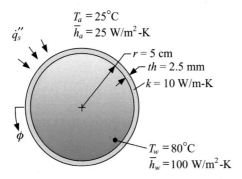

$T_a = 25°C$
$\overline{h}_a = 25$ W/m²-K
$\dot{q}_s''$
$r = 5$ cm
$th = 2.5$ mm
$k = 10$ W/m-K
ϕ
$T_w = 80°C$
$\overline{h}_w = 100$ W/m²-K

Figure P1-16: A solar collector.

The receiver tube is exposed to solar radiation that has been reflected from a concentrating mirror. The heat flux received by the tube is related to the position of the sun and the geometry and efficiency of the concentrating mirrors. For this problem, you may assume that all of the radiation heat flux is absorbed by the collector and neglect the radiation emitted by the collector to its surroundings. The flux received at the collector surface ($\dot{q}_s''$) is not circumferentially uniform but rather varies with angular position; the flux is uniform along the top of the collector, $\pi < \phi < 2\pi$ rad, and varies sinusoidally along the bottom, $0 < \phi < \pi$ rad, with a peak at $\phi = \pi/2$ rad.

$$\dot{q}_s''(\phi) = \begin{cases} \dot{q}_t'' + \left(\dot{q}_p'' - \dot{q}_t''\right)\sin(\phi) & \text{for } 0 < \phi < \pi \\ \dot{q}_t'' & \text{for } \pi < \phi < 2\pi \end{cases}$$

where $\dot{q}_t'' = 1000$ W/m² is the uniform heat flux along the top of the collector tube and $\dot{q}_p'' = 5000$ W/m² is the peak heat flux along the bottom. The receiver tube has an inner radius of $r = 5.0$ cm and thickness of $th = 2.5$ mm (because $th/r \ll 1$ it is possible to ignore the small difference in convection area on the inner and outer surfaces of the tube). The thermal conductivity of the tube material is $k = 10$ W/m-K. The solar collector is used to heat water, which is at $T_w = 80°C$ at the axial position of interest. The average heat transfer coefficient between the water and the internal surface of the collector is $\overline{h}_w = 100$ W/m²-K. The external surface of the collector is exposed to air at $T_a = 25°C$. The average heat transfer coefficient between the air and the external surface of the collector is $\overline{h}_a = 25$ W/m²-K.

a.) Can the collector be treated as an extended surface for this problem (i.e., can the temperature gradients in the radial direction in the collector material be neglected)?

b.) Develop an analytical model that will allow the temperature distribution in the collector wall to be determined as a function of circumferential position.

Analytical Solutions for Non-Constant Cross-Section Extended Surfaces

1–17 Figure P1-17 illustrates a disk brake for a rotating machine. The temperature distribution within the brake can be assumed to be a function of radius only. The brake is divided into two regions. In the outer region, from $R_p = 3.0$ cm to $R_d = 4.0$ cm, the stationary brake pads create frictional heating and the disk is not exposed to convection. The clamping pressure applied to the pads is $P = 1.0$ MPa and the coefficient of friction between the pad and the disk is $\mu = 0.15$. You may assume that the pads are not conductive and therefore all of the frictional heating is conducted into the disk. The disk rotates at $N = 3600$ rev/min and is $b = 5.0$ mm thick. The conductivity of the disk is $k = 75$ W/m-K and you may assume that the outer rim of the disk is adiabatic.

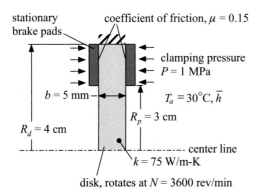

Figure P1-17: Disk brake.

The inner region of the disk, from 0 to R_p, is exposed to air at $T_a = 30°$C. The heat transfer coefficient between the air and disk surface depends on the angular velocity of the disk, ω, according to:

$$ h = 20 \left[\frac{\text{W}}{\text{m}^2\text{-K}} \right] + 1500 \left[\frac{\text{W}}{\text{m}^2\text{-K}} \right] \left(\frac{\omega}{100 \text{ [rad/s]}} \right)^{1.25} $$

a.) Develop an analytical model of the temperature distribution in the disk brake; prepare a plot of the temperature as a function of radius for $r = 0$ to $r = R_d$.
b.) If the disk material can withstand a maximum safe operating temperature of 750°C then what is the maximum allowable clamping pressure that can be applied? Plot the temperature distribution in the disk at this clamping pressure. What is the braking torque that results?
c.) Assume that you can control the clamping pressure so that as the machine slows down the maximum temperature is always kept at the maximum allowable temperature, 750°C. Plot the torque as a function of rotational speed for 100 rev/min to 3600 rev/min.

1–18 Figure P1-18 illustrates a fin that is to be used in the evaporator of a space conditioning system for a space-craft. The fin is a plate with a triangular shape. The thickness of the plate is $th = 1$ mm and the width of the fin at the base is $W_b = 1$ cm. The length of the fin is $L = 2$ cm. The fin material has conductivity $k = 50$ W/m-K. The average heat transfer coefficient between the fin surface and the air in the space-craft is $\bar{h} = 120$ W/m²-K. The air is at $T_\infty = 20°$C and the base of the fin is at $T_b = 10°$C. Assume that the temperature distribution in the fin is 1-D in x. Neglect convection from the edges of the fin.

a.) Obtain an analytical solution for the temperature distribution in the fin. Plot the temperature as a function of position.
b.) Calculate the rate of heat transfer to the fin.
c.) Determine the fin efficiency.

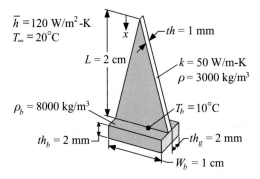

Figure P1-18: Fin on an evaporator.

The fin has density $\rho = 3000\,\text{kg/m}^3$ and is installed on a base material with thickness $th_b = 2$ mm and density $\rho_b = 8000\,\text{kg/m}^3$. The half-width of the gap between adjacent fins is $th_g = 2$ mm. Therefore, the volume of the base material associated with each fin is $th_b W_b (th + 2th_g)$.

d.) Determine the ratio of the absolute value of the rate of heat transfer to the fin to the total mass of material (fin and base material associated with the fin).

e.) Prepare a contour plot that shows the ratio of the heat transfer to the fin to the total mass of material as a function of the length of the fin (L) and the fin thickness (th).

f.) What is the optimal value of L and th that maximizes the absolute value of the fin heat transfer rate to the mass of material?

Numerical Solution of Extended Surface Problems

1–19 A fiber optic bundle (FOB) is shown in Figure P1-19 and used to transmit the light for a building application.

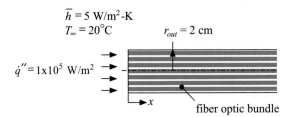

Figure P1-19: Fiber optic bundle used to transmit light.

The fiber optic bundle is composed of several, small-diameter fibers that are each coated with a thin layer of polymer cladding and packed in approximately a hexagonal close-packed array. The porosity of the FOB is the ratio of the open area of the FOB face to its total area. The porosity of the FOB face is an important characteristic because any radiation that does not fall directly upon the fibers will not be transmitted and instead contributes to a thermal load on the FOB. The fibers are designed so that any radiation that strikes the face of a fiber is "trapped" by total internal reflection. However, radiation that strikes the interstitial areas between the fibers will instead be absorbed in the cladding very close to the FOB face. The volumetric generation of thermal energy associated with this radiation can be represented by:

$$\dot{g}''' = \frac{\phi\,\dot{q}''}{L_{ch}}\exp\left(-\frac{x}{L_{ch}}\right)$$

where $\dot{q}'' = 1 \times 10^5$ W/m^2 is the energy flux incident on the face, $\phi = 0.05$ is the porosity of the FOB, x is the distance from the face, and $L_{ch} = 0.025$ m is the characteristic length for absorption of the energy. The outer radius of the FOB is $r_{out} = 2$ cm. The face of the FOB as well as its outer surface are exposed to air at $T_\infty = 20°$C with heat transfer coefficient $\bar{h} = 5$ W/m^2-K. The FOB is a composite structure and therefore conduction through the FOB is a complicated problem involving conduction through several different media. Section 2.9 discusses methods for computing the effective thermal conductivity for a composite. The effective thermal conductivity of the FOB in the radial direction is $k_{eff,r} = 2.7$ W/m-K. In order to control the temperature of the FOB near the face, where the volumetric generation of thermal energy is largest, it has been suggested that high conductivity filler material be inserted in the interstitial regions between the fibers. The result of the filler material is that the effective conductivity of the FOB in the axial direction varies with position according to:

$$k_{eff,x} = k_{eff,x,\infty} + \Delta k_{eff,x} \exp\left(-\frac{x}{L_k}\right)$$

where $k_{eff,x,\infty} = 2.0$ W/m-K is the effective conductivity of the FOB in the x-direction without filler material, $\Delta k_{eff,x} = 28$ W/m-K is the augmentation of the conductivity near the face, and $L_k = 0.05$ m is the characteristic length over which the effect of the filler material decays. The length of the FOB is effectively infinite. Assume that the volumetric generation is unaffected by the filler material.

a.) Is it appropriate to use a 1-D model of the FOB?

b.) Assume that your answer to (a) was yes. Develop a numerical model of the FOB.

c.) Overlay on a single plot the temperature distribution within the FOB for the case where the filler material is present ($\Delta k_{eff,x} = 28$ W/m-K) and the case where no filler material is present ($\Delta k_{eff,x} = 0$).

1–20 An expensive power electronics module normally receives only a moderate current. However, under certain conditions it might experience currents in excess of 100 amps. The module cannot survive such a high current and therefore, you have been asked to design a fuse that will protect the module by limiting the current that it can experience, as shown in Figure P1-20.

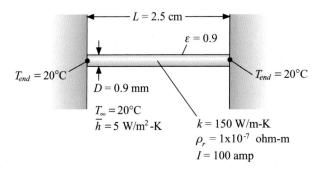

Figure P1-20: A fuse that protects a power electronics module from high current.

The space available for the fuse allows a wire that is $L = 2.5$ cm long to be placed between the module and the surrounding structure. The surface of the fuse wire is exposed to air at $T_\infty = 20°$C. The heat transfer coefficient between the surface of the fuse and the air is $\bar{h} = 5.0$ W/m^2-K. The fuse surface has an emissivity of $\varepsilon = 0.90$. The fuse is made of an aluminum alloy with conductivity

$k = 150$ W/m-K. The electrical resistivity of the aluminum alloy is $\rho_e = 1 \times 10^{-7}$ ohm-m and the alloy melts at approximately 500°C. Assume that the properties of the alloy do not depend on temperature. The ends of the fuse (i.e., at $x = 0$ and $x = L$) are maintained at $T_{end} = 20°C$ by contact with the surrounding structure and the module. The current passing through the fuse, I, results in a uniform volumetric generation within the fuse material. If the fuse operates properly, then it will melt (i.e., at some location within the fuse, the temperature will exceed 500°C) when the current reaches 100 amp. Your job will be to select the fuse diameter; to get your model started, you may assume a diameter of $D = 0.9$ mm. Assume that the volumetric rate of thermal energy generation due to ohmic dissipation is uniform throughout the fuse volume.

a.) Prepare a numerical model of the fuse that can predict the steady-state temperature distribution within the fuse material. Plot the temperature as a function of position within the wire when the current is 100 amp and the diameter is 0.9 mm.

b.) Verify that your model has numerically converged by plotting the maximum temperature in the wire as a function of the number of nodes in your model.

c.) Prepare a plot of the maximum temperature in the wire as a function of the diameter of the wire for $I = 100$ amp. Use your plot to select an appropriate fuse diameter.

REFERENCES

Cercignani, C., *Rarefied Gas Dynamics: From Basic Concepts to Actual Calculations*, Cambridge University Press, Cambridge, U.K., (2000).

Chen, G., *Nanoscale Energy Transport and Conversion: A Parallel Treatment of Electrons, Molecules, Phonons, and Photons*, Oxford University Press, Oxford, U.K., (2005).

Flynn, T. M., *Cryogenic Engineering, 2nd Edition, Revised and Expanded*, Marcel Dekker, New York, (2005).

Fried, E., *Thermal Conduction Contribution to Heat Transfer at Contacts*, in *Thermal Conductivity, Volume 2*, R. P. Tye, ed., Academic Press, London, (1969).

Iwasa, Y., *Case Studies in Superconducting Magnets: Design and Operational Issues*, Plenum Press, New York, (1994).

Izzo, F., "Other Thermal Ablation Techniques: Microwave and Interstitial Laser Ablation of Liver Tumors," *Annals of Surgical Oncology*, Vol. 10, pp. 491–497, (2003).

Keenan, J. H., "Adventures in Science," *Mechanical Engineering*, May, p. 79, (1958).

NIST Standard Reference Database Number 69, http://webbook.nist.gov/chemistry, June 2005 Release, (2005).

Que, L., *Micromachined Sensors and Actuators Based on Bent-Beam Suspensions*, Ph.D. Thesis, Electrical and Computer Engineering Dept., University of Wisconsin-Madison, (2000).

Schneider, P. J., *Conduction*, in *Handbook of Heat Transfer, 2nd Edition*, W.M. Rohsenow et al., eds., McGraw-Hill, New York, (1985).

Tien, C.-L., A. Majumdar, and F. M. Gerner, eds., *Microscale Energy Transport*, Taylor & Francis, Washington (1998).

Tompkins, D. T., *A Finite Element Heat Transfer Model of Ferromagnetic Thermoseeds and a Physiologically-Based Objective Function for Pretreatment Planning of Ferromagnetic Hypothermia*, Ph.D. Thesis, Mechanical Engineering, University of Wisconsin at Madison, (1992).

Walton, A. J., *Three Phases of Matter*, Oxford University Press, Oxford, U.K., (1989).

2 Two-Dimensional, Steady-State Conduction

Chapter 1 discussed the analytical and numerical solution of 1-D, steady-state problems. These are problems where the temperature within the material is independent of time and varies in only one spatial dimension (e.g., x). Examples of such problems are the plane wall studied in Section 1.2, which is truly a 1-D problem, and the constant cross section fin studied in Section 1.6, which is approximately 1-D. The governing differential equation for these problems is an ordinary differential equation and the mathematics required to solve the problem are straightforward.

In this chapter, more complex, 2-D steady-state conduction problems are considered where the temperature varies in multiple spatial dimensions (e.g., x and y). These can be problems where the temperature actually varies in only two coordinates or approximately varies in only two coordinates (e.g., the temperature gradient in the third direction is negligible, as justified by an appropriate Biot number). The governing differential equation is a partial differential equation and therefore the mathematics required to analytically solve these problems are more advanced and the bookkeeping required to solve these problems numerically is more cumbersome. However, many of the concepts that were covered in the context of 1-D problems continue to apply.

2.1 Shape Factors

There are many 2-D and 3-D conduction problems involving heat transfer between two well-defined surfaces (surface 1 and surface 2) that commonly appear in heat transfer applications and have previously been solved analytically and/or numerically. The solution to these problems is conveniently expressed in the form of a shape factor, S, which is defined as:

$$S = \frac{1}{kR} \tag{2-1}$$

where k is the conductivity of the material separating the surfaces and R is the thermal resistance between surfaces 1 and 2. Solving Eq. (2-1) for the thermal resistance leads to:

$$R = \frac{1}{kS} \tag{2-2}$$

Recall that the resistance of a plane wall, derived in Section 1.2, is given by:

$$R_{pw} = \frac{L}{kA_c} \tag{2-3}$$

where L is the length of conduction path and A_c is the area for conduction. Comparing Eqs. (2-2) and (2-3) suggests that:

$$S \approx \frac{A_c}{L} \tag{2-4}$$

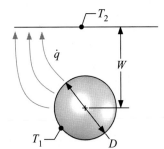

Figure 2-1: Sphere buried in a semi-infinite medium.

The shape factor has units of length and represents the ratio of the effective area for conduction to the effective length for conduction. Any shape factor solution should be checked against your intuition using Eq. (2-4). Given a problem, it should be possible to approximately identify the area and length that characterize the conduction process; the ratio of these quantities should have the same order of magnitude as the shape factor solution.

One example of a shape factor solution is for a sphere buried in a semi-infinite medium (i.e., a medium that extends forever in one direction but is bounded in the other) as shown in Figure 2-1. In this case, the surface of the sphere is surface 1 (assumed to be isothermal, at T_1) while the surface of the medium is surface 2 (assumed to be isothermal, at T_2).

The shape factor solution for a completely buried sphere in a semi-infinite medium is:

$$S = \frac{2\pi D}{1 - \dfrac{D}{4W}} \tag{2-5}$$

where D is the diameter of the sphere and W is the distance between the center of the sphere and the surface. The thermal resistance characterizing conduction between the surface of the sphere and the surface of the medium is:

$$R = \frac{1}{kS} = \frac{\left(1 - \dfrac{D}{4W}\right)}{2\pi D k} \tag{2-6}$$

The rate of conductive heat transfer between the sphere and the surface ($\dot{q}$) is:

$$\dot{q} = \frac{T_1 - T_2}{R} = \frac{2\pi D k (T_1 - T_2)}{\left(1 - \dfrac{D}{4W}\right)} \tag{2-7}$$

There are numerous formulae for shape factors that have been tabulated in various references; for example, Rohsenow et al. (1998). Table 2-1 summarizes a few shape factor solutions.

A library of shape factors, including those shown in Table 2-1 as well as others, has been integrated with EES. To access the shape factor library, select Function Information from the Options menu and then scroll through the list to Conduction Shape Factors, as shown in Figure 2-2. The shape factor functions that are available can be selected by moving the scroll bar below the picture.

Table 2-1: Shape factors.

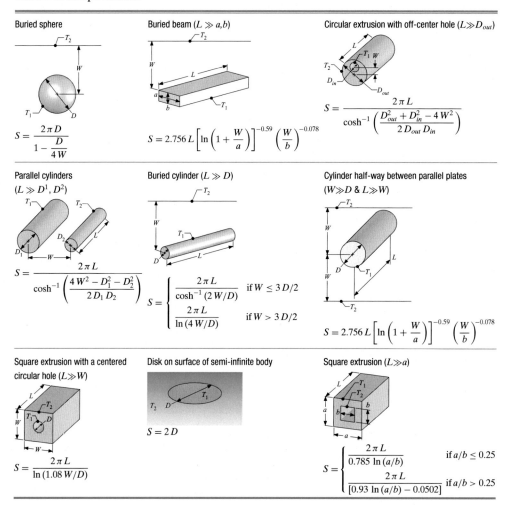

Buried sphere

$$S = \frac{2\pi D}{1 - \dfrac{D}{4W}}$$

Buried beam $(L \gg a,b)$

$$S = 2.756\,L\left[\ln\left(1 + \frac{W}{a}\right)\right]^{-0.59}\left(\frac{W}{b}\right)^{-0.078}$$

Circular extrusion with off-center hole $(L \gg D_{out})$

$$S = \frac{2\pi L}{\cosh^{-1}\left(\dfrac{D_{out}^2 + D_{in}^2 - 4W^2}{2\,D_{out}\,D_{in}}\right)}$$

Parallel cylinders $(L \gg D^1, D^2)$

$$S = \frac{2\pi L}{\cosh^{-1}\left(\dfrac{4W^2 - D_1^2 - D_2^2}{2\,D_1\,D_2}\right)}$$

Buried cylinder $(L \gg D)$

$$S = \begin{cases} \dfrac{2\pi L}{\cosh^{-1}(2W/D)} & \text{if } W \le 3D/2 \\[2ex] \dfrac{2\pi L}{\ln(4W/D)} & \text{if } W > 3D/2 \end{cases}$$

Cylinder half-way between parallel plates $(W \gg D\ \&\ L \gg W)$

$$S = 2.756\,L\left[\ln\left(1 + \frac{W}{a}\right)\right]^{-0.59}\left(\frac{W}{b}\right)^{-0.078}$$

Square extrusion with a centered circular hole $(L \gg W)$

$$S = \frac{2\pi L}{\ln(1.08\,W/D)}$$

Disk on surface of semi-infinite body

$$S = 2D$$

Square extrusion $(L \gg a)$

$$S = \begin{cases} \dfrac{2\pi L}{0.785\,\ln(a/b)} & \text{if } a/b \le 0.25 \\[2ex] \dfrac{2\pi L}{[0.93\,\ln(a/b) - 0.0502]} & \text{if } a/b > 0.25 \end{cases}$$

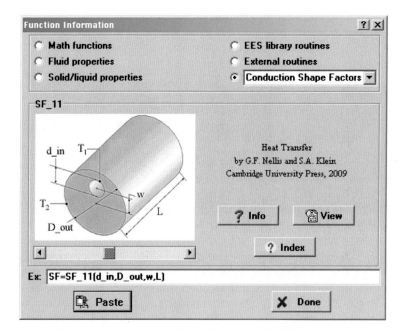

Figure 2-2: Accessing the shape factor library from EES.

EXAMPLE 2.1-1: MAGNETIC ABLATIVE POWER MEASUREMENT

This example revisits the magnetic ablation concept that was previously described in EXAMPLES 1.3-1 and 1.8-2. You want to measure the power generated by the thermoseed that is used for the ablation process. The radius of the thermoseed is $r_{ts} = 1.0$ mm. The sphere is placed $W = 5$ cm below the surface of a solution of agar, as shown in Figure 1. Agar is a material with well-known thermal properties that resembles gelatin and is sometimes used as a surrogate for tissue in biological experiments. The agar is allowed to solidify around the sphere and the container of agar is large enough to be considered semi-infinite. The surface of the agar is exposed to an ice-water bath in order to keep it at a constant temperature, $T_{ice} = 0°C$. The sphere is heated using an oscillating magnetic field and its surface temperature is measured using a thermocouple. The conductivity of agar is $k = 0.35$ W/m-K.

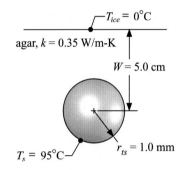

Figure 1: Test setup to measure the power generated by the thermoseed.

a) **If the measured surface temperature of the sphere is $T_s = 95°C$, how much energy is generated in the thermoseed?**

The inputs are entered in EES:

```
"EXAMPLE 2.1-1: Magnetic Ablative Power Measurement"

$UnitSystem SI MASS RAD PA K J
$Tabstops 0.2 0.4 0.6 3.5 in

"Inputs"
r_ts=1.0 [mm]*convert(mm,m)          "thermoseed radius"
W=5 [cm]*convert(cm,m)               "depth of sphere"
k=0.35 [W/m-K]                       "conductivity of agar"
T_ice=converttemp(C,K,0 [C])         "ice bath temperature"
T_s=converttemp(C,K,95 [C])          "surface temperature"
```

The shape factor associated with the buried sphere (S) is accessed from the EES library of shape factors:

```
S=SF_1(2*r_ts,W)                     "shape factor for buried sphere"
```

The thermal resistance between the surface of the sphere and the semi-infinite body (R) is:

$$R = \frac{1}{k\,S}$$

EXAMPLE 2.1-1: MAGNETIC ABLATIVE POWER MEASUREMENT

EXAMPLE 2.1-1: MAGNETIC ABLATIVE POWER MEASUREMENT

The heat transfer rate ($\dot{q}$) is computed using the thermal resistance and the known temperatures:

$$\dot{q} = \frac{T_s - T_{ice}}{R}$$

R=1/(k*S)   "thermal resistance"
q_dot=(T_s-T_ice)/R   "heat transfer rate"

which leads to a generation rate of 0.422 W.

b) **Estimate the uncertainty in your measurement of the power. Assume that your temperature measurements are accurate to $\delta T = 1.0°C$, the conductivity of agar is known to within 10% (i.e., $\delta k = 0.035$ W/m-K), the depth measurement has an uncertainty of $\delta W = 2.0$ mm, and the sphere radius is known to within $\delta r_{ts} = 0.1$ mm.**

It is possible to separately estimate the uncertainty introduced by each of the parameters listed above. For example, to evaluate the effect of the uncertainty in the conductivity, simply increase the conductivity by δk:

deltak=0.035 [W/m-K]   "uncertainty in conductivity"
k=0.35 [W/m-K]+deltak   "conductivity of agar"

and re-run the model. The result is a change in the generation rate from 0.422 W to 0.464 W which translates into an uncertainty in the power of 0.042 W or 10%. This process can be repeated for each of the independent variables in order to identify the uncertainty that is introduced into the dependent variable calculation. These contributions should be combined using the root-sum-square technique in order to obtain an overall uncertainty. This process can be carried out automatically in EES; select Uncertainty Propagation from the Calculate Menu to access the dialog shown in Figure 2.

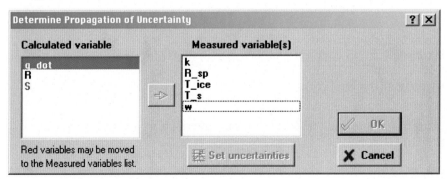

Figure 2: Propagation of uncertainty dialog.

The possible dependent (calculated) and independent (measured) variables are listed; highlight the independent variables that have some uncertainty (all of them for this problem). The calculated variable that you want to examine is **q_dot**. Select

EXAMPLE 2.1-1: MAGNETIC ABLATIVE POWER MEASUREMENT

the Set uncertainties button to reach the window shown in Figure 3. Set each of the uncertainties either in absolute or relative terms; note that each of these quantities can also be set using variables that are defined in the main equation window.

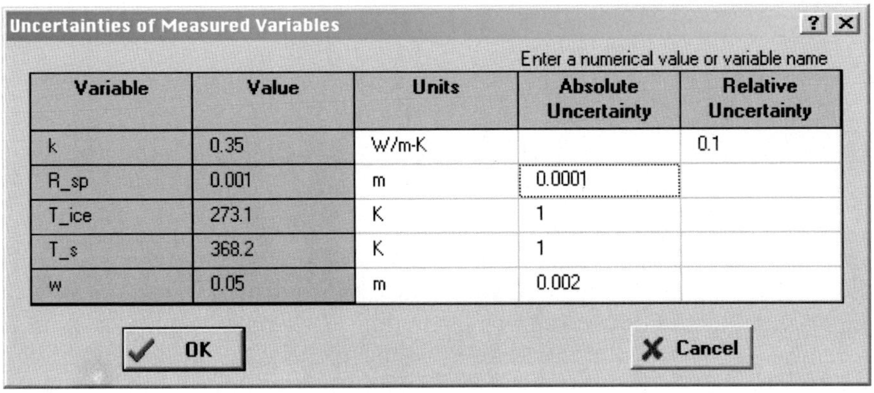

Uncertainties of Measured Variables				
			Enter a numerical value or variable name	
Variable	Value	Units	Absolute Uncertainty	Relative Uncertainty
k	0.35	W/m-K		0.1
R_sp	0.001	m	0.0001	
T_ice	273.1	K	1	
T_s	368.2	K	1	
w	0.05	m	0.002	

✓ OK   ✗ Cancel

Figure 3: Uncertainty of measured variables dialog.

Select OK twice in order to initiate the calculations; the results appear the Uncertainty Results window (Figure 4) which shows the total uncertainty in the variable q_dot (0.06 W or 14%) as well as a delineation of the source of the uncertainty.

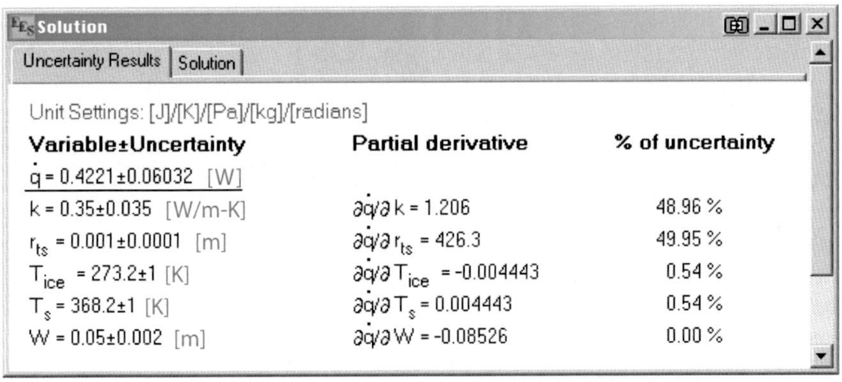

Solution

Uncertainty Results | Solution

Unit Settings: [J]/[K]/[Pa]/[kg]/[radians]

Variable±Uncertainty	Partial derivative	% of uncertainty
$\dot{q} = 0.4221 \pm 0.06032$ [W]		
$k = 0.35 \pm 0.035$ [W/m-K]	$\partial \dot{q}/\partial k = 1.206$	48.96 %
$r_{ts} = 0.001 \pm 0.0001$ [m]	$\partial \dot{q}/\partial r_{ts} = 426.3$	49.95 %
$T_{ice} = 273.2 \pm 1$ [K]	$\partial \dot{q}/\partial T_{ice} = -0.004443$	0.54 %
$T_s = 368.2 \pm 1$ [K]	$\partial \dot{q}/\partial T_s = 0.004443$	0.54 %
$W = 0.05 \pm 0.002$ [m]	$\partial \dot{q}/\partial W = -0.08526$	0.00 %

Figure 4: Uncertainty of Results window.

Notice that the dominant sources of uncertainty for this problem are the conductivity of agar and the radius of the sphere (each contributing approximately 50% of the total). The temperature measurements are adequate and the depth does not matter at all since the shape factor becomes nearly independent of the depth provided W is much larger than the diameter (see Eq. (2-5)).

2.2 Separation of Variables Solutions

2.2.1 Introduction

Two-dimensional steady-state conduction problems are governed by partial rather than ordinary differential equations; the analytical solution to partial differential equations

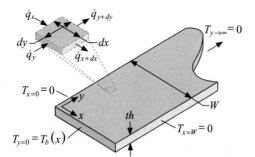

Figure 2-3: Plate.

is somewhat more involved. Separation of variables is a common technique that is used to solve the partial differential equations that arise in many areas of science and engineering. A complete understanding of separation of variables requires a substantial mathematical background; in this section, the technique is introduced and used to solve several problems. In the subsequent section, more difficult problems are solved using separation of variables. It is not possible to cover separation of variables thoroughly in this book and the interested reader is referred to the textbook by G. E. Myers (1998).

2.2.2 Separation of Variables

The method of separation of variables is most conveniently discussed in the context of a specific problem. In this section, the flat plate shown in Figure 2-3 is considered. The top and bottom surfaces of the plate are insulated and therefore there is no temperature variation in the z-direction; the problem is truly two-dimensional, as the temperature depends on x and y but not z. If the top and bottom surfaces were not insulated (e.g., they experienced convection to a surrounding fluid) but the plate was sufficiently thin and conductive, then it still might be possible to ignore temperature gradients in the z-direction and treat the problem as being two-dimensional. This assumption is equivalent to the extended surface assumption that was discussed in Section 1.6.2 and should be justified using an appropriately defined Biot number.

The plate in Figure 2-3 has conductivity k, thickness th, width (in the x-direction) W, and extends to infinity in the y-direction. The governing differential equation for the problem is derived in a manner that is analogous to the 1-D problems that have been previously considered. A differential control volume is defined (see Figure 2-3) and used to develop a steady-state energy balance. Note that the control volume must be differential in both the x- and y-directions because there are temperature gradients in both of these directions.

$$\dot{q}_x + \dot{q}_y = \dot{q}_{x+dx} + \dot{q}_{y+dy} \tag{2-8}$$

The $x + dx$ and $y + dy$ terms are expanded as usual:

$$\dot{q}_x + \dot{q}_y = \dot{q}_x + \frac{\partial \dot{q}_x}{\partial x} dx + \dot{q}_y + \frac{\partial \dot{q}_y}{\partial y} dy \tag{2-9}$$

Equation (2-9) can be simplified to:

$$\frac{\partial \dot{q}_x}{\partial x} dx + \frac{\partial \dot{q}_y}{\partial y} dy = 0 \tag{2-10}$$

Fourier's law is used to determine the conduction heat transfer rates in the x- and y-directions:

$$\dot{q}_x = -k\, th\, dy\, \frac{\partial T}{\partial x} \tag{2-11}$$

$$\dot{q}_y = -k\, th\, dx\, \frac{\partial T}{\partial y} \tag{2-12}$$

Equations (2-11) and (2-12) are substituted into Eq. (2-10):

$$\frac{\partial}{\partial x}\left[-k\, th\, dy\, \frac{\partial T}{\partial x}\right] dx + \frac{\partial}{\partial y}\left[-k\, th\, dx\, \frac{\partial T}{\partial y}\right] dy = 0 \tag{2-13}$$

If the thermal conductivity and plate thickness are both constant, then Eq. (2-13) can be simplified to:

$$\boxed{\frac{\partial^2 T}{\partial x^2} + \frac{\partial^2 T}{\partial y^2} = 0} \tag{2-14}$$

which is the governing partial differential equation for this problem. Equation (2-14) is called Laplace's equation. Equation (2-14) is second order in both the x- and y-directions and therefore two boundary conditions are required in each of these directions. The left and right edges of the plate have a temperature of zero:

$$\boxed{T_{x=0} = 0} \tag{2-15}$$

$$\boxed{T_{x=W} = 0} \tag{2-16}$$

The zero temperature boundaries are necessary here to ensure that the boundary conditions are homogeneous, as explained below. However, these boundary conditions are likely not of general interest. Techniques that allow the solution of problems with more realistic boundary conditions are presented in subsequent sections.

The edge of the plate at $y = 0$ has a specified temperature that is an arbitrary function of position x:

$$\boxed{T_{y=0} = T_b(x)} \tag{2-17}$$

The plate is infinitely long in the y-direction and the temperature approaches 0 as y becomes infinite:

$$\boxed{T_{y\to\infty} = 0} \tag{2-18}$$

Requirements for using Separation of Variables
The method of separation of variables will not work for every problem; there are some fairly restrictive conditions that limit where it can be applied. First, the governing equation must be linear; that is, the equation cannot contain any products of the dependent variable or its derivative. Equation (2-14) is certainly linear. An example of a non-linear equation might be:

$$T\frac{\partial^2 T}{\partial x^2} + \frac{\partial T}{\partial x}\frac{\partial^2 T}{\partial y^2} = 0 \tag{2-19}$$

The governing equation must also be homogeneous, which is a more restrictive condition. If T is a solution to a homogeneous equation then $C\,T$ is also a solution, where C

is an arbitrary constant. Equation (2-14) is homogeneous; to prove this, simply check if CT can be substituted into the equation and still satisfy the equality:

$$\frac{\partial^2 (CT)}{\partial x^2} + \frac{\partial^2 (CT)}{\partial y^2} = 0 \tag{2-20}$$

or

$$C \underbrace{\left(\frac{\partial^2 T}{\partial x^2} + \frac{\partial^2 T}{\partial y^2} \right)}_{=0 \text{ according to Eq. (2-14)}} = 0 \tag{2-21}$$

A non-homogeneous equation would result, for example, if the plate were exposed to a volumetric generation of thermal energy ($\dot{g}'''$); the governing differential equation for this situation would be:

$$\frac{\partial^2 T}{\partial x^2} + \frac{\partial^2 T}{\partial y^2} + \frac{\dot{g}'''}{k} = 0 \tag{2-22}$$

Substituting CT into Eq. (2-22) leads to:

$$C \underbrace{\left(\frac{\partial^2 T}{\partial x^2} + \frac{\partial^2 T}{\partial y^2} \right)}_{= -\frac{\dot{g}'''}{k} \text{ according to Eq. (2-22)}} + \frac{\dot{g}'''}{k} = 0 \tag{2-23}$$

Substituting Eq. (2-22) into Eq. (2-23) leads to:

$$C \left(-\frac{\dot{g}'''}{k} \right) + \frac{\dot{g}'''}{k} = 0 \tag{2-24}$$

Equation (2-24) shows that CT is a solution to Eq. (2-22) only if $\dot{g}''' = 0$. We will show how some non-homogeneous problems can be solved using separation of variables in Section 2.3.

In order to apply the separation of variables method, the boundary conditions must also be linear with respect to the dependent variable. Linear has the same definition for the boundary conditions that it does for the differential equation; i.e., the boundary condition cannot involve products of the dependent variable or its derivatives. The boundary conditions for the plate in Figure 2-3 are given by Eqs. (2-15) through (2-18) and are all linear. A non-linear boundary condition would result from, for example, radiation. If the right edge of the plate were radiating to surroundings at $T = 0$ then Eq. (2-15) should be replaced with:

$$-k \left. \frac{\partial T}{\partial x} \right|_{x=W} = \sigma \varepsilon T_{x=W}^4 \tag{2-25}$$

which is non-linear. Finally, both of the boundary conditions in one direction must be homogeneous (i.e., either both boundary conditions in the x-direction or both boundary conditions in the y-direction). Again, the meaning of homogeneity for a boundary condition is analogous to its meaning for the differential equation. If a boundary condition is homogeneous then any multiple of a solution also satisfies the boundary condition. Examination shows that all of the boundary conditions except for Eq. (2-17) are homogeneous. Therefore, both boundary conditions in the x-direction, Eqs. (2-15) and (2-16), are homogeneous. For this problem, x is therefore the homogeneous direction; this will be important to keep in mind as we solve the problem.

A final criterion for the use of separation of variables is that the computational domain must be simple; that is, it must have boundaries that lie along constant values of the coordinate axes. For a Cartesian coordinate system, we are restricted to rectangular problems. The problem shown in Figure 2-3 meets all of the criteria and should therefore be solvable using separation of variables.

Separate the Variables

The name of the technique, separation of variables, is related to the next step in the solution; it is assumed that the solution T, which is a function of both x and y, can be expressed as the product of two functions, TX which is only a function of x and TY which is only a function of y:

$$T(x, y) = TX(x) \; TY(y) \tag{2-26}$$

Substituting Eq. (2-26) into Eq. (2-14) leads to:

$$\frac{\partial^2}{\partial x^2}[TX \, TY] + \frac{\partial^2}{\partial y^2}[TX \, TY] = 0 \tag{2-27}$$

or

$$TY \frac{d^2 TX}{dx^2} + TX \frac{d^2 TY}{dy^2} = 0 \tag{2-28}$$

Dividing through by the product $TY \, TX$ leads to:

$$\underbrace{\frac{\dfrac{d^2 TX}{dx^2}}{TX}}_{\text{function of } x} + \underbrace{\frac{\dfrac{d^2 TY}{dy^2}}{TY}}_{\text{function of } y} = 0 \tag{2-29}$$

The first term in Eq. (2-29) is a function only of x while the second is a function only of y. Therefore, Eq. (2-29) can only be satisfied if both terms are equal and opposite and constant. To see this clearly, imagine moving along a line of constant y (i.e., across the plate in Figure 2-3 in the x-direction from one side to the other). If the first term were not constant then, by definition, its value would change as x changes; however, the second term is not a function of x and therefore it cannot change in response. Clearly then the sum of the two terms could not continue to be zero in this situation. Equation (2-29) can be expressed as two statements:

$$\frac{\dfrac{d^2 TX}{dx^2}}{TX} = \pm \lambda^2 \tag{2-30}$$

$$\frac{\dfrac{d^2 TY}{dy^2}}{TY} = \mp \lambda^2 \tag{2-31}$$

where λ^2 is a constant that must be positive. Notice that there is a choice that must be made at this point. The TX group can either be set equal to a positive constant (λ^2) or a negative constant $(-\lambda^2)$. Depending on this choice, the TY group must be set equal to a negative constant $(-\lambda^2)$ or a positive constant (λ^2), in order to satisfy Eq. (2-29). The choice at this point seems arbitrary but in fact it is important.

Recall that one condition for using separation of variables is that one of the coordinate directions must have homogeneous boundary conditions; this was referred to as the homogeneous direction. In this problem, the x-direction is the homogeneous direction because the boundary conditions at $x = 0$, Eq. (2-15), and at $x = W$, Eq. (2-16), are both homogeneous. It is necessary to choose the negative constant for the group associated with the homogeneous direction (i.e., for Eq. (2-30) in this problem). With this choice, rearranging Eqs. (2-30) and (2-31) leads to the two ordinary differential equations:

$$\frac{d^2 TX}{dx^2} + \lambda^2 \, TX = 0 \tag{2-32}$$

$$\frac{d^2 TY}{dy^2} - \lambda^2 \, TY = 0 \tag{2-33}$$

We have effectively converted our partial differential equation, Eq. (2-14), into two ordinary differential equations, Eqs. (2-32) and (2-33). The solutions to Eqs. (2-32) and (2-33) can be identified using Maple:

```
> restart;
> ODEX:=diff(diff(TX(x),x),x)+lambda^2*TX(x)=0;
```

$$ODEX = \left(\frac{d^2}{dx^2} TX(x)\right) + \lambda^2 TX(x) = 0$$

```
> Xs:=dsolve(ODEX);
```

$$Xs = TX(x) = \_C1 \sin(\lambda x) + \_C2 \cos(\lambda x)$$

```
> ODEY:=diff(diff(TY(y),y),y)-lambda^2*TY(y)=0;
```

$$ODEY := \left(\frac{d^2}{dy^2} TY(Y)\right) - \lambda^2 TY(Y) = 0$$

```
> Ys:=dsolve(ODEY);
```

$$Ys := TY(y) = \_C1 e^{(-\lambda Y)} + \_C2 e^{(\lambda Y)}$$

So the solution for TX (i.e., the solution in the homogeneous direction) is:

$$TX = C_1 \sin(\lambda x) + C_2 \cos(\lambda x) \tag{2-34}$$

where C_1 and C_2 are undetermined constants. The solution for TY (i.e., in the non-homogeneous direction) is:

$$TY = C_3 \exp(-\lambda y) + C_4 \exp(\lambda y) \tag{2-35}$$

where C_3 and C_4 are undetermined constants. Note that Eq. (2-35) could equivalently be expressed in terms of hyperbolic sines and cosines:

$$TY = C_3 \sinh(\lambda y) + C_4 \cosh(\lambda y) \tag{2-36}$$

where C_3 and C_4 are undetermined constants (different from those in Eq. (2-35)).

The choice of the negative value of the constant for the homogeneous direction (i.e., for Eq. (2-30)) has led directly to sine/cosine solutions in the homogeneous direction; this result is necessary in order to use separation of variables.

Solve the Eigenproblem
It is necessary to address the solution in the homogeneous direction (in this problem, TX) before moving on to the non-homogeneous direction. This portion of the problem is often called the eigenproblem and the solutions are referred to as eigenfunctions.

The boundary conditions for TX can be obtained by revisiting the original boundary conditions for the problem in the x-direction using the assumed, separated form of the solution. Equation (2-15) becomes:

$$TX_{x=0} \, TY = 0 \tag{2-37}$$

which can only be true at an arbitrary location y if:

$$TX_{x=0} = 0 \qquad (2\text{-}38)$$

The remaining boundary condition in the x-direction is given by Eq. (2-16) and leads to:

$$TX_{x=W} = 0 \qquad (2\text{-}39)$$

Substituting the solution to the ordinary differential equation in the homogeneous direction, Eq. (2-34), into Eq. (2-38) leads to:

$$C_1 \underbrace{\sin{(\lambda\,0)}}_{0} + C_2 \underbrace{\cos{(\lambda\,0)}}_{1} = 0 \qquad (2\text{-}40)$$

or

$$C_2 = 0 \qquad (2\text{-}41)$$

So that:

$$TX = C_1 \sin{(\lambda\,x)} \qquad (2\text{-}42)$$

Substituting Eq. (2-42) into Eq. (2-39) leads to:

$$C_1 \sin{(\lambda\,W)} = 0 \qquad (2\text{-}43)$$

Equation (2-43) could be satisfied if C_1 is 0, but that would lead to $TX = 0$ (and therefore $T = 0$) everywhere, which is not a useful solution. However, Eq. (2-43) is also satisfied whenever the sine function becomes zero; this occurs whenever the argument of sine is an integer multiple of π:

$$\lambda_i\,W = i\,\pi \quad \text{where } i = 0, 1, 2, \ldots \infty \qquad (2\text{-}44)$$

Equation (2-43) satisfies the eigenproblem (i.e., the ordinary differential equation in the x-direction, Eq. (2-32), and both boundary conditions in the x-direction, Eqs. (2-38) and (2-39)) for each value of λ_i identified by Eq. (2-44). Note that every negative integer will also satisfy the eigenproblem; however, the positive integers provide an infinite number of terms and therefore are sufficient.

$$TX_i = C_{1,i} \sin{(\lambda_i\,x)} \quad \text{where } \lambda_i = \frac{i\,\pi}{W} \quad i = 1, 2, \ldots \infty \qquad (2\text{-}45)$$

Note that the $i = 0$ case is not included in Eq. (2-45) because the sine of 0 is zero; therefore this solution does not provide any useful information. The functions TX_i given by Eq. (2-45) are referred to as the eigenfunctions that solve the linear, homogeneous problem for TX and the values λ_i are the eigenvalues associated with each eigenfunction. The function $\sin(i\,\pi x/W)$ is referred to as the ith eigenfunction and $\lambda_i = i\,\pi/W$ is the ith eigenvalue.

Solve the Non-homogeneous Problem for each Eigenvalue
With the eigenproblem solved, it is necessary to return to the non-homogeneous portion of the problem, TY. Each of the eigenvalues identified by Eq. (2-44) is associated with an ordinary differential equation in the y-direction according to Eq. (2-33):

$$\frac{d^2 TY_i}{dy^2} - \lambda_i^2\,TY_i = 0 \qquad (2\text{-}46)$$

These ordinary differential equations have either an exponential or hyperbolic solution according to Eqs. (2-35) or (2-36). The choice of one form over the other is arbitrary and either will lead to the same solution. Because one of the boundary conditions is at $y \to \infty$, the exponentials will provide a more concise solution for this problem.

However, in most other cases, the sinh and cosh solution will be easier to work with. The solution for TY_i is:

$$TY_i = C_{3,i} \exp(-\lambda_i y) + C_{4,i} \exp(\lambda_i y) \tag{2-47}$$

Obtain Solution for each Eigenvalue
According to Eq. (2-26), the solution for temperature associated with the ith eigenvalue is:

$$T_i = TX_i \, TY_i = C_{1,i} \sin(\lambda_i x) [C_{3,i} \exp(-\lambda_i y) + C_{4,i} \exp(\lambda_i y)] \tag{2-48}$$

The products of the undetermined constants $C_{1,i} C_{3,i}$ and $C_{1,i} C_{4,i}$ are also undetermined constants and therefore Eq. (2-48) can be written as:

$$T_i = \sin(\lambda_i x) [C_{3,i} \exp(-\lambda_i y) + C_{4,i} \exp(\lambda_i y)] \tag{2-49}$$

Equation (2-49) will, for any value of i, satisfy the governing differential equation, Eq. (2-14), and satisfy both of the boundary conditions in the homogeneous direction, Eqs. (2-15) and Eq. (2-16). This is a typical outcome of solving the eigenproblem: a set of solutions that each satisfy the governing partial differential equation and all of the boundary conditions in the homogeneous direction. It is worth checking that your solution has these properties using Maple.

First, it is necessary to let Maple know that i is an integer using the **assume** command.

```
> restart;
> assume(i,integer);
```

Next, define λ_i according to Eq. (2-44):

```
> lambda:=i*Pi/W;
```

$$\lambda := \frac{i \sim \pi}{W}$$

Note the use of Pi rather than pi in the Maple code; Pi indicates that π should be evaluated symbolically whereas pi is the numerical value of π. Create a function T in the independent variables x and y according to Eq. (2-49):

```
> T:=(x,y)->sin(lambda*x)*(C3*exp(-lambda*y)+C4*exp(lambda*y));
```

$$T := (x, y) \rightarrow \sin(\lambda x)(C3 e^{(-\lambda y)} + C4 e^{(\lambda y)})$$

You can verify that the two homogeneous direction boundary conditions, Eqs. (2-15) and (2-16), are satisfied:

```
> T(0,y);                                    0

> T(W,y);                                    0
```

and also that the partial differential equation, Eq. (2-14), is satisfied;

```
> diff(diff(T(x,y),x),x)+diff(diff(T(x,y),y),y);
```

$$-\frac{\sin\left(\dfrac{i\sim\pi x}{W}\right)i\sim^2\pi^2\left(C3e^{\left(-\frac{i\sim\pi y}{W}\right)}+C4e^{\left(\frac{i\sim\pi y}{W}\right)}\right)}{W^2}$$

$$+\sin\left(\frac{i\sim\pi x}{W}\right)\left(\frac{C3\,i\sim^2\pi^2 e^{\left(-\frac{i\sim\pi y}{W}\right)}}{W^2}+\frac{C4\,i\sim^2\pi^2 e^{\left(\frac{i\sim\pi y}{W}\right)}}{W^2}\right)$$

```
> simplify(%);
```

$$0$$

Create the Series Solution and Enforce the Remaining Boundary Conditions

Because the partial differential equation is linear and homogeneous, the sum of the solutions for each eigenvalue, T_i given by Eq. (2-49), is itself a solution:

$$T = \sum_{i=1}^{\infty} T_i = \sum_{i=1}^{\infty} \sin(\lambda_i x)\left[C_{3,i}\exp(-\lambda_i y) + C_{4,i}\exp(\lambda_i y)\right] \tag{2-50}$$

The final step of the solution selects the constants so that the boundary conditions in the non-homogeneous direction are satisfied. Equation (2-18) provides the boundary condition as y approaches infinity; substituting Eq. (2-50) into Eq. (2-18) leads to:

$$T_{y\to\infty} = \sum_{i=1}^{\infty} \sin(\lambda_i x)\left[C_{3,i}\underbrace{\exp(-\infty)}_{0} + C_{4,i}\underbrace{\exp(\infty)}_{\infty}\right] = 0 \tag{2-51}$$

or

$$\sum_{i=1}^{\infty} \sin(\lambda_i x)\,C_{4,i}\,\infty = 0 \tag{2-52}$$

Equation (2-52) can be solved by inspection; the equality can only be satisfied if $C_{4,i} = 0$ for all i.

$$T = \sum_{i=1}^{\infty} C_{3,i}\,\sin(\lambda_i x)\,\exp(-\lambda_i y) \tag{2-53}$$

Because only $C_{3,i}$ remains in our solution it is no longer necessary to designate it as the third undetermined constant:

$$T = \sum_{i=1}^{\infty} C_i\,\sin(\lambda_i x)\,\exp(-\lambda_i y) \tag{2-54}$$

Equation (2-17) provides the boundary condition at $y = 0$; substituting Eq. (2-54) into Eq. (2-17) leads to:

$$T_{y=0} = \sum_{i=1}^{\infty} C_i\,\sin(\lambda_i x)\,\underbrace{\exp(-\lambda_i 0)}_{1} = T_b(x) \tag{2-55}$$

or

$$\sum_{i=1}^{\infty} C_i \sin(\lambda_i x) = T_b(x) \qquad (2\text{-}56)$$

Equation (2-56) defines the constants in the solution; they are the Fourier coefficients of the non-homogeneous boundary condition. At first glance, it may seem like we have not really come very far. The solution to the problem is certainly provided by Eqs. (2-54) and (2-56), however an infinite number of unknown constants, C_i, are needed to evaluate this solution and it is not clear how Eq. (2-56) can be manipulated in order to evaluate these constants. Fortunately the eigenfunctions have the property of orthogonality, which makes it relatively easy to determine the constants C_i.

The meaning of orthogonality becomes evident when Eq. (2-56) is multiplied by a single eigenfunction, say the jth one, and then integrated in the homogeneous direction from one boundary to the other (i.e., from $x = 0$ to $x = W$):

$$\sum_{i=1}^{\infty} C_i \int_{0}^{W} \sin\left(\frac{i\pi}{W} x\right) \sin\left(\frac{j\pi}{W} x\right) dx = \int_{0}^{W} T_b(x) \sin\left(\frac{j\pi}{W} x\right) dx \qquad (2\text{-}57)$$

The property of orthogonality guarantees that the only term in the summation on the left side of Eq. (2-57) that will not integrate to zero is the one where $i = j$. We can verify this result by consulting a table of integrals. The integral of the product of two sine functions is:

$$\int_{0}^{W} \sin\left(\frac{i\pi}{W} x\right) \sin\left(\frac{j\pi}{W} x\right) dx = \frac{W}{2\pi} \left[\frac{\sin\left(\frac{\pi(j-i)x}{W}\right)}{(j-i)} - \frac{\sin\left(\frac{\pi(i+j)x}{W}\right)}{(i+j)} \right]_{0}^{W} \qquad (2\text{-}58)$$

This result can be obtained using Maple:

```
> assume(j,integer);
> assume(i,integer);
> int(sin(i*Pi*x/W)*sin(j*Pi*x/W),x);
```

$$\frac{1}{2} \frac{W \sin\left(\frac{\pi(-i+j)x}{W}\right)}{\pi(-i+j)} - \frac{1}{2} \frac{W \sin\left(\frac{\pi(i+j)x}{W}\right)}{\pi(i+j)}$$

Applying the limits of integration to Eq. (2-58) leads to:

$$\int_{0}^{W} \sin\left(\frac{i\pi}{W} x\right) \sin\left(\frac{j\pi}{W} x\right) dx = \frac{W}{2\pi} \left[\frac{\sin(\pi(j-i))}{(j-i)} - \frac{\sin(\pi(i+j))}{(i+j)} \right] \qquad (2\text{-}59)$$

The term involving $\sin(\pi(i+j))$ must always be zero for any positive integer values of i and j because the sine of any integer multiple of π is zero. The first term on the right side of Eq. (2-59) will also be zero provided $j \neq i$. However, if $j = i$ then both the numerator and denominator of this term are zero and the value of the integral is not obvious. In the limit that $j = i$, the value of the integral in Eq. (2-59) is:

$$\int_0^W \sin^2\left(\frac{i\pi}{W}x\right)dx = \frac{W}{2\pi}\underbrace{\lim_{(j-i)\to 0}\left[\frac{\sin(\pi(j-i))}{(j-i)}\right]}_{\pi} \tag{2-60}$$

The limit of the bracketed term in Eq. (2-60) can be evaluated using Maple:

```
> restart;
> limit(sin(Pi*x)/x,x=0);
                              π
```

which leads to:

$$\int_0^W \sin^2\left(\frac{i\pi}{W}x\right)dx = \frac{W}{2} \tag{2-61}$$

This behavior lies at the heart of orthogonality: the integral of the product of two different eigenfunctions between the homogeneous boundary conditions will always be zero while the integral of any eigenfunction multiplied by itself between the same boundary conditions will not be zero. It is possible to prove that this behavior is generally true for any solution in the homogeneous direction (see Myers (1987)). The orthogonality of the eigenfunctions simplifies the problem considerably because it allows the summation in Eq. (2-57) to be replaced by a single integration. (All of the terms on the left hand side where $j \neq i$ must integrate to zero and disappear.)

$$C_i \int_0^W \sin\left(\frac{i\pi}{W}x\right)\sin\left(\frac{i\pi}{W}x\right)dx = \int_0^W T_b(x)\sin\left(\frac{i\pi}{W}x\right)dx \quad \text{for } i = 1..\infty \tag{2-62}$$

The only term in the sum that has been retained is the integration of the eigenfunction $\sin(i\pi x/W)$ multiplied by itself; notice that Eq. (2-62) provides a single equation for each of the constants C_i and therefore completes the solution.

A more physical feel for the orthogonality of the eigenfunctions can be obtained by examining various eigenfunctions and their products. For example, Figure 2-4 shows the first eigenfunction, $\sin(\pi x/W)$, and the second eigenfunction, $\sin(2\pi x/W)$. Also shown in Figure 2-4 is the product of these two eigenfunctions; notice that the integral of the product must be zero as the areas above and below the axis are equal.

Figure 2-5 illustrates the behavior of the second and fourth eigenfunctions and their product; while their oscillations are more complex, it is clear that the product of these eigenfunctions must also integrate to zero. Finally, Figure 2-6 shows the behavior of the third eigenfunction and its value squared; the square of any real valued function is never negative and therefore it cannot integrate to zero.

The eigenfunctions that are appropriate for other problems with different boundary conditions will not be $\sin(i\pi x/W)$. However, they will be orthogonal functions. As a result, the sum that defines the constants in the series can always be reduced to one equation that allows each constant to be evaluated; the equivalent of Eq. (2-57) can always be reduced to the equivalent of Eq. (2-62).

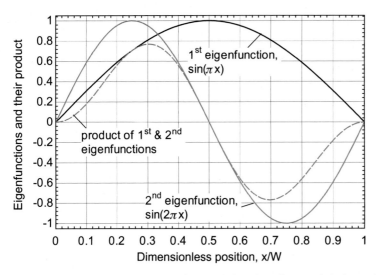

Figure 2-4: Behavior of the first and second eigenfunctions and their product.

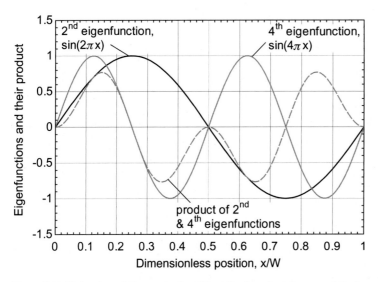

Figure 2-5: Behavior of the second and fourth eigenfunctions and their product.

Equation (2-62) provides an integral equation that can be used to evaluate each of the coefficients:

$$C_i = \frac{\int_0^W T_b(x) \sin\left(\frac{i\pi}{W}x\right) dx}{\int_0^W \sin^2\left(\frac{i\pi}{W}x\right) dx} \qquad i = 1..\infty \qquad (2\text{-}63)$$

Still, it is necessary to determine both of the integrals in Eq. (2-63) in order to evaluate each coefficient. In some cases, the integrals can be evaluated by inspection or by the use of mathematical tables; usually these integrals can be evaluated easily with the aid of Maple. The integral in the denominator of Eq. (2-63) was previously evaluated in

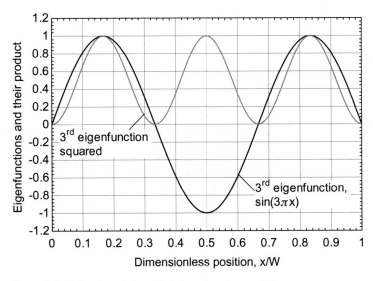

Figure 2-6: Behavior of the third eigenfunction and its square.

Eq. (2-61). The integral could also be evaluated with the aid of trigonometric identities and integral tables:

$$\int_0^W \sin^2\left(\frac{i\pi}{W}x\right) dx = \int_0^W \left[\frac{1}{2} - \cos\left(\frac{2i\pi}{W}x\right)\right] dx = \left[\frac{x}{2} - \frac{W}{2i\pi}\sin\left(\frac{2i\pi}{W}x\right)\right]_0^W = \frac{W}{2}$$

(2-64)

Maple makes this process much easier:

```
> int((sin(i*Pi*x/W))^2,x=0..W);
```

$$\frac{W}{2}$$

The ith coefficient is therefore:

$$C_i = \frac{2}{W}\int_0^W T_b(x)\sin\left(\frac{i\pi}{W}x\right) dx$$

(2-65)

The remaining integral in Eq. (2-65) depends on the functional form of the boundary condition. The simplest possibility is a constant temperature, T_b, which leads to:

$$C_i = \frac{2T_b}{W}\int_0^W \sin\left(\frac{i\pi}{W}x\right) dx = -\frac{2T_b}{i\pi}\left[\cos\left(\frac{i\pi}{W}x\right)\right]_0^W = \frac{2T_b}{i\pi}[1 - \cos(i\pi)]$$

(2-66)

or, using Maple:

```
> 2*Tb*int(sin(i*pi*x/W),x=0..W)/W;
```

$$-\frac{2Tb(-1 + \cos(i\pi))}{i\pi}$$

Substituting Eq. (2-66) into Eq. (2-54) leads to:

$$T = \sum_{i=1}^{\infty} \frac{2\,T_b}{i\,\pi}\left[1 - \cos\left(i\,\pi\right)\right] \sin\left(\frac{i\,\pi}{W}x\right) \exp\left(-\frac{i\,\pi}{W}y\right) \tag{2-67}$$

It is usually more convenient to let Maple carry out the symbolic math and then evaluate the solution using EES. The input parameters are entered in EES:

```
$UnitSystem SI MASS RAD PA K J
$Tabstops 0.2 0.4 0.6 3.5 in

"Inputs"
W=1.0 [m]                        "width of plate"
k=10 [W/m-K]                     "conductivity of plate"
T_b=1 [K]                        "base temperature"
```

The temperature is evaluated at an arbitrary location:

```
"position to evaluate temperature"
x=0.1 [m]                        "x-position"
y=0.25 [m]                       "y-position"
```

Each of the first N terms of the series solution in Eq. (2-67) are evaluated using a duplicate loop. The coefficient for each term is evaluated using the formula obtained in Maple by copying and pasting it into EES.

```
N=10 [-]                         "number of terms in the solution to evaluate"
duplicate i=1,N
  C[i]=-2*T_b*(-1+cos(i*pi))/(i*pi)    "constant for i'th term in the series"
  T[i]=C[i]*sin(i*pi*x/W)*exp(-i*pi*y/W)   "i'th term in the series"
end
```

The terms in the series are summed using the sum function in EES:

```
T=sum(T[1..N])                   "sum of N terms in the series"
```

A parametric table is created in order to examine the temperature as a function of position along the bottom of the plate, $y = 0$. The ability of the solution to match the imposed boundary condition depends on the number of terms that are used. Figure 2-7 illustrates the solution with 5, 10, and 100 terms.

A contour plot of the temperature distribution is generated by creating a parametric table containing the variables x, y, and T that includes 400 runs. Click on the arrow in the x-column header in order to bring up the dialog that allows values to be automatically entered into the table. Click the check box for the option to repeat the pattern of running x from 0 to 1 every 20 rows, as shown in Figure 2-8(a). Repeat the process for the y column, but in this case apply the pattern of running y from 0 to 1 every 20 rows, as shown in Figure 2-8(b).

When the table is solved, the solution is obtained over a 20×20 grid ranging from 0 to 1 in both x and y. Select X-Y-Z Plot from the New Plot Window selection under the Plots menu and select Isometric Lines to generate the contour plot shown in Figure 2-9.

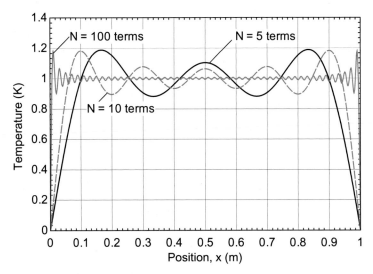

Figure 2-7: Temperature as a function of x at $y = 0$ for different values of N.

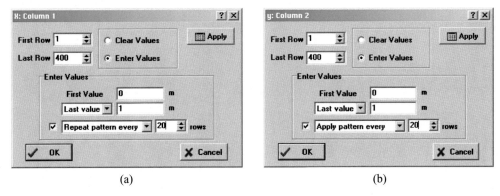

(a)   (b)

Figure 2-8: Automatically entering repeating values for (a) x and (b) y.

More complicated boundary conditions can be considered at $y = 0$. For example, the temperature may vary linearly from 0 to 1 K according to:

$$T_b(x) = T_b \frac{x}{W} \tag{2-68}$$

The coefficients in the general solution, Eq. (2-54) are obtained by substituting Eq. (2-68) into Eq. (2-65):

$$C_i = \frac{2 T_b}{W^2} \int_0^W x \, \sin\left(\frac{i\pi}{W} x\right) dx \tag{2-69}$$

which can be evaluated using Maple:

```
> restart;
> assume(i,integer);
> C[i]:=2*T_b*int(x*sin(i*pi*x/W),x=0..W)/W^2;
```

$$C_{i\sim} := -\frac{2\, T\_b(-\sin(i \sim \pi) + \cos(i \sim \pi)i \sim \pi)}{i \sim^2 \pi^2}$$

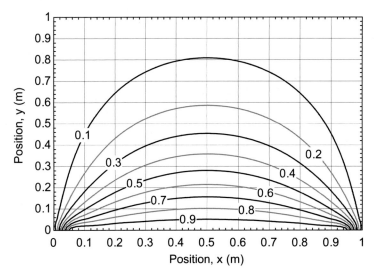

Figure 2-9: Contour plot of temperature distribution for constant temperature boundary.

The Maple result is copied and pasted into EES to achieve:

```
N=100 [-]                                        "number of terms in the solution to evaluate"
duplicate i=1,N
{   C[i]=-2*T_b*(-1+cos(i*pi))/(i*pi)
      "constant for i'th term in the series, constant temp. boundary"}
    C[i]=-2*T_b*(-sin(i*pi)+cos(i*pi)*i*pi)/i^2/pi^2
        "constant for i'th term in the series, linear variation in boundary temp."
    T[i]=C[i]*sin(i*pi*x/W)*exp(-i*pi*y/W)       "i'th term in the series"
end
T=sum(T[1..N])                                   "sum of N terms in the series"
```

It is obviously not possible to include an infinite number of terms in the solution and therefore a natural question is: how many terms are sufficient? The magnitude of the neglected terms can be assessed by considering the magnitude of the last term that was included. For example, Figure 2-10 shows the Arrays Table that results when the solution is evaluated at $x = 0.1$ m and $y = 0.25$ m with $N = 11$ terms. The size of the terms in the solution drop dramatically as the index of the term increases. The accuracy of the solution computed using 11 terms is within 3.2×10^{-6} K of the actual solution and therefore it is clear that only a few terms are required at this position.

However, the number of terms that are required depends on the position within the computational domain. More terms are typically required to resolve the solution near the boundary and, in particular, near boundaries where non-physical conditions are being enforced. For example, in this problem we are requiring that an edge at $T = 0$ (i.e., the left edge) intersect with an edge at $T = 1$ (i.e., the bottom edge) which results in an infinite temperature gradient at $x = y = 0$ that cannot physically exist.

Summary of Steps

The steps required to solve a problem using separation of variables are summarized below:

1. Verify that the problem satisfies all of the conditions that are required for separation of variables. The partial differential equation must be linear and homogeneous,

Sort	C_i	T_i
[1]	0.6366	0.0897
[2]	-0.3183	-0.03889
[3]	0.2122	0.01627
[4]	-0.1592	-0.006541
[5]	0.1273	0.002509
[6]	-0.1061	-0.0009065
[7]	0.09095	0.0003014
[8]	-0.07958	-0.00008735
[9]	0.07074	0.00001861
[10]	-0.06366	5.110E-18
[11]	0.05787	-0.000003165

Figure 2-10: Arrays Table containing the solution terms for $x = 0.1$ m and $y = 0.25$ m and $N = 11$ terms.

all boundary conditions must be linear, and both boundary conditions in one direction (the homogeneous direction) must be homogeneous. If the problem does not meet these requirements then it may be possible to apply a simple transformation to the boundary conditions (as discussed in Section 2.2.3), use superposition (as discussed in Section 2.4), or carefully divide the problem into its homogeneous and non-homogeneous parts (as discussed in Section 2.3.2).

2. Separate the variables; that is, express the solution (T) as the product of a function of x (TX) and a function of y (TY). Use this approach to split the partial differential equation into two ordinary differential equations; the ordinary differential equation in the homogeneous direction should be selected so that it is solved by a function involving sines and cosines.

3. Solve the ordinary differential equation in the homogeneous direction (the eigenproblem) and apply the boundary conditions in this direction in order to obtain the eigenfunctions and eigenvalues.

4. Solve the ordinary differential equation in the non-homogeneous direction for each eigenvalue.

5. Determine a solution for temperature associated with each eigenvalue, T_i, using the results from steps 3 and 4. This solution should satisfy the partial differential equation and both of the homogeneous direction boundary conditions. It is helpful to use Maple to check this solution.

6. Express your general solution as a series composed of the solutions for each eigenvalue that resulted from step 5.

7. Enforce the boundary conditions in the non-homogeneous direction in order to determine the constants in the series. Note that this step will require that the property of the orthogonality of the eigenfunctions be utilized at one or both of the two non-homogeneous direction boundary conditions. The property of orthogonality is utilized by multiplying the series solution by an arbitrary eigenfunction and integrating between the two homogeneous boundaries. This mathematical operation will reduce the series to a single equation involving the constants for only one of the terms in the series. The integration required to carry out this step can often be facilitated using Maple and the resulting equation can often be solved using EES.

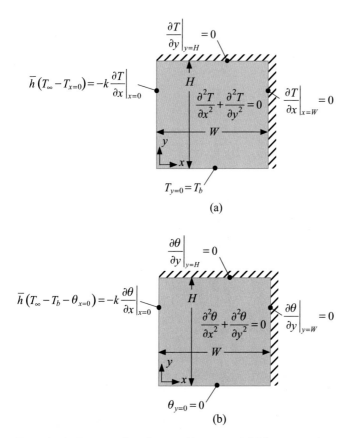

Figure 2-11: Rectangular plate problem stated (a) in terms of temperature, T, and (b) in terms of temperature difference, θ.

2.2.3 Simple Boundary Condition Transformations

The separation of variables technique discussed in Section 2.2.2 can be applied to a linear and homogeneous problem that has linear boundary conditions. In addition, both of the boundary conditions in one direction must be homogeneous; that is, any solution that satisfies the boundary condition must still satisfy the boundary condition if it is multiplied by a constant. Three types of linear boundary conditions are often encountered in heat transfer problems: (1) specified temperature, (2) specified heat flux, and (3) convection to a fluid of specified temperature. Direct application of any of these conditions generally results in a non-homogeneous boundary condition. In general, it is possible to deal with non-homogeneous boundary conditions through superposition, as discussed in Section 2.4, or by breaking a solution into its particular and homogeneous components, as discussed in Section 2.3.2. However, it is often possible to apply a relatively simple transformation in order to reduce the number of non-homogeneous boundary conditions by at least one, thereby (possibly) avoiding the need to use these advanced techniques.

Consider the rectangular plate shown in Figure 2-11(a). The governing differential equation (assuming that there is no convection from the top and bottom surfaces or thermal energy generation within the plate material) is:

$$\frac{\partial^2 T}{\partial x^2} + \frac{\partial^2 T}{\partial y^2} = 0 \tag{2-70}$$

The differential equation is linear and homogeneous. The plate has a specified temperature-type boundary condition at the bottom edge:

$$T_{y=0} = T_b \tag{2-71}$$

Equation (2-71) is not homogeneous unless $T_b = 0$. The plate has a convection-type boundary condition applied to the left edge:

$$\overline{h} \, (T_\infty - T_{x=0}) = -k \left. \frac{\partial T}{\partial x} \right|_{x=0} \tag{2-72}$$

Equation (2-72) is not homogeneous unless $T_\infty = 0$. The plate has specified heat flux-type boundary conditions applied to the remaining two edges; these boundaries are adiabatic and therefore the specified heat flux is equal to 0:

$$\left. \frac{\partial T}{\partial x} \right|_{x=W} = 0 \tag{2-73}$$

$$\left. \frac{\partial T}{\partial y} \right|_{y=H} = 0 \tag{2-74}$$

Because the specified heat flux is zero (i.e., the boundaries are adiabatic), Eqs. (2-73) and (2-74) are homogeneous.

The problem posed by Figure 2-11(a) cannot be directly solved using separation of variables as neither direction is characterized by two non-homogeneous boundary conditions. However, it is possible to reduce the number of non-homogeneous boundary conditions by one. The problem is transformed and solved for the temperature difference relative to a boundary temperature; that is, the problem is solved in terms of either $\theta = T - T_\infty$ or $\theta = T - T_b$ rather than T. The governing differential equation that results is unaffected by this modification (the derivatives of θ are the same as those of T) and it is easy to re-state the remaining boundary conditions in terms of θ rather than T. For example, transforming the problem shown in Figure 2-11(a) to solve for:

$$\theta = T - T_b \tag{2-75}$$

results in the problem shown in Figure 2-11(b). The problem posed in terms of θ can be solved directly by separation of variables because both boundary conditions in the y-direction are homogeneous.

EXAMPLE 2.2-1: TEMPERATURE DISTRIBUTION IN A 2-D FIN

In Section 1.6 the constant cross-section, straight fin shown in Figure 1 was analyzed under the assumption that it could be treated as an extended surface (i.e., temperature gradients in the y direction are neglected). In this example, the 2-D temperature distribution within the fin will be determined using separation of variables.

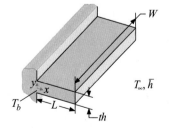

Figure 1: Straight, constant cross-sectional area fin.

Assume that the tip of the fin is insulated and that the width (W) is much larger than the thickness (th) so that convection from the edges can be neglected. The length of

EXAMPLE 2.2-1: TEMPERATURE DISTRIBUTION IN A 2-D FIN

EXAMPLE 2.2-1: TEMPERATURE DISTRIBUTION IN A 2-D FIN

the fin is L. The fin base temperature is T_b and the fin experiences convection with fluid at T_∞ with average heat transfer coefficient, $\overline{h}$.

a) Develop an analytical solution for the temperature distribution in the fin using separation of variables.

The upper and lower halves of the fin are symmetric; that is, there is no difference between the upper and lower portions of the fin and therefore no heat transfer across the mid-plane of the fin. The mid-plane of the fin (i.e., the surface at $y = 0$) can therefore be treated as if it were adiabatic. The computational domain including the boundary conditions is shown in Figure 2(a).

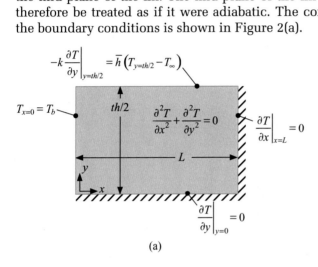

(a)

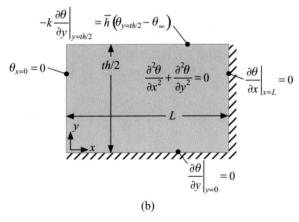

(b)

Figure 2: Problem statement posed in terms of (a) temperature, T, and (b) temperature difference, θ.

The governing equation within the fin can be derived using the process described in Section 2.2.2:

$$\frac{\partial^2 T}{\partial x^2} + \frac{\partial^2 T}{\partial y^2} = 0$$

Figure 2(a) indicates that the problem stated in terms of T has two non-homogeneous boundary conditions (the base and the top surface). However, the boundary condition at the base can be made homogeneous by defining:

$$\theta = T - T_b$$

EXAMPLE 2.2-1: TEMPERATURE DISTRIBUTION IN A 2-D FIN

so that the governing equation becomes:

$$\frac{\partial^2\theta}{\partial x^2} + \frac{\partial^2\theta}{\partial y^2} = 0 \tag{1}$$

The boundary conditions for the transformed problem, illustrated in Figure 2(b), are:

$$\theta_{x=0} = 0 \tag{2}$$

$$\left.\frac{\partial\theta}{\partial x}\right|_{x=L} = 0 \tag{3}$$

$$\left.\frac{\partial\theta}{\partial y}\right|_{y=0} = 0 \tag{4}$$

$$-k\left.\frac{\partial\theta}{\partial y}\right|_{y=th/2} = \overline{h}(\theta_{y=th/2} - \theta_\infty) \tag{5}$$

where

$$\theta_\infty = T_\infty - T_b$$

The problem stated in terms of θ satisfies all of the requirements discussed in Section 2.2.2 with x being the homogeneous direction. Therefore, the separation of variables solution proceeds using the steps laid out in Section 2.2.2. The solution for the temperature difference (θ) is expressed as the product of a function only of x (θX) and a function only of y (θY):

$$\theta(x, y) = \theta X(x)\ \theta Y(y) \tag{6}$$

Substitution of Eq. (6) into Eq. (1) leads to two ordinary differential equations, as shown in Section 2.2.2:

$$\frac{d^2\theta X}{dx^2} \pm \lambda^2\,\theta X = 0$$

$$\frac{d^2\theta Y}{dy^2} \mp \lambda^2\,\theta Y = 0$$

It is necessary to determine the sign of the constant λ^2 in the ordinary differential equations. Recall that it is necessary to have the sine/cosine eigenfunctions in the homogeneous direction. Therefore, it is necessary to select the positive sign for the ordinary differential equation for θX and the negative sign for the ordinary differential equation for θY:

$$\frac{d^2\theta X}{dx^2} + \lambda^2\,\theta X = 0 \tag{7}$$

$$\frac{d^2\theta Y}{dy^2} - \lambda^2\,\theta Y = 0 \tag{8}$$

The next step is to solve the eigenproblem (i.e., the problem for θX); the solution to the ordinary differential equation for θX, Eq. (7), is:

$$\theta X = C_1 \sin(\lambda x) + C_2 \cos(\lambda x) \tag{9}$$

EXAMPLE 2.2-1: TEMPERATURE DISTRIBUTION IN A 2-D FIN

The boundary conditions for θX are obtained by substituting Eq. (9) into Eqs. (2) and (3):

$$\theta X_{x=0} = 0 \qquad (10)$$

$$\left. \frac{d\theta X}{dx} \right|_{x=L} = 0 \qquad (11)$$

Substituting Eq. (9) into Eq. (10) leads to:

$$\theta X_{x=0} = C_1 \underbrace{\sin(\lambda\, 0)}_{0} + C_2 \underbrace{\cos(\lambda\, 0)}_{1} = 0$$

which can only be true if $C_2 = 0$. Substituting Eq. (9), with $C_2 = 0$, into Eq. (11) leads to:

$$\left. \frac{d\theta X}{dx} \right|_{x=L} = C_1 \lambda \cos(\lambda\, L) = 0$$

which can only be true if the argument of the cosine function is $\pi/2$, $3\pi/2$, $5\pi/2$, etc. Therefore, the argument of the cosine function must be:

$$\lambda_i L = \frac{(1+2\,i)}{2}\pi \quad \text{where } i = 0, 1, 2, \ldots$$

The eigenfunctions of the problem are:

$$\theta X_i = C_{1,i} \sin(\lambda_i x) \quad \text{where } i = 0, 1, 2, \ldots \qquad (12)$$

and the eigenvalues of the problem are:

$$\lambda_i = \frac{(1+2\,i)\pi}{2\,L} \qquad (13)$$

The next step is to solve the problem in the non-homogeneous direction. The ordinary differential equation in the y-direction that is associated with each eigenvalue is:

$$\frac{d^2\theta Y_i}{dy^2} - \lambda_i^2\, \theta Y_i = 0$$

which is solved by either

$$\theta Y_i = C_{3,i} \exp(\lambda_i\, y) + C_{4,i} \exp(-\lambda_i\, y)$$

or

$$\theta Y_i = C_{3,i} \cosh(\lambda_i\, y) + C_{4,i} \sinh(\lambda_i\, y) \qquad (14)$$

The choice of either exponentials or sinh and cosh is arbitrary in that both will lead to the correct solution. However, the proper choice often makes the solution process easier. The plate in Figure 2-3 extended to infinity where the temperature became zero. As a result, the constant multiplying the positive exponential was forced to be zero, which made the problem easier to solve. Looking ahead for this fin problem, we see that the gradient of temperature at $y = 0$ must be 0. This boundary condition would not eliminate either of the exponential terms. On the other hand, the boundary condition will force the constant $C_{4,i}$ in Eq. (14) to be zero and therefore the sinh term will be eliminated. Clearly then, Eq. (14) is the better choice; a little insight early in the problem can make the solution process easier.

EXAMPLE 2.2-1: TEMPERATURE DISTRIBUTION IN A 2-D FIN

The next step is to determine the temperature difference solution associated with each eigenvalue:

$$\theta_i = \theta X_i \, \theta Y_i = \sin(\lambda_i \, x)\,[C_{3,i}\cosh(\lambda_i \, y) + C_{4,i}\sinh(\lambda_i \, y)]$$

where the constant $C_{1,i}$ was absorbed into the constants $C_{3,i}$ and $C_{4,i}$. This solution should satisfy both of the homogeneous boundary conditions as well as the partial differential equation for all values of i; it is worthwhile using Maple to verify that this is true. Specify that i is an integer and enter the definition of the eigenvalues:

```
> restart;
> assume(i,integer);
> lambda:=(1+2*i)*Pi/(2*L);
```

$$\lambda := \frac{(1+2i\sim)\pi}{2L}$$

Enter the solution for each eigenvalue:

```
> T:=(x,y)->sin(lambda*x)*(C3*cosh(lambda*y)+C4*sinh(lambda*y));
```

$$T := (x,y) \to \sin(\lambda x)(C3\cosh(\lambda y) + C4\sinh(\lambda y))$$

Verify that the solution satisfies the two boundary conditions in the x-direction, Eqs. (2) and (3):

```
> T(0,y);
```

$$0$$

```
> eval(diff(T(x,y),x),x=L);
```

$$0$$

and the partial differential equation, Eq. (1):

```
> diff(diff(T(x,y),x),x)+diff(diff(T(x,y),y),y);
```

$$-\frac{1}{4}\frac{\sin\left(\frac{(1+2i\sim)\pi x}{2L}\right)(1+2i\sim)^2\pi^2\left(C3\cosh\left(\frac{(1+2i\sim)\pi y}{2L}\right)+C4\sinh\left(\frac{(1+2i\sim)\pi y}{2L}\right)\right)}{L^2}$$

$$+\sin\left(\frac{(1+2i\sim)\pi x}{2L}\right)\left(\frac{1}{4}\frac{C3\cosh\left(\frac{(1+2i\sim)\pi y}{2L}\right)(1+2i\sim)^2\pi^2}{L^2}\right.$$

$$\left.+\frac{1}{4}\frac{C3\sinh\left(\frac{(1+2i\sim)\pi y}{2L}\right)(1+2i\sim)^2\pi^2}{L^2}\right)$$

```
> simplify(%);
```

$$0$$

EXAMPLE 2.2-1: TEMPERATURE DISTRIBUTION IN A 2-D FIN

The sum of the solutions for each eigenvalue becomes the general solution to the problem:

$$\theta = \sum_{i=0}^{\infty} \theta_i = \theta X_i \theta Y_i = \sum_{i=0}^{\infty} \sin(\lambda_i x) \left[C_{3,i} \cosh(\lambda_i y) + C_{4,i} \sinh(\lambda_i y) \right] \qquad (15)$$

The boundary conditions in the non-homogeneous directions are enforced. Substituting Eq. (15) into Eq. (4) leads to:

$$\left. \frac{\partial \theta}{\partial y} \right|_{y=0} = \sum_{i=0}^{\infty} \sin(\lambda_i x) \left[C_{3,i}\lambda_i \underbrace{\sinh(\lambda_i 0)}_{0} + C_{4,i}\lambda_i \underbrace{\cosh(\lambda_i 0)}_{1} \right] = 0$$

The $\cosh(0) = 1$ and the $\sinh(0) = 0$ (much like the $\cos(0) = 1$ and the $\sin(0) = 0$) and therefore this boundary condition can be written as:

$$\sum_{i=0}^{\infty} \sin(\lambda_i x) C_{4,i} \lambda_i = 0$$

which can only be true if $C_{4,i} = 0$, therefore:

$$\theta = \sum_{i=0}^{\infty} C_i \sin(\lambda_i x) \cosh(\lambda_i y) \qquad (16)$$

where the subscript 3 has been removed from $C_{3,i}$ as it is the only remaining undetermined constant. Equation (16) is substituted into the boundary condition at $y = th/2$, Eq. (5):

$$-k \sum_{i=0}^{\infty} C_i \sin(\lambda_i x) \lambda_i \sinh\left(\lambda_i \frac{th}{2}\right) = \bar{h} \left(\sum_{i=0}^{\infty} C_i \sin(\lambda_i x) \cosh\left(\lambda_i \frac{th}{2}\right) - \theta_\infty \right)$$

which can be rearranged:

$$\sum_{i=0}^{\infty} C_i \sin(\lambda_i x) \left[\frac{k\lambda_i}{\bar{h}} \sinh\left(\lambda_i \frac{th}{2}\right) + \cosh\left(\lambda_i \frac{th}{2}\right) \right] = \theta_\infty \qquad (17)$$

The eigenfunctions must be orthogonal between $x = 0$ and $x = L$ (it is not necessary to prove this for each problem) and therefore Eq. (17) can be converted into an algebraic equation for each individual constant. Equation (17) is multiplied by one eigenfunction, $\sin(\lambda_j x)$, and integrated from $x = 0$ to $x = L$:

$$\sum_{i=0}^{\infty} C_i \left[\frac{k\lambda_i}{\bar{h}} \sinh\left(\lambda_i \frac{th}{2}\right) + \cosh\left(\lambda_i \frac{th}{2}\right) \right] \int_0^L \sin(\lambda_i x) \sin(\lambda_j x)\, dx = \theta_\infty \int_0^L \sin(\lambda_j x)\, dx$$

EXAMPLE 2.2-1: TEMPERATURE DISTRIBUTION IN A 2-D FIN

Orthogonality guarantees that the integral on the left side of this equation will be zero for every term in the summation except the one where $i = j$; therefore, the series equation can be rewritten as:

$$C_i \left[\frac{k \lambda_i}{\overline{h}} \sinh \left(\lambda_i \frac{th}{2} \right) + \cosh \left(\lambda_i \frac{th}{2} \right) \right] \int_0^L \sin^2 (\lambda_i x) \, dx = \theta_\infty \int_0^L \sin (\lambda_i x) \, dx$$

The coefficients are evaluated according to:

$$C_i = \frac{\theta_\infty \displaystyle\int_0^L \sin (\lambda_i x) \, dx}{\left[\dfrac{k \lambda_i}{\overline{h}} \sinh \left(\lambda_i \dfrac{th}{2} \right) + \cosh \left(\lambda_i \dfrac{th}{2} \right) \right] \displaystyle\int_0^L \sin^2 (\lambda_i x) \, dx} \tag{18}$$

The integrals in Eq. (18) can be evaluated either using math tables or, more easily, using Maple:

```
> restart;
> assume(i,integer);
> lambda:=(1+2*i)*Pi/(2*L);
```
$$\lambda := \frac{(1 + 2i\sim)\pi}{2L}$$
```
> int(sin(lambda*x),x=0..L);
```
$$\frac{2L}{(1 + 2i\sim)\pi}$$
```
> int(sin(lambda*x)*sin(lambda*x),x=0..L);
```
$$\frac{L}{2}$$

The constants can therefore be written as:

$$C_i = \frac{2 \theta_\infty}{L \lambda_i \left[\dfrac{k \lambda_i}{\overline{h}} \sinh \left(\lambda_i \dfrac{th}{2} \right) + \cosh \left(\lambda_i \dfrac{th}{2} \right) \right]} \tag{19}$$

Equations (16) and (19) together provide the analytical solution for the temperature distribution within the fin.

b) **Use the analytical solution to predict and plot the temperature distribution in a fin that is $L = 5.0$ cm long, $th = 4$ cm thick, with conductivity $k = 0.5$ W/m-K, and $\overline{h} = 100$ W/m²-K. The base temperature is $T_b = 200°C$ and the fluid temperature is $T_\infty = 20°C$.**

EXAMPLE 2.2-1: TEMPERATURE DISTRIBUTION IN A 2-D FIN

The inputs are entered in EES:

```
"EXAMPLE 2.2-1: 2-D Fin"

$UnitSystem SI MASS RAD PA C J
$Tabstops 0.2 0.4 0.6 0.8 3.5

"Inputs"
th_cm=4 [cm]                              "thickness of fin in cm"
th=th_cm*convert(cm,m)                    "thickness of fin"
L=5 [cm]*convert(cm,m)                    "length of fin"
k=0.5 [W/m-K]                             "thermal conductivity"
h_bar=100 [W/m^2-K]                       "heat transfer coefficient"
T_b=converttemp(C,K,200[C])              "base temperature"
T_infinity=converttemp(C,K,20[C])        "fluid temperature"
```

Dimensionless coordinates within the fin are defined in order to facilitate plotting the temperature distribution:

```
y_bar=0.5                                 "dimensionless y-position"
x_bar=0.5                                 "dimensionless x-position"
y=y_bar*th                                "y-position"
x=x_bar*L                                 "x-position"
```

The solution is implemented using a duplicate loop that calculates the first N terms of the series. The number of terms that is required for accuracy should be checked by exploring the sensitivity of the calculation to the number of terms in the same way that a numerical model should be checked for grid convergence.

```
N=100                                                   "number of terms in series"
duplicate i=0,N
   lambda[i]=(1+2*i)*pi/(2*L)                           "eigenvalues"
   C[i]=2*(T_infinity-T_b)/(L*lambda[i]*(k*lambda[i]*sinh(lambda[i]*th/2)/h_bar+cosh(lambda[i]*th/2)))
      "constants"
   theta[i]=C[i]*sin(lambda[i]*x)*cosh(lambda[i]*y)     "term in summation"
end
theta=sum(theta[0..N])                                  "temperature difference"
T=theta+T_b                                             "temperature"
T_C=converttemp(K,C,T)                                  "in C"
```

Figure 3 shows the temperature distribution as a function of x/L for various values of y/th. Notice that for these conditions, an extended surface (i.e., 1-D) model of the fin would not be justified because there is a substantial difference between the temperature at the center of the fin ($y/th = 0$) and the edge ($y/th = 0.5$). This is evident from the Biot number:

$$Bi = \frac{\overline{h}\,th}{2\,k}$$

EXAMPLE 2.2-1: TEMPERATURE DISTRIBUTION IN A 2-D FIN

Bi=h_bar*th/(2k)   "Biot number"

which leads to $Bi = 4.0$.

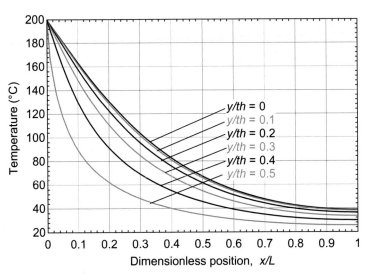

Figure 3: Temperature as a function of x/L for various values of y/th.

c) Use the analytical solution to predict the fin efficiency of the 2-D fin.

The rate of conductive heat transfer into the base of the fin is:

$$\dot{q}_{fin} = -2\, k\, W \int_0^{th/2} \left. \frac{\partial \theta}{\partial x} \right|_{x=0} dy \tag{20}$$

Substituting Eqs. (16) and (19) into Eq. (20) leads to:

$$\dot{q}_{fin} = -4\, \theta_\infty \frac{k\, W}{L} \sum_{i=0}^{\infty} \frac{\displaystyle\int_0^{th/2} \cosh\left(\lambda_i\, y\right) dy}{\left[\dfrac{k\,\lambda_i}{\bar{h}} \sinh\left(\lambda_i \dfrac{th}{2}\right) + \cosh\left(\lambda_i \dfrac{th}{2}\right) \right]}$$

The integral can be accomplished using Maple:

```
> restart;
> int(cosh(lambda*y),y=0..th/2);
```

$$\frac{\sinh\left(\dfrac{\lambda\, th}{2}\right)}{\lambda}$$

EXAMPLE 2.2-1: TEMPERATURE DISTRIBUTION IN A 2-D FIN

so that the rate of conductive heat transfer to the fin is:

$$\dot{q}_{fin} = -4\,\theta_\infty\,\frac{k\,W}{L}\sum_{i=0}^{\infty}\frac{\sinh\left(\lambda_i\frac{th}{2}\right)}{\lambda_i\left[\frac{k\,\lambda_i}{\overline{h}}\sinh\left(\lambda_i\frac{th}{2}\right)+\cosh\left(\lambda_i\frac{th}{2}\right)\right]} \tag{21}$$

The fin efficiency, discussed in Section 1.6.5, is defined as the ratio of the rate of heat transfer to the maximum possible rate of heat transfer rate that is obtained with an infinitely conductive fin:

$$\eta_{fin} = \frac{\dot{q}_{fin}}{2\,\overline{h}\,W\,L\,(T_b - T_\infty)} \tag{22}$$

Substituting Eq. (21) into Eq. (22) leads to the fin efficiency predicted by the 2-D analytical solution:

$$\eta_{fin,2D} = \frac{2\,k}{\overline{h}\,L^2}\sum_{i=0}^{\infty}\frac{\sinh\left(\lambda_i\frac{th}{2}\right)}{\lambda_i\left[\frac{k\,\lambda_i}{\overline{h}}\sinh\left(\lambda_i\frac{th}{2}\right)+\cosh\left(\lambda_i\frac{th}{2}\right)\right]}$$

which is evaluated in EES according to:

```
duplicate i=0,N
   eta_fin[i]=(2*k/(h_bar*lambda[i]*L^2))*sinh(lambda[i]*th/2)/(k*lambda[i]*sinh(lambda[i]*th/2)/h_bar+&
      cosh(lambda[i]*th/2))
end
eta_fin=sum(eta_fin[0..N])
```

d) Plot the fin efficiency predicted by the 2-D analytical solution as a function of the fin thickness and overlay on the plot the fin efficiency predicted using the extended surface approximation, developed in Section 1.6.

Figure 4 illustrates the fin efficiency predicted by the 2-D model as a function of fin thickness. Overlaid on Figure 4 is the solution from Section 1.6 that is listed in Table 1-4 for a fin with an adiabatic tip:

$$\eta_{fin,1D} = \frac{\tanh{(m\,L)}}{m\,L}$$

where

$$m\,L = \sqrt{\frac{\overline{h}\,2}{k\,th}}\,L$$

As the fin becomes thicker, the impact of the temperature gradients in the y-direction, neglected in the 1-D solution, become larger and therefore the 1-D

EXAMPLE 2.2-1: TEMPERATURE DISTRIBUTION IN A 2-D FIN

and 2-D solutions diverge, with the 1-D solution always over-predicting the performance.

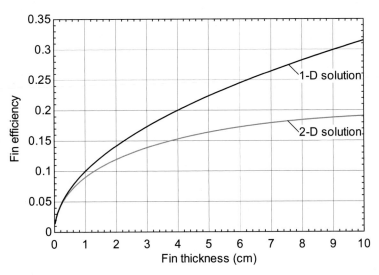

Figure 4: Fin efficiency as a function of the fin thickness predicted by the 2-D solution and the 1-D solution.

The ratio $\eta_{fin,2D}/\eta_{fin,1D}$ is shown in Figure 5 as a function of the Biot number; recall that the Biot number was used to justify the extended surface approximation in Section 1.6. Note that 1-D model is quite accurate (better than 2%) provided the Biot number is less than 0.1 and, surprisingly, remains reasonably accurate (10%) even up to a Biot number of 1.0.

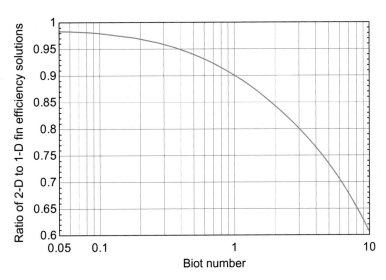

Figure 5: Ratio of the fin efficiency predicted by the 2-D solution to the fin efficiency predicted by the 1-D solution as a function of the Biot number.

EXAMPLE 2.2-2: CONSTRICTION RESISTANCE

EXAMPLE 2.2-2: CONSTRICTION RESISTANCE

Figure 1 illustrates the situation where energy is transferred by conduction through a structure that suddenly changes cross-sectional area. The conduction resistance associated with this structure can be computed using separation of variables. This problem also illustrates an issue that is often confusing for separation of variables problems; specifically, the zeroth term in a cosine series must often be treated separately from the rest of the series. The proper methodology for dealing with this situation is demonstrated in this example.

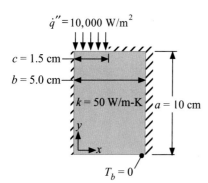

$\dot{q}'' = 10,000 \ \mathrm{W/m}^2$

$c = 1.5$ cm

$b = 5.0$ cm

$k = 50$ W/m-K $\quad a = 10$ cm

$T_b = 0$

Figure 1: A constriction in a conduction path.

The width of the larger cross-sectional area is $b = 5.0$ cm and its length is $a = 10$ cm. The heat flux, $\dot{q}'' = 10{,}000 \ \mathrm{W/m^2}$, is applied to the upper surface over a smaller width, $c = 1.5$ cm. The conductivity of the material is $k = 50$ W/m-K. The bottom surface of the object is maintained at some reference temperature, taken to be 0.

a) Develop a solution for the temperature distribution in the material.

The partial differential equation for the problem is:

$$\frac{\partial^2 T}{\partial x^2} + \frac{\partial^2 T}{\partial y^2} = 0 \tag{1}$$

The boundary conditions in the x-direction are:

$$\left.\frac{\partial T}{\partial x}\right|_{x=0} = 0 \tag{2}$$

$$\left.\frac{\partial T}{\partial x}\right|_{x=b} = 0 \tag{3}$$

and the boundary conditions in the y-direction are:

$$T_{y=0} = 0 \tag{4}$$

$$k\left.\frac{\partial T}{\partial y}\right|_{y=a} = \begin{cases} \dot{q}'' & x < c \\ 0 & x \geq c \end{cases} \tag{5}$$

The first step in the solution is to verify that separation of variables can be applied to the problem without transformation or superposition. The governing partial differential equation is linear and homogeneous, all of the boundary conditions are linear, and both boundary conditions in the x-direction are homogeneous. Therefore

EXAMPLE 2.2-2: CONSTRICTION RESISTANCE

the problem meets all of the requirements discussed in Section 2.2.2 and separation of variables can be applied, with x being the homogeneous direction.

The next step in the solution is to assume a separable solution:

$$T(x, y) = TX(x)\, TY(y) \tag{6}$$

which is substituted into Eq. (1) in order to achieve two ordinary differential equations for TX and TY, as discussed in Section 2.2.2:

$$\frac{d^2 TX}{dx^2} + \lambda^2\, TX = 0 \tag{7}$$

$$\frac{d^2 TY}{dy^2} - \lambda^2\, TY = 0 \tag{8}$$

Notice that the separation process was accomplished so that sine/cosine functions solve the ordinary differential equation for TX because x is the homogeneous direction. The next step in the solution is to solve the eigenproblem (i.e., the problem in the homogeneous direction). The solution to Eq. (7) is:

$$TX = C_1 \sin(\lambda x) + C_2 \cos(\lambda x) \tag{9}$$

The x-direction boundary conditions, Eqs. (2) and (3), expressed in terms of TX, become:

$$\left. \frac{dTX}{dx} \right|_{x=0} = 0 \tag{10}$$

$$\left. \frac{dTX}{dx} \right|_{x=b} = 0 \tag{11}$$

Substituting Eq. (9) into Eq. (10) leads to:

$$\left. \frac{dTX}{dx} \right|_{x=0} = C_1 \lambda \underbrace{\cos(0)}_{1} - C_2 \lambda \underbrace{\sin(0)}_{0} = 0$$

which can only be true if $C_1 = 0$. Substituting Eq. (9), with $C_1 = 0$, into Eq. (11) leads to:

$$\left. \frac{dTX}{dx} \right|_{x=b} = -C_2 \lambda \sin(\lambda b) = 0$$

which can only be true if the argument of the sine function is an integer multiple of π:

$$\lambda_i\, b = i\,\pi \quad \text{where } i = 0, 1, 2, \ldots$$

Therefore, the eigenfunctions for this problem are:

$$TX_i = C_{2,i} \cos(\lambda_i x) \quad \text{where } i = 0, 1, 2, \ldots \tag{12}$$

and the eigenvalues, λ_i, are:

$$\lambda_i = \frac{i\,\pi}{b} \tag{13}$$

Note that the zeroth eigenfunction is retained in Eq. (12) because TX_0 is not zero. The zeroth eigenfunction is a constant and it will be necessary to treat this term

EXAMPLE 2.2-2: CONSTRICTION RESISTANCE

separately from the others. The next step is to solve the problem in the non-homogeneous direction for each eigenvalue. The solution to Eq. (8) for each eigenvalue is:

$$TY_i = C_{3,i} \sinh(\lambda_i y) + C_{4,i} \cosh(\lambda_i y) \tag{14}$$

The solution associated with each eigenvalue is the product of Eqs. (12) and (14):

$$T_i = TX_i \, TY_i = C_{2,i} \cos(\lambda_i x) \, [C_{3,i} \sinh(\lambda_i y) + C_{4,i} \cosh(\lambda_i y)] \quad \text{where } i = 0, 1, 2, \ldots$$

or, absorbing the constant $C_{2,i}$ into the constants $C_{3,i}$ and $C_{4,i}$:

$$T_i = \cos(\lambda_i x) \, [C_{3,i} \sinh(\lambda_i y) + C_{4,i} \cosh(\lambda_i y)] \quad \text{where } i = 0, 1, 2, \ldots \tag{15}$$

The function T_i provided by Eq. (15) should satisfy both boundary conditions in the x-direction as well as the partial differential equation for any value of i; this should be checked using Maple before continuing.

The general solution is expressed as the sum of the solutions associated with each eigenvalue:

$$T = \sum_{i=0}^{\infty} T_i = \sum_{i=0}^{\infty} \cos(\lambda_i x) \, [C_{3,i} \sinh(\lambda_i y) + C_{4,i} \cosh(\lambda_i y)] \tag{16}$$

The final step forces the general solution to satisfy the boundary conditions in the non-homogeneous direction. Equation (16) is substituted into the boundary condition at $y = 0$, Eq. (4):

$$T_{y=0} = \sum_{i=0}^{\infty} \cos(\lambda_i x) \left[C_{3,i} \underbrace{\sinh(0)}_{=0} + C_{4,i} \underbrace{\cosh(0)}_{=1} \right] = 0$$

which leads to:

$$\sum_{i=0}^{\infty} \cos(\lambda_i x) \, C_{4,i} = 0$$

which can only be true if $C_{4,i} = 0$ for all i, therefore:

$$T = \sum_{i=0}^{\infty} C_i \cos(\lambda_i x) \sinh(\lambda_i y) \tag{17}$$

where the subscript 3 has been removed from $C_{3,i}$ since it is the only remaining undetermined constant associated with each eigenvalue. The zeroth term in the cosine series is a constant and it must be pulled out and treated separately; this is generally true for a cosine series where the zeroth eigenvalue is zero (i.e., the zeroth eigenfunction is a constant).

$$T = \lim_{i \to 0} \left[C_0 \cos\left(\frac{i\pi}{b} x\right) \sinh\left(\frac{i\pi}{b} y\right) \right] + \sum_{i=1}^{\infty} C_i \cos(\lambda_i x) \sinh(\lambda_i y)$$

or, recognizing that the cos(0) is 1.0:

$$T = \lim_{i \to 0} \left[C_0 \sinh\left(\frac{i\pi}{b} y\right) \right] + \sum_{i=1}^{\infty} C_i \cos(\lambda_i x) \sinh(\lambda_i y) \tag{18}$$

It is tempting to recognize that the sinh(0) $= 0$ and therefore if $i \to 0$ then the zeroth term will not contribute to the solution. This is true provided that C_0 is finite;

EXAMPLE 2.2-2: CONSTRICTION RESISTANCE

however, the product $C_0 \sinh(0)$ may not be zero and therefore the zeroth term must be retained.

The final, non-homogeneous boundary condition, Eq. (5), is used to compute the undetermined coefficients in Eq. (18). Equation (18) is substituted into Eq. (5):

$$k \left. \frac{\partial T}{\partial y} \right|_{y=a} = k \lim_{i \to 0} \left[C_0 \frac{i\pi}{b} \cosh\left(\frac{i\pi}{b} a\right) \right] + k \sum_{i=1}^{\infty} C_i \lambda_i \cos(\lambda_i x) \cosh(\lambda_i a) = \begin{cases} \dot{q}'' & x < c \\ 0 & x \geq c \end{cases}$$

or, recognizing that the $\cosh(0)$ is 1.0:

$$\frac{k\pi}{b} \lim_{i \to 0} [C_0 \, i] + k \sum_{i=1}^{\infty} C_i \lambda_i \cos(\lambda_i x) \cosh(\lambda_i a) = \begin{cases} \dot{q}'' & x < c \\ 0 & x \geq c \end{cases} \quad (19)$$

We take advantage of the orthogonality of the eigenfunctions to compute the constants in Eq. (19). First, we will deal with the zeroth term in the series. Both sides of the equation are multiplied by the zeroth eigenfunction, $\cos(\lambda_0 x)$ which is equal to 1, and the equation is integrated from $x = 0$ to $x = b$:

$$\frac{k\pi}{b} \lim_{i \to 0} [C_0 \, i] \int_0^b dx + k \sum_{i=1}^{\infty} C_i \lambda_i \cosh(\lambda_i a) \int_0^b \cos(\lambda_i x) \, dx = \int_0^c \dot{q}'' dx + \int_c^b 0 \, dx$$

The integral of any of the eigenfunctions (other than the zeroth one) from 0 to b is zero. Therefore, every term in the summation integrates to zero and we are left with:

$$k\pi \lim_{i \to 0} [C_0 \, i] = \dot{q}'' c$$

therefore:

$$\lim_{i \to 0} [C_0 \, i] = \frac{\dot{q}'' c}{\pi k} \quad (20)$$

Equation (20) shows why it is necessary to treat the zeroth term in the series separately. If the limit notation is removed and the term was treated in the normal way then we would conclude that C_0 was infinite. Substituting into Eq. (20) into Eq. (18) leads to:

$$T = \underbrace{\lim_{i \to 0} \left[\frac{\dot{q}'' c}{i\pi k} \sinh\left(\frac{i\pi}{b} y\right) \right]}_{\text{0th term in solution}} + \sum_{i=1}^{\infty} C_i \cos(\lambda_i x) \sinh(\lambda_i y)$$

Maple can be used to evaluate the zeroth term in the solution:

```
> limit(q_dot_flux*c*sinh(i*Pi*y/b)/(i*Pi*k),i=0);
```
$$\frac{y q\_dot\_flux \, c}{b k}$$

Substituting this result into Eq. (18) leads to:

$$T = \frac{\dot{q}'' c y}{b k} + \sum_{i=1}^{\infty} C_i \cos(\lambda_i x) \sinh(\lambda_i y) \quad (21)$$

Substituting Eq. (21) into Eq. (5) leads to:

$$\frac{\dot{q}'' c}{b} + k \sum_{i=1}^{\infty} C_i \lambda_i \cos(\lambda_i x) \cosh(\lambda_i a) = \begin{cases} \dot{q}'' & x < c \\ 0 & x \geq c \end{cases} \quad (22)$$

EXAMPLE 2.2-2: CONSTRICTION RESISTANCE

Next, we will deal with the non-zero terms in the series. Both sides of Eq. (22) are multiplied by $\cos(\lambda_j x)$ and integrated from $x = 0$ to $x = b$.

$$\frac{\dot{q}'' c}{b} \int_0^b \cos(\lambda_j x) dx + k \sum_{i=1}^{\infty} C_i \lambda_i \cosh(\lambda_i a) \int_0^b \cos(\lambda_i x) \cos(\lambda_j x) dx$$

$$= \int_0^c \dot{q}'' \cos(\lambda_j x) dx + \int_c^b 0 \cos(\lambda_j x) dx$$

The zeroth order term integrates to zero for any $j > 0$ and the only term of the summation that does not integrate to zero is the one where $i = j$:

$$k C_i \lambda_i \cosh(\lambda_i a) \int_0^b \cos^2(\lambda_i x) dx = \dot{q}'' \int_0^c \cos(\lambda_i x) dx \qquad (23)$$

The integrals in Eq. (23) are computed using Maple:

```
> int((cos(lambda*x))^2,x=0..b);
```

$$\frac{b}{2}$$

```
> int(cos(lambda*x),x=0..c);
```

$$\frac{b \sin\left(\frac{i \sim \pi c}{b}\right)}{i \sim \pi}$$

in order to obtain an equation for the undetermined coefficients:

$$k C_i \lambda_i \cosh(\lambda_i a) \frac{b}{2} = \dot{q}'' \frac{\sin(\lambda_i c)}{\lambda_i}$$

The solution is programmed in EES:

```
"EXAMPLE 2.2-2: Constriction Resistance"

$UnitSystem SI MASS RAD PA C J
$Tabstops 0.2 0.4 0.6 0.8 3.5

"Inputs"
q_dot_flux=10000 [W/m^2]              "Heat flux"
k=50 [W/m-K]                          "Conductivity"
c=1.5 [cm]*convert(cm,m)              "width of applied flux"
a=10 [cm]*convert(cm,m)              "length of object"
b=5.0 [cm]*convert(cm,m)             "width of object"
```

EXAMPLE 2.2-2: CONSTRICTION RESISTANCE

```
x_bar=0.75                          "dimensionless x-position"
y_bar=1                             "dimensionless y-position"
x=x_bar*b                           "x-position"
y=y_bar*a                           "y-position"

N=400                               "number of terms"
duplicate i=1,N                     "evaluate coefficients for N terms"
  lambda[i]=i*pi/b
  k*C[i]*lambda[i]*cosh(lambda[i]*a)*b/2=q_dot_flux*sin(lambda[i]*c)/lambda[i]
  T[i]=C[i]*cos(lambda[i]*x)*sinh(lambda[i]*y)
end
T=q_dot_flux*c*y/(k*b)+sum(T[1..N])
```

A parametric table is generated and used to generate the contour plot of temperature shown in Figure 2.

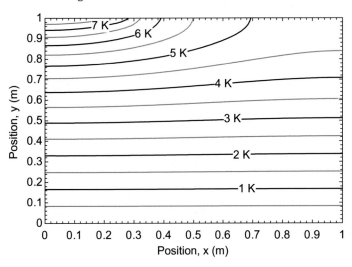

Figure 2: Contour plot of temperature distribution in constriction.

It is worth comparing the answer with physical intuition. The temperature elevation at the constriction relative to the base is approximately 8 K according to Figure 2; does this value make sense? In Section 2.8, methods for estimating the conduction resistance of 2-D geometries using 1-D models are discussed. However, it is clear that the resistance of the constriction cannot be greater than the resistance to conduction through the material if the heat flux is applied uniformly at the top surface:

$$R_{nc} = \frac{a}{b \, L \, k}$$

where L is the length of the material. The temperature elevation at the constriction in this limit is:

$$\Delta T_{nc} = R_{nc} \, \dot{q}'' \, c \, L = \frac{a \, \dot{q}'' \, c}{b \, k}$$

```
DeltaT_nc=a*q_dot_flux*c/(b*k)              "temperature rise without constriction"
```

This leads to $\Delta T_{nc} = 6.0$ K, which has the same magnitude as the observed temperature rise but is smaller, as expected.

2.3 Advanced Separation of Variables Solutions

This extended section of the book can be found on the website www.cambridge.org/ nellisandklein. Section 2.2 provides an introduction to the technique of separation of variables and discusses its application in the context of a few examples. The separation of variables method, as it is presented in Section 2.2, is rather limited as it does not allow, for example, non-homogeneous terms that might arise from effects such as volumetric generation or for problems that are in cylindrical coordinates. One technique for solving non-homogeneous partial differential equations is discussed in Section 2.3.2. The extension of separation of variables to cylindrical coordinates is presented in Section 2.3.3 and demonstrated in EXAMPLE 2.3-1.

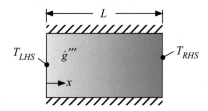

Figure 2-15: Plane wall with uniform volumetric generation and specified edge temperatures.

2.4 Superposition

2.4.1 Introduction

Many conduction heat transfer problems are governed by linear differential equations; in some cases, many different functions will all satisfy the differential equation. The sum of all of the functions that separately satisfy a linear homogeneous differential equation will itself be a solution. This property is used in separation of variables when the solutions associated with each of the eigenfunctions are added together in order to obtain a series solution to the problem. Superposition uses this property to determine the solution to a complex problem by breaking it into several, simpler problems that are solved individually and then added together. Care must be taken to ensure that the boundary conditions for the individual problems properly add to satisfy the desired boundary condition.

A series of 1-D steady-state problems appears in Chapter 1; although superposition was not used to solve these problems, it would have been possible to apply this methodology. For example, consider a plane wall with thickness (L) and conductivity (k) experiencing a constant volumetric rate of thermal energy generation ($\dot{g}'''$). The edges of the wall have specified temperatures, T_{LHS} and T_{RHS}, as shown in Figure 2-15.

The governing differential equation for this problem is:

$$\frac{d^2 T}{dx^2} = -\frac{\dot{g}'''}{k} \tag{2-126}$$

and the boundary conditions are:

$$T_{x=0} = T_{LHS} \tag{2-127}$$

$$T_{x=L} = T_{RHS} \tag{2-128}$$

Notice that the governing differential equation and both boundary conditions are linear but they are not homogeneous. It is possible to solve the problem posed by Eqs. (2-126) through (2-128) without resorting to superposition; indeed there is no advantage to using

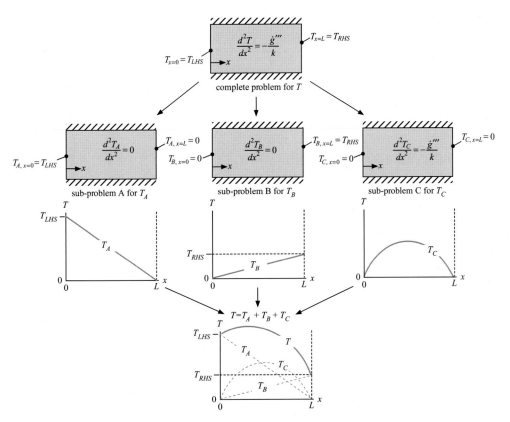

Figure 2-16: Principle of superposition applied to the plane wall of Figure 2-15.

superposition for such a simple problem. The solution was found in Section 1.3 to be:

$$T = \frac{\dot{g}''' L^2}{2k} \left[\frac{x}{L} - \left(\frac{x}{L} \right)^2 \right] - \frac{(T_{LHS} - T_{RHS})}{L} x + T_{LHS} \qquad (2\text{-}129)$$

Nevertheless, the problem shown in Figure 2-15 provides a useful introduction to superposition. The problem can be broken into three, simpler sub-problems each of which retains only one of the non-homogeneities that are inherent in the total problem.

The complete problem is solved by T and is broken into sub-problems A, B, and C which are solved by the functions T_A, T_B and T_C, respectively, as shown in Figure 2-16.

Sub-problem A retains the non-homogeneous boundary condition at $x = 0$, but uses the homogeneous version of the differential equation (i.e., the generation term is dropped) and the boundary condition at $x = L$.

$$\frac{d^2 T_A}{dx^2} = 0 \qquad (2\text{-}130)$$

$$T_{A,x=0} = T_{LHS} \qquad (2\text{-}131)$$

$$T_{A,x=L} = 0 \qquad (2\text{-}132)$$

The solution to sub-problem A, T_A, is linear from T_{LHS} to 0 as shown in Figure 2-16.

$$T_A = T_{LHS} \left(1 - \frac{x}{L} \right) \qquad (2\text{-}133)$$

Sub-problem B retains the non-homogeneous boundary condition at $x = L$, but uses the homogeneous version of the differential equation and the boundary condition at $x = 0$:

$$\frac{d^2 T_B}{dx^2} = 0 \tag{2-134}$$

$$T_{B,x=0} = 0 \tag{2-135}$$

$$T_{B,x=L} = T_{RHS} \tag{2-136}$$

The solution, T_B, is linear from 0 to T_{RHS}.

$$T_B = T_{RHS} \frac{x}{L} \tag{2-137}$$

Finally, sub-problem C retains the non-homogeneous differential equation but uses the homogeneous versions of both boundary conditions:

$$\frac{d^2 T_C}{dx^2} - \frac{\dot{g}'''}{k} \tag{2-138}$$

$$T_{C,x=0} = 0 \tag{2-139}$$

$$T_{C,x=L} = 0 \tag{2-140}$$

The solution, T_C, is a quadratic with a maximum at the center of the wall:

$$T_C = \frac{\dot{g}''' L^2}{2k} \left[\frac{x}{L} - \left(\frac{x}{L} \right)^2 \right] \tag{2-141}$$

The solution to the complete problem is the sum of the solutions to the three sub-problems:

$$T = T_A + T_B + T_C \tag{2-142}$$

or

$$T = \underbrace{T_{LHS} \left(1 - \frac{x}{L} \right)}_{T_A} + \underbrace{T_{RHS} \frac{x}{L}}_{T_B} + \underbrace{\frac{\dot{g}''' L^2}{2k} \left[\frac{x}{L} - \left(\frac{x}{L} \right)^2 \right]}_{T_C} \tag{2-143}$$

which is identical to Eq. (2-129), the solution obtained in Section 1.3. It is easy to see from Figure 2-16 that this process of superposition must work; the differential equations for the sub-problems, Eqs. (2-130), (2-134), and (2-138) can be added together to recover Eq. (2-126):

$$\frac{d^2 T_A}{dx^2} + \frac{d^2 T_B}{dx^2} + \frac{d^2 T_C}{dx^2} = -\frac{\dot{g}'''}{k} \rightarrow \frac{d^2 (T_A + T_B + T_C)}{dx^2} = -\frac{\dot{g}'''}{k} \rightarrow \frac{d^2 T}{dx^2} = -\frac{\dot{g}'''}{k}$$
$$\tag{2-144}$$

and the boundary conditions for the sub-problems can be added together to recover Eqs. (2-127) and (2-128):

$$T_{A,x=0} + T_{B,x=0} + T_{C,x=0} = T_{LHS} \rightarrow (T_A + T_B + T_C)_{x=0} = T_{LHS} \rightarrow T_{x=0} = T_{LHS}$$
$$\tag{2-145}$$

$$T_{A,x=L} + T_{B,x=L} + T_{C,x=L} = T_{RHS} \rightarrow (T_A + T_B + T_C)_{x=L} = T_{RHS} \rightarrow T_{x=L} = T_{RHS}$$
$$\tag{2-146}$$

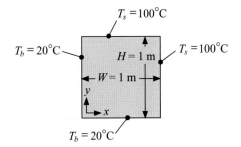

Figure 2-17: Rectangular plate used to illustrate superposition for 2-D problems.

2.4.2 Superposition for 2-D Problems

Superposition becomes much more useful for 2-D problems because the separation of variables technique is restricted to problems that have homogeneous boundary conditions in one direction. Most real problems will not satisfy this condition and therefore it is absolutely necessary to use superposition to solve these problems. The solution can be developed by superimposing several solutions, each constructed so that they are tractable using separation of variables. The process is only slightly more complex than the 1-D problem that is discussed in the previous section.

For example, consider the plate with height $H = 1$ m and width $W = 1$ m, shown in Figure 2-17. The top and right sides are kept at $T_s = 100°C$ while the bottom and left sides are kept at $T_b = 20°C$. The temperature distribution is a function only of x and y.

The complete problem is shown in Figure 2-18(a); notice that all four boundary conditions are non-homogeneous and even transforming the problem by subtracting T_b or T_s will not result in two homogeneous boundary conditions in either the x- or y-direction. Therefore, the problem cannot be solved directly using separation of variables. Figure 2-18(b) illustrates the problem transformed by defining the temperature difference relative to T_b:

$$\theta = T - T_b \qquad (2\text{-}147)$$

It is necessary to break the problem for θ into two sub-problems:

$$\theta = \theta_A + \theta_B \qquad (2\text{-}148)$$

Each sub-problem is characterized by a homogeneous direction, as shown in Figure 2-18(c). Note that for each sub-problem, the homogeneous boundary conditions that are selected are analogous to the original, non-homogeneous boundary conditions. In this problem, the specified temperature boundaries are replaced with a temperature of zero. A specified heat flux boundary should be replaced with an adiabatic boundary and a boundary with convection to fluid at T_∞ should be replaced by convection to fluid at zero temperature.

Each of the two sub-problems are solved using separation of variables, as discussed in Section 2.2. The governing partial differential equation for sub-problem A:

$$\frac{\partial^2 \theta_A}{\partial x^2} + \frac{\partial^2 \theta_A}{\partial y^2} = 0 \qquad (2\text{-}149)$$

is separated into two ordinary differential equations; note that the x-direction is homogeneous for sub-problem A and therefore the ordinary differential equation

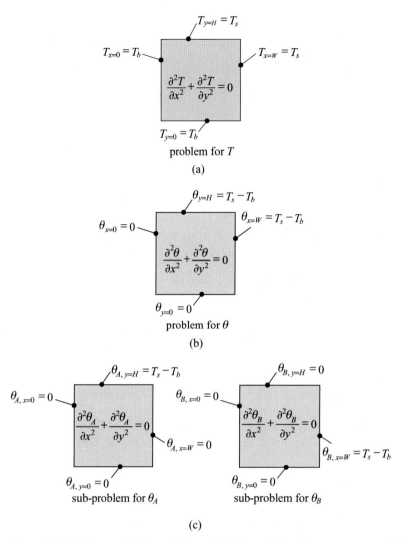

Figure 2-18: Mathematical description of (a) the problem for temperature T, (b) the problem for temperature difference θ, and (c) the two sub-problems θ_A and θ_B that can be solved using separation of variables.

for θX_A is selected so that it is solved by sines and cosines.

$$\frac{d^2\theta X_A}{dx^2} + \lambda_A^2\, \theta X_A = 0 \tag{2-150}$$

$$\frac{d^2\theta Y_A}{dx^2} - \lambda_A^2\, \theta Y_A = 0 \tag{2-151}$$

The eigenproblem is solved first; the solution to Eq. (2-150) is:

$$\theta X_A = C_{A,1} \sin(\lambda_A x) + C_{A,2} \cos(\lambda_A x) \tag{2-152}$$

The homogeneous boundary condition at $x = 0$ leads to:

$$\theta X_{A,x=0} = C_{A,1} \underbrace{\sin(\lambda_A 0)}_{0} + C_{A,2} \underbrace{\cos(\lambda_A 0)}_{1} = 0 \tag{2-153}$$

which can only be true if $C_{A,2} = 0$:

$$\theta X_A = C_{A,1} \sin (\lambda_A x) \tag{2-154}$$

The homogeneous boundary condition at $x = W$ leads to:

$$\theta X_{A,x=W} = C_{A,1} \sin (\lambda_A W) = 0 \tag{2-155}$$

which leads to the eigenvalues:

$$\lambda_{A,i} = \frac{i\pi}{W} \quad \text{for } i = 1, 2..\infty \tag{2-156}$$

The solution to the ordinary differential equation in the non-homogeneous direction, Eq. (2-151), is:

$$\theta Y_{A,i} = C_{A,3,i} \sinh (\lambda_{A,i} y) + C_{A,4,i} \cosh (\lambda_{A,i} y) \tag{2-157}$$

The solution for each eigenvalue is:

$$\theta_{A,i} = \theta X_{A,i} \theta Y_{A,i} = \sin (\lambda_{A,i} x) \left[C_{A,3,i} \sinh (\lambda_{A,i} y) + C_{A,4,i} \cosh (\lambda_{A,i} y) \right] \tag{2-158}$$

The general solution is the sum of the solutions for each eigenvalue:

$$\theta_A = \sum_{i=1}^{\infty} \theta_{A,i} = \sum_{i=1}^{\infty} \sin (\lambda_{A,i} x) \left[C_{A,3,i} \sinh (\lambda_{A,i} y) + C_{A,4,i} \cosh (\lambda_{A,i} y) \right] \tag{2-159}$$

The general solution is required to satisfy the boundary condition at $y = 0$:

$$\theta_{A,y=0} = \sum_{i=1}^{\infty} \sin (\lambda_{A,i} x) \left[C_{A,3,i} \underbrace{\sinh (\lambda_{A,i} 0)}_{=0} + C_{A,4,i} \underbrace{\cosh(\lambda_{A,i} 0)}_{=1} \right] = 0 \tag{2-160}$$

which can only be true if $C_{A,4,i} = 0$:

$$\theta_A = \sum_{i=1}^{\infty} C_{A,i} \sin (\lambda_{A,i} x) \sinh (\lambda_{A,i} y) \tag{2-161}$$

The solution is required to satisfy the boundary condition at $y = H$:

$$\theta_{A,y=H} = \sum_{i=1}^{\infty} C_{A,i} \sin (\lambda_{A,i} x) \sinh (\lambda_{A,i} H) = T_s - T_b \tag{2-162}$$

The orthogonality property of the eigenfunctions is used:

$$C_{A,i} \sinh (\lambda_{A,i} H) \int_0^W \sin^2 (\lambda_{A,i} x) \, dx = (T_s - T_b) \int_0^W \sin (\lambda_{A,i} x) \, dx \tag{2-163}$$

The integrals in Eq. (2-163) are evaluated in Maple:

```
> restart;
> assume(i,integer);
> lambda:=i*Pi/W;
```

$$\lambda := \frac{i \sim \pi}{W}$$

```
> int((sin(lambda*x))^2,x=0..W);
```

$$\frac{W}{2}$$

```
> int(sin(lambda*x),x=0..W);
```

$$-\frac{W(-1+(-1)^{i\sim})}{i\sim\pi}$$

which leads to:

$$C_{A,i}\sinh\left(\lambda_{A,i}\,H\right)\frac{W}{2}=-\left(T_s-T_b\right)\frac{W\left[-1+(-1)^i\right]}{i\,\pi} \qquad (2\text{-}164)$$

or:

$$C_{A,i}=-\left(T_s-T_b\right)\frac{2\left[-1+(-1)^i\right]}{i\,\pi\sinh\left(\lambda_{A,i}\,H\right)} \qquad (2\text{-}165)$$

When solving a problem using superposition, it is useful to separately implement and examine the solution to each of the sub-problems and verify that they separately satisfy the boundary conditions and satisfy our physical intuition. The inputs are entered in EES:

```
$UnitSystem SI MASS RAD PA C J
$Tabstops 0.2 0.4 0.6 0.8 3.5

"Inputs"
H=1 [m]                                    "height of plate"
W=1 [m]                                    "width of plate"
T_s=converttemp(C,K,100 [C])              "temperature of right and top of plate"
T_b=converttemp(C,K,20 [C])               "temperature of left and bottom of plate"
```

The position to evaluate the temperature is specified in terms of dimensionless variables:

```
x=x_bar*W                                  "x-position"
y=y_bar*H                                  "y-position"
```

The solution to sub-problem A is implemented in EES:

```
N=100 [-]                                                          "number of terms"
duplicate i=1,N
  lambda_A[i]=i*pi/W                                               "eigenvalue"
  C_A[i]=2*(T_s-T_b)*(-(-1+(-1)^i)/i/Pi)/sinh(lambda_A[i]*H)       "evaluate constants"
  theta_A[i]=C_A[i]*sin(lambda_A[i]*x)*sinh(lambda_A[i]*y)
end
theta_A=sum(theta_A[1..N])                                         "sub-problem A"
```

The solution to sub-problem A is shown in Figure 2-19(a).

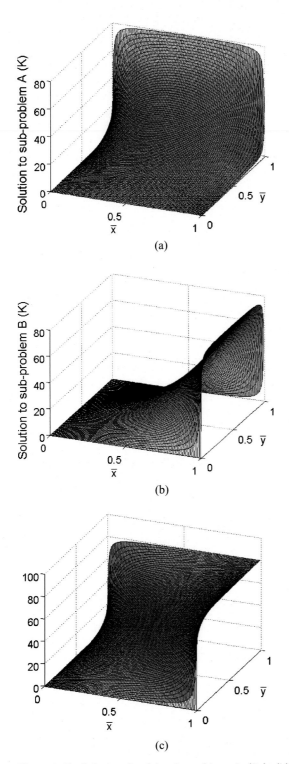

Figure 2-19: Solution for (a) sub-problem A (θ_A), (b) sub-problem B (θ_B), (c) and temperature difference (θ).

A similar process leads to the solution for sub-problem B:

$$\lambda_{B,i} = \frac{i\pi}{H} \quad \text{for } i = 1, 2..\infty \tag{2-166}$$

$$C_{B,i} = -(T_s - T_b)\frac{2\left[-1 + (-1)^i\right]}{i\pi \sinh(\lambda_{B,i} W)} \tag{2-167}$$

$$\theta_B = \sum_{i=1}^{\infty} C_{B,i} \sin(\lambda_{B,i} y) \sinh(\lambda_{B,i} x) \tag{2-168}$$

which is also implemented in EES:

```
duplicate i=1,N
  lambda_B[i]=i*pi/H                                        "eigenvalue"
  C_B[i]=2*(T_s-T_b)*(-(-1+(-1)^i)/i/Pi)/sinh(lambda_B[i]*W)   "evaluate constants"
  theta_B[i]=C_B[i]*sin(lambda_B[i]*y)*sinh(lambda_B[i]*x)
end
theta_B=sum(theta_B[1..N])                                  "sub-problem B"
```

The solution to sub-problem B is shown in Figure 2-19(b). The temperature difference solution is obtained by superposition, Eq. (2-148):

```
theta=theta_A+theta_B              "temperature difference, from superposition"
```

and shown in Figure 2-19(c). The temperature solution (in °C) is obtained with the following code:

```
T=theta+T_b                        "temperature"
T_C=converttemp(K,C,T)             "in C"
```

The process of superposition was illustrated in this section for a simple problem. However, it is possible to use superposition to break a relatively complicated problem with multiple, non-homogeneous and complex boundary conditions into a series of problems that are each tractable, allowing the problem to be solved and verified one sub-problem at a time.

2.5 Numerical Solutions to Steady-State 2-D Problems with EES

2.5.1 Introduction

Sections 2.1 through 2.4 present analytical techniques that are useful for solving 2-D conduction problems. These solution techniques have some fairly severe limitations. For example, the shape factors discussed in Section 2.1 can be used only in those situations where the problem can be represented using one of the limited set of shape factor solutions that are available. The separation of variables techniques coupled with superposition presented in Sections 2.2 through 2.4 can be applied to a more general set of problems; however, they are still limited to linear problems (e.g., radiation and temperature dependent properties cannot be explicitly considered) with simple boundaries. To consider a problem of any real complexity would require the superposition of many solutions and that would be somewhat time consuming. Also, the solution is specific to the problem; if any aspect of the problem changes then the solution must be re-derived. These analytical solutions are most useful for verifying numerical solutions or, in some cases, creating multi-scale models.

This section begins the discussion of numerical solutions to 2-D problems. There are two techniques that are used to solve 2-D problems: finite difference solutions and finite element solutions. Finite difference solutions are discussed in this section as well as in Section 2.6. The application of the finite difference approach to 2-D problems is a natural extension of the 1-D finite difference solutions that were studied in Chapter 1. Finite difference solutions are intuitive and powerful, but difficult to apply to complex geometries.

Finite element solutions are dramatically different from finite difference solutions and can be applied more easily to complex geometries. A complete description of the finite element technique is beyond the scope of this book; however, finite element solutions to heat transfer problems are extremely powerful and many commercial packages are available for this purpose. Section 2.7 provides a discussion of the finite element technique followed by an introduction to the finite element package FEHT. An academic version of FEHT can be downloaded from the website www.cambridge.org/nellisandklein.

Both finite difference and finite element techniques break a large computational domain into many smaller ones that are referred to as control volumes for the finite difference technique and elements for the finite element technique. The control volumes or elements are modeled approximately in order to generate a system of equations that can be efficiently solved using a computer. The approximate, numerical solution will approach the actual solution as the number of control volumes or elements is increased. It is important to remember that it is not sufficient to obtain a solution. Regardless of what technique you are using (including the use of a pre-packaged piece of software, such as FEHT), you must still:

1. verify that your solution has an adequately large number of control volumes or elements,
2. verify that your solution makes physical sense and obeys your intuition, and
3. verify your solution against an analytical solution in an appropriate limit.

These steps are widely accepted as being "best practice" when working with numerical solutions of any type.

2.5.2 Numerical Solutions with EES

Finite difference solutions to 1-D steady-state problems are presented in Sections 1.4 and 1.5. The steps required to set up a numerical solution to a 2-D problem are essentially the same; however, the bookkeeping process (i.e., the process of entering the algebraic equations into the computer) may be somewhat more cumbersome.

The first step is to define small control volumes that are distributed through the computational domain and to precisely define the locations at which the numerical model will compute the temperatures (i.e., the locations of the nodes). The control volumes are

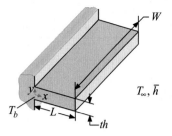

Figure 2-20: Straight, constant cross-sectional area fin.

small but finite; for the 1-D problems that were investigated in Chapter 1, the control volumes were small in a single dimension whereas they must be small in two dimensions for a 2-D problem. It is necessary to perform an energy balance on each differential control volume and provide rate equations that approximate each term in the energy balance based upon the nodal temperatures or other input parameters. The result of this step will be a set of equations (one for each control volume) in an equal number of unknown temperatures (one for each node). This set of equations can be solved in order to provide the numerical prediction of the temperature at each node. In this section, EES is used to solve the system of equations. In the next section, MATLAB is used to solve these types of problems.

In Section 1.6, the constant cross-section, straight fin shown in Figure 2-20 is analyzed under the assumption that it could be treated as an extended surface (i.e., temperature gradients in the y direction could be neglected). The fin is reconsidered using separation of variables in EXAMPLE 2.2-1 without making the extended surface approximation. In this section the problem is revisited again, this time using a finite difference technique to obtain a solution for the temperature distribution.

The tip of the fin is insulated and the width (W) is much larger than its thickness (th) so that convection from the edges of the fin can be neglected; therefore, the problem is 2-D. The length of the fin is $L = 5.0$ cm and its thickness is $th = 4.0$ cm. The fin base temperature is $T_b = 200°$C and it transfers heat to the surrounding fluid at $T_\infty = 20°$C with average heat transfer coefficient, $\bar{h} = 100$ W/m²-K. The conductivity of the fin material is $k = 0.5$ W/m-K.

The inputs are entered in EES:

```
$UnitSystem SI MASS RAD PA K J
$TABSTOPS 0.2 0.4 0.6 0.8 3.5 in

"Inputs"
L=5.0 [cm]*convert(cm,m)                    "length of fin"
th=4.0 [cm]*convert(cm,m)                   "width of fin"
k=0.5 [W/m-K]                               "thermal conductivity"
h_bar=100 [W/m^2-K]                         "heat transfer coefficient"
T_b=converttemp(C,K,200 [C])                "base temperature"
T_infinity=converttemp(C,K,20 [C])          "fluid temperature"
```

The computational domain associated with a half-symmetry model of the fin is shown in Figure 2-21.

The first step in obtaining a numerical solution is to position the nodes throughout the computational domain. A regularly spaced grid of nodes is uniformly distributed, with the first and last nodes in each dimension placed on the boundaries of the domain

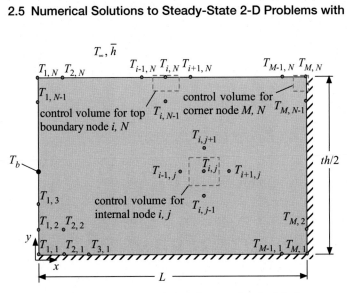

Figure 2-21: The computational domain associated with the constant cross-sectional area fin and the regularly spaced grid used to obtain a numerical solution.

as shown in Figure 2-21. The x- and y-positions of any node (i, j) are given by:

$$x_i = \frac{(i-1)\,L}{(M-1)} \tag{2-169}$$

$$y_j = \frac{(j-1)\,th}{2\,(N-1)} \tag{2-170}$$

where M and N are the number of nodes used in the x- and y-directions, respectively. The x- and y-distance between adjacent nodes (Δx and Δy, respectively) are:

$$\Delta x = \frac{L}{(M-1)} \tag{2-171}$$

$$\Delta y = \frac{th}{2\,(N-1)} \tag{2-172}$$

This information is entered in EES:

```
"Setup grid"
M=40 [-]                                    "number of x-nodes"
N=21 [-]                                    "number of y-nodes"
duplicate i=1,M
  x[i]=(i-1)*L/(M-1)                        "x-position of each node"
  x_bar[i]=x[i]/L                           "dimensionless x-position of each node"
end
DELTAx=L/(M-1)                              "x-distance between adjacent nodes"
duplicate j=1,N
  y[j]=(j-1)*th/(2*(N-1))                   "y-position of each node"
  y_bar[j]=y[j]/th                          "dimensionless y-position of each node"
end
DELTAy=th/(2*(N-1))                         "y-distance between adjacent nodes"
```

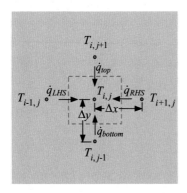

Figure 2-22: Energy balance for an internal node.

The next step in the solution is to write an energy balance for each node. Figure 2-22 illustrates a control volume and the associated energy transfers for an internal node (see Figure 2-21); each energy balance includes conduction from each side ($\dot{q}_{RHS}$ and $\dot{q}_{LHS}$), the top ($\dot{q}_{top}$), and the bottom ($\dot{q}_{bottom}$). Note that the direction associated with these energy transfers is arbitrary (i.e., they could have been taken as positive if energy leaves the control volume), but it is important to write the energy balance and rate equations in a manner that is consistent with the directions chosen in Figure 2-22. The energy balance suggested by Figure 2-22 is:

$$\dot{q}_{RHS} + \dot{q}_{LHS} + \dot{q}_{top} + \dot{q}_{bottom} = 0 \tag{2-173}$$

The next step is to approximate each of the terms in the energy balance. The material separating the nodes is assumed to behave as a plane wall thermal resistance. Therefore, $\Delta y\, W$ (where W is the width of the fin into the page) is the area for conduction between nodes (i, j) and $(i + 1, j)$ and Δx is the distance over which the conduction heat transfer occurs.

$$\dot{q}_{RHS} = \frac{k\, \Delta y\, W}{\Delta x}\, (T_{i+1,j} - T_{i,j}) \tag{2-174}$$

Note that the temperature difference in Eq. (2-174) is consistent with the direction of the arrow in Figure 2-22. The other conductive heat transfers are approximated using a similar model:

$$\dot{q}_{LHS} = \frac{k\, \Delta y\, W}{\Delta x}\, (T_{i-1,j} - T_{i,j}) \tag{2-175}$$

$$\dot{q}_{top} = \frac{k\, \Delta x\, W}{\Delta y}\, (T_{i,j+1} - T_{i,j}) \tag{2-176}$$

$$\dot{q}_{bottom} = \frac{k\, \Delta x\, W}{\Delta y}\, (T_{i,j-1} - T_{i,j}) \tag{2-177}$$

Substituting the rate equations, Eqs. (2-174) through (2-177), into the energy balance, Eq. (2-173) written for all of the internal nodes in Figure 2-21, leads to:

$$\frac{k\, \Delta y\, W}{\Delta x}\, (T_{i+1,j} - T_{i,j}) + \frac{k\, \Delta y\, W}{\Delta x}\, (T_{i-1,j} - T_{i,j}) + \frac{k\, \Delta x\, W}{\Delta y}\, (T_{i,j+1} - T_{i,j})$$
$$+ \frac{k\, \Delta x\, W}{\Delta y}\, (T_{i,j-1} - T_{i,j}) = 0 \quad \text{for } i = 2\ldots(M-1) \quad \text{and} \quad j = 2\ldots(N-1) \tag{2-178}$$

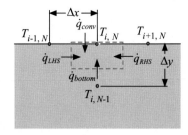

Figure 2-23: Energy balance for a node on the top boundary.

which can be simplified to:

$$\frac{\Delta y}{\Delta x}(T_{i+1,j} - T_{i,j}) + \frac{\Delta y}{\Delta x}(T_{i-1,j} - T_{i,j}) + \frac{\Delta x}{\Delta y}(T_{i,j+1} - T_{i,j}) + \frac{\Delta x}{\Delta y}(T_{i,j-1} - T_{i,j}) = 0$$

$$\text{for } i = 2 \dots (M-1) \quad \text{and} \quad j = 2 \dots (N-1) \quad (2\text{-}179)$$

These equations are entered in EES using nested duplicate loops:

```
"Internal node energy balances"
duplicate i=2,(M-1)
  duplicate j=2,(N-1)
    DELTAy*(T[i+1,j]-T[i,j])/DELTAx+DELTAy*(T[i-1,j]-T[i,j])/DELTAx&
      +DELTAx*(T[i,j+1]-T[i,j])/DELTAy+DELTAx*(T[i,j-1]-T[i,j])/DELTAy=0
  end
end
```

Note that each time the outer **duplicate** statement iterates once (i.e., i is increased by 1), the inner **duplicate** statement iterates $(N-2)$ times (i.e., j runs from 2 to $N-1$). Therefore, all of the internal nodes are considered with these two nested duplicate loops. Also note that the unknowns are placed in an array rather than a vector. The entries in the array T are accessed using two indices that are contained in square brackets.

Boundary nodes must be treated separately from internal nodes, just as they are in the 1-D problems that are considered in Section 1.4; however, 2-D problems have many more boundary nodes than 1-D problems. The left boundary ($x = 0$) is easy because the temperature is specified:

$$T_{1,j} = T_b \quad \text{for } j = 1 \dots N \quad (2\text{-}180)$$

where T_b is the base temperature. These equations are entered in EES:

```
"left boundary"
duplicate j=1,N
  T[1,j]=T_b
end
```

The remaining boundary nodes do not have specified temperatures and therefore must be treated using energy balances. Figure 2-23 illustrates an energy balance associated with a node that is located on the top boundary (at $y = th/2$, see Figure 2-21). The energy balance suggested by Figure 2-23 is:

$$\dot{q}_{RHS} + \dot{q}_{LHS} + \dot{q}_{bottom} + \dot{q}_{conv} = 0 \tag{2-181}$$

The conduction terms in the x-direction must be approximated slightly differently than for the internal nodes:

$$\dot{q}_{RHS} = \frac{k \, \Delta y \, W}{2 \Delta x} (T_{i+1,N} - T_{i,N}) \tag{2-182}$$

$$\dot{q}_{LHS} = \frac{k \, \Delta y \, W}{2 \, \Delta x} (T_{i-1,N} - T_{i,N}) \tag{2-183}$$

The factor of 2 in the denominator of Eqs. (2-182) and (2-183) appears because there is half the area available for conduction through the sides of the control volumes located on the top boundary. The conduction term in the y-direction is approximated as before:

$$\dot{q}_{bottom} = \frac{k \, \Delta x \, W}{\Delta y} (T_{i,N-1} - T_{i,N}) \tag{2-184}$$

The convection term is:

$$\dot{q}_{conv} = \overline{h} \, \Delta x \, W \, (T_{\infty} - T_{i,N}) \tag{2-185}$$

Substituting Eqs. (2-182) through (2-185) into Eq. (2-181) for all of the nodes on the upper boundary leads to:

$$\frac{k \, \Delta y \, W}{2 \Delta x} (T_{i+1,N} - T_{i,N}) + \frac{k \, \Delta y \, W}{2 \, \Delta x} (T_{i-1,N} - T_{i,N}) + \frac{k \, \Delta x \, W}{\Delta y} (T_{i,N-1} - T_{i,N})$$

$$+ W \, \Delta x \overline{h} \, (T_{\infty} - T_{i,N}) = 0 \quad \text{for } i = 2 \ldots (M-1) \tag{2-186}$$

which can be simplified to:

$$\frac{\Delta y}{2 \Delta x} (T_{i+1,N} - T_{i,N}) + \frac{\Delta y}{2 \, \Delta x} (T_{i-1,N} - T_{i,N}) + \frac{\Delta x}{\Delta y} (T_{i,N-1} - T_{i,N})$$

$$+ \frac{\Delta x \overline{h}}{k} (T_{\infty} - T_{i,N}) = 0 \quad \text{for } i = 2 \ldots (M-1) \tag{2-187}$$

These equations are entered in EES using a single **duplicate** statement:

```
"top boundary"
duplicate i=2,(m-1)
    DELTAy*(T[i+1,n]-T[i,n])/(2*DELTAx)+DELTAy*(T[i-1,n]-T[i,n])/(2*DELTAx)+&
        DELTAx*(T[i,n-1]-T[i,n])/DELTAy+DELTAx*h_bar*(T_infinity-T[i,n])/k=0
end
```

Notice that the control volume at the top left corner, node $(1, N)$, has already been specified by the equations for the left boundary, Eq. (2-180). It is important not to write an additional equation related to this node, or the problem will be over-specified. Therefore, the equations for the top boundary should only be written for $i = 2 \ldots (M-1)$.

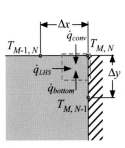

Figure 2-24: Energy balance for a node on the top right corner.

A similar procedure for the nodes on the lower boundary leads to:

$$\frac{\Delta y}{2\Delta x}\left(T_{i+1,1}-T_{i,1}\right)+\frac{\Delta y}{2\,\Delta x}\left(T_{i-1,1}-T_{i,1}\right)+\frac{\Delta x}{\Delta y}\left(T_{i,2}-T_{i,1}\right)=0 \quad \text{for } i=2\ldots(M-1)$$

(2-188)

Notice that there is no convection term in Eq. (2-188) because the lower boundary is adiabatic. These equations are entered into EES:

```
"bottom boundary"
duplicate i=2,(M-1)
   DELTAy*(T[i+1,1]-T[i,1])/(2*DELTAx)+DELTAy*(T[i-1,1]-T[i,1])/(2*DELTAx)&
      +DELTAx*(T[i,2]-T[i,1])/DELTAy=0
end
```

Energy balances for the nodes on the right-hand boundary $(x = L)$ lead to:

$$\frac{\Delta x}{2\Delta y}\left(T_{M,j+1}-T_{M,j}\right)+\frac{\Delta x}{2\,\Delta y}\left(T_{M,j-1}-T_{M,j}\right)+\frac{\Delta y}{\Delta x}\left(T_{M-1,j}-T_{M,j}\right)=0$$

(2-189)

$$\text{for } j=2\ldots(N-1)$$

```
"right boundary"
duplicate j=2,(n-1)
   DELTAx*(T[M,j+1]-T[M,j])/(2*DELTAy)+DELTAx*(T[M,j-1]-T[M,j])/(2*DELTAy)&
      +DELTAy*(T[M-1,j]-T[M,j])/DELTAx=0
end
```

The two corners (right upper and right lower) have to be considered separately. A control volume and energy balance for node (M, N), which is at the right upper corner (see Figure 2-21), is shown in Figure 2-24. The energy balance suggested by Figure 2-24 is:

$$\frac{k\,\Delta x\,W}{2\,\Delta y}\left(T_{M,N-1}-T_{M,N}\right)+\frac{k\,\Delta y\,W}{2\,\Delta x}\left(T_{M-1,N}-T_{M,N}\right)+\bar{h}\frac{\Delta x\,W}{2}\left(T_\infty-T_{M,N}\right)=0$$

(2-190)

which can be simplified to:

$$\frac{\Delta x}{2\,\Delta y}\left(T_{M,N-1}-T_{M,N}\right)+\frac{\Delta y}{2\,\Delta x}\left(T_{M-1,N}-T_{M,N}\right)+\frac{\bar{h}\,\Delta x}{2\,k}\left(T_\infty-T_{M,N}\right)=0 \quad \text{(2-191)}$$

and entered into EES:

```
"upper right corner"
DELTAx*(T[M,N-1]-T[M,N])/(2*DELTAy)+DELTAy*(T[M-1,N]-T[M,N])/(2*DELTAx)+&
   h_bar*DELTAx*(T_infinity-T[M,N])/(2*k)=0
```

The energy balance for the right lower boundary, node $(M, 1)$, leads to:

$$\frac{\Delta x}{2\,\Delta y}\,(T_{M,2} - T_{M,1}) + \frac{\Delta y}{2\,\Delta x}\,(T_{M-1,1} - T_{M,1}) = 0 \qquad (2\text{-}192)$$

```
"lower right corner"
DELTAx*(T[M,2]-T[M,1])/(2*DELTAy)+DELTAy*(T[M-1,1]-T[M,1])/(2*DELTAx)=0
```

We have derived a total of $M \times N$ equations in the $M \times N$ unknown temperatures; these equations completely specify the problem and they have now all been entered in EES. Therefore, a solution can be obtained by solving the EES code. The solution is contained in the Arrays window; each column of the table corresponds to the temperatures associated with one value of i and all of the values of j (i.e., the temperatures in a column are at a constant value of y and varying values of x). The temperature solution is converted from K to $^\circ$C with the following equations.

```
duplicate i=1,M
   duplicate j=1,N
      T_C[i,j]=converttemp(K,C,T[i,j])
   end
end
```

The solution is obtained for $N = 21$ and $M = 40$; the columns $T_{i,1}$ (corresponding to $y/th = 0$), $T_{i,5}$ (corresponding to $y/th = 0.10$), $T_{i,9}$ (corresponding to $y/th = 0.2$), etc. to $T_{i,21}$ (corresponding to $y/th = 0.50$) are plotted in Figure 2-25 as a function of the dimensionless x-position. The solution corresponds to our physical intuition as it exhibits temperature gradients in the x- and y-directions that correspond to conduction in these directions. The results from the analytical solution derived in EXAMPLE 2.2-1 are overlaid onto the plot and show nearly exact agreement with the numerical solution.

The fin efficiency will be used to verify that the grid is adequately refined. The fin efficiency is the ratio of the actual to the maximum possible heat transfer rates. The actual heat transfer rate per unit width ($\dot{q}'_{fin}$) is computed by evaluating the conductive heat transfer rate into the left hand side of each of the nodes that are located on the left boundary (i.e., all of the nodes where $i = 1$).

$$\dot{q}_{fin} = 2\left[k\frac{\Delta y}{2\,\Delta x}(T_{1,1} - T_{2,1}) + \sum_{j=2}^{m-1} k\frac{\Delta y}{\Delta x}(T_{1,j} - T_{2,j}) + k\frac{\Delta y}{2\,\Delta x}(T_{1,N} - T_{2,N}) \right]$$

$$(2\text{-}193)$$

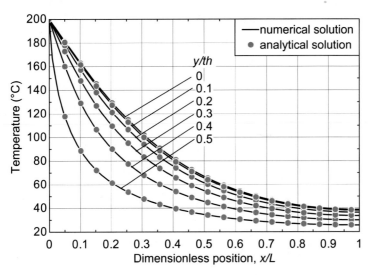

Figure 2-25: Temperature predicted by numerical model and analytical model (from EXAMPLE 2.2-1) as a function of x/L for various values of y/th.

while the maximum heat transfer rate per unit width ($\dot{q}'_{max}$) is associated with an isothermal fin:

$$\dot{q}'_{max} = 2\,L\,\overline{h}\,(T_b - T_\infty) \tag{2-194}$$

Note that the corner nodes must be considered outside of the duplicate loop as they have $\frac{1}{2}$ the cross-sectional area available for conduction. Also, the factor of 2 in Eq. (2-193) appears because the numerical model only considers one-half of the fin.

```
"calculate fin efficiency"
q_dot_fin[1]=(T[1,1]-T[2,1])*k*DELTAy/(2*DELTAx)
duplicate j=2,(N-1)
    q_dot_fin[j]=(T[1,j]-T[2,j])*k*DELTAy/DELTAx
end
q_dot_fin[N]=(T[1,N]-T[2,N])*k*DELTAy/(2*DELTAx)
q_dot_fin=2*sum(q_dot_fin[1..N])
q_dot_max=2*L*h_bar*(T_b-T_infinity)
eta_fin=q_dot_fin/q_dot_max
```

Figure 2-26 illustrates the fin efficiency as a function of the number of nodes in the x-direction (M) for various values of the number of nodes in the y direction (N). The solution is more sensitive to M than it is to N, but appears to converge (for the conditions considered here) when M is greater than 80 and N is greater than 10. (The solution for $N = 10$ is not shown in Figure 2-26, because it is nearly identical to the solution for $N = 20$.)

EES can solve up to 6,000 simultaneous equations, so a reasonably large problem can be considered using EES. However EES is not really the best tool for dealing with very large sets of equations. In the next section, we will look at how MATLAB can be used to solve this type of 2-D problem.

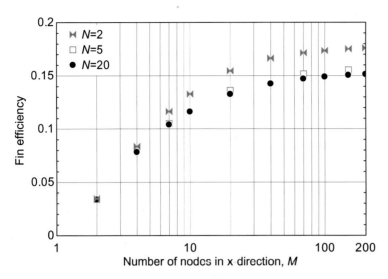

Figure 2-26: Fin efficiency as a function of M for various values of N predicted by the numerical model.

2.6 Numerical Solutions to Steady-State 2-D Problems with MATLAB

2.6.1 Introduction

Section 2.5 describes how 2-D, steady-state problems can be solved using a finite difference solution implemented in EES. This process is intuitive and easy because EES will automatically solve a set of implicit equations. However, EES is not well-suited for problems that involve very large numbers of equations. The finite difference method results in a system of algebraic equations that can be solved in a number of ways using different computer tools. Large problems will normally be implemented in a formal programming language such as C++, FORTRAN, or MATLAB. This section describes the methodology associated with solving the system of equations using MATLAB; however, the process is similar in any programming language.

2.6.2 Numerical Solutions with MATLAB

The system of equations that results from applying the finite difference technique to a steady-state problem can be solved by placing these equations into a matrix format:

$$\underline{\underline{A}}\,\underline{X} = \underline{b} \tag{2-195}$$

where the vector $\underline{X}$ contains the unknown temperatures. Each *row* of the $\underline{\underline{A}}$ matrix and $\underline{b}$ vector corresponds to an equation (for one of the control volumes in the computational domain) whereas each *column* of the $\underline{\underline{A}}$ matrix holds the coefficients that multiply the corresponding unknown (the nodal temperature) in that equation. To place a system of equations in matrix format, it is necessary to carefully define how the rows and energy balances are related and how the columns and unknown temperatures are related. This process is easy for the 1-D steady-state problems considered in Section 1.5, but it becomes somewhat more difficult for 2-D problems.

The basic steps associated with carrying out a 2-D finite difference solution using MATLAB remain the same as those discussed in Section 1.5 for a 1-D problem. The first step is to define the structure of the vector of unknowns, the vector $\underline{X}$ in Eq. (2-195). It doesn't really matter what order the unknowns are placed in $\underline{X}$, but the implementation of the solution is easier if a logical order is used. For a 2-D problem with M nodes in one dimension and N in the other, a logical technique for ordering the unknown temperatures in the vector $\underline{X}$ is:

$$\underline{X} = \begin{bmatrix} X_1 = T_{1,1} \\ X_2 = T_{2,1} \\ X_3 = T_{3,1} \\ \cdots \\ X_M = T_{M,1} \\ X_{M+1} = T_{1,2} \\ X_{M+2} = T_{2,2} \\ \cdots \\ X_{MN} = T_{M,N} \end{bmatrix} \tag{2-196}$$

Equation (2-196) indicates that temperature $T_{i,j}$ corresponds to element $X_{M(j-1)+i}$ of the vector $\underline{X}$; this mapping is important to keep in mind as you work towards implementing a solution in MATLAB.

The next step is to define how the control volume equations will be placed into each row of the matrix $\underline{\underline{A}}$. For a 2-D problem with M nodes in one dimension and N in the other, a logical technique is:

$$\underline{\underline{A}} = \begin{bmatrix} \text{row } 1 = \text{control volume equation for node } (1,1) \\ \text{row } 2 = \text{control volume equation for node } (2,1) \\ \text{row } 3 = \text{control volume equation for node } (3,1) \\ \cdots \\ \text{row } M = \text{control volume equation for node } (M,1) \\ \text{row } M+1 = \text{control volume equation for node } (1,2) \\ \text{row } M+2 = \text{control volume equation for node } (2,2) \\ \cdots \\ \text{row } MN = \text{control volume equation for node } (M,N) \end{bmatrix} \tag{2-197}$$

Equation (2-197) indicates that the equation for the control volume around node (i,j) is placed into row $M(j-1)+i$ of matrix $\underline{\underline{A}}$.

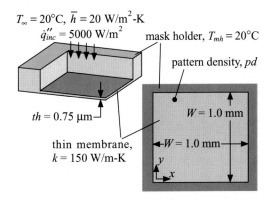

$T_\infty = 20°C$, $\bar{h} = 20$ W/m²-K
$\dot{q}''_{inc} = 5000$ W/m²   mask holder, $T_{mh} = 20°C$

pattern density, pd

$th = 0.75$ μm

$W = 1.0$ mm

thin membrane,
$k = 150$ W/m-K

$W = 1.0$ mm

y
x

Figure 2-27: EPL Mask.

The process of implementing a numerical solution in MATLAB is illustrated in the context of the problem shown in Figure 2-27. An electron projection lithography (EPL) mask may be used to generate the next generation of computer chips. The EPL mask is used to reflect an electron beam onto a resist-covered wafer. The EPL mask consists of a thin membrane that extends between the much larger struts of a mask holder. The membrane absorbs the electron beam in some regions and reflects it in others so that the resist on the wafer is developed (i.e., exposed to energy) only in certain locations. The developing process changes the chemical structure of the resist so that it is selectively etched away during subsequent processes; thus, the pattern on the mask is transferred to the wafer.

The portion of the membrane that absorbs the incident electron beam is heated. A precise calculation of the temperature rise in the mask is critical as any heating will result in thermally induced distortion that causes imaging errors in the printed features. Very small temperature rises can result in errors that are large relative to the printed features, as these features are themselves on the order of 100 nm or less.

A membrane that is $th = 0.75$ μm thick and $W = 1.0$ mm on a side contains the pattern to be written. The membrane is supported by the mask holder; this is a substantially thicker piece of material that can be assumed to be at a constant temperature, $T_{mh} = 20°$C. The thermal conductivity of the membrane is $k = 150$ W/m-K. The pattern density, pd (i.e., the fraction of the area of the mask that absorbs the incident radiation) can vary spatially across the EPL mask and therefore the thermal load applied to the surface of the mask area is non-uniform in x and y. The pattern density in this case is given by:

$$pd = 0.1 + 0.5\,\frac{xy}{W^2} \tag{2-198}$$

The thermal load on the mask per unit area at any location is equal to the product of the incident energy flux, $\dot{q}''_{inc} = 5000$ W/m^2, and the pattern density pd at that location. The mask is exposed to ambient air on both sides. The air temperature is $T_\infty = 20°$C and the average heat transfer coefficient is $\bar{h} = 20$ W/m^2-K.

The input parameters are entered in the MATLAB function **EPL_Mask**. The input arguments are M and N, the number of nodes in the x- and y-coordinates, while the output arguments are not specified yet.

```
function[]=EPL_Mask(M,N)
%[]=EPL_Mask
%
% This function determines the temperature distribution in an EPL_Mask
%
% Inputs:
% M - number of nodes in the x-direction (-)
% N - number of nodes in the y-direction (-)

%INPUTS
W=0.001;                          % width of membrane (m)
th=0.75e-6;                       % thickness of membrane (m)
q_dot_flux=5000;                  % incident energy (W/m^2)
k=150;                            % conductivity (W/m-K)
T_mh=20+273.2;                    % mask holder temperature (K)
T_infinity=20+273.2;              % ambient air temperature (K)
h_bar=20;                         % heat transfer coefficient (W/m^2-K)
```

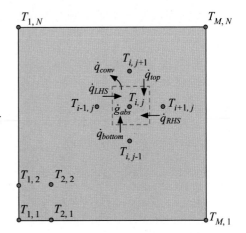

Figure 2-28: Numerical grid and control volume for an internal node (i, j).

A sub-function pd_f is created (at the bottom of the function EPL-Mask) in order to provide the pattern density:

```
function[pd]=pd_f(x,y,W)

% [pd]=pd_f(x,y,W)
%
% This sub-function returns the pattern density of the EPL mask
%
% Inputs:
% x - x-position (m)
% y - y-position (m)
% W - dimension of mask (m)
% Output:
% pd - pattern density (-)

  pd=0.1+0.5*x*y/W^2;
end
```

The problem is two-dimensional; the membrane is sufficiently thin that temperature gradients in the z-direction can be neglected. This assumption can be verified by calculating an appropriate Biot number:

$$Bi = \frac{\overline{h}\, th}{k} = 1 \times 10^{-7} \tag{2-199}$$

Therefore, a 2-D numerical model will be generated using the grid shown in Figure 2-28. The x- and y-coordinates of each node are provided by:

$$x_i = \frac{(i-1)\,W}{(M-1)} \quad \text{for } i = 1..M \tag{2-200}$$

$$y_i = \frac{(j-1)\,W}{(N-1)} \quad \text{for } j = 1..N \tag{2-201}$$

The distance between adjacent nodes is:

$$\Delta x = \frac{W}{(M-1)} \qquad (2\text{-}202)$$

$$\Delta y = \frac{W}{(N-1)} \qquad (2\text{-}203)$$

The grid is set up in the MATLAB function:

```
%Setup grid
for i=1:M
    x(i,1)=(i-1)*W/(M-1);
end
DELTAx=W/(M-1);
for j=1:N
    y(j,1)=(j-1)*W/(N-1);
end
DELTAy=W/(N-1);
```

The problem will be solved by placing the system of equations that result from considering each control volume into matrix format. A control volume for an internal node is shown in Figure 2-28. The energy balance for this control volume includes conduction from the left and right sides ($\dot{q}_{LHS}$ and $\dot{q}_{RHS}$) and the top and bottom ($\dot{q}_{top}$ and $\dot{q}_{bottom}$) as well as generation of thermal energy due to the absorbed illumination ($\dot{g}_{abs}$) and heat loss due to convection ($\dot{q}_{conv}$). The energy balance suggested by Figure 2-28 is:

$$\dot{q}_{RHS} + \dot{q}_{LHS} + \dot{q}_{top} + \dot{q}_{bottom} + \dot{g}_{abs} = \dot{q}_{conv} \qquad (2\text{-}204)$$

The conduction terms are approximated using the technique discussed in Section 2.5:

$$\dot{q}_{RHS} = \frac{k\,\Delta y\,th}{\Delta x}\,(T_{i+1,j} - T_{i,j}) \qquad (2\text{-}205)$$

$$\dot{q}_{LHS} = \frac{k\,\Delta y\,th}{\Delta x}\,(T_{i-1,j} - T_{i,j}) \qquad (2\text{-}206)$$

$$\dot{q}_{bottom} = \frac{k\,\Delta x\,th}{\Delta y}\,(T_{i,j-1} - T_{i,j}) \qquad (2\text{-}207)$$

$$\dot{q}_{top} = \frac{k\,\Delta x\,th}{\Delta y}\,(T_{i,j+1} - T_{i,j}) \qquad (2\text{-}208)$$

The absorbed energy is:

$$\dot{g}_{abs} = \dot{q}''_{inc}\,pd\,\Delta x\,\Delta y \qquad (2\text{-}209)$$

The rate of convection heat transfer is:

$$\dot{q}_{conv} = 2\,\overline{h}\,\Delta x\,\Delta y\,(T_{i,j} - T_\infty) \qquad (2\text{-}210)$$

Substituting Eqs. (2-205) to (2-210) into Eq. (2-204) for all internal nodes leads to:

$$\frac{k\,\Delta y\,th}{\Delta x}\,(T_{i-1,j} - T_{i,j}) + \frac{k\,\Delta y\,th}{\Delta x}\,(T_{i+1,j} - T_{i,j}) + \frac{k\,\Delta x\,th}{\Delta y}\,(T_{i,j-1} - T_{i,j}) + \frac{k\,\Delta x\,th}{\Delta y}\,(T_{i,j+1} - T_{i,j})$$

$$+ \dot{q}''_{inc}\,pd\,\Delta x\,\Delta y = 2\,\overline{h}\,\Delta x\,\Delta y\,(T_{i,j} - T_\infty) \quad \text{for } i = 2..(M-1) \quad \text{and} \quad j = 2..(N-1)$$

$$(2\text{-}211)$$

Equation (2-211) is rearranged to identify the coefficients that multiply each unknown temperature:

$$
T_{i,j}\underbrace{\left[-2\frac{k\,\Delta y\,th}{\Delta x}-2\frac{k\,\Delta x\,th}{\Delta y}-2\overline{h}\,\Delta x\,\Delta y\right]}_{A_{M(j-1)+i,M(j-1)+i}}+T_{i-1,j}\underbrace{\left[\frac{k\,\Delta y\,th}{\Delta x}\right]}_{A_{M(j-1)+i,M(j-1)+i-1}}+T_{i+1,j}\underbrace{\left[\frac{k\,\Delta y\,th}{\Delta x}\right]}_{A_{M(j-1)+i,M(j-1)+i+1}}
$$

$$
+T_{i,j-1}\underbrace{\left[\frac{k\,\Delta x\,th}{\Delta y}\right]}_{A_{M(j-1)+i,M(j-1-1)+i}}+T_{i,j+1}\underbrace{\left[\frac{k\,\Delta x\,th}{\Delta y}\right]}_{A_{M(j-1)+i,M(j+1-1)+i}}=\underbrace{-2\overline{h}\,\Delta x\,\Delta y\,T_{\infty}-\dot{q}''_{inc}\,pd\,\Delta x\,\Delta y}_{b_{M(j-1)+i}}\quad(2\text{-}212)
$$

$$
\text{for } i=2..(M-1)\quad\text{and}\quad j=2..(N-1)
$$

The control volume equations must be placed into the matrix equation:

$$
\underline{\underline{A}}\,\underline{X}=\underline{b}\tag{2-213}
$$

where the equation for the control volume around node (i,j) is placed into row $M(j-1)+i$ of $\underline{\underline{A}}$ and $T_{i,j}$ corresponds to element $X_{M(j-1)+i}$ in the vector $\underline{X}$, as required by Eqs. (2-197) and (2-196), respectively. Each coefficient in Eq. (2-212) (i.e., each term multiplying an unknown temperature on the left side of the equation) must be placed in the row of $\underline{\underline{A}}$ corresponding to the control volume being examined and the column of $\underline{\underline{A}}$ corresponding to the unknown in $\underline{X}$. The matrix assignments consistent with Eq. (2-212) are:

$$
A_{M(j-1)+i,M(j-1)+i}=-2\frac{k\,\Delta y\,th}{\Delta x}-2\frac{k\,\Delta x\,th}{\Delta y}-2\overline{h}\,\Delta x\,\Delta y\tag{2-214}
$$

$$
\text{for } i=2..(M-1)\quad\text{and}\quad j=2..(N-1)
$$

$$
A_{M(j-1)+i,M(j-1)+i-1}=\frac{k\,\Delta y\,th}{\Delta x}\quad\text{for } i=2..(M-1)\quad\text{and}\quad j=2..(N-1)\tag{2-215}
$$

$$
A_{M(j-1)+i,M(j-1)+i+1}=\frac{k\,\Delta y\,th}{\Delta x}\quad\text{for } i=2..(M-1)\quad\text{and}\quad j=2..(N-1)\tag{2-216}
$$

$$
A_{M(j-1)+i,M(j-1-1)+i}=\frac{k\,\Delta x\,th}{\Delta y}\quad\text{for } i=2..(M-1)\quad\text{and}\quad j=2..(N-1)\tag{2-217}
$$

$$
A_{M(j-1)+i,M(j+1-1)+i}=\frac{k\,\Delta x\,th}{\Delta y}\quad\text{for } i=2..(M-1)\quad\text{and}\quad j=2..(N-1)\tag{2-218}
$$

$$
b_{M(j-1)+i}=-2\overline{h}\,\Delta x\,\Delta y\,T_{\infty}-\dot{q}''_{inc}\,pd\,\Delta x\,\Delta y\quad\text{for } i=2..(M-1)\quad\text{and}\quad j=2..(N-1)\tag{2-219}
$$

A sparse matrix is allocated in MATLAB for $\underline{\underline{A}}$ and the equations derived above are implemented using nested for loops. The spalloc command requires the number of rows and columns and the maximum number of non-zero elements in the matrix. Note that there are at most five non-zero entries in each row of $\underline{\underline{A}}$, corresponding to Eqs. (2-214) through (2-218); thus the last argument in the spalloc command is 5 $M\,N$.

```
A=spalloc(M*N,M*N,5*M*N);          %allocate a sparse matrix for A
%energy balances for internal nodes
for i=2:(M-1)
  for j=2:(N-1)
    A(M*(j-1)+i,M*(j-1)+i)=-2*k*DELTAy*th/DELTAx-2*k*DELTAx*th/DELTAy-...
      2*h_bar*DELTAx*DELTAy;
    A(M*(j-1)+i,M*(j-1)+i-1)=k*DELTAy*th/DELTAx;
    A(M*(j-1)+i,M*(j-1)+i+1)=k*DELTAy*th/DELTAx;
    A(M*(j-1)+i,M*(j-1-1)+i)=k*DELTAx*th/DELTAy;
    A(M*(j-1)+i,M*(j+1-1)+i)=k*DELTAx*th/DELTAy;
    b(M*(j-1)+i,1)=-2*h_bar*DELTAx*DELTAy*T_infinity-...
      q_dot_flux*pd_f(x(i,1),y(j,1),W)*DELTAx*DELTAy;
  end
end
```

The boundary nodes have specified temperature:

$$\underbrace{T_{1,j}}_{A_{M(j-1)+1,M(j-1)+1}} \underbrace{[1]}_{} = \underbrace{T_{mh}}_{b_{M(j-1)+1}} \quad \text{for } j = 1..N \tag{2-220}$$

$$\underbrace{T_{M,j}}_{A_{M(j-1)+M,M(j-1)+M}} \underbrace{[1]}_{} = \underbrace{T_{mh}}_{b_{M(j-1)+M}} \quad \text{for } j = 1..N \tag{2-221}$$

$$\underbrace{T_{i,1}}_{A_{M(1-1)+i,M(1-1)+i}} \underbrace{[1]}_{} = \underbrace{T_{mh}}_{b_{M(1-1)+i}} \quad \text{for } i = 2..(M-1) \tag{2-222}$$

$$\underbrace{T_{i,N}}_{A_{M(N-1)+i,M(N-1)+i}} \underbrace{[1]}_{} = \underbrace{T_{mh}}_{b_{M(N-1)+i}} \quad \text{for } i = 2..(M-1) \tag{2-223}$$

Note that Eqs. (2-220) through (2-223) are written so that the corner nodes (e.g. node (1, 1)) are not specified twice. The matrix assignments suggested by Eqs. (2-220) through (2-223) are:

$$A_{M(j-1)+1,M(j-1)+1} = 1 \quad \text{for } j = 1..N \tag{2-224}$$

$$b_{M(j-1)+1} = T_{mh} \quad \text{for } j = 1..N \tag{2-225}$$

$$A_{M(j-1)+M,M(j-1)+M} = 1 \quad \text{for } j = 1..N \tag{2-226}$$

$$b_{M(j-1)+M} = T_{mh} \quad \text{for } j = 1..N \tag{2-227}$$

$$A_{M(1-1)+i,M(1-1)+i} = 1 \quad \text{for } i = 2..(M-1) \tag{2-228}$$

$$b_{M(1-1)+i} = T_{mh} \quad \text{for } i = 2..(M-1) \tag{2-229}$$

$$A_{M(N-1)+i,M(N-1)+i} = 1 \quad \text{for } i = 2..(M-1) \tag{2-230}$$

$$b_{M(N-1)+i} = T_{mh} \quad \text{for } i = 2..(M-1) \tag{2-231}$$

These assignments are implemented in MATLAB:

```
%specified temperatures around all edges
for j=1:N
    A(M*(j-1)+1,M*(j-1)+1)=1;
    b(M*(j-1)+1,1)=T_mh;
    A(M*(j-1)+M,M*(j-1)+M)=1;
    b(M*(j-1)+M,1)=T_mh;
end
for i=2:(M-1)
    A(M*(1-1)+i,M*(1-1)+i)=1;
    b(M*(1-1)+i,1)=T_mh;
    A(M*(N-1)+i,M*(N-1)+i)=1;
    b(M*(N-1)+i,1)=T_mh;
end
```

The vector $\underline{X}$ is obtained using MATLAB's backslash command and the temperature of each node in degrees Celsius is placed in the matrix T_C.

```
X=A\b;
for i=1:M
    for j=1:N
        T_C(i,j)=X(M*(j-1)+i)-273.2;
    end
end
end
```

The function header is modified so that running the MATLAB function provides the temperature prediction (the matrix T_C) as well as the vectors x and y that contain the *x*- and *y*-positions of each node in the matrix.

```
function[x,y,T_C]=EPL_Mask(M,N)

% [x,y,T_C]=EPL_Mask
%
% This function determines the temperature distribution in an EPL_Mask
%
% Inputs:
% M - number of nodes in the x-direction (-)
% N - number of nodes in the y-direction (-)
% Outputs:
% x - Mx1 vector of x-positions of each node (m)
% y - Nx1 vector of y-positions of each node (m)
% T_C - MxN matrix of temperature at each node (C)
```

The solution should be examined for grid convergence. Figure 2-29 illustrates the maximum temperature in the EPL mask as a function of M and N (the two parameters are set equal for this analysis). The analysis is carried out using the script varyM, below, which defines a vector Mv that contains a range of values of the number of nodes, M, and runs

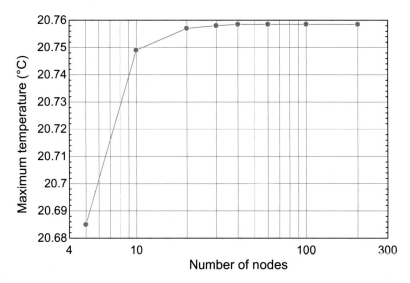

Figure 2-29: Maximum predicted temperature as a function of the number of nodes (M and N).

the function EPL_Mask for each value. The command max(max(T_C)) computes the maximum value of each column and then the maximum value of the resulting vector in order to obtain the maximum nodal temperature in the mask.

```
clear all;
Mv=[5;10;20;30;50;70;100;200];       % values of M to use
for i=1:8
    [x,y,T_C]=EPL_Mask(Mv(i),Mv(i));    % obtain temperature distribution
    T_Cmaxv(i,1)=max(max(T_C))          % obtain maximum temperature
end
```

There is a variety of 3-D plotting functions in MATLAB; these can be investigated by typing help graph3d at the command window. For example,

```
>> mesh(x,y,T_C');
>> colorbar;
```

produces a mesh plot indicating the temperature, as shown in Figure 2-30. Note that the matrix $\underline{\underline{T}}_C$ has to be transposed (by adding the ' character after the variable name) in order to match the dimensions of the $\underline{x}$ and $\underline{y}$ vectors.

2.6.3 Numerical Solution by Gauss-Seidel Iteration

This extended section of the book can be found on the website www.cambridge.org/ nellisandklein. A finite difference solution results in a system of algebraic equations that must be solved simultaneously. In Section 2.6.2, we looked at placing these equations into a matrix equation that was solved by a single matrix inversion (or the equivalent mathematical manipulation). An alternative technique, Gauss-Seidel iteration, can also be used to approximately solve the system of equations using an iterative technique that requires much less memory than the direct matrix solution method. In some cases

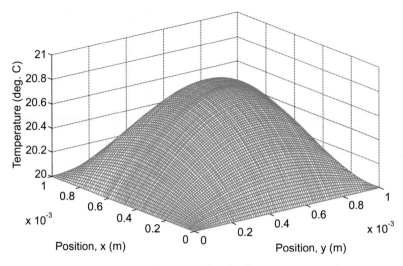

Figure 2-30: Mesh plot of temperature distribution.

it can require less computational effort as well. The Gauss-Seidel iteration process is illustrated using the EPL mask problem that was discussed in Section 2.6.2.

2.7 Finite Element Solutions

This extended section of the book can be found on the website www.cambridge.org/ nellisandklein. Sections 2.5 and 2.6 present the finite difference method for solving 2-D steady-state conduction problems. In Section 2.7.1, the FEHT (Finite Element Heat Transfer) program is discussed and used to solve EXAMPLE 2.7-1. FEHT implements the finite element technique to solve 2-D steady-state conduction problems. A version of FEHT that is limited to 1000 nodes can be downloaded from www.cambridge.org/ nellisandklein. In order to become familiar with FEHT it is suggested that the reader go through the tutorial provided in Appendix A.4 which can be found on the web site associated with the book (www.cambridge.org/nellisandklein). In Section 2.7.2, the theory behind finite element techniques is presented.

2.8 Resistance Approximations for Conduction Problems

2.8.1 Introduction

The resistance to conduction through a plane wall (R_{pw}) was derived in Section 1.2 and is given by:

$$R_{pw} = \frac{L}{k A_c} \tag{2-232}$$

The concept of a thermal resistance is a broadly useful and practical idea that goes beyond the simple situation for which Eq. (2-232) was derived. It is possible to understand conduction heat transfer in most situations if you can identify the appropriate distance that heat must be conducted (L) and the area through which that conduction occurs (A_c). EXAMPLE 2.8-1 illustrates this type of "back-of-the-envelope" calculation.

EXAMPLE 2.8-1: RESISTANCE OF A BRACKET

EXAMPLE 2.8-1: RESISTANCE OF A BRACKET

You may be faced with trying to understand the heat transfer through a complex, 2-D or 3-D geometry, such as the bracket illustrated in Figure 1. The bracket is made of steel having a thermal conductivity $k = 14$ W/m-K. One surface of the bracket is held at $T_H = 200°C$ and the other is at $T_C = 20°C$.

It is beyond the scope of any technique discussed in this book to analytically determine the heat flow through this geometry and therefore it will be necessary to use a finite element software package for this purpose. However, it is possible to use the resistance concept represented by Eq. (2-232) in order to bound and estimate the heat flow through the bracket. If you determine that the heat flow through the bracket cannot possibly be important to the larger application (whatever that is) then the time and money required to generate the finite element model can be saved. If a finite element model is generated, then the simple thermal resistance estimate can provide a sanity check on the results.

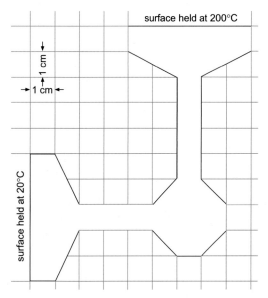

Figure 1: A bracket with a complex, 2-D geometry made of steel with $k = 14$ W/m-K and thickness 1 cm (into the page).

a) **Estimate the rate of heat transfer through the bracket using a resistance approximation.**

The length that heat must be conducted in order to go from the surface at T_H to the surface at T_C is approximately $L = 14$ cm and the area for conduction is approximately $A_c = 1$ cm². Clearly these are not exact values because the problem is two-dimensional; some energy must flow a longer distance to reach the more proximal regions of the bracket and there are several portions of the bracket where the area is larger than 1 cm². However, it is possible to estimate the resistance of the bracket with these approximations:

$$R_{bracket} \approx \frac{L}{k\,A} = \frac{14\,\text{cm}}{} \left| \frac{\text{m K}}{14\,\text{W}} \right| \frac{1}{1\,\text{cm}^2} \left\| \frac{100\,\text{cm}}{\text{m}} \right. = 100\,\frac{\text{K}}{\text{W}}$$

which provides an estimate of the heat flow:

$$\dot{q} \approx \frac{(T_H - T_C)}{R_{bracket}} = \frac{(200°\mathrm{C} - 20°\mathrm{C})}{} \left| \frac{\mathrm{W}}{100\,\mathrm{K}} \right. = 1.8\,\mathrm{W}$$

It may be that 1.8 W is a trivial rate of energy loss from whatever is being supported by the bracket and therefore the bracket does not require a more detailed analysis. However, if a more exact answer is required then a finite element solution is necessary.

b) Use FEHT to determine the rate of heat transfer through the bracket.

The geometry from Figure 1 can be entered in FEHT and solved, as discussed in Section 2.7.1 and Appendix A.4. Set a scale where 1 cm on the screen corresponds to 0.01 m and use the Outline selection from the Draw menu to approximately trace out the bracket. Then, right-click on each of the corner nodes and enter the exact position in the Node Information Dialog. The boundary conditions should be set as well as the material properties. Create a crude mesh and refine it.

The problem is solved and the solution is shown in Figure 2.

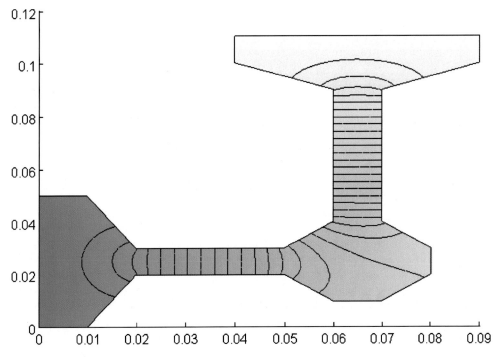

Figure 2: Solution.

The total heat flux at either the 200°C or the 20°C boundaries can be obtained by selecting Heat Flows from the View menu and then selecting all of the nodal boundaries along these boundaries. (Left-click and drag a selection rectangle.) At the T_H boundary, the total heat flow is reported as 249.8 W/m (Figure 3) or, for a 1 cm thick bracket, 2.5 W.

EXAMPLE 2.8-1: RESISTANCE OF A BRACKET

EXAMPLE 2.8-1: RESISTANCE OF A BRACKET

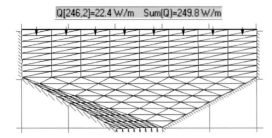

Figure 3: Heat flow along the top boundary.

The same calculation along the T_C boundary leads to 2.5 W. Therefore, the total heat flow is within 40% of the 1.8 W value predicted by the simple resistance approximation. The sanity check is valuable; if the finite element model had predicted 10's or 100's of W then it would be almost certain that there is an error in the solution (perhaps a unit conversion or a material property entered incorrectly). Furthermore, the 1-D solution was an underestimate of the heat transfer because it did not account for the regions of the bracket that have larger cross-section. In the next sections, several methods are presented that can be used to bound the thermal resistance of a multi-dimensional object using 1-D resistances that are calculated with specific assumptions.

2.8.2 Isothermal and Adiabatic Resistance Limits

Figure 2-31(a) illustrates a composite structure made from four materials (A through D). The composite structure experiences convection on its left and right sides to fluid temperatures $T_C = 0°C$ and $T_H = 100°C$, respectively, with average heat transfer coefficient $\bar{h} = 500\ \text{W/m}^2\text{-K}$. The other surfaces are insulated. The problem represented by Figure 2-31(a) is 2-D; to see this clearly, imagine the situation where material B has very low conductivity, $k_B = 1.0$ W/m-K, and material C has very high conductivity, $k_C = 100.0$ W/m-K. Material A and D both have intermediate conductivity, $k_A = k_D = 10$ W/m-K. In this limit, thermal energy will transfer primarily through material C with very little passing through material B. The problem was solved in FEHT and the resulting temperature distribution is shown in Figure 2-31(b); the heat transfer rate is 281.9 W (assuming unit width into the page).

The composite structure cannot be represented exactly with a 1-D resistance network due to the temperature gradients in the y-direction. In order to estimate the behavior of the system using 1-D resistance concepts, it is necessary to either allow unrestricted heat flow in the y-direction (referred to as the isothermal limit) or completely eliminate heat flow in the y-direction (referred to as the adiabatic limit). The temperature distribution is substantially simplified in these two limiting cases, as shown in Figure 2-32(a) and Figure 2-32(b), respectively. Note that Figure 2-32(a) and (b) can be obtained using FEHT; the material properties provides an Anisotropic ky/kx Type. For Figure 2-32(a), the ky/kx value is set to a large value (10,000) which effectively eliminates any resistance to heat flow in the y-direction. In Figure 2-32(b), the ky/kx value is set to a small value (0.0001) which essentially prevents any heat flow in the y-direction.

The 1-D resistance network that corresponds to the isothermal limit is shown in Figure 2-33. The isothermal limit implies that there are no temperature gradients in the y-direction and therefore the temperature at any axial location may be represented by a single node. Note that in Figure 2-33, A_c is the area of the composite (0.02 m^2 assuming

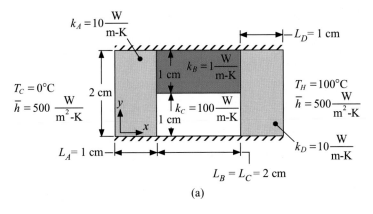

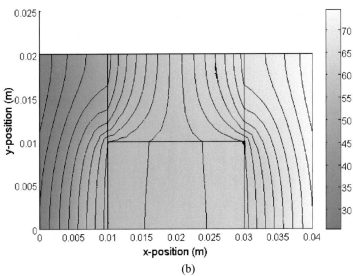

Figure 2-31: (a) A composite structure consisting of four materials with convection from each edge and (b) the temperature distribution (°C) that will occur if $k_C \gg k_B$.

a unit width into the page) and L_A, L_B, etc. are the thicknesses of the materials in the x-direction. The total resistance associated with the resistance network shown in Figure 2-33 is 0.32 K/W and therefore the rate of heat transfer through the composite structure predicted in the isothermal limit is 313 W. The isothermal limit corresponds to a lower bound on the thermal resistance, since the resistance to heat flow in the y-direction is neglected.

The adiabatic limit assumes that there is no heat transfer in the y-direction. Thermal energy can only pass through the composite axially and therefore the resistance network corresponding to the adiabatic limit consists of parallel paths for heat flow by convection and through materials B and C (these parallel paths are labeled 1 and 2), as shown in Figure 2-34. The total thermal resistance associated with the resistance network shown in Figure 2-34 is 0.50 K/W and the rate of heat transfer through the composite structure predicted in the adiabatic limit is 200 W. This adiabatic limit corresponds to an upper bound on the resistance since the thermal energy is prohibited from spreading in the y-direction. The true solution lies somewhere between the isothermal (313 W) and adiabatic (200 W) limits; the FEHT model in Figure 2-31 predicted 282 W. Thus, the limits

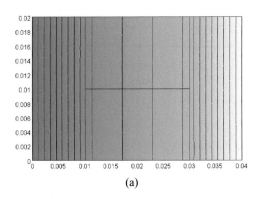

(a)

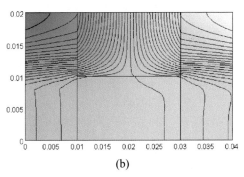

(b)

Figure 2-32: The temperature distribution in (a) the isothermal limit ($k_{A,y}$, $k_{B,y}$, $k_{C,y}$, and $k_{D,y} \rightarrow \infty$) and (b) the adiabatic limit ($k_{A,y}$, $k_{B,y}$, $k_{C,y}$, and $k_{D,y} \rightarrow 0$).

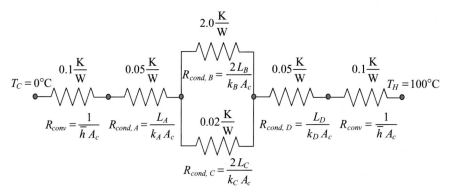

Figure 2-33: Resistance network representing the isothermal limit of the behavior of the composite structure.

Figure 2-34: Resistance network representing the adiabatic limit of the behavior of the composite structure.

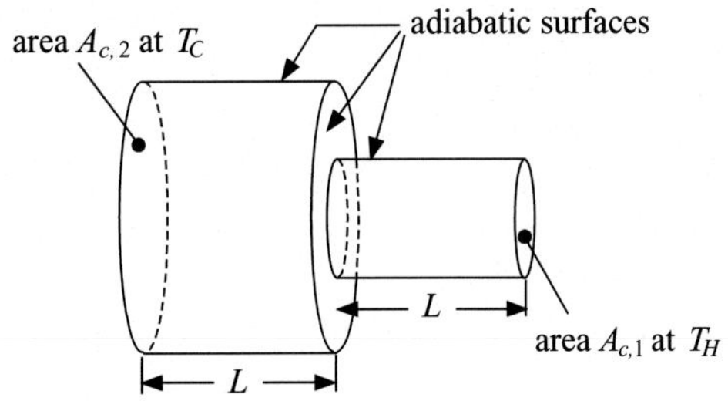

Figure 2-35: An example of a geometry with constant length and varying area.

are useful for determining the validity of a 2-D solution as well as bounding the problem without requiring a 2-D solution.

2.8.3 Average Area and Average Length Resistance Limits

There are conduction problems in which the length for conduction is known but the area of the conduction path varies along this length; a simple example is shown in Figure 2-35. The adiabatic approximation discussed in Section 2.8.2 suggests that the resistance of this shape is:

$$R_{ad} = \frac{2L}{k A_{c,1}} \tag{2-233}$$

where k is the conductivity of the material. The isothermal approximation for the resistance yields:

$$R_{iso} = \frac{L}{k A_{c,1}} + \frac{L}{k A_{c,2}} \tag{2-234}$$

An alternative technique for estimating the resistance in this situation is to use the average area for conduction:

$$R_{\overline{A}} = \frac{4L}{k (A_{c,1} + A_{c,2})} \tag{2-235}$$

The resistance based on the average area underestimates the actual resistance to a greater extent than even the isothermal approximation. To see that this is so, imagine the case where $A_{c,1}$ approaches zero; clearly the actual resistance will become infinite and both the adiabatic and isothermal approximations provided by Eqs. (2-233) and (2-234), respectively, predict this. However, the average area estimate remains finite and therefore substantially under-predicts the resistance.

The alternative situation may occur, where the area for conduction is essentially constant but the length varies (perhaps randomly, as in a contact resistance problem). In this case, it is natural to use an average length to compute the resistance as shown in Figure 2-36. The average length estimate of the resistance ($R_{\overline{L}}$) is:

$$R_{\overline{L}} = \frac{\overline{L}}{k A_c} \tag{2-236}$$

where $\overline{L}$ is the average length for conduction. The average length model overestimates the resistance to a greater extent than the adiabatic approximation. Consider the case where the length anywhere within the shape shown in Figure 2-36 approaches zero, which would cause the actual resistance to become zero. The adiabatic approximation will faithfully predict a zero resistance while the average length estimate will remain

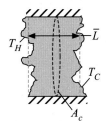

Figure 2-36: An example of a geometry with constant area and varying length.

finite. In terms of accuracy, the various 1-D estimates that have been discussed can be arranged in the following order:

$$R_{\overline{A}} \le R_{iso} \le R \le R_{ad} \le R_{\overline{L}} \tag{2-237}$$

where R is the actual thermal resistance.

EXAMPLE 2.8-2: RESISTANCE OF A SQUARE CHANNEL

Figure 1 illustrates a square channel. The inner and outer surfaces are held at different temperatures, $T_1 = 250°C$ and $T_2 = 50°C$, respectively.

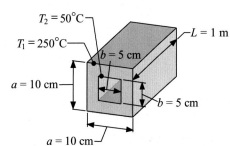

Figure 1: Square channel with heat transfer from inner to outer surface.

The outer dimension of the square channel is $a = 10$ cm and the inner dimension is $b = 5.0$ cm. The thermal conductivity of the material is $k = 100$ W/m-K and the length of the channel is $L = 1$ m.

a) **Using an appropriate shape factor, determine the actual rate of heat transfer through the square channel.**

The inputs are entered in EES:

```
"EXAMPLE 2.8-2: Resistance of a Square Channel"

$UnitSystem SI MASS RAD PA K J
$TABSTOPS 0.2 0.4 0.6 0.8 3.5 in

"Inputs"
a=10[cm]*convert(cm,m)              "outer dimension of square channel"
b=5[cm]*convert(cm,m)               "inner dimension"
L=1[m]                              "length"
k=100[W/m-K]                        "material conductivity"
T_1=converttemp(C,K,250 [C])        "inner wall temperature"
T_2=converttemp(C,K,50 [C])         "outer wall temperature"
```

EXAMPLE 2.8-2: RESISTANCE OF A SQUARE CHANNEL

This problem is a 2-D conduction problem; however, the solution for this 2-D problem is correlated in the form of a shape factor, S. Shape factors are discussed in Section 2.1. The shape factor (S) is defined according to:

$$\dot{q}_{cond} = S\,k\,(T_1 - T_2)$$

where $\dot{q}_{cond}$ is the rate of conductive heat transfer between the two surfaces at T_1 and T_2. The particular shape factor for a square channel can be accessed from EES' built-in shape factor library. Select Function Info from the Options menu and select Shape Factors from the lower-right pull down menu. Use the scroll-bar to select the shape factor function for a square channel. The function SF_7 is pasted into the Equation Window using the Paste button and used to calculate the actual heat transfer rate:

```
"Actual heat transfer rate"
SF=SF_7(b,a,L)                    "shape factor for square channel"
q_dot=SF*k*(T_1-T_2)             "heat transfer"
q_dot_kW=q_dot*convert(W,kW)     "heat transfer in kW"
```

The actual heat transfer rate predicted using a shape factor solution is 211.4 kW.

b) Provide a lower bound on the heat transfer through the square channel using an appropriate 1-D model.

According to Eq. (2-237), the adiabatic or average length models can be used to provide an upper bound on the resistance of the square channel. The adiabatic model does not allow the heat to spread as it moves across the channel, as shown in Figure 2(a).

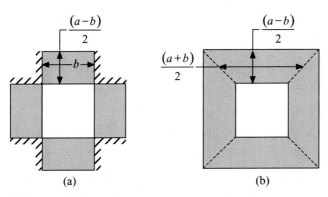

(a)   (b)

Figure 2: 1-D models based on the (a) adiabatic limit which allows no heat spreading and (b) the average area limit which uses the average area along the heat transfer path.

The resistance in the adiabatic limit is equal to the resistance of a plane wall with an area equal to the internal surface area of the channel and length equal to the channel thickness:

$$R_{ad} = \frac{(a-b)}{8\,k\,L\,b}$$

The rate of heat transfer predicted by the adiabatic limit is:

$$\dot{q}_{ad} = \frac{(T_1 - T_2)}{R_{ad}}$$

EXAMPLE 2.8-2: RESISTANCE OF A SQUARE CHANNEL

"Adiabatic limit"
R_ad=(a-b)/(8*k*L*b)   "thermal resistance in the adiabatic limit"
q_dot_ad=(T_1-T_2)/R_ad   "heat transfer in the adiabatic limit"
q_dot_ad_kW=q_dot_ad*convert(W,kW)

The adiabatic model predicts 160.0 kW and therefore leads to a 32% underestimate of the actual heat transfer rate (211.4 kW from part (a)).

c) Provide an upper bound on the heat transfer rate using an appropriate 1-D model.

Equation (2-237) indicates that either the isothermal or average area approach can be used to establish a lower bound on the thermal resistance and therefore an upper bound on the heat transfer. The average area along the heat transfer path is shown in Figure 2(b). The resistance calculated according to the average area model is:

$$R_{\overline{A}} = \frac{(a - b)}{4\,(a + b)\,L\,k}$$

The heat transfer in this limit is:

$$\dot{q}_{\overline{A}} = \frac{(T_1 - T_2)}{R_{\overline{A}}}$$

"Average area limit"
R_A_bar=(a-b)/(4*(a+b)*k*L)   "thermal resistance in the average area limit"
q_dot_A_bar=(T_1-T_2)/R_A_bar   "heat transfer in the average area limit"
q_dot_A_bar_kW=q_dot_A_bar*convert(W,kW)

The average area approximation predicts a heat transfer rate of 240 kW and is therefore a 12% overestimate of the actual heat transfer rate.

2.9 Conduction through Composite Materials

2.9.1 Effective Thermal Conductivity

Composite structures are made by joining different materials to create a structure with beneficial properties. Composites are often encountered in engineering applications; for example, motor laminations and windings, screens, and woven fabric composites. The length scale associated with the underlying structure of the composite material is often much smaller than the length scale associated with the overall problem of interest and therefore the details of the local energy flow through the materials that make up the structure are not important. In this case, the composite material can be modeled as a single, equivalent material with an effective conductivity that reflects the more complex behavior of the underlying structure. If the composite structure is not isotropic (i.e., it is anisotropic) then the effective conductivity may also be anisotropic; that is, the effective conductivity may depend on direction, reflecting some underlying characteristic of the composite structure that allows heat to flow more easily in certain directions.

The effective thermal conductivity of the composite structure must be determined by considering the details associated with heat transfer through the structure. The process of determining the effective conductivity involves (theoretically) imposing a temperature gradient in one direction and evaluating the resulting heat transfer rate. The

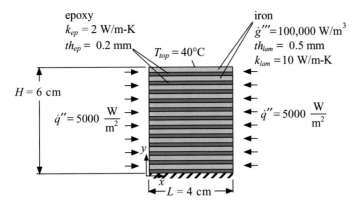

Figure 2-37: Motor pole.

effective conductivity in that direction is the conductivity of a homogeneous material that would yield the same heat transfer rate. It is often possible to determine the effective conductivity by inspection; however, for complex structures it will be necessary to generate a detailed, numerical model of a unit cell of the structure using a finite difference or finite element technique. The combination of a local model of the very small scale features of the underlying structure (in order to determine the effective conductivity) and a larger scale model of the global problem is sometimes referred to as multi-scale modeling.

Other effective characteristics of the composite may also be important. For example, an effective rate of volumetric generation or, for transient problems, an effective specific heat capacity and density. An effective property is the property that a homogenous material must have if it is to behave in the same way as the composite.

The process of estimating and using an effective conductivity to model a composite structure is illustrated in the context of the motor pole shown in Figure 2-37. The pole is composed of laminations of iron that are separated by an epoxy coating. Each iron lamination is $th_{lam} = 0.5$ mm thick and has conductivity $k_{lam} = 10$ W/m-K while the epoxy coating is approximately $th_{ep} = 0.2$ mm thick and has conductivity $k_{ep} = 2.0$ W/m-K. The motor pole is adiabatic on its bottom surface and experiences a heat flux of $\dot{q}'' = 5000$ W/m² from the windings on the sides. The top surface is maintained at a temperature of $T_{top} = 40°$C. The pole is $L = 4.0$ cm long and $H = 6.0$ cm high. The temperature distribution in the pole is 2-D in the x- and y-directions. The iron laminations are generating thermal energy due to eddy current heating at a volumetric rate of $\dot{g}''' = 100{,}000$ W/m³. There is no thermal energy generation in the epoxy.

The inputs required to determine the effective conductivity are entered into EES:

```
$UnitSystem SI MASS RAD PA K J
$TABSTOPS 0.2 0.4 0.6 0.8 3.5 in

"Inputs"
k_lam=10 [W/m-K]                        "lamination conductivity"
k_ep=2 [W/m-K]                          "epoxy conductivity"
th_lam=0.5 [mm]*convert(mm,m)           "lamination thickness"
th_ep=0.2 [mm]*convert(mm,m)           "epoxy thickness"
g'''_dot=100000 [W/m^3]                 "rate of volumetric generation in the laminations"
```

Heat transfer in the axial (x) direction occurs through the laminations and epoxy in parallel. The methodology for calculating the effective conductivity in the axial direction consists of imposing a temperature difference (ΔT) in the x-direction (i.e., across the width of the pole) and calculating the heat transfer rate through the two parallel paths. The effective conductivity in the x-direction ($k_{eff,x}$) is the conductivity of a homogeneous material that would provide the same heat transfer rate.

The rate of heat transfer through the iron laminations is:

$$\dot{q}_{lam} = \frac{k_{lam}\, th_{lam}\, H\, W}{(th_{lam} + th_{ep})\, L}\Delta T \tag{2-238}$$

where W is the width of the pole (into the page). The heat transfer rate through the epoxy is:

$$\dot{q}_{ep} = \frac{k_{ep}\, th_{ep}\, H\, W}{(th_{lam} + th_{ep})\, L}\Delta T \tag{2-239}$$

The total heat transfer rate in the x-direction through the equivalent material ($\dot{q}_{eff,x}$) must therefore be:

$$\dot{q}_{eff,x} = \dot{q}_{lam} + \dot{q}_{ep} = \frac{k_{lam}\, th_{lam}\, H\, W}{(th_{lam} + th_{ep})\, L}\Delta T + \frac{k_{ep}\, th_{ep}\, H\, W}{(th_{lam} + th_{ep})\, L}\Delta T = \frac{k_{eff,x}\, H\, W}{L}\Delta T \tag{2-240}$$

or, solving for $k_{eff,x}$:

$$k_{eff,x} = \frac{k_{lam}\, th_{lam}}{(th_{lam} + th_{ep})} + \frac{k_{ep}\, th_{ep}}{(th_{lam} + th_{ep})} \tag{2-241}$$

The effective conductivity in the x-direction is the thickness (or area) weighted average of the conductivity of the two parallel paths. The effective conductivity calculated using Eq. (2-241) depends only on the details of the microstructure (e.g., the conductivity of the laminations and their thickness) and not the macroscopic details of the problem (e.g., the size of the motor pole and the boundary conditions on the problem).

```
k_eff_x=k_lam*th_lam/(th_lam+th_ep)+k_ep*th_ep/(th_lam+th_ep)
```
"eff. conductivity in the x-direction"

which leads to $k_{eff,x} = 7.71$ W/m-K. Note that the effective conductivity must lie between the conductivity of the epoxy (2 W/m-K) and the laminations (10 W/m-K). If the thickness of the lamination becomes large relative to the thickness of the epoxy then the effective conductivity will approach the lamination conductivity. If the conductivity of one material (for example, the lamination) is substantially greater than the other (for example, the epoxy) then the effective conductivity in the x-direction will approach the product of the conductivity of the more conductive material and the fraction of the thickness that is occupied by that material. This behavior is typical of any parallel resistance network because the rate of heat transfer is more sensitive to the smaller resistance.

Heat transferred in the y-direction must pass through the laminations and epoxy in series. A temperature difference is applied across the pole in the y-direction. The heat transfer rate is calculated and used to establish the effective conductivity in the y-direction:

$$\dot{q}_{eff,y} = \frac{\Delta T}{\dfrac{H\, th_{lam}}{(th_{lam} + th_{ep})\, k_{lam}\, W\, L} + \dfrac{H\, th_{ep}}{(th_{lam} + th_{ep})\, k_{ep}\, W\, L}} = \frac{k_{eff,y}\, W\, L\, \Delta T}{H} \tag{2-242}$$

or, solving for $k_{eff,y}$:

$$k_{eff,y} = \cfrac{1}{\cfrac{th_{lam}}{(th_{lam} + th_{ep})\,k_{lam}} + \cfrac{th_{ep}}{(th_{lam} + th_{ep})\,k_{ep}}} \tag{2-243}$$

> k_eff_y=1/(th_lam/((th_lam+th_ep)*k_lam)+th_ep/((th_lam+th_ep)*k_ep))
> "eff. conductivity in the y-direction"

which leads to $k_{eff,y} = 4.67$ W/m-K. The effective conductivity is again bounded by k_{ep} and k_{lam}. The effective conductivity in the y-direction is dominated by the thicker and less conductive of the two materials. This behavior is typical of a series resistance network, where the larger resistance is the most important.

In addition to the effective conductivity, it is necessary to determine an effective rate of volumetric generation ($\dot{g}'''_{eff}$) that characterizes the motor pole. This is the volumetric generation rate for an equivalent piece of homogeneous material that produces the same total rate of generation. The total rate of energy generation in the motor pole is:

$$\dot{g} = \dot{g}''' \frac{H\,th_{lam}}{(th_{lam} + th_{ep})} W\,L = \dot{g}'''_{eff}\,H\,W\,L \tag{2-244}$$

and therefore:

$$\dot{g}'''_{eff} = \dot{g}''' \frac{th_{lam}}{(th_{lam} + th_{ep})} \tag{2-245}$$

> g'''_dot_eff=g'''_dot*th_lam/(th_lam+th_ep)   "effective rate of volumetric generation"

which leads to $\dot{g}'''_{eff} = 71,400$ W/m^3.

The effective properties of the motor pole, $k_{eff,x}$, $k_{eff,y}$, and $\dot{g}'''_{eff}$, can be used to generate a model of the motor pole that does not explicitly consider the microscale features of the composite structure but does capture the overall geometry and boundary conditions of the problem shown in Figure 2-37. The geometry is entered in FEHT as discussed in Section 2.7.1 and Appendix A.4. A grid is used where 1 cm of the screen corresponds to 1 cm and the corner nodes are approximately positioned using the Outline command in the Draw menu. The corner nodes are then precisely positioned by right-clicking on each in turn. The material properties are set by clicking on the outline and selecting Material Properties from the Specify menu. Create a new material (select "not specified" from the list of materials) and rename it lamination. Click on the box next to Type until the choice is Anisotropic; set the x-conductivity to 7.71 W/m-K and the ratio ky/kx to 0.605 (the ratio of $k_{eff,x}$ to $k_{eff,y}$). Set the effective rate of volumetric generation by clicking on the outline and selecting Generation from the Specify menu.

Set the boundary conditions according to Figure 2-37 and draw a crude grid (two triangles formed by a single element line across the pole will do). Refine the grid multiple times and solve. The temperature distribution in the pole is shown in Figure 2-38.

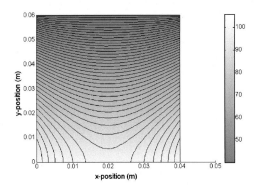

Figure 2-38: Temperature distribution in the motor pole ($^{\circ}$C).

Note that the 2-D temperature distribution is adequately captured using the finite element model with equivalent properties; however, the actual temperature distribution would include "ripples" corresponding to the effects of the individual laminations. Unless the characteristics of these very small-scale effects are important, it is convenient to consider the effect of the micro-structure on the larger-scale problem using the effective conductivity concept.

EXAMPLE 2.9-1: FIBER OPTIC BUNDLE

Lighting represents one of the largest uses of electrical energy in residential and commercial buildings; lighting loads are highest during on-peak hours when electrical energy is most costly. Also, the thermal energy deposited into conditioned space by electrical lighting adds to the air conditioning load on the building which, in turn, adds to the electrical energy required to run the air conditioning system.

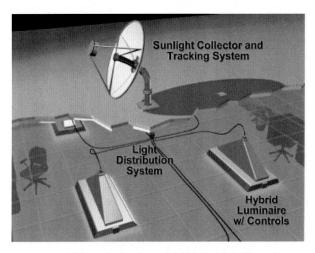

Figure 1: Hybrid lighting system (Cheadle, 2006).

A novel lighting system consists of a sunlight collector and a light distribution system, as shown in Figure 1. The sunlight collector tracks the sun and collects and concentrates solar radiation. The light distribution system receives the concentrated solar radiation and distributes it into a building where it is finally dispensed

EXAMPLE 2.9-1: FIBER OPTIC BUNDLE

in fixtures that are referred to as luminaires. Sunlight contains both visible and invisible energy; only the visible portion of the sunlight is useful for lighting and therefore the collector gathers the visible portion of the incident solar radiation while eliminating the invisible ultraviolet and infrared portions of the spectrum. (We will learn more about these characteristics of radiation in Chapter 10.) The fiber optic bundle used to transmit the visible light is composed of many, small diameter optical fibers that are packed in approximately a hexagonal close-packed array. A conductive filler material is wrapped around each fiber and the entire structure is simultaneously heated and compressed so that the fibers becomes hexagonal shaped with a thin layer of conductive filler separating each hexagon (Figure 2). The dimension of each face of the hexagon is $d = 1.0$ mm and the thickness of the filler that separates the hexagons is $a = 50$ μm thick. The fiber conductivity is $k_{fb} = 1.5$ W/m-K while the filler conductivity is $k_{fl} = 50.0$ W/m-K.

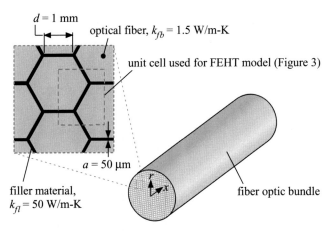

Figure 2: Array of optical fibers packed together and compressed in order to form a fiber optic bundle that has hexagonal units.

a) Determine the effective radial and axial conductivity associated with the bundle.

The inputs are entered in EES:

```
"EXAMPLE 2.9-1: Fiber Optic Bundle"

$UnitSystem SI MASS RAD PA K J
$TABSTOPS 0.2 0.4 0.6 0.8 3.5 in

"Composite Structure Inputs"
d=1 [mm]*convert(mm,m)                        "face dimension of hexagon"
a=0.05 [mm]*convert(mm,m)                      "filler thickness"
k_fl=50 [W/m-K]                                "conductivity of filler"
k_fb=1.5 [W/m-K]                               "conductivity of fiber"
```

The heat transfer in the axial direction can travel through two parallel paths, the filler and the fiber. The area of a single fiber (a hexagon with each side having length d) is:

$$A_{fb} = 2\,d^2\,\sin\left(\frac{\pi}{3}\right)\left[1 + \cos\left(\frac{\pi}{3}\right)\right]$$

A_fb=2*d^2*sin(pi/3)*(1+cos(pi/3))   "area of a fiber"

The heat transfer through the fiber in the axial direction for a given temperature difference (ΔT) is:

$$\dot{q}_{fb} = \frac{\Delta T\,k_{fb}\,A_{fb}}{L}$$

where L is the length of the fiber. The area of the filler material associated with a single fiber (which has thickness $a/2$ due to sharing of the filler with the neighboring fibers) is:

$$A_{fl} = 3\,d\,a$$

A_fl=3*d*a   "area of filler associated with a fiber"

The heat transfer through a unit length of the filler in the axial direction for the same temperature difference is:

$$\dot{q}_{fl} = \frac{\Delta T\,k_{fl}\,A_{fl}}{L}$$

The equivalent homogenous material will have conductivity $k_{eff,x}$ and area $A_{fl} + A_{fb}$; therefore, the heat transfer through the equivalent material $(\dot{q}_{eff,x})$ is:

$$\dot{q}_{eff,x} = \frac{\Delta T\,k_{eff,x}\,(A_{fl} + A_{fb})}{L}$$

The effective conductivity is defined so that the heat transfer through the equivalent material is equal to the sum of the heat transfer through the fiber and the filler:

$$\underbrace{\frac{\Delta T\,k_{eff,x}\,(A_{fl} + A_{fb})}{L}}_{\dot{q}_{eff,x}} = \underbrace{\frac{\Delta T\,k_{fb}\,A_{fb}}{L}}_{\dot{q}_{fb}} + \underbrace{\frac{\Delta T\,k_{fl}\,A_{fl}}{L}}_{\dot{q}_{fl}}$$

or, solving for $k_{eff,x}$:

$$k_{eff,x} = \frac{k_{fb}\,A_{fb} + k_{fl}\,A_{fl}}{A_{fl} + A_{fb}}$$

The effective conductivity in the axial direction is the area-weighted conductivity of the two parallel paths:

k_eff_x=(k_fb*A_fb+k_fl*A_fl)/(A_fl+A_fb)   "effective conductivity in the axial direction"

which leads to $k_{eff,x} = 4.1$ W/m-K.

The radial conductivity cannot be evaluated using a simple parallel or series resistance circuit because the heat flow across the bundle is complex and 2-D.

EXAMPLE 2.9-1: FIBER OPTIC BUNDLE

EXAMPLE 2.9-1: FIBER OPTIC BUNDLE

Therefore, a 2-D finite element model of a unit cell of the structure (shown in Figure 2) must be generated. Figure 3 illustrates the details of a unit cell and includes the coordinates of the points (in mm) that define the geometry.

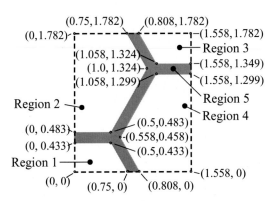

Figure 3: A unit cell of the fiber optic bundle structure; the points that define the structure are shown (dimensions are in mm).

The finite element model is generated using FEHT as discussed in Section 2.7.1 and Appendix A.4. A grid is specified where 1 cm of screen dimension corresponds to 0.2 mm. The five regions in Figure 3 are generated using five outlines; the points are initially placed approximately and then precisely positioned by double-clicking on each one in turn. The conductivity for regions 1 through 4 are set to 1.5 W/m-K (consistent with the optical fiber) and the conductivity for region 5 is set to 50.0 W/m-K (consistent with the filler material). The boundary conditions along the upper and lower edges are set as adiabatic. In order to set a temperature difference across the unit cell (from left to right) it would seem logical to set the temperature at the left hand side to 1.0°C and the right side to 0.0°C. However, due to the manner in which finite element techniques determine the heat flux at a surface, a more accurate answer is obtained if a convective boundary condition is set with a very high heat transfer coefficient; for example, 1×10^5 W/m^2-K.

A relatively crude mesh is generated and then refined, particularly in the filler material where most of the heat flow is expected. The result is shown in Figure 4(a). The finite element model is solved and the temperature contours are shown in Figure 4(b).

The heat flow through the unit cell can be determined by selecting Heat Flows from the View menu and then selecting all of the boundaries on either the left or right side. The selection process can be accomplished by using the mouse to 'drag' a selection rectangle around the lines on the boundary. The total heat flow is $\dot{q}_{eff,r}/L = 3.070$ W/m; this result can be used to compute the effective conductivity of the composite. The effective conductivity across the bundle ($k_{eff,r}$) is defined as the conductivity of a homogeneous material that would provide the same heat transfer per length into the page as the composite structure simulated by the finite element model.

$$\frac{\dot{q}_{eff,r}}{L} = k_{eff,r}\frac{H\,\Delta T}{W}$$

EXAMPLE 2.9-1: FIBER OPTIC BUNDLE

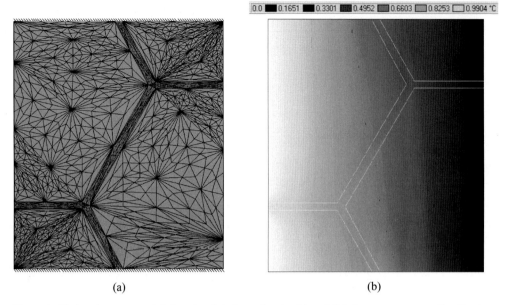

| 0.0 | ■ 0.1651 | ■ 0.3301 | ▦ 0.4952 | ▨ 0.6603 | ▥ 0.8253 | ☐ 0.9904 °C |

(a)   (b)

Figure 4: Finite element model showing (a) the refined grid and (b) the temperature distribution.

where H is the height of the unit cell (1.782 mm, from Figure 3), W is the width of the unit cell (1.558 mm, from Figure 3), and $\Delta T = 1.0$ K (the imposed temperature difference).

```
"k_eff_r calculation using inputs from FEHT"
H=1.782 [mm]*convert(mm,m)          "height of unit cell"
W=1.558 [mm]*convert(mm,m)          "width of unit cell"
q_dot_eff_r\L=3.070 [W/m]           "rate of heat transfer per unit length predicted by FE model"
DT=1 [K]                            "temperature difference imposed on FE model"
q_dot_eff_r\L=k_eff_r*H*DT/W        "effective conductivity in the radial direction"
```

which leads to $k_{eff,r} = 2.7$ W/m-K.

The effective conductivity in the x-direction is 4.1 W/m-K while the effective conductivity in the radial direction is 2.7 W/m-K. These values make sense. Both lie between the conductivity of the fiber and the filler and both are closer to the conductivity of the fiber because it occupies most of the space. Further, the conductivity in the x-direction is larger because the path through the high conductivity filler is more direct in this direction.

The advantage of the hexagonal pattern in Figure 2 is that is has the lowest possible porosity (ϕ). The porosity is defined as the fraction of the area of the face of the bundle that is occupied by the filler. The optical fibers are designed so that any radiation that strikes the face of a fiber is "trapped" by total internal reflection and transmitted without substantial loss. However, the radiation that strikes the opaque filler in the interstitial areas between the fibers will be absorbed and result in a thermal load on the bundle that manifests itself as a heat flux on the surface. This heat flux can lead to elevated temperatures and thermal failure.

Assume that the outer edge of the fiber optic bundle is exposed to air at $T_\infty = 20°C$ with a heat transfer coefficient $\bar{h}_{out} = 5.0$ W/m²-K. The front face of the bundle (at $x = 0$) is exposed to air at $T_\infty = 20°C$ with a heat transfer coefficient $\bar{h}_f =$

EXAMPLE 2.9-1: FIBER OPTIC BUNDLE

10.0 W/m²-K. The radius of the bundle is $r_{out} = 2.0$ cm and its length is $L = 2.5$ m. The end of the bundle at $x = L$ is maintained at a temperature of $T_L = 20°C$. The radiant heat flux incident on the face of the bundle is $\dot{q}''_{inc} = 1 \times 10^5$ W/m². A schematic of the problem is shown in Figure 5.

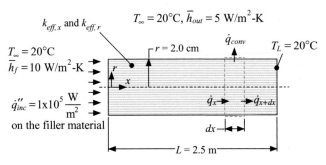

Figure 5: Schematic of the fiber optic bundle problem.

b) Is it appropriate to treat the bundle as an extended surface? Justify your answer.

The additional inputs for the problem are entered in EES:

```
"Problem Inputs"
r_out=2 [cm]*convert(cm,m)                    "outer radius"
L=2.5 [m]                                      "bundle length"
h_bar_out=5 [W/m^2-K]                          "heat transfer coefficient on outer surface"
T_infinity=converttemp(C,K,20[C])             "air temperature"
h_bar_f=10 [W/m^2-K]                           "heat transfer coefficient on the face"
T_L=converttemp(C,K,20[C])                     "temperature at x=L"
q_dot_flux_inc=100000 [W/m^2]                  "incident radiant heat flux on the face"
```

The extended surface approximation neglects temperature gradients in the radial direction within the bundle. The conduction resistance in the radial direction ($R_{cond,r}$) must be neglected in order to treat the bundle as an extended surface. This assumption is justified provided that $R_{cond,r}$ is small relative to the resistance that is being considered, convection from the outer surface of the bundle (R_{conv}). The appropriate Biot number is therefore:

$$Bi = \frac{R_{cond,r}}{R_{conv}}$$

It is not possible to precisely compute a resistance that characterizes conduction in the radial direction within the bundle; conduction from the center of the bundle is characterized by an infinite resistance. Instead, an approximate conduction length ($r_{out}/2$) and area ($\pi r_{out} L$) are used:

$$Bi = \underbrace{\left(\frac{r_{out}}{2 \, k_{eff,r} \, \pi \, r_{out} \, L} \right)}_{R_{cond,r}} \underbrace{\left(\frac{\overline{h}_{out} \, 2 \, \pi \, r_{out} \, L}{1} \right)}_{R_{conv}} = \frac{\overline{h}_{out} \, r_{out}}{k_{eff,r}}$$

```
Bi=h_bar_out*r_out/k_eff_r                    "Biot number"
```

EXAMPLE 2.9-1: FIBER OPTIC BUNDLE

The Biot number is 0.037 which is much less than 1 and therefore the extended surface approximation is justified.

c) Develop an analytical model for the temperature distribution in the bundle.

The differential control volume used to derive the governing equation is shown in Figure 5 and leads to the energy balance:

$$\dot{q}_x = \dot{q}_{x+dx} + \dot{q}_{conv}$$

which can be expanded and simplified:

$$0 = \frac{d\dot{q}_x}{dx}dx + \dot{q}_{conv} \tag{1}$$

The rate equations are:

$$\dot{q}_x = -k_{eff,x}\,\pi\,r_{out}^2\,\frac{dT}{dx} \tag{2}$$

$$\dot{q}_{conv} = 2\,\pi\,r_{out}\,\overline{h}_{out}\,dx\,(T - T_\infty) \tag{3}$$

Substituting Eqs. (1) and (2) into Eq. (3) leads to:

$$0 = \frac{d}{dx}\left[-k_{eff,x}\,\pi\,r_{out}^2\,\frac{dT}{dx}\right]dx + 2\,\pi\,r_{out}\,\overline{h}_{out}\,dx\,(T - T_\infty)$$

or

$$\frac{d^2T}{dx^2} - m^2\,T = -m^2\,T_\infty \tag{4}$$

where

$$m^2 = \frac{2\,\overline{h}_{out}}{k_{eff,x}\,r_{out}}$$

The governing differential equation, Eq. (4), is satisfied by exponential functions. The differential equation can also be entered in Maple and solved.

```
> restart;
> ODE:=diff(diff(T(x),x),x)-m^2*T(x)=-m^2*T_infinity;
```
$$ODE := \left(\frac{d^2}{dx^2}T(x)\right) - m^2 T(x) = -m^2 T\_infinity$$
```
> Ts:=dsolve(ODE);
```
$$Ts := T(x) = e^{(-mx)}\_C2 + e^{(mx)}\_C1 + T\_infinity$$

The solution is copied and pasted into EES.

```
"Solution"
m=sqrt(2*h_bar_out/(k_eff_x*r_out))        "solution parameter"
T=exp(-m*x)*C_2+exp(m*x)*C_1+T_infinity    "solution from Maple"
```

The first boundary condition is the specified temperature at $x = L$:

$$T_{x=L} = T_L$$

A symbolic expression for this boundary condition is obtained in Maple:

```
> rhs(eval(Ts,x=L))=T_L;
```
$$e^{(-mL)}\_C2 + e^{(mL)}\_C1 + T\_infinity = T\_L$$

and pasted into EES:

```
exp(-m*L)*C_2+exp(m*L)*C_1+T_infinity=T_L          "boundary condition at x=L"
```

The second boundary condition is obtained from an interface energy balance at $x = 0$. Recall that only the flux incident on the filler material results in a heat load. Therefore, the absorbed flux is the product of incident heat flux and the porosity (ϕ), which is the ratio of the area of the filler to the total area:

$$\phi = \frac{A_{fl}}{A_{fl} + A_{fb}}$$

```
phi=A_fl/(A_fl+A_fb)                               "porosity"
```

The absorbed flux must either be transferred by conduction to the bundle or convection to the air:

$$\dot{q}''_{inc}\, \phi = -k_{eff,x}\, \left.\frac{dT}{dx}\right|_{x=0} + \bar{h}_f\, (T_{x=0} - T_{air})$$

A symbolic expression for this boundary condition is obtained in Maple:

```
> q_dot_flux_inc*phi=-k_eff_x*rhs(eval(diff(Ts,x),x=0))+h_bar_f*(rhs(eval(Ts,x=0))-T_infinity);
```
$$q\_dot\_flux\_inc\, \phi = k\_eff\_x(-m\_C2 + m\_C1) + h\_bar\_f(\_C2 + \_C1)$$

and pasted into EES:

```
q_dot_flux_inc*phi = -k_eff_x*(-m*C_2+m*C_1)+h_bar_f*(C_2+C_1)
     "boundary condition at x=0"
```

The solution is converted to Celsius:

```
T_C=converttemp(K,C,T)                             "temperature in C"
```

Figure 6 illustrates the temperature distribution near the face of the fiber optic bundle.

EXAMPLE 2.9-1: FIBER OPTIC BUNDLE

EXAMPLE 2.9-1: FIBER OPTIC BUNDLE

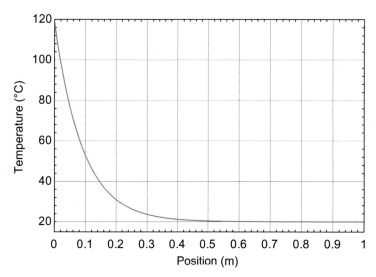

Figure 6: Temperature distribution in fiber optic bundle.

The model can be used to assess alternative methods of thermal management. For example, the heat transfer coefficient at the face might be increased by adding a fan or the conductivity of the filler material might be increased through material selection. Figure 7 illustrates the maximum temperature (the temperature at the face) as a function of $\bar{h}_f$ for various values of k_f.

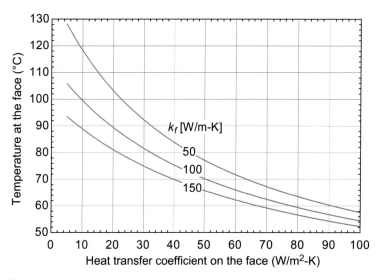

Figure 7: Temperature at the face as a function of the face heat transfer coefficient for various values of the filler material conductivity.

Chapter 2: Two-Dimensional, Steady-State Conduction

The website associated with this book (www.cambridge.org/nellisandklein) provides many more problems than are included here.

Shape Factors

2–1 Figure P2-1 illustrates two tubes that are buried in the ground behind your house and transfer water to and from a wood burner. The left tube carries hot water from the burner back to your house at $T_{w,h} = 135°F$ while the right tube carries cold water from your house to the burner at $T_{w,c} = 70°F$. Both tubes have outer diameter $D_o = 0.75$ inch and thickness $th = 0.065$ inch. The conductivity of the tubing material is $k_t = 0.22$ W/m-K. The heat transfer coefficient between the water and the tube internal surface (in both tubes) is $\bar{h}_w = 250$ W/m²-K. The center to center distance between the tubes is $w = 1.25$ inch and the length of the tubes is $L = 20$ ft (into the page). The tubes are buried in soil that has conductivity $k_s = 0.30$ W/m-K.

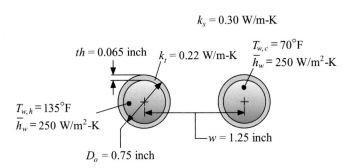

Figure P2-1: Tubes buried in soil.

a.) Estimate the heat transfer from the hot water to the cold water due to the proximity of the tubes to one another.

b.) To do part (a) you should have needed to determine a shape factor; calculate an approximate value of the shape factor and compare it to the accepted value.

c.) Plot the rate of heat transfer from the hot water to the cold water as a function of the center to center distance between the tubes.

2–2 A solar electric generation system (SEGS) employs molten salt as both the energy transport and storage fluid. The molten salt is heated to 500°C and stored in a buried hemispherical tank. The top (flat) surface of the tank is at ground level. The diameter of the tank before insulation is applied is 14 m. The outside surfaces of the tank are insulated with 0.30 m thick fiberglass having a thermal conductivity of 0.035 W/m-K. Sand having a thermal conductivity of 0.27 W/m-K surrounds the tank, except on its top surface. Estimate the rate of heat loss from this storage unit to the 25°C surroundings.

Separation of Variables Solutions

2–3 You are the engineer responsible for a simple device that is used to measure the heat transfer coefficient as a function of position within a tank of liquid (Figure P2-3). The heat transfer coefficient can be correlated against vapor quality, fluid composition, and other useful quantities. The measurement device is composed of many thin plates of low conductivity material that are interspersed with large, copper interconnects. Heater bars run along both edges of the thin plates. The heater bars are insulated and can only transfer energy to the plate; the heater bars are conductive and can therefore be assumed to come to a uniform temperature as a current is applied. This uniform temperature is assumed to be applied to the top and bottom edges of the plates. The copper interconnects are thermally well-connected to the fluid; therefore, the temperature of the left and right edges of each plate are equal to the fluid temperature. This is convenient because it isolates the effect of adjacent plates from one another, allowing each plate to measure the local heat transfer coefficient. Both surfaces of the plate are exposed to the fluid temperature via a heat transfer coefficient. It is possible to infer the heat transfer coefficient by measuring heat transfer required to elevate the heater bar temperature to a specified temperature above the fluid temperature.

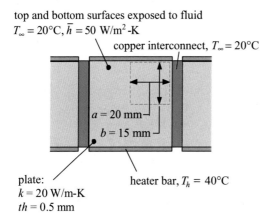

Figure P2-3: Device to measure local heat transfer coefficient.

The nominal design of an individual heater plate utilizes metal with $k = 20$ W/m-K, $th = 0.5$ mm, $a = 20$ mm, and $b = 15$ mm (note that a and b are defined as the half-width and half-height of the heater plate, respectively, and th is the thickness as shown in Figure P2-3). The heater bar temperature is maintained at $T_h = 40°$C and the fluid temperature is $T_\infty = 20°$C. The nominal value of the heat transfer coefficient is $\bar{h} = 50$ W/m²-K.

a.) Develop an analytical model that can predict the temperature distribution in the plate under these nominal conditions.

b.) The measured quantity is the rate of heat transfer to the plate from the heater ($\dot{q}_h$) and therefore the relationship between $\dot{q}_h$ and $\bar{h}$ (the quantity that is inferred from the heater power) determines how useful the instrument is. Determine the heater power and plot the heat transfer coefficient as a function of heater power.

c.) If the uncertainty in the measurement of the heater power is $\delta\dot{q}_h = 0.01$ W, estimate the uncertainty in the measured heat transfer coefficient ($\delta\bar{h}$).

2–4 A laminated composite structure is shown in Figure P2-4.

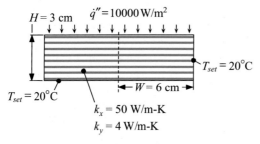

Figure P2-4: Composite structure exposed to a heat flux.

The structure is anisotropic. The effective conductivity of the composite in the x-direction is $k_x = 50$ W/m-K and in the y-direction it is $k_y = 4$ W/m-K. The top of the structure is exposed to a heat flux of $\dot{q}'' = 10,000$ W/m^2. The other edges are maintained at $T_{set} = 20°$C. The height of the structure is $H = 3$ cm and the half-width is $W = 6$ cm.

a.) Develop a separation of variables solution for the 2-D steady-state temperature distribution in the composite.

b.) Prepare a contour plot of the temperature distribution.

Advanced Separation of Variables Solutions

2–5 Figure P2-5 illustrates a pipe that connects two tanks of liquid oxygen on a spacecraft. The pipe is subjected to a heat flux, $\dot{q}'' = 8,000$ W/m^2, which can be assumed to be uniformly applied to the outer surface of the pipe and is entirely absorbed. Neglect radiation from the surface of the pipe to space. The inner radius of the pipe is $r_{in} = 6.0$ cm, the outer radius of the pipe is $r_{out} = 10.0$ cm, and the half-length of the pipe is $L = 10.0$ cm. The ends of the pipe are attached to the liquid oxygen tanks and therefore are at a uniform temperature of $T_{LOx} = 125$ K. The pipe is made of a material with a conductivity of $k = 10$ W/m-K. The pipe is empty and therefore the internal surface can be assumed to be adiabatic.

a.) Develop an analytical model that can predict the temperature distribution within the pipe. Prepare a contour plot of the temperature distribution within the pipe.

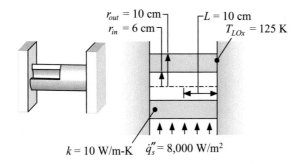

Figure P2-5: Cryogen transfer pipe connecting two liquid oxygen tanks.

2–6 Figure P2-6 illustrates a cylinder that is exposed to a concentrated heat flux at one end.

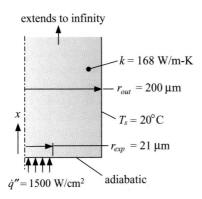

Figure P2-6: Cylinder exposed to a concentrated heat flux at one end.

The cylinder extends infinitely in the x-direction. The surface at $x = 0$ experiences a uniform heat flux of $\dot{q}'' = 1500$ W/cm^2 for $r < r_{exp} = 21$ μm and is adiabatic for $r_{exp} < r < r_{out}$ where $r_{out} = 200$ μm is the outer radius of the cylinder. The outer surface of the cylinder is maintained at a uniform temperature of $T_s = 20°$C. The conductivity of the cylinder material is $k = 168$ W/m-K.

a.) Develop a separation of variables solution for the temperature distribution within the cylinder. Plot the temperature as a function of radius for various values of x.

b.) Determine the average temperature of the cylinder at the surface exposed to the heat flux.

c.) Define a dimensionless thermal resistance between the surface exposed to the heat flux and T_s. Plot the dimensionless thermal resistance as a function of r_{out}/r_{in}.

d.) Show that your plot from (c) does not change if the problem parameters (e.g., T_s, k, etc.) are changed.

Superposition

2–7 The plate shown in Figure P2-7 is exposed to a uniform heat flux $\dot{q}'' = 1 \times 10^5$ W/m^2 along its top surface and is adiabatic at its bottom surface. The left side of the plate is kept at $T_L = 300$ K and the right side is at $T_R = 500$ K. The height and width of the plate are $H = 1$ cm and $W = 5$ cm, respectively. The conductivity of the plate is $k = 10$ W/m-K.

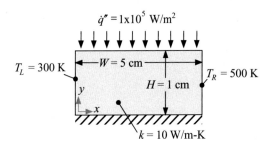

Figure P2-7: Plate.

a.) Derive an analytical solution for the temperature distribution in the plate.

b.) Implement your solution in EES and prepare a contour plot of the temperature.

Numerical Solutions to Steady-State 2-D Problems using EES

2–8 Figure P2-8 illustrates an electrical heating element that is affixed to the wall of a chemical reactor. The element is rectangular in cross-section and very long (into the page). The temperature distribution within the element is therefore two-dimensional, $T(x, y)$. The width of the element is $a = 5.0$ cm and the height is $b = 10.0$ cm. The three edges of the element that are exposed to the chemical (at $x = 0$, $y = 0$, and $x = a$) are maintained at a temperature $T_c = 200°$C while the upper edge (at $y = b$) is affixed to the well-insulated wall of the reactor and can therefore be considered adiabatic. The element experiences a uniform volumetric rate of thermal energy generation, $\dot{g}''' = 1 \times 10^6$ W/m^3. The conductivity of the material is $k = 0.8$ W/m-K.

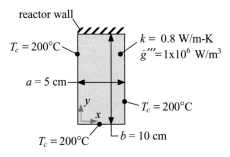

Figure P2-8: Electrical heating element.

a.) Develop a 2-D numerical model of the element using EES.
b.) Plot the temperature as a function of x at various values of y. What is the maximum temperature within the element and where is it located?
c.) Prepare a sanily check to show that your solution behaves according to your physical intuition. That is, change some aspect of your program and show that the results behave as you would expect (clearly describe the change that you made and show the result).

Finite-Difference Solutions to Steady-State 2-D Problems using MATLAB

2–9 Figure P2-9 illustrates a cut-away view of two plates that are being welded together. Both edges of the plate are clamped and held at temperature $T_s = 25°C$. The top of the plate is exposed to a heat flux that varies with position x, measured from joint, according to: $\dot{q}''_m(x) = \dot{q}''_j \exp(-x/L_j)$ where $\dot{q}''_j = 1 \times 10^6$ W/m² is the maximum heat flux (at the joint, $x = 0$) and $L_j = 2.0$ cm is a measure of the extent of

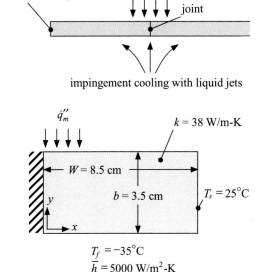

Figure P2-9: Welding process and half-symmetry model of the welding process.

the heat flux. The back side of the plates are exposed to liquid cooling by a jet of fluid at $T_f = -35°C$ with $\bar{h} = 5000$ W/m²-K. A half-symmetry model of the problem is shown in Figure P2-9. The thickness of the plate is $b = 3.5$ cm and the width of a single plate is $W = 8.5$ cm. You may assume that the welding process is steady-state and 2-D. You may neglect convection from the top of the plate. The conductivity of the plate material is $k = 38$ W/m-K.

a.) Develop a separation of variables solution to the problem. Implement the solution in EES and prepare a plot of the temperature as a function of x at $y = 0$, 1.0, 2.0, 3.0, and 3.5 cm.

b.) Prepare a contour plot of the temperature distribution.

c.) Develop a numerical model of the problem. Implement the solution in MATLAB and prepare a contour or surface plot of the temperature in the plate.

d.) Plot the temperature as a function of x at $y = 0$, $b/2$, and b and overlay on this plot the separation of variables solution obtained in part (a) evaluated at the same locations.

Finite Element Solutions to Steady-State 2-D Problems using FEHT

2–10 Figure P2-10(a) illustrates a double paned window. The window consists of two panes of glass each of which is $tg = 0.95$ cm thick and $W = 4$ ft wide by $H = 5$ ft high. The glass panes are separated by an air gap of $g = 1.9$ cm. You may assume that the air is stagnant with $k_a = 0.025$ W/m-K. The glass has conductivity $k_g = 1.4$ W/m-K. The heat transfer coefficient between the inner surface of the inner pane and the indoor air is $\bar{h}_{in} = 10$ W/m²-K and the heat transfer coefficient between the outer surface of the outer pane and the outdoor air is $\bar{h}_{out} = 25$ W/m²-K. You keep your house heated to $T_{in} = 70°F$.

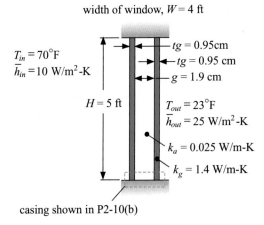

Figure P2-10(a): Double paned window.

The average heating season lasts about *time* = 130 days and the average outdoor temperature during this time is $T_{out} = 23°F$. You heat with natural gas and pay, on average, $ec = 1.415$ $/therm (a therm is an energy unit = 1.055×10^8 J).

a.) Calculate the average rate of heat transfer through the double paned window during the heating season using a 1-D resistance model.

b.) How much does the energy lost through the window cost during a single heating season?

There is a metal casing that holds the panes of glass and connects them to the surrounding wall, as shown in Figure P2-10(b). Because the metal casing has high conductivity, it seems likely that you could lose a substantial amount of heat by conduction through the casing (potentially negating the advantage of using a double paned window). The geometry of the casing is shown in Figure P2-10(b); note that the casing is symmetric about the center of the window.

All surfaces of the casing that are adjacent to glass, wood, or the air between the glass panes can be assumed to be adiabatic. The other surfaces are exposed to either the indoor or outdoor air.

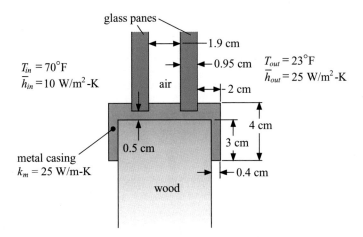

Figure P2-10(b): Metal casing.

c.) Prepare a 2-D thermal analysis of the casing using FEHT. Turn in a print out of your geometry as well as a contour plot of the temperature distribution. What is the rate of energy lost via conduction through the casing per unit length (W/m)?

d.) Show that your numerical model has converged by recording the rate of heat transfer per length for several values of the number of nodes.

e.) How much does the casing add to the cost of heating your house?

2–11 A radiator panel extends from a spacecraft; both surfaces of the radiator are exposed to space (for the purposes of this problem it is acceptable to assume that space is at 0 K); the emissivity of the surface is $\varepsilon = 1.0$. The plate is made of aluminum ($k = 200$ W/m-K and $\rho = 2700$ kg/m^3) and has a fluid line attached to it, as shown in Figure 2-11(a). The half-width of the plate is $a = 0.5$ m wide and the height of the plate is $b = 0.75$ m. The thickness of the plate is $th = 1.0$ cm. The fluid line carries coolant at $T_c = 320$ K. Assume that the fluid temperature is constant, although the fluid temperature will actually decrease as it transfers heat to the radiator. The combination of convection and conduction through the panel-to-fluid line mounting leads to an effective heat transfer coefficient of $h = 1,000$ W/m^2-K over the 3.0 cm strip occupied by the fluid line.

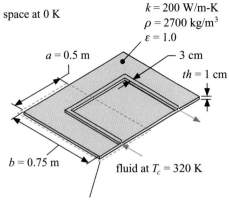

Figure 2-11(a): Radiator panel.

The radiator panel is symmetric about its half-width and the critical dimensions that are required to develop a half-symmetry model of the radiator are shown in Figure 2-11(b). There are three regions associated with the problem that must be defined separately so that the surface conditions can be set differently. Regions 1 and 3 are exposed to space on both sides while Region 2 is exposed to the coolant fluid one side and space on the other; for the purposes of this problem, the effect of radiation to space on the back side of Region 2 is neglected.

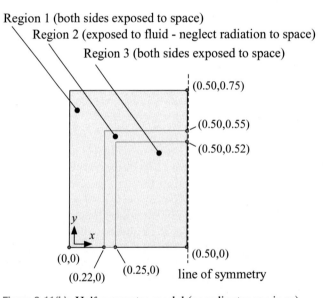

Figure 2-11(b): Half-symmetry model (coordinates are in m).

a.) Prepare a FEHT model that can predict the temperature distribution over the radiator panel.
b.) Export the solution to EES and calculate the total heat transferred from the radiator and the radiator efficiency (defined as the ratio of the radiator heat transfer to the heat transfer from the radiator if it were isothermal and at the coolant temperature).
c.) Explore the effect of thickness on the radiator efficiency and mass.

Resistance Approximations for Conduction Problems

2–12 There are several cryogenic systems that require a "thermal switch," a device that can be used to control the thermal resistance between two objects. One class of thermal switch is activated mechanically and an attractive method of providing mechanical actuation at cryogenic temperatures is with a piezoelectric stack. Unfortunately, the displacement provided by a piezoelectric stack is very small, typically on the order of 10 microns. A company has proposed an innovative design for a thermal switch, shown in Figure P2-12(a). Two blocks are composed of $th = 10\,\mu m$ laminations that are alternately copper ($k_{Cu} = 400$ W/m-K) and plastic ($k_p = 0.5$ W/m-K). The thickness of each block is $L = 2.0$ cm in the direction of the heat flow. One edge of each block is carefully polished and these edges are pressed together; the contact resistance associated with this joint is $R_c'' = 5 \times 10^{-4}$ K-m^2/W.

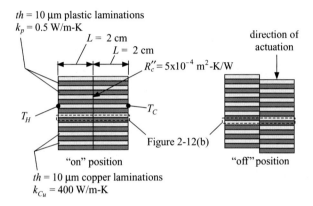

Figure P2-12(a): Thermal switch in the "on" and "off" positions.

Figure P2-12(a) shows the orientation of the two blocks when the switch is in the "on" position; notice that the copper laminations are aligned with one another in this configuration in order to provide a continuous path for heat through high conductivity copper (with the exception of the contact resistance at the interface). The vertical location of the right-hand block is shifted by 10 μm to turn the switch "off". In the "off" position, the copper laminations are aligned with the plastic laminations; therefore, the heat transfer is inhibited by low conductivity plastic. Figure P2-12(b) illustrates a closer view of half (in the vertical direction) of two adjacent laminations in the "on" and "off" configurations. Note that the repeating nature of the geometry means that it is sufficient to analyze a single lamination set and assume that the upper and lower boundaries are adiabatic.

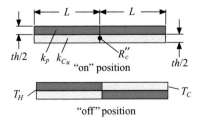

Figure P2-12(b): A single set consisting of half of two adjacent laminations in the "on" and "off" positions.

The key parameter that characterizes the performance of a thermal switch is the resistance ratio (RR) which is defined as the ratio of the resistance of the switch in the "off" position to its resistance in the "on" position. The company claims that they can achieve a resistance ratio of more than 100 for this switch.

a.) Estimate upper and lower bounds for the resistance ratio for the proposed thermal switch using 1-D conduction network approximations. Be sure to draw and clearly label the resistance networks that are used to provide the estimates. Use your results to assess the company's claim of a resistance ratio of 100.

b.) Provide one or more suggestions for design changes that would improve the performance of the switch (i.e., increase the resistance ratio). Justify your suggestions.

c.) Sketch the temperature distribution through the two parallel paths associated with the adiabatic limit of the switch's operation in the "off" position. Do not worry about the quantitative details of the sketch, just make sure that the qualitative features are correct.

d.) Sketch the temperature distribution through the two parallel paths associated with the adiabatic limit in the "on" position. Again, do not worry about the quantitative details of your sketch, just make sure that the qualitative features are correct.

2–13 Figure P2-13 illustrates a thermal bus bar that has width $W = 2$ cm (into the page).

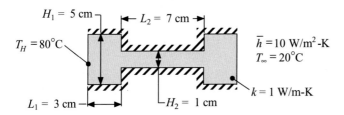

Figure P2-13: Thermal bus bar.

The bus bar is made of a material with conductivity $k = 1$ W/m-K. The middle section is $L_2 = 7$ cm long with thickness $H_2 = 1$ cm. The two ends are each $L_1 = 3$ cm long with thickness $H_1 = 3$ cm. One end of the bar is held at $T_H = 80°C$ and the other is exposed to air at $T_\infty = 20°C$ with $\bar{h} = 10$ W/m²-K.

a.) Use FEHT to predict the rate of heat transfer through the bus bar.

b.) Obtain upper and lower bounds for the rate of heat transfer through the bus bar using appropriately defined resistance approximations.

Conduction through Composite Materials

2–14 A laminated stator is shown in Figure P2-14. The stator is composed of laminations with conductivity $k_{lam} = 10$ W/m-K that are coated with a very thin layer of epoxy with conductivity $k_{epoxy} = 2.0$ W/m-K in order to prevent eddy current losses. The laminations are $th_{lam} = 0.5$ mm thick and the epoxy coating is 0.1 mm thick (the total amount of epoxy separating each lamination is $th_{epoxy} = 0.2$ mm). The inner radius of the laminations is $r_{in} = 8.0$ mm and the outer radius of the laminations is $r_{o,lam} = 20$ mm. The laminations are surrounded by a cylinder of plastic with conductivity $k_p = 1.5$ W/m-K that has an outer radius of $r_{o,p} = 25$ mm. The motor casing surrounds the plastic. The motor casing has an outer radius of $r_{o,c} = 35$ mm and is composed of aluminum with conductivity $k_c = 200$ W/m-K.

laminations, th_{lam} = 0.5 mm, k_{lam} = 10 W/m-K
epoxy coating, th_{epoxy} = 0.2 mm, k_{epoxy} = 2.0 W/m-K

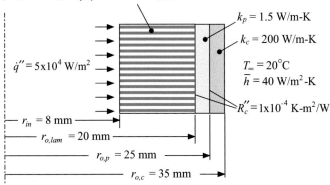

Figure P2-14: Laminated stator.

The heat flux due to the windage loss associated with the drag on the shaft is $\dot{q}'' = 5 \times 10^4$ W/m^2 and is imposed on the internal surface of the laminations. The outer surface of the motor is exposed to air at $T_\infty = 20°$C with a heat transfer coefficient $\bar{h} = 40$ W/m^2-K. There is a contact resistance $R_c'' = 1 \times 10^{-4}$ K-m^2/W between the outer surface of the laminations and the inner surface of the plastic and the outer surface of the plastic and the inner surface of the motor housing.

a.) Determine an upper and lower bound for the temperature at the inner surface of the laminations (T_{in}).

b.) You need to reduce the internal surface temperature of the laminations and there are a few design options available, including: (1) increase the lamination thickness (up to 0.7 mm), (2) reduce the epoxy thickness (down to 0.05 mm), (3) increase the epoxy conductivity (up to 2.5 W/m-K), or (4) increase the heat transfer coefficient (up to 100 W/m-K). Which of these options do you suggest and why?

REFERENCES

Cheadle, M., *A Predictive Thermal Model of Heat Transfer in a Fiber Optic Bundle for a Hybrid Solar Lighting System*, M.S. Thesis, University of Wisconsin, Dept. of Mechanical Engineering, (2006).

Moaveni, S., *Finite Element Analysis, Theory and Application with ANSYS, 2nd Edition*, Pearson Education, Inc., Upper Saddle River, (2003).

Myers, G. E., *Analytical Methods in Conduction Heat Transfer, 2nd Edition*, AMCHT Publications, Madison, WI, (1998).

Rohsenow, W. M., J. P. Hartnett, and Y. I. Cho, eds., *Handbook of Heat Transfer, 3rd Edition*, McGraw-Hill, New York, (1998).

3 Transient Conduction

3.1 Analytical Solutions to 0-D Transient Problems

3.1.1 Introduction

Chapters 1 and 2 discuss steady-state problems, i.e., problems in which temperature depends on position (e.g, x and y) but does not change with time (t). This chapter discusses transient conduction problems, where temperature depends on time.

3.1.2 The Lumped Capacitance Assumption

The simplest situation is zero-dimensional (0-D); that is, the temperature does not vary with position but only with time. This approximation is often referred to as the lumped capacitance assumption and it is appropriate for an object that is thin and conductive so that it can be assumed to be at a uniform temperature at any time. The lumped capacitance approximation is similar to the extended surface approximation that was discussed in Section 1.6. The resistance to conduction within the object is neglected as being small relative to the resistance to heat transfer from the surface of the object. Therefore, the lumped capacitance approximation is justified in the same way as the extended surface approximation, by defining an appropriate Biot number. For the case where only convection occurs from the surface of the object, the Biot number is:

$$Bi = \frac{R_{cond,int}}{R_{conv}} = \frac{L_{cond}\,\overline{h}}{k} \tag{3-1}$$

where L_{cond} is the conduction length within the object, $\overline{h}$ is the average heat transfer coefficient, and k is the conductivity of the material. Note that Eq. (3-1) is the simplest possible Biot number; it is often the case that heat transfer from the surface of the object will be resisted by mechanisms other than or in addition to convection (e.g., radiation or conduction through a thin insulating layer). These additional resistances should be included in the denominator of an appropriately defined Biot number.

The conduction length that characterizes an irregular shaped object can be ambiguous. Thermal energy will conduct out of the object along the easiest (i.e., shortest) path. For a thin plate, L_{cond} should be the half-width of the plate. For other shapes, L_{cond} should be selected so that it characterizes the minimum conduction length; the ratio of the volume of the object (V) to its surface area (A_s) is often used for this purpose:

$$L_{cond} = \frac{V}{A_s} \tag{3-2}$$

If the Biot number is much less than unity, then the lumped capacitance assumption is justified. A common criteria is that the Bi must be less than 0.1. However, this is certainly a case where engineering judgment is required based on the level of accuracy that is required for the model.

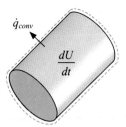

Figure 3-1: An object exposed to a time varying fluid temperature.

3.1.3 The Lumped Capacitance Problem

The lumped capacitance approximation reduces the spatial dimensionality of the problem to zero. Therefore, a control volume can be placed around the entire object (as all of the material is assumed to be at the same temperature at a given time). An energy balance will include all of the relevant energy transfers (e.g., convection and radiation) as well as the rate of energy storage. The result will be a first order differential equation that governs the temperature of the object. A single boundary condition, typically the initial temperature of the object, is required to obtain a solution using either analytical (as discussed in this section) or numerical (as discussed in Section 3.2) techniques.

Figure 3-1 illustrates an object with a small Biot number that can be considered to be lumped. The object is initially at a uniform temperature T_{ini} and is exposed to a time varying fluid temperature, T_∞, through a heat transfer coefficient, $\bar{h}$. The control volume in Figure 3-1 suggests the energy balance:

$$0 = \dot{q}_{conv} + \frac{dU}{dt} \tag{3-3}$$

where $\dot{q}_{conv}$ is the rate of convective heat transfer from the surface and U is the total energy stored in the object. The convective heat transfer is:

$$\dot{q}_{conv} = \bar{h} A_s \, (T - T_\infty) \tag{3-4}$$

where A_s is the surface area of the object exposed to the fluid. The rate of change of the total energy of the object can be expressed in terms of the specific energy of the material (u):

$$\frac{dU}{dt} = M \frac{du}{dt} \tag{3-5}$$

where M is the mass of the object. For solids or incompressible fluids, Eq. (3-5) can be written as:

$$\frac{dU}{dt} = M \underbrace{\frac{du}{dT}}_{c} \frac{dT}{dt} = M c \frac{dT}{dt} \tag{3-6}$$

where c is the specific heat capacity of the material. Substituting Eqs. (3-6) and (3-4) into Eq. (3-3) leads to:

$$0 = \bar{h} A_s \, (T - T_\infty) + M c \frac{dT}{dt} \tag{3-7}$$

which is the first order differential equation that governs the problem. Equation (3-7) can be rearranged:

$$\frac{dT}{dt} + \underbrace{\frac{\overline{h} A_s}{M c}}_{1/\tau_{lumped}} T = \frac{\overline{h} A_s}{M c} T_\infty \qquad (3\text{-}8)$$

Note that the group of variables that multiply the temperatures on the left and right sides of Eq. (3-8) must have units of inverse time; this group is used to define a lumped time constant, τ_{lumped}, that governs the problem:

$$\tau_{lumped} = \frac{M c}{\overline{h} A_s} \qquad (3\text{-}9)$$

Substituting Eq. (3-9) into Eq. (3-8) leads to:

$$\frac{dT}{dt} + \frac{T}{\tau_{lumped}} = \frac{T_\infty}{\tau_{lumped}} \qquad (3\text{-}10)$$

Equation (3-10) can be solved either analytically, using the same techniques discussed in Chapter 1 for 1-D steady-state conduction, or numerically (see Section 3.2). If additional heat transfer mechanisms or thermal loads are included in the problem (e.g., radiation or volumetric generation of thermal energy) then the governing differential equation will be different than Eq. (3-10). However, the steps associated with the derivation of the differential equation will remain the same and it will be possible to identify an appropriate lumped capacitance time constant.

3.1.4 The Lumped Capacitance Time Constant

The concept of a lumped capacitance time concept is useful even in the absence of an analytical solution to the problem. The lumped capacitance time constant is the product of the thermal resistance to heat transfer from the surface of the object (R) and the thermal capacitance of the object (C). For the object in Figure 3-1, the only resistance to heat transfer from the surface of the object is due to convection:

$$R = \frac{1}{\overline{h} A_s} \qquad (3\text{-}11)$$

The thermal capacitance of the object is the product of its mass and specific heat capacity. The thermal capacitance provides a measure of how much energy is required to change the temperature of the object:

$$C = M c \qquad (3\text{-}12)$$

The lumped time constant identified in Eq. (3-9) is the product of R, Eq. (3-11), and C, Eq. (3-12):

$$R C = \underbrace{\frac{1}{\overline{h} A_s}}_{R} \underbrace{M c}_{C} = \tau_{lumped} \qquad (3\text{-}13)$$

The lumped capacitance time constant for other situations can be calculated by modifying the resistance and capacitance terms in Eq. (3-13) appropriately. For example, if the object experiences both convection and radiation from its surface, then an

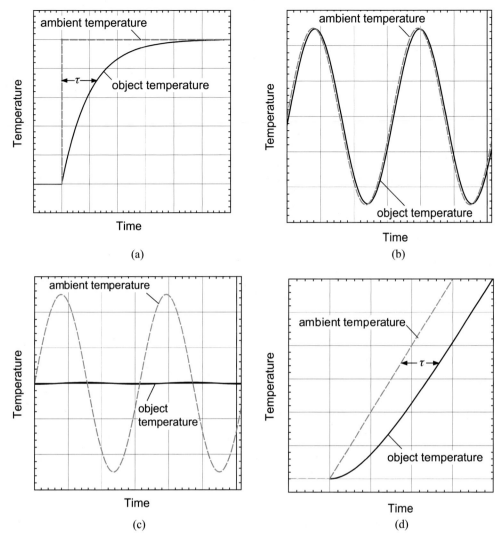

Figure 3-2: Approximate temperature response for an object subjected to (a) a step change in its ambient temperature, (b) and (c) an oscillatory ambient temperature where (b) the frequency of the oscillation is much less than the inverse of the time constant and (c) the frequency of the oscillation is much greater than the inverse of the time constant, and (d) a ramped ambient temperature.

appropriate thermal resistance is the parallel combination of a convective and radiative resistance:

$$R = \left(\overline{h} A_s + \sigma \varepsilon_s 4 \overline{T}^3 A_s\right)^{-1} \tag{3-14}$$

The lumped time constant is analogous to the electrical time constant associated with an R-C circuit and many of the concepts that may be familiar from electrical circuits can also be applied to transient heat transfer problems.

A quick estimate of the lumped time constant for a problem can provide substantial insight into the behavior of the system. The lumped time constant is, approximately,

Table 3-1: Summary of lumped capacitance solutions to some typical variations in the ambient temperature.

Situation	Solution	Nomenclature
Step change in ambient temperature	$$T = T_\infty + (T_{ini} - T_\infty)\exp\left(-\frac{t}{\tau_{lumped}}\right)$$	T_{ini} = initial temperature of environment and object T_∞ = temperature of environment after step τ_{lumped} = lumped time constant
Oscillatory ambient temperature (after initial transient has decayed)	$$T = \overline{T} + \frac{\Delta T}{\left[1 + (\omega\,\tau_{lumped})^2\right]}\sin(\omega t - \phi)$$ where $$\phi = \tan^{-1}(\omega\,\tau_{lumped})$$	$\overline{T}$ = average ambient temperature ΔT = amplitude of temperature oscillation ω = angular frequency of temperature oscillation (rad/s) τ_{lumped} = lumped time constant
Ramped ambient temperature	$$T = T_{ini} + \beta\,t + \beta\,\tau_{lumped}\left[\exp\left(-\frac{t}{\tau_{lumped}}\right) - 1\right]$$	T_{ini} = initial temperature of environment and object β = rate of ambient temperature change τ_{lumped} = lumped time constant

the amount of time that it will take the object to respond to any change in its thermal environment. For example, if the object is subjected to a step change in the ambient temperature, T_∞ in Eq. (3-10), then its temperature will be within 5% of the new temperature after 3 time constants, as shown in Figure 3-2(a). If the object is subjected to an oscillatory ambient temperature (e.g., within an engine cylinder or some other cyclic device) then the temperature of the object will follow the ambient temperature nearly exactly if the period of oscillation is much greater than the time constant (i.e., if the frequency is much less than the inverse of the time constant). In the opposite extreme, the object's temperature will be essentially constant if the period of oscillation is much less than the time constant (i.e., if the frequency is much greater than the inverse of the time constant). These extremes in behavior are shown in Figure 3-2(b) and (c). If the object is subjected to a ramped (i.e., linearly increasing) ambient temperature, then its temperature will tend to increase linearly as well, but its response will be delayed by approximately one time constant, as shown in Figure 3-2(d).

Some of the situations illustrated in Figure 3-2 will be investigated more completely in the following examples. The analytical solutions to these problems are summarized in Table 3-1. However, it is clear that simply knowing the time constant and its physical significance is sufficient in many cases.

EXAMPLE 3.1-1: DESIGN OF A CONVEYOR BELT

Plastic parts are formed in an injection mold and dropped (flat) onto a conveyor belt (Figure 1). The parts are disk-shaped with thickness $th = 2.0$ mm and diameter $D = 10.0$ cm. The plastic has thermal conductivity $k = 0.35$ W/m-K, density $\rho = 1100$ kg/m^3, and specific heat capacity $c = 1900$ J/kg-K. The side of the part that faces the conveyor belt is adiabatic. The top surface of the part is exposed to air at $T_\infty = 20°C$ with a heat transfer coefficient $\bar{h} = 15$ W/m^2-K. The temperature of the part immediately after it is formed is $T_{ini} = 180°C$. The part must be cooled to $T_{max} = 80°C$ before it can be stacked and packaged. The packaging system is positioned $L = 15$ ft away from the molding machine.

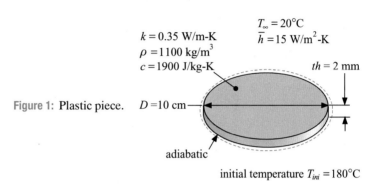

Figure 1: Plastic piece.

a) Is a lumped capacitance model of the part justified for this situation?

The inputs are entered in EES:

```
"EXAMPLE 3.1-1: Design of a Conveyor Belt"

$UnitSystem SI MASS RAD PA K J
$TABSTOPS 0.2 0.4 0.6 0.8 3.5 in

"Inputs"
th=2.0 [mm]*convert(mm,m)                "thickness"
k=0.35 [W/m-K]                           "conductivity"
D=10 [cm]*convert(cm,m)                  "diameter"
rho=1100 [kg/m^3]                        "density"
c=1900 [J/kg-K]                          "specific heat capacity"
h_bar=15 [W/m^2-K]                       "heat transfer coefficient"
T_ini=converttemp(C,K,180 [C])           "initial temperature of part"
T_infinity=converttemp(C,K,20 [C])       "ambient temperature"
T_max=converttemp(C,K,80 [C])            "maximum handling temperature"
L=15 [ft]*convert(ft,m)                  "conveyor length"
```

A lumped capacitance model of the part can be justified by examining the Biot number, the ratio of the resistance to internal conduction to the resistance to heat transfer from the surface of the object. In this problem, the resistance to heat transfer from the surface is due only to convection and therefore Eq. (3-1) is valid, where the conduction length is intuitively the thickness of the object (it would be the half-width if the conveyor side of the part were not adiabatic). Note that the characteristic

EXAMPLE 3.1-1: DESIGN OF A CONVEYOR BELT

length defined by Eq. (3-2) as the ratio of the volume (V) to the exposed area for heat transfer (A_s) is equal to the thickness of the part:

$$V = \frac{\pi D^2 th}{4}$$

$$A_s = \frac{\pi D^2}{4}$$

$$L_{cond} = \frac{V}{A_s} = \frac{\pi D^2 th}{4} \frac{4}{\pi D^2} = th$$

```
V=pi*D^2*th/4          "Volume"
As=pi*D^2/4            "Surface area exposed to cooling"
L_cond=V/As            "Conduction length"
Bi=h_bar*L_cond/k      "Biot number based on conduction length"
```

The Biot number predicted by EES is 0.09, which is sufficiently small to use the lumped capacitance model unless very high accuracy is required.

b) What is the maximum acceptable conveyor velocity so that the parts arrive at the packaging station below T_{max}?

The governing differential equation is obtained by considering a control volume that encloses the entire plastic part; the energy balance is:

$$0 = \dot{q}_{conv} + \frac{dU}{dt}$$

The rate of convection heat transfer is:

$$\dot{q}_{conv} = \overline{h} A_s (T - T_\infty)$$

and the rate of energy storage is:

$$\frac{dU}{dt} = \rho V c \frac{dT}{dt}$$

Combining these equations leads to:

$$0 = \overline{h} A_s (T - T_\infty) + \rho V c \frac{dT}{dt}$$

which can be rearranged:

$$\frac{dT}{dt} = -\frac{(T - T_\infty)}{\tau_{lumped}} \tag{1}$$

where τ_{lumped} is the time constant for this problem:

$$\tau_{lumped} = \frac{\rho V c}{\overline{h} A_s}$$

```
tau_lumped=V*rho*c/(h_bar*As)      "time constant"
```

The time constant for the part is 279 s. We should keep in mind that it will take on the order of 5 minutes to cool the plastic piece substantially and use this insight to check the more precise analytical solution that is obtained.

EXAMPLE 3.1-1: DESIGN OF A CONVEYOR BELT

Equation (1) is a first order differential equation with the boundary condition:

$$T_{t=0} = T_{ini} \qquad (2)$$

The differential equation is separable; that is, all of the terms involving the dependent variable, T, can be placed on one side while the terms involving the independent variable, t, can be placed on the other:

$$\frac{dT}{(T - T_\infty)} = -\frac{dt}{\tau_{lumped}}$$

The separated equation can be directly integrated:

$$\int_{T_{ini}}^{T} \frac{dT}{(T - T_\infty)} = -\int_{0}^{t} \frac{dt}{\tau_{lumped}} \qquad (3)$$

This integration is most easily accomplished by defining the temperature difference (θ):

$$\theta = T - T_\infty \qquad (4)$$

so that:

$$d\theta = dT \qquad (5)$$

Substituting Eqs. (4) and (5) into Eq. (3) leads to:

$$\int_{T_{ini}-T_\infty}^{T-T_\infty} \frac{d\theta}{\theta} = -\int_{0}^{t} \frac{dt}{\tau_{lumped}}$$

Carrying out the integration leads to:

$$\ln\left(\frac{T - T_\infty}{T_{ini} - T_\infty}\right) = -\frac{t}{\tau_{lumped}}$$

Solving for T leads to:

$$T = T_\infty + (T_{ini} - T_\infty)\exp\left(-\frac{t}{\tau_{lumped}}\right) \qquad (6)$$

which is equivalent to the entry in Table 3-1 for a step change in ambient temperature. The time required to cool the part from T_{ini} to T_{max} can be computed using Eq. (6):

```
T_max=T_infinity+(T_ini-T_infinity)*exp(-t_cool/tau_lumped)      "time required to cool part"
```

and is found to be $t_{cool} = 273$ s; note that this value is in good agreement with the previously calculated time constant. The linear velocity of the conveyor that is required so that it takes at least 273 s for the part to travel the 15 ft between the molding machine and the packaging station is:

$$u_c = \frac{L}{t_{cool}}$$

```
u_c=L/t_cool                                   "conveyor velocity"
u_c_fpm=u_c*convert(m/s,ft/min)                "conveyor velocity in ft/min"
```

The maximum allowable velocity u_c is found to be 0.0167 m/s (3.29 ft/min).

EXAMPLE 3.1-2: SENSOR IN AN OSCILLATING TEMPERATURE ENVIRONMENT

EXAMPLE 3.1-2: SENSOR IN AN OSCILLATING TEMPERATURE ENVIRONMENT

A temperature sensor is installed in a chemical reactor that operates in a cyclic fashion. The temperature of the fluid in the reactor varies in an approximately sinusoidal manner with a mean temperature $\overline{T}_\infty = 320°C$, an amplitude $\Delta T_\infty = 50°C$, and a frequency $f = 0.5$ Hz. The sensor can be modeled as a sphere with diameter $D = 1.0$ mm. The sensor is made of a material with conductivity $k_s = 50$ W/m-K, specific heat capacity $c_s = 150$ J/kg-K, and density $\rho_s = 16000$ kg/m^3. In order to provide corrosion resistance, the sensor has been coated with a thin layer of plastic; the coating is $th_c = 100$ μm thick with conductivity $k_c = 0.2$ W/m-K and has negligible heat capacity relative to the sensor itself. The heat transfer coefficient between the surface of the coating and the fluid is $\overline{h} = 500$ W/m^2-K. The sensor is initially at $T_{ini} = 260°C$.

a) Is a lumped capacitance model of the temperature sensor appropriate?

The inputs are entered in EES:

```
"EXAMPLE 3.1-2: Sensor in an Oscillating Temperature Environment"

$UnitSystem SI MASS RAD PA K J
$TABSTOPS 0.2 0.4 0.6 0.8 3.5 in

"Inputs"
T_infinity_bar=converttemp(C,K,320[C])      "average temperature of reactor"
T_ini=converttemp(C,K,260[C])               "initial temperature of sensor"
DT_infinity=50[K]                           "amplitude of reactor temperature change"
f=0.5 [Hz]                                  "frequency of reactor temperature change"
D=1.0 [mm]*convert(mm,m)                    "diameter of sensor"
k_s=50 [W/m-K]                              "conductivity of sensor material"
c_s=150 [J/kg-K]                            "specific heat capacity of sensor material"
rho_s=16000 [kg/m^3]                        "density of sensor material"
th_c=100 [micron]*convert(micron,m)         "thickness of coating"
k_c=0.2 [W/m-K]                             "conductivity of coating"
h_bar=500 [W/m^2-K]                         "heat transfer coefficient"
```

The Biot number is the ratio of the internal conduction resistance to the resistance to heat transfer from the surface of the object. In this problem, the resistance to heat transfer from the surface is the series combination of convection (R_{conv}):

$$R_{conv} = \frac{1}{\overline{h}\,4\,\pi\left(\dfrac{D}{2} + th_c\right)^2}$$

and the conduction resistance of the coating ($R_{cond,c}$, from Table 1-2):

$$R_{cond,c} = \frac{\left[\dfrac{2}{D} - \dfrac{2}{(D + 2\,th_c)}\right]}{4\,\pi\,k_c}$$

EXAMPLE 3.1-2: SENSOR IN AN OSCILLATING TEMPERATURE ENVIRONMENT

```
R_conv=1/(h_bar*4*pi*(D/2+th_c)^2)          "convective resistance"
R_cond_c=(1/(D/2)-1/(D/2+th_c))/(4*pi*k_c)  "conduction resistance of coating"
```

The resistance to internal conduction ($R_{cond,int}$) is approximated according to:

$$R_{cond,int} = \frac{L_{cond}}{k_s\,A_s}$$

where A_s is the surface area of the sensor:

$$A_s = 4\,\pi\left(\frac{D}{2}\right)^2$$

and L_{cond} is the conduction length, approximated according to Eq. (3-2):

$$L_{cond} = \frac{V}{A_s}$$

where:

$$V = \frac{4\,\pi}{3}\left(\frac{D}{2}\right)^3$$

```
V=4*pi*(D/2)^3/3              "volume of sensor"
A_s=4*pi*(D/2)^2             "surface area of sensor"
L_cond=V/A_s                 "approximate conduction length"
R_cond_int=L_cond/(k_s*A_s)  "internal conduction resistance"
```

The Biot number that characterizes this problem is therefore:

$$Bi = \frac{R_{cond}}{R_c + R_{conv}}$$

```
Bi=R_cond_int/(R_conv+R_cond_c)     "Biot number"
```

which leads to $Bi = 0.0018$; this is sufficiently less than 1 to justify a lumped capacitance model.

b) **What is the time constant associated with the sensor? Do you expect there to be a substantial temperature measurement error related to the dynamic response of the sensor?**

The time constant (τ_{lumped}) is the product of the resistance to heat transfer from the surface of the sensor (which is related to conduction through the coating and convection) and the thermal mass of the sensor (C):

$$\tau_{lumped} = (R_{cond,c} + R_{conv})C$$

EXAMPLE 3.1-2: SENSOR IN AN OSCILLATING TEMPERATURE ENVIRONMENT

where

$$C = V \, \rho_s \, c_s$$

C=V*rho_s*c_s   "capacitance of the sensor"
tau=(R_conv+R_cond_c)*C   "time constant of the sensor"

The time constant is 0.72 s and the time per cycle (the inverse of the frequency) is 2 s. These quantities are on the same order and therefore it is not likely that the temperature sensor will be able to faithfully follow the reactor temperature.

c) Develop an analytical model of the temperature response of the sensor.

The temperature sensor is exposed to a sinusoidally varying temperature:

$$T_\infty = \overline{T}_\infty + \Delta T_\infty \sin\left(2\,\pi\,f\,t\right) \tag{1}$$

The governing differential equation for the sensor balances heat transfer to ambient against energy storage:

$$0 = \frac{[T - T_\infty]}{R_{cond,c} + R_{conv}} + C\,\frac{dT}{dt} \tag{2}$$

Substituting Eq. (1) into Eq. (2) leads to:

$$0 = \frac{[T - \overline{T}_\infty - \Delta T_\infty \sin(2\,\pi\,f\,t)]}{\tau_{lumped}} + \frac{dT}{dt}$$

which is rearranged:

$$\frac{dT}{dt} + \frac{T}{\tau_{lumped}} = \frac{\overline{T}_\infty}{\tau_{lumped}} + \frac{\Delta T_\infty \sin(2\,\pi\,f\,t)}{\tau_{lumped}} \tag{3}$$

Equation (3) is a non-homogeneous ordinary differential equation. The solution is assumed to consist of a homogeneous and particular solution:

$$T = T_h + T_p \tag{4}$$

Substituting Eq. (4) into Eq. (3) leads to:

$$\underbrace{\frac{dT_h}{dt} + \frac{T_h}{\tau_{lumped}}}_{\substack{=0 \text{ for homogeneous} \\ \text{differential equation}}} + \underbrace{\frac{dT_p}{dt} + \frac{T_p}{\tau_{lumped}} = \frac{\overline{T}_\infty}{\tau_{lumped}} + \frac{\Delta T_\infty \sin(2\,\pi\,f\,t)}{\tau_{lumped}}}_{\text{particular differential equation}}$$

The homogeneous differential equation is:

$$\frac{dT_h}{dt} + \frac{T_h}{\tau_{lumped}} = 0$$

The solution to the homogeneous differential equation can be obtained by separating variables and integrating:

$$\int \frac{dT_h}{T_h} = -\int \frac{dt}{\tau_{lumped}}$$

Carrying out the indefinite integral leads to:

$$\ln\left(T_h\right) = -\frac{t}{\tau_{lumped}} + C_1 \tag{5}$$

EXAMPLE 3.1-2: SENSOR IN AN OSCILLATING TEMPERATURE ENVIRONMENT

where C_1 is a constant of integration. Equation (5) can be rearranged:

$$T_h = \underbrace{\exp(C_1)}_{C_1^*} \exp\left(-\frac{t}{\tau_{lumped}}\right) = C_1^* \exp\left(-\frac{t}{\tau_{lumped}}\right)$$

where C_1^* is an also undetermined constant that will subsequently be referred to as C_1:

$$T_h = C_1 \exp\left(-\frac{t}{\tau_{lumped}}\right) \tag{6}$$

Notice that the homogeneous solution provided by Eq. (6) dies off after about three time constants.

The particular solution (T_p) is obtained by identifying any function that satisfies the particular differential equation:

$$\frac{dT_p}{dt} + \frac{T_p}{\tau_{lumped}} = \frac{\overline{T}_\infty}{\tau_{lumped}} + \frac{\Delta T_\infty \sin(2\pi f t)}{\tau_{lumped}} \tag{7}$$

By inspection, the sum of a constant and a sine and cosine with the same frequency can be made to solve Eq. (7):

$$T_p = C_2 \sin(2\pi f t) + C_3 \cos(2\pi f t) + C_4 \tag{8}$$

Substituting Eq. (8) into Eq. (7) leads to:

$$C_2 \, 2\pi f \cos(2\pi f t) - C_3 \, 2\pi f \sin(2\pi f t) + \frac{C_2}{\tau_{lumped}} \sin(2\pi f t)$$

$$+ \frac{C_3}{\tau_{lumped}} \cos(2\pi f t) + \frac{C_4}{\tau_{lumped}} = \frac{\overline{T}_\infty}{\tau_{lumped}} + \frac{\Delta T_\infty}{\tau_{lumped}} \sin(2\pi f t) \tag{9}$$

Equation (9) can only be true if the constant, sine and cosine terms each separately add to zero:

$$\frac{C_4}{\tau_{lumped}} = \frac{\overline{T}_\infty}{\tau_{lumped}}$$

$$C_2 \, 2\pi f + \frac{C_3}{\tau_{lumped}} = 0$$

$$-C_3 \, 2\pi f + \frac{C_2}{\tau_{lumped}} = \frac{\Delta T_\infty}{\tau_{lumped}}$$

Solving for C_2, C_3, and C_4 leads to:

$$C_2 = \frac{\Delta T_\infty}{1 + (2\pi f \tau_{lumped})^2}$$

$$C_3 = -\frac{2\pi f \tau \Delta T_\infty}{1 + (2\pi f \tau_{lumped})^2}$$

$$C_4 = \overline{T}_\infty$$

so that the particular solution is:

$$T_p = \overline{T}_\infty + \frac{\Delta T_\infty}{1 + (2\pi f \tau_{lumped})^2}[\sin(2\pi f t) - (2\pi f \tau_{lumped}) \cos(2\pi f t)] \tag{10}$$

EXAMPLE 3.1-2: SENSOR IN AN OSCILLATING TEMPERATURE ENVIRONMENT

The solution is the sum of the particular and homogeneous solutions, Eqs. (6) and (10):

$$T = C_1 \exp\left(-\frac{t}{\tau_{lumped}}\right)$$

$$+\overline{T}_\infty + \frac{\Delta T_\infty}{1 + (2\pi f \tau_{lumped})^2}[\sin(2\pi f t) - (2\pi f \tau_{lumped})\cos(2\pi f t)] \tag{11}$$

Note that the same conclusion can be reached using two lines of Maple code; the governing differential equation, Eq. (3), is entered and solved:

```
> restart;
> ODE:=diff(T(t),t)+T(t)/tau=T_infinity_bar/tau+DT_infinity*sin(2*pi*f*t)/tau;
```

$$ODE := \left(\frac{d}{dt}T(t)\right) + \frac{T(t)}{\tau} = \frac{T\_infinity\_bar}{\tau} + \frac{DT\_infinity\,\sin(2\pi ft)}{\tau}$$

```
> Ts:=dsolve(ODE);
```

$$Ts := T(t) = e^{(-\frac{t}{\tau})}\_C1 + (T\_infinity\_bar + 4\,T\_infinity\_barf^2\pi^2\,\tau^2$$
$$-2\,DT\_infinity\,\cos(2\pi f t)f\,\pi\,\tau + DT\_infinity\,\sin(2\pi f t))/(1 + 4f^2\,\pi^2\,\tau^2)$$

The solution identified by Maple is the equivalent to Eq. (11); the solution is copied into EES, with minor editing:

```
"Solution"
Temp = exp(-1/tau*t)*C1-(-T_infinity_bar-4*T_infinity_bar*pi^2*f^2*tau^2+&
        2*DT_infinity*cos(2*pi*f*t)*pi*f*tau&
        -DT_infinity*sin(2*pi*f*t))/(1+4*pi^2*f^2*tau^2)
```

The constant C_1 must be selected so that the boundary condition is satisfied:

$$T_{t=0} = T_{ini} \tag{12}$$

Substituting Eq. (11) into Eq. (12) leads to:

$$T_{t=0} = C_1 + \overline{T}_\infty - \frac{\Delta T_\infty 2\pi f \tau_{lumped}}{1 + (2\pi f \tau_{lumped})^2} = T_{ini}$$

which leads to:

$$C_1 = \frac{\Delta T_\infty 2\pi f \tau_{lumped}}{1 + (2\pi f \tau_{lumped})^2} + T_{ini} - \overline{T}_\infty$$

The symbolic expression for the boundary condition can also be found using Maple:

```
> rhs(eval(Ts,t=0))=T_ini;
```

$$\_C1 + \frac{T\_infinity\_bar + 4T\_infinity\_barf^2\pi^2\,\tau^2 - 2\,DT\_infinityf\,\pi\,\tau}{1 + 4f^2\pi^2\,\tau^2} = T\_ini$$

which is copied and pasted into EES:

```
C_1+(T_infinity_bar+4*T_infinity_bar*f^2*pi^2*tau^2-2*DT_infinity*f*pi*tau)/(1+4*f^2*pi^2*tau^2) = T_ini
"initial condition"
```

The sensor temperature and fluid temperature are converted to Celsius.

```
T_infinity=T_infinity_bar+DT_infinity*sin(2*pi*f*t)        "ambient temperature"
T_C=converttemp(K,C,Temp)                                   "sensor temperature in C"
T_infinity_C=converttemp(K,C,T_infinity)                    "ambient temperature in C"
```

The temperatures are computed in a parametric table in which time is set so that it ranges from 0 to 10 s. The fluid and sensor temperature variation are shown as a function of time in Figure 1.

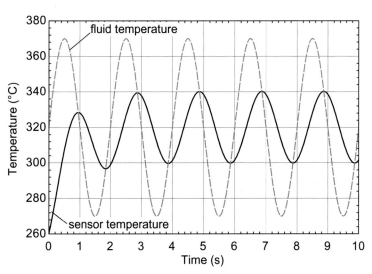

Figure 1: Temperature sensor and fluid temperature as a function of time.

Note that after approximately 3 seconds (i.e., a few time constants) the homogeneous solution has decayed to zero and the temperature response of the sensor is given entirely by the particular solution, Eq. (10):

$$T_p = \overline{T}_\infty + \frac{\Delta T_\infty}{1 + (2\pi f \tau_{lumped})^2} [\sin(2\pi f t) - (2\pi f \tau_{lumped}) \cos(2\pi f t)] \qquad (10)$$

Equation (10) can be rewritten in terms of an attenuation of the amplitude of the oscillation (Att) and a phase lag relative to the fluid temperature variation (ϕ)

$$T_p = \overline{T}_\infty + Att\, \Delta T_\infty \sin(2\pi f t - \phi) \qquad (13)$$

Equation (13) is rewritten using the trigonometric identity:

$$T_p = \overline{T}_\infty + Att\, \Delta T_\infty [\sin(2\pi f t)\cos(\phi) - \cos(2\pi f t)\sin(\phi)] \qquad (14)$$

Comparing Eq. (10) with Eq. (14) leads to:

$$Att\, \cos(\phi) = \frac{1}{1 + (2\pi f \tau)^2} \qquad (15)$$

EXAMPLE 3.1-2: SENSOR IN AN OSCILLATING TEMPERATURE ENVIRONMENT

EXAMPLE 3.1-2: SENSOR IN AN OSCILLATING TEMPERATURE ENVIRONMENT

and

$$Att \sin(\phi) = \frac{2\pi f \tau_{lumped}}{1 + (2\pi f \tau_{lumped})^2} \tag{16}$$

Dividing Eq. (16) by Eq. (15) leads to:

$$\tan(\phi) = 2\pi f \tau_{lumped}$$

so the phase (the lag) between the sensor and fluid temperature is:

$$\phi = \tan^{-1}(2\pi f \tau_{lumped})$$

and the attenuation is:

$$Att = \frac{1}{\sqrt{1 + (2\pi f \tau_{lumped})^2}}$$

Figure 2 shows the attenuation and phase angle as a function of the product of the frequency and the time constant ($f\tau_{lumped}$).

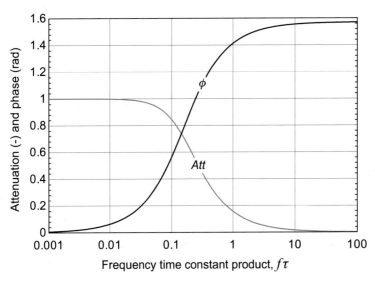

Figure 2: Attenuation and phase angle as a function of the product of the frequency and the time constant.

Notice that if either the frequency or the time constant is small, then the attenuation goes to unity and the phase goes to zero. In this limit, the temperature sensor will faithfully follow the fluid temperature with little error related to the transient response characteristics of the sensor; this is the situation that was illustrated earlier in Figure 3-2(b). In the other limit, if either the frequency or time constant of the sensor are very large then the attenuation will approach zero and the phase will approach $\pi/2$ rad (90 deg.). This situation corresponds to Figure 3-2(c) where the sensor cannot respond to the temperature oscillations. The dynamic characteristics of a temperature sensor are important in many applications and should be carefully considered when selecting an instrument for a transient temperature measurement.

3.2 Numerical Solutions to 0-D Transient Problems

3.2.1 Introduction

Section 3.1 discussed the lumped capacitance model, which neglects any spatial temperature gradients within an object and therefore approximates the temperature in a transient problem as being only a function of time. The analytical solution to such 0-D (or lumped capacitance) problems was examined in Section 3.1. In this section, lumped capacitance problems will be solved numerically.

The numerical solution to any transient problem begins with the derivation of the governing differential equation, which allows the calculation of the temperature rate of change as a function of the current temperature and time. The solution to the problem is therefore a matter of integrating the governing differential equation forward in time. There are a number of techniques available for numerical integration, each with its own characteristic level of accuracy, stability, and complexity. These numerical integration techniques are discussed in this section and revisited in Section 3.8 in order to solve 1-D transient problems and again in Chapter 5 in order to solve convection problems.

The governing differential equation that provides the rate of change of a variable (or several variables) given its own value (or values) is sometimes referred to as the state equation for the dynamic system. State equations characterize the transient behavior of problems in the areas of controls, dynamics, kinematics, fluids, electrical circuits, etc. Therefore, the numerical solution techniques provided in this section are generally relevant to a wide range of engineering problems.

The same caveats that were discussed previously in Sections 1.4 and 2.5 with regard to the numerical solution of steady-state conduction problems also apply to numerical solutions of transient conduction problems. Any numerical solution should be evaluated for numerical convergence (i.e., to ensure that you are using a sufficient number of time steps), examined against your physical intuition, and compared to an analytical solution in some limit.

3.2.2 Numerical Integration Techniques

In this section, the simplest lumped capacitance problem is solved numerically in order to illustrate the various options for numerical integration. An object initially in equilibrium with its environment at T_{ini} is subjected to a step change in the ambient temperature from T_{ini} to T_∞. The governing differential equation provides the temperature rate of change given the current value of the temperature; the governing differential equation is derived from an energy balance on the object (see EXAMPLE 3.1-1):

$$\frac{dT}{dt} = -\frac{(T - T_\infty)}{\tau_{lumped}} \tag{3-15}$$

where τ_{lumped} is the lumped time constant for the object. The analytical solution to the problem is derived in EXAMPLE 3.1-1, providing a basis for evaluating the accuracy of the numerical solutions that are derived in this section:

$$T_{an} = T_\infty + (T_{ini} - T_\infty) \exp\left(-\frac{t}{\tau_{lumped}}\right) \tag{3-16}$$

The solution can also be obtained numerically using one of several available numerical integration techniques. Each numerical technique requires that the total simulation time

(t_{sim}) be broken into small time steps. The simplest option is to use equal-sized steps, each with duration Δt:

$$\Delta t = \frac{t_{sim}}{(M-1)} \tag{3-17}$$

where M is the number of times at which the temperature will be evaluated. The temperature at each time step (T_j) is computed by the numerical model, where j indicates the time step (T_1 is the initial temperature of the object and T_M is the temperature at the end of a simulation). The time corresponding to each time step is therefore:

$$t_j = \frac{(j-1)}{(M-1)}t_{sim} \quad \text{for } j = 1..M \tag{3-18}$$

Euler's Method

The temperature at the end of each time step is computed based on the temperature at the beginning of the time step and the governing differential equation. The simplest (and generally the worst) technique for numerical integration is Euler's method. Euler's method approximates the rate of temperature change within the time step as being constant and equal to its value at the beginning of the time step. Therefore, for any time step j:

$$T_{j+1} = T_j + \left.\frac{dT}{dt}\right|_{T=T_j,t=t_j} \Delta t \tag{3-19}$$

Because the temperature at the end of the time step (T_{j+1}) can be calculated explicitly using information that is available at the beginning of the time step (T_j), Euler's method is referred to as an explicit numerical technique. The temperature at the end of the first time step (T_2) is given by:

$$T_2 = T_1 + \left.\frac{dT}{dt}\right|_{T=T_1,t=0} \Delta t \tag{3-20}$$

Substituting the state equation for our problem, Eq. (3-15), into Eq. (3-20) leads to:

$$T_2 = T_1 - \frac{(T_1 - T_\infty)}{\tau_{lumped}}\Delta t \tag{3-21}$$

Equation (3-21) is written for every time step, resulting in a set of explicit equations that can be solved sequentially for the temperature at each time.

As an example, consider the case where $\tau_{lumped} = 100\,\text{s}$, $T_{ini} = 300\,\text{K}$, and $T_\infty = 400\,\text{K}$. These inputs are entered in EES:

```
$UnitSystem SI MASS RAD PA K J
$TABSTOPS 0.2 0.4 0.6 0.8 3.5 in

"Inputs"
tau=100 [s]                        "time constant"
T_ini=300 [K]                      "initial temperature"
T_infinity=400 [K]                 "ambient temperature"
```

The time step duration and array of times for which the solution will be computed is specified using Eqs. (3-17) and (3-18):

```
t_sim=500 [s]                                     "simulation time"
M=501 [-]                                          "number of time steps"
DELTAt=t_sim/(M-1)                                 "duration of time step"
duplicate j=1,M
    time[j]=(j-1)*t_sim/(M-1)                      "time associated with each temperature"
end
```

For the values of t_{sim} and M entered above, $\Delta t = 1$ s and the temperature predicted by the numerical model at the end of the first timestep, T_2, computed using Eq. (3-21) is 301 K, as obtained from:

```
"Euler's Method"
T[1]=T_ini
T[2]=T[1]-(T[1]-T_infinity)*DELTAt/tau             "estimate of temperature at the 1st timestep"
```

The actual temperature at $t = 1$ s is obtained using the analytical solution to the problem, Eq. (3-16):

```
T_an[2]=T_infinity+(T_ini-T_infinity)*exp(-time[2]/tau)
                                                   "analytical solution for 1st time step"
```

and found to be $T_{an,t=1s} = 300.995$ K. The numerical technique is not exact; the error between the numerical and analytical solutions at the end of the first time step is $err = 0.005$ K. If the time step duration is increased by a factor of 10, to $\Delta t = 10$ s (by reducing M to 51), then the numerical solution at the end of the first time step becomes $T_2 = 310$ K and the analytical solution is $T_{an,t=10s} = 309.52°$C. Therefore, the error between the numerical and analytical solutions increases by approximately a factor of 100, to $err = 0.48$ K. This behavior is a characteristic of the Euler technique; the local error is proportional to the square of the size of the time step.

$$err \propto \Delta t^2 \tag{3-22}$$

This characteristic can be inferred by examination of Eq. (3-20), which is essentially the first two terms of a Taylor series expansion of the temperature about time $t = 0$:

$$T_2 = \underbrace{T_1 + \frac{dT}{dt}\bigg|_{T=T_1,t=0} \Delta t}_{\text{Euler's approximation}} + \underbrace{\frac{d^2T}{dt^2}\bigg|_{T=T_1,t=0} \frac{\Delta t^2}{2!} + \frac{d^3T}{dt^3}\bigg|_{T=T_1,t=0} \frac{\Delta t^3}{3!} + \cdots}_{err \approx \text{neglected terms}} \tag{3-23}$$

Examination of Eq. (3-23) shows that the error, corresponding to the neglected terms in the Taylor's series, is proportional to Δt^2 (neglecting the smaller, higher order terms). Therefore, Euler's technique is referred to as a first order method because it retains the first term in the Taylor's series approximation. The error associated with a first order technique is proportional to the second power of the time step duration. Most numerical techniques that are commonly used have higher order and therefore achieve higher accuracy.

The other drawback of Euler's technique (and any explicit numerical technique) is that it may become unstable if the duration of the time step becomes too large. For our example, if the time step duration is increased beyond the time constant of the object, $\tau_{lumped} = 100$ s, then the problem becomes unstable.

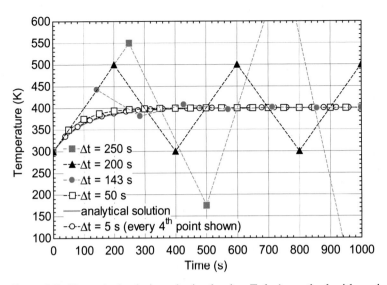

Figure 3-3: Numerical solution obtained using Euler's method with various values of Δt. Also shown is the analytical solution to the same problem.

The process of moving forward through all of the time steps may be automated using a duplicate loop:

```
"Euler's Method"
T[1]=T_ini                                          "initial temperature"
duplicate j=1,(M-1)
  T[j+1]=T[j]-(T[j]-T_infinity)*DELTAt/tau          "Euler solution"
  T_an[j+1]=T_infinity+(T_ini-T_infinity)*exp(-time[j+1]/tau)
                                                    "analytical solution"
  err[j+1]=abs(T_an[j+1]-T[j+1])    "error between numerical and analytical solution"
end
err_max=max(err[2..M])                              "maximum error"
```

Note that the analytical solution and the absolute value of the error between the numerical and analytical solutions are also obtained at each time step. The maximum error is computed at the conclusion of the duplicate loop.

Figure 3-3 illustrates the numerical solution for various values of the time step duration. The analytical solution provided by Eq. (3-16) is also shown in Figure 3-3. Notice that if the simulation is carried out with a time step duration that is less than $\Delta t = \tau_{lumped} = 100$ s, then the solution is stable and follows the analytical solution more precisely as the time step is reduced. If the numerical solution is carried out with a time step duration that is between $\Delta t = \tau_{lumped} = 100$ s and $\Delta t = 2\tau_{lumped} = 200$ s then the prediction oscillates about the actual temperature but remains bounded. For time step durations greater than $\Delta t = 2\tau_{lumped} = 200$ s, the solution oscillates in an unbounded manner.

Figure 3-3 shows that any solution in which $\Delta t > \tau_{lumped}$ is unstable. This threshold time step duration that governs the stability of the solution is called the critical time step, Δt_{crit}, and its value can be determined from the details of the problem. Substituting the

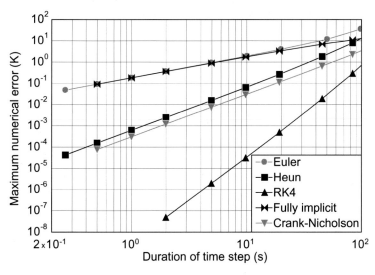

Figure 3-4: Maximum numerical error as a function of Δt for various numerical integration techniques.

governing equation, Eq. (3-15) into the definition of Euler's method, Eq. (3-19), leads to:

$$T_{j+1} = T_j - \frac{(T_j - T_\infty)}{\tau_{lumped}} \Delta t \tag{3-24}$$

Rearranging Eq. (3-24) leads to:

$$T_{j+1} = T_j \left[1 - \frac{\Delta t}{\tau_{lumped}} \right] + T_\infty \frac{\Delta t}{\tau_{lumped}} \tag{3-25}$$

The solution becomes unstable when the coefficient multiplying the temperature at the beginning of the time step (T_j) becomes negative; therefore:

$$\Delta t_{crit} = \tau_{lumped} \tag{3-26}$$

Not surprisingly, the numerical simulation of objects that have small time constants will require the use of small time steps.

The accuracy of the solution (i.e., the degree to which the numerical solution agrees with the analytical solution) improves as the duration of the time step is reduced. Figure 3-4 illustrates the global error, defined as the maximum error between the numerical and analytical solution, as a function of the duration of the time step. The global error is related to the local error associated with any one time step. Note that while the local error is proportional to Δt^2, the global error is proportional to Δt. Also shown in Figure 3-4 is the global accuracy of several alternative numerical integration techniques that are discussed in subsequent sections.

Any of the numerical integration techniques discussed in this section can be implemented in most software. As the problems become more complex (e.g., 1-D transient problems), the number of equations and the amount of data that must be stored increases and therefore it will be appropriate to utilize a formal programming language (e.g., MATLAB) for the solution.

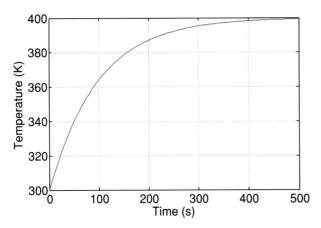

Figure 3-5: Temperature as a function of time predicted by MATLAB using Euler's method.

The numerical solution using Euler's method is implemented in MATLAB using a script with the inputs provided as the first lines:

```
clear all;

%INPUTS
tau=100;                                % time constant (s)
T_ini=300;                              % initial temperature (K)
T_infinity=400;                         % ambient temperature (K)
```

The array of times at which the solution will be determined is specified:

```
t_sim=500;                              % simulation time (s)
M=101;                                  % number of time steps (-)
DELTAt=t_sim/(M-1);                     % time step duration (s)
for j=1:M
    time(j)=(j-1)*t_sim/(M-1);          % time associated with each step
end
```

Because Euler's method is explicit, the solution is implemented in MATLAB using exactly the same technique that is used in EES; this is possible because each of the predictions for T_{j+1} require only knowledge of T_j, as shown by Eq. (3-25).

```
T(1)=T_ini;                             % initial temperature
for j=1:(M-1)
    T(j+1)=T(j)-(T(j)-T_infinity)*DELTAt/tau;    % Euler method
end
```

Running the script will place the vectors T and time in the command space. The numerical solution obtained by MATLAB is shown in Figure 3-5.

Heun's Method

Euler's method is the simplest example of a numerical integration technique; it is a first order explicit technique. In this section, Heun's method is presented. Heun's method is a second order explicit technique (but with the same stability characteristics as Euler's method).

Heun's method is an example of a predictor-corrector technique. In order to simulate any time step j, Heun's method begins with an Euler step to obtain an initial prediction for the temperature at the conclusion of the time step ($\hat{T}_{j+1}$). This first step in the solution is referred to as the predictor step and the details are essentially identical to Euler's method:

$$\hat{T}_{j+1} = T_j + \left. \frac{dT}{dt} \right|_{T=T_j, t=t_j} \Delta t \tag{3-27}$$

However, Heun's method uses the results of the predictor step to carry out a corrector step. The temperature predicted at the end of the time step ($\hat{T}_{j+1}$) is used to predict the temperature rate of change at the end of the time step ($\frac{dT}{dt}|_{T=\hat{T}_{j+1}, t=t_{j+1}}$). The corrector step predicts the temperature at the end of the time step (T_{j+1}) based on the average of the time rates of change at the beginning and end of the time step.

$$T_{j+1} = T_j + \left[\left. \frac{dT}{dt} \right|_{T=T_j, t=t_j} + \left. \frac{dT}{dt} \right|_{T=\hat{T}_{j+1}, t=t_{j+1}} \right] \frac{\Delta t}{2} \tag{3-28}$$

Heun's method is illustrated in the context of the same simple problem that was used to demonstrate Euler's technique. Initially, the technique will be implemented using EES:

```
$UnitSystem SI MASS RAD PA K J
$TABSTOPS 0.2 0.4 0.6 0.8 3.5 in

"Inputs"
tau=100 [s]                                        "time constant"
T_ini=300 [K]                                      "initial temperature"
T_infinity=400 [K]                                 "ambient temperature"

t_sim=500 [s]                                      "simulation time"
M=51 [-]                                           "number of time steps"
DELTAt=t_sim/(M-1)                                 "duration of time step"
duplicate j=1,M
    time[j]=(j-1)*t_sim/(M-1)                       "time associated with each temperature"
end
```

The process of moving through the first time step begins with the predictor step:

$$\hat{T}_2 = T_1 + \left. \frac{dT}{dt} \right|_{T=T_1, t=t_1} \Delta t \tag{3-29}$$

or, substituting Eq. (3-15) into Eq. (3-29):

$$\hat{T}_2 = T_1 - \frac{(T_1 - T_\infty)}{\tau_{lumped}} \Delta t \tag{3-30}$$

```
T[1]=T_ini                                         "initial temperature"
T_hat[2]=T[1]-(T[1]-T_infinity)*DELTAt/tau         "predictor step"
```

Note that for the same conditions considered previously (i.e., $\tau_{lumped} = 100\,\text{s}$, $T_{ini} = 300\,\text{K}$, and $T_\infty = 400\,\text{K}$), the predictor step using $\Delta t = 10\,\text{s}$ leads to $\hat{T}_2 = 310\,\text{K}$, which is about 0.5 K in error relative to the actual temperature calculated using Eq. (3-16), $T_{an,t=10s} = 309.52\,\text{K}$.

The corrector step follows:

$$T_2 = T_1 + \left[\frac{dT}{dt}\bigg|_{T=T_1, t=t_1} + \frac{dT}{dt}\bigg|_{T=\hat{T}_2, t=t_2} \right] \frac{\Delta t}{2} \tag{3-31}$$

or, substituting Eq. (3-15) into Eq. (3-31):

$$T_2 = T_1 - \left[\frac{(T_1 - T_\infty)}{\tau_{lumped}} + \frac{(\hat{T}_2 - T_\infty)}{\tau_{lumped}} \right] \frac{\Delta t}{2} \tag{3-32}$$

```
T[2]=T[1]-((T[1]-T_infinity)/tau+(T_hat[2]-T_infinity)/tau)*DELTAt/2          "corrector step"
```

The corrector step predicts $T_2 = 9.50°C$, which is only $0.02°C$ in error with respect to $T_{an,t=10s} = 9.52°C$. This result illustrates the power of the predictor/corrector process; a single-step correction (which approximately doubles the computational time) can lead to more than an order of magnitude increase in accuracy. The two-step predictor/corrector process is a second order method; the local error is proportional to the duration of the time step to the third power.

The EES code below implements a numerical solution to the problem using a duplicate loop and computes the numerical error. (Note that the EES code corresponding to the Euler solution must be commented out or deleted.)

```
"Heun's Method"
T[1]=T_ini
duplicate j=1,(M-1)
  T_hat[j+1]=T[j]-(T[j]-T_infinity)*DELTAt/tau              "predictor step"
  T[j+1]=T[j]-((T[j]-T_infinity)/tau+(T_hat[j+1]-T_infinity)/tau)*DELTAt/2
                                                           "corrector step"
  T_an[j+1]=T_infinity+(T_ini-T_infinity)*exp(-time[j+1]/tau)
                                                           "analytical solution"
  err[j+1]=abs(T_an[j+1]-T[j+1])                          "numerical error"
end
err_max=max(err[2..M])                                     "maximum numerical error"
```

Figure 3-6 illustrates the numerical solution obtained using Heun's method with time step durations of $\Delta t = 50$ s, $\Delta t = 83.3$ s, $\Delta t = 167.7$ s and $\Delta t = 250$ s. Also shown in Figure 3-6 are the analytical solution and the solution using Euler's method with $\Delta t = 50$ s. Notice that for the same time step duration ($\Delta t = 50$ s), the numerical result obtained with Heun's method is much closer to the analytical solution than the numerical result obtained with Euler's method. The maximum error between the numerical and analytical solutions is shown in Figure 3-4 for Heun's technique and the improvement relative to Euler's technique is obvious. Also note in Figure 3-6 that Heun's method is unstable for time step durations that are greater than $\tau_{lumped} = 100$ s and temperature actually decreases for $\Delta t > 2\tau_{lumped} = 200$ s. The instability is not oscillatory; however, the numerical result begins to diverge from the analytical results and provides a non-physical solution. Both Euler's and Heun's methods are explicit techniques and both will therefore become unstable when the time step duration exceeds the critical time step.

Heun's method is explicit and therefore its implementation in MATLAB is straightforward. The script begins by specifying the inputs and setting up the time steps and initial conditions:

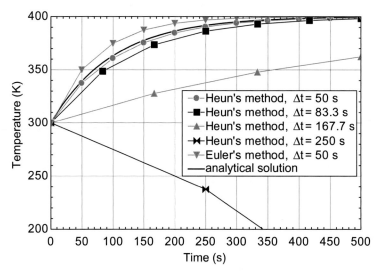

Figure 3-6: Numerical solution obtained using Heun's method for various values of Δt. Also shown is the analytical solution and the solution using Euler's method with $\Delta t = 50$ s.

```
clear all;

%INPUTS
tau=100;                                          % time constant (s)
T_ini=300;                                        % initial temperature (K)
T_infinity=400;                                   % ambient temperature (K)

t_sim=500;                                        % simulation time (s)
M=51;                                             % number of time steps (-)
DELTAt=t_sim/(M-1);                               % time step duration (s)
for j=1:M
    time(j)=(j-1)*t_sim/(M-1);
end
```

The integration follows the form provided by Eqs. (3-27) and (3-28). One advantage of using MATLAB over EES is that MATLAB utilizes assignment statements. Therefore, it is not necessary to store the intermediate variable $\hat{T}_{j+1}$ (i.e., the result of the predictor step) for each time step in an array. Instead, the value of the variable $\hat{T}_{j+1}$ (the variable T_hat) is over-written during each iteration of the for loop; this saves memory and time. EES uses equality statements rather than assignment statements and so it will not allow the value of a variable to be overwritten in the main body of the program. (It is possible to overwrite values in an EES function or procedure.)

```
T(1)=T_ini;                                       % initial temperature
for j=1:(M-1)
    T_hat=T(j)-(T(j)-T_infinity)*DELTAt/tau;      % predictor step
    T(j+1)=T(j)-((T(j)-T_infinity)/tau+(T_hat-T_infinity)/tau)*DELTAt/2;
                                                  % corrector step
end
```

Runge-Kutta Fourth Order Method

Heun's method is a two-step predictor/corrector technique that improves the order of accuracy by one (i.e., Heun's method is second order whereas Euler's method is first order). It is possible to carry out additional predictor/corrector steps and further improve the accuracy of the numerical solution. One of the most popular, higher order techniques is the fourth order Runge-Kutta method which involves four predictor/corrector steps.

The Runge-Kutta fourth order method (referred to subsequently as the RK4 method) estimates the time rate of change of the state variable four times; recall that Euler's method did this only once (at the beginning of time step) and Heun's method did this twice (at the beginning and end of the time step). The RK4 method begins by estimating the time rate of change at the beginning of the time step (referred to, for convenience, as aa):

$$aa = \frac{dT}{dt}\bigg|_{T=T_j,t=t_j} \tag{3-33}$$

The first estimate, aa, is used to predict the temperature half-way through the time step ($\hat{T}_{j+\frac{1}{2}}$):

$$\hat{T}_{j+\frac{1}{2}} = T_j + aa\frac{\Delta t}{2} \tag{3-34}$$

which is used to obtain the second estimate of the time rate of change, bb, at the midpoint of the time step:

$$bb = \frac{dT}{dt}\bigg|_{T=\hat{T}_{j+\frac{1}{2}},t=t_j+\frac{\Delta t}{2}} \tag{3-35}$$

The second estimate is used to re-compute the temperature half-way through the time step ($\hat{\hat{T}}_{j+\frac{1}{2}}$):

$$\hat{\hat{T}}_{j+\frac{1}{2}} = T_j + bb\frac{\Delta t}{2} \tag{3-36}$$

which is used to obtain a third estimate of the time rate of change, cc, also at the midpoint of the time step:

$$cc = \frac{dT}{dt}\bigg|_{T=\hat{\hat{T}}_{j+\frac{1}{2}},t=t_j+\frac{\Delta t}{2}} \tag{3-37}$$

The third estimate is used to predict the temperature at the end of the time step ($\hat{T}_{j+1}$):

$$\hat{T}_{j+1} = T_j + cc\,\Delta t \tag{3-38}$$

which is used to obtain the fourth and final estimate of the time rate of change, dd, at the end of the time step:

$$dd = \frac{dT}{dt}\bigg|_{T=\hat{T}_{j+1},t=t_{j+1}} \tag{3-39}$$

The integration is finally carried out using the weighted average of these four separate estimates of the time rate of change:

$$T_{j+1} = T_j + (aa + 2\,bb + 2\,cc + dd)\frac{\Delta t}{6} \tag{3-40}$$

The RK4 technique is implemented in EES in order to solve the problem discussed previously:

```
$UnitSystem SI MASS RAD PA K J
$TABSTOPS 0.2 0.4 0.6 0.8 3.5 in

"Inputs"
tau=100 [s]                              "time constant"
T_ini=300 [K]                            "initial temperature"
T_infinity=400 [K]                       "ambient temperature"

t_sim=500 [s]                            "simulation time"
M=4 [-]                                  "number of time steps"
DELTAt=t_sim/(M-1)                       "duration of time step"
duplicate j=1,M
    time[j]=(j-1)*t_sim/(M-1)            "time associated with each temperature"
end
```

The code below implements the RK4 technique:

```
"Runge-Kutta 4th Order Method"
T[1]=T_ini
duplicate j=1,(M-1)

    aa[j]=-(T[j]-T_infinity)/tau                "1st estimate of time-rate of change"
    T_hat1[j]=T[j]+aa[j]*DELTAt/2               "1st predictor step"
    bb[j]=-(T_hat1[j]-T_infinity)/tau           "2nd estimate of time-rate of change"
    T_hat2[j]=T[j]+bb[j]*DELTAt/2               "2nd predictor step"
    cc[j]=-(T_hat2[j]-T_infinity)/tau           "3rd estimate of time-rate of change"
    T_hat3[j]=T[j]+cc[j]*DELTAt                 "3rd predictor step"
    dd[j]=-(T_hat3[j]-T_infinity)/tau           "4th estimate of time-rate of change"
    T[j+1]=T[j]+(aa[j]+2*bb[j]+2*cc[j]+dd[j])*DELTAt/6
                                        "final integration uses all 4 estimates"
    T_an[j+1]=T_infinity+(T_ini-T_infinity)*exp(-time[j+1]/tau)
                                        "analytical solution"
    err[j+1]=abs(T_an[j+1]-T[j+1])             "numerical error"
end
err_max=max(err[2..M])                          "maximum numerical error"
```

Figure 3-4 illustrates the maximum numerical error for the RK4 solution as a function of the time step duration and clearly indicates the improvement that can be obtained by using a higher order method. If a certain level of accuracy is required, then much larger time steps (and therefore many fewer computations) can be used if a higher order technique is used. For example, if an accuracy of 0.1 K is required then $\Delta t = 0.4$ s must be used with Euler's method whereas $\Delta t = 70$ s can be used with the RK4 method. This difference represents more than two orders of magnitude of reduction in the number of time steps required while the number of computations per time step has only increased by a factor of four.

The slope of the curves shown in Figure 3-4 is related to the order of the technique with respect to the global error as opposed to the local error. The maximum error associated with Euler's method increases by an order of magnitude as the time step duration increases by an order of magnitude; therefore, while the local error associated with each time step has order two, the global error has order one. The maximum error associated with Heun's method changes by two orders of magnitude for every order of magnitude change in Δt and therefore Heun's method is order two with respect to the global error. The RK4 method is nominally order four with respect to global error.

The MATLAB code to implement the RK4 method follows naturally from the EES code because the method is explicit. Note that the intermediate variables (e.g., *aa*, *bb*, etc.) are not stored and therefore the MATLAB program is less memory intensive.

```
clear all;

%INPUTS
tau=100;                                    % time constant (s)
T_ini=300;                                  % initial temperature (K)
T_infinity=400;                             % ambient temperature (K)

t_sim=500;                                  % simulation time (s)
M=51;                                       % number of time steps (-)
DELTAt=t_sim/(M-1);                         % time step duration (s)
for j=1:M
    time(j)=(j-1)*t_sim/(M-1);
end

T(1)=T_ini;                                 % initial temperature
for j=1:(M-1)
    aa=-(T(j)-T_infinity)/tau;              % 1st estimate of time-rate of change
    T_hat=T(j)+aa*DELTAt/2;                 % 1st predictor step
    bb=-(T_hat-T_infinity)/tau;            % 2nd estimate of time-rate of change
    T_hat=T(j)+bb*DELTAt/2;                 % 2nd predictor step
    cc=-(T_hat-T_infinity)/tau;            % 3rd estimate of time-rate of change
    T_hat=T(j)+cc*DELTAt;                   % 3rd predictor step
    dd=-(T_hat-T_infinity)/tau;            % 4th estimate of time-rate of change
    T(j+1)=T(j)+(aa+2*bb+2*cc+dd)*DELTAt/6;

                                            % final integration uses all 4 estimates
end
```

Fully Implicit Method

The methods discussed thus far are explicit; they all therefore share the characteristic of becoming unstable when the time step exceeds a critical value. An implicit technique avoids this problem. The fully implicit method is similar to Euler's method in that the time rate of change is assumed to be constant throughout the time step. However, the time rate of change is computed at the end of the time step rather than the beginning. Therefore, for any time step j:

$$T_{j+1} = T_j + \frac{dT}{dt}\bigg|_{T=T_{j+1}, t=t_{j+1}} \Delta t \qquad (3\text{-}41)$$

The time rate of change at the end of the time step depends on the temperature at the end of the time step (T_{j+1}). Therefore, T_{j+1} cannot be calculated explicitly using information at the beginning of the time step (T_j) and instead an implicit equation is obtained for T_{j+1}. For the example problem that has been considered in previous sections, the implicit equation is obtained by substituting Eq. (3-15) into Eq. (3-41):

$$T_{j+1} = T_j - \frac{(T_{j+1} - T_\infty)}{\tau_{lumped}}\Delta t \qquad (3\text{-}42)$$

Notice that T_{j+1} appears on both sides of Eq. (3-42). Because EES solves implicit equations, it is not necessary to rearrange Eq. (3-42). The implicit solution is obtained using the following EES code:

```
$UnitSystem SI MASS RAD PA K J
$TABSTOPS 0.2 0.4 0.6 0.8 3.5 in

"Inputs"
tau=100 [s]                                         "time constant"
T_ini=300 [K]                                       "initial temperature"
T_infinity=400 [K]                                  "ambient temperature"

t_sim=500 [s]                                       "simulation time"
M=4 [-]                                             "number of time steps"
DELTAt=t_sim/(M-1)                                  "duration of time step"
duplicate j=1,M
    time[j]=(j-1)*t_sim/(M-1)                       "time associated with each temperature"
end

"Fully Implicit Method"
T[1]=T_ini                                          "initial temperature"
duplicate j=1,(M-1)
    T[j+1]=T[j]-(T[j+1]-T_infinity)*DELTAt/tau
                                                    "implicit equation for temperature"

    T_an[j+1]=T_infinity+(T_ini-T_infinity)*exp(-time[j+1]/tau)
                                                    "analytical solution"

    err[j+1]=abs(T_an[j+1]-T[j+1])                  "numerical error"
end
err_max=max(err[2..M])
```

Figure 3-7 illustrates the fully implicit solution for various values of the time step; also shown is the analytical solution, Eq. (3-16). The accuracy of the fully implicit solution is reduced as the duration of the time step increases. Figure 3-4 illustrates the maximum error associated with the fully implicit technique as a function of the time step duration; notice that the fully implicit technique is no more accurate than Euler's technique. However, Figure 3-7 shows that the fully implicit solution does not become unstable even when the duration of the time step is greater than the critical time step.

The implementation of the fully implicit method cannot be accomplished in the same way in MATLAB that it is in EES because Eq. (3-42) is not explicit for T_{j+1}; while EES can solve this implicit equation, MATLAB cannot. It is necessary to solve Eq. (3-42) for T_{j+1} in order to carry out the integration step in MATLAB:

$$T_{j+1} = \frac{T_j + T_\infty \dfrac{\Delta t}{\tau}}{\left(1 + \dfrac{\Delta t}{\tau}\right)} \qquad (3\text{-}43)$$

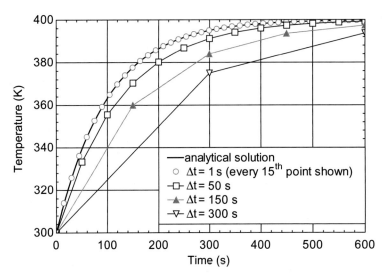

Figure 3-7: Fully implicit solution for various values of the time step duration as well as the analytical solution.

```
clear all;

%INPUTS
tau=100;                                                    % time constant (s)
T_ini=300;                                                  % initial temperature (K)
T_infinity=400;                                             % ambient temperature (K)

t_sim=500;                                                  % simulation time (s)
M=51;                                                       % number of time steps (-)
DELTAt=t_sim/(M-1);                                         % time step duration (s)
for j=1:M
    time(j)=(j-1)*t_sim/(M-1);
end

T(1)=T_ini;                                                 % initial temperature
for j=1:(M-1)
    T(j+1)=(T(j)+T_infinity*DELTAt/tau)/(1+DELTAt/tau);     % implicit step
end
```

Crank-Nicolson Method

The Crank-Nicolson method combines Euler's method with the fully implicit method. The time rate of change for the time step is estimated based on the average of its values at the beginning and end of the time step. Therefore, for any time step j:

$$T_{j+1} = T_j + \left[\left. \frac{dT}{dt} \right|_{T=T_j, t=t_j} + \left. \frac{dT}{dt} \right|_{T=T_{j+1}, t=t_{j+1}} \right] \frac{\Delta t}{2} \qquad (3\text{-}44)$$

Notice that the Crank-Nicolson is an implicit method because the solution for T_{j+1} involves a time rate of change that must be evaluated based on T_{j+1}. Therefore, the technique will have the stability characteristics of the fully implicit method. The solution also involves two estimates for the time rate of change and is therefore second order and more accurate than the fully implicit technique.

Substituting Eq. (3-15) into Eq. (3-44) leads to:

$$T_{j+1} = T_j - \left[\frac{(T_j - T_\infty)}{\tau_{lumped}} + \frac{(T_{j+1} - T_\infty)}{\tau_{lumped}} \right] \frac{\Delta t}{2} \qquad (3\text{-}45)$$

The EES code below implements the Crank-Nicolson solution for the example problem:

```
$UnitSystem SI MASS RAD PA K J
$TABSTOPS 0.2 0.4 0.6 0.8 3.5 in

"Inputs"
tau=100 [s]                                  "time constant"
T_ini=300 [K]                                "initial temperature"
T_infinity=400 [K]                           "ambient temperature"

t_sim=600 [s]                                "simulation time"
M=601 [-]                                    "number of time steps"
DELTAt=t_sim/(M-1)                           "duration of time step"
duplicate j=1,M
   time[j]=(j-1)*t_sim/(M-1)                 "time associated with each temperature"
end

"Crank Nicholson Method"
T[1]=T_ini
duplicate j=1,(M-1)
   T[j+1]=T[j]-((T[j]-T_infinity)/tau+(T[j+1]-T_infinity)/tau)*DELTAt/2
                                             "C-N equation for temperature"
   T_an[j+1]=T_infinity+(T_ini-T_infinity)*exp(-time[j+1]/tau)
                                             "analytical solution"
   err[j+1]=abs(T_an[j+1]-T[j+1])            "numerical error"
end
err_max=max(err[2..M])
```

The maximum numerical error associated with the Crank-Nicolson technique is shown in Figure 3-4. Notice that the fully implicit method has accuracy that is nominally equivalent to Euler's method while the Crank-Nicolson method has accuracy slightly better than Heun's method. The Crank-Nicolson method is a popular choice because it combines high accuracy with stability.

To implement the Crank-Nicolson technique using MATLAB, it is necessary to solve the implicit Eq. (3-45) for T_{j+1}:

$$T_{j+1} = \frac{T_j \left(1 - \dfrac{\Delta t}{2\tau}\right) + T_\infty \dfrac{\Delta t}{\tau}}{\left(1 + \dfrac{\Delta t}{2\tau}\right)} \qquad (3\text{-}46)$$

so the MATLAB script becomes:

```
clear all;

%INPUTS
tau=100;                                % time constant (s)
T_ini=300;                              % initial temperature (K)
T_infinity=400;                         % ambient temperature (K)

t_sim=500;                              % simulation time (s)
M=51;                                   % number of time steps (-)
DELTAt=t_sim/(M-1);                     % time step duration (s)
for j=1:M
    time(j)=(j-1)*t_sim/(M-1);
end

T(1)=T_ini;                             % initial temperature
for j=1:(M-1)
    T(j+1)=(T(j)*(1-DELTAt/(2*tau))+T_infinity*DELTAt/tau)/(1+DELTAt/(2*tau));    % C-N step

end
```

Adaptive Step-Size and EES' Integral Command

The implementation of the techniques that have been discussed to this point is accomplished using a fixed duration time step for the entire simulation. This implementation is often not efficient because there are regions of time during the simulation where the solution is not changing substantially and therefore large time steps could be taken with little loss of accuracy. Adaptive step-size solutions adjust the size of the time step used based on the local characteristics of the state equation. Typically, the absolute value of the local time rate of change or the second derivative of the time rate of change is used to set a step-size that guarantees a certain level of accuracy. For the current example, shown in Figure 3-6, smaller time steps would be used near $t = 0$ s because the rate of temperature change is largest at this time. Later in the simulation, for $t > 300$ s, the temperature is not changing substantially and therefore large time steps could be used to obtain a solution more rapidly.

The implementation of adaptive step-size solutions is beyond the scope of this book. However, a third order integration routine that optionally uses an adaptive step-size is provided with EES and can be accessed using the Integral command. EES' Integral command requires four arguments and allows an optional fifth argument:

F = Integral(Integrand,VarName,LowerLimit,UpperLimit,StepSize)

where Integrand is the EES variable or expression that must be integrated, VarName is the integration variable, LowerLimit and UpperLimit define the limits of integration, and StepSize provides the duration of the time step.

When using the Integral technique (or indeed any numerical integration technique) it is useful to first verify that, given temperature and time, the EES code is capable of computing the time rate of change of the temperature. Therefore, the first step is to implement Eq. (3-15) in EES for an arbitrary (but reasonable) value of T and t:

```
$UnitSystem SI MASS RAD PA K J
$TABSTOPS 0.2 0.4 0.6 0.8 3.5 in

"Inputs"
tau=100 [s]                          "time constant"
T_ini=300 [K]                        "initial temperature"
T_infinity=400 [K]                   "ambient temperature"
t_sim=600 [s]                        "simulation time"

"EES' Integral Method"
T=350 [K]                            "Temperature, arbitrary - used to test dTdt"
time = 0 [s]                         "Time, arbitrary - used to test dTdt"
dTdt=-(T-T_infinity)/tau             "Time rate of change"
```

which leads to $dTdt = 0.5$ K/s. The next step is to comment out the arbitrary values of temperature and time that are used to test the computation of the state equation and instead let EES' Integral function control these variables for the numerical integration. The temperature of the object is given by:

$$T = T_{ini} + \int_{0}^{t_{sim}} \frac{dT}{dt} dt \qquad (3\text{-}47)$$

Therefore, the solution to our example problem is obtained by calling the Integral function; Integrand is replaced with the variable dTdt, VarName with time, LowerLimit with 0, UpperLimit with the variable t_sim, and StepSize with DELTAt, the specified duration of the time step:

```
"EES' Integral Method"
{T=350 [K]                           "Temperature, arbitrary - used to test dTdt"
time=0 [s]                           "Time, arbitrary - used to test dTdt"}
dTdt=-(T-T_infinity)/tau             "Time rate of change"
DELTAt=1 [s]                         "Duration of the time step"
T=T_ini+INTEGRAL(dTdt,time,0,t_sim,DELTAt)    "Call EES' Integral function"
```

In order to accomplish the numerical integration, EES will adjust the value of the variable time from 0 to t_sim in increments of DELTAt. At each value of time, EES will iteratively solve all of the equations in the Equations window that depend on time. For the example above, this process will result in the variable T being evaluated at each value time.

When the solution converges, the value of the variable time is incremented and the process is repeated until time is equal to t_sim. The results shown in the Solution window will provide the temperature at the end of the process (i.e., the result of the integration, which is the temperature at time = t_sim). Often it is most interesting to know the temperature variation with time during the process. This information can be provided by including the $IntegralTable directive in the file. The format of the $IntegralTable directive is:

$IntegralTable VarName: Step, x,y,z

	time [s]	T [K]
Row 1	0	300
Row 2	1	301
Row 3	2	302
Row 4	3	303
Row 5	4	303.9

Figure 3-8: Integral Table generated by EES.

where VarName is the integration variable; the first column in the Integral Table will hold values of this variable. The colon followed by the parameter Step (which can either be a number or a variable name) and list of variables (x, y, z) are optional. If these values are provided, then the value of Step will be used as the output step size and the integration variables will be reported in the Integral Table at the specified output step size. The output step size may be a variable name, rather than a number, provided that the variable has been set to a constant value preceding the $IntegralTable directive. The step size that is used to report integration results is totally independent of the duration of the time step that is used in the numerical integration. If the numerical integration step size and output step size are not the same, then linear interpolation is used to determine the integrated quantities at the specified output steps. If an output step size is not specified, then EES will output all specified variables at every time step.

The variables x, y, z ... must correspond to variables in the EES program. Algebraic equations involving variables are not accepted. A separate column will be created in the Integral Table for each specified variable. The variables must be separated by a space or list delimiter (comma or semicolon).

Solving the EES code will result in the generation of an Integral Table that is filled with intermediate values resulting from the numerical integration. The values in the Integral Table can be plotted, printed, saved, and copied in exactly the same manner as for other tables. The Integral Table is saved when the EES file is saved and the table is restored when the EES file is loaded. If an Integral Table exists when calculations are initiated, it will be deleted if a new Integral Table is created.

Use the EES code below to generate an Integral Table containing the results of the numerical simulation and the analytical solution.

```
$IntegralTable time, T
```

After running the code with a time step duration of 1 s, the Integral Table shown in Figure 3-8 will be generated. The results of the integration can be plotted by selecting the Integral Table as the source of the data to plot.

The StepSize input to the Integral command is optional. If the parameter StepSize is not included or if it is set to a value of 0, then EES will use an adaptive step-size algorithm that maintains accuracy while maximizing computational speed. The parameters used to control the adaptive step-size algorithm can be accessed and adjusted by selecting Preferences from the Options menu and selecting the Integration Tab. A complete description of these parameters can be obtained from the EES Help menu.

To run the program using an adaptive step-size, remove the fifth argument from the Integral command (or set its value to 0):

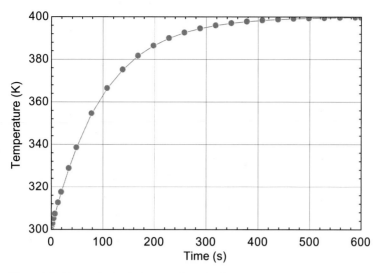

Figure 3-9: Numerical solution predicted using EES' internal Integration function with an adaptive time step.

```
"EES' Integral Method"
dTdt=-(T-T_infinity)/tau                    "Time rate of change"
T=T_ini+INTEGRAL(dTdt,time,0,t_sim)         "Call EES' Integral function"

$IntegralTable time, T
```

The results are shown in Figure 3-9. Note the concentration of time steps near $t = 0$ s. Also note that the Adaptive Step-Size parameters had to be adjusted in order for the algorithm to operate for such a simple problem; specifically, the minimum number of steps had to be reduced to 10 and the criteria for increasing the step-size had to be increased to 0.01.

MATLAB's Ordinary Differential Equation Solvers
MATLAB has a suite of solvers for initial value problems that implement advanced numerical integration algorithms (e.g., the functions ode45, ode23, etc.). These ordinary differential equation solvers all have a similar protocol; for example:

$$[\;\underbrace{\text{time,}}_{\substack{\text{vector} \\ \text{of time}}}\quad \underbrace{\text{T}}_{\substack{\text{temperature} \\ \text{at the times}}}\;] = \text{ode45(}\quad \underbrace{\text{'dTdt'}}_{\substack{\text{function that} \\ \text{returns the derivative} \\ \text{of temperature}}}\;,\quad \underbrace{\text{tspan}}_{\substack{\text{time span} \\ \text{to be integrated}}}\;,\quad \underbrace{\text{T0}}_{\substack{\text{initial} \\ \text{temperature}}}\;)$$

The ode solvers require three inputs, the name of the function that returns the derivative of temperature with respect to time given the current value of temperature and time (i.e., the function that implements the state equation for the system), the simulation time, and the initial temperature. The ode solver returns two vectors containing the solution times and the temperatures at the solution times. The ode solvers require that you have created a function that implements the state equation for the system. MATLAB assumes that this function will accept two inputs, corresponding to the current values of time and temperature, and return a single output, the rate at which temperature is changing with

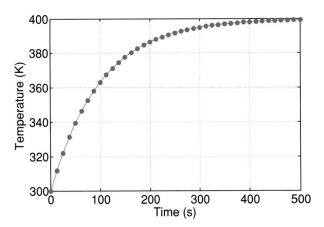

Figure 3-10: Predicted temperature as a function of time using MATLAB's ode45 solver.

time. The function dTdt_function, shown below, implements the state equation for the example considered previously:

```
function[dTdt]=dTdt_function(time,T)
  % this function computes the rate of change of temperature
  % Inputs
  % T - temperature (K)
  % time - time (s)
  %
  % Outputs
  % dTdt - time rate of change (K/s)
  tau=100;                              % time constant (s)
  T_infinity=500;                       % ambient temperature (K)
  dTdt=-(T-T_infinity)/tau;
end
```

The script that solves the problem calls the integration routine ode45 and specifies that the state equations are determined in the function dTdt_function, the simulation time is from 0 to t_sim and the initial condition is T_ini.

```
clear all;

%INPUTS
T_ini=300;                                 % initial temperature (K)
t_sim=500;                                 % simulation time (s)
[time,T]=ode45('dTdt_function',t_sim,T_ini);   % use ode45 to solve problem
```

The vectors time and T are the times and temperatures used to carry out the simulation. Figure 3-10 shows the temperature predicted using the ode45 solver as a function of time.

Figure 3-10 shows that the MATLAB solver is using an adaptive step-size algorithm because the duration of the time step varies throughout the computational domain. If the variable t_sim is specified as a vector of times, then MATLAB will return the solution at these specific times rather than the times that are actually used for the integration (although the integration process itself does not change). For example, in order to obtain a solution every 10 s the script should be changed as shown:

```
t_span=linspace(0,500,51);                        % specify times to return the solution at
[time,T]=ode45('dTdt_function',t_span,T_ini);
```

It is not convenient that the state equation function, dTdt_function, only allows two inputs. It is difficult to run parametric studies or optimization if the values of tau and T_infinity must be changed manually within the function. This requirement can be avoided by modifying the function dTdt_function so that it accepts all of the inputs of interest:

```
function[dTdt]=dTdt_function(time, T, tau, T_infinity)
  % this function computes the rate of change of temperature
  % Inputs
  % T - temperature (K)
  % time - time (s)
  % tau - time constant (s)
  % T_infinity - ambient temperature (K)
  %
  % Outputs
  % dTdt - time rate of change (K/s)

  dTdt=-(T-T_infinity)/tau;
end
```

The call to the function dTdt_function with the two inputs required by the function ode45 (time and T) is "mapped" to the full function call. This mapping has the form:

@(time, T) dTdt_function(time, T, tau, T_amb)

2 inputs that   the full function call, including the 2 required inputs
are required
by ode solver

The revised script becomes:

```
clear all;

%INPUTS
T_ini=300;                              % initial temperature (K)
t_sim=500;                              % simulation time (s)
tau=100;                                % time constant (s)
T_infinity=400;                         % ambient temperature (K)

[time,T]=ode45(@(time,T) dTdt_function(time,T, tau, T_infinity),[0,t_sim],T_ini);
```

A fourth argument (which is optional) can be added to the function ode45:

[time,T]=ode45('dTdt', tspan, T0, OPTIONS)

where OPTIONS is a vector that controls the integration properties. In the absence of a fourth argument, MATLAB will use the default settings for the integration properties. The most convenient way to adjust the integration settings is with the odeset function,

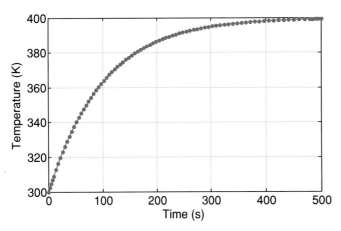

Figure 3-11: Predicted temperature as a function of time using MATLAB's ode45 solver with increased accuracy.

which has the format:

OPTIONS = odeset('property1', value1, 'property2', value2, . . .)

where property1 refers to one of the integration properties and value1 refers to its specified value, property2 refers to a different integration property and value2 refers to its specified value, etc. The properties that are not explicitly set in the odeset command remain at their default values. Type help odeset at the command line prompt in order to see a list and description of the integration properties with their default values.

The property RelTol refers to the relative error tolerance and has a default value of 0.001 (0.1% accuracy). The relative error tolerance of the integration process can be reduced to 1×10^{-6} by changing the script according to:

```
OPTIONS=odeset('RelTol',1e-6); % change the relative error tolerance used in the integration
[time,T]=ode45(@(time,T) dTdt_function(time,T, tau, T_infinity),[0,t_sim],T_ini,OPTIONS);
```

The results of the integration with the improved accuracy are shown in Figure 3-11. Notice that substantially more integration steps were required (compared with Figure 3-10) in order to obtain the improved accuracy.

The integration function ode45 was used in the discussion in this section. However, MATLAB provides other differential equation solvers that are specialized based on the stiffness and other aspects of the problem. To examine these solvers, type help funfun at the command line. The differential equation solvers have similar calling protocol, so it is easy to switch between them. For example, to use the ode15s rather than the ode45 function it is only necessary to change the name of the function call:

```
OPTIONS=odeset('RelTol',1e-6);
[time,T]=ode15s(@(time,T) dTdt_function(time,T, tau, T_infinity),[0,t_sim],T_ini,OPTIONS);
```

EXAMPLE 3.2-1(a): OVEN BRAZING (EES)

EXAMPLE 3.2-1(a): OVEN BRAZING (EES)

A brazing operation is carried out in an evacuated oven. The metal pieces to be brazed have a complex geometry; they are made of bronze (with $k = 50$ W/m-K, $c = 500$ J/kg-K, and $\rho = 8700$ kg/m^3) and have total volume $V = 10$ cm^3 and total surface area $A_s = 35$ cm^2. The pieces are heated by radiation heat transfer from the walls of the oven. A detailed presentation of radiation heat transfer is presented in Chapter 10. For this problem, assume that the emissivity of the surface of the piece is $\varepsilon = 0.8$ and that the wall of the oven is black. In this limit, the rate of radiation heat transfer from the wall to the piece ($\dot{q}_{rad}$) may be written as:

$$\dot{q}_{rad} = A_s \, \varepsilon \, \sigma \, \left(T_w^4 - T^4 \right)$$

where T_w is the temperature of the wall, T is the temperature of the surface of the piece, and σ is the Stefan-Boltzmann constant. The temperature of the oven wall is increased at a constant rate $\beta = 1$ K/s from its initial temperature $T_{ini} = 20°$C to its final temperature $T_f = 470°$C which is held for $t_{hold} = 1000$ s before the temperature of the wall is reduced at the same constant rate back to its initial temperature. The pieces and the oven are initially in thermal equilibrium at $T_{t=0} = T_{ini}$.

a) Is the lumped capacitance assumption appropriate for this problem?

The inputs are entered in EES.

```
"EXAMPLE 3.2-1(a): Oven Brazing (EES)"

$UnitSystem SI MASS RAD PA K J
$TABSTOPS 0.2 0.4 0.6 0.8 3.5 in

"Inputs"
V=10 [cm^3]*convert(cm^3,m^3)            "volume"
A_s=35 [cm^2]*convert(cm^2,m^2)          "surface area"
e=0.8 [-]                                 "emissivity of surface"
T_ini=converttemp(C,K,20 [C])            "initial temperature"
T_f=converttemp(C,K,470 [C])             "final oven temperature"
t_hold=1000 [s]                          "oven hold time"
beta=1 [K/s]                             "oven ramp rate"
c=500 [J/kg-K]                           "specific heat capacity"
k=50 [W/m-K]                             "conductivity"
rho=8700 [kg/m^3]                        "density"
```

The lumped capacitance assumption ignores the internal resistance to conduction as being small relative to the resistance to heat transfer from the surface of the object. The radiation heat transfer coefficient, defined in Section 1.2.6, is:

$$h_{rad} = \sigma \, \varepsilon \, \left(T_s^2 + T_w^2 \right) \left(T_s + T_w \right)$$

where T_s is the surface temperature of the object. The Biot number defined based on this radiation heat transfer coefficient is:

$$Bi = \frac{h_{rad} \, L_{cond}}{k}$$

where

$$L_{cond} = \frac{V}{A_s}$$

EXAMPLE 3.2-1(a): OVEN BRAZING (EES)

EXAMPLE 3.2-1(a): OVEN BRAZING (EES)

The Biot number is largest (and therefore the lumped capacitance model is least justified) when the value of h_{rad} is largest, which occurs if both T_s and T_w achieve their maximum possible value (T_f).

```
h_rad_max=sigma#*e*(T_f^2+T_f^2)*(T_f+T_f)      "maximum radiation coefficient"
L_cond=V/A_s                                     "characteristic length for conduction"
Bi=h_rad_max*L_cond/k                            "maximum Biot number"
```

The Biot number is calculated to be 0.004 at this upper limit, which indicates that the lumped capacitance assumption is valid.

b) Calculate a lumped capacitance time constant that characterizes the brazing process.

It is useful to calculate a time constant even when the problem is solved numerically. The value of the time constant provides guidance relative to the time step that will be required and it also allows a sanity check on your results. The time constant, discussed in Section 3.1.4, is the product of the thermal capacitance of the object and the net resistance from the surface. Using the concept of the radiation heat transfer coefficient allows the time constant to be written as:

$$\tau_{lumped} = \frac{V \rho c}{h_{rad} A_s}$$

The minimum value of the time-constant (again, corresponding to the object and the oven wall being at their maximum temperature) is computed in EES:

```
tau=rho*c*V/(h_rad_max*A_s)                      "time constant"
```

and found to be 167 s.

c) Develop a numerical solution based on Heun's method that predicts the temperature of the object for 3000 s after the oven is activated.

A function T_w is defined which returns the wall temperature as a function of time; the function is placed at the top of the EES file.

```
"Oven temperature function"
function T_w(time,T_ini,T_f,beta,t_hold)
   "INPUTS:
   time - time relative to initiation of process (s)
   T_ini - initial temperature (K)
   T_f - final temperature (K)
```

EXAMPLE 3.2-1(a): OVEN BRAZING (EES)

```
beta - ramp rate (K/s)
t_hold - hold time (s)
OUTPUTS
T_w - wall temperature (K)"

    T_w=T_ini+beta*time                          "temperature of wall during ramp up period"
    t_ramp=(T_f-T_ini)/beta                       "duration of ramp period"
    if (time>t_ramp) then
        T_w=T_f                                  "temperature of wall during hold period"
    endif
    if (time>(t_ramp+t_hold)) then
        T_w=T_f-beta*(time-t_ramp-t_hold)         "temperature during ramp down period"
    endif
    if (time>(2*t_ramp+t_hold)) then
        T_w=T_ini                                "temperature after ramp down period"
    endif
end
```

Note the use of the if-then-endif statements in the function to activate different equations for the wall temperature based on the time of the simulation. The function T_w be called from the Equation Window:

```
T_w=T_w(time,T_i,T_f,beta,tau_hold)
```

It is wise to check that the function is working correctly by setting up a parametric table that includes time and the wall temperature; the result is shown in Figure 1.

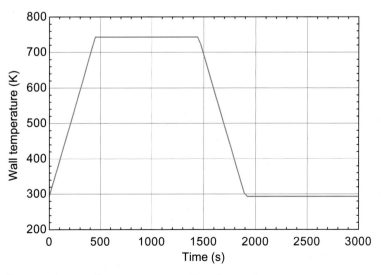

Figure 1: Oven wall temperature as a function of time.

EXAMPLE 3.2-1(a): OVEN BRAZING (EES)

The simulation time and number of time steps are defined and used to compute the time and wall temperature at each time step:

```
t_sim=3000 [s]                                    "simulation time"
M=101 [-]                                         "number of time-steps"
DELTAt=t_sim/(M-1)                               "time-step duration"
duplicate j=1,M
    time[j]=(j-1)*t_sim/(M-1)                    "time"
    T_w[j]=T_w(time[j],T_ini,T_f,beta,t_hold)   "wall temperature"
end
```

The governing differential equation is obtained from an energy balance on the piece:

$$\dot{q}_{rad} = \frac{dU}{dt}$$

or

$$A_s \varepsilon \sigma \left(T_w^4 - T^4\right) = V \rho c \frac{dT}{dt} \tag{1}$$

Rearranging Eq. (1) leads to the state equation that provides the time rate of change of the temperature:

$$\frac{dT}{dt} = \frac{A_s \varepsilon \sigma \left(T_w^4 - T^4\right)}{V \rho c} \tag{2}$$

The temperature at the beginning of the first time step is the initial condition:

$$T_1 = T_{ini}$$

```
T[1]=T_ini                                        "initial temperature"
```

Heun's method consists of an initial, predictor step:

$$\hat{T}_{j+1} = T_j + \left.\frac{dT}{dt}\right|_{T=T_j,t=t_j} \Delta t \tag{3}$$

Substituting the state equation, Eq. (2), into Eq. (3) leads to:

$$\left.\frac{dT}{dt}\right|_{T=T_j,t=t_j} = \frac{A_s \varepsilon \sigma \left(T_{w,t=t_j}^4 - T_j^4\right)}{V \rho c}$$

The corrector step is:

$$T_{j+1} = T_j + \left[\left.\frac{dT}{dt}\right|_{T=T_j,t=t_j} + \left.\frac{dT}{dt}\right|_{T=\hat{T}_{j+1},t=t_{j+1}}\right] \frac{\Delta t}{2}$$

where

$$\left.\frac{dT}{dt}\right|_{T=\hat{T}_{j+1},t=t_{j+1}} = \frac{A_s \varepsilon \sigma \left(T_{w,t=t_{j+1}}^4 - \hat{T}_{j+1}^4\right)}{V \rho c}$$

EXAMPLE 3.2-1(a): OVEN BRAZING (EES)

Heun's method is implemented in EES:

```
T[1]=T_ini                                          "initial temperature"
duplicate j=1,(M-1)
  dTdt[j]=e*A_s*sigma # *(T_w[j]^4-T[j]^4)/(rho*V*c)
    "Temperature rate of change at the beginning of the time-step"
  T_hat[j+1]=T[j]+dTdt[j]*DELTAt                     "Predictor step"
  dTdt_hat[j+1]=e*A_s*sigma #*(T_w[j+1]^4-T_hat[j+1]^4)/(rho*V*c)
    "Temperature rate of change at the end of the time-step"
  T[j+1]=T[j]+(dTdt[j]+dTdt_hat[j+1])*DELTAt/2       "Corrector step"
end
```

The solution for 100 time steps is shown in Figure 2.

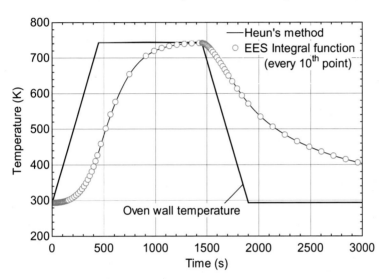

Figure 2: Oven wall and piece temperature as a function of time, predicted by Heun's method with 100 time steps and using EES' built-in Integral function.

Note that the piece temperature lags the oven wall temperature by 100's of seconds, which is consistent with the time constant calculated in (b).

d) Develop a numerical solution using EES' Integral function that predicts the temperature of the object for 3000 s after the oven is activated.

The adaptive time step algorithm is used within EES' Integral function and an integral table is created that holds the results of the integration:

```
"EES' Integral function"
dTdt=e*A_s*sigma #*(T_w(time,T_ini,T_f,beta,t_hold)^4-T^4)/(rho*V*c)
T=T_ini+INTEGRAL(dTdt,time,0,t_sim)                 "Call EES' Integral function"

$IntegralTable time, T
```

The results are included in Figure 2 and agree with Heun's method; note that only 1 out of every 10 points are included in the EES solution in order to show the results of the adaptive step-size.

EXAMPLE 3.2-1(b): OVEN BRAZING (MATLAB)

EXAMPLE 3.2-1(b): OVEN BRAZING (MATLAB)

Repeat EXAMPLE 3.2-1(a) using MATLAB rather than EES to do the calculations.

a) **Develop a numerical solution that is based on Heun's method and implemented in MATLAB that predicts the temperature of the object for 3000 s after the oven is activated.**

The solution will be obtained in a function called Oven_brazing that can be called from the command space. The function will accept the number of time steps (M) and the total simulation time (t_{sim}) and return three arrays that are the values of time, the wall temperature, and the surface temperature at each time step.

```
function [time, T_w, T]=Oven_Brazing(M, t_sim)
% EXAMPLE 3.2-1(b) Oven Brazing

% INPUTS
% M - number of time steps (-)
% t_sim - duration of simulation (s)

% OUTPUTS
% time - array with the values of time for each time step (s)
% T_w - array containing values of the wall temperature at each time (K)
% T - array containing values of the surface temperature at each time (K)
```

Next, the known information from the problem statement is entered into the function.

```
%Known information
V=10/100000;                  % volume (m^3)
A_s=35/1000;                  % surface area (m^2)
e=0.8;                        % emissivity of surface (-)
T_ini=293.15;                 % initial temperature (K)
T_f=743.15;                   % final oven temperature (K)
t_hold=1000;                  % oven hold temperature (s)
beta=1;                       % oven ramp rate (K/s)
c=500;                        % specific heat capacity (J/kg-K)
k=50;                         % conductivity (W/m-K)
rho=8700;                     % density (kg/m^3)
sigma=5.67e-8;                % Stefan-Boltzmann constant (W/m^2-K^4)
```

A function is needed to return the wall temperature as a function of time. The following code implements this function. Place this function at the bottom of the file, after an **end** statement that terminates the function Oven_Brazing.

```
function[T_w]=T_wf(time, T_ini, T_f, beta, t_hold)
% Oven temperature function
% Inputs
% time - current time value (s)
% T_ini - initial value of the wall temperature (K)
```

EXAMPLE 3.2-1(b): OVEN BRAZING (MATLAB)

```
% T_f - final value of the wall temperature (K)
% beta - rate of increase in temperature of the wall (K/s)
% t_hold - time period in which the wall temperature is held constant (s)

% Output
% T_w - wall temperature at specfied time (K)

    T_w=T_ini+beta*time;
    t_ramp=(T_f-T_ini)/beta;
    if (time>t_ramp)
        T_w=T_f;
    end
    if (time>(t_ramp+t_hold))
        T_w=T_f-beta*(time-t_ramp-t_hold);
    end
    if (time>(2*t_ramp+t_hold))
        T_w=T_ini;
    end
end
```

Note the use of the **if-end** clauses to activate different equations for the wall temperature based on the time of the simulation. The temperature of the wall at any time can be evaluated by a call to the function **T_w**. The following lines fill the **time** and **T_w** vectors with the values of time and the wall temperature at each time step.

```
DELTAt=t_sim/(M-1);                        % time step duration
for j=1:M
    time(j)=(j-1)*t_sim/(M-1);             % value of time for each step
    T_w(j)=T_wf(time(j),T_ini,T_f,beta,t_hold);
                                           % value of the wall temperature at each step
end
```

The last task is to enter the equations that implement Heun's method for solving the differential equation. The governing equation is obtained from an energy balance on the piece:

$$\dot{q}_{rad} = A_s\,\varepsilon\,\sigma\,\left(T_w^4 - T^4\right) = V\,\rho\,c\,\frac{dT}{dt}$$

or

$$\frac{dT}{dt} = \frac{A_s\,\varepsilon\,\sigma\,\left(T_w^4 - T^4\right)}{V\,\rho\,c} \tag{1}$$

Heun's method consists of an initial, predictor step:

$$\hat{T}_{j+1} = T_j + \left.\frac{dT}{dt}\right|_{T=T_j,t=t_j}$$

where:

$$\left.\frac{dT}{dt}\right|_{T=T_j,t=t_j} = \frac{A_s\,\varepsilon\,\sigma\,\left(T_{w,t=t_j}^4 - T_j^4\right)}{V\,\rho\,c}$$

EXAMPLE 3.2-1(b): OVEN BRAZING (MATLAB)

The corrector step is:

$$T_{j+1} = T_j + \left[\left. \frac{dT}{dt} \right|_{T=T_j, t=t_j} + \left. \frac{dT}{dt} \right|_{T=\hat{T}_{j+1}, t=t_{j+1}} \right] \frac{\Delta t}{2}$$

where, from Eq. (1),

$$\left. \frac{dT}{dt} \right|_{T=T^*_{j+1}, t=t_{j+1}} = \frac{A_s \, \varepsilon \, \sigma \left(T^4_{w, t=t_{j+1}} - \hat{T}^4_{j+1} \right)}{V \, \rho \, c}$$

The following equations implement Heun's method in MATLAB:

```
T(1)=T_ini;
for j=1:(M-1)
   dTdt=e*A_s*sigma*(T_w(j)^4-T(j)^4)/(rho*V*c);
                                          %Temp deriv. at the start of the time step
   T_hat=T(j)+dTdt*DELTAt;                %Predictor step
   dTdt_hat=e*A_s*sigma*(T_w(j+1)^4-T_hat^4)/(rho*V*c);   %deriv. at end of time step
   T(j+1)=T(j)+(dTdt+dTdt_hat)*DELTAt/2;  %Corrector step"
end
```

The Oven_Brazing function is terminated with an end statement and saved.

```
end
```

The function can now be called from the command window

```
>> [time, T_w, T]=Oven_Brazing(101, 3000);
```

The temperature of the wall and the piece is shown in Figure 1.

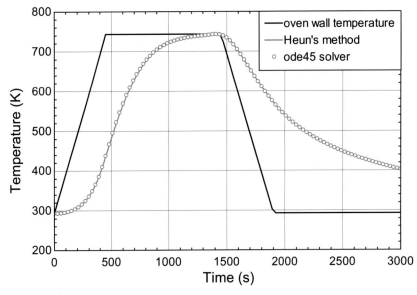

Figure 1: Temperature of the wall and work piece, predicted using Heun's method and the ode45 solver, as a function of time.

EXAMPLE 3.2-1(b): OVEN BRAZING (MATLAB)

b) Develop a numerical solution using MATLAB's ode45 function that predicts the temperature of the object for 3000 s after the oven is activated.

The ode solvers are designed to call a function the returns the derivative of dependent variable with respect to the independent variable. In our case, the dependent variable is the temperature of the work piece and time is the independent variable. A function must be provided that returns the time derivative of the temperature at a specified time and temperature. The MATLAB function dTdt_f accepts all of the input parameters that are needed to determine the derivative; according to Eq. (1), these include A_s, ε, σ, V, ρ, and c as well as parameters needed determine the wall temperature at any time. Place the function dTdt_f below the function T_wf. Note that the function dTdt_f calls the function T_wf in order to determine the wall temperature at a specified time.

```
function[dTdt]=dTdt_f(time,T,e,A_s,rho,V,c,T_f,T_ini,beta,t_hold)
% dTdt is called by the ode45 solver to evaluate the derivative dTdt
% INPUTS
% time - time relative to start of process (s)
% T - temperature of piece (K)
% e - emissivity of piece (-)
% A_s - surface area of piece (m^2)
% rho - density (kg/m^3)
% V - volume of piece (m^3)
% c - specific heat capacity of piece (J/kg-K)
% T_f - final oven temperature (K)
% T_ini - initial oven temperature (K)
% beta - oven ramp rate (K/s)
% t_hold - oven hold time (s)

% OUTPUTS
% dTdt - rate of change of temperature of the piece (K/s)


    sigma=5.67e-8;                              % Stefan-Boltzmann constant (W/m^2-K^4)
    T_w=T_wf(time, T_ini, T_f, beta, t_hold);   % wall temperature
    dTdt=e*A_s*sigma*(T_w^4-T^4)/(rho*V*c);     % energy balance
end
```

Now, all that is necessary is to comment out the code in the function Oven_brazing from part (a) and instead call the MATLAB ode45 solver function to determine the temperatures as a function of time.

```
% T(1)=T_ini;
% for j=1:(M-1)
%    dTdt=e*A_s*sigma*(T_w(j)^4-T(j)^4)/(rho*V*c);
%    T_hat=T(j)+dTdt*DELTAt;
%    dTdt_hat=e*A_s*sigma*(T_w(j+1)^4-T_hat^4)/(rho*V*c);
%    T(j+1)=T(j)+(dTdt+dTdt_hat)*DELTAt/2;
```

EXAMPLE 3.2-1(b)

% end

%Solution determined by ode solver
[time, T]=ode45(@(time,T) dTdt_f(time,T,e,A_s,rho,V,c,T_f,T_ini,beta,t_hold),time,T_ini);

The time span for the integration is provided by supplying the time array to the ode45 function. As a result, MATLAB will evaluate the temperatures at the same times as were used with Heun's method. A plot of the ode solver results (identified with circles) is superimposed onto the results obtained using Heun's method in Figure 1.

3.3 Semi-Infinite 1-D Transient Problems

3.3.1 Introduction

Sections 3.1 and 3.2 present analytical and numerical solutions to transient problems in which the spatial temperature gradients within the solid object can be neglected. Therefore, the problem is zero-dimensional; the transient solution is a function only of time. This section begins the discussion of transient problems where internal temperature gradients related to conduction are non-negligible (i.e., the Biot number is not much less than unity).

3.3.2 The Diffusive Time Constant

The simplest case for which the temperature gradients are one-dimensional (i.e., temperature varies in only one spatial dimension) occurs when the object itself is *semi-infinite*. A semi-infinite body is shown in Figure 3-12; semi-infinite means that the material is bounded on one edge (at $x = 0$) but extends to infinity in the other.

No object is truly semi-infinite although many approach this limit; the earth, for example, is semi-infinite relative to most surface phenomena. Furthermore, we will see that every object is essentially semi-infinite with respect to surface processes that occur over a sufficiently "small" time scale.

Figure 3-12 shows a semi-infinite body that is initially at a uniform temperature, T_{ini}, when at time $t = 0$ the temperature of the surface of the body (at $x = 0$) is raised to T_s. The temperature as a function of position is shown in Figure 3-12 for various times. The transient response can be characterized as a "thermal wave" that penetrates into the solid from the surface. The temperature of the solid is, at first, affected by the

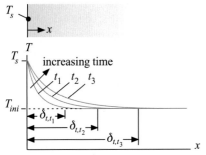

Figure 3-12: Semi-infinite body subjected to a sudden change in the surface temperature.

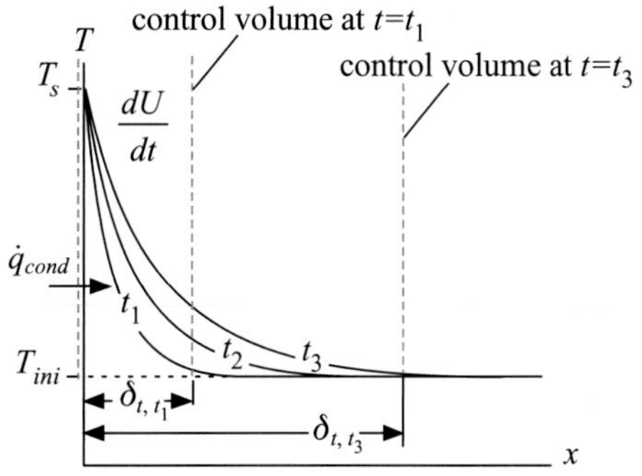

Figure 3-13: Control volume used to develop a simple model of the semi-infinite body.

surface change only at positions that are very near the surface. As time increases, the thermal wave penetrates deeper into the solid; the depth of the penetration (δ_t) grows and therefore the amount of material affected by the surface change increases.

The analytical and numerical methods discussed in this section as well as in Sections 3.4 and 3.8 can be used to provide the solution to the problem posed in Figure 3-12. However, before the problem is solved exactly, it is worthwhile to pause and understand the behavior of the thermal wave that characterizes this and all transient conduction problems. There are two phenomena occurring in Figure 3-12: (1) thermal energy is conducted from the surface into the body, and (2) the energy is stored by the temperature rise associated with the material that lies within the ever-growing thermal wave. By developing simple models for these two processes, it is possible to understand, to a first approximation, how the thermal wave behaves. A control volume is drawn from the surface to the outer edge of the thermal wave, as shown in Figure 3-13; the outer edge of the thermal wave is defined as the position within the material where the conduction heat transfer is small. The control volume must grow with time as the extent of the thermal wave increases.

An energy balance on the control volume shown in Figure 3-13 includes conduction into the surface ($\dot{q}_{cond}$) and energy storage:

$$\dot{q}_{cond} = \frac{dU}{dt} \tag{3-48}$$

The thermal resistance to conduction into the thermally affected region through the material that lies within the thermal wave (R_w) is approximated as a plane wall with the thickness of the thermal wave.

$$R_w \approx \frac{\delta_t}{k A_c} \tag{3-49}$$

where k is the conductivity of the material and A_c is the cross-sectional area of the material. This approximation is, of course, not exact as the temperature distribution within the thermal wave is not linear (see Figure 3-13). However, this approach is approximately correct; certainly it is more difficult to conduct heat into the solid as the thermal wave grows and Eq. (3-49) reflects this fact. The rate of conduction heat transfer is approximately:

$$\dot{q}_{cond} = \frac{T_s - T_{ini}}{R_w} \approx \frac{k A_c (T_s - T_{ini})}{\delta_t} \tag{3-50}$$

The thermal energy stored in the material (U) relative to its initial state is the product of the average temperature elevation of the material within the thermal wave ($\overline{\Delta T}$):

$$\overline{\Delta T} \approx \frac{(T_s - T_{ini})}{2} \tag{3-51}$$

and the heat capacity of the material within the control volume:

$$C \approx \rho c \, \delta_t A_c \tag{3-52}$$

where ρ and c are the density and specific heat capacity of the material, respectively. Note that Eqs. (3-51) and (3-52) are also only approximate.

$$U \approx \underbrace{\frac{(T_s - T_{ini})}{2}}_{\overline{\Delta T}} \underbrace{\rho c \, \delta_t A_c}_{C} \tag{3-53}$$

Substituting Eqs. (3-50) and (3-53) into Eq. (3-48) leads to:

$$\frac{k A_c (T_s - T_{ini})}{\delta_t} \approx \frac{d}{dt} \left[\frac{(T_s - T_{ini})}{2} \rho c \, \delta_t A_c \right] \tag{3-54}$$

Only the penetration depth (δ_t) varies with time in Eq. (3-54) and therefore Eq. (3-54) can be rearranged to provide an ordinary differential equation for δ_t:

$$\frac{k}{\delta_t} \approx \frac{\rho c}{2} \frac{d\delta_t}{dt} \tag{3-55}$$

Equation (3-55) is rearranged:

$$\frac{2 k}{\rho c} \approx \delta_t \frac{d\delta_t}{dt} \tag{3-56}$$

Equation (3-56) can be simplified somewhat using the definition of the thermal diffusivity (α):

$$\alpha = \frac{k}{\rho c} \tag{3-57}$$

Substituting Eq. (3-57) into Eq. (3-56) leads to:

$$2 \alpha \approx \delta_t \frac{d\delta_t}{dt} \tag{3-58}$$

Equation (3-58) is separated and integrated:

$$\int_0^t 2 \alpha \, dt \approx \int_0^{\delta_t} \delta_t \, d\delta_t \tag{3-59}$$

in order to obtain an approximate expression for the penetration of the thermal wave as a function of time:

$$2 \alpha t \approx \frac{\delta_t^2}{2} \tag{3-60}$$

or:

$$\boxed{\delta_t \approx 2\sqrt{\alpha\, t}} \tag{3-61}$$

Equation (3-61) indicates that the thermal wave will grow in proportion to $\sqrt{\alpha\, t}$. This is a very important and practical (but not an exact) result that governs transient conduction problems. The constant in Eq. (3-61) may change depending on the precise nature of the problem and the definition of the thermal penetration depth. However, Eq. (3-61) will be approximately correct for a large variety of transient conduction problems. For example, if you heat one side of a plate of stainless steel ($\alpha = 1.5 \times 10^{-5}$ m^2/s) that is $L = 1.0$ cm thick, then the temperature at the opposite side of the plate will begin to change in about $\tau_{diff} = 1.7$ s; this result follows directly from Eq. (3-61), rearranged to solve for time:

$$\boxed{\tau_{diff} \approx \frac{L^2}{4\,\alpha}} \tag{3-62}$$

The time required for the thermal wave to pass across the extent of a body is referred to as the diffusive time constant (τ_{diff}) and it is a broadly useful concept in the same way the lumped capacitance time constant (introduced in Section 3.1) is useful. The diffusive time constant characterizes, approximately, how long it takes for an object to equilibrate internally by conduction. The lumped capacitance time constant characterizes, approximately, how long it takes for an object to equilibrate externally with its environment. The first step in understanding any transient heat transfer problem involves the calculation of these two time constants.

EXAMPLE 3.3-1: TRANSIENT RESPONSE OF A TANK WALL

A metal wall (Figure 1) is used to separate two tanks of liquid at different temperatures, $T_{hot} = 500$ K and $T_{cold} = 400$ K. The thickness of the wall is $th = 0.8$ cm and its area is $A_c = 1.0$ m^2. The properties of the wall material are $\rho = 8000$ kg/m^3, $c = 400$ J/kg-K, and $k = 20$ W/m-K. The average heat transfer coefficient between the wall and the liquid in either tank is $\overline{h}_{liq} = 5000$ W/m^2-K.

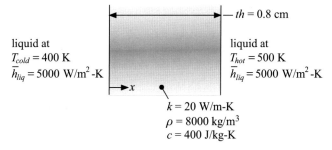

liquid at
$T_{cold} = 400$ K
$\overline{h}_{liq} = 5000$ W/m^2-K

$th = 0.8$ cm

liquid at
$T_{hot} = 500$ K
$\overline{h}_{liq} = 5000$ W/m^2-K

$k = 20$ W/m-K
$\rho = 8000$ kg/m^3
$c = 400$ J/kg-K

Figure 1: Tank wall exposed to fluid.

a) **Initially, the wall is at steady-state. That is, the wall has been exposed to the fluid in the tanks for a long time and therefore the temperature within the wall is not changing in time. What is the rate of heat transfer through the wall? What are the temperatures of the two surfaces of the wall (i.e., what is $T_{x=0}$ and $T_{x=th}$)?**

EXAMPLE 3.3-1: TRANSIENT RESPONSE OF A TANK WALL

EXAMPLE 3.3-1: TRANSIENT RESPONSE OF A TANK WALL

The known information is entered in EES:

```
"EXAMPLE 3.3-1: Transient Response of a Tank Wall"

$UnitSystem SI MASS RAD PA K J
$Tabstops 0.2 0.4 0.6 0.8 3.5

"Inputs"
k=20 [W/m-K]                            "thermal conductivity"
c=400 [J/kg-K]                          "specific heat capacity"
rho=8000 [kg/m^3]                       "density"
T_cold=400 [K]                          "cold fluid temperature"
T_hot=500 [K]                           "hot fluid temperature"
h_bar_liq=5000 [W/m^2-K]                "liquid-to-wall heat transfer coefficient"
th=0.8 [cm]*convert(cm,m)              "wall thickness"
A_c=1 [m^2]                             "wall area"
```

There are three thermal resistances governing this problem, convection from the surface on either side ($R_{conv,liq}$) and conduction through the wall (R_{cond}):

$$R_{conv,liq} = \frac{1}{\overline{h}_{liq} A_c}$$

$$R_{cond} = \frac{th}{k A_c}$$

```
"Steady-state solution, part (a)"
R_conv_liq=1/(h_bar_liq*A_c)           "convection resistance with liquid"
R_cond=th/(k*A_c)                      "conduction resistance"
```

The rate of heat transfer is:

$$\dot{q} = \frac{T_{hot} - T_{cold}}{2 R_{conv,liq} + R_{cond}}$$

and the temperatures at $x = 0$ and $x = th$ are:

$$T_{x=0} = T_{cold} + \dot{q} R_{conv,liq}$$

$$T_{x=th} = T_{hot} - \dot{q} R_{conv,liq}$$

```
q_dot=(T_hot-T_cold)/(2*R_conv_liq+R_cond)    "heat transfer"
T_0=T_cold+q_dot*R_conv_liq                   "temperature of cold side of wall"
T_L=T_hot-q_dot*R_conv_liq                    "temperature of hot side of wall"
```

The rate of heat transfer through the wall is $\dot{q} = 125$ kW and the edges of the wall are at $T_{x=0} = 425$ K and $T_{x=th} = 475$ K.

At time, $t = 0$, both tanks are drained and then both sides of the wall are exposed to gas at $T_{gas} = 300\,\text{K}$ (Figure 2). The average heat transfer coefficient between the walls and the gas is $\overline{h}_{gas} = 100$ W/m²-K. Assume that the process

EXAMPLE 3.3-1: TRANSIENT RESPONSE OF A TANK WALL

of draining the tanks and filling them with gas occurs instantaneously so that the wall has the linear temperature distribution from part (a) at time $t = 0$.

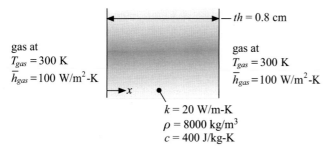

Figure 2: Tank wall exposed to gas.

b) **On the axes in Figure 3, sketch the temperature distribution in the wall (i.e., the temperature as a function of position) at $t = 0$ s (i.e., immediately after the process starts) and also at $t = 0.5$ s, 5 s, 50 s, 500 s, and 5000 s. Clearly label these different sketches. Be sure that you have the qualitative features of the temperature distribution drawn correctly.**

There are two processes that occur after the tank is drained. The tank material is not in equilibrium with itself due to its internal temperature gradient. Therefore, there is an internal equilibration process that will cause the wall material to come to a uniform temperature. Also, there is an external equilibration process as the wall transfers heat with its environment.

The internal equilibration process is governed by a diffusive time constant, τ_{diff}, discussed in Section 3.3.2 and provided, approximately, by Eq. (3-62):

$$\tau_{diff} = \frac{th^2}{4\,\alpha}$$

where α is the thermal diffusivity of the wall material:

$$\alpha = \frac{k}{\rho\,c}$$

```
"Internal equilibration process"
alpha=k/(rho*c)                              "thermal diffusivity"
tau_diff=th^2/(4*alpha)                      "diffusive time constant"
```

The diffusive time constant is about 2.6 seconds; this is, approximately, how long it will take for a thermal wave to pass from one side of the wall to the other and therefore this is the amount of time that is required for the wall to internally equilibrate. If the edges of the wall were adiabatic, then the wall will be at a nearly uniform temperature after a few seconds.

The external equilibration process is governed by a lumped time constant, discussed previously in Section 3.1 and defined for this problem as:

$$\tau_{lumped} = R_{conv,gas}\,C$$

EXAMPLE 3.3-1: TRANSIENT RESPONSE OF A TANK WALL

where $R_{conv,gas}$ is the thermal resistance to convection between the wall and the gas:

$$R_{conv,gas} = \frac{1}{2\,\overline{h}_{gas}\,A_c}$$

and C is the thermal capacitance of the wall:

$$C = A_c\,th\,\rho\,c$$

"External equilibration process"
h_bar_gas=100 [W/m^2-K]   "gas-to-wall heat transfer coefficient"
R_conv_gas=1/(2*h_bar_gas*A_c)   "convection resistance with gas"
C_total=A_c*th*rho*c   "capacity of wall"
tau_lumped=R_conv_gas*C_total   "lumped time constant"

The lumped time constant is about 130 s. Because the diffusive time constant is two orders of magnitude smaller than the lumped time constant, the wall will initially internally equilibrate rapidly and subsequently externally equilibrate more slowly. The initial internal equilibration process will be completed after about 5 to 10 s. Subsequently, the wall will equilibrate externally with the surrounding gas; this external equilibration process will be completed after 200 to 400 s. The temperature distributions sketched in Figure 3 are consistent with these time constants; they do not represent an exact solution, but they are consistent with the physical intuition that was gained through knowledge of the two time constants.

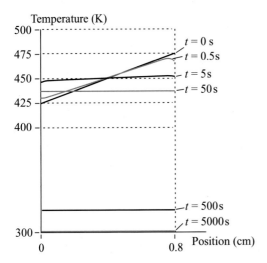

Figure 3: Temperature distributions in the wall as it equilibrates.

A more exact solution to this problem can be obtained using the techniques discussed in subsequent sections (see EXAMPLE 3.5-2). However, calculation of the diffusive and lumped time constants provide important physical intuition about the problem.

3.3.3 The Self-Similar Solution

The governing differential equation for the semi-infinite solid is derived using a control volume that is differential in x, as shown in Figure 3-14.

semi-infinite body with initial temperature T_{ini}

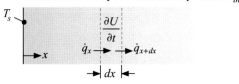

Figure 3-14: Differential control volume within semi-infinite body.

The energy balance suggested by Figure 3-14 is:

$$\dot{q}_x = \dot{q}_{x+dx} + \frac{\partial U}{\partial t} \tag{3-63}$$

Expanding the $x + dx$ term in Eq. (3-63) leads to:

$$\dot{q}_x = \dot{q}_x + \frac{\partial \dot{q}_x}{\partial x}dx + \frac{\partial U}{\partial t} \tag{3-64}$$

The conduction term is evaluated using Fourier's law:

$$\dot{q}_x = -k A_c \frac{\partial T}{\partial x} \tag{3-65}$$

where A_c is the area of the wall. The internal energy contained in the differential control volume is:

$$U = \rho c A_c \, dx \, T \tag{3-66}$$

where ρ and c are the density and specific heat capacity of the wall material. Assuming that the specific heat capacity is constant, the time rate of change of the internal energy is:

$$\frac{\partial U}{\partial t} = \rho c A_c \, dx \frac{\partial T}{\partial t} \tag{3-67}$$

Substituting Eqs. (3-65) and (3-67) into Eq. (3-64) leads to:

$$0 = \frac{\partial}{\partial x}\left[-k A_c \frac{\partial T}{\partial x} \right] dx + \rho c A_c \, dx \frac{\partial T}{\partial t} \tag{3-68}$$

which, for a constant thermal conductivity, can be simplified to:

$$\alpha \frac{\partial^2 T}{\partial x^2} = \frac{\partial T}{\partial t} \tag{3-69}$$

where α is the thermal diffusivity.

For the situation shown in Figure 3-14 in which the initial temperature of the solid is uniform (T_{ini}) and the surface temperature is suddenly elevated (to T_s), the boundary conditions are:

$$T_{x=0,t} = T_s \tag{3-70}$$

$$T_{x,t=0} = T_{ini} \tag{3-71}$$

$$T_{x\to\infty,t} = T_{ini} \tag{3-72}$$

The solution to the partial differential equation, Eq. (3-69), subject to the boundary conditions, Eqs. (3-70) through (3-72), is not obvious. Fortunately, the partial differential

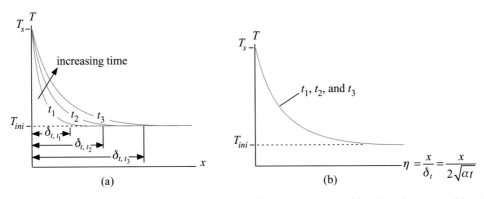

Figure 3-15: Temperature distribution in the semi-infinite body as (a) a function of position for various times, and (b) a function of position normalized by the thermal penetration depth for various times.

equation can be reduced to an ordinary differential by using the concept of the thermal penetration depth introduced in Section 3.3.2. Figure 3-12 shows the temperature distribution at different times; these distributions are repeated in Figure 3-15(a). Figure 3-15(b) shows that when the temperature distribution is plotted against x/δ_t where δ_t is $2\sqrt{\alpha t}$, according to Eq. (3-61), the temperature profiles for each time collapse onto a single curve.

The temperature in Figure 3-15(a) is a function of two independent variables x and t. In this problem, the two independent variables can be combined in order to express the temperature as a function of a single independent variable, defined as η:

$$\eta = \frac{x}{2\sqrt{\alpha t}} \qquad (3\text{-}73)$$

This process is often called combination of variables and the resulting solution is referred to as a self-similar solution.

The first step is to transform the governing partial differential equation in x and t to an ordinary differential equation in η. The similarity parameter η, given by Eq. (3-73), is substituted into Eq. (3-69). The temperature is expressed functionally as:

$$T\left(\eta\left(x, t\right)\right) \qquad (3\text{-}74)$$

Therefore, according to the chain rule, the partial derivative of temperature with respect to x is:

$$\frac{\partial T}{\partial x} = \frac{dT}{d\eta} \frac{\partial \eta}{\partial x} \qquad (3\text{-}75)$$

The partial derivative of η with respect to x is obtained by inspection of Eq. (3-73) or using Maple:

```
> eta:=x/(2*sqrt(alpha*t));
```

$$\eta := \frac{x}{2\sqrt{\alpha t}}$$

```
> diff(eta,x);
```

$$\frac{1}{2\sqrt{\alpha t}}$$

So that:

$$\frac{\partial \eta}{\partial x} = \frac{1}{2\sqrt{\alpha t}} \tag{3-76}$$

and therefore Eq. (3-75) becomes:

$$\frac{\partial T}{\partial x} = \frac{dT}{d\eta} \frac{1}{2\sqrt{\alpha t}} \tag{3-77}$$

The same process is used to calculate the second derivative of T with respect to x:

$$\frac{\partial^2 T}{\partial x^2} = \frac{d}{d\eta}\left(\frac{\partial T}{\partial x}\right)\frac{\partial \eta}{\partial x} \tag{3-78}$$

Substituting Eq. (3-77) into Eq. (3-78) leads to:

$$\frac{\partial^2 T}{\partial x^2} = \frac{d}{d\eta}\left(\frac{dT}{d\eta}\frac{1}{2\sqrt{\alpha t}}\right)\frac{\partial \eta}{\partial x} \tag{3-79}$$

Substituting Eq. (3-76) into Eq. (3-79) leads to:

$$\frac{\partial^2 T}{\partial x^2} = \frac{d^2 T}{d\eta^2}\frac{1}{4\alpha t} \tag{3-80}$$

The partial derivative of the temperature, Eq. (3-74), with respect to time is also obtained using the chain rule:

$$\frac{\partial T}{\partial t} = \frac{dT}{d\eta}\frac{\partial \eta}{\partial t} \tag{3-81}$$

where the partial derivative of η with respect to t is evaluated using Maple:

```
> diff(eta,t);
```

$$\frac{1}{4}\frac{x\alpha}{(\alpha t)^{(3/2)}}$$

So that:

$$\frac{\partial \eta}{\partial t} = -\frac{x}{4t\sqrt{\alpha t}} \tag{3-82}$$

and therefore Eq. (3-81) becomes:

$$\frac{\partial T}{\partial t} = -\frac{x}{4t\sqrt{\alpha t}}\frac{dT}{d\eta} \tag{3-83}$$

Substituting Eqs. (3-80) and (3-83) into Eq. (3-69) leads to:

$$\alpha\frac{d^2 T}{d\eta^2}\frac{1}{4\alpha t} = -\frac{x}{4t\sqrt{\alpha t}}\frac{dT}{d\eta} \tag{3-84}$$

which can be rearranged:

$$\frac{d^2T}{d\eta^2} = -2\underbrace{\frac{x}{2\sqrt{\alpha t}}}_{\eta}\frac{dT}{d\eta} \tag{3-85}$$

or

$$\boxed{\frac{d^2T}{d\eta^2} = -2\eta\frac{dT}{d\eta}} \tag{3-86}$$

which completes the transformation of the partial differential equation, Eq. (3-69), into an ordinary differential equation, Eq. (3-86). The boundary conditions must also be transformed. Equations (3-70) through (3-72) are transformed from expressions in x and t to expressions in η:

$$T_{x=0,t} = T_s \Rightarrow T_{\eta=0} = T_s \tag{3-87}$$

$$T_{x,t=0} = T_{ini} \Rightarrow T_{\eta\to\infty} = T_{ini} \tag{3-88}$$

$$T_{x\to\infty,t} = T_{ini} \Rightarrow T_{\eta\to\infty} = T_{ini} \tag{3-89}$$

Notice that Eqs. (3-88) and (3-89) are identical and so the partial differential equation has been transformed into a second order ordinary differential equation with two boundary conditions. It is not always possible to accomplish this transformation; if either x or t had been retained in the partial differential equation or any of the boundary conditions, then a self-similar solution would not be possible. In this case, an alternative analytical technique such as the Laplace transform (Section 3.4) or a numerical solution technique (Section 3.8) is required.

Equation (3-86) can be separated and solved. The variable w is defined as:

$$w = \frac{dT}{d\eta} \tag{3-90}$$

and substituted into Eq. (3-86):

$$\frac{dw}{d\eta} = -2\eta w \tag{3-91}$$

Equation (3-91) can be rearranged and integrated:

$$\int\frac{dw}{w} = -2\int\eta\,d\eta \tag{3-92}$$

which leads to:

$$\ln(w) = -\eta^2 + C_1 \tag{3-93}$$

where C_1 is a constant of integration that is necessary because Eq. (3-92) is an indefinite integral. Solving Eq. (3-93) for w leads to:

$$w = \exp(-\eta^2 + C_1) \tag{3-94}$$

Substituting Eq. (3-90) into Eq. (3-94) leads to:

$$\frac{dT}{d\eta} = \exp(-\eta^2 + C_1) \tag{3-95}$$

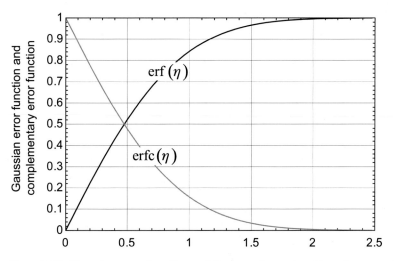

Figure 3-16: Gaussian error function and the complementary Gaussian error function.

or

$$\frac{dT}{d\eta} = C_2 \exp(-\eta^2) \tag{3-96}$$

where C_2 is another undetermined constant (equal to the exponential of C_1). Equation (3-96) can be integrated:

$$\int dT = \int C_2 \exp(-\eta^2)\, d\eta \tag{3-97}$$

or

$$T = C_2 \int_0^\eta \exp(-\eta^2)\, d\eta + C_3 \tag{3-98}$$

where C_3 is an undetermined constant. The integral in Eq. (3-98) cannot be evaluated analytically and yet it shows up often in engineering problems. The integral is defined in terms of the Gaussian error function (typically called the erf function, which is pronounced so that it rhymes with smurf). The Gaussian error function is defined as:

$$\mathrm{erf}(\eta) = \frac{2}{\sqrt{\pi}} \int_0^\eta \exp(-\eta^2)\, d\eta \tag{3-99}$$

and illustrated in Figure 3-16.

The complementary error function (erfc) is defined as:

$$\mathrm{erfc}\,(\eta) = 1 - \mathrm{erf}\,(\eta) \tag{3-100}$$

and is also illustrated in Figure 3-16. Equation (3-98) can be written in terms of the erf function:

$$T = C_2 \frac{\sqrt{\pi}}{2} \mathrm{erf}\,(\eta) + C_3 \tag{3-101}$$

The solution to the ordinary differential equation, Eq. (3-86), can also be obtained using Maple:

```
> GDE:=diff(diff(T(eta),eta),eta)=-2*eta*diff(T(eta),eta);
```

$$GDE := \frac{d^2}{d\eta^2} T(\eta) = -2\eta \left(\frac{d}{d\eta} T(\eta) \right)$$

```
> Ts:=dsolve(GDE);
```

$$Ts := T(\eta) = \_C1 + \mathrm{erf}(\eta)\_C2$$

The two constants in Eq. (3-101) are obtained using the boundary conditions. Substituting Eq. (3-101) into (3-87) leads to:

$$T_{\eta=0} = C_2 \frac{\sqrt{\pi}}{2} \underbrace{\mathrm{erf}(0)}_{0} + C_3 = T_s \tag{3-102}$$

Figure 3-16 shows that erf(0) = 0 so Eq. (3-102) becomes:

$$C_3 = T_s \tag{3-103}$$

Substituting Eq. (3-103) into Eq. (3-101) leads to:

$$T = C_2 \frac{\sqrt{\pi}}{2} \mathrm{erf}(\eta) + T_s \tag{3-104}$$

Substituting Eq. (3-104) into Eq. (3-88) leads to:

$$T_{\eta\to\infty} = C_2 \frac{\sqrt{\pi}}{2} \underbrace{\mathrm{erf}(\infty)}_{=1} + T_s = T_{ini} \tag{3-105}$$

Figure 3-16 shows that erf(∞) = 1 so Eq. (3-105) becomes:

$$C_2 = \frac{2}{\sqrt{\pi}} (T_{ini} - T_s) \tag{3-106}$$

The temperature distribution within the semi-infinite body is therefore:

$$\boxed{T = T_s + (T_{ini} - T_s)\,\mathrm{erf}\left(\frac{x}{2\sqrt{\alpha t}}\right)} \tag{3-107}$$

The heat transfer into the surface of the body, $\dot{q}_{x=0}$, is evaluated using Fourier's law:

$$\dot{q}_{x=0} = -k A_c \left.\frac{\partial T}{\partial x}\right|_{x=0} \tag{3-108}$$

which can be expressed in terms of η by substituting Eq. (3-77) into Eq. (3-108):

$$\dot{q}_{x=0} = -k A_c \frac{dT}{d\eta} \frac{1}{2\sqrt{\alpha t}} \tag{3-109}$$

Substituting Eq. (3-107) into Eq. (3-109) leads to:

$$\dot{q}_{x=0} = -k A_c \left.\frac{d}{d\eta}[T_s + (T_{ini} - T_s)\,\mathrm{erf}(\eta)]\right|_{\eta=0} \frac{1}{2\sqrt{\alpha t}} \tag{3-110}$$

or:

$$\dot{q}_{x=0} = -\frac{k A_c}{2\sqrt{\alpha t}} (T_{ini} - T_s) \frac{d}{d\eta} [\mathrm{erf}(\eta)]\bigg|_{\eta=0} \tag{3-111}$$

Substituting the definition of the erf function, Eq. (3-99), into Eq. (3-111) leads to:

$$\dot{q}_{x=0} = -\frac{k A_c}{2\sqrt{\alpha t}} (T_{ini} - T_s) \frac{d}{d\eta} \left[\frac{2}{\sqrt{\pi}} \int_0^\eta \exp(-\eta^2)\, d\eta \right]_{\eta=0} \tag{3-112}$$

The derivative of an integral is the integrand, and therefore Eq. (3-112) becomes:

$$\dot{q}_{x=0} = -\frac{k A_c}{2\sqrt{\alpha t}} (T_{ini} - T_s) \frac{2}{\sqrt{\pi}} [\exp(-\eta^2)]_{\eta=0} \tag{3-113}$$

or

$$\boxed{\dot{q}_{x=0} = \underbrace{\frac{k A_c}{\sqrt{\pi \alpha t}}}_{1/R_{semi-\infty}} (T_s - T_{ini})} \tag{3-114}$$

This result is extremely useful and intuitive; the material within the thermal wave in Figure 3-15 acts like a thermal resistance to heat transfer from the surface ($R_{semi-\infty}$):

$$\dot{q}_{x=0} = \frac{(T_s - T_{ini})}{R_{semi-\infty}} \tag{3-115}$$

where $R_{semi-\infty}$ increases with time as the thermal wave grows (and therefore the distance over which the conduction occurs increases) according to:

$$\boxed{R_{semi-\infty} = \frac{\sqrt{\pi \alpha t}}{k A_c}} \tag{3-116}$$

Equation (3-116) is similar to the thermal resistance to conduction through a plane wall with thickness that grows according to $\sqrt{\pi \alpha t}$. This concept is useful for understanding the physics associated with transient conduction problems. Of course, as soon as the thermal wave reaches a boundary, the problem is no longer semi-infinite and therefore the concepts of the semi-infinite resistance and thermal penetration wave are no longer valid.

3.3.4 Solution to other Semi-Infinite Problems

Analytical solutions to a semi-infinite body exposed to other surface boundary conditions have been developed and are summarized in Table 3-2. These analytical solutions can often be applied to various processes that occur over very short time-scales where the thermal penetration wave (δ_t) is small relative to the spatial extent of the object. Functions that return the temperature at a specified position and time for the semi-infinite body problems in Table 3-2 are available in EES. Select Function Info from the Options menu and select Transient Conduction from the pull-down menu at the lower right corner of the upper box, toggle to the Semi-Infinite Body library and scroll across to find the function of interest.

Table 3-2: Solutions to semi-infinite body problems.

Boundary Condition	Solution
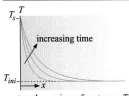 step change in surface temp.: $T_{x=0} = T_s$	$$\frac{T - T_{ini}}{T_s - T_{ini}} = 1 - \text{erf}\left(\frac{x}{2\sqrt{\alpha t}}\right)$$ $$\dot{q}''_{x=0} = \frac{k}{\sqrt{\pi \alpha t}}(T_s - T_{ini})$$
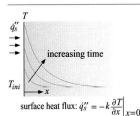 surface heat flux: $\dot{q}''_s = -k\frac{\partial T}{\partial x}\Big|_{x=0}$	$$T - T_{ini} = \frac{\dot{q}''_s}{k}\left[\sqrt{\frac{4\alpha t}{\pi}}\exp\left(-\frac{x^2}{4\alpha t}\right) - x\,\text{erfc}\left(\frac{x}{\sqrt{4\alpha t}}\right)\right]$$
convection to fluid: $\bar{h}(T_\infty - T_{x=0}) = -k\frac{\partial T}{\partial x}\Big|_{x=0}$	$$\frac{T - T_{ini}}{T_\infty - T_{ini}} = \text{erfc}\left(\frac{x}{2\sqrt{\alpha t}}\right)$$ $$- \exp\left(\frac{\bar{h}x}{k} + \frac{\bar{h}^2\alpha t}{k^2}\right)\text{erfc}\left(\frac{x}{2\sqrt{\alpha t}} + \frac{\bar{h}}{k}\sqrt{\alpha t}\right)$$
surface energy per unit area released at $t = 0$ wall adiabatic for $t>0$ surface energy pulse: $\lim\limits_{t,\Delta t\to 0}\dot{q}''_s\,\Delta t = E''$	$$T - T_{ini} = \frac{E''}{\rho c\sqrt{\pi \alpha t}}\exp\left(-\frac{x^2}{4\alpha t}\right)$$
$T_{x=0} = T_{ini} + \Delta T\sin(\omega t)$ periodic surface temperature	$$T - T_{ini} = \Delta T\exp\left(-x\sqrt{\frac{\omega}{2\alpha}}\right)\sin\left(\omega t - x\sqrt{\frac{\omega}{2\alpha}}\right)$$
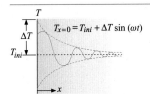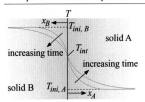 contact between two semi-infinite solids	$$\frac{T_{ini,A} - T_{int}}{T_{int} - T_{ini,B}} = \frac{\sqrt{k_B\rho_B c_B}}{\sqrt{k_A\rho_A c_A}}$$ $$\frac{T_A - T_{ini,A}}{T_{int} - T_{ini,A}} = 1 - \text{erf}\left(\frac{x_A}{2\sqrt{\alpha_A t}}\right),\quad \frac{T_B - T_{ini,B}}{T_{int} - T_{ini,B}} = 1 - \text{erf}\left(\frac{x_B}{2\sqrt{\alpha_B t}}\right)$$

T_{ini} = initial temperature t = time relative to surface disturbance k = conductivity
x = position from surface ρ = density c = specific heat capacity
α = thermal diffusivity

EXAMPLE 3.3-2: QUENCHING A COMPOSITE STRUCTURE

EXAMPLE 3.3-2: QUENCHING A COMPOSITE STRUCTURE

A laminated structure is fabricated by diffusion bonding alternating layers of high conductivity silicon ($k_s = 150$ W/m-K, $\rho_s = 2300$ kg/m³, $c_s = 700$ J/kg-K) and low conductivity pyrex ($k_p = 1.4$ W/m-K, $\rho_p = 2200$ kg/m³, $c_p = 800$ J/kg-K). The thickness of each layer is $th_s = th_p = 0.5$ mm, as shown in Figure 1.

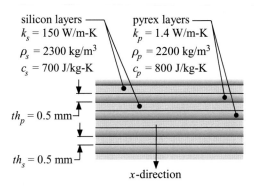

Figure 1: A composite structure formed from silicon and glass.

a) Determine an effective conductivity that can be used to characterize the composite structure with respect to heat transfer in the x-direction (see Figure 1).

The known information is entered in EES:

```
"EXAMPLE 3.3-2: Quenching a Composite Structure"

$UnitSystem SI MASS RAD PA K J
$Tabstops 0.2 0.4 0.6 0.8 3.5 in

"Inputs"
k_s=150 [W/m-K]                    "conductivity of silicon"
rho_s=2300 [kg/m^3]                "density of silicon"
c_s=700 [J/kg-K]                   "specific heat capacity of silicon"
k_p=1.4 [W/m-K]                    "conductivity of pyrex"
rho_p=2200 [kg/m^3]                "density of pyrex"
c_p=800 [J/kg-K]                   "specific heat capacity of pyrex"
th_s=0.5 [mm]*convert(mm,m)        "thickness of silicon lamination"
th_p=0.5 [mm]*convert(mm,m)        "thickness of pyrex lamination"
```

The method discussed in Section 2.9 is used to determine the effective thermal conductivity of the composite structure. The heat transfer through a thickness (L, in the x-direction) with cross-sectional area (A_c) of the composite structure when it is subjected to a given temperature difference (ΔT) is calculated as the series combination of the thermal resistance of the silicon and pyrex laminations according to:

$$\dot{q} = \frac{\Delta T}{\dfrac{L\,th_s}{(th_s + th_p)\,k_s\,A_c} + \dfrac{L\,th_p}{(th_s + th_p)\,k_p\,A_c}} \tag{1}$$

EXAMPLE 3.3-2: QUENCHING A COMPOSITE STRUCTURE

The effective conductivity in the x-direction (k_{eff}) is equal to the conductivity of a homogeneous material that would result in the same heat transfer rate:

$$\dot{q} = \frac{k_{eff}\, A_c\, \Delta T}{L} \qquad (2)$$

Setting Eq. (1) equal to Eq. (2) leads to:

$$\frac{\Delta T}{\dfrac{L\, th_s}{(th_s + th_p)\, k_s\, A_c} + \dfrac{L\, th_p}{(th_s + th_p)\, k_p\, A_c}} = \frac{k_{eff}\, A_c\, \Delta T}{L}$$

Solving for k_{eff} leads to:

$$k_{eff} = \left[\frac{th_s}{(th_s + th_p)\, k_s} + \frac{th_p}{(th_s + th_p)\, k_p} \right]^{-1}$$

"Part a: effective conductivity for heat transfer across laminations"
k_eff=1/(th_s/((th_s+th_p)*k_s)+th_p/((th_s+th_p)*k_p))   "effective conductivity"

The effective conductivity is $k_{eff} = 2.8$ W/m-K, which is between the conductivity of pyrex and silicon. Because the conductivity of silicon is high, the silicon laminations contribute essentially no thermal resistance; therefore, because the laminations are of equal size, the effective conductivity is twice that of pyrex.

b) Determine an effective heat capacity (c_{eff}) and density (ρ_{eff}) that can be used characterize the composite structure.

The process of determining an effective density and specific heat capacity is conceptually similar to the calculation of an effective thermal conductivity. The effective property is chosen so that the composite, when modeled as a homogeneous material using the effective property, behaves as the actual composite does.

The effective density is defined so that material has the same mass as the composite. The mass of the composite (with thickness L and cross-sectional area A_c) is:

$$M = \frac{th_s}{(th_p + th_s)}\, L\, A_c\, \rho_s + \frac{th_p}{(th_p + th_s)}\, L\, A_c\, \rho_p \qquad (3)$$

The mass of the homogeneous material with an effective density (ρ_{eff}) is:

$$M = L\, A_c\, \rho_{eff} \qquad (4)$$

Setting Eq. (3) equal to Eq. (4) and solving for ρ_{eff} leads to:

$$\rho_{eff} = \frac{th_s}{(th_p + th_s)}\, \rho_s + \frac{th_p}{(th_p + th_s)}\, \rho_p$$

"Part b: effective density and heat capacity"
rho_eff=th_s*rho_s/(th_p+th_s)+th_p*rho_p/(th_p+th_s)   "effective density"

The effective specific heat capacity is defined so that the homogeneous material model has the same total heat capacity as the composite. The total heat capacity of the composite (again with thickness L and area A_c) is:

$$C = \frac{th_s}{(th_p + th_s)}\, L\, A_c\, \rho_s\, c_s + \frac{th_p}{(th_p + th_s)}\, L\, A_c\, \rho_p\, c_p \qquad (5)$$

EXAMPLE 3.3-2: QUENCHING A COMPOSITE STRUCTURE

The total heat capacity of the homogeneous material with effective density (ρ_{eff}) and effective heat capacity (c_{eff}) is:

$$C = L\,A_c\,\rho_{eff}\,c_{eff} \tag{6}$$

Setting Eq. (5) equal to Eq. (6) and solving for c_{eff} leads to:

$$c_{eff} = \frac{th_s\,\rho_s}{(th_p + th_s)\,\rho_{eff}}\,c_s + \frac{th_p\,\rho_p}{(th_p + th_s)\,\rho_{eff}}\,c_p$$

c_eff=th_s*rho_s*c_s/((th_p+th_s)*rho_eff)_th_p*rho_p*c_p/((th_p+th_s)*rho_eff)
  "effective specific heat capacity"

The effective density and specific heat capacity of the composite structure are $\rho_{eff} = 2250$ kg/m³ and $c_{eff} = 749$ J/kg-K.

The diffusion bonding of the composite structure occurs at high temperature, $T_{bond} = 750°C$. When the bonding process is complete, the manufacturing process is terminated by quenching the composite structure from both sides with water at $T_w = 20°C$, as shown in Figure 2.

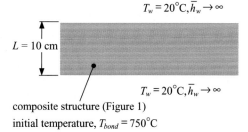

Figure 2: Quenching a composite structure with water.

$T_w = 20°C, \bar{h}_w \rightarrow \infty$

$L = 10$ cm

$T_w = 20°C, \bar{h}_w \rightarrow \infty$

composite structure (Figure 1)
initial temperature, $T_{bond} = 750°C$

The heat transfer coefficient between the water and the surface of the structure is very high because the water is vaporizing. Therefore the quenching process corresponds approximately to applying a step change in the surface temperature of the composite.

c) The laminated structure is $L = 10$ cm thick, for approximately how long will it be appropriate to model the composite as a semi-infinite body?

The thermal wave moves from the top and bottom surfaces of the structure approximately according to:

$$\delta_t = 2\sqrt{\alpha_{eff}\,t} \tag{7}$$

where the effective thermal diffusivity of the composite structure is:

$$\alpha_{eff} = \frac{k_{eff}}{\rho_{eff}\,c_{eff}}$$

"Part c"
alpha_eff=k_eff/(rho_eff*c_eff)   "effective diffusivity"

EXAMPLE 3.3-2: QUENCHING A COMPOSITE STRUCTURE

When the thermal wave reaches the half-thickness of the structure then it will no longer behave as a semi-infinite body but rather as a bounded, 1-D transient problem. Substituting $\delta_t = L/2$ into Eq. (7) leads to:

$$\frac{L}{2} = 2\sqrt{\alpha_{eff}\, t_{semi-\infty}}$$

where $t_{semi-\infty}$ is the time at which the semi-infinite model is no longer appropriate.

```
L=10[cm]*convert(cm,m)                          "width of structure"
2*sqrt(alpha_eff*t_semi_infinite)=L/2           "time that semi-infinite solution is valid"
```

The semi-infinite body solution is valid for approximately 380 s.

It has been observed that the laminations that are within $x_{fail} = 1.0$ cm of the two surfaces tend to de-bond during the quenching process. You suspect that the large spatial temperature gradients near the surface during the quench are causing thermally induced stresses that are responsible for this failure. In the presence of large temperature gradients, two laminations that are adjacent to one another will experience very different thermally induced expansions that result in a shear stress.

d) Prepare a plot of the temperature gradient as a function of time (up to the time at which the semi-infinite solution is no longer valid, from part (c)) at $x = x_{fail}$ as well as other positions that are greater than and less than x_{fail}. Determine the critical spatial temperature gradient that causes failure (i.e., the maximum spatial temperature gradient experienced at $x = x_{fail}$).

Table 3-2 or Eq. (3-107) provides the temperature within the semi-infinite body:

$$T = T_w + (T_{bond} - T_w)\,\mathrm{erf}\left(\frac{x}{2\sqrt{\alpha_{eff}\,t}}\right) \tag{8}$$

The spatial temperature gradient can be obtained by differentiating Eq. (8) using Maple:

```
> restart;
> T(x,time):=T_w+(T_bond-T_w)*erf(x/(2*sqrt(alpha_eff*time)));
```

$$T(x,\,time) := T\_w + (T\_bond - T\_w)\,\mathrm{erf}\left(\frac{x}{2\sqrt{alpha\_eff\,time}}\right)$$

```
> dTdx:=diff(T(x,time),x);
```

$$dTdx := \frac{(T\_bond - T\_w)\,e^{\left(-\frac{x^2}{4\,alpha\_eff\,time}\right)}}{\sqrt{\pi}\,\sqrt{alpha\_eff\,time}}$$

so the temperature gradient is:

$$\frac{\partial T}{\partial x} = \frac{(T_{bond} - T_w)}{\sqrt{\pi\,\alpha_{eff}\,t}}\exp\left(-\frac{x^2}{4\,\alpha_{eff}\,t}\right) \tag{9}$$

The result from Maple is copied and pasted into EES and then modified slightly for compatibility:

```
"Part d - analytical differentiation"
T_bond=converttemp(C,K,750 [C])          "bond temperature"
T_w=converttemp(C,K,20 [C])              "water temperature"
x_fail=1 [cm]*convert(cm,m)              "observed position of failure"
x=x_fail                                 "vary x from 0.5 to 4 * x_fail"
dTdx=(T_bond-T_w)/Pi^(1/2)*exp(-1/4*x^2/alpha_eff/time)/(alpha_eff*time)^(1/2)
```

A parametric table is used to generate Figure 3. The independent variable time is varied from 1×10^{-6} s to 380 s. A value of 1×10^{-6} s rather than 0 is used to avoid division by zero in Eq. (9). The value of x is changed in the Equation window in order to produce the different curves that are shown in Figure 3.

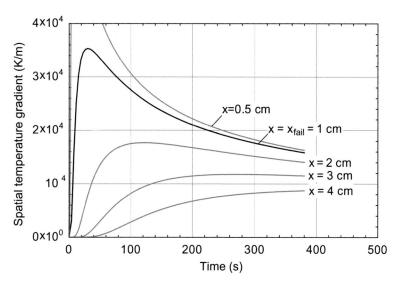

Figure 3: Spatial temperature gradient as a function of time for various locations in the composite structure.

Examination of Figure 3 suggests that the critical spatial temperature gradient for failure is about 3.5×10^4 K/m. Locations exposed to a spatial gradient greater than this value are likely to fail. A more exact solution can be obtained by determining the time at which the spatial gradient is maximized when $x = x_{fail}$. This result can be obtained by selecting Min/Max from the Calculate menu and maximizing dTdx by varying the independent variable time. Provide limits on time from slightly greater than 0 (to prevent a division by 0 error) to 380 s. The result will be dTdx = 3.53×10^4 K/m which occurs at 30.4 s.

In order to reduce the temperature gradients within the laminations, you suggest that the current water quenching process be replaced by a gas cooling process in which the surfaces of the composite are exposed to a gas at $T_{gas} = 20°C$ with a lower heat transfer coefficient, $\bar{h}_{gas} = 400$ W/m²-K.

EXAMPLE 3.3-2: QUENCHING A COMPOSITE STRUCTURE

EXAMPLE 3.3-2: QUENCHING A COMPOSITE STRUCTURE

e) Determine whether there will be any de-lamination using this process and, if so, what the thickness of the damaged region (x_{fail}) will be.

Rather than use the analytical solution for the temperature distribution within a semi-infinite body that is subjected to surface convection, from Table 3-2, the gradient will be evaluated numerically using the built-in function SemiInf3 in EES. The numerical derivative is obtained by adding and subtracting a very small value, Δx, to/from the nominal value of x according to:

$$\frac{\partial T}{\partial x} \approx \frac{T_{x+\Delta x} - T_{x-\Delta x}}{2\,\Delta x}$$

The value of Δx must be small in order for this approach to work, but not so small that numerical precision is exceeded. EES provides about 20 digits of numerical precision, so accurate determination of numerical derivatives is usually not a problem. The EES code to compute the temperature gradient for a given value of position and time is:

```
"Part e - gas quenching process"
T_bond=converttemp(C,K,750 [C])                "bond temperature"
T_gas=converttemp(C,K,20 [C])                  "temperature of gas"
h_bar_gas=400 [W/m^2-K]                         "heat transfer coefficient"
DELTAx=1e-5 [m]        "differential change used to evaluate numerical derivative"
x=2.0 [cm]*convert(cm,m)                        "position"
T_xplusdx=SemiInf3(T_bond,T_gas,h_bar_gas,k_eff,alpha_eff,x+DELTAx,time)
T_xminusdx=SemiInf3(T_bond,T_gas,h_bar_gas,k_eff,alpha_eff,x-DELTAx,time)
dTdx=(T_xplusdx-T_xminusdx)/(2*DELTAx)          "numerical temperature gradient"
```

In order for this code to run, it is necessary to comment out the EES code from (d). Figure 4 illustrates the spatial temperature gradient as a function of time for various values of position.

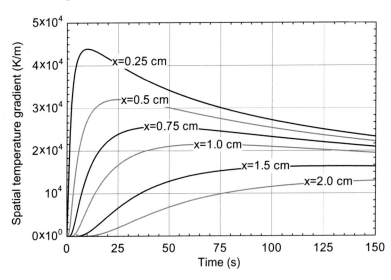

Figure 4: Spatial temperature gradient as a function of time for various values of position using the gas quenching process.

EXAMPLE 3.3-2: QUENCHING A COMPOSITE STRUCTURE

Figure 4 shows that there is a maximum temperature gradient experienced at each value of x that occurs as the thermal wave passes by that position. The maximum temperature gradient experienced as a function of position can be obtained by using the Min/Max Table option from the Calculate menu. Comment out the specified value of position and generate a parametric table that includes the variable x as well as the dependent variable to be maximized (the variable dTdx) and the independent variable to vary (the variable time). Vary x from 0.1 mm to 1.45 cm in the parametric table. Select Min/Max Table from the Calculate menu. Maximize the value of the temperature gradient by varying the independent variable time. The maximum temperature gradient as a function of position is shown in Figure 5.

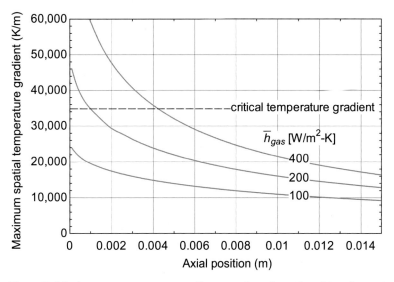

Figure 5: Maximum temperature gradient as a function of position for various values of the heat transfer coefficient.

Figure 5 suggests that for $\bar{h}_{gas} = 400$ W/m²-K, the extent of the damaged region will be reduced to only 0.42 cm. Figure 5 also shows that if $\bar{h}_{gas}$ is reduced to 100 W/m²-K then the spatial temperature gradient will not exceed 25,000 K/m anywhere in the composite. Because 25,000 K/m is less than the critical temperature gradient identified in part (d) (35,300 K/m), it is likely that no damage would occur for $\bar{h}_{gas} = 100$ W/m²-K.

3.4 The Laplace Transform

3.4.1 Introduction

Sections 3.1 and 3.3 present techniques for obtaining analytical solutions to 0-D (lumped) and 1-D transient conduction problems. This section presents an alternative method for obtaining an analytical solution to these problems: the Laplace transform. The Laplace transform is a mathematical technique that is used to solve differential equations that arise in many engineering disciplines.

The mathematical specification of a transient conduction problem will include a differential equation and boundary conditions that involve time. In the case of the lumped capacitance problems discussed in Section 3.1, an ordinary differential equation in time was obtained. The semi-infinite problems in Section 3.3 result in a partial differential equation in time and position. The Laplace transform maps a problem involving time,

t, onto a problem involving a different variable, typically called s. The attractive feature of the Laplace transform is that the mapping process removes time derivatives from the problem. Therefore, if an ordinary differential equation in t is Laplace-transformed, it becomes an algebraic equation in s. A partial differential equation in t and x becomes an ordinary differential equation in x that is algebraic in s. The complexity of the differential equation is reduced by one level and the problem is correspondingly easier to solve in the s domain than in the t domain.

The Laplace transform solution proceeds by obtaining the differential equation and boundary conditions as usual and transforming them into the Laplace (s) domain. The solution is obtained in the s domain and then transformed back to the time domain. The process of transforming between the s and t domains is facilitated by extensive tables that exist as well as symbolic software packages such as Maple.

This section is by no means a comprehensive review of Laplace transform theory and application; the interested reader is referred to more complete coverage which can be found in Arpaci (1966) and Myers (1998). However, the introduction provided here is sufficient to allow the solution of many interesting heat transfer problems and provide some insight into the technique. The Laplace transform is particularly useful for short time-scale and semi-infinite body problems.

3.4.2 The Laplace Transformation

The Laplace transform of a function of time, $T(t)$, is indicated by $\widehat{T}(s)$ and obtained according to:

$$\widehat{T}(s) = \langle T(t) \rangle = \underbrace{\int_0^\infty \exp(-st)\, T(t)\, dt}_{\text{the Laplace transform operation}} \qquad (3\text{-}117)$$

where the notation $\langle T(t) \rangle$ denotes the Laplace transform operation. As an example, consider the Laplace transform of a constant, C:

$$T(t) = C \qquad (3\text{-}118)$$

Substituting Eq. (3-118) into Eq. (3-117) leads to:

$$\widehat{T}(s) = \langle C \rangle = \int_0^\infty \exp(-st)\, C\, dt \qquad (3\text{-}119)$$

Carrying out the integral:

$$\widehat{T}(s) = -\frac{C}{s} \left[\exp(-st) \right]_0^\infty \qquad (3\text{-}120)$$

and evaluating the limits:

$$\widehat{T}(s) = \frac{C}{s} \qquad (3\text{-}121)$$

Thus the Laplace transform of C is C/s.

There are several techniques that can be used to transform a function from the time to the s domain. The integration expressed by Eq. (3-117) can be carried out explicitly. More commonly, one of the extensive tables of Laplace transforms that have been published (e.g., Abramowitz and Stegun, (1964)) is used to obtain the transform. Recently, symbolic software packages such as Maple have become available that are capable of identifying most Laplace transforms automatically.

Table 3-3: Some common Laplace transforms.

Function in t	Function in s	Function in t	Function in s
C	$\dfrac{C}{s}$	$\text{erfc}\left(\dfrac{C}{2\sqrt{t}}\right)$	$\dfrac{\exp\left(-C\sqrt{s}\right)}{s}$
Ct	$\dfrac{C}{s^2}$	$\dfrac{2\sqrt{t}}{\sqrt{\pi}}\exp\left(-\dfrac{C^2}{4t}\right) - C\,\text{erfc}\left(\dfrac{C}{2\sqrt{t}}\right)$	$\dfrac{\exp\left(-C\sqrt{s}\right)}{s\sqrt{s}}$
$\exp\left(Ct\right)$	$\dfrac{1}{s-C}$	$\left(t+\dfrac{C^2}{2}\right)\text{erfc}\left(\dfrac{C}{2\sqrt{t}}\right)$ $-\dfrac{C\sqrt{t}}{\sqrt{\pi}}\exp\left(-\dfrac{C^2}{4t}\right)$	$\dfrac{\exp\left(-C\sqrt{s}\right)}{s^2}$
$\sin\left(Ct\right)$	$\dfrac{C}{s^2+C^2}$	$\dfrac{C_1}{2\sqrt{\pi t^3}}\exp\left(-\dfrac{C_1^2}{4t}-C_2\,t\right)$	$\exp(-C_1\sqrt{s+C_2})$
$\cos\left(Ct\right)$	$\dfrac{s}{s^2+C^2}$	$\dfrac{\exp\left(-C_1\,t\right)-\exp\left(-C_2\,t\right)}{C_2-C_1}$	$\dfrac{1}{(s+C_1)(s+C_2)}$
$\sinh\left(Ct\right)$	$\dfrac{C}{s^2-C^2}$	$t\exp\left(-C_1\,t\right)$	$\dfrac{1}{(s+C_1)^2}$
$\cosh\left(Ct\right)$	$\dfrac{s}{s^2-C^2}$	$\dfrac{\left[\begin{array}{l}(C_3-C_2)\exp\left(-C_1\,t\right)\\+(C_1-C_3)\exp\left(-C_2\,t\right)\\+(C_2-C_1)\exp\left(-C_3\,t\right)\end{array}\right]}{(C_1-C_2)(C_2-C_3)(C_3-C_1)}$	$\dfrac{1}{(s+C_1)(s+C_2)(s+C_3)}$
$\dfrac{C}{2\sqrt{\pi t^3}}\exp\left(-\dfrac{C^2}{4t}\right)$	$\exp\left(-C\sqrt{s}\right)$	$\dfrac{\exp\left(-C_2\,t\right)-\exp\left(-C_1\,t\right)\left[1-(C_2-C_1)t\right]}{(C_2-C_1)^2}$	$\dfrac{1}{(s+C_1)^2(s+C_2)}$
$\dfrac{1}{\sqrt{\pi t}}\exp\left(-\dfrac{C^2}{4t}\right)$	$\dfrac{\exp\left(-C\sqrt{s}\right)}{\sqrt{s}}$	$\dfrac{t^2\exp\left(-Ct\right)}{2}$	$\dfrac{1}{(s+C)^3}$

Function in t	Function in s
$\dfrac{1}{\sqrt{\pi t}}\exp\left(-\dfrac{C^2}{4t}\right) - a\exp(a\,C)\exp(a^2\,t)\,\text{erfc}\left(a\sqrt{t}+\dfrac{C}{2\sqrt{t}}\right)$	$\dfrac{\exp(-C\sqrt{s})}{a+\sqrt{s}}$
$\text{erfc}\left(\dfrac{C}{2\sqrt{t}}\right) - \exp\left(a\,C\right)\exp(a^2\,t)\,\text{erfc}\left(a\sqrt{t}+\dfrac{C}{2\sqrt{t}}\right)$	$\dfrac{a\exp(-C\sqrt{s})}{s(a+\sqrt{s})}$
$\exp(a\,C)\exp(a^2\,t)\,\text{erfc}\left(a\sqrt{t}+\dfrac{C}{2\sqrt{t}}\right)$	$\dfrac{\exp(-C\sqrt{s})}{\sqrt{s}(a+\sqrt{s})}$

Laplace Transformations with Tables
Tables that provide Laplace transforms can be found in many mathematical references, a few common Laplace transforms are summarized in Table 3-3.

Laplace Transformations with Maple
The Laplace and inverse Laplace transforms can be obtained using Maple. To access the integral transform library in Maple, it is necessary to activate the inttrans package; this is accomplished using the with command:

```
> restart:
> with(inttrans):
```

The Laplace transform of an arbitrary function can be obtained using the laplace command. The laplace command requires three arguments; the first is the expression to be transformed, the second is the variable to transform from (typically t), and the third is the variable to transform to (typically s). To obtain the Laplace transform of a constant, as we did in Eqs. (3-118) through (3-121)

> laplace(C,t,s);

$$\frac{C}{s}$$

More complex functions can also be transformed:

> laplace(sin(C*t),t,s);

$$\frac{C}{s^2 + C^2}$$

> laplace(sin(C*t)*exp(-t/tau),t,s);

$$\frac{C}{\left(s + \dfrac{1}{\tau}\right)^2 + C^2}$$

which indicates that:

$$\langle \sin (C t) \rangle = \frac{C}{s^2 + C^2} \tag{3-122}$$

and

$$\left\langle \sin (C t) \exp \left(-\frac{t}{\tau}\right) \right\rangle = \frac{C}{\left(s + \dfrac{1}{\tau}\right)^2 + C^2} \tag{3-123}$$

3.4.3 The Inverse Laplace Transform

Once the solution to a problem has been obtained in terms of s, it is necessary to obtain the inverse Laplace transform of the solution in order to express the result in terms of t. There are several techniques for obtaining the inverse Laplace transform. The most general technique is to mathematically invert the Laplace transform, Eq. (3-117), but this operation requires integration in the complex plane. The computational effort required to obtain the inverse transform in this way defeats the purpose of using Laplace transforms to simplify the solution of heat transfer problems. However, many inverse transforms have been determined and tabulated. If the solution in the s domain appears in a table, such as Table 3-3, then it is a simple matter to obtain the inverse transform from the table (e.g., the inverse transform of C/s is C). More often, the particular transform that is needed will not be found in exactly the right form in a table and it will be necessary to break the solution into simpler pieces using the method of partial fractions.

The typical form of the solution in the s domain is a complicated fraction in s; for example:

$$\hat{T}(s) = \frac{s^2 - 3s + 4}{(s + 1)(s - 1)(s + 2)} \tag{3-124}$$

The inverse Laplace transform for Eq. (3-124) cannot be found in Table 3-3; however, the transform for simpler fractions are included in the table. The method of partial fractions is therefore required to effectively use tables of Laplace transforms.

Many common inverse Laplace transforms can be obtained automatically using Maple. Several additional Laplace transforms that are not normally available with Maple have been added to a file that can be downloaded from the text website (www.cambridge.org/nellisandklein), as discussed in Section 3.4.6.

Inverse Laplace Transform with Tables and the Method of Partial Fractions

Equation (3-124) can be reduced using the method of partial fractions, as discussed in Myers (1998). The method of partial fractions requires that the order of the numerator is less than that of the denominator.

Distinct Factors. For cases like Eq. (3-124) where there are distinct factors in the denominator, the fraction can be expressed as the sum of individual, lower order fractions each of which has one of the distinct factors in the denominator. For example, Eq. (3-124) can be written as:

$$\frac{s^2 - 3s + 4}{(s+1)(s-1)(s+2)} = \frac{C_1}{(s+1)} + \frac{C_2}{(s-1)} + \frac{C_3}{(s+2)} \tag{3-125}$$

where C_1, C_2, and C_3 are unknown constants. Both sides of Eq. (3-125) are multiplied by the denominator $(s+1)(s-1)(s+2)$:

$$s^2 - 3s + 4 = C_1(s-1)(s+2) + C_2(s+1)(s+2) + C_3(s+1)(s-1) \tag{3-126}$$

Eq. (3-126) must be valid for any value of s since s is the independent variable. If $s = -1$ is substituted into Eq. (3-126) then the terms involving C_2 and C_3 become zero (as they both involve $s + 1$) and an equation is obtained that involves only C_1:

$$(-1)^2 - 3(-1) + 4 = C_1((-1)-1)((-1)+2) \tag{3-127}$$

Equation (3-127) can easily be solved for C_1:

$$1 + 3 + 4 = C_1(-2)(1) \tag{3-128}$$

or $C_1 = -4$. A similar process can be used to obtain C_2 (i.e., eliminate C_1 and C_3 by substituting $s = 1$ into Eq. (3-126) and solve for C_2) and C_3. The result is $C_2 = 1/3$ (or 0.333) and $C_3 = 14/3$ (or 4.667). Therefore, Eq. (3-124) can be written as:

$$\hat{T}(s) = \frac{-4}{(s+1)} + \frac{0.333}{(s-1)} + \frac{4.667}{(s+2)} \tag{3-129}$$

Expressed in this form, the inverse transform of Eq. (3-129) can be obtained by inspection from Table 3-3:

$$T(t) = -4\exp(-t) + 0.333\exp(t) + 4.667\exp(-2t) \tag{3-130}$$

An alternative and more general method for obtaining C_1, C_2, and C_3 is to carry out the multiplications in Eq. (3-126):

$$s^2 - 3s + 4 = C_1(s^2 + s - 2) + C_2(s^2 + 3s + 2) + C_3(s^2 - 1) \tag{3-131}$$

and then require that the coefficients multiplying like powers of s on either side of Eq. (3-131) must be equal in order to obtain three equations (one for each power of s) in three unknowns (C_1, C_2, and C_3):

$$1 = C_1 + C_2 + C_3 \tag{3-132}$$

$$-3 = C_1 + 3\,C_2 \tag{3-133}$$

$$4 = -2\,C_1 + 2\,C_2 - C_3 \tag{3-134}$$

The solution of these simultaneous equations in EES:

```
1=C_1+C_2+C_3
-3=C_1+3*C_2
4=-2*C_1+2*C_2-C_3
```

yields $C_1 = -4$, $C_2 = 0.333$, and $C_3 = 4.667$.

Note that Maple can be used to quickly convert an expression to its partial fraction form using the convert command with the parfrac identifier. To convert Eq. (3-124) into its partial fraction form, enter the expression and then convert it using the convert command; note that the first argument is the expression to be converted, the second identifies the type of conversion, and the third identifies the name of the independent variable in the expression.

```
> restart;
> TS:=(s^2-3*s+4)/((s+1)*(s-1)*(s+2)); #expression to be converted
```
$$TS := \frac{s^2 - 3s + 4}{(s+1)\,(s-1)\,(s+2)}$$
```
> TSpf:=convert(TS,parfrac,s); #expression converted
```
$$TSpf := \frac{1}{3\,(s-1)} - \frac{4}{s+1} + \frac{14}{3\,(s+2)}$$

Notice that the coefficients identified by Maple (-4, $1/3$, and $14/3$) agree with the coefficients identified manually.

Repeated Factors. In the instance that the terms in the denominator of the expression to be converted are repeated, it is necessary to include each power of the term. For example, if the expression in the s domain is:

$$\widehat{T}(s) = \frac{s^2 - 3s + 4}{(s+1)^2\,(s-1)} \tag{3-135}$$

then the partial fraction form must include fractions with both $(s+1)$ and $(s+1)^2$:

$$\frac{s^2 - 3s + 4}{(s+1)^2\,(s-1)} = \frac{C_1}{(s+1)} + \frac{C_2}{(s+1)^2} + \frac{C_3}{(s-1)} \tag{3-136}$$

Otherwise, the solution proceeds as before. Both sides of Eq. (3-136) are multiplied by $(s+1)^2\,(s-1)$:

$$s^2 - 3s + 4 = C_1\,(s+1)\,(s-1) + C_2\,(s-1) + C_3\,(s+1)^2 \tag{3-137}$$

or

$$s^2 - 3s + 4 = C_1(s^2 - 1) + C_2(s - 1) + C_3(s^2 + 2s + 1) \tag{3-138}$$

which leads to:

$$1 = C_1 + C_3 \tag{3-139}$$

$$-3 = C_2 + 2C_3 \tag{3-140}$$

$$4 = -C_1 - C_2 + C_3 \tag{3-141}$$

The solution to these three equations:

```
1=C_1+C_3
-3=C_2+2*C_3
4=C_1-C_2+C_3
```

leads to $C_1 = 0.5$, $C_2 = -4$, and $C_3 = 0.5$. Therefore:

$$\widetilde{T}(s) = \frac{0.5}{(s+1)} + \frac{-4}{(s+1)^2} + \frac{0.5}{(s-1)} \tag{3-142}$$

Maple can provide the same result using the **convert** command:

```
> restart;
> TS:=(s^2-3*s+4)/((s+1)^2*(s-1)); #expression to be converted
```

$$TS := \frac{s^2 - 3s + 4}{(s+1)^2 (s-1)}$$

```
> TSpf:=convert(TS,parfrac,s); #expression converted
```

$$TSpf := \frac{1}{2(s-1)} + \frac{1}{2(s+1)} - \frac{4}{(s+1)^2}$$

The inverse transform of Eq. (3-142) can be obtained using Table 3-3:

$$\widehat{T}(t) = 0.5 \exp(t) + 0.5 \exp(-t) - 4t \exp(-t) \tag{3-143}$$

Polynomial Factors. In the instance that the terms in the denominator of the expression include a polynomial factor, it is necessary to include a polynomial numerator with the order of the polynomial reduced by 1. For example, if the expression in the s domain is:

$$\widehat{T}(s) = \frac{s^2 - 3s + 4}{(s^2 + 2)(s - 1)} \tag{3-144}$$

then the partial fraction form must include fractions:

$$\frac{s^2 - 3s + 4}{(s^2 + 2)(s - 1)} = \frac{C_1 s + C_2}{(s^2 + 2)} + \frac{C_3}{(s - 1)} \tag{3-145}$$

Otherwise, the solution proceeds as before. Both sides of Eq. (3-145) are multiplied by $(s^2 + 2)(s - 1)$:

$$s^2 - 3s + 4 = (C_1 s + C_2)(s - 1) + C_3(s^2 + 2) \tag{3-146}$$

or

$$s^2 - 3s + 4 = C_1 s^2 - C_1 s + C_2 s - C_2 + C_3 s^2 + 2 C_3 \tag{3-147}$$

which leads to:

$$1 = C_1 + C_3 \tag{3-148}$$

$$-3 = -C_1 + C_2 \tag{3-149}$$

$$4 = -C_2 + 2 C_3 \tag{3-150}$$

The solution to these three equations:

```
1=C_1+C_3
-3= -C_1+C_2
4= -C_2+2*C_3
```

leads to $C_1 = 0.333$, $C_2 = -2.667$, and $C_3 = 0.667$. Therefore:

$$\hat{T}(s) = \frac{0.333\, s - 2.667}{(s^2 + 2)} + \frac{0.667}{(s - 1)} \tag{3-151}$$

or

$$\hat{T}(s) = \frac{0.333\, s}{(s^2 + 2)} - \frac{2.667}{(s^2 + 2)} + \frac{0.667}{(s - 1)} \tag{3-152}$$

Maple can provide the same result:

```
> restart;
> TS:=(s^2-3*s+4)/((s^2+2)*(s-1)); #expression to be converted
```
$$TS := \frac{s^2 - 3s + 4}{(s^2 + 2)(s - 1)}$$
```
> TSpf:=convert(TS,parfrac,s); #expression converted
```
$$TSpf := \frac{s - 8}{3(s^2 + 2)} + \frac{2}{3(s - 1)}$$

The inverse transform of Eq. (3-152) can be obtained using Table 3-3:

$$\hat{T}(t) = 0.333 \cos(\sqrt{2}\,t) - \tfrac{2.667}{\sqrt{2}} \sin(\sqrt{2}\,t) + 0.667 \exp(t) \tag{3-153}$$

Inverse Laplace Transformation with Maple

Maple can also be used to transform from a function in the s domain to a function in the time domain using the invlaplace command. The invlaplace command has the same basic calling protocol as the laplace command. The first argument is the expression to be inverse transformed, the second argument is the variable to transform from (typically s), and the third argument is the variable to transform to (typically t). To obtain the inverse Laplace transform of C/s:

```
> restart:
> with(inttrans):
> invlaplace(C/s,s,t);
```

$$C$$

The inverse Laplace transforms that are obtained in Section 3.4.3 could also be obtained using the invlaplace command in Maple. For example, the inverse Laplace transform of Eq. (3-124) is:

```
> restart:
> with(inttrans):
> TS:=(s^2-3*s+4)/((s+1)*(s-1)*(s+2));
```

$$TS := \frac{s^2 - 3s + 4}{(s+1)(s-1)(s+2)}$$

```
> Tt:=invlaplace(TS,s,t);
```

$$Tt := \frac{11}{3}\cosh(t) + \frac{13}{3}\sinh(t) + \frac{14}{3}e^{(-2t)}$$

The solution identified by Maple looks different from Eq. (3-130), the solution identified using partial fractions and Table 3-3. However, if the hyperbolic cosine and sine terms are written in terms of exponentials then the result is the same. (This can be accomplished using the convert function with the exp identifier.)

```
> Tt:=convert(Tt,exp);
```

$$Tt := \frac{1}{3}e^t - \frac{4}{e^t} + \frac{14}{3}e^{(-2t)}$$

The inverse Laplace transforms of Eqs. (3-135) and (3-144) can also be identified using Maple:

```
> restart:
> with(inttrans):
> TS:=(s^2-3*s+4)/((s+1)^2*(s-1));
```

$$TS := \frac{s^2 - 3s + 4}{(s+1)^2(s-1)}$$

```
> Tt:=invlaplace(TS,s,t);
```

$$Tt := 4t\sinh(t) - \cosh(t)(-1+4t)$$

```
> Tt:=simplify(convert(Tt,exp));
```

$$Tt := 4te^{(-t)} + \frac{1}{2}e^t + \frac{1}{2}e^{(-t)}$$

```
> TS:=(s^2-3*s+4)/((s^2+2)*(s-1));
```
$$TS := \frac{s^2 - 3s + 4}{(s^2 + 2)(s - 1)}$$

```
> Tt:=invlaplace(TS,s,t);
```
$$Tt := \frac{1}{3} \cos(\sqrt{2}\,t) - \frac{4}{3}\sqrt{2}\,\sin(\sqrt{2}\,t) + \frac{2}{3}\,e^t$$

The library of inverse Laplace transforms that is available in Maple is limited and several transforms that are useful for solving heat transfer problems are not available. Additional inverse Laplace transforms are available from the website associated with this book (www.cambridge.org/nellisandklein). The file that includes these inverse Laplace transforms is titled **Inverse Laplace Transforms**. The top of this Maple file includes a section of code that adds a series of entries to the existing invlaplace table in Maple; these entries symbolically define some additional transforms that are commonly encountered in heat transfer problems and therefore augments Maple's functionality.

3.4.4 Properties of the Laplace Transformation

The Laplace transform is linear. Therefore, the transform of the sum of two functions is the sum of their individual transforms:

$$\langle T_1(t) + T_2(t) \rangle = \hat{T}_1(s) + \hat{T}_2(s) \tag{3-154}$$

and the transform of the product of a constant and a function is the product of the constant and the transform:

$$\langle C\,T(t) \rangle = C\,\hat{T}(s) \tag{3-155}$$

The transform of a time derivative of a function $T(t)$ is the product of the transform of the function and s less the initial condition in the time domain:

$$\left\langle \frac{dT(t)}{dt} \right\rangle = s\,\hat{T}(s) - T_{t=0} \tag{3-156}$$

This property of the Laplace transform is its primary feature and the reason it is useful for solving differential equations.

Equation (3-156) can be proven. Apply the Laplace transform to the time derivative of T:

$$\left\langle \frac{dT(t)}{dt} \right\rangle = \int_0^\infty \exp(-s\,t)\,\frac{dT}{dt}\,dt \tag{3-157}$$

Equation (3-157) can be simplified through integration by parts. We will encounter integration by parts again elsewhere in this book and therefore it is worth spending some time understanding the process. Integration by parts is based on the chain rule for differentiation; the differential of the product of two functions (u and v) is:

$$d(u\,v) = u\,dv + v\,du \tag{3-158}$$

which can be integrated:

$$\int_{(u\,v)_1}^{(u\,v)_2} d(u\,v) = \int_{v_1}^{v_2} u\,dv + \int_{u_1}^{u_2} v\,du \tag{3-159}$$

The left side of Eq. (3-159) is an exact differential and therefore:

$$(u\,v)_2 - (u\,v)_1 = \int_{v_1}^{v_2} u\,dv + \int_{u_1}^{u_2} v\,du \tag{3-160}$$

or, rearranging:

$$\int_{v_1}^{v_2} u\,dv = (u\,v)_2 - (u\,v)_1 - \int_{u_1}^{u_2} v\,du \tag{3-161}$$

Successful use of integration by parts requires that the functions u and v are identified and substituted into Eq. (3-161) in order to simplify the expression of interest, in this case Eq. (3-157):

$$\left\langle \frac{dT}{dt} \right\rangle = \int_0^{\infty} \underbrace{\exp(-s\,t)}_{u}\ \underbrace{\frac{dT}{}}_{dv} \tag{3-162}$$

By inspection of Eqs. (3-161) and (3-162), we will define:

$$u = \exp(-s\,t) \tag{3-163}$$

$$dv = dT \tag{3-164}$$

therefore

$$du = -s\,\exp(-s\,t)dt \tag{3-165}$$

$$v = T \tag{3-166}$$

Substituting Eqs. (3-163) through (3-166) into Eq. (3-161) leads to:

$$\left\langle \frac{dT}{dt} \right\rangle = \underbrace{[\exp(-s\,t)\,T]_{t=\infty}}_{(u\,v)_2} - \underbrace{[\exp(-s\,t)\,T]_{t=0}}_{(u\,v)_1} - \int_{t=0}^{t=\infty} \underbrace{T}_{v}\ \underbrace{(-s\,\exp(-s\,t)dt)}_{du} \tag{3-167}$$

or

$$\left\langle \frac{dT}{dt} \right\rangle = -T_{t=0} + s \underbrace{\int_0^{\infty} T\,\exp(-s\,t)dt}_{\widehat{T}\,(s)} \tag{3-168}$$

The final term in Eq. (3-168) is the product of s and the Laplace transform of T:

$$\left\langle \frac{dT}{dt} \right\rangle = s\,\widehat{T}\,(s) - T_{t=0} \tag{3-169}$$

Table 3-4: Useful properties of the Laplace transforms.

$\langle T_1(t) + T_2(t) \rangle = \widehat{T}_1(s) + \widehat{T}_2(s)$
$\langle C\,T(t) \rangle = C\,\widehat{T}(s)$
$\left\langle \dfrac{dT(t)}{dt} \right\rangle = s\,\widehat{T}(s) - T_{t=0}$
$\left\langle \dfrac{\partial T(x,t)}{\partial t} \right\rangle = s\,\widehat{T}(x,s) - T_{x,t=0}$
$\left\langle \dfrac{\partial^n T(x,t)}{\partial x^n} \right\rangle = \dfrac{\partial^n \widehat{T}(x,s)}{\partial x^n}$

The transform of a derivative is not, itself, a derivative. This is a very useful property, as it turns a differential equation in t into an algebraic equation in s.

It is possible to transform a partial derivative of temperature with respect to time (for example, when temperature depends on both position and time, $T(x,t)$):

$$\left\langle \frac{\partial T(x,t)}{\partial t} \right\rangle = \int_0^\infty \exp(-s\,t)\,\frac{\partial T}{\partial t}\,dt \qquad (3\text{-}170)$$

Carrying out the same process of integration by parts leads to a similar conclusion:

$$\left\langle \frac{\partial T(x,t)}{\partial t} \right\rangle = s\,\widehat{T}(x,s) - T_{x,t=0} \qquad (3\text{-}171)$$

The Laplace transform of a partial derivative of temperature with respect to position is the partial derivative of the transformed function with respect to position:

$$\left\langle \frac{\partial T(x,t)}{\partial x} \right\rangle = \frac{\partial \widehat{T}(x,s)}{\partial x} \qquad (3\text{-}172)$$

This is true for all higher derivatives as well:

$$\left\langle \frac{\partial^n T(x,t)}{\partial x^n} \right\rangle = \frac{\partial^n \widehat{T}(x,s)}{\partial x^n} \qquad (3\text{-}173)$$

These properties of the Laplace transform are summarized in Table 3-4.

3.4.5 Solution to Lumped Capacitance Problems

The Laplace transform can be used to obtain analytical solutions to the 0-D transient problems that are considered in Section 3.1. The process will be illustrated using the problem discussed in EXAMPLE 3.1-2. A temperature sensor is exposed to an oscillating temperature environment (T_∞):

$$T_\infty = \overline{T}_\infty + \Delta T_\infty \sin(2\pi f\,t) \qquad (3\text{-}174)$$

where $\overline{T}_\infty = 320°\mathrm{C}$ is the average temperature of the fluid and $\Delta T_\infty = 50\mathrm{K}$ and $f = 0.5\ \mathrm{Hz}$ are the amplitude and frequency of the temperature oscillation. The governing differential equation for the problem, determined in EXAMPLE 3.1-2, is:

$$\frac{dT}{dt} + \frac{T}{\tau_{lumped}} = \frac{\overline{T}_\infty}{\tau_{lumped}} + \frac{\Delta T_\infty \sin\left(2\pi f\, t\right)}{\tau_{lumped}} \qquad (3\text{-}175)$$

where $\tau_{lumped} = 0.72$ s is the lumped capacitance time constant of the sensor. The initial condition is:

$$T_{t=0} = T_{ini} \qquad (3\text{-}176)$$

where $T_{ini} = 260°C$ is the initial temperature of the sensor.

The known information is entered in EES:

```
$UnitSystem SI MASS RAD PA K J
$TABSTOPS 0.2 0.4 0.6 0.8 3.5 in

"Inputs"
T_infinity_bar=converttemp(C,K,320 [C])      "average environmental temperature"
T_ini=converttemp(C,K,260 [C])               "initial temperature"
DELTAT_infinity=50 [K]                        "amplitude of oscillation"
f=0.5 [Hz]                                    "frequency of oscillation"
tau=0.72 [s]                                  "time constant of temperature sensor"
```

In order to solve this problem using the Laplace transform approach it is necessary to transform the governing differential equation from the t domain to the s domain. The first term in the governing equation, Eq. (3-175), is transformed using Eq. (3-169):

$$\left\langle \frac{dT\,(t)}{dt} \right\rangle = s\,\widehat{T}\,(s) - T_{t=0} \qquad (3\text{-}177)$$

Substituting Eq. (3-176) into Eq. (3-177) leads to:

$$\left\langle \frac{dT}{dt} \right\rangle = s\,\widehat{T}\,(s) - T_{ini} \qquad (3\text{-}178)$$

The second term in Eq. (3-175) becomes:

$$\left\langle \frac{T\,(t)}{\tau_{lumped}} \right\rangle = \frac{\widehat{T}\,(s)}{\tau_{lumped}} \qquad (3\text{-}179)$$

which follows from the fact that the Laplace transform is linear, see Eq. (3-155). The final two terms in Eq. (3-175) are obtained from Table 3-3:

$$\left\langle \frac{\overline{T}_\infty}{\tau_{lumped}} \right\rangle = \frac{\overline{T}_\infty}{s\,\tau_{lumped}} \qquad (3\text{-}180)$$

and

$$\left\langle \frac{\Delta T_\infty}{\tau_{lumped}} \sin\left(2\pi f\, t\right) \right\rangle = \frac{\Delta T_\infty}{\tau_{lumped}} \left[\frac{2\pi f}{s^2 + (2\pi f)^2} \right] \qquad (3\text{-}181)$$

Substituting Eqs. (3-178) through (3-181) into Eq. (3-175) leads to the transformed governing equation:

$$s\,\widehat{T}\,(s) - T_{ini} + \frac{\widehat{T}\,(s)}{\tau_{lumped}} = \frac{\overline{T}_\infty}{s\,\tau_{lumped}} + \frac{\Delta T_\infty}{\tau_{lumped}}\left[\frac{2\,\pi f}{s^2 + (2\,\pi f)^2}\right]\qquad(3\text{-}182)$$

Notice that the differential equation in t, Eq. (3-175), has been transformed to an algebraic equation in s. The same result can be obtained using Maple. The governing differential equation is entered:

```
> restart:with(inttrans):
> ODEt:=diff(T(time),time)+T(time)/tau=T_infinity_bar/tau+DELTAT_infinity*sin (2*pi*f*time)/tau;
```

$$ODEt := \left(\frac{d}{dtime}\,T\,(time)\right)$$
$$+\ \frac{T\,(time)}{\tau} = \frac{T\_infinity\_bar}{\tau} + \frac{DELTAT\_infinity\,\sin\,(2\,\pi f\,time)}{\tau}$$

and the **laplace** command is used to obtain the Laplace transform of the entire differential equation:

```
> AEs:=laplace(ODEt,time,s);
```

$$AEs := s\,\text{laplace}(T(time),\ time,\ s) - T(0) + \frac{\text{laplace}(T(time),\ time,\ s)}{\tau}$$
$$= \frac{T\_infinity\_bar}{\tau s} + \frac{2\,DELTAT\_infinity\,\pi f}{\tau(s^2 + 4\pi^2 f^2)}$$

where **laplace(T(time),time,s)** indicates the Laplace transform of the function T. The transformed equation can be made more concise by using the **subs** command to substitute a single variable, T(s), for the **laplace()** result. Recall that the first argument of the **subs** command is the substitution you want to make while the second argument is the expression that you want to make the substitution into:

```
> AEs:=subs(laplace(T(time),time,s)=T(s),AEs);
```

$$AEs := s\,T\,(s) - T\,(0) + \frac{T\,(s)}{\tau} = \frac{T\_infinity\_bar}{\tau s} + \frac{2\,DELTAT\_infinity\,\pi f}{\tau\,(s^2 + 4\pi^2 f^2)}$$

The solution includes the initial condition, T(0), which can be eliminated using the **subs** command again:

```
> AEs:=subs(T(0)=T_ini,AEs);
```

$$AEs := s\,T\,(s) - T\_ini + \frac{T\,(s)}{\tau} = \frac{T\_infinity\_bar}{\tau s} + \frac{2\,DELTAT\_infinity\,\pi f}{\tau\,(s^2 + 4\pi^2 f^2)}$$

This result is identical to the result that was obtained manually, Eq. (3-182); the algebraic equation in the s domain, Eq. (3-182), can be solved for $\widehat{T}(s)$:

$$\widehat{T}(s) = \frac{T_{ini}}{\left(s + \dfrac{1}{\tau_{lumped}}\right)} \tag{3-183}$$

$$+ \frac{\overline{T}_\infty}{\tau_{lumped}\, s \left(s + \dfrac{1}{\tau_{lumped}}\right)} + \frac{2\pi f\,\Delta T_\infty}{\tau_{lumped}} \left[\frac{1}{(s^2 + (2\pi f)^2)\left(s + \dfrac{1}{\tau_{lumped}}\right)} \right]$$

The inverse Laplace transform of Eq. (3-183) does not appear in Table 3-3 and therefore it is necessary to use the method of partial fractions to reduce Eq. (3-183) to simpler fractions that do appear in Table 3-3. The second term in Eq. (3-183) can be written as:

$$\frac{\overline{T}_\infty}{\tau_{lumped}\, s \left(s + \dfrac{1}{\tau_{lumped}}\right)} = \frac{C_1}{s} + \frac{C_2}{\left(s + \dfrac{1}{\tau_{lumped}}\right)} \tag{3-184}$$

Multiplying through by $s\left(s + \dfrac{1}{\tau_{lumped}}\right)$ leads to:

$$\frac{\overline{T}_\infty}{\tau_{lumped}} = C_1 s + \frac{C_1}{\tau_{lumped}} + C_2 s \tag{3-185}$$

which leads to:

$$0 = C_1 + C_2 \tag{3-186}$$

and

$$\frac{\overline{T}_\infty}{\tau_{lumped}} = \frac{C_1}{\tau_{lumped}} \tag{3-187}$$

Solving Eq. (3-187) leads to:

$$C_1 = \overline{T}_\infty \tag{3-188}$$

Substituting Eq. (3-188) into Eq. (3-186) leads to:

$$C_2 = -\overline{T}_\infty \tag{3-189}$$

"coefficients from partial fraction expansion"
C_1=T_infinity_bar
C_2=-T_infinity_bar

The third term in Eq. (3-183) has a second order polynomial term (s^2) in the denominator and therefore it can be expressed as:

$$\frac{2\pi f\,\Delta T_\infty}{\tau_{lumped}} \left[\frac{1}{(s^2 + (2\pi f)^2)\left(s + \dfrac{1}{\tau_{lumped}}\right)} \right] = \frac{C_3 s + C_4}{(s^2 + (2\pi f)^2)} + \frac{C_5}{\left(s + \dfrac{1}{\tau_{lumped}}\right)}$$

$$\tag{3-190}$$

Multiplying Eq. (3-190) through by $(s^2 + (2\pi f)^2)(s + \frac{1}{\tau})$ leads to:

$$\frac{2\pi f\,\Delta T_\infty}{\tau_{lumped}} = (C_3 s + C_4)\left(s + \frac{1}{\tau_{lumped}}\right) + C_5(s^2 + (2\pi f)^2) \tag{3-191}$$

or

$$\frac{2\pi f\,\Delta T_\infty}{\tau_{lumped}} = C_3 s^2 + \frac{C_3 s}{\tau_{lumped}} + C_4 s + \frac{C_4}{\tau_{lumped}} + C_5 s^2 + C_5(2\pi f)^2 \tag{3-192}$$

which leads to:

$$0 = C_3 + C_5 \tag{3-193}$$

$$0 = \frac{C_3}{\tau_{lumped}} + C_4 \tag{3-194}$$

$$\frac{2\pi f\,\Delta T_\infty}{\tau_{lumped}} = \frac{C_4}{\tau_{lumped}} + C_5(2\pi f)^2 \tag{3-195}$$

Equations (3-193) through (3-195) are 3 equations in the 3 unknowns C_1, C_2, and C_3:

```
0=C_3+C_5
0=C_3/tau+C_4
2*pi*f*DELTAT_infinity/tau=C_4/tau+C_5*(2*pi*f)^2
```

With the coefficients for the partial fractions now identified, Eq. (3-183) can be expressed as:

$$\hat{T}(s) = \frac{T_{ini}}{\left(s + \dfrac{1}{\tau_{lumped}}\right)} + \frac{C_1}{s} + \frac{C_2}{\left(s + \dfrac{1}{\tau_{lumped}}\right)}$$

$$+ \frac{C_3 s}{(s^2 + (2\pi f)^2)} + \frac{C_4}{(s^2 + (2\pi f)^2)} + \frac{C_5}{\left(s + \dfrac{1}{\tau_{lumped}}\right)} \tag{3-196}$$

The inverse Laplace transform of each of the terms in Eq. (3-196) is obtained using Table 3-3.

$$T(t) = T_{ini}\exp\left(-\frac{t}{\tau_{lumped}}\right) + C_1 + C_2\exp\left(-\frac{t}{\tau_{lumped}}\right) + C_3\cos(2\pi f t)$$

$$+ \frac{C_4}{2\pi f}\sin(2\pi f t) + C_5\exp\left(-\frac{t}{\tau_{lumped}}\right) \tag{3-197}$$

The solution is programmed in EES:

```
T=T_ini*exp(-time/tau)+C_1+C_2*exp(-time/tau)+C_3*cos(2*pi*f*t+ &
    C_4*sin(2*pi*f*time)/(2*pi*f)+C_5*exp(-time/tau)        "sensor temp., obtained manually"
T_C=converttemp(K,C,T)                                      "in C"
T_infinity=T_infinity_bar+DELTAT_infinity*sin(2*pi*f*time)
                                                            "fluid temperature"
T_infinity_C=converttemp(K,C,T_infinity)                    "in C"
```

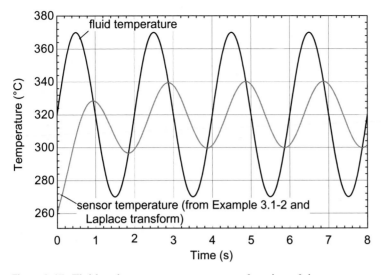

Figure 3-17: Fluid and sensor temperature as a function of time.

The solution obtained using the Laplace transform is identical to the solution obtained in EXAMPLE 3.1-2, as shown in Figure 3-17.

Maple could also be used to obtain the solution. The **solve** command can be used to carry out the algebra required to solve for $T(s)$:

```
> Ts:=solve(AEs,T(s));
```
$$Ts := (T\_ini\,\tau\,s^3 + 4\,T\_ini\,\tau\,s\,\pi^2 f^2 + T\_infinity\_bar\,s^2 + 4\,T\_infinity\_bar\,\pi^2 f^2$$
$$+ 2\,DELTAT\_infinity\,\pi\,f\,s)/(s(s^3\,\tau + 4\,s\,\tau\,\pi^2 f^2 + s^2 + 4\,\pi^2 f^2))$$

And the **invlaplace** command can be used to carry out the inverse Laplace transform:

```
> Tt:=invlaplace(Ts,s,time);
```
$$Tt := e^{\left(-\frac{time}{\tau}\right)} T\_ini + DELTAT\_infinity$$
$$\left(-\sinh(2\sqrt{-\pi^2 f^2}\,time)\sqrt{-\pi^2 f^2} + 2\,\tau\pi^2 f^2 \left(e^{\left(-\frac{time}{\tau}\right)} - \cosh(2\sqrt{-\pi^2 f^2}\,time)\right)\right) \Big/$$
$$(\pi f\,(1 + 4\pi^2 f^2\,\tau^2)) + T\_infinity\_bar\left(1 - e^{\left(-\frac{time}{\tau}\right)}\right)$$

Notice that the inverse Laplace transform identified by Maple includes the hyperbolic sine and cosine (rather than the sine and cosine); however, the argument of the sinh and cosh functions are complex (i.e., they involve $\sqrt{-1}$) and therefore these functions become sin and cos. It is possible to specify that the frequency, f, is positive using the **assume** command:

```
> assume(f,positive);
```

With this stipulation, Maple will correctly identify the solution in terms of sine and cosine:

```
> Tt:=invlaplace(Ts,s,time);
```

$$Tt := e^{\left(-\frac{time}{\tau}\right)} T\_ini$$
$$+ \frac{\left(\sin(2f \sim \pi\, time) + 2\tau\pi f \sim \left(-\cos(2f\tilde{\pi}\, time) + e^{\left(-\frac{time}{\tau}\right)}\right)\right) DELTAT\_infinity}{1 + 4\pi^2 f\sim^2 \tau^2}$$
$$+ T\_infinity\_bar\left(1 - e^{\left(-\frac{time}{\tau}\right)}\right)$$

The solution can be copied and pasted into EES:

```
T_maple=exp(-time/tau)*T_ini+(sin(2*f*pi*time)+2*tau*pi*f*(-cos(2*f*pi*time)+&
    exp(-time/tau)))*DELTAT_infinity/(1+4*pi^2*f^2*tau^2)+T_infinity_bar*(1-exp(-time/tau))
"solution from Maple"
T_maple_C=converttemp(K,C,T_maple)        "in C"
```

where it provides an identical solution to the one obtained manually (see Figure 3-17).

There is not usually a clear advantage associated with using the Laplace transform to solve the ordinary differential equations in time that result from lumped capacitance problems in heat transfer over the analytical techniques discussed in Section 3.1. However, the Laplace transform technique provides another useful tool and possibly a method for double-checking an important solution. The Laplace transform is very useful for certain types of 1-D transient problems, discussed in Section 3.4.6.

3.4.6 Solution to Semi-Infinite Body Problems

The Laplace transform can be used to obtain solutions to partial differential equations as well as to ordinary differential equations. The solution steps are basically the same; however, the Laplace transform directly incorporates the initial condition (see Eq. (3-169)) and therefore the initial condition does not have to be transformed. The Laplace transform of a partial differential equation in x and t will result in an ordinary differential equation in x in the s domain. Therefore, it is necessary to transform the boundary conditions involving x into the s domain so that the ordinary differential equation can be solved. The solution must be converted to a function of x and t using the inverse Laplace transform.

The process will be illustrated using the problem that was discussed in Section 3.3.3 in which a semi-infinite body that is initially at a uniform temperature T_{ini} is exposed to a step change in its surface temperature, from T_{ini} to T_s. The governing partial differential equation for this situation is:

$$\alpha \frac{\partial^2 T}{\partial x^2} = \frac{\partial T}{\partial t} \tag{3-198}$$

where α is the thermal diffusivity. The boundary conditions are provided by:

$$T_{x=0} = T_s \tag{3-199}$$

$$T_{t=0} = T_{ini} \tag{3-200}$$

$$T_{x\to\infty} = T_{ini} \tag{3-201}$$

The governing differential equation is transformed from the x, t domain to the x, s. The first term in Eq. (3-198) is transformed using Eq. (3-173):

$$\left\langle \alpha \frac{\partial^2 T\,(x,\,t)}{\partial x^2} \right\rangle = \alpha \frac{\partial^2 \,\widehat{T}\,(x,\,s)}{\partial x^2} \tag{3-202}$$

and the second term is transformed using Eq. (3-171):

$$\left\langle \frac{\partial T\,(x,\,t)}{\partial t} \right\rangle = s\,\widehat{T}\,(x,\,s) - T_{ini} \tag{3-203}$$

so that the transformed differential equation is:

$$\alpha \frac{\partial^2 \,\widehat{T}\,(x,\,s)}{\partial x^2} = s\,\widehat{T}\,(x,\,s) - T_{ini} \tag{3-204}$$

Maple can identify the same transformed governing differential equation using basically the same steps discussed Section 3.4.5:

> restart:with(inttrans):
> PDE:=alpha*diff(diff(T(x,time),x),x)=diff(T(x,time),time);

$$PDE := \alpha \left(\frac{\partial^2}{\partial x^2} T(x, time) \right) = \frac{\partial}{\partial time} T(x, time)$$

> ODE:=laplace(PDE,time,s);

$$ODE := \alpha \left(\frac{\partial^2}{\partial x^2} \text{laplace}(T(x, time),\ time,\ s) \right) = s\,\text{laplace}(T(x,\ time),\ time,\ s) - T(x,\ 0)$$

> ODE:=subs(T(x,0)=T_ini,ODE);

$$ODE := \alpha \left(\tfrac{\partial^2}{\partial x^2} \text{laplace}(T(x,\ time),\ time,\ s) \right) = s\,\text{laplace}(T(x,\ time),\ time,\ s) - T\_ini$$

> ODE:=subs(laplace(T(x,time),time,s)=Ts(x),ODE);

$$ODE := \alpha \left(\frac{d^2}{dx^2} Ts(x) \right) = s\,Ts(x) - T\_ini$$

Equation (3-204) does not involve any derivative with respect to s and therefore it is an ordinary differential equation in x; the partial differential in Eq. (3-204) can be changed to an ordinary differential:

$$\alpha \frac{d^2 \,\widehat{T}\,(s,\,x)}{dx^2} = s\,\widehat{T}\,(s,\,x) - T_{ini} \tag{3-205}$$

which can be rearranged:

$$\frac{d^2\widehat{T}}{dx^2} - \frac{s}{\alpha}\widehat{T} = -\frac{T_{ini}}{\alpha} \tag{3-206}$$

Equation (3-206) is a second order, non-homogeneous equation and therefore requires two boundary conditions; these are obtained from Eqs. (3-199) and (3-201), which must also be transformed to the s domain.

$$\widehat{T}_{x=0} = \frac{T_s}{s} \tag{3-207}$$

$$\widehat{T}_{x \to \infty} = \frac{T_{ini}}{s} \tag{3-208}$$

The second order differential equation is split into homogeneous and particular components:

$$\widehat{T} = \widehat{T}_h + \widehat{T}_p \tag{3-209}$$

Equation (3-209) substituted into Eq. (3-206):

$$\underbrace{\frac{d^2\widehat{T}_h}{dx^2} - \frac{s}{\alpha}\widehat{T}_h}_{\substack{=0 \text{ for homogeneous} \\ \text{differential equation}}} + \underbrace{\frac{d^2\widehat{T}_p}{dx^2} - \frac{s}{\alpha}\widehat{T}_p = -\frac{T_{ini}}{\alpha}}_{\text{particular differential equation}} \tag{3-210}$$

The homogeneous differential equation is:

$$\frac{d^2\widehat{T}_h}{dx^2} - \frac{s}{\alpha}\widehat{T}_h = 0 \tag{3-211}$$

which has the general solution:

$$\widehat{T}_h = C_1 \exp\left(\sqrt{\frac{s}{\alpha}}x\right) + C_2 \exp\left(-\sqrt{\frac{s}{\alpha}}x\right) \tag{3-212}$$

where C_1 and C_2 are undetermined constants. The solution to the particular differential equation

$$\frac{d^2\widehat{T}_p}{dx^2} - \frac{s}{\alpha}\widehat{T}_p = -\frac{T_{ini}}{\alpha} \tag{3-213}$$

is, by inspection:

$$\widehat{T}_p = \frac{T_{ini}}{s} \tag{3-214}$$

Substituting Eqs. (3-212) and (3-214) into Eq. (3-209) leads to:

$$\widehat{T} = C_1 \exp\left(\sqrt{\frac{s}{\alpha}}x\right) + C_2 \exp\left(-\sqrt{\frac{s}{\alpha}}x\right) + \frac{T_{ini}}{s} \tag{3-215}$$

Maple can be used to obtain the same solution:

```
> dsolve(ODE);
```

$$Ts(x) = e^{\left(\frac{\sqrt{s}x}{\sqrt{\alpha}}\right)} \_C2 + e^{\left(-\frac{\sqrt{s}x}{\sqrt{\alpha}}\right)} \_C1 + \frac{T\_ini}{s}$$

The constants C_1 and C_2 are obtained from the boundary conditions. The boundary condition at $x \to \infty$, Eq. (3-208), leads to:

$$\widehat{T}_{x\to\infty} = C_1 \exp\left(\sqrt{\frac{s}{\alpha}}\infty\right) + C_2 \exp\left(-\sqrt{\frac{s}{\alpha}}\infty\right) + \frac{T_{ini}}{s} = \frac{T_{ini}}{s} \tag{3-216}$$

or

$$C_1 \exp\left(\sqrt{\frac{s}{\alpha}}\infty\right) = 0 \tag{3-217}$$

which can only be true if $C_1 = 0$:

$$\hat{T} = C_2 \exp\left(-\sqrt{\frac{s}{\alpha}}x\right) + \frac{T_{ini}}{s} \tag{3-218}$$

The boundary condition at $x = 0$, Eq. (3-207), leads to:

$$\hat{T}_{x=0} = C_2 \exp\left(-\sqrt{\frac{s}{\alpha}}0\right) + \frac{T_{ini}}{s} = \frac{T_s}{s} \tag{3-219}$$

or

$$C_2 = \frac{(T_s - T_{ini})}{s} \tag{3-220}$$

Substituting Eq. (3-220) into Eq. (3-218) leads to the solution to the problem in the x, s domain:

$$\hat{T}(x, s) = \frac{(T_s - T_{ini})}{s} \exp\left(-\sqrt{\frac{s}{\alpha}}x\right) + \frac{T_{ini}}{s} \tag{3-221}$$

The solution in the x, t domain can be obtained using the inverse Laplace transforms contained in Table 3-3:

$$T(x, t) = (T_s - T_{ini}) \operatorname{erfc}\left(\frac{x}{2\sqrt{\alpha t}}\right) + T_{ini} \tag{3-222}$$

Recall that the complementary error function (erfc) is defined as:

$$\operatorname{erfc}(x) = 1 - \operatorname{erf}(x) \tag{3-223}$$

so Eq. (3-222) can be rewritten as:

$$T(x, t) = (T_s - T_{ini})\left[1 - \operatorname{erf}\left(\frac{x}{2\sqrt{\alpha t}}\right)\right] + T_{ini} \tag{3-224}$$

or

$$T(x, t) = T_s + (T_{ini} - T_s)\operatorname{erf}\left(\frac{x}{2\sqrt{\alpha t}}\right) \tag{3-225}$$

which is identical to the similarity solution obtained in Section 3.3.3.

Unfortunately, Maple is not able to automatically identify the inverse Laplace transform of Eq. (3-221):

```
> restart:with(inttrans):
> Ts:=(T_s-T_ini)*exp(-sqrt(s)*x/sqrt(alpha))/s+T_ini/s;
```

$$Ts := \frac{(T\_s - T\_ini)\,e^{\left(-\frac{\sqrt{s}x}{\alpha}\right)}}{s} + \frac{T\_ini}{s}$$

```
> invlaplace(Ts,s,time);
```

$$(T\_s - T\_ini)\,\operatorname{invlaplace}\left(\frac{e^{-\frac{\sqrt{s}x}{\sqrt{\alpha}}}}{s}, s, time\right) + T\_ini$$

The indicator invlaplace() is a placeholder that shows that the inverse Laplace transform was not found in Maple's library. It is possible to add entries to Maple's library, as discussed by Aziz (2006), in order to allow Maple to solve this and other semi-infinite body heat transfer problems. Several of the most common inverse transforms that are encountered for heat transfer problems have been added to the file Inverse Laplace Transforms, which can be downloaded from the website associated with the text (www.cambridge.org/nellisandklein). Rename the file after you download it and place the Maple code associated with this problem at the bottom of the file. Run the entire file to obtain:

```
> #Inverse Laplace transforms
> restart:with(inttrans):
> addtable(invlaplace,exp(-sqrt(s)*p::algebraic)/s,erfc(p/(2*sqrt(t))),s,t);
> addtable(invlaplace,exp(-sqrt(s)*p::algebraic)/s^(3/2),2*sqrt(t)*exp(-p^2/(4*t))/sqrt
  (pi)-p*erfc(p/(2*sqrt(t))),s,t);
> addtable(invlaplace,exp(-sqrt(s)*p::algebraic),p*exp(-p^2/(4*t))/(2*sqrt(pi*t^3)),s,t);
> addtable(invlaplace,exp(-sqrt(s)*p::algebraic)/sqrt(s),exp(-p^2/(4*t))/sqrt(pi*t^3),s,t);
> addtable(invlaplace,exp(-sqrt(s)*p::algebraic)/s^2,(t+p^2/2)*erfc(p/(2*sqrt(t)))-p*
  sqrt(t)*exp(-p^2/(4*t))/sqrt(pi),s,t);
> addtable(invlaplace,exp(-sqrt(s)*C1::algebraic)/(C2::algebraic+(C3::algebraic)*
  sqrt(s)),(exp(-C1^2/(4*t))/sqrt(Pi*t)-(C2/C3)*exp((C2/C3)*C1)*exp((C2/C3)^2*t)
  *erfc((C2/C3)*sqrt(t)+C1/(2*sqrt(t))))/C3,s,t);
> addtable(invlaplace,exp(-sqrt(s)*C1::algebraic)/(s*(C2::algebraic+(C3::algebraic)
  *sqrt(s))),erfc(1/2*C1/t^(1/2))-exp(C2/C3*C1)*exp(C2^2/C3^2*t)*erfc(C2/C3*t^(1/2)
  +1/2*C1/t^(1/2)),s,t;)
> addtable(invlaplace,exp(-sqrt(s)*C1::algebraic)/(sqrt(s)*(C2::algebraic+
  (C3::algebraic)*sqrt(s))),(exp((C2/C3)*C1)*exp((C2/C3)^2*t)*erfc((C2/C3)*sqrt(t)
  +C1/(2*sqrt(t))))/C3,s,t);
> #place your code below this line
> Ts:=(T_s-T_ini)*exp(-sqrt(s)*x/sqrt(alpha))/s+T_ini/s;
```

$$Ts := \frac{(T\_s - T\_ini)\, e^{\left(-\frac{\sqrt{s}\, x}{\sqrt{\alpha}}\right)}}{s} + \frac{T\_ini}{s}$$

```
> invlaplace(Ts,s,time);
```

$$T\_s + \operatorname{erf}\left(\frac{x}{2\sqrt{\alpha\, time}}\right)(-T\_s + T\_ini)$$

EXAMPLE 3.4-1: QUENCHING OF A SUPERCONDUCTOR

EXAMPLE 3.4-1: QUENCHING OF A SUPERCONDUCTOR

When a superconducting conductor "quenches," it goes from having no resistance (i.e., a superconducting state) to being resistive (i.e., a normal state). When the conductor is carrying current, the quenching process is accompanied by a step change in the rate of volumetric generation of thermal energy within the material (from nearly zero to a relatively high value, depending on the amount of current that is being carried). The result can be disastrous.

   This problem examines the temperature distribution in the conductor during the initial stages of a quenching process, shown schematically in Figure 1.

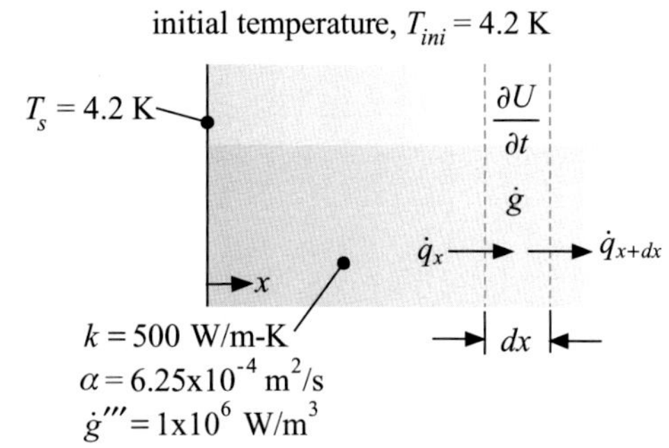

Figure 1: Quenching of a conductor.

The conductor is initially at a uniform temperature of $T_{ini} = 4.2$ K when the quench process occurs, resulting in a uniform rate of volumetric generation, $\dot{g}''' = 1 \times 10^6$ W/m^3. The conductor has conductivity $k = 500$ W/m-K and thermal diffusivity $\alpha = 0.000625$ m^2/s. The surface of the conductor is maintained at $T_s = 4.2$ K by boiling liquid helium.

a) **Develop an analytical model of the quench process that is valid for short times, while the conductor behaves as a semi-infinite body.**

The governing differential equation for the semi-infinite solid is derived by focusing on a differential control volume (see Figure 1). The energy balance suggested by Figure 1 is:

$$\dot{g} + \dot{q}_x = \dot{q}_{x+dx} + \frac{\partial U}{\partial t}$$

expanding the $x + dx$ term and simplifying leads to:

$$\dot{g} = \frac{\partial \dot{q}_x}{\partial x} dx + \frac{\partial U}{\partial t} \tag{1}$$

The conduction term is evaluated using Fourier's law:

$$\dot{q}_x = -k\, A_c \frac{\partial T}{\partial x} \tag{2}$$

where A_c is the cross-sectional area of the conductor perpendicular to the x-direction. The time rate of change of the internal energy of the material in the control volume is:

$$\frac{\partial U}{\partial t} = \rho\, c\, A_c\, dx \frac{\partial T}{\partial t} \tag{3}$$

The rate of generation of thermal energy in the control volume is:

$$\dot{g} = A_c\, dx\, \dot{g}''' \tag{4}$$

EXAMPLE 3.4-1: QUENCHING OF A SUPERCONDUCTOR

Substituting Eqs. (2) through (4) into Eq. (1) leads to:

$$A_c\,dx\,\dot{g}''' = \frac{\partial}{\partial x}\left[-k\,A_c\,\frac{\partial T}{\partial x}\right]dx + \rho\,c\,A_c\,dx\,\frac{\partial T}{\partial t}$$

or

$$\frac{\partial^2 T}{\partial x^2} - \frac{1}{\alpha}\frac{\partial T}{\partial t} = -\frac{\dot{g}'''}{k} \tag{5}$$

The boundary conditions for the problem include the initial condition:

$$T_{t=0} = T_{ini} \tag{6}$$

the surface temperature is specified:

$$T_{x=0} = T_s \tag{7}$$

and the temperature gradient must approach zero as you move away from the surface:

$$\left.\frac{dT}{dx}\right|_{x\to\infty} = 0 \tag{8}$$

To solve this problem using the Laplace transform it is necessary to transform the governing differential equation, Eq. (5):

$$\frac{d^2\widehat{T}}{dx^2} - \frac{1}{\alpha}\left(s\,\widehat{T} - T_{x,t=0}\right) = -\frac{\dot{g}'''}{k\,s} \tag{9}$$

Substituting Eq. (6) into Eq. (9) and rearranging:

$$\frac{d^2\widehat{T}}{dx^2} - \frac{s}{\alpha}\widehat{T} = -\frac{T_{ini}}{\alpha} - \frac{\dot{g}'''}{k\,s} \tag{10}$$

The spatial boundary conditions, Eqs. (7) and (8), are transformed to the s domain in order to provide the boundary conditions for the ordinary differential equation in the s domain, Eq. (10):

$$\widehat{T}_{x=0} = \frac{T_s}{s} \tag{11}$$

$$\left.\frac{d\widehat{T}}{dx}\right|_{x\to\infty} = 0 \tag{12}$$

The solution to the ordinary differential equation, Eq. (10), is:

$$\widehat{T} = C_1\exp\left(\sqrt{\frac{s}{\alpha}}x\right) + C_2\exp\left(-\sqrt{\frac{s}{\alpha}}x\right) + \frac{\dot{g}'''\,\alpha}{k\,s^2} + \frac{T_{ini}}{s}$$

The boundary condition at $x\to\infty$ leads to:

$$\left.\frac{d\widehat{T}}{dx}\right|_{x\to\infty} = C_1\sqrt{\frac{s}{\alpha}}\exp\left(\sqrt{\frac{s}{\alpha}}\infty\right) - C_2\sqrt{\frac{s}{\alpha}}\exp\left(-\sqrt{\frac{s}{\alpha}}\infty\right) = 0$$

which can only be true if $C_1 = 0$:

$$\widehat{T} = C_2\exp\left(-\sqrt{\frac{s}{\alpha}}x\right) + \frac{\dot{g}'''\,\alpha}{k\,s^2} + \frac{T_{ini}}{s}$$

The boundary condition at $x = 0$ leads to:

$$\hat{T}_{x=0} = C_2 \exp\left(-\sqrt{\frac{s}{\alpha}}0\right) + \frac{\dot{g}''' \alpha}{k\,s^2} + \frac{T_{ini}}{s} = \frac{T_s}{s}$$

which leads to:

$$C_2 = \frac{(T_s - T_{ini})}{s} - \frac{\dot{g}''' \alpha}{k\,s^2}$$

The solution in the s domain is:

$$\hat{T} = \left[\frac{(T_s - T_{ini})}{s} - \frac{\dot{g}''' \alpha}{k\,s^2}\right]\exp\left(-\sqrt{\frac{s}{\alpha}}x\right) + \frac{\dot{g}''' \alpha}{k\,s^2} + \frac{T_{ini}}{s}$$

which can be rearranged:

$$\hat{T} = \frac{(T_s - T_{ini})}{s}\exp\left(-\sqrt{\frac{s}{\alpha}}x\right) - \frac{\dot{g}''' \alpha}{k\,s^2}\exp\left(-\sqrt{\frac{s}{\alpha}}x\right) + \frac{\dot{g}''' \alpha}{k\,s^2} + \frac{T_{ini}}{s} \qquad (13)$$

The inverse Laplace transform can be obtained using Table 3-3:

$$T = (T_s - T_{ini})\,\mathrm{erfc}\left(\frac{x}{2\sqrt{\alpha\,t}}\right)$$
$$- \frac{\dot{g}''' \alpha}{k}\left[\left(t + \frac{x^2}{2\alpha}\right)\mathrm{erfc}\left(\frac{x}{2\sqrt{\alpha\,t}}\right) - x\sqrt{\frac{t}{\alpha\,\pi}}\exp\left(-\frac{x^2}{4\alpha\,t}\right) - t\right] + T_{ini} \qquad (14)$$

The known information is entered in EES:

```
"EXAMPLE 3.4-1: Quenching of a Superconductor"

$UnitSystem SI MASS RAD PA C J
$Tabstops 0.2 0.4 0.6 0.8 3.5

"Inputs"
T_ini=4.2 [K]                              "initial temperature of superconductor"
T_s=4.2 [K]                                "surface temperature"
g'''_dot=1e6 [W/m^3]                       "volumetric rate of generation"
k=500 [W/m-K]                              "conductivity"
alpha=0.000625 [m^2/s]                     "thermal diffusivity"
```

and the solution is programmed:

```
"Solution"
time=0.1 [s]                               "time"
x_mm=10 [mm]                               "position in mm"
x=x_mm*convert(mm,m)                       "position"
T=(T_s-T_ini)*erfc(x/(2*sqrt(alpha*time)))+(g'''_dot*alpha/k)*(time-(time+x^2/(2*alpha))* &
erfc(x(2*sqrt(alpha*time)))+x*sqrt(time/(pi*alpha))*exp(-x^2/(4*alpha*time)))+T_ini
```

Figure 2 illustrates the temperature as a function of position for various times relative to the start of the quench process.

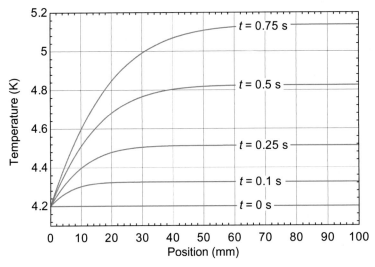

Figure 2: Temperature as a function of position at various times relative to the start of the quench process.

Notice that the portion of the conductor removed from the edge increases linearly in temperature. (This behavior corresponds to the term in Eq. (14) that is linear with time.) However, the effect of the cooled edge propagates into the conductor at a rate that increases with time. The solution can be checked against our physical intuition by verifying that the speed of the thermal wave agrees, approximately, with the diffusive time constant that was discussed in Section 3.3.2. At time = 0.5 s the thermal wave should have moved approximately:

$$\delta_t = 2\sqrt{\alpha\, t} \tag{15}$$

EES' calculator window can be used to carry out supplementary calculations such as this. EES' calculator window is accessed by selecting Calculator from the Windows menu (Figure 3).

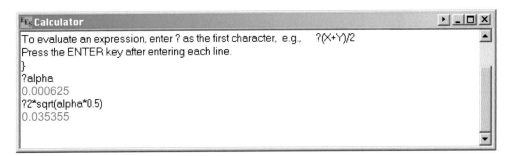

Figure 3: Calculator window.

To enter a command in the Calculator window, enter ? followed by the command and terminate the command with the enter key. The variables from the last time that the EES code was run are accessible from the Calculator; for example, typing ?alpha will provide the value of the thermal diffusivity of the superconductor. To determine the size of the thermal wave at 0.5 s using Eq. (15), type ?2*sqrt(alpha*0.5), as shown in Figure 3. Notice that the answer, 0.035 m or 35 mm, is consistent with the thermal wave thickness at 0.5 s in Figure 2.

EXAMPLE 3.4-1: QUENCHING OF A SUPERCONDUCTOR

The analytical solution of partial differential equations with the Laplace transform is convenient in some situations. For example, it is not possible to solve semi-infinite problems or (with some exceptions) problems with non-homogeneous boundary conditions in space using the method of separation of variables discussed in Section 3.5. Therefore, the Laplace transform solution or a self-similar solution (Section 3.3.3) may be the best alternative. However, the process of obtaining the inverse Laplace transform can be difficult and it is often easier to develop numerical solutions to these problems, as discussed in Section 3.8.

3.5 Separation of Variables for Transient Problems

3.5.1 Introduction

Section 3.3 discussed the behavior of a thermal wave within an un-bounded (i.e., semi-infinite) solid and introduced the concept of a diffusion time constant. The self-similar solution to a semi-infinite solid was presented in Section 3.3.3 and the solution to this type of problem using the Laplace transform approach was discussed in Section 3.4.6. This section examines the analytical solution to 1-D transient problems that are bounded using the method of separation of variables.

Figure 3-18 illustrates, qualitatively, the temperature distribution at various times that will be present in a plane wall that is initially at a uniform temperature, T_{ini}, when the temperature of one surface (at $x = L$) is suddenly increased to T_s while the other surface (at $x = 0$) is adiabatic.

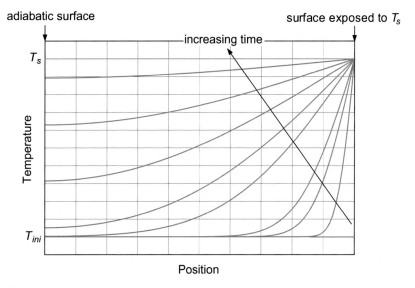

Figure 3-18: Temperature as a function of position at various times for a plane wall initially at T_{ini} that is subjected to a sudden change in the surface temperature to T_s.

The behavior illustrated in Figure 3-18 is initially consistent with the behavior of a semi-infinite body, discussed in Section 3.3. A thermal wave emanates from the surface at $x = L$ and penetrates into the solid; the depth of the thermal wave (δ_t) is approximately given by:

$$\delta_t = 2\sqrt{\alpha t} \qquad (3\text{-}226)$$

where α is the thermal diffusivity and t is time. When the thermal wave reaches the adiabatic edge at $x = 0$, it becomes bounded and cannot grow further. The character of the problem changes at this time. The temperature distributions will no longer collapse when plotted as a function of x/δ_t, as they did in Figure 3-15, and therefore a self-similar solution to the bounded problem is not possible.

The method of separation of variables can be used to analytically solve the problem shown in Figure 3-18. The solutions associated with some common shapes (a plane wall, cylinder, and sphere initially at a uniform temperature and exposed to a convective boundary condition) are presented in Section 3.5.2 without derivation. Sections 3.5.3 and 3.5.4 provide an introduction to the application of the method of separation of variables for 1-D transient problems in Cartesian and cylindrical coordinates, respectively. A more thorough discussion can be found in Myers (1998).

The concepts and steps used to obtain separation of variables solutions to 1-D transient problems are quite similar to those discussed for steady-state, 2-D problems in Section 2.2 and 2.3. The partial differential equation in position x (or r for cylindrical or spherical problems) and time t is transformed (by separation of variables) into a second order ordinary differential equation in position and a first order ordinary differential equation in time. The sub-problem involving x must be the eigenproblem and result in eigenfunctions. Therefore, both spatial boundary conditions must be homogeneous in order to apply separation of variables to a 1-D transient problem. Recall that a homogeneous boundary condition is one where any solution that satisfies the boundary condition must still satisfy the boundary condition if it is multiplied by a constant. Three types of homogeneous linear boundary conditions are encountered in heat transfer problems: (1) a specified temperature of zero, (2) an adiabatic boundary, and (3) convection to a fluid with a temperature of zero. Section 2.2.3 shows how it is possible to transform some non-homogeneous boundary conditions into homogeneous boundary conditions by subtracting the specified temperature or fluid temperature. Section 2.4 discusses the superposition of different solutions in order to accommodate multiple non-homogeneous boundary conditions. These techniques can also be used for 1-D transient problems.

3.5.2 Separation of Variables Solutions for Common Shapes

The separation of variables solutions for the plane wall, cylinder, and sphere are available in most textbooks as approximate formulae and in graphical format. The solutions are also available within EES both in dimensional and non-dimensional forms using the Transient Conduction library. This section provides, without derivation, the solutions to the basic problems associated with a plane wall, cylinder, and sphere initially at a uniform temperature when, at time $t = 0$, the surface is exposed via convection to a step change in the surrounding fluid temperature. These problems are summarized in Figure 3-19. The solutions to this set of problems are obtained using the separation of variables techniques discussed in Sections 3.5.3 and 3.5.4 and are widely applicable to 1-D transient problems.

The Plane Wall

Governing Equation and Boundary Conditions. Section 3.5.3 provides the solution to a plane wall subjected to a step change in the fluid temperature at a convective boundary. The partial differential equation associated with this problem is:

$$\frac{\partial^2 T}{\partial x^2} = \frac{1}{\alpha}\frac{\partial T}{\partial t} \tag{3-227}$$

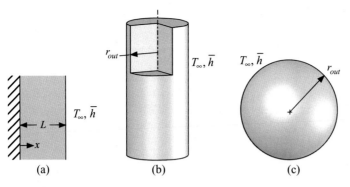

Figure 3-19: 1-D transient problems associated with (a) a plane wall, (b) a cylinder, and (c) a sphere, initially at a uniform temperature (T_{ini}) subjected to a step change in the convective boundary condition $(\bar{h}, T_\infty)$.

The boundary conditions for the problem are:

$$\left.\frac{\partial T}{\partial x}\right|_{x=0} = 0 \tag{3-228}$$

$$-k \left.\frac{\partial T}{\partial x}\right|_{x=L} = \bar{h}\left[T_{x=L} - T_\infty\right] \tag{3-229}$$

$$T_{t=0} = T_{ini} \tag{3-230}$$

Exact Solution. The solution derived in Section 3.5.3 for the plane wall problem shown in Figure 3-19(a) can be rearranged:

$$\tilde{\theta}(\tilde{x}, Fo) = \sum_{i=1}^{\infty} C_i \cos(\zeta_i \tilde{x}) \exp\left[-\zeta_i^2 Fo\right] \tag{3-231}$$

where $\tilde{x}$, Fo, and $\tilde{\theta}$ are the dimensionless position, Fourier number, and dimensionless temperature difference, defined as:

$$\tilde{x} = \frac{x}{L} \tag{3-232}$$

$$Fo = \frac{t\alpha}{L^2} \tag{3-233}$$

$$\tilde{\theta} = \frac{T - T_\infty}{T_{ini} - T_\infty} \tag{3-234}$$

The dimensionless eigenvalues, ζ_i in Eq. (3-231), correspond to the product of the dimensional eigenvalues λ_i and the wall thickness (L) and are therefore provided by the multiple roots of the eigencondition:

$$\tan(\zeta_i) = \frac{Bi}{\zeta_i} \tag{3-235}$$

where Bi is the Biot number:

$$Bi = \frac{\bar{h}L}{k} \tag{3-236}$$

The constants in Eq. (3-231) are:

$$C_i = \frac{2 \sin(\zeta_i)}{\zeta_i + \cos(\zeta_i) \sin(\zeta_i)} \tag{3-237}$$

The exact solution (the first 20 terms of Eq. (3-231)) is programmed in EES in its dimensionless format (i.e., the dimensionless temperature, $\tilde{\theta}$, as a function of the dimensionless independent variables Bi, Fo, and $\tilde{x}$) as the function planewall_T_ND. The solution can be accessed by selecting Function Info from the Options menu and selecting Transient Conduction from the pull-down menu. The different transient conduction functions that are available can be seen using the scroll-bar. The exact solutions for the cylinder and sphere are also programmed in EES as the functions cylinder_T_ND and sphere_T_ND.

Dimensional versions of each function are also available; these functions (planewall_T, cylinder_T, and sphere_T) return the temperature at a given position and time as a function of the dimensional independent variables (i.e., L, α, k, $\overline{h}$, T_i, T_∞).

It is often important to calculate the total amount of energy transferred to the wall. An energy balance on the wall indicates that the total energy transfer (Q) is the difference between the energy stored in the wall (U) and the energy stored in the wall at its initial condition ($U_{t=0}$):

$$Q = U - U_{t=0} \tag{3-238}$$

The total energy transfer is made dimensionless ($\tilde{Q}$) by normalizing it against the maximum amount of energy that could be transferred to the wall (Q_{max}):

$$\tilde{Q} = \frac{Q}{Q_{max}} \tag{3-239}$$

The maximum energy transfer would occur if the process continued to $t \to \infty$ and therefore the wall material equilibrates completely with the fluid temperature at T_∞.

$$Q_{max} = \rho c L A_c (T_\infty - T_{ini}) \tag{3-240}$$

The dimensionless energy transfer is related to the dimensionless, volume average temperature difference in the wall ($\bar{\bar{\theta}}$) according to:

$$\tilde{Q} = 1 - \bar{\bar{\theta}} \tag{3-241}$$

where

$$\bar{\bar{\theta}} = \int_0^1 \tilde{\theta} \, d\tilde{x} \tag{3-242}$$

Substituting the exact solution, Eq. (3-231), into the Eqs. (3-241) and (3-242) leads to:

$$\tilde{Q} = 1 - \sum_{i=1}^{\infty} C_i \frac{\sin(\zeta_i)}{\zeta_i} \exp(-\zeta_i^2 Fo) \tag{3-243}$$

The solution for the dimensionless total energy transfer for the plane wall, cylinder, and sphere have been programmed in EES as the functions planewall_Q_ND, cylinder_Q_ND, and sphere_Q_ND.

The exact solutions for the dimensionless temperature and heat transfer are often presented graphically in a form that was initially published by Heisler (1947) and is now referred to as the Heisler charts; these charts were convenient when access to computer solutions was not readily available. Heisler charts typically include the dimensionless center temperature difference ($\tilde{\theta}_{\tilde{x}=0}$) as a function of the Fourier number for various values of the inverse of the Biot number, Figure 3-20(a). The temperature information

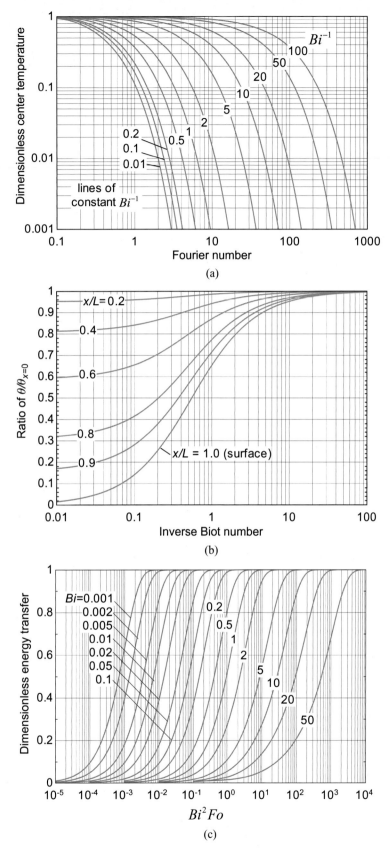

Figure 3-20: Heisler charts for a plane wall including (a) $\tilde{\theta}_{\tilde{x}=0}$ as a function of *Fo* for various values of Bi^{-1}, (b) $\tilde{\theta}/\tilde{\theta}_{\tilde{x}}$ as a function of Bi^{-1}, and (c) $\tilde{Q}$ as a function of $Bi^2\,Fo$ for various values of *Bi*.

presented in this figure corresponds to the position of the adiabatic boundary ($x = 0$). This is referred to as the center temperature because the center-line would be adiabatic if convection occured on both sides of the wall. The temperature at other locations can be obtained using Figure 3-20(b), which shows the ratio of the dimensionless temperature difference to the dimensionless center temperature difference ($\tilde{\theta}/\tilde{\theta}_{\tilde{x}=0}$) as a function of the inverse of the Biot number for various dimensionless positions; note that the Fourier number does not effect this ratio. Figure 3-20(c) illustrates the dimensionless total energy transfer, $\tilde{Q}$ as a function of $Bi^2 \, Fo$ for various values of the Biot number.

The Heisler charts shown in Figure 3-20 were generated using the solution programmed in EES. Figure 3-20(a) was generated by accessing the planewall_ND function using the following EES code (varying Fo for different values of invBi):

```
invBi=10                                       "Inverse Biot number"
theta_hat_0=planewall_T_ND(0,Fo,1/invBi)       "Dimensionless center temperature"
```

Figure 3-20(b) was generated by accessing planewall_ND twice using the following EES code (varying x_hat for different values of invBi):

```
x_hat=0.2                                       "Dimensionless position"
Fo=1                                            "Fourier No. (doesn't effect ratio)"
theta_hat_0=planewall_T_ND(0,Fo,1/invBi)       "Dimensionless center temperature"
thetatotheta_hat_0=planewall_T_ND(x_hat,Fo,1/invBi)/theta_hat_0
                                                "Dimensionless center temperature"
```

Figure 3-20(c) was generated by accessing the planewall_Q_ND function (varying Bi2Fo for different values of Bi):

```
Bi=50                                           "Biot number"
Fo=Bi2Fo/Bi^2                                   "Fourier number (doesn't change ratio)"
Q_hat=planewall_Q_ND(Fo, Bi)                    "Dimensionless energy transfer"
```

Approximate Solution for Large Fourier Number. When $Fo \gtrsim 0.2$, the series solution given by Eq. (3-231) can be adequately approximated using only the first term. Only the first eigenvalue (i.e., ζ_i, the first root of Eq. (3-235)) is required to implement this approximate solution and therefore many textbooks will tabulate the first eigenvalue as a function of the Biot number. The value of the first eigenvalue as a function Biot number is shown in Figure 3-21 for the plane wall, as well as the cylinder and sphere solutions, discussed subsequently.

Using ζ_1, the approximate solutions for the dimensionless temperature difference and dimensionless energy transfer become:

$$\tilde{\theta}(\tilde{x}, Fo) \approx \left[\frac{2 \sin(\zeta_1)}{\zeta_1 + \cos(\zeta_1) \sin(\zeta_1)} \right] \cos(\zeta_1 \tilde{x}) \exp\left[-\zeta_1^2 \, Fo \right] \qquad (3\text{-}244)$$

$$\tilde{Q} \approx 1 - \left[\frac{2 \sin(\zeta_1)}{\zeta_1 + \cos(\zeta_1) \sin(\zeta_1)} \right] \frac{\sin(\zeta_1)}{\zeta_1} \exp\left(-\zeta_1^2 \, Fo \right) \qquad (3\text{-}245)$$

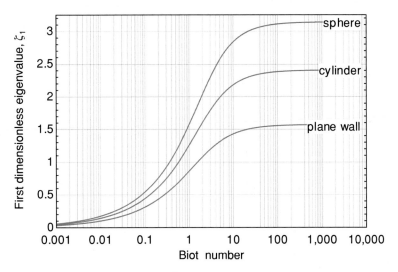

Figure 3-21: First eigenvalue for use in the approximate solution to the problems illustrated in Figure 3-19.

The Cylinder

Governing Equation and Boundary Conditions. The partial differential equation associated with the cylinder is derived by carrying out an energy balance on a control volume that is a differentially small cylindrical shell with thickness dr.

$$\dot{q}_r = \dot{q}_{r+dr} + \frac{\partial U}{\partial t} \tag{3-246}$$

or

$$0 = \frac{\partial \dot{q}_r}{\partial r} dr + \frac{\partial U}{\partial t} \tag{3-247}$$

The rate of conductive heat transfer is:

$$\dot{q}_r = -k \, 2 \, \pi \, r \, L \, \frac{\partial T}{\partial r} \tag{3-248}$$

where L is the length of the cylinder. The rate of energy storage is:

$$\frac{\partial U}{\partial t} = 2 \, \pi \, r \, L \, dr \, \rho \, c \, \frac{\partial T}{\partial t} \tag{3-249}$$

Substituting Eqs. (3-248) and (3-249) into Eq. (3-247) leads to:

$$0 = \frac{\partial}{\partial r} \left[-k \, 2 \, \pi \, r \, L \, \frac{\partial T}{\partial r} \right] dr + 2 \, \pi \, r \, L \, dr \, \rho \, c \, \frac{\partial T}{\partial t} \tag{3-250}$$

or with k constant,

$$\boxed{\frac{\alpha}{r} \frac{\partial}{\partial r} \left[r \frac{\partial T}{\partial r} \right] = \frac{\partial T}{\partial t}} \tag{3-251}$$

The initial condition is:

$$T_{t=0} = T_{ini} \tag{3-252}$$

and the spatial boundary conditions are:

$$\frac{\partial T}{\partial r}\bigg|_{r=0} = 0 \tag{3-253}$$

$$-k \frac{\partial T}{\partial r}\bigg|_{r=r_{out}} = \bar{h}\,(T_{r=r_{out}} - T_\infty) \tag{3-254}$$

Exact Solution. The exact solution for the cylinder problem shown in Figure 3-19(b) can be derived using the methods discussed in Section 3.5.4:

$$\tilde{\theta}(\tilde{r}, Fo) = \sum_{i=1}^{\infty} C_i\,\text{BesselJ}\,(0, \zeta_i\,\tilde{r}) \exp\left[-\zeta_i^2\,Fo\right] \tag{3-255}$$

where $\tilde{r}$, Fo, and $\tilde{\theta}$ are the dimensionless position, Fourier number, and dimensionless temperature difference, defined as:

$$\tilde{r} = \frac{r}{r_{out}} \tag{3-256}$$

$$Fo = \frac{t\,\alpha}{r_{out}^2} \tag{3-257}$$

$$\tilde{\theta} = \frac{T - T_\infty}{T_{ini} - T_\infty} \tag{3-258}$$

The dimensionless eigenvalues, ζ_i in Eq. (3-255) are the roots of the eigencondition:

$$\zeta_i\,\text{BesselJ}\,(1, \zeta_i) - Bi\,\text{BesselJ}\,(0, \zeta_i) = 0 \tag{3-259}$$

where Bi is the Biot number, defined as:

$$Bi = \frac{\bar{h}\,r_{out}}{k} \tag{3-260}$$

The constants in Eq. (3-255) are given by:

$$C_i = \frac{2\,\text{BesselJ}\,(1, \zeta_i)}{\zeta_i[\text{BesselJ}^2\,(0, \zeta_i) + \text{BesselJ}^2\,(1, \zeta_i)]} \tag{3-261}$$

The dimensionless energy transfer is related to the dimensionless, volume average temperature in the cylinder ($\bar{\tilde{\theta}}$) according to:

$$\tilde{Q} = 1 - \bar{\tilde{\theta}} \tag{3-262}$$

where

$$\bar{\tilde{\theta}} = 2 \int_0^1 \tilde{\theta}\,\tilde{r}\,d\tilde{r} \tag{3-263}$$

Substituting the exact solution, Eq. (3-255), into Eqs. (3-262) and (3-263) leads to:

$$\tilde{Q} = 1 - \sum_{i=1}^{\infty} C_i \frac{2\,\text{BesselJ}(1, \zeta_i)}{\zeta_i} \exp\left(-\zeta_i^2\,Fo\right) \tag{3-264}$$

Approximate Solution for Large Fourier Number. The series solution for the cylinder, like the plane wall, can be adequately approximated using only the first term when $Fo >$ 0.2. The value of the first eigenvalue as a function Biot number is shown in Figure 3-21. Using ζ_1, the approximate solutions for the dimensionless temperature difference and energy transfer become:

$$\tilde{\theta}(\tilde{r}, Fo) = \frac{2\,\text{BesselJ}(1, \zeta_1)\,\text{BesselJ}(0, \zeta_1\,\tilde{r})}{\zeta_1[\text{BesselJ}^2(0, \zeta_1) + \text{BesselJ}^2(1, \zeta_1)]}\,\exp[-\zeta_1^2\,Fo] \tag{3-265}$$

$$\tilde{Q} = 1 - \frac{4\,\text{BesselJ}^2(1, \zeta_1)}{\zeta_1^2[\text{BesselJ}^2(0, \zeta_1) + \text{BesselJ}^2(1, \zeta_1)]}\,\exp\left(-\zeta_i^2\,Fo\right) \tag{3-266}$$

The Sphere

Governing Equation and Boundary Conditions. The partial differential equation associated with the sphere is derived by carrying out an energy balance on a control volume that is a differentially small spherical shell with thickness r.

$$\dot{q}_r = \dot{q}_{r+dr} + \frac{\partial U}{\partial t} \tag{3-267}$$

or

$$0 = \frac{\partial \dot{q}_r}{\partial r}dr + \frac{\partial U}{\partial t} \tag{3-268}$$

The rate of conductive heat transfer is:

$$\dot{q}_r = -k\,4\,\pi\,r^2\,\frac{\partial T}{\partial r} \tag{3-269}$$

The rate of energy storage is:

$$\frac{\partial U}{\partial t} = 4\,\pi\,r^2\,dr\,\rho\,c\,\frac{\partial T}{\partial t} \tag{3-270}$$

Substituting Eqs. (3-269) and (3-270) into Eq. (3-268) leads to:

$$0 = \frac{\partial}{\partial r}\left[-k\,4\,\pi\,r^2\,\frac{\partial T}{\partial r}\right]dr + 4\,\pi\,r^2\,dr\,\rho\,c\,\frac{\partial T}{\partial t} \tag{3-271}$$

or

$$\boxed{\frac{\alpha}{r^2}\frac{\partial}{\partial r}\left[r^2\,\frac{\partial T}{\partial r}\right] = \frac{\partial T}{\partial t}} \tag{3-272}$$

The initial condition is:

$$T_{t=0} = T_{ini} \tag{3-273}$$

and the spatial boundary conditions are:

$$\left.\frac{\partial T}{\partial r}\right|_{r=0} = 0 \tag{3-274}$$

$$-k\left.\frac{\partial T}{\partial r}\right|_{r=r_{out}} = \bar{h}\,(T_{r=r_{out}} - T_\infty) \tag{3-275}$$

Exact Solution. The exact solution for the spherical problem shown in Figure 3-19(c) is:

$$\tilde{\theta}(\tilde{r}, Fo) = \sum_{i=1}^{\infty} C_i \frac{\sin(\zeta_i \tilde{r})}{\zeta_i \tilde{r}} \exp\left[-\zeta_i^2 Fo\right] \qquad (3\text{-}276)$$

where $\tilde{r}$, Fo, and $\tilde{\theta}$ are the dimensionless position, Fourier number, and dimensionless temperature difference, defined as:

$$\tilde{r} = \frac{r}{r_{out}} \qquad (3\text{-}277)$$

$$Fo = \frac{t\,\alpha}{r_{out}^2} \qquad (3\text{-}278)$$

$$\tilde{\theta} = \frac{T - T_\infty}{T_{ini} - T_\infty} \qquad (3\text{-}279)$$

The dimensionless eigenvalues, ζ_i in Eq. (3-276) are the roots of the eigencondition:

$$\zeta_i \cos(\zeta_i) + (Bi - 1)\sin(\zeta_i) = 0 \qquad (3\text{-}280)$$

where Bi is the Biot number, defined as:

$$Bi = \frac{\overline{h}\,r_{out}}{k} \qquad (3\text{-}281)$$

and the constants in Eq. (3-276) are:

$$C_i = \frac{2\left[\sin(\zeta_i) - \zeta_i \cos(\zeta_i)\right]}{\zeta_i - \sin(\zeta_i)\cos(\zeta_i)} \qquad (3\text{-}282)$$

The dimensionless energy transfer is related to the dimensionless, volume average temperature in the sphere $(\overline{\tilde{\theta}})$ according to:

$$\tilde{Q} = 1 - \overline{\tilde{\theta}} \qquad (3\text{-}283)$$

where

$$\overline{\tilde{\theta}} = 3\int_0^1 \tilde{\theta}\,\tilde{r}^2\,d\tilde{r} \qquad (3\text{-}284)$$

Substituting the exact solution, Eq. (3-276) into Eqs. (3-283) and (3-284) leads to:

$$\tilde{Q} = 1 - \sum_{i=1}^{\infty} C_i \frac{3\left[\sin(\zeta_i) - \zeta_i \cos(\zeta_i)\right]}{\zeta_i^3} \exp\left(-\zeta_i^2 Fo\right) \qquad (3\text{-}285)$$

Approximate Solution for Large Fourier Number. The series solution for the sphere can be adequately approximated using only the first term when Fo is > 0.2. The value of the first eigenvalue as a function Biot number is shown in Figure 3-21. Using ζ_1, the approximate solutions for the dimensionless temperature distribution and energy transfer become:

$$\tilde{\theta}(\tilde{r}, Fo) = \frac{2\left[\sin(\zeta_1) - \zeta_1 \cos(\zeta_1)\right]}{\left[\zeta_1 - \sin(\zeta_1)\cos(\zeta_1)\right]} \frac{\sin(\zeta_1 \tilde{r})}{\zeta_1 \tilde{r}} \exp\left[-\zeta_1^2 Fo\right] \qquad (3\text{-}286)$$

$$\tilde{Q} = 1 - \frac{6\left[\sin(\zeta_1) - \zeta_1 \cos(\zeta_1)\right]^2}{\zeta_1^3\left[\zeta_1 - \sin(\zeta_1)\cos(\zeta_1)\right]} \exp\left(-\zeta_1^2 Fo\right) \qquad (3\text{-}287)$$

EXAMPLE 3.5-1: MATERIAL PROCESSING IN A RADIANT OVEN

As part of a manufacturing process, long cylindrical pieces of material with radius $r_{out} = 5.0$ cm are placed into a radiant oven. The initial temperature of the material is $T_{ini} = 20°C$. The walls of the oven are maintained at a temperature of $T_{wall} = 750°C$ and the oven is evacuated so that the outer edge of the cylinder is exposed only to radiation heat transfer. The emissivity of the surface of the cylinder is $\varepsilon = 0.95$. The material is considered to be completely processed when the temperature everywhere is at least $T_p = 250°C$. The properties of the material are $k = 1.4$ W/m-K, $\rho = 2500$ kg/m^3, $c = 700$ J/kg-K.

a) Is a lumped capacitance model appropriate for this problem?

The known inputs are entered in EES:

```
"EXAMPLE 3.5-1: Material in a Radiant Oven"

$UnitSystem SI MASS RAD PA K J
$TABSTOPS 0.2 0.4 0.6 0.8 3.5 in

"Inputs"
r_out=5.0 [cm]*convert(cm,m)             "radius"
T_ini=converttemp(C,K,20[C])             "initial temperature"
T_wall=converttemp(C,K,750[C])           "wall temperature"
e=0.95                                    "emissivity"
T_p=converttemp(C,K,250[C])              "processing temperature"
k=1.4 [W/m-K]                            "conductivity"
rho=2500 [kg/m^3]                         "density"
c=700 [J/kg-K]                            "specific heat capacity"
alpha=k/(rho*c)                           "thermal diffusivity"
```

The effective heat transfer coefficient associated with radiation ($\bar{h}_{rad}$, discussed in Section 1.2.6) is:

$$\bar{h}_{rad} = \sigma\,\varepsilon\,\left(T_{wall}^2 + T_s^2\right)\left(T_{wall} + T_s\right)$$

where T_s is the surface temperature, which is not known but will certainly not be less then T_{ini} and will likely not be much higher than T_p. An average of these values is used to evaluate the radiation heat transfer coefficient.

```
T_s=(T_ini+T_p)/2                 "average temperature used for surface temp."
h_bar_rad=sigma# *e*(T_wall^2+T_s^2)*(T_wall+T_s)
                                  "effective heat transfer coefficient due to radiation"
```

Because T_{wall} is so much higher than T_s, the value of $\bar{h}_{rad}$ is not affected significantly by the choice of T_s. Using $\bar{h}_{rad}$, it is possible to calculate the Biot number:

$$Bi = \frac{\bar{h}_{rad}\,r_{out}}{k}$$

EXAMPLE 3.5-1: MATERIAL PROCESSING IN A RADIANT OVEN

EXAMPLE 3.5–1: MATERIAL PROCESSING IN A RADIANT OVEN

Bi=h_bar_rad*r_out/k   "Biot number"

which provides a Biot number of 3.3. Therefore, a lumped capacitance model is not valid.

b) **How long will the processing require? Use an effective, radiation heat transfer coefficient for this calculation.**

The dimensionless temperature difference at which the processing is complete can be calculated using Eq. (3-258):

$$\tilde{\theta}_p = \frac{T_p - T_{wall}}{T_{ini} - T_{wall}}$$

theta_hat_p=(T_p-T_wall)/(T_ini-T_wall)   "dimensionless temperature for processing"

The function cylinder_ND implements the exact solution for a cylinder, Eq. (3-255), and provides $\tilde{\theta}$ given $\tilde{r}$, Fo, Bi. In this case, we know $\tilde{\theta} = \tilde{\theta}_p$ at $\tilde{r} = 0$ (the center of the cylinder, which will be the lowest temperature part of the material), and the value of Bi is also known. We would like to solve for the corresponding value of Fo (and therefore time). EES will solve this implicit equation in order to determine Fo:

theta_hat_p=cylinder_T_ND(0, Fo, Bi)
                    "implements the dimensionless solution to a cylinder"

Initially you are likely to obtain an error in EES related to the value of Fo being negative. This error condition can be easily overcome by setting appropriate limits (e.g., 0.001 to 1000) on the value of Fo in the Variable Information window.

The Fourier number is used to calculate the processing time ($t_{process}$), according to Eq. (3-257):

$$t_{process} = Fo \frac{r_{out}^2}{\alpha}$$

t_process=Fo*r_out^2/alpha   "processing time"

which leads to a processing time of 678 s.

c) **What is the minimum amount of energy per unit length that is required to process the material? That is, how much energy would be required to bring the material to a uniform temperature of T_p?**

An energy balance on the material shows that the minimum amount of energy required to bring the material to a uniform temperature (Q_{min}) is:

$$Q_{min} = \pi \, r_{out}^2 \, L \, \rho \, c \, (T_p - T_{ini})$$

EXAMPLE 3.5-1: MATERIAL PROCESSING IN A RADIANT OVEN

```
L=1 [m]                                    "per unit length of material"
Q_min=pi*r_out^2*L*rho*c*(T_p-T_ini)       "minimum amount of energy transfer required"
```

which leads to Q_{min}= 3.16 MJ.

d) How much energy is actually required per unit length of material?

The dimensionless energy transfer to the cylinder ($\tilde{Q}$) is obtained using the cylin-der_Q_ND function in EES:

```
Q_hat=cylinder_Q_ND(Fo, Bi)                "dimensionless heat transfer"
```

The dimensionless energy transfer is defined according to Eq. (3-239)

$$\tilde{Q} = \frac{Q}{Q_{max}}$$

where Q_{max} is the energy transfer that occurs if the process is continued until the cylinder reaches equilibrium with the wall:

$$Q_{max} = \rho\,c\,L\,\pi\,r_{out}^2\,(T_{wall} - T_{ini})$$

```
Q_max=pi*r_out^2*L*rho*c*(T_wall-T_ini)
                          "maximum amount of energy transfer that could occur"
Q=Q_max*Q_hat             "actual energy transfer"
```

which leads to $Q = 5.62$ MJ.

e) The efficiency of the process (η) is defined as the ratio of the minimum possible energy transfer required to process the material (from part (b)) to the actual energy transfer (from part (c)). Plot the efficiency of the process as a function of the radius of the material for various values of T_{wall}.

The efficiency is defined as:

$$\eta = \frac{Q_{min}}{Q}$$

```
eta=Q_min/Q                                "Process efficiency"
```

The efficiency is computed over a range of radii, r_{out}, using a parametric table at several values of T_{wall} and the results are shown in Figure 1.

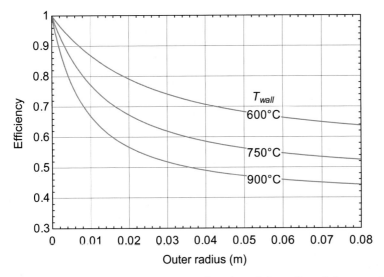

Figure 1: Efficiency of the process as a function of the radius of the material for various values of the oven wall temperature.

Notice that the efficiency improves with reduced radius because the temperature gradients within the material are reduced and so the amount of energy wasted in 'overheating' the material toward the outer radius of the cylinder is reduced. Also, reducing the oven temperature tends to improve the efficiency for the same reason, but at the expense of increased processing time. The most efficient process is associated with processing a very small amount of material (with small radius) very slowly (at low oven temperature); this is a typical result: very efficient processes are not usually very practical.

3.5.3 Separation of Variables Solutions in Cartesian Coordinates

The solution to 1-D transient problems using separation of variables is illustrated in the context of the problem shown in Figure 3-22. A plane wall is initially at a uniform temperature, $T_{ini} = 100$ K, when the surface is exposed to a step change in the surrounding fluid temperature to $T_\infty = 200$ K. The average heat transfer coefficient between the surface and the fluid is $\bar{h} = 200$ W/m²-K. The wall has thermal diffusivity $\alpha = 5 \times 10^{-6}$ m²/s

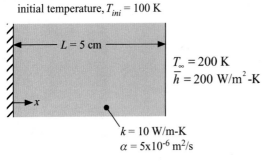

Figure 3-22: Plane wall initially at a uniform temperature that is exposed to convection at the surface.

and thermal conductivity $k = 10$ W/m-K. The thickness of the wall is $L = 5$ cm. The known information is entered in EES:

```
$UnitSystem SI MASS RAD PA K J
$TABSTOPS 0.2 0.4 0.6 0.8 3.5 in

"Inputs"
k=10 [W/m-K]                              "conductivity"
alpha=5e-6 [m^2/s]                        "thermal diffusivity"
T_ini=100 [K]                            "initial temperature"
T_infinity=200 [K]                        "fluid temperature"
h_bar=200 [W/m^2-K]                      "heat transfer coefficient"
L=5.0 [cm]*convert(cm,m)                  "thickness"
```

The governing partial differential equation for this situation is identical to the one derived for a semi-infinite body in Section 3.3.3:

$$\frac{\partial T}{\partial t} - \alpha \frac{\partial^2 T}{\partial x^2} = 0 \qquad (3\text{-}288)$$

The partial differential equation is first order in time and therefore it requires one boundary condition with respect to time; this is the initial condition:

$$T_{t=0} = T_{ini} \qquad (3\text{-}289)$$

Equation (3-288) is second order in space and therefore two boundary conditions are required with respect to x. At the adiabatic wall, the temperature gradient must be zero:

$$\left. \frac{\partial T}{\partial x} \right|_{x=0} = 0 \qquad (3\text{-}290)$$

An interface energy balance at the surface $(x = L)$ balances conduction with convection:

$$-k \left. \frac{\partial T}{\partial x} \right|_{x=L} = \bar{h} \left[T_{x=L} - T_\infty \right] \qquad (3\text{-}291)$$

The steps required to solve 1-D transient problems using separation of variables follow naturally from those presented in Section 2.2.2 for 2-D steady-state problems.

Requirements for using Separation of Variables

In order to apply separation of variables, it is necessary that the partial differential equation and all of the boundary condition be linear; these criteria are satisfied for the problem shown in Figure 3-22 by inspection of Eqs. (3-288) through (3-291). It is also necessary that the partial differential equation and both boundary conditions in space be homogeneous (there is only one direction associated with a 1-D transient problem and therefore it must be homogeneous). The partial differential equation, Eq. (3-288) is homogeneous and the boundary condition associated with the adiabatic wall, Eq. (3-290), is also homogeneous. However, the convective boundary condition at $x = L$, Eq. (3-291), is not homogeneous. Fortunately, it is possible to transform this problem, as discussed in Section 2.2.3. The temperature difference relative to the fluid temperature is defined:

$$\theta = T - T_\infty \qquad (3\text{-}292)$$

The transformed partial differential equation becomes:

$$\alpha \frac{\partial^2 \theta}{\partial x^2} = \frac{\partial \theta}{\partial t} \tag{3-293}$$

and the boundary conditions become:

$$\theta_{t=0} = T_{ini} - T_\infty \tag{3-294}$$

$$\left. \frac{\partial \theta}{\partial x} \right|_{x=0} = 0 \tag{3-295}$$

$$-k \left. \frac{\partial \theta}{\partial x} \right|_{x=L} = \overline{h}\, \theta_{x=L} \tag{3-296}$$

Notice that both spatial boundary conditions for the transformed problem, Eqs. (3-295) and (3-296), are homogeneous and therefore it will be possible to obtain a set of orthogonal eigenfunctions in x.

The initial condition, Eq. (3-294), is not homogeneous but it doesn't have to be. (Recall that the boundary conditions in one direction of the 2-D conduction problems, considered in Sections 2.2 and 2.3, did not have to be homogeneous.) As an aside, if the dimension of the problem were extended to infinity ($L \to \infty$) then the resulting spatial boundary condition, Eq. (3-296), would *not* be homogeneous and therefore the semi-infinite problem cannot be solved using separation of variables. One of the reasons that the Laplace transform is a valuable tool for heat transfer problems is that it can be used to analytically solve semi-infinite problems.

Separate the Variables

The temperature difference, θ, is a function of axial position and time. The separation of variables approach assumes that the solution can be expressed as the product of a function only of time, $\theta t(t)$, and a function only of position, $\theta X(x)$:

$$\theta(x, t) = \theta X(x)\, \theta t(t) \tag{3-297}$$

Substituting Eq. (3-297) into Eq. (3-293) leads to:

$$\alpha\, \theta t\, \frac{d^2 \theta X}{dx^2} = \theta X\, \frac{d\theta t}{dt} \tag{3-298}$$

Equation (3-298) is divided through by $\alpha\, \theta X \theta t$ in order to obtain:

$$\frac{\dfrac{d^2 \theta X}{dx^2}}{\theta X} = \frac{\dfrac{d\theta t}{dt}}{\alpha\, \theta t} \tag{3-299}$$

Both sides of Eq. (3-299) must be equal to a constant in order for the solution to be valid at an arbitrary time and position. We know from our experience with 2-D separation of variables that Eq. (3-299) will lead to two ordinary differential equations (one in x and the other in t). Furthermore, the ODE in x must lead to a set of eigenfunctions. This foresight motivates the choice of a negative constant $-\lambda^2$:

$$\frac{\dfrac{d^2 \theta X}{dx^2}}{\theta X} = \frac{\dfrac{d\theta t}{dt}}{\alpha\, \theta t} = -\lambda^2 \tag{3-300}$$

The ODE in x that is obtained from Eq. (3-300) is:

$$\frac{d^2 \theta X}{dx^2} + \lambda^2 \theta X = 0 \qquad (3\text{-}301)$$

which is solved by sines and cosines. The ODE in time suggested by Eq. (3-300) is:

$$\frac{d\theta t}{dt} + \lambda^2 \alpha \, \theta t = 0 \qquad (3\text{-}302)$$

Solve the Eigenproblem

It is always necessary to address the solution to the homogeneous sub-problem (in this problem, θX) before moving on to the non-homogeneous sub-problem (for θt). The homogeneous sub-problem is called the eigenproblem. The general solution to Eq. (3-301) is:

$$\theta X = C_1 \sin(\lambda x) + C_2 \cos(\lambda x) \qquad (3\text{-}303)$$

where C_1 and C_2 are unknown constants. Substituting Eq. (3-297) and Eq. (3-303) into the spatial boundary condition at $x = 0$, Eq. (3-295), leads to:

$$\left. \frac{\partial \theta}{\partial x} \right|_{x=0} = \theta t \left. \frac{d\theta X}{dx} \right|_{x=0} = \theta t \left[C_1 \lambda \underbrace{\cos(\lambda \, 0)}_{=1} - C_2 \lambda \underbrace{\sin(\lambda \, 0)}_{=0} \right] = 0 \qquad (3\text{-}304)$$

or

$$\theta t \, C_1 \lambda = 0 \qquad (3\text{-}305)$$

which can only be true (for a non-trivial solution) if $C_1 = 0$:

$$\theta X = C_2 \cos(\lambda x) \qquad (3\text{-}306)$$

Substituting Eq. (3-297) into the spatial boundary condition at $x = L$, Eq. (3-296), leads to:

$$-k \, \theta t \left. \frac{d\theta X}{dx} \right|_{x=L} = \overline{h} \, \theta t \, \theta X_{x=L} \qquad (3\text{-}307)$$

or

$$-k \left. \frac{d\theta X}{dx} \right|_{x=L} = \overline{h} \, \theta X_{x=L} \qquad (3\text{-}308)$$

Substituting Eq. (3-306) into Eq. (3-308) leads to:

$$k \, C_2 \lambda \sin(\lambda \, L) = \overline{h} \, C_2 \cos(\lambda \, L) \qquad (3\text{-}309)$$

Equation (3-309) provides the eigencondition for the problem, which defines multiple eigenvalues:

$$\frac{\sin(\lambda \, L)}{\cos(\lambda \, L)} = \frac{\overline{h}}{k \, \lambda} \qquad (3\text{-}310)$$

or, multiplying and dividing the right side of Eq. (3-310) by L:

$$\frac{\sin(\lambda L)}{\cos(\lambda L)} = \frac{\overline{h} L}{k \lambda L} \tag{3-311}$$

The dimensionless group $\overline{h} L/k$ is the Biot number (Bi) that was encountered in Section 1.6 and elsewhere. In this context, the Biot number represents the ratio of the resistance to conduction heat transfer within the wall to convective heat transfer from the surface. Writing Eq. (3-311) in terms of the Biot number leads to:

$$\tan(\lambda L) = \frac{Bi}{\lambda L} \tag{3-312}$$

Equation (3-312) is analogous to the eigenconditions for the problems considered in Section 2.2. There are an infinite number of values of λ that will satisfy Eq. (3-312). However, Eq. (3-312) provides an implicit rather than an explicit equation for these eigenvalues. This situation was encountered previously in EXAMPLE 2.3-1; the eigencondition involved the zeroes of the Bessel function. In EXAMPLE 2.3-1, the process of calculating the eigenvalues was automated by using EES to determine the roots of the eigencondition within specified ranges. We can use the same procedure for any problem with an implicit eigencondition, such as is given by Eq. (3-312). Figure 3-23 illustrates the left and right sides of Eq. (3-312) as a function of λL for the case where $Bi = 1.0$.

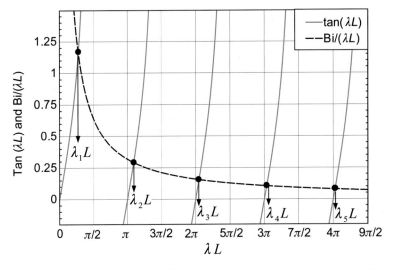

Figure 3-23: The left and right sides of the eigencondition equation for $Bi = 1.0$; the intersections correspond to eigenvalues for the problem, $\lambda_i L$.

Figure 3-23 shows that each successive value of $\lambda_i L$ can be found in a well-defined interval; $\lambda_1 L$ lies between 0 and $\pi/2$, $\lambda_2 L$ lies between π and $3\pi/2$, etc.; this will be true regardless of the value of the Biot number. The number of terms to use in the solution is specified and arrays of appropriate guess values and upper and lower bounds for each eigenvalue are generated.

```
Nterm=10 [-]                                    "number of terms to use in the solution"
"Setup guess values and lower and upper bounds for eigenvalues"
duplicate i=1,Nterm
   lowerlimit[i]=(i-1)*pi
   upperlimit[i]=lowerlimit[i]+pi/2
   guess[i]=lowerlimit[i]+pi/4
end
```

The eigencondition is programmed using a duplicate loop:

```
Bi=h_bar*L/k                                    "Biot number"
"Identify eigenvalues"
duplicate i=1,Nterm
   tan(lambdaL[i])=Bi/lambdaL[i]                "eigencondition"
   lambda[i]=lambdaL[i]/L                       "eigenvalue"
end
```

The solution obtained at this point will provide the same value for all of the eigenvalues. (Select Arrays from the Windows menu and you will see that each value of the array lambdaL[i] is the same, probably equal to whichever root of Eq. (3-312) lies closest to the default guess value of 1.0.) The interval for each eigenvalue can be controlled by selecting Variable Info from the Options menu. Deselect the Show array variables check box at the upper left so that the arrays are collapsed to a single entry and use the guess[], upperlimit[], and lowerlimit[] arrays to control the process of identifying the eigenvalues in the array lambdaL[].

The solution will now identify the first 10 eigenvalues of the problem (more can be obtained by changing the value of Nterm). The number of terms required depends on the time and position where you need the solution; this is discussed at the end of this section.

At this point, each of the eigenfunctions of the problem have been obtained. The ith eigenfunction is:

$$\theta X_i = C_{2,i} \cos(\lambda_i x) \tag{3-313}$$

where λ_i is the ith eigenvalue, identified by the eigencondition:

$$\tan(\lambda_i L) = \frac{Bi}{\lambda_i L} \tag{3-314}$$

Solve the Non-homogeneous Problem for each Eigenvalue

The solution to the non-homogeneous ordinary differential equation corresponding to the ith eigenvalue, Eq. (3-302):

$$\frac{d\theta t_i}{dt} + \lambda_i^2 \alpha \theta t_i = 0 \tag{3-315}$$

is

$$\theta t_i = C_{3,i} \exp(-\lambda_i^2 \alpha t) \tag{3-316}$$

where $C_{3,i}$ is an undetermined constant.

Obtain a Solution for each Eigenvalue

According to Eq. (3-297), the solution associated with the ith eigenvalue is:

$$\theta_i = \theta X_i \, \theta t_i = C_i \cos(\lambda_i x) \exp\left(-\lambda_i^2 \alpha t\right) \tag{3-317}$$

where the constants $C_{2,i}$ and $C_{3,i}$ have been combined to form a single undetermined constant C_i. Equation (3-317) will, for any value of i, satisfy the governing differential equation, Eq. (3-293), throughout the domain and satisfy all of the boundary conditions in the x-direction, Eqs. (3-295) and (3-296). It is worth checking that the solution has these properties using Maple before proceeding. Enter the solution as a function of x and t:

```
> restart;
> theta:=(x,t)->C*cos(lambda*x)*exp(-lambda^2*alpha*t);
```
$$\theta := (x, t) \rightarrow C\cos(\lambda x)e^{(-\lambda^2 \alpha t)}$$

Verify that it satisfies Eq. (3-295):

```
> eval(diff(theta(x,t),x),x=0);
```
$$0$$

and Eq. (3-296):

```
> -k*eval(diff(theta(x,t),x),x=L)-h_bar*theta(L,t);
```
$$kC\sin(\lambda L)\lambda \, e^{(-\lambda^2 \alpha t)} - h\_bar\, C\cos(\lambda L)\, e^{(-\lambda^2 \alpha t)}$$
```
> simplify(%);
```
$$Ce^{(-\lambda^2 \alpha t)}(k\sin(\lambda L)\lambda - h\_bar\cos(\lambda L))$$

Note that the eigencondition cannot be enforced in Maple using the assume command as it was in the problems in Section 2.2. However, it is clear that the eigencondition, Eq. (3-310), requires that the term within the parentheses must be zero and therefore the second spatial boundary condition will be satisfied for any eigenvalue. Finally, verify that the partial differential equation, Eq. (3-293), is satisfied:

```
> alpha*diff(diff(theta(x,t),x),x)-diff(theta(x,t),t);
```
$$0$$

Create the Series Solution and Enforce the Initial Condition

Because the partial differential equation is linear, the sum of the solution θ_i for each eigenvalue, Eq. (3-317), is itself a solution:

$$\theta = \sum_{i=1}^{\infty} \theta_i = \sum_{i=1}^{\infty} C_i \cos(\lambda_i x) \exp\left(-\lambda_i^2 \alpha t\right) \tag{3-318}$$

The final step of the problem selects the constants so that the series solution satisfies the initial condition, Eq. (3-294):

$$\theta_{t=0} = \sum_{i=1}^{\infty} C_i \cos(\lambda_i x) = T_{ini} - T_{\infty} \qquad (3\text{-}319)$$

We found in Section 2.2 that the eigenfunctions are orthogonal to one another. The property of orthogonality ensures that when any two, different eigenfunctions are multiplied together and integrated from one homogeneous boundary to the other, the result will necessarily be zero. Each side of Eq. (3-319) is multiplied by $\cos(\lambda_j x)$ and integrated from $x = 0$ to $x = L$:

$$\sum_{i=1}^{\infty} C_i \int_0^L \cos(\lambda_i x) \cos(\lambda_j x) dx = \int_0^L (T_{ini} - T_{\infty}) \cos(\lambda_j x) dx \qquad (3\text{-}320)$$

The property of orthogonality ensures that the only term on the left side of Eq. (3-320) that is not zero is the one for which $j = i$:

$$C_i \underbrace{\int_0^L \cos^2(\lambda_i x) dx}_{\text{Integral 1}} = (T_{ini} - T_{\infty}) \underbrace{\int_0^L \cos(\lambda_i x) \, dx}_{\text{Integral 2}} \qquad (3\text{-}321)$$

The integrals in Eq. (3-321) can be evaluated conveniently using either integral tables or Maple:

```
> Integral1:=int((cos(lambda*x))^2,x=0..L);
```
$$Integral1 := \frac{1}{2} \frac{\cos(\lambda L) \sin(\lambda L) + \lambda L}{\lambda}$$
```
> Integral2:=int(cos(lambda*x),x=0..L);
```
$$Integral2 := \frac{\sin(\lambda L)}{\lambda}$$

Substituting these results into Eq. (3-321) leads to:

$$C_i \frac{[\cos(\lambda_i L) \sin(\lambda_i L) + \lambda_i L]}{2\lambda_i} = \frac{(T_{ini} - T_{\infty}) \sin(\lambda_i L)}{\lambda_i} \qquad (3\text{-}322)$$

or

$$C_i = \frac{2(T_{ini} - T_{\infty}) \sin(\lambda_i L)}{[\cos(\lambda_i L) \sin(\lambda_i L) + \lambda_i L]} \qquad (3\text{-}323)$$

The result is used to evaluate each constant in EES:

```
"Evaluate constants"
duplicate i=1,Nterm
  C[i]=2*(T_ini-T_infinity)*sin(lambda[i]*L)/(cos(lambda[i]*L)*sin(lambda[i]*L)+lambda[i]*L)
end
```

The solution at a specific time and position is evaluated using Eq. (3-318):

```
x=0.01 [m]
time=1000 [s]
duplicate i=1,Nterm
   theta[i]=C[i]*cos(lambda[i]*x)*exp(-lambda[i]^2*alpha*time)
end
T=T_infinity+sum(theta[1..Nterm])
```

The solution to this bounded, 1-D transient problem is often expressed in terms of a dimensionless position ($\tilde{x}$), defined as:

$$\tilde{x} = \frac{x}{L} \tag{3-324}$$

A dimensionless time can also be defined. There is no characteristic time that appears in the problem statement. However, the diffusive time constant, discussed in Section 3.3.1, provides a convenient characteristic time for the problem that is related to the time required for a thermal wave to penetrate from the surface of the wall to the adiabatic boundary. The diffusive time constant is:

$$\tau_{diff} = \frac{L^2}{4\alpha} \tag{3-325}$$

and so an appropriate dimensionless time would be:

$$\tilde{t} = \frac{t}{\tau_{diff}} = \frac{4t\alpha}{L^2} \tag{3-326}$$

The dimensionless time is typically referred to as the Fourier number (Fo) and, in most textbooks, the factor of 4 is removed from the definition:

$$Fo = \frac{t\alpha}{L^2} \tag{3-327}$$

The dimensionless position and Fourier number can be used to more conveniently specify the position and time in the EES code:

```
x_hat=0.5 [-]                                               "dimensionless position"
x=x_hat*L                                                   "position"
Fo=0.2 [-]                                                  "Fourier number"
time=Fo*L^2/alpha                                           "time"
duplicate i=1,Nterm
   theta[i]=C[i]*cos(lambda[i]*x)*exp(-lambda[i]^2*alpha*time)
end
T=T_infinity+sum(theta[1..Nterm])
```

Figure 3-24 illustrates the temperature as a function of dimensionless position for various values of the Fourier number.

The transient response of a plane wall (as well as the response of a cylinder and sphere) that is initially at a uniform temperature and is subjected to a convective boundary condition can be accessed from the EES library of heat transfer functions. Select Function Info from the Options menu and select Transient Conduction from the

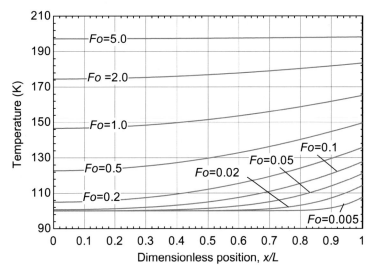

Figure 3-24: Temperature as a function of dimensionless position for various values of the Fourier number (dimensionless time).

pulldown menu in order to access these solutions. Note that the solutions programmed in EES are the first 20 terms of the infinite series solution (i.e., the first 20 terms of Eq. (3-318) for a plane wall). Access and use of these library functions is discussed in more detail in Section 3.5.2.

Limit Behaviors of the Separation of Variables Solution

It is worthwhile spending some time understanding the behavior of the separation of variables solution in order to reinforce some of the concepts that were introduced in earlier sections. Figure 3-24 shows that for Fo less than about 0.20, the wall behaves as a semi-infinite body. For $Fo < 0.2$ (approximately), the semi-infinite body solution listed in Table 3-2 or accessed from the EES function Semilnf3 will provide accurate results. Figure 3-25 illustrates the temperature at $\tilde{x} = 0.25$ as a function of Fo using the separation of variables solution developed in this section (100 terms are used to evaluate the series, so it is close to being exact) and obtained from the SemiInf3 function:

```
T_semiinf=SemiInf3(T_ini,T_infinity,h_bar,k,alpha,L-x,time)
    "temperature evaluated using the semi-infinite body assumption"
```

Note that the semi-infinite body solution is expressed in terms of the distance from the surface that is exposed to the fluid whereas the separation of variables solution is expressed in terms of the distance from the adiabatic surface; therefore, the coordinate transformation $(L\text{-}x)$ is required in the call to the SemiInf3 function.

Figure 3-25 shows that the 100 term separation of variables solution agrees extremely well with the semi-infinite body solution until Fo reaches about 0.20, at which point the thermal wave encounters the adiabatic wall and the problem becomes bounded.

The solution to the problem, Eq. (3-318), expressed in terms of $\tilde{x}$ and Fo, is:

$$\theta(\tilde{x}, Fo) = \sum_{i=1}^{\infty} C_i \cos(\lambda_i L \tilde{x}) \exp[-(\lambda_i L)^2 Fo] \qquad (3\text{-}328)$$

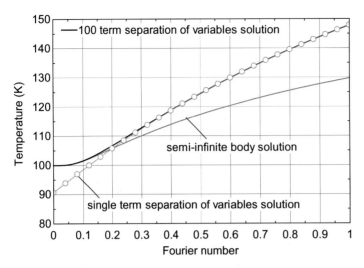

Figure 3-25: Temperature as a function of Fo at $\tilde{x} = 0.25$ evaluated using the separation of variables solution with 100 terms, with 1 term, and using the semi-infinite body function.

where

$$
C_i = \frac{2(T_{ini} - T_\infty) \overbrace{\sin(\lambda_i L)}^{\text{absolute value is } \leq 1}}{\underbrace{\cos(\lambda_i L)\sin(\lambda_i L)}_{\text{absolute value is } \leq 1 \text{ regardless of } i} + \underbrace{\lambda_i L}_{\text{grows with } i}}
\tag{3-329}
$$

It is interesting to look at how quickly the infinite series converges to a solution (i.e., how many terms are actually required?). The value of the constants will decrease in magnitude as they increase in index. Examination of Eq. (3-329) shows that the denominator increases as i increases. Therefore, regardless of $\tilde{x}$, Fo, and Bi, the terms in Eq. (3-328) corresponding to larger values of i will have a decreasing value. This can also be seen by examining the array theta[i] in the EES solution. Further, examination of Eq. (3-328) shows that the value of each term in the series decays in time as $\exp(-(\lambda_i L)^2 Fo)$. Because the value of $\lambda_i L$ increases with i, this decay will occur much more rapidly for the terms with larger values of i. As a result, as the Fourier number increases, fewer and fewer terms are required to achieve an accurate solution and eventually a single term will suffice. Many textbooks tabulate the first constant and first eigenvalue in Eq. (3-328) as a function of Bi and advocate using the "single-term approximation" for large values of Fourier number.

$$
\theta(\tilde{x}, Fo) \approx C_1 \cos(\lambda_1 L \tilde{x}) \exp[-(\lambda_1 L)^2 Fo] \quad \text{if } Fo > 0.2
\tag{3-330}
$$

The single term solution is also shown in Figure 3-25 and is clearly quite accurate for $Fo > 0.2$.

Finally, it is worth revisiting the concept of the Biot number. If the Biot number is very small, then we would expect that temperature gradients within the wall will be negligible and the lumped capacitance model, discussed in Section 3.1, will be accurate. The solution for a lumped capacitance subjected to a step-change in the fluid temperature is provided in Table 3-1:

$$
T = T_\infty + (T_{ini} - T_\infty) \exp\left(-\frac{t}{\tau_{lumped}}\right)
\tag{3-331}
$$

where the lumped capacitance time constant (τ_{lumped}) is:

$$\tau_{lumped} = R_{conv}\, C = \frac{L\,\rho\,c}{\overline{h}} \tag{3-332}$$

Substituting Eq. (3-332) into Eq. (3-331) leads to:

$$T = T_\infty + (T_{ini} - T_\infty)\exp\left(-\frac{\overline{h}\,t}{L\,\rho\,c}\right) \tag{3-333}$$

Multiplying and dividing the argument of the exponential by $k\,L$ allows it to be rewritten as:

$$T = T_\infty + (T_{ini} - T_\infty)\exp\left(-\underbrace{\frac{\overline{h}\,L}{k}}_{Bi}\,\underbrace{\frac{t\,k}{L^2\,\rho\,c}}_{Fo}\right) = T_\infty + (T_{ini} - T_\infty)\exp\left(-Bi\,Fo\right)$$

$$\tag{3-334}$$

The lumped capacitance solution as a function of the product of the Biot number and the Fourier number, Eq. (3-334), is implemented in EES:

```
T_lumped=T_infinity+(T_ini-T_infinity)*exp(-Bi*Fo)
   "temperature evaluated using the lumped capacitance assumption"
```

Figure 3-26 illustrates the temperature predicted by the lumped capacitance model, Eq. (3-334), as a function of the product $Fo\,Bi$. The temperature at the center of the wall ($\hat{x} = 0$) predicted by the separation of variables solution, Eq. (3-328), with 100 terms is also shown in Figure 3-26 for various values of the Biot number.

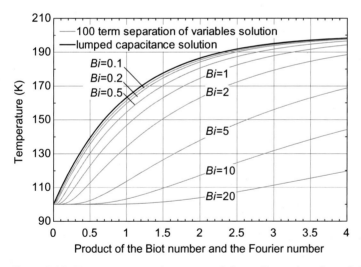

Figure 3-26: Temperature at the center of the wall as a function of the product $Fo\,Bi$ predicted by the lumped capacitance solution and by the separation of variables solution with 100 terms at $\tilde{x} = 0$.

Notice that the separation of variables solution approaches the lumped capacitance solution for $Bi < 0.2$. The separation of variables solution is an exact but complex method for analyzing a transient problem. However, Figure 3-25 and Figure 3-26 show that the solution limits to more simple expressions under certain conditions.

EXAMPLE 3.5-2: TRANSIENT RESPONSE OF A TANK WALL (REVISITED)

EXAMPLE 3.5-2: TRANSIENT RESPONSE OF A TANK WALL (REVISITED)

The separation of variables technique provides a tool that can be used to provide a more quantitative solution to the problem posed in EXAMPLE 3.3-1, which is re-stated here. Figure 1(a) shows a metal wall that separates two tanks of liquid at different temperatures, $T_{hot} = 500$ K and $T_{cold} = 400$ K; initially the wall is at steady state and therefore it has the linear temperature distribution shown in Figure 1(b). The thickness of the wall is $th = 0.8$ cm and its cross-sectional area is $A_c = 1.0\,\mathrm{m}^2$. The properties of the wall material are $\rho = 8000\,\mathrm{kg/m}^3$, $c = 400$ J/kg-K, and $k = 20$ W/m-K. The heat transfer coefficient between the wall and the liquid in either tank is $\bar{h}_{liq} = 5000\,\mathrm{W/m}^2$-K. At time, $t = 0$, both tanks are drained and then exposed to gas at $T_{gas} = 300$ K, as shown in Figure 1(c). The heat transfer coefficient between the walls and the gas is $\bar{h}_{gas} = 100$ W/m²-K. Assume that the process of draining the tanks and filling them with air occurs instantaneously so that the wall has the linear temperature distribution shown in Figure 1(b) at time $t = 0$.

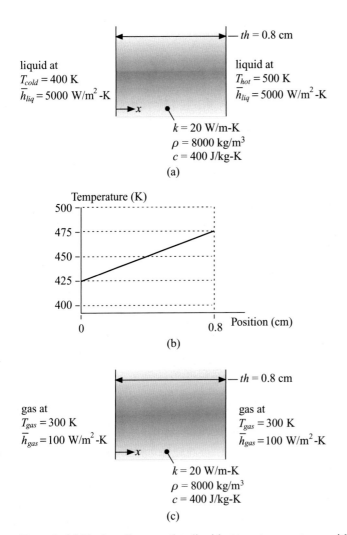

Figure 1: (a) Tank wall exposed to liquid at two temperatures with (b) a linear steady state initial temperature distribution, when (c) at time $t = 0$ both walls are exposed to low temperature gas.

EXAMPLE 3.5-2: TRANSIENT RESPONSE OF A TANK WALL (REVISITED)

a) In EXAMPLE 3.3-1, we calculated a diffusive and lumped time constant for this problem ($\tau_{diff} = 2.6$ s and $\tau_{lumped} = 130$ s) and used these values to sketch the expected temperature distribution at various times. Use the method of separation of variables to obtain a more precise solution.

Note that this problem can not be solved using the planewall_T function in EES because of the non-uniform initial temperature distribution. The separation of variables method must be applied. The known information is entered in EES:

```
"EXAMPLE 3.5-2: Transient Response of a Tank Wall (Revisited)"

$UnitSystem SI MASS RAD PA K J
$Tabstops 0.2 0.4 0.6 0.8 3.5

"Inputs"
k=20 [W/m-K]                          "thermal conductivity"
c=400 [J/kg-K]                        "specific heat capacity"
rho=8000 [kg/m^3]                     "density"
T_cold=400 [K]                        "cold fluid temperature"
T_hot=500 [K]                         "hot fluid temperature"
h_bar_liq=5000 [W/m^2-K]              "liquid-to-wall heat transfer coefficient"
th=0.8[cm]*convert(cm,m)             "wall thickness"
A_c=1 [m^2]                           "wall area"
h_bar_gas=100 [W/m^2-K]              "gas-to-wall heat transfer coefficient"
T_gas=300 [K]                         "gas temperature"
```

The steady state temperature distribution provides the initial condition for the transient problem:

$$T_{t=0} = T_{x=0,t=0} + (T_{x=th,t=0} - T_{x=0,t=0})\frac{x}{th} \tag{1}$$

where $T_{x=0,t=0}$ and $T_{x=th,t=0}$ are the steady-state temperatures at either edge of the wall (Figure 1(b)):

$$T_{x=0,t=0} = T_{cold} + \frac{(T_{hot} - T_{cold})\,R_{conv,liq}}{2\,R_{conv,liq} + R_{cond}}$$

and

$$T_{x=th,t=0} = T_{cold} + \frac{(T_{hot} - T_{cold})\,(R_{conv,liq} + R_{cond})}{2\,R_{conv,liq} + R_{cond}}$$

where $R_{conv,liq}$ and R_{cond} are the convective and conductive resistances:

$$R_{conv,liq} = \frac{1}{h_{liq}\,A_c}$$

$$R_{cond} = \frac{th}{k\,A_c}$$

EXAMPLE 3.5-2: TRANSIENT RESPONSE OF A TANK WALL (REVISITED)

"Initial condition"
R_conv_liq=1/(h_bar_liq*A_c)   "convection resistance with liquid"
R_cond=th/(k*A_c)   "conduction resistance"
T_x0_t0=T_cold+(T_hot-T_cold)*R_conv_liq/(2*R_conv_liq+R_cond)
                                               "initial temperature of cold side of wall"
T_xth_t0=T_cold+(T_hot-T_cold)*(R_conv_liq+R_cond)/(2*R_conv_liq+R_cond)
                                               "initial temperature of hot side of wall"

The governing differential equation is derived in Section 3.5.3:

$$\frac{\partial T}{\partial t} - \alpha \frac{\partial^2 T}{\partial x^2} = 0$$

The spatial boundary conditions are provided by interface balances at either edge of the wall:

$$\overline{h}_{gas} \left(T_{gas} - T_{x=0} \right) = -k \frac{\partial T}{\partial x}\bigg|_{x=0}$$

$$-k \frac{\partial T}{\partial x}\bigg|_{x=th} = \overline{h}_{gas} \left(T_{x=th} - T_{gas} \right)$$

The solution follows the steps that were outlined in Section 3.5.3. The problem as stated cannot be solved using separation of variables because neither of the two spatial boundary conditions are homogeneous. However, the problem can be transformed by defining the temperature difference relative to the gas temperature:

$$\theta = T - T_{gas}$$

The transformed partial differential equation becomes:

$$\alpha \frac{\partial^2 \theta}{\partial x^2} = \frac{\partial \theta}{\partial t} \tag{2}$$

and the initial and boundary conditions become:

$$\theta_{t=0} = \left(T_{x=0,t=0} - T_{gas} \right) + \left(T_{x=th,t=0} - T_{x=0,t=0} \right) \frac{x}{th} \tag{3}$$

$$\overline{h}_{gas} \theta_{x=0} = k \frac{\partial \theta}{\partial x}\bigg|_{x=0} \tag{4}$$

$$-k \frac{\partial \theta}{\partial x}\bigg|_{x=th} = \overline{h}_{gas} \theta_{x=th} \tag{5}$$

Both spatial boundary conditions, Eqs. (4) and (5), are homogeneous and therefore separation of variables can be used on the transformed problem. The solution is assumed to be the product of θX, a function of x, and θt, a function of t. The two ordinary differential equations that result are:

$$\frac{d^2 \theta X}{dx^2} + \lambda^2 \theta X = 0 \tag{6}$$

and

$$\frac{d\theta t}{dt} + \lambda^2 \alpha \theta t = 0 \tag{7}$$

The eigenproblem (the problem in x) will be solved first. The general solution to Eq. (6) is:

$$\theta X = C_1 \sin(\lambda x) + C_2 \cos(\lambda x) \tag{8}$$

Substituting Eq. (8) into the boundary condition at $x = 0$, Eq. (4), leads to:

$$\overline{h}_{gas} \left[C_1 \sin(\lambda 0) + C_2 \cos(\lambda 0) \right] = k \left[C_1 \lambda \cos(\lambda 0) - C_2 \lambda \sin(\lambda 0) \right]$$

or

$$C_2 \overline{h}_{gas} = k\, C_1 \lambda$$

Therefore, Eq. (8) can be written as:

$$\theta X = C_1 \left[\sin(\lambda x) + \frac{k\lambda}{\overline{h}_{gas}} \cos(\lambda x) \right] \tag{9}$$

Substituting Eq. (9) into the boundary condition at $x = th$, Eq. (5), leads to the eigencondition for the problem:

$$-k \left[\lambda \cos(\lambda\, th) - \frac{k\lambda^2}{\overline{h}_{gas}} \sin(\lambda\, th) \right] = \overline{h}_{gas} \left[\sin(\lambda\, th) + \frac{k\lambda}{\overline{h}_{gas}} \cos(\lambda\, th) \right] \tag{10}$$

The eigenvalues are identified automatically using EES, as discussed in Section 3.5.3. The left side of Eq. (10) is moved to the right side in order to identify a residual, *Res*, that must be zero at each eigenvalue:

$$Res = \sin(\lambda\, th) + \frac{k\,(\lambda\, th)}{\overline{h}_{gas}\, th} \cos(\lambda\, th) + \frac{k\,(\lambda\, th)}{\overline{h}_{gas}\, th} \left[\cos(\lambda\, th) - \frac{k\,(\lambda\, th)}{\overline{h}_{gas}\, th} \sin(\lambda\, th) \right] \tag{11}$$

Equation (11) can be simplified by defining the Biot number in the usual way:

$$Bi = \frac{\overline{h}_{gas}\, th}{k}$$

which leads to:

$$Res = \sin(\lambda\, th) + \frac{(\lambda\, th)}{Bi} \cos(\lambda\, th) + \frac{(\lambda\, th)}{Bi} \left[\cos(\lambda\, th) - \frac{(\lambda\, th)}{Bi} \sin(\lambda\, th) \right] \tag{12}$$

Equation (12) is programmed in EES:

```
"Eigenvalues"
Bi=h_bar_gas*th/k                          "Biot number"
Res=sin(lambdath)+lambdath*cos(lambdath)/Bi+lambdath*(cos(lambdath)&
    -lambdath*sin(lambdath)/Bi)/Bi         "Residual"
```

EXAMPLE 3.5-2: TRANSIENT RESPONSE OF A TANK WALL (REVISITED)

and used to generate Figure 2, which shows the residual as a function of $\lambda\,th$.

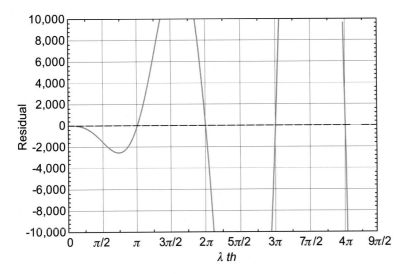

Figure 2: Residual as a function of $\lambda\,th$. Notice that the residual crosses zero between every interval of π.

Figure 2 shows that the roots of Eq. (12) lie in each interval of π. Therefore, upper and lower bounds and appropriate guess values can be identified for each eigenvalue and assigned to arrays that are subsequently used to constrain each value of $\lambda\,th$ (the entries in the array lambdath[ ]) in the Variable Information window:

```
"Eigenvalues"
Bi=h_bar_gas*th/k                                    "Biot number"
{Res=sin(lambdath)+lambdath*cos(lambdath)/Bi+lambdath*(cos(lambdath)&
  -lambdath*sin(lambdath)/Bi)/Bi                     "Residual"}
Nterm=10
duplicate i=1,Nterm
  lowerlimit[i]=(i-1)*pi                             "lower limit"
  upperlimit[i]=i*pi                                 "upper limit"
  guess[i]=(i-0.5)*pi                                "guess value"
  sin(lambdath[i])+lambdath[i]*cos(lambdath[i])/Bi+lambdath[i]*(cos(lambdath[i])&
    -lambdath[i]*sin(lambdath[i])/Bi)/Bi=0           "eigencondition"
end
```

The EES code will identify each of the eigenvalues. The solution for each eigenvalue is therefore:

$$\theta X_i = C_{1,i}\left[\sin\left(\lambda_i\,x\right) + \frac{k\,\lambda_i}{h_{gas}}\cos\left(\lambda_i\,x\right)\right] \qquad (13)$$

The solution to the ordinary differential equation in time, Eq. (7), for each eigenvalue is:

$$\theta t_i = C_{3,i}\,\exp\left(-\lambda_i^2\,\alpha\,t\right)$$

EXAMPLE 3.5-2: TRANSIENT RESPONSE OF A TANK WALL (REVISITED)

and therefore the solution for each eigenvalue is:

$$\theta_i = \theta x_i \, \theta t_i = C_i \left[\sin(\lambda_i \, x) + \frac{k\lambda_i}{h_{gas}} \cos(\lambda_i \, x) \right] \exp\left(-\lambda_i^2 \, \alpha \, t\right)$$

The sum of the solutions for each eigenvalue is the series solution to the problem:

$$\theta = \sum_{i=1}^{\infty} \theta_i = \sum_{i=1}^{\infty} C_i \left[\sin(\lambda_i \, x) + \frac{k\lambda_i}{h_{gas}} \cos(\lambda_i \, x) \right] \exp\left(-\lambda_i^2 \, \alpha \, t\right) \tag{14}$$

The series solution can be expressed in terms of dimensionless position ($\tilde{x}$) and Fourier number (Fo):

$$\theta = \sum_{i=1}^{\infty} C_i \left[\sin(\lambda_i \, th \, \tilde{x}) + \frac{\lambda_i \, th}{Bi} \cos(\lambda_i \, th \, \tilde{x}) \right] \exp\left[-(\lambda_i \, th)^2 \, Fo\right] \tag{15}$$

where $\tilde{x}$ and Fo are defined as:

$$\tilde{x} = \frac{x}{th}$$

$$Fo = \frac{\alpha \, t}{th^2}$$

The constants must be selected so that the initial condition, Eq. (3), is satisfied:

$$\theta_{t=0} = \sum_{i=1}^{\infty} C_i \left[\sin(\lambda_i \, th \, \hat{x}) + \frac{\lambda_i \, th}{Bi} \cos(\lambda_i \, th \, \tilde{x}) \right]$$

$$= \left(T_{x=0,t=0} - T_{gas}\right) + \left(T_{x=th} - T_{ss,x=0}\right)\hat{x} \tag{16}$$

The eigenfunctions must be orthogonal. Therefore, multiplying both sides of Eq. (16) by the jth eigenfunction and integrating from $x = 0$ to $x = th$ (or $\tilde{x} = 0$ to $\tilde{x} = 1$) leads to:

$$\sum_{i=1}^{\infty} \int_0^1 C_i \left[\sin(\lambda_i \, th \, \tilde{x}) + \frac{\lambda_i \, th}{Bi} \cos(\lambda_i \, th \, \tilde{x}) \right] \left[\sin(\lambda_j \, th \, \tilde{x}) + \frac{\lambda_j \, th}{Bi} \cos(\lambda_j \, th \, \tilde{x}) \right] d\tilde{x}$$

$$= \int_0^1 \left[T_{x=0,t=0} - T_{gas} + \left(T_{x=th,t=0} - T_{x=0,t=0}\right)\tilde{x} \right] \left[\sin(\lambda_j \, th \, \tilde{x}) + \frac{\lambda_j \, th}{Bi} \cos(\lambda_j \, th \, \tilde{x}) \right] d\tilde{x}$$

The only term on the left side of the equation that does not integrate to zero is $i = j$ and therefore:

$$\underbrace{C_i \int_0^1 \left[\sin(\lambda_i \, th \, \tilde{x}) + \frac{\lambda_i \, th}{Bi} \cos(\lambda_i \, th \, \tilde{x}) \right]^2 d\tilde{x}}_{(\text{Integral1})_i}$$

$$\tag{17}$$

$$= \underbrace{\int_0^1 \left[T_{x=0,t=0} - T_{gas} + \left(T_{x=th,t=0} - T_{x=0,t=0}\right)\tilde{x} \right] \left[\sin(\lambda_i \, th \, \tilde{x}) + \frac{\lambda_i \, th}{Bi} \cos(\lambda_i \, th \, \tilde{x}) \right] d\tilde{x}}_{(\text{Integral 2})_i}$$

EXAMPLE 3.5-2: TRANSIENT RESPONSE OF A TANK WALL (REVISITED)

EXAMPLE 3.5-2: TRANSIENT RESPONSE OF A TANK WALL (REVISITED)

The integrals in Eq. (17) seem imposing, but they can be accomplished relatively easily using Maple and then copied into EES for evaluation. Equation (17) is written in terms of the two integrals:

$$C_i \, (\text{Integral 1})_i = (\text{Integral 2})_i$$

where

$$(\text{Integral 1})_i = \int_0^1 \left[\sin(\lambda_i \, th \, \tilde{x}) + \frac{\lambda_i \, th}{Bi} \cos(\lambda_i \, th \, \tilde{x}) \right]^2 d\tilde{x}$$

and

$$(\text{Integral 2})_i = \int_0^1 \left[(T_{x=0,t=0} - T_{gas}) + (T_{x=th,t=0} - T_{x=0,t=0}) \tilde{x} \right]$$
$$\times \left[\sin(\lambda_i \, th \, \tilde{x}) + \frac{\lambda_i \, th}{Bi} \cos(\lambda_i \, th \, \tilde{x}) \right] d\tilde{x}$$

These integrals are entered in Maple:

```
> restart;
> Integral_1[i]:=int((sin(lambdath[i]*x_hat)+lambdath[i]*cos(lambdath[i]*x_hat)/Bi)^2,x_hat=0..1);
```

$Integral\_1_i := \frac{1}{2}(2\,Bi\,lambdath_i - Bi^2\cos(lambdath_i)\sin(lambdath_i) + Bi^2(lambdath_i)$
$\quad - 2\,Bi\,lambdath_i \cos(lambdath_i)^2 + lambdath_i^2 \cos(lambdath_i)\sin(lambdath_i)$
$\quad + lambdath_i^3)/(Bi^2\,lambdath_i)$

```
> Integral_2[i]:=int((T_x0_t0-T_gas+(T_xth_t0
T_x0_t0)*x_hat)*(sin(lambdath[i]*x_hat)+lambdath[i]*cos(lambdath[i]*x_hat)/Bi),x_hat=0..1);
```

$Integral\_2_i := -(-T\_x0\_t0\,Bi\,lambdath_i + T\_gas\,Bi\,lambdath_i + T\_xth\_t0\,lambdath_i$
$\quad - T\_x0\_t0\,lambdath_i - T\_gas\,Bi\cos(lambdath_i)\,lambdath_i$
$\quad + T\_gas\,lambdath_i^2 \sin(lambdath_i) - T\_xth\_t0\,Bi\sin(lambdath_i)$
$\quad + T\_xth\_t0\,Bi\,lambdath_i\cos(lambdath_i) - T\_xth\_t0\,lambdath_i\cos(lambdath_i)$
$\quad - T\_xth\_t0\,lambdath_i^2\sin(lambdath_i) + T\_x0\_t0\,Bi\sin(lambdath_i)$
$\quad + T\_x0\_t0\,lambdath_i\cos(lambdath_i))/(Bi\,lambdath_i^2)$

and copied into EES to evaluate each of the constants. The only modification required to the Maple results is to change the := to =

```
"Evaluate Constants"
duplicate i=1,Nterm
  Integral_1[i] = 1/2*(2*Bi*lambdath[i]-Bi^2*cos(lambdath[i])*sin(lambdath[i])&
    +Bi^2*lambdath[i]-2*Bi*lambdath[i]*cos(lambdath[i])^2+lambdath[i]^2*&
    cos(lambdath[i])*sin(lambdath[i])+lambdath[i]^3)/Bi^2/lambdath[i]
  Integral_2[i] = -(-T_x0_t0*Bi*lambdath[i]+T_gas*Bi*lambdath[i]+T_xth_t0*lambdath[i]&
    -T_x0_t0*lambdath[i]-T_gas*Bi*cos(lambdath[i])*lambdath[i]+T_gas*lambdath[i]^2&
    *sin(lambdath[i])-T_xth_t0*Bi*sin(lambdath[i])+T_xth_t0*Bi*lambdath[i]*cos(lambdath[i])&
    -T_xth_t0*lambdath[i]*cos(lambdath[i])-T_xth_t0*lambdath[i]^2*sin(lambdath[i])&
    +T_x0_t0*Bi*sin(lambdath[i])+T_x0_t0*lambdath[i]*cos(lambdath[i]))/Bi/lambdath[i]^2
  C[i]*Integral_1[i]=Integral_2[i]
end
```

EXAMPLE 3.5-2: TRANSIENT RESPONSE OF A TANK WALL (REVISITED)

The solution can be obtained at arbitrary values of position and time:

```
"Solution"
alpha=k/(rho*c)                          "thermal diffusivity"
x_hat=0                                  "dimensionless position"
x_hat=x/th
Fo=alpha*time/th^2                       "Fourier number"
time=5 [s]
duplicate i=1,Nterm
    theta[i]=C[i]*(sin(lambdath[i]*x_hat)+lambdath[i]*cos(lambdath[i]*x_hat)/Bi)*exp(-lambdath[i]^2*Fo)
end
T=T_gas+sum(theta[1..Nterm])
```

Figure 3 shows the temperature as a function of position in the wall for the same times ($t = 0$ s, 0.5 s, 5 s, 50 s, 500 s, and 5000 s) that were sketched in EXAMPLE 3.3-1.

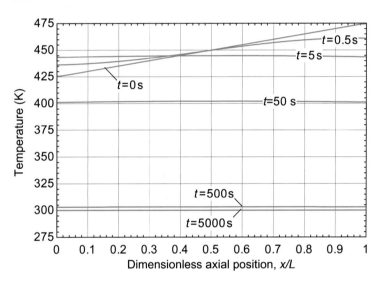

Figure 3: Temperature distribution within the wall at various times as it equilibrates, predicted by the separation of variables model. The times indicated are the same as those considered in EXAMPLE 3.3 -1.

Figure 3 shows the same trends that were discussed in EXAMPLE 3.3-1; there is an internal, conductive equilibration process that occurs very quickly (with a time scale that is similar to $\tau_{diff} = 2.6$ s). The Fourier number associated with this internal equilibration process will be on the order of 1.0, because the Fourier number is defined as the ratio of time to the time required for a thermal wave to move across the wall. There is subsequently an external, convective equilibration process that occurs much more slowly (with a time scale that is similar to $\tau_{lumped} = 130$ s). The Fourier-Biot number product associated with this external equilibration process will be on the order of 1.0 because the Fourier-Biot number product characterizes the ratio of time to the lumped time constant.

3.5.4 Separation of Variables Solutions in Cylindrical Coordinates

This extended section of the book can be found on the website (www.cambridge.org/nellisandklein). The separation of variables solution for transient problems can be applied in cylindrical coordinates. The solution for a cylinder exposed to a step-change in ambient conditions that is presented in Section 3.5.2 was derived using separation of variables in cylindrical coordinates. In this section, the techniques required to solve this problem are presented.

3.5.5 Non-homogeneous Boundary Conditions

This extended section of the book can be found on the website (www.cambridge.org/nellisandklein). The method of separation of variables can only be applied to 1-D transient problems where both spatial boundary conditions are homogeneous. In Sections 3.5.3 and 3.5.4, a single, obvious transformation is sufficient to make both spatial boundary conditions homogeneous. In many problems this will not be the case and more advanced techniques will be required. Section 2.3.2 discusses methods for breaking 2-D steady problems with non-homogeneous terms into sub-problems that can be solved either by separation of variables or by the solution of an ordinary differential equation. In Section 2.4, superposition for 2-D steady-state problems is discussed. These techniques for solving problems with non-homogeneous boundary conditions using separation of variables remain valid for 1-D transient problems and are presented in Section 3.5.5.

3.6 Duhamel's Theorem

This extended section of the book can be found on the website (www.cambridge.org/nellisandklein). The separation of variables technique discussed in Section 3.5 is not capable of solving problems with time-dependent spatial boundary conditions (e.g., an ambient temperature or heat flux that varies with time). However, problems with time dependent spatial boundary conditions are common. Duhamel's theorem provides one method of extending an analytical solution that is derived (for example, using separation of variables) assuming a time-invariant boundary condition in order to consider the temperature response to an arbitrary time variation of that boundary condition.

3.7 Complex Combination

This extended section of the book can be found on the website (www.cambridge.org/nellisandklein). Complex combination is a useful technique for solving problems that have periodic (i.e., oscillating) boundary conditions or forcing functions. This type of problem was encountered in EXAMPLE 3.1-2 where a temperature sensor (treated as a lumped capacitance) was exposed to an oscillating fluid temperature. In EXAMPLE 3.1-2, the problem is solved analytically and the temperature response of the sensor is found to be the sum of a homogeneous solution that decays to zero (as time became sufficiently greater than the time constant of the sensor), and a particular solution that is the sustained response. Complex combination is a convenient method for obtaining only this sustained solution, which is often the only portion of the solution that is of interest. Complex combination can be used for transient problems that are 0-D (i.e., lumped), 1-D, and even 2-D or 3-D.

3.8 Numerical Solutions to 1-D Transient Problems

3.8.1 Introduction

Sections 1.4 and 1.5 discuss numerical solutions to steady-state 1-D problems and Section 3.2 discusses the numerical solution to 0-D (lumped) transient problems. In this section, these concepts are extended in order to numerically solve 1-D transient problems.

The underlying methodology is the same; a set of equations is obtained from energy balances on small (but not differentially small) control volumes that are distributed throughout the computational domain. These equations will contain energy storage terms in addition to energy transfer terms (due to conduction, convection, and/or radiation) because of the transient nature of the problem. The energy storage terms involve the time rate of temperature change and therefore must be numerically integrated forward in time using one of the techniques that was discussed previously in Section 3.2 (e.g., the Crank-Nicolson technique). Most software applications, including EES and MATLAB, provide advanced numerical integration capabilities that can be applied to this problem.

3.8.2 Transient Conduction in a Plane Wall

The process of obtaining a numerical solution to a 1-D, transient conduction problem will be illustrated in the context of a plane wall subjected to a convective boundary condition on one surface, as shown in Figure 3-41.

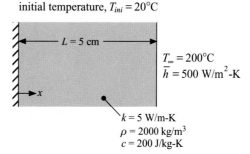

Figure 3-41: A plane wall exposed to a convective boundary condition at time $t = 0$.

The plane wall has thickness $L = 5.0$ cm and properties $k = 5.0$ W/m-K, $\rho = 2000$ kg/m³, and $c = 200$ J/kg-K. The wall is initially at $T_{ini} = 20°C$ when at time $t = 0$, the surface (at $x = L$) is exposed to fluid at $T_\infty = 200°C$ with average heat transfer coefficient $\bar{h} = 500$ W/m²-K. The wall at $x = 0$ is adiabatic. The known information is entered into EES.

```
$UnitSystem SI MASS RAD PA K J
$TABSTOPS 0.2 0.4 0.6 0.8 3.5 in

"inputs"
L=5 [cm]*convert(cm,m)              "wall thickness"
k=5.0 [W/m-K]                       "conductivity"
rho=2000 [kg/m^3]                   "density"
c=200 [J/kg-K]                      "specific heat capacity"
T_ini=converttemp(C,K,20 [C])      "initial temperature"
T_infinity=converttemp(C,K,200 [C]) "fluid temperature"
h_bar=500 [W/m^2-K]                "heat transfer coefficient"
```

The plane wall problem in Figure 3-41 corresponds to the problem that is solved using separation of variables in Section 3.5.3. This problem was selected so that the solution that is obtained numerically can be directly compared to the analytical solution accessed by the planewall_T function in EES that is described in Section 3.5.2. However, the major advantage of using a numerical technique is the capability to easily solve complex problems that may not have an analytical solution.

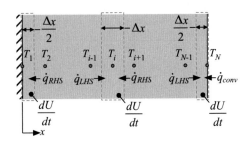

Figure 3-42: Nodes and control volumes distributed uniformly throughout computational domain.

The first step in developing the numerical solution is to partition the continuous medium into a large number of small volumes that are analyzed using energy balances in order to develop the set of equations that must be solved. The nodes (i.e., the positions at which the temperature will be obtained) are distributed uniformly through the wall, as shown in Figure 3-42.

For the uniform distribution of nodes that is shown in Figure 3-42, the location of each node (x_i) is:

$$x_i = \frac{(i-1)}{(N-1)} L \quad \text{for } i = 1..N \tag{3-495}$$

where N is the number of nodes used for the simulation. The distance between adjacent nodes (Δx) is:

$$\Delta x = \frac{L}{(N-1)} \tag{3-496}$$

This node spacing is specified in EES:

```
"Setup grid"
N=6 [-]                                     "number of nodes"
duplicate i=1,N
  x[i]=(i-1)*L/(N-1)                        "position of each node"
end
DELTAx=L/(N-1)                              "distance between adjacent nodes"
```

A control volume is defined around each node. The control surface bisects the distance between the nodes, as shown in Figure 3-42. An energy balance must be written for the control volume associated with every node. The control volume for an arbitrary, internal node experiences conduction heat transfer with the adjacent nodes as well as energy storage (as shown in Figure 3-42):

$$\dot{q}_{LHS} + \dot{q}_{RHS} = \frac{dU}{dt} \tag{3-497}$$

Each term in Eq. (3-497) must be approximated. The conduction terms from the adjacent nodes are modeled according to:

$$\dot{q}_{LHS} = \frac{k A_c (T_{i-1} - T_i)}{\Delta x} \tag{3-498}$$

$$\dot{q}_{RHS} = \frac{k A_c (T_{i+1} - T_i)}{\Delta x} \tag{3-499}$$

where A_c is the cross-sectional area of the wall. The rate of energy storage is the product of the time rate of change of the nodal temperature and the thermal mass of the control volume:

$$\frac{dU}{dt} = A_c \, \Delta x \, \rho \, c \, \frac{dT_i}{dt} \tag{3-500}$$

Substituting Eqs. (3-498) through (3-500) into Eq. (3-497) leads to:

$$A_c \, \Delta x \, \rho \, c \, \frac{dT_i}{dt} = \frac{k \, A_c \, (T_{i-1} - T_i)}{\Delta x} + \frac{k \, A_c \, (T_{i+1} - T_i)}{\Delta x} \quad \text{for } i = 2 ... (N-1) \tag{3-501}$$

Solving for the time rate of the temperature change:

$$\frac{dT_i}{dt} = \frac{k}{\Delta x^2 \, \rho \, c} (T_{i-1} + T_{i+1} - 2 \, T_i) \quad \text{for } i = 2 ... (N-1) \tag{3-502}$$

The control volumes at the boundaries must be treated separately because they have a smaller volume and experience different energy transfers. An energy balance on the control volume at the adiabatic wall (node 1 in Figure 3-42) leads to:

$$\dot{q}_{RHS} = \frac{dU}{dt} \tag{3-503}$$

or

$$\frac{A_c \, \Delta x \, \rho \, c}{2} \frac{dT_1}{dt} = \frac{k \, A_c \, (T_2 - T_1)}{\Delta x} \tag{3-504}$$

The factor of two on the left side of Eq. (3-504) results because the control volume around node 1 has half the width and thus half the heat capacity of the other nodes. Solving for the time rate of temperature change for node 1:

$$\frac{dT_1}{dt} = \frac{2k}{\rho \, c \, \Delta x^2} (T_2 - T_1) \tag{3-505}$$

An energy balance on the control volume for the node located at the outer surface (node N in Figure 3-42) leads to:

$$\frac{dU}{dt} = \dot{q}_{LHS} + \dot{q}_{conv} \tag{3-506}$$

or

$$\frac{A_c \, \Delta x \, \rho \, c}{2} \frac{dT_N}{dt} = \frac{k \, A_c \, (T_{N-1} - T_N)}{\Delta x} + \overline{h} \, A_c \, (T_\infty - T_N) \tag{3-507}$$

Solving for the time rate of temperature change for node N:

$$\frac{dT_N}{dt} = \frac{2 \, k}{\rho \, c \, \Delta x^2} (T_{N-1} - T_N) + \frac{2 \overline{h}}{\Delta x \, \rho \, c} (T_\infty - T_N) \tag{3-508}$$

Equations (3-502), (3-505), and (3-508) provide the time rate of change for the temperature of every node, given the temperatures of the nodes. This result is similar to the situation encountered in Section 3.2 when developing numerical solutions for lumped capacitance problems. In that case, the energy balance for the single control volume (around the object being studied) provided an equation for the time rate of change of the temperature in terms of the temperature. Here, the energy balance written for each of the lumped capacitances (i.e., each of the control volumes) has provided a set of equations for the time rates of change of each of the nodal temperatures. In order to solve the problem, it is necessary to integrate these time derivatives. All of the numerical integration techniques that were discussed in Section 3.2 to solve lumped capacitance problems can be applied here to solve 1-D transient problems.

The temperature of each node is a function both of position (x) and time (t). The index that specifies the node's position is i where $i = 1$ corresponds to the adiabatic wall and $i = N$ corresponds to the surface of the wall (see Figure 3-42). A second index, j, is added to each nodal temperature in order to indicate the time ($T_{i,j}$), where $j = 1$ corresponds to the beginning of the simulation and $j = M$ corresponds to the end of the simulation. The total simulation time, t_{sim}, is divided into M time steps. Most of the techniques discussed here will divide the simulation time into time steps of equal duration, Δt, although other distributions may be used.

$$\Delta t = \frac{t_{sim}}{(M - 1)} \tag{3-509}$$

The time associated with any time step is:

$$t_j = (j - 1)\,\Delta t \quad \text{for } j = 1...M \tag{3-510}$$

An array (time []) that provides the time for each step is defined:

```
"Setup time steps"
M=21 [-]                              "number of time steps"
t_sim=40 [s]                         "simulation time"
DELTAtime=t_sim/(M-1)               "time step duration"
duplicate j=1,M
  time[j]=(j-1)* DELTAtime
end
```

The initial conditions for this problem specifies that all of the temperatures at $t = 0$ are equal to T_{ini}.

$$T_{i,1} = T_{ini} \quad \text{for } i = 1...N \tag{3-511}$$

```
duplicate i=1,N
  T[i,1]=T_ini                       "initial condition"
end
```

Note that the variable T is a two-dimensional array (i.e., a matrix) and calculated results will be displayed in the Arrays Window. The use of a 2-D array simplifies the presentation of the methodology, but it requires unnecessary variable storage space in EES. MATLAB is a more appropriate tool for these types of simulations and the implementation of the numerical simulation in MATLAB will be discussed subsequently.

Euler's Method
The temperature of all of the nodes at the end of each time step (i.e., $T_{i,j+1}$ for all $i = 1$ to N) must be computed given the temperatures at the beginning of the time step (i.e., $T_{i,j}$ for all $i=1$ to N) and the algebraic equations derived from the energy balances (i.e., Eqs. (3-502), (3-505), and (3-508)). The simplest technique for numerical integration is

Euler's Method, which approximates the time rate of temperature change within each time step as being constant and equal to its value at the beginning of the time step. Therefore, for any node i during time step j:

$$T_{i,j+1} = T_{i,j} + \left. \frac{dT}{dt} \right|_{T=T_{i,j}, t=t_j} \Delta t \quad \text{for } i = 1...N \tag{3-512}$$

Note that it is often useful to develop a numerical simulation of a transient process by initially taking only a single step and then, once that works, automating the process of simulating all of the time steps. For example, the temperatures of all N nodes at the end of the first time step (i.e., $j = 2$) are determined from:

$$T_{i,2} = T_{i,1} + \left. \frac{dT}{dt} \right|_{T=T_{i,1}, t=t_1} \Delta t \quad \text{for } i = 1...N \tag{3-513}$$

Substituting Eqs. (3-502), (3-505), and (3-508), into Eq. (3-513) leads to:

$$T_{1,2} = T_{1,1} + \frac{2k}{\rho c \, \Delta x^2} (T_{2,1} - T_{1,1}) \, \Delta t \tag{3-514}$$

$$T_{i,2} = T_{i,1} + \frac{k}{\Delta x^2 \rho c} (T_{i-1,1} + T_{i+1,1} - 2 \, T_{i,1}) \, \Delta t \quad \text{for } i = 2...(N-1) \tag{3-515}$$

$$T_{N,2} = T_{N,1} + \left[\frac{2k}{\rho c \, \Delta x^2} (T_{N-1,1} - T_{N,1}) + \frac{2\overline{h}}{\Delta x \rho c} (T_f - T_{N,1}) \right] \Delta t \tag{3-516}$$

Equations (3-514) through (3-516) are programmed in EES:

```
"Take a single Euler step"
T[1,2]=T[1,1]+2*k*(T[2,1]-T[1,1])*DELTAtime/(rho*c*DELTAx^2)          "node 1"
duplicate i=2,(N-1)
  T[i,2]=T[i,1]+k*(T[i-1,1]+T[i+1,1]-2*T[i,1])*DELTAtime/(rho*c*DELTAx^2)
                                                                     "internal nodes"
end
T[N,2]=T[N,1]+(2*k*(T[N-1,1]-T[N,1])/(rho*c*DELTAx^2)+&
  2*h_bar*(T_infinity-T[N,1])/(rho*c*DELTAx))*DELTAtime               "node N"
```

Solving the program should provide a solution for the temperature at each node at the end of the first time step. The solution can be examined by selecting Arrays from the Windows menu. The equations used to move through any arbitrary time step j follow logically from Eqs. (3-514) through (3-516):

$$T_{1,j+1} = T_{1,j} + \frac{2k}{\rho c \, \Delta x^2} (T_{2,j} - T_{1,j}) \, \Delta t \tag{3-517}$$

$$T_{i,j+1} = T_{i,j} + \frac{k}{\Delta x^2 \rho c} (T_{i-1,j} + T_{i+1,j} - 2 \, T_{i,j}) \, \Delta t \quad \text{for } i = 2...(N-1) \tag{3-518}$$

$$T_{N,j+1} = T_{N,j} + \left[\frac{2k}{\rho c \, \Delta x^2} (T_{N-1,j} - T_{N,j}) + \frac{2\overline{h}}{\Delta x \rho c} (T_\infty - T_{N,j}) \right] \Delta t \tag{3-519}$$

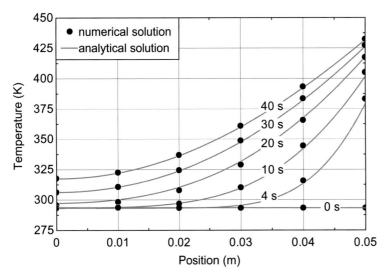

3-43: Temperature as a function of position predicted by Euler's method at various times.

Equations (3-517) through (3-519) are automated in order to simulate all of the time steps by placing the EES code within a second duplicate loop that steps from $j = 1$ to $j = (M - 1)$. This is accomplished by nesting the entire EES code associated with the first time step within an outer duplicate loop; wherever the second index was 2 (i.e., the end of time step 1) we replace it with $j + 1$ and wherever the second index was 1 (i.e., the beginning of time step 1) we replace it with j. The revised code is shown below; the additional or changed code is shown in bold:

```
"Move through all of the time steps"
duplicate j=1,(M-1)
    T[1,j+1]=T[1,j]+2*k*(T[2,j]-T[1,j])*DELTAtime/(rho*c*Deltax^2)          "node 1"
    duplicate i=2,(N-1)
    T[i,j+1]=T[i,j]+k*(T[i-1,j]+T[i+1,j]-2*T[i,j])*DELTAtime/(rho*c*Deltax^2)

                                                                            "internal nodes"

    end
    T[N,j+1]=T[N,j]+(2*k*(T[N-1,j]-T[N,j])/(rho*c*DELTAx^2)+&
        2*h_bar*(T_infinity-T[N,j])/(rho*c*DELTAx))*DELTAtime               "node N"
end
```

Figure 3-43 illustrates the temperature as a function of position at $t = 0$ s, $t = 4$ s, $t = 10$ s, $t = 20$ s, $t = 30$ s, and $t = 40$ s. Figure 3-43 was generated by selecting New Plot Window from the Plots menu and selecting the array x[ ] for the X-Axis and the arrays T[i,1], T[i,3], T[i,6], etc. for the Y-Axis (note that time[1] = 0, time[3] = 4 s, time[6] = 10 s, etc. You will be prompted that you are missing data in various rows because the time[ ] array is longer than the x[ ] or any of the T[i,j] arrays. Select Yes in response to this message in order to continue plotting the points.

The numerical solution can be compared to the analytical solution derived in Section 3.5.3 and accessed using the planewall_T function.

```
"Analytical solution"
alpha=k/(rho*c)
duplicate j=1,M
  duplicate i=1,N
    T_an[i,j]=planewall_T(x[i], time[j], T_ini, T_infinity, alpha, k, h_bar, L)
  end
end
```

The numerical and analytical solutions agree, as shown in Figure 3-43. If the duration of the time step is increased from 2.0 s to 4.0 s (by reducing M from 21 to 11) then the solution becomes unstable (try it and see; the solution will oscillate between large positive and negative temperatures). The existence of a stability limit is one of the key disadvantages associated with the Euler technique, as we saw in Section 3.2 for lumped capacitance problems. The maximum time step that can be used before the solution becomes unstable (i.e., the critical time step, Δt_{crit}) can be determined by examining the algebraic equations that are used to step through time. Rearranging Eq. (3-517), which governs the behavior of the node at the adiabatic edge, leads to:

$$T_{1,j+1} = T_{1,j}\left[1 - \frac{2k\,\Delta t}{\rho c\,\Delta x^2}\right] + \frac{2k\,\Delta t}{\rho c\,\Delta x^2}T_{2,j} \tag{3-520}$$

The solution will become unstable when the coefficient multiplying $T_{1,j}$ becomes negative. Therefore, Eq. (3-520) shows that the solution will tend to become unstable as Δt becomes larger or Δx becomes smaller. The critical time step associated with node 1 is:

$$\Delta t_{crit,1} = \frac{\rho c\,\Delta x^2}{2k} \tag{3-521}$$

Applying the same process to Eq. (3-518), which governs the behavior of the internal nodes, leads to:

$$T_{i,j+1} = T_{i,j}\left[1 - \frac{2k\,\Delta t}{\Delta x^2\,\rho c}\right] + \frac{k\,\Delta t}{\Delta x^2\,\rho c}\left(T_{i-1,j} + T_{i+1,j}\right) \quad \text{for } i = 2 \dots (N-1) \tag{3-522}$$

According to Eq. (3-522), the critical time step for the internal nodes is the same as for node 1:

$$\Delta t_{crit,i} = \frac{\rho c\,\Delta x^2}{2k} \quad \text{for } i = 2 \dots (N-1) \tag{3-523}$$

Equation (3-519), which governs the behavior of the node placed on the surface of the wall, is rearranged:

$$T_{N,j+1} = T_{N,j}\left[1 - \frac{2k\,\Delta t}{\rho c\,\Delta x^2} - \frac{2\bar{h}\,\Delta t}{\Delta x\,\rho c}\right] + \frac{2k\,\Delta t}{\rho c\,\Delta x^2}T_{N-1,j} + \frac{2h\,\Delta t}{\Delta x\,\rho c}T_f \tag{3-524}$$

Equation (3-524) leads to a different critical time step for node N:

$$\Delta t_{crit,N} = \frac{\Delta x\,\rho c}{2\left[\dfrac{k}{\Delta x} + \bar{h}\right]} \tag{3-525}$$

Comparing Eq. (3-525) with Eq. (3-523) indicates that the critical time step for node N will always be somewhat less than the critical time step for the other nodes and therefore Eq. (3-525) will govern the stability of the problem. The critical time step is calculated according to:

For the problem considered here, the critical time step $\Delta t_{crit,N} = 2.0$ s. Note that even a stable solution may not be sufficiently accurate.

We followed a similar process in Section 3.2 in order to identify the critical time step associated with the explicit integration of a lumped capacitance problem and found that the critical time step was related to the time constant of the object. Equations (3-521), (3-523), and (3-525) essentially restate this conclusion, but for an individual node rather than a lumped object. Recall that the time constant of an object is equal to the product of the heat capacity of the object and its thermal resistance to the environment. The heat capacity of an internal node (C_i) is:

$$C_i = \Delta x A_c \rho c \tag{3-526}$$

and the resistance (R_i) between the nodal temperature of an internal node and its environment (i.e., the adjacent nodes) is:

$$R_i = \left[\frac{k A_c}{\Delta x} + \frac{k A_c}{\Delta x}\right]^{-1} \tag{3-527}$$

or

$$R_i = \frac{\Delta x}{2 k A_c} \tag{3-528}$$

The lumped time constant of an internal node ($\tau_{lumped,i}$) is therefore:

$$\tau_{lumped,i} = R_i C_i = \frac{\Delta x}{2 k A_c} \Delta x A_c \rho c = \frac{\rho c \Delta x^2}{2 k} \tag{3-529}$$

and therefore the time constant of an internal node is exactly equal to its critical time step, Eq. (3-523). The time constant for the node located on the surface of the wall (node N) is smaller and therefore its critical time constant will be smaller. The heat capacity of node N (C_N) is:

$$C_N = \frac{\Delta x}{2} A_c \rho c \tag{3-530}$$

and R_N, the thermal resistance between node N and its environment (i.e., node $N-1$ and the fluid), is:

$$R_N = \left[\frac{k A_c}{\Delta x} + \bar{h} A_c\right]^{-1} \tag{3-531}$$

The time constant for node N is:

$$\tau_{lumped,N} = R_N C_N = \frac{\Delta x A_c \rho c}{2\left[\frac{k A_c}{\Delta x} + \bar{h} A_c\right]} = \frac{\rho c \Delta x}{2\left[\frac{k}{\Delta x} + \bar{h}\right]} \tag{3-532}$$

which is identical to the critical time step for node N, Eq. (3-525).

In Section 1.4 we found that the accuracy of a numerical solution is related to the size of the control volumes. Smaller values of Δx will provide more accurate solutions. However, smaller values of Δx will simultaneously reduce the thermal mass of each node as well as the resistance between the node and its adjacent nodes. Thus the time constant and therefore the critical time step duration for any node scales with Δx^2 (see Eq. (3-529)). As a result, reducing Δx will greatly increase the number of time steps required (by an explicit numerical technique) in order to maintain stability. It is often the case that the number of time steps required by stability is much larger than would be required for sufficient accuracy and therefore the use of explicit techniques can be computationally intensive.

Implementing Euler's Method within MATLAB is a straightforward extension of the EES code. The simulation in MATLAB is setup in a script. The inputs are provided as the first lines:

```
clear all;

% Inputs
L=0.05;                              % wall thickness (m)
k=5.0;                               % conductivity (W/m-K)
rho=2000;                            % density (kg/m^3)
c=200;                               % specific heat capacity (J/kg-K)
T_ini=293.2;                         % initial temperature (K)
T_infinity=473.2;                    % fluid temperature (K)
h_bar=500;                           % heat transfer coefficient (W/m^2-K)
```

The axial location of each node and the time steps are specified:

```
%Setup grid
N=11;                                % number of nodes (-)
for i=1:N
    x(i)=(i-1)*L/(N-1);              % position of each node (m)
end
DELTAx=L/(N-1);                      % distance between adjacent nodes (m)

%Setup time steps
M=81;                                % number of time steps (-)
t_sim=40;                            % simulation time (s)
DELTAtime=t_sim/(M-1);               % time step duration (s)
for j=1:M
   time(j)=(j-1)*DELTAtime;
end
```

The initial conditions for each node are inserted into the first column of the matrix T:

```
% Initial condition
for i=1:N
    T(i,1)=T_ini;
end
```

The remaining columns of T are filled in by stepping through time using the Euler approach:

```
% Step through time
for j=1:(M-1)
    T(1,j+1)=T(1,j)+2*k*(T(2,j)-T(1,j))*DELTAtime/(rho*c*DELTAx^2);
    for i=2:(N-1)
        T(i,j+1)=T(i,j)+k*(T(i-1,j)+T(i+1,j)-2*T(i,j))*DELTAtime/(rho*c*DELTAx^2);
    end
    T(N,j+1)=T(N,j)+(2*k*(T(N-1,j)-T(N,j))/(rho*c*DELTAx^2)+ ...
        2*h_bar*(T_infinity-T(N,j))/(rho*c*DELTAx))*DELTAtime;
end
```

One advantage of using MATLAB is that it is easy to manipulate the solution. For example, it is possible to plot the temperature as a function of time for various positions by entering plot(time,T) in the command window (Figure 3-44).

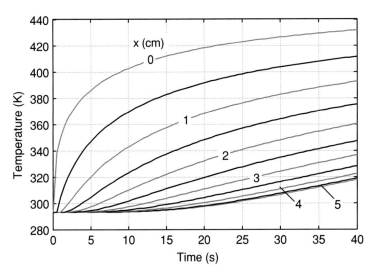

Figure 3-44: Temperature as a function of time for various positions.

Fully Implicit Method

Euler's method is an explicit technique. Therefore, it has the characteristic of becoming unstable when the time step exceeds a critical value. An implicit technique avoids this problem. The fully implicit method is similar to Euler's method in that the time rate of change is assumed to be constant throughout the time step. The difference is that the time rate of change is computed at the end of the time step rather than the beginning. Therefore, for any node i and time step j:

$$T_{i,j+1} = T_{i,j} + \left. \frac{dT}{dt} \right|_{T=T_{i,j+1}, t=t_{j+1}} \Delta t \quad \text{for } i = 1 \ldots N \tag{3-533}$$

The temperature rate of change at the end of the time step depends on the temperatures at the end of the time step $(T_{i,j+1})$. The temperatures $T_{i,j+1}$ cannot be calculated explicitly using information at the beginning of the time step $(T_{i,j})$ and instead Eq. (3-533) provides an implicit set of equations for $T_{i,j+1}$ where $i = 1$ to $i = N$. For the example problem shown in Figure 3-41, the implicit equations are obtained by substituting Eqs. (3-502), (3-505), and (3-508) into Eq.(3-533):

$$T_{1,j+1} = T_{1,j} + \frac{2k}{\rho c \, \Delta x^2} (T_{2,j+1} - T_{1,j+1}) \, \Delta t \qquad (3\text{-}534)$$

$$T_{i,j+1} = T_{i,j} + \frac{k}{\Delta x^2 \, \rho c} (T_{i-1,j+1} + T_{i+1,j+1} - 2\,T_{i,j+1}) \, \Delta t \quad \text{for } i = 2...(N-1) \qquad (3\text{-}535)$$

$$T_{N,j+1} = T_{N,j} + \left[\frac{2k}{\rho c \, \Delta x^2} (T_{N-1,j+1} - T_{N,j+1}) + \frac{2\overline{h}}{\Delta x \, \rho c} (T_f - T_{N,j+1}) \right] \Delta t \qquad (3\text{-}536)$$

Because EES solves implicit equations, it is not necessary to rearrange Eqs. (3-534) through (3-536) in order to solve them. The implicit solution is obtained using the following EES code. (Note that the modifications to the Euler method code are indicated in bold.)

```
$UnitSystem SI MASS RAD PA K J
$TABSTOPS 0.2 0.4 0.6 0.8 3.5 in

"Inputs"
L=5 [cm]*convert(cm,m)                      "wall thickness"
k=5.0 [W/m-K]                               "conductivity"
rho=2000 [kg/m^3]                           "density"
c=200 [J/kg-K]                              "specific heat capacity"
T_ini=converttemp(C,K,20 [C])              "initial temperature"
T_infinity=converttemp(C,K,200 [C])        "fluid temperature"
h_bar=500 [W/m^2-K]                         "heat transfer coefficient"

"Setup grid"
N=5 [-]                                      "number of nodes"
duplicate i=1,N
    x[i]=(i-1)*L/(N-1)                       "position of each node"
end
DELTAx=L/(N-1)                              "distance between adjacent nodes"

"Setup time steps"
M=21 [-]                                     "number of time steps"
t_sim=40 [s]                                "simulation time"
DELTAtime=t_sim/(M-1)                       "time step duration"
duplicate j=1,M
    time[j]=(j-1)* DELTAtime
end

duplicate i=1,N
    T[i,1]=T_ini                            "initial condition"
end
```

```
"Move through all of the time steps"
duplicate j=1,(M-1)
    T[1,j+1]=T[1,j]+2*k*(T[2,j+1]-T[1,j+1])*DELTAtime/(rho*c*Deltax^2) "node 1"
    duplicate i=2,(N-1)
        T[i,j+1]=T[i,j]+k*(T[i-1,j+1]+T[i+1,j+1]-2*T[i,j+1])*DELTAtime/(rho*c*Deltax^2)
                                                "internal nodes"
    end
    T[N,j+1]=T[N,j]+(2*k*(T[N-1,j+1]-T[N,j+1]))/(rho*c*DELTAx^2)+&
        2*h_bar*(T_infinity-T[N,j+1])/(rho*c*DELTAx))*DELTAtime "node N"
end
```

An attractive feature of the implicit technique is that it is possible to vary M and N independently in order to achieve sufficient accuracy without being constrained by stability considerations.

The implementation of the implicit technique in MATLAB is not a straightforward extension of the EES code because MATLAB cannot directly solve a set of implicit equations. Instead, Eqs. (3-534) through (3-536) must be placed in matrix format before they can be solved. This operation is similar to the solution of 1-D and 2-D steady-state problems using MATLAB, discussed in Sections 1.5, 1.9, and 2.6.

The inputs are entered in MATLAB and the grid and time steps are setup:

```
clear all;

% Inputs
L=0.05;                              % wall thickness (m)
k=5.0;                               % conductivity (W/m-K)
rho=2000;                            % density (kg/m^3)
c=200;                               % specific heat capacity (J/kg-K)
T_ini=293.2;                         % initial temperature (K)
T_infinity=473.2;                    % fluid temperature (K)
h_bar=500;                           % heat transfer coefficient (W/m^2-K)

% Setup grid
N=11;                                % number of nodes (-)
for i=1:N
    x(i)=(i-1)*L/(N-1);              % position of each node (m)
end
DELTAx=L/(N-1);                      % distance between adjacent nodes (m)

% Setup time steps
M=81;                                % number of time steps (-)
t_sim=40;                            % simulation time (s)
DELTAtime=t_sim/(M-1);               % time step duration (s)
for j=1:M
    time(j)=(j-1)*DELTAtime;
end

% Initial condition
for i=1:N
    T(i,1)=T_ini;
end
```

For any time step j, Eqs. (3-534) through (3-536) represent N linear equations for the N unknown temperatures $T_{i,j+1}$ for $i = 1$ to $i = N$. In order to solve these implicit equations, they must be placed in matrix format:

$$\underline{\underline{A}}\,\underline{X} = \underline{b} \qquad (3\text{-}537)$$

It is important to clearly specify the order that the equations are placed into the matrix $\underline{\underline{A}}$ and the order that the unknown temperatures are placed into the vector $\underline{X}$. The most logical method to setup the vector $\underline{X}$ is:

$$\underline{X} = \begin{bmatrix} X_1 = T_{1,j+1} \\ X_2 = T_{2,j+1} \\ \dots \\ X_N = T_{N,j+1} \end{bmatrix} \qquad (3\text{-}538)$$

so that $T_{i,j+1}$ corresponds to element i of $\underline{X}$. The most logical method for placing the equations into $\underline{\underline{A}}$ is:

$$\underline{\underline{A}} = \begin{bmatrix} \text{row 1} = \text{control volume 1 equation} \\ \text{row 2} = \text{control volume 2 equation} \\ \dots \\ \text{row } N = \text{control volume } N \text{ equation} \end{bmatrix} \qquad (3\text{-}539)$$

so that the equation derived based on the control volume for node i corresponds to row i of $\underline{\underline{A}}$. Equations (3-534) through (3-536) are rearranged so that the coefficients multiplying the unknowns and the constants for the linear equations are clear:

$$T_{1,j+1}\underbrace{\left[1 + \frac{2k\,\Delta t}{\rho c\,\Delta x^2}\right]}_{A_{1,1}} + T_{2,j+1}\underbrace{\left[-\frac{2k\,\Delta t}{\rho c\,\Delta x^2}\right]}_{A_{1,2}} = \underbrace{T_{1,j}}_{b_1} \qquad (3\text{-}540)$$

$$T_{i,j+1}\underbrace{\left[1 + \frac{2k\,\Delta t}{\Delta x^2\,\rho c}\right]}_{A_{i,i}} + T_{i-1,j+1}\underbrace{\left[-\frac{k\,\Delta t}{\Delta x^2\,\rho c}\right]}_{A_{i,i-1}} + T_{i+1,j+1}\underbrace{\left[-\frac{k\,\Delta t}{\Delta x^2\,\rho c}\right]}_{A_{i,i+1}} = \underbrace{T_{i,j}}_{b_i} \quad \text{for } i = 2\dots(N-1)$$

$$(3\text{-}541)$$

$$T_{N,j+1}\underbrace{\left[1 + \frac{2k\,\Delta t}{\rho c\,\Delta x^2} + \frac{2\bar{h}\,\Delta t}{\Delta x\,\rho c}\right]}_{A_{N,N}} + T_{N-1,j+1}\underbrace{\left[-\frac{2k\,\Delta t}{\rho c\,\Delta x^2}\right]}_{A_{N,N-1}} = \underbrace{T_{N,j} + \frac{2\bar{h}\,\Delta t}{\Delta x\,\rho c}T_\infty}_{b_N} \qquad (3\text{-}542)$$

The matrix $\underline{\underline{A}}$ and vector $\underline{b}$ are initialized:

```
A=spalloc(N,N,3*N);          % initialize A
b=zeros(N,1);                % initialize b
```

Note that the variable A is defined as being a sparse matrix with at most three nonzero entries per row. The maximum number of nonzero elements can be determined based on examination of Eqs. (3-540) through (3-542). The values of the elements in the matrix A are all initialized to zero. The matrix A does not change depending on the particular time step being considered (at least for this problem with constant properties). Therefore, the matrix A can be constructed just one time and used without modification for each time step, which saves computational effort.

```
% Setup A matrix
A(1,1)=1+2*k*DELTAtime/(rho*c*DELTAx^2);
A(1,2)=-2*k*DELTAtime/(rho*c*DELTAx^2);
for i=2:(N-1)
   A(i,i)=1+2*k*DELTAtime/(rho*c*DELTAx^2);
   A(i,i-1)=-k*DELTAtime/(rho*c*DELTAx^2);
   A(i,i+1)=-k*DELTAtime/(rho*c*DELTAx^2);
end
A(N,N)=1+2*k*DELTAtime/(rho*c*DELTAx^2)+2*h_bar*DELTAtime/(DELTAx*rho*c);
A(N,N-1)=-2*k*DELTAtime/(rho*c*DELTAx^2);
```

On the other hand, the vector b does change depending on the time step because it includes the current value of the temperatures. Therefore, the vector b will need to be reconstructed before the simulation of each time step.

```
for j=1:(M-1)
   % Setup b matrix
   for i=1:(N-1)
      b(i)=T(i,j);
   end
   b(N)=T(N,j)+2*h_bar*DELTAtime*T_infinity/(DELTAx*rho*c);
   % Simulate time step
   T(:,j+1)=A\b;
end
```

Note that T(:,j+1) is the MATLAB syntax for specifying the entire column j+1 in the matrix T. Running the script at this point will provide the matrix T, which contains the temperatures for each node (corresponding to each row of T) and each time step (corresponding to each column of T). This information can be plotted against either time or position by either typing:

```
plot(time,T);
```

or

```
plot(x,T);
```

in the command window, respectively.

Heun's Method
Heun's method was discussed previously in Section 3.2.2 as a simple example of a class of numerical integration methods referred to as predictor-corrector techniques. Predictor-corrector techniques take an initial predictor step (based on Euler's technique) followed by one or more corrector steps, in which the knowledge obtained from the predictor step is used to improve the integration process. Because the predictor-corrector techniques rely on an explicit predictor step, they suffer from the same limitations related to stability that were discussed in the context of Euler's method.

The implementation of Heun's method is illustrated using MATLAB. The problem is set up as before:

```
clear all;

% Inputs
L=0.05;                                    % wall thickness (m)
k=5.0;                                     % conductivity (W/m-K)
rho=2000;                                  % density (kg/m^3)
c=200;                                     % specific heat capacity (J/kg-K)
T_ini=293.2;                               % initial temperature (K)
T_infinity=473.2;                          % fluid temperature (K)
h_bar=500;                                 % heat transfer coefficient (W/m^2-K)

% Setup grid
N=11;                                      % number of nodes (-)
for i=1:N
    x(i)=(i-1)*L/(N-1);                    % position of each node (m)
end
DELTAx=L/(N-1);                            % distance between adjacent nodes (m)

% Setup time steps
M=81;                                      % number of time steps (-)
t_sim=40;                                  % simulation time (s)
DELTAtime=t_sim/(M-1);                     % time step duration (s)
for j=1:M
    time(j)=(j-1)*DELTAtime;
end

% Initial condition
for i=1:N
    T(i,1)=T_ini;
end
```

In order to simulate any time step j for any node i, Heun's method begins with an Euler step to obtain an initial prediction for the temperature at the conclusion of the time step ($\hat{T}_{i,j+1}$). This first step is referred to as the predictor step and the details are identical to Euler's method:

$$\hat{T}_{i,j+1} = T_{i,j} + \frac{dT}{dt}\bigg|_{T=T_{i,j},t=t_j} \Delta t \quad \text{for } i = 1\ldots N \tag{3-543}$$

The equations used to predict the time rate of change are specific to the problem being studied even though the methodology is broadly applicable. The results of the predictor step are used to carry out a subsequent corrector step. The predicted values of the temperatures at the end of the time step are used to predict the temperature rate of change at the end of the time step ($\frac{dT}{dt}|_{T=\hat{T}_{i,j+1},t=t_j+\Delta t}$). The corrector step predicts the temperature at the end of the time step ($T_{i,j+1}$) based on the average of $\frac{dT}{dt}|_{T=T_{i,j},t=t_j}$ and $\frac{dT}{dt}|_{T=\hat{T}_{i,j+1},t=t_j+\Delta t}$ according to:

$$T_{i,j+1} = T_{i,j} + \left[\frac{dT}{dt}\bigg|_{T=T_{i,j},t=t_j} + \frac{dT}{dt}\bigg|_{T=\hat{T}_{i,j+1},t=t_j+\Delta t}\right]\frac{\Delta t}{2} \quad \text{for } i = 1\ldots N \tag{3-544}$$

Heun's method is illustrated in the context of the problem that is illustrated in Figure 3-41. Equations (3-502), (3-505), and (3-508) are used to compute the time rate of change for each node at the beginning of a time step:

$$\left.\frac{dT}{dt}\right|_{T=T_{1,j},t=t_j} = \frac{2k}{\rho\,c\,\Delta x^2}\,(T_{2,j} - T_{1,j}) \tag{3-545}$$

$$\left.\frac{dT}{dt}\right|_{T=T_{i,j},t=t_j} = \frac{k}{\Delta x^2\,\rho\,c}\,(T_{i-1,j} + T_{i+1,j} - 2\,T_{i,j}) \quad \text{for } i = 2\ldots(N-1) \tag{3-546}$$

$$\left.\frac{dT}{dt}\right|_{T=T_{N,j},t=t_j} = \frac{2k}{\rho\,c\,\Delta x^2}\,(T_{N-1,j} - T_{N,j}) + \frac{2\overline{h}}{\Delta x\,\rho\,c}(T_\infty - T_{N,j}) \tag{3-547}$$

The time rate of change for the temperature of each node at the beginning of the time step is stored in a temporary vector, dTdt0, which is overwritten during each time step:

```
% Step through time
for j=1:(M-1)
    % compute time rates of change at the beginning of the time step
    dTdt0(1)=2*k*(T(2,j)-T(1,j))/(rho*c*DELTAx^2);
    for i=2:(N-1)
        dTdt0(i)=k*(T(i-1,j)+T(i+1,j)-2*T(i,j))/(rho*c*DELTAx^2);
    end
    dTdt0(N)=2*k*(T(N-1,j)-T(N,j))/(rho*c*DELTAx^2)+2*h_bar*(T_infinity-T(N,j))/(rho*c*DELTAx);
```

A first estimate of the temperatures at the end of the time step is obtained using Eq. (3-543); these values are also stored in a temporary vector, Ts, which is also not saved beyond the time step being simulated:

```
% compute first estimate of temperatures at the end of the time step
for i=1:N
    Ts(i)=T(i,j)+dTdt0(i)*DELTAtime;
end
```

The results of the predictor step are used to carry out a corrector step. The time rates of change for each temperature at the end of the time step are computed using Eqs. (3-502), (3-505), and (3-508) with the time derivative evaluated using the predicted temperatures.

$$\left.\frac{dT}{dt}\right|_{T=\hat{T}_{1,j+1},t=t_j+\Delta t} = \frac{2k}{\rho\,c\,\Delta x^2}\,(\hat{T}_{2,j+1} - \hat{T}_{1,j+1}) \tag{3-548}$$

$$\left.\frac{dT}{dt}\right|_{T=\hat{T}_{i,j+1},t=t_j+\Delta t} = \frac{k}{\Delta x^2\,\rho\,c}\,(\hat{T}_{i-1,j+1} + \hat{T}_{i+1,j+1} - 2\,\hat{T}_{i,j+1}) \quad \text{for } i = 2\ldots(N-1) \tag{3-549}$$

$$\left.\frac{dT}{dt}\right|_{T=\hat{T}_{N,j+1},t=t_j+\Delta t} = \frac{2k}{\rho\,c\,\Delta x^2}\,(\hat{T}_{N-1,j+1} - \hat{T}_{N,j+1}) + \frac{2\overline{h}}{\Delta x\,\rho\,c}\,(T_\infty - \hat{T}_{N,j+1}) \tag{3-550}$$

```
% compute time rates of change at the end of the time step
dTdts(1)=2*k*(Ts(2)-Ts(1))/(rho*c*DELTAx^2);
for i=2:(N-1)
   dTdts(i)=k*(Ts(i-1)+Ts(i+1)-2*Ts(i))/(rho*c*DELTAx^2);
end
dTdts(N)=2*k*(Ts(N-1)-Ts(N))/(rho*c*DELTAx^2)+2*h_bar*(T_infinity-Ts(N))/(rho*c*DELTAx);
```

The corrector step is based on the average of the two estimates of the time rate of change:

$$T_{i,j+1} = T_{i,j} + \left[\left. \frac{dT}{dt} \right|_{T=T_{i,j}, t=t_j} + \left. \frac{dT}{dt} \right|_{T=\hat{T}_{i,j+1}, t=t_j+\Delta t} \right] \frac{\Delta t}{2} \quad \text{for } i = 1 \ldots N \qquad (3\text{-}551)$$

```
% corrector step
for i=1:N
   T(i,j+1)=T(i,j)+(dTdt0(i)+dTdts(i))*DELTAtime/2;
end
end
```

The numerical error associated with Heun's method is substantially less than Euler's method or the fully implicit method because Heun's method is a second-order technique, as discussed in Section 3.2. Also note that if M is reduced from 21 to 11 (i.e., the time step is increased from 2.0 s to 4.0 s) then Heun's method becomes unstable, just as Euler's method did.

Runge-Kutta 4th Order Method

Heun's method is a two-step predictor/corrector technique, which improves the accuracy of the solution but not the stability characteristics. The fourth order Runge-Kutta method involves four predictor/corrector steps and therefore improves the accuracy to an even larger extent. Because the Runge-Kutta technique is explicit, the stability characteristics of the solution are not affected.

The Runge-Kutta fourth order method (or RK4 method) was previously discussed in Section 3.2.2. The RK4 technique estimates the time rate of change of the temperature of each node four times in order to simulate a single time step. (Contrast this method with Euler's method where the time rate of change was computed only once, at the beginning of time step.)

The implementation of the RK4 method is illustrated using MATLAB. The problem is setup as before:

```
clear all;

% Inputs
L=0.05;              % wall thickness (m)
k=5.0;               % conductivity (W/m-K)
rho=2000;            % density (kg/m^3)
c=200;               % specific heat capacity (J/kg-K)
T_ini=293.2;         % initial temperature (K)
T_infinity=473.2;    % fluid temperature (K)
h_bar=500;           % heat transfer coefficient (W/m^2-K)
```

```
% Setup grid
N=11;                              % number of nodes (-)
for i=1:N
   x(i)=(i-1)*L/(N-1);             % position of each node (m)
end
DELTAx=L/(N-1);                    % distance between adjacent nodes (m)

% Setup time steps
M=81;                              % number of time steps (-)
t_sim=40;                          % simulation time (s)
DELTAtime=t_sim/(M-1);             % time step duration (s)
for j=1:M
   time(j)=(j-1)*DELTAtime;
end

% Initial condition
for i=1:N
   T(i,1)=T_ini;
end
```

The RK4 method begins by estimating the time rate of change of the temperature of each node at the beginning of the time step (referred to as aa_i):

$$aa_i = \left.\frac{dT}{dt}\right|_{T=T_{i,j},t=t_j} \quad \text{for } i = 1\ldots N \tag{3-552}$$

or, for this problem:

$$aa_1 = \frac{2k}{\rho c\,\Delta x^2}\,(T_{2,j} - T_{1,j}) \tag{3-553}$$

$$aa_i = \frac{k}{\Delta x^2 \rho c}\,(T_{i-1,j} + T_{i+1,j} - 2\,T_{i,j}) \quad \text{for } i = 2\ldots(N-1) \tag{3-554}$$

$$aa_N = \frac{2\,k}{\rho c\,\Delta x^2}\,(T_{N-1,j} - T_{N,j}) + \frac{2\overline{h}}{\Delta x \rho c}\,(T_\infty - T_{N,j}) \tag{3-555}$$

```
% Step through time
for j=1:(M-1)
   % compute the 1st estimate of the time rate of change
   aa(1)=2*k*(T(2,j)-T(1,j))/(rho*c*DELTAx^2);
   for i=2:(N-1)
      aa(i)=k*(T(i-1,j)+T(i+1,j)-2*T(i,j))/(rho*c*DELTAx^2);
   end
   aa(N)=2*k*(T(N-1,j)-T(N,j))/(rho*c*DELTAx^2)+2*h_bar*(T_infinity-T(N,j))/(rho*c*DELTAx);
```

The first estimate of the time rate of change is used to predict the temperature of each node half-way through the time step ($\hat{T}_{i,j+\frac{1}{2}}$) in the first predictor step:

$$\hat{T}_{i,j+\frac{1}{2}} = T_{i,j} + aa_i\frac{\Delta t}{2} \tag{3-556}$$

```
% 1st predictor step
for i=1:N
   Ts(i)=T(i,j)+aa(i)*DELTAtime/2;
end
```

The estimated temperatures are used to obtain the second estimate of the time rate of change (bb_i), this time at the midpoint of the time step:

$$bb_i = \left.\frac{dT}{dt}\right|_{T=\hat{T}_{i,j+\frac{1}{2}},\, t=t_j+\frac{\Delta t}{2}} \qquad \text{for } i=1\ldots N \tag{3-557}$$

or, for this problem:

$$bb_1 = \frac{2k}{\rho c\, \Delta x^2}\left(\hat{T}_{2,j+\frac{1}{2}} - \hat{T}_{1,j+\frac{1}{2}}\right) \tag{3-558}$$

$$bb_i = \frac{k}{\Delta x^2\, \rho c}\left(\hat{T}_{i-1,j+\frac{1}{2}} + \hat{T}_{i+1,j+\frac{1}{2}} - 2\,\hat{T}_{i,j+\frac{1}{2}}\right) \qquad \text{for } i=2\ldots(N-1) \tag{3-559}$$

$$bb_N = \frac{2k}{\rho c\, \Delta x^2}\left(\hat{T}_{N-1,j+\frac{1}{2}} - \hat{T}_{N,j+\frac{1}{2}}\right) + \frac{2\overline{h}}{\Delta x \rho c}\left(T_\infty - \hat{T}_{N,j+\frac{1}{2}}\right) \tag{3-560}$$

```
% compute the 2nd estimate of the time rate of change
bb(1)=2*k*(Ts(2)-Ts(1))/(rho*c*DELTAx^2);
for i=2:(N-1)
   bb(i)=k*(Ts(i-1)+Ts(i+1)-2*Ts(i))/(rho*c*DELTAx^2);
end
bb(N)=2*k*(Ts(N-1)-Ts(N))/(rho*c*DELTAx^2)+2*h_bar*(T_infinity-Ts(N))/(rho*c*DELTAx);
```

The second estimate of the time rate of change is used to obtain a new prediction of the temperature of each node, again half-way through the time step ($\hat{T}_{i,j+\frac{1}{2}}$):

$$\hat{T}_{i,j+\frac{1}{2}} = T_{i,j} + bb_i\,\frac{\Delta t}{2} \qquad \text{for } i=1\ldots N \tag{3-561}$$

```
% 2nd predictor step
for i=1:N
   Tss(i)=T(i,j)+bb(i)*DELTAtime/2;
end
```

The third estimate of the time rate of change of each node (cc_i) is also obtained at the mid-point of the time step:

$$cc_i = \left.\frac{dT}{dt}\right|_{T=\hat{T}_{i,j+\frac{1}{2}},\, t=t_j+\frac{\Delta t}{2}} \qquad \text{for } i=1\ldots N \tag{3-562}$$

or, for this problem:

$$cc_1 = \frac{2k}{\rho c\, \Delta x^2}\left(\hat{T}_{2,j+\frac{1}{2}} - \hat{T}_{1,j+\frac{1}{2}}\right) \tag{3-563}$$

$$cc_i = \frac{k}{\Delta x^2 \rho c} \left(\hat{\hat{T}}_{i-1,j+\frac{1}{2}} + \hat{\hat{T}}_{i+1,j+\frac{1}{2}} - 2 \hat{\hat{T}}_{i,j+\frac{1}{2}} \right) \quad \text{for } i = 2 \dots (N-1) \qquad (3\text{-}564)$$

$$cc_N = \frac{2k}{\rho c \, \Delta x^2} \left(\hat{\hat{T}}_{N-1,j+\frac{1}{2}} - \hat{\hat{T}}_{N,j+\frac{1}{2}} \right) + \frac{2\overline{h}}{\Delta x \rho c} \left(T_\infty - \hat{\hat{T}}_{N,j+\frac{1}{2}} \right) \qquad (3\text{-}565)$$

```
% compute the 3rd estimate of the time rate of change
cc(1)=2*k*(Tss(2)-Tss(1))/(rho*c*DELTAx^2);
for i=2:(N-1)
  cc(i)=k*(Tss(i-1)+Tss(i+1)-2*Tss(i))/(rho*c*DELTAx^2);
end
cc(N)=2*k*(Tss(N-1)-Tss(N))/(rho*c*DELTAx^2)+2*h_bar*(T_infinity-Tss(N))/(rho*c*DELTAx);
```

The third estimate of the time rate of change is used to predict the temperature at the end of the time step ($\hat{T}_{i,j+1}$):

$$\hat{T}_{i,j+1} = T_{i,j} + cc_i \, \Delta t \quad \text{for } i = 1 \dots N \qquad (3\text{-}566)$$

```
% 3rd predictor step
for i=1:N
  Tsss(i)=T(i,j)+cc(i)*DELTAtime;
end
```

Finally, the fourth estimate of the time rate of change (dd_i) is obtained at the end of the time step:

$$dd_i = \left. \frac{dT}{dt} \right|_{\hat{T}_{i,j+1}, t = t_j + \Delta t} \quad \text{for } i = 1 \dots N \qquad (3\text{-}567)$$

or, for this problem:

$$dd_1 = \frac{2k}{\rho c \, \Delta x^2} \left(\hat{T}_{2,j+1} - \hat{T}_{1,j+1} \right) \qquad (3\text{-}568)$$

$$dd_i = \frac{k}{\Delta x^2 \rho c} \left(\hat{T}_{i-1,j+1} + \hat{T}_{i+1,j+1} - 2 \hat{T}_{i,j+1} \right) \quad \text{for } i = 2 \dots (N-1) \qquad (3\text{-}569)$$

$$dd_N = \frac{2k}{\rho c \, \Delta x^2} \left(\hat{T}_{N-1,j+1} - \hat{T}_{N,j+1} \right) + \frac{2\overline{h}}{\Delta x \rho c} \left(T_\infty - \hat{T}_{N,j+1} \right) \qquad (3\text{-}570)$$

```
% compute the 4th estimate of the time rate of change
dd(1)=2*k*(Tsss(2)-Tsss(1))/(rho*c*DELTAx^2);
for i=2:(N-1)
  dd(i)=k*(Tsss(i-1)+Tsss(i+1)-2*Tsss(i))/(rho*c*DELTAx^2);
end
dd(N)=2*k*(Tsss(N-1)-Tsss(N))/(rho*c*DELTAx^2)+2*h_bar*(T_infinity-Tsss(N))/(rho*c*DELTAx);
```

The integration through the time step is finally carried out using the weighted average of these four separate estimates of the time rate of change for each node:

$$T_{i,j+1} = T_{i,j} + (aa_i + 2\,bb_i + 2\,cc_i + dd_i)\,\frac{\Delta t}{6} \tag{3-571}$$

```
% corrector step
for i=1:N
   T(i,j+1)=T(i,j)+(aa(i)+2*bb(i)+2*cc(i)+dd(i))*DELTAtime/6;
   end
end
```

Crank-Nicolson Method

The Crank-Nicolson method combines Euler's method with the fully implicit method. The time rate of change for each time step is estimated based on the average of its values at the beginning and the end of the time step.

$$T_{i,j+1} = T_{i,j} + \left(\left.\frac{dT}{dt}\right|_{T=T_{i,j},t=t_j} + \left.\frac{dT}{dt}\right|_{T=T_{i,j+1},t=t_{j+1}}\right)\frac{\Delta t}{2} \quad \text{for } i = 1 \ldots N \tag{3-572}$$

Notice that Eq. (3-572) is not a predictor-corrector method, like Heun's method, because the temperature at the end of the time step is not predicted with an explicit predictor step (i.e., there is no $\hat{T}_{i,j+1}$ in Eq. (3-572)). Rather, the unknown solution for the temperatures at the end of the time step $(T_{i,j+1})$ is substituted directly into Eq. (3-572), resulting in a set of implicit equations for $T_{i,j+1}$ where $i = 1$ to $i = N$. Therefore, the Crank-Nicolson technique has stability characteristics that are similar to the fully implicit method. The solution involves two estimates for the time rate of change of each node and therefore has higher order accuracy than either Euler's method or the fully implicit technique.

The Crank-Nicolson method is illustrated using MATLAB. The problem is setup as before:

```
clear all;

% Inputs
L=0.05;                    % wall thickness (m)
k=5.0;                     % conductivity (W/m-K)
rho=2000;                  % density (kg/m^3)
c=200;                     % specific heat capacity (J/kg-K)
T_ini=293.2;               % initial temperature (K)
T_infinity=473.2;          % fluid temperature (K)
h_bar=500;                 % heat transfer coefficient (W/m^2-K)

% Setup grid
N=11;                      % number of nodes (-)
for i=1:N
   x(i)=(i-1)*L/(N-1);     % position of each node (m)
end
DELTAx=L/(N-1);            % distance between adjacent nodes (m)
```

```
% Setup time steps
M=81;                                         % number of time steps (-)
t_sim=40;                                     % simulation time (s)
DELTAtime=t_sim/(M-1);                        % time step duration (s)
for j=1:M
   time(j)=(j-1)*DELTAtime;
end

% Initial conditions
for i=1:N
   T(i,1)=T_ini;
end
```

Substituting Eqs. (3-502), (3-505), and (3-508) into Eq. (3-572) leads to:

$$T_{1,j+1} = T_{1,j} + \left[\frac{2k}{\rho c \, \Delta x^2} \left(T_{2,j} - T_{1,j} \right) + \frac{2k}{\rho c \, \Delta x^2} \left(T_{2,j+1} - T_{1,j+1} \right) \right] \frac{\Delta t}{2} \qquad (3\text{-}573)$$

$$T_{i,j+1} = T_{i,j} + \left[\frac{k}{\Delta x^2 \, \rho c} \left(T_{i-1,j} + T_{i+1,j} - 2\,T_{i,j} \right) + \frac{k}{\Delta x^2 \, \rho c} \left(T_{i-1,j+1} + T_{i+1,j+1} - 2\,T_{i,j+1} \right) \right] \frac{\Delta t}{2}$$
$$\text{for } i = 2 \ldots (N-1) \qquad (3\text{-}574)$$

$$T_{N,j+1} = T_{N,j} + \left[\begin{array}{l} \dfrac{2\,k}{\rho c \, \Delta x^2} \left(T_{N-1,j} - T_{N,j} \right) + \dfrac{2\,\overline{h}}{\Delta x \rho c} \left(T_\infty - T_{N,j} \right) + \\[4mm] \dfrac{2\,k}{\rho c \, \Delta x^2} \left(T_{N-1,j+1} - T_{N,j+1} \right) + \dfrac{2\,\overline{h}}{\Delta x \rho c} \left(T_\infty - T_{N,j+1} \right) \end{array} \right] \frac{\Delta t}{2} \qquad (3\text{-}575)$$

Equations (3-573) through (3-575) are a set of N equations in the unknown temperatures $T_{i,j+1}$ for $i = 1$ to $i = N$. These equations must be placed into matrix format in order to be solved in MATLAB. This is similar to the fully implicit method and the same process is used. Equations (3-573) through (3-575) are rearranged so that the coefficients multiplying the unknowns and the constants for the linear equations are clear:

$$T_{1,j+1} \underbrace{\left[1 + \frac{k\,\Delta t}{\rho c \, \Delta x^2} \right]}_{A_{1,1}} + T_{2,j+1} \underbrace{\left[-\frac{k\,\Delta t}{\rho c \, \Delta x^2} \right]}_{A_{1,2}} = \underbrace{T_{1,j} + \frac{k\,\Delta t}{\rho c \, \Delta x^2} \left(T_{2,j} - T_{1,j} \right)}_{b_1} \qquad (3\text{-}576)$$

$$T_{i,j+1} \underbrace{\left[1 + \frac{k\,\Delta t}{\Delta x^2 \, \rho c} \right]}_{A_{i,i}} + T_{i-1,j+1} \underbrace{\left[-\frac{k\,\Delta t}{2\,\Delta x^2 \, \rho c} \right]}_{A_{i,i-1}} + T_{i+1,j+1} \underbrace{\left[-\frac{k\,\Delta t}{2\,\Delta x^2 \, \rho c} \right]}_{A_{i,i+1}}$$

$$= \underbrace{T_{i,j} + \frac{k\,\Delta t}{2\,\Delta x^2 \, \rho c} \left(T_{i-1,j} + T_{i+1,j} - 2\,T_{i,j} \right)}_{b_i} \quad \text{for } i = 2 \ldots (N-1)$$

$$(3\text{-}577)$$

$$
T_{N,j+1}\underbrace{\left[1+\frac{k\,\Delta t}{\rho\,c\,\Delta x^2}+\frac{\bar h\,\Delta t}{\Delta x\,\rho\,c}\right]}_{A_{N,N}}+T_{N-1,j+1}\underbrace{\left[-\frac{k\,\Delta t}{\rho\,c\,\Delta x^2}\right]}_{A_{N,N-1}}
$$

$$
=\underbrace{T_{N,j}+\frac{k\,\Delta t}{\rho\,c\,\Delta x^2}\,(T_{N-1,j}-T_{N,j})+\frac{\bar h\,\Delta t}{\Delta x\,\rho\,c}\,(T_\infty-T_{N,j})+\frac{\bar h\,\Delta t}{\Delta x\,\rho\,c}T_\infty}_{b_N}
$$

(3-578)

The variables A and b are initialized:

```
A=spalloc(N,N,3*N);          % initialize A
b=zeros(N,1);                % initialize b
```

The matrix $\underline{\underline{A}}$ does not depend on the time step and can therefore be constructed just once:

```
% Setup A matrix
A(1,1)=1+k*DELTAtime/(rho*c*DELTAx^2);
A(1,2)=-k*DELTAtime/(rho*c*DELTAx^2);
for i=2:(N-1)
    A(i,i)=1+k*DELTAtime/(rho*c*DELTAx^2);
    A(i,i-1)=-k*DELTAtime/(2*rho*c*DELTAx^2);
    A(i,i+1)=-k*DELTAtime/(2*rho*c*DELTAx^2);
end
A(N,N)=1+k*DELTAtime/(rho*c*DELTAx^2)+h_bar*DELTAtime/(DELTAx*rho*c);
A(N,N-1)=-k*DELTAtime/(rho*c*DELTAx^2);
```

while the vector $\underline{b}$ must be reconstructed during each time step:

```
for j=1:(M-1)
    % Setup b matrix
    b(1)=T(1,j)+k*DELTAtime*(T(2,j)-T(1,j))/(rho*c*DELTAx^2);
    for i=2:(N-1)
        b(i)=T(i,j)+k*DELTAtime*(T(i-1,j)+T(i+1,j)-2*T(i,j))/(2*rho*c*DELTAx^2);
    end
    b(N)=T(N,j)+k*DELTAtime*(T(N-1,j)-T(N,j))/(rho*c*DELTAx^2)+...
        h_bar*DELTAtime*(T_infinity-T(N,j))/(DELTAx*rho*c)+...
        h_bar*DELTAtime*T_infinity/(DELTAx*rho*c);
    % Simulate time step
    T(:,j+1)=A\b;
end
```

The numerical error associated with the Crank-Nicolson technique is substantially less than either Euler's method or the fully implicit method and it retains the stability characteristics of the fully implicit technique. The Crank-Nicolson technique is likely to be the most attractive option for many problems due to its superior accuracy combined with its stability.

EES' Integral Command

The EES Integral function was introduced in Section 3.2.2 and used to solve lumped capacitance problems. EES is used in this section to illustrate Euler's method and the fully implicit method. However, these are not optimal methods for using EES to solve 1-D transient problems. Instead, the Integral command provides a more powerful and computationally efficient method to solve this class of problem. The Integral command implements a third order accurate integration scheme that uses an (optional) adaptive step-size in order to minimize computation time while maintaining accuracy.

The problem is set up in EES by entering the inputs:

```
$UnitSystem SI MASS RAD PA K J
$TABSTOPS 0.2 0.4 0.6 0.8 3.5 in

"Inputs"
L=5 [cm]*convert(cm,m)                "wall thickness"
k=5.0 [W/m-K]                         "conductivity"
rho=2000 [kg/m^3]                     "density"
c=200 [J/kg-K]                        "specific heat capacity"
T_ini=converttemp(C,K,20 [C])         "initial temperature"
T_infinity=converttemp(C,K,200 [C])   "fluid temperature"
h_bar=500 [W/m^2-K]                   "heat transfer coefficient"
t_sim=40 [s]                          "simulation time"
```

and the spatial grid is setup:

```
"Setup grid"
N=6 [-]                               "number of nodes"
duplicate i=1,N
    x[i]=(i-1)*L/(N-1)                "position of each node"
end
DELTAx=L/(N-1)                        "distance between adjacent nodes"
```

The EES Integral command requires four arguments (with an optional fifth argument to specify the integration step size):

```
F=INTEGRAL(Integrand,VarName,LowerLimit,UpperLimit, Stepsize)
```

where Integrand is the EES variable or expression that is be integrated (which, in this case, is each of the time derivatives of the nodal temperatures), VarName is the integration variable (time), LowerLimit and UpperLimit define the limits of integration (0 and t_sim), and Stepsize is the optional size of the time step.

The solution to our example problem is obtained by replacing Integrand with the variable that specifies the time rate of change of the temperature for each of the nodes; i.e., Eqs. (3-502), (3-505), and (3-508). To solve this problem using the Integral command, it is first necessary to set up the equations that calculate the time rates of temperature change within a one-dimensional array. Note that two-dimensional arrays are not needed when the problem is solved in this manner, since EES will automatically keep track of the integrated values as they change with time. The use of the Integral command requires many fewer variables and less computation time than the numerical schemes presented in earlier sections require when they are implemented in EES.

```
"time rate of change"
dTdt[1]=2*k*(T[2]-T[1])/(rho*c*Deltax^2)                                    "node 1"
duplicate i=2,(N-1)
    dTdt[i]=k*(T[i-1]+T[i+1]-2*T[i])/(rho*c*Deltax^2)                       "internal nodes"
end
dTdt[N]=2*k*(T[N-1]-T[N])/(rho*c*DELTAx^2)+2*h_bar*(T_infinity-T[N])/(rho*c*DELTAx)
                                                                            "node N"
```

The following code integrates the time rates of change for each node forward through time:

```
"integrate using the INTEGRAL command"
duplicate i=1,N
    T[i]=T_ini+INTEGRAL(dTdt[i],time,0,t_sim)
end
```

Note that after the calculations are completed, the array T[i] will provide the temperature of each node at the end of the time that is simulated, i.e., at time $t = t_{sim}$. The time variation in the temperature at each node can be recorded by including the $IntegralTable directive in the file. The format of the $IntegralTable directive is:

```
$IntegralTable VarName:Interval, x,y,z
```

where VarName is the integration variable (time), Interval is the output step size in the integration variable at which results will be placed in the integral table (which is completely independent of the integration step size), and the variables x, y, z are the EES variables that will be reported. Note that the range notation for an array can be used as a variable (i.e., T[1..N] indicates a list of all of the nodal temperatures, provided that the variable N is previously defined).

To obtain an integral table that contains the value of all of the nodal temperatures at 1 s intervals, add the following directive to the EES code:

```
$IntegralTable time:1, T[1..N]
```

After running the code (select Solve from the Calculate menu), an integral table will be generated. The results in the integral table can be used to create plots in the same way that results in a parametric table or an array table are used.

MATLAB's Ordinary Differential Equation Solvers
MATLAB's suite of integration routines were discussed in Section 3.2.2 in the context of lumped capacitance problems. The most basic method for calling these functions is:

$$[\underbrace{time}_{\substack{\text{vector} \\ \text{of time}}} , \underbrace{T}_{\substack{\text{temperatures} \\ \text{at the times}}}] = ode45(\underbrace{\text{'dTdt'}}_{\substack{\text{function that} \\ \text{returns the derivative} \\ \text{of temperatures}}} , \underbrace{tspan}_{\substack{\text{time span} \\ \text{to be integrated}}} , \underbrace{T0}_{\substack{\text{initial} \\ \text{temperatures}}})$$

In Section 3.2.2, the argument dTdt specified the name of a function that returns a single output, the time rate of change of temperature given the current temperature and time. For the 1-D transient problems considered in this section, an energy balance on each control volume leads to a system of equations that must be solved in order to provide the time rate of change of each nodal temperature given the current values of all of the nodal temperatures and time. Therefore, the function dTdt required to solve 1-D transient problems will return a vector that contains the time rate of change of each nodal temperature and it will require a vector of nodal temperatures and time as the input. The function dTdt_functionv is defined below with this calling protocol:

```
function[dTdt]=dTdt_functionv(time,T)
% Inputs:
% time – time in simulation (s)
% T – vector of nodal temperatures (K)
%
% Output:
% dTdt – vector of the time rate of temperature change for each node (K/s)
```

The problem inputs are entered in the body of the function for now. Eventually, this information will be passed to the function as input parameters.

```
% Inputs
L=0.05;               % wall thickness (m)
k=5.0;                % conductivity (W/m-K)
rho=2000;             % density (kg/m^3)
c=200;                % specific heat capacity (J/kg-K)
T_infinity=473.2;     % fluid temperature (K)
h_bar=500;            % heat transfer coefficient (W/m^2-K)
```

The function does not know the size of the vector T and therefore the size of the vector dTdt that must be returned. The number of nodes can be ascertained using the size command which returns the number of rows (N) and columns for the variable T.

```
[N,g]=size(T);        % determine size of T
```

The distance between adjacent nodes is computed:

```
DELTAx=L/(N-1);       % distance between adjacent nodes (m)
```

The vector containing the rate of temperature change for each node is initialized:

```
dTdt=zeros(N,1);      % initialize the dTdt vector
```

Equations (3-502), (3-505), and (3-508) are used to evaluate the temperature rate of change for every node:

```
    dTdt(1)=2*k*(T(2)-T(1))/(rho*c*DELTAx^2);
    for i=2:(N-1)
        dTdt(i)=k*(T(i-1)+T(i+1)-2*T(i))/(rho*c*DELTAx^2);
    end
    dTdt(N)=2*k*(T(N-1)-T(N))/(rho*c*DELTAx^2)+2*h_bar*(T_infinity-T(N))/(rho*c*DELTAx);
end
```

The function dTdt_functionv is integrated from a MATLAB script. The initial temperature, number of nodes and simulation time are specified:

```
clear all;
T_ini=293.2;                          % initial temperature (K)
N=11;                                 % number of nodes (-)
t_sim=40;                             % simulation time (s)
```

and the ode solver **ode45** is used to integrate the time rates of change:

```
[time,T]=ode45('dTdt_functionv',[0,t_sim],T_ini*ones(N,1));
```

Note that **ones(N,1)** provides an $N \times 1$ vector (which has the same size as the T vector) for which each element is set to 1. The solution is contained in the matrix T and includes temperatures at each node (corresponding to the columns) at each simulated time step (corresponding to each row). The time steps are returned in the vector **time**. (Note that the time steps at which the solution is returned can be specified by providing a vector of specific time values as the second argument, in place of [0, t_sim].)

It is not convenient that the function dTdt_functionv has only two arguments, **time** and **T**. Therefore, the argument list is expanded in order to include all of those parameters that are required to calculate the time rate of change of the nodes. This modification allows these parameters to be set just once (e.g., in the script) and passed through to the function.

```
function[dTdt]=dTdt_functionv(time,T,L,k,rho,c,T_infinity,h_bar)

% Inputs:
% time – time in simulation (s)
% T – vector of nodal temperatures (K)
% L – thickness of the wall (m)
% k – conductivity (W/m-K)
% rho – density (kg/m^3)
% c – specific heat capacity (J/kg-K)
% T_infinity – fluid temperature (K)
% h_bar – heat transfer coefficient (W/m^2-K)
%
% Output:
% dTdt – vector of the time rate of temperature change for each node (K/s)
```

```
    [N,g]=size(T);                              % determine size of T
    DELTAx=L/(N-1);                             % distance between adjacent nodes
    dTdt=zeros(N,1);                            % initialize the dTdt vector
    dTdt(1)=2*k*(T(2)-T(1))/(rho*c*DELTAx^2);
    for i=2:(N-1)
        dTdt(i)=k*(T(i-1)+T(i+1)-2*T(i))/(rho*c*DELTAx^2);
    end
    dTdt(N)=2*k*(T(N-1)-T(N))/(rho*c*DELTAx^2)+2*h_bar*(T_infinity-T(N))/(rho*c*DELTAx);
end
```

The parameters are specified within the script. The inputs time and T are mapped onto the new function dTdt_functionv, as described in Section 3.2.2.

```
clear all;

% Inputs
L=0.05;                                         % wall thickness (m)
k=5.0;                                          % conductivity (W/m-K)
rho=2000;                                       % density (kg/m^3)
c=200;                                          % specific heat capacity (J/kg-K)
T_ini=293.2;                                    % initial temperature (K)
T_infinity=473.2;                               % fluid temperature (K)
h_bar=500;                                      % heat transfer coefficient (W/m^2-K)

%Setup grid
N=11;                                           % number of nodes (-)
for i=1:N
    x(i)=(i-1)*L/(N-1);                         % position of each node (m)
end
DELTAx=L/(N-1);                                 % distance between adjacent nodes (m)

t_sim=40;                                       % simulation time (s)
[time,T]=ode45(@(time,T) dTdt_functionv(time,T,L,k,rho,c,T_infinity,h_bar),[0,t_sim],T_ini*ones(N,1));
```

The integration options (tolerance, etc.) can be controlled using the odeset function and an optional fourth argument, OPTIONS, provided to ode45. This capability was previously discussed in Section 3.2.2. For example, the relative error tolerance for the integration can be set:

```
OPTIONS=odeset('RelTol',1e-6);                  %set relative tolerance
[time,T]=ode45(@(time,T) dTdt_functionv(time,T,L,k,rho,c,...
    T_infinity,h_bar),[0,t_sim],T_ini*ones(N,1),OPTIONS);
```

EXAMPLE 3.8-1: TRANSIENT RESPONSE OF A BENT-BEAM ACTUATOR

EXAMPLE 3.8-1: TRANSIENT RESPONSE OF A BENT-BEAM ACTUATOR

The steady-state behavior of a micro-scale, lithographically fabricated device referred to as a bent-beam actuator was investigated in EXAMPLE 1.7-1. This example examines the transient response of the bent-beam actuator. A V-shaped structure (the bent-beam in Figure 1) is suspended between two pillars. The entire beam is initially at $T_{ini} = 20°C$ when, at time $t = 0$, a voltage difference is applied between the pillars causing current, $I = 10$ mA, to flow through the bent-beam structure. The volumetric generation of thermal energy associated with ohmic heating leads to a thermally induced expansion of both legs that causes the apex of the bent-beam to move outwards.

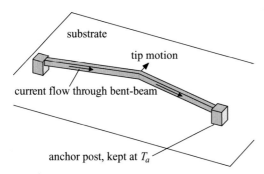

Figure 1: Bent-beam actuator.

The bent beam actuator considered here is identical to that in EXAMPLE 1.7-1. The anchors of the bent-beam actuator are placed $L_a = 1$ mm apart and the beam structure has a cross-section of $w = 10$ μm by $th = 5$ μm. The slope of the beams (with respect to a line connecting the two pillars) is $\theta = 0.5$ rad, as shown in Figure 2. The bent-beam material has conductivity $k = 80$ W/m-K, electrical resistivity $\rho_e = 1 \times 10^{-5}$ ohm-m, density $\rho = 2300$ kg/m^3, specific heat capacity $c = 700$ J/kg-K, and coefficient of thermal expansion $CTE = 3.5 \times 10^{-6}$ K^{-1}. Radiation from the beam surface can be neglected. All of the thermal energy that is generated in the beam is either convected to the surrounding air at temperature $T_\infty = 20°C$ with average heat transfer coefficient $\bar{h} = 100$ W/m^2-K or transferred conductively to the pillars that are maintained at $T_a = 20°C$.

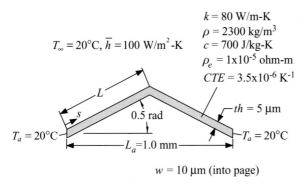

Figure 2: Dimensions and conditions associated with bent-beam actuator.

In EXAMPLE 1.7-1, an appropriate Biot number was used to show that the bent beam can be treated as an extended surface. Therefore, the temperature varies significantly only along the beam (i.e., in the s direction in Figure 2).

a) Develop a 1-D, transient simulation that predicts the time response of the bent beam.

The known information is entered in EES:

"EXAMPLE 3.8-1: Transient Response of a Bent-beam Actuator"

$UnitSystem SI MASS RAD PA K J
$Tabstops 0.2 0.4 0.6 3.5 in

"Inputs"
L_a=1 [mm]*convert(mm,m)   "distance between anchors"
w=10 [micron]*convert(micron,m)   "width of beam"
th=5 [micron]*convert(micron,m)   "thickness of beam"
Current=0.010 [Amp]   "current"
theta=0.5 [rad]   "slope of beam"
T_a=converttemp(C,K,20 [C])   "temperature of pillars"
T_ini=converttemp(C,K,20 [C])   "initial temperature of the beam"
T_infinity=converttemp(C,K,20 [C])   "temperature of air"
h_bar=100 [W/m^2-K]   "heat transfer coefficient"
k=80 [W/m-K]   "conductivity"
rho_e=1e-5 [ohm-m]   "electrical resistivity"
CTE=3.5e-6 [1/K]   "coefficient of thermal expansion"
c=700 [J/kg-K]   "specific heat capacity"
rho=2300 [kg/m^3]   "density"

A half-symmetry (around the apex of the V-shape) numerical model of the bent-beam is developed, with nodes positioned as shown in Figure 3.

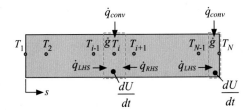

Figure 3: Nodes in a half-symmetry model of the bent-beam.

The length of each leg of the beam structure (L in Figure 2) is:

$$L = \frac{L_a}{2 \cos(\theta)}$$

The position of each node is given by:

$$s_i = \frac{(i-1)}{(N-1)}L \quad \text{for } i = 1..N$$

EXAMPLE 3.8-1: TRANSIENT RESPONSE OF A BENT-BEAM ACTUATOR

and the distance between adjacent nodes is:

$$\Delta s = \frac{L}{(N-1)}$$

```
L=L_a/(2*cos(theta))                    "length of one side of the beam"
N=6 [-]                                 "number of nodes"
DELTAs=L/(N-1)                          "distance between adjacent nodes"
duplicate i=1,N
    s[i]=L*(i-1)/(N-1)                  "position of each node"
    s_mm[i]=s[i]*convert(m,mm)          "in mm"
end
```

The equations that govern the behavior of each node must be obtained using energy balances. The control volume for an arbitrary internal node i (shown in Figure 3) leads to the energy balance:

$$\dot{q}_{LHS} + \dot{q}_{RHS} + \dot{g} + \dot{q}_{conv} = \frac{dU}{dt}$$

Each of the terms must be approximated. The conduction terms are:

$$\dot{q}_{LHS} = k\frac{w\,th}{\Delta s}(T_{i-1} - T_i)$$

$$\dot{q}_{RHS} = k\frac{w\,th}{\Delta s}(T_{i+1} - T_i)$$

The convection term is written as:

$$\dot{q}_{conv} = 2\,(w + th)\,\Delta s\,\overline{h}\,(T_\infty - T_i)$$

The generation term is:

$$\dot{g} = \rho_e \frac{\Delta s}{w\,th}I^2$$

The energy storage term is:

$$\frac{dU}{dt} = w\,th\,\Delta s\,\rho\,c\,\frac{dT_i}{dt}$$

These equations are combined to obtain:

$$k\frac{w\,th}{\Delta s}(T_{i-1} - T_i) + k\frac{w\,th}{\Delta s}(T_{i+1} - T_i) + \rho_e \frac{\Delta s}{w\,th}I^2 + 2\,(w + th)\,\Delta s\,\overline{h}\,(T_\infty - T_i)$$

$$= w\,th\,\Delta s\,\rho\,c\,\frac{dT_i}{dt} \quad \text{for } i = 2\ldots(N-1) \tag{1}$$

Equation (1) can be solved for the rate of temperature change for the internal nodes:

$$\frac{dT_i}{dt} = \frac{k}{\Delta s^2\,\rho\,c}(T_{i-1} + T_{i+1} - 2\,T_i) + \frac{\rho_e}{w^2\,th^2\,\rho\,c}I^2 + \frac{2\,(w + th)\,\overline{h}}{w\,th\,\rho\,c}(T_\infty - T_i)$$

$$\text{for } i = 2\ldots(N-1) \tag{2}$$

EXAMPLE 3.8-1: TRANSIENT RESPONSE OF A BENT-BEAM ACTUATOR

Node 1 is always maintained at T_a. Therefore, it is not necessary to carry out an energy balance on the control volume associated with node 1.

$$\frac{dT_1}{dt} = 0 \tag{3}$$

Node N is the half-node at the apex of the beam (i.e., at $s = L$); the energy balance for the control volume associated with node N (see Figure 3) leads to:

$$\dot{q}_{LHS} + \dot{g} + \dot{q}_{conv} = \frac{dU}{dt}$$

Notice that there is no $\dot{q}_{RHS}$ because the right side of node N is adiabatic according to the assumption of symmetry. The terms are approximated according to:

$$\dot{q}_{LHS} = k \frac{w \, th}{\Delta s} (T_{N-1} - T_N)$$

$$\dot{q}_{conv} - (w + th) \, \Delta s \, \bar{h} \, (T_\infty - T_N)$$

$$\dot{g} = \rho_e \frac{\Delta s}{2 \, w \, th} I^2$$

$$\frac{dU}{dt} = w \, th \frac{\Delta s}{2} \rho \, c \frac{dT_N}{dt}$$

Notice that the convection, generation, and storage terms have all changed by a factor of 2 because node N is a half-node. Combining these equations leads to:

$$k \frac{w \, th}{\Delta s} (T_{N-1} - T_N) + \rho_e \frac{\Delta s}{2 \, w \, th} I^2 + (w + th) \, \Delta s \, \bar{h} \, (T_\infty - T_N) = w \, th \frac{\Delta s}{2} \rho \, c \frac{dT_N}{dt}$$

or

$$\frac{dT_N}{dt} = \frac{2 \, k}{\Delta s^2 \, \rho \, c} (T_{N-1} - T_N) + \frac{\rho_e}{w^2 \, th^2 \, \rho \, c} I^2 + \frac{2 \, (w + th) \, \bar{h}}{w \, th \, \rho \, c} (T_\infty - T_N) \tag{4}$$

Equations (2) through (4) must be integrated forward in time using one of the techniques discussed in Section 3.8.2. Any of the techniques will work; here, the Integral command in EES is used.

Before implementing the solution, it is useful to estimate approximately how long the start up process will take in order to determine the simulation time and provide a sanity check on the solution. Transient conduction processes are characterized by the diffusive time constant (τ_{diff}, which was discussed in Section 3.3.1 in the context of a semi-infinite body) and a lumped capacitance time constant (τ_{lumped}, which was discussed in Section 3.1.3 in the context of 0-D transient problems). The fact that the bent-beam actuator cannot be treated as either a semi-infinite body or a lumped capacitance does not reduce the relevance of these time constants. The diffusive time constant is related to the amount of time required for a conduction wave to move through the material. For this problem, it is interesting to know approximately how long is required for energy to be conducted from the apex to the pillar:

$$\tau_{diff} = \frac{L^2}{\alpha}$$

The lumped time constant is related to the amount of time required for the beam material to equilibrate with the surrounding air:

$$\tau_{lumped} = C_{beam} R_{conv,beam}$$

EXAMPLE 3.8-1: TRANSIENT RESPONSE OF A BENT-BEAM ACTUATOR

where C_{beam} is the heat capacity of the beam:

$$C_{beam} = w\, th\, L\, \rho\, c$$

and $R_{conv,beam}$ is the convective resistance between the beam surface and the air:

$$R_{conv,beam} = \frac{1}{\bar{h}\, 2\, (w + th)\, L}$$

These equations are entered in EES:

```
C_beam=w*th*L*rho*c                          "heat capacity of beam"
R_conv_beam=1/(2*(w+th)*L*h_bar)             "convection resistance between beam and air"
tau_lumped=C_beam*R_conv_beam                "lumped time constant"
alpha=k/(rho*c)                              "thermal diffusivity"
tau_diff=L^2/alpha                           "diffusive time constant"
```

and lead to $\tau_{diff} = 7$ ms and $\tau_{lumped} = 27$ ms. Based on this result, we can expect that the process will be completed on the order of 27 ms (probably somewhat less than this, since the beam will not be able to come completely into thermal equilibrium with the air due to its conductive link with the pillars). Therefore, a simulation time $t_{sim} = 25$ ms is appropriate.

```
t_sim=0.025 [s]                              "simulation duration"
```

The time rate of change for each of the nodes are calculated using Eqs. (2) through (4):

```
dTdt[1]=0 [K/s]                              "pillar temperature never changes"
duplicate i=2,(N-1)
    dTdt[i]=k*(T[i-1]+T[i+1]-2*T[i])/(DELTAs^2*rho*c)+rho_e*Current^2/(w^2*th^2*rho*c)&
        +2*(w+th)*h_bar*(T_infinity-T[i])/(w*th*rho*c)      "internal nodes"
end
dTdt[N]=2*k*(T[N-1]-T[N])/(DELTAs^2*rho*c)+rho_e*Current^2/(w^2*th^2*rho*c)&
    +2*(w+th)*h_bar*(T_infinity-T[N])/(w*th*rho*c)          "node at apex"
```

and these are integrated forward in time using the **Integral** command:

```
duplicate i=1,N
    T[i]=T_ini+INTEGRAL(dTdt[i],time,0,t_sim)    "integrate forward in time"
end
```

The intermediate values of the temperature of each node are stored in an integral table at 0.5 ms intervals using the **$IntegralTable** directive:

```
$IntegralTable time:0.0005, T[1..N]
```

Figure 4 illustrates the temperature at various locations along the beam as a function of time.

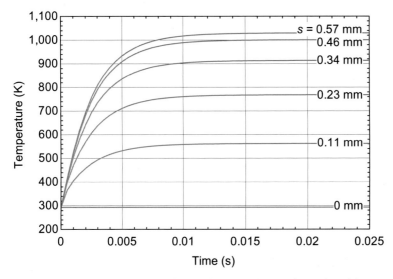

Figure 4: Temperature as a function of time at various values of position.

Notice that the solution has approached steady state after about 10 ms, which is in line with our physical intuition.

The unconstrained motion of the apex of the tip was discussed in EXAMPLE 1.7-1. The elongation of one leg of the beam (ΔL) is obtained by integrating the differential elongation dL along the beam:

$$\Delta L = \int_{0}^{L} dL$$

where dL is related to the product of the coefficient of thermal expansion (CTE) and the temperature change:

$$dL = CTE(T - T_{ini})\,ds$$

The integral can be approximated as a summation of the numerical results:

$$\Delta L_j = \sum_{i=2}^{(N-1)} CTE(T_{i,j} - T_a)\,\Delta s + CTE(T_{N,j} - T_a)\frac{\Delta s}{2} \quad \text{for} \quad j = 1..M$$

where node N is treated separately because it is half the length of the internal nodes and node 1 is not included because its temperature does not rise. The summation is accomplished using the **sum** command in EES:

```
DELTAL=sum(CTE*(T[i]-T_ini)*Deltas,i=2,(N-1))+Deltas*CTE*(T[N]-T_ini)/2 "elongation of beam"
```

Assuming that the joint associated with the apex does not provide a torque on either leg of the beam, the displacement of the apex can be estimated using trigonometry (Figure 5).

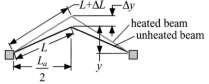

Figure 5: Trigonometry associated with apex motion.

The original position of the apex (y) is given by:

$$y = \sqrt{L^2 - \left(\frac{L_a}{2}\right)^2}$$

therefore, the motion of the apex (Δy) is:

$$\Delta y = \sqrt{(L + \Delta L)^2 - \left(\frac{L_a}{2}\right)^2} - \sqrt{L^2 - \left(\frac{L_a}{2}\right)^2}$$

```
DELTAy=sqrt((L+DELTAL)^2-(L_a/2)^2)-sqrt(L^2-(L_a/2)^2)     "displacement of apex"
DELTAy_micron=DELTAy*convert(m,micron)                      "in μm"
```

The variable **DELTAy_micron** is added to the **IntegralTable** directive:

```
$IntegralTable time:0.0005, T[1..N], DELTAy_micron
```

Figure 6 illustrates the actuator motion as a function of time.

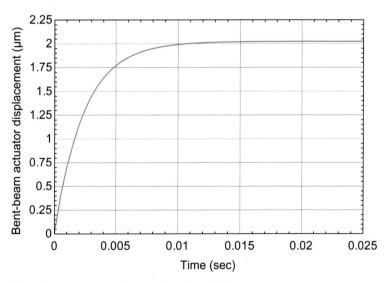

Figure 6: Actuator motion as a function of time.

3.8.3 Temperature-Dependent Properties

The simulation of steady-state 1-D problems with temperature-dependent properties is discussed in Section 1.4.3 (for EES models) and 1.5.6 (for MATLAB models). In either case, it is important that the equations are set up in a manner that ensures that there is no energy mismatch at the interface between adjacent nodes. This consideration is also important for transient simulations with temperature-dependent properties. It is easiest to simulate each time step by assuming that the properties have values that are consistent with the temperature distribution that exists at the beginning of the time step and then use the methods discussed in Section 3.8.2.

EXAMPLE 1.8-2 examines the steady state behavior of an ablative technique for heating cancerous tissue locally using small, conducting spheres that are exposed to magnetic waves. The proper method for treating temperature-dependent properties with a numerical model is illustrated in this section by examining the transient behavior of the ablative process. The conducting spheres experience a volumetric generation of thermal energy that causes their temperature and the temperature of the adjacent tissue to rise. The tissue surrounding the spheres experiences blood perfusion, which refers to the volumetric removal of thermal energy in the tissue by the blood flowing in the microvascular structure. Blood perfusion may be modeled as a volumetric heat sink (a negative volumetric rate of thermal energy generation, $\dot{g}_{bp}'''$) that is proportional to the difference between the local temperature and the normal body temperature, $T_b = 37°C$:

$$\dot{g}_{bp}''' = -\beta \left(T - T_b \right) \tag{3-579}$$

where the perfusion constant is $\beta = 20,000 \, \text{W/m}^3\text{-K}$. The sphere has a radius $r_{ts} = 1.0 \, \text{mm}$ and experiences the generation of thermal energy at the rate of $\dot{g}_{ts} = 1.0 \, \text{W}$. The temperature far from the sphere is $T_b = 37°C$. The density of tissue is $\rho = 1000 \, \text{kg/m}^3$ and the specific heat capacity of tissue is $c = 3500 \, \text{J/kg-K}$. The conductivity of tissue varies with temperature according to:

$$k = -0.621 \left[\frac{\text{W}}{\text{m-K}} \right] + 6.03 \times 10^{-3} \left[\frac{\text{W}}{\text{m-K}^2} \right] T - 7.87 \times 10^{-6} \left[\frac{\text{W}}{\text{m-K}^3} \right] T^2 \tag{3-580}$$

The tissue is initially at $T_b = 37°C$ when the thermoseed is activated. The heat capacity of the thermoseed itself is small and can be neglected. Therefore, the entire generation rate ($\dot{g}_{ts}$) is transferred to the tissue.

The inputs are entered into a MATLAB script:

```
clear all;

% Inputs
r_ts=0.001;              % thermoseed radius (m)
beta=20000;             % perfusion constant (W/m^3-K)
c=3500;                 % tissue specific heat capacity (J/kg-K)
rho=1000;               % tissue density (kg/m^3)
T_b=310.2;              % body temperature (K)
g_dot_ts=1.0;           % sphere generation (W)
```

In order to develop a numerical model for the tissue it is necessary to position nodes throughout the computational domain. However, the outer limit of the domain (r_{out}) is undefined. We will take $r_{out} = 10 \, r_{sp}$ based on the results of EXAMPLE 1.8-2 that showed that the temperature disturbance produced by the ablation process has died out at this distance. Without this insight, it would be necessary to increase r_{out} until our solution is unaffected by this choice.

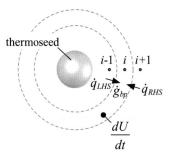

Figure 3-45: First law on a control volume defined around an internal node.

The nodes are uniformly positioned throughout the domain. The distance between adjacent nodes is given by:

$$\Delta r = \frac{(r_{out} - r_{ts})}{(N - 1)} \tag{3-581}$$

$$r_i = r_{ts} + (i - 1)\,\Delta r \quad \text{for } i = 1 \ldots N \tag{3-582}$$

where N is the number of nodes.

```
r_out=10*r_ts;                    % outer radius of computational domain (m)
N=51;                             % number of nodes spatially
DELTAr=(r_out-r_ts)/(N-1);        % distance between nodes
for i=1:N
  r(i,1)=r_ts+DELTAr*(i-1);       % radial location of each node
end
```

The initial temperature of each node is equal to the body temperature:

```
T_ini=T_b*ones(N,1);                    % initial temperature
```

A control volume is defined around each of the nodes and shown for node i in Figure 3-45. The energy balance suggested by Figure 3-45 is:

$$\dot{q}_{LHS} + \dot{q}_{RHS} = \frac{dU}{dt} + \dot{g}_{bp} \tag{3-583}$$

Note that the rate of heat removal due to blood perfusion ($\dot{g}_{bp}$) is placed on the outflow side of the energy balance. The rate equations for the conduction terms in the energy balance are:

$$\dot{q}_{LHS} = 4\pi \left(r_i - \frac{\Delta r}{2}\right)^2 k_{T=(T_i+T_{i-1})/2} \frac{(T_{i-1} - T_i)}{\Delta r} \tag{3-584}$$

$$\dot{q}_{RHS} = 4\pi \left(r_i + \frac{\Delta r}{2}\right)^2 k_{T=(T_i+T_{i+1})/2} \frac{(T_{i+1} - T_i)}{\Delta r} \tag{3-585}$$

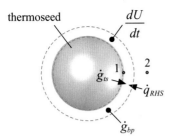

Figure 3-46: First law on the control volume defined around node 1.

where the conductivity is evaluated at the temperature of the interface (i.e., the average of adjacent node temperatures) rather than the temperature of the node in order to ensure that energy is not 'lost' between nodes. The rate of energy removed by blood perfusion within the control volume is given by:

$$\dot{g}_{bp} = 4\,\pi\,r_i^2\,\Delta r\,\beta\,(T_i - T_b) \tag{3-586}$$

The rate of energy storage within the control volume is given by:

$$\frac{dU}{dt} = 4\,\pi\,r_i^2\,\Delta r\,\rho\,c\,\frac{dT_i}{dt} \tag{3-587}$$

Substituting Eqs. (3-584) through (3-587) into Eq. (3-583) leads to:

$$4\,\pi\left(r_i - \frac{\Delta r}{2}\right)^2 k_{T=(T_i+T_{i-1})/2}\,\frac{(T_{i-1} - T_i)}{\Delta r} + 4\,\pi\left(r_i + \frac{\Delta r}{2}\right)^2 k_{T=(T_i+T_{i+1})/2}\,\frac{(T_{i+1} - T_i)}{\Delta r}$$
$$= 4\,\pi\,r_i^2\,\Delta r\,\rho\,c\,\frac{dT_i}{dt} + 4\,\pi\,r_i^2\,\Delta r\,\beta\,(T_i - T_b) \quad \text{for } i = 2\ldots(N-1) \tag{3-588}$$

which can be rearranged to provide the rate of temperature change for each internal node:

$$\frac{dT_i}{dt} = \frac{\left(r_i - \dfrac{\Delta r}{2}\right)^2}{r_i^2\,\Delta r^2\,\rho\,c}\,k_{T=(T_i+T_{i-1})/2}\,(T_{i-1} - T_i)$$

$$+ \frac{\left(r_i + \dfrac{\Delta r}{2}\right)^2}{r_i^2\,\Delta r^2\,\rho\,c}\,k_{T=(T_i+T_{i+1})/2}\,(T_{i+1} - T_i) - \frac{\beta\,(T_i - T_b)}{\rho\,c} \quad \text{for } i = 2\ldots(N-1) \tag{3-589}$$

The node N at $r = r_{out}$ is assumed to be at the body temperature:

$$T_N = T_b \tag{3-590}$$

and therefore its time rate of change is always zero:

$$\frac{dT_N}{dt} = 0 \tag{3-591}$$

An energy balance on node 1 at the surface of the sphere is shown in Figure 3-46. The energy balance suggested by Figure 3-46 is:

$$\dot{g}_{ts} + \dot{q}_{RHS} = \frac{dU}{dt} + \dot{g}_{bp} \tag{3-592}$$

where $\dot{g}_{ts}$ is the rate of energy transfer from the thermoseed. The rate equations for the terms in Eq. (3-592) are:

$$\dot{q}_{RHS} = 4\pi \left(r_1 + \frac{\Delta r}{2}\right)^2 k_{T=(T_1+T_2)/2} \frac{(T_2 - T_1)}{\Delta r} \tag{3-593}$$

$$\dot{g}_{bp} = 2\pi r_1^2 \Delta r \, \beta \, (T_1 - T_b) \tag{3-594}$$

$$\frac{dU}{dt} = 2\pi r_1^2 \Delta r \rho c \frac{dT_1}{dt} \tag{3-595}$$

Substituting Eqs. (3-593) through (3-595) into Eq. (3-592) leads to:

$$\dot{g}_{ts} + 4\pi \left(r_1 + \frac{\Delta r}{2}\right)^2 k_{T=(T_1+T_2)/2} \frac{(T_2 - T_1)}{\Delta r} = 2\pi r_1^2 \Delta r \rho c \frac{dT_1}{dt} + 2\pi r_1^2 \Delta r \beta \, (T_1 - T_b) \tag{3-596}$$

Solving for the time rate of change of the temperature of node 1 leads to:

$$\frac{dT_1}{dt} = \frac{\dot{g}_{ts}}{2\pi r_1^2 \Delta r \rho c} + \frac{2\left(r_1 + \frac{\Delta r}{2}\right)^2}{r_1^2 \Delta r^2 \rho c} k_{T=(T_1+T_2)/2} (T_2 - T_1) - \frac{\beta \, (T_1 - T_b)}{\rho c} \tag{3-597}$$

In order to solve this problem, Eqs. (3-589), (3-591), and (3-597) must be integrated forward through time using one of the techniques discussed in Section 3.8.2. Here, this is accomplished using MATLAB's native ODE solver ode45. The function dTdt_S3p8p3 is defined. The function returns the rate of temperature change for each of the nodes given the time and instantaneous temperature of each node as well as the other inputs to the problem:

```
function[dTdt]=dTdt_S3p8p3(time,T,r,g_dot_ts,T_b,c,rho,beta)
%
% Inputs:
% time – time in simulation (s)
% T – temperature of each node (K)
% r – radial position of each node (m)
% g_dot_ts – thermal energy generated by thermoseed (W)
% T_b – body temperature (K)
% c – specific heat capacity of tissue (J/kg-K)
% rho – density of tissue (kg/m^3)
% beta – perfusion constant (W/m^3-K)
[N,g]=size(T);                          % number of nodes
DELTAr=r(2)-r(1);                       % distance between nodes (m)
dTdt=zeros(N,1);                        % initialize temperature rate of change vector
dTdt(1)=g_dot_ts/(2*pi*r(1)^2*DELTAr*rho*c)+2*(r(1)+DELTAr/2)^2*k((T(1)+T(2))/2)*...
    (T(2)-T(1))/(r(1)^2*DELTAr^2*rho*c)-beta*(T(1)-T_b)/(rho*c);
for i=2:(N-1)
    dTdt(i)=(r(i)-DELTAr/2)^2*k((T(i)+T(i-1))/2)*(T(i-1)-T(i))/(r(i)^2*DELTAr^2*rho*c)...
    +(r(i)+DELTAr/2)^2*k((T(i)+T(i+1))/2)*(T(i+1)-T(i))/(r(i)^2*DELTAr^2*rho*c)-beta*(T(i)-T_b)/(rho*c);
end
dTdt(N)=0;
end
```

Note that dTdt_S3p8p3 implements Eqs. (3-589), (3-591), and (3-597). The function k is defined in order to provide the temperature-dependent conductivity according to Eq. (3-580):

```
function[k]=k(T)
% Input:
% T - temperature (K)
%
% Output:
% k - conductivity (W/m-K)
    k=-0.621+6.03e-3*T-7.87e-6*T^2;            % tissue conductivity (W/m-K)
end
```

The function k is located in the same M-file that contains dTdt_S3p8p3.

The simulation is carried out by calling the ode45 function from within the script S3p8p3. The simulation time is set to 300 sec and the relative tolerance for the integration is set to 1×10^{-6}. The ode45 integration routine is called and the results are converted from K to °C:

```
t_sim=300;          %simulation time (s)
OPTIONS=odeset('RelTol',1e-6);
[time,T]=ode45(@(time,T) dTdt_S3p8p3(time,T,r,g_dot_ts,T_b,c,rho,beta),[0,t_sim],T_ini,OPTIONS);
T_C=T-273.2;
```

The temperature as a function of radius at various times is shown in Figure 3-47.

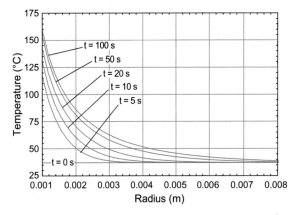

Figure 3-47: Temperature as a function of radius for various times.

3.9 Reduction of Multi-Dimensional Transient Problems

This extended section of the book can be found on the website (www.cambridge.org/nellisandklein). Section 3.5 discusses the solution to 1-D transient problems using separation of variables. In some cases, it is possible to solve a multidimensional transient problem using the product of 1-D transient solutions; this process is discussed in Section 3.9. It is not always possible to solve a multi-dimensional transient problem using this technique. The problem must be linear and completely homogeneous for this process to work; completely homogeneous indicates that: (1) the governing differential equation is homogeneous (e.g., there is no generation term), and (2) all of the spatial boundary conditions are homogeneous. The initial condition does not have to be homogeneous but it must be relatively simple. If the problem satisfies these conditions, then the steps outlined in Section 3.9 will indicate whether the multidimensional problem can be recast as several 1-D problems.

Chapter 3: Transient Conduction

The website associated with this book (www.cambridge.org/nellisandklein) provides many more problems than are included here.

Analytical Solution to 0-D Transient Problems

3–1 Your cabin is located close to a source of geothermal energy and therefore you have decided to heat it during the winter by lowering spheres of metal into the ground in the morning so that they are heated to a uniform temperature, $T_{gt} = 300°C$ during the day. In the evenings, you remove the spheres and carry them to your cabin; this trip requires about $\tau_{travel} = 30$ minutes. The spheres are placed in your cabin and give off heat during the night as they cool; the night is $\tau_{night} = 6$ hrs long. The heat transfer coefficient between a sphere and the surrounding air (outdoor or cabin) is $\bar{h} = 20\,\text{W/m}^2\text{-K}$ (neglect radiation) and the temperature of the surrounding air (outdoor or cabin) is $T_{amb} = 10°C$. You can carry about $M = 100\,\text{lb}_\text{m}$ of metal and are trying to decide what radius of sphere would work the best. You can carry a lot of spheres (as small as $r_{min} = 5.0\,\text{mm}$) or a single very large sphere. The thermal conductivity of the metal is $k = 80\,\text{W/m-K}$, density $\rho = 9000\,\text{kg/m}^3$, and $c = 1000\,\text{J/kg-K}$.

   a.) What is the largest sphere you could use, r_{max}? That is, what it is the size of a sphere with mass $M = 100\,\text{lb}_\text{m}$?

   b.) What is the Biot number associated with the maximum size sphere from (a)? Is a lumped capacitance model of the sphere appropriate for this problem?

   c.) Prepare a plot showing the amount of energy released from the metal (all of the spheres) during τ_{travel}, the period of time that is required to transport the metal back to your cabin, as a function of sphere radius. Explain the shape of your plot (that is, explain why it increases or decreases).

   d.) Prepare a plot showing the amount of energy released from the metal to your cabin during the night (i.e., from $t = \tau_{travel}$ to $t = \tau_{travel} + \tau_{night}$) as a function of sphere radius. Explain the shape of your plot (again, why does it look the way it does?).

   e.) Prepare a plot showing the efficiency of the heating process, η, as a function of radius. The efficiency is defined as the ratio of the amount of energy provided to your cabin to the maximum possible amount of energy you could get from the metal. (Note that this limit occurs if the metal is delivered to the cabin at T_{gt} and removed at T_{amb}.)

3–2 An instrument on a spacecraft must be cooled to cryogenic temperatures in order to function. The instrument has mass $M = 0.05$ kg and specific heat capacity $c = 300$ J/kg-K. The surface area of the instrument is $A_s = 0.02\,\text{m}^2$ and the emissivity of its surface is $\varepsilon = 0.35$. The instrument is exposed to a radiative heat transfer from surroundings at $T_{sur} = 300$ K. It is connected to a cryocooler that can provide $\dot{q}_{cooler} = 5$ W. The instrument is exposed to a solar flux that oscillates according to: $\dot{q}''_s = \overline{\dot{q}''_s} + \Delta\dot{q}''_s \sin(\omega t)$ where $\overline{\dot{q}''_s} = 100\,\text{W/m}^2$, $\Delta\dot{q}''_s = 100\,\text{W/m}^2$, and $\omega = 0.02094\,\text{rad/s}$. The initial temperature of the instrument is $T_{ini} = 300$ K. Assume that the instrument can be treated as a lumped capacitance. Model radiation using a constant radiation resistance.

a.) Develop an analytical model of the cool-down process and implement your model in EES.

b.) Plot the temperature as a function of time.

3–3 One technique for detecting chemical threats uses a laser to ablate small particles so that they can subsequently be analyzed using ion mobility spectroscopy. The laser pulse provides energy to a particle according to: $\dot{q}_{laser} = \dot{q}_{max} \exp[-\frac{(t-t_p)^2}{2\,t_d^2}]$ where $\dot{q}_{max} = 0.22$ W is the maximum value of the laser power, $t_p = 2\mu s$ is the time at which the peak laser power occurs, and $t_d = 0.5$ μs is a measure of the duration of the pulse. The particle has radius $r_p = 5\,\mu m$ and has properties $c = 1500$ J/kg-K, $k = 2.0$ W/m-K, and $\rho = 800$ kg/m^3. The particle is surrounded by air at $T_\infty = 20°C$. The heat transfer coefficient is $\bar{h} = 60000$ W/m^2-K. The particle is initially at T_∞.

a.) Is a lumped capacitance model of the particle justified?

b.) Assume that your answer to (a) is yes; develop an analytical model of the particle using Maple and EES. Plot the temperature of the particle as a function of time. Overlay on your plot (on a secondary axis) the laser power.

Numerical Solution to 0-D Transient Problems

3–4 Reconsider Problem 3-2 using a numerical model. The cooling power of the cryocooler is not constant but is a function of temperature:

$$\dot{q}_{cooler} = \begin{cases} -4.995\,[\text{W}] + 0.1013\,T\left[\dfrac{\text{W}}{\text{K}}\right] - 0.0001974\,T^2\left[\dfrac{\text{W}}{\text{K}^2}\right] & \text{if } T > 55.26\,\text{K} \\ 0 \quad \text{if } T < 55.26\,\text{K} \end{cases}$$

where T is the temperature of the instrument.

a.) Develop a numerical model in EES using Heun's method. Plot the temperature of the instrument as a function of time for 2000 s after the cryocooler is activated.

b.) Verify that your model from (a) limits to the analytical solution developed in Problem 3-2 in the limit that the cryocooler power is constant and radiation is treated using a constant, approximate radiation resistance. Overlay on the same plot the temperature of the instrument as a function of time predicted by the analytical and numerical models.

d.) Develop a numerical model in EES using the Integral command. Plot the temperature of the instrument as a function of time for 2000 s after the cryocooler is activated.

e.) Develop a numerical model in MATLAB using the ode solver. Plot the temperature of the instrument as a function of time for 2000 s after the cryocooler is activated.

3–5 Reconsider Problem 3-3.

a.) Develop a numerical model of the particle using the Euler technique implemented in either EES or MATLAB. Plot the temperature as a function of time and compare your answer with the analytical solution from Problem 3-3.

b.) Develop a numerical model of the particle using Heun's technique implemented in either EES or MATLAB. Plot the temperature as a function of time.

c.) Develop a numerical model of the particle using the fully implicit technique implemented in either EES or MATLAB. Plot the temperature as a function of time.

    d.) Develop a numerical model of the particle using the Crank-Nicolson technique implemented in either EES or MATLAB. Plot the temperature as a function of time.

    e.) Develop a numerical model of the particle using the Integral command in EES. Plot the temperature as a function of time.

    f.) Develop a numerical model of the particle using ode45 solver in MATLAB. Plot the temperature as a function of time.

3–6 You are interested in using a thermoelectric cooler to quickly reduce the temperature of a small detector from its original temperature of $T_{ini} = 295$ K to its operating temperature. As shown in Figure P3-6, the thermoelectric cooler receives power at a rate of $\dot{w} = 5.0$ W from a small battery and rejects heat at a rate of $\dot{q}_{rej}$ to ambient temperature $T_H = 305$ K. The cooler removes energy at a rate of $\dot{q}_{ref}$ from the detector which is at temperature T. (The detector temperature T will change with time, t). The detector has a total heat capacity, C, of 0.5 J/K. Despite your best efforts to isolate the detector from the ambient, the detector is subjected to a parasitic heat gain, $\dot{q}_p$, that can be modeled as occurring through a fixed resistance $R_p = 100$ K/W between T and T_H; this resistance represents the combined effect of radiation and conduction.

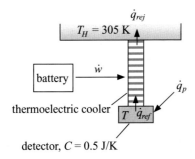

Figure P3-6: Detector cooled by a thermoelectric cooler.

The thermoelectric cooler has a second law efficiency $\eta_c = 10\%$ regardless of its operating temperatures. That is, the amount of refrigeration provided to the detector can be related to the input power provided to the thermoelectric cooler and its operating temperatures according to:

$$\dot{q}_{ref} = \frac{\dot{w}\,\eta_c}{\left(\dfrac{T_H}{T} - 1\right)}$$

    a.) Derive the governing differential equation that describes the temperature of the detector. Note that the result should be a symbolic equation for the rate of temperature change of the detector as a function of the quantities given in the problem (i.e., T_H, R_p, C, $\dot{w}$, η_c) and the instantaneous value of the detector temperature (T).

    b.) Develop an EES program that numerically solves this problem for the values given in the problem statement using a predictor-corrector technique (e.g., Heun's method). Using your program, prepare a plot showing the temperature of the detector as a function of time for 120 sec after the cooler is activated.

c.) Modify your program so that it accounts for the fact that your battery only has 100 J of energy storage capacity; once the 100 J of energy in the battery is depleted, then the power driving the thermoelectric cooler goes to zero. Prepare a plot showing the temperature of the detector as a function of time for 120 s after the cooler is activated.

d.) Assume that the objective of your cooler is to keep the detector at a temperature below 240 K for as long as possible, given that your battery only has 100 J of energy. What power ($\dot{w}$) would you use to run the thermoelectric cooler? Justify your answer with plots and an explanation.

Semi-infinite 1-D Transient Problems

3–7 A thin heater is sandwiched between two materials, A and B, as shown in Figure P3-7. Both materials are very thick and so they may be considered semi-infinite. Initially, both materials are at a uniform temperature of T_{ini}. The heater is activated at $t = 0$ and delivers a uniform heat flux, $\dot{q}''_{heater}$, to the interface; some of this energy will be conducted into material A ($\dot{q}''_A$) and some into material B ($\dot{q}''_B$). Materials A and B have the same thermal diffusivity, $\alpha_A = \alpha_B = \alpha$. and the same conductivity, $k_A = k_B = k$. There is no contact resistance anywhere in this problem and it is a 1-D, transient conduction problem.

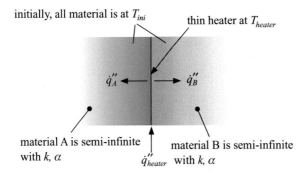

Figure P3-7: Thin heater sandwiched between two semi-infinite bodies.

a.) Draw a thermal resistance network that you could use to model this problem approximately. Your resistances should be written in terms of time, t, and the symbols in the problem statement. Clearly indicate on your network where $\dot{q}''_{heater}$ is added to the network and where the temperatures T_{ini} and T_{heater} are located.

b.) Use your resistance network from (a) to develop an equation for the heater temperature, T_{heater}, in terms of the symbols in the problem statement.

c.) Sketch the temperature distribution at $t = 0$ and two additional times after the heater has been activated (t_1 and t_2 where $t_2 > t_1$). Label your plots clearly. Focus on getting the qualitative features of your plot correct.

3–8 Figure P3-8 shows a slab of material that is $L = 5$ cm thick and is heated from one side ($x = 0$) by a radiant heat flux $\dot{q}''_s = 7500\ \text{W/m}^2$. The material has conductivity $k = 2.4\ \text{W/m-K}$ and thermal diffusivity $\alpha = 2.2 \times 10^{-4}\ \text{m}^2/\text{s}$. Both sides of the slab are exposed air at $T_\infty = 20°\text{C}$ with heat transfer coefficient $\bar{h} = 15\ \text{W/m}^2\text{-K}$. The initial temperature of the material is $T_{ini} = 20°\text{C}$.

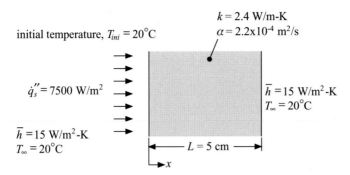

initial temperature, $T_{ini} = 20°C$

$k = 2.4$ W/m-K
$\alpha = 2.2 \times 10^{-4}$ m²/s

$\dot{q}_s'' = 7500$ W/m²

$\bar{h} = 15$ W/m²-K
$T_\infty = 20°C$

$\bar{h} = 15$ W/m²-K
$T_\infty = 20°C$

$L = 5$ cm

Figure P3-8: Slab of material heated at one surface.

a.) About how long do you expect it to take for the temperature of the material on the unheated side ($x = L$) to begin to rise?

b.) What do you expect the temperature of the material at the heated surface ($x = 0$) to be (approximately) at the time identified in (a)?

c.) Develop a simple and approximate model that can predict the temperature at the heated surface as a function of time for times that are less than the time calculated in (a). Plot the temperature as a function of time from $t = 0$ to the time identified in (a).

d.) Sketch the temperature as a function of position in the slab for several times less than the time identified in (a) and greater than the time identified in (a). Make sure that you get the qualitative features of the sketch correct. Also, sketch the temperature as a function of position in the slab at steady state; (make sure that you get the temperatures at either side correct).

3–9 A semi-infinite body has conductivity $k = 1.2$ W/m-K and thermal diffusivity $\alpha = 5 \times 10^{-4}$ m²/s. At time $t = 0$, the surface is exposed to fluid at $T_\infty = 90°C$ with heat transfer coefficient $\bar{h} = 35$ W/m² - K. The initial temperature of the material is $T_{ini} = 20°C$.

a.) Develop an approximate model that can provide the temperature of the surface and the rate of heat transfer into the surface as a function of time.

b.) Based on your model, develop an expression that provides a characteristic time related to how long will it take for the surface of the solid to approach T_∞?

c.) Compare the results of your model from (a) with the exact solution programmed in EES and accessed using the Semilnf3 function.

3–10 A rod with uniform cross-sectional area, $A_c = 0.1$ m² and perimeter $per = 0.05$ m is placed in a vacuum environment. The length of the rod is $L = 0.09$ m and the external surfaces of the rod can be assumed to be adiabatic. For a long time, a heat transfer rate of $\dot{q}_h = 100$ W is provided to the end of the rod at $x = 0$. The tip of the rod at $x = L$ is always maintained at $T_t = 20°C$. The rod material has density $\rho = 5000$ kg/m³, specific heat capacity $c = 500$ J/kg-K, and conductivity $k = 5$ W/m-K. The rod is at a steady state operating condition when, at time $t = 0$, the heat transfer rate at $x = 0$ becomes zero.

a.) About how long does it take for the rod to respond to the change in heat transfer?

b.) Sketch the temperature distribution you expect at $t = 0$ and $t \to \infty$. Make sure that you get the temperatures at either end of the rod and the shape of the temperature distributions correct.

c.) Overlay on your sketch from (b) the temperature distributions that you expect at the time that you calculated in (a) as well as half that time and twice that time.

d.) Sketch the heat transfer from the rod at $x = L$ (i.e., at the tip) as a function of time. Make sure that your sketch clearly shows the behavior before and after the time identified in (a). Make sure that you get the rate of heat transfer at $t = 0$ and $t \rightarrow \infty$ correct.

3–11 One technique that is being proposed for measuring the thermal diffusivity of a material is illustrated schematically in Figure P3-11.

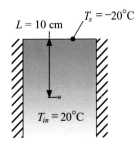

Figure P3-11: Test setup for measuring thermal diffusivity.

The material is placed in a long, insulated container and allowed to come to thermal equilibrium with its environment $T_{in} = 20°C$. A thermocouple is embedded in the material at a distance $L = 10$ cm below the surface. At time $t = 0$ the temperature of the surface is changed from T_{in} to $T_s = -20°C$ by applying a flow of chilled ethylene glycol to the surface. The time required for the thermocouple to change from T_{in} to $T_{target} = 0°C$ is found to be $t_{target} = 310.2$ s.

a.) What is the measured thermal diffusivity?

There is some error in your measurement from part (a) due to inaccuracies in your thermocouple and your measurement of time and position of the thermocouple. Assume that the following uncertainties characterize your experiment:

• the temperature measurements have an uncertainty of $\delta T_{in} = \delta T_{target} = \delta T_s = 0.2°C$
• the position measurement has an uncertainty of $\delta L = 0.1$ mm
• the time measurement has an uncertainty of $\delta t_{target} = 0.5$ s

b.) What is the uncertainty in your measured value of thermal diffusivity from part (a)? You can answer this question in a number of ways including using the built-in uncertainty propagation feature in EES.

The Laplace Transform

3–12 A disk shaped piece of material is used as the target of a laser, as shown in Figure P3-12. The laser target is $D = 5.0$ mm in diameter and $b = 2.5$ mm thick. The target is made of a material with $\rho = 2330$ kg/m^3, $k = 500$ W/m-K, and $c = 400$ J/kg-K. The target is mounted on a chuck with a constant temperature $T_c = 20°C$. The interface between the target and the chuck is characterized by a contact resistance, $R_c'' = 1 \times 10^{-4}$ K-m^2/W. The target is initially in thermal equilibrium with the chuck. You may neglect radiation and convection from the laser target. The laser flux is pulsed according to: $\dot{q}_{laser}'' = A t^2 \exp(-t/t_{pulse})$ where $A = 1 \times 10^7$ W/m^2-s^2 and $t_{pulse} = 0.10$ s.

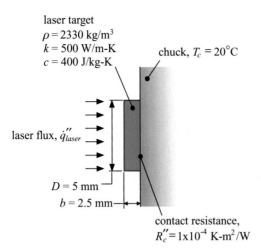

laser target
$\rho = 2330\ \text{kg/m}^3$
$k = 500\ \text{W/m-K}$
$c = 400\ \text{J/kg-K}$

chuck, $T_c = 20°\text{C}$

laser flux, $\dot{q}''_{laser}$

$D = 5\ \text{mm}$
$b = 2.5\ \text{mm}$

contact resistance,
$R''_c = 1\times10^{-4}\ \text{K-m}^2/\text{W}$

Figure P3-12: Laser target.

a.) Is a lumped capacitance model of the laser target appropriate? Justify your answer.

b.) Use Laplace transforms to determine the temperature of the laser target as a function of time. Prepare a plot of the temperature of the laser target as a function of time; overlay on this plot the laser heat flux as a function of time (on a secondary y-axis).

3–13 A semi-infinite piece of material with thermal diffusivity $\alpha = 1 \times 10^{-5}\ \text{m}^2/\text{s}$ and conductivity $k = 1\ \text{W/m-K}$ is initially (at $t = 0$) at $T_{ini} = 300\ \text{K}$ when the surface temperature (i.e., the temperature at $x = 0$) begins to increase linearly according to: $T_{x=0,t} = T_{in} + \beta t$ where $\beta = 1\ \text{K/s}$.

a.) For part (a), do not solve the problem exactly. Rather, use your conceptual knowledge of how a thermal wave moves through a semi-infinite body in order to obtain an approximate model for the heat flux at the surface as a function of time. Plot the approximate heat flux as a function of time for $t = 0$ to 5000 s.

b.) Use the Laplace transform technique to obtain an analytical solution to this problem. Implement your solution in EES and prepare a plot showing the temperature as a function of time (for $t = 0$ to 5000 s) at locations $x = 0$, 0.1 m, 0.2 m, and 0.3 m. Prepare another plot showing the temperature as a function of position (for $x = 0$ to 0.5 m) for $t = 0$ s, 300 s, 600 s, and 900 s.

c.) Overlay your exact solution onto the plot of your approximate solution from (a).

3–14 Solve Problem 3-8 using the Laplace transform technique for the period of time where the material can be treated as a semi-infinite body.

a.) Prepare a solution and implement your solution in EES. Plot the temperature as a function of position for several times.

b.) Compare the analytical solution obtained in (a) to the approximate model that you derived in (c) of Problem 3-8.

3–15 A sphere with radius $R = 1$ mm is composed of material with density $\rho = 9000\ \text{kg/m}^3$, specific heat capacity $c = 500\ \text{J/kg-K}$, and conductivity $k = 25\ \text{W/m-K}$. The surface is exposed to fluid at $T_\infty = 25°\text{C}$ with heat transfer coefficient

$\bar{h} = 1000 \, \text{W/m}^2\text{-K}$. The sphere is initially in equilibrium with the fluid when it experiences a time varying volumetric generation of thermal energy:

$$\dot{g}''' = \dot{g}'''_{max} \exp\left(-\frac{t}{a}\right)$$

where $\dot{g}'''_{max} = 1 \times 10^9 \, \text{W/m}^3$ and $a = 2$ s.

a.) Is a lumped capacitance solution appropriate for this problem? Justify your answer.

b.) Assume that your answer to (a) is yes. Determine an expression for the temperature as function of time using the Laplace transform technique and implement this solution in EES. Plot temperature as a function of time.

Section 3.5: Separation of Variables for Transient Properties

3–16 Ice cream containers are removed from a warehouse and loaded into a refrigerated truck. During this loading process, the ice cream may sit on the dock for a substantial amount of time. The dock temperature is higher than the warehouse temperature, which can cause two problems. First, the temperature of the ice cream near the surface can become elevated, resulting in a loss of food quality. Second, the energy absorbed by the ice cream on the dock must subsequently be removed by the equipment on the refrigerated truck, causing a substantial load on this relatively under-sized and inefficient equipment. The ice cream is placed in cylindrical cardboard containers. Assume that the containers are very long and therefore, the temperature distribution of the ice cream is one dimensional, as shown in Figure P3-16. The inner radius of the cardboard ice cream containers is $R_o = 10$ cm and the thickness of the wall is $th_{cb} = 2.0$ mm. The conductivity of cardboard is $k_{cb} = 0.08 \, \text{W/m-K}$. The ice cream comes out of the warehouse at $T_{ini} = 0°F$ and is exposed to the dock air at $T_{dock} = 45°F$ with heat transfer coefficient, $\bar{h} = 20 \, \text{W/m}^2\text{-K}$. The ice cream has properties $k_{ic} = 0.2 \, \text{W/m-K}$, $\rho_{ic} = 720 \, \text{kg/m}^3$, and $c_{ic} = 3200 \, \text{J/kg-K}$. (Assume that the ice cream does not melt.)

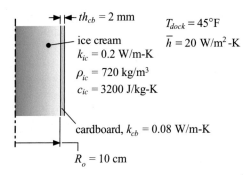

Figure P3-16: Ice cream containers.

a.) Determine an effective heat transfer coefficient, $\bar{h}_{eff}$, that can be used in conjunction with the analytical solutions for a cylinder subjected to a step change in fluid temperature but includes the effect of the conduction resistance associated with the cardboard as well as the convection to the air.

b.) If the ice cream remains on the dock for $t_{load} = 5$ minutes, what will the temperature of the surface of the ice cream be when it is loaded?

c.) How much energy must be removed from the ice cream (per unit length of container) after it is loaded in order to bring it back to a uniform temperature of $T_{ini} = 0°F$?

d.) What is the maximum amount of time that the ice cream can sit on the dock before the ice cream at the outer surface begins to melt (at $0°C$)?

3–17 A wall is exposed to a heat flux for a long time, as shown in Figure P3-17. The left side of the wall is exposed to liquid at $T_f = 20°C$ with a very high heat transfer coefficient; therefore, the left side of the wall ($T_{x=0}$) always has the temperature T_f. The right side of the wall is exposed to the heat flux and also convects to gas at $T_f = 20°C$ but with a heat transfer coefficient of $\bar{h} = 5000 \text{ W/m}^2\text{-K}$. The wall is $L = 0.5$ m thick and composed of a material with $k = 1.0 \text{ W/m-K}$, $\rho = 4000 \text{ kg/m}^3$, and $c = 700 \text{ J/kg-K}$. The wall is initially at steady state with the heat flux when, at time $t = 0$, the heat flux is suddenly shut off. The wall subsequently equilibrates with the liquid and gas, eventually it reaches a uniform temperature, equal to T_f.

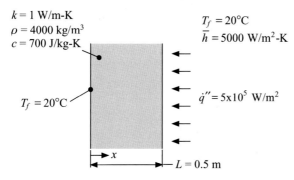

Figure 3-17: Wall exposed to a heat flux.

a.) Calculate the temperature of the right hand side of the wall at $t = 0$ (i.e., determine $T_{x=L,t=0}$).

b.) Sketch the temperature distribution at $t = 0$ and the temperature distribution as $t \to \infty$.

c.) Sketch the temperature distribution at $t = 10$ s, 100 s, 1×10^3 s, 1×10^4 s, and 1×10^5 s. Justify the shape of these sketches by calculating the characteristic time scales that govern the equilibration process.

d.) Prepare an analytical solution for the equilibration process using separation of variables. Implement your solution in EES and prepare a plot showing temperature as a function of position at the times requested in part (c).

3–18 A current lead carries 1000's of amps of current to a superconducting magnet, as shown in Figure P3-18.

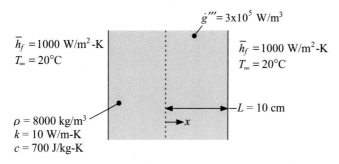

Figure P3-18: Current lead.

The edges of the current lead are cooled by flowing water at $T_\infty = 20°C$ with heat transfer coefficient $\bar{h}_f = 1000\,\text{W/m}^2\text{-K}$. The current lead material has density $\rho = 8000\,\text{kg/m}^3$, conductivity $k = 10\,\text{W/m-K}$, and specific heat capacity $c = 700\,\text{J/kg-K}$. The current causes a uniform rate of volumetric generation of thermal energy, $\dot{g}''' = 3 \times 10^5\,\text{W/m}^3$. The half-width of the current lead is $L = 10\,\text{cm}$.

a.) Determine the steady-state temperature distribution in the current lead, $T_{ss}(x)$. Plot the temperature distribution.

At time $t = 0$ the current is deactivated so that the rate of volumetric generation in the current lead goes to zero. The cooling water flow is also deactivated at $t = 0$, causing the heat transfer coefficient at the surface to be reduced to $\bar{h}_s = 100\,\text{W/m-K}$.

b.) Sketch the temperature distribution that you expect within the material at $t = 0\,\text{s}, t = 500\,\text{s}, t = 1000\,\text{s}, t = 5000\,\text{s}$, and $t = 10{,}000\,\text{s}$. Make sure that the qualitative characteristics of your sketch are correct and justify them by calculating the characteristic time scale that governs the problem.

c.) Sketch the rate of heat transfer per unit area to the cooling water as a function of time. Make sure that the qualitative characteristics of your sketch are correct and justify them. Include a rough sense of the scale on the t axis.

d.) Develop a separation of variables solution for the process. Prepare the plots requested in parts (b) and (c) using this model.

3–19 Reconsider Problem 3-10 using a separation of variables solution.

a.) Derive the governing differential equation, the boundary conditions, and the initial conditions for the problem.

b.) Does the mathematical problem statement derived in (a) satisfy all of the requirements for a separation of variables solution? If not, provide a simple transformation that can be applied so that the problem can be solved using separation of variables?

c.) Prepare a separation of variables solution to the transformed problem from (b) and implement your solution in EES.

d.) Prepare a plot of the temperature as a function of position for $t = 0$ and $t \to \infty$ as well as the times requested in Problem 3-10 part (c).

Duhamel's Theorem

3–20 An oscillating heat flux is applied to one side of a wall that is exposed to fluid on the other side, as shown in Figure P3-20.

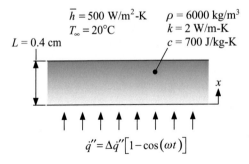

Figure P3-20: Wall exposed to an oscillating heat flux.

The wall thickness is $L = 0.4$ cm and the wall material has density $\rho = 6000\,\text{kg/m}^3$, conductivity $k = 2\,\text{W/m-K}$, and $c = 700\,\text{J/kg-K}$. The fluid temperature is $T_\infty = 20°C$ and the heat transfer coefficient is $\bar{h} = 500\ \text{W/m}^2\text{-K}$. Initially, the wall is in equilibrium with the fluid. The heat flux varies according to: $\dot{q}'' = \Delta\dot{q}''\ [1 - \cos(\omega t)]$ where $\Delta\dot{q}'' = 1000\ \text{W/m}^2$ and $\omega = 1\ \text{rad/s}$.

a.) Sketch the temperature as a function of time that you expect at $x = 0$ and $x = L$ for the first 10 oscillations ($0 < t < 62.8$ s). Try to get the qualitative characteristics of your sketch correct (e.g., the magnitude of the average temperature rise and temperature oscillations as well as the time scales involved).

b.) Use Duhamel's Theorem to develop an analytical model of the process. Plot the temperature as a function of time for the first 10 oscillations at $x = 0$, $x = L/2$, and $x = L$.

Complex Combination

3–21 Regenerative heat exchangers are discussed in Section 8.10. A regenerator operates in a cyclic fashion. Hot fluid passes across the regenerator material for half of a cycle, transferring energy to the material. Cold fluid passes across the regenerator material for the other half of the cycle, receiving energy from the material. After a sufficient number of cycles, the temperature distribution reaches a cyclic steady-state condition. Consider a regenerator matrix that consists of plates, one of which is shown in Figure P3-21.

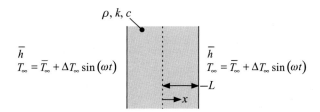

Figure P3-21: Plate regenerator matrix.

The half-thickness of the plate is L and the material properties are ρ, c, and k. The heat transfer coefficient between the surface of the plate and the fluid is $\bar{h}$ and the fluid temperature is assumed to vary sinusoidally with mean temperature $\overline{T}_\infty$, amplitude ΔT_∞, and frequency ω. In general, the temperature within the regenerator matrix is a function of both x and t.

a.) Using the method of complex combination, develop a solution for the sustained response of the temperature within the regenerator.

b.) Identify physically significant dimensionless parameters that can be used to correlate your solution. You should non-dimensionalize your solution and express it in terms of a dimensionless position and time as well as the Biot number and an additional dimensionless parameter that characterizes the frequency of oscillation.

c.) Prepare three plots of the dimensionless temperature as a function of dimensionless time for various values of the dimensionless position. Plot 1 should be for a large Biot number and large dimensionless frequency, plot 2 should be for a large Biot number and a small dimensionless frequency, and plot 3 should be for a small Biot number and a small dimensionless frequency. Explain why the behavior exhibited in each of these plots obeys your physical intuition.

Numerical Solution to 1-D Transient Problems

3–22 Prepare a numerical solution for the equilibration process discussed in Problem 3-17 using the Crank-Nicolson technique. Implement your solution in MATLAB and prepare a plot of the temperature as a function of position at $t = 10{,}000$ s; overlay the analytical solution derived in Problem 3-17 on this plot in order to demonstrate that the analytical and numerical solutions agree.

3–23 A pin fin is used as part of a thermal management system for a power electronics system, as shown in Figure P3-23.

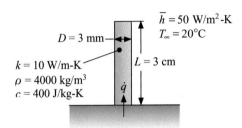

Figure P3-23: Pin fin subjected to a transient heat load.

The diameter of the fin is $D = 3$ mm and the length is $L = 3$ cm. The fin material has conductivity $k = 10$ W/m-K, $\rho = 4000$ kg/m^3, and $c = 400$ J/kg-K. The surface of the fin is exposed to air at $T_\infty = 20°C$ with heat transfer coefficient $\bar{h} = 50$ W/m^2-K. The tip of the fin can be assumed to be adiabatic. The power electronics system does not operate at steady state; rather, the load applied at the base of the fin cycles between a high and a low value with some angular frequency, ω. The average heat transfer rate is $\bar{q} = 0.5$ W and the amplitude of the fluctuation is $\Delta\dot{q} = 0.1$ W. The frequency of oscillation varies. The fin is initially at T_∞.

a.) Develop a 1-D transient model that can be used to analyze the startup and operating behavior of the pin fin. Use the ode solver in MATLAB.

b.) Plot the temperature as a function of time at various values of axial position for the start up assuming a constant heat load ($\omega = 0$).

c.) Calculate a diffusive time constant and a lumped capacitance time constant for the equilibration process. Is the plot from (b) consistent with these values?

d.) Adjust the diameter of the fin so that the lumped time constant is much greater than the diffusive time constant. Plot the temperature as a function of time at various values of axial position for the start up assuming a constant heat load ($\omega = 0$). Explain your result.

e.) Return the diameter of the fin to $D = 3$ mm and set the oscillation frequency to $\omega = 1$ rad/s. Prepare a contour plot showing the temperature of the fin as a function of position and time. You should see that the oscillation of the heat load causes a disturbance that penetrates only part-way along the axis of the fin. Explain this result.

f.) Is the maximum temperature experienced by the fin under oscillating conditions at cyclic steady-state (i.e., after the start-up transient has decayed) greater than or less than the maximum temperature experienced under steady-state conditions (i.e., with $\omega = 0$)?

g.) Plot the ratio of the maximum temperature under oscillating conditions to the maximum temperature under steady-state conditions as a function of frequency.

h.) Define a meaningful dimensionless frequency and plot the ratio of the maximum temperature under oscillating conditions to the maximum temperature under steady-state conditions as a function of this dimensionless frequency. Explain the shape of your plot.

Transient Conduction Problems using FEHT (FEHT can be downloaded from www.cambridge.org/nellisandklein.)

3–24 Figure P3-24(a) illustrates a disk brake that is used to bring a piece of rotating machinery to a smooth stop.

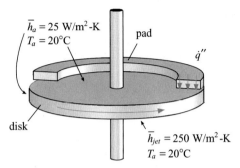

Figure P3-24(a): A disk brake on a rotating machine.

The brake pad engages the disk at its outer edge when the brake is activated. The outer edge and top surface of the disk (except under the pad) are exposed to air at $T_a = 20°C$ with $\bar{h}_a = 25\,W/m^2$-K. The bottom edge is exposed to air jets in order to control the disk temperature; the air jets have $T_a = 20°C$ and $\bar{h}_{jet} = 250\,W/m^2$-K. The problem can be modeled as a 2-D, radial problem as shown in Figure P3-24(b).

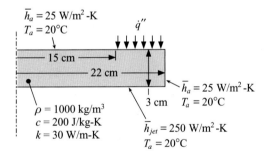

Figure P3-24(b): 2-D representation of disk brake.

The dimensions of the brake and boundary conditions are shown in Figure P3-24(b); the friction between the disk and the brake causes a spatially uniform heat flux that varies with time according to:

$$\dot{q}'' = 200000\,[W/m^2]\left(1 - \left(\frac{t\,[s]}{50\,[s]}\right)^2\right)$$

You may neglect the contribution of the shaft (i.e., assume that the brake is just a disk). The disk is initially at a uniform temperature of 20°C. The density of the disk material is $\rho = 1000\,kg/m^3$, the conductivity is $k = 30\,W/m$-K, and the specific heat capacity is $c = 200\,J/kg$-K.

a.) Develop a FEHT model that can predict the temperature distribution in the disk as a function of time during the 50 s that is required for the rotating machine to stop.

b.) Plot the maximum temperature in the disk as a function of time for two values of the number of nodes in order to demonstrate that your mesh is sufficiently refined. You will need to generate two plots and the comparison will be qualitative.

c.) Prepare a contour plot showing the temperature distribution at $t = 10$ s, $t = 25$ s, and $t = 50$ s. You may also want to animate your temperature contours by selecting Temperature Contours from the View menu and selecting From start to stop.

d.) Plot the temperature on the lower surface (the surface exposed to the jets of air) at various locations as a function of time. Explain the shape of the plot – does the result make physical sense to you based on any time constants that you can compute?

REFERENCES

Abramowitz, M., and I. A. Stegun, eds., *Handbook of Mathematical Functions with Formulas, Graphs, and Mathematical Tables*, Dover Publications, New York, NY, (1968).

Arpaci, V. S., *Conduction Heat Transfer*, Addison-Wesley Publishing Company, Reading, MA, (1966).

Aziz, A., *Heat Conduction with Maple*, Edwards Publishing, Philadelphia, PA, (2006).

Heisler, M. P., "Temperature charts for induction and constant temperature heating," *Transactions of the ASME*, Vol. 69, pp. 227–236, (1947).

Myers, G. E., *Analytical Methods in Conduction Heat Transfer, 2nd Edition*, AMCHT Publications, Madison, WI, (1998).

4 External Forced Convection

4.1 Introduction to Laminar Boundary Layers

4.1.1 Introduction

Chapters 1 through 3 consider conduction heat transfer in a stationary medium. Energy transport within the material of interest occurs entirely by conduction and is governed by Fourier's law. Convection is considered only as a boundary condition for the relatively simple ordinary or partial differential equations that govern conduction problems. Convection is the transfer of energy in a moving medium, most often a liquid or gas flowing through a duct or over an object. The transfer of energy in a flowing fluid is not only due to conduction (i.e., the interactions between micro-scale energy carriers) but also due to the enthalpy carried by the macro-scale flow. Enthalpy is the sum of the internal energy of the fluid and the product of its pressure and volume. The pressure-volume product is related to the work required to move the fluid across a boundary. You were likely introduced to this term in a thermodynamics course in the context of an energy balance on a system that includes flow across its boundary. The additional terms in the energy balance related to the fluid flow complicate convection problems substantially and link the heat transfer problem with an underlying fluid dynamics problem. The complete solution to many convection problems therefore requires sophisticated computational fluid dynamic (CFD) tools that are beyond the scope of this book.

The presentation of convection heat transfer that is provided in this book looks at convection processes at a conceptual level in order to build insight. In addition, the capabilities and tools that are required to solve typical convection heat transfer problems are presented. As engineers, we are most often interested in the interaction between a fluid and a surface; specifically the transport of momentum and energy between the surface and the fluid. The transport of momentum is related to the force exerted on the surface and it is usually represented in terms of a drag force or a shear stress. The transport of energy is expressed in terms of the heat transfer coefficient. These are the engineering quantities of interest and they are governed by the behavior of boundary layers, the thin layer of fluid that is adjacent to the surface and they are affected by its presence.

We will attempt to obtain physical intuition regarding the behavior of boundary layers and understand how the transport of momentum and energy are related. We will explore the equations that govern these transport processes and see how they can be simplified and non-dimensionalized. We will look at exact solutions to these simplified equations, where they exist, and develop some tools that provide approximate solutions. Most importantly, we will examine the correlations that are enabled by the non-dimensionalized equations and understand their proper use and the limits of their applicability. The convection heat transfer correlations included in this book are also built into EES, which simplifies their application. However, it is important that any solution be checked against physical intuition and understood at a deeper level than just "this is what the correlation predicts."

4.1.2 The Laminar Boundary Layer

Figure 4-1(a) and (b) illustrate, qualitatively, the laminar flow of a cold fluid over a heated plate that is at a uniform temperature (T_s). The flow approaching the plate (i.e., at $x < 0$) has a uniform velocity (u_∞) in the x-direction, no velocity in the y-direction ($v_\infty = 0$), and a uniform temperature (T_∞). The quantities u_∞ and T_∞ are referred to as the free-stream velocity and temperature, respectively. The difference between laminar and turbulent flow will be discussed in more detail in Section 4.5. For now, it is sufficient to understand that the laminar flow is steady, provided that the free-stream velocity and temperature do not change with time. An instrument placed in the flow would report a constant value of velocity and temperature with no fluctuations.

The presence of the plate affects the velocity and temperature for $x > 0$. The plate is stationary and therefore the fluid particles immediately adjacent to the plate (i.e., at $y = 0$) will have zero velocity. These particles exert a shear stress on those that are slightly farther from the plate, causing them to slow down. The result is the velocity distribution shown in Figure 4-1(a). A similar phenomena results in the temperature distribution shown in Figure 4-1(b). Those particles immediately adjacent to the warm plate approach the plate temperature, T_s, and transfer heat to the cooler particles that are farther from the plate.

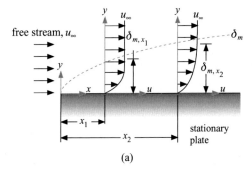

(a)

Figure 4-1: (a) The velocity distribution and (b) the temperature distribution associated with external flow over a flat plate.

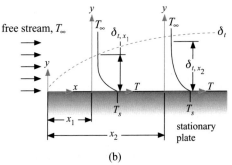

(b)

The velocity (or momentum) and temperature (or thermal) boundary layers refer to the regions of the flow that are affected by the presence of the plate (i.e., the region where $u < u_\infty$ and $T > T_\infty$, respectively). There are different specific definitions of the boundary layer thickness that can be used. A common definition of the momentum boundary layer thickness (δ_m) is the distance from the plate (y) where the velocity has recovered to 99% of its free-stream value, $u/u_\infty = 0.99$. The thermal boundary layer thickness (δ_t) is often defined in a similar manner; the distance from the plate where the fluid-to-plate temperature difference has achieved 99% of its free-stream value, $(T - T_s)/(T_\infty - T_s) = 0.99$.

The thickness of the momentum and thermal boundary layers will grow as the fluid moves downstream (i.e., towards larger x), as shown in Figure 4-1; notice that $\delta_{m,x_2} > \delta_{m,x_1}$ and $\delta_{t,x_2} > \delta_{t,x_1}$. The growth of a laminar boundary layer is almost exactly analogous to the penetration of a thermal wave by conduction into a semi-infinite body and can be understood using the concept of a diffusive time constant (discussed in Section 3.3.2). The fluid flow in Figure 4-1 is laminar. Therefore, there are no velocity fluctuations in the boundary layer and very little velocity in the y-direction. (We will solve the problem of laminar flow over a flat plate analytically in Section 4.4.2 and show that the velocity in the y-direction is small, but not zero.) The transport of thermal energy in the y-direction must therefore be primarily due to conduction, i.e., the diffusion of energy due to the interaction of the molecular scale energy carriers.

A Conceptual Model of the Laminar Boundary Layer

Conceptually, the free stream behaves much like a semi-infinite body that experiences a step-change in its surface temperature at the instant that the fluid encounters the leading edge of the plate (i.e., at $x = 0$ and $t = 0$). The disturbance associated with the change in the surface temperature diffuses as a thermal wave into the free stream (i.e., in the y-direction). This diffusion process takes time and, for the external flow problem, the fluid motion transports the wave downstream from the leading edge. Therefore, at $x = x_2$ the thermal wave has propagated farther into the free stream (i.e., in the y-direction) (in Figure 4-1) than it had at $x = x_1$. In Section 3.3.2, we learned that the motion of a thermal wave can be approximately represented by:

$$\delta_t \approx 2\sqrt{\alpha t} \tag{4-1}$$

where α is the thermal diffusivity of the material and t is the time relative to the disturbance at the surface. The transport of momentum by molecular diffusion is analogous to the transport of energy by conduction; a "momentum wave" will travel a distance δ_m according to:

$$\delta_m \approx 2\sqrt{\upsilon t} \tag{4-2}$$

where υ is the kinematic viscosity (the ratio of the dynamic viscosity of the fluid to its density, μ/ρ). Notice that both α and υ have units m^2/s.

Thermal diffusivity describes the ability of a fluid to transport energy by diffusion whereas kinematic viscosity describes the ability of a fluid to transport momentum by diffusion. For many fluids, α and υ have similar values because the transport of energy and momentum occur by basically the same mechanism. For example, the transport of energy and momentum in a gas both occur as a result of collisions of individual gas molecules, as discussed in Section 1.1.2. Room temperature ambient air, for example, has $\alpha = 2.2 \times 10^{-5}$ m^2/s and $\upsilon = 1.9 \times 10^{-5}$ m^2/s. The ratio of kinematic viscosity to thermal diffusivity is an important dimensionless parameter in convection heat transfer, referred to as the Prandtl number (Pr):

$$Pr = \frac{\upsilon}{\alpha} \tag{4-3}$$

The Prandtl number provides a measure of the relative ability of a fluid to transport momentum and energy; Figure 4-2 illustrates the Prandtl number as a function of temperature for a variety of fluids at atmospheric pressure.

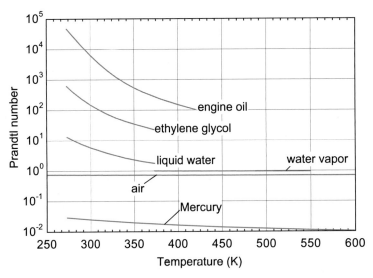

Figure 4-2: Prandtl number as a function of temperature for various fluids at atmospheric pressure.

A fluid with a large Prandtl number (e.g., engine oil) is viscous and non-conductive. Such a fluid will transport momentum very well but not thermal energy. At the other extreme, a fluid with a small Prandtl number (e.g., a liquid metal such as Mercury) is very conductive but inviscid. Such a fluid will transport thermal energy very well but not momentum.

Returning to the models of the momentum and thermal boundary layers, Eqs. (4-1) and (4-2), the transport time for momentum and thermal energy is approximately related to the distance from the leading edge and the free-stream velocity according to:

$$t \approx \frac{x}{u_\infty} \tag{4-4}$$

Substituting Eq. (4-4) into Eq. (4-2) provides a simple, but conceptually accurate, estimate for the momentum boundary layer thickness:

$$\delta_m \approx 2\sqrt{\frac{\upsilon x}{u_\infty}} \tag{4-5}$$

Equation (4-5) can be rearranged:

$$\frac{\delta_m}{x} \approx \frac{2}{x}\sqrt{\frac{\upsilon x}{u_\infty}} = 2\sqrt{\frac{\upsilon_{\text{,}}}{u_\infty x}} = \frac{2}{\sqrt{\dfrac{u_\infty x}{\upsilon}}} \tag{4-6}$$

Substituting the definition of kinematic viscosity into Eq. (4-6) leads to:

$$\frac{\delta_m}{x} \approx \frac{2}{\sqrt{\dfrac{u_\infty x \rho}{\mu}}} \tag{4-7}$$

The argument of the square root in the denominator of Eq. (4-7) ought to look familiar to anyone who has taken a fluids class; it is the definition of the Reynolds number based on x. The Reynolds number is defined differently for different flow configurations. In general, the Reynolds number is defined according to:

$$Re = \frac{u_{char} \, L_{char} \, \rho}{\mu} \tag{4-8}$$

where L_{char} is the characteristic length and u_{char} is the characteristic velocity associated with a problem. For flow over a flat plate, the characteristic length is the distance from the leading edge, x, and the characteristic velocity is the free-stream velocity. Therefore, the Reynolds number for flow over a flat plate is defined according to:

$$Re_x = \frac{u_\infty \, x \, \rho}{\mu} \tag{4-9}$$

Substituting Eq. (4-9) into Eq. (4-7) leads to:

$$\frac{\delta_m}{x} \approx \frac{2}{\sqrt{Re_x}} \tag{4-10}$$

The exact solution, discussed in Section 4.4.2, shows that the boundary layer thickness defined as the location where $u/u_\infty = 0.99$ is actually:

$$\frac{\delta_m}{x} = \frac{4.916}{\sqrt{Re_x}} \tag{4-11}$$

Therefore, the conceptual model is not perfect. However, it has certainly predicted the growth of the momentum boundary layer and its scale correctly. This should reinforce the idea that the laminar boundary layer is, to first order, a diffusive transport process that can be understood using the concepts developed in Chapters 1 through 3 as we studied conduction (the diffusive transport of energy).

Substituting Eq. (4-4) into Eq. (4-1) provides a similar conceptual model of the thermal boundary layer thickness:

$$\delta_t \approx 2\sqrt{\frac{\alpha \, x}{u_\infty}} \tag{4-12}$$

Equation (4-12) can be re-arranged:

$$\frac{\delta_t}{x} \approx \frac{2}{x}\sqrt{\frac{\alpha \, x}{u_\infty}} = 2\sqrt{\frac{\alpha}{u_\infty \, x}} = 2\sqrt{\frac{\alpha}{u_\infty \, x}\frac{\upsilon}{\upsilon}} = \frac{2}{\sqrt{\dfrac{u_\infty \, x \, \upsilon}{\upsilon \, \alpha}}} \tag{4-13}$$

Introducing the definition of the Reynolds number for flow over a flat plate, Eq. (4-9), and the Prandtl number, Eq. (4-3), into Eq. (4-13) leads to:

$$\frac{\delta_t}{x} \approx \frac{2}{\sqrt{Re_x \, Pr}} \tag{4-14}$$

The analytical solution to this problem, presented in Section 4.4.3, shows that over a large range of Prandtl number, the thermal boundary layer defined as the location where $(T - T_s)/(T_\infty - T_s) = 0.99$ is actually given by:

$$\frac{\delta_t}{x} = \frac{4.916}{Re_x^{1/2} \, Pr^{1/3}} \tag{4-15}$$

Again, the conceptual model has come close to the exact solution because the transport of thermal energy through the laminar boundary layer is similar to a conduction problem.

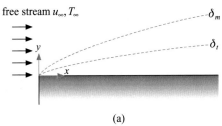

(a)

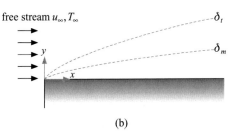

(b)

Figure 4-3: Sketch of the thermal and momentum boundary layers for (a) a fluid with $Pr \gg 1$ and (b) a fluid with $Pr \ll 1$.

Equations (4-10) and (4-14) can be used to estimate the relative thickness of the thermal and momentum boundary layers:

$$\frac{\delta_t}{\delta_m} \approx \frac{2}{\sqrt{Re_x \, Pr}} \frac{\sqrt{Re_x}}{2} \qquad (4\text{-}16)$$

or

$$\frac{\delta_t}{\delta_m} \approx \frac{1}{\sqrt{Pr}} \qquad (4\text{-}17)$$

The ratio of the boundary layers is only related to the Prandtl number. This is not surprising, given that the Prandtl number characterizes the ability of a fluid to transport momentum relative to its ability to transport thermal energy. Figure 4-3(a) illustrates the boundary layers for a fluid with a Prandtl number that is much higher than unity (e.g., engine oil). According to Eq. (4-17), δ_t will grow more slowly than δ_m because engine oil can transport momentum more efficiently than it can transfer energy. Figure 4-3(b) illustrates the converse situation, a fluid with a Prandtl number that is much less than unity (e.g., a liquid metal). According to Eq. (4-17), δ_t will be much greater than δ_m because the liquid metal can transport energy more efficiently than momentum.

Why should we care about the thickness of the boundary layers? The thermal boundary layer thickness does not appear to be directly useful for the design of a heat exchanger, for example. In the next sections, we will see that the interactions between the wall and the fluid that are of primary interest, shear stress and heat transfer, are directly related to the velocity gradient and temperature gradient at the wall, respectively. These gradients are, in turn, directly related to δ_m and δ_t. In fact, you could say that understanding convection heat and momentum transfer depends on your understanding of these boundary layer thicknesses.

A Conceptual Model of the Friction Coefficient and Heat Transfer Coefficient

Fourier's law relates the rate of diffusive transport of energy per unit area to the temperature gradient according to:

$$\dot{q}'' = -k \frac{\partial T}{\partial y} \qquad (4\text{-}18)$$

where k is the conductivity of the fluid. A similar phenomenological law in fluid dynamics relates the rate of diffusive transport of momentum per unit area (i.e., the shear stress τ) to the velocity gradient according to:

$$\tau = \mu \frac{\partial u}{\partial y} \tag{4-19}$$

where μ is the viscosity of the fluid. The similarity between Eqs. (4-18) and (4-19) is significant. Both energy and momentum are transported diffusively by virtue of a gradient in the potential that drives the transport process (i.e., by a gradient in temperature or velocity). We are most interested in the rate of transport at the surface of the plate ($\dot{q}''_{conv}$ and τ_s, at $y = 0$):

$$\dot{q}''_{conv} = -k \left. \frac{\partial T}{\partial y} \right|_{y=0} \tag{4-20}$$

$$\tau_s = \mu \left. \frac{\partial u}{\partial y} \right|_{y=0} \tag{4-21}$$

The temperature gradient and velocity gradient at the wall are directly related to the thermal and momentum boundary layer thicknesses. These gradients can be written approximately as:

$$\left. \frac{\partial T}{\partial y} \right|_{y=0} \approx \frac{(T_\infty - T_s)}{\delta_t} \tag{4-22}$$

$$\left. \frac{\partial u}{\partial y} \right|_{y=0} \approx \frac{u_\infty}{\delta_m} \tag{4-23}$$

Equations (4-22) and (4-23) are not exact because the gradients are not constant throughout the boundary layers (see Figure 4-1). However, Eq. (4-22) does correctly reflect the fact that the temperature of the fluid will change from T_s to T_∞ in the region between the surface of the plate and the top of the thermal boundary layer. Equation (4-23) indicates that the velocity will change from zero to u_∞ between the surface of the plate and the top of the momentum boundary layer. Substituting Eqs. (4-22) and (4-23) into Eqs. (4-20) and (4-21) shows that the thermal boundary layer thickness and the momentum boundary layer thickness govern the rate of heat transfer and shear at the plate surface, respectively:

$$\dot{q}''_{conv} \approx k \frac{(T_s - T_\infty)}{\delta_t} \tag{4-24}$$

$$\tau_s \approx \mu \frac{u_\infty}{\delta_m} \tag{4-25}$$

At the surface of the plate, the fluid motion is zero and so the heat transfer and shear represented by Eqs. (4-24) and (4-25) represent the total interaction between the plate and the fluid. Equations (4-24) and (4-25) suggest that the rates of heat transfer and shear are inversely proportional to the thermal and momentum boundary layer thicknesses, respectively. Accordingly, we expect that the rates of heat transfer and shear stress will be largest at the leading edge of the plate and decrease with x due to the thickening of the boundary layers, as shown in Figure 4-4.

Equations (4-24) and (4-25) show clearly one of the engineering challenges that faces the design of most energy conversion systems. It is often the case that you would

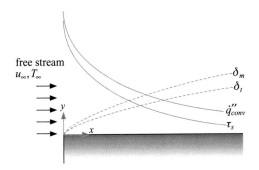

Figure 4-4: A sketch showing how the shear and heat transfer rate will vary with position.

like to maximize the convective heat flux, $\dot{q}''_{conv}$, in order to get the highest performance (for example, in a heat exchanger). However, you would also like to minimize the shear stress at the surface in order to minimize the pump or fan power required. Equations (4-24) and (4-25) show that it is usually not possible to increase the convective heat flux without simultaneously increasing the shear stress. Fluids with higher conductivity tend to also have higher viscosity and flow configurations with lower δ_t also tend to have lower δ_m. It is a kind of Murphy's law for the thermal engineer; you generally pay for improved heat transfer with increased shear stress.

Equation (4-24) can be written as a thermal resistance equation:

$$\dot{q}''_{conv} \approx \frac{(T_s - T_\infty)}{R_{bl}} \qquad (4\text{-}26)$$

where R_{bl} is the thermal resistance of the boundary layer:

$$R_{bl} \approx \frac{\delta_t}{k} \qquad (4\text{-}27)$$

Equations (4-26) and (4-27) should be familiar and show, again, that laminar flow can be understood as a conduction problem where the conduction length is the thermal boundary layer thickness. The laminar boundary layer can be thought of as a conduction resistance between the surface and the free stream.

Comparing Eq. (4-24) with Newton's law of cooling (i.e., the definition of the local heat transfer coefficient, h) leads to:

$$\dot{q}''_{conv} = h(T_s - T_\infty) \approx k\frac{(T_s - T_\infty)}{\delta_t} \qquad (4\text{-}28)$$

Rearranging Eq. (4-28) leads to:

$$\boxed{h \approx \frac{k}{\delta_t}} \qquad (4\text{-}29)$$

Equation (4-29) provides a physical understanding of the heat transfer coefficient. Fluids with high conductivity will, in general, provide a high heat transfer coefficient. For example, liquid water will almost always provide a higher heat transfer coefficient than air due, in part, to its much higher conductivity (0.60 W/m-K for water vs 0.026 W/m-K for air at room temperature and atmospheric pressure). Furthermore, flow situations where the boundary layer is thin will, in general, provide a high heat transfer coefficient. For example, a higher velocity flow will almost always lead to a higher heat transfer coefficient than a lower velocity flow; this result follows directly from Eq. (4-12).

Substituting the conceptual model of the thermal boundary layer thickness, Eq. (4-14), into Eq. (4-29) provides an approximate equation for the local heat transfer coefficient associated with laminar flow over a flat plate:

$$h \approx \frac{k}{\delta_t} = \frac{k}{2x}\sqrt{Re_x\, Pr} \qquad (4\text{-}30)$$

The heat transfer coefficient is typically made dimensionless using the Nusselt number (Nu). The Nusselt number is defined in general according to:

$$Nu = \frac{h\, L_{char}}{k} \qquad (4\text{-}31)$$

where L_{char} is the characteristic dimension of the problem. For a flat plate, the characteristic dimension is the distance from the leading edge, x. Therefore, the Nusselt number for flow over a flat plate is defined according to:

$$Nu_x = \frac{h\,x}{k} \qquad (4\text{-}32)$$

The major reason for using the dimensionless quantity (the Nusselt number) as opposed to the dimensional quantity (heat transfer coefficient) is the reduction in the number of independent variables required to describe the dimensionless problem. This same approach is used in fluid dynamics when, for example, a friction factor is used instead of the pressure gradient. We will see in Section 4.3 that the Nusselt number is often a function only of the Reynolds and Prandtl numbers. On the other hand, the heat transfer coefficient is a function of a larger number of dimensional parameters (e.g., velocity, viscosity, conductivity, plate length, etc.). It is convenient to correlate the Nusselt number as a function of the Reynolds number and the Prandtl number (based on either exact or approximate solutions or experimental data); such a correlation can be used for a wide range of problems that span different operating conditions, fluid properties, and length scales. These correlations will be discussed in more detail in Section 4.9.

Substituting Eq. (4-29) into Eq. (4-31) leads to:

$$Nu \approx \frac{L_{char}}{\delta_t} \qquad (4\text{-}33)$$

Equation (4-33) suggests that the Nusselt number is essentially a length ratio; the Nusselt number for a laminar flow can be thought of as the ratio of the characteristic length that is appropriate for the problem to the thermal boundary layer thickness (i.e., the distance through which energy must be conducted into the fluid).

For a flat plate, the characteristic length is the distance from the leading edge of the plate, substituting Eq. (4-29) into Eq. (4-32) leads to:

$$Nu_x \approx \frac{x}{\delta_t} \qquad (4\text{-}34)$$

For a flat plate under any practical operating condition, the boundary layer thickness will be much less than the length of the plate (this observation underlies the boundary layer simplifications that are discussed in Section 4.2). Therefore the Nusselt number will be much larger than unity. In other situations this will not be true. For example, internal laminar flow is discussed in Chapter 5. For an internal flow, the thermal boundary layer is confined (by the other side of the duct) and therefore cannot continue to grow. As a result, the boundary layer thickness is approximately equal to the radius of the duct ($\delta_t \approx R$). The characteristic length for an internal flow problem is the pipe diameter ($L_{char} = 2\,R$). Therefore, according to Eq. (4-33), the Nusselt number for fully developed laminar, internal flow through a round duct should be approximately 2.0. In fact, the Nusselt number ranges from 3.66 to 4.36 for this geometry, but you get the idea.

Substituting the conceptual model of the thermal boundary layer thickness for a flat plate, Eq. (4-14), into Eq. (4-34) provides an approximate equation for the Nusselt number associated with laminar flow over a flat plate:

$$Nu_x = \frac{hx}{k} \approx 0.5\sqrt{Re_x\,Pr}$$
(4-35)

Over a large range of Prandtl number, the exact solution, discussed in Section 4.4.3, leads to:

$$Nu_x = 0.332\,Re_x^{1/2}\,Pr^{1/3}$$
(4-36)

The shear stress for external flow over a flat plate is made dimensionless using the friction coefficient, C_f. The friction coefficient is defined according to:

$$C_f = \frac{2\,\tau_s}{\rho\,u_\infty^2}$$
(4-37)

Substituting Eq. (4-25) into Eq. (4-37) leads to:

$$C_f \approx \frac{2\,\mu}{\rho\,u_\infty\,\delta_m}$$
(4-38)

Substituting the approximate equation for the momentum boundary layer thickness, Eq. (4-10), into Eq. (4-38) leads to:

$$C_f \approx \frac{2\,\mu}{\rho\,u_\infty}\frac{\sqrt{Re_x}}{2\,x}$$
(4-39)

Equation (4-39) can be rearranged:

$$C_f \approx \frac{1.0}{\sqrt{Re_x}}$$
(4-40)

The exact solution, discussed in Section 4.4.2, leads to:

$$C_f = \frac{0.664}{\sqrt{Re_x}}$$
(4-41)

Without solving the complicated partial differential equations that describe the boundary layer behavior, we have been able to predict that the Nusselt number and friction coefficient vary with the square root of Reynolds number and, perhaps more importantly, develop some understanding of their behavior.

The Reynolds Analogy
The Reynolds analogy formalizes the similarity between the transport of heat and momentum. The Reynolds analogy is discussed more formally in Section 4.3.4, but it is worth introducing it here before diving into the details of the boundary layer equations in the next section. The Reynolds analogy simply states that, for many fluids, the Prandtl number is near unity and therefore the momentum and thermal boundary layer thickness will be approximately the same ($\delta_m \approx \delta_t$). In this limit, it is possible to relate (approximately) the hydrodynamic characteristics of the problem (i.e., the friction coefficient) to the thermal characteristics of the problem (i.e., the Nusselt number).

Equation (4-34) expresses the Nusselt number in terms of the thermal boundary layer thickness:

$$Nu_x \approx \frac{x}{\delta_t} \tag{4-42}$$

which can be solved for δ_t:

$$\delta_t \approx \frac{x}{Nu_x} \tag{4-43}$$

Equation (4-38) expressed the friction coefficient as a function of the momentum boundary layer thickness:

$$C_f \approx \frac{2\,\mu}{\rho\,u_\infty\,\delta_m} \tag{4-44}$$

which can be solved for δ_m:

$$\delta_m \approx \frac{2\,\mu}{\rho\,u_\infty\,C_f} \tag{4-45}$$

In the limit that the Prandtl number is near unity, Eq. (4-17) indicates that $\delta_t \approx \delta_m$ and therefore Eqs. (4-43) and (4-45) must be equal:

$$\frac{x}{Nu_x} \approx \frac{2\,\mu}{\rho\,u_\infty\,C_f} \tag{4-46}$$

or

$$Nu_x \approx \frac{C_f\,Re_x}{2} \tag{4-47}$$

Equation (4-47) is the Reynolds analogy and expresses the thermal solution (Nu_x) in terms of the hydrodynamic solution (C_f). This is a useful result because it is usually much easier to measure or model the shear stress than the heat transfer coefficient.

The Reynolds analogy can be extended so that it is valid over a wider range of Prandtl number. According to Eq. (4-17), the ratio of the boundary layer thicknesses is, approximately:

$$\frac{\delta_t}{\delta_m} \approx \frac{1}{\sqrt{Pr}} \tag{4-48}$$

Substituting Eqs. (4-43) and (4-45) into Eq. (4-48) leads to:

$$\frac{x}{Nu_x}\frac{\rho\,u_\infty\,C_f}{2\,\mu} \approx \frac{1}{\sqrt{Pr}} \tag{4-49}$$

or

$$Nu_x \approx \frac{Pr^{\frac{1}{2}}\,C_f\,Re_x}{2} \tag{4-50}$$

Typically, the exponent on the Prandtl number is taken to be $\frac{1}{3}$ rather than $\frac{1}{2}$ in order to provide slightly better results:

$$\boxed{Nu_x \approx \frac{Pr^{\frac{1}{3}}\,C_f\,Re_x}{2}} \tag{4-51}$$

The modified Reynolds analogy in Eq. (4-51) is referred to as the Chilton-Colburn analogy. These analogies should be applied with some caution. There are physical phenomena that can cause the momentum and thermal boundary layers to be substantially different even when the Prandtl number is near unity. For example large amounts of viscous dissipation or a strong pressure gradient will reduce the accuracy of the Reynolds or Chilton-Colburn analogies.

4.1.3 Local and Integrated Quantities

The discussion above was centered on the local value of the heat flux or shear stress at some particular position along the plate surface. These quantities are used to define a local heat transfer coefficient (h), shear stress (τ_s), friction coefficient (C_f), and Nusselt number (Nu_x). It is almost always more useful to know the average value of these quantities rather than the local ones. You may have noticed that the problems discussed in Chapters 1 through 3 almost always required an average heat transfer coefficient ($\bar{h}$) to specify their boundary conditions rather than a local heat transfer coefficient.

The average friction coefficient ($\overline{C_f}$) is defined based on the average shear stress experienced by the plate ($\bar{\tau}_s$) over its entire surface:

$$\overline{C_f} = \frac{2\,\bar{\tau}_s}{\rho\, u_\infty^2} \tag{4 52}$$

where the average shear stress is defined as:

$$\bar{\tau}_s = \frac{1}{A_s} \int_{A_s} \tau_s \, dA_s \tag{4-53}$$

and A_s is the surface area of the plate. If you are interested in calculating the total force on the plate, then the average friction coefficient is much more useful than the local one. For the flat plate, the shear stress varies only in the x-direction and therefore Eq. (4-53) can be written as:

$$\bar{\tau}_s = \frac{1}{L} \int_0^L \tau_s \, dx \tag{4-54}$$

where L is the length of the plate. While the local friction coefficient is correlated against the Reynolds number based on local position x (Re_x), the average friction factor will be correlated against the Reynolds number based on the plate length (Re_L):

$$Re_L = \frac{u_\infty L \rho}{\mu} \tag{4-55}$$

The total heat transfer from a plate is more conveniently expressed in terms of the average rather than the local heat transfer coefficient. The average heat transfer coefficient ($\bar{h}$) is defined according to:

$$\bar{h} = \frac{1}{A_s} \int_{A_s} h \, dA_s \tag{4-56}$$

and the average Nusselt number ($\overline{Nu}$) is defined as:

$$\overline{Nu} = \frac{\bar{h}\, L_{char}}{k} \tag{4-57}$$

For flow over a flat plate, the average heat transfer coefficient is given by:

$$\overline{h} = \frac{1}{L} \int_0^L h \, dx \qquad (4\text{-}58)$$

and the average Nusselt number $(\overline{Nu})$ is:

$$\overline{Nu}_L = \frac{\overline{h} \, L}{k} \qquad (4\text{-}59)$$

4.2 The Boundary Layer Equations

4.2.1 Introduction

In Section 4.1, laminar boundary layers are discussed at a conceptual level in order to obtain some physical feel for their behavior. In this section, the partial differential equations that govern the behavior of a laminar flow are derived and then simplified for the special case of flow inside of a boundary layer. In Section 4.3, the boundary layer equations are made dimensionless in order to identify the minimum set of parameters that govern the boundary layer problem. These steps are accomplished in order to justify the use of the correlations that are presented in Section 4.9 and are ubiquitous in the study of convection.

4.2.2 The Governing Equations for Viscous Fluid Flow

The governing equations are derived by enforcing the conservation of mass, momentum (in each direction), and thermal energy at every position within the fluid. In Chapters 1 through 3, the governing equations for conduction problems are derived by applying conservation of thermal energy to a differential control volume. The sequence of steps required to derive the mass and momentum equations are the same: a differentially small control volume is defined and used to write a conservation equation. The terms in the conservation equation are expanded and the first term in the Taylor series is retained. Finally, appropriate rate equations (e.g., Fourier's law) are substituted into the equation.

The Continuity Equation
The continuity equation enforces mass conservation for a differential control volume. A differential control volume is shown in Figure 4-5 for a 2-D flow in Cartesian coordinates.

Figure 4-5: A mass balance on a differential control volume.

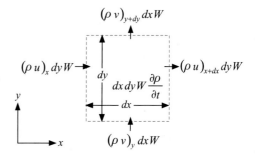

A mass balance on the differential control volume shown in Figure 4-5 leads to:

$$(\rho u)_x \, dy \, W + (\rho v)_y \, dx \, W = (\rho u)_{x+dx} \, dy \, W + (\rho v)_{y+dy} \, dx \, W + dx \, dy \, W \frac{\partial \rho}{\partial t} \qquad (4\text{-}60)$$

where W is the width of the control volume in the z-direction (into the page) and u and v are the velocity in the x- and y-directions, respectively. At this point, we will assume that the fluid is incompressible so that the density is constant. Therefore, Eq. (4-60) can be simplified to:

$$u_x \, dy \, W + v_y \, dx \, W = u_{x+dx} \, dy \, W + v_{y+dy} \, dx \, W \qquad (4\text{-}61)$$

The terms at $x + dx$ and $y + dy$ are expanded:

$$u_x \, dy \, W + v_y \, dx \, W = \left(u_x + \frac{\partial u}{\partial x} dx \right) dy \, W + \left(v_y + \frac{\partial v}{\partial y} dy \right) dx \, W \qquad (4\text{-}62)$$

Equation (4-62) can be simplified to:

$$\boxed{\frac{\partial u}{\partial x} + \frac{\partial v}{\partial y} = 0} \qquad (4\text{-}63)$$

In cylindrical coordinates (i.e., an x-r coordinate system), the continuity equation becomes:

$$\boxed{\frac{\partial u}{\partial x} + \frac{1}{r} \frac{\partial (r\,v)}{\partial r} = 0} \qquad (4\text{-}64)$$

where u and v are the components of velocity in the x- and r-directions, respectively.

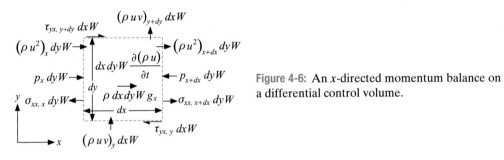

Figure 4-6: An x-directed momentum balance on a differential control volume.

The Momentum Conservation Equations

Momentum must be conserved in each of the coordinate directions. Therefore, there will be as many momentum conservation equations as there are dimensions to the problem. For the 2-D problem in Cartesian coordinates considered here, momentum conservation equations must be derived in both the x- and y-directions.

Figure 4-6 illustrates the x-directed momentum terms and forces for a differential control volume. It is worth discussing the source of the various terms that appear in Figure 4-6. The x-directed momentum per unit mass of fluid entering the control volume is u. Therefore, the momentum transfer terms are the product of the x-directed momentum per unit mass and the appropriate mass flow rates (shown in Figure 4-5). Similarly, the amount of momentum stored in the control volume is equal to the product of the momentum per unit mass (u) and the mass of fluid in the control volume. The rate of momentum storage is the time derivative of this quantity.

The x-directed momentum transfer and storage terms must be balanced by the forces that are exerted on the control volume in the x-direction. The forces shown in Figure 4-6 include pressure forces, viscous forces, and a gravitational force. The pressure forces in the x-direction are exerted on the left and right faces of the control volume and are the product of the area of these faces and the pressure (p) acting on these faces. The gravitational force is the product of the mass of fluid that is contained in the control volume and the component of gravity acting in the x-direction, g_x. If the x-direction is perpendicular to the gravity vector, then this force is zero. There are two additional forces considered in Figure 4-6 that are related to the viscous stresses, τ_{yx} and σ_{xx}; the first subscript indicates the face that the stress acts on and the second subscript indicates the direction that the stress acts in. The tangential stress, or shear, τ_{yx} acts on the bottom and top (i.e., the y-directed) faces while the normal stress (in excess of the pressure) σ_{xx} acts on the left and right (i.e., the x-directed) faces.

The momentum balance in the x-direction on the differential control volume shown in Figure 4-6 leads to:

$$\rho \, dx \, dy \, W \, g_x + (\rho u^2)_x \, dy \, W + (\rho u v)_y \, dx \, W + p_x \, dy \, W + \tau_{yx,y+dy} \, dx \, W$$
$$+ \sigma_{xx,x+dx} \, dy \, W = (\rho u^2)_{x+dx} \, dy \, W + (\rho u v)_{y+dy} \, dx \, W \qquad (4\text{-}65)$$
$$+ p_{x+dx} \, dy \, W + \tau_{yx,y} \, dx \, W + \sigma_{xx,x} \, dy \, W + dx \, dy \, W \, \frac{\partial(\rho u)}{\partial t}$$

The terms in Eq. (4-65) are expanded to achieve:

$$\underbrace{\rho g_x}_{\text{gravity force}} + \underbrace{\frac{\partial \tau_{yx}}{\partial y} + \frac{\partial \sigma_{xx}}{\partial x}}_{\text{viscous stresses}} = \underbrace{\rho \frac{\partial(u^2)}{\partial x} + \rho \frac{\partial(u v)}{\partial y}}_{\text{momentum transfer}} + \underbrace{\frac{\partial p}{\partial x}}_{\text{pressure force}} + \underbrace{\rho \frac{\partial u}{\partial t}}_{\substack{\text{rate of} \\ \text{momentum storage}}} \qquad (4\text{-}66)$$

The rate equations that govern the shear and normal stresses in a Newtonian fluid are substituted into Eq. (4-66) leading to (Bejan (1993)):

$$\rho g_x + \mu \frac{\partial^2 u}{\partial y^2} + \mu \frac{\partial^2 u}{\partial x^2} = \rho \frac{\partial(u^2)}{\partial x} + \rho \frac{\partial(u v)}{\partial y} + \frac{\partial p}{\partial x} + \rho \frac{\partial u}{\partial t} \qquad (4\text{-}67)$$

Equation (4-67) is rearranged to obtain:

$$\boxed{\rho \left[\frac{\partial u}{\partial t} + \frac{\partial(u^2)}{\partial x} + \frac{\partial(uv)}{\partial y} \right] = -\frac{\partial p}{\partial x} + \mu \left(\frac{\partial^2 u}{\partial y^2} + \frac{\partial^2 u}{\partial x^2} \right) + \rho g_x} \qquad (4\text{-}68)$$

Typically, the partial differentials on the left side of Eq. (4-68) are expanded:

$$\rho \left[\frac{\partial u}{\partial t} + 2u \frac{\partial u}{\partial x} + u \frac{\partial v}{\partial y} + v \frac{\partial u}{\partial y} \right] = -\frac{\partial p}{\partial x} + \mu \left(\frac{\partial^2 u}{\partial y^2} + \frac{\partial^2 u}{\partial x^2} \right) + \rho g_x \qquad (4\text{-}69)$$

Equation (4-69) can be rearranged:

$$\rho \left[\frac{\partial u}{\partial t} + u \underbrace{\left(\frac{\partial u}{\partial x} + \frac{\partial v}{\partial y} \right)}_{=0 \text{ by continuity}} + u \frac{\partial u}{\partial x} + v \frac{\partial u}{\partial y} \right] = -\frac{\partial p}{\partial x} + \mu \left(\frac{\partial^2 u}{\partial y^2} + \frac{\partial^2 u}{\partial x^2} \right) + \rho g_x \qquad (4\text{-}70)$$

The second term on the left side Eq. (4-70) must equal zero according to the continuity equation, Eq. (4-63). Therefore, the momentum balance in the x-direction can be written as:

$$\rho\left[\frac{\partial u}{\partial t} + u\frac{\partial u}{\partial x} + v\frac{\partial u}{\partial y}\right] = -\frac{\partial p}{\partial x} + \mu\left(\frac{\partial^2 u}{\partial y^2} + \frac{\partial^2 u}{\partial x^2}\right) + \rho g_x \qquad (4\text{-}71)$$

A differential momentum balance in the y-direction leads to a similar equation:

$$\rho\left[\frac{\partial v}{\partial t} + u\frac{\partial v}{\partial x} + v\frac{\partial v}{\partial y}\right] = -\frac{\partial p}{\partial y} + \mu\left(\frac{\partial^2 v}{\partial y^2} + \frac{\partial^2 v}{\partial x^2}\right) + \rho g_y \qquad (4\text{-}72)$$

where g_y is the component of gravity in the y-direction. Equations (4-71) and (4-72) are often referred to as the Navier-Stokes equation. In cylindrical coordinates, Eqs. (4-71) and (4-72) become:

$$\rho\left[\frac{\partial u}{\partial t} + u\frac{\partial u}{\partial x} + v\frac{\partial u}{\partial r}\right] = -\frac{\partial p}{\partial x} + \mu\left[\frac{1}{r}\frac{\partial}{\partial r}\left(r\frac{\partial u}{\partial r}\right) + \frac{\partial^2 u}{\partial x^2}\right] + \rho g_x \qquad (4\text{-}73)$$

$$\rho\left[\frac{\partial v}{\partial t} + u\frac{\partial v}{\partial x} + v\frac{\partial v}{\partial r}\right] = -\frac{\partial p}{\partial r} + \mu\left[\frac{\partial}{\partial r}\left(\frac{1}{r}\frac{\partial(rv)}{\partial r}\right) + \frac{\partial^2 u}{\partial r^2}\right] \qquad (4\text{-}74)$$

where u and v are the velocity components in the x- and r-directions, respectively.

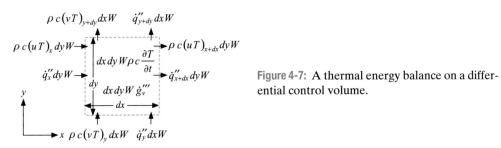

Figure 4-7: A thermal energy balance on a differential control volume.

The Thermal Energy Conservation Equation

The thermal energy equation is derived by requiring that thermal energy be conserved for a differential control volume, shown in Figure 4-7 for a fluid with a constant specific heat capacity (c).

The differential thermal energy balance shown in Figure 4-7 is nearly identical to one that would be used to analyze a 2-D, transient conduction problem with volumetric generation of thermal energy. There are a few differences, most notably the additional terms related to the enthalpy carried into the control volume by the flow through each of the faces. Also, the generation of thermal energy per unit volume is related to viscous dissipation ($\dot{g}_v'''$). Viscous dissipation is the rate at which the mechanical energy of the fluid is converted to thermal energy. The generation of thermal energy by viscous dissipation is proportional to the viscosity of the fluid and, like the shear stress, it is driven by gradients in the velocity. A total (rather than thermal) energy balance would necessarily include the kinetic and potential energy of the fluid as well as its thermal energy. There would be no viscous dissipation in a total energy equation because energy is conserved. The viscous dissipation term is the rate at which a fluid's mechanical energy is converted to thermal energy by viscous shear and it represents the linkage between the thermal and total energy equations. This linkage is similar to other problems that we've seen. For example, the volumetric thermal energy generation term in conduction problems that deal with ohmic dissipation represents the rate at which electrical energy is converted to thermal energy.

The energy balance suggested by Figure 4-7 is:

$$\rho c (uT)_x \, dy \, W + \rho c (vT)_y \, dx \, W + \dot{q}''_x \, dy \, W + \dot{q}''_y \, dx \, W + dx \, dy \, W \, \dot{g}'''_v$$

$$= \rho c (uT)_{x+dx} \, dy \, W + \rho c (vT)_{y+dy} \, dx \, W + \dot{q}''_{x+dx} \, dy \, W + \dot{q}''_{y+dy} \, dx \, W + dx \, dy \, W \rho c \frac{\partial T}{\partial t} \tag{4-75}$$

Expanding the terms in Eq. (4-75) and simplifying leads to:

$$\dot{g}'''_v = \rho c \frac{\partial (uT)}{\partial x} + \rho c \frac{\partial (vT)}{\partial y} + \frac{\partial \dot{q}''_x}{\partial x} + \frac{\partial \dot{q}''_y}{\partial y} + \rho c \frac{\partial T}{\partial t} \tag{4-76}$$

Fourier's law is used to compute the conductive heat flux in the x- and y-directions:

$$\dot{q}''_x = -k \frac{\partial T}{\partial x} \tag{4-77}$$

$$\dot{q}''_y = -k \frac{\partial T}{\partial y} \tag{4-78}$$

Substituting Eqs. (4-77) and (4-78) into Eq. (4-76) leads to:

$$\dot{g}'''_v = \rho c \frac{\partial (uT)}{\partial x} + \rho c \frac{\partial (vT)}{\partial y} + \frac{\partial}{\partial x} \left(-k \frac{\partial T}{\partial x} \right) + \frac{\partial}{\partial y} \left(-k \frac{\partial T}{\partial y} \right) + \rho c \frac{\partial T}{\partial t} \tag{4-79}$$

The thermal conductivity of the fluid is assumed to be constant. Therefore, Eq. (4-79) can be written as:

$$\rho c \left[\frac{\partial T}{\partial t} + \frac{\partial (uT)}{\partial x} + \frac{\partial (vT)}{\partial y} \right] = k \left(\frac{\partial^2 T}{\partial x^2} + \frac{\partial^2 T}{\partial y^2} \right) + \dot{g}'''_v \tag{4-80}$$

The derivatives on the left hand side of Eq. (4-80) are expanded:

$$\rho c \left[\frac{\partial T}{\partial t} + u \frac{\partial T}{\partial x} + v \frac{\partial T}{\partial y} + T \underbrace{\left(\frac{\partial u}{\partial x} + \frac{\partial v}{\partial y} \right)}_{=0 \text{ by continuity}} \right] = k \left(\frac{\partial^2 T}{\partial x^2} + \frac{\partial^2 T}{\partial y^2} \right) + \dot{g}'''_v \tag{4-81}$$

The last term on the left side of Eq. (4-81) must be zero according to the continuity equation, Eq. (4-63). Therefore, the thermal energy conservation equation is typically written as:

$$\boxed{\rho c \left[\frac{\partial T}{\partial t} + u \frac{\partial T}{\partial x} + v \frac{\partial T}{\partial y} \right] = k \left(\frac{\partial^2 T}{\partial x^2} + \frac{\partial^2 T}{\partial y^2} \right) + \dot{g}'''_v} \tag{4-82}$$

where the volumetric rate of generation associated with the viscous dissipation is (Bejan (1993)):

$$\boxed{\dot{g}'''_v = \mu \left\{ \left(\frac{\partial u}{\partial y} + \frac{\partial v}{\partial x} \right)^2 + 2 \left[\left(\frac{\partial u}{\partial x} \right)^2 + \left(\frac{\partial v}{\partial y} \right)^2 \right] \right\}} \tag{4-83}$$

In cylindrical coordinates, the thermal energy equation becomes:

$$\boxed{\rho c \left[\frac{\partial T}{\partial t} + u \frac{\partial T}{\partial x} + v \frac{\partial T}{\partial r} \right] = k \left[\frac{\partial^2 T}{\partial x^2} + \frac{1}{r} \frac{\partial}{\partial r} \left(r \frac{\partial T}{\partial r} \right) \right] + \dot{g}'''_v} \tag{4-84}$$

where u and v are the velocity components in the x- and r-directions, respectively. The volumetric rate of generation associated with viscous dissipation in radial coordinates is (Bejan (1993)):

$$\dot{g}_v''' = \mu \left\{ \left(\frac{\partial u}{\partial r} + \frac{\partial v}{\partial x} \right)^2 + 2 \left[\left(\frac{\partial u}{\partial x} \right)^2 + \left(\frac{v}{r} \right)^2 + \left(\frac{\partial v}{\partial r} \right)^2 \right] \right\} \qquad (4\text{-}85)$$

4.2.3 The Boundary Layer Simplifications

The boundary layer is very thin; this fact can be used to simplify the mass, momentum and thermal energy equations. The conceptual model discussed in Section 4.1 for a laminar boundary layer provided an estimate of the boundary layer thickness:

$$\frac{\delta_m}{L} \approx \frac{2}{\sqrt{Re_L}} \qquad (4\text{-}86)$$

For most practical flow situations, the Reynolds number will be very large. For example, the flow of air at room temperature and atmospheric pressure over an $L = 1$ m plate at $u_\infty = 10$ cm/s is characterized by $Re_L = 6000$. A flow of water at the same conditions and velocity is characterized by $Re_L = 1 \times 10^5$. The boundary layer will therefore be between $100\times$ and $1000\times$ smaller than the plate length under these conditions. This observation leads to the boundary layer simplifications, which are based fundamentally on the assumption that:

$$\delta_m, \delta_t \ll L \qquad (4\text{-}87)$$

Obviously, the boundary layer simplifications are not appropriate near the leading edge of the plate. Unless the Prandtl number is very large or very small, δ_m will have the same order of magnitude as δ_t. Therefore, for the purposes of the scaling arguments discussed in the section, we will refer only to a boundary layer thickness, δ, without specifying whether it is the thermal or momentum boundary layer thickness. Equation (4-86) provides the appropriate scaling for an order of magnitude analysis of the terms in the governing equations that were derived in Section 4.2.2.

The Continuity Equation
The continuity equation, Eq. (4-63), can be rearranged:

$$\frac{\partial u}{\partial x} = -\frac{\partial v}{\partial y} \qquad (4\text{-}88)$$

We are not interested in the exact values of the terms in Eq. (4-88) here, only their order of magnitude:

$$O \left(\frac{\partial u}{\partial x} \right) = O \left(\frac{\partial v}{\partial y} \right) \qquad (4\text{-}89)$$

where O indicates the order of magnitude of the argument. Equation (4-89) implies that the order of magnitude of the partial derivative of u with respect to x must be equal to the order of magnitude of the partial derivative of v with respect to y. If you move along the plate in the x-direction from the leading edge ($x = 0$) to the trailing edge ($x = L$), then the largest change in u that is possible is u_∞. Therefore, the partial differential of u with respect to x has order of magnitude u_∞/L.

$$O \left(\frac{\partial u}{\partial x} \right) = \frac{u_\infty}{L} \qquad (4\text{-}90)$$

There is no equally obvious scaling for the velocity in the y-direction. However, if you move in the y-direction from the plate surface ($y = 0$) to the edge of the boundary layer ($y = \delta$) then v will change from 0 to v_δ, where v_δ is the y-directed velocity at the outer edge of the boundary layer. Therefore, the partial differential of v with respect to y has order of magnitude v_δ/δ.

$$O\left(\frac{\partial v}{\partial y}\right) = \frac{v_\delta}{\delta} \tag{4-91}$$

Equations (4-90) and (4-91) are substituted into Eq. (4-89):

$$\frac{u_\infty}{L} = \frac{v_\delta}{\delta} \tag{4-92}$$

or

$$v_\delta = \frac{\delta}{L}u_\infty \tag{4-93}$$

Equation (4-93) indicates that when the boundary layer assumption, Eq. (4-87), is satisfied then the y-directed velocity in the boundary layer will be much less than the x-directed velocity. This result is intuitive. The x-directed velocity is driven by the free stream whereas the y-directed velocity originates because the boundary layer growth pushes against the free stream. The boundary layer is small and therefore the rate at which it grows is also small.

The x-Momentum Equation

Equation (4-93) provides a convenient scaling for the velocity in the y-direction (v) and allows an order of magnitude analysis of the x- and y-momentum equations. The x-directed momentum equation, Eq. (4-71), is simplified by assuming steady state and neglecting the gravitational force:

$$\rho\left(u\frac{\partial u}{\partial x} + v\frac{\partial u}{\partial y}\right) = -\frac{\partial p}{\partial x} + \mu\left(\frac{\partial^2 u}{\partial y^2} + \frac{\partial^2 u}{\partial x^2}\right) \tag{4-94}$$

The order of magnitude of each of the terms in Eq. (4-94) is estimated using the same type of scaling argument that was applied to the continuity equation.

$$\underbrace{O\left(\rho u\frac{\partial u}{\partial x}\right)}_{\rho\frac{u_\infty^2}{L}} + \underbrace{O\left(\rho v\frac{\partial u}{\partial y}\right)}_{\rho v_\delta\frac{u_\infty}{\delta}} = -\frac{\partial p}{\partial x} + \underbrace{O\left(\mu\frac{\partial^2 u}{\partial y^2}\right)}_{\mu\frac{u_\infty}{\delta^2}} + \underbrace{O\left(\mu\frac{\partial^2 u}{\partial x^2}\right)}_{\mu\frac{u_\infty}{L^2}} \tag{4-95}$$

The pressure change must scale with the inertial pressure drop based on the free-stream velocity; therefore:

$$O\left(\frac{\partial p}{\partial x}\right) = \frac{\rho u_\infty^2}{L} \tag{4-96}$$

Substituting Eqs. (4-93) and (4-96) into Eq. (4-95) leads to:

$$\underbrace{O\left(\rho u\frac{\partial u}{\partial x}\right)}_{\rho\frac{u_\infty^2}{L}} + \underbrace{O\left(\rho v\frac{\partial u}{\partial y}\right)}_{\rho\frac{u_\infty^2}{L}} = \underbrace{O\left(-\frac{\partial p}{\partial x}\right)}_{\rho\frac{u_\infty^2}{L}} + \underbrace{O\left(\mu\frac{\partial^2 u}{\partial y^2}\right)}_{\mu\frac{u_\infty}{\delta^2}} + \underbrace{O\left(\mu\frac{\partial^2 u}{\partial x^2}\right)}_{\mu\frac{u_\infty}{L^2}} \tag{4-97}$$

The order of magnitude of the first three terms in Eq. (4-97) are the same and equal to $\rho u_\infty^2/L$. The last two terms are therefore divided by $\rho u_\infty^2/L$ in order to evaluate their relative importance:

$$\rho \frac{u_\infty^2}{L} \left[O\left(\frac{\rho u \frac{\partial u}{\partial x}}{\rho \frac{u_\infty^2}{L}} \right) + O\left(\frac{\rho v \frac{\partial u}{\partial y}}{\rho \frac{u_\infty^2}{L}} \right) = O\left(\frac{-\frac{\partial p}{\partial x}}{\rho \frac{u_\infty^2}{L}} \right) + O\left(\frac{\mu \frac{\partial^2 u}{\partial y^2}}{\rho \frac{u_\infty^2}{L}} \right) + O\left(\frac{\mu \frac{\partial^2 u}{\partial x^2}}{\rho \frac{u_\infty^2}{L}} \right) \right]$$

$$\underbrace{\qquad}_{1} \qquad \underbrace{\qquad}_{1} \qquad \underbrace{\qquad}_{1} \qquad \underbrace{\qquad}_{\mu \frac{u_\infty}{\delta^2} \frac{L}{\rho u_\infty^2}} \qquad \underbrace{\qquad}_{\mu \frac{u_\infty}{L^2} \frac{L}{\rho u_\infty^2}}$$

$$(4\text{-}98)$$

or

$$\rho \frac{u_\infty^2}{L} \left[O\left(\frac{\rho u \frac{\partial u}{\partial x}}{\rho \frac{u_\infty^2}{L}} \right) + O\left(\frac{\rho v \frac{\partial u}{\partial y}}{\rho \frac{u_\infty^2}{L}} \right) = O\left(\frac{-\frac{\partial p}{\partial x}}{\rho \frac{u_\infty^2}{L}} \right) + O\left(\frac{\mu \frac{\partial^2 u}{\partial y^2}}{\rho \frac{u_\infty^2}{L}} \right) + O\left(\frac{\mu \frac{\partial^2 u}{\partial x^2}}{\rho \frac{u_\infty^2}{L}} \right) \right]$$

$$\underbrace{\qquad}_{1} \qquad \underbrace{\qquad}_{1} \qquad \underbrace{\qquad}_{1} \qquad \underbrace{\qquad}_{\frac{L^2}{\delta^2} \frac{1}{Re_L}} \qquad \underbrace{\qquad}_{\frac{1}{Re_L}}$$

$$(4\text{-}99)$$

The Reynolds number is large and therefore the last term will be small and can be neglected relative to the others. The scale of the fourth term is not clear; the Reynolds number is large but L^2/δ^2 is also large. Equation (4-86) suggests that the order of magnitude of the fourth term will be unity and therefore this term must be retained.

Two important conclusions have resulted from this analysis. First, the final term in Eq. (4-94) can be neglected relative to the others in the boundary layer:

$$\rho \left(u \frac{\partial u}{\partial x} + v \frac{\partial u}{\partial y} \right) = -\frac{\partial p}{\partial x} + \mu \frac{\partial^2 u}{\partial y^2} \qquad (4\text{-}100)$$

Second, the order of magnitude of the terms that have been retained in the x-directed momentum equation are all $\rho u_\infty^2/L$.

The y-Momentum Equation
The y-directed momentum equation, Eq. (4-72), is rewritten assuming steady state and neglecting the gravitational force:

$$\rho \left(u \frac{\partial v}{\partial x} + v \frac{\partial v}{\partial y} \right) = -\frac{\partial p}{\partial y} + \mu \left(\frac{\partial^2 v}{\partial y^2} + \frac{\partial^2 v}{\partial x^2} \right) \qquad (4\text{-}101)$$

The order of magnitude of each of the terms in Eq. (4-101) in the boundary layer is estimated:

$$O\left(\rho u \frac{\partial v}{\partial x} \right) + O\left(\rho v \frac{\partial v}{\partial y} \right) = O\left(-\frac{\partial p}{\partial y} \right) + O\left(\mu \frac{\partial^2 v}{\partial y^2} \right) + O\left(\mu \frac{\partial^2 v}{\partial x^2} \right) \qquad (4\text{-}102)$$

$$\underbrace{\qquad}_{\rho u_\infty \frac{v_\delta}{L}} \qquad \underbrace{\qquad}_{\rho v_\delta \frac{v_\delta}{\delta}} \qquad \underbrace{\qquad}_{\mu \frac{v_\delta}{\delta^2}} \qquad \underbrace{\qquad}_{\mu \frac{v_\delta}{L^2}}$$

Substituting Eq. (4-93) into Eq. (4-102) leads to:

$$\underbrace{O\left(\rho u \frac{\partial v}{\partial x}\right)}_{\rho u_\infty^2 \frac{\delta}{L^2}} + \underbrace{O\left(\rho v \frac{\partial v}{\partial y}\right)}_{\rho u_\infty^2 \frac{\delta}{L^2}} = O\left(-\frac{\partial p}{\partial y}\right) + \underbrace{O\left(\mu \frac{\partial^2 v}{\partial y^2}\right)}_{\mu \frac{u_\infty}{L\delta}} + \underbrace{O\left(\mu \frac{\partial^2 v}{\partial x^2}\right)}_{\mu \frac{\delta}{L^3} u_\infty} \qquad (4\text{-}103)$$

The order of the first two terms in Eq. (4-103) are $\rho u_\infty^2 \delta/L^2$ and so the last two terms are divided by this quantity in order to evaluate their importance:

$$\frac{\rho u_\infty^2}{L}\frac{\delta}{L}\left[\underbrace{O\left(\frac{\rho u \frac{\partial v}{\partial x}}{\rho u_\infty^2 \frac{\delta}{L}\frac{1}{L}}\right)}_{1} + \underbrace{O\left(\frac{\rho v \frac{\partial v}{\partial y}}{\rho u_\infty^2 \frac{\delta}{L}\frac{1}{L}}\right)}_{1} = O\left(\frac{-\frac{\partial p}{\partial y}}{\rho u_\infty^2 \frac{\delta}{L}\frac{1}{L}}\right) + \underbrace{O\left(\frac{\mu \frac{\partial^2 v}{\partial y^2}}{\rho u_\infty^2 \frac{\delta}{L}\frac{1}{L}}\right)}_{\frac{L^2}{\delta^2 Re_L}} + \underbrace{O\left(\frac{\mu \frac{\partial^2 v}{\partial x^2}}{\rho u_\infty^2 \frac{\delta}{L}\frac{1}{L}}\right)}_{\frac{1}{Re_L}}\right]$$

$$(4\text{-}104)$$

Equation (4-86) suggests that the order of the fourth term in Eq. (4-104) is unity. The last term will clearly be very small. The order of magnitude of the third term in Eq. (4-104), the pressure gradient in the y-direction, can be no larger than $\rho u_\infty^2 \delta/L^2$. Therefore, every term in the y-directed momentum equation is at least δ/L smaller than the terms in the x-directed momentum equation. The entire y-momentum equation can be neglected as being small when the boundary layer assumption is valid. This scaling analysis suggests that the pressure gradient in the y-direction is small and therefore the free stream pressure imposes itself through the boundary layer. The pressure in the boundary layer will only be a function of x. The partial derivative of pressure with respect to x in Eq. (4-100) can be replaced by the ordinary derivative of the free stream pressure:

$$\rho\left(u \frac{\partial u}{\partial x} + v \frac{\partial u}{\partial y}\right) = -\frac{dp_\infty}{dx} + \mu \frac{\partial^2 u}{\partial y^2} \qquad (4\text{-}105)$$

Equation (4-105) is the x-directed momentum equation in the boundary layer. In radial coordinates, if the flow is in the x-direction, it can be shown that the r-directed momentum equation, Eq. (4-74), is negligible and the x-directed momentum equation, Eq. (4-73) is simplified to:

$$\rho\left(u \frac{\partial u}{\partial x} + v \frac{\partial u}{\partial r}\right) = -\frac{dp_\infty}{dx} + \mu \frac{1}{r}\frac{\partial}{\partial r}\left(r \frac{\partial u}{\partial r}\right) \qquad (4\text{-}106)$$

where u and v are the components of velocity in the x and r directions, respectively.

The Thermal Energy Equation

The thermal energy equation, Eq. (4-82), is rewritten assuming steady state:

$$\rho c\left(u \frac{\partial T}{\partial x} + v \frac{\partial T}{\partial y}\right) = k\left(\frac{\partial^2 T}{\partial x^2} + \frac{\partial^2 T}{\partial y^2}\right) + \dot{g}_v''' \qquad (4\text{-}107)$$

where

$$\dot{g}_v''' = \mu \left\{ \left(\frac{\partial u}{\partial y} + \frac{\partial v}{\partial x} \right)^2 + 2 \left[\left(\frac{\partial u}{\partial x} \right)^2 + \left(\frac{\partial v}{\partial y} \right)^2 \right] \right\} \tag{4-108}$$

Initially, the order of magnitude of the viscous dissipation term is estimated. The order of magnitude of the terms in Eq. (4-108) are:

$$O(\dot{g}_v''') = O\left(\mu \left(\frac{\partial u}{\partial y} \right)^2 \right) + O\left(\mu \left(\frac{\partial u}{\partial y} \right) \left(\frac{\partial v}{\partial x} \right) \right) + O\left(\mu \left(\frac{\partial v}{\partial x} \right)^2 \right)$$
$$\underbrace{\phantom{O\left(\mu \left(\frac{\partial u}{\partial y} \right)^2 \right)}}_{\mu \frac{u_\infty^2}{\delta^2}} \qquad \underbrace{\phantom{O\left(\mu \left(\frac{\partial u}{\partial y} \right) \right)}}_{\mu \frac{u_\infty}{\delta} \frac{v_\delta}{L}} \qquad \underbrace{\phantom{O\left(\mu \left(\frac{\partial v}{\partial x} \right)^2 \right)}}_{\mu \frac{v_\delta^2}{L^2}}$$
$$+ O\left(2\mu \left(\frac{\partial u}{\partial x} \right)^2 \right) + O\left(2\mu \left(\frac{\partial v}{\partial y} \right)^2 \right) \tag{4-109}$$
$$\underbrace{\phantom{+ O\left(2\mu \left(\frac{\partial u}{\partial x} \right)^2 \right)}}_{\mu \frac{u_\infty^2}{L^2}} \qquad \underbrace{\phantom{+ O\left(2\mu \left(\frac{\partial v}{\partial y} \right)^2 \right)}}_{\mu \frac{v_\delta^2}{\delta^2}}$$

Substituting v_δ from Eq. (4-93) into Eq. (4-109) leads to:

$$O(\dot{g}_v''') = O\left(\mu \left(\frac{\partial u}{\partial y} \right)^2 \right) + O\left(\mu \left(\frac{\partial u}{\partial y} \right) \left(\frac{\partial v}{\partial x} \right) \right) + O\left(\mu \left(\frac{\partial v}{\partial x} \right)^2 \right)$$
$$\underbrace{\phantom{O\left(\mu \left(\frac{\partial u}{\partial y} \right)^2 \right)}}_{\mu \frac{u_\infty^2}{\delta^2}} \qquad \underbrace{\phantom{O\left(\mu \left(\frac{\partial u}{\partial y} \right) \right)}}_{\mu \frac{u_\infty^2}{L^2}} \qquad \underbrace{\phantom{O\left(\mu \left(\frac{\partial v}{\partial x} \right)^2 \right)}}_{\mu \frac{u_\infty^2}{L^2} \frac{\delta^2}{L^2}}$$
$$+ O\left(2\mu \left(\frac{\partial u}{\partial x} \right)^2 \right) + O\left(2\mu \left(\frac{\partial v}{\partial y} \right)^2 \right) \tag{4-110}$$
$$\underbrace{\phantom{+ O\left(2\mu \left(\frac{\partial u}{\partial x} \right)^2 \right)}}_{\mu \frac{u_\infty^2}{L^2}} \qquad \underbrace{\phantom{+ O\left(2\mu \left(\frac{\partial v}{\partial y} \right)^2 \right)}}_{\mu \frac{u_\infty^2}{L^2}}$$

The first term in Eq. (4-110) is of order $\mu\, u_\infty^2 / \delta^2$ and so the other terms in Eq. (4-110) are divided by this quantity:

$$O(\dot{g}_v''') = \mu \frac{u_\infty^2}{\delta^2} \left[O\left(\frac{\mu \left(\frac{\partial u}{\partial y} \right)^2}{\mu \frac{u_\infty^2}{\delta^2}} \right) + O\left(\frac{\mu \left(\frac{\partial u}{\partial y} \right) \left(\frac{\partial v}{\partial x} \right)}{\mu \frac{u_\infty^2}{\delta^2}} \right) + O\left(\frac{\mu \left(\frac{\partial v}{\partial x} \right)^2}{\mu \frac{u_\infty^2}{\delta^2}} \right) \right.$$
$$\underbrace{}_{1} \qquad\qquad \underbrace{}_{\frac{\delta^2}{L^2}} \qquad\qquad \underbrace{}_{\frac{\delta^4}{L^4}}$$
$$\left. + O\left(\frac{2\mu \left(\frac{\partial u}{\partial x} \right)^2}{\mu \frac{u_\infty^2}{\delta^2}} \right) + O\left(\frac{2\mu \left(\frac{\partial v}{\partial y} \right)^2}{\mu \frac{u_\infty^2}{\delta^2}} \right) \right] \tag{4-111}$$
$$\underbrace{}_{\frac{\delta^2}{L^2}} \qquad\qquad \underbrace{}_{\frac{\delta^2}{L^2}}$$

Equation (4-111) shows that only the first term in Eq. (4-108) is important in the boundary layer.

$$\dot{g}_v''' = \mu \left(\frac{\partial u}{\partial y}\right)^2 \tag{4-112}$$

Equation (4-112) should be intuitive. The most significant velocity gradient will be related to the change in the x-velocity in the y-direction since the x-velocity is the largest velocity and the y-direction is the smallest direction. The order of the viscous dissipation term is:

$$O(\dot{g}_v''') = \mu \frac{u_\infty^2}{\delta^2} \tag{4-113}$$

The order of magnitude of the thermal energy equation, Eq. (4-107), in the boundary layer are estimated, using Eq. (4-113):

$$\underbrace{O\left(\rho c u \frac{\partial T}{\partial x}\right)}_{\rho c u_\infty \frac{\Delta T}{L}} + \underbrace{O\left(\rho c v \frac{\partial T}{\partial y}\right)}_{\rho c v_\delta \frac{\Delta T}{\delta}} = \underbrace{O\left(k \frac{\partial^2 T}{\partial x^2}\right)}_{k \frac{\Delta T}{L^2}} + \underbrace{O\left(k \frac{\partial^2 T}{\partial y^2}\right)}_{k \frac{\Delta T}{\delta^2}} + \underbrace{O(\dot{g}_v''')}_{\mu \frac{u_\infty^2}{\delta^2}} \tag{4-114}$$

where ΔT is the surface-to-free stream temperature difference, $T_s - T_\infty$. Substituting Eqs. (4-93) into Eq. (4-114) leads to:

$$\underbrace{O\left(\rho c u \frac{\partial T}{\partial x}\right)}_{\rho c u_\infty \frac{\Delta T}{L}} + \underbrace{O\left(\rho c v \frac{\partial T}{\partial y}\right)}_{\rho c u_\infty \frac{\Delta T}{L}} = \underbrace{O\left(k \frac{\partial^2 T}{\partial x^2}\right)}_{k \frac{\Delta T}{L^2}} + \underbrace{O\left(k \frac{\partial^2 T}{\partial y^2}\right)}_{k \frac{\Delta T}{\delta^2}} + \underbrace{O(\dot{g}_v''')}_{\mu \frac{u_\infty^2}{\delta^2}} \tag{4-115}$$

The first two terms in Eq. (4-115) are of order $(\rho c u_\infty \Delta T/L^2)$ and so Eq. (4-115) is divided through by this quantity:

$$\rho c u_\infty \frac{\Delta T}{L} \left[\underbrace{O\left(\frac{\rho c u \frac{\partial T}{\partial x}}{\rho c u_\infty \frac{\Delta T}{L}}\right)}_{1} + \underbrace{O\left(\frac{\rho c v \frac{\partial T}{\partial y}}{\rho c u_\infty \frac{\Delta T}{L}}\right)}_{1} = \underbrace{O\left(\frac{k \frac{\partial^2 T}{\partial x^2}}{\rho c u_\infty \frac{\Delta T}{L}}\right)}_{k \frac{\Delta T}{L^2} \frac{L}{\rho c u_\infty \Delta T}} \right.$$

$$\left. + \underbrace{O\left(\frac{k \frac{\partial^2 T}{\partial y^2}}{\rho c u_\infty \frac{\Delta T}{L}}\right)}_{k \frac{\Delta T}{\delta^2} \frac{L}{\rho c u_\infty \Delta T}} + \underbrace{O\left(\frac{\dot{g}_v'''}{\rho c u_\infty \frac{\Delta T}{L}}\right)}_{\mu \frac{u_\infty^2}{\delta^2} \frac{L}{\rho c u_\infty \Delta T}} \right] \tag{4-116}$$

Introducing the Reynolds number and Prandtl number into Eq. (4-116):

$$
\rho c u_\infty \frac{\Delta T}{L} \left[\underbrace{O\left(\frac{\rho c u \frac{\partial T}{\partial x}}{\rho c u_\infty \frac{\Delta T}{L}} \right)}_{1} + \underbrace{O\left(\frac{\rho c v \frac{\partial T}{\partial y}}{\rho c u_\infty \frac{\Delta T}{L}} \right)}_{1} = \underbrace{O\left(\frac{k \frac{\partial^2 T}{\partial x^2}}{\rho c u_\infty \frac{\Delta T}{L}} \right)}_{\frac{1}{Re_L \, Pr}} \right.
$$

(4-117)

$$
\left. + \underbrace{O\left(\frac{k \frac{\partial^2 T}{\partial y^2}}{\rho c u_\infty \frac{\Delta T}{L}} \right)}_{\frac{L^2}{\delta^2} \frac{1}{Re_L \, Pr}} + \underbrace{O\left(\frac{\dot{g}_v'''}{\rho c u_\infty \frac{\Delta T}{L}} \right)}_{\frac{L^2}{\delta^2} \frac{u_\infty^2}{c \Delta T} \frac{1}{Re_L}} \right]
$$

The third term in Eq. (4-117) is clearly negligible in the boundary layer (recall that the Reynolds number will be large and the Prandtl number not too different from unity). Equation (4-86) suggests that the order of the fourth term in Eq. (4-117) will be near unity. The order of the last term is less clear and so it will be retained.

There are two important conclusions from this analysis. First, the axial conduction term (i.e., the term related to the second derivative of temperature with respect to x) can be neglected in the boundary layer. Second, the only significant term in the expression for the volumetric generation of thermal energy due to viscous dissipation is related to the gradient of the x-velocity in the y-direction. The thermal energy equation in the boundary layer is therefore:

$$
\rho c \left(u \frac{\partial T}{\partial x} + v \frac{\partial T}{\partial y} \right) = k \frac{\partial^2 T}{\partial y^2} + \mu \left(\frac{\partial u}{\partial y} \right)^2
$$

(4-118)

In radial coordinates, the thermal energy equation, Eqs. (4-84) and (4-85), may be simplified in the boundary layer to:

$$
\rho c \left(u \frac{\partial T}{\partial x} + v \frac{\partial T}{\partial r} \right) = k \frac{1}{r} \frac{\partial}{\partial r} \left(r \frac{\partial T}{\partial r} \right) + \mu \left(\frac{\partial u}{\partial r} \right)^2
$$

(4-119)

where u and v are the velocity components in the x- and r-directions, respectively.

Equations (4-105) and (4-118) are the momentum and thermal energy conservation equations in Cartesian coordinates, simplified for application in a boundary layer. These equations are used to examine convection problems in subsequent sections.

4.3 Dimensional Analysis in Convection

4.3.1 Introduction

In Section 4.2.3, it is shown that the governing equations within the boundary layer can be simplified relative to the general governing equations for an incompressible, viscous fluid derived in Section 4.2.2. The steady-state continuity, x-directed momentum, and thermal energy equations in a boundary layer are reduced to:

$$\frac{\partial u}{\partial x} + \frac{\partial v}{\partial y} = 0 \tag{4-120}$$

$$\rho \left(u \frac{\partial u}{\partial x} + v \frac{\partial u}{\partial y} \right) = -\frac{dp_\infty}{dx} + \mu \frac{\partial^2 u}{\partial y^2} \tag{4-121}$$

$$\rho c \left(u \frac{\partial T}{\partial x} + v \frac{\partial T}{\partial y} \right) = k \frac{\partial^2 T}{\partial y^2} + \mu \left(\frac{\partial u}{\partial y} \right)^2 \tag{4-122}$$

and the y-directed momentum equation was found to be negligible. Equations (4-120) through (4-122) may be simplified, but they are still pretty imposing and difficult to solve analytically in most practical cases. Consequently, researchers must carry out experiments or, more recently, develop computational fluid dynamic (CFD) models of the physical situation. Experiments and CFD models are relatively expensive and time consuming. Often it is not possible to build an experiment that has the same scale or operates under the same conditions as the physical device or situation of interest. In order to maximize the utility of a set of experimental results or CFD simulations, it is necessary to identify the minimum set of non-dimensional parameters that can be used to correlate the results.

Dimensional analysis provides a technique that can be used to reduce a very complicated problem to its simplest form in order to get the maximum possible use from the information that is available. The process of dimensional analysis has been the backbone of many scientific and engineering disciplines, as discussed by Bridgman (1922). Looking ahead, dimensional analysis provides the justification for correlating heat transfer data in the form that is encountered in most textbooks and handbooks, the Nusselt number as a function of the Reynolds number and the Prandtl number. We tend to take this presentation for granted and use these correlations without giving any real thought to the remarkable simplification that they represent. The same correlation can be used to estimate the heat transfer coefficient for a large cannonball traveling through air or a tiny spherical thermocouple mounted in a flowing liquid. The physical underpinnings and practical application of dimensional analysis are eloquently discussed by Sonin (1992) and others.

Given a physical problem of interest, there are at least two methods that can be used to identify the non-dimensional parameters that govern the problem. The classic technique is Buckingham's pi theorem (Buckingham (1914)). All of the independent physical quantities that are involved in the problem are listed. A complete and dimensionally independent subset of these quantities is selected and used to non-dimensionalize all of the remaining quantities. The result is one set of non-dimensional quantities that can be used to describe the problem.

An alternative approach is possible when additional information about the problem is available, for example the governing differential equation(s). In this situation, it is often more instructive to define physically meaningful, dimensionless quantities and substitute these into the governing equations which, through algebraic manipulation, are made dimensionless. The non-dimensional groups of quantities that result from this algebraic manipulation represent a more physically meaningful set of parameters than those arrived at using Buckingham's pi theorem because they can be directly attributed to each of the terms in the governing equation. In this section, the governing equations for a boundary layer are made dimensionless in order to identify the important non-dimensional quantities that govern a convective heat transfer problem.

4.3.2 Dimensionless Boundary Layer Equations

The process begins by defining the set of meaningful, non-dimensional quantities that can be identified by inspection. The coordinates x and y can be made dimensionless by normalizing them against the length of the plate (L), which is the only characteristic length in the problem:

$$\tilde{x} = \frac{x}{L} \tag{4-123}$$

$$\tilde{y} = \frac{y}{L} \tag{4-124}$$

The x- and y-components of velocity are normalized against the free-stream velocity (u_∞):

$$\tilde{u} = \frac{u}{u_\infty} \tag{4-125}$$

$$\tilde{v} = \frac{v}{u_\infty} \tag{4-126}$$

A dimensionless temperature difference relative to the plate temperature is defined by normalizing against the free stream to plate temperature difference:

$$\tilde{\theta} = \frac{T - T_s}{T_\infty - T_s} \tag{4-127}$$

The pressure is normalized against the fluid kinetic energy:

$$\tilde{p} = \frac{p}{\rho u_\infty^2} \tag{4-128}$$

These dimensionless parameters are substituted into the governing equations.

The Dimensionless Continuity Equation
Substituting the definitions for the dimensionless velocities and positions, Eqs. (4-123) through (4-126), into the continuity equation, Eq. (4-120), leads to:

$$\frac{\partial (\tilde{u} u_\infty)}{\partial (\tilde{x} L)} + \frac{\partial (\tilde{v} u_\infty)}{\partial (\tilde{y} L)} = 0 \tag{4-129}$$

or

$$\frac{u_\infty}{L} \frac{\partial \tilde{u}}{\partial \tilde{x}} + \frac{u_\infty}{L} \frac{\partial \tilde{v}}{\partial \tilde{y}} = 0 \tag{4-130}$$

Equation (4-130) is multiplied by L/u_∞ in order to obtain the dimensionless form of the continuity equation:

$$\boxed{\frac{\partial \tilde{u}}{\partial \tilde{x}} + \frac{\partial \tilde{v}}{\partial \tilde{y}} = 0} \tag{4-131}$$

The dimensionless continuity equation has the same form as the original continuity equation and has not resulted in the identification of any new non-dimensional groups.

The Dimensionless Momentum Equation in the Boundary Layer

Substituting the definitions of the dimensionless quantities into the x-momentum equation for the boundary layer, Eq. (4-121), leads to:

$$\rho\left(\tilde{u}\, u_\infty \frac{\partial\,(\tilde{u}\, u_\infty)}{\partial\,(\tilde{x}\, L)} + \tilde{v}\, u_\infty \frac{\partial\,(\tilde{u}\, u_\infty)}{\partial\,(\tilde{y}\, L)}\right) = -\frac{d\,(\tilde{p}_\infty\, \rho\, u_\infty^2)}{d\,(\tilde{x}\, L)} + \mu\frac{\partial^2\,(\tilde{u}\, u_\infty)}{\partial\,(\tilde{y}\, L)^2} \tag{4-132}$$

or

$$\frac{\rho\, u_\infty^2}{L}\left(\tilde{u}\frac{\partial \tilde{u}}{\partial \tilde{x}} + \tilde{v}\frac{\partial \tilde{u}}{\partial \tilde{y}}\right) = -\frac{\rho\, u_\infty^2}{L}\frac{d\tilde{p}_\infty}{d\tilde{x}} + \mu\frac{u_\infty}{L^2}\frac{\partial^2 \tilde{u}}{\partial \tilde{y}^2} \tag{4-133}$$

Equation (4-133) is divided by $\rho\, u_\infty^2/L$ (the scale of the inertial terms on the left side of the momentum equation) in order to obtain the dimensionless form of the x-momentum equation:

$$\tilde{u}\frac{\partial \tilde{u}}{\partial \tilde{x}} + \tilde{v}\frac{\partial \tilde{u}}{\partial \tilde{y}} = -\frac{d\tilde{p}_\infty}{d\tilde{x}} + \underbrace{\mu\frac{u_\infty}{L^2}\frac{L}{\rho\, u_\infty^2}}_{\substack{\text{viscous}\\ \text{inertial}}}\frac{\partial^2 \tilde{u}}{\partial \tilde{y}^2} \tag{4-134}$$

The group that multiplies the last term in Eq. (4-134) (the viscous shear term) must be dimensionless and it is an additional dimensionless group that governs the solution. Furthermore, the last term in Eq. (4-134) was obtained by dividing the viscous term by the inertial term. Therefore, the dimensionless parameter that has been identified represents the ratio of the viscous to the inertial force. Simplifying the last term leads to:

$$\frac{\text{viscous}}{\text{inertial}} = \mu\frac{u_\infty}{L^2}\frac{L}{\rho\, u_\infty^2} = \frac{\mu}{\rho\, u_\infty\, L} = \frac{1}{Re_L} \tag{4-135}$$

The Reynolds number must therefore represent the ratio of the inertial to the viscous forces. Equation (4-135) makes sense. As viscosity increases, the relative importance of the viscous forces will also increase. If the velocity or density increases, then the inertial forces will become more important. The transition from a laminar boundary layer to a turbulent boundary occurs when viscous forces are insufficient to suppress turbulent eddies, as discussed in Section 4.5. It makes sense that the transition to turbulence will occur at a critical value of the Reynolds number.

Substituting Eq. (4-135) into Eq. (4-134) leads to the dimensionless x-momentum equation for a boundary layer:

$$\boxed{\tilde{u}\frac{\partial \tilde{u}}{\partial \tilde{x}} + \tilde{v}\frac{\partial \tilde{u}}{\partial \tilde{y}} = -\frac{d\tilde{p}_\infty}{d\tilde{x}} + \frac{1}{Re_L}\frac{\partial^2 \tilde{u}}{\partial \tilde{y}^2}} \tag{4-136}$$

Notice that if the Reynolds number approaches infinity, then Eq. (4-136) will reduce to the governing equation for inviscid flow.

The Dimensionless Thermal Energy Equation in the Boundary Layer

Substituting the definitions of the dimensionless quantities into the thermal energy equation for the boundary layer, Eq. (4-122), leads to:

$$\rho c\left(\tilde{u}\, u_\infty \frac{\partial(\tilde{\theta}\,(T_\infty - T_s))}{\partial\,(\tilde{x}\, L)} + \tilde{v}\, u_\infty \frac{\partial(\tilde{\theta}\,(T_\infty - T_s))}{\partial\,(\tilde{y}\, L)}\right) = k\frac{\partial^2(\tilde{\theta}\,(T_\infty - T_s))}{\partial(\tilde{y}\, L)^2} + \mu\left(\frac{\partial(\tilde{u}\, u_\infty)}{\partial(\tilde{y}\, L)}\right)^2 \tag{4-137}$$

or

$$\frac{\rho c u_\infty (T_\infty - T_s)}{L} \left(\tilde{u} \frac{\partial \tilde{\theta}}{\partial \tilde{x}} + \tilde{v} \frac{\partial \tilde{\theta}}{\partial \tilde{y}} \right) = \frac{k (T_\infty - T_s)}{L^2} \frac{\partial^2 \tilde{\theta}}{\partial \tilde{y}^2} + \frac{\mu u_\infty^2}{L^2} \left(\frac{\partial \tilde{u}}{\partial \tilde{y}} \right)^2 \quad (4\text{-}138)$$

Equation (4-138) is made dimensionless by dividing through by the scale of the convective terms, $\rho c u_\infty (T_\infty - T_s) / L$:

$$\tilde{u} \frac{\partial \tilde{\theta}}{\partial \tilde{x}} + \tilde{v} \frac{\partial \tilde{\theta}}{\partial \tilde{y}} = \underbrace{\frac{k (T_\infty - T_s)}{L^2} \frac{L}{\rho c u_\infty (T_\infty - T_s)}}_{\substack{\text{conduction} \\ \text{convection}}} \frac{\partial^2 \tilde{\theta}}{\partial \tilde{y}^2} + \underbrace{\frac{\mu u_\infty^2}{L^2} \frac{L}{\rho c u_\infty (T_\infty - T_s)}}_{\substack{\text{dissipation} \\ \text{convection}}} \left(\frac{\partial \tilde{u}}{\partial \tilde{y}} \right)^2$$

$$(4\text{-}139)$$

The dimensionless group that appears in the first term on the right side of Eq. (4-139) must represent the ratio of conduction to convection. The dimensionless group that appears in the second term on the right side of Eq. (4-139) must represent the ratio of viscous dissipation to convection. Simplifying the first dimensionless group on the right side of Eq. (4-139) leads to:

$$\frac{\text{conduction}}{\text{convection}} = \frac{k (T_\infty - T_s)}{L^2} \frac{L}{\rho c u_\infty (T_\infty - T_s)} = \underbrace{\frac{k}{\rho c}}_{\alpha} \underbrace{\frac{1}{u_\infty L}}_{1/Re_L} \underbrace{\frac{\alpha}{v}}_{1/Pr} = \frac{1}{Re_L Pr} \quad (4\text{-}140)$$

Equation (4-140) makes sense. If the Prandtl number is very small (e.g., for a liquid metal) then the importance of conduction will increase. Also, as the Reynolds number increases, the importance of the convective terms (i.e., the transport of energy by the bulk motion of the fluid) will become more important.

Simplifying the second dimensionless group on the right side of Eq. (4-139) leads to:

$$\frac{\text{viscous dissipation}}{\text{convection}} = \frac{\mu u_\infty^2}{L^2} \frac{L}{\rho c u_\infty (T_\infty - T_s)} = \underbrace{\frac{u_\infty^2}{c (T_\infty - T_s)}}_{Ec} \underbrace{\frac{\mu}{L \rho u_\infty}}_{1/Re_L} = \frac{Ec}{Re_L} \quad (4\text{-}141)$$

where Ec is the Eckert number, defined as:

$$Ec = \frac{u_\infty^2}{c (T_\infty - T_s)} \quad (4\text{-}142)$$

A common simplification in convection problems is to neglect viscous dissipation. The majority of the correlations that are used to solve these problems have been developed in this limit. It is possible to evaluate the validity of this assumption using Eq. (4-141). If the ratio of the Eckert number to the Reynolds number is much less than unity, then very little error is introduced by neglecting viscous dissipation. As the value of Ec/Re_L increases, the effect of viscous dissipation becomes more important and the conventional correlations may no longer be valid. Equation (4-142) shows that this situation will occur for very high velocity flows with small temperature differences.

Substituting Eqs. (4-140) and (4-141) into Eq. (4-139) leads to the dimensionless form of the thermal energy equation for a boundary layer.

$$\boxed{\tilde{u} \frac{\partial \tilde{\theta}}{\partial \tilde{x}} + \tilde{v} \frac{\partial \tilde{\theta}}{\partial \tilde{y}} = \frac{1}{Re_L Pr} \frac{\partial^2 \tilde{\theta}}{\partial \tilde{y}^2} + \frac{Ec}{Re_L} \left(\frac{\partial \tilde{u}}{\partial \tilde{y}} \right)^2} \quad (4\text{-}143)$$

4.3.3 Correlating the Solutions of the Dimensionless Equations

In this section, the dimensionless equations that were derived in Section 4.3.2 are examined in order to understand how the important engineering quantities associated with boundary layer flows can be correlated.

The Friction and Drag Coefficients
The simultaneous solution of the dimensionless continuity and x-momentum equations for the boundary layer, Eqs. (4-131) and (4-136), would provide the dimensionless x- and y-directed velocities:

$$\tilde{u} = \tilde{u}\left(\tilde{x}, \tilde{y}, Re_L, \frac{d\tilde{p}_\infty}{d\tilde{x}}\right) \quad (4\text{-}144)$$

$$\tilde{v} = \tilde{v}\left(\tilde{x}, \tilde{y}, Re_L, \frac{d\tilde{p}_\infty}{d\tilde{x}}\right) \quad (4\text{-}145)$$

The solution will depend on a complete set of boundary conditions as well as the specification of the dimensionless free-stream pressure gradient. The free-stream pressure gradient in most flow situations will be dictated by the shape of the surface and the configuration being examined. For a flat plate exposed to a uniform free-stream velocity, the free-stream pressure gradient will be zero. For other objects, the free-stream pressure will vary based on the solution to the inviscid flow problem that exists away from the boundary layer. In this regard, the specified free-stream pressure gradient can be thought of as encapsulating information about the shape of object that is being considered.

The engineering quantity of interest is the shear stress at the surface of the plate:

$$\tau_s = \mu \left.\frac{\partial u}{\partial y}\right|_{y=0} \quad (4\text{-}146)$$

Equation (4-146) can be expressed in terms of the dimensionless quantities:

$$\tau_s = \mu \left.\frac{\partial\left(\tilde{u}\,u_\infty\right)}{\partial\left(\tilde{y}\,L\right)}\right|_{\tilde{y}=0} \quad (4\text{-}147)$$

or

$$\tau_s = \frac{\mu\,u_\infty}{L} \left.\frac{\partial\tilde{u}}{\partial\tilde{y}}\right|_{\tilde{y}=0} \quad (4\text{-}148)$$

Equation (4-148) could be made dimensionless by dividing through by $\mu u_\infty/L$ in order to obtain one possible dimensionless form of the shear stress; however, the more common (and equally valid) dimensionless form of the shear stress is the friction coefficient, C_f, introduced in Section 4.1.2:

$$C_f = \frac{2\,\tau_s}{\rho\,u_\infty^2} \quad (4\text{-}149)$$

Substituting Eq. (4-148) into Eq. (4-149) leads to:

$$C_f = \frac{2}{\rho\,u_\infty^2}\frac{\mu\,u_\infty}{L}\left.\frac{\partial\tilde{u}}{\partial\tilde{y}}\right|_{\tilde{y}=0} \quad (4\text{-}150)$$

or

$$C_f = \frac{2}{Re_L}\left.\frac{\partial\tilde{u}}{\partial\tilde{y}}\right|_{\tilde{y}=0} \quad (4\text{-}151)$$

Equation (4-151) shows that the solution for the friction factor depends on the solution for $\tilde{u}$. However, only the gradient of $\tilde{u}$ with respect to $\tilde{y}$ at $\tilde{y} = 0$ is required. Therefore, while $\tilde{u}$ depends on $\tilde{y}$, C_f does not:

$$C_f = C_f \left(\tilde{x}, Re_L, \frac{d\tilde{p}_\infty}{d\tilde{x}} \right) \tag{4-152}$$

The friction coefficient defined by Eq. (4-151) is the local friction coefficient that characterizes the shear at any location on the surface. Typically, the average friction coefficient ($\overline{C}_f$) is more useful. The average friction coefficient was introduced in Section 4.1.3 and is defined as:

$$\overline{C}_f = \frac{2 \overline{\tau}_s}{\rho u_\infty^2} \tag{4-153}$$

where the average shear stress for the 2-D problems considered here is defined as:

$$\overline{\tau}_s = \frac{1}{L} \int_0^L \tau_s \, dx \tag{4-154}$$

Substituting Eq. (4-148) into Eq. (4-154) and rearranging leads to:

$$\overline{\tau}_s = \frac{\mu \, u_\infty}{L} \int_0^1 \left. \frac{\partial \tilde{u}}{\partial \tilde{y}} \right|_{\tilde{y}=0} d\tilde{x} \tag{4-155}$$

Substituting Eq. (4-155) into Eq. (4-153) leads to:

$$\overline{C}_f = \frac{2}{\rho u_\infty^2} \frac{\mu \, u_\infty}{L} \int_0^1 \left. \frac{\partial \tilde{u}}{\partial \tilde{y}} \right|_{\tilde{y}=0} d\tilde{x} \tag{4-156}$$

or

$$\overline{C}_f = \frac{2}{Re_L} \int_0^1 \left. \frac{\partial \tilde{u}}{\partial \tilde{y}} \right|_{\tilde{y}=0} d\tilde{x} \tag{4-157}$$

Equation (4-157) shows that the solution for the average friction factor depends on the solution for $\tilde{u}$. However, only the gradient of $\tilde{u}$ with respect to $\tilde{y}$ at $\tilde{y} = 0$ is required. Furthermore, the solution for $\overline{C}_f$ does not depend on $\left. \frac{\partial \tilde{u}}{\partial \tilde{y}} \right|_{\tilde{y}=0}$ at any single value of $\tilde{x}$ but rather the quantity integrated over the entire range of $0 < \tilde{x} < 1$. Therefore, while $\tilde{u}$ depends on $\tilde{x}$ and $\tilde{y}$, $\overline{C}_f$ does not depend on either of these parameters:

$$\overline{C}_f = \overline{C}_f \left(Re_L, \frac{d\tilde{p}_\infty}{d\tilde{x}} \right) \tag{4-158}$$

For a given shape (e.g., for a flat plate), the pressure gradient is not an independent parameter. Therefore, the average friction coefficient for a particular shape will depend only on the Reynolds number:

$$\overline{C}_f = \overline{C}_f (Re_L) \quad \text{for a given shape} \tag{4-159}$$

Equation (4-159) is remarkable in that it indicates that a single dimensionless parameter can be used to correlate the functional behavior of a particular geometry. Equation (4-159) allows us to use small scale and therefore affordable wind tunnel tests to design large aircraft and also allows us to use large scale and therefore manageable and observable flow tests to characterize the flow around microscopic objects.

The same concepts can be applied to the external flow over shapes other than a flat plate (e.g., a cylinder or sphere). For these flow configurations, average friction coefficient is replaced by the drag coefficient, C_D. The drag coefficient is defined according to the drag force (F) exerted by the flow rather than the average shear:

$$C_D = \frac{2\,F}{\rho\,u_\infty^2\,A_p} \tag{4-160}$$

where A_p is the projected area of the object in the direction of the flow. The drag coefficient can be correlated against the Reynolds number defined based on the characteristic dimension of the problem, L_{char} (for example, the cylinder diameter):

$$C_D = C_D\,(Re_{L_{char}}) \quad \text{for a given shape} \tag{4-161}$$

The Nusselt Number

The solution of the dimensionless thermal energy equation would provide the dimensionless temperature difference:

$$\tilde{\theta} = \tilde{\theta}\,(\tilde{x},\,\tilde{y},\,Re_L,\,Pr,\,Ec,\,\tilde{u},\,\tilde{v}) \tag{4-162}$$

Note that the solution for $\tilde{\theta}$ depends on the dimensionless velocity solutions, Eqs. (4-144) and (4-145); therefore:

$$\tilde{\theta} = \tilde{\theta}\left(\tilde{x},\,\tilde{y},\,Re_L,\,\frac{d\tilde{p}_\infty}{d\tilde{x}},\,Pr,\,Ec\right) \tag{4-163}$$

Even though the dimensionless free stream pressure gradient does not appear explicitly in the dimensionless thermal energy equation, it will influence the solution through the velocity solution. The solution for $\tilde{\theta}$ depends on the same parameters as $\tilde{u}$ and $\tilde{v}$, as well as two additional parameters (the Prandtl number and Eckert number) that appear in the dimensionless thermal energy equation but were not present in the dimensionless momentum or continuity equations. The functional dependence of the temperature is somewhat more complex than that of the velocity. Said differently, two situations may be hydrodynamically similar if they have the same Reynolds number and $\frac{d\tilde{p}_\infty}{d\tilde{x}}$ (or shape) and therefore identical solutions for $\tilde{u}$ and $\tilde{v}$ according to Eqs. (4-144) and (4-145). However, the two situations will not also be thermally similar unless they also have the same Prandtl number and Eckert number, according to Eq. (4-162).

The engineering quantity of interest is the heat flux at the surface of the plate:

$$\dot{q}_s'' = -k\,\left.\frac{\partial T}{\partial y}\right|_{y=0} \tag{4-164}$$

which can be expressed in terms of the dimensionless quantities:

$$\dot{q}_s'' = -k\,\left.\frac{\partial\,(\tilde{\theta}\,(T_\infty - T_s))}{\partial\,(\tilde{y}\,L)}\right|_{\tilde{y}=0} \tag{4-165}$$

or

$$\dot{q}_s'' = \frac{k\,(T_s - T_\infty)}{L}\,\left.\frac{\partial\tilde{\theta}}{\partial\tilde{y}}\right|_{\tilde{y}=0} \tag{4-166}$$

Equation (4-166) can be made dimensionless by dividing through by $k(T_s - T_\infty)/L$ in order to obtain the dimensionless form of the heat flux:

$$\underbrace{\frac{\dot{q}_s''}{(T_s - T_\infty)} \frac{L}{k}}_{h} = \left.\frac{\partial \tilde{\theta}}{\partial \tilde{y}}\right|_{\tilde{y}=0} \tag{4-167}$$

Equation (4-167) can be rewritten in terms of the heat transfer coefficient:

$$\underbrace{\frac{h\,L}{k}}_{Nu} = \left.\frac{\partial \tilde{\theta}}{\partial \tilde{y}}\right|_{\tilde{y}=0} \tag{4-168}$$

The left side of Eq. (4-167) is equal to the Nusselt number, introduced in Section 4.1.2.

$$Nu = \left.\frac{\partial \tilde{\theta}}{\partial \tilde{y}}\right|_{\tilde{y}=0} \tag{4-169}$$

Equation (4-169) shows that the solution for the Nusselt number depends on the solution for $\tilde{\theta}$. However, the Nusselt number is only a function of the gradient in the dimensionless temperature difference at $\tilde{y} = 0$ and therefore the solution for the Nusselt number does not depend on $\tilde{y}$:

$$Nu = Nu\left(\tilde{x}, Re_L, \frac{d\tilde{p}_\infty}{d\tilde{x}}, Pr, Ec\right) \tag{4-170}$$

Also, if the average Nusselt number $(\overline{Nu})$ is required, then the local Nusselt number will be integrated from $\tilde{x} = 0$ to $\tilde{x} = 1$. Therefore, the solution for the average Nusselt number does not depend on $\tilde{x}$:

$$\overline{Nu} = \overline{Nu}\left(Re_L, \frac{d\tilde{p}_\infty}{d\tilde{x}}, Pr, Ec\right) \tag{4-171}$$

For a particular shape, the average Nusselt number will depend on the Reynolds number, Prandtl number, and Eckert number.

$$\overline{Nu} = \overline{Nu}\,(Re_L, Pr, Ec) \quad \text{for a given shape} \tag{4-172}$$

In most situations of general engineering interest, the effect of viscous dissipation is small and therefore the effect of Ec can be neglected:

$$\overline{Nu} = \overline{Nu}\,(Re_L, Pr) \quad \text{for a given shape if viscous dissipation is negligible} \tag{4-173}$$

Equation (4-173) is as remarkable as Eq. (4-159). It shows that only two parameters are required to correlate the thermal behavior of a given shape under most conditions. The experimental and theoretical results obtained by many researchers are usually presented in terms of the Nusselt number as a function of Reynolds number and Prandtl number. Indeed handbooks of heat transfer are filled with figures and correlations cast in terms of these parameters.

EXAMPLE 4.3-1: SUB-SCALE TESTING OF A CUBE-SHAPED MODULE

EXAMPLE 4.3-1: SUB-SCALE TESTING OF A CUBE-SHAPED MODULE

In a mine-detection and detonation application, large ($L = 0.5$ m on a side) cube-shaped modules are to be towed behind a ship with a velocity of $u_\infty = 1.0$ m/s. The water temperature is approximately $T_\infty = 20°C$ and the operation of the module dissipates energy at a rate of $\dot{q} = 10$ kW. You need to know the force that will be exerted on the ship by the tow cable (F) and the steady-state surface temperature of the module (T_s) under these conditions.

You have not been able to locate any correlations in the literature for the external flow over cubes that are oriented with respect to the flow in the same way as the tow module. Therefore an experimental test facility has been developed. Because you have a limited budget, experiments must be carried out on a much smaller cube ($L_{test} = 2$ cm on a side) that is placed in a pipe and mounted in the correct orientation relative to the flow.

A flow of water at $T_{\infty,test} = 20°C$ and $p_{test} = 1$ atm is initiated in the pipe. The pipe is quite large and flow straighteners are placed upstream and downstream of the test section in order to create a nearly uniform free-stream velocity in the pipe, $u_{\infty,test}$. A heater is embedded in the test cube and the heater power is adjusted so that the surface temperature ($T_{s,test}$) is exactly 10°C warmer than the water temperature. The heater power ($\dot{q}_{test}$) and the force exerted on the cube (F_{test}) are measured over a range of free-stream velocity. The results are summarized (with no attempt at quantifying experimental error, which is clearly a very important part of any real test of this type) in Table 1.

Table 1: Data from sub-scale testing.

Free-stream velocity, $u_{\infty,test}$ (m/s)	Drag force, F (N)	Heater power, $\dot{q}_{test}$ (W)
4.2	4	284
8.2	15	419
12.5	35	545
16.1	55	629
20.4	90	720
24.8	120	807
28.5	150	872
32.2	180	950
36.2	220	1023
40.1	250	1071

a) Use the results in Table 1 to estimate the force exerted on the full size module.

The discussion in Section 4.3.3 provides guidance relative to correlating these data. The drag force that was measured is a function of many variables, including the properties of the water, the free-stream velocity, the size of the cube, etc. On the other hand, the drag coefficient is a function of only the Reynolds number according to Eq. (4-161). Therefore, if we identify the data point with the same Reynolds number as the full scale module then we can be assured that the measured drag coefficient will apply to the full scale situation. In order for the data in Table 1 to be useful, it must be reduced to its dimensionless form: drag coefficient (C_D) and

EXAMPLE 4.3-1: SUB-SCALE TESTING OF A CUBE-SHAPED MODULE

Reynolds number (*Re*). This will be done in EES. The known information is entered in EES:

```
"EXAMPLE 4.3-1: Sub-Scale Testing of a Cube-Shaped Module"

$UnitSystem SI MASS RAD PA K J
$TabStops 0.2 3.5 in

"Inputs"
L=0.5 [m]                               "length of cube"
u_infinity=1.0 [m/s]                    "ship velocity"
T_infinity=converttemp(C,K,20)          "water temperature"
q_dot=10 [kW]*convert(kW,W)             "dissipation"
L_test=2 [cm]*convert(cm,m)             "length of test cube"
T_infinity_test=converttemp(C,K,20)     "temperature of test water"
T_s_test=T_infinity_test+10 [K]         "surface temperature during test"
p_test=1 [atm]*convert(atm,Pa)          "pressure of test"
```

The data in Table 1 are made available in EES as a Lookup Table. A Lookup Table provides a means of using tabular information in the solution of the equations. Select New Lookup Table from the Tables menu. Provide a title for the table as well as the number of columns and rows. Name the table Data and specify 3 columns and 10 rows. Select OK and a table will be created. Right-click on the first column heading and select Properties; the first column should be named u_infinity_test and have units of m/s. Repeat this process for the remaining columns so that the Lookup Table is ready to accept the data. Enter the data from Table 1 into the Lookup Table (Figure 1). Note that tabular data can be imported from various types of data files (e.g., text files or comma separated files) by selecting Open Lookup Table from the Tables menu.

Data			
Paste Special	$u_{\infty,test}$ [m/s]	F_{test} [N]	q_{test} [W]
Row 1	4.2	4	284
Row 2	8.2	15	419
Row 3	12.5	35	545
Row 4	16.1	55	629
Row 5	20.4	90	720
Row 6	24.8	120	807
Row 7	28.5	150	872
Row 8	32.2	180	950
Row 9	36.2	220	1023
Row 10	40.1	250	1071

Figure 1: Lookup Table.

The data in the lookup table can be accessed from the Equations window using the **Lookup** command. The **Lookup** command requires 3 arguments:

```
Value = Lookup(Table,Row,Column)
```

The first argument is a string constant (in single quotes) or string variable containing the name of the table that is being accessed. (Variables having names that end with $ are string variables in EES.) The second argument is the row number and the third is the column. The column can be specified as either a number or by using a string that contains the name of the column. The test data for the first data point are accessed using the following EES code:

```
i=1                                              "data point"
u_infinity_test=Lookup('Data',i,'u_infinity_test')   "test velocity"
F_test=Lookup('Data',i,'F_D_test')               "test force"
q_dot_test=Lookup('Data',i,'q_dot_test')         "test heat transfer rate"
```

The properties of liquid water at the test conditions (ρ_{test}, μ_{test}, and k_{test}) are obtained using EES' built-in functions for fluid properties:

```
rho_test=density(Water,T=T_infinity_test,P=P_test)     "test water density"
mu_test=viscosity(Water,T=T_infinity_test,P=P_test)    "test water viscosity"
k_test=conductivity(Water,T=T_infinity_test,P=P_test)  "test water conductivity"
```

The Reynolds number that characterizes each data point is:

$$Re_{L,test} = \frac{\rho_{test}\, L_{test}\, u_{\infty,test}}{\mu_{test}}$$

and is computed according to:

```
Re_test=L_test*rho_test*u_infinity_test/mu_test          "test Reynolds number"
```

The drag coefficient that characterizes each data point is:

$$C_{D,test} = \frac{2\,F_{test}}{\rho_{test}\, u_{\infty,test}^2\, L_{test}^2}$$

and is computed according to:

```
C_D_test=F_test*2/(rho_test*u_infinity_test^2*L_test^2)   "test drag coefficient"
```

The Reynolds number and drag coefficient for the first data point are $Re_{L,test} = 84{,}000$ and $C_{D,test} = 0.994$. A parametric table can be generated to carry out the calculations for each of the data points. Select New Parametric Table from the Tables menu and include the variables i, C_D_test, and Re_test.

Vary the column i from 1 to 10 in the parametric table and comment out the set value of i in the Equations Window. Solve the parametric table (select Solve Table from the Calculate menu) in order to obtain the drag coefficient and Reynolds

EXAMPLE 4.3-1: SUB-SCALE TESTING OF A CUBE-SHAPED MODULE

EXAMPLE 4.3-1: SUB-SCALE TESTING OF A CUBE-SHAPED MODULE

number at each data point. Figure 2 shows the measured drag coefficient as a function of the Reynolds number.

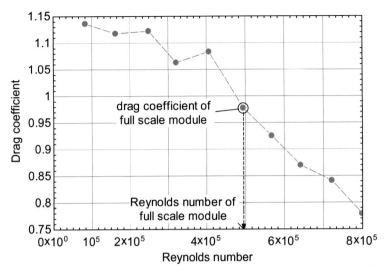

Figure 2: Drag coefficient as a function of Reynolds number.

The information in Figure 2 can be used to estimate the force on the full scale module. The Reynolds number associated with the full scale module is:

$$Re_L = \frac{\rho \, L \, u_\infty}{\mu}$$

where ρ and μ are the density and viscosity of the sea water:

```
rho=density(Water,T=T_infinity,P=1 [atm]*convert(atm,Pa))     "water density"
mu=viscosity(Water,T=T_infinity,P=1 [atm]*convert(atm,Pa))    "water viscosity"
k=conductivity(Water,T=T_infinity,P=1 [atm]*convert(atm,Pa))  "water conductivity"
Re=rho*L*u_infinity/mu                          "Reynolds number of full scale module"
```

which leads to a Reynolds number of 5×10^5 and therefore, from Figure 2, a drag coefficient of about 0.97. The drag force expected on the module will be:

$$F = \frac{C_D \, \rho \, u_\infty^2}{2} L^2$$

which is calculated using EES:

```
C_D=0.97                          "drag coefficient (from data)"
F=C_D*rho*u_infinity^2*L^2/2      "drag force"
```

and found to be 121 N.

b) Use the results in Table 1 to estimate the surface temperature of the full size module.

The average Nusselt number for each data point is defined as:

$$\overline{Nu}_{test} = \frac{\overline{h}_{test}\, L_{test}}{k_{test}}$$

where k_{test} is the conductivity of the water and $\overline{h}_{test}$ is the measured heat transfer coefficient. The measured heat transfer coefficient is calculated according to:

$$\overline{h}_{test} = \frac{\dot{q}_{test}}{6\, L_{test}^2\, (T_{s,test} - T_{\infty,test})}$$

h_bar_test=q_dot_test/(6*L_test^2*(T_s_test-T_infinity_test))

                                "heat transfer coefficient during the test"

Nusselt_bar_test=h_bar_test*L_test/k_test   "Nusselt number"

Add a new column to the parametric table (right click on one of the columns and select Insert Column to the Right and then add the variable Nusselt_bar_test to the parametric table). Solve the table and plot the variation of average Nusselt number with Reynolds number (Figure 3).

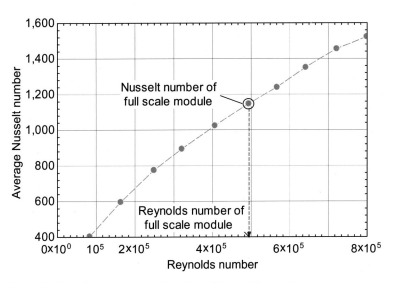

Figure 3: Nusselt number as a function of Reynolds number.

The average Nusselt number is a function of both the Reynolds number and Prandtl number. However, the testing was carried out for a single fluid at one temperature and pressure and therefore at a constant value of Prandtl number. The data shown in Figure 3 correspond to the average Nusselt number for the shape of interest as a function of Reynolds number at a constant Prandtl number of 7.15 (the Prandtl number of water at 20°C and 1 atm). If the full scale modules were immersed in a fluid with a markedly different Prandtl number, then the results in Figure 3 would not be as helpful. Fortunately, the test fluid has approximately the same Prandtl number as the sea water that will surround the full scale module and therefore the matching of the Prandtl numbers is taken care of.

EXAMPLE 4.3-1: SUB-SCALE TESTING OF A CUBE-SHAPED MODULE

EXAMPLE 4.3-1

Figure 3 indicates that the Nusselt number that can be expected for the full scale module is nominally $\overline{Nu} = 1150$. The heat transfer coefficient for the full scale module will therefore be:

$$\overline{h} = \overline{Nu}\frac{k}{L}$$

and the surface temperature of the full scale module will be:

$$T_s = T_\infty + \frac{\dot{q}}{6\,L^2\,\overline{h}}$$

```
Nusselt_bar=1150                        "Nusselt number of full scale module (from data)"
h_bar=Nusselt_bar*k/L                   "heat transfer coefficient"
T_s=T_infinity+q_dot/(6*L^2*h_bar)      "surface temperature"
T_s_C=converttemp(K,C,T_s)              "surface temperature in C"
```

which leads to a surface temperature of $T_s = 24.9°C$.

4.3.4 The Reynolds Analogy (revisited)

The Reynolds analogy is discussed in Section 4.1.2 using a simple conceptual model of the boundary layer that is based on the diffusive transport of momentum and energy into the free stream. It is informative to revisit the Reynolds analogy using the dimensionless governing equations that are derived in Section 4.3.2 in order to understand its limitations. The dimensionless momentum and energy equations for the boundary layer are:

$$\tilde{u}\frac{\partial \tilde{u}}{\partial \tilde{x}} + \tilde{v}\frac{\partial \tilde{u}}{\partial \tilde{y}} = -\frac{d\tilde{p}_\infty}{d\tilde{x}} + \frac{1}{Re_L}\frac{\partial^2 \tilde{u}}{\partial \tilde{y}^2} \tag{4-174}$$

$$\tilde{u}\frac{\partial \tilde{\theta}}{\partial \tilde{x}} + \tilde{v}\frac{\partial \tilde{\theta}}{\partial \tilde{y}} = \frac{1}{Re_L\,Pr}\frac{\partial^2 \tilde{\theta}}{\partial \tilde{y}^2} + \frac{Ec}{Re_L}\left(\frac{\partial \tilde{u}}{\partial \tilde{y}}\right)^2 \tag{4-175}$$

In the limit that (1) $Pr = 1$, (2) there is a negligible free-stream pressure gradient, and (3) viscous dissipation is not important, Eqs. (4-174) and (4-175) reduce to:

$$\tilde{u}\frac{\partial \tilde{u}}{\partial \tilde{x}} + \tilde{v}\frac{\partial \tilde{u}}{\partial \tilde{y}} = \frac{1}{Re_L}\frac{\partial^2 \tilde{u}}{\partial \tilde{y}^2} \tag{4-176}$$

$$\tilde{u}\frac{\partial \tilde{\theta}}{\partial \tilde{x}} + \tilde{v}\frac{\partial \tilde{\theta}}{\partial \tilde{y}} = \frac{1}{Re_L}\frac{\partial^2 \tilde{\theta}}{\partial \tilde{y}^2} \tag{4-177}$$

The boundary conditions for the dimensionless x-velocity include no-slip at the wall:

$$\tilde{u}_{\tilde{y}=0} = 0 \tag{4-178}$$

The free-stream velocity must be recovered as $\tilde{y}$ becomes large:

$$\tilde{u}_{\tilde{y}\to\infty} = 1 \tag{4-179}$$

There is a uniform approach velocity at the leading edge of the plate:

$$\tilde{u}_{\tilde{x}=0} = 1 \tag{4-180}$$

The boundary conditions for the dimensionless temperature difference are the same. At the wall, the temperature must be the surface temperature:

$$\tilde{\theta}_{\tilde{y}=0} = 0 \tag{4-181}$$

The free-stream temperature must be recovered as $\tilde{y}$ becomes large:

$$\tilde{\theta}_{\tilde{y}\to\infty} = 1 \tag{4-182}$$

There is a uniform approach temperature at the leading edge of the plate:

$$\tilde{\theta}_{\tilde{x}=0} = 1 \tag{4-183}$$

Therefore, under the limiting conditions discussed above (i.e., $Pr \approx 1$, $\frac{d\tilde{p}_\infty}{d\tilde{x}} \approx 0$, and $Ec \approx 0$) the dimensionless temperature and velocity are governed by the same partial differential equation with the same boundary conditions. They must have the same solutions. This is a more formal statement of the Reynolds analogy and it provides a clearer picture of its limitations.

Equations (4-151) and (4-169) expressed the friction factor and Nusselt number in terms of the dimensionless velocity and the dimensionless temperature difference, respectively:

$$C_f = \frac{2}{Re_L} \frac{\partial \tilde{u}}{\partial \tilde{y}}\bigg|_{\tilde{y}=0} \tag{4-184}$$

$$Nu = \frac{\partial \tilde{\theta}}{\partial \tilde{y}}\bigg|_{\tilde{y}=0} \tag{4-185}$$

If the Reynolds analogy holds, then the gradients of $\tilde{\theta}$ and $\tilde{u}$ must be identical and therefore:

$$\frac{\partial \tilde{u}}{\partial \tilde{y}}\bigg|_{\tilde{y}=0} = \frac{\partial \tilde{\theta}}{\partial \tilde{y}}\bigg|_{\tilde{y}=0} = \frac{C_f Re_L}{2} = Nu \tag{4-186}$$

which is the same statement of the Reynolds analogy obtained in Section 4.1.2.

4.4 Self-Similar Solution for Laminar Flow over a Flat Plate

4.4.1 Introduction

The behavior of a boundary layer was considered in Section 4.1 using a simple model that treated boundary layer growth over a flat-plate as being entirely related to the diffusion of momentum and energy. In fact, this is an over-simplification due to the two-dimensional velocity distribution that is generated by the interaction of the free stream with the plate. The boundary layer equations are derived in Section 4.2 and simplified for application in a boundary layer. The coupled set of partial differential equations that result are not easily solved in most situations. Therefore, in Section 4.3 these equations are made dimensionless in order to identify a minimal set of dimensionless parameters that can be used to generalize experimental measurements and numerical solutions.

The specific case of laminar flow over a flat plate is considered in this section. With some simplifications, this is one situation that can be solved analytically and the solution provides insight into the behavior of the boundary layer.

4.4.2 The Blasius Solution

The Blasius solution (as discussed in Schlichting (2000)), provides the velocity distribution in the laminar boundary layer by transforming the coupled partial differential equations associated with continuity and x-momentum into a single ordinary differential equation through the use of the stream function and a clever choice of a similarity variable. This technique resembles the process of obtaining a self-similar solution for transient conduction through a semi-infinite solid, discussed in Section 3.3.2. However, the Blasius solution is mathematically more complex.

The Problem Statement
The continuity equation for an incompressible, 2-D steady flow is:

$$\frac{\partial u}{\partial x} + \frac{\partial v}{\partial y} = 0 \tag{4-187}$$

The corresponding x-momentum equation for a flat plate (i.e., with no pressure gradient), simplified according to the boundary layer assumptions is:

$$u\frac{\partial u}{\partial x} + v\frac{\partial u}{\partial y} = \upsilon\frac{\partial^2 u}{\partial y^2} \tag{4-188}$$

The no-slip condition at the wall provides:

$$u_{y=0} = 0 \tag{4-189}$$

$$v_{y=0} = 0 \tag{4-190}$$

As y becomes large, the free-stream x-velocity must be recovered:

$$u_{y\to\infty} = u_\infty \tag{4-191}$$

At the leading edge of the plate, the velocity of the fluid approaching the plate is the free-stream velocity:

$$u_{x=0} = u_\infty \tag{4-192}$$

The Similarity Variables
The growth of the velocity and thermal boundary layers occur primarily due to the diffusive transport of momentum and energy. Therefore, the momentum boundary layer (δ_m) will grow approximately according to:

$$\delta_m \approx 2\sqrt{\upsilon t} \tag{4-193}$$

where υ is the kinematic viscosity. In a boundary layer, the time available for the diffusion of momentum is related to the distance from the leading edge (x) and the free-stream velocity (u_∞) according to:

$$t = \frac{x}{u_\infty} \tag{4-194}$$

Substituting Eq. (4-194) into Eq. (4-193) leads to:

$$\delta_m \approx 2\sqrt{\frac{\upsilon x}{u_\infty}} \tag{4-195}$$

In Section 3.3.3, the solution for the temperature in a semi-infinite body is obtained by defining a similarity parameter, η, as the position normalized by the depth of the thermal wave. This definition caused the temperature distributions at all times to collapse onto a

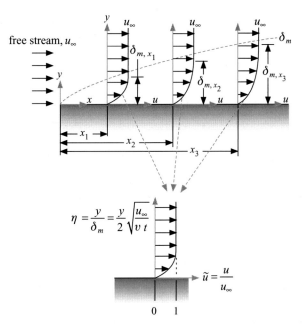

Figure 4-8: Velocity distributions at every position x collapse when expressed in terms of η.

single curve and transformed the partial differential equation that governed the problem into an ordinary differential equation. The Blasius solution takes the same approach. The similarity variable, η, is defined as the ratio of the distance from the plate surface (y) to the thickness of the momentum boundary layer:

$$\eta = \frac{y}{\delta_m} \tag{4-196}$$

The boundary layer grows as the fluid moves downstream. Therefore, η is a function of both x and y. Substituting Eq. (4-195) into Eq. (4-196) leads to:

$$\eta = \frac{y}{\delta_m} = \frac{y}{2}\sqrt{\frac{u_\infty}{\nu x}} \tag{4-197}$$

This definition of the similarity parameter is entered in Maple:

```
> restart;
> eta(x,y):=y*sqrt(u_infinity/(nu*x))/2;
```

$$\eta(x, y) := \frac{y\sqrt{\frac{u\_infinity}{\nu x}}}{2}$$

The motivation behind this choice of η is the anticipation that the dimensionless velocity:

$$\tilde{u} = \frac{u}{u_\infty} \tag{4-198}$$

at any position x will collapse onto a single curve when expressed in terms of η, as shown in Figure 4-8:

$$\tilde{u} = \tilde{u}(x, y) = \tilde{u}(\eta) \tag{4-199}$$

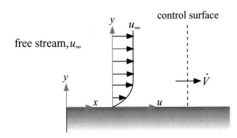

Figure 4-9: Volumetric flow rate through a surface placed in the boundary layer.

The stream function (Ψ) is a scalar function of position (x and y) that is defined according to its derivatives in order to automatically satisfy the continuity equation, Eq. (4-187).

$$u = \left(\frac{\partial \Psi}{\partial y}\right)_x \tag{4-200}$$

$$v = -\left(\frac{\partial \Psi}{\partial x}\right)_y \tag{4-201}$$

Substituting Eqs. (4-200) and (4-201) into Eq. (4-187) leads to:

$$\frac{\partial u}{\partial x} + \frac{\partial v}{\partial y} = 0 \Rightarrow \frac{\partial^2 \Psi}{\partial x \partial y} - \frac{\partial^2 \Psi}{\partial y \partial x} = 0 \tag{4-202}$$

The stream function Ψ is a continuous function of x and y and therefore the order in which the partial derivatives are taken does not matter. The continuity equation, Eq. (4-187), will automatically be satisfied by any continuous stream function. Therefore, it is not necessary to consider the continuity equation in our solution, provided that we work with Ψ rather than with u and v directly.

The stream function has some physical significance. The volumetric flow rate ($\dot{V}$) through a surface that extends vertically from the plate surface to a position y (see Figure 4-9) is given by:

$$\dot{V} = W \int_0^y u \, dy \tag{4-203}$$

where W is the width into the page.

Substituting Eq. (4-200) into Eq. (4-203) leads to:

$$\dot{V} = W \int_0^y \frac{\partial \Psi}{\partial y} \, dy \tag{4-204}$$

Integrating Eq. (4-204) leads to:

$$\dot{V} = W(\Psi - \Psi_{y=0}) \tag{4-205}$$

The stream function is defined in terms of its derivatives; therefore, a constant, reference value can be added to any stream function and it will still satisfy Eqs. (4-200) and (4-201). The reference value of the stream function is defined so that the stream function is 0 at $y = 0$. Therefore, Eq. (4-205) becomes:

$$\dot{V} = W \Psi \tag{4-206}$$

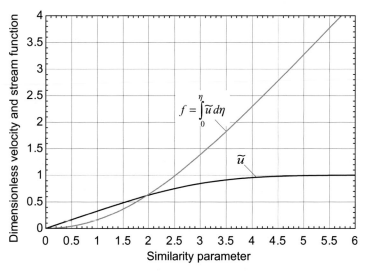

Figure 4-10: The anticipated qualitative form of the dimensionless velocity, $\tilde{u}$, and the dimensionless stream function, f.

Equation (4-206) shows that the value of the stream function at any position x, y represents the volumetric flow rate per unit width (with units m^3/s-m) between that position and the plate surface.

The volumetric flow rate through the surface in Figure 4-9, Eq. (4-203), can be expressed in terms of the dimensionless variables ($\tilde{u}$ and η):

$$\dot{V} = W\, u_\infty\, \delta_m \int_0^\eta \tilde{u}\, d\eta \tag{4-207}$$

Substituting Eq. (4-206) into Eq. (4-207) leads to:

$$\Psi = u_\infty\, \delta_m \underbrace{\int_0^\eta \tilde{u}\,(\eta)\, d\eta}_{f(\eta)} \tag{4-208}$$

The term labeled $f(\eta)$ in Eq. (4-208) can be thought of as a dimensionless form of the stream function.

$$f(\eta) = \frac{\Psi}{u_\infty\, \delta_m} = \int_0^\eta \tilde{u}\,(\eta)\, d\eta \tag{4-209}$$

The anticipated variation of $\tilde{u}$ and f with η are shown qualitatively in Figure 4-10. The value of f starts at 0 at $\eta = 0$ and grows with an increasing rate (as the dimensionless velocity increases) until finally it grows linearly with a slope of 1.0 at large η because the dimensionless velocity is equal to 1.0 outside of the boundary layer.

Substituting Eq. (4-195) into Eq. (4-208) leads to a working definition of the stream function:

$$\Psi = u_\infty\, 2\sqrt{\frac{\upsilon x}{u_\infty}}\, f(\eta) \tag{4-210}$$

The stream function definition is entered in Maple:

```
> Psi(x,y):=2*sqrt(u_infinity*nu*x)*f(eta(x,y));
```

$$\Psi(x, y) := 2\sqrt{u\_infinity\, v\, x}\, f\left(\frac{y\sqrt{\dfrac{u\_infinity}{vx}}}{2}\right)$$

The Problem Transformation

The stream function expressed in terms of the similarity variable, Eq. (4-210), is substituted into the governing x-momentum equation, Eq. (4-188), and the boundary conditions, Eqs. (4-189) through (4-192). The process transforms the partial differential equation that governs the problem into an ordinary differential equation. The substitutions required to transform the problem are discussed in this section. The symbolic manipulations are shown explicitly and Maple is used in parallel to accomplish these same steps.

The x-velocity is expressed in terms of the similarity variables by substituting Eq. (4-210) into Eq. (4-200):

$$u = \left(\frac{\partial \Psi}{\partial y}\right)_x = \frac{\partial}{\partial y}[2\sqrt{u_\infty\, v\, x}\, f(\eta)]_x = 2\sqrt{u_\infty\, v\, x}\frac{\partial}{\partial y}[f(\eta)]_x \qquad (4\text{-}211)$$

Applying the chain rule to the partial derivative of f in Eq. (4-211) leads to:

$$\frac{\partial}{\partial y}[f(\eta(x, y))]_x = \frac{df}{d\eta}\left(\frac{\partial \eta}{\partial y}\right)_x \qquad (4\text{-}212)$$

Substituting Eq. (4-212) into Eq. (4-211) leads to:

$$u = 2\sqrt{u_\infty\, v\, x}\frac{df}{d\eta}\left(\frac{\partial \eta}{\partial y}\right)_x \qquad (4\text{-}213)$$

The partial derivative of η, Eq. (4-197), with respect to y at constant x is:

$$\left(\frac{\partial \eta}{\partial y}\right)_x = \frac{1}{2}\sqrt{\frac{u_\infty}{vx}} \qquad (4\text{-}214)$$

Substituting Eq. (4-214) into Eq. (4-213) leads to:

$$u = 2\sqrt{u_\infty\, v\, x}\frac{df}{d\eta}\frac{1}{2}\sqrt{\frac{u_\infty}{vx}} \qquad (4\text{-}215)$$

or, after simplification:

$$\boxed{u = u_\infty \frac{df}{d\eta}} \qquad (4\text{-}216)$$

So the dimensionless velocity (u/u_∞) is the gradient of the dimensionless stream function. This result is consistent with Figure 4-10 and Eq. (4-209). Maple can be used to obtain the equivalent result:

```
> u:=diff(Psi(x,y),y);
```

$$u := \sqrt{u\_infinity\, v\, x}\, \mathrm{D}\,(f) \left(\frac{y\sqrt{\frac{u\_infinity}{v\, x}}}{2} \right) \sqrt{\frac{u\_infinity}{v\, x}}$$

The D operator in the Maple result indicates the derivative of the argument. The y-velocity is obtained by substituting Eq. (4-210) into Eq. (4-201)

$$v = -\frac{\partial}{\partial x} \left[2\sqrt{u_\infty\, v\, x}\, f\, (\eta\,(x,y)) \right]_y \qquad (4\text{-}217)$$

It is important to recognize that η is a function of x and y and, as a result, f is itself a function of x and y. Therefore, the partial derivative in Eq. (4-217) can be expanded according to:

$$v = -f\frac{\partial}{\partial x}[2\sqrt{u_\infty\, v\, x}] - 2\sqrt{u_\infty\, v\, x}\frac{df}{d\eta}\frac{\partial \eta}{\partial x} \qquad (4\text{-}218)$$

Substituting Eq. (4-197) into Eq. (4-218) leads to:

$$v = -f\frac{\partial}{\partial x}[2\sqrt{u_\infty\, v\, x}] - 2\sqrt{u_\infty\, v\, x}\frac{df}{d\eta}\frac{\partial}{\partial x}\left[\frac{y}{2}\sqrt{\frac{u_\infty}{v\, x}}\right] \qquad (4\text{-}219)$$

or

$$v = -f\sqrt{\frac{u_\infty\, v}{x}} + \sqrt{u_\infty\, v\, x}\frac{df}{d\eta}\frac{y}{2x^{3/2}}\sqrt{\frac{u_\infty}{v}} \qquad (4\text{-}220)$$

Equation (4-220) can be rearranged:

$$\boxed{v = \sqrt{\frac{u_\infty\, v}{x}}\left(\eta\frac{df}{d\eta} - f\right)} \qquad (4\text{-}221)$$

The equivalent result can be obtained using Maple:

```
> v:=-simplify(diff(Psi(x,y),x));
```

$$v := \frac{1}{2}u\_infinity$$

$$\left(-2f\left(\frac{y\sqrt{\frac{u\_infinity}{v\, x}}}{2} \right) v\, x\sqrt{\frac{u\_infinity}{v\, x}} + u\_infinity\, \mathrm{D}\,(f)\left(\frac{y\sqrt{\frac{u\_infinity}{v\, x}}}{2} \right) y \right) \Bigg/ \left(x\left(\sqrt{u\_infinity\, v\, x}\sqrt{\frac{u\_infinity}{v\, x}} \right) \right)$$

Examination of the momentum equation, Eq. (4-188), indicates that the partial derivatives of u must be obtained in terms of f and η in order complete the problem transformation. If the velocity distributions collapse, as shown in Figure 4-8, then u is a function only of η where η which is a function of x and y. Therefore, the partial derivative of u with respect to x is:

$$\frac{\partial}{\partial x}[u\,(\eta\,(x,y))] = \frac{du}{d\eta}\left(\frac{\partial \eta}{\partial x}\right)_y \qquad (4\text{-}222)$$

Substituting Eqs. (4-216) and (4-197) into Eq. (4-222) leads to:

$$\frac{\partial u}{\partial x} = \frac{d}{d\eta}\left[u_\infty \frac{df}{d\eta}\right]\frac{\partial}{\partial x}\left[\frac{y}{2}\sqrt{\frac{u_\infty}{\nu x}}\right]_y \tag{4-223}$$

or

$$\frac{\partial u}{\partial x} = u_\infty \frac{d^2 f}{d\eta^2}\left[-\frac{y}{4x^{3/2}}\sqrt{\frac{u_\infty}{\nu}}\right] \tag{4-224}$$

Equation (4-224) can be rearranged:

$$\frac{\partial u}{\partial x} = -\frac{u_\infty}{2x}\eta \frac{d^2 f}{d\eta^2} \tag{4-225}$$

The equivalent result can be obtained using Maple:

```
> dudx:=simplify(diff(u,x));
```

$$dudx := -\frac{1}{4}\frac{u\_infinity^2 (D^{(2)})(f)\left(\frac{y\sqrt{\frac{u\_infinity}{\nu x}}}{2}\right)y}{x\sqrt{u\_infinity\,\nu x}}$$

The partial derivative of u with respect to y is:

$$\frac{\partial}{\partial y}[u(\eta(x,y))] = \frac{du}{d\eta}\left(\frac{\partial\eta}{\partial y}\right)_x \tag{4-226}$$

Substituting Eqs. (4-216) and (4-197) into Eq. (4-226) leads to:

$$\frac{\partial u}{\partial y} = \frac{d}{d\eta}\left[u_\infty \frac{df}{d\eta}\right]\frac{\partial}{\partial y}\left[\frac{y}{2}\sqrt{\frac{u_\infty}{\nu x}}\right]_x \tag{4-227}$$

or

$$\frac{\partial u}{\partial y} = \frac{u_\infty}{2}\frac{d^2 f}{d\eta^2}\sqrt{\frac{u_\infty}{\nu x}} \tag{4-228}$$

The equivalent result can be obtained using Maple:

```
> dudy:=simplify(diff(u,y));
```

$$dudy := -\frac{1}{2}\frac{\sqrt{u\_infinity\,\nu x}\,(D^{(2)})(f)\left(\frac{y\sqrt{\frac{u\_infinity}{\nu x}}}{2}\right)u\_infinity}{\nu x}$$

Finally, the second derivative of u with respect to y is:

$$\frac{\partial^2 u}{\partial y^2} = \frac{\partial}{\partial y}\left(\frac{\partial u(\eta(x,y))}{\partial y}\right) = \frac{\partial}{\partial\eta}\left(\frac{\partial u}{\partial y}\right)_x\left(\frac{\partial\eta}{\partial y}\right)_x \tag{4-229}$$

Substituting Eqs. (4-228) and (4-197) into Eq. (4-229) leads to:

$$\frac{\partial^2 u}{\partial y^2} = \frac{\partial}{\partial \eta}\left[\frac{u_\infty}{2}\frac{d^2 f}{d\eta^2}\sqrt{\frac{u_\infty}{\upsilon x}}\right]\frac{\partial}{\partial y}\left[\frac{y}{2}\sqrt{\frac{u_\infty}{\upsilon x}}\right]_x \tag{4-230}$$

or

$$\frac{\partial^2 u}{\partial y^2} = \frac{u_\infty^2}{4\upsilon x}\frac{d^3 f}{d\eta^3} \tag{4-231}$$

The equivalent result can be obtained using Maple:

> d2udy2:=simplify(diff(dudy,y));

$$d2udy2 := -\frac{1}{4}\frac{\sqrt{u\_infinity\,\upsilon x}\,(D^{(3)})(f)\left(\dfrac{y\sqrt{\dfrac{u\_infinity}{\upsilon x}}}{2}\right)\sqrt{\dfrac{u\_infinity}{\upsilon x}}\,u\_infinity}{\upsilon x}$$

Substituting Eqs. (4-216), (4-221), (4-225), (4-228), and (4-231) into the *x*-momentum equation for the boundary layer, Eq. (4-188) leads to:

$$\underbrace{u_\infty \frac{df}{d\eta}}_{u}\underbrace{\left(-\frac{u_\infty}{2x}\eta\frac{d^2 f}{d\eta^2}\right)}_{\frac{\partial u}{\partial x}}+\underbrace{\sqrt{\frac{u_\infty \upsilon}{x}}\left(\eta\frac{df}{d\eta}-f\right)}_{\upsilon}\underbrace{\frac{u_\infty}{2}\frac{d^2 f}{d\eta^2}\sqrt{\frac{u_\infty}{\upsilon x}}}_{\frac{\partial u}{\partial y}}=\upsilon\underbrace{\frac{u_\infty^2}{4\upsilon x}\frac{d^3 f}{d\eta^3}}_{\frac{\partial^2 u}{\partial y^2}} \tag{4-232}$$

Notice that the quantity u_∞^2/x can be cancelled from every term on both sides of Eq. (4-232) in order to obtain:

$$-\frac{\eta}{2}\frac{d^2 f}{d\eta^2}\frac{df}{d\eta}+\frac{\eta}{2}\frac{df}{d\eta}\frac{d^2 f}{d\eta^2}-\frac{f}{2}\frac{d^2 f}{d\eta^2}=\frac{1}{4}\frac{d^3 f}{d\eta^3} \tag{4-233}$$

Simplification of Eq. (4-233) leads to an ordinary differential equation for the dimensionless stream function, f:

$$\frac{d^3 f}{d\eta^3}+2f\frac{d^2 f}{d\eta^2}=0 \tag{4-234}$$

The same ordinary differential equation is obtained using Maple:

> simplify(u*dudx+v*dudy=nu*d2udy2);

$$-\frac{1}{2}\frac{u\_infinity^2(D^{(2)})(f)\left(\dfrac{y\sqrt{\dfrac{u\_infinity}{\upsilon x}}}{2}\right)f\left(\dfrac{y\sqrt{\dfrac{u\_infinity}{\upsilon x}}}{2}\right)}{x}$$

$$=-\frac{1}{4}\frac{\sqrt{u\_infinity\,\upsilon x}\,(D^{(3)})(f)\left(\dfrac{y\sqrt{\dfrac{u\_infinity}{\upsilon x}}}{2}\right)\sqrt{\dfrac{u\_infinity}{\upsilon x}}\,u\_infinity}{x}$$

Three boundary conditions must be obtained for this third-order ordinary differential equation. Equation (4-216) is substituted into Eq. (4-189):

$$u_{y=0} = u_\infty \left. \frac{df}{d\eta} \right|_{\eta=0} = 0 \qquad (4\text{-}235)$$

or

$$\boxed{\left. \frac{df}{d\eta} \right|_{\eta=0} = 0} \qquad (4\text{-}236)$$

Equation (4-216) is also substituted into Eq. (4-191) (the same result could be obtained using Eq. (4-192)):

$$u_{y\to\infty} = u_\infty \left. \frac{df}{d\eta} \right|_{\eta\to\infty} = u_\infty \qquad (4\ 237)$$

or

$$\boxed{\left. \frac{df}{d\eta} \right|_{\eta\to\infty} = 1} \qquad (4\text{-}238)$$

The same result can be obtained by considering Eq. (4-192). Equation (4-221) is substituted into Eq. (4-190):

$$v_{y=0} = \frac{1}{2}\sqrt{\frac{u_\infty \upsilon}{x}} \left(\underbrace{\eta\frac{df}{d\eta}}_{=0} - f \right)_{\eta=0} = 0 \qquad (4\text{-}239)$$

or

$$\boxed{f_{\eta=0} = 0} \qquad (4\text{-}240)$$

Equation (4-234) together with Eqs. (4-236), (4-238), and (4-240) represent a well-defined but nonlinear third-order ordinary differential equation.

Numerical Solution

Any of the numerical integration techniques discussed in Chapter 3 in the context of transient conduction problems can be applied to this problem as well. Recall that these numerical techniques are based on a system of equations that compute the rate of change of a set of state variables given their current values. For this problem, the state variables are f, $\frac{df}{d\eta}$, and $\frac{d^2f}{d\eta^2}$. The rates of change of the first two state variables are obtained from the other state variables:

$$\frac{d}{d\eta}[f] = \frac{df}{d\eta} \qquad (4\text{-}241)$$

$$\frac{d}{d\eta}\left[\frac{df}{d\eta}\right] = \frac{d^2f}{d\eta^2} \qquad (4\text{-}242)$$

The rate of change of the final state variable is obtained from the governing ordinary differential equation, Eq. (4-234):

$$\frac{d}{d\eta}\left[\frac{d^2f}{d\eta^2}\right] = \frac{d^3f}{d\eta^3} = -2f\frac{d^2f}{d\eta^2} \tag{4-243}$$

Given the values of the state variables at the wall ($\eta = 0$) it is possible to numerically integrate Eqs. (4-241) through (4-243) in order to obtain the solution. However, only two of the three boundary conditions for the problem, Eqs. (4-236) and (4-240), are specified at the wall. The third boundary condition, Eq. (4-238), is specified at $\eta \to \infty$. In other words, the value of two state variables at $\eta = 0$, $f_{\eta=0}$ and $\frac{df}{d\eta}\big|_{\eta=0}$ are known but the value of the third state variable at $\eta = 0$ is not. Therefore, it is necessary to implement an implicit numerical integration technique that guesses a value of $\frac{d^2f}{d\eta^2}\big|_{\eta=0}$ and adjusts this guess until the final boundary condition, $\frac{df}{d\eta}\big|_{\eta\to\infty} = 1$, is satisfied. This process is referred to as using a "shooting method" because it is analogous to taking multiple shots with a gun and adjusting your aim between each shot.

The shooting method for this type of two-point boundary problem can be accomplished using any of the numerical techniques that are introduced in Section 3.2. Here, the Crank-Nicolson technique is used to integrate from $\eta = 0$ to a position far from the wall (but not at infinity) where the boundary condition given in Eq.(4-238) must be satisfied. Recall that η is defined in Eq. (4-196) as the ratio of the vertical distance from the plate to the thickness of the velocity boundary layer. Therefore, it should be reasonable to terminate the numerical integration at a value of η that is much larger than unity, for example at $\eta_\infty = 10$, and enforce the final boundary condition at η_∞. The computational domain ($0 < \eta < \eta_\infty$) is divided into steps of size $\Delta\eta$:

$$\Delta\eta = \frac{\eta_\infty}{(N-1)} \tag{4-244}$$

and the location of the nodes is provided by:

$$\eta_i = \eta_\infty \frac{(i-1)}{(N-1)} \quad \text{for } i = 1...N \tag{4-245}$$

```
$UnitSystem SI MASS RAD PA K J
$TabStops 0.2 3.5 in

eta_infinity=10 [-]              "outer edge of the computational domain"
N=101 [-]                        "number of steps in the numerical integration"
DELTAeta=eta_infinity/(N-1)      "size of the integration steps"
duplicate i=1,N
  eta[i]=(i-1)*eta_infinity/(N-1)   "position of integration steps"
end
```

The initial conditions for the integration process are specified; note that the value of $\frac{d^2f}{d\eta^2}\big|_{\eta=0}$ (the value of variable **d2fdeta2[1]**) is assumed and will be adjusted to complete the problem.

```
f[1]=0 [-]                "f at eta = 0"
dfdeta[1]=0 [-]           "dfdeta at eta = 0"
d2fdeta2[1]=0.3 [-]       "d2fdeta2 at eta = 0, this is a guess"
```

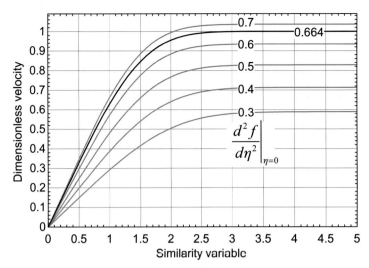

Figure 4-11: Dimensionless velocity, $\frac{df}{d\eta} = u/u_\infty$, as a function of η for several values of $\left.\frac{d^2f}{d\eta^2}\right|_{\eta=0}$; note that $\left.\frac{d^2f}{d\eta^2}\right|_{\eta=0} = 0.664$ causes $\frac{df}{d\eta} = 1$ and this curve is the self-similar solution.

The Crank-Nicolson technique uses the average of the rates of change evaluated at the beginning and end of the integration step in order to simulate each step:

$$f_{i+1} = f_i + \left[\left.\frac{df}{d\eta}\right|_i + \left.\frac{df}{d\eta}\right|_{i+1}\right]\frac{\Delta\eta}{2} \quad \text{for } i = 1..(N-1) \tag{4-246}$$

$$\left.\frac{df}{d\eta}\right|_{i+1} = \left.\frac{df}{d\eta}\right|_i + \left[\left.\frac{d^2f}{d\eta^2}\right|_i + \left.\frac{d^2f}{d\eta^2}\right|_{i+1}\right]\frac{\Delta\eta}{2} \quad \text{for } i = 1..(N-1) \tag{4-247}$$

$$\left.\frac{d^2f}{d\eta^2}\right|_{i+1} = \left.\frac{d^2f}{d\eta^2}\right|_i - \left[2f_i\left.\frac{d^2f}{d\eta^2}\right|_i + 2f_{i+1}\left.\frac{d^2f}{d\eta^2}\right|_{i+1}\right]\frac{\Delta\eta}{2} \quad \text{for } i = 1..(N-1) \tag{4-248}$$

```
"Crank-Nicolson integration"
duplicate i=1,(N-1)
  f[i+1]=f[i]+(dfdeta[i]+dfdeta[i+1])*DELTAeta/2
  dfdeta[i+1]=dfdeta[i]+(d2fdeta2[i]+d2fdeta2[i+1])*DELTAeta/2
  d2fdeta2[i+1]=d2fdeta2[i]-(2*f[i]*d2fdeta2[i]+2*f[i+1]*d2fdeta2[i+1])*DELTAeta/2
end
```

Figure 4-11 illustrates the dimensionless velocity distribution ($\tilde{u} = \frac{df}{d\eta}$, according to Eq. (4-216)) as a function of η and shows that the initial guess $\left.\frac{d^2f}{d\eta^2}\right|_{\eta=0} = 0.3$ is not appropriate because the solution does not satisfy the boundary condition given by Eq. (4-238), $\left.\frac{df}{d\eta}\right|_{\eta\to\infty} = 1$. The solutions for various values of $\left.\frac{d^2f}{d\eta^2}\right|_{\eta=0}$ are also shown in Figure 4-11 and it is clear that a value of $\left.\frac{d^2f}{d\eta^2}\right|_{\eta=0}$ that is between 0.6 and 0.7 will satisfy this boundary condition. The process of determining the correct value of $\left.\frac{d^2f}{d\eta^2}\right|_{\eta=0}$ can be automated by commenting out the guess for variable d2fdeta2[1] and replacing it with the requirement that $\left.\frac{df}{d\eta}\right|_{\eta\to\infty} = 1$ by specifying that dfdeta[N] = 1.0.

Table 4-1: Self-similar solution for laminar flow over a flat plate.

$\eta = \dfrac{y}{2}\sqrt{\dfrac{u_\infty}{\nu x}}$	$f = \dfrac{\dot{V}}{2\,W\,\sqrt{u_\infty\,\nu\,x}}$	$\dfrac{df}{d\eta} = \dfrac{u}{u_\infty}$	$\dfrac{d^2f}{d\eta^2}$
0.0	0.0000	0.0000	0.6640
0.2	0.0133	0.1327	0.6627
0.4	0.0530	0.2645	0.6544
0.6	0.1189	0.3934	0.6327
0.8	0.2099	0.5162	0.5929
1.0	0.3246	0.6291	0.5332
1.2	0.4605	0.7282	0.4562
1.4	0.6147	0.8107	0.3683
1.6	0.7835	0.8754	0.2787
1.8	0.9635	0.9228	0.1967
2.0	1.1515	0.9552	0.1289
2.2	1.3447	0.9757	0.0783
2.4	1.5411	0.9877	0.0439
2.458	**1.6000**	**0.9900**	**0.0365**
2.6	1.7394	0.9942	0.0227
2.8	1.9386	0.9975	0.0108
3.0	2.1382	0.9990	0.0048

```
{d2fdeta2[1]=0.3 [-]   "d2fdeta2 at eta = 0, this is a guess"}
dfdeta[N]=1.0          "specify that the free stream velocity is reached"
```

Solving the problem shows that the correct value of the third boundary condition is 0.664 (i.e., this is the value of the variable d2fdeta2[1] that leads to dfdeta[N]=1). The dimensionless velocity distribution for $\left.\dfrac{d^2f}{d\eta^2}\right|_{\eta=0} = 0.664$ is shown in Figure 4-11. The self-similar solution is summarized in Table 4-1.

The velocity boundary layer thickness defined based on achieving 99% of the free-stream velocity ($\delta_{m,99\%}$) can be obtained from the self-similar solution. Examining Table 4-1 shows that the dimensionless velocity reaches 0.99 at $\eta = 2.458$. Substituting $\eta = 2.458$ into Eq. (4-197) leads to:

$$2.458 = \frac{\delta_{m,99\%}}{2}\sqrt{\frac{u_\infty}{\nu x}} \tag{4-249}$$

or

$$\delta_{m,99\%} = \frac{4.916}{\sqrt{\dfrac{u_\infty}{\nu x}}} \tag{4-250}$$

which can be rearranged to provide:

$$\boxed{\frac{\delta_{m,99\%}}{x} = \frac{4.916}{\sqrt{Re_x}}} \tag{4-251}$$

The shear at the surface of the plate is:

$$\tau_s = \mu\left.\frac{\partial u}{\partial y}\right|_{y=0} \tag{4-252}$$

Substituting Eq. (4-228) into Eq. (4-252) leads to:

$$\tau_s = \mu \frac{u_\infty}{2} \sqrt{\frac{u_\infty}{\upsilon x}} \frac{d^2 f}{d\eta^2}\bigg|_{\eta=0} \tag{4-253}$$

or

$$\tau_s = 0.332 \, \mu \, u_\infty \sqrt{\frac{u_\infty}{\upsilon x}} \tag{4-254}$$

The local friction coefficient is defined as:

$$C_f = \frac{2 \, \tau_s}{\rho \, u_\infty^2} \tag{4-255}$$

Substituting Eq. (4-254) into Eq. (4-255) leads to:

$$C_f = \frac{2}{\rho \, u_\infty^2} 0.332 \, \mu \, u_\infty \sqrt{\frac{u_\infty}{\upsilon x}} \tag{4-256}$$

or

$$\boxed{C_f = \frac{0.664}{\sqrt{Re_x}}} \tag{4-257}$$

It is interesting to use the self-similar solution in order to determine the y-velocity in the boundary layer; Eq. (4-221) indicates that an appropriately scaled dimensionless y-velocity can be defined according to:

$$\frac{\upsilon}{\sqrt{\dfrac{u_\infty \upsilon}{x}}} = \eta \frac{df}{d\eta} - f \tag{4-258}$$

The dimensionless y-velocity defined in Eq. (4-258) is evaluated in EES:

```
v_bar[1]=0                              "y-velocity at plate"
duplicate i=1,(N-1)
   v_bar[i+1]=eta[i+1]*dfdeta[i+1]-f[i+1]   "y-velocity"
end
```

Figure 4-12 illustrates the dimensionless y-velocity as a function of dimensionless position. Note that the y-velocity is nonzero away from the wall. Therefore, energy is carried toward the free stream by the fluid flow. It is this additional energy transfer that prevents the convective heat transfer problem from being a simple conduction problem. Also notice that the y-velocity does not become 0 as η becomes large because the boundary layer is growing and therefore pushing the free stream away from the plate. Figure 4-12 shows that the y-velocity in the free stream will be:

$$\boxed{\upsilon_{y\to\infty} = 0.862 \sqrt{\frac{u_\infty \upsilon}{x}}} \tag{4-259}$$

The y-velocity at the edge of the boundary layer is related to the rate at which the boundary layer is growing. A result that is close to the exact answer provided by Eq. (4-259) can be obtained by taking the time derivative of the boundary layer thickness as the fluid

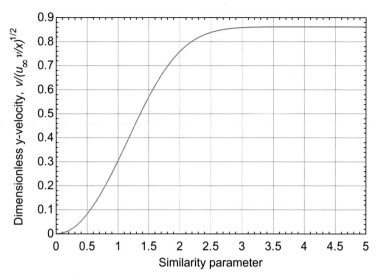

Figure 4-12: Dimensionless y-velocity as a function of dimensionless position.

moves along the plate:

$$v_{y \to \infty} \approx \frac{d\delta_m}{dt} \tag{4-260}$$

Substituting Eq. (4-193) into Eq. (4-260) leads to:

$$v_{y \to \infty} \approx \frac{d}{dt}\left[2\sqrt{\upsilon t}\right] \tag{4-261}$$

or

$$v_{y \to \infty} \approx \sqrt{\frac{\upsilon}{t}} \tag{4-262}$$

The time in Eq. (4-262) is the ratio of the distance from the leading edge to the free stream velocity and so:

$$v_{y \to \infty} \approx \sqrt{\frac{u_\infty \upsilon}{x}} \tag{4-263}$$

which is very close to the exact solution provided by Eq. (4-259). The similarity of Eqs. (4-259) and (4-263) should reinforce the idea that the value of v far from the plate is essentially equivalent to the rate at which the boundary layer is growing.

4.4.3 The Temperature Solution

In this section, the self-similar velocity distribution is used in the thermal energy equation in order to obtain an exact solution for the temperature distribution.

The Problem Statement
The steady-state thermal energy equation, simplified for application within the boundary layer, is derived in Section 4.2.3. If viscous dissipation is neglected, then the thermal energy equation is:

$$u\frac{\partial T}{\partial x} + v\frac{\partial T}{\partial y} = \alpha\frac{\partial^2 T}{\partial y^2} \tag{4-264}$$

The plate temperature (T_s) must be obtained at the plate surface:

$$T_{y=0} = T_s \tag{4-265}$$

The free stream temperature is recovered far from the plate:

$$T_{y\to\infty} = T_\infty \tag{4-266}$$

At the leading edge of the plate, the temperature of the fluid approaching the plate is:

$$T_{x=0} = T_\infty \tag{4-267}$$

The Similarity Variables

The dimensionless temperature difference is defined as in Section 4.3.2:

$$\tilde{\theta} = \frac{T - T_s}{T_\infty - T_s} \tag{4-268}$$

Substituting Eq. (4-268) into Eq. (4-264) leads to:

$$u\frac{\partial\tilde{\theta}}{\partial x} + v\frac{\partial\tilde{\theta}}{\partial y} = \alpha\frac{\partial^2\tilde{\theta}}{\partial y^2} \tag{4-269}$$

Note that if $\alpha = \upsilon$ (i.e., if $Pr = 1.0$) then Eq. (4-269) for $\tilde{\theta}$ is the same as Eq. (4-188) for u and we would expect that the solutions would be the same; this is the Reynolds analogy. Even if Pr is not equal to unity, the same similarity parameter (η, defined in Eq. (4-197)) can be used to transform the problem provided that the velocity and temperature boundary layers both develop from the leading edge of the plate so that the ratio of the momentum and thermal boundary layer thickness is constant (and related to the value of the Prandtl number). In this case, the dimensionless temperature difference $\tilde{\theta}$ at any location x will collapse when plotted against either y/δ_m or y/δ_t. It is convenient to continue to use $\eta = y/\delta_m$ so that the solutions for u and v obtained in Section 4.4.2 can be used directly in Eq. (4-269).

The boundary conditions, Eqs. (4-265) through (4-267), can be expressed in terms of $\tilde{\theta}$:

$$\tilde{\theta}_{\eta=0} = 0 \tag{4-270}$$

$$\tilde{\theta}_{\eta\to\infty} = 1 \tag{4-271}$$

The Problem Transformation

The first derivative of $\tilde{\theta}$ with respect to x is obtained using the chain rule, recognizing that $\tilde{\theta}$ is a function only of η, which is a function of x and y:

$$\frac{\partial}{\partial x}\left[\tilde{\theta}\left(\eta\left(x, y\right)\right)\right] = \frac{d\tilde{\theta}}{d\eta}\left(\frac{\partial\eta}{\partial x}\right)_y \tag{4-272}$$

or

$$\frac{\partial\tilde{\theta}}{\partial x} = \frac{d\tilde{\theta}}{d\eta}\frac{\partial}{\partial x}\left[\frac{y}{2}\sqrt{\frac{u_\infty}{\upsilon x}}\right]_y = -\left[\frac{y}{4x^{3/2}}\sqrt{\frac{u_\infty}{\upsilon}}\right]\frac{d\tilde{\theta}}{d\eta} \tag{4-273}$$

Equation (4-273) can be expressed as:

$$\boxed{\frac{\partial\tilde{\theta}}{\partial x} = -\frac{\eta}{2x}\frac{d\tilde{\theta}}{d\eta}} \tag{4-274}$$

An equivalent expression can be obtained from Maple:

```
> dthetadx:=diff(theta(eta(x,y)),x);
```

$$dthetadx := -\frac{1}{4}\frac{D(\theta)\left(\dfrac{y\sqrt{\dfrac{u\_infinity}{v\,x}}}{2}\right)y\,u\_infinity}{\sqrt{\dfrac{u\_infinity}{v\,x}}\,v\,x^2}$$

The first derivative of $\tilde\theta$ with respect to y is obtained in a similar manner:

$$\frac{\partial\tilde\theta}{\partial y} = \frac{d\tilde\theta}{d\eta}\left(\frac{\partial\eta}{\partial y}\right)_x \tag{4-275}$$

or

$$\boxed{\frac{\partial\tilde\theta}{\partial y} = \frac{d\tilde\theta}{d\eta}\frac{1}{2}\sqrt{\frac{u_\infty}{vx}}} \tag{4-276}$$

An equivalent expression can be obtained using Maple:

```
> dthetady:=diff(theta(eta(x,y)),y);
```

$$dthetady := \frac{1}{2}D(\theta)\left(\dfrac{y\sqrt{\dfrac{u\_infinity}{v\,x}}}{2}\right)\sqrt{\dfrac{u\_infinity}{v\,x}}$$

The second derivative of $\tilde\theta$ with respect to y is:

$$\frac{\partial^2\tilde\theta}{\partial y^2} = \frac{\partial}{\partial y}\left(\frac{\partial\tilde\theta}{\partial y}\right)_x = \frac{\partial}{\partial\eta}\left[\frac{\partial\tilde\theta}{\partial y}\right]\left(\frac{\partial\eta}{\partial y}\right)_x \tag{4-277}$$

Substituting Eq. (4-276) into Eq. (4-277) leads to:

$$\frac{\partial^2\tilde\theta}{\partial y^2} = \frac{\partial}{\partial\eta}\left[\frac{d\tilde\theta}{d\eta}\frac{1}{2}\sqrt{\frac{u_\infty}{vx}}\right]\left(\frac{\partial\eta}{\partial y}\right)_x \tag{4-278}$$

or

$$\boxed{\frac{\partial^2\tilde\theta}{\partial y^2} = \frac{\partial^2\tilde\theta}{\partial\eta^2}\frac{u_\infty}{4vx}} \tag{4-279}$$

An equivalent expression can be obtained using Maple:

```
> d2thetady2:=diff(diff(theta(eta(x,y)),y),y);
```

$$d2thetady2 := \frac{1}{4}\frac{(D^{(2)})(\theta)\left(y\dfrac{\sqrt{\dfrac{u\_infinity}{v\,x}}}{2}\right)u\_infinity}{v\,x}$$

Since it is assumed that properties are not temperature dependent, the solutions for u and v in terms of the similarity variables, Eqs. (4-216) and (4-221), remain valid as temperature changes. These solutions are substituted into the energy equation, Eq. (4-269), together with Eqs. (4-274), (4-276), and (4-279), in order to obtain:

$$
\underbrace{u_\infty \frac{df}{d\eta}}_{u} \underbrace{\left[-\frac{\eta}{2x} \frac{d\tilde{\theta}}{d\eta} \right]}_{\frac{\partial\tilde{\theta}}{\partial x}} + \underbrace{\sqrt{\frac{u_\infty \upsilon}{x}} \left(\eta \frac{df}{d\eta} - f \right) \frac{d\tilde{\theta}}{d\eta}}_{v} \underbrace{\frac{1}{2}\sqrt{\frac{u_\infty}{\upsilon x}}}_{\frac{\partial\tilde{\theta}}{\partial y}} = \alpha \underbrace{\frac{d^2\tilde{\theta}}{d\eta^2} \frac{u_\infty}{4\upsilon x}}_{\frac{\partial^2\tilde{\theta}}{\partial y^2}} \tag{4-280}
$$

Notice that Eq. (4-280) can be divided through by u_∞/x in order to make it dimensionless:

$$
-\frac{\eta}{2} \frac{d\tilde{\theta}}{d\eta} \frac{df}{d\eta} + \frac{\eta}{2} \frac{d\tilde{\theta}}{d\eta} \frac{df}{d\eta} - \frac{f}{2} \frac{d\tilde{\theta}}{d\eta} = \frac{1}{4\,Pr} \frac{d^2\tilde{\theta}}{d\eta^2} \tag{4-281}
$$

Equation (4-281) can be simplified in order to obtain the governing ordinary differential equation for $\tilde{\theta}$:

$$
\boxed{\frac{d^2\tilde{\theta}}{d\eta^2} + 2f\,Pr\,\frac{d\tilde{\theta}}{d\eta} = 0} \tag{4-282}
$$

An equivalent expression can be obtained using Maple:

```
> simplify(u*dthetadx+v*dthetady=alpha*d2thetady2);
```

$$
-\frac{1}{2} \frac{D(\theta)\left(y\sqrt{\dfrac{u\_infinity}{\upsilon x}{2}} \right) u\_infinity \sqrt{\dfrac{u\_infinity}{\upsilon x}} f\left(y\sqrt{\dfrac{u\_infinity}{\upsilon x}{2}} \right) v}{\sqrt{u\_infinity\,\upsilon x}}
$$

$$
= \frac{1}{4} \frac{\alpha(D^{(2)})(\theta)\left(y\sqrt{\dfrac{u\_infinity}{\upsilon x}{2}} \right) u\_infinity}{\upsilon x}
$$

Numerical Solution

The function f was obtained previously from the Blasius solution and reflects the influence of the velocity distribution on the temperature distribution in terms of η. Therefore, Eq. (4-282) is an ordinary differential equation for $\tilde{\theta}$ that can be solved numerically. The boundary conditions for Eq. (4-282) are specified at $\eta = 0$ and $\eta \to \infty$ and therefore this solution will also require a shooting method. The problem is solved here using the same approach discussed in Section 4.4.2. The additional EES code that is required is appended to the original file containing the Blasius solution.

Notice that Eq. (4-282) includes the Prandtl number, which must be specified in order to obtain the dimensionless temperature distribution:

```
"Temperature solution"
Pr=0.7 [-]   "Prandtl number"
```

The computational domain continues to be $0 < \eta < \eta_\infty$. However, it is important to note that it is no longer sufficient for η_∞ to be simply much greater than unity. The edge of the computational domain should be substantially greater than both δ_m and δ_t in order for the second boundary condition, Eq. (4-271), to be enforced. The dimensionless position η was defined as y/δ_m and therefore $\eta \gg 1$ ensures that $y \gg \delta_m$ but it does not guarantee that we have passed beyond the thermal boundary layer. Fluids with a very low Prandtl number may have $\delta_t \gg \delta_m$. Therefore it is important to adjust the value of η_∞ in order to ensure that $y \gg \delta_t$ and $y \gg \delta_m$ at the outer edge of the computational domain. Based on the conceptual model of laminar boundary layers discussed in Section 4.1, we expect that:

$$\frac{\delta_t}{\delta_m} \approx \frac{1}{\sqrt{Pr}} \tag{4-283}$$

and therefore:

$$\eta \approx \frac{y}{\delta_m} \frac{\delta_t}{\delta_t} \approx \frac{y}{\delta_t} \frac{1}{\sqrt{Pr}} \tag{4-284}$$

The outer edge of the computational domain will be selected in order to ensure that both $\frac{y}{\delta_m} = \eta \gg 1$ and $\frac{y}{\delta_t} = \eta \sqrt{Pr} \gg 1$:

$$\eta_\infty = \mathrm{Max}\left(10, \frac{10}{\sqrt{Pr}}\right) \tag{4-285}$$

This assignment is accomplished in EES using the Max command, which returns the maximum of the arguments:

```
{eta_infinity=10 [-]}                "outer edge of the computational domain"
eta_infinity=Max(10,10/sqrt(Pr))     "outer edge of computational domain"
```

The nodes are setup as in Section 4.4.2. The state variables are $\tilde{\theta}, \frac{d\tilde{\theta}}{d\eta}$. The rate of change of the first state variable is obtained from:

$$\frac{d}{d\eta}[\tilde{\theta}] = \frac{d\tilde{\theta}}{d\eta} \tag{4-286}$$

The rate of change of the second state variable is obtained from the governing ordinary differential equation, Eq. (4-282):

$$\frac{d}{d\eta}\left[\frac{d\tilde{\theta}}{d\eta}\right] = \frac{d^2\tilde{\theta}}{d\eta^2} = -2f\,Pr\,\frac{d\tilde{\theta}}{d\eta} \tag{4-287}$$

The initial conditions for the integration process are specified. Note that the value of $\frac{d\tilde{\theta}}{d\eta}\Big|_{\eta=0}$ (the value of variable dthetadeta[1]) is assumed, and will be adjusted so that the boundary condition at large η is met.

```
theta[1]=0 [-]            "theta at eta = 0"
dthetadeta[1]=0.5 [-]     "dthetadeta at eta = 0, this is a guess"
```

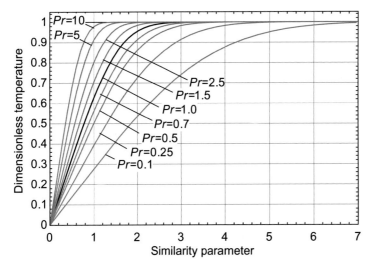

Figure 4-13: Dimensionless temperature difference as a function of η for various values of Pr.

The Crank-Nicolson technique is used to integrate the state equations from $\eta = 0$ to $\eta = \eta_\infty$:

$$\tilde{\theta}_{i+1} = \tilde{\theta}_i + \left[\left. \frac{d\tilde{\theta}}{d\eta} \right|_i + \left. \frac{d\tilde{\theta}}{d\eta} \right|_{i+1} \right] \frac{\Delta\eta}{2} \quad \text{for } i = 1..(N-1) \qquad (4\text{-}288)$$

$$\left. \frac{d\tilde{\theta}}{d\eta} \right|_{i+1} = \left. \frac{d\tilde{\theta}}{d\eta} \right|_i + \left[-2 f_i\, Pr \left. \frac{d\tilde{\theta}}{d\eta} \right|_i - 2 f_{i+1}\, Pr \left. \frac{d\tilde{\theta}}{d\eta} \right|_{i+1} \right] \frac{\Delta\eta}{2} \quad \text{for } i = 1..(N-1) \quad (4\text{-}289)$$

```
"Crank-Nicolson integration"
duplicate i=1,(N-1)
  theta[i+1]=theta[i]+(dthetadeta[i]+dthetadeta[i+1])*DELTAeta/2
  dthetadeta[i+1]=dthetadeta[i]+(-2*f[i]*Pr*dthetadeta[i]-2*f[i+1]*Pr*dthetadeta[i+1])*DELTAeta/2
end
```

The solution is obtained using a reasonable value of **dthetadeta[1]** and then the guess values are updated. The assumed value for $\left. \frac{d\tilde{\theta}}{d\eta} \right|_{\eta=0}$ is commented out and replaced with the boundary condition, Eq. (4-271):

```
{dthetadeta[1]=0.5 [-]   "this is a guess"}
theta[N]=1
```

Figure 4-13 illustrates dimensionless temperature difference as a function of η for various values of the Prandtl number. Note that the solution for $Pr = 1.0$ is identical to the solution for the dimensionless velocity shown in Figure 4-13 because the Reynolds analogy holds in this limit. As the Prandtl number is reduced, the temperature distribution extends further from the wall.

The local Nusselt number is defined as:

$$Nu_x = \frac{h\,x}{k} \tag{4-290}$$

where the heat transfer coefficient is the ratio of the local heat flux to the plate-to-free stream temperature difference:

$$Nu_x = \frac{\dot{q}_s''\, x}{k\,(T_s - T_\infty)} \tag{4-291}$$

The heat flux is related to the temperature gradient in the fluid at $y = 0$ according to:

$$Nu_x = \frac{-k\left(\dfrac{\partial T}{\partial y}\right)_{y=0} x}{k\,(T_s - T_\infty)} \tag{4-292}$$

According to Eq. (4-268), the partial derivative of temperature with respect to y is:

$$\frac{\partial T}{\partial y} = (T_\infty - T_s)\,\frac{\partial \tilde{\theta}}{\partial y} \tag{4-293}$$

Substituting Eq. (4-293) into Eq. (4-292) leads to:

$$Nu_x = x\left(\frac{\partial \tilde{\theta}}{\partial y}\right)_{y=0} \tag{4-294}$$

Substituting Eq. (4-276) into Eq. (4-294) leads to:

$$Nu_x = \frac{1}{2}\sqrt{\frac{u_\infty\, x}{\upsilon}}\,\left.\frac{d\tilde{\theta}}{d\eta}\right|_{\eta=0} \tag{4-295}$$

or

$$Nu_x = \sqrt{Re_x}\,\underbrace{\frac{1}{2}\left.\frac{d\tilde{\theta}}{d\eta}\right|_{\eta=0}}_{\text{function of } Pr} \tag{4-296}$$

The quantity $\frac{1}{2}\left.\frac{d\tilde{\theta}}{d\eta}\right|_{\eta=0}$ in Eq. (4-296) is a function only of the Prandtl number and is obtained from the self-similar solution for the temperature distribution. Figure 4-13 shows that $\frac{1}{2}\left.\frac{d\tilde{\theta}}{d\eta}\right|_{\eta=0}$ is an increasing function of Pr. The specified value for the variable Pr is commented out in the Equations Window and a parametric table is generated that includes $\frac{1}{2}\left.\frac{d\tilde{\theta}}{d\eta}\right|_{\eta=0}$ and Pr. The value of $\frac{1}{2}\left.\frac{d\tilde{\theta}}{d\eta}\right|_{\eta=0}$ as a function of Pr is shown in Figure 4-14. Also shown in Figure 4-14 is the curve fit to this solution that has been suggested by Churchill and Ozoe (1973):

$$\frac{1}{2}\left.\frac{d\tilde{\theta}}{d\eta}\right|_{\eta=0} = \frac{0.3387\,Pr^{1/3}}{\left[1 + \left(\dfrac{0.0468}{Pr}\right)^{2/3}\right]^{1/4}} \tag{4-297}$$

For $Pr > 0.6$, a simpler curve fit can be used:

$$\frac{1}{2}\left.\frac{d\tilde{\theta}}{d\eta}\right|_{\eta=0} = 0.332\,Pr^{1/3} \tag{4-298}$$

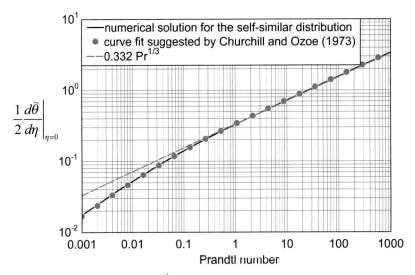

Figure 4-14: The quantity $\frac{1}{2}\frac{d\tilde{\theta}}{d\eta}\Big|_{\eta=0}$ as a function of Pr provided by the numerical solution for the self-similar temperature distribution. Also shown are curve fits to the solution.

4.4.4 The Falkner-Skan Transformation

This extended section of the book can be found on the website www.cambridge.org/nellisandklein. The self-similar solutions provided in Sections 4.4.2 and 4.4.3 do not allow the consideration of any variation in the free-stream velocity or surface temperature. The Falkner-Skan transformation can be used to develop self-similar solutions to a class of problems where the velocity and temperature vary according to a power law. This solution is useful for a number of interesting problems, including flow over a wedge and flow over a plate with a constant heat flux.

4.5 Turbulent Boundary Layer Concepts

4.5.1 Introduction

Section 4.1 demonstrated that the existence of a laminar boundary layer is the result of the diffusive transport of momentum and energy into the free stream. In Section 4.4, this concept was used to identify a similarity solution to the partial differential equations that were derived in Section 4.2 for laminar flow in a boundary layer. Laminar flow will only persist when the action of viscosity is sufficiently strong relative to inertial forces. In Section 4.3, the ratio of inertial to viscous forces is defined as the Reynolds number. For flow over a flat plate (Figure 4-15), the Reynolds number is defined as:

$$Re_x = \frac{\rho\, u_\infty\, x}{\mu} \qquad (4\text{-}299)$$

Flow over a flat plate will transition from laminar to turbulent at a location on the plate (x_{crit}) that is defined by a critical Reynolds number, $Re_{x,crit}$.

$$x_{crit} = \frac{Re_{x,crit}\,\mu}{\rho\, u_\infty} \qquad (4\text{-}300)$$

In fact, the transition to turbulent flow is neither sudden nor precise. The conditions under which the flow will transition to turbulence depend on the shape of the leading

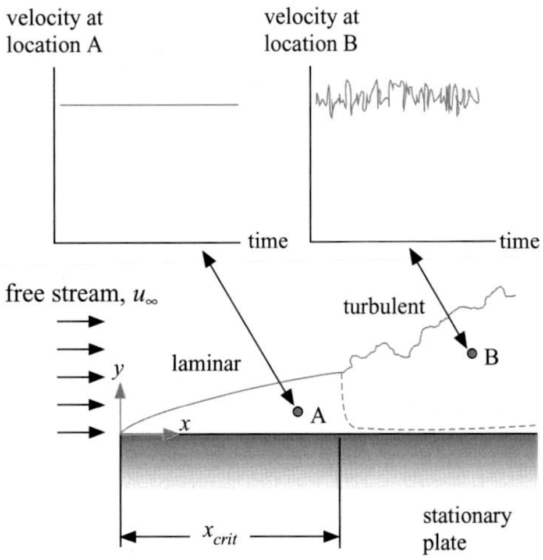

Figure 4-15: Flow over a flat plate transitioning from laminar to turbulent conditions.

edge, the presence of destabilizing oscillations (e.g., structural vibrations), the roughness of the surface, the character of the free stream, etc. The flow will not transition from completely laminar to completely turbulent flow at a particular location. Instead, there will be a region that is characterized by local, intermittent bursts of turbulence that progressively become more intense as the flow moves downstream; eventually the flow becomes a fully turbulent boundary layer. The critical Reynolds number will typically lie between 3×10^5 and 6×10^5, although in well-controlled experiments the critical Reynolds number can be as high as 4×10^6.

The characteristics of a turbulent flow are discussed in this section so that the differences between a laminar and a turbulent flow are clear. In Sections 4.6 and 4.7, turbulent flows are treated somewhat more rigorously; however, a complete discussion of turbulent flow is beyond the scope of this book. There are several good references that provide a more thorough presentation of turbulent heat transfer; for example, Tennekes and Lumley (1972), Hinze (1975), and Schlichting (2000).

4.5.2 A Conceptual Model of the Turbulent Boundary Layer

A laminar boundary layer is highly ordered. The trajectories of fluid particles (streamlines) are smooth and an instrument placed in a steady laminar flow will report a constant value of velocity, temperature, etc., as shown in Figure 4-15 (for location A). A turbulent boundary layer is characterized by vortices or eddies that randomly exist at many time and length scales. Streamlines cannot be easily identified because fluid particles, while always moving on average in the x-direction with the mean flow, will also move randomly up and down in the y-direction under the influence of turbulent eddies. The fluid particles may even move in the opposite direction of the mean flow in some locations. An instrument placed in a turbulent flow will report an oscillating value of velocity, temperature, etc., as shown in Figure 4-15 (for location B), even if the overall behavior of the problem is otherwise at steady-state. Turbulent eddies exist throughout the flow except in a region very near the wall that is referred to as the viscous sublayer.

In laminar flow, the transport of momentum and energy in the y-direction (i.e., perpendicular to the free stream flow) occurs primarily by diffusion at the molecular level because the y-directed velocity is quite small. Molecules of liquid with high energy collide with those at lower energy and this interaction leads to a diffusive transfer process. Thermal conductivity and viscosity are fluid properties because they characterize this

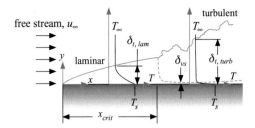

Figure 4-16: The temperature distribution in the laminar and turbulent boundary layers.

micro-scale process; fluid at a given state will always possess the same conductivity and viscosity.

In a turbulent flow, the transport of momentum and energy occurs due to the same micro-scale diffusive processes that are responsible for viscous shear and conduction heat transfer. However, the lateral motion of macro-scale packets of fluid under the influence of the turbulent eddies provides an additional and very efficient mechanism for transporting momentum and energy. Much more energy is transported when a large packet of fluid is moved away from the wall by a turbulent eddy than would be transported by conduction alone in the same amount of time. The presence of the turbulent eddies increases the effective viscosity and conductivity of the fluid tremendously everywhere that these eddies exist; that is, everywhere except in the viscous sublayer. The turbulent eddies are suppressed in the viscous sublayer by the viscous action of the wall itself and therefore the flow in the viscous sublayer behaves as if it were laminar. The thickness of the viscous sublayer (δ_{vs}) is, however, much thinner than the thickness of a laminar boundary layer under comparable conditions.

This description of a turbulent boundary layer is sufficient to provide a conceptual, albeit qualitative, model of a turbulent flow that explains many of its characteristics. Figure 4-16 illustrates a sketch of the temperature distribution that exists within the laminar and turbulent boundary layers. The temperature varies from T_s at the plate surface ($y = 0$) to T_∞ at the edge of the thermal boundary layer ($y = \delta_t$). The temperature distribution in the laminar boundary layer is quite different than in the turbulent boundary layer. While the temperature distribution in the laminar boundary layer is not linear, it does change smoothly and gradually over the entire extent of the boundary layer. This behavior is consistent with the conceptual model of the laminar boundary layer discussed in Section 4.1, in which heat transfer is treated, approximately, as conduction through a layer of fluid with thickness $\delta_{t,lam}$ (the laminar boundary layer thickness). The heat flux through a laminar boundary ($\dot{q}''_{lam}$) can be expressed approximately as:

$$\dot{q}''_{lam} \approx \frac{(T_s - T_\infty)}{\left(\dfrac{\delta_{t,lam}}{k}\right)} = h_{lam}\,(T_s - T_\infty) \qquad (4\text{-}301)$$

where k is the conductivity of the fluid. The area-specific thermal resistance of a laminar boundary layer is, approximately, $\delta_{t,lam}/k$, and therefore the heat transfer coefficient for a laminar flow (h_{lam}) is, approximately:

$$h_{lam} \approx \frac{k}{\delta_{t,lam}} \qquad (4\text{-}302)$$

Energy that is transported through a turbulent boundary layer to the free stream must pass through two regions of fluid. The viscous sublayer is a very thin layer of fluid with no turbulent eddies. Therefore, the area-specific thermal resistance of the viscous sublayer

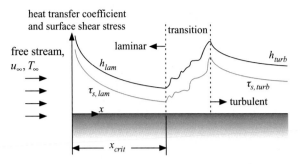

Figure 4-17: Qualitative variation of the heat transfer coefficient and shear stress along a flat plate.

is, approximately, δ_{vs}/k. The bulk of the boundary layer (the turbulent core) is characterized by large turbulent eddies and therefore has a large, effective thermal conductivity (k_{turb}). The area-specific thermal resistance of the turbulent core is $\delta_{t,turb}/k_{turb}$, where $\delta_{t,turb}$ is the turbulent boundary layer thickness. It is important to recognize that k_{turb} is not a fluid property, but rather a property of the flow itself; k_{turb} is sometimes referred to as the eddy conductivity of the fluid. The heat flux through the turbulent boundary layer ($\dot{q}''_{turb}$) can be expressed, approximately, as:

$$\dot{q}''_{turb} \approx \frac{(T_s - T_\infty)}{\dfrac{\delta_{vs}}{k} + \dfrac{\delta_{t,turb}}{k_{turb}}} = h_{turb}(T_s - T_\infty) \tag{4-303}$$

A turbulent boundary layer therefore behaves somewhat like a composite wall consisting of a thin, low conductivity material (the viscous sublayer) beneath a thicker, but extremely high conductivity material (the turbulent core). The thermal resistance of the viscous sublayer is much larger than that of the turbulent core (i.e., the first term in the denominator of Eq. (4-303) is much larger than the second term). In such a series combination of thermal resistances, the largest resistance will dominate the problem. Therefore, the majority of the temperature change will occur across the viscous sublayer, as shown in the sketch of the temperature distribution in Figure 4-16.

The characteristics of the viscous sublayer will dictate the behavior of a turbulent boundary layer. According to Eq. (4-303), the heat transfer coefficient for a turbulent boundary layer can be written approximately as:

$$h_{turb} \approx \frac{1}{\dfrac{\delta_{vs}}{k} + \dfrac{\delta_{t,turb}}{k_{turb}}} \tag{4-304}$$

Because the thermal resistance of the viscous sublayer is much larger than that of the turbulent core, Eq. (4-304) can be approximated by:

$$h_{turb} \approx \frac{k}{\delta_{vs}} \tag{4-305}$$

Comparing Eqs. (4-302) and (4-305) and recalling that $\delta_{vs} \ll \delta_{t,turb}$ suggests that $h_{turb} \gg h_{lam}$. Figure 4-17 illustrates, qualitatively, the variation of the heat transfer coefficient and shear stress with distance along the plate. There is a transition region between a fully laminar and fully turbulent boundary layer where the variation of heat transfer coefficient is not clear. However, the heat transfer coefficient will increase substantially when the flow becomes turbulent.

The shear stress at the surface of the plate in the laminar boundary layer ($\tau_{s,lam}$) is the product of the fluid viscosity and the velocity gradient at the plate surface and is given, approximately, by:

$$\tau_{s,lam} = \mu \left. \frac{\partial u}{\partial y} \right|_{y=0} \approx \mu \frac{u_\infty}{\delta_{m,lam}} \tag{4-306}$$

where $\delta_{m,lam}$ is the momentum boundary layer thickness in a laminar flow. The shear stress at the surface of the plate in the turbulent boundary layer ($\tau_{s,turb}$) is also equal to the product of the molecular viscosity and the velocity gradient at the plate. However, in a turbulent flow the velocity gradient will be much steeper as the velocity change will occur primarily across the viscous sublayer (see Figure 4-16); the shear stress at the plate surface in a turbulent flow is given, approximately, by:

$$\tau_{s,turb} = \mu \left. \frac{\partial u}{\partial y} \right|_{y=0} \approx \mu \frac{u_\infty}{\delta_{vs}} \tag{4-307}$$

Comparing Eqs. (4-306) and (4-307) suggests that the shear stress will increase substantially when the flow transitions from laminar to turbulent, as shown in Figure 4-17.

Although the viscous sublayer is very thin, the thickness of a turbulent boundary layer will be substantially larger than thickness of a laminar boundary layer. Recall that the thermal boundary layer grows due to the penetration of energy into the free stream. In a laminar boundary layer, the thermal boundary layer grows approximately according to:

$$\delta_{t,lam} \approx 2\sqrt{\alpha t} = 2\sqrt{\frac{k\,x}{\rho\,c\,u_\infty}} \tag{4-308}$$

and the momentum boundary layer grows according to:

$$\delta_{m,lam} \approx 2\sqrt{\upsilon t} = 2\sqrt{\frac{\mu\,x}{\rho\,u_\infty}} \tag{4-309}$$

The ratio of these laminar boundary layer thicknesses is equal, approximately, to the square root of the fluid Prandtl number:

$$\frac{\delta_{m,lam}}{\delta_{t,lam}} \approx \frac{2\sqrt{\upsilon t}}{2\sqrt{\alpha t}} = \sqrt{\frac{\upsilon}{\alpha}} = \sqrt{Pr} \tag{4-310}$$

A turbulent flow has a higher effective thermal conductivity (k_{turb}) and viscosity (μ_{turb}) and so it is reasonable to expect that it will grow faster and be much thicker than a laminar boundary layer. Figure 4-18 illustrates the qualitative variation of the thermal and momentum boundary layer along the plate for a substance with a Prandtl number that is greater than unity. Notice that the momentum boundary layer grows faster than the thermal boundary layer when the flow is laminar, but the momentum and thermal boundary layers grow at about the same rate for turbulent flow.

A turbulent thermal boundary layer will grow according to:

$$\delta_{t,turb} \approx 2\sqrt{\frac{k_{turb}\,x}{\rho\,c\,u_\infty}} \tag{4-311}$$

and a turbulent momentum boundary layer will grow according to:

$$\delta_{m,turb} \approx 2\sqrt{\frac{\mu_{turb}\,x}{\rho\,u_\infty}} \tag{4-312}$$

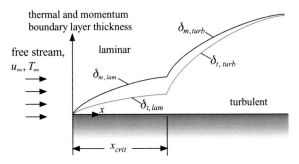

Figure 4-18: Qualitative variation of the thermal and momentum boundary layer thickness for a fluid with $Pr > 1$.

The ratio of these turbulent boundary layer thicknesses is:

$$\frac{\delta_{m,turb}}{\delta_{t,turb}} \approx 2\sqrt{\frac{\mu_{turb}\,x}{\rho\,u_\infty}}\,\frac{1}{2}\sqrt{\frac{\rho\,c\,u_\infty}{k_{turb}\,x}} = \sqrt{\underbrace{\frac{\mu_{turb}}{\rho}}_{\upsilon_{turb}}\,\underbrace{\frac{\rho\,c}{k_{turb}}}_{1/\alpha_{turb}}} \qquad (4\text{-}313)$$

An effective turbulent thermal diffusivity and kinematic viscosity (α_{turb} and υ_{turb}) are defined:

$$\alpha_{turb} = \frac{k_{turb}}{\rho\,c} \qquad (4\text{-}314)$$

$$\upsilon_{turb} = \frac{\mu_{turb}}{\rho} \qquad (4\text{-}315)$$

Substituting Eqs. (4-314) and (4-315) into Eq. (4-313) leads to:

$$\frac{\delta_{m,turb}}{\delta_{t,turb}} \approx \sqrt{\underbrace{\frac{\upsilon_{turb}}{\alpha_{turb}}}_{Pr_{turb}}} \qquad (4\text{-}316)$$

The turbulent Prandtl number is defined as:

$$Pr_{turb} = \frac{\upsilon_{turb}}{\alpha_{turb}} \qquad (4\text{-}317)$$

Substituting Eq. (4-317) into Eq. (4-316) leads to:

$$\frac{\delta_{m,turb}}{\delta_{t,turb}} \approx \sqrt{Pr_{turb}} \qquad (4\text{-}318)$$

Upon transitioning, Figure 4-18 shows that the turbulent momentum and thermal boundary layers will grow at approximately the same rate because the turbulent eddies transport both energy and momentum into the free stream. The transport mechanism is the same and thus the rate of growth of both boundary layers will be approximately the same, regardless of the Prandtl number of the fluid itself. Said differently, the turbulent Prandtl number in Eq. (4-318) is near unity and approximately independent of the Prandtl number of the fluid, Pr, which is sometimes referred to as the molecular Prandtl number. The molecular Prandtl number is a property of the fluid while the turbulent Prandtl number depends on the flow conditions (just as k and μ are properties of the fluid while k_{turb} and μ_{turb} are properties of the flow).

Finally, this simple conceptual model provides some understanding of the effect of surface roughness (e.g., imperfections on the surface of the plate related to machining).

In order to affect the shear stress or heat transfer coefficient, the surface roughness must be sufficiently large that it disrupts the velocity and temperature gradient that exists at $y = 0$. Figure 4-16 suggests that the size of the roughness required to change the characteristics of a turbulent boundary layer is much less than the size that is required to disrupt a laminar boundary layer. Therefore, the behavior of a turbulent flow is much more sensitive to surface roughness. Indeed, correlations for the friction coefficient or Nusselt number in a laminar flow will almost never include any provision for specifying the surface roughness whereas turbulent flow correlations often will.

4.6 The Reynolds Averaged Equations

4.6.1 Introduction

The governing continuity, momentum, and thermal energy equations for a boundary layer were derived in Section 4.2.3:

$$\frac{\partial u}{\partial x} + \frac{\partial v}{\partial y} = 0 \tag{4-319}$$

$$\rho \left(\frac{\partial u}{\partial t} + u \frac{\partial u}{\partial x} + v \frac{\partial u}{\partial y} \right) = -\frac{dp_\infty}{dx} + \mu \frac{\partial^2 u}{\partial y^2} \tag{4-320}$$

$$\rho c \left(\frac{\partial T}{\partial t} + u \frac{\partial T}{\partial x} + v \frac{\partial T}{\partial y} \right) = k \frac{\partial^2 T}{\partial y^2} + \mu \left(\frac{\partial u}{\partial y} \right)^2 \tag{4-321}$$

Note that the unsteady terms in the momentum and thermal equations have been retained because turbulent flows are inherently unsteady. Figure 4-19 illustrates, conceptually, that the velocity at a particular location in the bulk of a turbulent boundary layer fluctuates even if the flow is externally steady (i.e., there is no time variation in free stream velocity); the velocity in the x-direction can fluctuate by as much as 5–10% of its mean value, $\bar{u}$. The fluctuations are very rapid (with time scales, τ_{turb}, that are on the order of milliseconds) and therefore the flow is highly unsteady. As a result, even though Eqs. (4-319) through (4-321) govern both laminar and turbulent flows, they are much more difficult to solve in a turbulent flow due to the extreme unsteadiness that characterizes the turbulent eddies. Direct Numerical Simulation (DNS) of a turbulent

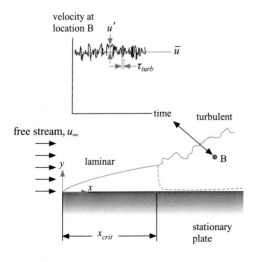

Figure 4-19: Velocity in the turbulent boundary layer that forms above a flat plate.

flow refers to the technique where these equations are solved numerically. However, an accurate solution can only be obtained if the entire range of spatial and temporal scales that characterize the turbulence are resolved numerically. Because the smallest turbulence scales are very small relative to the macro-scale features of the problem, the computational cost of a DNS turbulent model is extreme and such simulations have only recently been attempted.

The solution of the complete governing equations is not practical for turbulent flows and therefore alternative engineering approaches have been developed. Many of these engineering models of turbulent flows are based on the Reynolds averaged equations. The Reynolds averaged equations are derived by integrating the continuity, momentum, and thermal energy equations for a period of time (t_{int}) that is much longer than the turbulent time scales (τ_{turb} in Figure 4-19). The turbulence-induced fluctuations will integrate to zero over t_{int} and therefore the result of this integration is a set of equations that are steady. The integration process leaves behind some additional terms that are related to the transport of momentum and energy by the turbulent fluctuations. Therefore, it remains impossible to solve the Reynolds averaged equations without having some model of these terms. However, the process of deriving the Reynolds averaged equations and their final form provides some insight into turbulent flow. Furthermore, when the Reynolds averaged equations are coupled to a turbulence model it is possible to obtain useful solutions. For example, in Section 4.7 the famous "law of the wall" and the "temperature law of the wall" that describe the mean velocity and temperature distributions, respectively, in a turbulent flow are derived using the mixing length model of turbulence within the Reynolds averaged equations.

4.6.2 The Averaging Process

Each of the independent quantities in Eqs. (4-319) through (4-321) (i.e., u, v, and T) are written as the sum of an averaged and fluctuating component:

$$u(x, y, t) = \bar{u}(x, y) + u'(x, y, t) \tag{4-322}$$

$$v(x, y, t) = \bar{v}(x, y) + v'(x, y, t) \tag{4-323}$$

$$T(x, y, t) = \bar{T}(x, y) + T'(x, y, t) \tag{4-324}$$

where the average component is obtained by averaging over time period t_{int} which is much larger than the time scale associated with any turbulent fluctuations (i.e., $t_{int} > \tau_{turb}$) but much less than the time scale associated with any gross unsteadiness in the problem itself (i.e., the time scale associated with an imposed change in the flow conditions).

$$\bar{u}(x, y) = \frac{1}{t_{int}} \int_0^{t_{int}} u(x, y, t) \, dt \tag{4-325}$$

$$\bar{v}(x, y) = \frac{1}{t_{int}} \int_0^{t_{int}} v(x, y, t) \, dt \tag{4-326}$$

$$\bar{T}(x, y) = \frac{1}{t_{int}} \int_0^{t_{int}} T(x, y, t) \, dt \tag{4-327}$$

The fluctuating component of each dependent quantity (e.g., $u'(x, y, t)$) must integrate to zero over the time scale t_{int}. To see this for u, substitute Eq. (4-322) into Eq. (4-325):

$$\bar{u}(x, y) = \frac{1}{t_{int}} \int_0^{t_{int}} \bar{u}(x, y) \, dt + \frac{1}{t_{int}} \int_0^{t_{int}} u'(x, y, t) \, dt \qquad (4\text{-}328)$$

The first term in Eq. (4-328) is not a function of time, therefore:

$$\bar{u}(x, y) = \bar{u}(x, y) + \frac{1}{t_{int}} \int_0^{t_{int}} u'(x, y, t) \, dt \qquad (4\text{-}329)$$

or

$$\frac{1}{t_{int}} \int_0^{t_{int}} u'(x, y, t) \, dt = 0 \qquad (4\text{-}330)$$

All of the fluctuating quantities possesses this property; they will oscillate about their mean value and therefore integrate to zero over a sufficiently long time period. It is easy to show that the product of a fluctuating quantity and one or more non-fluctuating quantities will also integrate to zero. For example, the product of the averaged and fluctuating components of the x-component must integrate to zero:

$$\frac{1}{t_{int}} \int_0^{t_{int}} \bar{u}(x, y) \, u'(x, y, t) \, dt = \frac{1}{t_{int}} \bar{u}(x, y) \underbrace{\int_0^{t_{int}} u'(x, y, t) \, dt}_{=0} = 0 \qquad (4\text{-}331)$$

Finally, it should be intuitive that the spatial derivatives of a fluctuating quantity will also integrate to zero. With these results in mind, it is possible to derive the Reynolds averaged continuity, momentum, and thermal energy equations.

The Reynolds Averaged Continuity Equation
The continuity equation is:

$$\frac{\partial u}{\partial x} + \frac{\partial v}{\partial y} = 0 \qquad (4\text{-}319)$$

Equations (4-322) and (4-323) are substituted into Eq. (4-319) and the result is integrated from $t = 0$ to $t = t_{int}$.

$$\frac{1}{t_{int}} \int_0^{t_{int}} \left[\frac{\partial (\bar{u} + u')}{\partial x} + \frac{\partial (\bar{v} + v')}{\partial y} = 0 \right] dt \qquad (4\text{-}332)$$

or

$$\frac{1}{t_{int}} \left[\int_0^{t_{int}} \frac{\partial \bar{u}}{\partial x} dt + \int_0^{t_{int}} \frac{\partial u'}{\partial x} dt + \int_0^{t_{int}} \frac{\partial \bar{v}}{\partial y} dt + \int_0^{t_{int}} \frac{\partial v'}{\partial y} dt \right] \qquad (4\text{-}333)$$

The spatial gradient of the fluctuating velocities in Eq. (4-333) must integrate to zero while the spatial gradients of the average velocities are independent of time. Therefore, the Reynolds averaged continuity equation is:

$$\boxed{\frac{\partial \overline{u}}{\partial x} + \frac{\partial \overline{v}}{\partial y} = 0} \qquad (4\text{-}334)$$

which is identical in form to the original continuity equation with the instantaneous values of the velocities replaced by their mean values.

The Reynolds Averaged Momentum Equation
The x-momentum equation is:

$$\frac{\partial u}{\partial t} + u\frac{\partial u}{\partial x} + v\frac{\partial u}{\partial y} = -\frac{1}{\rho}\frac{dp_\infty}{dx} + \upsilon\frac{\partial^2 u}{\partial y^2} \qquad (4\text{-}320)$$

The product of the continuity equation, Eq. (4-319), and the x-velocity is added to the left hand side of Eq. (4-320):

$$u\underbrace{\left(\frac{\partial u}{\partial x} + \frac{\partial v}{\partial y}\right)}_{=0,\text{ by continuity}} = 0 \qquad (4\text{-}335)$$

Note that according to continuity, Eq. (4-335) must be identically zero. Therefore, adding Eq. (4-335) to Eq. (4-320) does not change the validity of the equation.

$$\frac{\partial u}{\partial t} + u\frac{\partial u}{\partial x} + v\frac{\partial u}{\partial y} + \underbrace{u\left(\frac{\partial u}{\partial x} + \frac{\partial v}{\partial y}\right)}_{=0} = -\frac{1}{\rho}\frac{dp_\infty}{dx} + \upsilon\frac{\partial^2 u}{\partial y^2} \qquad (4\text{-}336)$$

Eq. (4-336) is rearranged:

$$\frac{\partial u}{\partial t} + \underbrace{2\,u\frac{\partial u}{\partial x}}_{\frac{\partial(u^2)}{\partial x}} + \underbrace{v\frac{\partial u}{\partial y} + u\frac{\partial v}{\partial y}}_{\frac{\partial(u\,v)}{\partial x}} = -\frac{1}{\rho}\frac{dp_\infty}{dx} + \upsilon\frac{\partial^2 u}{\partial y^2} \qquad (4\text{-}337)$$

$$\frac{\partial u}{\partial t} + \frac{\partial(u^2)}{\partial x} + \frac{\partial(u\,v)}{\partial y} = -\frac{1}{\rho}\frac{dp_\infty}{dx} + \upsilon\frac{\partial^2 u}{\partial y^2} \qquad (4\text{-}338)$$

Equation (4-338) is similar to the original governing differential equation for a viscous flow, Eq. (4-68), that is naturally obtained by carrying out a momentum balance on a differential control volume. Equations (4-322) and (4-323) are substituted into Eq. (4-338) and the result is integrated from $t = 0$ to t_{int}.

$$\frac{1}{t_{int}}\int_0^{t_{int}}\left[\frac{\partial(\overline{u}+u')}{\partial t} + \frac{\partial((\overline{u}+u')^2)}{\partial x} + \frac{\partial((\overline{u}+u')(\overline{v}+v'))}{\partial y} = -\frac{1}{\rho}\frac{dp_\infty}{dx} + \upsilon\frac{\partial^2(\overline{u}+u')}{\partial y^2}\right]dt$$

$$(4\text{-}339)$$

or

$$\frac{1}{t_{int}}\int_0^{t_{int}}\left[\frac{\partial\overline{u}}{\partial t} + \cancel{\frac{\partial u'}{\partial t}} + \frac{\partial(\overline{u}^2)}{\partial x} + \cancel{\frac{\partial(2\overline{u}u')}{\partial x}} + \frac{\partial((u')^2)}{\partial x} + \frac{\partial(\overline{u}\,\overline{v})}{\partial y} + \cancel{\frac{\partial(\overline{u}v')}{\partial y}} + \cancel{\frac{\partial(u'\overline{v})}{\partial y}} + \frac{\partial(u'\,v')}{\partial y}\right]dt$$

$$(4\text{-}340)$$

$$= \frac{1}{t_{int}}\int_0^{t_{int}}\left[-\frac{1}{\rho}\frac{dp_\infty}{dx} + \upsilon\frac{\partial^2\overline{u}}{\partial y^2} + \upsilon\cancel{\frac{\partial^2(u')}{\partial y^2}}\right]dt$$

average x-velocity, $\overline{u}$

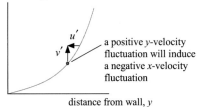

a positive y-velocity
fluctuation will induce
a negative x-velocity
fluctuation

Figure 4-20: Average x-velocity as a function of y.

distance from wall, y

Several of the terms in Eq. (4-340) are proportional to fluctuating terms and therefore must integrate to zero; these have been crossed out. Further, many of the terms in Eq. (4-340) are related only to average quantities and therefore will be unchanged after integration. Equation (4-340) can be written as:

$$\underbrace{\frac{\partial u}{\partial t} + \frac{\partial \left(\overline{u}^2\right)}{\partial x} + \frac{\partial \left(\overline{u}\,\overline{v}\right)}{\partial y} = -\frac{1}{\rho}\frac{dp_\infty}{dx} + \upsilon\frac{\partial^2 \overline{u}}{\partial y^2}}_{\text{original terms in momentum equation}} \quad \underbrace{-\frac{1}{t_{int}}\int_0^{t_{int}}\frac{\partial((u')^2)}{\partial x}dt - \frac{1}{t_{int}}\int_0^{t_{int}}\frac{\partial(u'\,v')}{\partial y}dt}_{\text{additional terms related to fluctuating nature of turbulent flow}}$$

$$(4\text{-}341)$$

There are two terms in Eq. (4-341) that are related to the product of fluctuating components of velocity that do not integrate to zero. It is clear that u' is negative as much as it is positive and therefore u' will integrate to zero. However, $(u')^2$ must always be positive and it therefore does not integrate to zero. The sign of the term related to $u'\,v'$ is less obvious. If the fluctuating components of the velocities in the x- and y-directions are truly random, then the term $u'\,v'$ should be positive and negative an equal amount of time. However, in a turbulent flow, the fluctuations in the x- and y-velocities at any position are correlated; that is, when u' is positive then v' will tend to be negative and vice versa. Thus the product $u'\,v'$ will tend to be negative and will not integrate to zero.

Figure 4-20 illustrates, qualitatively, the average x-velocity ($\overline{u}$) as a function of y, the distance from the wall. The result of a positive fluctuation in the y-velocity ($v' > 0$) is shown conceptually in Figure 4-20. A packet of fluid will be transported away from the wall; from a region of low $\overline{u}$ to a region of higher $\overline{u}$. The effect of the introduction of this low momentum fluid is to induce a reduction in the instantaneous value of the x-velocity (i.e., to induce $u' < 0$). The opposite would happen if there were a negative fluctuation in the y-velocity ($v' < 0$); fluid with high $\overline{u}$ will be transported toward the wall inducing $u' > 0$. The discussion above suggests that v' and u' are negatively correlated and therefore the term related to their product will be exclusively negative and cannot be eliminated from Eq. (4-341).

In Section 4.2.3, the fact that the boundary layer thickness is much less than the plate length is used to justify neglecting gradients in x as being small relative to gradients in y. Physically, we expect that the velocity fluctuations in the x- and y-directions should have similar magnitude. Therefore, using a similar reasoning, the gradient of $(u')^2$ with respect to x should be small relative to the gradient of $u'\,v'$ with respect to y. The boundary layer assumption suggests that:

$$\frac{1}{t_{int}}\int_0^{t_{int}}\frac{\partial((u')^2)}{\partial x}dt \ll \frac{1}{t_{int}}\int_0^{t_{int}}\frac{\partial(u'\,v')}{\partial y}dt \qquad (4\text{-}342)$$

and therefore the Reynolds averaged momentum equation in the boundary layer is:

$$\frac{\partial \bar{u}}{\partial t} + \frac{\partial (\bar{u}^2)}{\partial x} + \frac{\partial (\bar{u}\,\bar{v})}{\partial y} = -\frac{1}{\rho}\frac{dp_\infty}{dx} + \upsilon \frac{\partial^2 \bar{u}}{\partial y^2} - \frac{1}{t_{int}} \int_0^{t_{int}} \frac{\partial (u'\,v')}{\partial y} dt \qquad (4\text{-}343)$$

The unsteady term in Eq. (4-343) reflects any time variation in the average flow. This term accounts for changes in the flow that occur with a time scale that is much larger than t_{int} and therefore also much larger than τ_{turb}. Equation (4-343) can be rearranged to make it correspond more closely to the momentum equation that was derived for laminar flow in a boundary layer. The derivatives on the left side are expanded:

$$\frac{\partial \bar{u}}{\partial t} + 2\bar{u}\frac{\partial \bar{u}}{\partial x} + \bar{u}\frac{\partial \bar{v}}{\partial y} + \bar{v}\frac{\partial \bar{u}}{\partial y} = -\frac{1}{\rho}\frac{dp_\infty}{dx} + \upsilon \frac{\partial^2 \bar{u}}{\partial y^2} - \frac{1}{t_{int}} \int_0^{t_{int}} \frac{\partial (u'\,v')}{\partial y} dt \qquad (4\text{-}344)$$

or

$$\frac{\partial \bar{u}}{\partial t} + \bar{u}\frac{\partial \bar{u}}{\partial x} + \bar{v}\frac{\partial \bar{u}}{\partial y} + \underbrace{\bar{u}\left[\frac{\partial \bar{u}}{\partial x} + \frac{\partial \bar{v}}{\partial y}\right]}_{=0 \text{ by continuity}} = -\frac{1}{\rho}\frac{dp_\infty}{dx} + \upsilon \frac{\partial^2 \bar{u}}{\partial y^2} - \frac{1}{t_{int}} \int_0^{t_{int}} \frac{\partial (u'\,v')}{\partial y} dt \qquad (4\text{-}345)$$

The terms that add to zero according to the Reynolds averaged continuity equation, Eq. (4-334), are removed:

$$\underbrace{\frac{\partial \bar{u}}{\partial t} + \bar{u}\frac{\partial \bar{u}}{\partial x} + \bar{v}\frac{\partial \bar{u}}{\partial y}}_{\text{inertia of averaged flow}} = \underbrace{-\frac{1}{\rho}\frac{dp_\infty}{dx}}_{\text{pressure force}} + \underbrace{\upsilon \frac{\partial^2 \bar{u}}{\partial y^2}}_{\text{viscous shear}} + \underbrace{\left[-\frac{1}{t_{int}} \int_0^{t_{int}} \frac{\partial (u'\,v')}{\partial y} dt\right]}_{\substack{\text{momentum transport due} \\ \text{to turbulent eddies}}} \qquad (4\text{-}346)$$

Equation (4-346) indicates that the Reynolds averaged momentum equation is identical to the non-averaged momentum equation, Eq. (4-320), with the addition of the last term that is related to the transport of momentum due to mixing induced by turbulent eddies. (Recall that the product $u'\,v'$ is negative and therefore the sign of the last term is positive.) The turbulent momentum transport occurs in addition to momentum transport due to viscous shear. Notice that if the fluctuating components of velocity are zero, then Eq. (4-346) reduces to Eq. (4-320). The analogy between the viscous and "turbulent" momentum transport can be made clearer by rearranging the last two terms in Eq. (4-346):

$$\frac{\partial \bar{u}}{\partial t} + \bar{u}\frac{\partial \bar{u}}{\partial x} + \bar{v}\frac{\partial \bar{u}}{\partial y} = -\frac{1}{\rho}\frac{dp_\infty}{dx} + \frac{1}{\rho}\frac{\partial}{\partial y}\overbrace{\left[\underbrace{\mu\frac{\partial \bar{u}}{\partial y}}_{\substack{\text{viscous} \\ \text{stress}}} + \underbrace{\left[-\frac{\rho}{t_{int}} \int_0^{t_{int}} u'\,v'\,dt\right]}_{\text{Reynolds stress}}\right]}^{\text{total stress}} \qquad (4\text{-}347)$$

The last term in Eq. (4-347) is the gradient of the total (or apparent) shear stress, which has two components. The viscous stress is related to the molecular transport of momentum and is captured by the fluid property viscosity. The Reynolds stress is related to the turbulent mixing and depends on the local flow conditions. As discussed in Section 4.5, the momentum transport due to turbulent mixing tends to dominate the momentum transport due to molecular diffusion over the bulk of the turbulent boundary layer. Therefore, we expect the Reynolds stress to dominate the viscous stress everywhere except in the viscous sublayer.

The Reynolds Averaged Thermal Energy Equation

The thermal energy equation is:

$$\rho c \left(\frac{\partial T}{\partial t} + u \frac{\partial T}{\partial x} + v \frac{\partial T}{\partial y} \right) = k \frac{\partial^2 T}{\partial y^2} \tag{4-348}$$

where the viscous dissipation term in Eq. (4-321) has been neglected. The product of the continuity equation and the temperature is added to the left side of Eq. (4-348).

$$\frac{\partial T}{\partial t} + u \frac{\partial T}{\partial x} + v \frac{\partial T}{\partial y} + T \underbrace{\left(\frac{\partial u}{\partial x} + \frac{\partial v}{\partial y} \right)}_{=0, \text{ by continuity}} = \alpha \frac{\partial^2 T}{\partial y^2} \tag{4-349}$$

Equation (4-349) can be rearranged:

$$\frac{\partial T}{\partial t} + \frac{\partial (u\,T)}{\partial x} + \frac{\partial (v\,T)}{\partial y} = \alpha \frac{\partial^2 T}{\partial y^2} \tag{4-350}$$

Equations (4-322) through (4-324) are substituted into Eq. (4-350) and the result is integrated from $t = 0$ to t_{int}.

$$\frac{1}{t_{int}} \int_0^{t_{int}} \left[\frac{\partial (\overline{T} + T')}{\partial t} + \frac{\partial ((\overline{u} + u')(\overline{T} + T'))}{\partial x} + \frac{\partial ((\overline{v} + v')(\overline{T} + T'))}{\partial y} = \alpha \frac{\partial^2 (\overline{T} + T')}{\partial y^2} \right] dt \tag{4-351}$$

or

$$\frac{1}{t_{int}} \int_0^{t_{int}} \left[\frac{\partial \overline{T}}{\partial t} + \frac{\partial T'}{\partial t} + \frac{\partial (\overline{u}\,\overline{T})}{\partial x} + \frac{\partial (\overline{u}T')}{\partial x} + \frac{\partial (u'\overline{T})}{\partial x} + \frac{\partial (u'\,T')}{\partial x} \right] dt$$

$$\times \frac{1}{t_{int}} \int_0^{t_{int}} \left[\frac{\partial (\overline{v}\,\overline{T})}{\partial y} + \frac{\partial (\overline{v}T')}{\partial y} + \frac{\partial (v'\overline{T})}{\partial y} + \frac{\partial (v'\,T')}{\partial y} \right] dt = \frac{1}{t_{int}} \int_0^{t_{int}} \left[\alpha \frac{\partial^2 \overline{T}}{\partial y^2} + \alpha \frac{\partial^2 T'}{\partial y^2} \right] dt \tag{4-352}$$

Those terms in Eq. (4-352) that integrate to zero have been crossed out and the result is:

$$\frac{\partial \overline{T}}{\partial t} + \frac{\partial (\overline{u}\,\overline{T})}{\partial x} + \frac{\partial (\overline{v}\,\overline{T})}{\partial y} = \alpha \frac{\partial^2 \overline{T}}{\partial y^2} - \frac{1}{t_{int}} \int_0^{t_{int}} \frac{\partial (u'\,T')}{\partial x} dt - \frac{1}{t_{int}} \int_0^{t_{int}} \frac{\partial (v'\,T')}{\partial y} dt \tag{4-353}$$

The product of fluctuating terms will not integrate to zero because the velocity and temperature fluctuations are correlated. The boundary layer approximation suggests that gradients in the y-direction (perpendicular to the flow) are much greater than gradients in the x-direction so:

$$\left| \frac{1}{t_{int}} \int_0^{t_{int}} \frac{\partial\,(u'\,T')}{\partial x}\,dt \right| \ll \left| \frac{1}{t_{int}} \int_0^{t_{int}} \frac{\partial\,(v'\,T')}{\partial y}\,dt \right| \qquad (4\text{-}354)$$

With this understanding, Eq. (4-353) becomes:

$$\frac{\partial \overline{T}}{\partial t} + \frac{\partial(\overline{u}\,\overline{T})}{\partial x} + \frac{\partial(\overline{v}\,\overline{T})}{\partial y} = \alpha \frac{\partial^2 \overline{T}}{\partial y^2} - \frac{1}{t_{int}} \int_0^{t_{int}} \frac{\partial\,(v'\,T')}{\partial y}\,dt \qquad (4\text{-}355)$$

The values v' and T' are correlated in the same manner that the values v' and u' are correlated. The result of a positive fluctuation in the y-velocity ($v' > 0$) will transport a packet of fluid away from the wall. If the wall is hotter than the free stream, then hot fluid will be transported to a region of colder fluid and a positive fluctuation in temperature, $T' > 0$, will be induced. If the wall is colder than the free stream, then the opposite will occur. Either way, v' and T' will tend to have the same sign and therefore the product $v'\,T'$ will tend to be exclusively positive and will not integrate to zero. The Reynolds averaging process again leaves behind a term that is related to the effect of the turbulence in the flow.

Equation (4-355) can be rearranged to make it correspond more closely to the boundary layer thermal energy equation that we have seen previously. The derivatives on the left side are expanded:

$$\frac{\partial \overline{T}}{\partial t} + \overline{u}\frac{\partial \overline{T}}{\partial x} + \overline{v}\frac{\partial \overline{T}}{\partial y} + \overline{T}\underbrace{\left[\frac{\partial \overline{u}}{\partial x} + \frac{\partial \overline{v}}{\partial y}\right]}_{=0\text{ by continuity}} = \alpha \frac{\partial^2 \overline{T}}{\partial y^2} - \frac{1}{t_{int}} \int_0^{t_{int}} \frac{\partial\,(v'\,T')}{\partial y}\,dt \qquad (4\text{-}356)$$

The terms that sum to zero according to the Reynolds averaged continuity equation are removed:

$$\frac{\partial \overline{T}}{\partial t} + \overline{u}\frac{\partial \overline{T}}{\partial x} + \overline{v}\frac{\partial \overline{T}}{\partial y} = \alpha \frac{\partial^2 \overline{T}}{\partial y^2} + \underbrace{\left[-\frac{1}{t_{int}} \int_0^{t_{int}} \frac{\partial\,(v'\,T')}{\partial y}\,dt \right]}_{\text{energy transport due to turbulent eddies}} \qquad (4\text{-}357)$$

The Reynolds averaged thermal energy equation is identical to the non-averaged equation, Eq. (4-348), with the addition of one term that is related to the transport of energy due to mixing induced by turbulent eddies. Notice that the turbulent energy transport occurs in addition to the normal molecular level diffusion of thermal energy (conduction). The analogy between the molecular and "turbulent" transport can be made clearer by rearranging the last two terms in Eq. (4-357):

$$\frac{\partial \overline{T}}{\partial t} + \overline{u}\frac{\partial \overline{T}}{\partial x} + \overline{v}\frac{\partial \overline{T}}{\partial y} = -\frac{1}{\rho c}\frac{\partial}{\partial y}\underbrace{\left[\underbrace{-k\frac{\partial \overline{T}}{\partial y}}_{\substack{\text{diffusive}\\\text{heat flux}}} + \underbrace{\frac{\rho c}{t_{int}} \int_0^{t_{int}} v'\,T'\,dt}_{\substack{\text{heat flux due to}\\\text{turbulent mixing}}} \right]}_{\text{total heat flux}} \qquad (4\text{-}358)$$

The last term is the gradient of the total (or apparent) heat flux which has two components. The diffusive heat flux is related to the molecular transport of energy and is captured by the fluid property thermal conductivity. The turbulent heat flux is related to the turbulent mixing and depends on the local flow conditions. As discussed in Section 4.5, the turbulent mixing term tends to dominate the molecular diffusion term everywhere except in the viscous sublayer.

4.7 The Laws of the Wall

4.7.1 Introduction

The qualitative differences between a turbulent and laminar flow are discussed in Section 4.5. In Section 4.6, the governing differential equations for a turbulent flow are averaged over a time scale that is much larger than the fluctuating time scale associated with the turbulence. The result is the Reynolds averaged equations, which describe the average velocity ($\bar{u}$ and $\bar{v}$) and temperature ($\bar{T}$).

$$\frac{\partial \bar{u}}{\partial x} + \frac{\partial \bar{v}}{\partial y} = 0 \tag{4-359}$$

$$\frac{\partial \bar{u}}{\partial t} + \bar{u}\frac{\partial \bar{u}}{\partial x} + \bar{v}\frac{\partial \bar{u}}{\partial y} = -\frac{1}{\rho}\frac{dp_\infty}{dx} + \frac{1}{\rho}\frac{\partial}{\partial y}\underbrace{\left[\mu\frac{\partial \bar{u}}{\partial y} - \frac{\rho}{t_{int}}\int_0^{t_{int}} u'\, v'\, dt\right]}_{\text{apparent shear stress, } \tau_{app}} \tag{4-360}$$

$$\frac{\partial \bar{T}}{\partial t} + \bar{u}\frac{\partial \bar{T}}{\partial x} + \bar{v}\frac{\partial \bar{T}}{\partial y} = -\frac{1}{\rho c}\frac{\partial}{\partial y}\underbrace{\left[-k\frac{\partial \bar{T}}{\partial y} + \frac{\rho c}{t_{int}}\int_0^{t_{int}} v'\, T'\, dt\right]}_{\text{apparent heat flux, } \dot{q}_{app}} \tag{4-361}$$

The unsteadiness related to the turbulence has been removed from the Reynolds averaged equations and therefore they are more amenable to solution. However, it is important to notice that the averaging process does not remove the terms related to the correlated fluctuations in Eqs. (4-360) and (4-361). These terms arise due to the transport of momentum and energy as a result of turbulent mixing and add to the transport of momentum and energy due to molecular diffusion. The sum of the molecular and turbulent transport of momentum is referred to as the apparent shear stress, τ_{app}, and the sum of the molecular and turbulent transport of energy is referred to as the apparent heat flux, $\dot{q}''_{app}$.

The attempt to model these turbulent transport terms and therefore solve the Reynolds equations is referred to as the turbulence closure problem. First order models of the turbulent transport terms are empirical and express the turbulent fluxes as being analogous to laminar fluxes. The mixing length model presented in this section is an example of a first order turbulence model. More advanced turbulence models are available, but they are beyond the scope of this book. For example, the k-ε model is often used in numerical solutions to flow problems; k refers the turbulent kinetic energy and ε to the turbulent dissipation. These quantities are computed over the flow field using a set of coupled transport and conservation equations and subsequently used to model the transport of momentum and energy in the flow due to turbulent mixing.

The correct scaling for a turbulent flow is discussed in Section 4.7.2. Meaningful scaling parameters are related to the magnitude of the turbulent velocity and temperature fluctuations and the size of the viscous sublayer. These quantities are estimated and used to define the "inner variables" that are the dimensionless position, velocity, and temperature difference appropriate for a turbulent flow. Turbulent flow results are often presented in terms of these inner variables.

In Section 4.7.3, the eddy diffusivity of momentum is defined so that it is analogous to the kinematic viscosity of the fluid; the eddy diffusivity of momentum represents turbulent rather than molecular diffusion of momentum. The mixing length model of the eddy diffusivity is presented in Section 4.7.4 and used in Section 4.7.5 to derive the universal velocity profile (sometimes called the law of the wall). Section 4.7.6 presents some additional, more advanced models for the eddy diffusivity of momentum. The law of the wall describes the velocity in a turbulent boundary layer in the "inner" region; the region very near the wall where the inertial terms on the left side of Eq. (4-360) can be neglected. The wake region, which is the region further from the wall where the inertial terms are non-negligible, is discussed in Section 4.7.7. In Section 4.7.8, a similar process is applied to the transport of thermal energy; the eddy diffusivity for heat transfer is defined and related to the eddy diffusivity of momentum through a turbulent Prandtl number. The mixing length model is used to derive the thermal law of the wall.

The material presented in this section provides only an overview of the simplest possible analysis of turbulent boundary layers. The objective is to provide some insight that can be used to understand the behavior of turbulent flows and the correlations that are still used to solve most engineering problems. Some of these correlations are presented in Section 4.9 and these are also provided as functions in EES.

4.7.2 Inner Variables

It is useful to establish the correct length, velocity, and temperature scaling relations for a turbulent flow. The discussion in Section 4.5 showed that the critical length scale relative to the important engineering quantities (e.g., shear and heat transfer coefficient) for a turbulent flow is not the boundary layer thickness (δ_{turb}) but rather the size of the viscous sublayer, which is much smaller than the boundary layer thickness. Furthermore, the turbulent transport of momentum and energy dominates the viscous diffusion of these quantities. Turbulent transport processes are driven by the fluctuations in the velocity and temperature; therefore, it is useful to obtain an approximate scale for these velocity and temperature fluctuations. The scale of the velocity fluctuation is referred to as the eddy velocity, u^*, and the scale of the temperature fluctuation is referred to as the eddy temperature fluctuation, T^*. The position (y), velocity (u), and temperature (T) normalized against these approximate scaling values for the viscous sublayer thickness ($L_{char,vs}$), eddy velocity (u^*) and eddy temperature fluctuation (T^*) are referred to as the inner position (y^+), inner velocity (u^+), and inner temperature difference (θ^+). These inner variables are the most appropriate and widely used dimensionless parameters for a turbulent flow.

The Reynolds averaged momentum equation for a steady flow (i.e., a flow in which the average velocity distribution is not changing in time) with no pressure gradient is:

$$\overline{u}\frac{\partial \overline{u}}{\partial x} + \overline{v}\frac{\partial \overline{u}}{\partial y} = \frac{1}{\rho}\frac{\partial}{\partial y}\left[\mu \frac{\partial \overline{u}}{\partial y} - \frac{\rho}{t_{int}}\int_0^{t_{int}} u'\, v'\, dt \right] \qquad (4\text{-}362)$$

Equation (4-362) can be simplified further in the region very near the wall (within approximately the first 20% of the boundary layer thickness) where the inertial terms on the left side can be neglected relative to the molecular and turbulent transport of momentum:

$$\frac{d}{dy}\left[\underbrace{\mu\frac{d\overline{u}}{dy} - \frac{\rho}{t_{int}}\int_0^{t_{int}} u'\,v'\,dt}_{\text{apparent shear stress, } \tau_{app}}\right] \approx 0 \tag{4-363}$$

The simplification of Eq. (4-362) to Eq. (4-363) is consistent with assuming that the apparent shear stress is constant and is referred to as the Couette flow approximation because the viscous shear in a laminar Couette flow is constant and the apparent shear is constant in an appropriate portion of turbulent boundary layers. Integration of Eq. (4-363) shows that the apparent shear stress is constant in the "inner" region where the Couette flow approximation is valid:

$$\tau_{app} = \mu\frac{d\overline{u}}{dy} - \frac{\rho}{\tau_{int}}\int_0^{\tau_{int}} u'\,v'\,dt = \text{constant} \tag{4-364}$$

If the apparent shear stress is constant near the wall, then it must be equal to the shear stress at the wall, τ_s:

$$\tau_s = \mu\frac{d\overline{u}}{dy} - \frac{\rho}{\tau_{int}}\int_0^{\tau_{int}} u'\,v'\,dt \tag{4-365}$$

Outside of the viscous sublayer, the turbulent transport of momentum, the second term in Eq. (4-365), will dominate the transport of momentum by molecular diffusion, the first term. Therefore:

$$\tau_s \approx -\frac{\rho}{t_{int}}\int_0^{t_{int}} u'\,v'\,dt \quad\text{outside of viscous sublayer} \tag{4-366}$$

The time-averaged integral of the velocity fluctuation product in Eq. (4-366) will scale according to the second power of the eddy velocity:

$$\tau_s \approx \rho\,(u^*)^2 \tag{4-367}$$

Equation (4-367) provides the definition of the eddy velocity, sometimes referred to as the friction velocity:

$$u^* = \sqrt{\frac{\tau_s}{\rho}} \tag{4-368}$$

The inner velocity is the ratio of the average velocity to the eddy velocity:

$$\boxed{u^+ = \frac{\overline{u}}{u^*} = \overline{u}\sqrt{\frac{\rho}{\tau_s}} = \frac{\overline{u}}{u_\infty}\sqrt{\frac{2}{C_f}}} \tag{4-369}$$

where the definition of the local friction factor, C_f, has been substituted into the final term of Eq. (4-369).

The Reynolds averaged energy equation for a steady flow is:

$$\overline{u}\frac{\partial\overline{T}}{\partial x} + \overline{v}\frac{\partial\overline{T}}{\partial y} = \frac{1}{\rho c}\frac{\partial}{\partial y}\left[k\frac{\partial\overline{T}}{\partial y} - \frac{\rho c}{t_{int}}\int_0^{t_{int}} v'\,T'\,dt\right] \tag{4-370}$$

The convective terms on the left side of Eq. (4-370) can be neglected near the wall according to the Couette flow approximation, leaving:

$$\frac{d}{dy}\left[\underbrace{k\frac{\partial \overline{T}}{\partial y} - \frac{\rho c}{t_{int}}\int_0^{t_{int}} v'\,T'\,dt}_{\text{apparent heat flux, }\dot{q}''_{app}}\right] \approx 0 \qquad (4\text{-}371)$$

Integration of Eq. (4-371) shows that the apparent heat flux is constant and must therefore be equal to the heat flux at the surface of the wall, $\dot{q}''_s$:

$$\dot{q}''_{app} = k\frac{\partial \overline{T}}{\partial y} - \frac{\rho c}{t_{int}}\int_0^{t_{int}} v'\,T'\,dt = \text{constant} = \dot{q}''_s \qquad (4\text{-}372)$$

The turbulent transport of energy will dominate the transport of momentum by molecular diffusion outside of the viscous sublayer, therefore:

$$\dot{q}''_s \approx -\frac{\rho c}{t_{int}}\int_0^{t_{int}} v'\,T'\,dt \quad \text{outside of viscous sublayer} \qquad (4\text{-}373)$$

Equation (4-373) will scale according to:

$$\dot{q}''_s \approx \rho c\,u^*\,T^* \qquad (4\text{-}374)$$

where T^* is the eddy temperature fluctuation. Equation (4-374) provides the definition of the eddy temperature fluctuation:

$$T^* = \frac{\dot{q}''_s}{\rho c\,u^*} \qquad (4\text{-}375)$$

The inner temperature difference is the ratio of the wall-to-average temperature difference to the eddy temperature fluctuation:

$$\theta^+ = \frac{T_s - \overline{T}}{T^*} = \frac{T_s - \overline{T}}{\dfrac{\dot{q}''_s}{\rho c\,u^*}} = \frac{T_s - \overline{T}}{\dfrac{\dot{q}''_s}{\rho c}\sqrt{\dfrac{\rho}{\tau_s}}} = \frac{T_s - \overline{T}}{\dfrac{\dot{q}''_s}{\rho c\,u_\infty}\sqrt{\dfrac{2}{C_f}}} \qquad (4\text{-}376)$$

where T_s is the wall temperature.

Within the viscous sublayer, the transport of momentum by molecular diffusion will dominate the turbulent transport of momentum. Therefore, according to Eq. (4-364):

$$\tau_s \approx \mu\frac{\partial \overline{u}}{\partial y} \quad \text{inside viscous sublayer} \qquad (4\text{-}377)$$

Integrating Eq. (4-377) leads to:

$$\overline{u} \approx \frac{\tau_s}{\mu}y \quad \text{inside viscous sublayer} \qquad (4\text{-}378)$$

An estimate of the length scale that characterizes the viscous sublayer ($L_{char,vs}$) is obtained by determining the location at which the average velocity of the fluid ($\overline{u}$)

approaches the eddy velocity (u^*); at this location, the stream has sufficient kinetic energy to produce turbulent eddies.

$$u^* = \frac{\tau_s}{\mu} L_{char,vs} \tag{4-379}$$

Combining Eqs. (4-368) and (4-379) leads to:

$$L_{char,vs} = \frac{\mu}{\tau_s} \sqrt{\frac{\tau_s}{\rho}} \tag{4-380}$$

or

$$L_{char,vs} = \frac{\rho\,\mu}{\rho\,\tau_s} \sqrt{\frac{\tau_s}{\rho}} = \upsilon \sqrt{\frac{\rho}{\tau_s}} = \frac{\upsilon}{u^*} \tag{4-381}$$

The inner position (y^+) is the ratio of position to $L_{char,vs}$:

$$\boxed{y^+ = \frac{y}{L_{char,vs}} = \frac{y\,u^*}{\upsilon} = \frac{y}{\upsilon} \sqrt{\frac{\tau_s}{\rho}} = \frac{y\,u_\infty}{\upsilon} \sqrt{\frac{C_f}{2}}} \tag{4-382}$$

4.7.3 Eddy Diffusivity of Momentum

The turbulent transport of momentum (i.e., the Reynolds stress) is associated with the integral of the correlated x- and y-velocity fluctuations in the Reynolds averaged momentum equation, Eq. (4-360). The velocity fluctuations are correlated because a positive y-velocity fluctuation ($v' > 0$) will transport fluid with low $\bar{u}$ (closer to the wall) to a region with higher $\bar{u}$ (further from the wall) and thus induce a negative x-velocity fluctuation ($u' < 0$); this argument was presented in Section 4.6. The correlation of u' and v', and therefore the existence of the turbulent transport of momentum, relies on the presence of a gradient in the average velocity in the y-direction. It is therefore reasonable to expect that the magnitude of the Reynolds stress is proportional to this velocity gradient:

$$\underbrace{-\frac{\rho}{t_{int}} \int_0^{t_{int}} u'\,v'\,dt}_{\text{Reynolds stress}} \propto \frac{\partial \bar{u}}{\partial y} \tag{4-383}$$

The constant of proportionality that makes Eq. (4-383) an equality is used to define the eddy diffusivity of momentum (ε_M):

$$-\frac{\rho}{t_{int}} \int_0^{t_{int}} u'\,v'\,dt = \rho\,\varepsilon_M \frac{\partial \bar{u}}{\partial y} \tag{4-384}$$

Substituting Eq. (4-384) into Eq. (4-360) leads to the Reynolds averaged momentum equation, expressed in terms of the eddy diffusivity of momentum:

$$\frac{\partial \bar{u}}{\partial t} + \bar{u}\frac{\partial \bar{u}}{\partial x} + \bar{v}\frac{\partial \bar{u}}{\partial y} = -\frac{1}{\rho}\frac{dp_\infty}{dx} + \frac{1}{\rho}\frac{\partial}{\partial y}\underbrace{\left[(\mu + \rho\,\varepsilon_M)\frac{\partial \bar{u}}{\partial y}\right]}_{\text{apparent shear stress, }\tau_{app}} \tag{4-385}$$

The final term in Eq. (4-385) is related to the gradient in the apparent shear stress, τ_{app}:

$$\tau_{app} = (\mu + \rho\,\varepsilon_M)\frac{\partial \bar{u}}{\partial y} \tag{4-386}$$

   The apparent shear stress is the sum of the molecular diffusion of momentum and the transport of momentum by turbulent mixing. The molecular diffusion of momentum is characterized by the molecular viscosity, μ. The turbulent diffusion of momentum is characterized by a turbulent viscosity, $\mu_{turb} = \rho \varepsilon_M$.

4.7.4 The Mixing Length Model

The underlying meaning of the material property thermal conductivity is presented in Section 1.1. Thermal conductivity is related to the characteristics of the micro-scale energy carriers that are present in the substance. Specifically, the thermal conductivity is proportional to the product of the number of energy carriers per unit volume (n_{ms}), their average velocity (v_{ms}), the mean distance between their interactions (L_{ms}), and the ratio of the amount of energy carried by each energy carrier to their temperature (c_{ms}).

$$k \approx n_{ms}\, v_{ms}\, c_{ms}\, L_{ms} \qquad (4\text{-}387)$$

The transport of momentum is analogous to transport of energy. A similar discussion of viscosity would show that the viscosity of a fluid is related to the product of the number density of the micro-scale momentum carriers (molecules), the mass per momentum carrier (m_{ms}), their velocity, and the distance between their interactions:

$$\mu \approx n_{ms}\, m_{ms}\, v_{ms}\, L_{ms} \qquad (4\text{-}388)$$

Both μ and $\rho\, \varepsilon_M$ in Eq. (4-386) have units of Pa-s and both quantities characterize the transport of momentum, albeit by very different mechanisms. However, the transport of momentum by turbulent eddies is analogous to the transport of momentum by molecular scale momentum carriers. The mass density of the turbulent momentum carriers (analogous to $n_{ms}\, m_{ms}$ in Eq. (4-391)) is related to the density of the fluid, ρ. The velocity of the momentum carriers (analogous to v_{ms}) is taken to be the magnitude of the velocity fluctuations ($|u'|$). The distance that the turbulent eddies move (analogous to L_{ms}) is referred to as the mixing length, L_{ml}. Using Eq. (4-388) as a guide, the "turbulent viscosity" can therefore be written as:

$$\rho\, \varepsilon_M \approx \rho |u'|\, L_{ml} \qquad (4\text{-}389)$$

The magnitude of the velocity fluctuations and the mixing length should be related by the gradient in the average velocity. If a fluid packet moves a distance L_{ml} in a velocity field, then the magnitude of the velocity fluctuation that is induced in the fluid will be approximately equal to the product of the mixing length and the gradient in the average velocity:

$$|u'| = \frac{\partial \overline{u}}{\partial y} L_{ml} \qquad (4\text{-}390)$$

This is shown conceptually in Figure 4-21. Substituting Eq. (4-390) into Eq. (4-389) leads to:

$$\rho\, \varepsilon_M \approx \rho\, L_{ml}^2 \frac{\partial \overline{u}}{\partial y} \qquad (4\text{-}391)$$

Prandtl's mixing length model provides an empirical expression for the mixing length and therefore a model for the turbulent transport term in the Reynolds averaged momentum equation. The model assumes that eddies are constrained by the presence of the wall and therefore the mixing length decreases linearly to 0 at $y = 0$ according to:

$$L_{ml} = \kappa\, y \qquad (4\text{-}392)$$

average x-velocity, $\overline{u}$

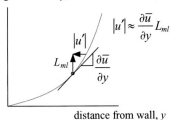

Figure 4-21: Relationship between velocity gradient, velocity fluctuation, and mixing length.

where κ is referred to as the von Kármán constant. The von Kármán constant is typically taken to be 0.41 based on data. Note that the mixing length model provided in Eq. (4-392) is not valid within the viscous sublayer where turbulent eddies are suppressed. More advanced mixing length models are applicable all the way to the wall.

4.7.5 The Universal Velocity Profile

The Couette flow approximation neglects the inertial terms in the Reynolds averaged momentum equation.

$$\frac{1}{\rho}\frac{d}{dy}\underbrace{\left[(\mu + \rho\varepsilon_M)\frac{d\overline{u}}{dy}\right]}_{\text{apparent shear stress, }\tau_{app}} = 0 \qquad (4\text{-}393)$$

Equation (4-393) indicates that the apparent shear stress is constant with y and therefore must be equal to the wall shear stress:

$$\tau_s = (\mu + \rho\varepsilon_M)\frac{d\overline{u}}{dy} \qquad (4\text{-}394)$$

Equation (4-394) can also be written in terms of the inner variables defined in Section 4.7.2:

$$1 = \left(1 + \frac{\varepsilon_M}{\upsilon}\right)\frac{du^+}{dy^+} \qquad (4\text{-}395)$$

Given a model of the eddy diffusivity of momentum, it is possible to use either Eq. (4-394) or Eq. (4-395) to obtain the velocity or inner velocity distribution, respectively, in the inner region. One model of the eddy diffusivity is Prandtl's mixing length model, discussed in Section 4.7.4. Substituting the mixing length model of the turbulent viscosity, Eq. (4-391), into Eq. (4-394) leads to:

$$\tau_s = \mu\frac{d\overline{u}}{dy} + \rho L_{ml}^2\left(\frac{d\overline{u}}{dy}\right)^2 \qquad (4\text{-}396)$$

Recall that Eq. (4-392) for the mixing length is not valid in the viscous sublayer; therefore, a two layer model must be used. The molecular diffusion of momentum (the first term in Eq. (4-396)) is assumed to dominate the turbulent diffusion of momentum (the second term in Eq. (4-396)) in the viscous sublayer:

$$\tau_s = \mu\frac{d\overline{u}}{dy} \quad \text{inside viscous sublayer} \qquad (4\text{-}397)$$

Integrating Eq. (4-397) from the surface of the wall to a position y that is inside the viscous sublayer leads to:

$$\int_0^{\bar{u}} d\bar{u} = \frac{\tau_s}{\mu} \int_0^y dy \tag{4-398}$$

Carrying out the integration in Eq. (4-398) shows that the velocity distribution in the viscous sublayer is linear:

$$\bar{u} = \frac{\tau_s}{\mu} y \tag{4-399}$$

Equation (4-399) is multiplied by $\sqrt{\rho/\tau_s}$ in order to express the result in terms of the inner coordinates:

$$\underbrace{\bar{u}\sqrt{\frac{\rho}{\tau_s}}}_{1/u^*} = y \underbrace{\frac{\rho}{\mu}\sqrt{\frac{\tau_s}{\rho}}}_{1/L_{char,vs}} \tag{4-400}$$

so

$$u^+ = y^+ \tag{4-401}$$

Experimental data have shown that the linear velocity gradient provided by Eq. (4-401) persists out to $y^+ \approx 6$. Recall that $y^+ = y/L_{char,vs}$ and so this result indicates that the viscous sublayer actually extends about six times further from the surface than our estimate from Section 4.7.2 would suggest.

Further from the wall, the turbulent transport of momentum dominates the molecular transport. Therefore, Eq. (4-396) reduces to:

$$\tau_s = \rho L_{ml}^2 \left(\frac{d\bar{u}}{dy}\right)^2 \quad \text{outside of viscous sublayer} \tag{4-402}$$

According to Prandtl's mixing length model, $L_{ml} = \kappa y$, where κ is the von Kármán constant:

$$\tau_s = \rho \kappa^2 y^2 \left(\frac{d\bar{u}}{dy}\right)^2 \tag{4-403}$$

The average velocity gradient is therefore:

$$\frac{d\bar{u}}{dy} = \sqrt{\frac{\tau_s}{\rho}} \frac{1}{\kappa y} \tag{4-404}$$

Dividing Eq. (4-404) by $\sqrt{\tau_s/\rho}$ allows it to be expressed in terms of inner coordinates:

$$\frac{du^+}{dy^+} = \frac{1}{\kappa y^+} \tag{4-405}$$

Equation (4-405) can be integrated:

$$\int du^+ = \int \frac{dy^+}{\kappa y^+} \tag{4-406}$$

to obtain:

$$u^+ = \frac{1}{\kappa} \ln(y^+) + C \tag{4-407}$$

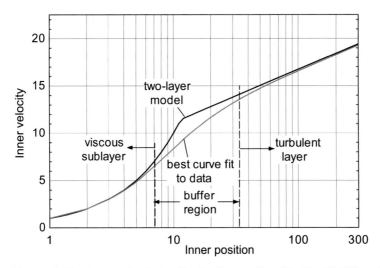

Figure 4-22: Universal velocity distribution predicted by Eq. (4-408) and based on data.

Equation (4-407) is sometimes referred to as the log-law and has been shown to apply for y^+ greater than approximately 30 (i.e. far from the viscous sublayer) and y/δ_{turb} less than approximately 0.2 (i.e., in the region where the Couette assumption is valid, note that δ_{turb} is the boundary layer thickness). The variable C in Eq. (4-407) is a constant of integration that appears because there are no limits on the integrals in Eq. (4-406). The constant of integration is typically taken to be 5.5 in order to provide a best fit to the experimentally measured velocity distribution in a turbulent flow. The parameters $C = 5.5$ and $\kappa = 0.41$ cause the log-law, Eq. (4-407), to intersect the linear velocity distribution, Eq. (4-401), at $y^+ = 11.5$. This simple form of the law of the wall (sometimes referred to as the universal velocity distribution) is therefore:

$$u^+ = \begin{cases} y^+ & \text{for } 0 < y^+ < 11.5 \\ 2.44 \ln (y^+) + 5.5 & \text{for } y^+ > 11.5 \end{cases} \tag{4-408}$$

Equation (4-408) is illustrated in Figure 4-22.

Also shown in Figure 4-22 is the best fit curve to experimental measurements of the velocity distribution. The Couette approximation is valid out to approximately $y^+ = 300$ and even further in the absence of a pressure gradient. Notice that the inner region (i.e., the region where the Couette approximation is valid and the wall shear stress dominates the velocity distribution, sometimes called the near-wall region) can be approximately broken into three regions. The viscous sublayer is the region between $0 < y^+ < \approx 6$. In the viscous sublayer, the molecular diffusion of momentum dominates the turbulent mixing and therefore the linear velocity distribution provides an excellent fit to the data. The turbulent layer is the region from $\approx 30 < y^+ < \approx 300$. In the turbulent region, the turbulent diffusion of momentum dominates the molecular diffusion and the Couette flow approximation still holds. Therefore the log-law provides an excellent fit to the data. The buffer region extends from $\approx 6 < y^+ < \approx 30$. In the buffer region, both molecular and turbulent diffusion of momentum are important and therefore neither the linear nor the log-law distributions are adequate.

Table 4-2: Selected universal velocity distribution models

Model	Expression	Reference
Prandtl-Taylor	$u^+ = \begin{cases} y^+ & \text{for } 0 < y^+ < 11.5 \\ 2.44 \ln(y^+) + 5.5 & \text{for } y^+ > 11.5 \end{cases}$	Rubesin et al. (1998)
von Kármán	$u^+ = \begin{cases} y^+ & \text{for } 0 < y^+ < 5.0 \\ 5 \ln(y^+) - 3.05 & \text{for } 5 < y^+ < 30 \\ 2.5 \ln(y^+) + 5.5 & \text{for } 30 < y^+ \end{cases}$	von Kármán (1939)
Spalding	$y^+ = u^+ + 0.11408 \left[\exp(\kappa u^+) - 1 - \kappa u^+ - \dfrac{(\kappa u^+)^2}{2} - \dfrac{(\kappa u^+)^3}{6} - \dfrac{(\kappa u^+)^4}{24} \right]$	Spalding (1961)
van Driest	$\dfrac{du^+}{dy^+} = \dfrac{2}{1 + \sqrt{1 + 4\,(\kappa y^+)^2 \left[1 - \exp\left(-\dfrac{y^+}{26} \right) \right]^2}}$	van Driest (1956)

4.7.6 Eddy Diffusivity of Momentum Models

There have been several proposed models for the average velocity distribution within the inner region; some of these are summarized in Table 4-2. Any of the velocity distributions proposed in Table 4-2 can be used to provide models for the eddy diffusivity that are more accurate than Prandtl's mixing length model. In the Couette flow region, Eq. (4-394) can be solved for the ratio of the eddy diffusivity of momentum to its molecular analog, the kinematic viscosity:

$$\frac{\varepsilon_M}{\upsilon} = \frac{\tau_s}{\mu \dfrac{d\overline{u}}{dy}} - 1 \qquad (4\text{-}409)$$

Equation (4-395) can be solved for the same quantity in terms of inner coordinates:

$$\frac{\varepsilon_M}{\upsilon} = \frac{1}{\dfrac{du^+}{dy^+}} - 1 \qquad (4\text{-}410)$$

Equations (4-409) and (4-410) relate information about the average velocity to the eddy diffusivity of momentum. For example substituting the log-law, Eq. (4-407), into Eq. (4-410) shows that:

$$\frac{\varepsilon_M}{\upsilon} = \kappa y^+ - 1 \qquad (4\text{-}411)$$

in the turbulent layer.

There are several published models for the eddy diffusivity of momentum. These models are useful in that they provide a simple mechanism for modeling turbulent flows. The molecular viscosity used in laminar flows may be replaced by the total viscosity, $\mu + \rho \varepsilon_M$, in order to model turbulent flows. The eddy diffusivity models that are associated with the velocity distributions listed in Table 4-2 are summarized in Table 4-3 and illustrated in Figure 4-23. Several of these eddy viscosity models are used in Section 5.5.3 in order to numerically simulate an internal turbulent flow.

Table 4-3: Selected eddy diffusivity models

Model	Expression	Reference
Prandtl-Taylor	$\dfrac{\varepsilon_M}{\upsilon} = \begin{cases} 0 & \text{for } y^+ < 11.5 \\ \kappa y^+ - 1 & \text{for } y^+ > 11.5 \end{cases}$	Rubesin et al. (1998)
von Kármán	$\dfrac{\varepsilon_M}{\upsilon} = \begin{cases} 0 & \text{for } y^+ < 5 \\ \dfrac{y^+}{5} - 1 & \text{for } 5 < y^+ < 30 \\ \kappa y^+ - 1 & \text{for } y^+ > 30 \end{cases}$	von Kármán (1939)
Spalding	$\dfrac{\varepsilon_M}{\upsilon} = 0.0526\left[\exp(\kappa u^+) - 1 - \kappa u^+ - \dfrac{(\kappa u^+)^2}{2} - \dfrac{(\kappa u^+)^3}{6}\right]$	Spalding (1961)
van Driest	$\dfrac{\varepsilon_M}{\upsilon} = \left\{\kappa y^+ \left[1 - \exp\left(-\dfrac{y^+}{24.7}\right)\right]\right\}^2 \left|\dfrac{du^+}{dy^+}\right|$	van Driest (1956)

4.7.7 Wake Region

Outside of the inner layer, the Couette flow assumption is not valid and therefore the shear stress is no longer constant. As a result, the velocity distribution no longer collapses when expressed in terms of the inner coordinates. The portion of the boundary layer beyond $y^+ = 300$ is referred to as the wake region. The behavior in the wake region must be correlated in terms of both an inner and an "outer" coordinate, where the outer coordinate is the ratio of the position (y) to the boundary layer thickness (δ_{turb}). Coles (1956) suggests that the velocity profile in the wake region is:

$$u^+ = \frac{1}{\kappa}\ln(y^+) + C + \frac{2}{\kappa}\Pi \sin^2\left(\frac{\pi}{2}\frac{y}{\delta_{turb}}\right) \tag{4-412}$$

where $\kappa = 0.41$ and C is approximately 5.5. The parameter Π is approximately 0.50 for a turbulent flow with no pressure gradient.

The "power law" velocity distributions is simpler, but less accurate than Eq. (4-412). The power law velocity distribution is given by:

$$\frac{u}{u_\infty} = \left(\frac{y}{\delta_{turb}}\right)^{1/7} \tag{4-413}$$

or, in terms of the inner variables:

$$u^+ = 8.75(y^+)^{1/7} \tag{4-414}$$

The mixing length and eddy diffusivity of momentum in the wake region cannot be ascertained directly from the velocity distribution using Eq. (4-410) because the shear stress is not constant. However, relatively simple correlations for these quantities are adequate. The mixing length in the wake region can be taken to be a constant:

$$L_{ml} = 0.09\,\delta_{turb} \quad \text{for } y > 0.2\,\delta_{turb} \tag{4-415}$$

The eddy diffusivity of momentum associated with Eq. (4-415) can be obtained using Eq. (4-391).

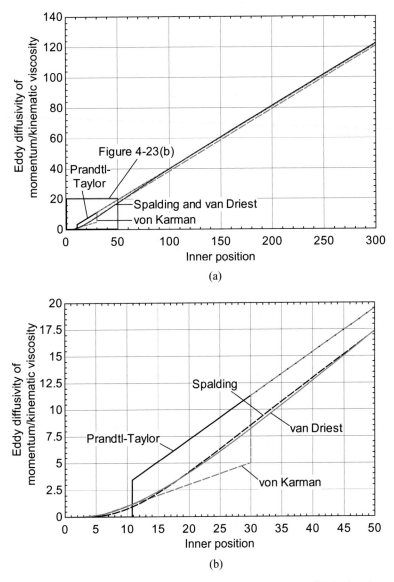

Figure 4-23: Eddy viscosity models (a) in the inner region and (b) in the viscous sublayer and buffer region.

4.7.8 Eddy Diffusivity of Heat Transfer

The correlation between v' and T' that leads to the turbulent transport of energy in the Reynolds averaged thermal energy equation, Eq. (4-361), is fundamentally linked to the magnitude of the average temperature gradient in the same way that the correlation between v' and u' is related to the magnitude of the average velocity gradient. Therefore, the magnitude of the integral of the product $v'\,T'$ is proportional to the average temperature gradient:

$$-\frac{1}{t_{int}}\int_{0}^{t_{int}} v'\,T'\,dt \propto \frac{\partial \overline{T}}{\partial y} \qquad (4\text{-}416)$$

The constant of proportionality that makes Eq. (4-416) an equality is referred to as the eddy diffusivity of heat transfer (ε_H):

$$-\frac{1}{t_{int}} \int_0^{t_{int}} v'\, T'\, dt = \varepsilon_H\, \frac{\partial \overline{T}}{\partial y} \tag{4-417}$$

Substituting Eq. (4-417) into Eq. (4-361) leads to the Reynolds averaged thermal energy conservation equation expressed in terms of the eddy diffusivity of heat transfer:

$$\frac{\partial \overline{T}}{\partial t} + \overline{u}\frac{\partial \overline{T}}{\partial x} + \overline{v}\frac{\partial \overline{T}}{\partial y} = \frac{1}{\rho c}\frac{\partial}{\partial y}\underbrace{\left[(k + \rho c\,\varepsilon_H)\frac{\partial \overline{T}}{\partial y}\right]}_{\text{apparent heat flux}} \tag{4-418}$$

The final term in Eq. (4-418) is the gradient in the apparent heat flux, $\dot{q}''_{app}$:

$$\dot{q}''_{app} = (k + \rho c\,\varepsilon_H)\frac{\partial \overline{T}}{\partial y} \tag{4-419}$$

The apparent heat flux is the sum of the molecular diffusion of energy and the transport of energy by turbulent mixing. The molecular diffusion of energy is characterized by the molecular conductivity, k. The turbulent diffusion of energy is characterized by a turbulent conductivity, $\rho c\,\varepsilon_H$.

4.7.9 The Thermal Law of the Wall

The Reynolds averaged energy equation for a steady flow is:

$$\overline{u}\frac{\partial \overline{T}}{\partial x} + \overline{v}\frac{\partial \overline{T}}{\partial y} = \frac{1}{\rho c}\frac{\partial}{\partial y}\underbrace{\left[(k + \rho c\,\varepsilon_H)\frac{\partial \overline{T}}{\partial y}\right]}_{\text{apparent heat flux}} \tag{4-420}$$

Equation (4-420) is simplified by neglecting the convective terms on the left side of Eq. (4-420) in the inner region ($y^+ < 300$).

$$\frac{d}{dy}\left[(k + \rho c\,\varepsilon_H)\frac{d\overline{T}}{dy}\right] = 0 \tag{4-421}$$

This is analogous to the Couette approximation that was used to derive the velocity law of the wall, Eq. (4-408). Integrating Eq. (4-421) once leads to:

$$(k + \rho c\,\varepsilon_H)\frac{d\overline{T}}{dy} = -\dot{q}''_s \tag{4-422}$$

which can be rearranged:

$$(\alpha + \varepsilon_H)\frac{d\overline{T}}{dy} = -\frac{\dot{q}''_s}{\rho c} \tag{4-423}$$

Equation (4-423) can be expressed in terms of the inner temperature difference, Eq. (4-376), and inner position, Eq. (4-382):

$$(\alpha + \varepsilon_H)\left(-\frac{\dot{q}''_s}{\rho c\, u^*}\right)\frac{u^*}{v}\frac{d\theta^+}{dy^+} = -\frac{\dot{q}''_s}{\rho c} \tag{4-424}$$

or

$$\frac{(\alpha + \varepsilon_H)}{\upsilon} \frac{d\theta^+}{dy^+} = 1 \tag{4-425}$$

Rearranging Eq. (4-425) leads to:

$$\frac{d\theta^+}{dy^+} = \frac{1}{\left(\dfrac{\alpha}{\upsilon} + \dfrac{\varepsilon_H}{\upsilon}\right)} \tag{4-426}$$

The molecular Prandtl number is the ratio of the kinematic viscosity to the thermal diffusivity of a fluid and reflects the relative ability of that fluid to transfer momentum and energy by molecular diffusion:

$$Pr = \frac{\upsilon}{\alpha} \tag{4-427}$$

In a similar manner, the turbulent Prandtl number is defined as the ratio of the eddy diffusivity of momentum to the eddy diffusivity of heat transfer:

$$Pr_{turb} = \frac{\varepsilon_M}{\varepsilon_H} \tag{4-428}$$

Because the same mechanism (i.e., turbulent eddies) is responsible for both ε_M and ε_H, it seems reasonable to expect that the turbulent Prandtl number will be near 1.0. In fact, researchers have found that the turbulent Prandtl number is approximately 0.90 for many flows. Substituting Eqs. (4-427) and (4-428) into Eq. (4-426) leads to:

$$\frac{d\theta^+}{dy^+} = \frac{1}{\left(\dfrac{1}{Pr} + \dfrac{\varepsilon_M}{Pr_{turb}\,\upsilon}\right)} \tag{4-429}$$

Equation (4-429) can be integrated from the surface of the wall through the boundary layer by substituting any of the models for the eddy diffusivity of momentum that were presented in Section 4.7.6.

$$\int_{\theta^+=0}^{\theta^+} d\theta^+ = \int_{y^+=0}^{y^+} \frac{dy^+}{\left(\dfrac{1}{Pr} + \dfrac{\varepsilon_M}{Pr_{turb}\,\upsilon}\right)} \tag{4-430}$$

For example, the Prandtl-Taylor model of the eddy diffusivity of momentum (from Table 4-3) is:

$$\frac{\varepsilon_M}{\upsilon} = \begin{cases} 0 & \text{for } y^+ < 11.5 \\ \kappa\, y^+ - 1 & \text{for } y^+ > 11.5 \end{cases} \tag{4-431}$$

The thermal law of the wall corresponding to the Prandtl-Taylor model is obtained by substituting Eq. (4-431) into Eq. (4-430). For $y^+ < 11.5$, the eddy diffusivity of heat transfer is zero, according to the Prandtl-Taylor model:

$$\int_{\theta^+=0}^{\theta^+} d\theta^+ = \int_{y^+=0}^{y^+} Pr\, dy^+ \quad \text{for } y^+ < 11.5 \tag{4-432}$$

$$\theta^+ = Pr\, y^+ \quad \text{for } y^+ < 11.5 \tag{4-433}$$

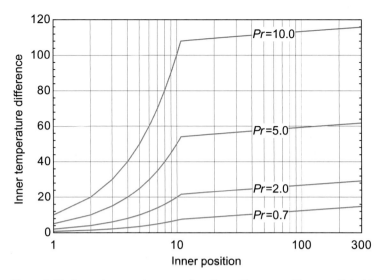

Figure 4-24: Inner temperature as a function of inner position predicted by Eq. (4-438) for various values of Pr with $Pr_{turb} = 0.9$ and $\kappa = 0.41$.

For $y^+ > 11.5$, Eq. (4-430) becomes:

$$\int_{\theta^+ = 11.5\,Pr}^{\theta^+} d\theta^+ = \int_{y^+ = 11.5}^{y^+} \frac{dy^+}{\left[\dfrac{1}{Pr} + \dfrac{1}{Pr_{turb}}(\kappa\,y^+ - 1)\right]} \quad \text{for } y^+ > 11.5 \qquad (4\text{-}434)$$

The integral on the right side of Eq. (4-434) can be determined by making the substitution:

$$w = \frac{1}{Pr} + \frac{1}{Pr_{turb}}(\kappa\,y^+ - 1) \qquad (4\text{-}435)$$

$$dw = \frac{\kappa}{Pr_{turb}}\,dy^+ \qquad (4\text{-}436)$$

Substituting Eqs. (4-435) and (4-436) into Eq. (4-434) leads to:

$$\theta^+ - 11.5\,Pr = \int_{\frac{1}{Pr} + \frac{1}{Pr_{turb}}(11.5\,\kappa - 1)}^{\frac{1}{Pr} + \frac{1}{Pr_{turb}}(\kappa\,y^+ - 1)} \frac{Pr_{turb}}{\kappa}\frac{dw}{w} \qquad (4\text{-}437)$$

so that

$$\theta^+ - 11.5\,Pr = \frac{Pr_{turb}}{\kappa}\,\ln\left(\frac{\dfrac{1}{Pr} + \dfrac{1}{Pr_{turb}}(\kappa\,y^+ - 1)}{\dfrac{1}{Pr} + \dfrac{1}{Pr_{turb}}(11.5\,\kappa - 1)}\right) \quad \text{for } y^+ > 11.5 \qquad (4\text{-}438)$$

Figure 4-24 illustrates the inner temperature as a function of inner position predicted by Eq. (4-438) for various values of the molecular Prandtl number, assuming $\kappa = 0.41$ and $Pr_{turb} = 0.90$.

4.8 Integral Solutions

4.8.1 Introduction

The analytical solution for external flow over a flat plate is presented in Section 4.4. Using advanced techniques, such as the Falkner-Skan transformation, it is possible to obtain analytical solutions for other shapes. These solutions are exact, but they are limited to relatively simple shapes and boundary conditions. The integral techniques introduced in this section are approximate, but much more flexible.

Integral techniques begin with the integral form of the boundary layer equations (the x-directed momentum and energy equations). In this section, the integral equations for momentum and thermal energy are derived in their most general form. These integral equations are written in terms of the integrals of the velocity and temperature distribution across the momentum and thermal boundary layers.

Integral techniques do not attempt to determine an exact solution for the temperature or velocity distribution. Rather, "reasonable" functional forms for the velocity and temperature are assumed based on whether the flow is laminar or turbulent; these distributions are written in terms of the momentum and thermal boundary layer thicknesses and the free stream and surface conditions. The functions are integrated across the boundary layer and forced to satisfy the integral form of the momentum and energy equations. This operation results in differential equations that describe how the boundary layers grow with position. These ordinary differential equations may be solved analytically or numerically in order to predict the boundary layer thicknesses as a function of position. Knowledge of the boundary layer thicknesses can be used to infer the engineering quantities of interest (i.e., shear and heat flux or, more generally, friction coefficient and Nusselt number).

4.8.2 The Integral Form of the Momentum Equation

In this section, the integral form of the momentum equation is derived in its most general form. The result is an integral equation that can be used as the starting point for most integral solutions. In a typical application, many of the terms in the integral equation will be negligible; these can easily be removed in order to proceed with the solution. The integral technique is subsequently applied to a benchmark problem, flow over a flat plate, in order to demonstrate and evaluate the accuracy of the technique.

Derivation of the Integral Form of the Momentum Equation
The steady state, x-momentum equation simplified based on the boundary layer assumptions is derived in Section 4.2.3:

$$u\frac{\partial u}{\partial x} + v\frac{\partial u}{\partial y} = -\frac{1}{\rho}\frac{dp_\infty}{dx} + \upsilon\frac{\partial^2 u}{\partial y^2} \tag{4-439}$$

The flow outside of the boundary layer (i.e., the free stream) can be considered inviscid and therefore the free-stream pressure (p_∞) can be related to the free-stream velocity (u_∞) according to Bernoulli's equation:

$$p_\infty + \frac{\rho u_\infty^2}{2} = \text{constant} \tag{4-440}$$

The derivative of Eq. (4-440) is:

$$\frac{dp_\infty}{dx} + \rho u_\infty\frac{du_\infty}{dx} = 0 \tag{4-441}$$

Equation (4-441) substituted into Eq. (4-439), leading to:

$$u\frac{\partial u}{\partial x} + v\frac{\partial u}{\partial y} = u_\infty \frac{du_\infty}{dx} + v\frac{\partial^2 u}{\partial y^2} \tag{4-442}$$

The final term in Eq. (4-442) can be expanded:

$$u\frac{\partial u}{\partial x} + v\frac{\partial u}{\partial y} = u_\infty \frac{du_\infty}{dx} + \frac{1}{\rho}\frac{\partial}{\partial y}\underbrace{\left[\mu\frac{\partial u}{\partial y}\right]}_{\tau} \tag{4-443}$$

The term within the brackets is the shear stress, τ:

$$u\frac{\partial u}{\partial x} + v\frac{\partial u}{\partial y} = u_\infty \frac{du_\infty}{dx} + \frac{1}{\rho}\frac{\partial \tau}{\partial y} \tag{4-444}$$

Equation (4-444) is integrated across the momentum boundary layer:

$$\int_{y=0}^{y=\delta_m} u\frac{\partial u}{\partial x}dy + \int_{y=0}^{y=\delta_m} v\frac{\partial u}{\partial y}dy = \int_{y=0}^{y=\delta_m} u_\infty \frac{du_\infty}{dx}dy + \frac{1}{\rho}\int_{y=0}^{y=\delta_m} \frac{\partial \tau}{\partial y}dy \tag{4-445}$$

The integration of the last term is straightforward because τ is a continuous function of x and y and therefore the last term is the integral of an exact derivative:

$$\int_{y=0}^{y=\delta_m} u\frac{\partial u}{\partial x}dy + \int_{y=0}^{y=\delta_m} v\frac{\partial u}{\partial y}dy = \int_{y=0}^{y=\delta_m} u_\infty \frac{du_\infty}{dx}dy + \frac{1}{\rho}[\tau_{y=\delta_m} - \tau_s] \tag{4-446}$$

where τ_s is the wall shear stress.

The sequence of mathematical manipulations that follows is not obvious. However, the motivation for these steps can be understood if we recall the underlying objective: obtain an integral form of the equation that requires only the integration of an assumed form of the u velocity distribution. With this in mind, the second term in Eq. (4-446) is clearly a problem as it includes the v velocity. However, this problem can be addressed using integration by parts and the continuity equation.

Integration by parts starts with the chain rule; the differential of the product of the two functions u and v is:

$$d(uv) = u\,dv + v\,du \tag{4-447}$$

Equation (4-447) is integrated from $y = 0$ to $y = \delta_m$:

$$\int_{y=0}^{y=\delta_m} d(uv) = \int_{y=0}^{y=\delta_m} u\frac{\partial v}{\partial y}dy + \int_{y=0}^{y=\delta_m} v\frac{\partial u}{\partial y}dy \tag{4-448}$$

or

$$\left[\underbrace{u_{y=\delta_m}v_{y=\delta_m}}_{u_\infty} - u_{y=0}v_{y=0}\right] = \int_{y=0}^{y=\delta_m} u\frac{\partial v}{\partial y}dy + \underbrace{\int_{y=0}^{y=\delta_m} v\frac{\partial u}{\partial y}dy}_{\text{term of interest}} \tag{4-449}$$

Solving Eq. (4-449) for the second term in Eq. (4-446) leads to:

$$\int_{y=0}^{y=\delta_m} v\frac{\partial u}{\partial y}dy = [u_\infty v_{y=\delta_m} - u_{y=0}v_{y=0}] - \int_{y=0}^{y=\delta_m} u\underbrace{\frac{\partial v}{\partial y}}_{=-\frac{\partial u}{\partial x}}dy \qquad (4\text{-}450)$$

where the free stream velocity, u_∞, has been substituted for $u_{y=\delta_m}$. At first glance, it seems as if Eq. (4-450) has traded one problematic term for two others. The y-velocity at the boundary layer edge $(v_{y=\delta_m})$ is nonzero (the boundary layer is growing, as discussed in Section 4.4.2) and the last term in Eq. (4-450) also involves v. However, the continuity equation derived in Section 4.2,

$$\frac{\partial v}{\partial y} = -\frac{\partial u}{\partial x} \qquad (4\text{-}451)$$

can be used to eliminate the last term in Eq. (4-450):

$$\int_{y=0}^{y=\delta_m} v\frac{\partial u}{\partial y}dy = [u_\infty v_{y=\delta_m} - u_{y=0}v_{y=0}] + \int_{y=0}^{y=\delta_m} u\frac{\partial u}{\partial x}dy \qquad (4\text{-}452)$$

The continuity equation, Eq. (4-451), can also be integrated across the boundary layer in order to express $v_{y=\delta_m}$ in terms of u and its derivatives:

$$\int_{y=0}^{y=\delta_m} \frac{\partial v}{\partial y}dy = -\int_{y=0}^{y=\delta_m} \frac{\partial u}{\partial x}dy \qquad (4\text{-}453)$$

or

$$v_{y=\delta_m} = v_{y=0} - \int_{y=0}^{y=\delta_m} \frac{\partial u}{\partial x}dy \qquad (4\text{-}454)$$

Substituting Eq. (4-454) into Eq. (4-452) leads to:

$$\int_{y=0}^{y=\delta_m} v\frac{\partial u}{\partial y}dy = u_\infty \left[v_{y=0} - \int_{y=0}^{y=\delta_m} \frac{\partial u}{\partial x}dy \right] - u_{y=0}v_{y=0} + \int_{y=0}^{y=\delta_m} u\frac{\partial u}{\partial x}dy \qquad (4\text{-}455)$$

Substituting $\int_{y=0}^{y=\delta_m} v\frac{\partial u}{\partial y}dy$ from Eq. (4-455) into the original integrated momentum equation, Eq. (4-446), leads to:

$$\boxed{\int_{y=0}^{y=\delta_m} u\frac{\partial u}{\partial x}dy} + u_\infty \left[v_{y=0} - \int_{y=0}^{y=\delta_m} \frac{\partial u}{\partial x}dy \right] - u_{y=0}v_{y=0} + \boxed{\int_{y=0}^{y=\delta_m} u\frac{\partial u}{\partial x}dy}$$

$$(4\text{-}456)$$

$$= \int_{y=0}^{y=\delta_m} u_\infty \frac{du_\infty}{dx}dy + \frac{1}{\rho}[\tau_{y=\delta_m} - \tau_s]$$

Equation (4-456) is simplified by combining the two terms that are enclosed in boxes:

$$\int_{y=0}^{y=\delta_m} \frac{\partial (u^2)}{\partial x} dy + u_\infty v_{y=0} - u_{y=\delta_m} \int_{y=0}^{y=\delta_m} \frac{\partial u}{\partial x} dy - u_{y=0}\, v_{y=0} = \int_{y=0}^{y=\delta_m} u_\infty \frac{du_\infty}{dx} dy + \frac{1}{\rho}[\tau_{y=\delta_m} - \tau_s]$$

(4-457)

Noting that $u_{y=\delta_m} = u_\infty$, Eq (4-457) is rearranged:

$$\int_{y=0}^{y=\delta_m} \frac{\partial (u^2)}{\partial x} dy - u_\infty \int_{y=0}^{y=\delta_m} \frac{\partial u}{\partial x} dy - \int_{y=0}^{y=\delta_m} u_\infty \frac{du_\infty}{dx} dy + u_\infty v_{y=0} - u_{y=0}\, v_{y=0} = \frac{1}{\rho}[\tau_{y=\delta_m} - \tau_s]$$

(4-458)

The second term in Eq. (4-458) can be simplified. The partial derivative of the product of the x velocity and the free stream velocity with respect to x is:

$$\frac{\partial (u\, u_\infty)}{\partial x} = u \frac{du_\infty}{dx} + u_\infty \frac{\partial u}{\partial x}$$

(4-459)

Equation (4-459) is integrated:

$$\int_{y=0}^{y=\delta_m} \frac{\partial (u\, u_\infty)}{\partial x} dy = \int_{y=0}^{y=\delta_m} u \frac{du_\infty}{dx} dy + u_\infty \int_{y=0}^{y=\delta_m} \frac{\partial u}{\partial x} dy$$

(4-460)

and rearranged:

$$u_\infty \int_{y=0}^{y=\delta_m} \frac{\partial u}{\partial x} dy = \int_{y=0}^{y=\delta_m} \frac{\partial (u\, u_\infty)}{\partial x} dy - \int_{y=0}^{y=\delta_m} u \frac{du_\infty}{dx} dy$$

(4-461)

Substituting Eq. (4-461) into Eq. (4-458) leads to:

$$\int_{y=0}^{y=\delta_m} \frac{\partial (u^2)}{\partial x} dy - \int_{y=0}^{y=\delta_m} \frac{\partial (u\, u_\infty)}{\partial x} dy + \int_{y=0}^{y=\delta_m} u \frac{du_\infty}{dx} dy - \int_{y=0}^{y=\delta_m} u_\infty \frac{du_\infty}{dx} dy + u_\infty v_{y=0} - u_{y=0}\, v_{y=0}$$

(4-462)

$$= \frac{1}{\rho}[\tau_{y=\delta_m} - \tau_s]$$

The first and second and the third and fourth terms in Eq. (4-462) are combined:

$$\int_{y=0}^{y=\delta_m} \frac{\partial (u^2 - u\, u_\infty)}{\partial x} dy + \int_{y=0}^{y=\delta_m} (u - u_\infty) \frac{du_\infty}{dx} dy + u_\infty v_{y=0} - u_{y=0}\, v_{y=0} = \frac{1}{\rho}[\tau_{y=\delta_m} - \tau_s]$$

(4-463)

Liebniz' rule is used to simplify the first term in Eq. (4-463). Liebniz' rule states that:

$$\frac{d}{dx}\left[\int_{y=a}^{y=b} F(x, y)\, dy\right] = \int_{y=a}^{y=b} \frac{\partial F}{\partial x} dy + F_{x,y=b} \frac{db}{dx} - F_{x,y=a} \frac{da}{dx}$$

(4-464)

Replacing the function F in Eq. (4-464) with the quantity $(u^2 - uu_\infty)$ and the limits with 0 and δ_m leads to:

$$\frac{d}{dx}\left[\int_{y=0}^{y=\delta_m} (u^2 - u\,u_\infty)\,dy\right]$$

$$= \int_{y=0}^{y=\delta_m} \frac{\partial(u^2 - u\,u_\infty)}{\partial x}\,dy + \underbrace{(u_\infty^2 - u_\infty\,u_\infty)}_{=0}\frac{d\delta_m}{dx} - \left(u_{y=0}^2 - u_{y=0}\,u_\infty\right)\underbrace{\frac{d0}{dx}}_{=0} \quad (4\text{-}465)$$

The last two terms in Eq. (4-465) are zero. Therefore:

$$\frac{d}{dx}\left[\int_{y=0}^{y=\delta_m} (u^2 - u\,u_\infty)\,dy\right] = \int_{y=0}^{y=\delta_m} \frac{\partial(u^2 - u\,u_\infty)}{\partial x}\,dy \quad (4\text{-}466)$$

Substituting Eq. (4-466) into Eq. (4-463) leads to:

$$\underbrace{\frac{d}{dx}\left[\int_{y=0}^{y=\delta_m} (u^2 - u\,u_\infty)\,dy\right]}_{\text{momentum change}} + \underbrace{\frac{du_\infty}{dx}\int_{y=0}^{y=\delta_m}(u - u_\infty)\,dy}_{\text{pressure force}} + \underbrace{v_{y=0}(u_\infty - u_{y=0})}_{\text{momentum injected at surface}}$$

$$= \underbrace{\frac{1}{\rho}[\tau_{y=\delta_m} - \tau_s]}_{\text{shear force}}$$

$\quad (4\text{-}467)$

Equation (4-467) is the working form of the momentum integral technique. In most problems, one or more of the terms in Eq. (4-467) can be neglected. For example, there will rarely be a shear force at the outer edge of the boundary layer or a y-velocity at the surface of the wall. However, these terms have been retained in the derivation in order to produce the most broadly useful result.

Application of the Integral Form of the Momentum Equation
In order to carry out the integrations in Eq. (4-467), it is necessary to assume a functional form for the velocity distribution (i.e., a function with undetermined coefficients). The assumed function can range from the very simple (e.g., a linear velocity distribution) to more complex (e.g., a third order polynomial). The more elaborate the function, the more accurate the solution will be because more sophisticated functions are better able to capture the characteristics of the velocity distribution. The undetermined coefficients for the velocity distribution function are selected so that:

1. The no-slip condition is satisfied at the plate surface:

$$u_{y=0} = 0 \quad (4\text{-}468)$$

2. The free-stream velocity is recovered at the edge of the boundary layer:

$$u_{y=\delta_m} = u_\infty \quad (4\text{-}469)$$

Table 4-4: Forms of the velocity distribution that are appropriate for use in the momentum integral equation for a laminar momentum boundary layer.

Form	Velocity distribution	Wall shear stress
linear	$\dfrac{u}{u_\infty} = \dfrac{y}{\delta_m}$	$\tau_s = \mu \dfrac{u_\infty}{\delta_m}$
2nd order polynomial	$\dfrac{u}{u_\infty} = \left[2\,\dfrac{y}{\delta_m} - \dfrac{y^2}{\delta_m^2}\right] + \dfrac{\tau_{y=\delta_m}\,\delta_m}{\mu\,u_\infty}\left[\dfrac{y^2}{\delta_m^2} - \dfrac{y}{\delta_m}\right]$	$\tau_s = 2\mu\,\dfrac{u_\infty}{\delta_m} - \tau_{y=\delta_m}$
3rd order polynomial	$\dfrac{u}{u_\infty}\left[1 + \dfrac{v_{y=0}\,\rho\,\delta_m}{4\,\mu}\right] = \left[\dfrac{3}{2}\dfrac{y}{\delta_m} - \dfrac{1}{2}\dfrac{y^3}{\delta_m^3}\right]$ $+ \left(\dfrac{\delta_m^2}{\mu}\rho\,\dfrac{du_\infty}{dx}\right)\left[\dfrac{1}{4}\dfrac{y}{\delta_m} - \dfrac{1}{2}\dfrac{y^2}{\delta_m^2} + \dfrac{1}{4}\dfrac{y^3}{\delta_m^3}\right]$ $+ \left(\dfrac{\delta_m\,\tau_{y=\delta_m}}{u_\infty\,\mu}\right)\left[-\dfrac{1}{2}\dfrac{y}{\delta_m} + \dfrac{1}{2}\dfrac{y^3}{\delta_m^3}\right]$ $+ \left(\dfrac{\rho\,v_{y=0}\,\delta_m}{\mu}\right)\left[\dfrac{3}{4}\dfrac{y^2}{\delta_m^2} - \dfrac{1}{2}\dfrac{y^3}{\delta_m^3}\right]$ $+ \left(\dfrac{\rho\,v_{y=0}\,\tau_{y=\delta_m}\,\delta_m^2}{u_\infty\,\mu^2}\right)\left[-\dfrac{1}{4}\dfrac{y^2}{\delta_m^2} + \dfrac{1}{4}\dfrac{y^3}{\delta_m^3}\right]$	$\tau_s\left[1 + \dfrac{v_{y=0}\,\rho\,\delta_m}{4\,\mu}\right] = \dfrac{3}{2}\mu\,\dfrac{u_\infty}{\delta_m}$ $+ \dfrac{1}{4}\delta_m\,\rho\,u_\infty\,\dfrac{du_\infty}{dx} - \dfrac{1}{2}\tau_{y=\delta_m}$

3. The shear stress at the edge of the boundary layer is recovered:

$$\mu\,\frac{\partial u}{\partial y}\bigg|_{y=\delta_m} = \tau_{y=\delta_m} \tag{4-470}$$

4. The differential x-momentum equation is satisfied at the wall:

$$v_{y=0}\,\frac{\partial u}{\partial y}\bigg|_{y=0} = u_\infty\,\frac{du_\infty}{dx} + \upsilon\,\frac{\partial^2 u}{\partial y^2}\bigg|_{y=0} \tag{4-471}$$

Depending on the complexity of the assumed form of the velocity distribution, it may not be possible to satisfy all of the conditions listed above. In general, it is best to satisfy these conditions in the order that they are listed using as many undetermined coefficients as there are available in the selected functional form.

Table 4-4 summarizes some appropriate choices for the velocity distribution within a laminar, momentum boundary layer and includes the shear stress at the wall evaluated according to each function. Turbulent boundary layers can also be considered using integral techniques; however, the assumed velocity distribution must be selected based on the discussion in Section 4.7. The velocity distributions provided in Table 4-4 include many terms that will not be considered in most problems (e.g., $v_{y=0}$ and $\tau_{y=\delta_m}$ are usually zero). However, these terms can be removed more easily than they can be added and so Table 4-4 together with Eq. (4-467) provide a useful starting point for most problems.

The velocity distributions summarized in Table 4-4 are illustrated in Figure 4-25 in the limit corresponding to flow over a flat plate with no transpiration ($\tau_{y=\delta_m} = 0$, $v_{y=0} = 0$, and $\frac{du_\infty}{dx} = 0$). Also shown in Figure 4-25 is the Blasius solution for flow over a flat plate, derived in Section 4.4.2. The higher order velocity profiles closely approach the Blasius solution.

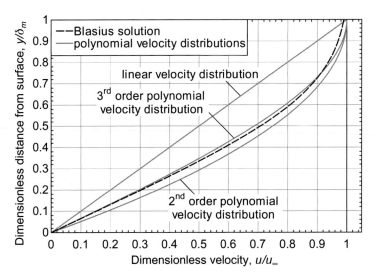

Figure 4-25: Velocity distributions summarized in Table 4-4 in the limit that $\tau_{y=\delta_m} = 0$, $v_{y=0} = 0$, and $\frac{du_\infty}{dx} = 0$. Also shown is the Blasius solution derived in Section 4.4.2.

The integral technique is applied to a benchmark problem, laminar external flow over a flat plate with no transpiration. The integral form of the momentum equation, Eq. (4-467), is simplified for this situation:

$$\frac{d}{dx}\left[\int_{y=0}^{y=\delta_m} (u^2 - u\,u_\infty)\,dy\right] = -\frac{\tau_s}{\rho} \tag{4-472}$$

A third order polynomial is assumed for the velocity distribution. The entry in Table 4-4 is simplified for these conditions:

$$u = \frac{3\,u_\infty}{2}\,\frac{y}{\delta_m} - \frac{u_\infty}{2}\,\frac{y^3}{\delta_m^3} \tag{4-473}$$

The shear stress at the wall is:

$$\tau_s = \mu\,\frac{3\,u_\infty}{2\,\delta_m} \tag{4-474}$$

Equations (4-473) and (4-474) are substituted into Eq.(4-472):

$$\frac{d}{dx}\left[\int_{y=0}^{y=\delta_m}\left(\left(\frac{3\,u_\infty}{2}\,\frac{y}{\delta_m} - \frac{u_\infty}{2}\,\frac{y^3}{\delta_m^3}\right)^2 - \left(\frac{3\,u_\infty}{2}\,\frac{y}{\delta_m} - \frac{u_\infty}{2}\,\frac{y^3}{\delta_m^3}\right)u_\infty\right)dy\right] = -\frac{\mu}{\rho}\,\frac{3\,u_\infty}{2\,\delta_m}$$
$$\tag{4-475}$$

Expanding the integrand leads to:

$$\frac{d}{dx}\left[u_\infty^2\int_{y=0}^{y=\delta_m}\left(\frac{9\,y^2}{4\,\delta_m^2} - \frac{3}{2}\,\frac{y^4}{\delta_m^4} + \frac{1}{4}\,\frac{y^6}{\delta_m^6} - \frac{3}{2}\,\frac{y}{\delta_m} + \frac{1}{2}\,\frac{y^3}{\delta_m^3}\right)dy\right] = -\frac{\mu}{\rho}\,\frac{3\,u_\infty}{2\,\delta_m} \tag{4-476}$$

The integration of Eq. (4-476) leads to:

$$\frac{d}{dx}\left[u_\infty^2\left(\frac{9\,y^3}{12\,\delta_m^2}-\frac{3}{10}\frac{y^5}{\delta_m^4}+\frac{1}{28}\frac{y^7}{\delta_m^6}-\frac{3}{4}\frac{y^2}{\delta_m}+\frac{1}{8}\frac{y^4}{\delta_m^3}\right)_{y=0}^{y=\delta_m}\right]=-\frac{\mu}{\rho}\frac{3\,u_\infty}{2\,\delta_m} \qquad (4\text{-}477)$$

Evaluating the limits leads to:

$$\frac{d}{dx}\left[u_\infty^2\,\delta_m\left(\frac{9}{12}-\frac{3}{10}+\frac{1}{28}-\frac{3}{4}+\frac{1}{8}\right)\right]=-\frac{\mu}{\rho}\frac{3\,u_\infty}{2\,\delta_m} \qquad (4\text{-}478)$$

or

$$\frac{d}{dx}\left[-0.140\,u_\infty^2\,\delta_m\right]=-\frac{\mu}{\rho}\frac{3\,u_\infty}{2\,\delta_m} \qquad (4\text{-}479)$$

In this problem, the free-stream velocity is not a function of x and therefore Eq. (4-479) can be simplified:

$$\delta_m\frac{d\delta_m}{dx}=10.762\frac{\mu}{\rho\,u_\infty} \qquad (4\text{-}480)$$

Equation (4-480) is the typical result of the momentum integral solution technique: an ordinary differential equation that describes the growth of the momentum boundary layer thickness and can be solved either analytically or numerically. Equation (4-480) is separated and integrated:

$$\int_{\delta_m=0}^{\delta_m}\delta_m\,d\delta_m=10.762\frac{\mu}{\rho\,u_\infty}\int_{x=0}^{x}dx \qquad (4\text{-}481)$$

where the limits of the integral are consistent with a boundary layer that grows from the leading edge of the plate. Carrying out the integral in Eq. (4-481) provides an expression for the boundary layer thickness as a function of x:

$$\frac{\delta_m^2}{2}=10.762\frac{\mu\,x}{\rho\,u_\infty} \qquad (4\text{-}482)$$

or

$$\delta_m=4.640\sqrt{\frac{\mu\,x}{\rho\,u_\infty}} \qquad (4\text{-}483)$$

Note that the analysis carried out in Eqs. (4-475) through (4-483) can be facilitated using Maple. The assumed velocity distribution and associated shear stress, Eqs. (4-473) and (4-474), are entered in Maple:

```
> u:=3*u_infinity*y/(2*delta_m(x))-u_infinity*y^3/(2*delta_m(x)^3);
```

$$u:=\frac{3}{2}\frac{u\_infinity\,y}{delta\_m(x)}-\frac{1}{2}\frac{u\_infinity\,y^3}{delta\_m(x)^3}$$

```
> tau_s:=mu*3*u_infinity/(2*delta_m(x));
```

$$tau\_s:=\frac{3}{2}\frac{\mu\,u\_infinity}{delta\_m(x)}$$

Notice that it is necessary to specify that the momentum boundary layer thickness is a function of x by entering delta_m(x). The ordinary differential equation that governs the growth of the momentum boundary layer is obtained using Eq. (4-472):

```
> ODE:=diff(int(u^2-u*u_infinity,y=0..delta_m(x)),x)=-tau_s/rho;
```

$$ODE := -\frac{39}{280} \, u\_infinity^2 \left(\frac{d}{dx} \, \text{delta\_m}(x)\right) = -\frac{3}{2} \frac{\mu \, u\_infinity}{\text{delta\_m}(x) \, \rho}$$

which is the same result as Eq. (4-480). The solution is obtained using the dsolve command in Maple, with the initial condition specified:

```
> dsolve({ODE,delta_m(0)=0});
```

$$\text{delta\_m}(x) = \frac{2\sqrt{910}\,\sqrt{u\_infinity\,\rho\,\mu\,x}}{13\,u\_infinity\,\rho}, \quad \text{delta\_m}(x) = -\frac{2\sqrt{910}\,\sqrt{u\_infinity\,\rho\,\mu\,x}}{13\,u\_infinity\,\rho}$$

There are two roots to Eq. (4-482). Both a positive and negative boundary layer thickness will satisfy the ordinary differential equation and Maple has identified both solutions. The positive solution is physical and identical to Eq. (4-483).

The viscous shear experienced at the plate surface is obtained by substituting Eq. (4-483) into Eq. (4-474):

$$\tau_s = \mu \, \frac{3 \, u_\infty}{2} \frac{1}{4.640} \sqrt{\frac{\rho \, u_\infty}{\mu \, x}} \tag{4-484}$$

or

$$\tau_s = 0.323 \, u_\infty \sqrt{\frac{\rho \, \mu \, u_\infty}{x}} \tag{4-485}$$

The shear stress is used to evaluate the local friction coefficient:

$$C_f = \frac{2 \, \tau_s}{\rho \, u_\infty^2} = \frac{(2) \, 0.323 \, u_\infty}{\rho \, u_\infty^2} \sqrt{\frac{\rho \, \mu \, u_\infty}{x}} \tag{4-486}$$

which can be rearranged and expressed in terms of the Reynolds number:

$$C_f = \frac{0.647}{\sqrt{\dfrac{\rho \, u_\infty \, x}{\mu}}} = \frac{0.647}{\sqrt{Re_x}} \tag{4-487}$$

The exact, analytical solution for a flat plate boundary layer was discussed in Section 4.4.2. The local friction coefficient predicted by the Blasius solution is:

$$C_f = \frac{0.664}{\sqrt{Re_x}} \tag{4-488}$$

Comparing Eq. (4-487) with Eq. (4-488) indicates that the approximate solution obtained using the integral technique is within 2.5% of the exact solution. The agreement would not be as good if a lower order velocity distribution were used. For example, if a linear velocity distribution were assumed in place of a third order polynomial then the result would have been:

$$C_f = \frac{0.578}{\sqrt{Re_x}} \tag{4-489}$$

which is 13% in error relative to the analytical solution.

EXAMPLE 4.8-1: PLATE WITH TRANSPIRATION

EXAMPLE 4.8-1: PLATE WITH TRANSPIRATION

Transpiration is one technique for insulating a surface from an adjacent flowing fluid (e.g., a turbine blade exposed to high temperature gas). Fluid is blown through small holes in the surface so that there is a specified y-directed velocity at the plate surface ($v_{y=0} = v_b$).

a) Determine the ordinary differential equation that governs the growth of the momentum boundary layer for a flat plate experiencing transpiration. Use the momentum integral technique with an assumed second order velocity distribution.

The second order velocity distribution in Table 4-4 is used and the terms associated with shear at the edge of the boundary layer are neglected. The resulting velocity distribution and surface shear stress are:

$$\frac{u}{u_\infty} = \left[2\frac{y}{\delta_m} - \frac{y^2}{\delta_m^2} \right] \tag{1}$$

and

$$\tau_s = 2\mu \frac{u_\infty}{\delta_m} \tag{2}$$

The momentum integral equation, Eq. (4-467), can be simplified for this problem:

$$\frac{d}{dx} \left[\int_{y=0}^{y=\delta_m} (u^2 - u u_\infty)\, dy \right] + v_b u_\infty = -\frac{\tau_s}{\rho} \tag{3}$$

Substituting Eqs. (1) and (2) into Eq. (3) leads to:

$$u_\infty^2 \frac{d}{dx} \left[\int_{y=0}^{y=\delta_m} \left(\left(2\frac{y}{\delta_m} - \frac{y^2}{\delta_m^2}\right)^2 - 2\frac{y}{\delta_m} + \frac{y^2}{\delta_m^2} \right) dy \right] + v_b u_\infty = -2\frac{\mu u_\infty}{\rho \delta_m}$$

or

$$u_\infty^2 \frac{d}{dx} \left[\int_{y=0}^{y=\delta_m} \left(-2\frac{y}{\delta_m} + 5\frac{y^2}{\delta_m^2} - 4\frac{y^3}{\delta_m^3} + \frac{y^4}{\delta_m^4} \right) dy \right] + v_b u_\infty = -2\frac{\mu u_\infty}{\rho \delta_m}$$

Carrying out the integration leads to:

$$u_\infty^2 \frac{d}{dx} \left[\left(-\frac{y^2}{\delta_m} + \frac{5}{3}\frac{y^3}{\delta_m^2} - \frac{y^4}{\delta_m^3} + \frac{1}{5}\frac{y^5}{\delta_m^4} \right)_0^{\delta_m} \right] + v_b u_\infty = -2\frac{\mu u_\infty}{\rho \delta_m}$$

Applying the limits leads to:

$$-\frac{2}{15}u_\infty^2 \frac{d\delta_m}{dx} + v_b u_\infty = -2\frac{\mu u_\infty}{\rho \delta_m} \tag{4}$$

Solving for the rate of change of the momentum boundary layer:

$$\frac{d\delta_m}{dx} = \frac{15}{2 u_\infty} \left(2\frac{\mu}{\rho \delta_m} + v_b \right) \tag{5}$$

which is the ordinary differential equation that governs the boundary layer growth.

EXAMPLE 4.8-1: PLATE WITH TRANSPIRATION

This derivation can also be accomplished using Maple. The assumed velocity distribution and shear stress, Eqs. (1) and (2), are entered:

```
> restart;
> u:=u_infinity*(2*y/delta_m(x)-y^2/delta_m(x)^2);
```
$$u := u\_infinity \left(\frac{2y}{delta\_m(x)} - \frac{y^2}{delta\_m(x)^2} \right)$$

```
> tau_s:=2*mu*u_infinity/delta_m(x);
```
$$tau\_s := \frac{2\,\mu\,u\_infinity}{delta\_m(x)}$$

The momentum integral equation, Eq. (3), is entered.

```
> ODE:=diff(int(u^2-u*u_infinity,y=0..delta_m(x)),x)+v_b*u_infinity=-tau_s/rho;
```
$$ODE := -\frac{2}{15}u\_infinity^2 \left(\frac{d}{dx} delta\_m(x) \right) + v\_bu\_infinity = -\frac{2\mu\,u\_infinity}{delta\_m(x)\,\rho}$$

The result identified by Maple is identical to Eq. (4). The rate of change of the momentum boundary layer is obtained using the **solve** command.

```
> ddeltamdx:=solve(ODE,diff(delta_m(x),x));
```
$$ddeltamdx := \frac{15}{2} \frac{v\_b\,delta\_m(x)\,\rho + 2\,\mu}{u\_infinity\,delta\_m(x)\,\rho}$$

The result identified by Maple is identical to Eq. (5)

b) **Use a numerical method to obtain a solution for the local friction coefficient as a function of Reynolds number. The plate is $L = 0.2$ m long and the fluid has properties $\rho = 10$ kg/m^3 and $\mu = 0.0005$ Pa-s. The free stream velocity is $u_\infty = 10$ m/s. The transpiration velocity is $v_b = 0.1$ m/s.**

The inputs are entered in EES:

```
"EXAMPLE 4.8-1: Plate with Transpiration"

$UnitSystem SI MASS RAD PA K J
$TABSTOPS 0.2 0.4 0.6 0.8 3.5 in

"Inputs"
L=0.2 [m]                          "length of plate"
v_b=0.1 [m/s]                      "blowing velocity"
rho=10 [kg/m^3]                    "density of fluid"
mu=0.0005 [Pa-s]                   "viscosity of fluid"
u_infinity=10 [m/s]                "velocity of free stream"
```

EXAMPLE 4.8-1: PLATE WITH TRANSPIRATION

The variation of the boundary layer thickness with position is the solution to the ordinary differential equation, Eq. (5), subject to the initial condition:

$$\delta_{m,x=0} = 0 \tag{6}$$

The rate of change of the boundary layer thickness with position is infinite at $x = 0$ (to see this, substitute Eq. (6) into Eq. (5)). Therefore, it is difficult to start the numerical integration. One approach is to specify a small, but non-zero, boundary layer thickness at the leading edge of the plate and integrate from that initial condition. A more sophisticated and more reliable technique recognizes that the first term in Eq. (5) dominates near the leading edge of the plate. Therefore, very near the leading edge the ordinary differential equation, Eq. (5), becomes:

$$\frac{d\delta_m}{dx} \approx 15 \frac{\mu}{u_\infty \, \rho \, \delta_m}$$

which can be integrated analytically rather than numerically from $x = 0$ to $x = x_{si}$, where x_{si} is a position sufficiently removed from the leading edge that both terms in Eq. (5) must be considered:

$$\int_0^{\delta_{m,si}} \delta_m \, d\delta_m = 15 \frac{\mu}{u_\infty \, \rho} \int_0^{x_{si}} dx$$

which leads to:

$$\delta_{m,si} = \sqrt{\frac{30 \, \mu \, x_{si}}{u_\infty \, \rho}} \tag{7}$$

The numerical integration will therefore start from $x = x_{si}$ and $\delta_m = \delta_{m,si}$ rather than at $x = 0$ and $\delta_m = 0$. The starting point for the numerical integration, x_{si}, should be selected based on the location where the second term in Eq. (5) becomes significant in relation to the first term. The ratio of the second to the first terms of Eq. (5) is:

$$\frac{\text{second term in Eq. (5)}}{\text{first term in Eq. (5)}} = \frac{v_b \, \rho \, \delta_m}{2 \, \mu} \tag{8}$$

The starting point for the integration will be selected so that the ratio in Eq. (8) reaches a value of 0.01 at x_{si}. Substituting Eq. (7) into Eq. (8) leads to:

$$\frac{v_b \, \rho \, \delta_{m,si}}{2 \, \mu} = \frac{v_b \, \rho}{2 \, \mu} \sqrt{\frac{30 \, \mu \, x_{si}}{u_\infty \, \rho}} = 0.01 \tag{9}$$

Solving Eq. (9) for x_{si} leads to:

$$x_{si} = \frac{4 \, (0.01)^2}{30} \frac{\mu \, u_\infty}{v_b^2 \, \rho}$$

x_si=4*(0.01)^2*mu*u_infinity/(30*v_b^2*rho)   "starting point for integration"
delta_m_si=sqrt(30*mu*x_si/(u_infinity*rho))   "boundary layer thickness at the starting point"

EXAMPLE 4.8-1: PLATE WITH TRANSPIRATION

Equation (5) is entered in EES and the Integral command is used to integrate from $x = x_{si}$ to $x = L$:

```
ddelta_mdx=15*(2*mu/(rho*delta_m)+v_b)/(2*u_infinity)
            "rate of change of boundary layer with position"
delta_m=delta_m_si+integral(ddelta_mdx,x,x_si,L)     "integral solution"
```

The surface shear stress (τ_s) is computed according to Eq. (2):

```
tau_s=2*mu*u_infinity/delta_m                "shear stress"
```

The Reynolds number is calculated according to:

$$Re_x = \frac{\rho \, u_\infty \, x}{\mu}$$

and the friction factor is calculated according to:

$$C_f = \frac{2 \, \tau_s}{\rho \, u_\infty^2}$$

```
Re_x=rho*u_infinity*x/mu                "Reynolds number"
C_f=2*tau_s/(rho*u_infinity^2)          "friction factor"
```

The analytical solution for the friction factor over a flat plate without transpiration was obtained from the Blasius solution in Section 4.4.2.

$$C_{f,bs} = \frac{0.664}{\sqrt{Re_x}}$$

```
C_f_bs=0.664/sqrt(Re_x)                 "friction factor from Blasius solution"
```

The quantities are included in an integral table.

```
DELTAx=L/500                            "spacing in integral table"
$integraltable x:DELTAx,delta_m,tau_s,Re_x,C_f,C_f_bs
```

The spacing in the integral table is specified by the variable DELTAx; however, DELTAx has no impact on the spatial step used in the numerical integration.

Figure 1 illustrates C_f and $C_{f,bs}$ as a function of Re_x for various values of the blowing velocity. Notice that the integral solution agrees well with the Blasius

EXAMPLE 4.8-1: PLATE WITH TRANSPIRATION

solution when v_b approaches zero and that the friction coefficient is reduced by transpiration because the boundary layer thickness is increased.

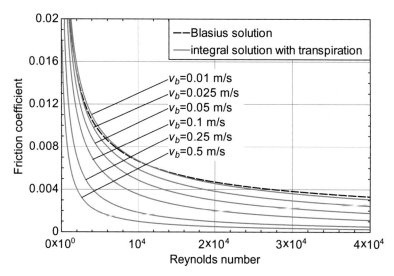

Figure 1: Friction coefficient as a function of the Reynolds number for various values of the blowing velocity; also shown is the Blasius solution.

4.8.3 The Integral Form of the Energy Equation

In this section, the integral form of the energy equation is derived in its most general form. The result will be similar to Eq. (4-467), an integral equation that can be used as the starting point for an integral solution. The integral form of the energy equation is subsequently applied to the same benchmark problem considered in Section 4.8.2, the flow over a flat plate, in order to demonstrate and evaluate the accuracy of the technique.

Derivation of the Integral Form of the Energy Equation
The steps required to derive the integral form of the energy equation are similar to those discussed in Section 4.8.2 to obtain the integral form of the momentum equation. The steady-state energy equation, simplified based on the boundary layer assumptions, is derived in Section 4.2.3:

$$u\frac{\partial T}{\partial x} + v\frac{\partial T}{\partial y} = \frac{k}{\rho c}\frac{\partial^2 T}{\partial y^2} + \frac{\mu}{\rho c}\left(\frac{\partial u}{\partial y}\right)^2 \tag{4-490}$$

The third term in Eq. (4-490) can expanded so that the equation can be written as:

$$u\frac{\partial T}{\partial x} + v\frac{\partial T}{\partial y} = -\frac{1}{\rho c}\frac{\partial}{\partial y}\underbrace{\left[-k\frac{\partial T}{\partial y}\right]}_{\dot{q}''} + \frac{\mu}{\rho c}\left(\frac{\partial u}{\partial y}\right)^2 \tag{4-491}$$

The term inside the brackets is the heat flux ($\dot{q}''$):

$$u\frac{\partial T}{\partial x} + v\frac{\partial T}{\partial y} = -\frac{1}{\rho c}\frac{\partial \dot{q}''}{\partial y} + \frac{\mu}{\rho c}\left(\frac{\partial u}{\partial y}\right)^2 \tag{4-492}$$

Equation (4-492) is integrated across the thermal boundary layer (from $y = 0$ to $y = \delta_t$):

$$\int_0^{\delta_t} u\frac{\partial T}{\partial x}\,dy + \int_0^{\delta_t} v\frac{\partial T}{\partial y}\,dy = -\frac{1}{\rho c}\int_0^{\delta_t}\frac{\partial \dot{q}''}{\partial y}\,dy + \frac{\mu}{\rho c}\int_0^{\delta_t}\left(\frac{\partial u}{\partial y}\right)^2 dy \qquad (4\text{-}493)$$

The integration of the third term is straightforward because it is an exact derivative:

$$\int_0^{\delta_t} u\frac{\partial T}{\partial x}\,dy + \int_0^{\delta_t} v\frac{\partial T}{\partial y}\,dy = -\frac{1}{\rho c}[\dot{q}''_{y=\delta_t} - \dot{q}''_s] + \frac{\mu}{\rho c}\int_0^{\delta_t}\left(\frac{\partial u}{\partial y}\right)^2 dy \qquad (4\text{-}494)$$

where $\dot{q}''_s$ is the heat flux at the wall.

A series of manipulations is required in order obtain an integral equation that requires only the integration of the assumed functions for u and T. The terms in Eq. (4-494) that involve the v velocity must be eliminated through integration by parts and application of the continuity equation.

Integration by parts is applied to the second term in Eq. (4-494):

$$d(vT) = T\,dv + v\,dT \qquad (4\text{-}495)$$

Integrating Eq. (4-495) from the wall to the edge of the thermal boundary layer leads to:

$$\int_0^{\delta_t} d(vT) = \int_0^{\delta_t} T\,dv + \int_0^{\delta_t} v\,dT \qquad (4\text{-}496)$$

or

$$v_{y=\delta_t}\,T_\infty - v_{y=0}\,T_s = \int_0^{\delta_t} T\frac{\partial v}{\partial y}\,dy + \int_0^{\delta_t} v\frac{\partial T}{\partial y}\,dy \qquad (4\text{-}497)$$

where T_∞ is the free-stream temperature and T_s is the surface temperature. The second term on the right side of Eq. (4-497) can be written as:

$$\int_0^{\delta_t} v\frac{\partial T}{\partial y}\,dy = v_{y=\delta_t}\,T_\infty - v_{y=0}\,T_s - \int_0^{\delta_t} T\underbrace{\frac{\partial v}{\partial y}}_{-\frac{\partial u}{\partial x}}\,dy \qquad (4\text{-}498)$$

The last term in Eq. (4-498) can be rewritten using the continuity equation:

$$\frac{\partial v}{\partial y} = -\frac{\partial u}{\partial x} \qquad (4\text{-}499)$$

so that:

$$\int_0^{\delta_t} v\frac{\partial T}{\partial y}\,dy = v_{y=\delta_t}\,T_\infty - v_{y=0}\,T_s + \int_0^{\delta_t} T\frac{\partial u}{\partial x}\,dy \qquad (4\text{-}500)$$

The y-directed velocity at the edge of the boundary layer was previously considered in Eq. (4-454) and this equation is substituted into Eq. (4-500):

$$\int_0^{\delta_t} v\frac{\partial T}{\partial y}\,dy = \left[v_{y=0} - \int_{y=0}^{y=\delta_m}\frac{\partial u}{\partial x}\,dy\right]T_\infty - v_{y=0}\,T_s + \int_0^{\delta_t} T\frac{\partial u}{\partial x}\,dy \qquad (4\text{-}501)$$

Substituting Eq. (4-501) into Eq. (4-494) leads to:

$$\int_0^{\delta_t} u \frac{\partial T}{\partial x} \, dy + \left[v_{y=0} - \int_{y=0}^{y=\delta_m} \frac{\partial u}{\partial x} dy \right] T_\infty - v_{y=0} \, T_s + \int_0^{\delta_t} T \frac{\partial u}{\partial x} \, dy$$

$$= -\frac{1}{\rho c} [\dot{q}''_{y=\delta_t} - \dot{q}''_s] + \frac{\mu}{\rho c} \int_0^{\delta_t} \left(\frac{\partial u}{\partial y} \right)^2 dy \qquad (4\text{-}502)$$

which is rearranged:

$$\int_0^{\delta_t} u \frac{\partial T}{\partial x} \, dy - v_{y=0} \, (T_s - T_\infty) + \int_0^{\delta_t} (T - T_\infty) \frac{\partial u}{\partial x} \, dy$$

$$= -\frac{1}{\rho c} [\dot{q}''_{y=\delta_t} - \dot{q}''_s] + \frac{\mu}{\rho c} \int_0^{\delta_t} \left(\frac{\partial u}{\partial y} \right)^2 dy \qquad (4\text{-}503)$$

If T_∞ is constant then the first term in Eq. (4-503) can be written in terms of the temperature difference relative to the free stream temperature:

$$\int_0^{\delta_t} u \frac{\partial (T - T_\infty)}{\partial x} \, dy - v_{y=0} \, (T_s - T_\infty) + \int_0^{\delta_t} (T - T_\infty) \frac{\partial u}{\partial x} \, dy$$

$$= -\frac{1}{\rho c} [\dot{q}''_{y=\delta_t} - \dot{q}''_s] + \frac{\mu}{\rho c} \int_0^{\delta_t} \left(\frac{\partial u}{\partial y} \right)^2 dy \qquad (4\text{-}504)$$

and combined with the third term in Eq. (4-504):

$$\int_0^{\delta_t} \frac{\partial [u \, (T - T_\infty)]}{\partial x} \, dy - v_{y=0} \, (T_s - T_\infty) = -\frac{1}{\rho c} [\dot{q}''_{y=\delta_t} - \dot{q}''_s] + \frac{\mu}{\rho c} \int_0^{\delta_t} \left(\frac{\partial u}{\partial y} \right)^2 dy \quad (4\text{-}505)$$

Using Liebniz' rule, Eq. (4-464), to simplify the first term in Eq. (4-505) leads to the working version of the integral form of the energy equation:

$$\underbrace{\frac{d}{dx} \left[\int_0^{\delta_t} u \, (T - T_\infty) \, dy \right]}_{\text{enthalpy change}} = \underbrace{v_{y=0} \, (T_s - T_\infty)}_{\substack{\text{energy due to fluid} \\ \text{injected at surface}}} + \underbrace{\frac{1}{\rho c} [\dot{q}''_s - \dot{q}''_{y=\delta_t}]}_{\text{conduction}} + \underbrace{\frac{\mu}{\rho c} \int_0^{\delta_t} \left(\frac{\partial u}{\partial y} \right)^2 dy}_{\text{viscous dissipation}}$$

$$(4\text{-}506)$$

In most problems, one or more of the terms in Eq. (4-506) can be neglected. For example, there will rarely be a heat flux at the outer edge of the boundary layer or a y-velocity at the surface of the wall. However, these terms have been retained in the derivation in order to produce the most broadly useful result. The velocity distributions provided in Table 4-4 may be substituted into Eq. (4-506) together with an assumed form of the temperature distribution, as discussed in the subsequent section. Note that an underlying assumption in the application of the integral form of the energy equation is that the temperature distribution does not affect the velocity distribution and therefore the integral form of the momentum equation can be solved before the integral form of the energy equation.

Application of the Integral Form of the Energy Equation

In order to carry out the integration in Eq. (4-506), it is necessary to assume a functional form for both the temperature distribution and the velocity distribution. Again, these functions can range in complexity. The undetermined coefficients for the temperature distribution function should be selected so that:

1. The wall temperature is retained at the plate surface:

$$T_{y=0} = T_s \tag{4-507}$$

2. The free-stream temperature is recovered at the edge of the thermal boundary layer:

$$T_{y=\delta_t} = T_\infty \tag{4-508}$$

3. The specified heat flux at the edge of the thermal boundary layer is recovered:

$$-k\frac{\partial T}{\partial y}\bigg|_{y=\delta_t} = \dot{q}''_{y=\delta_t} \tag{4-509}$$

4. The differential thermal energy equation is satisfied at the wall,

$$v_{y=0}\frac{\partial T}{\partial y}\bigg|_{y=0} = \alpha\frac{\partial^2 T}{\partial y^2}\bigg|_{y=0} + \frac{\mu}{\rho c}\left(\frac{\partial u}{\partial y}\bigg|_{y=0}\right)^2 \tag{4-510}$$

It may not be possible to satisfy all of the conditions listed above. However, it is best to satisfy the conditions in the order that they are listed. Table 4-5 summarizes some appropriate choices for the temperature distribution within a laminar, thermal boundary layer and includes the associated heat flux at the wall. The analysis of a turbulent flow with the integral technique requires the use of a temperature distribution that is based on the discussion in Section 4.7. The temperature distributions provided in Table 4-5 are as general as possible and include many terms that will not typically be considered. These terms can be easily removed so that Table 4-5 and Eq. (4-506) provide a useful starting point for most problems.

The temperature distributions summarized in Table 4-5 are illustrated in Figure 4-26 in the limit that $\dot{q}''_{y=\delta_t} = 0$, $v_{y=0} = 0$, and viscous dissipation is ignored (i.e., the $\frac{\partial u}{\partial y}|_{y=0}$ term in the third order polynomial is ignored). Also shown in Figure 4-26 is the self-similar solution for the temperature distribution that was discussed in Section 4.4.3 (with $Pr = 1$). Notice that the higher order velocity profiles approach the self-similar solution.

The integral technique is again illustrated by application to a benchmark problem: laminar flow over a flat plate with no pressure gradient, no transpiration, no externally applied heat flux, and negligible viscous dissipation. A third order polynomial velocity distribution (from Table 4-4):

$$u = \frac{3}{2}\frac{u_\infty}{\delta_m}\frac{y}{\delta_m} - \frac{u_\infty}{2}\frac{y^3}{\delta_m^3} \tag{4-511}$$

Table 4-5: Forms of the temperature distribution that are appropriate for use in the energy integral equation for a laminar thermal boundary layer.

Form	Temperature distribution	Wall heat flux
linear	$\dfrac{T - T_s}{T_\infty - T_s} = \dfrac{y}{\delta_t}$	$\dot{q}''_s = k\,\dfrac{(T_s - T_\infty)}{\delta_t}$
2nd order polynomial	$\dfrac{T - T_s}{T_\infty - T_s} = \left[2\,\dfrac{y}{\delta_t} - \dfrac{y^2}{\delta_t^2}\right]$ $+ \dfrac{\dot{q}''_{y=\delta_t}\,\delta_t}{k\,(T_\infty - T_s)}\left[\dfrac{y}{\delta_t} - \dfrac{y^2}{\delta_t^2}\right]$	$\dot{q}''_s = 2k\,\dfrac{(T_s - T_\infty)}{\delta_t} - \dot{q}''_{y=\delta_t}$
3rd order polynomial	$\left[\dfrac{T - T_s}{T_\infty - T_s}\right]\left(1 + \dfrac{\delta_t\,v_{y=0}\,\rho\,c}{4\,k}\right) = \left[\dfrac{3}{2}\dfrac{y}{\delta_t} - \dfrac{1}{2}\dfrac{y^3}{\delta_t^3}\right]$ $+ \dfrac{\dot{q}''_{y=\delta_t}\,\delta_t}{k\,(T_\infty - T_s)}\left[\dfrac{1}{2}\dfrac{y}{\delta_t} - \dfrac{1}{2}\dfrac{y^3}{\delta_t^3}\right]$ $+ \dfrac{\delta_t^2\,\mu\left(\dfrac{\partial u}{\partial y}\Big|_{y=0}\right)^2}{k\,(T_\infty - T_s)}\left[\dfrac{1}{4}\dfrac{y}{\delta_t} - \dfrac{1}{2}\dfrac{y^2}{\delta_t^2} + \dfrac{1}{4}\dfrac{y^3}{\delta_t^3}\right]$ $+ \dfrac{\rho\,c\,v_{y=0}\,\delta_t}{k}\left[\dfrac{3}{4}\dfrac{y^2}{\delta_t^2} - \dfrac{1}{2}\dfrac{y^3}{\delta_t^3}\right]$ $+ \dfrac{\rho\,c\,v_{y=0}\,\delta_t^2\,\dot{q}''_{y=\delta_t}}{k^2\,(T_\infty - T_s)}\left[\dfrac{1}{4}\dfrac{y^2}{\delta_t^2} - \dfrac{1}{4}\dfrac{y^3}{\delta_t^3}\right]$	$\dot{q}''_s\left(1 + \dfrac{\delta_t\,v_{y=0}\,\rho\,c}{4\,k}\right) = \dfrac{3}{2}k\,\dfrac{(T_s - T_\infty)}{\delta_t}$ $- \dfrac{\dot{q}''_{y=\delta_t}}{2} - \dfrac{\delta_t\,\mu}{4}\left(\dfrac{\partial u}{\partial y}\Big|_{y=0}\right)^2$

is substituted into the integral form of the momentum equation and used to determine the variation of the momentum boundary layer thickness with position:

$$\delta_m = 4.640\sqrt{\dfrac{\mu\,x}{\rho\,u_\infty}} \tag{4-512}$$

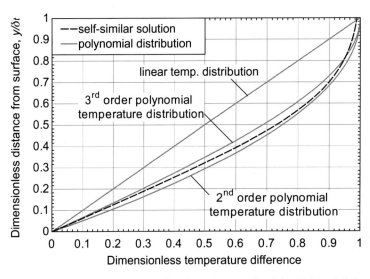

Figure 4-26: Temperature distributions summarized in Table 4-5 in the limit that $\dot{q}''_{y=\delta_t} = 0$, $v_{y=0} = 0$, and viscous dissipation is ignored. Also shown is the self-similar solution (with $Pr = 1$).

Here, a third order polynomial temperature distribution is assumed. The appropriate entry in Table 4-5 is simplified for this situation:

$$\frac{T - T_s}{T_\infty - T_s} = \frac{3}{2}\frac{y}{\delta_t} - \frac{1}{2}\frac{y^3}{\delta_t^3} \tag{4-513}$$

The heat flux experienced at the plate surface that is consistent with Eq. (4-513) is also obtained from Table 4-5:

$$\dot{q}_s'' = \frac{3}{2}k\frac{(T_s - T_\infty)}{\delta_t} \tag{4-514}$$

The integral form of the energy equation, Eq. (4-506) is simplified for this problem:

$$\frac{d}{dx}\left[\int_0^{\delta_t} u\,(T - T_\infty)\,dy\right] = \frac{\dot{q}_s''}{\rho\,c} \tag{4-515}$$

or

$$\frac{d}{dx}\left[\int_0^{\delta_t} u\,[(T - T_s) - (T_\infty - T_s)]\,dy\right] = \frac{\dot{q}_s''}{\rho\,c} \tag{4-516}$$

Substituting Eqs. (4-511), (4-513), and (4-514) into Eq. (4-516) leads to:

$$\frac{d}{dx}\left[\int_0^{\delta_t} u_\infty\,(T_\infty - T_s)\left(\frac{3}{2}\frac{y}{\delta_m} - \frac{1}{2}\frac{y^3}{\delta_m^3}\right)\left(\frac{3}{2}\frac{y}{\delta_t} - \frac{1}{2}\frac{y^3}{\delta_t^3} - 1\right)dy\right] = \frac{3}{2}\frac{k}{\rho\,c}\frac{(T_s - T_\infty)}{\delta_t} \tag{4-517}$$

Note that Eq. (4-517) is only strictly valid provided that the thermal boundary layer thickness is less than or approximately equal to the momentum boundary layer thickness. If this is not the case then the velocity distribution provided by Eq. (4-511) is not valid throughout the entire thermal boundary layer. This requirement implies that Eq. (4-517) is only valid if the fluid has a Prandtl number that is greater than or approximately equal to 1.0. If this condition is not met then the integration must be split into two parts; from 0 to δ_m, where the velocity distribution is given by Eq. (4-511), and from δ_m to δ_t, where the velocity is constant and equal to u_∞. Carrying out the multiplication in the integrand of Eq. (4-517) leads to:

$$\frac{d}{dx}\left[\int_0^{\delta_t}\left(\frac{3}{2}\frac{y}{\delta_m} - \frac{9}{4}\frac{y^2}{\delta_t\,\delta_m} + \frac{3}{4}\frac{y^4}{\delta_t^3\,\delta_m} - \frac{1}{2}\frac{y^3}{\delta_m^3} + \frac{3}{4}\frac{y^4}{\delta_t\,\delta_m^3} - \frac{1}{4}\frac{y^6}{\delta_t^3\,\delta_m^3}\right)dy\right] = \frac{3}{2\delta_t}\frac{k}{\rho\,c\,u_\infty} \tag{4-518}$$

Carrying out the integration leads to:

$$\frac{d}{dx}\left[\left(\frac{3}{4}\frac{y^2}{\delta_m} - \frac{9}{12}\frac{y^3}{\delta_t\,\delta_m} + \frac{3}{20}\frac{y^5}{\delta_t^3\,\delta_m} - \frac{1}{8}\frac{y^4}{\delta_m^3} + \frac{3}{20}\frac{y^5}{\delta_t\,\delta_m^3} - \frac{1}{28}\frac{y^7}{\delta_t^3\,\delta_m^3}\right)_0^{\delta_t}\right] = \frac{3}{2\delta_t}\frac{k}{\rho\,c\,u_\infty} \tag{4-519}$$

Applying the limits of integration leads to:

$$\frac{d}{dx}\left[\frac{3}{4}\frac{\delta_t^2}{\delta_m} - \frac{9}{12}\frac{\delta_t^2}{\delta_m} + \frac{3}{20}\frac{\delta_t^2}{\delta_m} - \frac{1}{8}\frac{\delta_t^4}{\delta_m^3} + \frac{3}{20}\frac{\delta_t^4}{\delta_m^3} - \frac{1}{28}\frac{\delta_t^4}{\delta_m^3}\right] = \frac{3}{2\delta_t}\frac{k}{\rho\,c\,u_\infty} \tag{4-520}$$

or

$$\frac{d}{dx}\left[0.15\frac{\delta_t^2}{\delta_m} - 0.01\frac{\delta_t^4}{\delta_m^3}\right] = \frac{3}{2\,\delta_t}\frac{k}{\rho\,c\,u_\infty} \tag{4-521}$$

Here, we are looking at the case where the thermal and the momentum boundary layers develop together from the leading edge of the plate. Therefore, we can expect that:

$$\frac{\delta_t}{\delta_m} = Pr^{-\frac{1}{3}} \tag{4-522}$$

based on the analysis and concepts discussed in Section 4.1. Substituting Eq. (4-522) into Eq. (4-521) leads to:

$$\left(0.15\,Pr^{-\frac{1}{3}} - 0.01Pr^{-1}\right)\frac{d\delta_t}{dx} = \frac{3}{2\,\delta_t}\frac{k}{\rho\,c\,u_\infty} \tag{4-523}$$

The second term in parentheses is neglected as being small relative to the first, leading to:

$$\delta_t\frac{d\delta_t}{dx} = 10\frac{k}{\rho\,c\,u_\infty}Pr^{\frac{1}{3}} \tag{4-524}$$

Equation (4-524) is integrated from the leading edge of the plate:

$$\int_0^{\delta_t}\delta_t\,d\delta_t = \int_0^x 10\frac{\alpha}{u_\infty}Pr^{\frac{1}{3}}dx \tag{4-525}$$

which leads to:

$$\delta_t^2 = 20\frac{\alpha}{u_\infty}Pr^{\frac{1}{3}}x \tag{4-526}$$

Equation (4-526) is solved for the thermal boundary layer thickness:

$$\delta_t = 4.47\sqrt{\frac{\alpha x}{u_\infty}}Pr^{\frac{1}{6}} \tag{4-527}$$

Substituting Eq. (4-527) into Eq. (4-514) leads to:

$$\dot{q}_s'' = \frac{3\,k\,(T_s - T_\infty)}{2\,(4.47)\,Pr^{\frac{1}{6}}}\sqrt{\frac{u_\infty}{\alpha\,x}} \tag{4-528}$$

Equation (4-528) can be expressed in terms of the local Nusselt number:

$$Nu_x = \frac{\dot{q}_s''\,x}{k\,(T_s - T_\infty)} = 0.336\frac{1}{Pr^{\frac{1}{6}}}\sqrt{\frac{x\,u_\infty}{\alpha}} \tag{4-529}$$

Multiplying the numerator and denominator of the argument of the square root in Eq. (4-529) by the kinematic viscosity leads to:

$$Nu_x = 0.336\frac{1}{Pr^{\frac{1}{6}}}\sqrt{\frac{x\,u_\infty\,\upsilon}{\alpha\,\upsilon}} = 0.336\frac{1}{Pr^{\frac{1}{6}}}\sqrt{\frac{Pr\,x\,u_\infty}{\upsilon}} = 0.336\,Pr^{\frac{1}{3}}Re_x^{\frac{1}{2}} \tag{4-530}$$

which is within 1% of the self-similar solution discussed in Section 4.4.

4.8.4 Integral Solutions for Turbulent Flows

The characteristics of turbulent flows are discussed in Sections 4.5 through 4.7 without providing any techniques that can be applied to solve specific problems. The integral solution technique described in this section can be applied to a turbulent flow problem. However, the functional forms of the velocity and temperature distributions and the associated shear stress and heat flux relationships that are presented in Table 4-4 and Table 4-5 for a laminar flow must be modified appropriately. The velocity distribution throughout the majority of the turbulent boundary layer can be described by the power law relationship given in Section 4.7.7 by Eq. (4-413):

$$\frac{u}{u_\infty} = \left(\frac{y}{\delta_m}\right)^{1/7} \tag{4-531}$$

Very near the wall, in the viscous sublayer, the simple power law velocity distribution is not adequate. Therefore, it is not appropriate to relate the shear stress at the wall (τ_s) to the gradient of Eq. (4-531) at $y = 0$. Rather, the wall shear stress must be separately related to the momentum boundary layer using a correlation based on experimental data. For example, the correlation provided by Kays et al. (2005):

$$\tau_s = 0.0225 \, \rho \, u_\infty^2 \left(\frac{\upsilon}{u_\infty \, \delta_m}\right)^{1/4} \tag{4-532}$$

The integral form of momentum equation, Eq. (4-467), is simplified for application to a flat plate with no transpiration or pressure gradient:

$$\frac{d}{dx}\left[\int_{y=0}^{y=\delta_m} (u^2 - u \, u_\infty) \, dy\right] = -\frac{\tau_s}{\rho} \tag{4-533}$$

Substituting Eqs. (4-531) and (4-532) into Eq. (4-533) leads to:

$$\frac{d}{dx}\left[\int_{y=0}^{y=\delta_m} \left[\left(\frac{y}{\delta_m}\right)^{2/7} - \left(\frac{y}{\delta_m}\right)^{1/7}\right] dy\right] = -0.0225 \left(\frac{\upsilon}{u_\infty}\delta_m\right)^{1/4} \tag{4-534}$$

Note that the velocity distribution in Eq. (4-531) is integrated from the wall surface ($y = 0$) to the edge of the boundary layer ($y = \delta_m$), even though the power law relationship is not valid in the viscous sublayer. This does not introduce a significant error because the amount of momentum carried by the viscous sublayer is very small.

Carrying out the integration in Eq. (4-534) leads to:

$$\frac{d}{dx}\left[\left(\frac{7}{9}\frac{y^{9/7}}{\delta_m^{2/7}} - \frac{7}{8}\frac{y^{8/7}}{\delta_m^{1/7}}\right)_0^{\delta_m}\right] = -0.0225 \left(\frac{\upsilon}{u_\infty \, \delta_m}\right)^{1/4} \tag{4-535}$$

Applying the limits of integration leads to:

$$-0.0972\frac{d\delta_m}{dx} = -0.0225 \left(\frac{\upsilon}{u_\infty \, \delta_m}\right)^{1/4} \tag{4-536}$$

which is the ordinary differential equation that governs δ_m Equation (4-536) is separated:

$$\delta_m^{1/4} d\delta_m = 0.2315 \left(\frac{v}{u_\infty}\right)^{1/4} dx \tag{4-537}$$

and integrated from the leading edge of the plate:

$$\int_0^{\delta_m} \delta_m^{1/4} d\delta_m = 0.232 \left(\frac{v}{u_\infty}\right)^{1/4} \int_0^x dx \tag{4-538}$$

which leads to:

$$\frac{4}{5}\delta_m^{5/4} = 0.232 \left(\frac{v}{u_\infty}\right)^{1/4} x \tag{4-539}$$

Solving Eq. (4-539) for the momentum boundary layer thickness leads to:

$$\delta_m = 0.371 \left(\frac{v}{u_\infty}\right)^{1/5} x^{4/5} \tag{4-540}$$

Substituting the definition of the Reynolds number into Eq. (4-540) leads to:

$$\delta_m = \frac{0.371}{Re_x^{0.2}} x \tag{4-541}$$

The friction coefficient is defined as:

$$C_f = \frac{2\,\tau_s}{\rho\,u_\infty^2} \tag{4-542}$$

Substituting Eq. (4-532) into Eq. (4-542) leads to:

$$C_f = 0.045 \left(\frac{v}{u_\infty\,\delta_m}\right)^{1/4} \tag{4-543}$$

Substituting Eq. (4-541) into Eq. (4-543) leads to:

$$C_f = 0.045 \left(\frac{v\,Re_x^{0.2}}{u_\infty\,0.371\,x}\right)^{1/4} \tag{4-544}$$

which can be simplified to:

$$C_f = 0.057\,Re_x^{-0.2} \tag{4-545}$$

Equation (4-545) is within 5% of the correlation for the friction factor for turbulent flow over a smooth plate suggested by Schlichting (2000), discussed in Section 4.9.2.

According to the modified Reynolds analogy, discussed in Section 4.1.2, the Nusselt number is related to the friction coefficient according to:

$$Nu_x \approx \frac{Pr^{1/3}\,C_f\,Re_x}{2} \tag{4-546}$$

Substituting Eq. (4-545) into Eq. (4-546) leads to:

$$Nu_x = 0.0285\,Pr^{1/3}\,Re_x^{0.8} \tag{4-547}$$

which is within 5% of the correlation for the Nusselt number for turbulent flow over a smooth plate that is presented in Section 4.9.2.

4.9 External Flow Correlations

4.9.1 Introduction

Sections 4.1 and 4.5 discuss the behavior of laminar and turbulent boundary layers, respectively, at a conceptual level without presenting either analytical or experimental information that could be used to solve an external flow problem. In Section 4.2, the boundary layer equations are derived and Section 4.3 shows how, with some limitations, the engineering results associated with any analytical, numerical, or experimental solution to an external flow problem can be correlated using a limited set of nondimensional parameters (specifically the Reynolds number, Prandtl number, and Nusselt number). Sections 4.4 and 4.8 present some methods for solving external convection problems; however, these tools are limited to only very simple problems.

This section presents a series of useful correlations for external flow that can be used to solve a wide range of engineering problems. These results are based on very careful experimental and theoretical work accomplished by many researchers. The correlations are examined as they are presented in order to verify that they agree with the physical insight that is developed in Sections 4.1 and 4.5. A relatively complete and useful set of these correlations has been provided as libraries of functions that are available in EES and the use of these functions to solve practical engineering problems is illustrated by example.

4.9.2 Flow over a Flat Plate

Flow over a flat plate has been discussed in previous sections. The most common correlations apply to a plate with a constant surface temperature, T_s, exposed to a free stream with a uniform velocity and temperature, u_∞ and T_∞. The Reynolds number for this configuration is defined based on the position with respect to the leading edge, x:

$$Re_x = \frac{\rho u_\infty x}{\mu} \tag{4-548}$$

where ρ and μ are the density and viscosity of the fluid evaluated at the film temperature, T_{film}, which is the average of the free stream and surface temperature:

$$T_{film} = \frac{T_\infty + T_s}{2} \tag{4-549}$$

The local shear stress experienced by the plate (τ_s) is correlated by the local friction coefficient:

$$C_f = \frac{2 \tau_s}{\rho u_\infty^2} \tag{4-550}$$

and the local heat transfer coefficient (h) is correlated by the local Nusselt number, also based on x:

$$Nu_x = \frac{h x}{k} \tag{4-551}$$

where k is the conductivity of the fluid evaluated at the film temperature.

Friction Coefficient

The flow is laminar for $Re_x < Re_{crit}$ where Re_{crit} is the critical Reynolds number. The critical Reynolds number may vary based on the flow situation, but a typical value is taken to be 5×10^5. If the Reynolds number is less than Re_{crit}, then the self-similar solution discussed in Section 4.4 is valid and the local friction coefficient is given by:

$$C_f = \frac{0.664}{\sqrt{Re_x}} \quad \text{for } Re_x < Re_{crit}$$

(4-552)

The thickness of the laminar momentum boundary layer, defined as the position where the velocity reaches 99% of the free stream velocity, is:

$$\delta_{m,lam} = \frac{4.916\,x}{\sqrt{Re_x}} \quad \text{for } Re_x < Re_{crit}$$

(4-553)

Recognizing that the Reynolds number is proportional to x and u_∞, Eq. (4-553) suggests that the laminar momentum boundary layer will grow as $\sqrt{x/u_\infty} = \sqrt{t}$ where t is the time that the fluid has been in contact with the plate. This result is consistent with a diffusive penetration of momentum into the free-stream, as discussed in Section 4.1.

The laminar thermal boundary layer thickness is:

$$\delta_{t,lam} = \frac{4.916\,x}{\sqrt{Re_x}} Pr^{-\frac{1}{3}} \quad \text{for } Re_x < Re_{crit}$$

(4-554)

Correlations for turbulent flow are based on experimental data. Several alternative correlations are available in the literature. For a smooth plate, the local friction coefficient in a turbulent flow has been correlated according to:

$$C_f = 0.0592\,Re_x^{-0.20} \quad \text{for } Re_{crit} < Re_x < 1 \times 10^8$$

(4-555)

for a Reynolds number ranging from Re_{crit} to 1×10^8 (Schlichting (2000)).

The thickness of the turbulent boundary layer ($\delta_{m,turb}$) can be estimated approximately using the concepts discussed in Sections 4.5 through 4.7. The turbulent boundary layer grows as the fluid moves along the plate surface. This growth is not due to the viscous penetration of momentum but rather due to the turbulent fluctuations at the edge of the boundary layer. Therefore, it is reasonable to expect that the rate that the turbulent boundary layer grows will be given approximately by:

$$\frac{d\delta_{m,turb}}{dt} \approx u^*$$

(4-556)

where u^* is the eddy velocity. Time, t in Eq. (4-556), is related to the distance along the plate and the free stream velocity:

$$u_\infty \frac{d\delta_{m,turb}}{dx} \approx u^*$$

(4-557)

Substituting the definition of the eddy velocity, Eq. (4-368), into Eq. (4-557) leads to:

$$u_\infty \frac{d\delta_{m,turb}}{dx} \approx \sqrt{\frac{\tau_s}{\rho}}$$

(4-558)

Substituting the definition of the local friction coefficient, Eq. (4-550), into Eq. (4-558) leads to:

$$\frac{d\delta_{m,turb}}{dx} \approx \sqrt{\frac{C_f}{2}}$$

(4-559)

Substituting the correlation for C_f provided by Eq. (4-555) into Eq. (4-559) leads to:

$$\frac{d\delta_{m,turb}}{dx} \approx \sqrt{0.0296}\,Re_x^{-0.10}$$

(4-560)

If the laminar region is ignored, then Eq. (4-560) can be integrated from the leading edge of the plate in order to obtain an estimate of the turbulent momentum boundary thickness:

$$\int_0^{\delta_{m,turb}} d\delta_{m,turb} \approx \sqrt{0.0296} \left(\frac{\mu}{\rho u_\infty}\right)^{0.1} \int_0^x x^{-0.1} dx \qquad (4\text{-}561)$$

or

$$\delta_{m,turb} \approx \frac{0.19\,x}{Re_x^{0.1}} \qquad (4\text{-}562)$$

According to White (2003), the turbulent momentum boundary thickness is approximately:

$$\boxed{\delta_{m,turb} \approx \frac{0.16\,x}{Re_x^{1/7}} \quad \text{for } Re_x > Re_{crit}} \qquad (4\text{-}563)$$

Equations (4-562) and (4-563) are consistent with each other, as well as with the solution for $\delta_{m,turb}$ that is obtained using the integral technique in Section 4.8.4, Eq. (4-541). The turbulent Prandtl number, Pr_{turb}, is near unity and therefore Eq. (4-562) or Eq. (4-563) can also be used to estimate the thermal boundary layer thickness in a turbulent flow.

The size of the viscous sublayer can be estimated based on concepts discussed in Sections 4.5 through 4.7. In Section 4.7, we found that the viscous sublayer extends approximately to an inner position, $y^+ = 6$, where y^+ is given by:

$$y^+ = \frac{y\,u_\infty}{\upsilon}\sqrt{\frac{C_f}{2}} \qquad (4\text{-}564)$$

Substituting $y = \delta_{vs}$ at $y^+ = 6$ into Eq. (4-564) leads to:

$$6 \approx \frac{\delta_{vs}\,u_\infty}{\upsilon}\sqrt{\frac{C_f}{2}} \qquad (4\text{-}565)$$

Solving for δ_{vs} leads to:

$$\delta_{vs} \approx \frac{6\,\upsilon}{u_\infty}\sqrt{\frac{2}{C_f}} = \frac{6\sqrt{2}\,x}{Re_x\sqrt{C_f}} \qquad (4\text{-}566)$$

Substituting the correlation for C_f, Eq. (4-555), into Eq. (4-566) leads to:

$$\delta_{vs} \approx \frac{6\sqrt{2}\,x}{Re_x\sqrt{0.0592\,Re_x^{-0.20}}} \qquad (4\text{-}567)$$

or

$$\boxed{\delta_{vs} \approx \frac{34.9\,x}{Re_x^{0.9}}} \qquad (4\text{-}568)$$

Figure 4-27 illustrates the momentum boundary layer thickness and the viscous sublayer thickness as a function of position for a 1 m smooth flat plate exposed to a flow of water at room temperature and pressure with a free stream velocity of $u_\infty = 10$ m/s. Equations (4-553), (4-563), and (4-568) were used to generate Figure 4-27. Notice that the flow transitions to turbulence at approximately $x_{crit} = 5.5$ mm and that the turbulent momentum boundary layer thickness is much thicker and grows more rapidly. Equation (4-563) suggests that the turbulent boundary layer thickness will grow as $x^{6/7}$, whereas the laminar

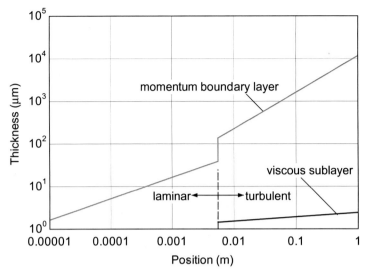

Figure 4-27: Momentum boundary layer and viscous sublayer thickness for water flowing over a 1 m long flat plate at $u_\infty = 10$ m/s.

boundary layer will grow as $x^{1/2}$, according to Eq. (4-553). The increased growth rate is related to the action of the turbulent eddies and the very effective transport of momentum that they provide. The viscous sublayer is several orders of magnitude smaller than either the laminar or turbulent momentum boundary layer. Equation (4-568) suggests that the viscous sublayer grows only as $x^{0.1}$; the viscous sublayer is therefore not only much smaller than the turbulent boundary layer but it also grows much more slowly along the plate.

The smooth plate correlation, Eq. (4-555), is valid provided that the roughness on the plate surface (e) is less than the size of the viscous sublayer. Such a surface is referred to as aerodynamically (or hydrodynamically) smooth; the effect of larger roughness elements on the flow over a flat plate is discussed subsequently in this section.

Figure 4-28 illustrates the local friction factor as a function of the Reynolds number using Eqs. (4-552) and (4-555). As expected, there is a substantial jump in the friction coefficient at the critical Reynolds number as the flow transitions from laminar to turbulent.

The average friction coefficient was discussed in Section 4.1 and is defined according to:

$$\overline{C}_f = \frac{2\,\overline{\tau}_s}{\rho\,u_\infty^2} \tag{4-569}$$

where $\overline{\tau}_s$ is the average shear stress:

$$\overline{\tau}_s = \frac{1}{L} \int_0^L \tau_s\, dx \tag{4-570}$$

Substituting the definition of the local friction coefficient, Eq. (4-550), into Eq. (4-570) leads to:

$$\overline{\tau}_s = \frac{\rho\,u_\infty^2}{2\,L} \int_0^L C_f\, dx \tag{4-571}$$

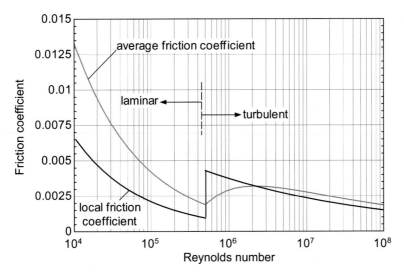

Figure 4-28: Local and average friction coefficient as a function of Reynolds number.

Substituting Eq. (4-571) into Eq. (4-569) leads to:

$$\overline{C_f} = \frac{2}{\rho u_\infty^2} \frac{\rho u_\infty^2}{2L} \int\limits_0^L C_f \, dx \tag{4-572}$$

or

$$\overline{C_f} = \frac{1}{L} \int\limits_0^L C_f \, dx \tag{4-573}$$

so the average friction factor is the average of the local friction factor over the plate surface. The local friction factor, C_f, is correlated in terms of Re_x and therefore it is useful to transform the coordinate of integration in Eq. (4-573) from x to Re_x:

$$x = \frac{\mu \, Re_x}{\rho u_\infty} \tag{4-574}$$

so that

$$dx = \frac{\mu}{\rho u_\infty} dRe_x \tag{4-575}$$

Substituting Eqs. (4-574) and (4-575) into Eq. (4-573) leads to:

$$\overline{C_f} = \frac{1}{L} \int\limits_0^{Re_L} C_f \frac{\mu}{\rho u_\infty} dRe_x = \frac{\mu}{\rho u_\infty L} \int\limits_0^{Re_L} C_f \, dRe_x \tag{4-576}$$

or

$$\overline{C_f} = \frac{1}{Re_L} \int\limits_0^{Re_L} C_f \, dRe_x \tag{4-577}$$

where Re_L is the Reynolds number evaluated at the trailing edge of the plate. Equation (4-577) is valid regardless of the specific correlation(s) that are used to express

C_f as a function of Re_x. In the laminar region, Eq. (4-552) can be substituted into Eq. (4-577) to provide:

$$\overline{C}_f = \frac{1}{Re_L} \int_0^{Re_L} \frac{0.664}{\sqrt{Re_x}} dRe_x = \frac{0.664}{Re_L} \left[2\,Re_x^{0.5} \right]_0^{Re_L} \tag{4-578}$$

or

$$\boxed{\overline{C}_f = \frac{1.328}{\sqrt{Re_L}} \quad \text{for } Re_L < Re_{crit}} \tag{4-579}$$

The average friction coefficient including both the laminar and the turbulent regions is obtained by integrating Eq. (4-577) in two parts, corresponding to a laminar region with Eq. (4-552) for $Re_x < Re_{crit}$ and a turbulent region with Eq. (4-555) for $Re_{crit} < Re_x < Re_L$:

$$\overline{C}_f = \frac{1}{Re_L} \left[\int_0^{Re_{crit}} \frac{0.664}{\sqrt{Re_x}} dRe_x + \int_{Re_{crit}}^{Re_L} 0.0592\,Re_x^{-0.20} dRe_x \right] \tag{4-580}$$

or

$$\boxed{\overline{C}_f = \frac{1}{Re_L} \left[1.328\,Re_{crit}^{0.5} + 0.0740 \left(Re_L^{0.8} - Re_{crit}^{0.8} \right) \right] \quad \text{for } Re_{crit} < Re_L < 1 \times 10^8} \tag{4-581}$$

The average friction coefficient evaluated using Eqs. (4-579) and (4-581) is also shown in Figure 4-28.

Nusselt Number

When the Reynolds number is less than Re_{crit}, then the self-similar solution discussed in Section 4.4 is valid. The self-similar solution is not explicit, but the local Nusselt number has been correlated by Churchill and Ozoe (1973) according to:

$$\boxed{Nu_x = \frac{0.3387\,Re_x^{1/2}\,Pr^{1/3}}{\left[1 + \left(\dfrac{0.0468}{Pr} \right)^{2/3} \right]^{1/4}} \quad \text{for } Re_x < Re_{crit}} \tag{4-582}$$

where Pr is the Prandtl number (which should be evaluated at the film temperature).

For turbulent flow, correlations are based on data and, again, several are available in the literature. For a smooth plate, the local Nusselt number can be computed using the modified Reynolds analogy, discussed in Sections 4.1.2 and 4.3.4, together with Eq. (4-555):

$$\boxed{Nu_x = 0.0296\,Re_x^{4/5}\,Pr^{1/3} \quad \text{for } Re_{crit} < Re_x < 1 \times 10^8 \quad \text{and} \quad 0.6 < Pr < 60} \tag{4-583}$$

Figure 4-29 illustrates the local Nusselt number in the laminar and turbulent regions for various values of the Prandtl number and shows the large increase in the local Nusselt number that occurs at the transition from laminar to turbulent flow conditions.

Figure 4-29 shows that the Nusselt number increases with Prandtl number by an amount that is essentially the same in both the laminar and turbulent regions. The conceptual model of the turbulent boundary layer presented in Section 4.5 indicates that

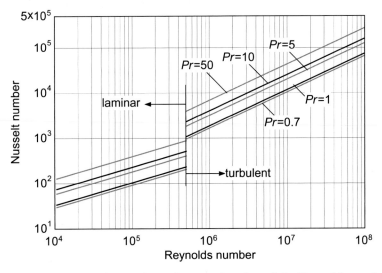

Figure 4-29: Local Nusselt number as a function of the Reynolds number for various values of the Prandtl number.

the transport of energy is primarily due to turbulent eddies throughout the boundary layer. Therefore, the substantial improvement in the Nusselt number with the molecular Prandtl number (which is a property of the fluid rather than the turbulent eddies) in the turbulent region may be non-intuitive. However, the dominant resistance to heat transfer in a turbulent boundary layer is associated with conduction through the viscous sublayer, and the thermal resistance of the viscous sublayer is substantially affected by the molecular Prandtl number.

Figure 4-29 can be misleading; the local Nusselt number increases with the Reynolds number (which is proportional to the position along the plate) even though the heat transfer coefficient tends to decrease with x due to the thickening of the boundary layer. The apparent discrepancy is related to the fact that the characteristic length used to define the Nusselt number for a flat plate is position, x:

$$Nu_x = \frac{h\,x}{k} \tag{4-584}$$

Therefore, even if the heat transfer coefficient were constant the Nusselt number would increase with x and with Re_x. Figure 4-30 illustrates the heat transfer coefficient for the flow of water with a free stream velocity of 10 m/s at ambient conditions as a function of position on a 1 m long flat plate. Figure 4-30 was generated using Eqs. (4-582) and (4-583). Notice that the heat transfer coefficient does not decrease as quickly in the turbulent region. This behavior is consistent with the earlier observation from Figure 4-27 that the viscous sublayer does not grow very quickly.

The average Nusselt number ($\overline{Nu}$) was discussed in Section 4.1.3 and defined according to:

$$\overline{Nu}_L = \frac{\overline{h}\,L}{k} \tag{4-585}$$

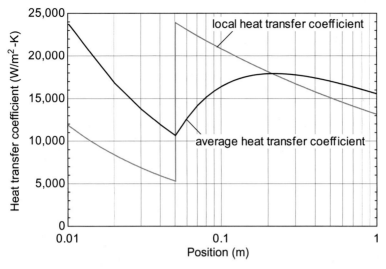

Figure 4-30: The local and average heat transfer coefficient as a function of position for a flow of water at ambient conditions with $u_\infty = 10$ m/s over a 1 m long flat plate.

where $(\overline{h})$ is the average heat transfer coefficient, defined according to:

$$\overline{h} = \frac{1}{L} \int_0^L h \, dx \tag{4-586}$$

Substituting Eqs. (4-586) and (4-584) into Eq. (4-585) leads to:

$$\overline{Nu_L} = \frac{L}{k} \frac{1}{L} \int_0^L \frac{Nu_x k}{x} \, dx \tag{4-587}$$

or

$$\overline{Nu_L} = \int_0^L \frac{Nu_x}{x} \, dx \tag{4-588}$$

Because Nusselt number correlations are expressed in terms of the Reynolds number, it is convenient to change the coordinate of integration in Eq. (4-588) from x to Re_x by substituting Eqs. (4-574) and (4-575) into Eq. (4-588):

$$\overline{Nu_L} = \int_0^{Re_L} \frac{Nu_x}{Re_x} \, dRe_x \tag{4-589}$$

Equation (4-589) can be integrated using any correlation or set of correlations for the local Nusselt number. In the laminar region, Eq. (4-582) is substituted into Eq. (4-589) in order to obtain:

$$\overline{Nu}_L = \frac{0.3387\,Pr^{1/3}}{\left[1 + \left(\dfrac{0.0468}{Pr}\right)^{2/3}\right]^{1/4}} \int_0^{Re_L} Re_x^{-1/2}\,dRe_x = \frac{0.3387\,Pr^{1/3}}{\left[1 + \left(\dfrac{0.0468}{Pr}\right)^{2/3}\right]^{1/4}} \left[2\,Re_x^{1/2}\right]_0^{Re_L}$$

(4-590)

or

$$\boxed{\overline{Nu}_L = \frac{0.6774\,Pr^{1/3}\,Re_L^{1/2}}{\left[1 + \left(\dfrac{0.0468}{Pr}\right)^{2/3}\right]^{1/4}} \quad \text{for } Re_L < Re_{crit}}$$

(4-591)

The average Nusselt number for the combined laminar and turbulent regions is obtained by integrating Eq. (4-589) in two parts; Eq. (4-582) is used in the laminar region, from $Re_x < Re_{crit}$, and Eq. (4-583) is used in the turbulent region, for $Re_{crit} < Re_x < Re_L$:

$$\overline{Nu}_L = \frac{0.3387\,Pr^{1/3}}{\left[1 + \left(\dfrac{0.0468}{Pr}\right)^{2/3}\right]^{1/4}} \int_0^{Re_{crit}} Re_x^{-1/2}\,dRe_x + 0.0296\,Pr^{1/3} \int_{Re_{crit}}^{Re_L} Re_x^{-0.2}\,dRe_x$$

(4-592)

or

$$\boxed{\overline{Nu}_L = \frac{0.6774\,Pr^{1/3}\,Re_{crit}^{1/2}}{\left[1 + \left(\dfrac{0.0468}{Pr}\right)^{2/3}\right]^{1/4}} + 0.037\,Pr^{1/3}\left(Re_L^{0.8} - Re_{crit}^{0.8}\right)}$$

(4-593)

The average heat transfer coefficient for a flow of water at ambient conditions with free stream velocity $u_\infty = 10$ m/s over a 1 m long flat plate, predicted using Eqs. (4-591) and (4-593), is overlaid onto Figure 4-30.

The correlations for flow over a flat plate with constant surface temperature are summarized in Table 4-6. The average friction coefficient and Nusselt number correlations are available for a smooth flat plate as a built-in procedure in EES. To access this procedure, select Function Info from the Options menu and then select convection from the pull down menu at the lower right corner of the upper window. Select External Flow – Non-dimensional and scroll to the flat plate. The External_Flow_Plate_ND code is a procedure rather than a function. Procedures may return multiple outputs as opposed to functions, which can return only one. The arguments to the left of the colon are inputs; these include the Reynolds number and the Prandtl number for the procedure External_Flow_Plate_ND. The arguments to the right of the colon are outputs; these include the average Nusselt number and average friction coefficient.

Table 4-6: Summary of correlations for a smooth isothermal flat plate.

Flow condition	Parameter	Local value	Average value
laminar, $Re_x < Re_{crit}$	friction coefficient	$C_f = \dfrac{0.664}{\sqrt{Re_x}}$	$\overline{C}_f = \dfrac{1.328}{\sqrt{Re_L}}$
	Nusselt number	$Nu_x = \dfrac{0.3387\, Re_x^{1/2}\, Pr^{1/3}}{\left[1 + \left(\dfrac{0.0468}{Pr}\right)^{2/3}\right]^{1/4}}$	$\overline{Nu}_L = \dfrac{0.6774\, Pr^{1/3}\, Re_L^{1/2}}{\left[1 + \left(\dfrac{0.0468}{Pr}\right)^{2/3}\right]^{1/4}}$
turbulent*, $Re_x > Re_{crit}$	friction coefficient	$C_f = 0.0592\, Re_x^{-0.20}$	$\overline{C}_f = \dfrac{1}{Re_L}\left[1.328\, Re_{crit}^{0.5} + 0.0740\left(Re_L^{0.8} - Re_{crit}^{0.8}\right)\right]$
	Nusselt number	$Nu_x = 0.0296\, Re_x^{4/5}\, Pr^{1/3}$	$\overline{Nu}_L = \dfrac{0.6774\, Pr^{1/3}\, Re_{crit}^{1/2}}{\left[1 + \left(\dfrac{0.0468}{Pr}\right)^{2/3}\right]^{1/4}} + 0.037\, Pr^{1/3}\left(Re_L^{0.8} - Re_{crit}^{0.8}\right)$

* note that the average correlations in the turbulent region include integration of the local correlations for laminar flow through the laminar region.

EXAMPLE 4.9-1: PARTIALLY SUBMERGED PLATE

A plate is pulled through water at a velocity $u_\infty = 5$ m/s (Figure 1). The upper half of the plate is exposed to air at $T_a = 30°C$ and the lower half is exposed to water at $T_w = 5°C$. The water and air are both stagnant (i.e., there is no wind or current). The length of the plate is $L = 1$ m and the width is $W = 1$ m. The plate is submerged exactly half way into the water.

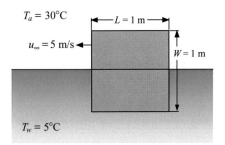

Figure 1: Plate partially submerged in water.

The plate is thin (and therefore form drag can be neglected) and has high conductivity so that it is isothermal.

a) **Determine the average heat transfer coefficient between the plate and the water and between the plate and the air. Determine the temperature of the plate.**

The known information is entered in EES:

```
"EXAMPLE 4.9-1: Partially Submerged Plate"

$UnitSystem SI MASS RAD PA K J
$Tabstops 0.2 0.4 0.6 3.8 in

"Inputs"
u_infinity=5.0 [m/s]                    "plate velocity"
T_a=converttemp(C,K,30 [C])             "air temperature"
T_w=converttemp(C,K,5 [C])              "water temperature"
L=1 [m]                                 "length"
W=1 [m]                                 "width"
```

The Reynolds number associated with the air flowing over the plate is:

$$Re_{L,a} = \frac{\rho_a \, L \, u_\infty}{\mu_a}$$

where ρ_a and μ_a are the density and viscosity of air, evaluated at the film temperature:

$$T_{film,a} = \frac{T_a + T_p}{2}$$

where T_p is the plate temperature. The plate temperature is not known but is required to obtain the properties. Therefore, it is convenient to provide a reasonable guess for the plate temperature (something between T_a and T_w) and use the guess

EXAMPLE 4.9-1: PARTIALLY SUBMERGED PLATE

EXAMPLE 4.9-1: PARTIALLY SUBMERGED PLATE

value to compute the film temperature and the required properties (including the conductivity, k_a, and Prandtl number, Pr_a).

```
T_p=290 [K]                                              "guess for plate temperature"
T_p_C=converttemp(K,C,T_p)                              "in C"
T_film_a=(T_p+T_a)/2                                     "air-side film temperature"
rho_a=density(Air,P=1[atm]*convert(atm,Pa),T=T_film_a)  "air density"
mu_a=viscosity(Air,T=T_film_a)                          "air viscosity"
k_a=conductivity(Air,T=T_film_a)                        "air conductivity"
Pr_a=Prandtl(Air,T=T_film_a)                            "air Prandtl number"
Re_a=rho_a*u_infinity*L/mu_a                            "air Reynolds number"
```

The average Nusselt number ($\overline{Nu}_a$) and average friction coefficient ($\overline{C}_{f,a}$) associated with the air flow are obtained using the procedure **External_Flow_Plate ND** from EES' convection library.

```
Call External_Flow_Plate_ND(Re_a,Pr_a: Nusselt_bar_a,C_f_bar_a)
        "obtain average Nusselt number and friction coefficient"
```

The average Nusselt number is used to compute the average heat transfer coefficient on the air side:

$$\overline{h}_a = \frac{\overline{Nu}_a\, k_a}{L}$$

```
h_bar_a=Nusselt_bar_a*k_a/L    "average air-side heat transfer coefficient"
```

The Reynolds number associated with the water flowing over the plate is:

$$Re_{L,w} = \frac{\rho_w\, L\, u_\infty}{\mu_w}$$

where ρ_w and μ_w are the density and viscosity of water, respectively, evaluated at the film temperature on the water side:

$$T_{film,w} = \frac{T_w + T_p}{2}$$

The water properties (including conductivy, k_w, and Prandtl number, Pr_w) are evaluated using EES' built-in property routines.

```
T_film_w=(T_p+T_w)/2                                       "water-side film temperature"
rho_w=density(Water,P=1[atm]*convert(atm,Pa),T=T_film_w)   "water density"
mu_w=viscosity(Water,P=1[atm]*convert(atm,Pa),T=T_film_w)  "water viscosity"
k_w=conductivity(Water,P=1[atm]*convert(atm,Pa),T=T_film_w)"water conductivity"
Pr_w=Prandtl(Water,P=1[atm]*convert(atm,Pa),T=T_film_w)    "water Prandtl number"
Re_w=rho_w*u_infinity*L/mu_w                               "water Reynolds number"
```

EXAMPLE 4.9-1: PARTIALLY SUBMERGED PLATE

The water-side Reynolds number and Prandtl number are used to evaluate the average Nusselt number ($\overline{Nu}_w$) and friction coefficient ($\overline{C}_{f,w}$) for the submerged portion of the plate using the procedure External_Flow_Plate_ND. The average Nusselt number is used to calculate the average heat transfer coefficient:

$$\overline{h}_w = \frac{\overline{Nu}_w\, k_w}{L}$$

Call External_Flow_Plate_ND(Re_w,Pr_w: Nusselt_bar_w,C_f_bar_w)
  "obtain average Nusselt number and friction coefficient"
h_bar_w=Nusselt_bar_w*k_w/L   "average water-side heat transfer coefficient"

An energy balance on the plate balances convection from the air with convection to the water:

$$\overline{h}_a\, L\frac{W}{2}(T_a - T_p) = \overline{h}_w\, L\frac{W}{2}(T_p - T_w) \tag{1}$$

The solution of Eq. (1) provides the plate temperature and therefore will over-specify the problem. Solve the problem and then update the guess values (select Update Guesses from the Calculate menu). Comment out the assignment of the guessed value of the plate temperature:

{T_p=290 [K]}   "guess for plate temperature"

and replace the equation with the energy balance, Eq. (1):

h_bar_a*L*W*(T_a-T_p)/2=h_bar_w*L*W*(T_p T_w)/2   "energy balance on plate"

The result is $\overline{h}_a = 8.5$ W/m²-K, $\overline{h}_w = 6520$ W/m²-K, and $T_p = 278.2$ K (5.03°C). Note that the plate will come very close to the water temperature because the average heat transfer coefficient on the water side is so much higher than on the air side.

b) Determine the total force exerted on the plate. How much of this force is related to the drag from the water and how much to the drag from the air?

The force exerted on the air-side and the water-side of the plate can be obtained using the average friction coefficient:

$$F_a = \overline{C}_{f,a}\frac{\rho_a u_\infty^2}{2} W\, L$$

$$F_w = \overline{C}_{f,w}\frac{\rho_w u_\infty^2}{2} W\, L$$

F_a=C_f_bar_a*0.5*rho_a*u_infinity^2*W*L   "air-side force"
F_w=C_f_bar_w*0.5*rho_w*u_infinity^2*W*L   "water-side force"

The result is $F_a = 0.035$ N and $F_w = 39.4$ N for a total force of about 39.4 N.

free stream, u_∞, T_∞

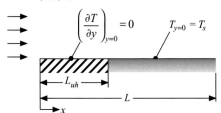

$$\left(\frac{\partial T}{\partial y}\right)_{y=0} = 0 \qquad T_{y=0} = T_s$$

L_{uh}

L

x

Figure 4-31: Plate with an unheated starting length.

Unheated Starting Length

The integral techniques discussed in Section 4.8 can be used to provide the approximate solution for a flat plate that has an adiabatic section at its leading edge so that the flow develops hydrodynamically before it begins to develop thermally, as shown in Figure 4-31.

The solution to to the problem under laminar conditions is:

$$Nu_x = \frac{Nu_{x,L_{uh}=0}}{\left[1 - \left(\frac{L_{uh}}{x}\right)^{0.75}\right]^{1/3}} \tag{4-594}$$

and under turbulent conditions, the solution is:

$$Nu_x = \frac{Nu_{x,L_{uh}=0}}{\left[1 - \left(\frac{L_{uh}}{x}\right)^{0.90}\right]^{1/9}} \tag{4-595}$$

where $Nu_{x,L_{uh}=0}$ is the Nusselt number at the same position, x, that would be obtained with $L_{uh} = 0$. That is, $Nu_{x,L_{uh}=0}$ is the result obtained Eq. (4-582) for laminar flow and Eq. (4-583) for turbulent flow. The average Nusselt number coefficient ($\overline{Nu}_L$) (averaged over the constant surface temperature portion of the plate) is (Ameel (1997)):

$$\overline{Nu}_L = \overline{Nu}_{L,L_{uh}=0} \frac{L}{(L - L_{uh})} \left[1 - \left(\frac{L_{uh}}{x}\right)^{\frac{(p+1)}{(p+2)}}\right]^{\frac{p}{(p+1)}} \tag{4-596}$$

where $\overline{Nu}_{L,L_{uh}=0}$ is the average Nusselt number obtained if $L_{uh} = 0$ (i.e., the result obtained using Eq. (4-591) or Eq. (4-593)). The parameter p in Eq. (4-596) is 2 for laminar flow and 8 for turbulent flow.

Constant Heat Flux

The local Nusselt number for a flat plate with a constant heat flux, $\dot{q}_s''$, under laminar conditions is (Kays et al. (2005)):

$$Nu_x = 0.453 \, Re_x^{0.5} \, Pr^{1/3} \quad \text{for } Pr > 0.6 \tag{4-597}$$

The local Nusselt number for turbulent flow over a flat plate with constant heat flux is:

$$Nu_x = 0.0308 \, Re_x^{0.8} \, Pr^{1/3} \quad \text{for } 0.6 < Pr < 60 \tag{4-598}$$

In the case of a uniform heat flux, the total rate of heat transfer from the surface is known directly from the product of the area and the heat flux. Therefore, the average surface temperature $(\overline{T}_s)$ is used to define an average heat transfer coefficient:

$$\overline{h} = \frac{\dot{q}_s''}{(\overline{T}_s - T_\infty)} \tag{4-599}$$

and an average Nusselt number:

$$\overline{Nu}_L = \frac{\overline{h}\,L}{k} \tag{4-600}$$

The average Nusselt number associated with the laminar flow solution, Eq. (4-597), is:

$$\boxed{\overline{Nu}_L = 0.680\,Re_x^{0.5}\,Pr^{1/3}} \tag{4-601}$$

Equation (4-601) is very close to the result associated with a constant surface temperature, Eq. (4-591). The average Nusselt number for a turbulent flow agrees even more closely with the constant temperature result, Eq. (4-593).

Flow over a Rough Plate

The thickness of the momentum and thermal boundary layers for laminar flow are substantially larger than the roughness that will be encountered on most engineering surfaces. Therefore, laminar flows tend to be insensitive to the roughness of the surface. However, the gradients of velocity and temperature at the wall in a turbulent flow are controlled by the viscous sublayer, which may be extremely thin (see Figure 4-27) and can easily be on the same scale as the roughness that exists on many engineering surface. Therefore, the behavior of a turbulent flow can be substantially affected by the surface characteristics. The presence of roughness will tend to disrupt the viscous sublayer and therefore increase both the friction coefficient and the heat transfer coefficient. The heat transfer surfaces in heat exchangers are sometimes roughened specifically to improve the heat transfer. This improvement will come at the expense of increased fluid friction and therefore additional pump or fan power.

In Section 4.4, it was shown that the viscous sublayer extends until the inner position, y^+ reaches a value of approximately 6; this leads to the estimate for the viscous sublayer that is provided by Eq. (4-568). The ratio of the size of the roughness elements (e) to the size of the viscous sublayer (δ_{vs}) provides the appropriate measure of the importance of surface roughness for a turbulent flow:

$$\frac{e}{\delta_{vs}} = \frac{e}{x}\frac{Re_x\sqrt{C_f}}{6\sqrt{2}} \tag{4-602}$$

If e/δ_{vs} is less than 1, then the correlations provided for smooth plates are valid. However, if e/δ_{vs} is between 1 and 12, then the surface is referred to as transitionally rough and the roughness will substantially affect the friction and heat transfer performance. Reliable correlations for the transitionally rough regime are not available. If e/δ_{vs} is greater than 12, then the surface is fully rough. The roughness elements on a fully rough surface are much larger than the viscous sublayer; you can imagine that the roughness acts like a series of cylinders that penetrate far into the fully turbulent core. The force experienced by the fully rough plate is primarily due to form drag on these cylinders rather than viscous shear at the surface. The friction coefficient is therefore no longer a function of Reynolds number. To understand why this is so, recall the definition of the

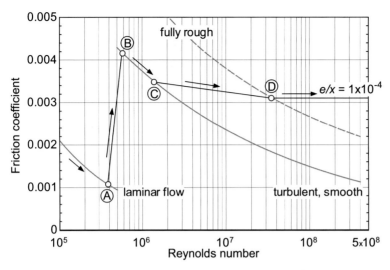

Figure 4-32: Friction coefficient as a function of Reynolds number for flow over a rough flat plate.

friction coefficient:

$$C_f = \frac{2\,\tau_s}{\rho\,u_\infty^2} \qquad (4\text{-}603)$$

For a fully rough surface, τ_s in Eq. (4-603) is no longer the viscous shear at the plate surface but rather the force on the plate per unit area. The force exerted on each roughness element, and therefore τ_s, will be proportional to the pressure difference between the stagnation point and the wake region of the element. This pressure difference is proportional to $\rho\,u_\infty^2/2$. (This behavior of flow across a cylinder is discussed in more detail in Section 4.9.3.)

$$\tau_s \propto \frac{\rho\,u_\infty^2}{2} \qquad (4\text{-}604)$$

Substituting Eq. (4-604) into Eq. (4-603) suggests that the friction factor will be constant for a specified roughness. The local friction coefficient in the fully rough regime has been correlated in terms of the roughness as (Mills and Hang (1983)):

$$C_f = \frac{1}{\left[3.476 + 0.707\,\ln\left(\frac{x}{e}\right)\right]^{2.46}} \quad \text{for } 150 < \frac{x}{e} < 1.5 \times 10^7 \qquad (4\text{-}605)$$

The average friction coefficient in the fully rough regime is given by:

$$\overline{C}_f = \frac{1}{\left[2.635 + 0.618\,\ln\left(\frac{L}{e}\right)\right]^{2.57}} \quad \text{for } 150 < \frac{L}{e} < 1.5 \times 10^7 \qquad (4\text{-}606)$$

Figure 4-32 illustrates the approximate progression of the local friction factor for a plate with roughness e at a position x where $e/x = 1 \times 10^{-4}$ as a function of the Reynolds number. The flow is initially laminar and the friction coefficient is predicted by Eq. (4-552) until the critical Reynolds number is reached, at point A, where the flow transitions to turbulent flow, point B. The friction coefficient is predicted by Eq. (4-555), the correlation for smooth plates, until the viscous sublayer becomes comparable to the size of the roughness at point C. The Reynolds number at which this occurs can be ascertained by setting the ratio e/δ_{vs}, Eq. (4-602), equal to 1 and using Eq. (4-555) for the friction coefficient. This is accomplished in the following EES code:

```
eoverx=1e-4 [-]                          "relative roughness"
Cf=0.0592*Re^(-0.2)                      "friction coefficient for a smooth plate"
1=eoverx*Re*sqrt(Cf)/(6*sqrt(2))         "ratio of the viscous sublayer thickness to the roughness"
```

The result of this calculation indicates that the flow will enter the transitionally rough regime when $Re_x \approx 1.4 \times 10^6$ (note that appropriate limits on C_f and Re must be applied in order for EES to converge to this solution). From point C to point D, the flow is transitioning to the fully rough regime; no correlations were presented for this transition, but you might expect that the behavior will resemble the transition from smooth to fully rough behavior for internal flow that is seen on the Moody diagram (this is discussed in Section 5.2.3). Finally, the flow becomes fully rough at point D. The Reynolds number at which this occurs can be determined by setting the ratio e/δ_{vs}, Eq. (4-602), equal to 12 and using Eq. (4-605) for the friction coefficient in the fully rough region.

```
eoverx=1e-4 [-]                              "relative roughness"
Cf=1/(3.476+0.707*ln(1/eoverx))^2.46         "friction coefficient for a fully rough plate"
12=eoverx*Re*sqrt(Cf)/(6*sqrt(2))            "ratio of the viscous sublayer thickness to the roughness"
```

The result of the calculation indicates that the flow will become fully rough when $Re_x \approx 1.7 \times 10^7$. In the fully rough regime, the friction coefficient will be independent of Reynolds number, as shown in Figure 4-32.

The local Nusselt number in the fully rough regime is (Dipprey and Sabersky (1963)):

$$Nu_x = \frac{Re_x\, Pr\, (C_f/2)}{\left\{ 0.9 + \sqrt{C_f/2}\left[24 \left(\frac{e}{\delta_{vs}}\right)^{0.2} Pr^{0.44} - 7.65 \right] \right\}} \qquad (4\text{-}607)$$

4.9.3 Flow across a Cylinder

The boundary layer formed by external flow across a cylinder, or any other shape that has a geometry that is more complex than a flat plate, will be subjected to pressure gradients that are generated as the free-stream flow decelerates and accelerates. For example, Figure 4-33 illustrates the flow over a cylinder that is exposed to a uniform upstream velocity, u_f. The free-stream velocity that is experienced at the outer edge of the boundary layer that forms at the cylinder surface, u_∞, is not constant and therefore the free-stream pressure, p_∞, varies as well. The pressure gradient term in the momentum equation is not zero as it was for the flat plate. The free-stream pressure variation will affect the boundary layer growth and will likely cause the boundary layer to separate from the surface, resulting in a wake region.

Figure 4-33(a) illustrates, qualitatively, the streamlines associated with flow around a cylinder at very low velocity. A stagnation point will occur at $\theta = 0°$, where the fluid is initially brought to rest. The reduction in the fluid velocity is accompanied by an increase in the pressure at the stagnation point. The velocity of the fluid in the free stream that

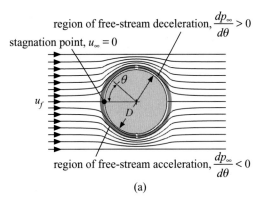

(a)

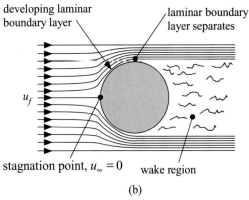

(b)

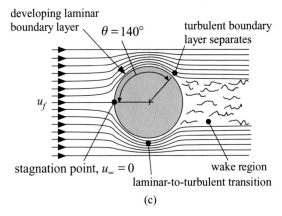

(c)

Figure 4-33: Flow over a cylinder, (a) at very low velocity, (b) under conditions where the boundary layer is laminar and separates, and (c) under conditions where the boundary layer transitions to turbulence and then separates.

is adjacent to the cylinder surface will then increase from $\theta = 0°$ to $\theta = 90°$ (notice that the streamlines are getting closer together) resulting in a decrease in pressure. The pressure gradient experienced by the boundary layer in this region is negative; pressure decreases in the direction of flow from $\theta = 0°$ to $\theta = 90°$. Because fluid likes to flow from high pressure to low pressure, a negative pressure gradient is referred to as a "favorable" pressure gradient. The boundary layer develops from the stagnation point and thickens in the downstream direction. The favorable pressure gradient results in a very stable boundary layer that is somewhat thinner than it would be at the same distance from the leading edge of a flat plate.

The flow will decelerate from $\theta = 90°$ to $\theta = 180°$ and therefore the pressure will increase, the pressure gradient experienced by the boundary layer in this region will be positive. Fluid does not easily flow from low pressure to high pressure and therefore a positive pressure gradient is referred to as an "adverse" pressure gradient. The adverse pressure gradient results in a very unstable boundary layer and therefore the flow is likely to separate from the cylinder's surface, as shown in Figure 4-33(b). The effect of the adverse pressure gradient and the probability of separation cause the behavior of external flow around a cylinder to deviate substantially from the behavior of a boundary layer on a flat plate. However, the behavior can be correlated using the essentially the same set of non-dimensional parameters that were used for a flat plate.

The Reynolds number for flow over a cylinder is defined according to:

$$Re_D = \frac{\rho\, u_f\, D}{\mu} \tag{4-608}$$

where D is the diameter of the cylinder, u_f is the upstream velocity, and ρ and μ are the density and viscosity of the fluid, respectively, evaluated at the film temperature, T_{film}, which is the average of the free stream and surface temperature.

The drag force experienced by the cylinder (F) is correlated by the drag coefficient (C_D):

$$C_D = \frac{2\,F}{\rho\, u_f^2\, L\, D} \tag{4-609}$$

where L is the length of the cylinder. The local heat transfer coefficient (h) is correlated by the Nusselt number:

$$Nu_D = \frac{h\, D}{k} \tag{4-610}$$

where k is the conductivity of the fluid, also evaluated at the film temperature. The local Nusselt number is a function of θ, the position on the surface of the cylinder. The average Nusselt number, $\overline{Nu_D}$, correlates the average heat transfer coefficient.

$$\overline{Nu_D} = \frac{\overline{h}\, D}{k} \tag{4-611}$$

Drag Coefficient

Figure 4-34 illustrates the drag coefficient for a cylinder in cross flow as a function of the Reynolds number (Schlichting (2000)). The behavior exhibited in Figure 4-34 can be understood by returning to Figure 4-33. At a very low Reynolds number (less than about unity), the flow does not separate and therefore the drag force is primarily related to viscous shear at the surface. This is the same phenomenon that provides a shear force on the flat plate. The force associated with viscous shear will scale according to the area exposed to shear and the product of the viscosity and the velocity gradient at the wall. To first order, the shear force is given by:

$$F \approx \mu \frac{u_f}{\delta_m} D\, L \tag{4-612}$$

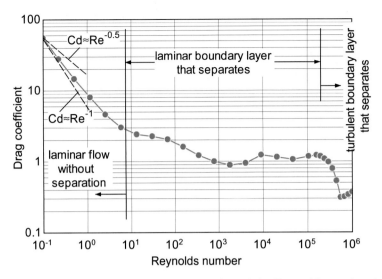

Figure 4-34: Drag coefficient as a function of the Reynolds number for a cylinder (based on Schlichting (2000)).

and so the drag coefficient will scale according to:

$$C_D \approx \frac{\mu \frac{u_f}{\delta_m} D L}{\rho u_f^2 D L} = \frac{\mu}{\rho u_f \delta_m} = \frac{\mu}{\rho u_f D} \frac{D}{\delta_m} = \frac{1}{Re_D} \frac{D}{\delta_m} \qquad (4\text{-}613)$$

The ratio of the momentum boundary layer thickness to the diameter should scale approximately as $Re_D^{-0.5}$ according to flat plate theory, Eq. (4-553), and therefore the drag coefficient should scale approximately as:

$$C_D \approx \frac{1}{\sqrt{Re_D}} \qquad (4\text{-}614)$$

In fact, Figure 4-34 shows that the drag coefficient scales somewhere between $Re_D^{-0.5}$ and Re_D^{-1} at a very low Reynolds number.

When the Reynolds number reaches approximately 10, the laminar boundary layer detaches and a wake region forms, as shown in Figure 4-33(b). The laminar boundary layer is thick and has relatively low momentum throughout. Therefore, it is particularly susceptible to adverse pressure gradients and it will detach almost immediately, at the apex of the cylinder. The wake region covers essentially the entire downstream side of the cylinder.

The drag force on the cylinder after the laminar boundary layer separates is primarily due to form drag. The pressure associated with the fluid deceleration exerted on the upstream side of the cylinder is, approximately, $\rho u_f^2/2$ greater than ambient pressure, while the pressure in the wake region is approximately ambient. Therefore, the drag force experienced by the cylinder is approximately:

$$F \approx \frac{\rho u_f^2}{2} D L \qquad (4\text{-}615)$$

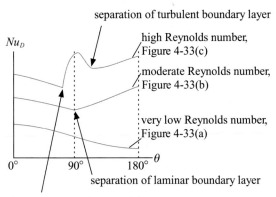

Figure 4-35: The qualitative behavior of the local Nusselt number as a function of angular position for very low Reynolds number, moderate Reynolds number, and very high Reynolds number.

Substituting Eq. (4-615) into Eq. (4-609) leads to:

$$C_D \approx \frac{\dfrac{\rho u_f^2}{2} D L}{\dfrac{\rho u_f^2}{2} D L} = 1 \qquad (4\text{-}616)$$

Figure 4-34 shows that the drag coefficient becomes relatively insensitive to the Reynolds number and remains close to unity for a wide range of Reynolds number. This behavior persists until the Reynolds number approaches the critical Reynolds number for transition to turbulence ($Re_{crit} \approx 5 \times 10^5$ according to flat plate theory); notice the sharp drop in the drag coefficient that occurs at this point. The turbulent boundary layer has a much higher momentum throughout because the region of low velocity, the viscous sublayer, is very thin. As a result, a turbulent boundary layer has sufficient momentum to overcome the adverse pressure gradient and separation is delayed until approximately $\theta = 140°$, as shown in Figure 4-33(c). A smaller area of the cylinder is exposed to the low pressure wake region and the form drag is substantially smaller. A similar phenomenon is observed for other smooth bodies, such as spheres. The reason that golf balls are dimpled is to promote the transition to turbulence at lower values of the Reynolds number and therefore reduce the drag force that is experienced by the golf ball during flight.

The data provided in Figure 4-34 have been entered in EES and are accessible as the built-in procedure External_Flow_Cylinder_ND that also returns the average Nusselt number, discussed in the subsequent section.

Nusselt Number

The behavior of the local Nusselt for flow past a cylinder is shown qualitatively in Figure 4-35 for various values of the Reynolds number. Detailed data for this situation can be found in Giedt (1949) and elsewhere.

At a very low Reynolds number, the local Nusselt number decreases with angular position as the laminar thermal boundary layer thickens in the downstream direction and never detaches. This situation corresponds to Figure 4-33(a). At a higher Reynolds number, the local Nusselt number again decreases from $\theta = 0°$ to 90° as the laminar thermal boundary layer thickens. At approximately 90°, the boundary layer separates and the

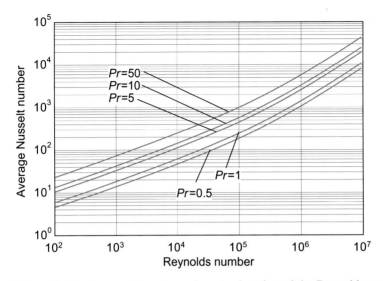

Figure 4-36: Average Nusselt number as a function of the Reynolds number for various values of the Prandtl number for a cylinder in cross-flow.

local heat transfer coefficient increases in the wake region. This situation corresponds to Figure 4-33(b). For a still higher Reynolds number, the Nusselt number decreases as the laminar boundary layer thickens and then, at some critical value of θ (similar to x_{crit} on a flat plate), the boundary layer transitions to turbulence with a corresponding jump in the local Nusselt number. The local Nusselt number associated with the turbulent boundary layer decreases in the downstream direction until finally the turbulent boundary layer detaches and a second jump in the local Nusselt number occurs in the wake region. This situation corresponds to Figure 4-33(c).

The behavior of the local Nusselt number in Figure 4-35 is quite complex. However, the average value of the Nusselt number increases monotonically with Reynolds number even as the characteristics of the flow change substantially. The average Nusselt number is correlated over a wide range of Reynolds numbers and Prandtl numbers by Churchill and Bernstein (1977).

$$\overline{Nu}_D = 0.3 + \frac{0.62 \, Re_D^{0.5} \, Pr^{\frac{1}{3}}}{\left[1 + \left(\frac{0.4}{Pr}\right)^{\frac{2}{3}}\right]^{0.25}} \left[1 + \left(\frac{Re_D}{2.82 \times 10^5}\right)^{0.625}\right]^{0.80}$$

$$\text{for } 1 \times 10^2 < Re_D < 1 \times 10^7 \quad \text{and} \quad Re_D \, Pr > 0.2$$

(4-617)

The average Nusselt number predicted by Eq. (4-617) is shown in Figure 4-36 and can be accessed using the procedure External_Flow_Cylinder_ND in EES.

EXAMPLE 4.9-2: HOT WIRE ANEMOMETER

EXAMPLE 4.9-2: HOT WIRE ANEMOMETER

A wire with diameter $D = 0.5$ mm and length $L = 1.0$ cm is used to measure the velocity of air flowing in a duct. The local air temperature is $T_\infty = 20°C$. The electrical resistance of the wire depends on temperature according to:

$$R_e = 0.2 \left[\frac{\Omega}{K} \right] T \text{ [K]} \tag{1}$$

where R_e is the resistance (in ohm) and T is the wire temperature (in K). The wire is provided with a current $I = 0.1$ Ampere and the electrical dissipation is transferred as heat to the free stream. The temperature of the wire depends on the average heat transfer coefficient and therefore on the velocity of the air. The voltage across the wire (V) is measured and can be related to the local velocity (u_f).

Neglect radiation from the wire surface and assume that the wire comes to a uniform temperature.

a) Estimate the voltage that will be measured if the velocity is $u_f = 20$ m/s.

The known information is entered in EES:

```
"EXAMPLE 4.9-2: Hot Wire Anemometer"

$UnitSystem SI MASS RAD PA K J
$Tabstops 0.2 0.4 0.6 3.5 in

"Inputs"
D=0.5 [mm]*convert(mm,m)          "diameter of wire"
L=1.0 [cm]*convert(cm,m)          "length of wire"
T_infinity=converttemp(C,K,20 [C])  "air temperature"
I=0.1 [amp]                       "current"
u_f=20 [m/s]                      "air velocity"
```

In order to evaluate the film temperature (required for the properties) and the resistance (required for the ohmic dissipation) it is necessary to know the wire temperature (T). This situation is typical of most convection problems; it is almost always necessary to iterate so that an assumed value of the surface temperature that is used to evaluate the required properties matches a final value that is determined from a complete solution. In this case, it is convenient to provide a reasonable guess for the wire temperature and use that temperature to evaluate the resistance of the wire based on Eq. (1).

```
T=300 [K]              "assumed temp., used to calculate properties"
R_e=0.2 [ohm/K]*T      "wire resistance"
```

The assumed value of temperature is used to compute the film temperature:

$$T_{film} = \frac{T + T_\infty}{2}$$

and the air properties, including k, ρ, μ, and Pr (using EES' built-in properties for air).

EXAMPLE 4.9-2: HOT WIRE ANEMOMETER

```
T_film=(T+T_infinity)/2                          "film temperature"
k=conductivity(Air,T=T_film)                     "air conductivity"
rho=density(Air,T=T_film,P=1[atm]*convert(atm,Pa))   "air density"
mu=viscosity(Air,T=T_film)                       "air viscosity"
Pr=Prandtl(Air,T=T_film)                         "air Prandtl number"
```

The Reynolds number characterizing the flow across the wire is computed:

$$Re_D = \frac{\rho\, u_f\, D}{\mu}$$

and the average Nusselt number $(\overline{Nu_D})$ is obtained using EES' built-in procedure for external flow over a cylinder, External_Flow_Cylinder_ND:

```
Re=rho*u_f*D/mu                                  "Reynolds number"
Call External_Flow_Cylinder_ND(Re,Pr:Nusselt_bar,C_D_bar)
                                 "access correlations for Nu and Cd"
```

The average Nusselt number is used to compute the average heat transfer coefficient:

$$\overline{h} = \frac{\overline{Nu_D}\, k}{D}$$

```
h_bar=Nusselt_bar*k/D                            "average heat transfer coefficient"
```

At steady state, an energy balance on the wire balances ohmic dissipation against convection to the air:

$$I^2 R_e = \overline{h}\, \pi\, D\, L\, (T - T_\infty) \tag{2}$$

which provides a solution for T and therefore over-specifies the problem. Before Eq. (2) is entered, the problem should be solved and the guess values updated. (Select Update Guess Values from the Calculate menu.) Then the initial assignment of the guess temperature should be commented out:

```
{T=300 [K]}    "assumed temp., used to calculate properties"
```

and the energy balance, Eq. (2), inserted in its place:

```
I^2*R_e=h_bar*pi*D*L*(T-T_infinity)    "energy balance on the wire"
```

The problem should solve without any problems. Note that if you skip the step of updating the guess values, then the problem is not likely to converge because the default guess value for T is 1 K, which is far from any reasonable solution and out of the range where the property correlations for air are valid.

The voltage measured across the wire is obtained from Ohm's law:

$$V = I\, R_e$$

V=I*R_e   "voltage"

which leads to $V = 7.27$ V.

b) Prepare a plot of voltage vs air velocity for velocities between 5 m/s and 100 m/s. Where is the instrument most useful?

The line in the EES code that assigns the velocity of the air is commented out and a parametric table is setup to run the code over a range of the variable u_f. The voltage as a function of the air velocity is shown in Figure 1.

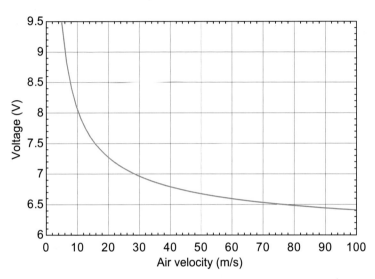

Figure 1: Voltage as a function of air velocity.

The instrument is most sensitive (i.e., the change in voltage associated with a given change in velocity is largest) at relatively low values of air velocity. Depending on the resolution of the voltage measurement, Figure 1 suggests that the instrument may not be very useful for velocities greater than about 50 m/s.

Flow across a Bank of Cylinders
External flow across a bank of cylinders is encountered quite often in the analysis of engineering devices such as shell-and-tube heat exchangers. Procedures that return the average heat transfer and pressure drop associated with flow across inline and staggered tube banks are available in EES. These procedures are based on the correlations provided by Zukauskas (1972) and are only available in dimensional form.

Non-Circular Extrusions
Cylinders are circular extrusions; non-circular extrusions (e.g., square channels) appear often in engineering applications and their behavior can be correlated using a similar set of dimensionless parameters. The average Nusselt number and the Reynolds number are defined according to:

$$\overline{Nu}_W = \frac{\overline{h}\,W}{k} \qquad (4\text{-}618)$$

EXAMPLE 4.9-2: HOT WIRE ANEMOMETER

and

$$Re_W = \frac{\rho u_f W}{\mu} \tag{4-619}$$

where W is the width of the extrusion as viewed from the direction of the frontal velocity, u_f.

Jakob (1949) has correlated the average Nusselt number for flow across several common extrusions as a function of the Reynolds number and the Prandtl number according to:

$$\boxed{\overline{Nu}_W = C Re_W^n Pr^{1/3}} \tag{4-620}$$

where C and n are dimensionless constants that are specific to the shape of the extrusion. The constants C and n as well as the correct definition of W are provided in Table 4-7. Care should be taken not to apply the correlations beyond their range of validity (also indicated in Table 4-7).

Table 4-7: Average Nusselt number correlations for external flow over non-circular extrusions.

Geometry	Reynolds number range	Constants for Eq. (4-620)	
		C	n
	5×10^3 to 1×10^5	0.246	0.588
	5×10^3 to 1×10^5	0.102	0.675
	5×10^3 to 1.95×10^4	0.160	0.668
	1.95×10^4 to 1×10^5	0.0385	0.782
	5×10^3 to 1×10^5	0.153	0.638
	4×10^3 to 1.5×10^4	0.228	0.731

4.9.4 Flow Past a Sphere

Figure 4-37 illustrates external flow past a sphere. The behavior of flow past a sphere is correlated in terms of the Prandtl number and the Reynolds number defined based on the sphere diameter:

$$Re_D = \frac{\rho u_f D}{\mu}, \tag{4-621}$$

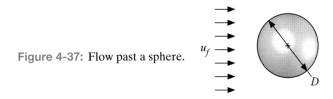

Figure 4-37: Flow past a sphere.

The drag force, F, is correlated with a drag coefficient:

$$C_D = \frac{8\,F}{\rho\,u_f^2\,\pi\,D^2},$$ (4-622)

and the average heat transfer coefficient is correlated with an average Nusselt number

$$\overline{Nu}_D = \frac{\overline{h}\,D}{k},$$ (4-623)

Figure 4-38 illustrates the drag coefficient as a function of the Reynolds number for a sphere (Schlichting (2000)); note that the drag coefficient exhibits essentially the same behavior that is discussed in Section 4.9.3 in the context of flow across a cylinder.

Whitaker (1972), as provided by Mills (1995), has proposed the correlation for average Nusselt number:

$$\overline{Nu}_D = 2 + \left(0.4\,Re_D^{0.5} + 0.06\,Re_D^{\frac{2}{3}}\right) Pr^{0.4}$$ (4-624)

The drag coefficient for a sphere based on interpolation of data shown in Figure 4-38 and the average Nusselt number predicted by Eq. (4-624) can be accessed using the built-in EES function External_Flow_Sphere_ND.

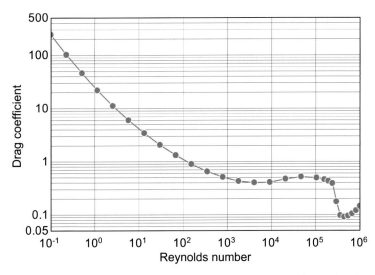

Figure 4-38: Drag coefficient as a function of the Reynolds number for flow past a sphere.

EXAMPLE 4.9-3: BULLET TEMPERATURE

EXAMPLE 4.9-3: BULLET TEMPERATURE

You have been asked to develop a model that can determine the relationship between the temperature of a bullet at impact and the distance that it traveled after it was fired. Such a model may be useful for forensic science by allowing investigators to ascertain details of the crime from the characteristics of the entrance wound. The bullet can be modeled (approximately) as a sphere with diameter $D = 0.25$ inch. Develop the model assuming that the velocity of the bullet as it leaves a gun is $u_{ini} = 1150$ ft/s and the initial temperature of the bullet is $T_{ini} = 513°F$; these parameters can be adjusted depending on the model of the gun. The bullet travels through still air at $T_\infty = 70°F$. The bullet can be modeled as a lumped capacitance and the bullet material has density $\rho = 0.30$ lb$_m$/in^3 and $c = 0.10$ Btu/lb$_m$-R. Neglect the effects of radiation and gravity in this analysis.

a) Develop a model that can relate temperature to distance traveled.

The known information is entered in EES:

```
"EXAMPLE 4.9-3: Bullet Temperature"

$UnitSystem SI MASS RAD PA K J
$Tabstops 0.2 0.4 0.6 3.5 in

"Inputs"
D=0.25 [inch]*convert(inch,m)                    "bullet diameter"
u_ini=1150 [ft/s]*convert(ft/s,m/s)              "initial velocity"
T_ini=converttemp(F,K,513 [F])                   "initial temperature"
T_infinity=converttemp(F,K,70 [F])               "air temperature"
rho=0.30 [lbm/inch^3]*convert(lbm/inch^3,kg/m^3) "density"
c=0.10 [Btu/lbm-R]*convert(Btu/lbm-R,J/kg-K)     "specific heat capacity"
```

The mass of the bullet, M, is:

$$M = \frac{4\pi}{3}\left(\frac{D}{2}\right)^3 \rho$$

```
M=4*pi*rho*(D/2)^3/3                             "mass of bullet"
```

The governing differential equations for the distance traveled by the bullet (x) and the bullet velocity (u) as a function of time (t) are obtained by carrying out a force balance on the bullet. The bullet acceleration is balanced by the drag force (F):

$$M\frac{du}{dt} = -F \tag{1}$$

Equation (1) can be rearranged to provide the time rate of change of the bullet velocity given its instantaneous velocity (which can be used to determine the drag force):

$$\frac{du}{dt} = -\frac{F}{M} \tag{2}$$

The time rate of change of the distance traveled by the bullet is equal to the velocity:

$$\frac{dx}{dt} = u \qquad (3)$$

The temperature of the bullet is governed by an energy balance:

$$M c \frac{dT}{dt} = \bar{h} \pi D^2 (T_\infty - T) \qquad (4)$$

where $\bar{h}$ is the average heat transfer coefficient. Equation (4) is rearranged to provide the instantaneous temperature rate of change:

$$\frac{dT}{dt} = \frac{\bar{h} \pi D^2}{mc} (T_\infty - T) \qquad (5)$$

Equations (2), (3), and (5) are the state equations for the problem. Given the value of the state variables (x, u, and T), Eqs. (2), (3) and (5) allow the calculation of their time derivative. Therefore, it is possible to solve the problem through numerical integration using any of the techniques discussed in Sections 3.2 or 3.8. Here, the Integral command in EES is used.

This is a complex problem and it is important to solve it in a logical and sequential manner in order to avoid frustrating errors and problems with the computer solution. The easiest way to solve this problem is to start by setting a value for each of the state variables and, for these arbitrary values, calculating the value of their time derivatives. Once you are sure that your program can compute these time derivatives, the next step is to numerically integrate them through time. Here, we will start by specifying that the state variables have their initial values:

```
"arbitrary specification of the state variables – used to calculate their derivatives"
T=T_ini                                          "temperature"
u=u_ini                                          "velocity"
x=0 [m]                                          "position"
```

In order to obtain the drag coefficient and heat transfer coefficient, it is necessary to compute the properties of air (ρ_a, μ_a, k_a, and Pr_a) at the film temperature:

$$T_{film} = \frac{T + T_\infty}{2}$$

```
T_film=(T+T_infinity)/2                          "film temperature"
rho_a=density(Air,T=T_film,P=1[atm]*convert(atm,Pa))   "density of air"
mu_a=viscosity(Air,T=T_film)                     "viscosity of air"
k_a=conductivity(Air,T=T_film)                   "conductivity of air"
Pr_a=Prandtl(Air,T=T_film)                       "Prandtl number of air"
```

The Reynolds number is computed:

$$Re_D = \frac{\rho_a u D}{\mu_a}$$

```
Re=rho_a*u*D/mu_a                                "Reynolds number"
```

EXAMPLE 4.9-3: BULLET TEMPERATURE

EXAMPLE 4.9-3: BULLET TEMPERATURE

The procedure External_Flow_Sphere_ND in EES is used to access the correlations that return the average Nusselt number $(\overline{Nu}_D)$ and the drag coefficient (C_D):

```
Call External_Flow_Sphere_ND(Re,Pr_a:Nusselt_bar, C_D)
                                                   "access correlation for a sphere"
```

The drag coefficient is used to compute the drag force on the bullet:

$$F = C_D \frac{\rho_a u^2}{2} \frac{\pi D^2}{4}$$

```
F=C_D*(rho_a*u^2/2)*(pi*D^2/4)                    "drag force"
```

The average Nusselt number is used to compute the average heat transfer coefficient:

```
h_bar=Nusselt_bar*k_a/D                            "average heat transfer coefficient"
```

Equations (2), (3) and (5) are used to compute the time derivatives of the velocity, position, and temperature of the bullet, respectively:

```
dudt=-F/M                            "velocity rate of change"
dxdt=u                               "position rate of change"
dTdt=h_bar*pi*D^2*(T_infinity-T)/(M*c)    "temperature rate of change"
```

Once you are sure that the program is capable of calculating these derivatives (and the answers make sense, the units check, etc.) it is relatively easy to integrate them forward with time. The velocity, position, and temperature of the bullet are obtained according to the integrals:

$$u = u_{ini} + \int_0^t \frac{du}{dt} dt \tag{6}$$

$$x = \int_0^t \frac{dx}{dt} dt \tag{7}$$

$$T = T_{ini} + \int_0^t \frac{dT}{dt} dt \tag{8}$$

The specification of the arbitrary values of the state variables u, x, and T are removed:

```
"arbitrary specification of the state variables - used to calculate their derivatives"
{T=T_ini                             "temperature"
u=u_ini                              "velocity"
x=0 [m]                              "position"}
```

EXAMPLE 4.9-3: BULLET TEMPERATURE

and these quantities are calculated by the Integral command in EES according to Eqs. (6) through (8). An integral table is obtained using the $IntegralTable directive.

```
u=u_ini+Integral(dudt,time,0,t_sim)      "velocity integral"
x=Integral(dxdt,time,0,t_sim)            "position integral"
T=T_ini+Integral(dTdt,time,0,t_sim)      "temperature integral"
$IntegralTable time:0.01,u,x,T
```

Figure 1 illustrates the velocity, position, and temperature of the bullet as a function of time.

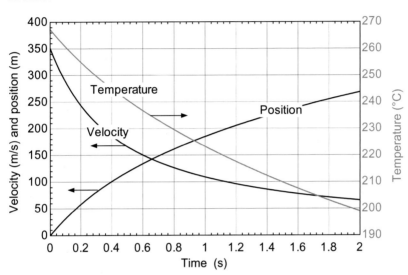

Figure 1: Velocity, position, and temperature of the bullet as a function of time.

Figure 2 illustrates the distance that the bullet has traveled as a function of its temperature and shows that these quantities are strongly related.

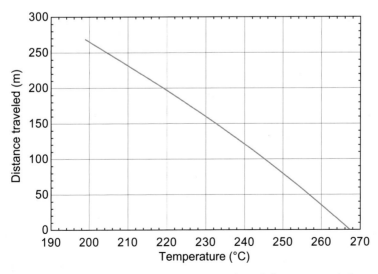

Figure 2: Temperature of the bullet as a function of distance traveled.

Chapter 4: External Convection

The instructors' resource website associated with this book (www.cambridge.org/nellisandklein) provides many more problems than are included here.

Introduction to Laminar Boundary Layers

4–1 Water at atmospheric pressure, free stream velocity $u_\infty = 1.0$ m/s and temperature $T_\infty = 25°C$ flows over a flat plate with a surface temperature $T_s = 90°C$. The plate is $L = 0.15$ m long. Assume that the flow is laminar over the entire length of the plate.
   a.) Estimate, using your knowledge of how boundary layers grow, the size of the momentum and thermal boundary layers at the trailing edge of the plate (i.e., at $x = L$). Do not use a correlation from your book, instead use the approximate model for boundary layer growth.
   b.) Use your answer from (a) to estimate the shear stress at the trailing edge of the plate and the heat transfer coefficient at the trailing edge of the plate.
   c.) You measure a shear stress of $\tau_{s,meas} = 1.0$ Pa at the trailing edge of the plate; use the Modified Reynolds Analogy to predict the heat transfer coefficient at this location.

4–2 Figure P4-2 illustrates the flow of a fluid with $T_\infty = 0°C$, $u_\infty = 1$ m/s over a flat plate.

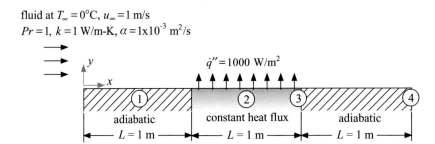

fluid at $T_\infty = 0°C$, $u_\infty = 1$ m/s
$Pr = 1$, $k = 1$ W/m-K, $\alpha = 1 \times 10^{-3}$ m²/s

$\dot{q}'' = 1000$ W/m²

adiabatic   constant heat flux   adiabatic
$L = 1$ m $L = 1$ m $L = 1$ m

Figure P4-2: Flow over a flat plate.

The flat plate is made up of three sections, each with length $L = 1$ m. The first and last sections are insulated and the middle section is exposed to a constant heat flux, $\dot{q}'' = 1000$ W/m². The properties of the fluid are Prandtl number $Pr = 1$, conductivity $k = 1$ W/m-K, and thermal diffusivity $\alpha = 1 \times 10^{-3}$ m²/s. Assume that the flow is laminar over the entire surface.
   a.) Sketch the momentum and thermal boundary layers as a function of position, x. Do not worry about the qualitative characteristics of your sketch – get the quantitative characteristics correct.
   b.) Sketch the temperature distribution (the temperature as a function of distance from the plate y) at the 4 locations indicated in Figure P4-2. Location 1 is half-way through the first adiabatic region, Location 2 is half-way through the heated region, Location 3 is at the trailing edge of the heated region (in the heated region), and Location 4 is at the trailing edge of the final adiabatic region. Again, focus on getting as many of the qualitative characteristics of your sketch correct as you can.
   c.) Sketch the temperature of the surface of the plate as a function of position, x. Get the qualitative features of your sketch correct.

d.) Predict, approximately, the temperature of the surface at locations 1, 2, 3, and 4 in Figure P4-2. Do not use a correlation. Instead, use your conceptual understanding of how boundary layers behave to come up with very approximate estimates of these temperatures.

The Boundary Layer Equations and Dimensional Analysis in Convection

4–3 You have fabricated a 1000× scale model of a microscale feature that is to be used in a microchip. The device itself is only 1 µm in size and is therefore too small to test accurately. However, you'd like to know the heat transfer coefficient between the device and an air flow that has a velocity of 10 m/s.
   a.) What velocity should you use for the test and how will the measured heat transfer coefficient be related to the actual one?

4–4 Your company has come up with a randomly packed fibrous material that could be used as a regenerator packing. Currently there are no correlations available that would allow the prediction of the heat transfer coefficient for the packing. Therefore, you have carried out a series of tests to measure the heat transfer coefficient. A $D_{bed} = 2$ cm diameter bed is filled with these fibers; the diameter of the individual fibers is $d_{fiber} = 200$ µm. The nominal temperature and pressure used to carry out the test is $T_{nom} = 20°C$ and $p_{nom} = 1$ atm, respectively. The mass flow rate of the test fluid, $\dot{m}$, is varied and the heat transfer coefficient is measured. Several fluids, including air, water, and ethanol, are used for testing. The data are shown in Table P4-4; the data can be downloaded from the website www.cambridge.org/nellisandklein as EES lookup tables (P4-4_air.lkt, P4-4_ethanol.lkt, and P4-4_water.lkt).

Table P4-4: Heat transfer data.

Air		Water		Ethanol	
Mass flow rate (kg/s)	Heat transfer coefficient (W/m²-K)	Mass flow rate (kg/s)	Heat transfer coefficient (W/m²-K)	Mass flow rate (kg/s)	Heat transfer coefficient (W/m²-K)
0.0001454	170.7	0.00787	8464	0.009124	4162
0.0004073	311.9	0.02204	15470	0.02555	7607
0.0006691	413.7	0.0362	20515	0.04197	10088
0.0009309	491.8	0.05037	24391	0.05839	11993
0.001193	572.7	0.06454	28399	0.07481	13964
0.001454	631.1	0.0787	31296	0.09124	15388

   a.) Plot the heat transfer coefficient as a function of mass flow rate for the three different different test fluids.
   b.) Plot the Nusselt number as a function of the Reynolds number for the three different test fluids. Use the fiber diameter as the characteristic length and the free-flow velocity (i.e., the velocity in the bed if it were empty) as the characteristic velocity.
   c.) Correlate the data for all of the fluids using a function of the form: $Nu = a\,Re^b\,Pr^c$. Note that you will want to transform the results using a natural logarithm and use the Linear Regression option from the Tables menu to determine a, b, and c.
   d.) Use your correlation to estimate the heat transfer coefficient for 20 kg/s of oil passing through a 50 cm diameter bed composed of fibers with 2 mm diameter.

The oil has density 875 kg/m³, viscosity 0.018 Pa-s, conductivity 0.14 W/m-K, and Prandtl number 20.

4–5 Your company makes an extrusion that can be used as a lightweight structural member; the extrusion is long and thin and has an odd cross-sectional shape that is optimized for structural performance. This product has been used primarily in the aircraft industry; however, your company wants to use the extrusion in an application where it will experience a cross-flow of water at $V = 10$ m/s rather than air. There is some concern that the drag force experienced by the extrusion will be larger than it can handle. Because the cross-section of the extrusion is not simple (e.g., circular or square) you cannot look up a correlation for the drag coefficient in the same way that you could for a cylinder. However, because of the extensive use of the extrusion in the air-craft industry you have a large amount of data relating the drag force on the extrusion to velocity when it is exposed to a cross-flow of air. These data have been collated and are shown graphically in Figure P4-5.

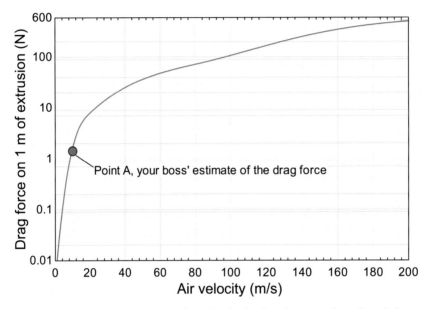

Figure P4-5: Drag force as a function of velocity for the extrusion when it is exposed to a cross-flow of air.

Your boss insists that the drag force for the extrusion exposed to water can be obtained by looking at Figure P4-5 and picking off the data at the point where $V = 10$ m/s (Pt. A in Figure P4-5); this corresponds to a drag force of about 1.7 N/m of extrusion.

a.) Is your boss correct? Explain why or why not.

b.) If you think that your boss is not correct, then use the data in Figure P4-5 to estimate the drag force that will be experienced by the extrusion for a water cross-flow velocity of 10 m/s.

The Self-Similar Solution for Laminar Flow over a Flat Plate

4–6 The momentum and thermal boundary layer can be substantially affected by either injecting or removing fluid at the plate surface. For example, Figure P4-6 shows the surface of a turbine blade exposed to the free stream flow of a hot combustion gas with velocity u_∞ and temperature T_∞. The surface of the blade is protected by blowing gas through pores in the surface in a process called transpiration cooling.

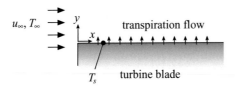

Figure P4-6: Transpiration cooled turbine blade.

The velocity of the injected gas is a function of x: $v_{y=0} = C\sqrt{\frac{u_\infty v}{x}}$, where C is a dimensionless constant and v is the kinematic viscosity of the fluid. The gas is injected at the same temperature as the surface of the plate, T_s. The Prandtl number of the combustion gas is $Pr = 0.7$.

a.) Develop a self-similar solution to the momentum equation for this problem using a Crank-Nicolson numerical integration implemented in EES.

b.) Plot the dimensionless velocity (u/u_∞) as a function of the similarity parameter, η, for various values of C.

c.) The boundary layer will "blow-off" of the plate at the point where the shear stress at the plate surface becomes zero. What is the maximum value of C that can be tolerated before the boundary layer becomes unstable?

d.) Plot the ratio of the friction factor experienced by the plate with transpiration to the friction factor experienced by a plate without transpiration as a function of the parameter C.

e.) Develop a self-similar solution to the thermal energy equation for this problem using a Crank-Nicolson numerical integration implemented in EES.

f.) Plot the dimensionless temperature difference, $(T - T_s)/(T_\infty - T_s)$, as a function of the similarity parameter, η, for various values of C.

g.) Plot the ratio of the Nusselt number experienced by the plate with transpiration to the Nusselt number experienced by a plate without transpiration as a function of the parameter C.

4–7 Develop a self-similar solution for the flow over a flat plate that includes viscous dissipation. The ordinary differential equation governing the dimensionless temperature difference should include an additional term that is related to the Eckert number.

a.) Plot the dimensionless temperature difference, $(T - T_s)/(T_\infty - T_s)$, as a function of the similarity parameter, η, for various values of Ec with $Pr = 10$.

b.) Plot the ratio of the Nusselt number to the Nusselt number neglecting viscous dissipation (i.e., with $Ec = 0$) as a function of the Eckert number for various values of the Prandtl number.

Turbulence

4–8 Use the Spalding model to obtain a velocity and temperature law of the wall (your temperature law of the wall should be obtained numerically using the EES Integral command). Compare your result with the Prandtl-Taylor model. Use a molecular Prandtl number of $Pr = 0.7$ and a turbulent Prandtl number of $Pr_{turb} = 0.9$.

4–9 Use the van Driest model to obtain a velocity and temperature law of the wall. (Both of these results should be obtained numerically using the EES Integral command). Compare your result with the Prandtl-Taylor model. Use a molecular Prandtl number of $Pr = 0.7$ and a turbulent Prandtl number of $Pr_{turb} = 0.9$.

4–10 In Section 4.5, a conceptual model of a turbulent flow was justified based on the fact that the thermal resistance of the viscous sublayer (δ_{vs}/k) is larger than the thermal resistance of the turbulent boundary layer (δ_{turb}/k_{turb}). Estimate the magnitude of each of these terms for a flow of water over a smooth flat plate and evaluate the validity of this simplification. The free stream velocity is $u_\infty = 10$ m/s and the plate is $L = 1$ m long. The water is at 20°C and 1 atm.

Integral Solutions

4–11 Figure P4-11 illustrates a flat plate that has an unheated starting length (ε); the hydrodynamic boundary layer grows from the leading edge of the plate while the thermal boundary layer grows from $x > L_{uh}$. Assume that the plate has a constant surface temperature, T_s, for $x > L_{uh}$.

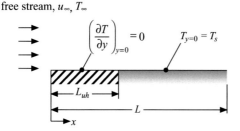

Figure P4-11: Plate with an unheated starting length.

Determine a correlation for the local Nusselt number in this situation using the integral technique. Use a third order velocity and temperature distribution. Neglect viscous dissipation. You may find it useful to solve for the ratio of the thermal to momentum boundary layer thickness.

4–12 A flow of a liquid metal over a flat plate, shown in Figure P4-12, is being considered during the design of an advanced nuclear reactor.

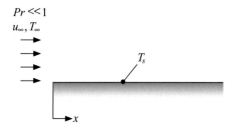

Figure P4-12: Flow of a low Prandtl number liquid metal over a flat plate.

You are to develop a solution to this problem using an integral technique. Because the Prandtl number is much less than one, it is appropriate to assume that $\delta_m \ll \delta_t$ and therefore the velocity is constant and equal to u_∞ throughout the thermal boundary layer. Use a second order temperature distribution and neglect viscous dissipation.

a.) If the temperature of the plate is constant and equal to T_s then derive a solution for the local Nusselt number on the plate surface as a function of the Reynolds number and the Prandtl number.

b.) Plot your result from (a) as a function of Re for $Pr = 0.001$. Overlay on your plot the correlation for the local Nusselt number for flow over a constant temperature flat plate found in Section 4.9.

c.) Plot your result from (a) and the correlation from Section 4.9 as a function of Pr for $Re = 1 \times 10^4$. Your Prandtl number range should be from 0.001 to 100 and it should be clear from your plot that the solutions begin to diverge as the Prandtl number approaches unity.

d.) If the heat flux at the surface of the plate is constant and equal to q_s'' then derive a solution for the local Nusselt number on the plate surface as a function of the Reynolds number and the Prandtl number.

e.) If the heat flux at the surface of the plate varies linearly with position and is zero at the leading edge of the plate, then derive a solution for the local Nusselt number on the plate surface as a function of the Reynolds number and the Prandtl number.

f.) Plot the solutions from (a), (d), and (e) as a function of Reynolds number for $Pr = 0.001$.

4–13 Determine the local friction coefficient as a function of Reynolds number for laminar flow over a flat plate using the momentum integral technique. Assume a velocity distribution of the form: $u/u_\infty = a \sin(b y/\delta_m + c)$ where a, b, and c are undetermined constants. Compare your answer to the Blasius solution from Section 4.4.

4–14 A flat plate that is $L = 0.2$ m long experiences a heat flux given by:

$$\dot{q}_s'' = \dot{q}_{max}'' \cos\left(\frac{\pi x}{4 L}\right)$$

where $\dot{q}_{max}'' = 2 \times 10^4$ W/m^2. The free stream velocity is $u_\infty = 20$ m/s and the free stream temperature is $T_\infty = 20°$C. The fluid passing over the plate has thermal diffusivity $\alpha = 1 \times 10^{-4}$ m^2/s, conductivity $k = 0.5$ W/m-K, and Prandtl number $Pr = 2.0$.

a.) Use a linear temperature distribution and a linear velocity distribution in order to obtain an ordinary differential equation for the thermal boundary layer thickness.

b.) Solve the ordinary differential equation from (a) numerically. At the leading edge of the plate there is a singularity; obtain an analytical solution in this region and start your numerical solution at the x position where the analytical solution is no longer valid.

c.) Plot the surface temperature of the plate as a function of axial position.

d.) Overlay on your plot from (c) the surface temperature calculated using the correlation for the local heat transfer coefficient on an isothermal flat plate.

External Flow Correlations

4–15 You and your friend are looking for an apartment in a high-rise building. You have your choice of 4 different south-facing units (units #2 through #5) in a city where the wind is predominately from west-to-east, as shown in Figure P4-15. You are responsible for paying the heating bill for your apartment and you have noticed that the exterior wall is pretty cheaply built. Your friend has taken heat transfer and therefore is convinced that you should take unit #5 in order to minimize the cost of heating the unit because the boundary layer will be thickest and heat transfer coefficient smallest for the exterior wall of that unit. Prepare an analysis that can predict the cost of heating each of the 4 units so that you can (a) decide whether the difference is worth considering, and (b) if it is, choose the optimal unit. Assume that the heating season is *time* = 4 months long (120 days), the average outdoor air temperature during that time is $T_\infty = 0°C$ and the average wind velocity is $u_\infty = 5$ mph. The dimensions of the external walls are provided in Figure P4-15; assume that no heat loss occurs except through the external walls. Further, assume that the walls have a total thermal resistance on a unit area basis (not including convection) of $R''_w = 1$ K-m^2/W. The internal heat transfer coefficient is $\bar{h}_{in} = 10$ W/m^2-K. You like to keep your apartment at $T_{in} = 22°C$ and use electric heating at a cost of $ec = 0.15\$/$kW-hr. You may use the properties of air at T_∞ for your analysis and neglect the effect of any windows.

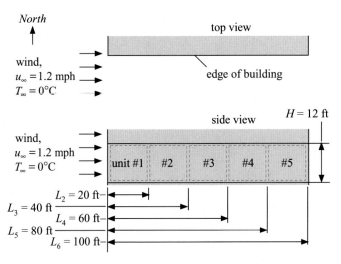

Figure P4-15: Location of the external wall of units 2 through 5 relative to the wind direction.

a.) Determine the average yearly heating cost for each of the 4 units and discuss which apartment is best and why.
b.) Prepare a plot showing the heating cost for unit #2 and unit #5 as a function of the wind velocity for the range 0.5 mph to 5.0 mph. Explain any interesting characteristics that you observe.

4–16 A solar photovoltaic panel is mounted on a mobile traffic sign in order to provide power without being connected to the grid. The panel is $W = 0.75$ m wide by $L = 0.5$ m long. The wind blows across the panel with velocity $u_\infty = 5$ miles/hr and temperature $T_\infty = 90°F$, as shown in Figure P4-16. The back side of the panel is insulated. The panel surface has an emissivity of $\varepsilon = 1.0$ and radiates to surroundings at T_∞. The PV panel receives a solar flux of $\dot{q}''_s = 490$ W/m^2. The electricity generated by the panel is quantified with an efficiency η, defined as the ratio of the electrical energy produced by the panel to the incident solar radiation. The electrical generation is the product of the efficiency, the solar flux, and the panel area. The efficiency of the panel is a function of surface temperature; at 20°C the efficiency is 15% and the efficiency drops by 0.25%/K as the surface temperature increases (i.e., if the panel surface is at 40°C then the efficiency has been reduced to 10%). All of the solar radiation absorbed by the panel and not transformed into electrical energy must be either radiated or convected to its surroundings.

a.) Determine the panel surface temperature, T_s, and the amount of electrical energy generated by the panel.

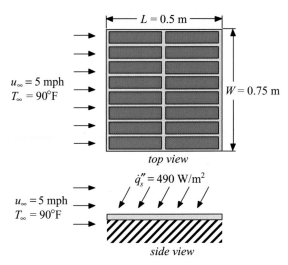

Figure P4-16: Solar panel.

b.) Prepare a plot of the electrical energy generated by the panel as a function of the solar flux for $\dot{q}_s''$ ranging from 100 W/m² to 700 W/m². Your plot should show that there is an optimal value for the solar flux – explain this result.

c.) Prepare a plot of the electrical energy generated by the panel as a function of the wind velocity (with $\dot{q}_s'' = 490$ W/m²) for u_∞ ranging from 5 mph to 50 mph – explain any interesting aspects of your plot.

d.) Prepare a plot of the shear force experienced by the panel due to the wind as a function of wind velocity for u_∞ ranging from 5 mph to 50 mph – explain any interesting aspects of your plot.

4–17 The wind chill temperature is loosely defined as the temperature that it "feels like" outside when the wind is blowing. More precisely, the wind chill temperature is the temperature of still air that will produce the same bare skin temperature that you experience on a windy day. If you are alive, then you are always transferring thermal energy (at rate $\dot{q}$) from your skin (at temperature T_{skin}). On a windy day, this heat loss is resisted by a convection resistance where the heat transfer coefficient is related to forced convection ($R_{conv,fc}$), as shown in Figure P4-17(a). The skin temperature is therefore greater than the air temperature (T_{air}). On a still day, this heat loss is resisted by a larger convection resistance because the heat transfer coefficient is related to natural convection ($R_{conv,nc}$), as shown in Fig. P4-17(b). For a given heat loss, air temperature, and wind velocity, the wind chill temperature (T_{WC}) is the temperature of still air that produces the same skin temperature.

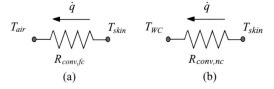

Figure P4-17: Resistance network for a body losing heat on (a) a windy and (b) a still day.

It is surprisingly complicated to compute the wind chill temperature because it requires that you know the rate at which the body is losing heat and the heat transfer coefficient between a body and air on both a windy and still day. At the same time, the wind chill temperature is important and controversial because it affects winter tourism in many places. The military and other government agencies that deploy personnel in extreme climates are also very interested in the wind chill temperature in order to establish allowable exposure limits. This problem looks at the wind chill temperature using your heat transfer background. It has been shown that the heat transfer coefficient for most animals can be obtained by treating them as if they were spherical with an equivalent volume.

a.) What is the diameter of a sphere that has the same volume as a man weighing $M = 170$ lb$_m$ (assume that the density of human flesh is $\rho_f = 64$ lb$_m$/ft^3)?

b.) Assuming that the man can be treated as a sphere, compute the skin temperature for the man on a day when the wind blows at $V = 10$ mph and the air temperature is $T_{air} = 0°$F. Assume that the metabolic heat generation for the man is $\dot{q} = 150$ W.

c.) Assume that the natural convection heat transfer coefficient that would occur on a day with no wind is $h_{nc} = 8.0$ W/m^2-K. What is the wind chill temperature?

According to the National Weather Service (http://www.weather.gov/om/windchill/), the wind chill temperature can be computed according to:

$$T_{WC} = 35.74 + 0.6215\, T_{air} - 35.75\, V^{0.16} + 0.4275\, T_{air}\, V^{0.16}$$

where T_{air} is the air temperature in $°$F and V is the wind velocity in mph.

d.) Use the National Weather Service equation to compute T_{WC} on a day when $T_{air} = 0°$F and $V = 10$ mph.

e.) Plot the wind chill temperature on a day with $T_{air} = 0°$F as a function of the wind velocity; show the value predicted by your model and by the National Weather Service equation for wind velocities ranging from 5 to 30 mph.

4–18 A soldering iron tip can be approximated as a cylinder of metal with radius $r_{out} = 5.0$ mm and length $L = 20$ mm. The metal is carbon steel; assume that the steel has constant density $\rho = 7854$ kg/m^3 and constant conductivity $k = 50.5$ W/m-K, but a specific heat capacity that varies with temperature according to:

$$c = 374.9\left[\frac{J}{\text{kg-K}}\right] + 0.0992\left[\frac{J}{\text{kg-K}^2}\right] T + 3.596 \times 10^{-4}\left[\frac{J}{\text{kg-K}^3}\right] T^2$$

The surface of the iron radiates and convects to surroundings that have temperature $T_{amb} = 20°$C. Radiation and convection occur from the sides of the cylinder (the top and bottom are insulated). The soldering iron is exposed to an air flow (across the cylinder) with a velocity $V = 3.5$ m/s at T_{amb} and $P_{amb} = 1$ atm. The surface of the iron has an emissivity $\varepsilon = 1.0$. The iron is heated electrically by ohmic dissipation; the rate at which electrical energy is added to the iron is $\dot{g} = 35$ W.

a.) Assume that the soldering iron tip can be treated as a lumped capacitance. Develop a numerical model using the Euler technique that can predict the temperature of the soldering iron as a function of time after it is activated. The tip is at ambient temperature at the time of activation. Be sure to account for the fact that the heat transfer coefficient, the radiation resistance, and the heat capacity of the soldering iron tip are all a function of the temperature of the tip.

b.) Plot the temperature of the soldering iron as a function of time. Make sure that your plot covers sufficient time that your soldering iron has reached steady state.

c.) Verify that the soldering iron tip can be treated as a lumped capacitance.

4–19 Molten metal droplets must be injected into a plasma for an extreme ultraviolet radiation source, as shown in Figure P4-19.

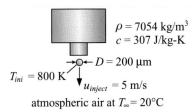

$\rho = 7054 \ \mathrm{kg/m^3}$
$c = 307 \ \mathrm{J/kg\text{-}K}$
$\longleftarrow D = 200 \ \mu\mathrm{m}$
$T_{ini} = 800 \ \mathrm{K}$
$u_{inject} = 5 \ \mathrm{m/s}$
atmospheric air at $T_\infty = 20°\mathrm{C}$

Figure P4-19: Injection of molten metal droplets.

The fuel droplets have a diameter of $D = 200 \ \mu\mathrm{m}$ and are injected at a velocity $u_{inject} = 5 \ \mathrm{m/s}$ with temperature $T_{ini} = 800 \ \mathrm{K}$. The density of the droplet $\rho = 7054 \ \mathrm{kg/m^3}$ and the specific heat capacity is $c = 307 \ \mathrm{J/kg\text{-}K}$. You may assume that the droplet can be treated as a lumped capacitance. The droplet is exposed to still air at $T_\infty = 20°\mathrm{C}$.

a.) Develop a numerical model in EES using the Integral command that can predict the velocity, temperature, and position of the droplet as a function of time.

b.) Plot the velocity as a function of time and the temperature as a function of time.

c.) Plot the temperature as a function of position. If the temperature of the droplet must be greater than 500 K when it reaches the plasma then what is the maximum distance that can separate the plasma from the injector?

4–20 Figure 4-34 in your text illustrates the drag coefficient for a cylinder as a function of Reynolds number.

a.) Using Figure 4-34 of your text, discuss briefly (1-2 sentences) why it might make sense to add dimples to a baseball bat.

b.) Using Figure 4-34 of your text, estimate how fast you would have to be able to swing a bat in order for it to make sense to think about adding dimples (the estimate can be rough, but should be explained well). Assume that a bat has diameter $D = 0.04 \ \mathrm{m}$ and air has properties $\rho = 1 \ \mathrm{kg/m^3}$ and $\mu = 0.00002 \ \mathrm{Pa\text{-}s}$.

REFERENCES

Ameel, T. A., "Average effects of forced convection over a flat plate with an unheated starting length," *Int. Comm. Heat Mass Transfer*, Vol. 24, No. 8, pp. 1113–1120, (1997).

Bejan, A., *Heat Transfer*, John Wiley & Sons, New York, (1993).

Bridgman, P. W., *Dimensional Analysis*, Yale University Press, New Haven, (1922).

Buckingham, E., "On Physically Similar Systems; Illustrations of the Use of Dimensional Analysis," *Physical Review*, Vol. 4, pp. 345–376, (1914).

Churchill, S. W., and H. Ozoe, "Correlations for Laminar Forced Convection with Uniform Heating in Flow over a Plate and in Developing and Fully Developed Flow in a Tube," *Journal of Heat Transfer*, Vol. 95, pp. 78, (1973).

Churchill, S. W. and M. Bernstein, "A Correlating Equation for Forced Convection from Gases and Liquids to a Circular Cylinder in Crossflow," *J. Heat Transfer*, Vol. 99, pp. 300–306, (1977).

Coles, D., "The law of the wall in the turbulent boundary layer," *J. Fluid Mechanics*, Vol. 1, Part 2, pp. 191–226, (1956).

Dipprey, D. F. and R. H. Sabersky, "Heat and momentum transfer in smooth and rough tubes at various Prandtl numbers," *Int. J. Heat Mass Transfer*, Vol. 6, pp. 329–353, (1963).

Giedt, W. H., "Investigation of Variation of Point Unit-Heat Transfer Coefficient around a Cylinder Normal to an Air Stream," *Trans. ASME*, Vol. 71, pp. 375–381, (1949).

Hinze, J. O., *Turbulence*, 2nd Edition, McGraw-Hill, New York, (1975).

Jakob, M., *Heat Transfer*, John Wiley and Sons, New York, (1949).

Kays, W. M., M. E. Crawford, and B. Weigand, *Convective Heat and Mass Transfer*, 4th Ed., McGraw-Hill, Boston, (2005).

Mills, A. F. and X. Hang, "On the skin friction coefficient for a fully rough flat plate," *J. Fluids Engineering*, Vol. 105, pp. 364–365, (1983).

Mills, A. F., *Basic Heat and Mass Transfer*, Irwin, Inc., Chicago, (1995).

Rubesin, M. W., M. Inouye, and P. G. Parikh, *Forced Convection External Flows*, in the *Handbook of Heat Transfer*, W. M. Rohsenow, J. P. Hartnett, and Y. I. Cho, eds., McGraw-Hill, Boston, (1998).

Schlichting, H., *Boundary Layer Theory*, Springer Publishing, New York, (2000).

Sonin, A. A., *The Physical Basis of Dimensional Analysis*, Class notes for Advanced Fluid Mechanics in the Department of Mechanical Engineering at the Massachusetts Institute of Technology, Cambridge, MA, (1992).

Spalding, D. B., "Heat Transfer to a Turbulent Stream from a Surface with a Stepwise Discontinuity in Wall Temperature," *Proc. Conf. Int. Dev. Heat Transfer, Part 2*, pp. 439–446, ASME, New York, (1961).

Tennekes, H. and J. L. Lumley, *A First Course in Turbulence*, The MIT Press, Cambridge, (1972).

van Driest, E. R., "On Turbulent Flow Near a Wall," *J. Aeronaut. Sci.*, Vol. 23, pp. 1007–1011, (1956).

von Kármán, T., "The Analogy Between Fluid Friction and Heat Transfer," *Trans. ASME*, Vol. 61, pp. 705–710, (1939).

Whitaker, S., "Forced convection heat-transfer coerrelations for flow in pipes, past flat plates, single cylinders, single spheres, and flow in packed beds and tube bundles," *AIChE J.*, Vol. 18, p. 361, (1972).

White, F. M., *Fluid Mechanics*, 5th Edition, McGraw-Hill, New York, (2003).

Zukauskas, A., "Heat Transfer from Tubes in Cross-Flow," in J. P. Hartnett and T. F. Irvine, Jr., *Advances in Heat Transfer*, Vol. 8, Academic Press, New York, (1972).

5 Internal Forced Convection

5.1 Internal Flow Concepts

5.1.1 Introduction

Chapter 4 discusses the behavior of the momentum and thermal boundary layers associated with an external flow. An external flow is broadly defined as one where the boundary layer can grow without bound; for the flat plate considered in Section 4.1, the boundary layer was never confined by the presence of another object. An internal flow is defined as one where the growth of the boundary layer is confined. Internal flows are often encountered in engineering applications (e.g., the flow through tubes or ducts) and this section discusses the qualitative behavior of internal flows. Many of the concepts that are discussed in Section 4.1 for an external flow can also be applied to internal flows in order to provide a physical understanding of their behavior.

5.1.2 Momentum Considerations

Figure 5-1(a) illustrates laminar external flow over a plate and shows, qualitatively, the momentum boundary layer and velocity distribution that results. Figure 5-1(b) illustrates laminar flow through a passage that is formed between two parallel plates; notice that at some location, the momentum boundary layer becomes bounded.

The momentum boundary layers growing from the upper and lower plates in Figure 5-1(b) meet at some distance from the inlet; this distance is referred to as the hydrodynamic entry length, $x_{fd,h}$. The momentum boundary layer thickness will remain constant as the fluid moves further down the flow passage (i.e., for $x > x_{fd,h}$). The region where the boundary layers are growing in an internal flow (i.e., for $x < x_{fd,h}$) is referred to as the hydrodynamically developing region. The behavior of an internal flow in the developing region is similar to an external flow.

Recall that the shear stress at the wall, τ_s, is proportional to the velocity gradient at the wall:

$$\tau_s = \mu \left. \frac{\partial u}{\partial y} \right|_{y=0} \qquad (5\text{-}1)$$

For a laminar flow, the shear stress can be expressed, approximately, as:

$$\tau_s \approx \mu \frac{u_\infty}{\delta_m} \qquad (5\text{-}2)$$

where δ_m is the momentum boundary layer thickness. Figure 5-1(c) illustrates, qualitatively, the variation of the shear stress as a function of position for the external and internal flow cases shown in Figure 5-1(a) and Figure 5-1(b), respectively.

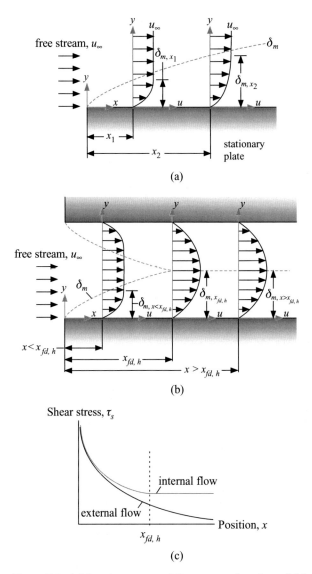

Figure 5-1: (a) Laminar external flow over a flat plate, (b) laminar internal flow between two parallel plates, and (c) the shear stress as a function of position for the external and internal flow situations.

The shear stress associated with an external flow was discussed in Section 4.1. The shear stress continues to decrease with position as the boundary layer grows. Eventually, the viscous shear will become sufficiently small that the flow will become turbulent. In the developing region of the internal flow, the shear stress at the plate surface will also decrease as the boundary layer grows and the velocity gradient at the wall is reduced. The shear stress in the developing region of the internal flow is not quite identical to the shear stress at the corresponding position in the external flow case because the velocity at the outer edge of the boundary layer must increase to satisfy continuity for an internal flow; the total flow through the channel at any position is constant and so the velocity at the center of the duct must increase as the flow near the edges is retarded. The region after the boundary layers have joined (i.e., for $x > x_{fd,h}$) is referred to as the hydrodynamically fully developed region. Because the momentum boundary layer thickness does not change with x in the hydrodynamically fully developed region, the velocity gradient at the wall (and therefore the shear stress) will also be independent of position in this region, as shown in Figure 5-1(c). In some sense, the behavior of a laminar internal flow is easier to understand and predict than an external flow since you have a good idea of the boundary layer thickness based on the geometry of the duct. If the duct has a constant cross-sectional area, then the velocity distribution for an internal flow will not change with x in the hydrodynamically fully developed region.

The Mean Velocity

The mean velocity (u_m, sometimes referred to as the bulk velocity) is used to characterize an internal flow. The mean velocity is a single velocity that represents the mass flow rate ($\dot{m}$) carried by the actual velocity distribution that exists within the duct:

$$u_m = \frac{\dot{m}}{\rho A_c} \tag{5-3}$$

where ρ is the density of the flow and A_c is the cross-sectional area of the flow. Equation (5-3) can be written in terms of the volumetric flow rate ($\dot{V}$):

$$u_m = \frac{\dot{V}}{A_c} \tag{5-4}$$

Both Eqs. (5-3) and (5-4) assume that the flow is incompressible (i.e., that ρ is constant). Figure 5-1(b) shows that the velocity changes across the cross-section of the duct and therefore the volumetric flow rate must be obtained by integration of the velocity distribution across the duct cross-sectional area, A_c:

$$\dot{V} = \int_{A_c} u \, dA_c \tag{5-5}$$

where u is the axial velocity. Substituting Eq. (5-5) into Eq. (5-4) leads to:

$$u_m = \frac{1}{A_c} \int_{A_c} u \, dA_c \tag{5-6}$$

The mass or volume rate of flow in a duct is typically known and therefore Eqs. (5-3) or (5-4) can be used to compute the bulk velocity. The bulk velocity provides the reference velocity for an internal flow. Notice that continuity requires that u_m be constant with position for an internal incompressible flow, provided that the cross-sectional area of the duct remains constant. Therefore, u_m must be identical to u_∞, the free-stream velocity at the entrance to the duct.

The Reynolds number that characterizes an internal flow is based on u_m and the hydraulic diameter, D_h, which is the characteristic dimension of the duct cross-section.

$$Re_{D_h} = \frac{\rho D_h u_m}{\mu} \tag{5-7}$$

The hydraulic diameter is defined according to:

$$D_h = \frac{4 A_c}{per} \tag{5-8}$$

where *per* is the wetted perimeter of the duct. For a circular duct, the hydraulic diameter is equal to the diameter of the duct (D):

$$D_h = \frac{4 \pi D^2}{4 \pi D} = D \tag{5-9}$$

The Laminar Hydrodynamic Entry Length

The hydrodynamic entry length is the distance required for the momentum boundary layers to join. The momentum boundary layer growth for a laminar external flow is discussed in Section 4.1. A simple model of the boundary layer based on the diffusion of momentum into the free stream leads to:

$$\delta_{m,lam} = \frac{2x}{\sqrt{\dfrac{\rho u_\infty x}{\mu}}} \tag{5-10}$$

The laminar hydrodynamic entry length ($x_{fd,h,lam}$) is equal to the axial position at which the momentum boundary layer first spans the half-width of the duct. Substituting $\delta_{m,lam} = D_h/2$ and $x = x_{fd,h,lam}$ into Eq. (5-10) leads to:

$$\frac{D_h}{2} \approx \frac{2 x_{fd,h,lam}}{\sqrt{\dfrac{\rho u_\infty x_{fd,h,lam}}{\mu}}} \tag{5-11}$$

where $x_{fd,h,lam}$ is the hydrodynamic entry length for a laminar internal flow. Squaring both sides of Eq. (5-11) leads to:

$$\frac{D_h^2}{4} \approx \frac{4 x_{fd,h,lam} \, \mu}{\rho u_\infty} \tag{5-12}$$

Equation (5-12) is solved for the ratio of the hydrodynamic entry length to the hydraulic diameter:

$$\frac{x_{fd,h,lam}}{D_h} \approx \frac{\rho u_\infty D_h}{16 \, \mu} \tag{5-13}$$

Recognizing the u_∞ is equal to u_m when density and cross-sectional area are constant allows Eq. (5-13) to be rewritten as:

$$\frac{x_{fd,h,lam}}{D_h} \approx 0.06 \, Re_{D_h} \tag{5-14}$$

Equation (5-14) is the accepted correlation for the hydrodynamic entry length of a laminar internal flow (White (1991)). Clearly, our understanding of the characteristics of external flows translates well to the behavior of internal flows.

Turbulent Internal Flow

The internal flow will be turbulent provided that it transitions to turbulence before it becomes hydrodynamically fully developed. Once the boundary layers join, the viscous shear stress is constant and therefore the flow is not likely to become turbulent unless it experiences a disturbance or there is a change in the cross-sectional area or fluid properties that causes an increase in the Reynolds number. We can use our understanding of external flow behavior to approximately identify the flow conditions that will lead to a turbulent internal flow.

The critical Reynolds number based on position in the flow direction ($Re_{x,crit}$) at which an external flow becomes turbulent depends significantly on the free-stream conditions and other characteristics of the flow, but typically ranges from 3×10^5 to 6×10^6, as discussed in Section 4.5. A value of $Re_{x,crit} = 5 \times 10^5$ is often used.

$$Re_{x,crit} = \frac{\rho\, u_\infty\, x_{crit}}{\mu} \tag{5-15}$$

Solving Eq. (5-15) for x_{crit} leads to:

$$x_{crit} = Re_{x,crit}\frac{\mu}{\rho\, u_\infty} \tag{5-16}$$

Equation (5-10) is used to model the boundary layer thickness at the transition from laminar to turbulent flow, $\delta_{m,crit}$:

$$\delta_{m,crit} \approx 2\sqrt{\frac{x_{crit}\,\mu}{\rho\, u_\infty}} \tag{5-17}$$

Substituting Eq. (5-16) into Eq. (5-17) leads to an expression for the critical boundary layer thickness:

$$\delta_{m,crit} \approx \frac{2\,\mu}{\rho\, u_\infty}\sqrt{Re_{x,crit}} \tag{5-18}$$

If the critical boundary layer thickness is less than the approximate half-width of the duct, $D_h/2$, then the flow will transition to turbulence before becoming fully developed. Therefore, the condition at which an internal flow will become turbulent is, approximately:

$$\frac{D_h}{2} \approx \frac{2\,\mu}{\rho\, u_\infty}\sqrt{Re_{x,crit}} \tag{5-19}$$

Rearranging Eq. (5-19) and recognizing that $u_\infty = u_m$ leads to:

$$\underbrace{\frac{\rho\, u_m\, D_h}{\mu}}_{Re_{D_h,crit}} \approx 4\sqrt{Re_{x,crit}} \tag{5-20}$$

Assuming that $Re_{x,crit} = 5 \times 10^5$ and recognizing that the left side of Eq. (5-20) is the Reynolds number based on the hydraulic diameter, Eq. (5-7), leads to:

$$Re_{D_h,crit} \approx 2800 \tag{5-21}$$

Equation (5-21) suggests that an internal flow will be turbulent provided that the Reynolds number based on the hydraulic diameter (Re_{D_h}) is greater than a critical value ($Re_{D_h,crit}$). The critical Reynolds number is typically assumed to be $Re_{D_h,crit} = 2300$, but can be as low as 2100 or as high as 10,000 for a carefully controlled experiment (Lienhard and Lienhard (2005)).

Figure 5-2 illustrates flow through the passage formed by two parallel plates. A laminar boundary layer develops from the leading edge of the plate and the boundary layer grows until it reaches a critical value, $\delta_{m,crit}$ given by Eq. (5-18), at which point the flow transitions to turbulence. The turbulent, internal flow has many of the same characteristics as turbulent external flow that were discussed in Section 4.5. Throughout the bulk of the flow, turbulent eddies provide an efficient mechanism for momentum transport and therefore the fluid has a very large, effective viscosity. As a result, the turbulent boundary layer grows quickly and the flow becomes fully developed soon after it transitions to turbulence. The turbulent eddies are suppressed in the thin viscous sublayer

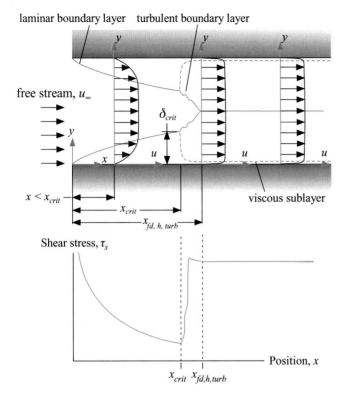

Figure 5-2: Qualitative characteristics of a turbulent internal flow.

that exists next to the wall. As a result, the velocity gradient occurs primarily across the viscous sublayer and so the shear stress jumps dramatically upon transitioning to turbulence. The viscous sublayer thickens very slowly in the downstream direction and therefore the shear stress drops slightly until the flow becomes fully developed. The velocity distribution and therefore the shear stress remain unchanged after the flow is fully developed.

The Turbulent Hydrodynamic Entry Length
The thickness of the turbulent boundary layer for an external flow is discussed in Section 4.9.2:

$$\delta_{m,turb} \approx \frac{0.16\,x}{Re_x^{1/7}} \tag{5-22}$$

Equation (5-22) is rearranged:

$$\delta_{m,turb} \approx 0.16\,x \left(\frac{\rho\,u_\infty\,x}{\mu} \right)^{-1/7} \tag{5-23}$$

A turbulent flow will become fully developed when the boundary layer extends approximately across the half-width of the duct. Substituting $\delta_{m,turb} \approx D_h/2$ and $x = x_{fd,h,turb}$ into Eq. (5-23) leads to:

$$\frac{D_h}{2} \approx 0.16\,x_{fd,h,turb} \left(\frac{\rho\,u_\infty\,x_{fd,h,turb}}{\mu} \right)^{-1/7} \tag{5-24}$$

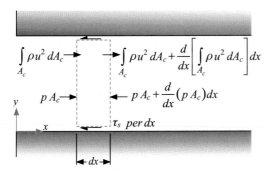

Figure 5-3: A differential momentum balance on an internal flow.

where $x_{fd,h,turb}$ is the hydrodynamic entry length for a turbulent flow. Equation (5-24) is rearranged:

$$\left[\frac{D_h}{2} \approx 0.16 \, x_{fd,h,turb}^{6/7}\left(\frac{\rho u_\infty}{\mu}\right)^{-1/7}\right]\frac{D_h^{1/7}}{D_h^{1/7}} \quad \rightarrow \quad \left(\frac{x_{fd,h,turb}}{D_h}\right)^{-6/7} \approx 0.16 \, (2)\left(\underbrace{\frac{\rho u_\infty D_h}{\mu}}_{Re_{D_h}}\right)^{-1/7} \tag{5-25}$$

or

$$\frac{x_{fd,h,turb}}{D_h} \approx 3.8 \, Re_{D_h}^{1/6} \tag{5-26}$$

Equation (5-26) shows that $x_{fd,h,turb}/D_h$ is a weak function of the Reynolds number. For Reynolds numbers ranging from 2300 and 11×10^7, the value of $x_{fd,h,turb}/D_h$ predicted by Eq. (5-26) goes from 14 to 56. Therefore, the turbulent hydrodynamic entry length will be relatively short for any reasonable Reynolds number.

The Friction Factor

The shear stress at the wall of an internal flow is related to the pressure gradient in the flow direction. The pressure gradient is the more critical engineering quantity, because the pressure drop experienced by the fluid must be overcome by a fan or pump that consumes energy and therefore costs money to operate. The shear stress and pressure gradient are related by a momentum balance on a differential element of the duct (in x), shown in Figure 5-3.

The momentum balance suggested by Figure 5-3 is:

$$p A_c + \int_{A_c} \rho u^2 \, dA_c = \tau_s \, per \, dx + p A_c + \frac{d\,(p A_c)}{dx} dx + \int_{A_c} \rho u^2 \, dA_c + \frac{d}{dx}\left[\int_{A_c} \rho u^2 \, dA_c\right] dx \tag{5-27}$$

or

$$-\frac{d\,(p A_c)}{dx} = \tau_s \, per + \frac{d}{dx}\left[\int_{A_c} \rho u^2 \, dA\right] \tag{5-28}$$

Equation (5-28) shows that the pressure force must balance the shear force at the wall as well as any change in the momentum of the fluid. For a constant cross-sectional area

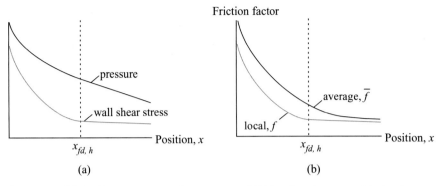

Figure 5-4: (a) Wall shear stress and pressure and (b) local and average friction factor as a function of position in the entrance region of an internal flow.

duct, Eq. (5-28) reduced to:

$$-\frac{dp}{dx} = \tau_s \frac{per}{A_c} + \frac{d}{dx}\left[\frac{1}{A_c}\int_{A_c} \rho u^2\, dA\right] \tag{5-29}$$

Equation (5-29) suggests that the pressure gradient will be largest in the developing region, both because the shear stress is higher (see Figure 5-1(c)) as well as because the core of the flow is accelerating. The second term on the right side of Eq. (5-29) is related to the change in the momentum of the flow. The flow enters the duct with a uniform velocity that becomes non-uniform due to the effect of the shear applied by the wall. The velocity at the center of the duct increases while the velocity at the edge of the duct decreases; the integral of the velocity squared must therefore increase as the flow develops hydrodynamically. This effect also contributes to an increased pressure gradient in the developing region.

In the hydrodynamically fully developed region (i.e., $x > x_{fd,h}$), the velocity distribution does not change with x and therefore the last term in Eq. (5-29) will be zero (i.e., the momentum of the flow does not change in the x direction). Furthermore, in the hydrodynamically fully developed region the momentum boundary layer does not grow (see Figure 5-1(b)) and therefore the shear stress at the wall does not change in the x-direction (see Figure 5-1(c)). The pressure gradient in the hydrodynamically fully developed region will be constant:

$$\left(-\frac{dp}{dx}\right)_{x>x_{fd,h}} = \tau_s \frac{per}{A_c} \tag{5-30}$$

Figure 5-4(a) illustrates, qualitatively, the variation of the shear stress and pressure in the entrance region of an internal flow. The pressure initially falls sharply due to the larger shear stress and, to a lesser extent, the fact that the core of the flow is being accelerated. However, for $x > x_{fd,h}$ the pressure gradient is constant and the pressure decreases linearly with position.

For an external flow over a plate, the shear stress was correlated with the friction coefficient, C_f. For an internal flow, the pressure gradient is correlated using the Moody (or Darcy) friction factor, f, defined according to:

$$f = -\frac{dp}{dx}\frac{2\,D_h}{\rho\,u_m^2} \tag{5-31}$$

The friction factor will be large in the hydrodynamically developing region and then become constant in the fully developed region, as shown in Figure 5-4(b). It is important that the Moody friction factor not be confused with the Fanning friction factor. The Fanning friction factor is defined based on the wall shear stress. The Darcy or Moody friction factor is four times larger than the Fanning friction factor and care should be taken so that these friction factor definitions are not accidentally interchanged. This book deals exclusively with the Moody friction factor, defined in Eq. (5-31).

The friction factor defined by Eq. (5-31) is a local friction factor that is based on the local pressure gradient. An average or apparent friction factor $(\bar{f})$ is defined based on the total change in the pressure:

$$\bar{f} = (p_{x=0} - p_{x=L}) \frac{2 D_h}{L \rho u_m^2} \tag{5-32}$$

where L is the length of the pipe. It is almost always neccesary to calculate the average rather than the local friction factor in order to solving an engineering problem because it is usually necessary to know the total pressure change across a duct that must be overcome by a pump or fan. Equation (5-32) can be rewritten in terms of the local friction factor:

$$\bar{f} = -\frac{2 D_h}{L \rho u_m^2} \int_0^L \frac{dp}{dx} dx = \frac{2 D_h}{L \rho u_m^2} \int_0^L \frac{\rho u_m^2}{2 D_h} f\, dx = \frac{1}{L} \int_0^L f\, dx \tag{5-33}$$

The average friction factor will approach the local friction factor in the fully developed region. However, because the average friction factor has some memory of the developing region it will always be somewhat larger than the local value, as shown in Figure 5-4(b).

Specific correlations for the behavior of the friction factor in an internal flow are provided in Section 5.2. However, some of the characteristics of these correlations can be anticipated based upon our knowledge of laminar and turbulent flows. For example, the velocity gradient in a laminar flow encompasses the entire boundary layer which, in a fully developed internal flow, corresponds to the entire cross-section of the duct. Therefore, we should expect that the friction factor for a laminar internal flow will be quite sensitive to the shape of the duct (e.g., round vs square) but insensitive to small-scale roughness at the duct surface.

The velocity gradient in a turbulent flow is primarily confined to the viscous sub-layer and therefore does not encompass the entire duct. The shear stress at the wall of a turbulent duct will therefore be insensitive to the shape of the duct. Substituting Eq. (5-30) into (5-31) and recalling the definition of the hydraulic diameter leads to:

$$f = \tau_s \frac{per}{A_c} \frac{2 D_h}{\rho u_m^2} = \tau_s \frac{per}{A_c} \frac{2}{\rho u_m^2} \frac{4 A_c}{per} \tag{5-34}$$

or

$$f = \tau_s \frac{8}{\rho u_m^2} \tag{5-35}$$

The shear stress in a turbulent flow is related to the velocity gradient across the viscous sublayer. Therefore, Eq. (5-35) suggests that the friction factor for a turbulent flow does not depend on the characteristics of the duct (A_c or per). However, the turbulent friction factor will be strongly affected by the presence of roughness at the duct surface because the velocity gradient is confined to the region very near the wall. In Section 5.2, we will see that these intuitive characteristics of internal flow are reflected in the specific correlations that are used to carry out engineering calculations.

5.1.3 Thermal Considerations

Figure 5-5(a) illustrates laminar external flow over a plate and shows, qualitatively, the thermal boundary layer and temperature distribution that results. Figure 5-5(b) illustrates the laminar flow through a passage that is formed between two parallel plates with constant surface temperature; notice that at some location the thermal boundary layer becomes bounded in the same way that the momentum boundary layer does.

The position where the thermal boundary layers that are growing from the upper and lower plates in Figure 5-5(b) meet is referred to as the thermal entry length $x_{fd,t}$. The thermal boundary layer does not grow further as the fluid moves downstream (i.e., for $x > x_{fd,t}$). The region where the thermal boundary layers are growing (i.e., for $x < x_{fd,t}$) is referred to as the thermally developing region. The region where the thermal boundary layers are bounded (i.e., for $x > x_{fd,t}$) is referred to as the thermally fully developed region.

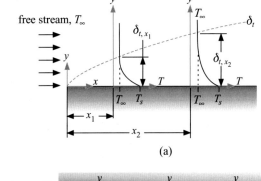

(a)

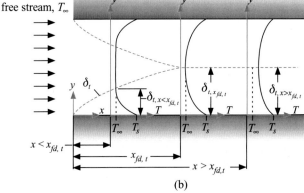

(b)

Figure 5-5: (a) Laminar external flow over a flat plate, and (b) laminar internal flow between two parallel plates.

The Mean Temperature

The velocity distribution for an internal flow does not change with x in the hydrodynamically fully developed region (see Figure 5-1(b)) if the shape of the duct does not change. However, Figure 5-5(b) illustrates that the temperature distribution in the

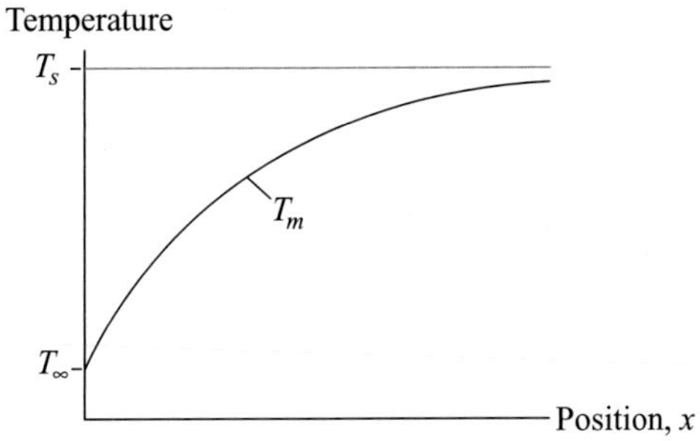

Temperature

Figure 5-6: Mean temperature as a function of position in a duct with a constant surface temperature.

thermally fully developed region does continue to change even though the thermal boundary layer thickness does not. In general, energy continues to be added or removed from the fluid in the thermally fully developed region and therefore the fluid temperature must change in order to conserve energy. As a result, the free-stream temperature at the duct inlet, T_∞ in Figure 5-5(b), is not a useful reference temperature for the flow. Instead, the mean temperature (T_m, sometimes referred to as the bulk temperature) is defined. The mean temperature is the single temperature that represents the thermal energy carried by the flow at a particular axial position:

$$T_m = \frac{\rho}{\dot{m}} \int_{A_c} T \, u \, dA_c \tag{5-36}$$

where T is the local temperature of the flow. Equation (5-36) can be written in terms of the volumetric flow rate ($\dot{V}$):

$$T_m = \frac{1}{\dot{V}} \int_{A_c} T \, u \, dA_c \tag{5-37}$$

Note that both Eqs. (5-36) and (5-37) are only valid for incompressible fluids with constant specific heat capacity. The mean temperature is the single temperature that can be used in an energy balance on the flow, as discussed in Section 5.3. Figure 5-6 illustrates the variation of the mean temperature of the fluid with position in the constant surface temperature duct that is shown in Figure 5-5.

The Heat Transfer Coefficient and Nusselt Number

The mean temperature represents the local rate at which the flow carries thermal energy (or enthalpy) and it is convenient for carrying out energy balances within a heat exchanger or other heat transfer device. The mean temperature is therefore the most meaningful reference temperature for the flow and is used to define the local heat transfer coefficient for an internal flow:

$$h = \frac{\dot{q}_s''}{(T_s - T_m)} \tag{5-38}$$

where $\dot{q}_s''$ is the surface heat flux and T_s and T_m are the local surface and mean temperatures, respectively. The local heat transfer coefficient is correlated using a local Nusselt number that is based on the hydraulic diameter:

$$Nu_{D_h} = \frac{h \, D_h}{k} \tag{5-39}$$

Heat transfer coefficient
and Nusselt number

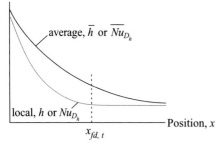

Figure 5-7: Local and average heat transfer coefficient or Nusselt number as a function of position.

where k is the thermal conductivity of the fluid. For a laminar flow, the thermal boundary layer can be thought of approximately as a conduction resistance to heat transfer. Therefore, the heat flux at the wall can be written approximately as:

$$\dot{q}_s'' \approx \frac{k}{\delta_t}(T_s - T_m) \qquad (5\text{-}40)$$

Comparing Eqs. (5-38) and (5-40) indicates that:

$$h \approx \frac{k}{\delta_t} \qquad (5\text{-}41)$$

As with an external flow, the heat transfer coefficient for a laminar internal flow will be inversely proportional to the boundary layer thickness. Therefore, the heat transfer coefficient (and Nusselt number) will be large in the thermally developing region and decrease until it reaches a constant value in the thermally fully developed region, as shown in Figure 5-7.

The heat transfer coefficient and Nusselt number defined by Eqs. (5-38) and (5-39) are local values. An average heat transfer coefficient ($\bar{h}$) is often more useful:

$$\bar{h} = \frac{1}{L}\int_0^L h\,dx \qquad (5\text{-}42)$$

The average Nusselt number for an internal flow is defined according to:

$$\overline{Nu}_{D_h} = \frac{\bar{h}\,D_h}{k} \qquad (5\text{-}43)$$

Substituting Eqs. (5-42) and (5-39) into Eq. (5-43) leads to:

$$\overline{Nu}_{D_h} = \frac{D_h}{k}\frac{1}{L}\int_0^L \frac{k\,Nu_{D_h}}{D_h}\,dx = \frac{1}{L}\int_0^L Nu_{D_h}\,dx \qquad (5\text{-}44)$$

Figure 5-7 illustrates the average heat transfer coefficient and Nusselt number for an internal flow. The average Nusselt number will approach the local Nusselt number in the thermally fully developed region. However, because the average Nusselt number has some memory of the developing region, it will always be somewhat larger than the local value.

The Laminar Thermal Entry Length
The thermal entry length is the distance required for the thermal boundary layers to join. The thermal boundary layer growth for a laminar, external flow is discussed conceptually

in Section 4.1. A simple model based on the diffusion of energy leads to:

$$\delta_{t,lam} \approx \frac{2x}{\sqrt{Re_x\,Pr}} \tag{5-45}$$

where Re_x is the Reynolds number based on x:

$$\delta_{t,lam} \approx \frac{2x}{\sqrt{\dfrac{\rho\,u_\infty\,x}{\mu}\,Pr}} \tag{5-46}$$

The thermal entry length is the position at which the thermal boundary layer extends across the half-width of the duct. Substituting $\delta_{t,lam} = D_h/2$ and $x = x_{fd,t,lam}$ into Eq. (5-46) leads to:

$$\frac{D_h}{2} \approx \frac{2x_{fd,t,lam}}{\sqrt{\dfrac{\rho\,u_\infty\,x_{fd,t,lam}}{\mu}\,Pr}} \tag{5-47}$$

where $x_{fd,t,lam}$ is the thermal entry length for a laminar internal flow. Squaring both sides of Eq. (5-47) leads to:

$$\frac{D_h^2}{4} \approx \frac{4\,x_{fd,t,lam}\,\mu}{\rho\,u_\infty\,Pr} \tag{5-48}$$

Equation (5-48) is solved for the ratio of the thermal entry length to the hydraulic diameter:

$$\frac{x_{fd,t,lam}}{D_h} \approx \underbrace{\frac{\rho\,u_\infty\,D_h}{\mu}}_{Re_{D_h}}\frac{Pr}{16} \tag{5-49}$$

Recognizing the u_∞ is equal to u_m for an incompressible fluid allows Eq. (5-49) to be rewritten as:

$$\frac{x_{fd,t,lam}}{D_h} \approx 0.06\,Re_{D_h}\,Pr \tag{5-50}$$

The ratio of the thermal to the hydrodynamic laminar entry lengths $(x_{fd,t,lam}/x_{fd,h,lam})$ is obtained by dividing Eq. (5-50) by Eq. (5-14):

$$\frac{x_{fd,t,lam}}{x_{fd,h,lam}} \approx Pr \tag{5-51}$$

Equation (5-51) matches our intuition. The relative rate of the momentum to thermal boundary layer development in a laminar external flow was previously shown to be related to the Prandtl number; this continues to be true in an internal flow. The relative growth of the boundary layers for a fluid such as engine oil is shown in Figure 5-8(a). Engine oil is viscous but not conductive and therefore it has a high Prandtl number (much greater than unity). The momentum boundary layer will grow more quickly than the thermal boundary layer and the flow will become hydrodynamically fully developed much sooner than it will become thermally developed. The ratio of $x_{fd,t,lam}$ to $x_{fd,h,lam}$ should be much greater than unity, as predicted by Eq. (5-51).

The opposite behavior occurs for a fluid such as a liquid metal that has a Prandtl number that is much less than unity. The thermal boundary layer develops quickly due to the conductive nature of the liquid metal and therefore $x_{fd,t,lam}$ is small. The momentum boundary layer takes longer to develop and so the ratio of $x_{fd,t,lam}$ to $x_{fd,h,lam}$ should be much less than unity, as predicted by Eq. (5-51) and shown in Figure 5-8(b).

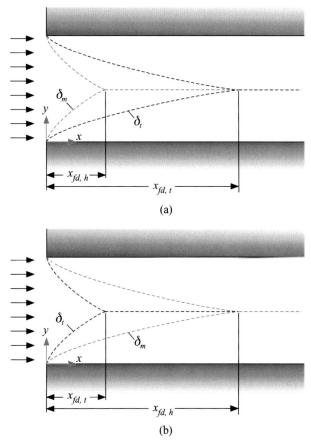

(a)

(b)

Figure 5-8: Thermal and momentum boundary layer growth for (a) a high Prandtl number, and (b) a low Prandtl number fluid.

Turbulent Internal Flow

As discussed in Section 5.1.2, an internal flow will be turbulent provided that it transitions before it becomes hydrodynamically fully developed. This condition is consistent with a critical Reynolds number based on the hydraulic diameter, Re_{D_h}, that is nominally equal to $Re_{D_h, crit} = 2300$. A turbulent internal flow will exhibit the same characteristics that were previously discussed for a turbulent external flow. For example, Figure 5-9 illustrates, qualitatively, the temperature distribution expected for a laminar and turbulent flow with the same mean temperature and surface temperature.

The energy transport in the laminar internal flow shown in Figure 5-9(a) is primarily diffusive across the entire cross-section because no turbulent eddies are present. The energy transport is completely diffusive in the hydrodynamically fully developed region where there is no velocity component in the y-direction. As a result, the temperature gradient extends over the entire cross-section. The energy transport in the turbulent internal flow shown in Figure 5-9(b) is primarily due to the macroscopic fluid motion induced by turbulent eddies. The effective conductivity associated with this turbulent condition, k_{turb}, is much higher than the conductivity of the fluid itself. Only in the viscous sublayer very near the wall will the energy transport be diffusive. The temperature gradient is primarily confined to the viscous sublayer because $k_{turb} \gg k$.

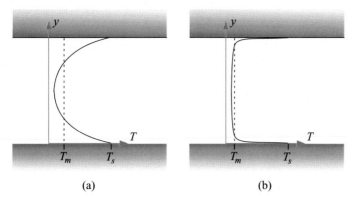

Figure 5-9: Qualitative temperature distribution expected in (a) a laminar and (b) a turbulent internal flow.

The heat transfer coefficients for the laminar and turbulent flows are, approximately:

$$h_{lam} \approx \frac{k}{\delta_{t,lam}} \tag{5-52}$$

$$h_{turb} \approx \frac{k}{\delta_{vs}} \tag{5-53}$$

where δ_{vs} is the thickness of the viscous sublayer. Comparing Eqs. (5-52) and (5-53) indicates that the heat transfer coefficient and therefore the Nusselt number will be substantially higher in a turbulent flow because δ_{vs} is much smaller than $\delta_{t,lam}$.

Specific correlations for the Nusselt number in an internal flow are provided in Section 5.2; however, some of the characteristics of these correlations can be anticipated based on the discussion above. For example, the temperature gradient in a laminar flow encompasses the entire boundary layer which, in a fully developed internal flow, corresponds to the entire cross-section of the duct. Therefore the Nusselt number for a laminar internal flow will be quite sensitive to the shape of the duct (e.g., round vs square) as well as the boundary conditions on the flow (e.g., constant temperature or constant heat flux) but insensitive to small-scale roughness. The temperature gradient in a turbulent flow is confined to the viscous sublayer. Therefore the turbulent Nusselt number will be insensitive to the shape of the duct or the boundary conditions but will be strongly affected by the presence of surface roughness.

5.2 Internal Flow Correlations

5.2.1 Introduction

Section 5.1 discusses the behavior of laminar and turbulent internal flow at a conceptual level without providing specific information that could be used to solve an internal flow problem. This section presents a set of useful correlations that are based on analytical solutions and experimental data. These correlations can be used to solve a wide range of engineering problems. A relatively complete set of these correlations has been provided in a library of functions that are available in EES. The use of these functions to solve engineering problems is illustrated in this section. The correlations are examined as they are presented in order to verify that they agree with our physical understanding of internal flow processes.

5.2.2 Flow Classification

In order to select the appropriate correlation, it is necessary to classify the salient features of the flow. The most critical classification is the flow condition, which is determined by the Reynolds number based on the hydraulic diameter (D_h):

$$Re_{D_h} = \frac{\rho u_m D_h}{\mu} \tag{5-54}$$

The hydraulic diameter is defined according to:

$$D_h = \frac{4 A_c}{per} \tag{(5-55)}$$

where A_c is the cross-sectional area of the duct and *per* is the wetted perimeter. The flow will be laminar if the Reynolds number based on the hydraulic diameter is less than approximately 2300, otherwise it will be turbulent.

Both the Nusselt number and friction factor depend on axial position in the developing region, as discussed in Section 5.1. The behavior of an internal flow is very different in the developing as opposed to the fully developed regions. The Nusselt number depends on whether the flow is developing both thermally and hydrodynamically (referred to as simultaneously developing) or is thermally developing after it has previously become hydrodynamically fully developed (referred to as thermally developing/hydrodynamically developed). The flow will be simultaneously developing at the inlet to a pipe whereas it will be thermally developing when heating or cooling is applied at some location on a pipe surface that is removed from the inlet.

The heat transfer coefficient and friction factor for a laminar flow will be affected by the large-scale features of the flow, for example the shape of the duct. Correlations are presented in this section for circular, rectangular, and annular passages. The Nusselt number for a laminar flow also depends on the thermal boundary conditions. The two most common boundary conditions are constant wall temperature and constant heat flux; these two boundary conditions provide lower and upper bounds, respectively, for more realistic boundary conditions that are not as well-defined. The roughness at the duct surface will affect the Nusselt number and friction factor for a turbulent flow but not a laminar flow.

The engineer must carefully classify an internal flow problem according to the criteria discussed above in order to identify the most appropriate correlation. The internal flow convection library in EES automatically accomplishes this classification and implements the correct correlation from among those presented in this section. However, it is incumbent on the engineer to understand this process so that the result can be critically assessed and also so that correlations for geometries and conditions that are not considered in this section or implemented in EES can be correctly applied.

5.2.3 Friction Factor

The local friction factor in the fully developed region is shown in Figure 5-10 as a function of the Reynolds number for various duct shapes (in the laminar region) and wall roughness (in the turbulent region). The wall roughness is quantified as the ratio of the average height of the surface disparities (e) to the hydraulic diameter of the duct. The specific correlations that correspond to the information in Figure 5-10, as well as the increase in friction factor that occurs in the developing region, are discussed in this section.

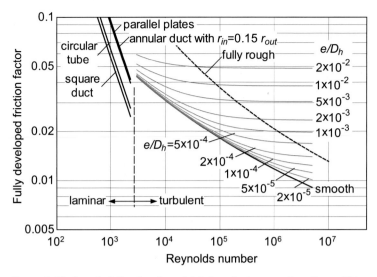

Figure 5-10: Local, fully developed friction factor as a function of Reynolds number for various duct shapes (in the laminar region) and relative roughness, e/D_h (in the turbulent region).

Laminar Flow

As shown in Figure 5-10 and discussed in Section 5.1, the friction for a laminar flow is affected by the duct shape but not surface roughness.

Circular Tubes. The local friction factor for a laminar, hydrodynamically fully developed flow in a circular tube is given by:

$$f_{fd,h} = \frac{64}{Re_{D_h}}$$
(5-56)

This result is obtained by solving the momentum equation in the fully developed region, as discussed in Section 5.4.2. The average friction factor (sometimes referred to as the apparent friction factor) that includes the developing region for a circular tube (defined in Section 5.1.2) is given by Shah and London (1978):

$$\bar{f} = \frac{4}{Re_{D_h}} \left[\frac{3.44}{\sqrt{L^+}} + \frac{\frac{1.25}{4\,L^+} + \frac{64}{4} - \frac{3.44}{\sqrt{L^+}}}{1 + \frac{0.00021}{(L^+)^2}} \right]$$
(5-57)

where L^+ is the appropriate dimensionless length for a hydrodynamically developing internal flow, defined as:

$$L^+ = \frac{L}{D_h\,Re_{D_h}}$$
(5-58)

Notice that in the limit that $L^+ \to \infty$, Eq. (5-57) approaches Eq. (5-56), as it should. Also, the ratio of the apparent friction factor to the fully developed friction factor

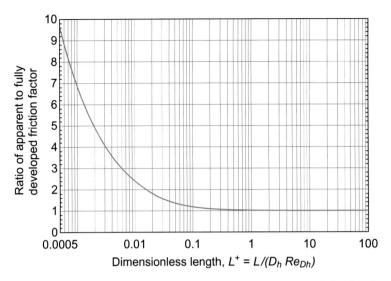

Figure 5-11: The ratio of the apparent friction factor to the fully developed friction factor as a function of $L^+ = L/D_h\,Re_{D_h}$.

(i.e., $\overline{f}/f_{fd,h}$) is independent of the Reynolds number and depends only on L^+:

$$\frac{\overline{f}}{f_{fd,h}} = \frac{1}{16}\left[\frac{3.44}{\sqrt{L^+}} + \frac{\dfrac{1.25}{4\,L^+} + \dfrac{64}{4} - \dfrac{3.44}{\sqrt{L^+}}}{1 + \dfrac{0.00021}{(L^+)^2}}\right] \tag{5-59}$$

Figure 5-11 illustrates the ratio of the apparent friction factor to the fully developed friction factor as a function of the dimensionless position, L^+. Notice that the apparent friction factor approaches the fully developed friction factor when L^+ reaches a value of approximately 0.1; according to Eq. (5-58), this corresponds to $L \approx 0.1\,Re_{D_h}\,D_h$. This behavior is consistent with the discussion in Section 5.1.2 and Eq. (5-14), which predict that a laminar flow will become hydrodynamically fully developed at $x_{fd,h,lam} \approx 0.06\,Re_{D_h}\,D_h$.

Rectangular Ducts. In the laminar, fully developed region, the product of the Reynolds number and the friction factor will be a constant that depends on the shape of the duct. For example, in a circular duct, this constant is 64 according to Eq. (5-56). For a rectangular duct, the fully developed friction factor is a function of the aspect ratio (AR, defined as the ratio of the minimum to the maximum dimensions of the duct) according to:

$$f_{fd,h} = \frac{96}{Re_{D_h}}(1 - 1.3553\,AR + 1.9467\,AR^2 - 1.7012\,AR^3 + 0.9564\,AR^4 - 0.2537\,AR^5)$$

$$\tag{5-60}$$

As the aspect ratio approaches 0, the solution provided by Eq. (5-60) limits to $96/Re_{D_h}$. This is the correct solution for flow between two infinite parallel plates, derived in Section 5.4.2. The apparent friction factor for laminar flow in rectangular ducts can be approximately computed using the dimensionless position L^+ and the solution given by Eq. (5-57) for circular ducts with the 64 in the numerator of the second term replaced

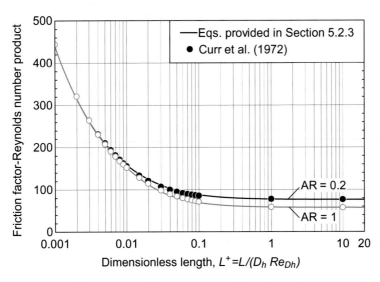

Figure 5-12: The product of the apparent friction factor in a rectangular duct and the Reynolds number as a function of L^+ for two values of the aspect ratio. The results predicted by Eqs. (5-60) and (5-61) are shown, as well as the more exact solution provided by Curr et al. (1972).

by the appropriate friction factor-Reynolds number product for the duct shape being considered:

$$\overline{f} \approx \frac{4}{Re_{D_h}} \left[\frac{3.44}{\sqrt{L^+}} + \frac{\frac{1.25}{4L^+} + \frac{\overbrace{(f_{fd,h} Re_{D_h})}^{\text{depends on duct shape}}}{4} - \frac{3.44}{\sqrt{L^+}}}{1 + \frac{0.00021}{(L^+)^2}} \right] \tag{5-61}$$

where $f_{fd,h}$ should be obtained from Eq. (5-60) for a rectangular duct. Figure 5-12 illustrates the product of the apparent friction factor and the Reynolds number as a function of L^+ for two values of the aspect ratio. Also shown are the more exact results obtained by Curr et al. (1972).

Annular Duct. Flow in the space between concentric inner and outer tubes is often encountered in engineering applications. For an annular duct, the fully developed friction factor is a function of the radius ratio (RR, defined as the ratio of the inner to the outer radii of the flow passage).

$$f_{fd,h} = \frac{64}{Re_{D_h}} \sqrt{\frac{(1 - RR^2)}{1 + RR^2 - \left(\frac{1 - RR^2}{\ln(RR^{-1})} \right)}} \tag{5-62}$$

Notice that as the radius ratio approaches 0, the solution provided by Eq. (5-62) limits to $64/Re_{D_h}$, which is the correct solution for flow through a circular duct. It is also true that the solution provided by Eq. (5-62) limits to $96/Re_{D_h}$ as RR approaches unity, which is the correct solution for flow through parallel plates. The apparent friction factor can be obtained using the same approximation previously presented for a rectangular duct,

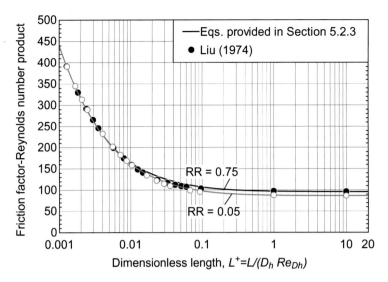

Figure 5-13: The product of the apparent friction factor for an annular duct and the Reynolds number as a function of L^+ for two values of the radius ratio. The results predicted by Eqs. (5-62) and Eq. (5-61) are shown, as well as the more exact solution provided by Liu (1974).

Eq. (5-61), with $f_{fd,h}$ obtained from Eq. (5-62). Figure 5-13 illustrates the product of the apparent friction factor and the Reynolds number as a function of L^+ for two values of the radius ratio. Also shown are the more exact results obtained by Liu (1974).

Turbulent Flow

Figure 5-10 shows that the friction factor for a turbulent flow is not significantly affected by the duct shape, but it is sensitive to the scale of the surface roughness (e). The friction factor for fully developed turbulent flow in an aerodynamically smooth duct is provided by Petukhov (1970):

$$f_{fd,h,e=0} = \frac{1}{[0.790 \ \ln (Re_{D_h}) - 1.64]^2} \quad \text{for } 3000 < Re_{D_h} < 5.0 \times 10^6 \qquad (5\text{-}63)$$

Colebrook (1939) presents an implicit expression for the fully developed, turbulent friction factor in a duct with surface roughness, e:

$$\frac{2}{\sqrt{f_{fd,h}}} = 3.48 - 1.7373 \ \ln \left(\frac{2e}{D_h} + \frac{2(9.35)}{Re_{D_h} \sqrt{f_{fd,h}}} \right) \qquad (5\text{-}64)$$

Zigrang and Sylvester (1982) present an explicit and therefore more convenient correlation for the fully developed, turbulent friction factor in a rough duct:

$$f_{fd,h} = \left\{ -2.0 \ \log_{10} \left[\frac{2e}{7.54 \, D_h} - \frac{5.02}{Re_{D_h}} \log_{10} \left(\frac{2e}{7.54 \, D_h} + \frac{13}{Re_{D_h}} \right) \right] \right\}^{-2} \qquad (5\text{-}65)$$

Because the turbulent entry length is so short, as discussed in Section 5.1.2, most correlations ignore the additional pressure drop incurred in the hydrodynamically developing region and assume that the apparent friction factor for a turbulent flow is equal to the fully developed value, $\bar{f} \approx f_{fd,h}$. However, the effect of the developing region on the apparent friction factor can be approximately accounted for using:

$$\overline{f} \approx f_{fd,h} \left(1 + \left(\frac{D_h}{L} \right)^{0.7} \right) \tag{5-66}$$

Equation (5-66) has the same form as is used to account for the effect of the developing region on the average turbulent Nusselt number in Section 5.2.4. Notice that the apparent friction factor is only slightly higher than the fully developed friction factor and quickly approaches $f_{fd,h}$ as L/D_h increases.

The effect of surface roughness on a turbulent flow is consistent with its impact on the flow over a plate, discussed in Section 4.9.2. The viscous sublayer is the critical dimension that governs the behavior of a turbulent flow and therefore the impact of surface roughness is related to the value of e/δ_{vs}, where δ_{vs} is the viscous sublayer thickness. If e/δ_{vs} is <1 then Petukhov's correlation for a smooth tube, Eq. (5-63), is valid. In the range $1 < e/\delta_{vs} < 12$, the surface is transitionally rough and the roughness will substantially affect the friction factor and heat transfer coefficient. If $e/\delta_{vs} > 12$ then the surface is fully rough and the friction factor becomes insensitive to the Reynolds number.

It is possible to estimate the transition to fully rough behavior based on our understanding of turbulent boundary layers. In Section 4.7, we saw that the thickness of the viscous sublayer is approximately consistent with an inner coordinate, $y^+ \approx 6$, and therefore can be estimated according to:

$$\delta_{vs} \approx 6 \, \upsilon \sqrt{\frac{\rho}{\tau_s}} \tag{5-67}$$

where υ is the kinematic viscosity and τ_s is the wall shear stress. According to the momentum balance, presented in Section 5.1.2, the shear stress at the wall in the fully developed region is related to the pressure gradient according to:

$$\tau_s = \left(-\frac{dp}{dx} \right) \frac{D_h}{4} \tag{5-68}$$

Substituting the definition of the friction factor, Eq. (5-31), into Eq. (5-68) leads to:

$$\tau_s = f \frac{\rho u_m^2}{2 D_h} \frac{D_h}{4} = f \frac{\rho u_m^2}{8} \tag{5-69}$$

Substituting Eq. (5-69) into Eq. (5-67) leads to an estimate of the viscous sublayer thickness in an internal flow:

$$\delta_{vs} \approx 6 \, \upsilon \sqrt{\frac{8 \rho}{f \rho u_m^2}} = \frac{6 \upsilon}{u_m} \sqrt{\frac{8}{f}} \tag{5-70}$$

The fully rough region will therefore occur when:

$$\frac{e}{\delta_{vs}} \approx \frac{e \, u_m}{6 \, \upsilon} \sqrt{\frac{f}{8}} \approx 12 \tag{5-71}$$

Multiplying and dividing Eq. (5-71) by D_h leads to:

$$\underbrace{\frac{e}{D_h}}_{\substack{\text{relative}\\\text{roughness}}} \underbrace{\frac{u_m D_h}{\upsilon}}_{Re_{D_h}} \frac{1}{6} \sqrt{\frac{f}{8}} \approx 12 \tag{5-72}$$

or

$$\frac{e}{D_h} Re_{D_h} \sqrt{f} \approx 204 \tag{5-73}$$

Table 5-1: Typical roughness of commercial pipes (various sources).

Material	Roughness, e
Riveted steel	0.9 to 9.0 mm
Galvanized steel	0.15 mm
Forged steel	0.045 mm
New cast iron	0.26 to 0.80 mm
Rusty cast iron	1.5 to 2.5 mm
Drawn tubing	0.0015 mm
Concrete	0.3 to 3.0 mm
Glass	smooth
PVC and plastic pipes	0.0015 to 0.007 mm
Wood	0.5 mm
Rubber	0.01 mm

If a correlation for the friction factor is used in conjunction with Eq. (5-73), then two equations (Eq. (5-73) and, for example, Eq. (5-65)) are available in three unknowns (f, e/D_h, and Re_{D_h}). Therefore, it is possible to develop a curve in the space of f vs Re_{D_h} that delineates the fully rough region. This is accomplished using the EES code below:

```
f=(-2*log10(2*Relrough/7.4-5.02*log10(2*Relrough/7.4+13/Re)/Re))^(-2)
                                              "Zigrang & Sylvester correlation"
Relrough*Re*sqrt(f)=204                        "e/delta_vs = 12"
```

A parametric table is setup containing the variables Re and f and used to generate the dashed line in Figure 5-10 that delineates the fully rough region from the transitionally rough region. Notice that in the fully rough region, the friction factor is independent of Reynolds number but strongly dependent on the size of the roughness. The friction factor in the fully rough region can be obtained from (Nikuradse (1950)):

$$f = \frac{1}{\left[1.74 + 2.0 \log_{10}\left(\frac{D_h}{2e}\right)\right]^2} \tag{5-74}$$

In the transitionally rough region both roughness and the Reynolds number affect the pressure drop. Typical roughness values for various surfaces are listed in Table 5-1.

EES' Internal Flow Convection Library

The friction factor correlations discussed in Section 5.4.3 have been implemented in EES and can be accessed by selecting Function Information from the Options menu and then selecting Convection from the list of options in the Heat Transfer pulldown menu. Select Internal Flow – Non-dimensional and scroll through the available duct shapes. The internal flow library implements correlations for the apparent friction factor in ducts of various shapes (as well as the Nusselt number for constant temperature and constant heat flux boundary conditions, discussed in Section 5.2.4). The companion library, Internal Flow – Non-dim (local), includes procedures that return the local friction factor (and Nusselt numbers) for the same duct geometries. Procedures are accessed using the Call command. For example, to access the correlations for the average friction factor for circular ducts, it is necessary to call the procedure PipeFlow_N. For example:

```
$UnitSystem SI MASS RAD PA K J
$TABSTOPS 0.2 0.4 0.6 0.8 3.5 in

Re=1000 [-]                                              "Reynolds number"
Pr=1 [-]                                                 "Prandtl number"
LoverD=100 [-]                                           "length to diameter ratio"
RelRough=1e-4 [-]                                        "relative roughness"
call PipeFlow_N(Re,Pr,LoverD,RelRough: Nusselt_T_bar,Nusselt_H_bar,f_bar)
                                                         "access correlation"
```

which leads to $\overline{f} = 0.0758$; this result is consistent with the value indicated in Figure 5-10 (it is slightly higher, due to the effect of the hydrodynamically developing region).

The EES library procedures identify whether the flow is laminar or turbulent and which of the correlations should be used. The arguments to the left of the colon in the procedure are inputs and those to the right are outputs; however, it is possible to specify an output, such as the friction factor, and have the procedure determine one of inputs (provided that the guess values and limits for the variables are appropriately set). The inputs to the procedure include the Reynolds number, Prandtl number, length to diameter ratio (L/D_h), and the relative roughness (e/D_h). The outputs are the average Nusselt numbers assuming constant wall temperature and heat flux (which provide lower and upper bounds, as discussed in the subsequent section), and the average friction factor. The procedure models the transition from laminar to turbulence (i.e., the region between a Reynolds number of 2300 and 3000) by determining the fully laminar and fully turbulent results and interpolating between these values; this gradual rather than abrupt transition will prevent instability and convergence problems in many numerical solutions. Further details regarding the procedure can be obtained by examining the Help information in EES.

EXAMPLE 5.2-1: FILLING A WATERING TANK

You are filling a watering tank for livestock from a stream using a small pump. The watering tank holds $V_{tank} = 600$ gallons of water and is approximately at the same elevation as the stream, therefore there is no hydrostatic head to be overcome. The water is pumped through a PVC pipe that is $L = 20$ ft long and has an inner diameter of $D = 1.0$ inch. Several points on the manufacturer's pump curve are provided in Table 1.

Table 1: Data from the manufacturer's pump curve.

Head (ft of water)	Flow (gpm)
60 ft	0 gpm
60 ft	20 gpm
57 ft	40 gpm
52 ft	60 gpm
44 ft	80 gpm
33 ft	100 gpm
18 ft	120 gpm
0 ft	140 gpm

EXAMPLE 5.2-1: FILLING A WATERING TANK

EXAMPLE 5.2-1: FILLING A WATERING TANK

a) Estimate how long it will take to fill the tank using the pump and the PVC pipe.

The known information is entered in EES:

```
"EXAMPLE 5.2-1: Filling a Watering Tank"

$UnitSystem SI MASS RAD PA K J
$Tabstops 0.2 0.4 0.6 3.5 in

"Inputs"
V_tank=600 [gal]*convert(gal,m^3)          "tank volume"
L=20[ft]*convert(ft,m)                      "pipe length"
D=1.0 [inch]*convert(inch,m)               "pipe diameter"
```

The pump curve data points are entered into a lookup table (select New Lookup Table from the Tables menu and specify a table with 2 columns and 8 rows). The pump curve data are plotted in Figure 1.

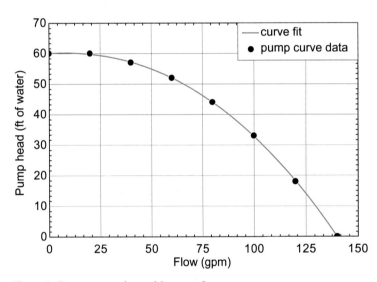

Figure 1: Pump curve data with curve fit.

The data in the plot are used to generate a curve fit that represents the pump performance. Select Curve Fit from the Plots menu and fit a third order polynomial to the Head vs Flow data; select Fit and then plot to overlay the curve fit on the plot (Figure 2). Select Copy equation to Clipboard and then paste the equation into the Equations Window:

```
Head=60.0606061+0.0398809524*Flow - 0.00255681818*Flow^2 - 0.00000568181818*Flow^3
        "Curve fit for the pump curve"
```

Change the variable names in the equation above in order to reflect the non-SI units used in the pump curve. Include units for the constants in the pump curve formula:

EXAMPLE 5.2-1: FILLING A WATERING TANK

Head_ftH2O=60.0606061 [ftH2O] + 0.0398809524 [ftH2O/gpm]*Flow_gpm &
   - 0.00255681818[ftH2O/gpm^2]*Flow_gpm^2 &
   - 0.00000568181818[ftH2O/gpm^3]*Flow_gpm^3

"Curve fit for the pump curve"

The pressure rise and flow rate are converted to SI units.

DeltaP_pump=Head_ftH2O*convert(ftH2O,Pa)   "pressure provided by pump"
V_dot=Flow_gpm*convert(gpm,m^3/s)   "flow provided by the pump"

The pressure rise produced by the pump at a particular flow rate, for example $\dot{V} = 0.005$ m^3/s, can be obtained:

V_dot=0.005 [m^3/s]   "flow rate"

which leads to $\Delta p_{pump} = 132500$ Pa (44.3 ft H$_2$O). Comment out the specification of the variable V_dot and generate a pump curve in SI units (Figure 2) using a parametric table in which the variable V_dot is varied from 0 to the maximum flow rate that the pump can produce (140 gal/min).

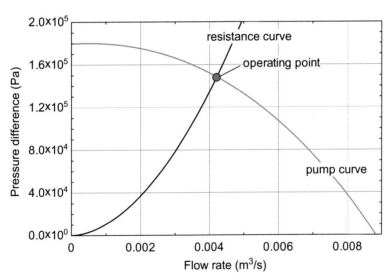

Figure 2: Pump curve and resistance curve in SI units.

The operating point is the intersection of the pump curve and a resistance curve that characterizes the system that the pump is connected to, in this case the PVC pipe. If the pump exit is completely closed (i.e., "dead-headed") then there will be no flow regardless of the pressure. This is an infinitely resistive system; the pressure provided by the pump will be approximately 1.8×10^5 Pa according to Figure 2. (This is sometimes referred to as the dead-head pressure.) If the pump exit is wide open (i.e., the pump is not connected to anything) then there will be no pressure

EXAMPLE 5.2-1: FILLING A WATERING TANK

rise across the pump regardless of flow. This is a system with no resistance; the pump will provide 0.0089 m³/s according to Figure 2. (This is sometimes referred to as the open-circuit flow rate.) The resistance of the PVC pipe will fall somewhere between these limits and can be estimated using the friction factor correlations discussed in Section 5.2.3.

The properties of water (ρ and μ) are computed at ambient conditions:

```
"Properties of water"
rho=density(Water,T=converttemp(C,K,20[C]),p=1 [atm]*convert(atm,Pa))    "density"
mu=viscosity(Water,T=converttemp(C,K,20[C]),p=1 [atm]*convert(atm,Pa))   "viscosity"
```

Given a volumetric flow rate, $\dot{V}$, the mean velocity is computed according to:

$$u_m = \frac{4\,\dot{V}}{\pi\,D^2}$$

```
u_m=V_dot/(pi*D^2/4)              "mean velocity"
```

The mean velocity is used to calculate the Reynolds number:

$$Re = \frac{\rho\,u_m\,D}{\mu}$$

```
Re=rho*u_m*D/mu                  "Reynolds number"
```

The roughness of the surface of a PVC pipe is at most 0.007 mm, according to Table 5-1; this value is used to compute the relative roughness.

```
e=0.007 [mm]*convert(mm,m)       "roughness"
relrough=e/D                     "relative roughness"
```

The function PipeFlow_N is used to determine the average friction factor ($\bar{f}$); note that the Nusselt number results are not required and so the value of the Prandtl number used in the call to the prodedure is arbitrary.

```
call PipeFlow_N(Re,1 [-],L/D,relrough: Nusselt_T_bar,Nusselt_H_bar,f_bar)
                                 "access correlations"
```

The apparent friction factor is used to compute the pressure drop. (Note that the calculated pressure loss neglects any entrance or exit effects as well as inertial losses at bends, etc.)

$$\Delta p = \bar{f}\,\frac{L\,\rho\,u_m^2}{2\,D}$$

```
DeltaP_pipe=f_bar*L*rho*u_m^2/(2*D)       "pressure drop across pipe"
```

A resistance curve is obtained by commenting out the specified value of the variable V_dot and using a parametric table to vary V_dot from 0 to the open-circuit flow rate

EXAMPLE 5.2-1: FILLING A WATERING TANK

of the pump. The resistance curve is overlaid on the pump curve (see Figure 2) and their intersection determines the operating point (approximately 0.0043 m³/s at a pressure drop of 1.5×10^5 Pa). The operating point can be found exactly by requiring that the pressure drop across the pipe be equal to the pressure rise provided by the pump. Update the guess values, comment out the assumed value of the volumetric flow rate and specify that the pressure drop and pressure rise are equal:

```
{V_dot=0.005 [m^3/s]              "flow rate"}
DeltaP_pipe=DeltaP_pump          "find operating point"
```

The time required to fill the tank (t_{fill}) is:

$$t_{fill} = \frac{V_{tank}}{\dot{V}}$$

```
t_fill=V_tank/V_dot                    "time required to fill tank"
t_fill_min=t_fill*convert(s,min)       "in minute"
```

The result is 539 s or 9.0 min. If the length is commented out and varied in a parametric table then it is possible to generate Figure 3, which shows the time required to fill the tank as a function of length for various values of the pipe diameter.

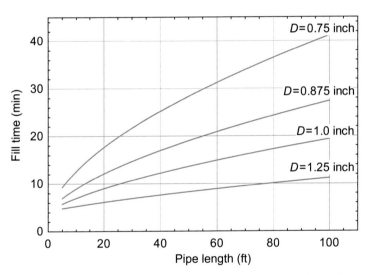

Figure 3: Fill time as a function of pipe length for various values of pipe diameter.

5.2.4 The Nusselt Number

The local Nusselt number in the fully developed region is shown in Figure 5-14 as a function of Reynolds number for various duct shapes and boundary conditions (in the laminar region) and relative roughness and Prandtl number (in the turbulent region). Notice that the fully developed Nusselt number in the laminar region is independent of Reynolds number, Prandtl number, and surface roughness. The laminar thermal boundary layer, which dictates the thermal resistance to convection and therefore the heat transfer coefficient and Nusselt number, is not affected by the Reynolds number and is much larger than the surface roughness. The fully developed Nusselt number in the

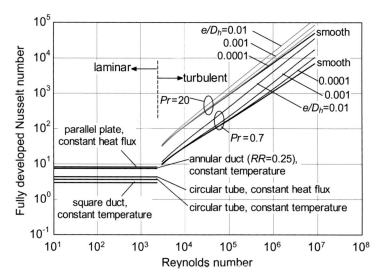

Figure 5-14: Local, fully developed Nusselt Number as a function of the Reynolds number for various duct shapes and boundary conditions (in the laminar region) and relative roughness and Prandtl number (in the turbulent region).

turbulent region is affected by the Reynolds number and Prandtl number because the viscous sublayer, which dictates the thermal resistance to convection, becomes thinner at a higher Reynolds number. The viscous sublayer is on the same scale as the surface roughness and therefore the turbulent Nusselt number is affected by the presence of roughness.

Laminar Flow
Figure 5-14 shows that the Nusselt number (and thus the heat transfer coefficient) for a fully developed laminar flow is affected by the duct shape and boundary conditions but not the surface roughness, Prandtl number, or even the Reynolds number. This behavior is consistent with the discussion of the heat transfer coefficient in a laminar flow in Section 5.1.3. The heat transfer coefficient is approximately equal to the ratio of the thermal conductivity of the fluid to the thermal boundary layer thickness: $h \approx k/\delta_t$. In the fully developed, laminar region of a duct, the thermal boundary layer is, approximately, half the duct-width: $\delta_t \approx D_h/2$. Substituting these approximate relationships into the definition of the Nusselt number leads to:

$$Nu_{D_h} = \frac{h\,D_h}{k} \approx \frac{2\,k}{D_h}\frac{D_h}{k} \approx 2 \tag{5-75}$$

In fact, the fully developed Nusselt number can range anywhere from 3.66 (for a circular tube with a constant temperature) to 8.24 (for flow between infinite parallel plates with constant heat flux). However, Eq. (5-75) explains why the Nusselt number is independent of the Reynolds number and Prandtl number.

Circular Tubes. The local Nusselt number for a laminar, hydrodynamically and thermally fully developed flow in a circular tube depends on the thermal boundary condition. For a uniform heat flux (indicated by the subscript H), the Nusselt number is:

$$\boxed{Nu_{D_h,H,fd} = 4.36} \tag{5-76}$$

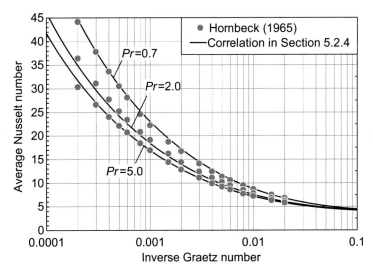

Figure 5-15: Average Nusselt number for developing flow with constant wall temperature as a function of L^* for various values of the Prandtl number. The graphical results presented by Hornbeck (1965) are shown, as well as the correlation provided by Eq. (5-80).

This result is derived in Section 5.4.3. For a uniform wall temperature (indicated by the subscript T), the fully developed Nusselt number is (Kays and Crawford, (1993)):

$$\boxed{Nu_{D_h,T,fd} = 3.66}$$
(5-77)

The average Nusselt number depends on the entrance conditions. Figure 5-15 illustrates the average Nusselt number for simultaneously developing flow with a constant wall temperature for various values of the Prandtl number as a function of L^*, where L^* is the dimensionless length appropriate for a thermally developing flow:

$$\boxed{L^* = \frac{L^+}{Pr} = \frac{L}{D_h\, Re_{D_h}\, Pr}}$$
(5-78)

The dimensionless length L^* is sometimes referred to as the inverse of the Graetz number, Gz:

$$\boxed{Gz = \frac{1}{L^*} = \frac{D_h\, Re_{D_h}\, Pr}{L}}$$
(5-79)

The symbols in Figure 5-15 correspond to graphical results presented by Hornbeck (1965). The correlation provided by Eq. (5-80) is also shown in Figure 5-15 and provides an adequate fit to the results of Hornbeck. Notice that the correlation limits to the fully developed Nusselt number as Gz approaches 0 (i.e., L^* becomes large):

$$\boxed{\overline{Nu}_{D_h,T} = 3.66 + \frac{\left[0.049 + \dfrac{0.020}{Pr}\right] Gz^{1.12}}{[1 + 0.065\, Gz^{0.7}]}}$$
(5-80)

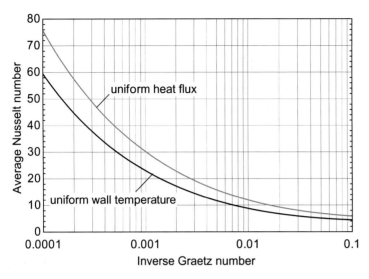

Figure 5-16: Average Nusselt number as a function of L^* for a simultaneously developing flow subjected to a constant wall temperature and constant heat flux for $Pr = 0.70$.

The correlation:

$$\overline{Nu}_{D_h,H} = 4.36 + \frac{\left[0.1156 + \dfrac{0.08569}{Pr^{0.4}}\right] Gz}{\left[1 + 0.1158 \, Gz^{0.6}\right]} \tag{5-81}$$

provides the average Nusselt number for simultaneously developing flow in a duct that is exposed to a uniform heat flux. Equation (5-81) is based on integrating and fitting data provided by Hornbeck (1965) and presented in Shah and London (1978).

The average heat transfer coefficient in the case where the heat flux is specified is not typically required to solve an internal flow problem. However, the average heat transfer coefficient for the constant temperature and constant heat flux conditions provide natural bounding cases and therefore it is generally useful to compute both values. Figure 5-16 illustrates the average Nusselt number as a function of L^* calculated using Eqs. (5-80) and (5-81) for a Prandtl number of 0.70. The actual average Nusselt number is likely to fall between these bounds.

Figure 5-17 shows the average Nusselt number for a thermally developing/hydrodynamically developed flow as a function of L^*. Both the constant temperature and constant heat flux boundary condition solutions are shown, based on the results presented in Shah (1975). Notice that the Prandtl number does not affect the solution because the momentum boundary does not change with position in a hydrodynamically fully developed flow. In the limit that the Prandtl number is very large, the simultaneously developing solution will approach the thermally developing/hydrodynamically developed result because the momentum boundary layer grows much faster than the thermal boundary layer for a high Prandtl number fluid. The average Nusselt number predicted using Eqs. (5-80) and (5-81) for a simultaneously developing flow in the limit that $Pr \rightarrow \infty$ are also shown in Figure 5-17.

The EES library of internal forced convection procedures provides the Nusselt number for simultaneously developing flow under conditions of both constant heat flux and temperature; these provide upper and lower bounds on the result, respectively. The

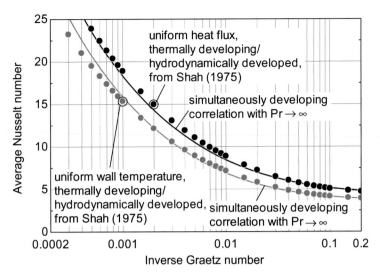

Figure 5-17: Average Nusselt number as a function of L^* for a thermally developing/hydrodynamically developed flow from Shah (1975). Also shown are the results predicted by the correlations for the simultaneously developing flow, Eqs. (5-80) and (5-81), in the limit that $Pr \to \infty$.

average Nusselt number for a circular duct may be accessed using the PipeFlow_N procedure and the local Nusselt number may be obtained using the PipeFlow_N_local procedure. The local Nusselt number is obtained by numerically differentiating Eqs. (5-80) and (5-81) according to Eq. (5-44). It is possible to obtain results that are consistent with a thermally developing/hydrodynamically developed flow by calling EES' internal convection procedures with a very large Prandtl number.

Rectangular Ducts. Shah and London (1978) provide the local Nusselt number for a laminar, hydrodynamically and thermally fully developed flow in a rectangular duct that is exposed to a uniform heat flux and uniform wall temperature. The constant temperature result is correlated by:

$$Nu_{D_h,T,fd} = 7.541(1 - 2.610\,AR + 4.970\,AR^2 - 5.119\,AR^3 + 2.702\,AR^4 - 0.548\,AR^5)$$

$$(5\text{-}82)$$

and the constant heat flux result is correlated by:

$$Nu_{D_h,H,fd} = 8.235(1 - 2.042\,AR + 3.085\,AR^2 - 2.477\,AR^3 + 1.058\,AR^4 - 0.186\,AR^5)$$

$$(5\text{-}83)$$

where AR is the aspect ratio of the duct (the ratio of the minimum to the maximum dimensions). Figure 5-18 illustrates the average Nusselt number for a simultaneously developing flow in a rectangular duct exposed to a constant wall temperature and constant heat flux with $Pr = 0.72$ for various values of the aspect ratio as a function of dimensionless position L^* (Wibulswas, (1966)). Also shown in Figure 5-18 are the predictions obtained from EES' procedure DuctFlow_N, which interpolates a table of data provided by Kakaç et al. (1987) and corrects for the Prandtl number effects by applying the correction associated with a square duct, also from Kakaç et al. (1987).

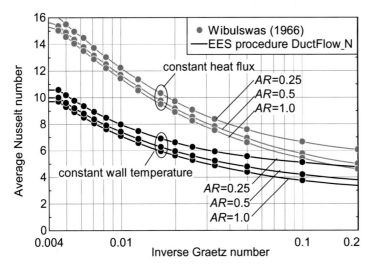

Figure 5-18: Average Nusselt number in a rectangular duct with simultaneously developing flow as a function of L^* for various values of aspect ratio and $Pr = 0.72$. Results are shown for uniform heat flux and uniform wall temperature from Wibulswas (1966) and the EES procedure DuctFlow_N.

Figure 5-19 illustrates the average Nusselt number in a rectangular duct exposed to a constant heat flux that is thermally developing/hydrodynamically developed for various values of the aspect ratio (Wibulswas (1966)). Also shown in Figure 5-19 are the results from the DuctFlow_N procedure in EES called with a large Prandtl number; note that the results do not match exactly, but are sufficiently accurate for most engineering calculations.

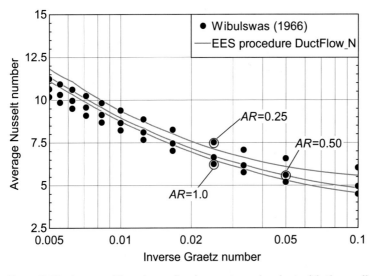

Figure 5-19: Average Nusselt number in a rectangular duct with thermally developing/hydrodynamically developed flow as a function of L^* for various values of aspect ratio; results are shown for uniform heat flux from Wibulswas (1966) and the EES procedure DuctFlow_N called with $Pr \to \infty$.

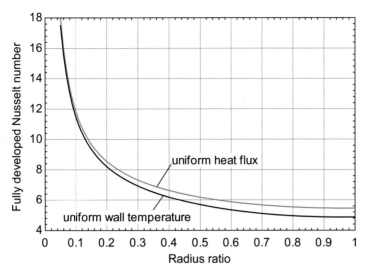

Figure 5-20: Fully developed Nusselt number in an annular duct with an adiabatic external surface and a constant temperature and constant heat flux applied to the internal surface as a function of the radius ratio.

More complex boundary conditions can be considered in which one or more of the duct walls are adiabatic. Solutions for alternative boundary conditions are presented in various references including Shah and London (1978) and Rohsenow et al. (1998).

Annular Ducts. The fully developed Nusselt number for an annular duct with an adiabatic external surface and an internal surface that is subjected to a constant temperature and a constant heat flux is provided by Rohsenow et al. (1998) and shown in Figure 5-20 as a function of the radius ratio (RR, the ratio of the inner to the outer radii of the duct).

The AnnularFlow_N procedure provides the average Nusselt number for simultaneously developing flow in an annular duct with a uniform heat flux and uniform wall temperature boundary conditions. The procedure interpolates the solution for $Pr = 0.72$ and applies a Prandtl based number correction using the solution for a square duct.

Turbulent Flow

Figure 5-14 shows that the Nusselt number for a turbulent flow is not affected by the duct shape or boundary conditions, but it is sensitive to the scale of the surface roughness (e). The Nusselt number for fully developed turbulent flow is provided by Gnielinski (1976):

$$Nu_{D_h,fd} = \frac{\left(\dfrac{f_{fd}}{8}\right)(Re_{D_h} - 1000)Pr}{1 + 12.7(Pr^{2/3} - 1)\sqrt{\dfrac{f_{fd}}{8}}} \quad \text{for } 0.5 < Pr < 2000 \text{ and } 2300 < Re_{D_h} < 5 \times 10^6$$

(5-84)

where f_{fd} is the fully developed friction factor, obtained as discussed in Section 5.2.3. The effect of roughness on the Nusselt number in Eq. (5-84) is included through its effect on the friction factor. The average Nusselt number can be approximately computed according to (Kakaç et al. (1987)):

$$\overline{Nu_{D_h}} \approx Nu_{D_h,fd}\left[1 + C\left(\frac{x}{D_h}\right)^{-m}\right] \tag{5-85}$$

where reasonable values of the constants C and m are 1.0 and 0.7.

EXAMPLE 5.2-2: DESIGN OF AN AIR HEATER

EXAMPLE 5.2-2: DESIGN OF AN AIR HEATER

You are designing a device to heat a flow of air from an inlet temperature of $T_{in} = 20°C$ to a mean outlet temperature of $T_{out} = 80°C$. The inlet pressure is $p_{in} = 100$ psia and the mass flow rate is $\dot{m} = 0.01$ kg/s. Your preliminary design concept is a copper tube wrapped with a heater that provides a uniform heat flux, $\dot{q}_s''$. The thickness of the tube is $th = 0.035$ inch and the tube/heater assembly is insulated. You must specify the tube outer diameter, D_o, and length, L. There are a few design constraints: the pressure drop through the tube must be no larger than $\Delta p_{max} = 10$ psi, the heat flux provided by the heater must be no larger than $\dot{q}_{s,max}'' = 9000\,\text{W/m}^2$, and the surface temperature of the tube (which is approximately equal to the heater temperature) must be no larger than $T_{s,max} = 100°C$.

a) **Develop a model that can analyze a particular design. Use the model to determine whether a heater that is $L = 4$ ft long with outer diameter $D_o = 0.25$ inch meets the design criteria.**

The inputs and the initial design are entered in EES:

```
"EXAMPLE 5.2-2: Design of an Air Heater"

$UnitSystem SI MASS RAD PA K J
$Tabstops 0.2 0.4 0.6 3.5 in

"Inputs"
T_in=converttemp(C,K,20[C])                    "inlet air temperature"
T_out=converttemp(C,K,80[C])                   "outlet mean air temperature"
m_dot=0.01 [kg/s]                              "mass flow rate"
p_in=100 [psi]*convert(psi,Pa)                 "inlet pressure"
th=0.035 [inch]*convert(inch,m)                "thickness"
DELTAp_max=10 [psi]*convert(psi,Pa)            "maximum pressure drop"
q"_dot_s_max=9000 [W/m^2]                      "maximum heat flux"
T_s_max_max=converttemp(C,K,100[C])
                            "maximum allowable maximum surface temperature"

"Initial design"
L_ft=4 [ft]                                    "tube length, in ft"
L=L_ft*convert(ft,m)                           "tube length"
D_o_inch=0.25 [inch]                           "tube diameter, in inch"
D_o=D_o_inch*convert(inch,m)                   "tube diameter"
```

The properties of air (ρ, c, μ, k, and Pr) are obtained at the average air temperature and inlet air pressure using EES' internal property functions:

EXAMPLE 5.2-2: DESIGN OF AN AIR HEATER

```
T_bar=(T_in+T_out)/2                    "average air temperature"
rho=density(Air,T=T_bar,p=p_in)         "density of air"
mu=viscosity(Air,T=T_bar)               "viscosity of air"
c=cP(Air,T=T_bar)                       "specific heat capacity of air"
k=conductivity(Air,T=T_bar)             "conductivity of air"
Pr=mu*c/k                               "Prandtl number of air"
```

An energy balance on the tube provides the rate of heat transfer that must be provided by the heater:

$$\dot{q} = \dot{m} c \, (T_{out} - T_{in})$$

The heat flux is the ratio of the heater power to the external surface area of the tube:

$$\dot{q}_s'' = \frac{\dot{q}}{\pi \, D_o \, L}$$

```
q_dot=m_dot*c*(T_out-T_in)              "heater power required"
q"_dot_s=q_dot/(pi*D_o*L)               "heat flux"
```

which leads to $\dot{q}_s'' = 24{,}820$ W/m^2; therefore, the initial design does not satisfy the design constraint related to the heat flux.

The bulk velocity of the air flow is:

$$u_m = \frac{4 \, \dot{m}}{\rho \, \pi \, D_i^2}$$

where D_i is the inner diameter of the tube:

$$D_i = D_o - 2 \, th$$

```
D_i=D_o-2*th                            "inner diameter of the tube"
u_m=m_dot/(rho*pi*D_i^2/4)              "mean velocity"
```

The Reynolds number is:

$$Re_{D_h} = \frac{\rho \, u_m \, D_i}{\mu}$$

```
Re=rho*u_m*D_i/mu                       "Reynolds number"
```

The roughness of a drawn copper tube can be obtained by consulting with the manufacturer. According to Table 5-1, $e \approx 0.0015$ mm. The correlations for the average friction factor and Nusselt numbers ($\overline{f}$, $\overline{Nu}_{D_h,T}$, and $\overline{Nu}_{D_h,H}$) for internal flow through a circular tube are accessed using the procedure **PipeFlow_N**.

```
e=0.0015 [mm]*convert(mm,m)             "roughness of tube"
call PipeFlow_N(Re,Pr,L/D_i,e/D_i: Nusselt_bar_T,Nusselt_bar_H,f_bar)
                    "correlations for average friction factor and Nusselt numbers"
```

EXAMPLE 5.2-2: DESIGN OF AN AIR HEATER

The average friction factor is used to compute the pressure drop across the tube:

$$\Delta p = \frac{\rho u_m^2}{2}\left(\bar{f}\frac{L}{D_i}\right)$$

```
DELTAp=(rho*u_m^2/2)*(f_bar*L/D_i)          "pressure drop"
DELTAp_psi=DELTAp*convert(Pa,psi)           "in psi"
```

which leads to $\Delta p = 127{,}000$ Pa (18.4 psi); therefore, the initial design also does not satisfy the pressure drop requirement.

The tube surface temperature will be highest at the outlet, where the air temperature is highest and the heat transfer coefficient lowest. The definition of the local heat transfer coefficient is:

$$h = \frac{\dot{q}_s''}{(T_s - T_m)}$$

where T_m is the local bulk temperature of the air. The maximum surface temperature of the tube is therefore:

$$T_{s,max} = T_{out} + \frac{\dot{q}_s''}{h_{x=L}} \tag{1}$$

The local heat transfer coefficient at the tube exit can be obtained using the local Nusselt number at the tube exit. The correlations for the local friction factor and Nusselt number (f, $Nu_{D_h,T}$, and $Nu_{D_h,H}$) for internal flow through a circular tube are accessed using the procedure PipeFlow_N_local:

```
call PipeFlow_N_local(Re,Pr,L/D_i,e/D_i: Nusselt_T,Nusselt_H,f)
       "correlations for local friction factor and Nusselt numbers"
```

The local heat transfer coefficient at the tube outlet is computed according to:

$$h_{x=L} = \frac{Nu_{D_h,H}\,k}{D_i}$$

The Nusselt number based on a constant heat flux boundary condition is more appropriate than a constant temperature boundary condition for this problem. However, because the flow is turbulent, $Nu_{D_h,T}$ and $Nu_{D_h,H}$ are the same.

```
h=Nusselt_H*k/D_i                    "local heat transfer coefficient at the outlet"
T_s_max=T_out+q"_dot_s/h             "maximum tube temperature"
T_s_max_C=converttemp(K,C,T_s_max)   "in C"
```

which leads to $T_{s,max} = 368.4$ K (95.3°C); therefore, the initial design does satisfy the surface temperature requirement.

b) Use your model to obtain a heater design that satisfies all of the design criteria.

Because your EES solution is robust with respect to the inputs, it is possible to use it for design studies. This is the advantage of solving the problem using a computer

EXAMPLE 5.2-2: DESIGN OF AN AIR HEATER

program rather than by hand. Two of the three design constraints are not met by the initial design and there are two geometric parameters that can be adjusted (tube diameter and length). Solve the problem with the initial design geometry and update guess values (select Update Guesses from the Calculate menu). Then comment out the specified value of the heater length and substitute in an equation that enforces the heat flux design constraint:

```
{L_ft=4 [ft]}                    "tube length, in ft"
q"_dot_s=q"_dot_s_max            "heat flux criteria"
```

After solving the problem, you should see that EES has adjusted the length of the tube in order to satisfy the heat flux requirement. The pressure drop requirement is still not satisfied. Therefore, comment out the specified value of the diameter and substitute in an equation that enforces the pressure drop design constraint:

```
{D_o_inch=0.25 [inch]}           "tube diameter, in inch"
DELTAp=DELTAp_max                "pressure drop criteria"
```

Again solve the problem and you should see that EES has simultaneously adjusted the length and diameter of the tube in order to satisfy both design constraints. The maximum tube surface temperature is 90°C which is still sufficiently low. Therefore, the final heater design is $L = 2.72$ m (8.9 ft) and $D_o = 0.0079$ m (0.31 inch).

5.3 The Energy Balance

5.3.1 Introduction

Unlike the free-stream temperature in an external flow, the mean temperature of an internal flow will vary with position as energy is added to or removed from the flow (see Figure 5-5 and Figure 5-6). Therefore, in order to solve an internal flow problem, it is necessary to determine the variation in the mean temperature with position. This is accomplished using an energy balance on the fluid. For a single internal flow interacting with a prescribed boundary condition (e.g., a constant temperature duct surface), the energy balance will result in an ordinary differential equation that can be solved analytically or numerically. In Chapter 8, we will consider heat exchangers where two (or more) flow streams interact. Multiple energy balances are required to solve a heat exchanger problem, leading to a set of coupled ordinary or partial differential equations.

5.3.2 The Energy Balance

Figure 5-21 illustrates the general steady-state energy balance on a control volume that is differential in the flow direction, x.

The energy balance shown in Figure 5-21 balances heat flux from the duct wall to the fluid ($\dot{q}''_s$), volumetric generation due to viscous dissipation ($\dot{g}'''_v$), externally imposed volumetric generation in the fluid ($\dot{g}'''$, for example related to an electric current passing through the fluid or chemical reactions), enthalpy carried by the flow, and axial

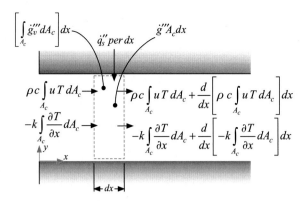

Figure 5-21: Steady-state energy balance on a differential control volume.

conduction in the fluid. The energy balance suggested by Figure 5-21 is:

$$\rho c \int_{A_c} u\, T\, dA_c - k \int_{A_c} \frac{\partial T}{\partial x}\, dA_c + \left[\int_{A_c} \dot{g}'''_v A_c \right] dx + \dot{g}''' A_c\, dx + \dot{q}''_s\, per\, dx$$

$$= \rho c \int_{A_c} u\, T\, dA_c + \rho c\, \frac{d}{dx} \left[\int_{A_c} u\, T\, dA_c \right] dx - k \int_{A_c} \frac{\partial T}{\partial x}\, dA_c - k \frac{d}{dx} \left[\int_{A_c} \frac{\partial T}{\partial x}\, dA_c \right] dx$$

(5-86)

Substituting the definition of the mean temperature, Eq. (5-36), into Eq. (5-86) and simplifying leads to:

$$\underbrace{\left[\int_{A_c} \dot{g}'''_v A_c \right]}_{\substack{\text{viscous}\\ \text{dissipation}}} + \underbrace{\dot{g}''' A_c}_{\substack{\text{imposed}\\ \text{volumetric}\\ \text{generation}}} + \underbrace{\dot{q}''_s\, per}_{\substack{\text{heat transfer}\\ \text{from duct}\\ \text{surface}}} = \underbrace{\dot{m} c\, \frac{dT_m}{dx}}_{\substack{\text{enthalpy carried}\\ \text{by flow}}} + \underbrace{\frac{d}{dx} \left[\int_{A_c} -k \frac{\partial T}{\partial x}\, dA_c \right]}_{\text{axial conduction}}$$

(5-87)

In most situations, $\dot{g}'''$ will be zero and the volumetric generation related to viscous dissipation and axial conduction can be neglected. (The conditions under which these assumptions are justified are discussed in Section 5.4.3.) Therefore, Eq. (5-87) can usually be simplified to a balance between energy input by convection at the wall with the change in the mean temperature of the fluid:

$$\dot{q}''_s\, per = \dot{m} c\, \frac{dT_m}{dx}$$

(5-88)

If the heat flux at the wall is prescribed, then Eq. (5-88) can be integrated directly to obtain T_m as a function of x. If the wall temperature, T_s, is prescribed, then the definition of the heat transfer coefficient, Eq. (5-38), must be substituted into Eq. (5-88):

$$per\, h\, (T_s - T_m) = \dot{m} c\, \frac{dT_m}{dx}$$

(5-89)

The energy balances provided by either Eq. (5-88) or Eq. (5-89) are 1-D ordinary differential equations for T_m whereas the actual flow situation in the duct is clearly 2-D or 3-D. The use of the 1-D energy balance is facilitated by the definition of the mean temperature and the associated definition of the heat transfer coefficient in terms of the mean temperature. This is an engineering approach that simplifies the solution to internal flow problems. The mean temperature represents the average thermal energy level of the fluid while the heat transfer coefficient encompasses the details of the local temperature distribution and flow conditions. The heat transfer coefficient for many practical problems is correlated in the form of the Nusselt number as a function of the Reynolds number and the Prandtl number (as discussed in Section 5.2) and therefore many engineering problems can be solved without ever considering the 2-D or 3-D details of the flow. Sections 5.4 and 5.5 present analytical and numerical solution techniques for internal flow problems. These techniques do consider the details of the flow and can be used to obtain exact solutions for the temperature distribution and the heat transfer coefficient. However, in most problems we rely on correlations that approximately fit the flow situation under consideration.

Two common wall conditions that are encountered in practice are the constant wall temperature and constant surface temperature conditions. Analytical solutions to the energy balance can be obtained for these limiting cases. However, the energy balance equations, Eqs. (5-88) and (5-89), are not limited to these simple wall conditions.

5.3.3 Prescribed Heat Flux

In a duct with a specified heat flux (e.g., resulting from electric dissipation of a heater wire) the energy balance provided by Eq. (5-88) can be separated and integrated to determine the variation in the mean temperature with position:

$$\int_{T_{in}}^{T_m} dT_m = \frac{per}{\dot{m}\,c} \int_0^x \dot{q}_s'' \, dx \tag{5-90}$$

or

$$T_m = T_{in} + \frac{per}{\dot{m}\,c} \int_{x=0}^x \dot{q}_s'' \, dx \tag{5-91}$$

where T_{in} is the mean temperature of the flow at the inlet (T_∞ in Figure 5-5(b) and Figure 5-6). Equation (5-91) shows that the mean temperature of the flow increases at a rate that is proportional to the applied heat flux and inversely proportional to the total capacitance rate of the fluid. The total capacitance rate, $\dot{C}$, is the product of the mass flow rate and specific heat capacity ($\dot{m}\,c$) and represents the rate at which energy must be added to the fluid in order to increase its temperature. Equation (5-91) can be solved either analytically (for simple variations in the heat flux) or numerically using any of the numerical integration techniques discussed in Section 3.2.

Equation (5-91) indicates the variation in the mean temperature of the fluid is insensitive to the details of the flow and the local heat transfer coefficient for situations where the heat flux is specified. The surface temperature (the temperature of the wall) depends on the local heat transfer coefficient according to Eq. (5-38):

$$T_s = T_m + \frac{\dot{q}_s''}{h} \tag{5-92}$$

Temperature

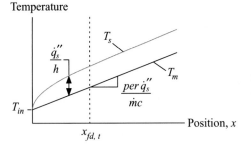

Figure 5-22: The mean fluid temperature and surface temperature as a function of position for a duct exposed to a constant heat flux.

Constant Heat Flux

In the limit that a constant heat flux is applied at the surface of the duct, Eq. (5-91) provides:

$$T_m = T_{in} + \frac{per\,\dot{q}_s''}{\dot{m}\,c}x \tag{5-93}$$

Figure 5-22 illustrates the mean fluid temperature and wall temperature as a function of position for a duct with a constant heat flux. Notice that the wall-to-mean temperature difference becomes constant in the thermally fully developed region because the heat transfer coefficient is constant in this region. Further, the surface temperature approaches the mean temperature at the inlet because the heat transfer coefficient is very large in the developing region.

5.3.4 Prescribed Wall Temperature

In a duct with a prescribed wall temperature, the energy balance provided by Eq. (5-89) must be used:

$$\frac{dT_m}{dx} + \frac{per\,h}{\dot{m}\,c}T_m = \frac{per\,h}{\dot{m}\,c}T_s \tag{5-94}$$

If the heat transfer coefficient, perimeter, mass flow rate, and specific heat capacity are constant, then Eq. (5-94) is a linear, first-order ordinary differential equation that may be solved analytically (for simple variations in the duct surface temperature) by dividing the solution into homogeneous and particular components. It is possible to solve Eq. (5-94) numerically for more complicated situations using any of the integration techniques previously discussed in Section 3.2.

Constant Wall Temperature

In the limit that the duct has a constant wall temperature, the energy balance, Eq. (5-89), can be separated and integrated:

$$\int_{T_{in}}^{T_m} \frac{dT_m}{(T_s - T_m)} = \frac{per}{\dot{m}\,c}\int_0^x h\,dx \tag{5-95}$$

Substituting the definition of the average heat transfer coefficient, Eq. (5-42), into Eq. (5-95) leads to:

$$\int_{T_{in}}^{T_m} \frac{dT_m}{(T_s - T_m)} = \frac{per\,x\,\bar{h}}{\dot{m}\,c} \tag{5-96}$$

Equation (5-96) is integrated:

$$\ln\left(\frac{T_s - T_m}{T_s - T_{in}}\right) = -\frac{per\,x\,\overline{h}}{\dot{m}\,c} \tag{5-97}$$

or

$$T_m = T_s - (T_s - T_{in})\exp\left(-\frac{per\,x\,\overline{h}}{\dot{m}\,c}\right) \tag{5-98}$$

Equation (5-98) shows that the mean fluid temperature will approach the wall temperature approximately exponentially (exactly exponentially if $\overline{h}$ is constant). This behavior is similar to the transient problem associated with an object that can be treated a lumped capacitance and is subjected to a step change in the ambient temperature. The solution to this problem is studied in Section 3.1 and presented in Table 3-1:

$$T = T_\infty + (T_{ini} - T_\infty)\exp\left(-\frac{\overline{h}\,A_s}{M\,c}t\right) \tag{5-99}$$

Equation (5-99) is the temperature experienced by a differential mass of fluid as a function of time as it moves through the tube (i.e., the temperature expressed in a Lagrangian frame of reference where the observer moves with the fluid) whereas Eq. (5-98) is the temperature expressed as a function of position (i.e., the temperature expressed in an Eulerian frame of reference where the observer is stationary). The equivalence of these solutions can be demonstrated by replacing time, t in Eq. (5-99), with the time that the differential mass of fluid has been in the tube (x/u_m), replacing mass, M in Eq. (5-99), with the differential mass of fluid in the tube ($A_c\,dx\,\rho$), and surface area, A_s in Eq. (5-99), with the surface area of the differential segment of tube ($per\,dx$). The ambient temperature (T_∞) and initial temperature (T_{ini}) in Eq. (5-99) are replaced by the tube surface temperature (T_s) and inlet temperature (T_{in}), respectively.

$$T = T_s + (T_{in} - T_s)\exp\left(-\frac{x\,\overline{h}\,per\,dx}{u_m\,A_c\,dx\,\rho\,c}\right) \tag{5-100}$$

or

$$T = T_s + (T_{in} - T_s)\exp\left(-\frac{x\,\overline{h}\,per}{\dot{m}\,c}\right) \tag{5-101}$$

which is equivalent to Eq. (5-98).

The mean temperature of the fluid leaving the duct, T_{out}, is obtained by evaluating Eq. (5-98) at $x = L$:

$$T_{out} = T_s - (T_s - T_{in})\exp\left(-\frac{per\,L\,\overline{h}}{\dot{m}\,c}\right) \tag{5-102}$$

5.3.5 Prescribed External Temperature

It is more often the case that a temperature external to the duct will be prescribed, rather than the temperature of the duct wall itself. For example, Figure 5-23 illustrates fluid running through a tube with length L and inner and outer diameters D_{in} and D_{out}, respectively. The tube is placed in an air flow at T_∞. The fluid enters the tube at T_{in}. Although the tube is exposed to a constant external air temperature, the temperature of the duct wall itself is not constant.

In this situation, it is necessary to write a differential energy balance on the fluid in terms of the total thermal resistance that separates the prescribed external temperature (T_∞) from the local mean temperature of the fluid (T_m) evaluated on a unit length

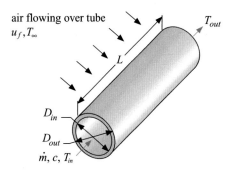

Figure 5-23: Fluid in a tube exposed to a flow of air.

basis (R'_{tot}). For the situation shown in Figure 5-23, the total thermal resistance will include a resistance associated with convection from the internal surface of the tube wall to the fluid as well as the resistance to conduction through the tube wall and convection from the outside surface of the tube:

$$R'_{tot} = \underbrace{\frac{1}{h_{in}\,\pi\,D_{in}}}_{\text{internal convection}} + \underbrace{\frac{\ln\left(\dfrac{D_{out}}{D_{in}}\right)}{2\,\pi\,k_{tube}}}_{\text{conduction through tube}} + \underbrace{\frac{1}{h_{out}\,\pi\,D_{out}}}_{\text{external convection}} \tag{5-103}$$

Additional resistances could be added in order to represent fins, contact resistance, radiation, or fouling (the build-up of scale or deposit on the inside or outside surfaces of a tube subjected to a flowing fluid, discussed in Section 8.1.5).

The differential energy balance is shown in Figure 5-24 for a duct that is exposed to a constant external temperature with negligible viscous dissipation and axial conduction. The energy balance suggested by Figure 5-24 is:

$$\dot{m}\,c\,T_m + \frac{(T_\infty - T_m)}{\dfrac{R'_{tot}}{dx}} = \dot{m}\,c\,T_m + \dot{m}\,c\,\frac{dT_m}{dx}dx \tag{5-104}$$

where R'_{tot}/dx is the thermal resistance that is contained within the control volume. Equation (5-104) can be simplified:

$$\frac{(T_\infty - T_m)}{R'_{tot}} = \dot{m}\,c\,\frac{dT_m}{dx} \tag{5-105}$$

The difference between Eq. (5-104) and Eq. (5-89), which was derived for a constant tube surface temperature, is that the heat transfer rate per unit length of tube is written in terms of total resistance on a unit length basis rather than the internal heat transfer coefficient. Equation (5-105) can be solved numerically using any of the techniques discussed in Section 3.2 for a situation in which the external temperature varies in a

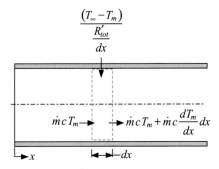

Figure 5-24: Differential control volume for fluid subjected to a constant external temperature rather than a constant wall temperature.

prescribed way or the value of R'_{tot} is not constant. However, if T_∞ and R'_{tot} are constant (or can be reasonably approximated as being constant), then Eq. (5-105) can be separated and integrated to provide:

$$T_{out} = T_\infty - (T_\infty - T_{in}) \exp\left(-\frac{L}{R'_{tot}\, \dot{m}\, c}\right) \tag{5-106}$$

The parameter R'_{tot}/L in Eq. (5-106) represents the total thermal resistance between the fluid temperature and the external temperature over the entire length of the duct, R_{tot}:

$$R_{tot} = \frac{R'_{tot}}{L} = \frac{1}{\overline{h}_{in}\, \pi\, D_{in}\, L} + \frac{\ln\left(\dfrac{D_{out}}{D_{in}}\right)}{2\, L\, \pi\, k_{tube}} + \frac{1}{\overline{h}_{out}\, \pi\, D_{out}\, L} \tag{5-107}$$

The inverse of the total resistance is referred to as the total conductance (UA):

$$UA = \frac{1}{R_{tot}} \tag{5-108}$$

Substituting the definition of conductance, Eq. (5-108), into Eq. (5-106) leads to:

$$T_{out} = T_\infty - (T_\infty - T_{in}) \exp\left(-\frac{UA}{\dot{m}\, c}\right) \tag{5-109}$$

which is the same as the solution for the constant duct temperature, Eq. (5-102), with the terms $per\, L\overline{h}$ replaced by UA and T_s replaced by T_∞. It is often the case that Eq. (5-109) is used to obtain an engineering solution to a problem even when it is clear that R'_{tot} varies substantially (for example, due to the temperature dependent conductivity of the fluid).

In Chapter 8, heat exchangers are introduced. In a heat exchanger, two fluids interact thermally and therefore two energy balances of the type shown in Figure 5-24 must be written. The analysis becomes more complex. However, the idea is the same and the concept of a total conductance is very useful for a heat exchanger problem.

EXAMPLE 5.3-1: ENERGY RECOVERY WITH AN ANNULAR JACKET

One concept for recovering energy from the exhaust of a large truck is to run water through an annular jacket that surrounds the exhaust pipe, as shown in Figure 1.

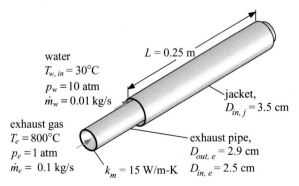

water
$T_{w, in} = 30°C$
$p_w = 10$ atm
$\dot{m}_w = 0.01$ kg/s

$L = 0.25$ m

jacket,
$D_{in, j} = 3.5$ cm

exhaust gas
$T_e = 800°C$
$p_e = 1$ atm
$\dot{m}_e = 0.1$ kg/s

exhaust pipe,
$D_{out, e} = 2.9$ cm
$D_{in, e} = 2.5$ cm

$k_m = 15$ W/m-K

Figure 1: An annular duct placed around the exhaust pipe to recovery energy.

The exhaust pipe is a drawn tube with an inner diameter $D_{in,e} = 2.5$ cm and an outer diameter $D_{out,e} = 2.9$ cm. The jacket surrounding the pipe has an inner diameter

EXAMPLE 5.3-1: ENERGY RECOVERY WITH AN ANNULAR JACKET

EXAMPLE 5.3-1: ENERGY RECOVERY WITH AN ANNULAR JACKET

$D_{in,j} = 3.5$ cm. The outer surface of the jacket is well-insulated. The conductivity of the exhaust pipe is $k_m = 15$ W/m-K. The length of the jacket is $L = 0.25$ m. The exhaust gas has a mass flow rate of $\dot{m}_e = 0.1$ kg/s and enters the exhaust pipe with a temperature $T_e = 800°C$ and pressure $p_e = 1$ atm. The temperature and pressure of the exhaust gas are approximately constant as it passes through the pipe and the exhaust gas has properties that are consistent with air. Water runs through the jacket with a mass flow rate of $\dot{m}_w = 0.01$ kg/s. The water enters at temperature $T_{w,in} = 30°C$ and pressure $p_w = 10$ atm.

a) **Assume that the water does not change phase in the jacket. Develop a numerical model of the flow through the jacket that can predict the amount of energy transferred to the water.**

The known information is entered in EES:

```
"EXAMPLE 5.3-1: Energy Recovery with an Annular Jacket"

$UnitSystem SI MASS RAD PA K J
$Tabstops 0.2 0.4 0.6 3.5 in

"Inputs"
m_dot_e=0.1 [kg/s]                          "exhaust mass flow rate"
T_e=converttemp(C,K,800 [C])                "exhaust gas temperature"
p_e=1 [atm]*convert(atm,Pa)                 "exhaust gas pressure"
m_dot_w=0.01 [kg/s]                         "water mass flow rate"
T_w_in=converttemp(C,K,30 [C])             "water inlet temperature"
p_w=10 [atm]*convert(atm,Pa)               "water pressure"
D_in_e=2.5 [cm] *convert(cm,m)             "inner diameter of exhaust pipe"
D_out_e=2.9 [cm]*convert(cm,m)             "outer diameter of exhaust pipe"
D_in_j=3.5 [cm]*convert(cm,m)              "inner diameter of jacket"
L=0.25 [m]                                  "length of jacket"
k_m=15 [W/m-K]                              "conductivity of tubes"
```

A differential energy balance on the water in the jacket is shown in Figure 2.

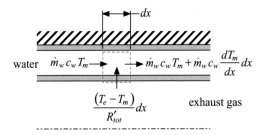

Figure 2: A differential energy balance.

The energy balance suggested by Figure 2 is:

$$\dot{m}_w\, c_w\, T_m + \frac{(T_e - T_m)}{R'_{tot}}dx = \dot{m}_w\, c_w\, T_m + \dot{m}_w\, c_w\, \frac{dT_m}{dx}dx \qquad (1)$$

where R'_{tot} is the total thermal resistance per unit length between the water and the exhaust gas and c_w is the specific heat capacity of the water. Equation (1) is simplified to:

$$\frac{(T_e - T_m)}{R'_{tot}} = \dot{m}_w\, c_w\, \frac{dT_m}{dx} \tag{2}$$

Equation (2) is integrated numerically along the length of the pipe. The state equation for the numerical integration is obtained by solving Eq. (2) for the rate of change of the state variable, T_m, the mean temperature of the water:

$$\frac{dT_m}{dx} = \frac{(T_e - T_m)}{\dot{m}_w\, c_w\, R'_{tot}} \tag{3}$$

The thermal resistance between T_m and T_e includes convection between the water and the exhaust pipe, conduction through the exhaust pipe, and convection between the exhaust pipe and the exhaust gas:

$$R'_{tot} = \underbrace{\frac{1}{h_w\, \pi\, D_{out,e}}}_{\substack{\text{convection between}\\\text{the water and the}\\\text{exhaust pipe}}} + \underbrace{\frac{\ln\left(\dfrac{D_{out,e}}{D_{in,e}}\right)}{2\,\pi\, k_m}}_{\substack{\text{conduction through}\\\text{the exhaust pipe}}} + \underbrace{\frac{1}{h_e\, \pi\, D_{in,e}}}_{\substack{\text{convection between}\\\text{the exhaust pipe and the}\\\text{exhaust gas}}} \tag{4}$$

The first step in carrying out a numerical integration is to set up the program so that the state equation can be computed, given an arbitrary value of the state variable (T_m) and the integration variable (x). Once this is accomplished, it is not difficult to integrate the state equation. Arbitrary values of T_m and x are used to start this process; these values will later be commented out to complete the problem.

```
T_m=600 [K]        "an arbitrary value of the mean water temperature"
x=0.1 [m]          "and position"
```

Equation (4) indicates that two convection problems must be solved in order to compute R'_{tot}. The properties of the exhaust gas are obtained using the film temperature on the exhaust gas side:

$$T_{film,\, e} = \frac{T_e + T_{s,e}}{2}$$

where $T_{s,e}$ is the temperature of the inner surface of the exhaust pipe. This temperature is between T_e and T_m, but it cannot be calculated until the resistances in Eq. (4) are determined. Therefore, a reasonable value of $T_{s,e}$ is assumed; this equation will be commented out later.

```
"Properties"
T_s_e=T_m        "assumed value of the inner surface temperature of the exhaust pipe"
T_film_e=(T_s_e+T_e)/2    "film temperature for exhaust gas"
```

The properties of the exhaust gas (ρ_e, μ_e, k_e, and Pr_e) are computed using EES' built-in property routines for air:

EXAMPLE 5.3-1: ENERGY RECOVERY WITH AN ANNULAR JACKET

```
mu_e=viscosity(Air,T=T_film_e)              "exhaust viscosity"
rho_e=density(Air,T=T_film_e,P=P_e)         "exhaust density"
k_e=conductivity(Air,T=T_film_e)            "exhaust conductivity"
Pr_e=Prandtl(Air,T=T_film_e)                "exhaust Prandtl number"
```

The properties of water (ρ_w, μ_w, k_w, Pr_w, and c_w) are also obtained using EES' built-in property routines using the film temperature on the water side:

$$T_{film,\ w} = \frac{T_m + T_{s,w}}{2}$$

where $T_{s,w}$ is the temperature of the outer surface of the exhaust pipe. Again, a reasonable value of $T_{s,w}$ is assumed to start the process.

```
T_s_w=T_m                                   "assumed outer surface temp. of the exhaust pipe"
T_film_w=(T_s_w+T_m)/2                       "film temperature for water"
mu_w=viscosity(Water,T=T_film_w,P=P_w)      "water viscosity"
rho_w=density(Water,T=T_film_w,P=P_w)       "water density"
k_w=conductivity(Water,T=T_film_w,P=P_w)    "water conductivity"
c_w=cP(Water,T=T_film_w,P=P_w)              "water specific heat capacity"
Pr_w=(mu_w/rho_w)/(k_w/(rho_w*c_w))         "water Prandtl number"
```

The water-side cross-sectional area ($A_{c,w}$) and wetted perimeter (per_w) are:

$$A_{c,w} = \frac{\pi}{4}\left(D_{in,j}^2 - D_{out,e}^2\right)$$

$$per_w = \pi\left(D_{in,j} + D_{out,e}\right)$$

The hydraulic diameter for the annular duct is:

$$D_{h,w} = \frac{4\,A_{c,w}}{per_w}$$

```
"Water-side convection"
Ac_w=pi*(D_in_j^2-D_out_e^2)/4              "cross-sectional area of annular duct"
per_w=pi*(D_in_j+D_out_e)                   "wetted perimeter of annular duct"
D_h_w=4*Ac_w/per_w                          "hydraulic diameter of annular duct"
```

The mean velocity of the water is:

$$u_w = \frac{\dot{m}_w}{\rho_w\,A_{c,w}}$$

The Reynolds number of the water is:

$$Re_w = \frac{u_w\,D_{h,w}\,\rho_w}{\mu_w}$$

```
u_w=m_dot_w/(rho_w*Ac_w)                    "mean velocity of water"
Re_w=u_w*D_h_w*rho_w/mu_w                   "water-side Reynolds number"
```

The roughness of drawn tubes is approximately 0.0015 mm according to Table 5-1; this is sufficient information to obtain the local Nusselt number of the flow using the AnnularFlow_N_local procedure.

```
e_w=0.0015 [mm]*convert(mm,m)                "roughness of drawn tubing"
call AnnularFlow_N_local(Re_w, Pr_w, x/D_h_w, D_out_e/D_in_j, e_w/D_h_w:&
   Nusselt_T_w,Nusselt_H_w, f_w)
                          "access correlations for local heat transfer coefficient in an annular duct"
```

The constant temperature Nusselt number $(Nu_{w,T})$ is used to compute the heat transfer coefficient; this is a lower bound and therefore conservative:

$$h_w = \frac{Nu_{w,T}\, k_w}{D_{h,w}}$$

The resistance per unit length related to water-side convection is:

$$R'_{conv,w} = \frac{1}{h_w\, \pi\, D_{out,e}}$$

```
h_w=Nusselt_T_w*k_w/D_h_w                    "water-side heat transfer coefficient"
R'_conv_w=1/(h_w*pi*D_out_e)                 "resistance per unit length on water-side"
```

The cross-sectional area on the exhaust side is:

$$A_{c,e} = \frac{\pi\, D_{in,e}^2}{4}$$

The mean velocity of the exhaust gas is:

$$u_e = \frac{\dot{m}_e}{\rho_e\, A_{c,e}}$$

The exhaust gas Reynolds number is:

$$Re_e = \frac{u_w\, D_{in,e}\, \rho_e}{\mu_e}$$

```
"Exhaust gas convection"
Ac_e=pi*D_in_e^2/4                           "cross sectional area of exhaust pipe"
u_e=m_dot_e/(rho_e*Ac_e)                     "mean velocity of exhaust"
Re_e=u_e*D_in_e*rho_e/mu_e                   "exhaust-side Reynolds number"
```

The procedure PipeFlow_N_local is used to obtain the local characteristics of the exhaust gas flow:

```
e_e=0.0015 [mm]*convert(mm,m)               "roughness of drawn tube"
call PipeFlow_N_local(Re_e,Pr_e,x/D_in_e,e_e/D_in_e: Nusselt_T_e,Nusselt_H_e,f_e)
                          "access correlations for local heat transfer coefficient in a circular duct"
```

EXAMPLE 5.3-1: ENERGY RECOVERY WITH AN ANNULAR JACKET

EXAMPLE 5.3-1: ENERGY RECOVERY WITH AN ANNULAR JACKET

The constant temperature Nusselt number $(Nu_{e,T})$ is used to estimate the exhaust-side heat transfer coefficient:

$$h_e = \frac{Nu_{e,T}\, k_e}{D_{in,e}}$$

The resistance per unit length related to exhaust-side convection is:

$$R'_{conv,e} = \frac{1}{h_e\, \pi\, D_{in,e}}$$

```
h_e=Nusselt_T_e*k_e/D_in_e          "exhaust-side heat transfer coefficient"
R'_conv_e=1/(h_e*pi*D_in_e)         "resistance per unit length on exhaust-side"
```

The resistance per unit length due to conduction is:

$$R'_{cond} = \frac{\ln\left(\dfrac{D_{out,e}}{D_{in,e}}\right)}{2\,\pi\, k_m}$$

```
R'_cond=ln(D_out_e/D_in_e)/(2*pi*k_m)    "resistance per unit length due to conduction"
```

The problem should be solved and the guess values updated at this point. The assumed values of $T_{s,w}$ and $T_{s,e}$ can then be commented out and instead computed using the resistances:

$$T_{s,w} = T_m + \frac{(T_e - T_m)\, R'_{conv,w}}{\left(R'_{conv,w} + R'_{cond} + R'_{conv,e}\right)}$$

$$T_{s,e} = T_m + \frac{(T_e - T_m)\left(R'_{conv,w} + R'_{cond}\right)}{\left(R'_{conv,w} + R'_{cond} + R'_{conv,e}\right)}$$

```
{T_s_e=T_m}    "assumed value of the inner surface temperature of the exhaust pipe"
{T_s_w=T_m}    "assumed value of the outer surface temperature of the exhaust pipe"
T_s_w=T_m+(T_e-T_m)*R'_conv_w/(R'_conv_w+R'_cond+R'_conv_e)
               surface temperature on water side"
T_s_e=T_m+(T_e-T_m)*(R'_conv_w+R'_cond)/(R'_conv_w+R'_cond+R'_conv_e)
               surface temperature on exhaust side"
```

The total resistance to heat transfer between the exhaust and water per unit length is:

$$R'_{tot} = R'_{conv,w} + R'_{cond} + R'_{conv,e}$$

and the rate of change of the mean temperature of the water can be computed using Eq. (3).

```
R'_tot=R'_conv_w+R'_cond+R'_conv_e          "total resistance per unit length"
dTmdx=(T_e-T_m)/(m_dot_w*c_w*R'_tot)        "rate of change of T_m"
```

The problem should be solved at this point and the guess values again updated; the numerical integration of the state equation is accomplished by commenting out the assumed values of T_m and x and instead varied using the Integral command in EES:

```
{T_m=600 [K]                          "an arbitrary value of the mean water temperature"
x=0.1 [m]                             "and position"}
T_m=T_w_in+Integral(dTmdx,x,L/100,L)
                                      "integrate rate equation to get outlet temperature"
T_m_C=converttemp(K,C,T_m)            "outlet temperature in C"
T_w_out=T_m                           "water outlet temperature"
```

Note that the integration cannot start at $x = 0$ because the correlations for the local heat transfer coefficient are not defined in this limit (in the same way that the heat transfer coefficient at the leading edge of a plate is not well-defined). Instead, the integration begins at a very small value of x. The exit temperature is $T_{w,out} = 436.3$ K (163.2°C), which is less than the saturation temperature at the water pressure, T_{sat}, obtained using EES' internal property routine for water. Therefore, no phase change occurs.

```
T_sat=temperature(Water,x=0,p=p_w)    "saturation temperature"
T_sat_C=converttemp(K,C,T_sat)        "in C"
```

The total rate of heat transfer is:

$$\dot{q} = \dot{m}_w\, c_w (T_{w,out} - T_{w,in})$$

```
q_dot=m_dot_w*c_w*(T_m-T_w_in)        "heat recovered"
q_dot_kW=q_dot*convert(W,kW)          "in kW"
```

which leads to $\dot{q}$=2840 W (2.84 kW).

b) **Sanity check your numerical solution using the analytical solution presented in Section 5.3.5 with an approximate, constant value of the total resistance.**

The previous solution is commented out. The properties of the exhaust gas (ρ_e, μ_e, k_e, and Pr_e) and water (ρ_w, μ_w, k_w, Pr_w, and c_w) are obtained at the average of the inlet exhaust gas and inlet water temperature:

```
"Sanity check"
T_bar=(T_e+T_w_in)/2                  "average temperature used to evaluate properties"
mu_e=viscosity(Air,T=T_bar)           "exhaust viscosity"
rho_e=density(Air,T=T_bar,P=P_e)      "exhaust density"
k_e=conductivity(Air,T=T_bar)         "exhaust conductivity"
Pr_e=Prandtl(Air,T=T_bar)             "exhaust Prandtl number"
mu_w=viscosity(Water,T=T_bar,P=P_w)   "water viscosity"
rho_w=density(Water,T=T_bar,P=P_w)    "water density"
k_w=conductivity(Water,T=T_bar,P=P_w) "water conductivity"
c_w=cP(Water,T=T_bar,P=P_w)           "water specific heat capacity"
Pr_w=(mu_w/rho_w)/(k_w/(rho_w*c_w))   "water Prandtl number"
```

EXAMPLE 5.3-1: ENERGY RECOVERY WITH AN ANNULAR JACKET

EXAMPLE 5.3-1: ENERGY RECOVERY WITH AN ANNULAR JACKET

The cross-sectional area ($A_{c,w}$), perimeter (per_w), hydraulic diameter ($D_{h,w}$), mean velocity (u_w), and Reynolds number (Re_w) on the water-side are computed as before.

```
"Water-side convection"
Ac_w=pi*(D_in_j^2-D_out_e^2)/4        "cross-sectional area of annular duct"
per_w=pi*(D_in_j+D_out_e)             "wetted perimeter of annular duct"
D_h_w=4*Ac_w/per_w                    "hydraulic diameter of annular duct"
u_w=m_dot_w/(rho_w*Ac_w)              "mean velocity of water"
Re_w=u_w*D_h_w*rho_w/mu_w             "water-side Reynolds number"
```

The average (rather than local) Nusselt number on the water-side is computed using the **Annular_Flow_N** procedure:

```
e_w=0.0015 [mm]*convert(mm,m)         "roughness of drawn tubing"
call AnnularFlow_N(Re_w, Pr_w, L/D_h_w, D_out_e/D_in_j, e_w/D_h_w:&
    Nusselt_bar_T_w,Nusselt_bar_H_w, f_bar_w)
            "access correlations for average heat transfer coefficient in an annular duct"
```

The average heat transfer coefficient on the water-side is estimated based on the constant temperature Nusselt number ($\overline{Nu}_{T,w}$):

$$\overline{h}_w = \frac{\overline{Nu}_{T,w}\,k_w}{D_{h,w}}$$

The water-side resistance to convection is:

$$R_{conv,w} = \frac{1}{\overline{h}_w\,\pi\,D_{out,e}\,L}$$

```
h_bar_w=Nusselt_bar_T_w*k_w/D_h_w     "water-side heat transfer coefficient"
R_conv_w=1/(h_bar_w*pi*D_out_e*L)     "resistance on water-side"
```

The cross-sectional area ($A_{c,e}$), mean velocity (u_e), and Reynolds number (Re_e) on the exhaust-side are computed as before.

```
"Exhaust gas convection"
Ac_e=pi*D_in_e^2/4                    "cross sectional area of exhaust pipe"
u_e=m_dot_e/(rho_e*Ac_e)             "mean velocity of exhaust"
Re_e=u_e*D_in_e*rho_e/mu_e           "exhaust-side Reynolds number"
```

The average Nusselt number on the exhaust-side is computed using the **Pipe_Flow_N** procedure:

```
e_e=0.0015 [mm]*convert(mm,m)         "roughness of drawn tube"
call PipeFlow_N(Re_e,Pr_e,L/D_in_e,e_e/D_in_e: Nusselt_bar_T_e,Nusselt_bar_H_e, f_bar_e)
            "access correlations for local heat transfer coefficient in a circular duct"
```

The average heat transfer coefficient on the exhaust-side is estimated based on the constant temperature Nusselt number $(\overline{Nu}_{T,e})$:

$$\overline{h}_e = \frac{\overline{Nu}_{T,e}\, k_e}{D_{in,e}}$$

The exhaust-side resistance to convection is:

$$R_{conv,e} = \frac{1}{\overline{h}_e\, \pi\, D_{in,e}\, L}$$

EXAMPLE 5.3-1: ENERGY RECOVERY WITH AN ANNULAR JACKET

h_bar_e=Nusselt_bar_T_e*k_e/D_in_e   "exhaust-side heat transfer coefficient"
R_conv_e=1/(h_bar_e*pi*D_in_e*L)   "resistance on exhaust-side"

The total resistance to conduction is:

$$R_{cond} = \frac{\ln\left(\dfrac{D_{out,e}}{D_{in,e}}\right)}{2\,\pi\, k_m\, L}$$

R_cond=ln(D_out_e/D_in_e)/(2*pi*k_m*L)   "resistance due to conduction"

The total resistance is:

$$R_{tot} = R_{conv,w} + R_{cond} + R_{conv,e}$$

The total conductance is:

$$UA = \frac{1}{R_{tot}}$$

R_tot=R_conv_e+R_cond+R_conv_w   "total resistance"
UA=1/R_tot   "total conductance"

The outlet temperature is computed using Eq. (5-109)

$$T_{w,out} = T_e - (T_e - T_{in})\exp\left(-\frac{UA}{\dot{m}_w\, c_w}\right)$$

and the heat transfer rate is calculated as before:

$$\dot{q} = \dot{m}_w\, c_w \left(T_{w,\,out} - T_{w,in}\right)$$

T_w_out=T_e-(T_e-T_w_in)*exp(-UA/(m_dot_w*c_w))   "outlet water temperature"
T_w_out_C=converttemp(K,C,T_w_out)   "in C"
q_dot=m_dot_w*c_w*(T_w_out-T_w_in)   "heat recovered"
q_dot_kW=q_dot*convert(W,kW)   "in kW"

which leads to $\dot{q} = 2920$ W (2.92 kW). This answer is within 5% of the more detailed, numerical solution. Note that this is a heat exchanger problem and most heat exchanger problems rely on computing a total conductance based on average properties rather than carrying out a detailed numerical simulation that accounts for changing fluid properties and flow conditions.

c) **Any substantial back-pressure on the engine cylinder will reduce the perfor-mance of the engine and potentially negate any advantage offered by the energy recovery. Estimate the pressure drop in the exhaust gas.**

The apparent friction factor of the exhaust gas $(\bar{f}_e)$ was returned by the function PipeFlow_N. The definition of the apparent friction factor is used to estimate the pressure drop:

$$\Delta p_e = \frac{\rho_e u_e^2}{2} \bar{f}_e \frac{L}{D_{in,e}}$$

"Pressure loss"
DP_e=f_bar_e*0.5*rho_e*u_e^2*L/D_in_e   "pressure loss in exhaust"
DP_e_psid=DP_e*convert(Pa,psi)   "pressure loss in psi"

The pressure loss in the exhaust gas is 8250 Pa (1.2 psi).

5.4 Analytical Solutions for Internal Flows

5.4.1 Introduction

In Section 5.1, internal convection is discussed and some important concepts are intro-duced, including the hydrodynamic and thermal entry length, the bulk velocity and tem-perature, and the friction factor and Nusselt number. In Section 5.2, correlations for the friction factor and Nusselt number based on analytical solutions and experimental data are presented. In this section, the governing momentum and thermal energy equations for an internal flow are investigated in order to identify important dimensionless param-eters that allow the equation to be simplified and solved under some conditions. These solutions are limited to the fully developed flows with simple boundary conditions. More general solutions can be obtained using the numerical techniques discussed in Section 5.5.

5.4.2 The Momentum Equation

The x-direction momentum equation for steady-state, laminar flow within a boundary layer was derived in Section 4.2 in Cartesian coordinates:

$$\rho \underbrace{\left(u \frac{\partial u}{\partial x} + v \frac{\partial u}{\partial y} \right)}_{\text{inertia}} = \underbrace{-\frac{dp}{dx}}_{\text{pressure gradient}} + \underbrace{\mu \frac{\partial^2 u}{\partial y^2}}_{\text{viscous shear}} \qquad (5\text{-}110)$$

and in radial coordinates:

$$\rho \underbrace{\left(u \frac{\partial u}{\partial x} + v \frac{\partial u}{\partial r} \right)}_{\text{inertia}} = \underbrace{-\frac{dp}{dx}}_{\text{pressure gradient}} + \underbrace{\mu \frac{1}{r} \frac{\partial}{\partial r} \left(r \frac{\partial u}{\partial r} \right)}_{\text{viscous shear}} \qquad (5\text{-}111)$$

In the hydrodynamically fully developed region of a constant cross-sectional area duct, the x-velocity distribution does not change with axial position x, as discussed in Section 5.1.2. Therefore, the y-velocity or r-velocity must be zero and both of the inertial terms on the left side of Eq. (5-110) or Eq. (5-111) are zero. However, it is sometimes possi-ble to neglect the inertial terms relative to the viscous shear even if the cross-sectional area of the duct does change. A situation that can be simplified in this way is sometimes referred to as an inertia-free flow. The conditions under which it is appropriate to assume inertia-free flow can be determined by scaling each of the terms in Eq. (5-110) appropriately for an internal flow through a duct. The x-velocity is scaled by the mean velocity of the flow, u_m. Gradients in the y-direction or r-direction are scaled by the hydraulic diameter of the channel, D_h. Gradients in the x-direction are scaled by a characteristic length, L_{char}, which is the axial distance over which the characteristics of the flow change. Typically, this is the length of the channel (L). The y-velocity or r-velocity is scaled as it was in Section 4.2.3, where examination of the continuity equation showed that:

$$v \approx \frac{\delta}{L} u_\infty \qquad (5\text{-}112)$$

For an internal flow, the analogous scaling is:

$$v \approx \frac{D_h}{L_{char}} u_m \qquad (5\text{-}113)$$

With these scaling assumptions, the order of magnitude of each term in Eq. (5-110) and Eq. (5-111) can be ascertained:

$$\rho \underbrace{\frac{u_m^2}{L_{char}}}_{\text{inertia}} \approx \underbrace{\frac{\Delta p}{L_{char}}}_{\text{pressure gradient}} + \underbrace{\frac{\mu\, u_m}{D_h^2}}_{\text{viscous shear}} \qquad (5\text{-}114)$$

The ratio of inertial to viscous forces in an internal flow is therefore:

$$\frac{\text{inertia}}{\text{viscous shear}} \approx \frac{\rho\, D_h\, u_m}{\mu} \frac{D_h}{L_{char}} \qquad (5\text{-}115)$$

Equation (5-115) is sometimes referred to as the modified Reynolds number. If the modified Reynolds number is much less than unity (i.e., the flow is inertia-free) or the flow is hydrodynamically fully developed, then the x-momentum equation for internal flow in Cartesian coordinates, Eq. (5-110), reduces to:

$$\frac{dp}{dx} = \mu \frac{\partial^2 u}{\partial y^2} \qquad (5\text{-}116)$$

In radial coordinates, the x-momentum equation for fully developed or inertia free flow becomes:

$$\frac{dp}{dx} = \frac{\mu}{r} \frac{\partial}{\partial r}\left(r \frac{\partial u}{\partial r} \right) \qquad (5\text{-}117)$$

Fully Developed Flow between Parallel Plates

Equation (5-116) is used to investigate fully developed laminar flow between two parallel plates that are separated by a distance H. In this case, the x-velocity is only a function of y and therefore Eq. (5-116) can be written as:

$$\frac{dp}{dx} = \mu \frac{d^2 u}{dy^2} \qquad (5\text{-}118)$$

The x-velocity distribution can be expressed in terms of the local pressure gradient by separating Eq. (5-118) and integrating in y:

$$\int d\frac{du}{dy} = \frac{1}{\mu} \frac{dp}{dx} \int dy \qquad (5\text{-}119)$$

or

$$\frac{du}{dy} = \frac{1}{\mu}\frac{dp}{dx}y + C_1 \tag{5-120}$$

where C_1 is a constant of integration. Equation (5-120) is integrated again:

$$\int du = \int \left(\frac{1}{\mu}\frac{dp}{dx}y + C_1\right) dy \tag{5-121}$$

which leads to:

$$u = \frac{dp}{dx}\frac{y^2}{2\mu} + C_1 y + C_2 \tag{5-122}$$

where C_2 is another constant of integration. The constants of integration are determined by requiring that the no-slip condition is satisfied at $y = 0$ and $y = H$:

$$u_{y=0} = 0 \rightarrow C_2 = 0 \tag{5-123}$$

$$u_{y=H} = 0 \rightarrow \frac{dp}{dx}\frac{H^2}{2\mu} + C_1 H = 0 \tag{5-124}$$

so

$$u = \frac{1}{2\mu}\frac{dp}{dx}(y^2 - Hy) \tag{5-125}$$

The mean velocity is obtained by substituting the velocity distribution, Eq. (5-125), into Eq. (5-6):

$$u_m = \frac{1}{H}\int_0^H u\,dy = \frac{1}{2H\mu}\frac{dp}{dx}\int_0^H (y^2 - Hy)dy \tag{5-126}$$

which leads to:

$$\boxed{u_m = -\frac{dp}{dx}\frac{H^2}{12\mu}} \tag{5-127}$$

Substituting Eq. (5-127) into Eq. (5-125) leads to:

$$\boxed{u = 6u_m\left[\frac{y}{H} - \left(\frac{y}{H}\right)^2\right]} \tag{5-128}$$

Substituting Eq. (5-127) into the definition of the friction factor, Eq. (5-31), leads to:

$$f = \underbrace{\frac{12\mu u_m}{H^2}}_{-\frac{dp}{dx}}\frac{2D_h}{\rho u_m^2} \tag{5-129}$$

The hydraulic diameter for a passage formed by parallel plates is:

$$D_h = 2H \tag{5-130}$$

Substituting Eq. (5-130) into Eq. (5-129) leads to:

$$f = \frac{48\mu}{H\rho u_m} \tag{5-131}$$

or

$$f = \frac{96}{Re_{D_h}}$$ (5-132)

where Re_{D_h} is the Reynolds number based on hydraulic diameter:

$$Re_{D_h} = \frac{2\,H\,\rho\,u_m}{\mu}$$ (5-133)

Notice that Eq. (5-132) is consistent with the friction factor for fully developed flow through a rectangular duct with an aspect ratio that approaches zero, presented in Section 5.2.3.

The Reynolds Equation
This extended section can be found on the website associated with this textbook (www.cambridge.org/nellisandklein). The x-momentum equation for inertia-free flow, Eq. (5-116), provides the basis for the Reynolds equation, which is derived in this section. The Reynolds equation is used to analyze the flow of viscous fluids through small gaps (i.e., lubrication problems).

Fully Developed Flow in a Circular Tube
This extended section can be found on the website associated with this textbook (www.cambridge.org/nellisandklein). Equation (5-117) is the appropriate x-momentum equation for fully developed laminar flow in a circular tube. In this section, Maple is used to carry out the same steps that were applied to flow between parallel plates in order to obtain the velocity distribution and friction factor.

5.4.3 The Thermal Energy Equation

In this section, the thermal energy equation for a laminar internal flow is examined in order to identify dimensionless parameters that can be used to simplify the governing equation. Analytical solutions for the temperature distribution associated with hydrodynamically and thermally fully developed flow through a circular tube and between parallel plates are developed for a constant surface heat flux.

The thermal energy equation for laminar flow is derived in Section 4.2.2. Under steady conditions, the time derivative can be neglected and in the hydrodynamically fully developed region of an internal flow, the velocity perpendicular to the axis of the duct (v, the velocity in the y- or r-directions) is zero. Therefore, the thermal energy equation in Cartesian coordinates simplifies to:

$$\underbrace{\rho\,c\,u\frac{\partial T}{\partial x}}_{\substack{\text{enthalpy carried}\\\text{by flow}}} = \underbrace{k\frac{\partial^2 T}{\partial x^2}}_{\text{axial conduction}} + \underbrace{k\frac{\partial^2 T}{\partial y^2}}_{\text{lateral conduction}} + \underbrace{\mu\left(\frac{\partial u}{\partial y}\right)^2}_{\text{viscous dissipation}}$$ (5-157)

and, in radial coordinates:

$$\underbrace{\rho\,c\,u\frac{\partial T}{\partial x}}_{\substack{\text{enthalpy carried}\\\text{by flow}}} = \underbrace{k\frac{\partial^2 T}{\partial x^2}}_{\text{axial conduction}} + \underbrace{\frac{k}{r}\frac{\partial}{\partial r}\left(r\frac{\partial T}{\partial r}\right)}_{\text{radial conduction}} + \underbrace{\mu\left(\frac{\partial u}{\partial r}\right)^2}_{\text{viscous dissipation}}$$ (5-158)

Equations (5-157) and (5-158) represent a balance between the enthalpy carried by the fluid, conduction in the direction of the flow (i.e., the x-direction), conduction in the direction perpendicular to the flow (i.e., the y- or r-directions), and viscous dissipation. These equations cannot be solved analytically for most conditions. However, it is often reasonable to ignore both axial conduction and viscous dissipation. Equations (5-157) and (5-158) are examined in more detail in order to identify the conditions where these effects can be neglected. Each of the terms are scaled based on an internal flow situation in order to evaluate their relative size. The temperature gradients are scaled by the surface-to-mean temperature difference, $T_s - T_m$. Axial velocity is scaled by the mean velocity of the flow, u_m. Gradients in the y- or r-direction are scaled by the hydraulic diameter of the duct, D_h. Gradients in the x-direction are scaled by a characteristic length, L_{char}, that represents the axial distance over which the flow properties change.

$$\rho c u_m \frac{(T_s - T_m)}{L_{char}} = k \frac{(T_s - T_m)}{L_{char}^2} + k \frac{(T_s - T_m)}{D_h^2} + \mu \frac{u_m^2}{D_h^2} \tag{5-159}$$

Most internal convection problems are a balance between the enthalpy carried by the flow, the first term in Eq. (5-159), and conduction perpendicular to the flow, the third term in Eq. (5-159). Therefore, it is appropriate to define a characteristic axial length scale that ensures that these two terms are of the same order of magnitude:

$$\rho c u_m \frac{(T_s - T_m)}{L_{char}} \approx k \frac{(T_s - T_m)}{D_h^2} \tag{5-160}$$

or

$$L_{char} = \frac{u_m}{\alpha} D_h^2 \tag{5-161}$$

The characteristic length identified by Eq. (5-161) is approximately the length required for energy to diffuse from the edge to the center of the passage by conduction through the fluid. Substituting Eq. (5-161) into Eq. (5-159) leads to:

$$\rho c u_m \alpha \frac{(T_s - T_m)}{u_m D_h^2} = k \alpha^2 \frac{(T_s - T_m)}{u_m^2 D_h^4} + k \frac{(T_s - T_m)}{D_h^2} + \mu \frac{u_m^2}{D_h^2} \tag{5-162}$$

or

$$\underbrace{k \frac{(T_s - T_m)}{D_h^2}}_{\substack{\text{enthalpy carried} \\ \text{by flow}}} = \underbrace{k \alpha^2 \frac{(T_s - T_m)}{u_m^2 D_h^4}}_{\text{axial conduction}} + \underbrace{k \frac{(T_s - T_m)}{D_h^2}}_{\text{radial conduction}} + \underbrace{\mu \frac{u_m^2}{D_h^2}}_{\text{viscous dissipation}} \tag{5-163}$$

Equation (5-163) allows us to examine the relative magnitude of axial conduction with respect to radial conduction (or the enthalpy carried by the flow, these terms have the same scale due to our definition of L_{char}).

$$\frac{\text{axial conduction}}{\text{radial conduction \& convection}} \approx \frac{\alpha^2}{u_m^2 D_h^2} \tag{5-164}$$

Equation (5-164) can be written in terms of the Reynolds number and the Prandtl number:

$$\frac{\text{axial conduction}}{\text{radial conduction \& convection}} \approx \underbrace{\frac{\nu^2}{u_m^2 D_h^2}}_{\frac{1}{Re_{D_h}^2}} \underbrace{\frac{\alpha^2}{\nu^2}}_{\frac{1}{Pr^2}} = \frac{1}{(Re_{D_h} Pr)^2} \tag{5-165}$$

The product of the Reynolds number and the Prandtl number is referred to as the Peclet number (Pe_{D_h}):

$$Pe_{D_h} = Re_{D_h}\, Pr = \frac{u_m\, D_h}{\alpha} \tag{5-166}$$

Substituting Eq. (5-166) into Eq. (5-165) leads to:

$$\frac{\text{axial conduction}}{\text{radial conduction \& convection}} \approx \frac{1}{Pe_{D_h}^2} \tag{5-167}$$

Typically, the Prandtl number will be near unity and the Reynolds number will be larger than unity. Therefore, the Peclet number is usually large and, according to Eq. (5-167), axial conduction can safely be ignored. Axial conduction will become important when the Peclet number is small. According to Eq. (5-166), this will occur for very conductive fluids (e.g., liquid metals where Pr is small) or flows with very low velocity.

Using Eq. (5-163), the significance of viscous dissipation in an internal flow problem can be evaluated according to:

$$\frac{\text{viscous dissipation}}{\text{radial conduction \& convection}} \approx \frac{\mu\, u_m^2}{k\,(T_s - T_m)} \tag{5-168}$$

Equation (5-168) can be written in terms of the Prandtl number and the Eckert number:

$$\frac{\text{viscous dissipation}}{\text{radial conduction \& convection}} \approx \underbrace{\frac{\mu\, c}{k}}_{Pr}\ \underbrace{\frac{u_m^2}{c\,(T_s - T_m)}}_{Ec} = Pr\, Ec \tag{5-169}$$

The product of the Eckert number and the Prandtl number is referred to as the Brinkman number (Br):

$$Br = Pr\, Ec = \frac{\mu\, u_m^2}{k\,(T_s - T_m)} \tag{5-170}$$

Substituting Eq. (5-170) into Eq. (5-169) leads to:

$$\frac{\text{viscous dissipation}}{\text{radial conduction \& convection}} \approx Br \tag{5-171}$$

The Brinkman number will be small for most internal flow problems, allowing viscous dissipation to be ignored. Viscous dissipation will become important when the Brinkman number is large. According to Eq. (5-170), this situation corresponds to a very viscous or high velocity flow or to a flow where the characteristic temperature difference is small.

Fully Developed Flow through a Round Tube with a Constant Heat Flux

In the typical limits where the Brinkman number is small (i.e., viscous dissipation can be ignored) and the Peclet number is large (i.e., axial conduction can be ignored), the thermal energy equation in cylindrical coordinates for a thermally and hydrodynamically fully developed flow, Eq. (5-158), reduces to:

$$\rho\, c\, u \frac{\partial T}{\partial x} = \frac{k}{r} \frac{\partial}{\partial r}\left(r \frac{\partial T}{\partial r} \right) \tag{5-172}$$

In Section 5.3.3, the energy balance for flow through a tube with a constant heat flux ($\dot{q}_s''$) was considered and it was shown that the mean temperature of the fluid increases

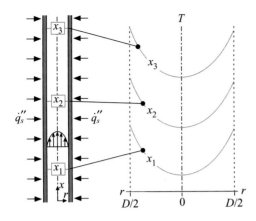

Figure 5-26: Temperature as a function of radius at various axial positions within a fully developed flow through a round tube subjected to a uniform heat flux.

linearly according to:

$$T_m = T_{in} + \frac{per \, \dot{q}_s''}{\dot{m} \, c} x \qquad (5\text{-}173)$$

where *per* is the perimeter of the channel, $\dot{m}$ is the mass flow rate, and c is the specific heat capacity. According to Eq. (5-173), the mean temperature gradient is:

$$\frac{dT_m}{dx} = \frac{per \, \dot{q}_s''}{\dot{m} \, c} \qquad (5\text{-}174)$$

For a round tube, Eq. (5-174) can be written as:

$$\frac{dT_m}{dx} = \frac{4 \, \dot{q}_s''}{D \, \rho \, u_m \, c} \qquad (5\text{-}175)$$

where D is the inner diameter of the tube, u_m is the mean velocity of the flow, and ρ is the density of the fluid.

The fluid temperature distribution in the fully developed region of a round tube exposed to a constant heat flux is shown qualitatively in Figure 5-26. The shape of the temperature distribution in Figure 5-26 does not change once the flow becomes thermally fully developed. However, the absolute value of the temperature at every radial position increases at the same rate. Therefore, the temperature gradient at any radial location is equal to the mean temperature gradient given by Eq. (5-175):

$$\frac{\partial T}{\partial x} = \frac{4 \, \dot{q}_s''}{D \, \rho \, u_m \, c} \qquad (5\text{-}176)$$

Substituting Eqs. (5-176) and the velocity distribution for fully developed flow in a round tube, Eq. (5-153), into Eq. (5-172) leads to:

$$\rho \, c \, 2 \, u_m \left[1 - \left(\frac{2r}{D} \right)^2 \right] \frac{4 \, \dot{q}_s''}{D \, \rho \, u_m \, c} = \frac{k}{r} \frac{\partial}{\partial r} \left(r \frac{\partial T}{\partial r} \right) \qquad (5\text{-}177)$$

or

$$\left[1 - \left(\frac{2r}{D} \right)^2 \right] \frac{8 \, \dot{q}_s''}{D} = \frac{k}{r} \frac{\partial}{\partial r} \left(r \frac{\partial T}{\partial r} \right) \qquad (5\text{-}178)$$

Note that even though temperature is a function of both x and r, the differential equation, Eq. (5-178), includes only derivatives of temperature with respect to radius and is therefore essentially an ordinary differential equation for temperature. The same ordinary differential equation can be obtained using Maple:

```
> restart;
> dTdx:=4*qs/(D*rho*u_m*c);
```

$$dT\_dx := \frac{4\,qs}{D\rho u\_mc}$$

```
> u:=2*u_m*(1-(2*r/D)^2);
```

$$u := 2\,u\_m\left(1 - \frac{4\,r^2}{D^2}\right)$$

```
> ODE:=rho*c*u*dTdx=k*diff(r*diff(T(r),r),r)/r;
```

$$ODE := \frac{8\left(1 - \frac{4\,r^2}{D^2}\right)qs}{D} = \frac{k\left(\left(\frac{d}{dr}T(r)\right) + r\left(\frac{d^2}{dr^2}T(r)\right)\right)}{r}$$

Rearranging Eq. (5-178) and integrating once in r leads to:

$$\int \partial\left(r\frac{\partial T}{\partial r}\right) = \frac{8\,\dot{q}_s''}{k\,D}\int\left[r - \frac{4\,r^3}{D^2}\right]\partial r \tag{5-179}$$

or

$$r\frac{\partial T}{\partial r} = \frac{8\,\dot{q}_s''}{k\,D}\left[\frac{r^2}{2} - \frac{r^4}{D^2}\right] + C_1 \tag{5-180}$$

where C_1 is a constant of integration. Rearranging Eq. (5-180) and integrating again in r leads to:

$$\int \partial T = \int\left\{\frac{8\,\dot{q}_s''}{k\,D}\left[\frac{r}{2} - \frac{r^3}{D^2}\right] + \frac{C_1}{r}\right\}\partial r \tag{5-181}$$

or

$$T = \frac{8\,\dot{q}_s''}{k\,D}\left[\frac{r^2}{4} - \frac{r^4}{4\,D^2}\right] + C_1\ln(r) + C_2 \tag{5-182}$$

where C_2 is another constant of integration. This solution can be obtained using Maple:

```
> T_sol:=dsolve(ODE);
```

$$T\_sol := T(r) = -\frac{2\,qs\,r^4}{D^3\,k} + \frac{2\,qs\,r^2}{D\,k}\_C1\ln(r) + \_C2$$

The constants of integration are determined by applying appropriate boundary conditions. The temperature must remain bounded at $r = 0$; this requires that $C_1 = 0$ so that:

$$T = \frac{8\,\dot{q}_s''}{k\,D}\left[\frac{r^2}{4} - \frac{r^4}{4\,D^2}\right] + C_2 \tag{5-183}$$

```
> T_sol:=subs(_C1=0,T_sol);
```

$$T\_sol := T(r) = -\frac{2\,qs\,r^4}{D^3\,k} + \frac{2\,qs\,r^2}{D\,k} + \_C2$$

The tube surface temperature (T_s) is obtained at $r = D/2$:

$$T_{r=\frac{D}{2}} = T_s \tag{5-184}$$

Note that the tube surface temperature is a function of x. Substituting Eq. (5-183) into Eq. (5-184) leads to:

$$\frac{8\,\dot{q}_s''}{k\,D}\left[\frac{D^2}{16} - \frac{D^4}{64\,D^2}\right] + C_2 = T_s \tag{5-185}$$

Solving Eq. (5-185) for C_2:

$$C_2 = T_s - \frac{3\,\dot{q}_s''\,D}{8\,k} \tag{5-186}$$

Substituting Eq. (5-186) into Eq. (5-183):

$$\boxed{T = \frac{8\,\dot{q}_s''}{k\,D}\left[\frac{r^2}{4} - \frac{r^4}{4\,D^2}\right] + T_s - \frac{3\,\dot{q}_s''\,D}{8\,k}} \tag{5-187}$$

or in Maple:

```
> T_sol:=subs(_C2=solve(eval(rhs(T_sol),r=D/2)=T_s,_C2),T_sol);
```

$$T\_sol := T(r) = -\frac{2\,qs\,r^4}{D^3\,k} + \frac{2\,qs\,r^2}{D\,k} - \frac{3\,D\,qs - 8\,T\_s\,k}{8\,k}$$

The definition of the mean temperature for an incompressible flow is discussed in Section 5.1.3:

$$T_m = \frac{1}{A_c\,u_m}\int_{A_c} T\,u\,dA_c \tag{5-188}$$

The mean temperature is obtained by substituting the temperature and velocity distributions, Eqs (5-187) and (5-153), respectively, into Eq. (5-188):

$$T_m = \frac{16}{D^2}\int_{r=0}^{r=\frac{D}{2}}\left\{\frac{8\,\dot{q}_s''}{k\,D}\left[\frac{r^2}{4} - \frac{r^4}{4\,D^2}\right] + T_s - \frac{3\,\dot{q}_s''\,D}{8\,k}\right\}\left[1 - \left(\frac{2r}{D}\right)^2\right]r\,dr \tag{5-189}$$

Carrying out the integral and applying the limits leads to:

$$T_m = T_s - \frac{11\,D\,\dot{q}_s''}{48\,k} \tag{5-190}$$

or in Maple:

```
> T_m:=simplify(int(rhs(T_sol)*u*2*pi*r,r=0..D/2)/(pi*D^2*u_m/4));
```

$$T\_m := -\frac{11\,D\,qs - 48\,T\_s\,k}{48\,k}$$

The Nusselt number for an internal flow is defined in Section 5.1.3:

$$Nu = \frac{h D}{k} = \frac{\dot{q}_s'' D}{(T_s - T_m) k} \tag{5-191}$$

Substituting Eq. (5-190) into Eq. (5-191) leads to:

$$Nu = \frac{\dot{q}_s'' D}{k} \frac{48 k}{11 D \dot{q}_s''} = \frac{48}{11} \tag{5-192}$$

or

$$\boxed{Nu = 4.36} \tag{5-193}$$

This solution can be obtained in Maple:

```
> Nusselt:=simplify(qs*D/(k*(T_s-T_m)));
```

$$Nusselt := \frac{48}{11}$$

The Nusselt number for laminar, fully developed flow through a round tube subjected to a constant heat flux is 4.36; this is consistent with the result presented in Section 5.2.4.

Fully Developed Flow between Parallel Plates with a Constant Heat Flux

The solution for fully developed flow between parallel plates subjected to a constant heat flux (at both surfaces) is derived using the steps presented in Section 5.4.3 for flow in a circular tube. In this section, the derivation of the solution is accomplished using Maple. The governing differential equation is obtained by simplifying the thermal energy equation, Eq. (5-157), by neglecting viscous dissipation and axial conduction:

$$\rho c u \frac{\partial T}{\partial x} = k \frac{\partial^2 T}{\partial y^2} \tag{5-194}$$

According to the energy balance for a flow with a constant heat flux, the rate of temperature change is given by:

$$\frac{\partial T}{\partial x} = \frac{per \, \dot{q}_s''}{\dot{m} c} \tag{5-195}$$

or, for flow between parallel plates:

$$\frac{\partial T}{\partial x} = \frac{2 \, \dot{q}_s''}{u_m H \rho c} \tag{5-196}$$

where H is the distance between the plates. Substituting Eqs. (5-196) and the fully developed velocity distribution for flow between parallel plates, Eq. (5-128), into Eq. (5-194) leads to the differential equation for the problem:

```
> restart;
> dTdx:=2*qs/(u_m*H*rho*c);
```

$$dT \, dx := \frac{2 \, qs}{u\_m H \rho c}$$

```
> u:=6*u_m*(y/H-(y/H)^2);
```

$$u := 6 \, u\_m \left(\frac{y}{H} - \frac{y^2}{H^2} \right)$$

```
> ODE:=rho*c*u*dTdx=k*diff(diff(T(y),y),y);
```

$$ODE := \frac{12 \left(\dfrac{y}{H} - \dfrac{y^2}{H^2} \right) qs}{H} = k \left(\frac{d^2}{dy^2} T(y) \right)$$

The differential equation is solved to obtain the general solution for the temperature distribution:

```
> T_sol:=dsolve(ODE);
```

$$T\_sol := T(y) = -\frac{qs\,y^4}{H^3\,k} + \frac{2\,qs\,y^3}{H^2\,k} + \_C1\,y + \_C2$$

The two boundary conditions are:

$$T_{y=0} = T_s \tag{5-197}$$

$$T_{y=H} = T_s \tag{5-198}$$

```
> BC1:=rhs(eval(T_sol,y=0))=T_s;
```

$$BC1 := \_C2 = T\_s$$

```
> BC2:=rhs(eval(T_sol,y=H))=T_s;
```

$$BC2 := \frac{qsH}{k} + \_C1H + \_C2 = T\_s$$

These equations are solved simultaneously and substituted into the general solution:

```
> constants:=solve({BC1,BC2},{_C1,_C2});
```

$$constants := \left\{ \_C2 = T\_s, \_C1 = -\frac{qs}{k} \right\}$$

```
> T_sol:=subs(constants,T_sol);
```

$$T\_sol := T(y) = -\frac{qs\,y^4}{H^3\,k} + \frac{2\,qs\,y^3}{H^2\,k} - \frac{qs\,y}{k} + T\_s$$

The temperature distribution is therefore:

$$T = T_s + \frac{\dot{q}_s'' H}{k} \left(-\frac{y^4}{H^4} + 2\frac{y^3}{H^3} - \frac{y}{H} \right) \tag{5-199}$$

The mean temperature is obtained according to:

$$T_m = \frac{1}{H\,u_m} \int_0^H u\,T\,dy \tag{5-200}$$

```
> T_m:=simplify(int(rhs(T_sol)*u,y=0..H)/(H*u_m));
```

$$T\_m := -\frac{17\,qs\,H - 70\,T\_s\,k}{70\,k}$$

The Nusselt number is obtained from:

$$Nu = \frac{h\,D_h}{k} = \frac{\dot{q}_s''\,2\,H}{(T_s - T_m)\,k} \tag{5-201}$$

```
> Nusselt:=simplify(qs*2*H/(k*(T_s-T_m)));
```
$$Nusselt := \frac{140}{17}$$

The Nusselt number for laminar, fully developed flow between parallel plates subjected to a constant heat flux is 8.235; this is consistent with the result presented in Section 5.2.4 for a rectangular duct in the limit that the aspect ratio approaches zero. Substituting Eq. (5-201) and the result for the Nusselt number into Eq. (5-199) leads to:

$$T = T_m + \frac{\dot{q}_s''\,H}{k}\left(-\frac{y^4}{H^4} + 2\,\frac{y^3}{H^3} - \frac{y}{H} + 0.243\right) \tag{5-202}$$

5.5 Numerical Solutions to Internal Flow Problems

5.5.1 Introduction

Numerical solution to convection problems is an extensive area of research that has seen tremendous growth in the past decades. A detailed discussion of computational fluid dynamic models is beyond the scope of this book. However, there are a few cases where the numerical solution to internal flow heat transfer problems follows naturally from the techniques discussed in Section 3.8 to solve 1-D transient problems. The cases considered in this book are restricted to those where the velocity distribution is fully developed and prescribed so that the fluid dynamics problem is straightforward. Furthermore, the transfer of heat by conduction in the flow direction is neglected. According to the discussion in Section 5.4.3, this assumption implies that the Peclet number is large. In this limit, the thermal energy equation for an internal flow in Cartesian coordinates, Eq. (5-157), reduces to:

$$\rho\,c\,u\frac{\partial T}{\partial x} = k\frac{\partial^2 T}{\partial y^2} + \mu\left(\frac{\partial u}{\partial y}\right)^2 \tag{5-203}$$

Equation (5-203) should be compared to the partial differential equation that governs the 1-D transient process associated with a plane wall experiencing volumetric generation of thermal energy, considered in Chapter 3.

$$\rho\,c\,\frac{\partial T}{\partial t} = k\frac{\partial^2 T}{\partial y^2} + \dot{g}_v''' \tag{5-204}$$

The ratio x/u in Eq. (5-203) plays the same role that time (t) does in Eq. (5-204) and viscous dissipation in Eq. (5-203) is analogous to the volumetric generation term in Eq. (5-204). The removal of the axial conduction term from the thermal energy conservation equation for fully developed flow has made Eq. (5-203) a parabolic partial differential equation. Mathematically, Eq. (5-203) is analogous to Eq. (5-204). Indeed, the process of a flow moving through a slot shaped passage is analogous to the thermal equilibration process associated with a plane wall, as discussed previously in Section 5.1. As a result, the numerical techniques that were used in Section 3.8 for 1-D transient conduction problems can also be employed to solve this type of internal flow problem.

5.5.2 Hydrodynamically Fully Developed Laminar Flow

Section 3.8 presents several methods that can be used to integrate a system of state equations (i.e., equations that provide the time rate of change of the nodal temperatures) forward through time. These integration techniques were implemented in both EES and MATLAB. The presentation in this section will mirror that previous discussion. However, the state equations for an internal flow problem will provide the axial rate of change of the nodal temperature. The state equations can be integrated in the flow direction using any of the numerical techniques that were previously introduced.

The solution method will be illustrated in the context of the internal flow problem that is illustrated in Figure 5-27.

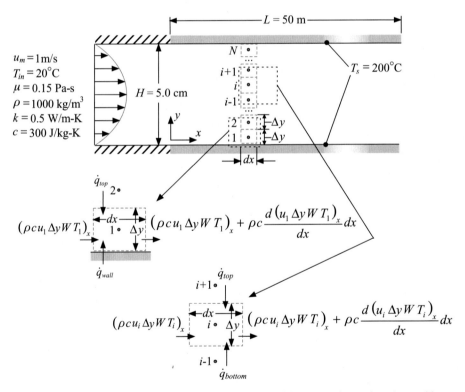

Figure 5-27: Hydrodynamically fully developed internal flow entering a duct formed by two parallel plates with uniform wall temperature.

A laminar, hydrodynamically fully developed flow with mean velocity $u_m = 1.0$ m/s flows between parallel plates. The flow has a uniform temperature, $T_{in} = 20°C$ when it enters a region of the duct where the wall temperature is maintained at a constant value, $T_s = 200°C$. The height of the duct is $H = 5.0$ cm and the length of the heated region is

$L = 50$ m. The fluid has properties $\rho = 1000\,\text{kg/m}^3$, $\mu = 0.15$ Pa-s, $k = 0.5$ W/m-K, and $c = 300$ J/kg-K. These inputs are entered in EES:

```
$UnitSystem SI MASS RAD PA K J
$TABSTOPS 0.2 0.4 0.6 0.8 3.5 in

"Inputs"
u_m=1 [m/s]                              "mean velocity"
H=5.0 [cm]*convert(cm,m)                 "duct height"
T_in=converttemp(C,K,20 [C])             "inlet fluid temperature"
rho=1000 [kg/m^3]                        "density"
mu=0.15 [Pa-s]                           "viscosity"
k=0.5 [W/m-K]                            "conductivity"
c=300 [J/kg-K]                           "specific heat capacity"
T_s=converttemp(C,K,200 [C])            "wall temperature"
L=50 [m]                                 "length of duct"
W=1.0 [m]                                "unit width into page"
alpha=k/(rho*c)                          "thermal diffusivity"
nu=mu/rho                                "kinematic viscosity"
Pr=nu/alpha                              "Prandtl number"
```

The numerical solution proceeds by distributing N nodes uniformly in the y-direction across the passage, as shown in Figure 5-27. The placement of the nodes is slightly different than was used in Section 3.8. There are no nodes placed on the walls; rather, the nodes are positioned in the center of N full sized control volumes. The reason for this adjustment becomes clear when the state equations are derived, but is related to the fact that the velocity at the wall is zero according to the no slip condition. Therefore, the axial rate of change of the temperature of a node that is placed on the surface of the wall will become unbounded according to Eq. (5-203).

The distance between adjacent nodes in Figure 5-27 is:

$$\Delta y = \frac{H}{N} \tag{5-205}$$

and the location of each of the nodes is given by:

$$y_i = \Delta y \left(i - \frac{1}{2} \right) \quad \text{for } i = 1..N \tag{5-206}$$

```
N=21 [-]                                 "number of nodes"
DELTAy=h/N                               "distance between nodes"
duplicate i=1,N
   y[i]=DELTAy*(i-1/2)                   "position of each node"
end
```

The velocity distribution in the duct is parabolic, as derived in Section 5.4.2. Therefore, the velocity at each nodal location is:

$$u_i = 6\,u_m \left[\frac{y_i}{H} - \left(\frac{y_i}{H} \right)^2 \right] \quad \text{for } i = 1..N \tag{5-207}$$

```
duplicate i = 1,N
  u[i]=6*u_m*(y[i]/H-(y[i]/H)^2)                    "velocity at each node"
end
```

The hydraulic diameter associated with the duct is:

$$D_h = 2\,H \qquad\qquad (5\text{-}208)$$

and the Reynolds number that characterizes the flow is:

$$Re_{D_h} = \frac{\rho\,D_h\,u_m}{\mu} \qquad\qquad (5\text{-}209)$$

```
D_h=2*H                                             "hydraulic diameter"
Re=rho*u_m*D_h/mu                                   "Reynolds number"
```

The Reynolds number is $Re_{D_h} = 667$; this is sufficiently low that the flow is laminar, as discussed in Section 5.1.2. Therefore, the conductive heat transfer terms may be approximated using the molecular conductivity, k, rather than by using an effective turbulent conductivity.

The Peclet and Brinkman numbers that characterize the flow are:

$$Pe = Pr\,Re_{D_h} \qquad\qquad (5\text{-}210)$$

$$Br = \frac{\mu\,u_m^2}{k\,(T_s - T_{in})} \qquad\qquad (5\text{-}211)$$

```
Pe=Re*Pr                                            "Peclet number"
Br=mu*u_m^2/(k*(T_s-T_in))                          "Brinkman number"
```

The Peclet number is $Pe = 6 \times 10^4$ and the Brinkman number is 1.7×10^{-3}. Therefore, according to the discussion in Section 5.4.3, conduction in the flow direction and viscous dissipation can both be neglected.

The state equations are derived by defining a control volume around each of the nodes. The control volumes extend a finite spatial extent in the y-direction (Δy) but a differentially small spatial extent in the x-direction (dx). This definition is consistent with the approach that was used in Section 3.8 to derive the state equations for the time rate of change of temperature; the control volumes were finite in space but differentially small in time.

The internal nodes are treated separately from the boundary nodes. A control volume for an internal node is shown in Figure 5-27. The node experiences conduction in the y-direction from adjacent nodes and energy is carried by the fluid entering the control volume at x and leaving at $x + dx$. The energy balance suggested by the control volume in Figure 5-27 is:

$$(\rho\,c\,u_i\,\Delta y\,W\,T_i)_x + \dot{q}_{top} + \dot{q}_{bottom} = (\rho\,c\,u_i\,\Delta y\,W\,T_i)_x + \rho\,c\,\frac{d(u_i\,\Delta y\,W\,T_i)}{dx}dx \qquad (5\text{-}212)$$

where W is the depth of the channel into the page. The conduction heat transfer rates are approximated according to:

$$\dot{q}_{top} = \frac{k\,dx\,W}{\Delta y}(T_{i+1} - T_i) \tag{5-213}$$

$$\dot{q}_{bottom} = \frac{k\,dx\,W}{\Delta y}(T_{i-1} - T_i) \tag{5-214}$$

Equations (5-213) and (5-214) are substituted into Eq. (5-212):

$$\frac{k\,dx\,W}{\Delta y}(T_{i+1} - T_i) + \frac{k\,dx\,W}{\Delta y}(T_{i-1} - T_i) = \rho c \frac{d\,(u_i\,\Delta y\,W\,T_i)}{dx}dx \tag{5-215}$$

$$\text{for } i = 2..\,(N-1)$$

Note that the only term in the x-derivative of Eq. (5-215) that changes with x is the temperature (provided that the flow is incompressible, u_i is independent of x in the hydrodynamically fully developed region) and therefore Eq. (5-215) can be rewritten as:

$$\frac{k\,dx\,W}{\Delta y}(T_{i+1} - T_i) + \frac{k\,dx\,W}{\Delta y}(T_{i-1} - T_i) = \rho c\,u_i\,\Delta y\,W\frac{dT_i}{dx}\,dx \quad \text{for } i = 2..\,(N-1) \tag{5-216}$$

Solving for the rate of change of T_i with respect to x:

$$\frac{dT_i}{dx} = \frac{k}{\rho c\,\Delta y^2\,u_i}(T_{i+1} + T_{i-1} - 2\,T_i) \quad \text{for } i = 2..\,(N-1) \tag{5-217}$$

An energy balance for the control volume around node 1 is also shown in Figure 5-27 and leads to:

$$\dot{q}_{top} + \dot{q}_{wall} = \rho c\,u_1\,\Delta y\,W\frac{dT_1}{dx}dx \tag{5-218}$$

The conductive heat transfer from node 2 is approximated according to:

$$\dot{q}_{top} = \frac{k\,dx\,W}{\Delta y}(T_2 - T_1) \tag{5-219}$$

and the conductive heat transfer from the wall is approximated according to:

$$\dot{q}_{wall} = \frac{2\,k\,dx\,W}{\Delta y}(T_s - T_1) \tag{5-220}$$

The factor of two that appears in the numerator of Eq. (5-220) is related to the fact that the energy must only be conducted a distance $\Delta y/2$ in order to reach node 1 from the wall. Substituting Eqs. (5-219) and (5-220) into Eq. (5-218) leads to:

$$\frac{k\,dx\,W}{\Delta y}(T_2 - T_1) + \frac{2\,k\,dx\,W}{\Delta y}(T_s - T_1) = \rho c\,u_1\,\Delta y\,W\frac{dT_1}{dx}dx \tag{5-221}$$

Solving for the rate of change of the temperature of node 1 leads to:

$$\frac{dT_1}{dx} = \frac{k}{\rho c\,\Delta y^2\,u_1}(T_2 + 2\,T_s - 3\,T_1) \tag{5-222}$$

A similar process applied to node N leads to:

$$\frac{dT_N}{dx} = \frac{k}{\rho c\,\Delta y^2\,u_N}(T_{N-1} + 2\,T_s - 3\,T_N) \tag{5-223}$$

Equations (5-217), (5-222), and (5-223) are the N state equations for the problem; these equations must be integrated from the entrance of the heated region downstream. Several of the more powerful techniques that were previously discussed in Section 3.8 are applied to this problem.

EES' Integral Command

The temperature rates of change for each node, Eqs. (5-217), (5-222), and (5-223) are programmed in EES:

```
dTdx[1]=k*(T[2]+2*T_s-3*T[1])/(rho*c*DELTAy^2*u[1])
                                    "node 1 temperature rate of change"
duplicate i=2,(N-1)
  dTdx[i]=k*(T[i+1]+T[i-1]-2*T[i])/(rho*c*DELTAy^2*u[i])
                                    "internal node temperature rate of change"
end
dTdx[N]=k*(T[N-1]+2*T_s-3*T[N])/(rho*c*DELTAy^2*u[N])
                                    "node N temperature rate of change"
```

The EES Integral function is used to integrate these coupled differential equations along the length of the duct:

```
duplicate i=1,N
  T[i]=T_in+Integral(dTdx[i],x,0,L)          "integrate temperatures through position"
end
```

An integral table is used to record the temperature of each node every 0.5 m.

```
$IntegralTable: x: 0.5, T[1..N]
```

Figure 5-28 illustrates the temperature as a function of position along the duct for various values of y.

The mean temperature can be computed at any position according to:

$$T_m = \frac{1}{A_c \, u_m} \int_{A_c} T \, u \, dA_c \qquad (5\text{-}224)$$

For the geometry considered here, Eq. (5-224) becomes:

$$T_m = \frac{1}{H \, u_m} \int_0^H u \, T \, dy \qquad (5\text{-}225)$$

The integrand of Eq. (5-225) is evaluated for each node and the integral is approximated as a sum, which is implemented using the sum command:

```
duplicate i=1,N
  TuA[i]=T[i]*u[i]*DELTAy
               "product of temperature and volumetric flow rate for each control volume"
end
T_m=sum(TuA[1..N])/(H*u_m)              "mean temperature"
```

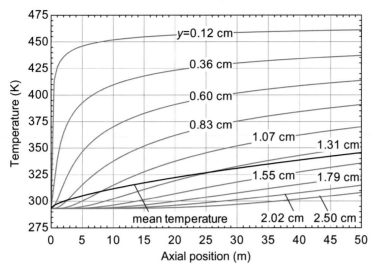

Figure 5-28: Temperature as a function of x for various values of y predicted using EES' Integral command.

The mean temperature, the variable T_m, is added to the integral table:

```
$IntegralTable: x: 0.5, T[1..N], T_m
```

and plotted in Figure 5-28. The heat flux from the bottom surface of the duct to the fluid is obtained according to:

$$\dot{q}_s'' = \frac{2\,k(T_s - T_1)}{\Delta y} \tag{5-226}$$

Again, note the factor of two in the numerator of Eq. (5-226) indicating that the distance required to conduct from the edge of the duct to the center of node 1 is only $\Delta y/2$.

```
q"_s=k*(T_s-T[1])/(DELTAy/2)            "heat flux"
```

The heat transfer coefficient is defined as:

$$h = \frac{\dot{q}_s''}{(T_s - T_m)} \tag{5-227}$$

and the Nusselt number is computed according to:

$$Nu = \frac{h\,D_h}{k} = \frac{\dot{q}_s''\,D_h}{k(T_s - T_m)} \tag{5-228}$$

```
Nusselt=q"_s*D_h/(k*(T_s-T_m))          "Nusselt number"
```

The Nusselt number is added to the integral table. The Nusselt number as a function of axial position is illustrated in Figure 5-29. Notice that the Nusselt number approaches 7.54 as the flow becomes thermally fully developed; this is consistent with the correlation provided by Shah and London (1978) for a constant temperature duct (discussed in Section 5.2.4).

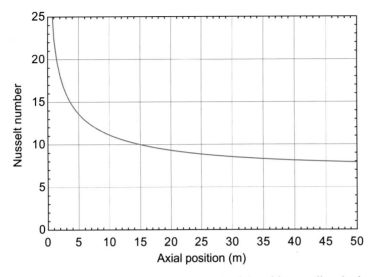

Figure 5-29: Nusselt number as a function of axial position predicted using EES' Integral command.

The Euler Technique

The temperature rates of change for each node, Eqs. (5-217), (5-222), and (5-223) can also be numerically integrated using the Euler technique. This technique is illustrated in MATLAB. The inputs are entered in a script:

```
clear all;

u_m=1;                          % mean velocity (m/s)
H=0.05;                         % duct height (m)
T_in=293.2;                     % inlet fluid temperature (K)
rho=1000;                       % density (kg/m^3)
mu=0.15;                        % viscosity (Pa-s)
k=0.5;                          % conductivity (Pa-s)
c=300;                          % specific heat capacity (J/kg-K)
T_s=473.2;                      % wall temperature (K)
L=50;                           % length of duct (m)
W=1.0;                          % unit width into page (m)
```

The *y*-position and the velocity at each node are set up:

```
% Setup y grid
N=21;                           % number of nodes in y direction (-)
Dy=H/N;                         % distance between nodes (m)
for i=1:N
    y(i)=Dy*(i-1/2);            % position of each node (m)
end

% mean velocity
for i=1:N
    u(i)=6*u_m*(y(i)/H-(y(i)/H)^2);   % velocity at each node (m/s)
end
```

In order to implement Euler's technique, the total length of the duct must be broken into length steps of size:

$$\Delta x = \frac{L}{(M-1)} \qquad (5\text{-}229)$$

where M is the number of length steps. The position of each node in the x-direction is:

$$x_j = (j-1)\,\Delta x \quad \text{for } j = 1..M \qquad (5\text{-}230)$$

```
% setup x grid
M=1001;                                    % number of length steps
Dx=L/(M-1);                                % size of length steps
for j=1:M
  x(j)=(j-1)*Dx;
end
```

The initial temperatures of each node are assigned:

```
% Initial condition
for i=1:N
  T(i,1)=T_in;
end
```

Equations (5-217), (5-222), and (5-223) are integrated using Euler's technique:

$$T_{1,j+1} = T_{1,j} + \frac{k\,\Delta x}{\rho c\,\Delta y^2\,u_1}\,(T_{2,j} + 2\,T_s - 3\,T_{1,j}) \qquad (5\text{-}231)$$

$$T_{i,j+1} = T_{i,j} + \frac{k\,\Delta x}{\rho c\,\Delta y^2\,u_i}\,(T_{i+1,j} + T_{i-1,j} - 2\,T_{i,j}) \quad \text{for } i = 2..(N-1) \qquad (5\text{-}232)$$

$$T_{N,j+1} = T_{N,j} + \frac{k\,\Delta x}{\rho c\,\Delta y^2\,u_N}\,(T_{N-1,j} + 2\,T_s - 3\,T_{N,j}) \qquad (5\text{-}233)$$

```
% step through axial position, x
for j=1:(M-1)
  T(1,j+1)=T(1,j)+k*(T(2,j)+2*T_s-3*T(1,j))*Dx/(rho*c*Dy^2*u(1));
  for i=2:(N-1)
    T(i,j+1)=T(i,j)+k*(T(i+1,j)+T(i-1,j)-2*T(i,j))*Dx/(rho*c*Dy^2*u(i));
  end
  T(N,j+1)=T(N,j)+k*(T(N-1,j)+2*T_s-3*T(N,j))*Dx/(rho*c*Dy^2*u(N));
end
```

Figure 5-30 illustrates the temperature predicted by Euler's method as a function of axial position at various values of y.

The Euler technique has the stability issues discussed previously in the context of 1-D transient conduction problems. Reducing the number of axial steps, M, to a value

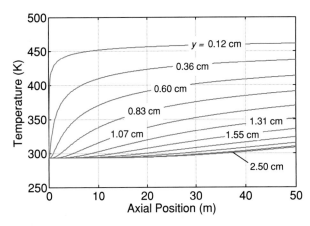

Figure 5-30: Temperature as a function of axial position at various y-locations predicted using the Euler technique with $M = 1001$.

that is less than about 250 leads to oscillations near the wall, as shown in Figure 5-31. Reducing the value of M to less than 160 will lead to a completely unstable solution.

The process of determining the mean temperature, heat flux at the duct surface, heat transfer coefficient, and Nusselt number follows from the discussion in Section 5.5.2:

```
for j=1:M
  T_mean(j)=sum(T(:,j).*u')*Dy/(H*u_m);
  qf(j)=k*(T_s-T(1,j))/(Dy/2);
  htc(j)=qf(j)/(T_s-T_mean(j));
  Nusselt(j)=htc(j)*2*H/k;
end
```

The Crank-Nicolson Technique
In order to mitigate the stability problem illustrated in Figure 5-31, the temperature rates of change for each node, Eqs. (5-217), (5-222), and (5-223) can be integrated using

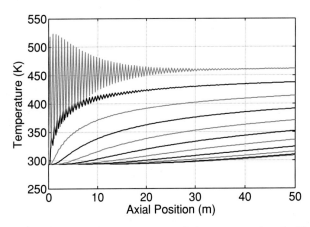

Figure 5-31: Instability near the wall that occurs using the Euler technique with $M = 170$.

the Crank-Nicolson technique. The problem inputs, grid, mean velocity, and initial conditions are setup without change:

```
clear all;

u_m=1;                                  % mean velocity (m/s)
H=0.05;                                 % duct height (m)
T_in=293.2;                             % inlet fluid temperature (K)
rho=1000;                               % density (kg/m^3)
mu=0.15;                                % viscosity (Pa-s)
k=0.5;                                  % conductivity (Pa-s)
c=300;                                  % specific heat capacity (J/kg-K)
T_s=473.2;                              % wall temperature (K)
L=50;                                   % length of duct (m)
W=1.0;                                  % unit length into page (m)

% Setup y grid
N=21;                                   % number of nodes in y direction (-)
Dy=H/N;                                 % distance between nodes (m)
for i=1:N
    y(i)=Dy*(i-1/2);                    % position of each node (m)
end

% mean velocity
for i=1:N
    u(i)=6*u_m*(y(i)/H-(y(i)/H)^2);     % velocity at each node (m/s)
end

% setup x grid
M=170;                                  % number of length steps
Dx=L/(M-1);                             % size of length steps
for j=1:M
    x(j)=(j-1)*Dx;
end

% Initial condition
for i=1:N
    T(i,1)=T_in;
end
```

The Crank-Nicolson method accomplishes each integration step using the rate of change estimated based on the average of its values at the beginning and the end of the length step. The temperature rate of change at the end of the step depends on the temperature at this point, which is not yet known. Consequently, the formula for taking a Crank-Nicolson step is implicit rather than explicit.

$$T_{i,j+1} = T_{i,j} + \left[\frac{dT}{dx}\bigg|_{T=T_{i,j}, x=x_j} + \frac{dT}{dx}\bigg|_{T=T_{i,j+1}, x=x_{j+1}} \right] \frac{\Delta x}{2} \quad \text{for } i = 1...N \quad (5\text{-}234)$$

Substituting Eqs. (5-217), (5-222), and (5-223) into Eq. (5-234) leads to:

$$T_{1,j+1} = T_{1,j} + \frac{k}{\rho c\,\Delta y^2\,u_1}[(T_{2,j} + 2\,T_s - 3\,T_{1,j}) + (T_{2,j+1} + 2\,T_s - 3\,T_{1,j+1})]\frac{\Delta x}{2}$$

(5-235)

$$T_{i,j+1} = T_{i,j} + \frac{k}{\rho c\,\Delta y^2\,u_i}[(T_{i+1,j} + T_{i-1,j} - 2\,T_{i,j}) + (T_{i+1,j+1} + T_{i-1,j+1} - 2\,T_{i,j+1})]\frac{\Delta x}{2}$$

for $i = 2 \ldots (N-1)$ (5-236)

$$T_{N,j+1} = T_{N,j} + \frac{k}{\rho c\,\Delta y^2\,u_N}[(T_{N-1,j} + 2\,T_s - 3\,T_{N,j}) + (T_{N-1,j+1} + 2\,T_s - 3\,T_{N,j+1})]\frac{\Delta x}{2}$$

(5-237)

Equations (5-235) through (5-237) are a set of N linear equations in the unknown temperatures $T_{i,j+1}$ where $i = 1..N$. These equations must be placed into matrix format in order to move forward a length step:

$$\underline{\underline{A}}\,\underline{X} = \underline{b}$$ (5-238)

It is important to clearly specify the order in which the equations are placed into the matrix $\underline{\underline{A}}$ and the unknown temperatures are placed into the vector $\underline{X}$. The most logical method for placing the unknowns into $\underline{X}$ is:

$$\underline{X} = \begin{bmatrix} X_1 = T_{1,j+1} \\ X_2 = T_{2,j+1} \\ \ldots \\ X_N = T_{N,j+1} \end{bmatrix}$$ (5-239)

Therefore $T_{i,j+1}$ corresponds to element i of the vector $\underline{X}$. The most logical method for placing the equations into $\underline{\underline{A}}$ is:

$$\underline{\underline{A}} = \begin{bmatrix} \text{row } 1 = \text{control volume 1 equation} \\ \text{row } 2 = \text{control volume 2 equation} \\ \text{row } 3 = \text{control volume 3 equation} \\ \ldots \\ \text{row } N = \text{control volume } N \text{ equation} \end{bmatrix}$$ (5-240)

Therefore, the equation derived based on the control volume for node i corresponds to row i of the matrix $\underline{\underline{A}}$. Equations (5-235) through (5-237) are rearranged so that the coefficients multiplying the unknowns and the constants for the linear equations are clear:

$$T_{1,j+1}\underbrace{\left[1 + \frac{3\,k\,\Delta x}{2\,\rho c\,\Delta y^2\,u_1}\right]}_{A_{1,1}} + T_{2,j+1}\underbrace{\left[-\frac{k\,\Delta x}{2\,\rho c\,\Delta y^2\,u_1}\right]}_{A_{1,2}} = \underbrace{T_{1,j} + \frac{k\,\Delta x}{2\,\rho c\,\Delta y^2\,u_1}[T_{2,j} + 4\,T_s - 3\,T_{1,j}]}_{b_1}$$

(5-241)

$$T_{i,j+1} \underbrace{\left[1 + \frac{k\,\Delta x}{\rho\,c\,\Delta y^2\,u_i}\right]}_{A_{i,i}} + T_{i+1,j+1} \underbrace{\left[-\frac{k\,\Delta x}{2\,\rho\,c\,\Delta y^2\,u_i}\right]}_{A_{i,i+1}} + T_{i-1,j+1} \underbrace{\left[-\frac{k\,\Delta x}{2\,\rho\,c\,\Delta y^2\,u_i}\right]}_{A_{i,i-1}}$$

(5-242)

$$= \underbrace{T_{i,j} + \frac{k\,\Delta x}{2\,\rho\,c\,\Delta y^2\,u_i}[T_{i+1,j} + T_{i-1,j} - 2\,T_{i,j}]}_{b_i} \quad \text{for } i = 2 \ldots (N-1)$$

$$T_{N,j+1} \underbrace{\left[1 + \frac{3\,k\,\Delta x}{2\,\rho\,c\,\Delta y^2\,u_N}\right]}_{A_{N,N}} + T_{N-1,j+1} \underbrace{\left[-\frac{k\,\Delta x}{2\,\rho\,c\,\Delta y^2\,u_N}\right]}_{A_{N,N-1}}$$

(5-243)

$$= \underbrace{T_{N,j} + \frac{k\,\Delta x}{2\,\rho\,c\,\Delta y^2\,u_N}[T_{N-1,j} + 4\,T_s - 3\,T_{N,j}]}_{b_N}$$

The matrix $\underline{\underline{A}}$ and vector $\underline{b}$ are initialized:

```
A=spalloc(N,N,3*N);                    % initialize A
b=zeros(N,1);                          % initialize b
```

For this problem, the entries in the matrix $\underline{\underline{A}}$ do not depend on the length step or include any temperature dependent properties. Therefore, $\underline{\underline{A}}$ can be constructed just once and used without modification to move through each length step:

```
% setup A matrix
A(1,1)=1+3*k*Dx/(2*rho*c*Dy*2*u(1));
A(1,2)=-k*Dx/(2*rho*c*Dy*2*u(1));
for i=2:(N-1)
   A(i,i)=1+k*Dx/(rho*c*Dy*2*u(i));
   A(i,i+1)=-k*Dx/(2*rho*c*Dy*2*u(i));
   A(i,i-1)=-k*Dx/(2*rho*c*Dy*2*u(i));
end
A(N,N)=1+3*k*Dx/(2*rho*c*Dy*2*u(N));
A(N,N-1)=-k*Dx/(2*rho*c*Dy*2*u(N));
```

while the vector $\underline{b}$ must be reconstructed during each step because the elements of $\underline{b}$ depend on the temperatures:

```
%step through space
for j=1:(M-1)
   b(1)=T(1,j)+k*Dx*(T(2,j)+4*T_s-3*T(1,j))/(2*rho*c*Dy^2*u(1));
   for i=2:(N-1)
      b(i)=T(i,j)+k*Dx*(T(i+1,j)+T(i-1,j)-2*T(i,j))/(2*rho*c*Dy^2*u(i));
   end
   b(N)=T(N,j)+k*Dx*(T(N-1,j)+4*T_s-3*T(N,j))/(2*rho*c*Dy^2*u(N));
   T(:,j+1)=A/b;
end
```

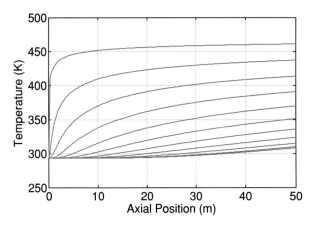

Figure 5-32: Temperature as a function of x at various values of y predicted using the Crank-Nicolson technique with $M = 170$.

The temperature as a function of axial position at various values of y is shown in Figure 5-32. Figure 5-32 was generated with $M = 170$ length steps, the value of M that caused Euler's technique to become unstable in Figure 5-31.

MATLAB's Ordinary Differential Equation Solvers

The use of MATLAB's suite of integration routines to integrate a system of state equations through time is discussed in Section 3.8.2. These same functions can be used to integrate Eqs. (5-217), (5-222), and (5-223) along the length of the duct; length replaces time as the independent variable for this process. A function dTdx_functionv must be setup in order to compute the state equations that are required for the integration. This function returns a vector that contains the rate of change of each of the nodal temperatures. The first two inputs to the function dTdx_functionv are the position x (a scalar) and a vector of nodal temperatures, T. The function dTdt_functionv is defined below:

```
function[dTdx]=dTdx_functionv(x,T,Dy,k,rho,c,u,T_s)
    % Inputs:
    % x − position (m)
    % T − vector of nodal temperatures (K)
    % Dy − vertical spacing (m)
    % k − thermal conductivity (W/m-K)
    % rho − density (kg/m^3)
    % c − specific heat capacity (J/kg-K)
    % u − vector of nodal velocities (m/s)
    % T_s − surface temperature (K)

    [N,g]=size(T);                          % determine number of nodes
    dTdx=zeros(N,1);                        % initialize dTdx
    dTdx(1)=k*(T(2)+2*T_s-3*T(1))/(rho*c*Dy^2*u(1));
    for i=2:(N-1)
     dTdx(i)=k*(T(i+1)+T(i-1)-2*T(i))/(rho*c*Dy^2*u(i));
    end
    dTdx(N)=k*(T(N-1)+2*T_s-3*T(N))/(rho*c*Dy^2*u(N));
end
```

The problem is solved in a MATLAB script; the inputs, grid, and velocity distribution are entered:

```
clear all;

u_m=1;                                          % mean velocity (m/s)
H=0.05;                                         % duct height (m)
T_in=293.2;                                     % inlet fluid temperature (K)
rho=1000;                                       % density (kg/m^3)
mu=0.15;                                        % viscosity (Pa-s)
k=0.5;                                          % conductivity (Pa-s)
c=300;                                          % specific heat capacity (J/kg-K)
T_s=473.2;                                      % wall temperature (K)
L=50;                                           % length of duct (m)
W=1.0;                                          % unit length into page (m)

% Setup y grid
N=21;                                           % number of nodes in y direction (-)
Dy=H/N;                                         % distance between nodes (m)
for i=1:N
    y(i)=Dy*(i-1/2);                            % position of each node (m)
end

% velocity
for i=1:N
    u(i)=6*u_m*(y(i)/H-(y(i)/H)^2);            % velocity at each node (m/s)
end
```

The **odeset** function is used to specify the **OPTIONS** vector that controls the integration process:

```
OPTIONS=odeset('RelTol',1e-6);
```

The **ode45** function is used to integrate the state equations forward through the duct. The call to the function **dTdx_functionv** is mapped to the two inputs required by the **ode45** solver, x and T, as discussed in Section 3.2.2:

```
[x,T]=ode45(@(x,T) dTdx_functionv(x,T,Dy,k,rho,c,u,T_s),[0,L],T_in*ones(N,1),OPTIONS);
```

The process of computing the mean temperature, heat flux, heat transfer coefficient and Nusselt number using the numerical solution follows from the discussion in Section 5.5.2. First, the number of length steps used for the integration is determined:

```
[M,g]=size(T);               % determine number of length steps used
```

The mean temperature is the velocity-weighted average temperature at each axial position:

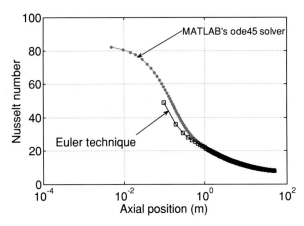

Figure 5-33: Nusselt number as a function of axial position predicted using Euler's technique and MATLAB's native ode solver.

```
for j=1:M
  T_mean(j)=sum(T(j,:).*u)*Dy/(H*u_m);
```

The heat flux is obtained from the thermal resistance of the node at the wall:

```
qf(j)=k*(T_s-T(j,1))/(Dy/2);
```

The heat transfer and Nusselt number follow:

```
  htc(j)=qf(j)/(T_s-T_mean(j));
  Nusselt(j)=htc(j)*2*H/k;
end
```

Figure 5-33 illustrates the Nusselt number as a function of x predicted using Euler's technique, discussed earlier in this section, and using the solver ode45. Notice that Figure 5-33 has a logarithmic x-axis in order to clearly illustrate the problems associated with Euler's technique near the entrance to the heated region of the duct. The MATLAB ode solver is able to select very small length steps in this region and therefore maintain its high accuracy.

5.5.3 Hydrodynamically Fully Developed Turbulent Flow

Several models for the eddy diffusivity of momentum are presented in Section 4.7; these models can be used to account for turbulent transport in numerical models. The process of developing a numerical model of a turbulent internal flow is illustrated in the context of the circular pipe flow problem shown in Figure 5-34.

Fluid with properties $\rho = 1000 \, \text{kg/m}^3$, $\mu = 0.001$ Pa-s, $k = 0.6$ W/m-K, and $c = 4200$ J/kg-K flows through the pipe with a mean velocity of $u_m = 2.5$ m/s. The flow is hydrodynamically fully developed and has a uniform temperature, $T_{in} = 20°$C, when it enters

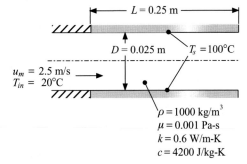

Figure 5-34: Turbulent internal flow through a pipe.

a section that is $L = 0.25$ m long with a constant surface temperature $T_s = 100°C$. The inner diameter of the pipe is $D = 0.025$ m and the pipe surface is smooth.

The inputs are entered in EES:

```
$UnitSystem SI MASS RAD PA K J
$Tabstops 0.2 0.4 0.6 3.5 in

rho=1000 [kg/m^3]                        "density"
mu=0.001 [Pa-s]                          "viscosity"
k=0.6 [W/m-K]                            "conductivity"
c=4200 [J/kg-K]                          "specific heat capacity"
u_m=2.5 [m/s]                            "mean velocity"
D=0.025 [m]                              "diameter"
T_in=converttemp(C,K,20 [C])            "inlet temperature"
T_s=converttemp(C,K,100 [C])            "surface temperature"
L=0.25 [m]                               "length of heated section"
nu=mu/rho                                "kinematic viscosity"
alpha=k/(rho*c)                          "thermal diffusivity"
Pr=nu/alpha                              "Prandtl number"
```

The Reynolds number is calculated:

$$Re_D = \frac{\rho D u_m}{\mu} \tag{5-244}$$

```
Re=rho*D*u_m/mu                          "Reynolds number"
```

which leads to $Re_D = 6.25 \times 10^4$; therefore, the flow is turbulent. The velocity profile is given, approximately, by the universal velocity profile discussed in Section 4.7.5. The universal velocity profile is provided in terms of inner coordinates. Therefore the friction velocity, u^*, and the length scale characterizing the viscous sublayer, $L_{char,vs}$, must be computed in order to obtain the universal velocity profile in dimensional form. The friction velocity is based on the surface shear stress, τ_s, which is related to the pressure gradient according to the momentum balance discussed in Section 5.1.2 and given by Eq. (5-30) for hydrodynamically fully developed flow.

$$\tau_s = \frac{A_c}{per}\left(-\frac{dp}{dx}\right) \tag{5-245}$$

where A_c and *per* are the cross-sectional area and perimeter of the duct:

$$A_c = \pi \frac{D^2}{4} \tag{5-246}$$

$$per = \pi D \tag{5-247}$$

```
A_c=pi*D^2/4                              "cross-sectional area"
per=pi*D                                  "perimeter"
```

The pressure gradient is related to the friction factor, f, according to:

$$\left(-\frac{dp}{dx}\right) = \frac{\rho u_\infty^2}{2} \frac{f}{D} \tag{5-248}$$

Substituting Eq. (5-248) into Eq. (5-245) leads to:

$$\tau_s = \frac{A_c}{per} \frac{\rho u_\infty^2}{2} \frac{f}{D} \tag{5-249}$$

The friction factor can be estimated using the internal flow correlations presented in Section 5.2.3 and accessed using the **PipeFlow_N** procedure:

```
call PipeFlow_N(Re,Pr,9999 [-],0 [-]: Nusselt_T,Nusselt_H,f)     "estimate friction factor"
tau_s=A_c*f*rho*u_m^2/(2*per*D)                                  "surface shear stress"
```

Note that the flow is assumed to be hydrodynamically fully developed and therefore an arbitrary, large value of L/D_h (9999) is used in the call to **PipeFlow_N**. The values of u^* and $L_{char,vs}$ are computed:

$$u^* = \sqrt{\frac{\tau_s}{\rho}} \tag{5-250}$$

$$L_{char,vs} = \frac{\upsilon}{u^*} \tag{5-251}$$

```
u_star=sqrt(tau_s/rho)         "friction velocity"
L_char_vs=nu/u_star            "length scale characterizing the viscous sublayer"
```

The nodes and control volumes are defined in the computational domain as shown in Figure 5-35. The universal velocity distribution is expressed in terms of the distance from the wall, y, rather than the radius, r. Because the velocity and temperature gradients associated with a turbulent flow are so sharp near the wall, the nodes must be concentrated in this region. One methodology for accomplishing this places the first node at $y^+ = 1$. The inner position of each subsequent node is distributed logarithmically:

$$y_i^+ = MF^{i-1} \quad \text{for } i = 1..N \tag{5-252}$$

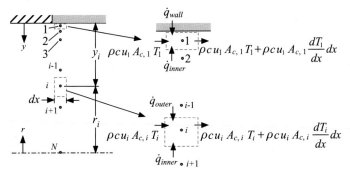

Figure 5-35: Control volumes used for numerical simulation of internal turbulent flow.

where MF is a multiplicative factor that is greater than 1. The last node (node N) should be placed at the center line of the duct, which corresponds to an inner position:

$$y_N^+ = R^+ = \frac{D}{2\,L_{char,vs}} \tag{5-253}$$

Substituting Eq. (5-253) into Eq. (5-252) leads to:

$$\frac{D}{2\,L_{char,vs}} = MF^{N-1} \tag{5-254}$$

Solving Eq. (5-254) for MF leads to:

$$MF = \left(\frac{D}{2\,L_{char,vs}}\right)^{1/(N-1)} \tag{5-255}$$

```
N=32 [-]                                "number of nodes distributed across the pipe"
MF=(D/(2*L_char_vs))^(1/(N-1))          "multiplication factor"
duplicate i=1,N
  y_plus[i]=MF^(i-1)                     "inner position of each node"
end
```

The actual distance of each node from the wall, y_i, is computed:

$$y_i = L_{char,vs}\, y_i^+ \quad \text{for } i = 1..N \tag{5-256}$$

and the radial location of each node is computed:

$$r_i = \frac{D}{2} - y_i \quad \text{for } i = 1..N \tag{5-257}$$

```
duplicate i=1,N
  y[i]=y_plus[i]*L_char_vs              "distance from the wall"
  r[i]=(D/2)-y[i]                        "radial position"
end
```

The inner velocity at each node, u^+, can be estimated using any of the universal velocity distributions that are presented in Table 4-2. The Prandtl-Taylor two-layer model derived in Section 4.7.5 is the simplest (and least accurate) option:

$$u_i^+ = \begin{cases} y_i^+ & \text{if } y^+ < 11.5 \\ \\ 2.44 \ln\left(y_i^+\right) + 5.5 & \text{if } y^+ > 11.5 \end{cases} \qquad \text{for } i = 1..N \qquad (5\text{-}258)$$

The velocity at the node is the product of the inner velocity and the friction velocity:

$$u_i = u_i^+ \, u^* \quad \text{for } i = 1..N \qquad (5\text{-}259)$$

The logic associated with deciding between the two choices in Eq. (5-258) is accomplished using the if command in EES; the if command has the following protocol:

f=if(a,b,c,d,e)

The variable f is assigned the value c if a < b, the value d if a = b, and the value e if a > b.

```
duplicate i=1,N
  u_plus[i]=if(y_plus[i],11.5,y_plus[i],y_plus[i],2.44*ln(y_plus[i])+5.5)
                            "Prandtl-Taylor form of the universal velocity distribution"
  u[i]=u_plus[i]*u_star          "velocity at each node"
end
```

The friction factor estimated using the correlations for turbulent flow is not completely consistent with the universal velocity distribution. That is, the mean velocity associated with integrating the universal velocity distribution across the pipe cross-section will not be exactly the same as the mean velocity specified in the problem and used to compute the friction velocity (although they should be close). The mean velocity of the flow associated with the universal velocity distribution can be obtained from:

$$u_m = \frac{1}{A_c} \int\limits_{A_c} u \, dA_c \qquad (5\text{-}260)$$

or, numerically:

$$u_m = \frac{1}{A_c} \sum_{i=1}^{N} u_i \, A_{c,i} \qquad (5\text{-}261)$$

where $A_{c,i}$ is the cross-sectional area for flow associated with each control volume. For the control volume surrounding node 1 (adjacent to the pipe wall, see Figure 5-35), the cross-sectional area is:

$$A_{c,1} = \pi \left[\left(\frac{D}{2}\right)^2 - \left(\frac{r_1 + r_2}{2}\right)^2 \right] \qquad (5\text{-}262)$$

The cross-sectional area for each of the internal control volumes is:

$$A_{c,i} = \pi \left[\left(\frac{r_i + r_{i-1}}{2}\right)^2 - \left(\frac{r_i + r_{i+1}}{2}\right)^2 \right] \quad \text{for } i = 2..(N-1) \qquad (5\text{-}263)$$

and the cross-sectional area of the control volume surrounding node N (at the pipe center line) is:

$$A_{c,N} = \pi \left(\frac{r_{N-1}}{2}\right)^2 \qquad (5\text{-}264)$$

```
A_c[1]=pi*((D/2)^2-((r[1]+r[2])/2)^2)                      "cross-sectional area for node 1"
duplicate i=2,(N-1)
  A_c[i]=pi*(((r[i]+r[i-1])/2)^2-((r[i]+r[i+1])/2)^2)
                                                           "cross-sectional area for inner nodes"
end
A_c[N]=pi*(r[N-1]/2)^2                                      "cross-sectional area for node N"
duplicate i=1,N
  uA[i]=u[i]*A_c[i]                                         "volume flow through each control volume"
end
u_m_c=sum(uA[i],i=1,N)/A_c                                  "calculated mean velocity"
```

The result of the calculation is $u_m = 2.462$ m/s, which is close to the specified mean velocity in the problem statement (2.5 m/s) but not equal to it. In order to obtain a self-consistent velocity distribution, update the guess values and comment out the calculation of the friction factor. Instead, let EES iterate to determine the friction factor that provides the appropriate mean velocity:

```
{call PipeFlow_N(Re,Pr,9999 [-],0 [-]: Nusselt_T,Nusselt_H,f)    "estimate friction factor"}
u_m=u_m_c       "require that the calculated and specified mean velocities are equal"
```

With the velocity distribution established, it is possible to use any of the eddy diffusivity models listed in Table 4-3 to model the energy transport associated with turbulent eddies. Here, the model for the eddy diffusivity of momentum that is related to the Prandtl-Taylor velocity distribution is used:

$$\left(\frac{\varepsilon_M}{\upsilon}\right)_i = \begin{cases} 0 & \text{if } y_i^+ < 11.5 \\ 0.41\, y_i^+ - 1 & \text{if } y_i^+ > 11.5 \end{cases} \quad \text{for } i = 1..N \tag{5-265}$$

```
duplicate i=1,N
  epsilonMovernu[i]=if(y_plus[i],11.5,0,0,0.41*y_plus[i]-1)
    "eddy diffusivity of momentum based on Prandtl-Taylor model"
end
```

The eddy diffusivity of momentum coupled with a turbulent Prandtl number is used to determine the effective conductivity of the fluid (k_{eff}). The effective conductivity represents the simultaneous transport of energy by molecular diffusion (k) and turbulent eddies (k_{turb}).

$$k_{eff,i} = k + \underbrace{\frac{\upsilon \rho c}{Pr_{turb}} \left(\frac{\varepsilon_M}{\upsilon}\right)_i}_{k_{turb,i}} \quad \text{for } i = 1..N \tag{5-266}$$

where Pr_{turb} is taken to be 0.9. Note that k_{eff} is a function of radial location because it depends on the local flow conditions.

```
Pr_turb=0.9 [-]                                            "turbulent Prandtl number"
duplicate i=1,N
  k_eff[i]=k+nu*rho*c*epsilonMovernu[i]/Pr_turb            "effective conductivity"
end
```

The state equations are derived using the same technique discussed in Section 5.5.2 for a laminar internal flow. A control volume for node 1, adjacent to the pipe wall, is shown in Figure 5-35 and suggests the energy balance:

$$\rho c u_1 A_{c,1} T_1 + \dot{q}_{inner} + \dot{q}_{wall} = \rho c u_1 A_{c,1} T_1 + \rho c u_1 A_{c,1} \frac{dT_1}{dx} dx \qquad (5\text{-}267)$$

The heat transfer rate from node 2 is approximated according to:

$$\dot{q}_{inner} = \frac{(k_{eff,1} + k_{eff,2})}{2} 2\pi \frac{(r_1 + r_2)}{2} dx \frac{(T_2 - T_1)}{(r_1 - r_2)} \qquad (5\text{-}268)$$

Note that the cross-sectional area for conduction and effective conductivity are evaluated at the interface between control volumes 1 and 2 in order to avoid artificial destruction/generation of energy. The heat transfer rate from the wall is approximated according to:

$$\dot{q}_{wall} = k_{eff,1} \pi D \, dx \frac{(T_s - T_1)}{\left(\dfrac{D}{2} - r_1\right)} \qquad (5\text{-}269)$$

Substituting Eqs. (5-268) and (5-269) into Eq. (5-267) leads to:

$$k_{eff,1} \pi D \, dx \frac{(T_s - T_1)}{\left(\dfrac{D}{2} - r_1\right)} + \frac{(k_{eff,1} + k_{eff,2})}{2} 2\pi \frac{(r_1 + r_2)}{2} dx \frac{(T_2 - T_1)}{(r_1 - r_2)} = \rho c u_1 A_{c,1} \frac{dT_1}{dx} dx$$
$$(5\text{-}270)$$

Solving for the rate of change of T_1:

$$\frac{dT_1}{dx} = \left[k_{eff,1} \pi D \frac{(T_s - T_1)}{\left(\dfrac{D}{2} - r_1\right)} + \frac{(k_{eff,1} + k_{eff,2})}{2} 2\pi \frac{(r_1 + r_2)}{2} \frac{(T_2 - T_1)}{(r_1 - r_2)} \right] \frac{1}{\rho c u_1 A_{c,1}}$$
$$(5\text{-}271)$$

```
dTdx[1]=(((k_eff[1]+k_eff[2])/2)*2*pi*((r[1]+r[2])/2)*(T[2]-T[1])/(r[1]-r[2])&
    +(k_eff[1])*pi*D*(T_s-T[1])/(D/2-r[1]))/(A_c[1]*rho*c*u[1])
```
"state eq. for node at wall"

A control volume for an arbitrary internal is shown in Figure 5-35 and suggests the energy balance:

$$\rho c u_i A_{c,i} T_i + \dot{q}_{inner} + \dot{q}_{outer} = \rho c u_i A_{c,i} T_i + \rho c u_i A_{c,i} \frac{dT_i}{dx} dx \qquad (5\text{-}272)$$

The heat transfer rate from node $i + 1$ ($\dot{q}_{inner}$) is approximated according to:

$$\dot{q}_{inner} = \frac{(k_{eff,i} + k_{eff,i+1})}{2} 2\pi \frac{(r_i + r_{i+1})}{2} dx \frac{(T_{i+1} - T_i)}{(r_i - r_{i+1})} \qquad (5\text{-}273)$$

The heat transfer rate from node $i - 1$ ($\dot{q}_{outer}$) is approximated according to:

$$\dot{q}_{outer} = \frac{(k_{eff,i} + k_{eff,i-1})}{2} 2\pi \frac{(r_i + r_{i-1})}{2} dx \frac{(T_{i-1} - T_i)}{(r_{i-1} - r_i)} \qquad (5\text{-}274)$$

Substituting Eqs. (5-273) and (5-274) into Eq. (5-272) leads to:

$$\frac{(k_{eff,i} + k_{eff,i+1})}{2} \, 2\pi \, \frac{(r_i + r_{i+1})}{2} dx \frac{(T_{i+1} - T_i)}{(r_i - r_{i+1})}$$

$$+ \frac{(k_{eff,i} + k_{eff,i-1})}{2} \, 2\pi \, \frac{(r_i + r_{i-1})}{2} dx \frac{(T_{i-1} - T_i)}{(r_{i-1} - r_i)} = \rho \, c \, u_i \, A_{c,i} \, \frac{dT_i}{dx} dx$$

(5-275)

Solving for the rate of change of T_i:

$$\frac{dT_i}{dx} = \left[\frac{(k_{eff,i} + k_{eff,i+1})}{2} \, 2\pi \, \frac{(r_i + r_{i+1})}{2} \frac{(T_{i+1} - T_i)}{(r_i - r_{i+1})} + \frac{(k_{eff,i} + k_{eff,i-1})}{2} \right.$$

$$\left. \times 2\pi \, \frac{(r_i + r_{i-1})}{2} \frac{(T_{i-1} - T_i)}{(r_{i-1} - r_i)} \right] \frac{1}{\rho \, c \, u_i \, A_{c,i}} \quad \text{for } i = 2..(N-1) \quad (5\text{-}276)$$

```
duplicate i=2,(N-1)
  dTdx[i]=(((k_eff[i]+k_eff[i+1])/2)*2*pi*((r[i]+r[i+1])/2)*(T[i+1]-T[i])/(r[i]-r[i+1]))&
    +((k_eff[i]+k_eff[i-1])/2)*2*pi*((r[i-1]+r[i])/2)*(T[i-1]-T[i])/(r[i-1]-r[i]))&
    /(A_c[i]*rho*c*u[i])                                    "state eq. for internal nodes"
end
```

A similar process for node N (at the center line) leads to:

$$\frac{dT_N}{dx} = \left[\frac{(k_{eff,N} + k_{eff,N-1})}{2} \, 2\pi \, \frac{(r_N + r_{N-1})}{2} \frac{(T_{N-1} - T_N)}{(r_{N-1} - r_N)} \right] \frac{1}{\rho \, c \, u_N \, A_{c,N}} \quad (5\text{-}277)$$

```
dTdx[N]=(((k_eff[N]+k_eff[N-1])/2)*2*pi*((r[N]+r[N-1])/2)*(T[N-1]-T[N])/(r[N-1]-r[N]))&
  /(A_c[N]*rho*c*u[N])                                    "state eq. for node at center"
```

The state equations are integrated forward along the length of pipe using EES' Integral command:

```
duplicate i=1,N
  T[i]=T_in+Integral(dTdx[i],x,0,L,0)                      "integrate rate equations"
end
```

and the results are saved in an integral table:

```
DELTAx_table=L/250                                         "interval to save results"
$IntegralTable x:DELTAx_table,T[1..N]
```

Figure 5-36 illustrates the temperature at various values of radius (or inner position) as a function of axial position.

The mean temperature of the flow can be computed according to:

$$T_m = \frac{1}{A_c \, u_m} \int_{A_c} u \, T \, dA_c \quad (5\text{-}278)$$

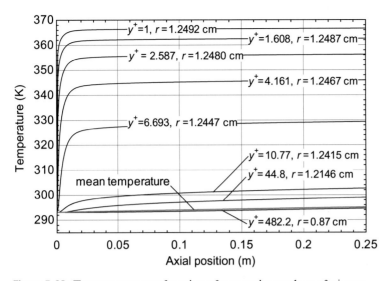

Figure 5-36: Temperature as a function of x at various values of y^+ or r.

or numerically:

$$T_m = \frac{1}{A_c\,u_m}\sum_{i=1}^{N} u_i\,T_i\,A_{c,i} \qquad (5\text{-}279)$$

```
duplicate i=1,N
   uAT[i]=u[i]*A_c[i]*T[i]                                    "energy carried by flow"
end
T_m=sum(uAT[i],i=1,N)/(A_c*u_m)                               "mean temperature"
$IntegralTable x:DELTAx_table,T[1..N],T_m
```

The mean temperature is also shown in Figure 5-36. The heat flux at the wall is computed according to:

$$\dot{q}_s'' = k_{eff,1}\frac{(T_s - T_1)}{\left(\dfrac{D}{2} - r_1\right)} \qquad (5\text{-}280)$$

and used to compute the local Nusselt number:

$$Nu = \frac{h\,k}{D} = \frac{\dot{q}_s''\,k}{(T_s - T_m)\,D} \qquad (5\text{-}281)$$

```
q"_s=k_eff[1]*(T_s-T[1])/(D/2-r[1])                           "surface heat flux"
Nusselt=q"_s*D/(k*(0.001 [K]+T_s-T_m))                        "Nusselt number"
$IntegralTable x:DELTAx_table,T[1..N],T_m,Nusselt
```

Figure 5-37 illustrates the local Nusselt number as a function of axial position in the duct.

The temperature distribution predicted by the numerical model can be compared to the thermal law of the wall that is discussed in Section 4.7.9. The eddy temperature

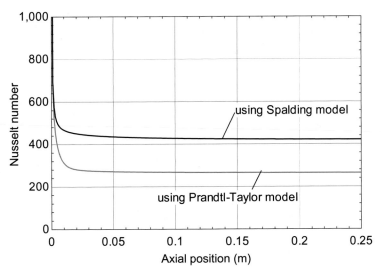

Figure 5-37: Nusselt number as a function of position obtained using the Prandtl-Taylor and Spalding models for the universal velocity distribution and eddy diffusivity of momentum.

fluctuation, T^*, is computed:

$$T^* = \frac{\dot{q}_s''}{\rho c u^*} \qquad (5\text{-}282)$$

and used to compute the inner temperature difference at each node:

$$\theta_i^+ = \frac{(T_s - T_i)}{T^*} \qquad (5\text{-}283)$$

```
T_star=q"_s/(rho*c*u_star)                          "eddy temperature fluctuation"
duplicate i=1,N
    theta_plus[i]=(T_s-T[i])/T_star                 "inner temperature difference"
end
```

The thermal law of the wall derived in Section 4.7.9 is given by:

$$\theta_i^+ = \begin{cases} Pr\, y^+ & \text{if } y^+ < 11.5 \\[2ex] 11.5\,Pr + \dfrac{Pr_{turb}}{\kappa} \ln\left(\dfrac{\dfrac{1}{Pr} + \dfrac{1}{Pr_{turb}}(\kappa\, y^+ - 1)}{\dfrac{1}{Pr} + \dfrac{1}{Pr_{turb}}(11.5\,\kappa - 1)} \right) & \text{if } y^+ > 11.5 \end{cases} \quad \text{for } i = 1..N$$

$$(5\text{-}284)$$

```
duplicate i=1,N
    theta_plus_TLW[i]=IF(y_plus[i],11.5,Pr*y_plus[i],Pr*y_plus[i],&
        11.5*Pr+Pr_turb*ln(abs(1/Pr+(0.41*y_plus[i]-1)/Pr_turb)/(1/Pr+(11.5*0.41-1)/Pr_turb))/0.41)
        "thermal law of the wall"
end
```

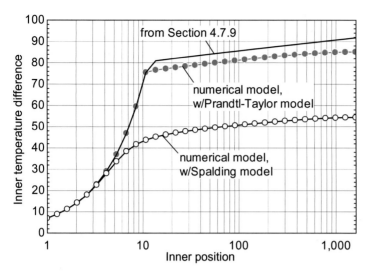

Figure 5-38: Inner temperature difference as a function of inner position. Also shown is the result from Section 4.7.9.

The inner temperature difference predicted by the numerical model and Eq. (5-284) are shown as a function of inner position in Figure 5-38. Note the good agreement between the results from the numerical model and the solution derived in Section 4.7.9; this is because both techniques use the Prandtl-Taylor model.

It is relatively easy to use a more sophisticated universal velocity distribution and therefore a more accurate model of the eddy diffusivity of momentum within the numerical model. For example, the Spalding model, presented in Tables 4-2 and 4-3, can be used:

$$y^+ = u^+ + 0.11408 \left[\exp(0.41\,u^+) - 1 - 0.41\,u^+ - \frac{(0.41\,u^+)^2}{2} - \frac{(0.41\,u^+)^3}{6} - \frac{(0.41\,u^+)^4}{24} \right]$$

$$(5\text{-}285)$$

$$\frac{\varepsilon_M}{\upsilon} = 0.0526 \left[\exp(0.41\,u^+) - 1 - 0.41\,u^+ - \frac{(0.41\,u^+)^2}{2} - \frac{(0.41\,u^+)^3}{6} \right] \qquad (5\text{-}286)$$

The Prandtl-Taylor model is commented out and Eqs. (5-285) and (5-286) are used in their place:

```
duplicate i=1,N
    {u_plus[i]=IF(y_plus[i],11.5,y_plus[i],y_plus[i],2.44*ln(y_plus[i])+5.5)
    "Prandtl-Taylor form of the universal velocity distribution"}
    y_plus[i]=u_plus[i]+0.11408*(exp(0.41*u_plus[i])-1-0.41*u_plus[i]-&
    (0.41*u_plus[i])^2/2-(0.41*u_plus[i])^3/6-(0.41*u_plus[i])^4/24)
        "Spalding universal velocity distribution"
    u[i]=u_plus[i]*u_star                                     "velocity at each node"
end

duplicate i=1,N
{   epsilonMovernu[i]=IF(y_plus[i],11.5,0,0,0.41*y_plus[i]-1)
    "eddy diffusivity of momentum based on Prandtl-Taylor model"}
```

```
      epsilonMovernu[i]=0.0526*(exp(0.41*u_plus[i])-1-0.41*u_plus[i]-&
         (0.41*u_plus[i])^2/2-(0.41*u_plus[i])^3/6)            "Spalding model"
end
```

The Nusselt number and inner temperature distribution obtained using the Spalding model are shown in Figure 5-37 and Figure 5-38, respectively. The results obtained with the Spalding model are much more accurate. The fully developed Nusselt number associated with the Gnielinski correlation (obtained using EES' convection library):

```
call PipeFlow_N(Re,Pr,9999 [-], 0 [-]: Nusselt_T_fd,Nusselt_H_fd,f_fd)
   "use correlations to determine fully developed Nusselt number"
```

is equal to $Nu_T = 399.9$ which agrees well with the result obtained using the numerical model with Spalding's eddy diffusivity model. The numerical model that uses the simpler Prandtl-Taylor eddy diffusivity model is 25% in error relative to the Gnielinski correlation (see Figure 5-37).

Chapter 5: Internal Convection

Many more problems can be found on the website (www.cambridge.org/nellisandklein).

Internal Flow Concepts

5–1 You have been asked to help interpret some measured data for flow through a tube with inner diameter D. Figure P5-1 illustrates the heat transfer coefficient h measured in the thermally fully developed region of the tube as a function of the tube diameter; note that the mass flow rate of fluid ($\dot{m}$), the type of fluid, and all other aspects of the experiment are not changed for these measurements.

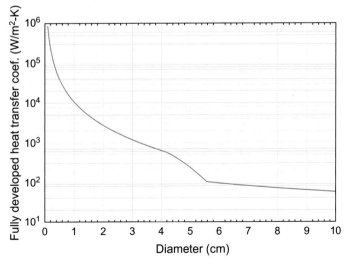

Figure P5-1: Heat transfer coefficient as a function of diameter (for a constant fluid mass flow rate and fluid type and all other aspects of the problem held constant.)

a.) Explain in a few sentences the abrupt change in the heat transfer coefficient observed that occurs at approximately $D = 5.5$ cm.

b.) Explain in a few sentences why the heat transfer is inversely proportional to diameter for diameters above about $D = 5.5$ cm; that is, why is it true that $h \propto D^{-1}$ for $D > 5.5$ cm?

c.) Sketch on your expectation for how Figure P5-1 would change if the roughness of the tube wall is increased dramatically.

5–2 Figure P5-2 shows the flow of a fluid with a low Prandtl number, $Pr < 1$, through a pipe.

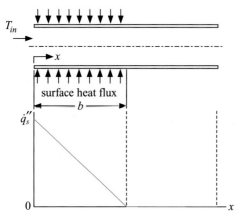

Figure P5-2: Pipe with a surface heat flux that depends on position.

The fluid becomes thermally fully developed at location $x = b$. The flow of the fluid is laminar.

a.) Sketch the thermal and momentum boundary layer thickness as a function of position (δ_t and δ_m – be sure to clearly show which is which). Label the hydrodynamic and thermal entry lengths, $x_{fd,t}$ and $x_{fd,h}$, in your sketch. Show the location $x = b$ in your sketch.

b.) Sketch the local and average heat transfer coefficient, h and $\overline{h}$, as a function of x; indicate on your sketch the location $x = b$.

c.) Sketch the local and average friction factor, f and $\overline{f}$, as a function of x; indicate on your sketch the location $x = b$.

Figure P5-2 shows that a non-uniform heat flux is applied to the surface of the pipe. The heat flux decreases linearly from $x = 0$ to $x = b$ and remains at 0 for all subsequent x. The fluid enters the pipe with mean temperature, T_{in}.

d.) Sketch the mean temperature of the fluid as a function of position.

e.) Sketch the surface temperature of the pipe as a function of position.

Internal Flow Correlations and the Energy Balance

5–3 Dismounted soldiers and emergency response personnel are routinely exposed to high temperature/humidity environments as well as external energy sources such as flames, motor heat or solar radiation. The protective apparel required by chemical, laser, biological, and other threats tend to have limited heat removal capability. These and other factors can lead to severe heat stress. One solution is a portable, cooling system integrated with an encapsulating garment to provide metabolic heat removal. A portable metabolic heat removal system that is acceptable for use by a dismounted soldier or emergency response personnel must satisfy a unique set of criteria. The key requirement for such a system is that it be extremely low mass and very compact in order to ensure that any gain in performance due to active cooling is not offset by fatigue related to an increase in pack load. In order to allow operation for an extended period of time, a system must either be passive (require no consumable energy source), very efficient (require very little consumable energy), or draw energy from a high energy density power source.

One alternative for providing portable metabolic heat removal is with an ice pack, as shown in Figure P5-3.

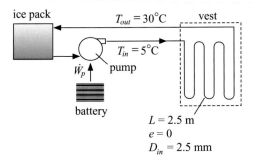

ice pack

$T_{out} = 30°C$ vest

$T_{in} = 5°C$

$\dot{W}_p$ pump

battery

$L = 2.5$ m
$e = 0$
$D_{in} = 2.5$ mm

Figure P5-3: Schematic of a portable metabolic heat removal system that utilizes an ice pack.

The pump forces a liquid antifreeze solution to flow through plastic tubes in the vest in order to transfer the cooling from the ice to the person. Assume that the surface of the plastic is completely smooth, $e = 0$, the total length of the tube is $L = 2.5$ m and the inner diameter of the tube is $D_{in} = 2.5$ mm. There are $N_b = 20$ bends in the vest; the loss coefficient associated with each bend is $C_b = 1.0$. The fluid that is being circulated through the vest has properties $\rho_f = 1110$ kg/m³, $c_f = 2415$ J/kg-K, $\mu_f = 0.0157$ Pa-s, and $Pr_f = 151$. The fluid enters the vest at $T_{in} = 5.0°C$ and leaves the vest at $T_{out} = 30°C$. You may assume that the pressure drop associated with the vest is much greater than the pressure drop associated with any other part of the system.

a.) Assume that the bulk velocity in the tube is $u_m = 1.0$ m/s. Determine the pressure drop required to circulate the fluid through the vest.

A miniature diaphragm pump is used to circulate the fluid. Assume that the pressure rise produced by the pump (Δp_p) varies linearly from the dead head pressure rise $\Delta p_{p,dh} = 30$ psi at no flow ($\dot{V}_p = 0$) to zero at the maximum unrestricted flow rate $\dot{V}_{p,open} = 650$ mL/min.

$$\Delta p_p = \Delta p_{p,dh}\left(1 - \frac{\dot{V}_p}{\dot{V}_{p,open}}\right)$$

b.) Determine the fluid flow rate through the vest that is consistent with the pump curve given by the equation above; that is, vary the value of the mean velocity, u_m, until the pressure drop across the vest and flow rate through the vest falls on the pump curve.

c.) How much cooling is provided by the vest?

d.) If the pump efficiency is $\eta_p = 0.20$ then how much power is consumed by the pump?

e.) If the system is run for *time* $= 1$ hour then what is the mass of ice that is consumed? (Assume that the latent heat of fusion associated with melting ice is $i_{fs} = 3.33 \times 10^5$ J/kg and that the only energy transfer to the ice is from the fluid.) What is the mass of batteries that are consumed, assuming that the energy density of a lead acid battery is $ed_b = 0.05$ kW-hr/kg?

5–4 One concept for rapidly launching small satellites involves a rocket boosted, expendable launch vehicle that is dropped from the cargo bay of a military cargo aircraft. The launch vehicle is propelled by self-pressurized tanks of liquid oxygen and liquid propane. The liquid oxygen fuel tank (referred to as the propellant tank) is at elevated pressure and must be kept full while the aircraft sits on the runway, flies to the launch coordinates, and potentially holds position in order to wait for a strategically appropriate launch time; the design requires that the propellant tank remain full for $time_{wait} = 12$ hours. The propellant tank contains saturated liquid oxygen at $p_{tank} = 215$ psia. Saturated liquid oxygen at this pressure has a temperature of $T_{tank} = 126.8$ K. Because the tank is so cold, it is subjected to a large heat leak, $\dot{q}_{tank}$. Without external cooling, it would be necessary to vent the liquid oxygen that boils off in order to maintain the proper pressure and therefore the tank would slowly be emptied. It is not possible to place a cryogenic refrigerator in the propellant tank in order to re-liquefy the oxygen. Rather, an adjacent dewar of liquid oxygen (referred to as the conditioning tank) is used to remove the parasitic heat transfer and prevent any oxygen in the propellant tank from boiling away. Figure P5-4 illustrates the proposed system. A pump is used to circulate liquid oxygen from the propellant tank through a cooling coil that is immersed in the conditioning tank. The pump and conditioning tank can be quickly removed from the launch vehicle when it is time for launch. The conditioning tank is maintained at $p_{ct} = 14.7$ psia and contains saturated liquid oxygen; any oxygen that evaporates due to the heat added by the cooling coil is allowed to escape. The cooling coil is a coiled up tube with total length $L = 10$ m, inner diameter $D_i = 0.8$ cm and outer diameter $D_o = 1.0$ cm. The internal surface of the tube has roughness $e = 50$ μm and the conductivity of the tube material is $k_{tube} = 2.5$ W/m-K. The mass flow rate provided by the pump is $\dot{m} = 0.25$ kg/s and the pump efficiency is $\eta_{pump} = 0.45$. The heat transfer coefficient associated with the evaporation of the liquid oxygen in the conditioning tank from the external surface of the tube is $h_o = 2 \times 10^4$ W/m²-K. You may assume that the liquid oxygen that is pumped through the cooling coil has constant properties that are consistent with saturated liquid oxygen at the tank pressure.

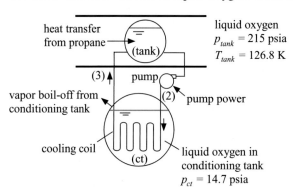

Figure P5-4: Liquid from the propellant tank is pumped through a coil immersed in the conditioning tank.

a.) What is the pressure drop associated with forcing the liquid oxygen through the cooling coil?

b.) What is the power required by the pump?

c.) If all of the pump power is ultimately transferred to the liquid oxygen that is being pumped then what is the temperature of the liquid oxygen leaving the pump (T_2 in Figure P5-4)?

d.) What is the heat transfer coefficient between the liquid oxygen flowing through the cooling coil and the internal surface of the tube?

e.) What is the total conductance associated with the cooling coil?

f.) What is the temperature of the liquid oxygen leaving the cooling coil (T_3 in Figure P5-4)?

g.) How much cooling is provided to the propellant tank?

h.) Plot the cooling provided to the conditioning tank and the pump power as a function of the mass flow rate. If the parasitic heat leak to the propellant tank is $\dot{q}_{tank} = 10\,\text{kW}$ then suggest the best mass flow rate to use for the system.

5–5 You have been asked to help with the design of a water source heat pump, as shown in Figure P5-5. During the cooling season, the water source heat pump is, essentially, an air conditioner that rejects heat to a water source rather than to air. The building is located next to a lake and therefore you intend to reject heat by running a plastic tube through the lake. Currently, you have selected a tube with an outer diameter, $D_{out} = 0.50$ inch and a wall thickness $th = 0.065$ inch. You measure the temperature of the water in the lake to be $T_{lake} = 50°F$ and estimate that the heat transfer coefficient between the external surface of the pipe and the water is $\overline{h}_o = 450\,\text{W/m}^2\text{-K}$. The conductivity of the tube material is $k_{tube} = 1.5\,\text{W/m-K}$.

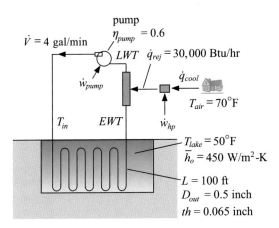

Figure P5-5: Water source heat pump rejecting heat to a lake.

The manufacturer's sheet for the particular heat pump that has been purchased lists many characteristics of the heat pump as a function of the entering water temperature, EWT. The manufacturer recommends a fixed flow rate of water through the pipe of $\dot{V} = 4.0$ gal/min, and so you have found an appropriate fixed displacement pump to provide this constant volumetric flow rate of water; the pump has an efficiency, $\eta_{pump} = 0.60$. The data from the manufacturer's sheet have been used to correlate the heat pump power consumption as a function of the entering water temperature according to:

$$\dot{w}\,[\text{kW}] = 0.8513\,[\text{kW}] + 1.347 \times 10^{-3}\left[\frac{\text{kW}}{°F}\right]EWT\,[°F]$$
$$+ 9.901 \times 10^{-5}\left[\frac{\text{kW}}{°F^2}\right](EWT\,[°F])^2$$

You have been asked to determine the length of tube, L, that should be run through the lake in order to maximize the efficiency of the system (defined as the coefficient of performance, COP, which is the ratio of the cooling provided to the power consumed by both the pump and the heat pump). This is not a straightforward problem because it is difficult to see where to start. We'll tackle it in small steps here. We'll start by making a couple of assumptions that will eventually be relaxed; the assumptions are just to get the solution going – it is easier to accomplish a meaningful analysis when you have a working model. Assume that the leaving water temperature is $LWT = 40°C$ and that the length of the tube is $L = 100$ ft.

a.) Calculate the pressure drop required to force the water through the tube in the lake.

b.) Predict the temperature of the water leaving the pump (T_{in} in Figure P5-5); assume all of the pump energy goes into the water.

c.) Predict the temperature of the water leaving the lake and entering the heat pump (EWT in Figure P5-5) by considering the heat transfer coefficient associated with the flow of water in the tube and the energy balance for this flow.

d.) Using your model, adjust the leaving water temperature (that you initially assumed to be 40°C) until the heat rejected to the water is equal to the heat rejection required by the heat pump (i.e., $\dot{q}_{rej} = 30 \times 10^3$ Btu/hr).

e.) Using the manufacturer's data provided by the curve fit, calculate the power required by the heat pump and, from that, the cooling provided to the cabin and the total COP (including both the heat pump and the pump).

f.) Use your model to prepare a single plot that shows how the COP and cooling capacity vary with length of tube. You should see an optimal length of tube that maximizes the COP; explain why this optimal value exists.

5–6 Figure P5-6 illustrates a cold plate that is used as the heat sink for an array of diodes in a power supply.

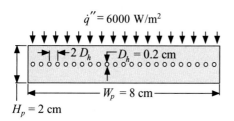

Figure P5-6: Cold plate.

The operation of the diodes provides a uniform heat flux $\dot{q}'' = 6000$ W/m^2 over the top surface of the cold plate. The plate is cooled by the flow of a coolant with density $\rho_c = 1090$ kg/m^3, conductivity $k_c = 0.8$ W/m-K, viscosity $\mu_c = 0.01$ Pa-s, and specific heat capacity $c_c = 1500$ J/kg-K. The mass flow rate of coolant is $\dot{m} = 0.1$ kg/s and the inlet temperature is $T_{c,in} = 30°C$. The coolant flows along the length of the cold plate through holes that are $D_h = 0.2$ cm in diameter. The length of the cold plate (in the flow direction) is $L_p = 15$ cm, the width is $W_p = 8$ cm, and the thickness is $H_p = 2$ cm. The conductivity of the cold plate is $k_p = 650$ W/m-K. The distance between the centers of two adjacent holes is twice the hole diameter. All of the surfaces of the cold plate that are not exposed to the heat flux are adiabatic. Your initial model should assume that the resistance to conduction through the cold plate from the surface where the heat flux is applied to the surface of the holes is negligible. Further, your model should assume that the resistance to conduction along the length of the cold plate is infinite.

a.) Plot the coolant temperature and plate temperature as a function of position, x.

b.) Plot the maximum coolant temperature and maximum plate temperature as a function of mass flow rate for mass flow rates varying between 0.0005 kg/s and 10 kg/s. Use a log-scale for the mass flow rate. Overlay on your plot the difference between the maximum plate temperature and maximum coolant temperature. You should see three distinct types of behavior as you increase the mass flow rate. Explain these behaviors.

c.) Assess the validity of neglecting the resistance to conduction through the cold plate from the surface where the heat flux is applied to the surface of the holes for the nominal mass flow rate ($\dot{m} = 0.1$ kg/s).

d.) Assess the validity of assuming that the resistance to conduction along the length of the cold plate is infinite for the nominal mass flow rate ($\dot{m} = 0.1$ kg/s). Refine your model so that it includes the effect of conduction along the length of the cold plate. This refined model should continue to assume that the resistance to conduction through the cold plate from the surface where the heat flux is applied to the surface of the holes is negligible.

e.) Using differential energy balances on the plate material and the coolant, derive the state equations that govern this problem. For this problem, the state variables include the coolant temperature (T_c), the plate temperature (T_p), and the gradient of the plate temperature ($\frac{dT_p}{dx}$).

f.) Assume that the temperature of the plate material at $x = 0$ is $T_{p,x=0} = 310$ K. The coolant temperature and plate temperature gradient at $x = 0$ are both specified. Use the Crank-Nicolson technique to integrate the state equations from $x = 0$ to $x = L$. Do not attempt to enforce the fact that the plate is adiabatic at $x = L$ during this step. Your Crank-Nicolson technique should be implicit in the temperatures but explicit in the heat transfer coefficient (i.e., the heat transfer coefficient can be calculated at the beginning of the length step). Plot the temperature of the coolant and the plate as a function of position.

g.) Adjust the assumed value of the plate temperature at $x = 0$, $T_{p,x=0}$, until the temperature gradient in the plate material at $x = L$ is zero (i.e., the end of the plate is adiabatic). Overlay on your plot from (a) the temperature of the coolant and the conductor as a function of position.

Analytical Solutions to Internal Flow Problems

5-7 Figure P5-7 illustrates a simple slider bearing used to provide support against thrust loads.

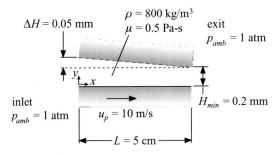

Figure P5-7: Thrust bearing.

The slider is a close clearance, converging gap formed between a moving surface (e.g., the surface of a rotating shaft) and a stationary surface. The velocity of the moving surface is $u_p = 10$ m/s. The length of the gap is $L = 5.0$ cm and the minimum clearance in the gap is at the exit (i.e., at $x = L$) is $H_{min} = 0.2$ mm. The maximum clearance of the gap is at the inlet (i.e., at $x = 0$) is $H_{min} + \Delta H$ where $\Delta H/H_{min} = 0.25$. The clearance varies linearly with position according to: $H = H_{min} + \Delta H \frac{(L-x)}{L}$. The pressure at the inlet and exit of the gap is ambient, $p_{amb} = 1$ atm. The properties of the oil that flows through the gap are $\rho = 800$ kg/m^3 and $\mu = 0.5$ Pa-s.

a.) Is it appropriate to model the flow through the gap as inertia-free flow using the Reynolds equation?

b.) Use the Reynolds equation to obtain an analytical solution for the pressure distribution within the gap.

c.) Determine the force per unit width provided by the thrust bearing.

d.) Determine an appropriate scaling relation for the force per unit width and use it to define a non-dimensional thrust force. Plot the dimensionless thrust force as a function of the parameter $\Delta H/H_{min}$.

5-8 A very viscous fluid is pumped through a circular tube at a rate of $\dot{V} = 15$ liter/min. The tube is thin walled and made of metal; the thickness of the tube and its resistance to conduction can be neglected. The tube diameter is $D = 0.5$ inch. The tube is covered with insulation that is $th_{ins} = 0.25$ inch with conductivity $k_{ins} = 0.5$ W/m-K. The external surface of the tube is exposed to air at $T_\infty = 20°$C with heat transfer coefficient $\bar{h} = 120$ W/m^2-K. The viscosity of the fluid is $\mu = 0.6$ Pa-s and its conductivity is $k = 0.15$ W/m-K.

a.) Prepare an analytical solution for the radial temperature distribution within the fluid at a location where the fluid temperature is not changing in the x-direction (i.e., in the direction of the flow). Include the effect of viscous dissipation. You may neglect axial conduction. Assume that the fluid is hydrodynamically fully developed.

b.) Plot the temperature as a function of radial position. Overlay on your plot the temperature distribution for $\dot{V} = 5, 10,$ and 20 liter/min.

Numerical Solutions to Internal Flow Problems

5-9 Immersion lithography is a technique that will potentially allow optical lithography (the manufacturing technique used to fabricate computer chips) to create smaller features. A liquid is inserted into the space between the last optical element (the lens) and the wafer that is being written in order to increase the index of refraction in this volume relative to the air that would otherwise fill this gap. A simplified version of this concept is shown in Figure P5-9.

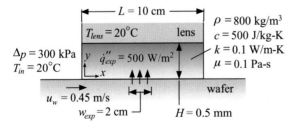

Figure P5-9: An immersion lithography tool.

It is important to predict the temperature distribution in the fluid during this process; even very small temperature changes will result in imaging problems associated with changes in the properties of the fluid or thermally induced distortions of the wafer. The fluid enters the gap at the left hand side ($x = 0$) with a uniform temperature, $T_{in} = 20°C$ and flows from left to right through the lens/wafer gap; the total length of the gap is $L = 10$ cm and the height of the gap is $H = 0.5$ mm. The fluid is driven by the viscous shear as the wafer is moved under the lens with velocity $u_w = 0.45$ m/s. The fluid is also driven by a pressure gradient; the pressure at the left side of the gap ($x = 0$) is elevated relative to the pressure at the right side ($x = L$) by an amount, $\Delta p = 300$ kPa. Assume that the flow is laminar and that the problem is two-dimensional (i.e., the slot extends a long way into the page). In this case, the liquid has a fully developed velocity distribution when it enters the gap:

$$u = u_w \left(1 - \frac{y}{H}\right) + \frac{H^2 \, \Delta p}{2 \mu L}\left[\left(\frac{y}{H}\right) - \left(\frac{y}{H}\right)^2\right]$$

where y is the distance from the wafer and μ is the liquid viscosity.

The next generation of immersion lithography tools will use advanced liquids with very high viscosity and so you have been asked to generate a model that can evaluate the impact of viscous dissipation on the temperature distribution. The liquid has density $\rho = 800$ kg/m^3, specific heat capacity $c = 500$ J/kg-K, thermal conductivity $k = 0.1$ W/m-K, and viscosity $\mu = 0.1$ Pa-s. The energy required to develop the resist layer and therefore carry out the lithography process passes through the lens and the water and is deposited into the wafer; the energy can be modeled as a heat flux at the wafer surface into the liquid (assume all of the heat flux will go to the liquid rather than the wafer). The heat flux is concentrated in a small strip ($w_{exp} = 2.0$ cm wide) at the center of the lens, as shown in Figure P5-9 and given by:

$$\dot{q}_s''(x) = \begin{cases} 0 & \text{for } x < (L - w_{exp})/2 \\ \dot{q}_{exp}'' & \text{for } (L - w_{exp})/2 < x < (L + w_{exp})/2 \\ 0 & \text{for } x > (L + w_{exp})/2 \end{cases}$$

where $\dot{q}_{exp}'' = 500$ W/m^2. The lens is maintained at a constant temperature, $T_{lens} = 20°C$.

a.) What is the mean velocity and the Reynolds number that characterizes the flow through the lens/wafer gap?

b.) You'd like to calculate a Brinkman number in order to evaluate the relative impact of viscous dissipation for the process but you don't have a convenient reference temperature difference to use. Use the heat flux to come up with a meaningful reference temperature difference and from that temperature difference determine a Brinkman number. Comment on the importance of viscous dissipation for this problem.

c.) Is axial conduction important for this problem? Justify your answer.

d.) Develop a numerical model of the thermal behavior of the flow through the gap that accounts for viscous dissipation but not axial conduction. Use the native ODE solver in MATLAB.

e.) Prepare a contour plot that shows the temperature distribution in the lens/wafer gap.

f.) Prepare a contour plot that shows the temperature distribution in the absence of any applied heat flux (i.e., what is the heating caused by the viscous dissipation?).

5–10 Figure P5-10 shows a thin-wall tube with radius $R = 5.0$ mm carrying a flow of liquid with density $\rho = 1000\,\text{kg/m}^3$, specific heat capacity $c = 1000$ J/kg-K, conductivity $k = 0.5$ W/m-K, and viscosity $\mu = 0.017$ Pa-s. The fluid is fully developed hydrodynamically with a bulk velocity $u_m = 0.2$ m/s and has a uniform temperature $T_{ini} = 80°C$ when it enters a section of the tube that is exposed to air at temperature $T_\infty = 20°C$ with heat transfer coefficient $\bar{h}_a$.

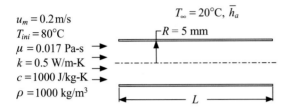

Figure P5-10: Thin-wall tube carrying fluid exposed to air.

The Reynolds number for this flow is around 100 and so this is a laminar flow. Typically, the heat transfer between the fluid and the air is modeled with a laminar flow heat transfer coefficient that is calculated using correlations that are based on a constant tube surface temperature. In fact, the surface temperature of the tube is not constant for this process. The objective of this problem is to understand how much this approximation affects the solution.

a.) Develop a numerical model of this situation using MATLAB. Prepare a plot showing the temperature at various radii as a function of axial position for the case where $\bar{h}_a = 100$ W/m²-K and $L = 5.0$ m. Prepare a plot of the Nusselt number as a function of axial position for this situation.

b.) Verify your solution by comparing it with an appropriate analytical model.

c.) Investigate the effect of the external convection coefficient; in the limit that $\bar{h}_a$ is very large your solution should limit (at long length) to $Nu = 3.66$. What is the effect of a finite $\bar{h}_a$? Present your conclusions in a logical and systematic manner.

5–11 Figure P5-11 illustrates a flow of liquid in a passage formed between two parallel plates.

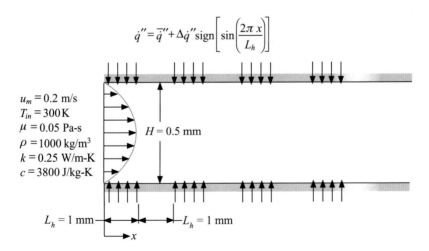

Figure P5-11: Flow between two parallel plates.

The flow enters the duct having been exposed to a uniform heat flux at $\overline{q''_s} = 9500$ W/m^2 for a long time. Therefore, the flow is both thermally and hydrodynamically fully developed. The velocity distribution is:

$$u = 6\,u_m \left(\frac{y}{H} - \frac{y^2}{H^2} \right)$$

where $u_m = 0.2$ m/s is the bulk velocity and $H = 0.5$ mm is the plate-to-plate spacing. The temperature distribution at the inlet is:

$$T = T_{in} + \frac{\overline{q''_s} H}{k} \left[-\left(\frac{y}{H}\right)^4 + 2 \left(\frac{y}{H}\right)^3 - \left(\frac{y}{H}\right) + 0.243 \right]$$

where $T_{in} = 300$ K is the mean temperature of the fluid at the inlet. The properties of the fluid are density $\rho = 1000$ kg/m^3, viscosity $\mu = 0.05$ Pa-s, conductivity $k = 0.25$ W/m-K, and specific heat capacity $c = 3800$ J/kg-K. The heat flux applied to the surfaces of the channel is non-uniform and you need to evaluate the impact of the non-uniform heat flux on the surface temperature of the duct. The heat flux at both the upper and lower surfaces of the duct varies according to:

$$\dot{q}'' = \overline{\dot{q}''} + \Delta\dot{q}'' \, \text{sign} \left[\sin \left(\frac{2\pi x}{L_h} \right) \right]$$

where $\Delta\dot{q}'' = 9500$ W/m^2 is the amplitude of the heat flux variation and $L_h = 1$ mm is the width of the heated regions and the function sign returns +1 if the argument is positive and −1 if it is negative. This equation, with the specified inputs, leads to a wall that alternates between having a heat flux of 19,000 W/m^2 and then being adiabatic.

a.) Is the flow laminar or turbulent?

b.) Is viscous dissipation important?

c.) Is axial conduction important?

d.) Develop a 2-D numerical model of the flow in the gap using the Crank-Nicolson solution technique implemented in MATLAB. Plot the temperature as a function of position x at various values of y for $0 < x < 1$ cm.

e.) Determine the surface temperature of the duct and the mean temperature of the fluid at each axial position. Plot T_s and T_m as a function of position x.

f.) Determine the Nusselt number at each axial position. Plot the Nusselt number as a function of x. Explain the shape of your plot.

g.) Verify that your model is working correctly by setting $\Delta\dot{q}'' = 0$ and showing that the Nusselt number in the duct is consistent with the Nusselt number for fully developed flow between parallel plates subjected to a constant heat flux.

Return the value of $\Delta\dot{q}''$ to 9500 W/m^2. You are interested in studying the impact of the non-uniform heat flux on the surface temperature of the duct. There are two natural limits to this problem.

h.) Calculate the surface-to-mean temperature difference experienced when the average heat flux is applied at the wall and the flow is fully developed.

i.) Calculate the surface-to-mean temperature difference experienced when the peak heat flux is applied at the wall and the flow is fully developed.

j.) Define a meaningful dimensionless spatial period, $\tilde{L}_h$, and plot the surface-to-mean temperature difference as a function of dimensionless spatial period. Show that when $\tilde{L}_h$ is small, the solution limits to your answer from (h) and when $\tilde{L}_h$ is large then your solution limits to your solution from (i). Explain this result.

REFERENCES

Colebrook, C. F., "Turbulent Flow in Pipes with Particular Reference to the Transitional Region between the Smooth and the Rough Pipe Laws," *J. Inst. Civil Eng.*, Vol. 11, pp. 133–156, (1939).

Curr, R. M., D. Sharma, and D.G. Tatchell, "Numerical predictions of some three-dimensional boundary layers in ducts," *Comput. Methods Appl. Mech. Eng.*, Vol. 1, pp. 143–158, (1972).

Gnielinski, V., "New Equations for Heat and Mass Transfer in Turbulent Pipe and Channel Flow," *Int. Chem. Eng.*, Vol. 16, pp. 359–368, (1976).

Hornbeck, R. W., "An all-numerical method for heat transfer in the inlet of a tube," *Am. Soc. Mech. Eng.,* Paper 65-WA/HT-36, (1965).

Kakaç, S., R. K. Shah, and W. Aung, eds., *Handbook of Single-Phase Convective Heat Transfer*, Wiley-Interscience, (1987).

Kays, W. M. and M. E. Crawford, *Convective Heat and Mass Transfer, 3rd Edition*, McGraw-Hill, New York, (1993).

Lienhard, J. H., IV and J. H. Lienhard V, *A Heat Transfer Textbook, 3rd Edition*, Phlogiston Press, Cambridge, MA, (2005).

Liu, J., *Flow of a Bingham Fluid in the Entrance Region of an Annular Tube*, M. S. Thesis, University of Wisconsin at Milwaukee, (1974).

Nikuradse, J., *Laws of flow in rough pipes*, Technical report: NACA Technical Memo 1292, National Advisory Commission for Aeronautics, Washington, D.C., (1950).

Petukhov, B. S., in *Advances in Heat Transfer*, Vol. 6, T. F. Irvine and J. P. Hartnett eds., Academic Press, New York, (1970).

Rohsenow, W. M., J. P. Hartnett, and Y. I. Cho, eds., *Handbook of Heat Transfer, 3rd Edition*, McGraw-Hill, New York, (1998).

Shah, R.K., "Thermal entry length solutions for the circular tube and parallel plates," *Proc. Natl. Heat Mass Transfer Conf., 3rd* Indian Inst. Technology., Bombay, Vol. I, Paper No. HMT-11-75, (1975).

Shah, R. K. and A. L. London, *Laminar Flow Forced Convection in Ducts*, Academic Press, New York, (1978).

White, F. M., *Viscous Fluid Flow*, McGraw-Hill, New York, (1991).

Wibulswas, P., *Laminar Flow Heat Transfer in Non-Circular Ducts*, Ph.D. thesis, London University, London, (1966).

Zigrang, D. J. and N. D. Sylvester, "Explicit approximations to the solution of Colebrook's friction factor equation," *AIChE Journal*, Vol. 28, pp. 514–515, (1982).

6 Natural Convection

6.1 Natural Convection Concepts

6.1.1 Introduction

Chapters 4 and 5 focus on forced convection problems in which the fluid motion is driven externally, for example by a fan or a pump. However, even in the limit of no externally driven fluid motion, a solid surrounded by a fluid may not reduce to a conduction problem because the fluid adjacent to a heated or cooled surface will usually not be stagnant. Natural (or free) convection refers to convection problems in which the fluid is not driven mechanically but rather thermally; that is, fluid motion is driven by density gradients that are induced in the fluid as it is heated or cooled. The velocities induced by these density gradients are typically small and therefore the absolute magnitude of natural convection heat transfer coefficients is also typically small.

The flow patterns induced by heating or cooling can be understood intuitively; hot fluid tends to have lower density and therefore rise (flow against gravity) while cold fluid with higher density tends to fall (flow with gravity). The existence of a temperature gradient does not guarantee fluid motion. Figure 6-1(a) illustrates fluid between two plates oriented horizontally (i.e., perpendicular to the gravity vector g) where the lower plate is heated (to T_H) and the upper plate is cooled (to T_C). The heated fluid will tend to rise and the cooled fluid fall, resulting in the natural convection "cells" that are shown in Figure 6-1(a). Figure 6-1(b) illustrates fluid between horizontal plates where the lower plate is cooled and the upper one heated. This situation is stable; the cold fluid cannot fall further and the hot fluid cannot rise further. The heat transfer rate between the two plates shown in Figure 6-1(a) will be substantially higher than in Figure 6-1(b).

This section discusses the natural set of dimensionless parameters that are used to correlate the solutions for free convection problems. In Section 6.2, several commonly encountered configurations are examined and correlations are presented that can be used to solve engineering problems. A relatively comprehensive set of correlations are included in EES; the use of these correlations is illustrated with examples.

6.1.2 Dimensionless Parameters for Natural Convection

Section 4.3 shows that the average Nusselt number for most forced convection problems can be correlated using the Reynolds number and the Prandtl number. The appropriate analogous set of dimensionless parameters for a natural convection problem may be obtained either by physical reasoning or through the more rigorous process of making the governing differential equations, including the gravitational term, dimensionless. Both methods are presented in the subsequent sections.

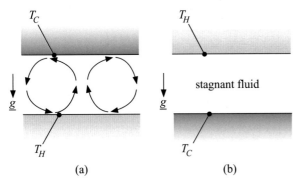

Figure 6-1: Flow patterns induced between horizontal plates in which (a) the upper plate is cooled and the lower plate heated, and (b) the upper plate is heated and the lower plate cooled.

Identification from Physical Reasoning

Figure 6-2(a) illustrates a vertically oriented (i.e., parallel to the gravity vector, g) plate with a surface that is heated to temperature T_s in an environment of fluid that would otherwise be stagnant at temperature T_∞.

The governing equation for a laminar boundary layer was derived in Section 4.2. The momentum conservation equation in the x-direction (vertical, see Figure 6-2(a)) is:

$$\rho \left[\frac{\partial u}{\partial t} + u\frac{\partial u}{\partial x} + v\frac{\partial u}{\partial y} \right] = -\frac{\partial p}{\partial x} + \mu \left(\frac{\partial^2 u}{\partial y^2} + \frac{\partial^2 u}{\partial x^2} \right) - \rho g \qquad (6\text{-}1)$$

The gravitational term in Eq. (6-1) is negative because the gravity vector is in the negative x-direction. In the absence of any fluid motion (i.e., $u = v = 0$), Eq. (6-1) reduces to:

$$\frac{\partial p}{\partial x} + \rho g = 0 \qquad (6\text{-}2)$$

If the density of the fluid is everywhere constant, then the pressure in the stagnant fluid will only vary with x (i.e., due to hydrostatic effects) and the pressure at any y-position will be the same. However, if density is a function of temperature, then the pressure gradient immediately adjacent to the heated plate (along line A-B in Figure 6-2(a)) will be different than the pressure gradient far from the plate (along line C-D in Figure 6-2(a)).

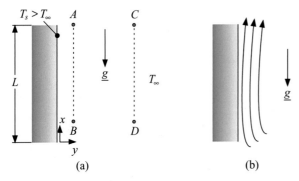

Figure 6-2: (a) A plate heated to T_H in an environment of fluid at T_∞ and (b) the associated flow pattern.

The pressure gradient away from the plate will be:

$$\left(\frac{\partial p}{\partial x}\right)_{C-D} = -\rho_{T=T_\infty}\, g \tag{6-3}$$

whereas the pressure gradient adjacent to the plate will be:

$$\left(\frac{\partial p}{\partial x}\right)_{A-B} = -\rho_{T=T_s}\, g \tag{6-4}$$

If the pressures at points A and C are the same, then Eqs. (6-3) and (6-4) imply that, in the absence of any fluid motion, a pressure difference will be induced between points B and D in Figure 6-2(a):

$$p_D - p_B = \Delta p = g\, L\left(\rho_{T=T_\infty} - \rho_{T=T_s}\right) \tag{6-5}$$

The y-directed pressure difference predicted by Eq. (6-5) will not exist; instead, fluid will be pushed towards the plate where it is heated and rises against gravity, as shown in Figure 6-2(b). The pressure difference in Eq. (6-5) provides the driving force for fluid motion and allows the definition of a characteristic velocity for the natural convection problem, $u_{char,nc}$. The pressure difference will induce a consistent fluid momentum change, according to:

$$g\, L\left(\rho_{T=T_\infty} - \rho_{T=T_s}\right) = \rho\, u_{char,nc}^2 \tag{6-6}$$

Therefore, the characteristic velocity is:

$$u_{char,nc}^2 = \frac{g\, L}{\rho}\left(\rho_{T=T_\infty} - \rho_{T=T_s}\right) \tag{6-7}$$

where ρ is the average density of the fluid. In natural convection problems, the density difference is driven by a temperature difference. Density depends approximately linearly on temperature, for moderate temperature changes in most fluids, according to:

$$\rho_{T=T_\infty} - \rho_{T=T_s} = \left(\frac{\partial \rho}{\partial T}\right)_p (T_\infty - T_s) \tag{6-8}$$

The volumetric thermal expansion coefficient (β) is defined as:

$$\beta = -\frac{1}{\rho}\left(\frac{\partial \rho}{\partial T}\right)_p \tag{6-9}$$

Substituting Eq. (6-9) into Eq. (6-8) leads to:

$$\rho_{T=T_\infty} - \rho_{T=T_s} = -\beta\,\rho\,(T_\infty - T_s) \tag{6-10}$$

where ρ is the nominal density of the fluid (i.e., the average density). Substituting Eq. (6-10) into Eq. (6-7) leads to the definition of a characteristic velocity for a natural convection problem:

$$\boxed{u_{char,nc} = \sqrt{g\, L\, \beta\,(T_s - T_\infty)}} \tag{6-11}$$

Equation (6-11) shows that the magnitude of the induced velocity will increase with the temperature difference and the volumetric thermal expansion coefficient. Figure 6-3 illustrates the volumetric thermal expansion coefficient of several fluids as a function of temperature.

It is worth noting that the magnitude of the velocity that will be induced by a temperature difference under most conditions is quite small. For example, a 20 cm long plate

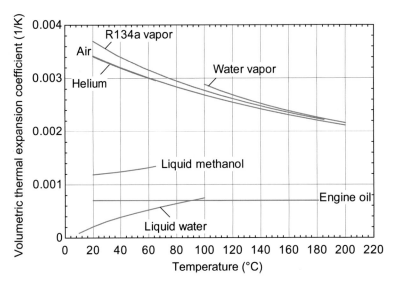

Figure 6-3: Volumetric thermal expansion coefficient of several fluids as a function of temperature at atmospheric pressure.

heated to 100°C in room temperature air leads to $u_{char,nc} \approx 0.7$ m/s; a fan or blower can easily provide air flow velocities that are an order of magnitude higher than this value.

Notice that the gases shown in Figure 6-3 tend to collapse onto a single line, particularly at higher temperature. This observation can be explained by substituting the ideal gas law into the definition of the volumetric thermal expansion coefficient. The ideal gas law is:

$$\rho = \frac{p}{RT} \tag{6-12}$$

where R is the gas constant. Substituting Eq. (6-12) into Eq. (6-9) leads to:

$$\beta = -\frac{1}{\rho}\left(\frac{\partial \rho}{\partial T}\right)_p = -\frac{RT}{p}\frac{\partial}{\partial T}\left(\frac{p}{RT}\right) = -\frac{RT}{p}\left(-\frac{p}{RT^2}\right) = \frac{1}{T} \tag{6-13}$$

The volumetric thermal expansion coefficient of an ideal gas is the inverse of its absolute temperature. The small differences that can be seen in Figure 6-3 result because these gases do not exactly behave according to the ideal gas law at the conditions used to construct the plot.

The external and internal forced convection results are correlated using the Nusselt number, Prandtl number and a Reynolds number that is based on the free-stream velocity (u_∞) or mean velocity (u_m), respectively. Natural convection correlations are based on the Reynolds number associated with the characteristic velocity induced by the driving temperature difference, Eq. (6-11):

$$Re_L = \frac{\rho L u_{char,nc}}{\mu} = \frac{\rho L}{\mu}\sqrt{g L \beta (T_s - T_\infty)} \tag{6-14}$$

In fact, natural convection correlations are often presented in terms of the Grashof number (Gr_L), which is the Reynolds number defined in Eq. (6-14) squared:

$$Gr_L = \left(\frac{\rho L u_{char,nc}}{\mu}\right)^2 = \frac{g L^3 \beta (T_s - T_\infty)}{\upsilon^2} \tag{6-15}$$

Alternatively, the Rayleigh number (Ra_L) is sometimes used to correlate the natural convection results. The Rayleigh number is defined as the product of the Grashof number and the Prandtl number:

$$Ra_L = Gr_L \, Pr = \frac{g \, L^3 \, \beta \, (T_s - T_\infty)}{\nu \alpha} \tag{6-16}$$

Identification from the Governing Equations
The governing differential equations for a natural convection problem are based on the conservation of mass, momentum (in the x- and y-directions, see Figure 6-2) and thermal energy. These equations were derived in Section 4.2 and are repeated below:

$$\frac{\partial u}{\partial x} + \frac{\partial v}{\partial y} = 0 \tag{6-17}$$

$$\rho \left[\frac{\partial u}{\partial t} + u \frac{\partial u}{\partial x} + v \frac{\partial u}{\partial y} \right] = -\frac{\partial p}{\partial x} + \mu \left(\frac{\partial^2 u}{\partial y^2} + \frac{\partial^2 u}{\partial x^2} \right) - \rho g \tag{6-18}$$

$$\rho \left[\frac{\partial v}{\partial t} + u \frac{\partial v}{\partial x} + v \frac{\partial v}{\partial y} \right] = -\frac{\partial p}{\partial y} + \mu \left(\frac{\partial^2 v}{\partial y^2} + \frac{\partial^2 v}{\partial x^2} \right) \tag{6-19}$$

$$\rho c \left[\frac{\partial T}{\partial t} + u \frac{\partial T}{\partial x} + v \frac{\partial T}{\partial y} \right] = k \left(\frac{\partial^2 T}{\partial x^2} + \frac{\partial^2 T}{\partial y^2} \right) + \dot{g}_v''' \tag{6-20}$$

Note that the gravitational term is included in the x-momentum equation, Eq. (6-18). Equations (6-17) through (6-20) are simplified by employing the boundary layer simplifications discussed in Section 4.2. Also, steady state is assumed and the viscous dissipation term in Eq. (6-20) is neglected.

$$\frac{\partial u}{\partial x} + \frac{\partial v}{\partial y} = 0 \tag{6-21}$$

$$\rho \left[u \frac{\partial u}{\partial x} + v \frac{\partial u}{\partial y} \right] = -\frac{dp}{dx} + \mu \frac{\partial^2 u}{\partial y^2} - \rho g \tag{6-22}$$

$$\rho c \left[u \frac{\partial T}{\partial x} + v \frac{\partial T}{\partial y} \right] = k \frac{\partial^2 T}{\partial y^2} \tag{6-23}$$

Far away from the plate, the fluid everywhere is stagnant. Setting all of the velocities in Eq. (6-22) to zero leads to:

$$\left(\frac{dp}{dx} \right)_{y \to \infty} = -\rho_{T=T_\infty} \, g \tag{6-24}$$

According to the boundary layer simplifications, the pressure gradient in the y-direction is negligibly small. Therefore, Eq. (6-24) should hold approximately at any value of y. Substituting Eq. (6-24) into Eq. (6-22) leads to:

$$\rho \left[u \frac{\partial u}{\partial x} + v \frac{\partial u}{\partial y} \right] = \rho_{T=T_\infty} \, g + \mu \frac{\partial^2 u}{\partial y^2} - \rho g \tag{6-25}$$

or

$$\rho \left[u \frac{\partial u}{\partial x} + v \frac{\partial u}{\partial y} \right] = g \left(\rho_{T=T_\infty} - \rho \right) + \mu \frac{\partial^2 u}{\partial y^2} \tag{6-26}$$

Dividing through by ρ leads to:

$$u\frac{\partial u}{\partial x} + v\frac{\partial u}{\partial y} = \frac{g\left(\rho_{T=T_\infty} - \rho\right)}{\rho} + v\frac{\partial^2 u}{\partial y^2} \tag{6-27}$$

The first term on the right side of Eq. (6-27) can be written in terms of the volumetric thermal expansion coefficient:

$$\rho_{T=T_\infty} - \rho = -\beta\rho\left(T_\infty - T\right) \tag{6-28}$$

Substituting Eq. (6-28) into Eq. (6-27) leads to:

$$u\frac{\partial u}{\partial x} + v\frac{\partial u}{\partial y} = g\beta\left(T - T_\infty\right) + v\frac{\partial^2 u}{\partial y^2} \tag{6-29}$$

The simplifications that lead to Eq. (6-29) are referred to as the Boussinesq approximation and refer specifically to the assumption that all of the properties of the fluid are constant except for the density, and density is assumed to only depend linearly on temperature. These simplifications are justified to some extent by examining the magnitude of β for most common substances (Figure 6-3). The product of β and the driving temperature difference is the maximum fractional change in fluid density. Clearly, for reasonable driving temperature differences the fractional change in the density will be quite small. Therefore although the density change is sufficient to induce a buoyancy force that drives fluid motion, it is not enough to otherwise change the character of the governing equations. The governing equations typically used to analyze a laminar natural convection problem are:

$$\boxed{\frac{\partial u}{\partial x} + \frac{\partial v}{\partial y} = 0} \tag{6-30}$$

$$\boxed{u\frac{\partial u}{\partial x} + v\frac{\partial u}{\partial y} = g\beta(T - T_\infty) + v\frac{\partial^2 u}{\partial y^2}} \tag{6-31}$$

$$\boxed{\rho c\left[u\frac{\partial T}{\partial x} + v\frac{\partial T}{\partial y}\right] = k\frac{\partial^2 T}{\partial y^2}} \tag{6-32}$$

Notice that the momentum and energy equations are inherently coupled because temperature appears in the momentum equation and the velocities appear in the energy equation. Therefore, it is not possible to solve the continuity and momentum equations in isolation (as we did for forced convection over a plate) and subsequently solve the energy equation.

The governing equations are non-dimensionalized by defining non-dimensional positions ($\tilde{x}$ and $\tilde{y}$), velocities ($\tilde{u}$ and $\tilde{v}$), and temperature difference ($\tilde{\theta}$):

$$\tilde{x} = \frac{x}{L} \tag{6-33}$$

$$\tilde{y} = \frac{y}{L} \tag{6-34}$$

$$\tilde{u} = \frac{u}{u_{char,nc}} \tag{6-35}$$

$$\tilde{v} = \frac{v}{u_{char,nc}} \tag{6-36}$$

$$\tilde{\theta} = \frac{T - T_s}{T_\infty - T_s} \tag{6-37}$$

where $u_{char,nc}$ is the characteristic velocity identified in Eq. (6-11). Substituting Eqs. (6-33) through (6-37) into Eqs. (6-30) through (6-32) produces nearly the same results that were obtained in Section 4.3 for forced convection:

$$\frac{\partial \tilde{u}}{\partial \tilde{x}} + \frac{\partial \tilde{v}}{\partial \tilde{y}} = 0 \tag{6-38}$$

$$\tilde{u}\frac{\partial \tilde{u}}{\partial \tilde{x}} + \tilde{v}\frac{\partial \tilde{u}}{\partial \tilde{y}} = \frac{L\,g\,\beta}{u_{char,nc}^2}(T - T_\infty) + \frac{\upsilon}{L\,u_{char,nc}}\frac{\partial^2 \tilde{u}}{\partial \tilde{y}^2} \tag{6-39}$$

$$\tilde{u}\frac{\partial \tilde{\theta}}{\partial \tilde{x}} + \hat{v}\frac{\partial \tilde{\theta}}{\partial \tilde{y}} = \frac{\mu}{\rho\,u_{char,nc}\,L\,Pr}\frac{\partial^2 \tilde{\theta}}{\partial \tilde{y}^2} \tag{6-40}$$

Substituting the reference velocity definition, Eq. (6-11), into Eqs. (6-39) and (6-40) leads to:

$$\tilde{u}\frac{\partial \tilde{u}}{\partial \tilde{x}} + \tilde{v}\frac{\partial \tilde{u}}{\partial \tilde{y}} = \frac{(T - T_\infty)}{(T_s - T_\infty)} + \underbrace{\frac{\upsilon}{L\sqrt{g\,L\,\beta\,(T_s - T_\infty)}}}_{=Gr_L^{-1/2}}\frac{\partial^2 \tilde{u}}{\partial \tilde{y}^2} \tag{6-41}$$

$$\tilde{u}\frac{\partial \tilde{\theta}}{\partial \tilde{x}} + \tilde{v}\frac{\partial \tilde{\theta}}{\partial \tilde{y}} = \underbrace{\frac{\mu}{\rho\sqrt{g\,L\,\beta\,(T_s - T_\infty)}\,L\,Pr}}_{=Gr_L^{-1/2}\,Pr^{-1}}\frac{\partial^2 \tilde{\theta}}{\partial \tilde{y}^2} \tag{6-42}$$

or

$$\tilde{u}\frac{\partial \tilde{u}}{\partial \tilde{x}} + \tilde{v}\frac{\partial \tilde{u}}{\partial \tilde{y}} = \tilde{\theta} + \frac{1}{Gr_L^{1/2}}\frac{\partial^2 \tilde{u}}{\partial \tilde{y}^2} \tag{6-43}$$

$$\tilde{u}\frac{\partial \tilde{\theta}}{\partial \tilde{x}} + \tilde{v}\frac{\partial \tilde{\theta}}{\partial \tilde{y}} = \frac{1}{Gr_L^{1/2}\,Pr}\frac{\partial^2 \tilde{\theta}}{\partial \tilde{y}^2} \tag{6-44}$$

The two correlating parameters are identified in Eqs. (6-43) and (6-44) as the Grashof and Prandtl numbers (or equivalently the Rayleigh and Prandtl numbers).

6.2 Natural Convection Correlations

6.2.1 Introduction

Section 6.1 provides an introduction to natural convection and identifies the important correlating parameters. This section presents correlations that can be used to examine several commonly encountered natural convection situations.

6.2.2 Plate

A heated or cooled plate placed in an effectively infinite environment is often encountered in various engineering applications and this geometry has been thoroughly studied. The behavior of the natural convection flow that is induced depends strongly on the orientation of the plate with respect to gravity. The correlations provided in this section are from Raithby and Hollands (1998).

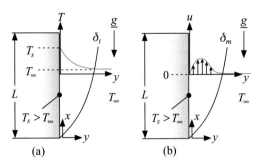

Figure 6-4: The (a) temperature and (b) velocity distribution associated with natural convection from a vertical plate.

Heated or Cooled Vertical Plate

Figure 6-4 illustrates, qualitatively, the velocity and temperature distribution associated with natural convection from a heated vertical plate; vertical refers to a plate that is parallel to the gravity vector.

Notice that both the momentum and thermal boundary layers develop from the lower edge of a heated plate in a manner that is similar to external forced convection over a plate. However, the velocity distribution is somewhat different in that the velocity is zero both at $y = 0$ and as $y \to \infty$. Also, the temperature rise associated with the thermal boundary layer drives the velocity. Therefore, the momentum and thermal boundary layers are related and will have approximately the same magnitude.

The solution for natural convection from a vertical plate is correlated using a Rayleigh number and average Nusselt number that are defined based on the length of the plate in the vertical direction (see Figure 6-4):

$$Ra_L = \frac{g\,L^3\,\beta(T_s - T_\infty)}{\upsilon\,\alpha} \tag{6-45}$$

$$\overline{Nu}_L = \frac{\bar{h}\,L}{k} \tag{6-46}$$

The boundary layer will become turbulent at a critical Rayleigh number of approximately $Ra_{crit} \approx 1 \times 10^9$ and the average Nusselt number will subsequently increase substantially. Note that the square root of this critical Rayleigh number is not too different from the critical Reynolds number associated with the transition to turbulence for forced flow over a flat plate.

The average Nusselt number associated with an isothermal, vertical heated plate is provided by asymptotically weighting the Nusselt numbers for laminar and turbulent flow ($\overline{Nu}_{L,lam}$ and $\overline{Nu}_{L,turb}$, respectively) using the empirical formula:

$$\overline{Nu}_L = \left(\overline{Nu}_{L,lam}^6 + \overline{Nu}_{L,turb}^6\right)^{1/6} \tag{6-47}$$

The laminar Nusselt number is:

$$\overline{Nu}_{L,lam} = \frac{2.0}{\ln\left(1 + \dfrac{2.0}{C_{lam}\,Ra_L^{0.25}}\right)} \tag{6-48}$$

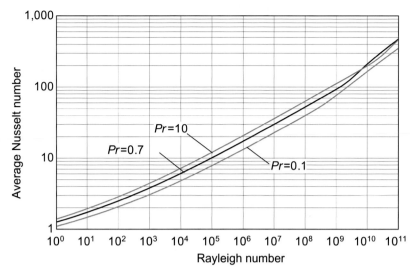

Figure 6-5: The average Nusselt number for a heated vertical plate as a function of Rayleigh number for various values of the Prandtl number.

where

$$C_{lam} = \frac{0.671}{\left[1 + \left(\dfrac{0.492}{Pr}\right)^{9/16}\right]^{4/9}} \qquad (6\text{-}49)$$

The turbulent Nusselt number is:

$$\overline{Nu}_{L,turb} = \frac{C_{turb,V}\, Ra_L^{1/3}}{1 + \left(1.4 \times 10^9\right)\dfrac{Pr}{Ra_L}} \qquad (6\text{-}50)$$

where

$$C_{turb,V} = \frac{0.13\, Pr^{0.22}}{\left(1 + 0.61\, Pr^{0.81}\right)^{0.42}} \qquad (6\text{-}51)$$

The correlations associated with Eqs. (6-47) through (6-51) are valid for $0.1 < Ra_L < 1 \times 10^{12}$. These correlations are implemented by the procedure FC_plate_vertical_ND in EES. Figure 6-5 illustrates the average Nusselt number as a function of the Rayleigh number for various values of the Prandtl number.

Figure 6-6 illustrates the average Nusselt number for natural convection from a vertical flat plate as a function of the square root of the Grashof number. Also shown in Figure 6-6 is the average Nusselt number for external flow over a flat plate at an equivalent Reynolds number based on $u_{char,nc}$ (i.e., at $Re_L = \sqrt{Gr_L}$). The results are shown for a few values of Prandtl number. Figure 6-6 emphasizes the similarity between external forced convection over a plate and natural convection from a vertical plate. The forced convection and natural convection results are similar when presented in this way. At an equivalent flow condition (i.e., at $Re_L = \sqrt{Gr_L}$ with the same Prandtl number), the natural and forced convection Nusselt numbers are similar. This is particularly true for

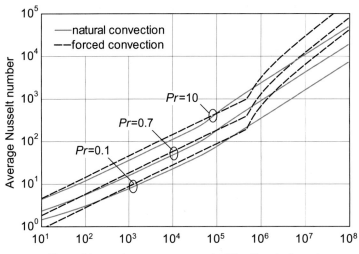

Reynolds number or square root of the Grashof number

Figure 6-6: The average Nusselt number for a heated vertical plate as a function of $\sqrt{Gr_L}$ and the average Nusselt number for external flow over a plate as a function of Re_L. Note that $\sqrt{Gr_L}$ is equal to the Reynolds number based on $u_{char,nc}$ and therefore $Re_L = \sqrt{Gr_L}$ represents an approximately equivalent flow condition. Results are shown for several values of the Prandtl number.

laminar flow (i.e., when $Re_L = \sqrt{Gr_L} < \approx 5 \times 10^5$). The difference between the natural and forced convection correlations becomes more substantial for turbulent flow.

If the vertical plate were cooled rather than heated (i.e., if $T_s < T_\infty$), then the boundary layers shown in Figure 6-4 would initiate at the upper edge of the plate and grow in the downward direction. However, the average Nusselt number for a cooled vertical plate can be computed using the correlations presented in this section (note that the Rayleigh and Grashof numbers must be defined based on the absolute value of the temperature difference in this instance).

Specific correlations exist for a plate with a uniform heat flux and other thermal conditions. However, reasonable accuracy can be obtained using the correlations for an isothermal plate with a Rayleigh number based on the average temperature difference between the plate and the ambient air.

Horizontal Heated Upward Facing or Cooled Downward Facing Plate

Figure 6-7 illustrates, qualitatively, the flow induced by a heated plate oriented horizontally with the heated surface facing up and no restriction to flow at the edges of the plate.

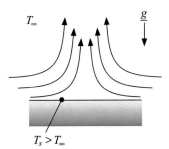

Figure 6-7: Flow induced by a horizontal heated plate facing upwards.

The solution to the problem shown in Figure 6-7 has been correlated using a Rayleigh number and the average Nusselt number that are based on the characteristic length scale of the plate, L_{char}:

$$Ra_{L_{char}} = \frac{g\,L_{char}^3\,\beta(T_s - T_\infty)}{\upsilon\,\alpha} \qquad (6\text{-}52)$$

$$\overline{Nu}_{L_{char}} = \frac{\overline{h}\,L_{char}}{k} \qquad (6\text{-}53)$$

The characteristic length scale of the plate is defined as the ratio of the plate surface area to its perimeter:

$$L_{char} = \frac{A_s}{per} \qquad (6\text{-}54)$$

For a square plate, the characteristic length is one quarter the side length and for a circular plate, the characteristic length is one quarter the diameter. The average Nusselt number is the weighted average of the laminar and turbulent Nusselt numbers:

$$\overline{Nu}_{L_{char}} = \left(\overline{Nu}_{L_{char},lam}^{10} + \overline{Nu}_{L_{char},turb}^{10}\right)^{1/10} \qquad (6\text{-}55)$$

The laminar Nusselt number is:

$$\overline{Nu}_{L_{char},lam} = \frac{1.4}{\ln\left(1 + \dfrac{1.4}{0.835\,C_{lam}\,Ra_{L_{char}}^{0.25}}\right)} \qquad (6\text{-}56)$$

where C_{lam} is given by Eq. (6-49). The turbulent Nusselt number is:

$$\overline{Nu}_{L_{char},turb} = C_{turb,U}\,Ra_{L_{char}}^{1/3} \qquad (6\text{-}57)$$

where $C_{turb,U}$ is calculated according to:

$$C_{turb,U} = 0.14\left(\frac{1 + 0.0107\,Pr}{1 + 0.01\,Pr}\right) \qquad (6\text{-}58)$$

The correlation associated with Eqs. (6-55) through (6-58) is valid for $1.0 < Ra_{L_{char}} < 1 \times 10^{10}$ and is implemented by the procedure FC_plate_horizontal1_ND in EES. Figure 6-8 illustrates the average Nusselt number as a function of the Rayleigh number for various values of the Prandtl number.

If the plate were cooled (i.e., $T_s < T_\infty$) and facing downward, then the flow pattern would appear as shown in Figure 6-7 but flipped over so that the fluid tends to fall downwards after it is cooled by the plate. The average Nusselt number for a cooled downward facing plate can be computed using the correlations presented in this section; note that the Rayleigh and Grashof numbers must be defined based on the absolute value of the temperature difference in this instance.

Horizontal Heated Downward Facing or Cooled Upward Facing Plate
Figure 6-9 illustrates, qualitatively, the flow induced by a heated plate oriented horizontally with the heated surface facing down and no restriction to flow from the edges. The heated fluid tends to escape from the side of the plate, inducing further fluid flow from the ambient.

The solution to the problem shown in Figure 6-9 is correlated using a Rayleigh number and an average Nusselt number that are based on the characteristic length scale L_{char}

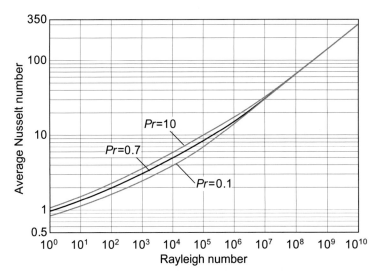

Figure 6-8: The average Nusselt number for a heated horizontal vertical plate facing upward as a function of Rayleigh number for various values of the Prandtl number.

defined previously in Eq. (6-54). The flow velocity that is induced in this configuration is substantially less than for the vertical or horizontal upward facing plate because the configuration is nearly stable. As a result, the flow is laminar even at high Rayleigh numbers. The average Nusselt number is therefore taken to be the laminar Nusselt number:

$$\overline{Nu}_{L_{char}} = \frac{2.5}{\ln\left\{1 + \dfrac{2.5}{0.527\,Ra_{L_{char}}^{0.20}\left[1 + \left(\dfrac{1.9}{Pr}\right)^{0.9}\right]^{2/9}}\right\}} \qquad (6\text{-}59)$$

The correlation provided by Eq. (6-59) is valid for $1 \times 10^3 < Ra_{L_{char}} < 1 \times 10^{10}$ and is implemented by the procedure **FC_plate_horizontal2_ND** in EES. Figure 6-10 illustrates the average Nusselt number as a function of the Rayleigh number for various values of the Prandtl number.

If the plate were cooled (i.e., $T_s < T_\infty$) and facing upward, then the flow pattern would appear as shown in Figure 6-9, but flipped over so that the fluid tends to fall downwards after it is cooled by the plate. The average Nusselt number for an upward facing cooled plate can be computed using the correlations presented in this section; note that the Rayleigh and Grashof numbers must be defined based on the absolute value of the temperature difference in this instance.

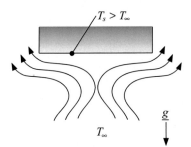

Figure 6-9: Flow induced by a horizontal heated plate facing downward.

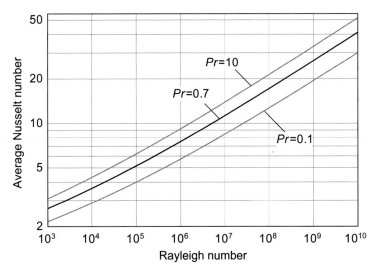

Figure 6-10: The average Nusselt number for a heated horizontal vertical plate facing downward as a function of Rayleigh number for various values of the Prandtl number.

Plate at an Arbitrary Tilt Angle

The correlations presented thus far can be used to calculate the heat transfer coefficient for a heated plate that is horizontal upward facing ($\zeta = 0$ rad), vertical ($\zeta = \pi/2$ rad), and horizontal downward facing ($\zeta = \pi$ rad), as shown in Figure 6-11(a). These correlations are implemented in EES as functions FC_plate_horizontal1_ND,

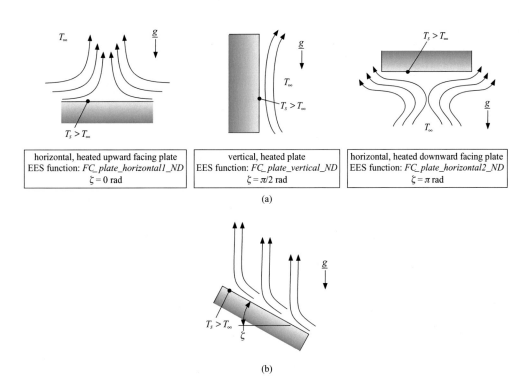

Figure 6-11: Heated plate (a) in a horizontal upward facing, vertical, and horizontal downward facing configuration, and (b) at an arbitrary angle ζ.

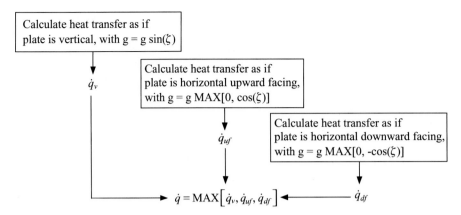

Figure 6-12: Methodology used to calculate the heat transfer rate from a heated plate at arbitrary angle.

FC_plate_vertical_ND, and FC_plate_horizontal2_ND, respectively. Raithby and Hollands (1998) present a methodology that uses these correlations to estimate the heat transfer from a heated plate that is inclined at an arbitrary angle, $0 < \zeta < \pi$ rad, as shown in Figure 6-11(b).

The procedure is illustrated in Figure 6-12 and requires that each of the three functions shown in Figure 6-11(a) is called using an appropriate projection of the gravity vector. The maximum of the three heat transfer rates that are calculated is taken as the best estimate of the actual heat transfer rate. (Note that it is not correct to take the maximum of the three Nusselt numbers, since different length scales are used to define the Nusselt number for the various orientations.) This methodology is implemented in the EES dimensional function FC-plate-tilted. For a cooled plate, the same procedure applies but the tilt angle should be adjusted to be $(\pi - \zeta)$ rad.

EXAMPLE 6.2-1: AIRCRAFT FUEL ULLAGE HEATER

EXAMPLE 6.2-1: AIRCRAFT FUEL ULLAGE HEATER

A rectangular plate heater is placed in the ullage space of a fuel tank on a military aircraft, as shown in Figure 1.

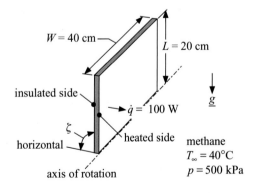

Figure 1: Ullage heater plate.

One side of the heater is insulated and the other is heated. The heater is normally oriented vertically with respect to gravity and achieves a nearly uniform temperature. The length of the heater is $L = 20$ cm and the width is $W = 40$ cm. The plate is exposed to fuel that has properties consistent with methane at $T_\infty = 40°C$ and $p = 500$ kPa. The aircraft undergoes maneuvers and so the orientation of the heater

EXAMPLE 6.2-1: AIRCRAFT FUEL ULLAGE HEATER

with respect to gravity may change dramatically. The plate will rotate about the axis of rotation shown in Figure 1. The heater power is $\dot{q} = 100$ W regardless of the orientation and so the surface temperature of the plate (T_s) changes depending on the natural convection heat transfer coefficient between the plate and the fuel.

a) If the tilt of the plate relative to horizontal is $\zeta = 60°$ (where $0°$ is horizontal with the heated surface facing upward) then determine the surface temperature of the heater.

The known information is entered in EES:

```
"EXAMPLE 6.2-1: Aircraft Fuel Ullage Heater"

$UnitSystem SI MASS RAD PA K J
$Tabstops 0.2 0.4 0.6 3.5 in

"Inputs"
L=20 [cm]*convert(cm,m)              "length of the plate (perpendicular to axis of tilt)"
W=40 [cm]*convert(cm,m)              "width of plate (parallel to axis of tilt)"
Fluid$='Methane'                     "fuel"
T_infinity=converttemp(C,K,40 [C])   "fuel temperature"
p=500 [kPa]*convert(kPa,Pa)          "fuel pressure"
tilt=60 [degree]*convert(degree,rad) "tilt angle"
q_dot=100 [W]                        "heat transfer rate"
```

The fluid properties of the methane (β, ρ, μ, k, c, ν, α, and Pr) are obtained using the built-in EES property functions. The properties should be evaluated at the film temperature, which is the average of the fluid and surface temperatures:

$$T_{film} = \frac{(T_s + T_\infty)}{2}$$

The surface temperature is not yet known. The best method for proceeding with the solution is to assume a value of the surface temperature. After the calculations have been successfully completed and an estimate of the surface temperature, T_s, is available, then the guess values can be updated and the assumed value of the surface temperature will be commented out.

```
T_s=400 [K]                          "assumed surface temperature"

"Fluid properties"
T_film=(T_s+T_infinity)/2            "film temperature"
beta=VolExpCoef(Fluid$,T=T_film,P=p) "volumetric thermal expansion coefficient"
rho=density(Fluid$,T=T_film,P=p)     "density"
mu=viscosity(Fluid$,T=T_film,P=p)    "viscosity"
k=conductivity(Fluid$,T=T_film,P=p)  "conductivity"
c=cP(Fluid$,T=T_film,P=p)            "specific heat capacity"
nu=mu/rho                            "kinematic viscosity"
alpha=k/(rho*c)                      "thermal diffusivity"
Pr=nu/alpha                          "Prandtl number"
```

The heat transfer rate is computed using the technique illustrated in Figure 6-12.

EXAMPLE 6.2-1: AIRCRAFT FUEL ULLAGE HEATER

The heat transfer coefficient is first estimated using the correlation for a vertical plate, with the Rayleigh number computed using the projection of gravity onto the plate surface:

$$Ra_{L,v} = \frac{g \sin(\zeta)\, L^3\, \beta\, (T_s - T_\infty)}{\nu\, \alpha}$$

The procedure FC_plate_vertical_ND is used to determine the average Nusselt number for the vertical plate calculation ($\overline{Nu}_{L,v}$) and the associated heat transfer coefficient:

$$\overline{h}_v = \frac{\overline{Nu}_{L,v}\, k}{L}$$

Note that the characteristic length used to compute the Rayleigh number is L.

```
"vertical plate calculation"
Ra_L_v=g#*sin(tilt)*L^3*beta*(T_s-T_infinity)/(nu*alpha)          "Raleigh number"
Call FC_plate_vertical_ND(Ra_L_v, Pr: Nusselt_L_v)               "Nusselt number"
h_v=Nusselt_L_v*k/L                                              "heat transfer coefficient"
```

The heat transfer coefficient is next estimated using the correlation for a horizontal heated upward facing plate with modified gravity, as shown in Figure 6-12. The characteristic length of the upward facing plate is calculated according to:

$$L_{char} = \frac{L\,W}{2\,(L+W)}$$

and used to compute a Raleigh number:

$$Ra_{L_{char},uf} = \frac{g\, \text{MAX}\,[0, \cos(\zeta)]\, L_{char}^3\, \beta(T_s - T_\infty)}{\nu\, \alpha}$$

The procedure FC_plate_horizontal1_ND is used to determine the average Nusselt number for a horizontal, upward facing plate ($\overline{Nu}_{L_{char},uf}$) and the associated heat transfer coefficient:

$$\overline{h}_{uf} = \frac{\overline{Nu}_{L_{char},uf}\, k}{L_{char}}$$

```
"horizontal upward calculation"
L_char=(L*W)(2*L+2*W)                                            "characteristic length"
Ra_uf=g# *MAX(0,cos(tilt))*L_char^3*beta*(T_s-T_infinity)/(nu*alpha)
                                                                "Raleigh number"
Call FC_plate_horizontal1_ND(Ra_uf, Pr: Nusselt_uf)             "Nusselt number"
h_uf=Nusselt_uf*k/L_char                                         "heat transfer coefficient"
```

The heat transfer coefficient is finally estimated using the correlation for a horizontal heated downward facing plate with modified gravity, as shown in Figure 6-12. The Raleigh number is:

$$Ra_{L_{char},uf} = \frac{g\, \text{MAX}\,[0, -\cos(\zeta)]\, L_{char}^3\, \beta(T_s - T_\infty)}{\nu\, \alpha}$$

The procedure FC_plate_horizontal2_ND is used to determine the average Nusselt number for the horizontal, downward facing plate $(\overline{Nu}_{L_{char},df})$ and the associated heat transfer coefficient:

$$\overline{h}_{df} = \frac{\overline{Nu}_{L_{char},df}\, k}{L_{char}}$$

```
"horizontal downward calculation"
Ra_df=g#*MAX(0.01,cos(-tilt))*L_char^3*beta*(T_s-T_infinity)/(nu*alpha)
Call FC_plate_horizontal2_ND(Ra_df, Pr: Nusselt_df)          "Nusselt number"
h_df=Nusselt_df*k/L_char                                     "heat transfer coefficient"
```

The maximum of the three heat transfer coefficients is used to compute the rate of heat transfer from the plate:

$$\overline{h} = \text{MAX}\left(\overline{h}_v, \overline{h}_{uf}, \overline{h}_{df}\right)$$

```
h=max(h_v,h_uf,h_df)                    "actual heat transfer coefficient"
```

At this point, it is best to update the guess values and comment out the assumed value of the surface temperature. The problem is fully specified by computing the surface temperature according to:

$$\dot{q} = \overline{h}\,L\,W\,(T_s - T_\infty)$$

```
{T_s=400 [K]}                   "assumed surface temperature"
q_dot=h*L*W*(T_s-T_infinity)    "heat transfer rate"
T_s_C=converttemp(K,C,T_s)      "surface temperature in C"
```

The result leads to a surface temperature of $T_s = 379.2\,\text{K}$ (106.1°C).

b) **Prepare a plot of the surface temperature as a function of tilt angle (where 0 is horizontal upward facing and π radian is horizontal downward facing, as shown in Figure 1). If the maximum allowable plate temperature is 125°C, then what are the limits on the tilt angle that the plane can experience?**

The tilt specification is commented out of the Equations Window and a parametric table is generated in which tilt is varied from near 0 radian to near π radian. Note that a tilt angle of exactly 0 radian and π radian will lead to singularities in the correlations. Figure 2 shows the plate surface temperature as a function of the tilt angle.

EXAMPLE 6.2-1: AIRCRAFT FUEL ULLAGE HEATER

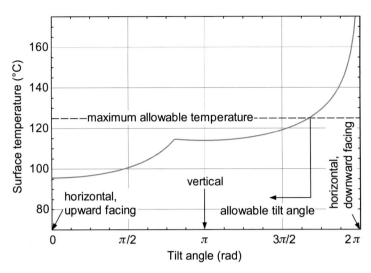

Figure 2: The plate temperature as a function of the tilt angle.

The maximum temperature occurs when the plate is horizontal facing downward. This geometry provides the smallest heat transfer coefficient because the least fluid motion is induced (i.e., the situation is most stable). Figure 2 shows that the tilt can range between approximately 0 rad (0°) and 2.6 radian (149°) without exceeding the maximum temperature limit.

6.2.3 Sphere

Natural convection from a heated or cooled sphere is correlated using a Rayleigh number and average Nusselt number that are based on the diameter of the sphere:

$$Ra_D = \frac{g\,D^3\,\beta\,(T_s - T_\infty)}{\upsilon\,\alpha} \tag{6-60}$$

$$\overline{Nu_D} = \frac{\overline{h}\,D}{k} \tag{6-61}$$

Churchill (1983) recommends the following correlation:

$$\overline{Nu_D} = 2 + \frac{0.589\,Ra_D^{0.25}}{\left[1 + \left(\frac{0.469}{Pr}\right)^{9/16}\right]^{4/9}} \quad \text{for } Pr > 0.5 \quad \text{and} \quad Ra_D < 1 \times 10^{11} \tag{6-62}$$

Note that the Nusselt number limits to a value of two at very low Rayleigh number; this is not accidental. In the limit that the Rayleigh number approaches zero, there will be no thermally induced fluid motion and therefore the problem is reduced to conduction through a stationary medium. One dimensional, steady-state conduction through a spherical shell is characterized by the spherical conduction resistance derived in Section 1.2.5 and listed in Table 1-2:

$$R_{sph} = \frac{1}{4\,\pi\,k}\left[\frac{1}{r_{in}} - \frac{1}{r_{out}}\right] \tag{6-63}$$

If the surface of the sphere (at $r_{in} = D/2$) is at T_s and the surrounding medium (removed from the proximity of the sphere, at $r_{out} \to \infty$) is at T_∞, then the rate of heat transfer from the sphere will be:

$$\dot{q} = \frac{(T_s - T_\infty)}{R_{sph}} = \frac{4\pi k (T_s - T_\infty)}{\left[\dfrac{2}{D} - \dfrac{1}{\infty}\right]} \tag{6-64}$$

or

$$\dot{q} = 2\pi k D(T_s - T_\infty) \tag{6-65}$$

The heat transfer rate may also be written in terms of the average heat transfer coefficient and the surface area of the sphere:

$$\dot{q} = \overline{h} A_s (T_s - T_\infty) \tag{6-66}$$

Expressing the average heat transfer coefficient in Eq. (6-66) in terms of an average Nusselt number and substituting the surface area for a sphere leads to:

$$\dot{q} = \underbrace{\overline{Nu}_D \frac{k}{D}}_{\overline{h}} \underbrace{\pi D^2}_{A_s} (T_s - T_\infty) = \underbrace{\overline{Nu}_D}_{=2} \pi k D(T_s - T_\infty) \tag{6-67}$$

Comparing Eqs. (6-65) and (6-67) shows that the average Nusselt number for a sphere in a stagnant medium must be two.

The correlation provided by Eq. (6-62) is implemented in EES in the procedure FC_sphere_ND. The average Nusselt number for a cooled sphere may be computed using the correlations provided in this section provided that the absolute value of the temperature difference is used.

EXAMPLE 6.2-2: FRUIT IN A WAREHOUSE

The quality of frozen fruit is particularly susceptible to changes in its storage temperature. A large warehouse full of fruit is cooled by several evaporators. The nominal temperature of the air in the warehouse is $T_{\infty,ini} = -5°C$ and the fruit is initially at this temperature. However, each of the evaporators must be periodically defrosted by running hot gas through the refrigeration tubes. Because the remaining evaporators do not have sufficient capacity to meet the freezer load, the defrost process leads to an increase in the warehouse air temperature. The air temperature in the freezer following the initiation of a defrost is given by:

$$T_\infty = \begin{cases} T_{\infty,ini} + \Delta T_\infty \sin\left(\frac{\pi t}{t_{defrost}}\right) & \text{for } t < t_{defrost} \\ T_{\infty,ini} & \text{for } t > t_{defrost} \end{cases} \tag{1}$$

where $\Delta T_\infty = 7\,K$ is the temperature rise and $t_{defrost} = 60$ min. The fruit can be modeled as spheres with diameter $D = 2.0$ cm and properties $\rho = 1000$ kg/m^3, $c = 4000$ J/kg-K, and $k = 0.9$ W/m-K.

a) Can the fruit be treated using a lumped capacitance model?

The known information is entered in EES.

EXAMPLE 6.2-2: FRUIT IN A WAREHOUSE

EXAMPLE 6.2-2: FRUIT IN A WAREHOUSE

"EXAMPLE 6.2-2: Fruit in a Warehouse"

```
$UnitSystem SI MASS RAD PA K J
$Tabstops 0.2 0.4 0.6 3.5 in

"Inputs"
D=2 [cm]*convert(cm,m)                          "diameter of fruit"
rho=1000 [kg/m^3]                               "fruit density"
c=4000 [J/kg-K]                                 "specific heat capacity"
k=0.9 [W/m-K]                                   "conductivity"
T_infinity_ini=converttemp(C,K,-5 [C])          "initial freezer air temperature"
DT_infinity=7 [K]                               "freezer air temperature change"
t_defrost=60[min]*convert(min,s)                "defrost time"
```

The Biot number that characterizes the relative importance of temperature gradients within the fruit is:

$$Bi = \frac{\bar{h}\,V}{A_s\,k} = \frac{\bar{h}\,D}{6\,k} \qquad (2)$$

where $\bar{h}$ is the average heat transfer coefficient. In order to use a lumped capacitance model, the Biot number must be much less than unity. The Biot number will be largest for a natural convection problem when the temperature difference is largest. Therefore, the largest possible temperature difference should be used to compute $\bar{h}$ in Eq. (2). The largest possible temperature difference occurs if $T = T_{\infty,ini}$ and $T_\infty = T_{\infty,ini} + \Delta T_\infty$.

```
T_infinity=T_infinity_ini+DT_infinity           "maximum ambient temperature"
T=T_infinity_ini                                "minimum fruit temperature"
```

The properties of air (ρ_a, k_a, c_a, μ_a, and β_a) are evaluated at the film temperature:

$$T_{film} = \frac{T_\infty + T}{2}$$

using EES' built-in property routines for air:

```
T_film=(T_infinity+T)/2                         "film temperature"
rho_a=density(Air,T=T_film,P=1[atm]*convert(atm,Pa))
                                                "density of air"
k_a=conductivity(Air,T=T_film)                  "conductivity of air"
c_a=cP(Air,T=T_film)                            "specific heat capacity of air"
mu_a=viscosity(Air,T=T_film)                    "viscosity of air"
beta_a=volexpcoef(Air,T=T_film)                 "volumetric expansion coefficient of air"
```

The air properties are used to evaluate the thermal diffusivity (α_a), kinematic viscosity (ν_a), and Prandtl number (Pr_a).

```
nu_a=mu_a/rho_a                                 "kinematic viscosity of air"
alpha_a=k_a/(rho_a*c_a)                          "thermal diffusivity of air"
Pr_a=nu_a/alpha_a                               "Prandtl number of air"
```

EXAMPLE 6.2-2: FRUIT IN A WAREHOUSE

The Rayleigh number based on diameter is:

$$Ra_D = \frac{g\, D^3\, \beta (T_\infty - T)}{\upsilon\, \alpha}$$

```
Ra_D=g# *D^3*beta_a*abs(T-T_infinity)/(nu_a*alpha_a)        "Rayleigh number"
```

Notice the use of the **abs** function in the Rayleigh number calculation; this ensures that even if the fruit temperature is larger than ambient (as it will be during the latter parts of the defrost process) the Rayleigh number will still be calculated appropriately. The function **FC_sphere_ND** is used to determine the average Nusselt number ($\overline{Nu}_D$) and the heat transfer coefficient:

$$\overline{h} = \frac{\overline{Nu}_D\, k_a}{D}$$

The heat transfer coefficient is used to compute the Biot number according to Eq. (2).

```
Call FC_sphere_ND(Ra_D, Pr_a: Nusselt_bar_D)        "Nusselt number"
h_bar=Nusselt_bar_D*k_a/D                           "heat transfer coefficient"
Biot=h_bar*D/(6*k)                                  "Biot number"
```

The maximum value of the Biot number will be 0.03; this is sufficiently less than unity to justify a lumped capacitance model for an engineering estimate of the fruit behavior.

b) Determine the temperature variation of the fruit during a defrost process using a numerical model that employs the lumped capacitance assumption.

Comment out the values of T and T_∞ that were used in part (a) to estimate the Biot number:

```
{T_infinity=T_infinity_ini+DT_infinity        "maximum ambient temperature"}
{T=T_infinity_ini                             "minimum fruit temperature"}
```

An energy balance on the fruit balances convection with energy storage:

$$\overline{h}\, A_s\, (T_\infty - T) = M c \frac{dT}{dt} \qquad (3)$$

where A_s is the surface area:

$$A_s = \pi D^2$$

and M is the mass:

$$M = \frac{4\,\pi}{3} \left(\frac{D}{2}\right)^3 \rho$$

```
M=rho*4*pi*(D/2)^3/3        "mass"
A_s=4*pi*(D/2)^2            "surface area"
```

EXAMPLE 6.2-2: FRUIT IN A WAREHOUSE

Solving Eq. (3) for the temperature rate of change provides the state equation for the problem:

$$\frac{dT}{dt} = \frac{\overline{h}A_s}{Mc}(T_\infty - T) \tag{4}$$

In order to numerically integrate Eq. (4) forward in time, it is necessary to compute the rate of change of the state variable (T) given an arbitrary value of the state variable and the integration variable (t). The air temperature is entered as a function of time using the If command; the If command allows conditional assignment statements in the equations window and has the protocol:

If(A, B, X, Y, Z)

The If command returns X if A < B, Y if A = B, and Z if A > B. Therefore, the air temperature variation provided by Eq. (1) can be programmed according to:

```
T_infinity=IF(time,t_defrost,T_infinity_ini+DT_infinity*sin(pi*time/t_defrost), &
    T_infinity_ini,T_infinity_ini)
        "ambient air temperature"
```

The state equation is calculated using Eq. (4):

```
dTdt=h_bar*A_s*(T_infinity-T)/(M*c)            "rate of change of fruit temperature"
```

At this point, the variable dTdt can be computed for any values of the variables T and time; you can check that this is so by entering arbitrary values of temperature and time:

```
T=300 [K]                    "arbitrary temperature – to check state eq. calc."
time=500 [s]                 "arbitrary time – to check state eq. calc"
```

which leads to dTdt = −0.021 K/s. Comment out the arbitary values of the variables T and time and allow the Integral function to vary these values.

```
{T=300 [K]                   "arbitrary temperature – to check state eq. calc."
time=500 [s]                 "arbitrary time – to check state eq. calc"}
T=T_infinity_ini+Integral(dTdt,time,0,2*t_defrost)
                             "integration of the state equation"
```

The temperatures are converted to Celsius for presentation.

```
T_infinity_C=converttemp(K,C,T_infinity)      "ambient temperature in C"
T_C=converttemp(K,C,T)                        "fruit temperature in C"
```

An integral table is used to store the results.

```
$IntegralTable time:10, T_infinity_C, T_C
```

EXAMPLE 6.2-2: FRUIT IN A WAREHOUSE

The variation of the air temperature and the fruit temperature with time is shown in Figure 1.

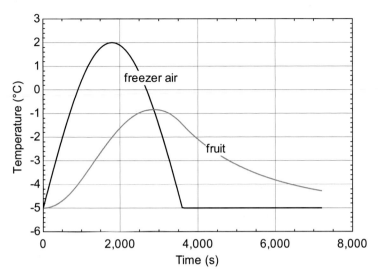

Figure 1: Freezer air and fruit temperature as a function of time during a defrost process.

6.2.4 Cylinder

Horizontal Cylinder

Natural convection from a heated or cooled cylinder (neglecting edge effects) that is horizontal with respect to gravity (i.e., gravity is perpendicular to the axis of the cylinder) is correlated using a Rayleigh number and average Nusselt number that are based on the diameter of the cylinder:

$$Ra_D = \frac{g\, D^3\, \beta (T_s - T_\infty)}{\upsilon\,\alpha} \tag{6-68}$$

$$\overline{Nu}_D = \frac{\overline{h}\, D}{k} \tag{6-69}$$

The correlation recommended by Raithby and Hollands (1998) is similar in form to the correlation for a plate and valid from $1 \times 10^{-10} < Ra_D < 1 \times 10^7$. The laminar and turbulent Nusselt numbers are calculated separately and the average Nusselt number is computed using an asymptotically weighted average of the two results. The laminar Nusselt number is given by:

$$\overline{Nu}_{D,lam} = \frac{2\, C_{cyl}}{\ln\left(1 + \dfrac{2\, C_{cyl}}{0.772\, C_{lam}\, Ra_D^{0.25}}\right)} \tag{6-70}$$

where C_{lam} is the laminar coefficient given previously by Eq. (6-49) and C_{cyl} is given by:

$$C_{cyl} = \begin{cases} 1 - \dfrac{0.13}{\left(0.772\, C_{lam}\, Ra_D^{0.25}\right)^{0.16}} & \text{for } Ra_D < 1 \times 10^{-4} \\[2ex] 0.8 & \text{for } Ra_D > 1 \times 10^{-4} \end{cases} \tag{6-71}$$

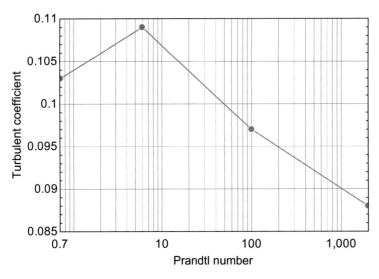

Figure 6-13: Turbulent coefficient as a function of Prandtl number.

The turbulent Nusselt number is given by:

$$\overline{Nu}_{D,turb} = C_{turb}\, Ra_D^{1/3} \tag{6-72}$$

where the turbulent coefficient C_{turb} is a weak function of the Prandtl number, as shown in Figure 6-13, and therefore can be taken to be approximately 0.1.

The average Nusselt number is calculated according to:

$$\overline{Nu}_D = \left(\overline{Nu}_{D,lam}^{10} + \overline{Nu}_{D,turb}^{10}\right)^{1/10} \tag{6-73}$$

For Rayleigh numbers greater than 1×10^7, the correlation provided by Churchill and Chu (1975) should be used:

$$\overline{Nu}_D = \left\{0.60 + \frac{0.387\, Ra_D^{1/6}}{\left[1 + \left(\dfrac{0.559}{Pr}\right)^{9/16}\right]^{8/27}}\right\}^2 \tag{6-74}$$

Figure 6-14 illustrates the average Nusselt number for a horizontal cylinder as a function of the Rayleigh number for various values of the Prandtl number.

The correlations discussed in this section are implemented in EES as the procedure FC_horizontal_cylinder_ND and they can be used for either a heated or cooled cylinder (provided that the absolute value of the temperature difference is used).

Vertical Cylinder

Natural convection from a heated or cooled cylinder that is vertical with respect to gravity (i.e., gravity is parallel to the axis of the cylinder) leads to the development of a boundary layer that starts at one end (the lower end of a heated cylinder and upper edge of a cooled cylinder) and grows in the direction of the induced fluid motion. The

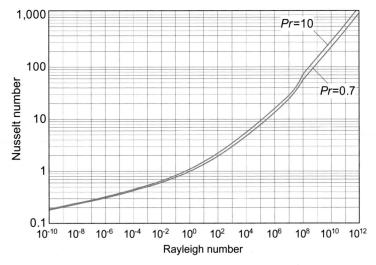

Figure 6-14: Nusselt number for a horizontal cylinder as a function of the Rayleigh number for various values of the Prandtl number.

flow pattern is therefore similar to natural convection from a vertical plate (Figure 6-4). Therefore, the Rayleigh number and average Nusselt number should be based on the length of the cylinder rather than its diameter.

$$Ra_L = \frac{g\,L^3\,\beta(T_s - T_\infty)}{\upsilon\,\alpha} \tag{6-75}$$

$$\overline{Nu_L} = \frac{\overline{h}\,L}{k} \tag{6-76}$$

where L is the length of the cylinder. A vertical cylinder can be treated as a flat plate provided that the boundary layer thickness (δ_m) is much smaller than the cylinder diameter (D). Section 4.1 shows that the momentum boundary layer thickness for laminar flow over a flat plate in forced convection is approximately:

$$\delta_{m,lam} \approx \frac{5\,x}{\sqrt{Re_x}} \tag{6-77}$$

Recall from the discussion in Section 6.1 that in natural convection problems, the square root of the Grashof number is analogous to the Reynolds number for a forced convection problem. Therefore, the boundary layer thickness at the upper edge of a heated cylinder will be approximately:

$$\delta_{m,lam} = \frac{5\,L}{Gr_L^{0.25}} \tag{6-78}$$

We will require that the ratio of the diameter to the boundary layer thickness be greater than 10 in order for the curvature of the cylinder to be insignificant:

$$\frac{D}{\delta_{m,lam}} > 10 \tag{6-79}$$

Substituting Eq. (6-78) into Eq. (6-79) leads to the criteria that:

$$\frac{D}{L} > \frac{50}{Gr_L^{0.25}} \tag{6-80}$$

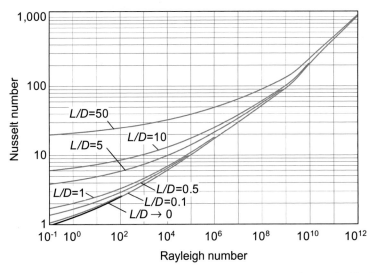

Figure 6-15: Nusselt number for a vertical cylinder as a function of the Rayleigh number for various values of L/D and $Pr = 0.7$.

Sparrow and Gregg (1956) suggest that natural convection from a vertical cylinder can be treated as a vertical flat plate when the diameter to length ratio meets the criteria:

$$\frac{D}{L} > \frac{35}{Gr_L^{0.25}} \tag{6-81}$$

For smaller diameter to length ratios, the Nusselt number is augmented by the effect of the curvature of the surface and can be obtained approximately from:

$$\overline{Nu}_L = \overline{Nu}_{L,nc} \frac{\zeta}{\ln(1+\zeta)} \tag{6-82}$$

where $\overline{Nu}_{L,nc}$ is the average Nusselt number neglecting curvature (i.e., obtained using the correlation for a vertical flat plate) and ζ is given by:

$$\zeta = \frac{1.8}{\overline{Nu}_{L,nc}} \frac{L}{D} \tag{6-83}$$

Figure 6-15 illustrates the average Nusselt number for a vertical cylinder as a function of the Rayleigh number for various values of the length-to-diameter ratio with $Pr = 0.7$. The augmentation associated with slender cylinders (i.e., cylinders with small diameters and long lengths) is most apparent at low Rayleigh numbers where the boundary layer is the thickest.

The correlation for a vertical cylinder are implemented in EES by the procedure FC_vertical_cylinder_ND. The correlations in this section can be used for either a heated or a cooled cylinder (provided that the absolute value of the temperature difference is used).

6.2.5 Open Cavity

Sections 6.2.2 through 6.2.4 explored natural convection problems associated with flow induced around geometries that are immersed in an infinitely large medium of otherwise stagnant fluid. These problems are analogous to external flow forced convection

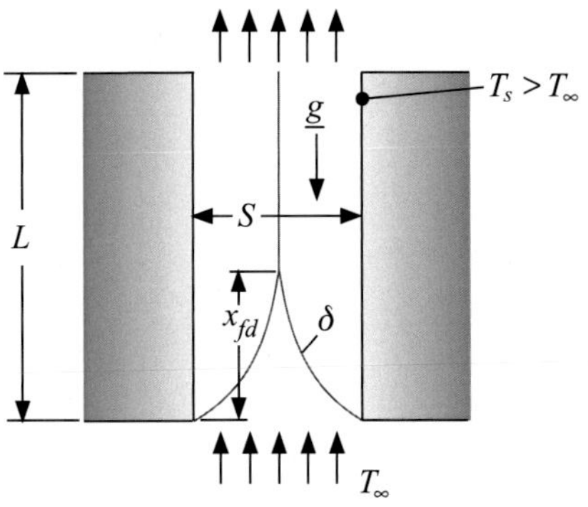

Figure 6-16: Open channel flow induced in a vertical channel formed by two vertical plates.

problems in that the boundary layers that form are unbounded. This section presents some correlations for free convection in cavities, a flow situation that is analogous to internal forced convection because the boundary layers that form are inherently bounded by the cavity walls.

Vertical Parallel Plates

A typical geometry encountered in engineering applications is an array of channels formed by parallel plates that are oriented parallel to gravity (vertical channels), as shown in Figure 6-16. The flow within these channels will resemble a forced convection internal flow problem. There is a developing region where the boundary layers are growing, followed by a fully developed region where the boundary layers are bounded. For $L < x_{fd}$, the flow can be adequately modeled using the correlations for a vertical plate (from Section 6.2.2). However, if the ratio of the channel length to spacing (L/S) is large, then the flow will become fully developed and the correlations provided in this section must be used.

The Rayleigh number and average Nusselt number for flow between parallel plates is defined based on the plate spacing, S:

$$\overline{Nu}_S = \frac{\overline{h}\,S}{k} \tag{6-84}$$

$$Ra_S = \frac{g\,S^3\,\beta(T_s - T_\infty)}{\upsilon\alpha} \tag{6-85}$$

The average heat transfer coefficient in Eq. (6-84) is defined as:

$$\overline{h} = \frac{\dot{q}}{A_s(T_s - T_\infty)} \tag{6-86}$$

where $\dot{q}$ is the total rate of heat transfer from one of the plates and A_s is the surface area of the plate. Note that the average heat transfer coefficient in Eq. (6-86) is based upon the difference between the plate surface temperature (T_s) and the temperature of the fluid that is being pulled into the channel (T_∞) rather than the temperature difference between the wall and the local mean temperature (as was the case for forced, internal flow). This difference in definition leads to substantial differences between the behavior of the Nusselt number defined for an open channel, natural convection problem as compared with the behavior of a forced convection, internal Nusselt number.

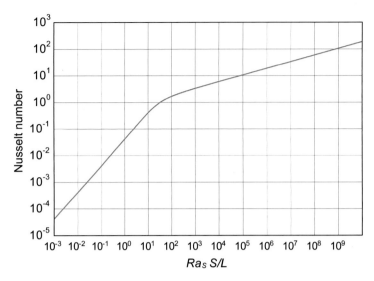

Figure 6-17: Nusselt number for open channel flow between parallel plates as a function of $Ra_S\,S/L$.

The Nusselt number is correlated against the Rayleigh number and the ratio of the channel length to spacing (L/S) according to Elenbaas (1942):

$$\overline{Nu}_S = \frac{Ra_S}{24}\frac{S}{L}\left[1 - \exp\left(-\frac{35}{Ra_S}\frac{L}{S}\right)\right]^{0.75} \tag{6-87}$$

Figure 6-17 illustrates the average Nusselt number as a function of $Ra_S\,S/L$. Notice that even for very long channels and therefore fully developed conditions ($S/L \to 0$) the Nusselt number does not approach a constant value as it does for internal flow; instead, the Nusselt number continues to decrease with the length of the channel. This behavior is a result of the definition of the average heat transfer coefficient based on the plate-to-inlet temperature difference rather than the plate-to-mean fluid temperature difference. Substituting Eq. (6-86) into Eq. (6-84) leads to:

$$\overline{Nu}_S = \frac{\dot{q}\,S}{k\,A_s(T_s - T_\infty)} \tag{6-88}$$

As the length increases, the heat transfer rate decreases because the fluid temperature will eventually approach the plate temperature. However, the temperature difference in the denominator of Eq. (6-88) remains constant and therefore the Nusselt number continues to decrease with length.

The correlation provided in this section is implemented in the EES procedure FC_Vertical_Channel_ND. Correlations for parallel plate channels under other boundary conditions (e.g., uniform heat flux or with one channel insulated) are presented by various researchers including Bar-Cohen and Rohsenow (1984). Correlations for channels with other cross-sections (e.g., circular channels) and arrays of extended surfaces (such as are commonly encountered in heat sink applications) are also available and have been compiled by Raithby and Hollands (1998).

EXAMPLE 6.2-3: HEAT SINK DESIGN

EXAMPLE 6.2-3: HEAT SINK DESIGN

You are designing a passive heat sink for an electronic component. Plates of copper with thickness $th_c = 1.5$ mm extend between two parallel surfaces, as shown in Figure 1. The space between adjacent plates forms an open channel. The copper is sufficiently conductive that these plates can be assumed to be isothermal. (This assumption could be confirmed by computing the fin efficiency of the plates.) The copper plates, and therefore the channels, are $L = 25$ cm long in the direction of gravity. The surfaces are $W = 10$ cm wide and separated by a distance $H = 10$ cm.

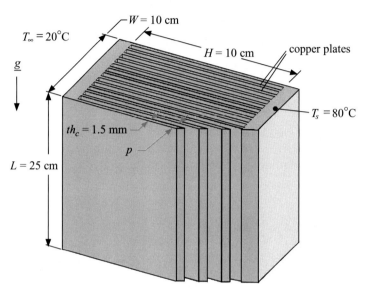

Figure 1: A heat sink fabricated using copper plates that extend between heated surfaces.

The heat sink is placed in stagnant air at a nominal temperature $T_\infty = 20°C$ and the plates are maintained at a temperature $T_s = 80°C$.

a) Determine the plate-to-plate pitch (p in Figure 1) that maximizes the rate of heat transfer from the heat sink.

The known information is entered in EES:

```
"EXAMPLE 6.2-3: Heat Sink Design"

$UnitSystem SI MASS RAD PA K J
$Tabstops 0.2 0.4 0.6 3.5 in

"Inputs"
th_c=1.5 [mm]*convert(mm,m)              "plate thickness"
L=25 [cm]*convert(cm,m)                  "length of channel in the gravity direction"
W=10 [cm]*convert(cm,m)                  "width of plates"
H=10 [cm]*convert(cm,m)                  "distance between plates"
T_s=converttemp(C,K,80 [C])             "plate surface temperature"
T_infinity=converttemp(C,K,20 [C])      "air temperature"
```

EXAMPLE 6.2-3: HEAT SINK DESIGN

The properties of air (β_a, k_a, μ_a, ρ_a, c_a, ν_a, α_a, and Pr_a) are evaluated at the film temperature using EES' internal property functions for air:

$$T_{film} = \frac{T_\infty + T_s}{2}$$

```
T_film=(T_s+T_infinity)/2              "film temperature"
beta_a=volexpcoef(Air,T=T_film)        "volumetric coefficient of thermal expansion"
k_a=conductivity(Air,T=T_film)         "conductivity"
mu_a=viscosity(Air,T=T_film)           "viscosity"
rho_a=density(Air,T=T_film,P=1[atm]*convert(atm,Pa))
                                       "density"
c_a=cP(Air,T=T_film)                   "specific heat capacity"
nu_a=mu_a/rho_a                        "kinematic viscosity"
alpha_a=k_a/(rho_a*c_a)                "thermal diffusivity"
Pr_a=nu_a/alpha_a                      "Prandtl number"
```

In order to develop the model, a plate-to-plate pitch of $p = 5.0$ mm is used; this value will be parametrically varied in order to determine the optimal value. The channel spacing (S) is equal to the plate-to-plate pitch less the plate thickness:

$$S = p - th_c$$

```
p=5 [mm]*convert(mm,m)                 "plate-to-plate spacing"
S=p-th_c                               "channel spacing"
```

The Rayleigh number based on the channel spacing is:

$$Ra_S = \frac{g\, S^3\, \beta(T_s - T_\infty)}{\nu\, \alpha}$$

The average Nusselt number ($\overline{Nu_S}$) is computed using the correlation provided by Eq. (6-87).

```
Ra_S=g#*S^3*beta_a*(T_s-T_infinity)/(nu_a*alpha_a)     "Rayleigh number"
Nusselt_bar=Ra_s*S*(1-exp(-35*L/(Ra_S*S)))^(0.75)/(24*L)   "Nusselt number"
```

The average heat transfer coefficient is computed according to:

$$\overline{h} = \frac{\overline{Nu_S}\, k}{S}$$

According to the definition of the heat transfer coefficient for an open channel flow, Eq. (6-86), the heat transfer for a single plate ($\dot{q}_{plate}$) is:

$$\dot{q}_{plate} = 2\,\overline{h}\, L\, H(T_s - T_\infty)$$

```
h_bar=k_a*Nusselt_bar/S                "heat transfer coefficient"
q_dot_plate=2*H*L*h_bar*(T_s-T_infinity)   "per-plate heat transfer rate"
```

EXAMPLE 6.2-3: HEAT SINK DESIGN

The number of plates that can be installed in the heat sink is:

$$N_{plate} = \frac{W}{p}$$

and the total heat transfer rate for the heat sink is:

$$\dot{q}_{total} = N_{plate}\, \dot{q}_{plate}$$

```
N_plate=W/p                                "number of plates"
q_dot=q_dot_plate*N_plate                   "total heat transfer rate"
```

The solution for $p = 5$ mm is $\dot{q}_{total} = 47.7$ W. The value of p is commented out and a parametric table is created that includes the variables p, q_dot_plate, and q_dot_total. The value of p is varied from 1.6 mm (just greater than the plate thickness, which leads to very thin channel spacing) to 5.0 cm (i.e., two very large channels). Figure 2 illustrates the total heat transfer rate and the rate of heat transfer per plate as a function of the plate-to-plate pitch.

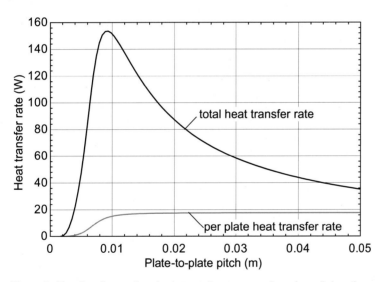

Figure 2: Total and per-plate heat transfer rate as a function of the plate-to-plate pitch.

Figure 2 shows that there is an optimal plate-to-plate pitch (around 9.0 mm) that maximizes the heat transfer rate from the heat sink. The per-plate heat transfer rate tends to increase with pitch because the developing region extends further into the channel and the fluid in the channel will warm up less. However, the number of plates that can be used decreases with pitch; the optimal plate-to-plate pitch balances these effects.

A more exact prediction of the optimal spacing can be obtained using EES' built-in optimization capability. Select Min/Max from the Calculate menu and then indicate that the variable to be maximized is q_dot_total and the independent variable is p. Reasonable bounds (0.0016 m to 0.05 m) and a guess (0.01 m) must be provided for p. The result is an optimal pitch of $p = 9.2$ mm which leads to a total heat transfer rate of 153.7 W.

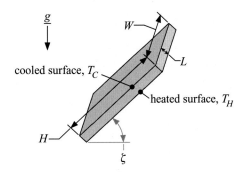

Figure 6-18: A high aspect ratio cavity tilted at angle ζ relative to horizontal.

6.2.6 Enclosures

The fluid in an enclosed volume may not remain stagnant if it is heated and cooled at different surfaces in the presence of gravity. The rectangular enclosure is encountered often in engineering applications and it has been extensively studied. Rectangular enclosures with large aspect ratios in both directions are common; such an enclosure is shown in Figure 6-18, notice that the separation distance between two walls (L) is much less than either of the other dimensions of the enclosures (H or W). Examples of such high aspect ratio enclosures include solar collectors and multi-pane windows.

The enclosure in Figure 6-18 is tilted with angle ζ relative to the horizontal. If ζ is equal to 0, then the cooled surface (at temperature T_C) lies above the heated surface (at temperature T_H). When the tilt angle reaches $\pi/2$ radian (90°), then the cooled and heated surfaces are both parallel to the gravity vector. If ζ reaches π radian (180°), then the enclosure is again horizontal but with the heated surface over the cooled surface. The free convection flows that are induced within the enclosure depend strongly on the angle of tilt. Regardless of tilt angle, the results are correlated using a Rayleigh number and an average Nusselt number that are defined based on the separation distance, L:

$$Ra_L = \frac{g\,L^3\,\beta(T_H - T_C)}{\upsilon\,\alpha} \tag{6-89}$$

$$\overline{Nu}_L = \frac{\overline{h}\,L}{k} \tag{6-90}$$

where the heat transfer coefficient is defined based on the imposed temperature difference across the enclosure:

$$\overline{h} = \frac{\dot{q}}{W\,H(T_H - T_C)} \tag{6-91}$$

and $\dot{q}$ is the total rate of heat transfer from the heated surface to the cooled surface.

At a tilt angle of $\zeta = \pi/2$ radian (vertical), the fluid adjacent to the heated wall tends to rise until it reaches the top of the cavity and comes into contact with the cooled wall. However, if the Rayleigh number is less than a critical value of $Ra_{L,crit} \approx 1000$, then the buoyancy force is insufficient to overcome the viscous force and the fluid remains stagnant. In this limit, the free convection problem reduces to a conduction problem and the heat transfer can be calculated according to:

$$\dot{q} = \frac{k\,W\,H}{L}(T_H - T_C) \tag{6-92}$$

Substituting Eq. (6-92) into Eq. (6-91) leads to:

$$\bar{h} = \frac{k}{L} \tag{6-93}$$

and therefore, according to Eq. (6-90), the Nusselt number will be 1 if $Ra_L < Ra_{L,crit}$. MacGregor and Emery (1969) recommend the following correlation for an enclosure at $\zeta = \pi/2$ radian.

$$\overline{Nu}_{L,\zeta=\pi/2} = 0.42\,Ra_L^{0.25} \left(\frac{H}{L}\right)^{-0.3} \quad \text{for } 10 < \frac{H}{L} < 40,\, 1\times 10^4 < Ra_L < 1\times 10^7 \tag{6-94}$$

and for higher Rayleigh numbers:

$$\overline{Nu}_{L,\zeta=\pi/2} = 0.046\,Ra_L^{1/3} \quad \text{for } 1 < \frac{H}{L} < 40,\, 1 < Pr < 20,\, 1\times 10^6 < Ra_L < 1\times 10^9 \tag{6-95}$$

When the tilt angle is reduced below a critical tilt angle, ζ_{crit}, then the regularly spaced, convective cells shown in Figure 6-1(a) tend to form, provided that the Rayleigh number exceeds a critical value, $Ra_{L,crit}$:

$$Ra_{L,crit} = \frac{1708}{\cos(\zeta)} \tag{6-96}$$

The critical tilt angle is a function of the aspect ratio, H/L; however, for high aspect ratio enclosures ($H/L > 12$), the critical angle is approximately $\zeta_{crit} = 1.22$ rad (or 70°). Below the critical Rayleigh number, the buoyancy force is not sufficient to overcome the viscous force that suppresses fluid motion and therefore the fluid remains stagnant. Above the critical Rayleigh number, the Nusselt number has been correlated by Hollands et al. (1976) according to:

$$\overline{Nu}_L = 1 + 1.44\,\text{MAX}\left[0,\, 1 - \frac{1708}{Ra_L\,\cos(\zeta)}\right]\left[1 - \frac{1708\,[\sin(1.8\,\zeta)]^{1.6}}{Ra_L\,\cos(\zeta)}\right] \tag{6-97}$$

$$+ \text{MAX}\left[0,\, \left(\frac{Ra_L\,\cos(\zeta)}{5830}\right)^{1/3} - 1\right] \quad \frac{H}{L},\, \frac{W}{L} > 12 \quad \text{and} \quad 0 < \zeta < 1.22 \text{ rad}$$

Note that Eq. (6-97) will reduce to 1 when $Ra_L < Ra_{L,crit}$. For tilt angles between the critical angle and vertical (i.e., 1.22 radian $< \zeta < \pi/2$ radian), Ayyaswamy and Catton (1973) recommend:

$$\overline{Nu}_L = \overline{Nu}_{L,\zeta=\pi/2}\,[\sin(\zeta)]^{0.25} \tag{6-98}$$

For tilt angles greater than vertical (i.e., $\pi/2$ radian $< \zeta < \pi$ radian), Arnold et al. (1975) recommend:

$$\overline{Nu}_L = 1 + \left[\overline{Nu}_{L,\zeta=\pi/2} - 1\right]\sin(\zeta) \tag{6-99}$$

where $\overline{Nu}_{L,\zeta=\pi/2}$ is given by either Eq. (6-94) or Eq. (6-95) depending on the conditions. Notice that if $\zeta = \pi$ radian, then the enclosure is horizontal with the heated side up. This is an unconditionally stable situation and therefore the Nusselt number approaches unity regardless of the Rayleigh number.

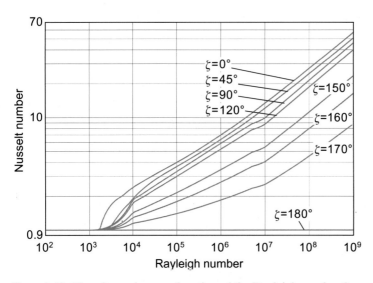

Figure 6-19: Nusselt number as a function of the Rayleigh number for various values of tilt angle.

The correlations discussed in this section are implemented in EES by the procedure Tilted_Rect_Enclosure_ND. Figure 6-19 illustrates the Nusselt number as a function of the Rayleigh number for various values of the tilt angle. Notice that the Nusselt number increases with the Rayleigh number and decreases with tilt. As the cooled edge of the enclosure moves from being at the top of the enclosure ($\zeta = 0$) to the bottom of the enclosure ($\zeta = \pi$ radian), the free convection situation becomes progressively more stable and therefore the Nusselt number is reduced.

6.2.7 Combined Free and Forced Convection

In Chapters 4 and 5, we examined forced convection as if the only fluid motion is related to the flow induced by a fan or pump. However, the discussion in this chapter indicates that fluid motion can occur without a mechanical input. Whenever an unstable temperature gradient is present in a fluid in the presence of gravity, there will be some buoyancy induced fluid motion. The relative importance of the buoyancy induced fluid motion as compared to the forced fluid motion is quantified by the ratio of the Grashof number to the Reynolds number squared. This ratio provides an index that is consistent with the discussion in Section 6.1.2, where the Grashof number is identified as the square of the Reynolds number based on the buoyancy induced velocity. In Chapters 4 and 5, we implicitly assumed that the buoyancy induced flow can be neglected in favor of forced flow. This is consistent with:

$$\frac{Gr}{Re^2} \ll 1 \rightarrow \text{ consider only forced convection effects} \qquad (6\text{-}100)$$

In this chapter, forced flow has been neglected and only buoyancy induced flow is considered. This assumption is valid provided that:

$$\frac{Gr}{Re^2} \gg 1 \rightarrow \text{ consider only free convection effects} \qquad (6\text{-}101)$$

However, there will be situations where:

$$\frac{Gr}{Re^2} \approx 1 \rightarrow \text{ both free and forced convection are important} \qquad (6\text{-}102)$$

There is a substantial body of literature that addresses these mixed convection problems. However, an approximate method for dealing with a mixed convection problem is to separately calculate the heat transfer coefficients using the appropriate forced convection correlation ($\overline{h}_{fc}$) and free convection correlation ($\overline{h}_{nc}$). A rough approximation of the actual heat transfer coefficient ($\overline{h}$) may be obtained by taking the maximum of these two values:

$$\overline{h} \approx \text{MAX}\left[\overline{h}_{fc}, \overline{h}_{nc}\right] \qquad (6\text{-}103)$$

A slightly more sophisticated methodology combines the two estimates of the heat transfer coefficient according to:

$$\overline{h} = \left[\left(\text{MAX}\left[\overline{h}_{fc}, \overline{h}_{nc}\right]\right)^m \pm \left(\text{MIN}\left[\overline{h}_{fc}, \overline{h}_{nc}\right]\right)^m\right]^{1/m} \qquad (6\text{-}104)$$

where m is typically taken to be 3. The positive sign in Eq. (6-104) is used if the natural and forced convection flows augment one another (i.e., the velocities are in the same direction or perpendicular to one another) and the negative sign is used if the natural and forced convection flows suppress one another (i.e., the velocities are in opposite directions).

EXAMPLE 6.2-4: SOLAR FLUX METER

You have been asked to evaluate a simple device for measuring the solar flux, $\dot{q}_s''$, that is incident on the side of a building, as shown in Figure 1.

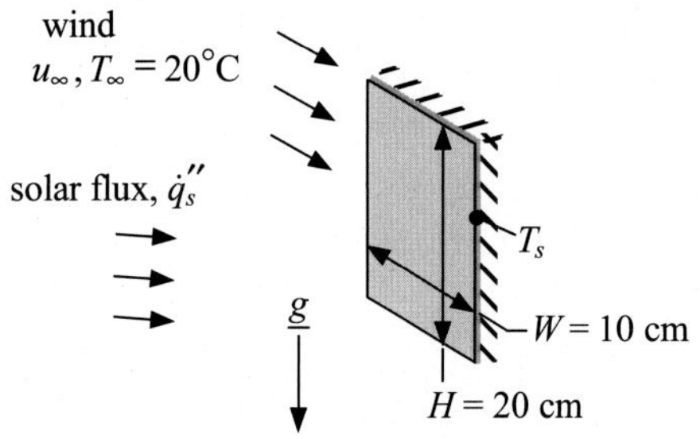

Figure 1: Simple device for measuring solar flux.

A rectangular plate made of conductive material is placed on the side of the building and the surface temperature of the plate, T_s, is related to the solar flux, $\dot{q}_s''$. However, the instrument is sensitive to the wind velocity, u_∞. The plate is surrounded by ambient air at $T_\infty = 20°C$. The plate has height $H = 20$ cm vertically (i.e., in the direction of gravity) and width $W = 10$ cm horizontally (i.e., in the wind direction). The plate is black and absorbs all of the solar flux (i.e., the emissivity is $\varepsilon = 1$). The solar flux is normal to the surface of the plate and the plate is insulated on its back surface.

a) **Develop a model of the plate and use it to prepare a plot showing the surface temperature as a function of the solar flux for various values of the wind velocity. Based on this plot, comment on the usefulness of this solar flux meter.**

EXAMPLE 6.2-4: SOLAR FLUX METER

The known information is entered in EES; note that a wind velocity and solar flux are required in order to develop the model. We will assume that the solar flux is $\dot{q}''_s = 500 \, \text{W/m}^2$ initially and parametrically vary this value once the model is complete. The wind velocity is initially assumed to be $u_\infty = 0.1$ m/s, which represents nearly still air.

```
"EXAMPLE 6.2-4: Solar flux meter"

$UnitSystem SI MASS RAD PA K J
$Tabstops 0.2 0.4 0.6 3.5 in

"Inputs"
q"_s=500 [W/m^2]                              "solar flux"
T_infinity=converttemp(C,K,20 [C])           "air temperature"
W=10 [cm]*convert(cm,m)                       "width of the plate"
H=20 [cm]*convert(cm,m)                       "height of the plate"
u_infinity=0.1 [m/s]                          "wind velocity"
```

The properties of the air are obtained at the film temperature which is related to the surface temperature of the plate. The natural convection heat transfer coefficient is also a function of the surface temperature through the Rayleigh number. Therefore, to proceed with the solution, it is convenient to assume a surface temperature; this is typical of a natural convection problem.

```
T_s=330 [K]                 "assumed surface temperature – used to get started"
```

The film temperature, T_{film}, is the average of the surface temperature and the air temperature:

$$T_{film} = \frac{T_s + T_\infty}{2}$$

The film temperature is used to compute the properties of air (ρ, k, μ, α, ν, and Pr):

```
T_film=(T_s+T_infinity)/2                     "film temperature"
rho=density(Air,T=T_film,P=1[atm]*convert(atm,Pa))   "density"
k=conductivity(Air,T=T_film)                  "conductivity"
mu=viscosity(Air,T=T_film)                    "viscosity"
c=cP(Air,T=T_film)                            "specific heat capacity"
beta=volexpcoef(Air,T=T_film)                 "volumetric coefficient of thermal expansion"
nu=mu/rho                                     "kinematic viscosity"
alpha=k/(rho*c)                               "thermal diffusivity"
Pr=nu/alpha                                   "Prandtl number"
```

The Reynolds number is based on the length of the plate in the direction of the wind flow:

$$Re = \frac{\rho \, u_\infty \, W}{\mu}$$

EXAMPLE 6.2-4: SOLAR FLUX METER

and the average Nusselt number associated with forced convection $(\overline{Nu_{fc}})$ is computed using the procedure External_Flow_Plate_ND. The average forced convection heat transfer coefficient $(\overline{h}_{fc})$ is:

$$\overline{h}_{fc} = \frac{\overline{Nu_{fc}}\,k}{W}$$

"Forced flow correlation"
Re=rho*W*u_infinity/mu "Reynolds number"
Call External_Flow_Plate_ND(Re,Pr: Nusselt_fc,C_f)
"Nusselt number for forced flow"
h_fc=Nusselt_fc*k/W "forced convection heat transfer coefficient"

The Rayleigh number is based on the length of the plate in the direction of gravity:

$$Ra = \frac{g\,H^3\,\beta(T_s - T_\infty)}{\nu\,\alpha}$$

The average Nusselt number for free convection from a vertical plate $(\overline{Nu_{nc}})$ is obtained using the procedure FC_plate_vertical_ND. The average natural convection heat transfer coefficient $(\overline{h}_{nc})$ is obtained according to:

$$\overline{h}_{nc} = \frac{\overline{Nu_{nc}}\,k}{H}$$

"Natural convection"
Ra=g#*H^3*beta*(T_s-T_infinity)/(nu*alpha) "Rayleigh number"
Call FC_plate_vertical_ND(Ra, Pr: Nusselt_nc)

"Nusselt number for free convection"
h_nc=Nusselt_nc*k/H "free convection heat transfer coefficient"

The free and forced convection flows are perpendicular to one another and therefore they augment one another. The mixed heat transfer coefficient is given by:

$$\overline{h} = \left[\overline{h}_{fc}^{\,m} + \overline{h}_{nc}^{\,m}\right]^{1/m}$$

where $m = 3$.

"Mixed convection"
h_bar=(h_fc^m+h_nc^m)^(1/m) "mixed convection heat transfer coefficient"
m=3 [-] "exponent"

An energy balance written for the plate balances solar flux against the mixed convection heat loss:

$$\dot{q}_s'' = \overline{h}\,(T_s - T_\infty)$$

The guess values for the problem are updated (Update Guesses from the Calculate menu). The assumed value for the surface temperature is commented out and replaced by the relation for the heat flux:

EXAMPLE 6.2-4: SOLAR FLUX METER

```
{T_s=330 [K]}                    "assumed surface temperature – used to get started"
q"_s=h_bar*(T_s-T_infinity)      "energy balance"
T_s_C=converttemp(K,C,T_s)       "surface temperature in C"
```

A parametric table is generated containing the variables T_s_C and q"_s. The parametric table is run several times, for various values of the variable u_infinity, in order to generate the plot requested by the problem statement, Figure 2.

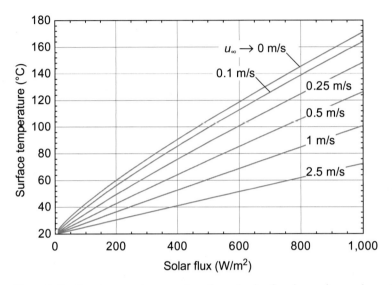

Figure 2: Surface temperature as a function of solar flux for various values of the wind velocity.

The surface temperature increases with solar flux and decreases with wind velocity, as expected. The surface temperature is a strong function of both the wind velocity and the solar flux and therefore the proposed instrument will not work unless some other method is used to measure the wind velocity.

6.3 Self-Similar Solution

This extended section can be found on the website (www.cambridge.org/nellisandklein). In this section, the development of the self-similar solution for natural convection from a vertical, isothermal heated plate is presented using the methodology provided by Ostrach (1953) and discussed in Kays and Crawford (1993). The steps leading to a solution are nearly identical to those presented in Section 4.4.2 to derive the self-similar solution for laminar forced convection flow over a flat plate. However, the definitions of the similarity variable and stream function are slightly different and the inclusion of the buoyancy force in the x-momentum equation complicates the solution.

6.4 Integral Solution

This extended section can be found on the website (www.cambridge.org/nellisandklein). Section 4.8 presents integral techniques for laminar and turbulent external forced convection flows. In this section, these techniques are extended to study problems where buoyancy induced flow is important by including the buoyancy force in the integral form of the momentum equation.

Chapter 6: Natural Convection

Many more problems can be found on the website (www.cambridge.org/nellisandklein).

Natural Convection Correlations

6–1 A pipe that is 3 m long has an outer diameter $D_{out} = 0.1$ m and is bent in the center to form an "L" shape, as shown in Figure P6-1. One leg is vertical and the other leg is horizontal. The pipe is made of thin-walled copper and saturated steam at atmospheric pressure is circulating through the pipe. The pipe is in a large room and the air temperature far from the pipe is at $T_\infty = 30°$C and atmospheric pressure. The conduction resistance associated with the pipe wall and the convection resistance associated with steam can be neglected.

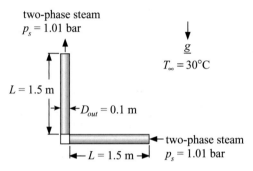

Figure P6-1: An "L"-shaped pipe.

   a.) Determine the Grashof, Rayleigh, and Nusselt numbers and the corresponding average heat transfer coefficient for the horizontal section of the pipe.
   b.) Determine the Grashof, Rayleigh, and Nusselt numbers and the corresponding average heat transfer coefficient for the vertical section of the pipe.
   c.) Calculate the total rate of heat transfer to the air.

6–2 A resistance temperature detector (RTD) is inserted into a methane pipeline to measure the gas temperature. The sensor is spherical with diameter $D = 5.0$ mm and is exposed to methane at $p_f = 10$ atm with temperature $T_f = 20°$C. The resistance of the sensor is related to its temperature; the resistance is measured by passing a known current through the RTD and measuring the associated voltage drop. The current causes an ohmic dissipation of $\dot{q} = 5.0$ milliW. You have been asked to estimate the associated self-heating error as a function of the velocity of the methane in the pipe, V_f. Focus on the very low velocity operation (e.g., 0 to 0.1 m/s) where self-heating might be large. The self-heating error is the amount that the temperature sensor surface must rise relative to the surrounding fluid in order to transfer the heat associated with ohmic dissipation. You may neglect radiation for this problem. Assume that the pipe is mounted horizontally.

   a.) Assume that only forced convection is important and prepare a plot showing the self-heating error as a function of the methane velocity for velocities ranging from 0 to 0.1 m/s.
   b.) Assume that only natural convection is important and determine the self-heating error in this limit. Overlay this value on your plot from (a).
   c.) Prepare a plot that shows your prediction for the self heating error as a function of velocity considering both natural convection and forced convection effects.

6–3 Figure P6-3 illustrates a flat plate solar collector that is mounted at an angle of $\tau = 45$ degrees on the roof of a house. The collector is used to heat water; a series of tubes are soldered to the back-side of a black plate. The collector plate is contained in a case with a glass cover.

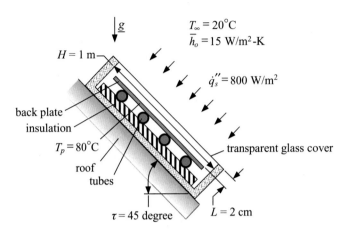

Figure P6-3: Flat plate solar collector.

Assume that the solar collector is $H = 1$ m wide by $W = 1$ m long (into the page) and the distance between the heated plate and the glass covering is $L = 2$ cm. The collector receives a solar flux $\dot{q}''_s = 800$ W/m^2 and the collector plate can be assumed to absorb all of the solar energy. The collected energy is either transferred to the water in the pipe (in which case the energy is used to provide useful water heating) or lost due to heat transfer with the environment (either by radiation, which will be neglected in this problem, or convection). The collector plate temperature is $T_p = 80°$C and the ambient temperature is $T_\infty = 20°$C. The heat transfer coefficient on the external surface of the glass is due to forced convection (there is a slight breeze) and equal to $\bar{h}_o = 15$ W/m^2-K. The glass is thin and can be neglected from the standpoint of providing any thermal resistance between the plate and ambient.

a.) Determine the rate of heat loss from plate due to convection; you may assume that the insulation on the back of the tubes is perfect so that no heat is conducted to the roof and that radiation from the plate is negligible.

b.) What is the efficiency of the solar collector, $\eta_{collector}$, defined as the ratio of the energy delivered to the water to the energy received from the sun?

c.) Prepare a plot showing the collector efficiency as a function of the plate to glass spacing, L. Explain the shape of the plot.

6–4 Figure P6-4 illustrates a single pane glass window that is $L = 6$ ft high and $W = 4$ ft wide; the glass is $th_g = 0.25$ inch thick and has conductivity $k_g = 1.4$ W/m-K.

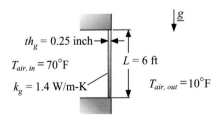

Figure P6-4: Single pane glass window.

On a typical winter day, the outdoor temperature is $T_{air,out} = 10°F$ and you keep the indoor temperature at $T_{air,in} = 70°F$.

a.) On a still winter day, estimate the rate of heat loss from the window.

b.) Winter lasts $t_{winter} = 90$ days and you are heating with electrical resistance heaters. Electricity costs $e_{cost} = \$0.12/kW\text{-hr}$. How much does the heat loss through the window cost you over the course of 1 winter?

c.) Assume that 50% of the heat loss in your house is through your windows and that you have $N_{window} = 10$ single paned windows in your house. Prepare a plot showing the cost of heating your house as a function of the thermostat set point (i.e., the indoor air temperature).

6–5 The single-glazed window in Problem 6-4 is replaced with a double-glazed window. Both glass panes are 0.25 inch thick and the gap between the panes is 0.5 inch. The gap contains dry air at atmospheric pressure. All other information is the same as in Problem 6-4. Neglect heat transfer by radiation.

a.) Repeat the calculations requested in parts (a) and (b) of Problem 6-4.

b.) Summarize and explain the benefits of the double-glazed window.

6–6 You have seen an advertisement for argon-filled windows. These windows arc similar in construction to the window described in Problem 6-5, except that argon, rather than air, is contained in the gap. Neglect heat transfer by radiation.

a.) Repeat Problem 6-5 assuming that the gap contains argon.

b.) Are the claims that argon reduces heat loss valid? If so, why does this behavior occur?

c.) Would nitrogen (which is cheaper) work as well? Why or why not? Can you suggest another gas that would work better than argon?

6–7 You are involved in a project to design a solar collector for heating air. Two competing designs are shown in Figure P6-7.

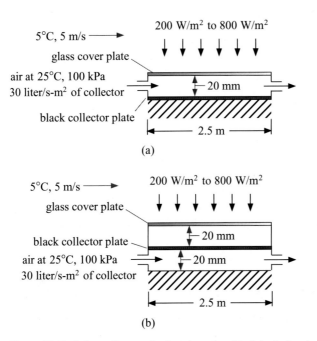

Figure P6-7: Solar collector for heating air with (a) air flowing above the collector plate and (b) air flowing below the collector plate.

Both designs employ a transparent glass cover plate and a thin metal opaque black collector plate upon which solar radiation is completely absorbed. The glazing is standard safety glass with a thickness of 6 mm. In the first design, shown in Figure P6-7(a), air is blown through the gap between the cover and plate. In the second design, shown in Figure P6-7(b), the air flows in a second gap that is below the collector plate and free convection occurs between the collector plate and the glass cover plate. The collector is 1 meter wide (into the page) and 2.5 m long (in the air flow direction) and oriented horizontally. In both designs, the gaps are 20 mm wide. Air at 25°C, 100 kPa enters the flow passage at a flow rate of 30 liters/sec per square meter of collector area (area exposed to solar radiation) in both cases. The outdoor temperature (above the glass cover plate) is 5°C and there is a wind that may be represented as a forced convective flow with a free-stream velocity of 5 m/s in the flow direction. Calculate and plot the efficiency of the two collector designs as a function of the solar radiation absorbed on the plate for values between 200 and 800 W/m^2. Assume that the insulation is adiabatic and neglect radiation (other than the absorbed solar flux) in these calculations

Self-Similar Solution

6–8 Reconsider Problem 6-4. In Problem 6-4, the glass was assumed to be isothermal and the correlations for the average heat transfer coefficient were used. In this problem, account for the variation of the local heat transfer coefficient on either side of the window using the self-similar solution. Neglect conduction along the length of the glass, but allow the glass temperature to vary with position due to the variation of the heat transfer coefficient with position.
   a.) Determine the total rate of heat transfer through the window. Compare your answer with the solution for Problem 6-4.
   b.) Plot the inner and outer temperature of the glass as a function of position.
   c.) If the relative humidity of the indoor air is 35%, will condensate form on the window? If so, at what location will the condensate end?
6–9 A self-similar solution can be obtained for the free convection problem where a heated vertical plate has a surface temperature (T_s) that varies with position according to: $T_s - T_\infty = A x^n$ where x is measured from the bottom of the plate.
   a.) Transform the governing partial differential equations for momentum conservation in the x-direction and thermal energy conservation into ordinary differential equations for f and $\tilde{\theta}$.
   b.) Transform the boundary conditions for u, v, and T into boundary conditions for f and $\tilde{\theta}$.
   c.) Develop a numerical solution for this problem.
   d.) Plot the dimensionless temperature and velocity ($\tilde{\theta}$ and $\frac{df}{d\eta}$) as a function of dimensionless position (η) for the case where $Pr = 1$ and $n = 0.5$.
   e.) Plot the product of the local Nusselt number and the Grashof number (based on the local plate temperature to the $-1/4$ power) as a function of Pr for various values of n.
   f.) Plot the product of the average Nusselt number and the Grashof number (based on the average plate temperature to the $-1/4$ power) as a function of Pr for various values of n.
   g.) Plot the average Nusselt number as a function of the Grashof number (based on the average plate temperature to the $-1/4$ power) for a plate with a constant heat flux for $Pr = 0.7$.

REFERENCES

Arnold, J. N., I. Catton, and D. K. Edwards, "Experimental Investigation of Natural Convection in Inclined Rectangular Regions of Differing Aspect Ratios," ASME Paper 75-HT-62, (1975).

Ayyaswamy, P. S. and I. Catton, *J. Heat Transfer*, Vol. 95, pp. 543, (1973).

Bar-Cohen, A., and W. M. Rohsenow, "Thermally Optimum Spacing of Vertical Natural Convection Cooled, Parallel Plates," *J. Heat Transfer*, Vol. 106, pp. 116, (1984).

Churchill, S. W., and H. H. S. Chu, "Correlating Equations for Laminar and Turbulent Free Convection from a Horizontal Cylinder," *Int. J. Heat Mass Transfer*, Vol. 18, pp. 1049, (1975).

Churchill, S. W., "Free Convection Around Immersed Bodies," in E. U. Schlünder, Ed., *Heat Exchanger Design Handbook*, Hemisphere Publishing, New York, (1983).

Elenbaas, W., "Heat Dissipation of Parallel Plates by Free Convection," *Physica*, Vol. 9, pp. 1, (1942).

Hollands, K. G. T., S. E. Unny, G. D. Raithby, and L. Konicek, *J. Heat Transfer*, Vol. 98, pp. 189, (1976).

Kays, W. M. and M. E. Crawford, *Convective Heat and Mass Transfer*, 3rd Edition, McGraw-Hill, New York, (1993).

LeFevre, E. J., "Laminar Free Convection from a Vertical Plane Surface," *Proc. Ninth Int. Congr. Appl. Mech.*, Brussels, Vol. 4, pp. 168, (1956).

MacGregor, R. K. and A. P. Emery, *J. Heat Transfer*, Vol. 91, pp. 391, (1969).

Ostrach, S., *An Analysis of Laminar Free Convection Flow and Heat Transfer about a Flat Plate Parallel to the Direction of the Generating Body Force*, National Advisory Committee for Aeronautics, Report 1111, (1953).

Raithby, G. D. and K. G. T. Hollands, *Natural Convection* in *The Handbook of Heat Transfer, 3rd Edition*, W. M. Rohsenow, J. P. Hartnett, and Y. I. Cho eds., McGraw-Hill, New York, (1998).

Sparrow, E. M, and J. L. Gregg, "Laminar Free Convection Heat Transfer from the Outer Surface of a Vertical Circular Cylinder," *Trans. ASME*, Vol. 78, pp. 1823, (1956).

7 Boiling and Condensation

7.1 Introduction

Chapters 4 through 6 discuss convection involving single-phase fluids. The thermodynamic state of single-phase fluids is sufficiently far from their vapor dome so that even though temperature variations may be present, only one phase exists (vapor or liquid). In this chapter, two-phase convection processes are examined. Two-phase processes occur when the fluid is experiencing heat transfer near the vapor dome so that vapor and liquid are simultaneously present. If the fluid is being transformed from liquid to vapor through heat addition, then the process is referred to as boiling or evaporation. If vapor is being transformed to liquid by heat removal, then the process is referred to as condensation.

Chapter 6 showed that temperature-induced density variations in a single-phase fluid may have a substantial impact on a heat transfer problem because they drive buoyancy induced fluid motion. However, the temperature-induced density gradients that are present in a typical single-phase fluid are small and so the resulting buoyancy-induced fluid velocity is also small. As a result, the heat transfer coefficients that characterize natural convection processes are usually much lower than those encountered in forced convection processes. The density difference between a vapor and a liquid is typically quite large. For example, saturated liquid water at 1 atm has a density of 960 kg/m^3 while saturated water vapor at 1 atm has a density of 0.60 kg/m^3. Large differences in density lead to correspondingly large buoyancy-induced fluid velocities and heat transfer coefficients. The heat transfer coefficients that characterize boiling and condensation processes are often much larger than those encountered in either natural convection or forced convection heat transfer with single phase fluids.

Most power and refrigeration cycles operate using thermodynamic cycles that rely on both boiling and condensation. The Rankine cycle for power systems includes a boiler where steam (water vapor) is generated through heat addition and a condenser where steam is returned to liquid through heat rejection. The vapor compression refrigeration cycle includes an evaporator where refrigerant turns from liquid to vapor at low pressure (accomplishing the heat extraction from the cooled space) and a condenser where the vapor is returned to liquid at a higher pressure (accomplishing the heat rejection process). Despite the fact that boiling and condensation are present in so many of our thermal energy conversion systems, the fundamental physics associated with these two-phase convection processes are not as well understood as they are for single-phase convection. This lack of understanding is certainly due to the complexity of the phase change process; surface tension and other forces that are not important for single-phase convection may play a dominant role in boiling and condensation processes. However, another reason that boiling and condensation are less well understood is that boiling and condensation heat transfer coefficients are generally high. As a result, the thermal resistance that limits the performance of these power and refrigeration devices is usually

778

not related to the boiling and condensation processes. For example, the condenser in a home air conditioner must transfer heat from condensing refrigerant (e.g., R134a) to outdoor air. The total thermal resistance that characterizes the condenser (the inverse of the total conductance of the heat exchanger, which is mentioned in Section 5.3.5 and will be studied in more detail in Chapter 8), is likely to be dominated by the convection to the outdoor air rather than by convection to the condensing refrigerant. Thus, an engineer focused on improving the performance of this device will be motivated to study and understand the single-phase heat transfer coefficient associated with air flow over a finned surface rather than forced condensation of the refrigerant.

Because the heat transfer coefficients that characterize two-phase heat transfer are high, engineers also use boiling and condensation processes to accomplish heat removal or addition in applications where equipment must be compact. For example, two-phase heat transfer is used for electronics cooling in high performance devices (e.g., supercomputers) and other high heat flux removal applications. The need for the high heat transfer rates provided by phase-change heat transfer processes is expected to increase in order to allow the size of this type of equipment to be reduced.

The study of two-phase heat transfer processes is an extremely rich and interesting field of research. The complexity of the processes that are involved makes them difficult to model and therefore careful experimental studies and visualization efforts are important. There are several excellent reviews that discuss these topics in detail, including Collier and Thome (1996), Thome (2006), Carey (1992), and Whalley (1987). In this chapter, a qualitative description of the boiling and condensation processes is presented and a few useful correlations are discussed; these correlations are also available as built-in functions in EES. It should be noted that there is substantially more uncertainty associated with the use of correlations for two-phase heat transfer than is associated with the single-phase heat transfer correlations discussed in Chapters 4 through 6. Correlations are typically derived from particular, limited data sets; while these correlations may be appropriate beyond the test conditions associated with the data, they should be applied with some caution. If accurate two-phase heat transfer coefficients are required, then the engineer is advised to obtain data either from the literature or through testing at conditions that are similar to the application of interest.

7.2 Pool Boiling

7.2.1 Introduction

A surface that is heated to a temperature greater than the saturation temperature of the surrounding liquid may result in evaporation (boiling). Pool boiling is analogous to natural convection in that there is no external mechanism to cause fluid motion. However, vigorous fluid motion occurs during pool boiling due to the dramatic difference in the density of the vapor that is generated by the evaporation process in comparison to the density of the surrounding liquid. Flow boiling (which is analogous to forced convection) is discussed in Section 7.3. If the temperature of the surrounding liquid is lower than the saturation temperature, then the process is referred to as sub-cooled pool boiling. If the liquid is at its saturation temperature, then the process is referred to as saturated pool boiling. The general behavior observed for pool boiling is discussed in this section and some correlations are presented that predict the behavior of the most commonly encountered mode of pool boiling, nucleate boiling, and the limit of this mode, the critical heat flux.

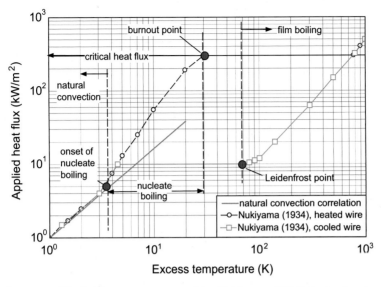

Figure 7-1: The boiling curve measured by Nukiyama (1934) during heating and cooling.

7.2.2 The Boiling Curve

A famous experiment is presented by Nukiyama (1934) in which a platinum wire submerged in a pool of saturated liquid water at ambient pressure is subjected to a controlled level of power (i.e., a controlled heat flux, $\dot{q}_s''$, from the wire surface). The heat flux is measured as a function of the excess temperature, ΔT_e, defined as the temperature difference between the surface of the wire (T_s) and the saturation temperature of the fluid (T_{sat}):

$$\Delta T_e = T_s - T_{sat} \tag{7-1}$$

The results of the experiment are shown in Figure 7-1; the applied heat flux is shown as a function of the excess temperature difference. Notice that the behavior that is observed as the wire is heated (i.e., as $\dot{q}_s''$ is increased) is substantially different than the behavior observed when the wire is cooled (i.e., as $\dot{q}_s''$ is reduced).

At low power (low excess temperature), there is no evaporation and therefore heat transfer from the wire is due to single-phase natural convection. Heated liquid near the wire surface tends to rise due to its lower density and cooler liquid from the pool flows in to take its place. The single-phase natural convection correlations presented in Section 6.2 are sufficient to predict this portion of the boiling curve. For example, the heat flux as a function of the surface-to-fluid temperature predicted by the correlation for natural convection from a horizontal cylinder presented in Section 6.2.4 and implemented using the EES function FC_horizontal_cylinder_ND is shown in Figure 7-1 and it predicts the boiling curve at low excess temperature quite well.

As the surface heat flux increases, nucleate boiling begins approximately at the point labeled 'onset of nucleate boiling' in Figure 7-1. Nucleate boiling is initially characterized by vapor bubbles that form at nucleation sites on the surface and grow until the buoyancy force is sufficient to cause them to detach from the surface and rise against gravity, as shown in Figure 7-2(a). As the heat flux increases further, more nucleation sites are activated and vapor is generated at a higher rate. Eventually, the bubbles may coalesce and form jets and columns of vapor, as shown in Figure 7-2(b).

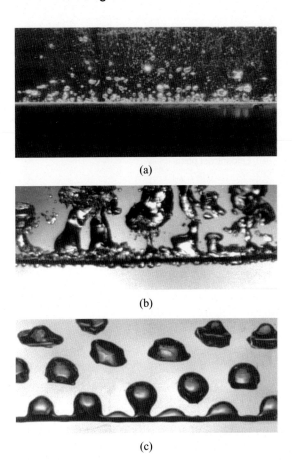

(a)

(b)

(c)

Figure 7-2: Photographs of pool boiling at (a) low temperature difference, (b) moderate temperature difference, and (c) high temperature difference. Photographs are from Lienhard and Lienhard (2005), available at http://web.mit.edu/lienhard/www/ahtt.html.

Notice that the heat flux in the nucleate boiling region is considerably higher than would be present for natural convection in a single-phase fluid at the same temperature difference. The enhancement is due to the vigorous fluid motion that is induced by the vapor bubbles leaving the surface; as a vapor bubble leaves, relatively cold liquid rushes in to take its place. Therefore, the surface is continuously being exposed to cold fluid. We learned in Chapters 4 and 5 that the heat transfer coefficient can be thought of as the ratio of the conductivity of the fluid to the distance that energy has to be conducted into the fluid. Because the cold fluid is continuously being pulled into contact with the surface, the distance that energy must be conducted is very small and the heat transfer coefficient for boiling therefore tends to be very high. The boiling curve data shown in Figure 7-1 are presented in Figure 7-3 in terms of the associated heat transfer coefficient as a function of the excess temperature; the heat transfer coefficient is defined for pool boiling in the usual way:

$$h = \frac{\dot{q}_s''}{\Delta T_e} \qquad (7\text{-}2)$$

Figure 7-3 shows the sharp increase in the heat transfer coefficient that occurs at the onset of nucleate boiling; the heat transfer coefficient rises above the value that would be expected for natural convection from a horizontal cylinder (also shown in Figure 7-3). The heat transfer coefficient continues to increase as the rate of vapor generation

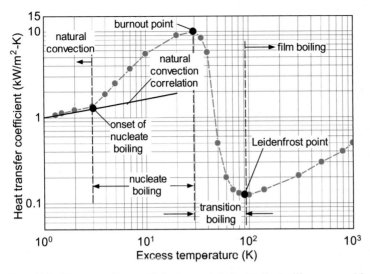

Figure 7-3: Heat transfer coefficient associated with the boiling curve. Also shown is the heat transfer coefficient predicted for natural convection from a horizontal cylinder.

increases due to the aforementioned fluid motion that is produced as vapor escapes from the wire surface.

Eventually, vapor is produced at a rate that is so high that it begins to interfere with the ability of the liquid to re-wet the surface, as shown Figure 7-2(c). The vapor phase has a substantially lower conductivity than the liquid phase (e.g., for water at 1 atm, the conductivity of saturated vapor is 0.025 W/m-K while the conductivity of saturated liquid is 0.67 W/m-K). When low conductivity vapor interferes with the heat transfer path, it causes the excess temperature to become larger for a specified heat flux. This situation is unstable because larger excess temperatures results in more vapor generation which further interferes with the flow of liquid and further increases the excess temperature. The result is a very dramatic increase in the excess temperature at the burnout point, which is indicated in Figure 7-1 and Figure 7-3. Exceeding the burnout point is often referred to as the boiling crisis because the large increase in the excess temperature that results from a small increase in the applied heat flux will tend to damage or melt most materials. For water at 1 atm, Figure 7-1 indicates that the excess temperature will rise from approximately 30 K to almost 800 K when the burnout point is reached. The heat flux at the burnout point is referred to as the peak heat flux or critical heat flux ($\dot{q}''_{s,crit}$). In most devices, it is important that the heat flux be kept below the critical heat flux so that the boiling crisis is avoided and the device operates safely in the nucleate boiling region. For this reason, the correlations available in the literature and discussed in Section 7.2.3 focus on predicting the heat transfer behavior in the nucleate boiling regime and predicting the critical heat flux.

After the wire has experienced the boiling crisis (assuming it has not melted), the excess temperature will be very high because the wire will be completely coated with vapor and heat transfer will therefore occur by conduction and radiation through the low conductivity vapor layer. If the wire is subsequently cooled, the excess temperature remains very high even as the heat flux decreases below the critical heat flux. The excess temperature during the cooling process is much higher than the excess temperature at the same heat flux during the heating process; see the data for the cooled wire in Figure 7-1. Film boiling, the condition at which the surface is completely coated with a

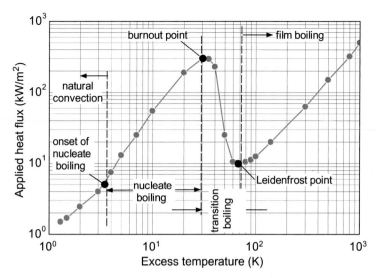

Figure 7-4: Boiling curve measured by Drew and Mueller (1937) by controlling the excess temperature.

vapor blanket, persists until the Leidenfrost point is reached. If the heat flux is decreased still further, then the excess temperature drops dramatically (from approximately 70 K to 5 K in Figure 7-1) and nucleate boiling is observed again.

The hysteresis exhibited by the boiling curve shown in Figure 7-1 is a consequence of the fact that the surface heat flux is controlled and the surface temperature measured. In this situation, a single value of the surface heat flux can be associated with two different values of the surface temperature. For example, Figure 7-1 shows that it is possible to provide 100 kW/m² under nucleate boiling conditions with the modest excess temperature of 14 K. The same heat flux of 100 kW/m² can also be provided under film boiling conditions, but only with the much larger excess temperature of 400 K. Therefore, the excess temperature that is measured when the experiment is run in a heat flux controlled mode depends on the history of the test.

Later experiments carried out by Drew and Mueller (1937) control the surface temperature and measure the heat flux; they are therefore able to measure the complete boiling curve shown in Figure 7-4. The heat transfer coefficient in Figure 7-3 is based on data of this nature. There is a unique value of the surface heat flux corresponding to each value of the excess temperature and so the boiling curve measured by Drew and Mueller does not exhibit the same hysteresis that is evident in the data collected by Nukiyama (1934).

In addition to the natural convection, nucleate boiling and film boiling modes already discussed, there is a transition boiling regime that joins the burnout point to the Leidenfrost point. Notice that the surface heat flux tends to go down as the excess temperature increases in the transition boiling regime. This behavior occurs because the amount of the surface that is completely coated by vapor increases with excess temperature so that the applied heat flux (or equivalently, the heat transfer coefficient) decreases. Notice in Figure 7-3 that the heat transfer coefficient associated with transition and film boiling is actually lower than would be expected with single-phase natural convection; these are typically undesirable operating regimes for engineering equipment.

Table 7-1: Values of the coefficient C_{nb} in Eq. (7-3) for various surface/fluid combinations, from Rohsenow (1952), Collier and Thome (1994), Vachon et al. (1968).

Fluid	Surface	C_{nb}
water	polished copper	0.0127
	lapped copper	0.0147
	scored copper	0.0068
	ground & polished stainless steel	0.0080
	teflon-pitted stainless steel	0.0058
	chemically etched stainless steel	0.0133
	mechanically polished stainless steel	0.0132
	brass	0.0060
	nickel	0.0060
	platinum	0.0130
n-pentane	polished copper	0.0154
	polished nickel	0.0127
	lapped copper	0.0049
	emery-rubbed copper	0.0074
carbon tetrachloride	polished copper	0.0070
benzene	chromium	0.0101
ethyl alcohol	chromium	0.0027
isopropyl alcohol	copper	0.0023
n-butyl alcohol	copper	0.0030

7.2.3 Pool Boiling Correlations

The heat flux ($\dot{q}_s''$) in the nucleate boiling region has been correlated by Rohsenow (1952) to excess temperature (ΔT_e), fluid properties, and an empirical constant that is related to the surface-fluid combination:

$$\dot{q}_s'' = \mu_{sat,l}\, \Delta i_{vap} \sqrt{\frac{g(\rho_{sat,l} - \rho_{sat,v})}{\sigma}} \left(\frac{c_{sat,l}\, \Delta T_e}{C_{nb}\, \Delta i_{vap}\, Pr_{sat,l}^n}\right)^3 \tag{7-3}$$

where $\mu_{sat,l}$, $\rho_{sat,l}$, $c_{sat,l}$, and $Pr_{sat,l}$ are the viscosity, density, specific heat capacity, and Prandtl number of saturated liquid, $\rho_{sat,v}$ is the density of saturated vapor, Δi_{vap} is the latent heat of vaporization (the difference between the enthalpy of the saturated vapor and the enthalpy of saturated liquid), g is the acceleration of gravity, and σ is the liquid-vapor surface tension. The dimensionless constant C_{nb} in Eq. (7-3) is an experimentally determined coefficient that depends on the surface-fluid combination. It makes sense that the characteristics of the surface play a role in the nucleate boiling behavior because the number of nucleation sites that are active depends on the surface preparation. The dimensionless exponent n on Prandtl number is equal to 1.0 for water and 1.7 for other fluids. Some values of C_{nb} are listed in Table 7-1; note that it is typical to assume that $C_{nb} = 0.013$ if there are no data available for the surface-fluid combination of interest. According to Eq. (7-3), $\dot{q}_s'' \propto \Delta T_e^3$; therefore, large errors (as much as 100%) can occur when the correlation is used to estimate the heat flux given the temperature difference.

Table 7-2: Values of C_{crit} for Eq. (7-4) for various heater geometries, from Mills (1992).

Geometry	C_{crit}	Characteristic length	Range of $\tilde{L}$
large flat plate	0.15	width or diameter	$\tilde{L} > 27$
small flat plate	$0.15 \dfrac{12 \pi L_{char,nb}^2}{A_s}$	width or diameter	$9 < \tilde{L} < 20$
large horizontal cylinder	0.12	cylinder radius	$\tilde{L} > 1.2$
small horizontal cylinder	$0.12\,\tilde{L}^{-0.25}$	cylinder radius	$0.15 < \tilde{L} < 1.2$
large sphere	0.11	sphere radius	$4.26 < \tilde{L}$
small sphere	$0.227\,\tilde{L}^{-0.5}$	sphere radius	$0.15 < \tilde{L} < 4.26$
large, finite body	≈ 0.12		

On the other hand, $\Delta T_e \propto \dot{q}_s''^{1/3}$ so the error is much smaller when Eq. (7-3) is used to estimate the excess temperature associated with a particular heat flux. The nucleate boiling correlation provided by Eq. (7-3) is programmed in EES as the function **Nucleate_Boiling**.

The critical heat flux (i.e., the heat flux at the burnout point) is also an important engineering quantity for many applications. Lienhard and Dhir (1973) suggest the following correlation for the critical heat flux ($\dot{q}_{s,crit}''$):

$$\dot{q}_{s,crit}'' = C_{crit}\, \Delta i_{vap}\, \rho_{v,sat} \left[\frac{\sigma\, g\, (\rho_{l,sat} - \rho_{v,sat})}{\rho_{v,sat}^2} \right]^{1/4} \qquad (7\text{-}4)$$

where C_{crit} is a dimensionless constant that does not depend on the surface characteristics but does depend weakly on the surface geometry. The size of the surface is characterized by a dimensionless length ($\tilde{L}$) that is defined as the ratio of the characteristic length of the surface (L, for example the radius of a cylindrical heater) to the characteristic length associated with the nucleate boiling process ($L_{char,nb}$):

$$\tilde{L} = \frac{L}{L_{char,nb}} \qquad (7\text{-}5)$$

where

$$L_{char,nb} = \sqrt{\frac{\sigma}{g(\rho_{l,sat} - \rho_{v,sat})}} \qquad (7\text{-}6)$$

Table 7-2 summarizes the values of C_{crit} for various heater geometries. This correlation for critical heat flux is implemented in the **Critical_Heat_Flux** EES library function. A correlation for film boiling is also provided in the EES boiling heat transfer library.

EXAMPLE 7.2-1: COOLING AN ELECTRONICS MODULE USING NUCLEATE BOILING

An electronics module is immersed in a pool of saturated liquid R134a at $p = 550$ kPa, as shown in Figure 1. The module is in the form of a long plate with width $W = 3.5$ cm. The heat flux from the surface of the plate is $\dot{q}''_s = 20$ W/cm^2.

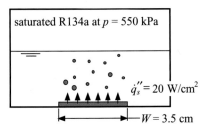

saturated R134a at $p = 550$ kPa

$\dot{q}''_s = 20$ W/cm^2

$W = 3.5$ cm

Figure 1: Electronics module immersed in a pool of saturated liquid R134a.

a) Estimate the surface temperature of the module.

The inputs are entered in EES:

```
"EXAMPLE 7.2-1"

$UnitSystem SI MASS RAD PA K J
$Tabstops 0.2 0.4 0.6 3.5 in

F$='R134a'                                          "fluid"
p=550 [kPa]*convert(kPa,Pa)                         "pressure"
W=3.5 [cm]*convert(cm,m)                            "width of module"
q"_s_Wcm2=20 [W/cm^2]                               "surface heat flux in W/cm^2"
q"_s=q"_s_Wcm2*convert(W/cm^2,W/m^2)                "surface heat flux"
```

The temperature of the pool of R134a is the saturation temperature associated with the stated pressure:

```
T_sat=temperature(F$,p=p,x=0)                       "saturation temperature"
T_sat_C=converttemp(K,C,T_sat)                      "in C"
```

which leads to $T_{sat} = 18.7°C$. The properties of saturated R134a liquid and vapor required for the correlations discussed in Section 7.2.3 ($\rho_{l,sat}$, $c_{l,sat}$, $\mu_{l,sat}$, $Pr_{l,sat}$, $\rho_{v,sat}$, σ, and Δi_{vap}) are evaluated using EES' internal property routines:

```
mu_l_sat=viscosity(F$,p=p,x=0)                      "saturated liquid viscosity"
c_l_sat=cP(F$,p=p,x=0)                              "saturated liquid specific heat capacity"
Pr_l_sat=Prandtl(F$,p=p,x=0)                        "saturated liquid Prandtl number"
rho_l_sat=density(F$,p=p,x=0)                       "saturated liquid density"
rho_v_sat=density(F$,p=p,x=1)                       "saturated vapor density"
sigma=SurfaceTension(F$,T=T_sat)                    "surface tension"
DELTAi_vap=enthalpy(F$,p=p,x=1)-enthalpy(F$,p=p,x=0)
                                                    "enthalpy of vaporization"
```

EXAMPLE 7.2-1: COOLING AN ELECTRONICS MODULE USING NUCLEATE BOILING

The Rohsenow correlation presented in Section 7.2.3 is used to evaluate the relationship between heat flux and excess temperature

$$\dot{q}''_s = \mu_{sat,l}\,\Delta i_{vap} \sqrt{\frac{g(\rho_{sat,l} - \rho_{sat,v})}{\sigma}} \left(\frac{c_{sat,l}\,\Delta T_e}{C_{nb}\,\Delta i_{vap}\,Pr^n_{sat,l}}\right)^3$$

Because the fluid-surface combination associated with this problem is not included in Table 7-1, an assumed value of $C_{nb} = 0.013$ is used for the surface coefficient. The value of the exponent n is taken to be 1.7 because the fluid is not water.

```
C_nb=0.013 [-]                                          "estimate for the coefficient"
q"_s=mu_l_sat*DELTAi_vap*sqrt(g#*(rho_l_sat-rho_v_sat)/sigma)*&
    (c_l_sat*DELTAT_e/(C_nb*DELTAi_vap*Pr_l_sat^1.7))^3
                                                        "Rohsenow's correlation for heat flux"
T_s=T_sat+DELTAT_e                                      "surface temperature"
T_s_C=converttemp(K,C,T_s)                              "in C"
```

which leads to $T_s = 41.4°C$. Note that the function Nucleate_Boiling could be used to carry out this calculation.

```
{q"_s=mu_l_sat*DELTAi_vap*sqrt(g# *(rho_l_sat-rho_v_sat)/sigma)*&
    (c_l_sat*DELTAT_e/(C_nb*DELTAi_vap*Pr_l_sat^1.7))^3}
                                                        "Rohsenow's correlation for heat flux"
q"_s= Nucleate_Boiling(F$, T_sat, T_s, C_nb)           "using the Nucleate_Boiling function"
```

which also leads to $T_s = 41.4°C$.

b) **What is the critical heat flux? That is, what is the maximum heat flux that can be applied before burnout will occur?**

The characteristic length of the nucleate boiling process is calculated according to:

$$L_{char,nb} = \sqrt{\frac{\sigma}{g(\rho_{l,sat} - \rho_{v,sat})}}$$

and used to evaluate the dimensionless size of the module:

$$\tilde{L} = \frac{W}{L_{char,nb}}$$

```
L_char_nb=sqrt(sigma/(g#*(rho_l_sat-rho_v_sat)))       "characteristic length"
L_bar=W/L_char_nb                                       "dimensionless size of the module"
```

which leads to $\tilde{L} = 40.2$. According to Table 7-2, $C_{crit} = 0.15$ for a large flat plate. The critical heat flux is estimated according to Eq. (7-4):

$$\dot{q}''_{s,crit} = C_{crit}\,\Delta i_{vap}\,\rho_{v,sat} \left[\frac{\sigma\,g(\rho_{l,sat} - \rho_{v,sat})}{\rho^2_{v,sat}}\right]^{1/4}$$

EXAMPLE 7.2-1: COOLING AN ELECTRONICS MODULE USING NUCLEATE BOILING

```
C_crit=0.15 [-]                                "coefficient for the critical heat flux equation"
q"_s_crit=C_crit*DELTAi_vap*rho_v_sat*(sigma*g# *(rho_l_sat-rho_v_sat)/&
    rho_v_sat^2)^0.25                          "critical heat flux"
q"_s_crit_Wcm2=q"_s_crit*convert(W/m^2,W/cm^2)
                                               "in W/cm^2"
```

which leads to $\dot{q}''_{s,crit} = 45.6$ W/cm^2.

c) Prepare a plot showing the surface temperature of the module as a function of the heat flux for values of heat flux up to the critical heat flux calculated in (b).

Figure 2 illustrates the surface temperature of the module as a function of the heat flux.

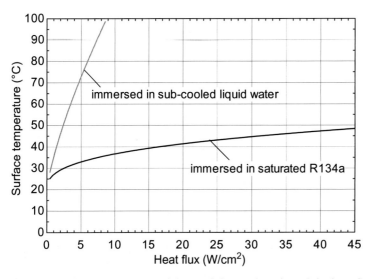

Figure 2: Surface temperature of the module as a function of the heat flux.

d) Overlay on your plot from (c) the surface temperature of the module as a function of the heat flux that would be expected if the module were immersed in a pool of sub-cooled liquid water at $p_w = 1$ atm and $T_w = 20°C$. Do not extend your plot beyond the point where the water begins to boil.

The module immersed in sub-cooled water experiences natural convection and so the problem requires the correlations for a horizontal upward facing heated plate, presented in Section 6.3.2. The specified temperature and pressure are entered in EES:

```
"free convection in water"
p_w=1 [atm]*convert(atm,Pa)                    "pressure of water pool"
T_w=converttemp(C,K,20 [C])                    "temperature of water pool"
```

In order to carry out the natural convection problem (i.e., calculate properties at the film temperature and the Raleigh number) the problem follows logically if a surface temperature is assumed and then finally calculated to complete the problem. A

EXAMPLE 7.2-1: COOLING AN ELECTRONICS MODULE USING NUCLEATE BOILING

reasonable surface temperature ($T_{s,nc}$) is assumed and used to compute the film temperature:

$$T_{film} = \frac{T_{s,nc} + T_w}{2}$$

```
T_s_nc=T_w+10 [K]                          "guess for the surface temperature"
T_film=(T_s+T_w)/2                         "film temperature"
```

The film temperature is used to evaluate the properties of water that are required for a natural convection problem ($\rho_w, c_w, \beta_w, \mu_w,$ and k_w):

```
rho_w=density(Water,p=p_w,T=T_film)        "density of water"
c_w=cP(Water,p=p_w,T=T_film)               "specific heat capacity of water"
beta_w=VolExpCoef(Water,p=p_w,T=T_film)    "volumetric expansion coefficient of water"
mu_w=viscosity(Water,p=p_w,T=T_film)       "viscosity of water"
k_w=conductivity(Water,p=p_w,T=T_film)     "conductivity of water"
```

The kinematic viscosity is:

$$v_w = \frac{\mu_w}{\rho_w}$$

The thermal diffusivity is:

$$\alpha_w = \frac{k_w}{\rho_w c_w}$$

The Prandtl number is:

$$Pr_w = \frac{v_w}{\alpha_w}$$

```
nu_w=mu_w/rho_w                            "kinematic viscosity"
alpha_w=k_w/(rho_w*c_w)                     "thermal diffusivity"
Pr_w=nu_w/alpha_w                          "Prandtl number"
```

The Rayleigh number for an upward heated flat plate is based on the characteristic length, L_{char}, defined by:

$$L_{char} = \frac{A_{plate}}{per_{plate}}$$

For a long flat plate with width W, the characteristic length is therefore:

$$L_{char} = \frac{W}{2}$$

and so the Rayleigh number is:

$$Ra = \frac{(T_{s,nc} - T_w) g \beta_w \left(\frac{W}{2}\right)^3}{v_w \alpha_w}$$

The correlation for the average Nusselt number associated with natural convection from an upward heated flat plate ($\overline{Nu}_{nc}$) is accessed using the **FC_plate_horizontal1_ND** function.

```
Ra=(T_s_nc-T_w)*g#*beta_w*(W/2)^3/(nu_w*alpha_w)
```
"Rayleigh number"
```
Call FC_plate_horizontal1_ND(Ra, Pr_w: Nusselt_bar_nc)
```
"correlation for upward facing heated plate"

The Nusselt number is used to compute the average heat transfer coefficient:

$$\bar{h}_{nc} = \frac{\overline{Nu}\,k_w}{L_{char}} = \frac{\overline{Nu}\,k_w}{\left(\dfrac{W}{2}\right)}$$

```
h_bar_nc=Nusselt_bar_nc*k_w/(W/2)
```
"natural convection heat transfer coefficient"

The guess values for the problem should be updated (select Update Guesses from the Calculate menu). The assumed surface temperature should be commented out and Newton's law of cooling used to compute the actual surface temperature:

$$\dot{q}_s'' = \bar{h}_{nc}\left(T_{s,nc} - T_w\right)$$

```
{T_s_nc=T_w+10 [K]}
T_s_nc=T_w+q"_s/h_bar_nc
T_s_nc_C=converttemp(K,C,T_s_nc)
```
"guess for the surface temperature"
"surface temperature in water pool"
"in C"

The surface temperature as a function of the heat flux is shown in Figure 2. Notice that nucleate boiling provides a substantial improvement in performance over natural convection (i.e., a substantial reduction in surface temperature at a given heat flux). This fact has motivated engineers to develop thermal management systems for electronics cooling that are based on two-phase heat transfer.

7.3 Flow Boiling

7.3.1 Introduction

Most power generation and refrigeration systems rely on evaporation of the working fluid while the fluid is flowing within a tube or annulus. This process is referred to as flow boiling. Flow boiling occurs in the evaporator of a vapor compression refrigeration cycle and in the boiler of Rankine-type power cycles. In a direct expansion evaporator of a refrigeration cycle, for example, the refrigerant enters a heat exchanger as a low quality, two-phase mixture and exits as a saturated or possibly super-heated vapor as a result of heat transfer through the tube walls from another fluid. The purpose of this section is to provide a method of estimating the heat transfer coefficient for flow boiling processes.

The physical processes involved in the evaporation of a flowing fluid are much more complicated than those associated with the single-phase forced convection heat transfer processes that are examined in Chapter 5. Figure 7-5 shows qualitatively the behavior of a sub-cooled liquid flowing through a horizontal tube in which the wall is heated to a temperature above the saturation temperature of the fluid. Near the inlet, fluid at the wall is heated and undergoes nucleate boiling, even though the bulk average temperature may be below the saturation temperature. The vapor produced by this

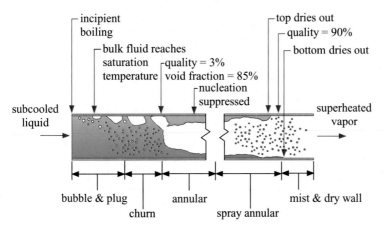

Figure 7-5: Flow Regimes occurring during flow boiling in a smooth horizontal tube.

boiling process coalesces into vapor bubbles that are distributed in the fluid. The specific volume of the vapor is usually much higher than the specific volume of liquid so that even when the vapor represents a relatively small mass fraction of the fluid (i.e., the two-phase mixture has a low quality), it may still represent a large volume fraction. The bubbles tend to concentrate near the center of the tube, forcing the liquid to towards the wall. Eventually, the flow may enter what is called an annular flow regime where the walls are coated with a liquid film and there is a vapor core. The liquid film in contact with the walls continues to produce vapor by nucleate boiling. However, vapor is also produced by evaporation of the liquid at the liquid-vapor interface in a process called convective boiling. Both the nucleate and convective boiling processes contribute to high heat transfer coefficients.

As boiling continues, the liquid film is thinned, reducing the amount of nucleate boiling. At some point, the amount of liquid is no longer sufficient to wet the entire perimeter of the tube. Due to the force of gravity, the top part of a horizontal tube will tend to dry out first. The liquid-wall interface has a much higher heat transfer coefficient than the vapor-wall interface that takes its place as dry out progresses. Therefore, the heat transfer coefficient tends to drop precipitously when dry out occurs. The fraction of the perimeter that is dry increases as the remaining liquid is vaporized, resulting in a continuous decrease in the heat transfer coefficient until single-phase conditions exist with the fluid becoming entirely vapor.

This discussion of flow boiling is simplistic and there continue to be large numbers of researchers who are working toward a more complete understanding of this complex process. The interested reader is directed towards books such as Collier and Thome (1996).

7.3.2 Flow Boiling Correlations

Correlations for the heat transfer coefficient for a single-phase fluid (either all liquid or all vapor) flowing in a tube are provided in Section 5.2. The physical situation occurring in two-phase heating processes is more complicated and, consequently, the correlations for estimating flow-boiling heat transfer coefficients are also more complicated. These correlations are almost always based on fitting some set of measurements rather than a complete model of the physical situation. Therefore, the predicted heat transfer coefficients also have larger uncertainty bands, particularly when the correlations are applied outside of the range of conditions associated with the data. It is fortunate that the heat transfer coefficients resulting from evaporation are ordinarily large and therefore the thermal resistance associated with flow boiling is not typically the limiting thermal resistance in a heat exchanger. In this situation, a large uncertainty in the heat transfer coefficient that contributes to the smallest thermal resistance will not have a significant effect on the performance of the heat exchanger.

There have been literally hundreds of correlations proposed for flow boiling heat transfer coefficients. A review of the most well-accepted correlations has been prepared by Shah (2006). The review concludes that the correlation proposed by Shah (1976, 1982) provides the most consistent agreement with the available experimental data, with a mean deviation of less than 20%. The Shah correlation was developed for saturated flow boiling at sub-critical heat fluxes and it is applicable for horizontal or vertical flow situations. The correlation can be used for a wide range of vapor qualities, ranging from saturated liquid ($x = 0$) to the liquid-deficient and dry-out regimes that occur at qualities of 0.8 or higher.

The Shah correlation was selected from the available correlations because it is applicable to any fluid in horizontal and vertical tubes and it has been compared to a large data base. The Shah correlation correlates the dimensionless heat transfer coefficient, $\tilde{h}$, in terms of the three dimensionless parameters:

$$\tilde{h} = \tilde{h}\,(Co,\ Bo,\ Fr) \tag{7-7}$$

The dimensionless heat transfer coefficient is defined as the ratio of the local heat transfer coefficient for flow boiling (h) to the local heat transfer coefficient that would occur if only the liquid-phase of the two-phase flow were present (h_l, referred to as the superficial heat transfer coefficient of the liquid phase).

$$\tilde{h} = \frac{h}{h_l} \tag{7-8}$$

The superficial heat transfer coefficient of the liquid phase is determined using the Gnielinski correlation (Gnielinski (1976)), presented in Section 5.2. The Gnielinski correlation predicts the Nusselt number and thus the heat transfer coefficient for fully developed single-phase flow under turbulent conditions:

$$h_l = \left[\frac{\left(\dfrac{f_l}{8}\right) (Re_{D_h,l} - 1000)\, Pr_{l,sat}}{1 + 12.7 \left(Pr_{l,sat}^{2/3} - 1\right)\sqrt{\dfrac{f_l}{8}}} \right] \frac{k_{l,sat}}{D_h} \tag{7-9}$$

The Reynolds number appearing in Eq. (7-9) is the liquid superficial Reynolds number which is based on the hydraulic diameter D_h of the tube and evaluated using the mass flow rate of the liquid only:

$$Re_{D_h,l} = \frac{G\,(1 - x)\,D_h}{\mu_{l,sat}} \tag{7-10}$$

where G is the mass velocity of the flow. The mass velocity of the flow is equal to the total mass flow rate of the two-phase flow ($\dot{m}$, the sum of the liquid and vapor flow rates) divided by the cross-sectional area of the tube (A_c):

$$G = \frac{\dot{m}}{A_c} \tag{7-11}$$

The properties $\mu_{l,sat}$, $k_{l,sat}$, and $Pr_{l,sat}$ in Eqs. (7-9) and (7-10) are the dynamic viscosity, thermal conductivity, and Prandtl number, respectively, of the saturated liquid. The parameter f_l in Eq. (7-9) is the friction factor associated with the flow of the liquid alone.

The Petukhov correlation (Petukhov (1970)) for fully developed single-phase flow under turbulent conditions in a smooth passage is used to evaluate f_l:

$$f_l = \frac{1}{[0.790 \ \ln{(Re_{D_h,l})} - 1.64]^2} \tag{7-12}$$

The dimensionless parameter Co is the convection number, defined according to:

$$Co = \left(\frac{1}{x} - 1\right)^{0.8} \sqrt{\frac{\rho_{v,sat}}{\rho_{l,sat}}} \tag{7-13}$$

where $\rho_{l,sat}$ and $\rho_{v,sat}$ are the densities of saturated liquid and vapor, respectively, and x is the quality.

The dimensionless parameter Bo is the boiling number, defined as the ratio of the heat flux at the wall ($\dot{q}_s''$) to the heat flux required to completely vaporize the fluid.

$$Bo = \frac{q_s''}{G \, \Delta i_{vap}} \tag{7-14}$$

where Δi_{vap} is the enthalpy of vaporization (the difference between the specific enthalpies of saturated vapor and liquid, $i_{v,sat} - i_{l,sat}$).

The dimensionless parameter Fr is the Froude number, defined as the ratio of the inertial force of the fluid to the gravitational force:

$$Fr = \frac{G^2}{\rho_{l,sat}^2 \, g \, D_h} \tag{7-15}$$

where g is the acceleration of gravity. Note that, according to Shah (1982), the Reynolds number in Eq. (7-10) should be evaluated using the liquid mass velocity, i.e., $G(1-x)$, while the Froude number in Eq. (7-15) should be evaluated using the total mass velocity, G.

The correlation for $\tilde{h}$ in terms of Co, Bo, and Fr is facilitated by defining one additional dimensionless parameter, N, in terms of the others:

$$N = \begin{cases} Co & \text{for vertical tubes or horizontal tubes with } Fr > 0.04 \\ 0.38 \, Co \, Fr^{-0.3} & \text{for horizontal tubes with } Fr \leq 0.04 \end{cases} \tag{7-16}$$

The Shah correlation is expressed by Eqs. (7-17) through (7-21):

$$\tilde{h}_{cb} = 1.8 N^{-0.8} \tag{7-17}$$

$$\tilde{h}_{nb} = \begin{cases} 230\sqrt{Bo} & \text{if } Bo \geq 0.3 \times 10^{-4} \\ 1 + 46\sqrt{Bo} & \text{if } Bo < 0.3 \times 10^{-4} \end{cases} \tag{7-18}$$

$$\tilde{h}_{bs,1} = \begin{cases} 14.70 \sqrt{Bo} \, \exp(2.74 N^{-0.1}) & \text{if } Bo \geq 11 \times 10^{-4} \\ 15.43 \sqrt{Bo} \, \exp(2.74 N^{-0.1}) & \text{if } Bo < 11 \times 10^{-4} \end{cases} \tag{7-19}$$

$$\tilde{h}_{bs,2} = \begin{cases} 14.70 \sqrt{Bo} \, \exp(2.47 N^{-0.15}) & \text{if } Bo \geq 11 \times 10^{-4} \\ 15.43 \sqrt{Bo} \, \exp(2.47 N^{-0.15}) & \text{if } Bo < 11 \times 10^{-4} \end{cases} \tag{7-20}$$

$$\tilde{h} = \begin{cases} \text{MAX}(\tilde{h}_{cb}, \tilde{h}_{bs,2}) & \text{if } N \leq 0.1 \\ \text{MAX}(\tilde{h}_{cb}, \tilde{h}_{bs,1}) & \text{if } 0.1 < N \leq 1.0 \\ \text{MAX}(\tilde{h}_{cb}, \tilde{h}_{nb}) & \text{if } N > 0.1 \end{cases} \tag{7-21}$$

The Shah correlation is implemented in the **Flow_Boiling** procedure in EES. There are a few implementation details used in the procedure **Flow_Boiling** that are not directly addressed by Shah (1982). For example, as the quality increases to 1, the Reynolds number calculated using Eq. (7-10) tends toward 0 while the Gnielinski correlation, Eq. (7-9), requires that the Reynolds number be greater than about 2300. In addition, it is necessary to ensure that the heat transfer coefficient provided by the procedure for saturated vapor ($x = 1$) is consistent with the heat transfer coefficient for single-phase vapor (i.e., the heat transfer coefficient predicted by Eq. (7-9) evaluated using the properties of the saturated vapor). Therefore, if the Reynolds number determined by Eq. (7-10) is less than 2300, then the Flow_Boiling procedure determines the quality at which Reynolds number is 2300 and computes the heat transfer coefficient at this quality. The heat transfer coefficient at a quality of 1 (i.e., for pure vapor) is also computed. The heat transfer coefficient is found by linear interpolation between these two values.

EXAMPLE 7.3-1: CARBON DIOXIDE EVAPORATING IN A TUBE

Carbon dioxide is being considered as the refrigerant for an automotive air-conditioning application because it is a natural working fluid that results in no harmful environmental effects if it released to the environment. Tests are being conducted to determine the heat transfer coefficient that should be used in a refrigeration cycle analysis of the CO_2 system. The test facility is shown in Figure 1. Saturated liquid carbon dioxide at $p_{sat} = 3.2$ MPa enters horizontal tubes having an inner diameter $D = 2.5$ mm and length $L = 2.0$ m. A constant heat flux is applied to the external surface of the tube by an electrical heating tape. The heat flux is adjusted so that the carbon dioxide exits the tubes as a saturated vapor.

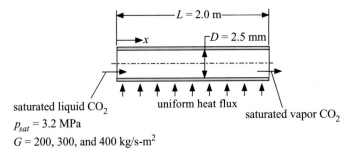

saturated liquid CO_2
$p_{sat} = 3.2$ MPa
$G = 200, 300,$ and 400 kg/s-m^2

Figure 1: Carbon dioxide flow boiling in a horizontal tube.

a) Calculate and plot the local convective boiling heat transfer coefficient predicted by the Shah correlation for this test as a function of the quality, x, for mass velocities, $G = 200, 300,$ and 400 kg/s-m^2.

The known information is entered into EES.

```
"EXAMPLE 7.3-1 "

$UnitSystem SI MASS RAD PA K J
$Tabstops 0.2 0.4 0.6 3.5 in

F$ = 'CarbonDioxide'                          "fluid type"
D=2.5 [mm]*convert(mm,m)                       "tube inner diameter"
L=2 [m]                                        "tube length"
p_sat=3.2 [MPa]*convert(MPa,Pa)                "boiling saturation pressure"
```

EXAMPLE 7.3-1: CARBON DIOXIDE EVAPORATING IN A TUBE

The calculations are carried out at a particular value of x and G; these quantities will later be varied in order to prepare the plot requested in the problem statement:

```
x=0.5 [-]                                    "quality"
G=200 [kg/s-m^2]                             "mass velocity"
```

The saturation temperature (T_{sat}) is determined using EES' built-in property routine for carbon dioxide. It is assumed that the pressure drop is negligible and therefore the saturation temperature is constant in the tube.

```
T_sat=T_sat(F$,P=p_sat)                      "saturation temperature"
```

The total rate of heat transfer to the fluid, $\dot{q}$, is the amount that is required to vaporize the flow. An energy balance on the fluid leads to:

$$\dot{q} = \underbrace{G \frac{\pi\, D^2}{4}}_{\dot{m}}\ \underbrace{\Delta i_{vap}}_{(i_{v,sat}-i_{l,sat})}$$

where the Δi_{vap} is the heat of vaporization:

$$\Delta i_{vap} = i_{v,sat} - i_{l,sat}$$

and $i_{v,sat}$ and $i_{l,sat}$ are the specific enthalpies of saturated carbon dioxide vapor and liquid, respectively, evaluated using EES' internal property routine.

```
i_v_sat=enthalpy(F$,T=T_sat,x=1)             "specific enthalpy of saturated vapor"
i_l_sat=enthalpy(F$,T=T_sat,x=0)             "specific enthalpy of saturated liquid"
DELTAi_vap=i_v_sat-i_l_sat                   "enthalpy change of vaporization"
q_dot=G*pi*D^2/4*DELTAi_vap                  "rate of heat transfer"
```

The heat flux at the tube inner surface, $\dot{q}_s''$, is the ratio of the total rate of heat transfer to the internal surface area of the tube:

$$\dot{q}_s'' = \frac{\dot{q}}{\pi\, D\, L}$$

```
q"_dot_s=q_dot/(pi*D*L)                       "heat flux"
```

The heat transfer coefficient and corresponding heat flux can be determined using the **Flow_Boiling** procedure in EES.

```
call Flow_Boiling(F$, T_sat, G, D, x, q"_dot_s, 'Horizontal': h, T_w)
                                             "heat transfer coefficient"
```

A parametric table is generated in which the quality is varied from 0 to 1. The heat transfer coefficient as a function of the quality for the mass velocities specified in the problem statement is shown in Figure 2.

EXAMPLE 7.3-1: CARBON DIOXIDE EVAPORATING IN A TUBE

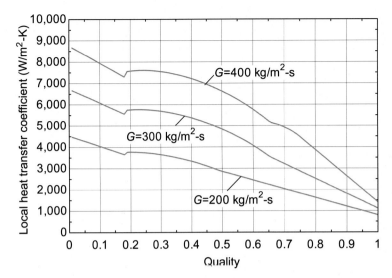

Figure 2: Local heat transfer coefficient as a function of quality for carbon dioxide evaporating in a tube at various values of the mass velocity.

b) A single, average heat transfer coefficient is to be used in the refrigeration cycle analysis of the evaporator. Calculate the average heat transfer coefficient for the quality range 0 to 1 and plot the average heat transfer coefficient as a function of the applied heat flux.

The average heat transfer coefficient over the length of the tube is defined as:

$$\bar{h} = \frac{1}{L} \int_0^L h \, ds \tag{1}$$

where L is the length of the tube and s is the position along the tube; s is used rather than x in order to avoid confusion with quality. It is more convenient in this problem to integrate with respect to quality than position because the heat transfer coefficient is an explicit function of quality. The quality (x) and position along the tube (s) are related by a differential energy balance on the CO_2, as shown in Figure 3.

$$\left[G\pi \frac{D^2}{4}(i_L + x\Delta i_{vap}) \right]_s \longrightarrow \longrightarrow \left[G\pi \frac{D^2}{4}(i_L + x\Delta i_{vap}) \right]_{s+ds}$$

$$\dot{q}_s'' \pi \, D \, ds$$

Figure 3: Differential energy balance.

The differential energy balance suggested by Figure 3 is:

$$\left[G\pi \frac{D^2}{4}(i_l + x\,\Delta i_{vap}) \right]_s + \dot{q}_s'' \pi \, D \, ds = \left[G\pi \frac{D^2}{4}(i_l + x\,\Delta i_{vap}) \right]_{s+ds}$$

The $s + ds$ term can be expanded:

$$\left[G \pi \frac{D^2}{4}(i_l + x \, \Delta i_{vap}) \right]_s + \dot{q}''_s \pi D \, ds$$

$$= \left[G \pi \frac{D^2}{4}(i_l + x \, \Delta i_{vap}) \right]_s + G \pi \frac{D^2}{4} \frac{d}{ds}(i_l + x \, \Delta i_{vap}) \, ds$$

which can be simplified:

$$\dot{q}''_s \pi D \, ds = G \pi \frac{D^2}{4} \Delta i_{vap} \, dx \qquad (2)$$

Equation (2) relates ds and dx; substituting Eq. (2) into Eq. (1) transforms the integration with respect to position (s) into an integration with respect to quality (x):

$$\bar{h} = \frac{G \, D \, \Delta i_{vap}}{4 \, L \, \dot{q}''_s} \int_0^1 h \, dx \qquad (3)$$

The specified quality is commented out and the Integral command is used to carry out the integral in Eq. (3):

```
{x=0.5 [-]}                                    "quality"
h_bar=G*D*DELTAi_vap*Integral(h,x,0,1)/(4*L*q"_dot_s)
                                    "average heat transfer coefficient"
```

A parametric table is generated that includes the mass flux, heat flux, and average heat transfer coefficient. The mass flux is varied from 200 to 500 kg/m²-s in the table and the average heat transfer coefficient is plotted a function of heat flux in Figure 4.

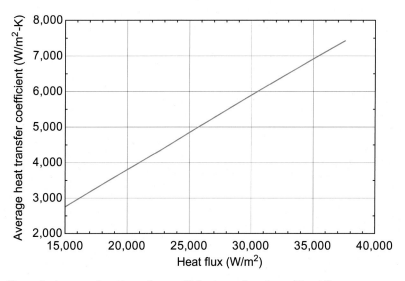

Figure 4: Average heat transfer coefficient as a function of heat flux.

EXAMPLE 7.3-1: CARBON DIOXIDE EVAPORATING IN A TUBE

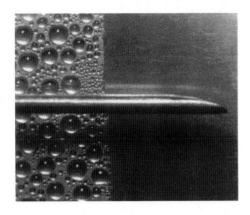

Figure 7-6: Photograph of drop condensation (on left) and film condensation (on right). Photograph from J.F. Welch and J.W. Westwater, Dept. of Chemical Engineering, University of Illinois, Urbana.

7.4 Film Condensation

7.4.1 Introduction

The condensation processes referred to as film and drop condensation processes are analogous to pool boiling for evaporation in that fluid motion is induced by density differences between the liquid and vapor. A surface is maintained at a temperature T_s, that is below the saturation temperature of a surrounding vapor, T_{sat}; therefore, liquid condenses onto the surface. Gravity causes the liquid to drain away from the surface, if possible. There are no other external forces acting on the fluid; only gravity causes the fluid motion and therefore film and drop condensation, like pool boiling, are analogous to natural convection albeit with much larger density differences and therefore larger heat transfer coefficients.

Film condensation occurs when the liquid wets the wall and therefore forms a contiguous film that is pulled downwards, as shown in Figure 7-6 (on the right). If the condensate does not wet the wall then it will bead up and form droplets that grow from nucleation sites. Eventually, the droplets break off under the force of gravity and then roll down the wall, as shown in Figure 7-6 (on the left).

In film condensation, the thermal resistance between the surface of the wall and the surrounding vapor is related to conduction through the thin film of liquid condensate on the wall. If the thickness of the film is δ, then the rate of heat transfer to the wall from the vapor is, approximately:

$$\dot{q}_s'' \approx \frac{k_{l,sat}}{\delta}\left(T_{sat} - T_s\right) \qquad (7\text{-}22)$$

where $k_{l,sat}$ is the conductivity of saturated liquid. Comparing Eq. (7-22) with Newton's law of cooling:

$$\dot{q}_s'' = h\left(T_{sat} - T_s\right) \qquad (7\text{-}23)$$

suggests that the heat transfer coefficient for film condensation is, approximately:

$$h \approx \frac{k_{l,sat}}{\delta} \qquad (7\text{-}24)$$

Equation (7-24) indicates that the heat transfer coefficient is inversely proportional to the film thickness. Since the film thickness grows in the direction of flow, it is desirable to limit the length of the surface using short vertical surfaces. Therefore, horizontal tubes are often used for condensers.

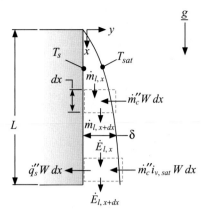

Figure 7-7: Film condensation on a vertical wall.

Because of the sweeping action of the droplets that occurs in drop condensation, the average thickness of the liquid film on the wall tends to be smaller than it is in film condensation. Also, the droplets themselves are small and well-mixed due to their motion. Therefore, the heat transfer coefficient for drop condensation processes tends to be higher than for film condensation processes. This observation has led heat exchanger designers to strive for drop condensation in heat exchangers by applying coatings and treatments that make the condensation surfaces hydrophobic. However, it is difficult to maintain drop condensation for long periods of time because surface treatments tend to lose their effectiveness over prolonged periods of operation. Therefore, while drop condensation is desirable, most design calculations assume film condensation in order to capture the conservative, long-term performance of the heat exchanger. The correlations presented in Section 7.4.3 are consistent with film condensation in a few geometries.

7.4.2 Solution for Inertia-Free Film Condensation on a Vertical Wall

Figure 7-7 illustrates steady-state film condensation on a vertical wall with a uniform surface temperature, T_s. Nusselt (1916) derived an analytical solution for the falling film problem in the limit that the inertia of the liquid can be neglected. This solution provides the basis for many of the heat transfer correlations that are presented in Section 7.4.3.

The x-momentum conservation equation, simplified for application within a boundary layer but including a buoyancy term, is derived in Chapter 6 in order to analyze natural convection problems:

$$\rho \underbrace{\left[u\frac{\partial u}{\partial x} + v\frac{\partial u}{\partial y}\right]}_{\text{inertia force}} = \underbrace{-g(\rho_{y\to\infty} - \rho)}_{\text{buoyancy force}} + \underbrace{\mu \frac{\partial^2 u}{\partial y^2}}_{\text{viscous force}} \tag{7-25}$$

Note that the negative sign associated with the buoyancy term in Eq. (7-25) is due to the fact that x is defined in the same direction as gravity in Figure 7-7. Within the condensate film, Eq. (7-25) becomes:

$$\rho_{l,sat}\left[u\frac{\partial u}{\partial x} + v\frac{\partial u}{\partial y}\right] = -g\left(\rho_{v,sat} - \rho_{l,sat}\right) + \mu_{l,sat}\frac{\partial^2 u}{\partial y^2} \tag{7-26}$$

where $\rho_{l,sat}$ and $\mu_{l,sat}$ are the density and viscosity of saturated liquid, respectively, and $\rho_{v,sat}$ is the density of saturated vapor. (The liquid properties are assumed to be constant and equal to their saturation values.) In most film condensation problems, the inertia force is small and can be neglected relative to the buoyancy and viscous forces. A scaling analysis of Eq. (7-26) leads to:

$$\underbrace{\rho_{l,sat}\frac{u_{char}^2}{L}}_{\text{inertia force}} \approx \underbrace{g(\rho_{l,sat}-\rho_{v,sat})}_{\text{buoyancy force}}+\underbrace{\mu_{l,sat}\frac{u_{char}}{\delta^2}}_{\text{viscous force}} \tag{7-27}$$

where u_{char} is a characteristic velocity. According to Eq. (7-27), the inertia force can be neglected as being small relative to the viscous force if:

$$\left(\frac{\rho_{l,sat}\,u_{char}\,\delta}{\mu_{l,sat}}\right)\left(\frac{\delta}{L}\right)\ll 1 \tag{7-28}$$

Equation (7-28) is equivalent to the modified Reynolds number condition that is derived in Section 5.4.2 in order to justify the assumption of inertia-free flow. The condition provided by Eq. (7-28) must be verified for any solution that is based on this simplification. An appropriate characteristic velocity for film condensation is identified by balancing buoyancy and viscous forces in Eq. (7-27):

$$u_{char}=\frac{g(\rho_{l,sat}-\rho_{v,sat})\delta^2}{\mu_{l,sat}} \tag{7-29}$$

If the inertia force term in Eq. (7-26) is neglected, then the simplified governing momentum equation is:

$$\mu_{l,sat}\frac{\partial^2 u}{\partial y^2}=-g(\rho_{l,sat}-\rho_{v,sat}) \tag{7-30}$$

The no-slip condition provides the boundary condition at the wall:

$$u_{y=0}=0 \tag{7-31}$$

The viscous shear must be balanced at the liquid-vapor interface. Typically, the viscosity of the vapor is much less than the viscosity of the liquid. Therefore, the shear exerted by the vapor at the interface is negligible:

$$\left.\frac{\partial u}{\partial y}\right|_{y=\delta}=0 \tag{7-32}$$

Rearranging and integrating Eq. (7-30):

$$\int \partial\left(\frac{\partial u}{\partial y}\right)=-\int\frac{g}{\mu_{l,sat}}(\rho_{l,sat}-\rho_{v,sat})\,\partial y \tag{7-33}$$

Carrying out the integration in Eq. (7-33) leads to:

$$\frac{\partial u}{\partial y}=-\frac{g}{\mu_{l,sat}}(\rho_{l,sat}-\rho_{v,sat})\,y+C_1 \tag{7-34}$$

where C_1 is a constant of integration. Integration of Eq. (7-34) results in:

$$\int \partial u=\int\left(-\frac{g}{\mu_{l,sat}}(\rho_{l,sat}-\rho_{v,sat})\,y+C_1\right)\partial y \tag{7-35}$$

or:

$$u=-\frac{g}{2\,\mu_{l,sat}}(\rho_{l,sat}-\rho_{v,sat})\,y^2+C_1\,y+C_2 \tag{7-36}$$

where C_2 is a second constant of integration. Substituting the no-slip boundary condition, Eq. (7-31), into Eq. (7-36) leads to:

$$u_{y=0} = C_2 = 0 \qquad (7\text{-}37)$$

so that Eq. (7-36) becomes:

$$u = -\frac{g}{2\,\mu_{l,sat}}(\rho_{l,sat} - \rho_{v,sat})\,y^2 + C_1\,y \qquad (7\text{-}38)$$

Substituting the boundary condition at the liquid-vapor interface, Eq. (7-32), into Eq. (7-38) leads to

$$\left.\frac{\partial u}{\partial y}\right|_{y=\delta} = -\frac{g}{\mu_{l,sat}}(\rho_{l,sat} - \rho_{v,sat})\,\delta + C_1 = 0 \qquad (7\text{-}39)$$

Solving Eq. (7-39) for C_1 and substituting into Eq. (7-38) leads to:

$$u = \underbrace{\frac{g\,\delta^2}{\mu_{l,sat}}(\rho_{l,sat} - \rho_{v,sat})}_{u_{char}}\left(-\frac{1}{2}\frac{y^2}{\delta^2} + \frac{y}{\delta}\right) \qquad (7\text{-}40)$$

Substituting the characteristic velocity for film condensation, Eq. (7-29), into Eq. (7-40) leads to:

$$u = u_{char}\left(-\frac{1}{2}\frac{y^2}{\delta^2} + \frac{y}{\delta}\right) \qquad (7\text{-}41)$$

The mass flow rate of liquid in the film ($\dot{m}_l$) is obtained by integrating the velocity distribution, Eq. (7-41), across the thickness of the film:

$$\dot{m}_l = \int_0^{\delta} u\,\rho_{l,sat}\,W\,dy = \rho_{l,sat}\,W\,u_{char}\int_0^{\delta}\left(-\frac{1}{2}\frac{y^2}{\delta^2} + \frac{y}{\delta}\right)dy \qquad (7\text{-}42)$$

where W is the width of the wall (into the page). Carrying out the integration in Eq. (7-42) leads to:

$$\dot{m}_l = \frac{\rho_{l,sat}\,W\,\delta\,u_{char}}{3} = \frac{\rho_{l,sat}\,W\,g(\rho_{l,sat} - \rho_{v,sat})\,\delta^3}{3\,\mu_{l,sat}} \qquad (7\text{-}43)$$

The steady state thermal energy conservation equation in a boundary layer is derived in Section 4.2. In the liquid film, the thermal energy conservation equation (neglecting viscous dissipation) becomes:

$$\rho_{l,sat}\,c_{l,sat}\underbrace{\left[u\frac{\partial T}{\partial x} + v\frac{\partial T}{\partial y}\right]}_{\text{energy carried by fluid flow}} = \underbrace{k_{l,sat}\frac{\partial^2 T}{\partial y^2}}_{\text{conduction}} \qquad (7\text{-}44)$$

where $c_{l,sat}$ is the specific heat capacity of the liquid. In most film condensation problems, the energy carried by the fluid flow (i.e., the term on the left hand side of Eq. (7-44)) is small and can be neglected relative to conduction. A scaling analysis of Eq. (7-44) leads to:

$$\underbrace{\frac{\rho_{l,sat}\,c_{l,sat}\,u_{char}(T_{sat} - T_s)}{L}}_{\text{energy carried by fluid flow}} = \underbrace{\frac{k_{l,sat}(T_{sat} - T_s)}{\delta^2}}_{\text{conduction}} \qquad (7\text{-}45)$$

According to Eq. (7-45), the energy carried by the flow can be neglected as being small relative to conduction provided that:

$$\left(\frac{\rho_{l,sat}\, u_{char}\, \delta}{\mu_{l,sat}} \right) \left(\frac{\delta}{L} \right) Pr_{l,sat} \ll 1 \tag{7-46}$$

Note that unless the Prandtl number of the liquid ($Pr_{l,sat}$) is very large, Eq. (7-46) will necessarily be satisfied when Eq. (7-28) is satisfied. Therefore, the energy carried by the fluid can be neglected for most situations where the inertia of the fluid flow can be neglected. With this simplification, Eq. (7-44) becomes:

$$k_{l,sat} \frac{\partial^2 T}{\partial y^2} = 0 \tag{7-47}$$

The temperature at the wall is specified:

$$T_{y=0} = T_s \tag{7-48}$$

The temperature at the liquid/vapor interface is the saturation temperature:

$$T_{y=\delta} = T_{sat} \tag{7-49}$$

Integrating Eq. (7-47) twice leads to:

$$T = C_3\, y + C_4 \tag{7-50}$$

where C_3 and C_4 are constants of integration. Substituting Eq. (7-50) into Eqs. (7-48) and (7-49) leads to:

$$T = T_s + (T_{sat} - T_s)\frac{y}{\delta} \tag{7-51}$$

The enthalpy of the liquid defined relative to the enthalpy of saturated liquid is given by:

$$i_l = c_{l,sat}\, (T - T_{sat}) \tag{7-52}$$

Substituting Eq. (7-51) into Eq. (7-52) leads to:

$$i_l = c_{l,sat}\, (T_{sat} - T_s) \left(\frac{y}{\delta} - 1 \right) \tag{7-53}$$

The total rate of energy carried by the liquid ($\dot{E}_l$) at any axial location is obtained by integrating the product of the enthalpy and velocity across the film:

$$\dot{E}_l = \int_0^\delta W\, u\, \rho_{l,sat}\, i_l\, dy \tag{7-54}$$

Substituting the velocity distribution, Eq. (7-41), and enthalpy distribution, Eq. (7-53), into Eq. (7-54) leads to:

$$\dot{E}_l = W u_{char}\, \rho_{l,sat}\, c_{l,sat}\, (T_{sat} - T_s) \int_0^\delta \left(-\frac{1}{2}\frac{y^2}{\delta^2} + \frac{y}{\delta} \right) \left(\frac{y}{\delta} - 1 \right) dy \tag{7-55}$$

Equation (7-55) is rearranged:

$$\dot{E}_l = W u_{char}\, \rho_{l,sat}\, c_{l,sat}(T_{sat} - T_s) \int_0^\delta \left(-\frac{1}{2}\frac{y^3}{\delta^3} + \frac{3}{2}\frac{y^2}{\delta^2} - \frac{y}{\delta} \right) dy \tag{7-56}$$

and integrated:

$$\dot{E}_l = W\, u_{char}\, \rho_{l,sat}\, c_{l,sat}\, (T_{sat} - T_s) \left(-\frac{1}{8}\frac{y^4}{\delta^3} + \frac{y^3}{2\,\delta^2} - \frac{y^2}{2\,\delta} \right)\Bigg|_0^{\delta} \qquad (7\text{-}57)$$

Applying the limits of integration leads to:

$$\dot{E}_l = \frac{W u_{char}\, \rho_{l,sat}\, c_{l,sat}(T_s - T_{sat})\,\delta}{8} \qquad (7\text{-}58)$$

Substituting the definition of the characteristic velocity, Eq. (7-29), into Eq. (7-58) leads to:

$$\dot{E}_l = \frac{W g\, (\rho_{l,sat} - \rho_{v,sat})\, \rho_{l,sat}\, c_{l,sat}\, \delta^3\, (T_s - T_{sat})}{8\,\mu_{l,sat}} \qquad (7\text{-}59)$$

A mass balance on a differential segment of the liquid film (see Figure 7-7) leads to:

$$\dot{m}_{l,x} + \dot{m}_c''\, W\, dx = \dot{m}_{l,x+dx} \qquad (7\text{-}60)$$

where $\dot{m}_c''$ is the mass flux of the vapor that is condensing to liquid at the interface. Expanding the $x + dx$ term in Eq. (7-60) leads to:

$$\dot{m}_c''\, W = \frac{d\dot{m}_l}{dx} \qquad (7\text{-}61)$$

Substituting $\dot{m}_l$ from Eq. (7-43) into Eq. (7-61) leads to:

$$\dot{m}_c'' = \frac{\rho_{l,sat}\, g\, (\rho_{l,sat} - \rho_{v,sat})\, \delta^2}{\mu_{l,sat}}\, \frac{d\delta}{dx} \qquad (7\text{-}62)$$

An energy balance on a differential segment of the liquid film (see Figure 7-7) leads to:

$$\dot{E}_{l,x} + \dot{m}_c''\, i_{v,sat}\, W\, dx = \dot{E}_{l,x+dx} + \dot{q}_s''\, W\, dx \qquad (7\text{-}63)$$

where $i_{v,sat}$ is the enthalpy of the saturated vapor that is condensing and $\dot{q}_s''$ is the heat flux removed from the wall surface. Because enthalpy has been defined relative to the enthalpy of saturated liquid, the enthalpy of saturated vapor in Eq. (7-63) is equal to the latent heat of vaporization:

$$i_{v,sat} = \Delta i_{vap} \qquad (7\text{-}64)$$

The heat flux into the wall is:

$$\dot{q}_s'' = k_{l,sat}\, \frac{\partial T}{\partial y}\bigg|_{y=0} \qquad (7\text{-}65)$$

Substituting the temperature distribution, Eq. (7-51), into Eq. (7-65) leads to:

$$\dot{q}_s'' = k_{l,sat}\, \frac{(T_{sat} - T_s)}{\delta} \qquad (7\text{-}66)$$

Substituting Eqs. (7-64) and (7-66) into Eq. (7-63) and expanding the $x + dx$ term leads to:

$$\dot{m}_c''\, \Delta i_{vap}\, W = \frac{d\dot{E}_l}{dx} + k_{l,sat}\, \frac{(T_{sat} - T_s)}{\delta}\, W \qquad (7\text{-}67)$$

Substituting Eqs. (7-62) and (7-59) into Eq. (7-67) leads to:

$$\begin{aligned}
&\frac{\rho_{l,sat}\, g\, (\rho_{l,sat} - \rho_{v,sat})\, \delta^2}{\mu_{l,sat}}\, \frac{d\delta}{dx}\, \Delta i_{vap} \\
&= \frac{3\, g\, (\rho_{l,sat} - \rho_{v,sat})\, \rho_{l,sat}\, c_{l,sat}\, (T_s - T_{sat})\, \delta^2}{8\,\mu_{l,sat}}\, \frac{d\delta}{dx} + k_{l,sat}\, \frac{(T_{sat} - T_s)}{\delta}
\end{aligned} \qquad (7\text{-}68)$$

Rearranging Eq. (7-68) leads to:

$$\frac{\rho_{l,sat}\, g\,(\rho_{l,sat}-\rho_{v,sat})\,\delta^2}{\mu_{l,sat}}\frac{d\delta}{dx}\left[\Delta i_{vap}+\frac{3\,c_{l,sat}\,(T_{sat}-T_s)}{8}\right]=k_{l,sat}\frac{(T_{sat}-T_s)}{\delta} \quad (7\text{-}69)$$

Equation (7-69) can be separated and integrated from $x=0$ where $\delta=0$:

$$\frac{\rho_{l,sat}\, g\,(\rho_{l,sat}-\rho_{v,sat})}{k_{l,sat}\,\mu_{l,sat}\,(T_{sat}-T_s)}\left[\Delta i_{vap}+\frac{3\,c_l\,(T_{sat}-T_s)}{8}\right]\int_0^\delta \delta^3\, d\delta=\int_0^x dx \quad (7\text{-}70)$$

Carrying out the integration in Eq. (7-70) leads to:

$$\frac{\rho_{l,sat}\, g(\rho_{l,sat}-\rho_{v,sat})}{k_{l,sat}\,\mu_{l,sat}(T_{sat}-T_s)}\left[\Delta i_{vap}+\frac{3\,c_{l,sat}\,(T_{sat}-T_s)}{8}\right]\frac{\delta^4}{4}=x \quad (7\text{-}71)$$

Solving Eq. (7-71) for the film thickness leads to:

$$\delta=\left\{\frac{4\,x\,k_{l,sat}\,\mu_{l,sat}\,(T_{sat}-T_s)}{\rho_{l,sat}\, g\,(\rho_{l,sat}-\rho_{v,sat})\left[\Delta i_{vap}+\frac{3\,c_{l,sat}\,(T_{sat}-T_s)}{8}\right]}\right\}^{1/4} \quad (7\text{-}72)$$

In most cases, the latent heat of vaporization is much larger than the sensible heat capacity associated with the sub-cooling at the wall:

$$\frac{c_{l,sat}(T_{sat}-T_s)}{\Delta i_{vap}}\ll 1 \quad (7\text{-}73)$$

and therefore the film thickness can be written approximately as:

$$\delta\approx\left[\frac{4\,x\,k_{l,sat}\,\mu_{l,sat}\,(T_{sat}-T_s)}{\rho_{l,sat}\, g\,(\rho_{l,sat}-\rho_{v,sat})\,\Delta i_{vap}}\right]^{1/4} \quad (7\text{-}74)$$

Comparing Eq. (7-66) to Newton's Law of Cooling shows that the local heat transfer coefficient is:

$$h=\frac{k_{l,sat}}{\delta} \quad (7\text{-}75)$$

which is consistent with our physical of understanding of film condensation, discussed in Section 7.4.1. Substituting Eq. (7-72) into Eq. (7-75) leads to:

$$h=\left\{\frac{\rho_{l,sat}\, g\,(\rho_{l,sat}-\rho_{v,sat})\,k_l\left[\Delta i_{vap}+\frac{3\,c_{l,sat}\,(T_{sat}-T_s)}{8}\right]}{4\,x\,\mu_{l,sat}(T_{sat}-T_s)}\right\}^{1/4} \quad (7\text{-}76)$$

If Eq. (7-73) is satisfied, then it is appropriate to substitute Eq. (7-74) into Eq. (7-75), which leads to:

$$h\approx\left[\frac{\rho_{l,sat}\, g\,(\rho_{l,sat}-\rho_{v,sat})\,k_{l,sat}\,\Delta i_{vap}}{4\,x\,\mu_{l,sat}\,(T_{sat}-T_s)}\right]^{1/4} \quad (7\text{-}77)$$

The average heat transfer coefficient, $\overline{h}$, is obtained according to:

$$\overline{h}=\frac{1}{L}\int_0^L h\, dx \quad (7\text{-}78)$$

Substituting Eq. (7-76) into Eq. (7-78) leads to:

$$\overline{h} = \frac{1}{L} \left\{ \frac{\rho_{l,sat}\, g(\rho_{l,sat} - \rho_{v,sat})\, k_{l,sat} \left[\Delta i_{vap} + \dfrac{3\, c_{l,sat}\, (T_{sat} - T_s)}{8} \right]}{4\, \mu_{l,sat}(T_{sat} - T_s)} \right\}^{1/4} \int_0^L x^{-1/4}\, dx \qquad (7\text{-}79)$$

or:

$$\overline{h} = \frac{4}{3} \left\{ \frac{\rho_{l,sat}\, g\, (\rho_{l,sat} - \rho_{v,sat})\, k_{l,sat} \left[\Delta i_{vap} + \dfrac{3\, c_{l,sat}\, (T_{sat} - T_s)}{8} \right]}{4\, \mu_{l,sat}\, L(T_{sat} - T_s)} \right\}^{1/4} \qquad (7\text{-}80)$$

If Eq. (7-73) is satisfied, then Eq. (7-80) can be written approximately as:

$$\overline{h} \approx \frac{4}{3} \left[\frac{\rho_{l,sat}\, g(\rho_{l,sat} - \rho_{v,sat}) k_{l,sat}\, \Delta i_{vap}}{4\, \mu_{l,sat}\, L(T_{sat} - T_s)} \right]^{1/4} \qquad (7\text{-}81)$$

Finally, we can evaluate the conditions for which the underlying assumption of inertia free flow holds. Substituting Eq. (7-29) into Eq. (7-28) leads to an expression in terms of the film thickness:

$$\frac{\rho_{l,sat}\, \delta^4 g\, (\rho_{l,sat} - \rho_{v,sat})}{\mu_{l,sat}^2\, L} \ll 1 \qquad (7\text{-}82)$$

According to Eq. (7-74), the largest film thickness (i.e., the film thickness at the trailing edge of the plate) is:

$$\delta_{x=L} \approx \left[\frac{4\, L\, k_{l,sat}\, \mu_{l,sat}(T_{sat} - T_s)}{\rho_{l,sat}\, g(\rho_{l,sat} - \rho_{v,sat})\Delta i_{vap}} \right]^{1/4} \qquad (7\text{-}83)$$

Substituting Eq. (7-83) into Eq. (7-82) leads to:

$$\frac{4\, k_{l,sat}\, (T_{sat} - T_s)}{\mu_{l,sat}\, \Delta i_{vap}} \ll 1 \qquad (7\text{-}84)$$

7.4.3 Correlations for Film Condensation

This section provides correlations for film condensation on a variety of geometries. These correlations are also implemented as EES procedures.

Vertical Wall

Correlations for condensation from a vertical wall aligned with gravity (Figure 7-7) are typically expressed in terms of a condensate film Reynolds number, Re_c, that is defined based on the mean velocity of the liquid in the film (u_m) and the hydraulic diameter associated with flow through the film (D_h). The mean velocity of the liquid in the film is:

$$u_m = \frac{\dot{m}_l}{\rho_{l,sat}\, W\, \delta} \qquad (7\text{-}85)$$

where W is the width of the wall, $\dot{m}_l$ is the mass flow rate of liquid condensate, $\rho_{l,sat}$ is the density of saturated liquid, and δ is the film thickness. The hydraulic diameter of the film is twice the film thickness:

$$D_h = \frac{4\,A_c}{per} = \frac{4\,W\,\delta}{2\,W} = 2\,\delta \qquad (7\text{-}86)$$

Therefore, the film Reynolds number is:

$$Re_c = \frac{\rho_{l,sat}\,u_m\,D_h}{\mu_{l,sat}} = \frac{2\,\dot{m}_l}{W\,\mu_{l,sat}} \qquad (7\text{-}87)$$

where $\mu_{l,sat}$ is the viscosity of saturated liquid. The mass flow rate of condensate per unit width of plate, $\dot{m}_l/W$ in Eq. (7-87), is usually not specified but rather calculated. Therefore, it is necessarily an iterative process to compute the condensate film Reynolds number and the heat transfer coefficient. If the Reynolds number is sufficiently small ($Re_c < 30$), then the solution derived in Section 7.4.2 is valid. With some algebra, the solution can be expressed in terms of the film Reynolds number:

$$\text{if } Re_c < 30 \quad \text{then } \frac{\overline{h}}{k_{l,sat}} \left[\frac{\mu_{l,sat}^2}{\rho_{l,sat}\,(\rho_{l,sat} - \rho_{v,sat})\,g} \right]^{1/3} = 1.47\,Re_c^{1/3} \qquad (7\text{-}88)$$

where $\rho_{v,sat}$ is the density of saturated vapor and g is the acceleration of gravity. As the film Reynolds number increases, the film becomes unstable and waves appear at the liquid/vapor interface. At very high values of the film Reynolds number, the film may become turbulent. For higher Reynolds numbers, Butterworth (1981) suggests the correlations:

$$\text{if } 30 < Re_c < 1600 \text{ then } \frac{\overline{h}}{k_{l,sat}} \left[\frac{\mu_{l,sat}^2}{\rho_{l,sat}(\rho_{l,sat} - \rho_{v,sat})g} \right]^{1/3} = \frac{Re_c}{1.08\,Re_c^{1.22} - 5.2} \qquad (7\text{-}89)$$

$$Re_c > 1600 \text{ then } \frac{\overline{h}}{k_{l,sat}} \left[\frac{\mu_{l,sat}^2}{\rho_{l,sat}\,(\rho_{l,sat} - \rho_{v,sat})\,g} \right]^{1/3} = \frac{Re_c}{8750 + \dfrac{58}{\sqrt{Pr_{l,sat}}}\,(Re_c^{0.75} - 253)}$$

$$(7\text{-}90)$$

These correlations for film condensation from a vertical wall are implemented in the EES procedure Cond_vertical_plate.

EXAMPLE 7.4-1: WATER DISTILLATION DEVICE

A very simple water purification system consists of a pressure vessel that is partially filled with brackish water, as shown in Figure 1. A polished copper heater element is submerged in the water and causes it to evaporate. The heater element is $L_{htr} = 8.0$ cm long and $W = 10$ cm wide (into the page). The heater element transfers $\dot{q}_{htr} = 4000$ W to the water.

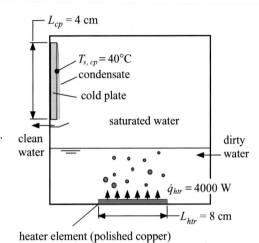

Figure 1: Simple water purification system.

A cold plate affixed to the side of the vessel is cooled to a uniform temperature of $T_{s,cp} = 40°C$. The water vapor condenses on the cold plate and is collected for drinking. The length of the cold plate is $L_{cp} = 4.0$ cm long and $W = 10$ cm wide (into the page). Ignore the effect of non-condensable air in the vessel.

a) Determine the steady-state rate that the device produces clean liquid water.

The inputs are entered in EES:

```
"EXAMPLE 7.4-1"

$UnitSystem SI MASS RAD PA K J
$Tabstops 0.2 0.4 0.6 3.5 in

"Inputs"
q_dot_htr=4000 [W]                    "heat transfer to heating element"
W=10 [cm]*convert(cm,m)               "width of heater and cool plate"
L_htr=8.0 [cm]*convert(cm,m)          "length of heater"
L_cp=4.0 [cm]*convert(cm,m)           "length of cool plate"
T_s_cp=converttemp(C,K,40 [C])        "cool plate surface temperature"
```

At steady state, the rate at which vapor is produced by the heating element must be balanced by the rate at which condensate is produced at the cold plate. The pressure, and therefore saturation temperature, within the pressure vessel will adjust itself to achieve this balance. To get started, a saturation temperature (T_{sat}) will be assumed and the rate of condensation and evaporation will be calculated; the saturation temperature will then be adjusted in order to balance these rates. The pressure in

EXAMPLE 7.4-1: WATER DISTILLATION DEVICE

the vessel corresponding to the assumed saturation temperature (p) is computed using EES' internal property routine for water:

```
T_sat=converttemp(C,K,100 [C])              "guess for saturation temperature"
p=pressure(Water,T=T_sat,x=0)               "pressure in vessel"
p_atm=p*convert(Pa,atm)                      "in atm"
```

The correlations for condensation on a flat plate are accessed using the Cond_Vertical_Plate procedure. Note that the Cond_Vertical_Plate procedure returns the average condensation heat transfer coefficient ($\bar{h}_c$), the film Reynolds number (Re_c), the heat transfer to the plate ($\dot{q}_c$), and the mass flow rate of condensation ($\dot{m}_c$).

```
Call Cond_vertical_plate('Water', L_cp, W, T_s_cp, T_sat :h_bar_c, Re_c, q_dot_cp, m_dot_c)
                              "call correlation for condensation on a vertical plate"
```

The predicted mass flow rate of condensate is $\dot{m}_c = 0.00086$ kg/s. The density of liquid water ($\rho_{l,sat}$) is calculated using EES' internal property routine and used to compute the volumetric flow rate of condensate:

$$\dot{V}_c = \frac{\dot{m}_c}{\rho_{l,sat}}$$

```
rho_l_sat=density('Water',p=p,x=0)          "liquid density"
V_dot_c=m_dot_c/rho_l_sat                    "volumetric flow rate of distilled water"
V_dot_c_lph=V_dot_c*convert(m^3/s,liter/hr)  "in liter/hr"
```

which leads to $\dot{V}_c = 8.9 \times 10^{-7}$ m^3/s (3.21 liter/hr).

The latent heat of vaporization (Δi_{vap}) is the difference between the enthalpies of saturated vapor and liquid and is computed using EES' built-in property routines. The mass flow rate at which vapor is produced by the heater is given by:

$$\dot{m}_{htr} = \frac{\dot{q}_{htr}}{\Delta i_{vap}} \tag{1}$$

Note that Eq. (1) neglects the energy required to heat the brackish water from its entering temperature to the saturation temperature; however, this energy is likely to be small relative to the amount of energy required to vaporize the water.

```
DELTAi_vap=enthalpy('Water',p=p,x=1)-enthalpy('Water',p=p,x=0)
                                             "enthalpy of vaporization"
m_dot_htr=q_dot_htr/DELTAi_vap               "rate of vapor production"
```

which leads to $\dot{m}_{htr} = 0.0018$ kg/s. The rate of vapor production is higher than the rate of condensation for the assumed value of the saturation temperature. This would lead to an increase in the pressure in the closed vessel and an increase in the saturation temperature. Select Update Guesses from the Calculate menu and then replace the assumed value of T_{sat} with the constraint $\dot{m}_c = \dot{m}_{htr}$.

```
{T_sat=converttemp(C,K,100 [C])}            "guess for saturation temperature"
m_dot_htr=m_dot_c                            "mass balance"
```

EXAMPLE 7.4-1: WATER DISTILLATION DEVICE

The solution indicates that $\dot{m}_c = 0.0020$ kg/s and $\dot{V}_c = 2.3 \times 10^{-6}$ m^3/s (8.19 liter/hr) with $T_{sat} = 458.4$ K. The pressure in the vessel is $p = 1.13 \times 10^6$ Pa (11.1 atm).

b) Determine the temperature of the heating element surface.

The heat flux at the surface of the heater is:

$$\dot{q}''_{htr} = \frac{\dot{q}_{htr}}{L_{htr}\,W}$$

```
q"_dot_htr=q_dot_htr/(L_htr*W)                    "heat flux on the heater plate"
```

The nucleate boiling constant for polished copper with water is $C_{nb} = 0.013$ according to Table 7-1. The Nucleate_Boiling function is used to evaluate the heater surface temperature, $T_{s,htr}$:

```
C_nb=0.013 [-]                                    "nucleate boiling constant"
q"_dot_htr= Nucleate_Boiling('Water', T_sat, T_s_htr, C_nb)
                                                  "nucleate boiling correlation"
T_s_htr_C=converttemp(K,C,T_s_htr)               "surface temperature in C"
```

which leads to $T_{s,htr} = 467.3$ K (194.1°C).

c) Estimate the critical heat flux for the heating element and compare this with the operating heat flux.

The critical heat flux is estimated using the correlation presented in Section 7.2.3. The vapor density ($\rho_{v,sat}$) and surface tension (σ) are computed using EES' internal property routine.

```
rho_v_sat=density('Water',p=p,x=1)               "saturated vapor density"
sigma=SurfaceTension('Water',T=T_sat)            "surface tension"
```

The characteristic length for the nucleate boiling process is determined according to:

$$L_{char,nb} = \sqrt{\frac{\sigma}{g\,(\rho_{l,sat} - \rho_{v,sat})}}$$

and used to evaluate the dimensionless size of the heater:

$$\tilde{L} = \frac{L_{htr}}{L_{char,nb}}$$

```
L_char_nb=sqrt(sigma/(g#*(rho_l_sat-rho_v_sat)))
                                                  "characteristic length"
L_bar=L_htr/L_char_nb                             "dimensionless size of the element"
```

which leads to $\tilde{L} = 37$. According to Table 7-2, the appropriate critical heat flux constant for a large flat plate is $C_{crit} = 0.15$. The critical heat flux is estimated

EXAMPLE 7.4-1: WATER DISTILLATION DEVICE

according to:

$$\dot{q}''_{s,crit} = C_{crit}\,\Delta i_{vap}\,\rho_{v,sat}\left[\frac{\sigma g(\rho_{l,sat} - \rho_{v,sat})}{\rho^2_{v,sat}}\right]^{1/4}$$

C_crit=0.15 [-]   "coefficient for the critical heat flux equation"
q"_dot_s_chf=C_crit*DELTAi_vap*rho_v_sat*(sigma*g# *(rho_l_sat-rho_v_sat)/rho_v_sat^2)^0.25
                                                   "critical heat flux"

which leads to $\dot{q}''_{s,crit} = 3.1 \times 10^6$ W/m². This is approximately 6 × the heat flux applied to the heating element, $\dot{q}''_{htr} = 5 \times 10^5$ W/m². Therefore, it is not likely that the heating element will experience a boiling crisis. The EES function Critical_Heat_Flux could also be used to do this calculation.

d) **Plot the rate of condensation and the pressure in the vessel as a function of the power applied to the heating element.**

Figure 2 illustrates the rate of condensation and pressure in the vessel as a function of the heater power.

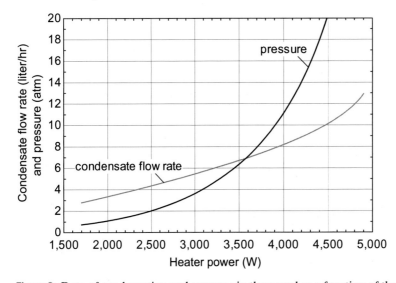

Figure 2: Rate of condensation and pressure in the vessel as a function of the heater power.

Horizontal, Downward Facing Plate
Gerstmann and Griffith (1967) present correlations for condensation on a horizontal, downward facing plate where the condensate is removed in the form of droplets that form, grow, and detach:

$$\text{if } 1 \times 10^6 < Ra < 1 \times 10^8 \quad \text{then} \quad \frac{\overline{h}}{k_{l,sat}}\sqrt{\frac{\sigma}{(\rho_{l,sat} - \rho_{v,sat})\,g}} = 0.69\,Ra^{0.20} \quad (7\text{-}91)$$

$$\text{if } 1 \times 10^8 < Ra < 1 \times 10^{10} \quad \text{then} \quad \frac{\overline{h}}{k_{l,sat}}\sqrt{\frac{\sigma}{(\rho_{l,sat} - \rho_{v,sat})\,g}} = 0.81\,Ra^{0.193} \quad (7\text{-}92)$$

where

$$Ra = \frac{g\,\rho_{l,sat}(\rho_{l,sat} - \rho_{v,sat})\Delta i_{vap}}{\mu_{l,sat}(T_{sat} - T_s)k_{l,sat}}\left[\frac{\sigma}{(\rho_{l,sat} - \rho_{v,sat})g}\right]^{1/3} \quad (7\text{-}93)$$

For slightly inclined surfaces (less than 20°), it is possible to use Eqs. (7-91) through (7-93) provided that g is replaced by $g\cos(\theta)$ where θ is the angle of inclination of the plate from horizontal. The correlation for condensation from a horizontal, downward facing plate is implemented in the EES procedure Cond_Horizontal_Down.

Horizontal, Upward Facing Plate
Nimmo and Leppert (1970) present a correlation for condensation on a horizontal, upward facing plate which is infinite in one direction and has length L in the other. The condensate drains from the side under the influence of the hydrostatic pressure gradient related to the film thickness variation between the side and center of the plate.

$$\frac{\overline{h}\,L}{k_{l,sat}} = 0.82\left[\frac{\rho_{l,sat}^2\, g\,\Delta i_{vap}\, L^3}{\mu_{l,sat}\,(T_{sat}-T_s)\,k_{l,sat}}\right]^{1/5} \tag{7-94}$$

The correlation for condensation on a horizontal, upward facing plate is implemented in the EES procedure Cond_horizontal_up.

Single Horizontal Cylinder
Marto (1998) presents a solution for laminar film condensation on the outer surface of a single cylinder of diameter D:

$$\frac{\overline{h}\,D}{k_{l,sat}} = 0.728\left[\frac{\rho_{l,sat}\,(\rho_{l,sat}-\rho_{v,sat})\,g\,\Delta i_{vap}\,D^3}{\mu_{l,sat}\,(T_{sat}-T_s)\,k_{l,sat}}\right]^{1/4} \tag{7-95}$$

The correlation for condensation on a single horizontal cylinder is implemented in the EES procedure Cond_horizontal_Cylinder.

Bank of Horizontal Cylinders
Film condensation in bundles of tubes is common in industrial condenser applications. The heat transfer coefficient associated with the tubes inside the bundle may be substantially less than the heat transfer coefficient calculated for a single tube using Eq. (7-95). Kern (1958), as presented in Kakaç and Liu (1998), suggests that the average heat transfer coefficient for a single tube ($\overline{h}_{1\text{-tube}}$) should be modified based on the number of rows of tubes in the vertical direction ($N_{tube,vert}$) according to:

$$\overline{h} = \overline{h}_{1\text{-tube}}\, N_{tube,vert}^{-1/6} \tag{7-96}$$

The correlation for condensation on a bank of horizontal cylinders is implemented in the procedure Cond_horizontal_N_Cylinders.

Single Horizontal Finned Tube
Film condensation on a horizontal finned tube (see Figure 7-8) was studied by Beatty and Katz (1948); the following correlation is suggested:

$$\overline{h} = 0.689\left[\frac{\rho_{l,sat}^2\, k_{l,sat}^3\, g\,\Delta i_{vap}}{\mu_{l,sat}\,(T_{sat}-T_s)\,D_{eff}}\right]^{1/4} \tag{7-97}$$

where D_{eff} is defined according to:

$$D_{eff} = \left(1.30\,\eta_{fin}\,\frac{A_f}{A_{eff}\,\overline{L}^{1/4}} + \frac{A_{uf}}{A_{eff}\,D_r^{1/4}}\right)^{-4} \tag{7-98}$$

where D_r is the root diameter of the finned tube (see Figure 7-8) and η_{fin} is the fin efficiency, calculated using the solution derived in Section 1.8.3 and implemented using

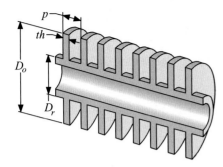

Figure 7-8: A finned tube.

the EES function eta_fin_annular_rect. The surface area of the flanks of a single fin, A_f in Eq. (7-98), is calculated according to:

$$A_f = \frac{\pi}{2}\left(D_o^2 - D_r^2\right) \qquad (7\text{-}99)$$

where D_o is the outer diameter of the fins. The area of the exposed tube between adjacent fins, A_{uf} in Eq. (7-98), is calculated according to:

$$A_{uf} = \pi D_o (p - th) \qquad (7\text{-}100)$$

where p is the fin pitch and th is the fin thickness (see Figure 7-8). The effective area of a fin, A_{eff} in Eq. (7-98), is calculated according to:

$$A_{eff} = \eta_f A_f + A_{uf} \qquad (7\text{-}101)$$

The parameter $\tilde{L}$ in Eq. (7-98) is calculated according to:

$$\tilde{L} = \frac{\pi(D_o^2 - D_r^2)}{4 D_o} \qquad (7\text{-}102)$$

Note that the fin efficiency in Eqs. (7-98) and (7-101) depends on the heat transfer coefficient and therefore implementation of the correlation is necessarily implicit and iterative. The correlation for condensation from a finned tube is implemented by the EES function Cond_finned_tube.

7.5 Flow Condensation

7.5.1 Introduction

Flow condensation processes are analogous to the flow boiling processes that were discussed in Section 7.3. When the condensation phase change process occurs while the fluid is flowing within a duct, then it is referred to as flow condensation. Flow condensation occurs within the condenser component of many vapor compression refrigeration cycles. A flow condensation process is likely to move through several different flow regimes that are similar to those shown in Figure 7-5 for flow boiling, but occur in reverse. Flow condensation, like flow boiling, is complicated and there are a large number of researchers working toward understanding this complex process. The interested reader is directed towards books such as Collier and Thome (1996). This section provides correlations that can be used to estimate the heat transfer coefficient for flow condensation processes.

7.5.2 Flow Condensation Correlations

The correlation suggested by Dobson and Chato (1998) presented in this section has been implemented in EES. This correlation has been experimentally validated by Smit et al. (2002) and others. However, it should be noted that the correlation strongly overpredicts the heat transfer coefficient for very high pressure refrigerants (e.g., R125, R32, and R410a).

The correlation is divided by flow regime into wavy or annular depending on the mass flux and the modified Froude number. The mass flux is:

$$G = \frac{4\dot{m}}{\pi D^2} \tag{7-103}$$

where D is the internal diameter of the tube and $\dot{m}$ is the total mass flow rate of the fluid (vapor and liquid). If the mass flux is greater than 500 kg/m²-s, then the flow is assumed to be annular regardless of the modified Froude number and the local heat transfer coefficient is computed according to:

$$\text{if } G > 500 \text{ kg/m}^2 \text{ then } h = \underbrace{\frac{k_{l,sat}}{D} 0.023 \, Re_{D,l}^{0.8} \, Pr_{l,sat}^{0.4}}_{\text{Dittus-Boelter equation}} \underbrace{\left(1 + \frac{2.22}{X_{tt}^{0.89}}\right)}_{\text{two-phase multiplier}} \tag{7-104}$$

where X_{tt} is the Lockhart Martinelli parameter and $Re_{D,l}$ is the superficial liquid Reynolds number. The Lockhart Martinelli parameter is computed according to:

$$X_{tt} = \sqrt{\frac{\rho_{v,sat}}{\rho_{l,sat}}} \left(\frac{\mu_{l,sat}}{\mu_{v,sat}}\right)^{0.1} \left[\frac{(1-x)}{x}\right]^{0.9} \tag{7-105}$$

where x is the local quality. The superficial liquid Reynolds number is the Reynolds number that is consistent with the liquid flowing alone in the tube:

$$Re_{D,l} = \frac{GD(1-x)}{\mu_{l,sat}} \tag{7-106}$$

Equation (7-104) can be compared to the Dittus-Boelter equation presented in Section 5.2 for fully developed turbulent flow. The parameter in Eq. (7-104) that is enclosed in parentheses is not present in the Dittus-Boelter equation and is referred to as the two-phase multiplier.

If the mass flux is less than 500 kg/m²-s, then the flow is either annular or wavy depending on the modified Froude number. The modified Froude number is computed according to:

$$\text{if } Re_{D,l} \leq 1250 \quad \text{then } Fr_{mod} = \frac{0.025 \, Re_{D,l}^{1.59}}{Ga^{0.5}} \left(\frac{1 + 1.09 \, X_{tt}^{0.039}}{X_{tt}}\right)^{1.5} \tag{7-107}$$

$$\text{if } Re_{D,l} > 1250 \quad \text{then } Fr_{mod} = \frac{1.26 \, Re_{D,l}^{1.04}}{Ga^{0.5}} \left(\frac{1 + 1.09 \, X_{tt}^{0.039}}{X_{tt}}\right)^{1.5} \tag{7-108}$$

where Ga is the Galileo number, defined as:

$$Ga = \frac{g \, \rho_{l,sat}(\rho_{l,sat} - \rho_{v,sat})D^3}{\mu_{l,sat}^2} \tag{7-109}$$

If the modified Froude number is greater than 20, then the flow is assumed to be annular and the local heat transfer coefficient is computed according to:

$$\text{if } G < 500 \text{ kg/m}^2 \quad \text{and} \quad Fr_{mod} > 20 \quad \text{then } h = \frac{k_{l,sat}}{D} 0.023 \, Re_{D,l}^{0.8} \, Pr_{l,sat}^{0.4} \left(1 + \frac{2.22}{X_{tt}^{0.89}}\right)$$
(7-110)

If the modified Froude number is less than 20, then the flow is assumed to be wavy and the local heat transfer coefficient is computed according to:

$$\text{if } G < 500 \text{ kg/m}^2 \quad \text{and} \quad Fr_{mod} < 20 \quad \text{then}$$

$$h = \left(\frac{k_{l,sat}}{D}\right) \left[\left(\frac{0.23}{1 + 1.11 \, X_{tt}^{0.58}}\right) \left(\frac{G D}{\mu_{v,sat}}\right)^{0.12} \left(\frac{\Delta i_{vap}}{c_{l,sat} \, (T_{sat} - T_s)}\right)^{0.25} Ga^{0.25} \, Pr_{l,sat}^{0.25} + A \, Nu_{fc}\right]$$
(7-111)

The parameter A in Eq. (7-111) is related to the angle from the top of the tube to the liquid level:

$$A = \frac{\arccos{(2 \, vf - 1)}}{\pi}$$
(7-112)

where vf is the void fraction (the fraction of the volume occupied by vapor), evaluated using the correlation provided by Zivi (1964):

$$vf = \left[1 + \frac{(1-x)}{x} \left(\frac{\rho_{v,sat}}{\rho_{l,sat}}\right)^{2/3}\right]^{-1}$$
(7-113)

The parameter Nu_{fc} in Eq. (7-111) is a Nusselt number related to forced convection in the bottom pool, evaluated according to:

$$Nu_{fc} = 0.0195 \, Re_{D,l}^{0.8} \, Pr_{l,sat}^{0.4} \sqrt{1.376 + \frac{C_1}{X_{tt}^{C_2}}}$$
(7-114)

The parameters C_1 and C_2 in Eq. (7-114) are evaluated based on the Froude number (Fr, note that this is not the modified Froude number):

$$Fr = \frac{G^2}{\rho_{l,sat}^2 \, g \, D}$$
(7-115)

If the Froude number is greater than 0.7 then:

$$\text{if } Fr > 0.7 \quad \text{then } C_1 = 7.242 \quad \text{and} \quad C_2 = 1.655$$
(7-116)

otherwise:

$$\text{if } Fr \leq 0.7 \quad \text{then}$$
$$C_1 = 4.172 + 5.48 \, Fr - 1.564 \, Fr^2 \quad \text{and} \quad C_2 = 1.773 - 0.169 \, Fr$$
(7-117)

The Dobson and Chato correlation for condensation in a horizontal tube is implemented in the procedure Cond_HorizontalTube. Note that the procedure slightly modifies the correlation described in this section in order to smooth out the sharp transition and the associated discontinuity that otherwise occurs at $Fr_{mod} = 20$ according to Eqs. (7-110) and (7-111). Also, the Cond_HorizontalTube procedure provides the film condensation correlation recommended by Chato (1962) as reported in Incropera and DeWitt (2002) when the mass flow rate of fluid in the tube is set to zero.

Chapter 7: Boiling and Condensation

The website associated with this book (www.cambridge.org/nellisandklein) provides additional problems.

Pool Boiling

7–1 One method of removing water and other contamination from a gas is to pass it through a cooled tube so that contaminants with high freezing and liquefaction points (e.g., water) tend to be collected at the wall. A quick and easy liquid nitrogen trap for methane is constructed by placing a tube in a Styrofoam cooler that is filled with liquid nitrogen, as shown in Figure P7-1.

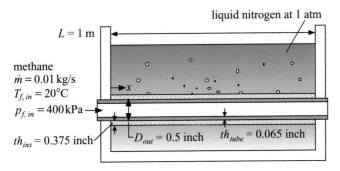

Figure P7-1: Liquid nitrogen trap.

The length of the tube is $L = 1$ m. The outer diameter of the tube is $D_{out} = 0.5$ inch and the tube thickness is $th_{tube} = 0.065$ inch. The tube conductivity is $k_{tube} = 150$ W/m-K. The tube is wrapped in insulation. The thickness of the insulation is $th_{ins} = 0.375$ inch and the insulation conductivity is $k_{ins} = 1.5$ W/m-K. Methane enters the tube at $\dot{m} = 0.01$ kg/s with temperature $T_{f,in} = 20°$C and pressure $p_{f,in} = 400$ kPa. The liquid nitrogen that fills the container is at 1 atm and is undergoing nucleate boiling on the external surface of the insulation. You may neglect axial conduction through the tube.

a.) Set up an EES program that can evaluate the state equations for this problem. That is, given a value of position, x, methane temperature, T_f, and methane pressure, p_f, your program should be able to compute $\frac{dT_f}{dx}$ and $\frac{dp_f}{dx}$.

b.) Use the Integral command in EES to integrate the state equations from $x = 0$ to $x = L$. Plot the fluid temperature and pressure of the methane as a function of position.

c.) Plot the heat flux at the insulation surface and the critical heat flux as a function of position.

d.) Plot the temperature of the methane at the surface of the tube as a function of position.

e.) Plot the lowest temperature experienced by the methane in the trap as a function of the insulation thickness. If the methane temperature must be maintained at or above its liquefaction point (131.4 K at 400 kPa) then what should the insulation thickness be?

7–2 An industrial boiler generates steam by heat exchanging combustion gases with saturated water at 125 kPa through mechanically polished AISI 302 stainless steel tubing having an inside diameter of 5.48 cm with a wall thickness of 2.7 mm and a total submerged length of 10 m. The combustion gases enter the tubing at 750°C with a mass flow rate of 0.0115 kg/s. The gases exhaust at ambient pressure. Assume that the combustion gases have the same thermodynamic properties as air.

  a.) Identify the state equation for this problem; the differential equation that can be used to determine the rate of change of the temperature of the combustion gas with respect to position.

  b.) Integrate the state equations developed in part (a) in order to determine the outlet temperature of the combustion gases

  c.) Calculate the rate at which steam is generated in this boiler.

7–3 You are preparing a spaghetti dinner for guests when you realize that your heat transfer training can be used to answer some fundamental questions about the process. The pot you are using holds four liters of water. The atmospheric pressure is 101 kPa. When on its high setting, the electric stove heating unit consumes 1.8 kW of electrical power of which 20% is transferred to the surroundings, rather than to the water. The pot is made of 4 mm thick polished AISI 304 stainless steel and it has a diameter of 0.25 m. The burner diameter is also 0.25 m.

  a.) How much time is required to heat the water from 15°C to its boiling temperature?

  b.) What are the temperatures of the outside and inside surfaces of the bottom of pot while the water is boiling?

  c.) What would the burner electrical power have to be in order to achieve the critical heat flux? Compare the actual heat flux during the boiling process to the critical heat flux.

  d.) How much water is vaporized during the 10 minutes required to cook the spaghetti?

7–4 A tungsten wire having a diameter of 1 mm and a length of 0.45 m is suspended in saturated carbon dioxide liquid maintained at 3.25 MPa. The fluid-surface coefficient needed in the nucleate boiling relation, C_{nb}, is estimated to be 0.01 and the emissivity of the tungsten wire is 0.4. Prepare a plot of the electrical power dissipated in the wire versus the excess temperature for power levels ranging from 10 W to the power corresponding to the critical heat flux for the nucleate boiling regime. What is your estimate of the excess temperature at the burnout point?

7–5 A cross-section of one type of evacuated solar collector is shown in Figure P7-5. The collector consists of a cylindrical glass tube with an outer diameter of 7.5 cm and wall thickness of 5 mm. In the center of the tube is a heat pipe, which is a copper tube with an outer diameter of 2 cm and wall thickness of 1.5 mm. The heat pipe contains a small amount of water at a pressure of 200 kPa that experiences nucleate boiling as solar radiation is incident on the outside surface of the copper tube at a rate of 745 W/m². You may assume that the glass is transparent to solar radiation and that the absorptivity of the copper tube with respect to solar radiation is 1.0. The surface of the copper tube has an emissivity of 0.13 with respect to its radiative interaction with the inner surface of the glass tube. The glass may be assumed to be opaque to thermal radiation from the copper tube with an emissivity of 1.0 on both its inner and outer surfaces. The outside surface of the glass interacts with the 25°C, 101.3 kPa surroundings through radiation and free convection.

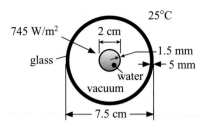

Figure P7-5: Cross-section of an evacuated tubular collector.

a.) Calculate the net rate of energy transfer to the water per unit length.
b.) Calculate the efficiency of the solar collector, defined as the ratio of the rate of energy transfer to the water in the heat pipe to the incident solar radiation.

Flow Boiling

7–6 When one fluid is changing phase in a heat exchanger, it is commonly assumed to be at a uniform temperature. However, there is a pressure drop in the evaporating fluid, which affects its saturation temperature. In a particular case, a 2 m long horizontal concentric tube heat exchanger made of copper is used to evaporate 0.028 kg/s of refrigerant R134a from an entering state of 300 kPa with a quality of 0.35. Heat transfer is provided by a flow of water that enters the heat exchanger at 12°C, 1.10 bar with a mass flow rate of 0.20 kg/s. The refrigerant passes through the central tube of the heat exchanger, which has an inner diameter of 1.25 cm and a wall thickness of 2 mm. The water flows through the annulus; the inner diameter of the outer tube is 2.5 cm.
 a.) Estimate the outlet temperature of the water and the outlet temperature and quality of the refrigerant.

7–7 A circular finned tube evaporator designed for cooling air is made from aluminum. The outer diameter of the tubes is 10.21 mm with a tube wall thickness of 1 mm. The evaporator is plumbed such that there are 12 parallel circuits of tubes with each circuit having a length of 0.6 m. Refrigerant R134a enters the evaporator at a mass flowrate of 0.15 kg/s. The refrigerant enters the throttle valve upstream of the evaporator as 35°C saturated liquid. The pressure in the evaporator is 240 kPa. The refrigerant exits as a saturated vapor.
 a.) What is the rate of heat transfer to the air for this evaporator?
 b.) Estimate of the average heat transfer coefficient between the R134a and the tube wall.
 c.) Estimate the pressure drop of the R134a as it passes through the evaporator. Does this pressure drop significantly effect the saturation temperature?

7–8 Repeat EXAMPLE 7.3-1 using the Flow_Boiling_avg function rather than integrating in order to determine the average heat transfer coefficient. Compare the two methods of obtaining the average heat transfer coefficient.

7–9 A computer manufacturer is reviewing alternative ways to remove heat from electronic components. The electronic circuit board can be assumed to be a thin horizontal plate with a width of 8 cm and a length of 16 cm. Currently, air is blown over the top of the circuit board at a velocity of 10 m/s. Additional cooling could be obtained by a higher air velocity, but the increased noise associated with the larger fan required is judged to be unacceptable. Another alternative is to immerse the board in a fluid at atmospheric pressure that is undergoing nucleate boiling.

The fluid R245fa has been chosen as a possibility. The surface tension of R245fa at atmospheric pressure is 0.0153 N/m. Other thermodynamic and transport properties are available from EES. Prepare a plot that shows the surface temperature of the plate as a function of the heat flux using air at 10 m/s and nucleate boiling at atmospheric pressure with R245fa for heat fluxes ranging from 100 to 10000 W/m^2.

Film Condensation

7–10 Evacuated tubular solar collectors often employ a heat pipe to transfer collected solar energy for water heating. Heat transfer between the water that is being heated and the solar collector occurs at the condenser of the heat pipe, which is a thin-walled cylinder made of copper with a length of 6 cm and a diameter of 1 cm, as shown in Figure P7-10. Water at 40°C and 1 atm flows past the condenser at a velocity that can be specified by the flow rate and duct diameter. The fluid inside the heat pipe is also water and it condenses at a pressure of 100 kPa. The heat transfer situation associated with the condensing water within the heat pipe is not known, but will here be assumed that it can be represented with the same relations that are used for film condensation on the inside surface of a cylinder. This heat transfer coefficient for film condensation on the inner surface of a cylinder is provided by the Cond_HorizontalTube procedure when the mass flow rate is set to 0. Plot the rate of heat transfer from the solar collector to the water that is being heated as a function of the flow velocity of the water for velocities between 1 and 10 m/s.

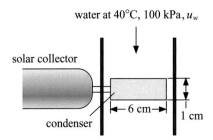

Figure P7-10: Condenser of evacuated solar collector.

7–11 The condenser in a steam power cycle utilizes a shell and tube heat exchanger that consists of 1200 nominal 1.5 inch schedule 40 tubes each made of brass. Each tube is 8 ft long and internally smooth. Cooling water enters each of the tubes at 68°F and exits at 74°F. Saturated steam at 1 psia having a quality of 91% enters the condenser at a low velocity and is condensed on the tubes. Estimate the rate of condensate formation and the associated water flowrate at steady state conditions.

7–12 Problem 7-9 described an electronics cooling system that removes the heat dissipated in an electronic circuit board by submersing the board in R245fa. The circuit board is maintained at a relatively low and uniform temperature over a range of heat fluxes by boiling R245fa. However, a problem now arises in dealing with the vapor produced by the evaporation. One possibility is to condense the vapor on the bottom side of a plate that is cooled by chilled water on its top side in a sealed container, as shown in Figure P7-12. The top of the enclosure is made of metal and can be considered to be isothermal. The chilled water is at 1 atm and has a free stream velocity of 10 m/s and a free stream temperature of 10°C. The circuit board is 8 cm wide and 16 cm long. The saturation pressure (and thus temperature) of the R245fa in the enclosure should vary with the heat flux.

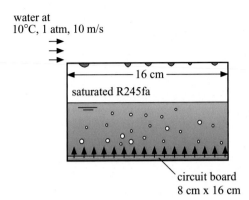

Figure P7-12: Sealed container full of evaporating and condensing R245fa.

    a.) Prepare a plot of the saturation pressure as a function of the heat flux for heat fluxes ranging from 100 to 10000 W/m^2.

    b.) How sensitive are the results to the velocity of the chilled water?

7–13 Recycled refrigerant R134a is purified in a simple distillation process in which a heater that is submerged in the liquid refrigerant heats the liquid, which causes it to vaporize. The distillation unit is a container with a square base that is 25 cm on a side. Piping and a float valve (not shown) are provided to maintain a constant liquid level of refrigerant in the container as condensate is removed. The vapor condenses on the bottom side of a copper plate that is placed at the top of the device, as shown in Figure P7-13. Liquid water at 25°C with free stream velocity 3 m/s flows over the top of the copper plate. The plate is slightly inclined so that the condensed refrigerant travels to the left side of the bottom side of the plate and drips into a collection gutter.

    a.) Calculate the saturation pressure and temperature of the refrigerant as a function of heater power for a range of heater powers from 100 W to 1000 W.

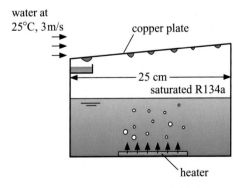

Figure P7-13: Refrigerant recycling apparatus.

7–14 Calculate the heat flux for a square plate that is 1 m on each side and is used for condensing steam at 6 kPa. Consider three plate orientations: (1) horizontal facing downward; (2) horizontal facing upward; and (3) vertical (one-side only is active).

a.) Plot the heat flux for each orientation as a function of plate surface temperature.

b.) Which geometry provides the highest rate of condensation per unit surface area?

c.) How do the answers to (a) and (b) change if the plate dimensions are reduced to 0.5 m per side?

7–15 A heat pipe has been instrumented in order to test its ability to transfer thermal energy. The heat pipe consists of a sealed vertical thin-walled copper tube that is 1.5 m in length and 2.5 cm in diameter. The heat pipe contains liquid and vapor toluene. The bottom 5 cm of the tube are wrapped with heater tape that provides 100 W of heat input to the toluene. The toluene evaporates at the lower end of the tube and the vapor rises to the top where it is condensed by contact with the cold top surface of the tube. The top 6 cm of the heat pipe are maintained at 29°C by a flow of liquid water at 25°C and 1 atm. Toluene condensate flows back to the bottom of the tube; the flow is assisted by surface tension due to the presence of a wicking material on the inner surface of the copper tube. The heat pipe tube is well-insulated except for the bottom part that is in contact with the heater and the top part that is in contact with the water.

a.) Estimate the saturation temperature and pressure of the toluene in the heat pipe.

b.) Estimate the surface temperature of the tube that is in contact with the heater.

c.) Estimate the velocity of the cooling water provided at 25°C needed to maintain the top surface of the heat pipe at 29°C.

d.) Compare the heat transfer rate provided by the heat pipe to the heat transfer rate that would occur if the tube were replaced with 2.5 cm diameter solid copper rod with the same temperatures imposed at the hot and cold ends.

e.) What is the effective thermal conductivity of the heat pipe? What do you see as advantages of the heat pipe?

7–16 A vertical cylindrical container is made of aluminum having a wall thickness of 2.5 mm. The cylinder is 0.24 m in height and it has an outer diameter of 7.5 cm. Liquid water is placed in the cylinder and the bottom is heated, evaporating the liquid. The vapor that is produced escapes through a vent at the top of the cylinder. The flow of vapor drives out air that was originally in the cylinder. When all of the liquid has been boiled, the heating is stopped and the vent at the top of the cylinder is closed. The aluminum surfaces are nearly at a uniform temperature of 100°C. The cylinder is allowed to stand in a large room and it transfers energy by free convection to the 25°C air.

a.) Calculate and plot the pressure inside the cylinder as a function of time for a 5 minute period after the vent is closed. State and justify any assumptions that you employ. (Note that the heat transfer coefficient for film condensation on the inside surfaces of the cylinder can be estimated using the Cond_HorizontalTube procedure with a mass flow rate of zero.)

Flow Condensation

7–17 You have fabricated an inexpensive condenser for your air conditioner by running the refrigerant through a plastic tube that you have submerged in a lake. The outer diameter of the tube is $D_o = 7.0$ mm and the inner diameter is $D_i = 5.0$ mm. The tube conductivity is $k_{tube} = 1.4$ W/m-K. The refrigerant is R134a and enters the tube with quality $x = 0.97$, temperature $T_{sat} = 35°C$ and mass flow rate $\dot{m}_r = 0.01$ kg/s. The water in the lake has temperature $T_\infty = 10°C$.

    a.) Determine the heat transfer rate per unit length from the refrigerant to the lake at the tube inlet.

    b.) If the length of the tube is $L = 6$ m, then what is the quality of the refrigerant leaving the tube? Plot the quality and refrigerant heat transfer coefficient as a function of position s in the tube.

7–18 Condensation and boiling are analogous processes in that both a involve phase change. Heat exchangers that provide condensation and boiling are often designed in a similar manner. In a particular case, a phase change of R134a takes place within horizontal tubes having an inner diameter of 1 cm. The mass velocity is 300 kg/s-m^2.

    a.) Prepare a plot of the heat transfer coefficient for condensation and boiling as a function of quality at a saturation temperature of 10°C for heat fluxes of 5,000 and 10,000 W/m^2.

    b.) Plot the excess temperature for condensation and boiling as a function of quality.

    c.) What conclusion can you draw from the results?

7–19 Absorption refrigeration cycles are often used to operate small refrigerators in hotel rooms because they do not require compressors and thus operate quietly. In a particular case, an absorption refrigeration system uses an ammonia-water mixture. The ammonia is separated from the water and passes through a condenser where it is isobarically changed from saturated vapor at 76°C to subcooled liquid at 40°C. The condenser consists of a single unfinned thin-walled copper tube with a 1 cm inner diameter. The thermal energy released from the ammonia in this process is transferred to the 25°C room air by free convection. The subcooled ammonia is throttled to a saturation temperature of −5° and evaporated to saturated vapor to produce a refrigeration capacity of 110 W.

    a.) Determine the mass flow rate of ammonia through the evaporator and condenser.

    b.) Estimate the length of piping required to condense the ammonia from saturated vapor to saturated liquid at 76°C.

    c.) Estimate the additional length of piping required to subcool the ammonia to 40°C.

REFERENCES

Beatty, K. O. and D. L. Katz, "Condensation of Vapors on Outside of Finned Tubes," *Chem. Eng. Prog.*, Vol. 44, pp. 55–70, (1948).

Butterworth, D., "Condensers: Basic Heat Transfer and Fluid Flow," in *Heat Exchanger*, S. Kakac, A. E. Bergles, and F. Mayinger, eds., Hemisphere Publishing Corp., New York, pp. 289–314, (1981).

Carey, V. P., *Liquid Vapor Phase Change Phenomena*, Hemisphere Publishing Corporation, Washington, D.C., (1992).

Chato, J. C., "Laminar Condensation Inside Horizontal and Inclined Tubes," *ASHRAE J.*, 4, pp. 52–60, 1962.

Collier, J. G. and J. R. Thome, *Convective Boiling and Condensation, 3rd Edition*, Oxford University Press, United Kingdom, (1996).

Dobson, M. K. and J. C. Chato, "Condensation in Smooth Horizontal Tubes," *Journal of Heat Transfer*, Vol. 120, pp. 193–213, (1998).

Drew, T. B. and C. Mueller, "Boiling," *Trans. AIChE*, Vol. 33, pp. 449, (1937).

Gerstmann, J. and P. Griffith, "Laminar Film Condensation on the Underside of Horizontal and Inclined Surfaces," *Int. J. Heat Mass Transfer*, Vol. 10, pp. 567–580, (1967).

Gnielinski, V., "New Equations for Heat and Mass Transfer in Turbulent Pipe and Channel Flow," *Int. Chem. Eng.*, Vol. 16, pp. 359–368, (1976).

Kakaç, S. and H. Liu, *Heat Exchangers: Selection, Rating, and Thermal Design*, CRC Press, New York, (1998).

Kern, D. Q., "Mathematical development of loading in horizontal condensers," *AIChE J.*, Vol. 4, pp. 157–160, (1958).

Incropera, F. P and DeWitt, D. P. *Introduction to Heat Transfer*, 4th edition, John Wiley & Sons, New York, 2002.

Lienhard, J. H. and V. K. Dhir, "Hydrodynamic prediction of peak pool-boiling heat fluxes from finite bodies," *Journal of Heat Transfer*, Vol. 95, pp. 152–158, (1973).

Lienhard, J. H., IV and J. H. Lienhard V, *A Heat Transfer Textbook, 3rd Edition*, Phlogiston Press, Cambridge, MA, (2005).

Marto, P. J, "Condensation," in *Handbook of Heat Transfer, 3rd Edition*, W. M. Rohsenow, J. P. Hartnett, and Y. I. Cho, eds., McGraw-Hill, New York, (1998).

Mills, A. F., *Heat Transfer*, Irwin Publishing, Homewood, IL, (1992).

Nimmo, B. G. and G. Leppert, "Laminar Film Condensation on a Finite Horizontal Surface," *Proc. 4th Int. Heat Transfer Conf.*, Paris, Vol. 6, Cs2.2, (1970).

Nukiyama, S., "The Maximum and Minimum Values of Heat Transmitted from Metal to Boiling Water Under Atmospheric Pressure," *J. Japan Soc. Mech. Eng.*, Vol. 37, pp. 367–374, (1934) (in Japanese); translated in *Int. J. Heat Mass Transfer*, Vol. 9, pp. 1419–1433, (1966).

Nusselt, W., "Die Oberflachenkondensation des Wasserdampfes," *Z. Ver. D-Ing.*, Vol. 60, pp. 541–546, (1916).

Petukhov, B. S., in *Advances in Heat Transfer*, Vol. 6, T. F. Irvine and J. P. Hartnett eds., Academic Press, New York, (1970).

Rohsenow, W. M., "A Method of Correlating Heat Transfer Data for Surface Boiling of Liquids," *Transactions of the ASME*, Vol. 74, pp. 969, (1952).

Shah, M. M., "A New Correlation for Heat Transfer during Boiling Flow Through Pipes," *ASHRAE Trans.*, Vol. 82(2), pp. 66–86, (1976).

Shah, M. M., "Chart Correlation for Saturated Boiling Heat Transfer: Equations and Further Study," *ASHRAE Trans.*, Vol. 88(1), pp. 185–186, (1982).

Shah, M. M., "Evaluation of General Correlations for Heat Transfer During Boiling of Saturated Liquids in Tubes and Annuli," *HVAC&R Journal*, Vol. 12, No. 4, pp. 1047–1065, (2006).

Smit, F. J., J. R. Thome, and J. P. Meyer, "Heat Transfer Coefficients During Condensation of the Zeotropic Refrigerant Mixture HCFC-22/HCFC-142b," *Journal of Heat Transfer*, Vol. 124, pp. 1137–1146, (2002).

Thome, J. R., *Engineering Data Book III*, Wolverine Tube, Huntsville, AL, (2006); accessible online at http://www.wlv.com/products/databook/db3/DataBookIII.pdf.

Vachon, R. I., G. H. Nix, and G. E. Tanger, *Journal of Heat Transfer*, Vol. 90, pp. 239, (1968).

Whalley, P. B., *Boiling, Condensation and Gas-Liquid Flow*, Clarendon Press, Oxford United Kingdom, (1987).

Zivi, S. M., "Estimation of Steady-State Steam Void-Fraction by Means of the Principle of Minimum Entropy Production," *Journal of Heat Transfer*, Vol. 86, pp. 247–252, (1964).

8 Heat Exchangers

8.1 Introduction to Heat Exchangers

8.1.1 Introduction

A heat exchanger is a device that is designed to transfer thermal energy from one fluid to another. The term "heat exchanger" like "heat transfer" is inconsistent with the thermodynamic definition of heat; these devices would be more appropriately called thermal energy exchangers. However, the term "heat exchanger" is ubiquitous. Heat exchangers are also ubiquitous; nearly all thermal systems employ at least one and usually several heat exchangers.

The background on conduction and convection, presented in Chapters 1 through 7, is required to analyze and design heat exchangers. This section reviews the applications and types of heat exchangers that are commonly encountered. Subsequent sections provide the theory and tools required to determine the performance of these devices.

8.1.2 Applications of Heat Exchangers

You may be unaware of just how common heat exchangers are in both residential and industrial applications. For example, you live in a residence that is heated to a comfortable temperature in winter and possibly cooled in the summer. Heating is usually accomplished by combusting a fuel (e.g., natural gas, propane, wood, or oil) that provides the desired thermal energy, but also produces combustion gases that can be harmful. Therefore, your furnace includes a heat exchanger that transfers thermal energy from the combustion gases to an air stream that can be safely circulated through the building.

You shower and wash clothes and dishes using water that is much warmer than the temperature of the water supplied by the city or a well. Some water heaters use electrical heaters, but many rely on combustion of fuel. The hot combustion gases are heat exchanged with the city or well water in an insulated tank in order to produce domestic hot water that may be stored for later use.

Your food is preserved in a refrigerator that includes at least two heat exchangers. One of the heat exchangers, the evaporator, transfers thermal energy from the evaporating refrigerant (typically an organic fluid such as R134a contained in a hermetically-sealed circuit) to the air inside the refrigerator. The second heat exchanger, the condenser, transfers the thermal energy removed from the refrigerator (plus the thermal equivalent of the required mechanical work added by the compressor) from the condensing refrigerant to the surroundings. Additional heat exchangers are used in some refrigerators to improve their efficiency.

The electricity that is provided to your home and used to run the refrigerator as well as other appliances is produced using a thermal power cycle. The common Rankine cycle uses hot combustion gases to boil water in a heat exchanger called the boiler. After it passes through a turbine and produces work, the spent steam is condensed in another heat exchanger called the condenser. A variety of additional heat exchangers are employed in the power plant to pre-heat the air used for combustion and to pre-heat the water returning to the boiler.

You may drive to work in an automobile that requires a variety of heat exchangers for thermal management. Heat exchangers are provided for window defrosting and cabin comfort. Also, the thermal energy that is transferred to the engine by the combustion process must be discharged to the surroundings in order to avoid over-heating. This heat exchange process is accomplished by circulating an anti-freeze solution of water and ethylene glycol through passages in the engine block and then to a heat exchanger that is located in the front of the vehicle and cooled by ambient air. This heat exchanger is often called the 'radiator,' but this is a misnomer as radiation plays a small role in its operation. The vehicle may also provide additional heat exchangers to cool the transmission fluid and the lubricating oil with the glycol solution.

The efficiency of most energy systems is controlled by the performance of its heat exchangers. The heat exchangers are often physically large and expensive. Therefore, the design of the heat exchangers is critical to the economic success of the energy system. The heat exchanger analysis tools that are provided in this chapter will be helpful in identifying an optimum design for a heat exchanger application. Optimal designs are usually defined in economic terms. Investing in a larger heat exchanger will cost you money now, but save you money over time due to the saving in energy associated with its higher performance. In order to carry out a meaningful design it is therefore necessary to have at least a rudimentary knowledge of economic analysis and the time value of money. For this reason, an introduction to economic analysis is available in Appendix A.5, which can be downloaded at www.cambridge.org/nellisandklein.

8.1.3 Heat Exchanger Classifications and Flow Paths

Heat exchangers can be broadly classified as either direct transfer or indirect transfer devices. A direct transfer heat exchanger transfers energy directly from one stream of fluid to another, typically through a separating wall that forms a pressure boundary. A direct transfer heat exchanger will usually operate at steady state or at least under quasi-steady conditions. Sections 8.2 through 8.9 discuss the analysis of direct transfer heat exchangers of various types. Indirect transfer heat exchangers do not transfer thermal energy directly from one stream to the other but rather utilize a secondary, intermediate medium to accomplish this process. A common type of indirect transfer heat exchanger is the regenerator. In a regenerator, the fluid flow oscillates through a matrix of material with high heat capacity (e.g., a packed bed of lead spheres). When the fluid flows in one direction, energy is transferred from the fluid to the matrix. When fluid flows in the opposite direction, the energy is transferred from the matrix to the fluid. Therefore, the matrix is a secondary medium in an indirect transfer device. Regenerators operate in a periodic rather than a steady-state manner and are considered in Section 8.10.

Many different types of heat exchanger configurations exist in order to accommodate different fluid properties and operating requirements. One of the simplest configurations is the concentric tube heat exchanger. The concentric tube heat exchanger shown in Figure 8-1 is operating in a parallel-flow arrangement. In a parallel-flow arrangement, the two fluids flow in the same direction (parallel to each other, from left to right). An alternative (and usually more effective) flow arrangement is counter-flow. In a counter-flow arrangement, the two fluids flow in opposite directions.

The plate heat exchanger is another common type of heat exchanger often used for thermal energy exchange between two liquid streams. A plate heat exchanger consists of

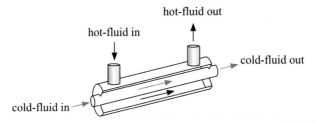

Figure 8-1: Concentric tube heat exchanger operating in parallel-flow.

many individual plates that are stacked together, as shown in Figure 8-2. The plates are corrugated in order to form flow channels between adjacent plates. The stack of plates is manifolded (via the large holes at the ends) so that one fluid flows on one side of each plate while a second fluid flows (usually in the opposite direction) on the other side. The plates are bolted together and sealed with gaskets. Plate heat exchangers offer several advantages. They are compact, easy to disassemble for cleaning and it is relatively easy to increase or decrease their size as needed by adding or removing plates.

Figure 8-3(a) is a cutaway of a shell-and-tube heat exchanger and Figure 8-3(b) is a schematic of the shell-and-tube heat exchanger configuration. One fluid flows through a bank of tubes that is situated within a large shell while the other fluid flows within the shell and around the outer surface of the tubes. Baffles are typically placed in the shell in order to force the shell-side flow to pass across the tube bundle as it makes its way through the shell in a serpentine pattern (see Figure 8-3(b)).

Figure 8-3(b) shows a shell-and-tube heat exchanger with a single tube pass (i.e., the fluid in the tubes flows along the length of the heat exchanger just one time) and a single shell pass (i.e., the fluid in the shell passes along the length of the heat exchanger just one time). Shell-and-tube heat exchangers can be configured in a number of ways, depending on how the fluid is directed through the tubes and the shell. For example, Figure 8-4(a) shows a shell-and-tube heat exchanger with two tube passes and a single shell pass; a manifold forces the fluid to flow one way (from right to left) through half of the tubes before turning around and flowing the opposite way (from left to

Figure 8-2: Plate heat exchanger (Turns (2006)).

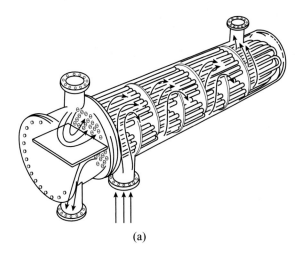

(a)

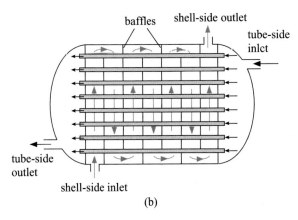

(b)

Figure 8-3: (a) Cutaway view of a shell-and-tube heat exchanger (Turns (2006)) and (b) a schematic of a shell-and-tube heat exchanger with a single tube pass and a single shell pass.

right) through the other half of the tubes. Multiple shell-passes require the fluid in the shell to pass back-and-forth through the heat exchanger. For example, Figure 8-4(b) shows a shell-and-tube heat exchanger with a single tube pass and two shell passes. Note that the multi-pass flow configurations shown in Figure 8-4(a) and Figure 8-4(b) are neither counter-flow nor parallel-flow configurations, but rather have some characteristics of both. Shell-and-tube heat exchangers commonly employ many tube and/or shell passes, as these multiple-pass flow configurations tend to increase the performance of the heat exchanger. Heat exchange processes in which one of the fluids changes phase, such as during boiling or condensation, often use shell-and-tube heat exchanger designs.

Fins are often placed on the gas side of a gas-to-liquid heat exchanger in order to increase the heat exchanger surface area and therefore compensate for the low convection heat transfer coefficients that are typical for forced convection with a gas. Gas-to-liquid heat exchangers are usually configured in a cross-flow arrangement in which the direction of the gas flow is perpendicular to that of the liquid flow. An automobile 'radiator,' which is designed to transfer unusable thermal energy from the

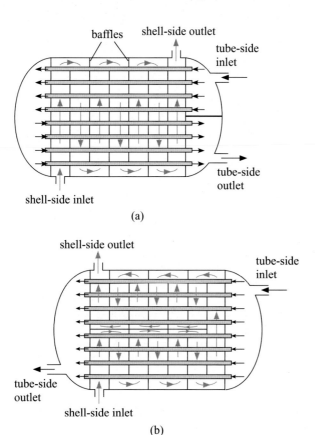

Figure 8-4: Schematic of a shell-and-tube heat exchanger with (a) two tube passes and one shell pass, and (b) one tube pass and two shell passes.

engine to the surrounding air, is a familiar example of a cross-flow, finned gas-to-liquid heat exchanger. Another example of a cross-flow heat exchanger is the evaporator in a refrigeration system. Two photographs of cross-flow heat exchangers are shown in Figure 8-5.

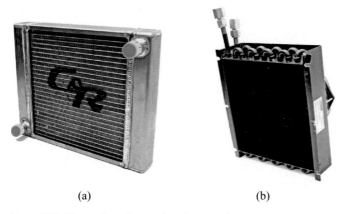

(a)   (b)

Figure 8-5: Examples of cross-flow heat exchangers.

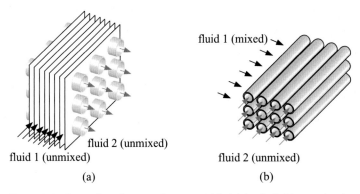

Figure 8-6: Cross-flow heat exchangers with (a) both fluids unmixed and (b) one fluid mixed.

The behavior of a cross-flow heat exchanger depends on how well the fluid at any position along the flow path has mixed together; that is, how well the fluid can mix in the direction perpendicular to the main flow. There are two limiting behaviors in this regard, referred to as mixed and unmixed. The actual heat exchanger behavior is likely somewhere between these limits, however the behavior can approach one of these limits depending on the geometric configuration. For example, the cross-flow heat exchanger shown in Figure 8-6(a) is likely to behave as if both fluids are unmixed since the tubes (for fluid 2) and the fins (for fluid 1) limit the amount of mixing that can occur in transverse direction. Figure 8-6(b) shows a cross-flow heat exchanger consisting of an un-finned bank of tubes. Fluid 1 in Figure 8-6(b) will approach mixed behavior (as there are no geometric barriers to flow laterally across the tubes) while fluid 2 will be un-mixed (as the tubes themselves prevent flow laterally).

There are many other heat exchanger geometries and flow configurations. This section has only reviewed the most common types. Additional information can be found in reference texts on this subject, such as Kakaç and Liu (1998) and Rohsenow et al. (1998).

8.1.4 Overall Energy Balances

Consider a heat exchanger operating in a counter-flow arrangement, shown in Figure 8-7. In Figure 8-7, the hot-fluid (indicated by subscript H) flows continuously with a mass flow rate of $\dot{m}_H$. The hot-fluid enters with a mean temperature $T_{H,in}$ and leaves

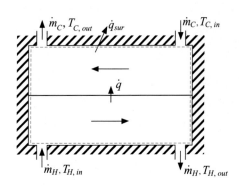

Figure 8-7: Schematic of a heat exchanger in a counter-flow arrangement.

with a mean temperature of $T_{H,out}$. The cold-fluid (indicated by subscript C) flows in the opposite direction with a mass flow rate of $\dot{m}_C$. The cold-fluid enters with a mean temperature $T_{C,in}$ and leaves with a mean temperature of $T_{C,out}$. The purpose of any heat exchanger analysis is to determine $\dot{q}$, the rate of heat transfer from the hot-fluid to the cold-fluid as well as the outlet temperatures of the hot and cold streams.

Regardless of the flow configuration or geometry, an overall energy balance can be written for a control volume that encloses the heat exchanger (shown in Figure 8-7). The kinetic and potential energy of the fluid streams are generally negligible and the heat exchanger is assumed to be at steady state. Therefore, the energy balance simplifies to:

$$\dot{m}_H i_{H,in} + \dot{m}_C i_{C,in} - \dot{m}_H i_{H,out} - \dot{m}_C i_{C,out} + \dot{q}_{sur} = 0 \tag{8-1}$$

where i is the specific enthalpy and $\dot{q}_{sur}$ is the rate of heat loss to the surroundings. Specific enthalpy is (nearly) independent of pressure for liquids and totally independent of pressure for gases that conform to the ideal gas law. In either case, if the specific heat capacity (c) is constant, then the enthalpy of the fluid can be written as the product of the constant pressure specific heat capacity and the temperature (T) relative to an arbitrary reference temperature (T_{ref}).

$$i = c(T - T_{ref}) \tag{8-2}$$

The specific heat capacity for most fluids is a function of temperature. However, in many cases the temperature dependence of specific heat capacity is small and can be neglected over the temperature range in the heat exchanger, as it is in this analysis. The specific heat capacity for each fluid used in the energy balance should be the average value over the range of temperatures encountered in the heat exchanger. If the temperature-dependent specific heat capacity is to be considered rigorously then a numerical analysis of the heat exchanger is required, as described in Section 8.6.

It is further assumed that the rate of heat transfer to the surroundings through the heat exchanger jacket, is negligible (i.e., the heat exchanger is well-insulated). With these simplifications, the overall energy balance, Eq. (8-1), reduces to:

$$\dot{m}_H c_H (T_{H,in} - T_{H,out}) = \dot{m}_C c_C (T_{C,out} - T_{C,in}) \tag{8-3}$$

Equation (8-3) cannot be solved since both of the fluid outlet temperatures, $T_{H,out}$ and $T_{C,out}$, are not known. However, if either of these temperatures are known, then $\dot{q}$, the rate of heat transfer from the hot to the cold-fluid, can be determined from an energy balance on either the hot or cold stream:

$$\dot{q} = \dot{m}_H c_H (T_{H,in} - T_{H,out}) = \dot{m}_C c_C (T_{C,out} - T_{C,in}) \tag{8-4}$$

The product of the mass flow rate and specific heat capacity appear in the energy balance and will continue to appear in heat exchanger analyses. The mass flow rate-heat capacity product is referred to as the capacitance rate of the hot and cold-fluid, $\dot{C}_H$ and $\dot{C}_C$, respectively:

$$\dot{C}_H = \dot{m}_H c_H \tag{8-5}$$

$$\dot{C}_C = \dot{m}_C c_C \tag{8-6}$$

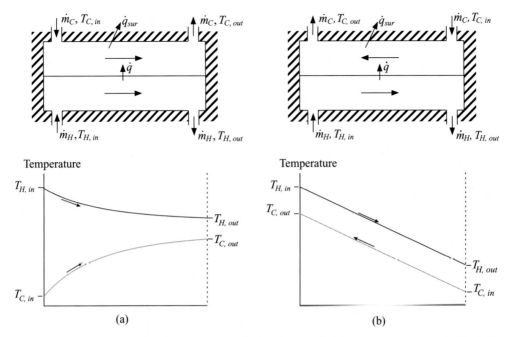

Figure 8-8: Typical temperature distribution within (a) a parallel-flow arrangement and (b) within a counter-flow arrangement.

The capacitance rate has units of power per temperature change (e.g., W/K) and can be thought of as the rate of heat transfer required to change the temperature of the stream by 1 unit. A large heat transfer rate is required to change the temperature of a stream with a large capacitance rate. In heat exchanger problems, it is the capacitance rate, rather than the mass flow rate or specific heat capacity, that matters. Substituting Eqs. (8-5) and (8-6) into Eq. (8-4) leads to:

$$\dot{q} = \dot{C}_H \left(T_{H,in} - T_{H,out}\right) = \dot{C}_C \left(T_{C,out} - T_{C,in}\right) \tag{8-7}$$

Additional information is required to solve Eq. (8-7). The geometry of the heat exchanger and the convection coefficients between the fluids and their heat exchange surfaces must be considered in order to establish the total thermal resistance that couples the fluids as they pass through the heat exchanger. The total thermal resistance is the inverse of the heat exchanger conductance, discussed in the next section.

Even after the heat exchanger conductance is calculated, it is not straightforward to determine $\dot{q}$ because the temperatures of the fluids change as they pass through the heat exchanger. The variation in the fluid temperature depends on the flow rates and conductance as well as the configuration. For example, Figure 8-8(a) and (b) illustrate the temperature distributions that are consistent with a parallel-flow and counter-flow arrangement, respectively.

The performance of the heat exchanger can only be predicted by deriving and solving the governing differential equation that accounts for the local heat transfer rate between the streams and the associated temperature change of each stream. The solution to this equation provides a second independent relation between the inlet and outlet temperatures of the hot and cold-fluids. Fortunately, the solutions to the governing differential equations that are associated with the heat exchanger configurations that are commonly encountered have been obtained. These solutions are available in two forms: the log-mean temperature difference (LMTD) and the effectiveness-NTU (ε-NTU) equations. These alternative methods for representing these heat exchanger solutions are described in Sections 8.2 and 8.3, respectively.

8.1.5 Heat Exchanger Conductance

All heat exchanger analyses require an estimate of the thermal resistance between the fluids. The inverse of the total thermal resistance is commonly referred to as the heat exchanger conductance, UA. The determination of UA is a straightforward application of the resistance concepts introduced in Section 1.2.3 and, when fins are used, Section 1.6.5, as well as the correlations for the internal flow convection heat transfer coefficients discussed in Chapter 5. The conductance can be estimated by considering all of the thermal resistances between the two fluids. In the absence of any fluid temperature change (e.g., for condensing or evaporating flows or flows with very high mass flow rate or specific heat capacity), the conductance could be used directly to determine the heat transfer rate. In most heat exchangers the temperature change of one or both fluids is significant (this is, after all, the purpose of most heat exchangers) and so the conductance must be used within one of the heat exchanger solutions discussed in Sections 8.2 and 8.3.

Fouling Resistance

An additional resistance is often encountered for heat exchangers that operate for prolonged periods of time; heat exchange surfaces that are clean when the heat exchanger is installed become "fouled." Fouling refers to any type of build-up or contamination that has the effect of increasing the thermal resistance between the underlying surface and the adjacent fluid; for example rust or scale. Fouling of heat exchange surfaces can cause a dramatic reduction in the performance of a heat exchanger.

The effect of fouling can be represented by the addition of a fouling resistance to the thermal network separating the two fluids. The fouling resistance (R_f) is calculated using a fouling factor (R_f'') according to:

$$R_f = \frac{R_f''}{A_s} \tag{8-8}$$

where A_s is the area of the surface being fouled. Notice the similarity between the fouling resistance and the contact resistance discussed in Section 1.2.6. The fouling factor, like the area-specific contact resistance, depends on a variety of factors and therefore is tabulated for specific situations based on experiments. Extensive tables of fouling factor values are provided in Kakaç and Liu (1998) and Rohsenow (1998). Much of the information in these sources has been compiled in the EES FoulingFactor function.

EXAMPLE 8.1-1: CONDUCTANCE OF A CROSS-FLOW HEAT EXCHANGER

EXAMPLE 8.1-1: CONDUCTANCE OF A CROSS-FLOW HEAT EXCHANGER

A finned, circular tube cross-flow heat exchanger is shown in Figure 1. The width and height of the front face of the heat exchanger are $W = 0.2$ m and $H = 0.26$ m, respectively. The fins are made of copper with a thickness $th_{fin} = 0.33$ mm and a fin pitch $p_{fin} = 3.18$ mm. Ten rows of tubes ($N_{t,row} = 10$) in two columns ($N_{t,col} = 2$) are connected in series. The vertical and horizontal spacing between adjacent tubes is $s_v = 25.4$ mm and $s_h = 22$ mm, respectively. The length of the heat exchanger in the direction of the air flow is $L = 0.06$ m. The tubes are made of copper with an outer diameter $D_{out} = 1.02$ cm and a wall thickness $th = 0.9$ mm. The roughness of the inner surface of the tube is $e = 1.0$ µm.

Treated water enters the tube with mass flow rate $\dot{m}_H = 0.03$ kg/s and inlet temperature $T_{H,in} = 60°C$. Clean dry air is forced to flow through the heat exchanger perpendicular to the tubes (i.e., in cross-flow) with a volumetric flow rate $\dot{V}_C = 0.06$ m³/s. The inlet temperature of the air is $T_{C,in} = 20°C$ and the air is at atmospheric pressure.

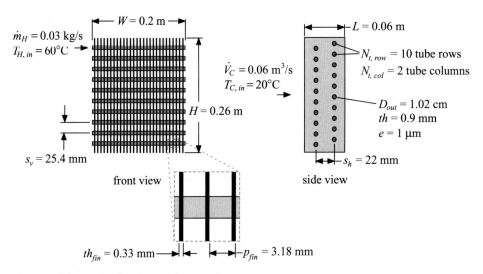

Figure 1: Schematic of a plate fin heat exchanger.

a) Determine the conductance of the heat exchanger.

The inputs are entered in EES:

```
"EXAMPLE 8.1-1: Conductance of a Cross-Flow Heat Exchanger"

$UnitSystem SI MASS RAD PA K J
$Tabstops 0.2 0.4 0.6 3.5 in

"Inputs"
D_out=1.02 [cm]*convert(cm,m)          "outer diameter of tube"
th=0.9 [mm]*convert(mm,m)              "tube wall thickness"
N_t_row=10 [-]                        "number of tube rows"
N_t_col=2 [-]                         "number of tube columns"
H=0.26 [m]                           "height of heat exchanger face"
W=0.2 [m]                            "width of heat exchanger face"
```

EXAMPLE 8.1-1: CONDUCTANCE OF A CROSS-FLOW HEAT EXCHANGER

```
L=0.06 [m]                                "length of heat exchanger in air flow direction"
V_dot_C=0.06 [m^3/s]                       "volumetric flow rate of air"
P=1 [atm]*convert(atm,Pa)                  "atmospheric pressure"
T_C_in=convertTemp(C,K,20 [C])             "inlet air temperature"
T_H_in=convertTemp(C,K,60 [C])             "inlet water temperature"
m_dot_H=0.03 [kg/s]                        "water flow rate"
s_v=25.4 [mm]*convert(mm,m)                "vertical separation distance between tubes"
s_h=22 [mm]*convert(mm,m)                  "horizontal separation distance between tubes"
th_fin=0.33 [mm]*convert(mm,m)             "fin thickness"
p_fin=3.18 [mm]*convert(mm,m)             "fin pitch"
e=1.0 [micron]*convert(micron,m)           "roughness of tube internal surface"
```

The total thermal resistance between the water and the air, R_{tot}, is the inverse of the conductance (UA). The total resistance can be found by summing all of the resistances in series:

$$R_{tot} = \frac{1}{(UA)} = R_{in} + R_{f,in} + R_{cond} + R_{out} \tag{1}$$

where R_{in} is the convection resistance between the water and the inner surface of the tube, $R_{f,in}$ is the fouling resistance that occurs on the internal surface of the tube as a result of deposits that accumulate from the flowing fluid. (The fouling on the external surface is expected to be negligible since there should be no build-up associated with clean dry air.) R_{cond} is the resistance to conduction through the tube wall, and R_{out} is the resistance between the air and the surface of the plate fins and the outer tube surface. (This resistance is due to both convection and the conduction resistance of the fins.)

The resistance between the liquid and the inner surface of the tube can be represented as:

$$R_{in} = \frac{1}{\overline{h}_{in} \pi D_{in} L_{tube}}$$

where $\overline{h}_{in}$ is the average heat transfer coefficient between the water and the tube wall, D_{in} is the inner diameter of the tube

$$D_{in} = D_{out} - 2\,th,$$

and L_{tube} is the total length of all of the tubes:

$$L_{tube} = N_{t,col} N_{t,row} W$$

```
"Internal flow through the tube"
D_in=D_out-2*th                            "tube inner diameter"
L_tube=N_t_row*N_t_col*W                    "total tube length"
```

The average convective heat transfer coefficient on the water-side can be determined using an internal forced convection flow correlation, as explained in Chapter 5. The process is simplified by the use of the **PipeFlow** procedure. The **PipeFlow** procedure is the dimensional form of the **PipeFlow_ND** procedure that was introduced in

EXAMPLE 8.1-1: CONDUCTANCE OF A CROSS-FLOW HEAT EXCHANGER

Chapter 5. The use of the **PipeFlow** procedure frees us from having to compute the fluid properties, Reynolds number, etc., that would be necessary to use the dimensionless version of the function.

```
T_avg=(T_H_in+T_C_in)/2                        "average temperature"
call PipeFlow('Water',T_avg,P,m_dot_H,D_in,L_tube,e/D_in:h_bar_T_H, &
    h_bar_H_H ,DELTAP_H, Nusselt_bar_T_H, f_bar_H, Re_H)
                                                "access correlations for internal flow through a tube"
h_bar_in=h_bar_T_H                             "average heat transfer coefficient on water side"
R_in=1/(pi*D_in*L_tube*h_bar_in)              "resistance to convection on water-side"
```

Note that the **PipeFlow** procedure provides outputs other than $\bar{h}_{in}$, but they are not needed for this calculation. The value of $\bar{h}_{in}$ is taken to be the average heat transfer coefficient predicted for a constant temperature (as opposed to constant heat flux) boundary condition. The heat transfer coefficient for a constant temperature boundary condition is generally smaller than for a constant heat flux boundary condition, leading to a conservative estimate of UA. If the flow is turbulent, then the two answers are the same.

Also note that the determination of the heat exchanger conductance is necessarily an iterative process when the outlet fluid temperatures are not known. The heat transfer coefficients depend on the outlet temperatures as a result of the temperature dependent properties of the fluids. For example, the temperature that should be provided to the **PipeFlow** procedure is an average of the inlet and outlet water temperatures. The methods required to completely solve the heat exchanger problem and therefore predict the outlet fluid temperatures are presented in Sections 8.2 and 8.3. As a reasonable first guess, the average water temperature is taken to be the average of the inlet water and inlet air temperatures.

The fouling resistance on the inner surface of the tube can be expressed in terms of its fouling factor, $R''_{f,in}$:

$$R_{f,in} = \frac{R''_{f,in}}{\pi \, D_{in} \, L_{tube}}$$

The fouling factor can be estimated using an appropriate handbook reference or, more simply, with the **FoulingFactor** function.

```
"Fouling resistance"
R"_f_in=FoulingFactor('Closed-loop treated water')   "fouling factor on inner surface of tube"
R_f_in=R"_f_in/(pi*D_in*L_tube)                       "fouling resistance on inner surface of tube"
```

The resistance of the tube wall is probably not worth calculating because it is small in comparison with the others in Eq. (1). However, it is easy to include. The resistance for a cylindrical tube was derived in Section 1.2.4.

$$R_{cond} = \frac{\ln\left(\dfrac{D_{out}}{D_{in}}\right)}{2 \, k_m \, \pi \, L_{tube}}$$

EXAMPLE 8.1-1: CONDUCTANCE OF A CROSS-FLOW HEAT EXCHANGER

where k_m is the conductivity of the tube, obtained using EES' built-in property routine.

```
"Conduction resistance"
k_m=k_('Copper',T_avg)                                "tube conductivity"
R_cond=ln(D_out/D_in)/(2*pi*k_m*L_tube)               "tube resistance"
```

The resistance between the air and the outer surface of the finned tube can be expressed in terms of an overall surface efficiency, η_o, as discussed in Section 1.6.6.

$$R_{out} = \frac{1}{\eta_o \bar{h}_{out} A_{tot,out}} \quad (2)$$

where $A_{tot,out}$ is the sum of the total surface area of the fins ($A_{s,fin,tot}$) and the un-finned tube wall surface ($A_{s,unfin}$) and $\bar{h}_{out}$ is the average heat transfer coefficient between the air and these surfaces. The overall surface efficiency is related to the fin efficiency, η_{fin}, as discussed in Section 1.6.6:

$$\eta_o = 1 - \frac{A_{s,fin,tot}}{A_{tot}}(1 - \eta_{fin}) \quad (3)$$

The total fin area is the total surface area of the plates (both sides) less the area that is occupied by the tubes.

$$A_{s,fin,tot} = 2\frac{W}{p_{fin}}\left(H L - N_{t,row} N_{t,col}\frac{\pi D_{out}^2}{4}\right)$$

where dimensions H, L, and W are shown in Figure 1. The total un-finned tube wall surface is:

$$A_{s,unfin} = \pi D_{out} L_{tube}\left(1 - \frac{th_{fin}}{p_{fin}}\right)$$

The total surface area is:

$$A_{tot} = A_{s,fin,tot} + A_{s,unfin}$$

```
"External resistance"
A_s_fin_tot=2*(W/p_fin)*(H*L-N_t_row*N_t_col*pi*D_out^2/4)   "total fin area"
A_s_unfin=pi*D_out*L_tube*(1-th_fin/p_fin)                  "total un-finned area"
A_tot=A_s_fin_tot+A_s_unfin                                 "total air-side surface area"
```

In order to determine the fin efficiency, it is first necessary to estimate the heat transfer coefficient on the air-side. The best method to determine $\bar{h}_{out}$ is not apparent. The flow of the air through the heat exchanger core is actually very complex, combining aspects of internal flow through the passages formed between adjacent fins with external flow over the tubes. The heat transfer coefficient $\bar{h}_{out}$ can be calculated using the techniques discussed for external flow over a bare cylinder, as presented in Section 4.9.3. On the other hand, the fins provide channels for the air flow, so perhaps $\bar{h}_{out}$ should be calculated using the techniques discussed in Section 5.2.4 for internal flow in a rectangular channel. Here, we will estimate $\bar{h}_{out}$ both ways

EXAMPLE 8.1-1: CONDUCTANCE OF A CROSS-FLOW HEAT EXCHANGER

and compare the results. In EXAMPLE 8.1-2, the value of $\bar{h}_{out}$ will be determined using a compact heat exchanger correlation that is based on experimental data for this particular geometry.

The average velocity of the air in the core (u_m) is determined by dividing the volumetric flow rate by the cross-sectional area that is available for the air flow (A_c).

$$u_m = \frac{\dot{V}_C}{A_c}$$

where

$$A_c = (H - N_{t,row}\, D_{out})\, W \left(1 - \frac{th_{fin}}{p_{fin}}\right)$$

The External_Flow_Cylinder procedure, is used to estimate $\bar{h}_{out}$ based on external flow over a cylinder. Note that External_Flow_Cylinder provides additional outputs that are not used here.

```
A_c=(H-N_t_row*D_out)*W*(1-th_fin/p_fin)        "cross-sectional area available for flow"
u_m=V_dot_C/A_c                                  "frontal velocity for external flow calculation"
"Heat transfer coefficient with external flow over tube"
Call External_Flow_Cylinder('Air', T_C_in, T_avg, P, u_m, D_out: &
    F_d\L, h_bar_out_ext, C_d, Nusselt_bar_out_ext, Re_out_ext)
                                                 "external flow correlation"
```

This calculation provides one estimate for $\bar{h}_{out}$ that is based on modeling the flow as external flow over tubes, $\bar{h}_{out,ext} = 47.7$ W/m^2-K.

The air flow through the core can also be modeled as an internal flow through rectangular channels. The effective channel width $(W_{ch,eff})$ is taken to be the space between the adjacent fins:

$$W_{ch,eff} = \frac{W - N_{fin}\, th_{fin}}{N_{fin}}$$

where N_{fin} is the number of fins:

$$N_{fin} = \frac{W}{p_{fin}}$$

The effective channel height $(H_{ch,eff})$ is taken to be the height of the heat exchanger:

$$H_{ch,eff} = H$$

The effective hydraulic diameter of the channel is:

$$D_{h,ch,eff} = \frac{W_{ch,eff}\, H_{ch,eff}}{2(W_{ch,eff} + H_{ch,eff})}$$

```
"Heat transfer coefficient with internal flow through duct"
N_fin=W/p_fin                                          "number of fins"
W_ch_eff=(W-N_fin*th_fin)/N_fin                        "effective channel width"
H_ch_eff=H                                             "effective channel height"
D_h_ch_eff=W_ch_eff*H_ch_eff/(2*(W_ch_eff+H_ch_eff))  "hydraulic diameter of channel"
```

The air mass flow rate through an individual channel is determined based on the density, volumetric flow and the number of passages.

$$\dot{m}_{ch} = \frac{\dot{V}_C \, \rho_{air}}{N_{fin}}$$

where ρ_{air} is the density of the air, evaluated at the average temperature using EES' internal property function.

```
rho_air=density(Air,T=T_avg,P=P)          "density of air"
m_dot_ch=rho_air*V_dot_C/N_fin            "mass flow rate per channel"
```

The temperature provided to the DuctFlow procedure is the average of the inlet water and air temperatures. This estimate could be improved after calculating the outlet air temperature as described in Sections 8.2 and 8.3.

```
call DuctFlow('Air',T_avg,P,m_dot_ch,W_ch_eff,H_ch_eff,L,e/D_h_ch_eff:&
    h_bar_out_int, h_bar_H_out_int ,DELTAP_C, Nusselt_bar_T_out_int, f_C, Re_out_int)
                                          "internal flow correlation"
```

This calculation provides another estimate for $\overline{h}_{out}$ that is based on modeling the air flow as internal flow through rectangular channels, $\overline{h}_{out,int} = 39.0 \, \text{W/m}^2\text{-K}$.

Initially, we will use the value of $\overline{h}_{out,ext}$ to determine R_{out} and therefore R_{tot} and UA.

```
h_bar_out=h_bar_out_ext          "set the air-side heat transfer coefficient"
```

Methods for calculating the fin efficiency are discussed in Chapter 1. However, the plate fins in the heat exchanger core shown in Figure 1 are not considered in Chapter 1. The plate material acts approximately like annular fins that are connected to the tubes. Each fin has an effective fin radius ($r_{fin,eff}$) that is defined so that the fictitious annular fins have the same surface area as the plates.

$$A_{s,fin,tot} = 2 \frac{L_{tube}}{p_{fin}} \pi \left[r_{fin,eff}^2 - \left(\frac{D_{out}}{2} \right)^2 \right]$$

```
"Fin efficiency"
A_s_fin_tot=2*(L_tube/p_fin)*pi*(r_fin_eff^2-(D_out/2)^2)    "effective fin radius"
```

EXAMPLE 8.1-1: CONDUCTANCE OF A CROSS-FLOW HEAT EXCHANGER

The fin efficiency is obtained using the function eta_fin_annular_rect.

eta_fin=eta_fin_annular_rect(th_fin, D_out/2, r_fin_eff, h_bar_out, k_m)   "fin efficiency"

Finally, the surface efficiency and the resistance, R_{out}, are calculated using Eqs. (2) and (3).

eta_o=1-A_s_fin_tot*(1-eta_fin)/A_tot   "overall surface efficiency"
R_out=1/(eta_o*h_bar_out*A_tot)   "resistance on air-side"

The overall thermal resistance and conductance are calculated.

"Overall resistance and conductance"
R_tot=R_in+R_f_in+R_cond+R_out   "total thermal resistance"
UA=1/R_tot   "conductance"

which leads to $UA = 62.1$ W/K.

If the air-side heat transfer estimate based on the internal flow calculation ($\bar{h}_{out,int}$) is used in place of the external flow calculation ($\bar{h}_{out,ext}$):

h_bar_out=h_bar_out_int   "set the air-side heat transfer coefficient"

then the conductance is $UA = 53.8$ W/K. EXAMPLE 8.1-2 compares these results with a compact heat transfer correlation, which is likely to be more accurate since it is based on experimental data for this specific core geometry. Notice that the heat transfer coefficient on the water-side, $\bar{h}_{in} = 3496$ W/m^2-K, is more than an order of magnitude larger than the heat transfer coefficient on the air-side, so that the air-side thermal resistance dominates the problem.

8.1.6 Compact Heat Exchanger Correlations

The results of EXAMPLE 8.1-1 revealed the following information.

1. The heat transfer coefficient for the gas-side of a gas-to-liquid heat exchanger is often orders of magnitude smaller than the heat transfer coefficient on the liquid-side. Therefore, even with the additional surface area added by the high-efficiency fins, the heat transfer resistance between the gas and the tube and fin surfaces (R_{out}) is usually the major heat transfer resistance. If the heat exchanger conductance must be predicted accurately, then it is necessary to focus on the gas-side heat exchanger coefficient.

2. The flow through a typical heat exchanger core is a complicated combination of internal flow between fins and external flow over tubes. The gas-side heat transfer coefficient calculated by considering the gas flow as an external flow differs significantly from the result obtained by assuming that it is an internal flow. The calculated air-side heat transfer coefficients, $\bar{h}_{out,int}$ and $\bar{h}_{out,ext}$, differ significantly in EXAMPLE 8.1-1.

This is a situation where experimental data are required. Cross-flow heat exchangers are common and there exists a great deal of experimental data for these heat exchanger cores. Experimental measurements of gas-side heat transfer coefficient and pressure drop data have been correlated by Kays and London (1984) for many heat exchanger core geometries. These heat transfer data are typically presented in terms of the Colburn j_H factor, defined as:

$$j_H = St\, Pr^{(2/3)} \tag{8-9}$$

where St is the Stanton number and Pr is the Prandtl number. The Stanton number is defined as:

$$St = \frac{\overline{h}_{out}}{G\,c} \tag{8-10}$$

where $\overline{h}_{out}$ is the gas-side heat transfer coefficient, c is the gas specific heat and G is the gas mass flux. The mass flux is defined as the mass flow rate per unit of flow area (A_{min}):

$$G = \frac{\dot{m}}{A_{min}} \tag{8-11}$$

Note that A_{min} in Eq. (8-11) is the minimum flow area in the core. The Colburn j_H factor is correlated in terms of the Reynolds and Prandtl numbers, defined as:

$$Re = \frac{G\,D_h}{\mu} \tag{8-12}$$

$$Pr = \frac{c\,\mu}{k} \tag{8-13}$$

where μ is the kinematic viscosity and k is the thermal conductivity of the gas. The hydraulic diameter of the flow channels (D_h) as well as other details of the geometry such as the outer diameter of the tube (D_{out}), number of fins per length (fpl), fin thickness (th_{fin}), ratio of free-flow to frontal area (σ), ratio of gas-side heat transfer area to core volume ($A_{s,out}/V$), and the ratio of finned to total surface area on the gas-side ($A_{s,fin}/A_{s,out}$) are provided with each correlation for a specific core geometry.

A gas-side friction factor, f, is also provided in the Kays and London compact heat exchanger data base. Using the friction factor, Kakaç and Liu (1997) show that the gas-side pressure drop, Δp, for finned heat exchangers can be estimated according to:

$$\Delta p = \frac{G^2}{2\rho_{in}}\left[f\,\frac{4L_{flow}}{D_h}\,\frac{\rho_{in}}{\overline{\rho}} + (1+\sigma^2)\left(\frac{\rho_{in}}{\rho_{out}} - 1\right)\right] \tag{8-14}$$

where L_{flow} is the length of flow passage in the direction of the gas flow, ρ_{in} and ρ_{out} are the gas densities at the inlet and outlet of the heat exchanger, respectively, and $\overline{\rho}$ is the average gas density, defined as:

$$\frac{1}{\overline{\rho}} = \frac{1}{2}\left(\frac{1}{\rho_{in}} + \frac{1}{\rho_{out}}\right) \tag{8-15}$$

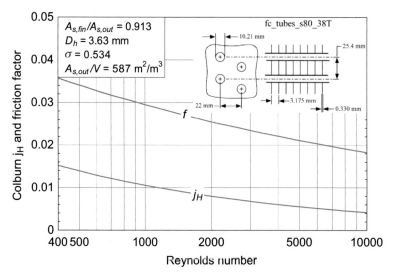

Figure 8-9: Colburn j_H and friction factors for heat exchanger surface 8.0-3/8T; based on data from Kays and London (1984).

For plate-fin exchangers, Kakaç and Liu recommend including additional terms for entrance and exit effects, so that the total pressure drop is calculated according to:

$$\Delta p = \frac{G^2}{2\rho_{in}}\left[(k_c + 1 - \sigma^2) + 2\left(\frac{\rho_{in}}{\rho_{out}} - 1\right) + f\frac{4L_{flow}}{D_h}\frac{\rho_{in}}{\overline{\rho}} - (1 - k_e - \sigma^2)\left(\frac{\rho_{in}}{\rho_{out}}\right)\right]$$

(8-16)

where k_c and k_e are the contraction and expansion loss coefficients. The frictional term in Eq. (8-16) is ordinarily responsible for 90% of the total pressure drop. The entrance and exit losses are important only for heat exchangers with short flow lengths.

Kays and London (1984) provide the heat exchanger correlations in graphical form; an example is shown in Figure 8-9. A library of EES procedures have been developed to provide the information contained in these graphs for many common heat exchanger geometries. There are four categories of compact heat exchanger procedures: Geometry, Non-dimensional, Coefficient of heat transfer, and Pressure drop. Within each category, different heat exchanger core configurations are available (e.g., Finned circular tubes) and within each core configuration it is possible to use the scroll bar to select a particular geometry (e.g., surface CF-8.8-1.0J). The Geometry procedures return the geometric characteristics of the core (e.g., the hydraulic diameter) that allow the user to compute the Reynolds number and heat transfer area. The non-dimensional functions provide the Colburn j_H and friction factors as a function of the Reynolds number (i.e., the information provided in charts like Figure 8-9). The Coefficient of heat transfer and Pressure drop procedures carry out the additional calculations required to determine the dimensional heat transfer coefficient and pressure drop.

EXAMPLE 8.1-2: CONDUCTANCE OF A CROSS-FLOW HEAT EXCHANGER (REVISITED)

EXAMPLE 8.1-2: CONDUCTANCE OF A CROSS-FLOW HEAT EXCHANGER (REVISITED)

a) Calculate the total conductance for the crossflow heat exchanger described in EXAMPLE 8.1-1 using the compact heat exchanger library in EES. Also, estimate the air-side pressure drop across this heat exchanger.

The EES code for this example is appended to the code from EXAMPLE 8.1-1. The heat exchanger geometry provided in EXAMPLE 8.1-1 is consistent with the finned circular tube surface 8.0-3/8T which has the identifier 'fc-tubes_s80-38T'. The heat transfer coefficient predicted by the compact heat exchanger library is obtained using the CHX_h_finned_tube procedure:

```
"Compact heat exchanger correlation"
TypeHX$='fc_tubes_s80-38T'                "heat exchanger identifier name"
Call CHX_h_finned_tube(TypeHX$, V_dot_C*rho_air, W*H, 'Air',T_avg, P:h_bar_out_CHX)
                                "access compact heat exchanger procedure"
```

which leads to $\bar{h}_{out,CHX} = 43.7$ W/m²-K. This estimate of the heat transfer coefficient compares favorably to the two estimates obtained in EXAMPLE 8.1-1, $\bar{h}_{out,ext} = 47.7$ W/m²-K and $\bar{h}_{out,int} = 39.0$ W/m²-K. The heat transfer coefficient is used to predict the total conductance (note that the calculation of the water-side resistance, fouling resistance, etc., remains as discussed in EXAMPLE 1.8-1).

```
h_bar_out=h_bar_out_CHX                "set the air-side heat transfer coefficient"
```

which leads to $UA = 58.4$ W/m²-K.

The air-side pressure drop for this finned circular tube crossflow heat exchanger can be estimated using the procedure CHX_DELTAp_finned_tube, which obtains the friction factor and uses Eq. (8-14) to calculate the associated pressure drop:

```
Call CHX_DELTAp_finned_tube(TypeHX$, V_dot_C*rho_air, W*H,L, 'Air', T_avg, T_avg, P: DELTAp)
                                "access compact heat exchanger procedure"
```

which leads to $\Delta p = 6.0$ Pa.

8.2 The Log-Mean Temperature Difference Method

8.2.1 Introduction

An overall energy balance on a heat exchanger alone is not sufficient to predict its performance. Additional information is required that relates the heat exchanger configuration, the capacity rates of the flows, and the conductance (discussed in Section 8.1) to the performance.

Two methods are typically used for simple heat exchanger calculations; the log-mean temperature difference method (*LMTD*) and the effectiveness-NTU (*ε-NTU*) method. Although these methods appear to be different, they are actually algebraically identical and represent different presentations of the same information. This section shows how the *LMTD* method is developed and applied for heat exchangers of different types.

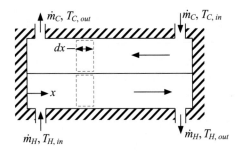

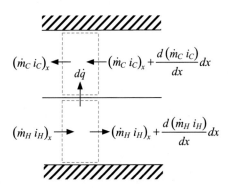

Figure 8-10: A counter-flow heat exchanger; the differential control volumes used to derive the governing differential equation are also shown.

8.2.2 *LMTD* Method for Counter-Flow and Parallel-Flow Heat Exchangers

The *LMTD* method expresses the heat transfer rate between the two fluid streams in a heat exchanger as the product of a temperature difference (later referred to as the log-mean temperature difference, ΔT_{lm}) and the conductance of the heat exchanger (UA):

$$\dot{q} = UA\,\Delta T_{lm} \tag{8-17}$$

The major consideration in applying Eq. (8-17) is the definition of the temperature difference ΔT_{lm}. Both the hot and cold-fluid temperatures will change by different amounts (and usually subtantially) as the fluid passes through the heat exchanger as seen, for example, in Figure 8-8. Therefore, it is not clear what temperature difference should be used in Eq. (8-17). Because the temperature distribution in the heat exchanger depends on flow configuration, the necessary form for ΔT_{lm} also depends upon the flow configuration. An analytical form for ΔT_{lm} can be derived for counter-flow and parallel-flow configurations.

Consider a differential control volume within a counter-flow heat exchanger, shown in Figure 8-10. The jacket of the heat exchanger is insulated and assumed to be adiabatic. The hot-fluid is flowing with mass flow rate $\dot{m}_H$ and the cold-fluid is flowing in the opposite direction with mass flow rate $\dot{m}_C$. The entering fluid temperatures are known ($T_{H,in}$ and $T_{C,in}$) and it is necessary to determine the exiting fluid temperatures ($T_{H,out}$ and $T_{C,out}$) and the heat transfer rate $\dot{q}$ between the two fluid streams.

An energy balance on the hot-fluid for the differential control volume shown in Figure 8-10 is:

$$(\dot{m}_H\,i_H)_x = (\dot{m}_H\,i_H)_x + \frac{d\,(\dot{m}_H\,i_H)}{dx}\,dx + d\dot{q} \tag{8-18}$$

where i_H is the specific enthalpy of the hot stream. The mass flow rate of the fluid must be spatially uniform for an incompressible fluid with no leakage. Therefore, Eq. (8-18) can be simplified:

$$0 = \dot{m}_H \frac{di_H}{dx} dx + d\dot{q} \qquad (8\text{-}19)$$

In general, the change in enthalpy may be driven by both pressure and temperature changes in the fluid. Therefore, Eq. (8-19) may be written as:

$$0 = \dot{m}_H \left[\underbrace{\left(\frac{\partial i_H}{\partial T} \right)_P}_{c_H} \frac{dT_H}{dx} + \underbrace{\left(\frac{\partial i_H}{\partial P} \right)_T \frac{dp_H}{dx}}_{\text{neglected}} \right] dx + d\dot{q} \qquad (8\text{-}20)$$

The partial derivative of enthalpy with respect to temperature at constant pressure (the first term in brackets in Eq. (8-20)) is the constant pressure specific heat capacity of the hot-fluid, c_H. The pressure driven change in the enthalpy (the second term in brackets) is typically neglected because both the pressure gradient and the partial derivative of enthalpy with respect to pressure at constant temperature are usually small.

$$0 = \dot{m}_H c_H \frac{dT_H}{dx} dx + d\dot{q} \qquad (8\text{-}21)$$

The quantity $d\dot{q}$ is the differential rate of energy transfer from the hot-fluid that occurs within segment dx. Rearranging Eq. (8-21) leads to:

$$d\dot{q} = -\dot{m}_H c_H \frac{dT_H}{dx} dx \qquad (8\text{-}22)$$

A similar energy balance for the differential control volume on the cold-fluid (see Figure 8-10) leads to:

$$d\dot{q} = -\dot{m}_C c_C \frac{dT_C}{dx} dx \qquad (8\text{-}23)$$

where, c_C is the specific heat capacity of the cold-fluid.

The differential energy transfer between the streams within the control volume is related to the conductance according to:

$$d\dot{q} = \underbrace{(T_H - T_C)}_{\substack{\text{local temp.} \\ \text{difference}}} \underbrace{UA \frac{dx}{L}}_{\substack{\text{amount of} \\ \text{conductance} \\ \text{in segment } dx}} \qquad (8\text{-}24)$$

where L is the total length of the heat exchanger. Note that Eq. (8-24) is the product of the local driving temperature difference and the amount of the total conductance that is contained in the differential segment. Substituting Eq. (8-24) into Eqs. (8-22) and (8-23) leads to:

$$UA (T_H - T_C) \frac{dx}{L} = -\dot{m}_H c_H \frac{dT_H}{dx} dx \qquad (8\text{-}25)$$

$$UA (T_H - T_C) \frac{dx}{L} = -\dot{m}_C c_C \frac{dT_C}{dx} dx \qquad (8\text{-}26)$$

Solving Eqs. (8-25) and (8-26) for the temperature gradients leads to:

$$\frac{dT_H}{dx} = -\frac{UA}{L\,\dot{m}_H\,c_H}(T_H - T_C) \tag{8-27}$$

$$\frac{dT_C}{dx} = -\frac{UA}{L\,\dot{m}_C\,c_C}(T_H - T_C) \tag{8-28}$$

Equations (8-27) and (8-28) are the state equations for this problem; they provide the rate of change of the state variables, T_H and T_C, in terms of the state variables and therefore can be integrated using any of the numerical integration techniques that are discussed in Section 3.1 and elsewhere. If the temperature-dependent specific heat capacity must be considered, then a numerical solution is required. Methodologies for obtaining numerical solutions to heat exchanger problems are discussed in Section 8.5.

Here, the analytical solution to the problem is derived for the case where c_C and c_H are constant. Equation (8-28) is subtracted from Eq. (8-27) in order to obtain:

$$\frac{d(T_H - T_C)}{dx} = -\frac{UA}{L}(T_H - T_C)\left(\frac{1}{\dot{m}_H c_H} - \frac{1}{\dot{m}_C c_C}\right) \tag{8-29}$$

which is a single ordinary differential equation in terms of the local temperature difference, θ, defined as:

$$\theta = T_H - T_C \tag{8-30}$$

Substituting Eq. (8-30) into Eq. (8-29) leads to:

$$\frac{d\theta}{dx} = -\frac{UA}{L}\theta\left(\frac{1}{\dot{m}_H c_H} - \frac{1}{\dot{m}_C c_C}\right) \tag{8-31}$$

Equation (8-31) can be separated:

$$\frac{d\theta}{\theta} = -\frac{UA}{L}\left(\frac{1}{\dot{m}_H c_H} - \frac{1}{\dot{m}_C c_C}\right)dx \tag{8-32}$$

and integrated from $x = 0$ to $x = L$:

$$\int_{\theta_{x=0}}^{\theta_{x=L}} \frac{d\theta}{\theta} = -\frac{UA}{L}\left(\frac{1}{\dot{m}_H c_H} - \frac{1}{\dot{m}_C c_C}\right)\int_0^L dx \tag{8-33}$$

Carrying out the integration leads to:

$$\ln\frac{\theta_{x=L}}{\theta_{x=0}} = -UA\left(\frac{1}{\dot{m}_H c_H} - \frac{1}{\dot{m}_C c_C}\right) \tag{8-34}$$

Substituting $\theta_{x=L} = T_{H,out} - T_{C,in}$ and $\theta_{x=0} = T_{H,in} - T_{C,out}$ (see Figure 8-10) into Eq. (8-34) leads to:

$$\ln\left(\frac{T_{H,out} - T_{C,in}}{T_{H,in} - T_{C,out}}\right) = -UA\left(\frac{1}{\dot{m}_H c_H} - \frac{1}{\dot{m}_C c_C}\right) \tag{8-35}$$

Notice that the product of the mass flow rate and specific heat capacity appears in the heat exchanger solution; this product was defined as the capacitance rate in Section 8.1.4:

$$\ln\left(\frac{T_{H,out} - T_{C,in}}{T_{H,in} - T_{C,out}}\right) = -UA\left(\frac{1}{\dot{C}_H} - \frac{1}{\dot{C}_C}\right) \tag{8-36}$$

Equation (8-36) is a fundamental relationship between the exit temperatures, conductance, and capacitance rates for a counter-flow heat exchanger; this equation provides the missing piece of information that can be used in conjunction with the overall energy balance, Eq. (8-7), in order to solve a counter-flow heat exchanger problem.

The fundamental relationship can be expressed in either log-mean temperature difference or effectiveness-NTU form. These two relationships are algebraically equivalent. To express Eq. (8-36) in log-mean temperature difference form, shown in Eq. (8-17), it is necessary to solve the overall heat exchanger energy balances:

$$\dot{q} = \dot{C}_H \left(T_{H,in} - T_{H,out} \right) \tag{8-37}$$

$$\dot{q} = \dot{C}_C \left(T_{C,out} - T_{C,in} \right) \tag{8-38}$$

for the capacitance rates:

$$\dot{C}_H = \frac{\dot{q}}{\left(T_{H,in} - T_{H,out} \right)} \tag{8-39}$$

$$\dot{C}_C = \frac{\dot{q}}{\left(T_{C,out} - T_{C,in} \right)} \tag{8-40}$$

Equations (8-39) and (8-40) are substituted into Eq. (8-36):

$$\ln \left[\frac{\left(T_{H,out} - T_{C,in} \right)}{\left(T_{H,in} - T_{C,out} \right)} \right] = -UA \left[\frac{\left(T_{H,in} - T_{H,out} \right) - \left(T_{C,out} - T_{C,in} \right)}{\dot{q}} \right] \tag{8-41}$$

Equation (8-41) can be rearranged so that it resembles Eq. (8-17):

$$\dot{q} = -UA \left[\frac{\left(T_{H,in} - T_{H,out} \right) - \left(T_{C,out} - T_{C,in} \right)}{\ln \left[\frac{\left(T_{H,out} - T_{C,in} \right)}{\left(T_{H,in} - T_{C,out} \right)} \right]} \right] \tag{8-42}$$

or

$$\dot{q} = UA \underbrace{\left[\frac{\left(T_{H,out} - T_{C,in} \right) - \left(T_{H,in} - T_{C,out} \right)}{\ln \left[\frac{\left(T_{H,out} - T_{C,in} \right)}{\left(T_{H,in} - T_{C,out} \right)} \right]} \right]}_{\Delta T_{lm,cf}} \tag{8-43}$$

Comparing Eq. (8-42) with Eq. (8-17) shows that the log-mean temperature difference for a counter-flow heat exchanger, $\Delta T_{lm,cf}$, is:

$$\boxed{\Delta T_{lm,cf} = \frac{\left(T_{H,out} - T_{C,in} \right) - \left(T_{H,in} - T_{C,out} \right)}{\ln \left[\frac{\left(T_{H,out} - T_{C,in} \right)}{\left(T_{H,in} - T_{C,out} \right)} \right]}} \tag{8-44}$$

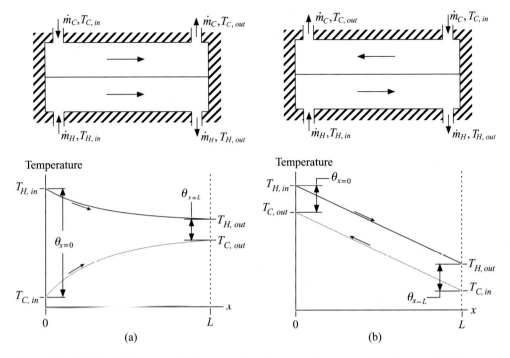

Figure 8-11: (a) Parallel-flow and (b) counter-flow heat exchangers and their associated temperature distributions and temperature differences that are required to compute the log-mean temperature difference.

The log-mean temperature difference for a parallel-flow heat exchanger is derived using a similar procedure:

$$\Delta T_{lm,pf} = \frac{(T_{H,in} - T_{C,in}) - (T_{H,out} - T_{C,out})}{\ln\left[\dfrac{(T_{H,in} - T_{C,in})}{(T_{H,out} - T_{C,out})}\right]} \tag{8-45}$$

Note that the log-mean temperature difference equations for counter-flow and parallel-flow configurations are identical if they are expressed in terms of the temperature difference at the two ends of the heat exchanger (see Figure 8-11):

$$\Delta T_{lm,pf} = \Delta T_{lm,cf} = \frac{\theta_{x=0} - \theta_{x=L}}{\ln\left[\dfrac{\theta_{x=0}}{\theta_{x=L}}\right]} \tag{8-46}$$

Also notice that it does not matter which end of the heat exchanger is defined as being $x = 0$ and $x = L$; changing the order in the numerator and denominator of Eq. (8-46) both result in a negative sign, which cancels:

$$\Delta T_{lm,pf} = \Delta T_{lm,cf} = \frac{\theta_{x=0} - \theta_{x=L}}{\ln\left[\dfrac{\theta_{x=0}}{\theta_{x=L}}\right]} = \frac{-(\theta_{x=L} - \theta_{x=0})}{-\ln\left[\dfrac{\theta_{x=L}}{\theta_{x=0}}\right]} = \frac{\theta_{x=L} - \theta_{x=0}}{\ln\left[\dfrac{\theta_{x=L}}{\theta_{x=0}}\right]} \tag{8-47}$$

8.2.3 *LMTD* Method for Shell-and-Tube and Cross-flow Heat Exchangers

The appropriate form of the log-mean temperature difference to use with Eq. (8-17) for shell-and-tube and cross-flow arrangements is more difficult to derive. Although an analytical solution for ΔT_{lm} can be obtained for some cases, the resulting expression is algebraically complicated and therefore inconvenient. An alternative approach recognizes that ΔT_{lm} for these heat exchanger arrangements will always be less than the log-mean temperature difference for the counter-flow arrangement, Eq. (8-44). Therefore, ΔT_{lm} for any configuration can be expressed as

$$\Delta T_{lm} = F \, \Delta T_{lm,cf} \tag{8-48}$$

where $\Delta T_{lm,cf}$ is the log-mean temperature difference for a counter-flow arrangement, given by Eq. (8-44), and F is a correction factor that has a value that is less than unity. For a given heat exchanger configuration, the value of F depends on the capacitance rates and the heat exchanger conductance. The effect of these factors can be expressed in terms of two nondimensional numbers, P and R, defined according to (for a cross-flow heat exchanger with both fluids unmixed):

$$P = \frac{(T_{C,out} - T_{C,in})}{(T_{H,in} - T_{C,in})} \tag{8-49}$$

$$R = \frac{\dot{C}_C}{\dot{C}_H} = \frac{T_{H,in} - T_{H,out}}{T_{C,out} - T_{C,in}} \tag{8-50}$$

The definition of P and R depend on the configuration. The quantity P is sometimes referred to as the *LMTD* effectiveness and R is as the *LMTD* capacitance ratio. These quantities are related to, but different from, the effectiveness (ε) and capacitance ratio (C_R) presented in Section 8.3 for the ε-NTU method.

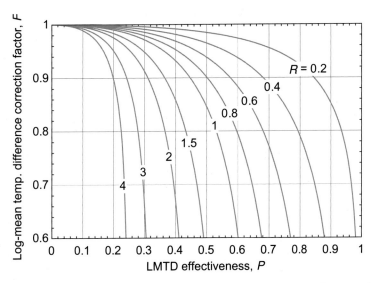

Figure 8-12: The correction parameter for a crossflow heat exchanger with both fluids unmixed as a function of P for various values of R.

The correction factor F has been derived for most common heat exchanger configurations. For example, Figure 8-12 provides the correction factor for a cross-flow heat exchanger with both fluids unmixed as a function of P for several values of R. Libraries of functions have been developed and integrated in EES in order to provide F as a function of P and R for a variety of heat exchanger geometries; these functions are accessed from the Function Information window by selecting the category Heat Exchangers and the sub-category F for $LMTD$ and then scrolling to find the configuration of interest. The definition of P and R for each configuration can be found in the help information associated with these functions.

Notice in Figure 8-12 that the value of the parameter P is limited to values between 0 and an upper bound that depends on R and is less than 1. For example, if $R = 3$ then P cannot exceed about 0.31. The $LMTD$ effectiveness, P, is the ratio of the temperature change of the cold-fluid to the maximum possible temperature change that the cold-fluid could experience; this maximum possible change occurs if the cold-fluid is heated to the hot inlet temperature. The ratio of the capacitance rates of the two streams, the $LMTD$ capacity ratio R, determines how closely the cold stream can approach the hot stream and therefore dictates the allowable values of P and the associated value of F. It is difficult to accurately determine values of F in terms of P and R when F is less than about 0.70 because of the strong sensitivity of F to P, as shown in Figure 8-12. In general, the effectiveness-NTU method, discussed in Section 8.3, is superior to the $LMTD$ method.

EXAMPLE 8.2-1: PERFORMANCE OF A CROSS-FLOW HEAT EXCHANGER

The cross-flow heat exchanger investigated in EXAMPLE 8.1-1 and EXAMPLE 8.1-2 is used to heat air with hot water. Water enters the heat exchanger tube with a mass flow rate, $\dot{m}_H = 0.03$ kg/s and temperature, $T_{H,in} = 60°C$. Air at $T_{C,in} = 20°C$ and atmospheric pressure is blown across the heat exchanger with a volumetric flowrate of $\dot{V}_C = 0.06$ m^3/s. The conductance of this heat exchanger has been calculated using several different techniques in Section 8.1; the best estimate of the conductance is $UA = 58.4$ W/K, obtained using the compact heat exchanger correlations.

a) Determine the outlet temperatures of the water and air and the heat transfer rate using the $LMTD$ method.

To solve this problem, we could add code to the EES program developed for EXAMPLE 8.1-2. Instead, a new program will be generated so that the calculations needed to implement the $LMTD$ method are clear. The conductance calculated in EXAMPLE 8.1-2 is an input to this program. It would appear to be straightforward to use the information provided together with the $LMTD$ heat exchanger formulation in Eq. (8-48) to determine the outlet temperatures. However, this example will show that the $LMTD$ method is not as easy to use as the ε-NTU technique discussed in Section 8.3.

The known information is entered into EES.

```
"EXAMPLE 8.2-1: Performance of a Cross-Flow Heat Exchanger"

$UnitSystem SI MASS RAD PA K J
$Tabstops 0.2 0.4 0.6 3.5 in

"Inputs"
V_dot_C=0.06 [m^3/s]                    "volumetric flow rate of air"
p=1 [atm]*convert(atm,Pa)               "atmospheric pressure"
T_C_in=convertTemp(C,K,20 [C])          "inlet air temperature"
T_H_in=convertTemp(C,K,60 [C])          "inlet water temperature"
m_dot_H=0.03 [kg/s]                     "water flow rate"
UA=58.4 [W/K]                           "conductance (from EXAMPLE 8.1-2)"
```

EXAMPLE 8.2-1: PERFORMANCE OF A CROSS-FLOW HEAT EXCHANGER

EXAMPLE 8.2-1: PERFORMANCE OF A CROSS-FLOW HEAT EXCHANGER

The density of air (ρ_C) is calculated at the inlet condition using EES' built-in property routine for air. The air-side mass flow rate is calculated according to:

$$\dot{m}_C = \rho_C \dot{V}_C$$

```
rho_C=density(Air,T=T_C_in,P=p)          "density of air"
m_dot_C=rho_C*V_dot_C                     "air mass flow rate"
```

The specific heat capacities should be evaluated at the average of the inlet and outlet temperatures for each fluid stream. However, these temperatures are not yet known. Notice that the values of $\Delta T_{lm,cf}$, P and R in Eqs. (8-44), (8-49) and (8-50) all require the outlet temperatures as well; this is the disadvantage of the *LMTD* method. The log-mean temperature difference is easy to use if the outlet temperatures are known and you want to solve for the required conductance (i.e., a design-type of problem). However, for problems where the conductance is known and you would like to know the outlet temperatures (i.e., a simulation-type of problem) then the *LMTD* method is inconvenient. Here, we will take the approach that has been used throughout this text; reasonable values for $T_{C,out}$ and $T_{H,out}$ are guessed so that the problem can be solved sequentially and then these values are adjusted by EES in order to complete the problem.

```
T_C_out=convertTemp(C,K,25 [C])          "guess for the cold stream exit temp."
T_H_out=convertTemp(C,K,50 [C])          "guess for the hot stream exit temp."
```

The specific heat capacities are evaluated at the average of the inlet and outlet temperatures:

```
c_C=cP(Air,T=(T_C_in+T_C_out)/2)         "specific heat capacity of air"
c_H=cP(Water,T=(T_H_in+T_H_out)/2,P=p)   "specific heat capacity of water"
```

The capacitance rates of the fluids are calculated:

$$\dot{C}_C = \dot{m}_C \, c_C$$

$$\dot{C}_H = \dot{m}_H \, c_H$$

```
C_dot_C=m_dot_C*c_C                       "capacitance rate of the air"
C_dot_H=m_dot_H*c_H                       "capacitance rate of the water"
```

The log-mean temperature difference that would result if the heat exchanger were in a counter-flow configuration is computed according to:

$$\Delta T_{lm,cf} = \frac{(T_{H,out} - T_{C,in}) - (T_{H,in} - T_{C,out})}{\ln\left[\dfrac{(T_{H,out} - T_{C,in})}{(T_{H,in} - T_{C,out})}\right]}$$

EXAMPLE 8.2-1: PERFORMANCE OF A CROSS-FLOW HEAT EXCHANGER

```
DELTAT_lm_cf=((T_H_out-T_C_in)-(T_H_in-T_C_out))/ln((T_H_out-T_C_in)/(T_H_in-T_C_out))
     "log-mean temp. difference for counter-flow configuration"
```

The *LMTD F* factor is required because the heat exchanger is cross-flow rather than parallel-flow or counter-flow. The value of *P* and *R* are computed according to:

$$P = \frac{(T_{C,out} - T_{C,in})}{(T_{H,in} - T_{C,in})}$$

$$R = \frac{\dot{C}_C}{\dot{C}_H} = \frac{T_{H,in} - T_{H,out}}{T_{C,out} - T_{C,in}}$$

```
P_HX=(T_C_out-T_C_in)/(T_H_in-T_C_in)                      "LMTD effectiveness"
R_HX=(T_H_in-T_H_out)/(T_C_out-T_C_in)                     "LMTD capacitance ratio"
```

The *LMTD F* factor is obtained from the appropriate EES function. To review the available functions, select the Function Info menu item in the Options menu and then select the Heat Exchangers option and F for LMTD options from the pull-down menus in the dialog. Scroll down to select the correct heat exchanger geometry. For this case, the appropriate heat exchanger is a cross-flow heat exchanger with both fluids unmixed since the plate fins prevent the air from mixing in the direction perpendicular to the air flow.

```
F_HX=LMTD_CF('crossflow_both_unmixed',P_HX,R_HX)          "LMTD correction factor"
```

The log-mean temperature difference in the heat exchanger is computed according to:

$$\Delta T_{lm} = F \, \Delta T_{lm,cf}$$

and the heat transfer rate is:

$$\dot{q} = (UA)\Delta T_{lm}$$

```
DELTAT_lm=DELTAT_lm_cf*F_HX                                "log-mean temp. difference"
q_dot=UA*DELTAT_lm                                        "heat transfer rate"
```

The solution that is obtained is clearly not correct as it was based on assumed fluid exit temperatures. Update the guess values (select Update Guesses from the Calculate menu) and then comment out the assumed outlet temperatures:

```
{T_C_out=convertTemp(C,K,25 [C])                          "guess for the cold stream exit temp."
T_H_out=convertTemp(C,K,50 [C])                           "guess for the hot stream exit temp."}
```

EXAMPLE 8.2-1: PERFORMANCE OF A CROSS-FLOW HEAT EXCHANGER

The solution is completed by calculating the outlet temperatures using energy balances on the two sides of the heat exchanger:

$$T_{C,out} = T_{C,in} + \frac{\dot{q}}{\dot{C}_C}$$

$$T_{H,out} = T_{H,in} - \frac{\dot{q}}{\dot{C}_H}$$

T_C_out=T_C_in+q_dot/C_dot_C	"cold-side fluid exit temperature"
T_C_out_C=converttemp(K,C,T_C_out)	"in C"
T_H_out=T_H_in-q_dot/C_dot_H	"hot-side fluid exit temperature"
T_H_out_C=converttemp(K,C,T_H_out)	"in C"

Select Solve from the Calculate menu and you are likely to be confronted with an error message. Even after carefully setting up the problem in a way that guaranteed a reasonable starting point for the iterative calculations, the *LMTD* method will often have problems converging; these problems are avoided using the ε-*NTU* method presented in Section 8.3. It is possible to force the problem to converge by setting appropriate limits in the Variable Information window. For example, specify that the outlet temperatures must lie between the inlet temperatures and that the values of R and P must be in a reasonable range. With the limits set, it should be possible to obtain the solution $\dot{q} = 1371$ W with $T_{C,out} = 38.9°C$ and $T_{H,out} = 49.1°C$.

There are a few sanity checks that you should use at this point to ensure that your answer makes physical sense. The outlet temperature of the hot-fluid cannot be less than the inlet temperature of the cold-fluid (and the outlet temperature of the cold-fluid cannot exceed the inlet temperature of the hot-fluid). The rate of heat transfer must always be less than the rate of heat transfer that would occur in the limit that neither fluid stream changed temperature, which provides the maximum possible driving temperature difference; that is, $\dot{q}$ must be less than UA $(T_{H,in} - T_{C,in})$ which for this problem is 2336 W.

You may be somewhat disappointed that EES was not able to solve the equations with the default guess values. The nature of the *LMTD* method makes this problem computationally difficult when the outlet fluid temperatures are not known. Figure 8-12 shows that there may be no solution for F for what appear to be reasonable outlet temperatures; this same problem occurs if you try to solve this problem by hand through manual iterations. The effectiveness method presented in Section 8.3 is much better in this respect.

8.3 The Effectiveness-*NTU* Method

8.3.1 Introduction

This section presents the effectiveness-*NTU* method for solving heat exchanger problems. The ε-*NTU* method is more flexible and easy to use than the *LMTD* method presented in Section 8.2. For example, the effectiveness-*NTU* method can be used to directly determine the outlet temperatures of a heat exchanger when the heat exchanger conductance is known (i.e., carry out a simulation-type of problem) or directly determine the conductance if the desired outlet temperatures are known (i.e., carry out a

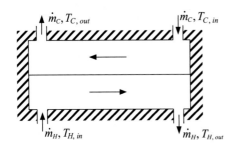

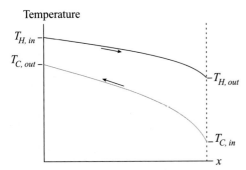

Figure 8-13: Counter-flow heat exchanger and the associated temperature distribution.

design-type of problem). As shown in Section 8.2, the *LMTD* method is difficult to apply to a simulation-type of problem where the outlet temperatures are unknown because it requires iteratively solving a set of non-linear equations that involve these temperatures. The effectiveness-*NTU* method is algebraically equivalent to the *LMTD* method and provides exactly the same results. However, the formulation is much better suited for a wide variety of heat exchanger problems.

8.3.2 The Maximum Heat Transfer Rate

In the *LMTD* method, the rate of heat transfer in the heat exchanger ($\dot{q}$) is expressed in terms of the conductance and an effective driving temperature difference (the log-mean temperature difference):

$$\dot{q} = UA \, \Delta T_{lm} \qquad (8\text{-}51)$$

The ε-*NTU* method expresses the rate of heat transfer in terms of the maximum possible heat transfer rate ($\dot{q}_{max}$) and the effectiveness (ε), which is the dimensionless heat exchanger performance:

$$\dot{q} = \varepsilon \, \dot{q}_{max} \qquad (8\text{-}52)$$

The value of the maximum possible heat transfer rate, $\dot{q}_{max}$, is not immediately obvious. An appropriate value of $\dot{q}_{max}$ is identified by considering the counter-flow heat exchanger shown in Figure 8-13.

The expected temperature distributions of the hot and cold-fluids as they progress through the heat exchanger are sketched in Figure 8-13 for the case where the capacitance rate (i.e., the mass flow rate – specific heat product) of the hot-fluid is somewhat larger than the capacity rate of the cold-fluid. (Recall, a fluid with a large capacitance rate will change temperature less than one with a smaller capacitance rate.)

The hot-fluid outlet temperature approaches the cold inlet temperature and the cold outlet temperature approaches the hot inlet temperature. However, the temperature distributions can never cross because doing so would violate the second law of thermo-dynamics (i.e., heat must transfer *from* hot *to* cold). The outlet temperatures of the two streams are related by an energy balance. Assuming that the heat exchanger is well-insulated, energy balances on the hot and cold streams lead to:

$$\dot{q} = \dot{C}_H \left(T_{H,in} - T_{H,out} \right) = \dot{C}_C \left(T_{C,out} - T_{C,in} \right) \tag{8-53}$$

where $\dot{C}_H$ and $\dot{C}_C$ are the capacitance rates of the hot and cold streams, respectively.

As the conductance (UA) of the heat exchanger increases, the temperature difference between the two fluids at either end of the heat exchanger decreases and the rate of heat transfer between the fluids increases. The quantity $\dot{q}_{max}$ is the rate of heat transfer that results if UA were to become infinitely large. In a counter-flow heat exchanger with an infinite UA, one fluid (but not necessarily both) must exit at the inlet temperature of the other fluid. For example, in the flow situation shown in Figure 8-13, it appears that the cold-fluid outlet temperature will approach the hot-fluid inlet temperature. However, in this limit the hot-fluid outlet temperature will not be equal to the cold-fluid inlet temperature. This behavior occurs because the capacitance rate of the hot-fluid is larger than the capacitance rate of the cold-fluid. If the capacitance rate of the one stream is larger than the other, the temperature change of the stream with the larger capacitance rate must be smaller in order to be consistent with the energy balance in Eq. (8-53). If the capacitance rates of the two streams were equal, then the temperature differences at both ends of the heat exchanger (and throughout the entire heat exchanger) would approach zero as UA becomes infinite.

The fluid having the smaller capacitance rate will experience the larger temperature change; as the conductance approaches infinity, the fluid with the minimum capacitance rate will experience a temperature change that is equal to $T_{H,in} - T_{C,in}$ and the maximum heat transfer rate can be written as:

$$\dot{q}_{max} = \begin{cases} \dot{C}_C \left(T_{H,in} - T_{C,in} \right) & \text{if } \dot{C}_C \le \dot{C}_H \\ \dot{C}_H \left(T_{H,in} - T_{C,in} \right) & \text{if } \dot{C}_C \ge \dot{C}_H \end{cases} \tag{8-54}$$

A more compact manner of expressing $\dot{q}_{max}$ is:

$$\dot{q}_{max} = \dot{C}_{min} \left(T_{H,in} - T_{C,in} \right) \tag{8-55}$$

where $\dot{C}_{min}$ is the minimum of the fluid capacitance rates.

$$\dot{C}_{min} = \text{MIN} \left(\dot{C}_C, \dot{C}_H \right) \tag{8-56}$$

Substituting Eq. (8-55) into Eq. (8-52) provides the definition of effectiveness:

$$\boxed{\dot{q} = \varepsilon \, \dot{C}_{min} \left(T_{H,in} - T_{C,in} \right)} \tag{8-57}$$

8.3.3 Heat Exchanger Effectiveness

The ε-*NTU* method is nothing more than an alternative presentation of the general solution for a particular heat exchanger configuration. In Section 8.2.2, the general solution for a counter-flow heat exchanger is derived:

$$\ln \left(\frac{T_{H,out} - T_{C,in}}{T_{H,in} - T_{C,out}} \right) = -UA \left(\frac{1}{\dot{C}_H} - \frac{1}{\dot{C}_C} \right) \tag{8-58}$$

The *LMTD* method rearranged Eq. (8-58) by using the energy balances to eliminate the capacitance rates; this led to:

$$\dot{q} = UA \underbrace{\frac{(T_{H,out} - T_{C,in}) - (T_{H,in} - T_{C,out})}{\ln\left[\dfrac{(T_{H,out} - T_{C,in})}{(T_{H,in} - T_{C,out})}\right]}}_{\Delta T_{lm,cf}}$$

(8-59)

The ε-*NTU* relationship for a counter-flow heat exchanger is an alternative arrangement of Eq. (8-58) that leads to a relationship between the effectiveness introduced in Section 8.3.2 and two additional parameters that are referred to as the number of transfer units (*NTU*) and the capacity ratio (C_R). The energy balances on the two streams are used to eliminate the fluid outlet temperatures:

$$T_{C,out} = T_{C,in} + \frac{\dot{q}}{\dot{C}_C}$$

(8-60)

$$T_{H,out} = T_{H,in} - \frac{\dot{q}}{\dot{C}_H}$$

(8-61)

Substituting the definition of effectiveness, Eq. (8-57), into Eqs. (8-60) and (8-61) leads to:

$$T_{C,out} = T_{C,in} + \frac{\varepsilon_{cf} \dot{C}_{min} (T_{H,in} - T_{C,in})}{\dot{C}_C}$$

(8-62)

$$T_{H,out} = T_{H,in} - \frac{\varepsilon_{cf} \dot{C}_{min} (T_{H,in} - T_{C,in})}{\dot{C}_H}$$

(8-63)

where ε_{cf} is the effectiveness of a counter-flow heat exchanger. Substituting Eqs. (8-62) and (8-63) into Eq. (8-58) leads to:

$$\ln\left(\frac{T_{H,in} - \varepsilon_{cf}\dfrac{\dot{C}_{min}}{\dot{C}_H}(T_{H,in} - T_{C,in}) - T_{C,in}}{T_{H,in} - T_{C,in} - \varepsilon_{cf}\dfrac{\dot{C}_{min}}{\dot{C}_C}(T_{H,in} - T_{C,in})}\right) = -UA\left(\frac{1}{\dot{C}_H} - \frac{1}{\dot{C}_C}\right)$$

(8-64)

which can be rearranged:

$$\ln\left(\frac{(T_{H,in} - T_{C,in})\left(1 - \varepsilon_{cf}\dfrac{\dot{C}_{min}}{\dot{C}_H}\right)}{(T_{H,in} - T_{C,in})\left(1 - \varepsilon_{cf}\dfrac{\dot{C}_{min}}{\dot{C}_C}\right)}\right) = -UA\left(\frac{1}{\dot{C}_H} - \frac{1}{\dot{C}_C}\right)$$

(8-65)

and simplified:

$$\ln\left(\frac{1 - \varepsilon_{cf}\dfrac{\dot{C}_{min}}{\dot{C}_H}}{1 - \varepsilon_{cf}\dfrac{\dot{C}_{min}}{\dot{C}_C}}\right) = -UA\left(\frac{1}{\dot{C}_H} - \frac{1}{\dot{C}_C}\right)$$

(8-66)

To proceed with the derivation, we will assume that $\dot{C}_{min} = \dot{C}_C$ and therefore $\dot{C}_{max} = \dot{C}_H$; this is consistent with the sketch in Figure 8-13. However, the final result is general and can be applied regardless of which stream has the minimum capacitance rate. With this assumption, Eq. (8-66) becomes:

$$\ln \left(\frac{1 - \varepsilon_{cf} \dfrac{\dot{C}_{min}}{\dot{C}_{max}}}{1 - \varepsilon_{cf} \dfrac{\dot{C}_{min}}{\dot{C}_{min}}} \right) = -UA \left(\frac{1}{\dot{C}_{max}} - \frac{1}{\dot{C}_{min}} \right) \tag{8-67}$$

Equation (8-67) can be rearranged:

$$\ln \left(\frac{1 - \varepsilon_{cf} \overbrace{\dfrac{\dot{C}_{min}}{\dot{C}_{max}}}^{C_R}}{1 - \varepsilon_{cf}} \right) = - \underbrace{\frac{UA}{\dot{C}_{min}}}_{NTU} \left(\underbrace{\frac{\dot{C}_{min}}{\dot{C}_{max}}}_{C_R} - 1 \right) \tag{8-68}$$

The dimensionless number $UA/\dot{C}_{min}$ in Eq. (8-68) is referred to as the number of transfer units or *NTU*.

$$NTU = \frac{UA}{\dot{C}_{min}} \tag{8-69}$$

The number of transfer units represents the dimensionless size of the heat exchanger. If the conductance of the heat exchanger increases, then the heat exchanger is physically larger. If the minimum capacitance rate decreases, then the heat exchanger must process less fluid. In either case, its dimensionless size, *NTU*, will increase.

The dimensionless number $\dot{C}_{min}/\dot{C}_{max}$ in Eq. (8-68) is referred to as the capacity ratio and reflects how well-balanced the heat exchanger is:

$$C_R = \frac{\dot{C}_{min}}{\dot{C}_{max}} \tag{8-70}$$

If the capacity ratio is unity, then the heat exchanger is operating in a balanced condition; the capacitance rates of the two fluids and therefore the temperature change experienced by the two fluids are the same. If the capacity ratio is very small, then the heat exchanger is operating in an unbalanced condition; the capacitance rate of one fluid is very large compared to the other and the temperature of this fluid will not change substantially. This situation occurs when one fluid is changing phase.

Substituting Eqs. (8-69) and (8-70) into Eq. (8-68) leads to:

$$\ln \left(\frac{1 - \varepsilon_{cf} C_R}{1 - \varepsilon_{cf}} \right) = -NTU \left(C_R - 1 \right) \tag{8-71}$$

It was assumed that $\dot{C}_{max} = \dot{C}_H$ and $\dot{C}_{min} = \dot{C}_C$ in order to derive Eq. (8-71). However, if we had instead assumed that $\dot{C}_{max} = \dot{C}_C$ and $\dot{C}_{min} = \dot{C}_H$, then the same result would be obtained. Equation (8-71) is the fundamental effectiveness-*NTU* relationship for a counter-flow heat exchanger that relates the three dimensionless parameters ε, *NTU*, and C_R. Equation (8-71) can be rearranged to provide the effectiveness in terms of *NTU* and C_R (which is useful for simulation-type problems) or the number of transfer units in terms of the ε and C_R (which is useful for design-type problems). Solving Eq. (8-71) for the effectiveness leads to:

$$\boxed{\varepsilon_{cf} = \frac{1 - \exp \left[-NTU \left(1 - C_R \right) \right]}{1 - C_R \exp \left[-NTU \left(1 - C_R \right) \right]} \quad \text{for } C_R < 1} \tag{8-72}$$

Table 8-1: Effectiveness-NTU relations for various heat exchanger configurations in the form effectiveness as a function of number of transfer units and capacity ratio.

Flow arrangement		$\varepsilon\,(NTU, C_R)$
One fluid (or any configuration with $C_R = 0$)		$\varepsilon = 1 - \exp(-NTU)$
Counter-flow		$\varepsilon = \begin{cases} \dfrac{1 - \exp\left[-NTU\left(1 - C_R\right)\right]}{1 - C_R \exp\left[-NTU\left(1 - C_R\right)\right]} & \text{for } C_R < 1 \\[3mm] \dfrac{NTU}{1 + NTU} & \text{for } C_R = 1 \end{cases}$
Parallel-flow		$\varepsilon = \dfrac{1 - \exp\left[-NTU\left(1 + C_R\right)\right]}{1 + C_R}$
Cross-flow	both fluids unmixed	$\varepsilon = 1 - \exp\left[\dfrac{NTU^{0.22}}{C_R}\left\{\exp(-C_R\, NTU^{0.78}) - 1\right\}\right]$
	both fluids mixed	$\varepsilon = \left[\dfrac{1}{1 - \exp(-NTU)} + \dfrac{C_R}{1 - \exp(-C_R\, NTU)} - \dfrac{1}{NTU}\right]^{-1}$
	$\dot{C}_{max}$ mixed & $\dot{C}_{min}$ unmixed	$\varepsilon = \dfrac{1 - \exp\left[C_R\left\{\exp(-NTU) - 1\right\}\right]}{C_R}$
	$\dot{C}_{min}$ mixed & $\dot{C}_{max}$ unmixed	$\varepsilon = 1 - \exp\left[-\dfrac{1 - \exp(-C_R\, NTU)}{C_R}\right]$
Shell-and-tube	one shell pass & an even # of tube-passes	$\varepsilon_1 = 2\left[1 + C_R + \sqrt{1 + C_R^2}\,\dfrac{1 + \exp\left(-NTU_1\sqrt{1 + C_R^2}\right)}{1 - \exp\left(-NTU_1\sqrt{1 + C_R^2}\right)}\right]^{-1}$
	N shell passes & $2N$, $4N, \ldots$ tube-passes	$\varepsilon = \dfrac{\left(\dfrac{1 - \varepsilon_1 C_R}{1 - \varepsilon_1}\right)^N - 1}{\left(\dfrac{1 - \varepsilon_1 C_R}{1 - \varepsilon_1}\right)^N - C_R}$ where ε_1 and NTU_1 is for one shell pass

Equation (8-72) is indeterminate if $C_R = 1$. However, taking the limit of Eq. (8-72) as C_R approaches 1 using Maple leads to:

```
> restart;
> eff:=(1-exp(-NTU*(1-CR)))/(1-CR*exp(-NTU*(1-CR)));
```

$$eff := \frac{1 - e^{(-NTU(1-CR))}}{1 - CR\,e^{(-NTU(1-CR))}}$$

```
> limit(eff,CR=1);
```

$$\frac{NTU}{NTU + 1}$$

$$\boxed{\varepsilon_{cf} = \frac{NTU}{1 + NTU} \quad \text{for } C_R = 1}$$

(8-73)

Table 8-2: Effectiveness-*NTU* relations for various heat exchanger configurations in the form of number transfer units as a function of effectiveness and capacity ratio.

Flow arrangement		$NTU\,(\varepsilon,\,C_R)$
One fluid (or any configuration with $C_R = 0$)		$NTU = -\ln(1-\varepsilon)$
Counter-flow		$NTU = \begin{cases} \dfrac{\ln\left[\dfrac{1-\varepsilon\,C_R}{1-\varepsilon}\right]}{1-C_R} & \text{for } C_R < 1 \\[1em] \dfrac{\varepsilon}{1-\varepsilon} & \text{for } C_R = 1 \end{cases}$
Parallel-flow		$NTU = \dfrac{\ln[1-\varepsilon\,(1+C_R)]}{1+C_R}$
Cross-flow	$\dot{C}_{max}$ mixed & $\dot{C}_{min}$ unmixed	$NTU = -\ln\left[1 + \dfrac{\ln(1-\varepsilon\,C_R)}{C_R}\right]$
	$\dot{C}_{min}$ mixed & $\dot{C}_{max}$ unmixed	$NTU = -\dfrac{\ln[C_R\,\ln(1-\varepsilon)+1]}{C_R}$
Shell-and-tube	one shell pass & an even # of tube-passes	$NTU_1 = \dfrac{\ln\left(\dfrac{E+1}{E-1}\right)}{\sqrt{1+C_R^2}}$ where $E = \dfrac{2-\varepsilon_1\,(1+C_R)}{\varepsilon_1\sqrt{1+C_R^2}}$
	N shell passes & 2N, 4N, ... tube-passes	use solution for one shell pass with: $\varepsilon_1 = \dfrac{F-1}{F-C_R}$ with $F = \left(\dfrac{\varepsilon\,C_R-1}{\varepsilon-1}\right)^{1/N}$

A similar analysis carried out for a parallel–flow configuration leads to:

$$\boxed{\varepsilon_{pf} = \dfrac{1-\exp\left[-NTU\,(1+C_R)\right]}{1+C_R}} \tag{8-74}$$

Effectiveness-*NTU* relations (i.e., algebraic equations that relate the quantities ε, *NTU*, and C_R) have been derived for the most common heat exchanger configurations; analytical or graphical presentations of these solutions are available in Kays and London (1984) and elsewhere. The ε-*NTU* solutions are summarized in Table 8-1 in the form of $\varepsilon(NTU, C_R)$ and in Table 8-2 in the form of $NTU\,(\varepsilon, C_R)$.

EES functions are available that implement the solutions listed in Table 8-1 and Table 8-2 as well as other solutions that are less easily expressed analytically. These functions are implemented in the two forms that are most useful: $\varepsilon\,(NTU, C_R)$ and *NTU* (ε, C_R). They can be accessed from the Function Information window by selecting Heat Exchangers and then either NTU-> Effectiveness or Effectiveness -> NTU, respectively. The first form assumes that *UA* and the capacitance rates of the two streams are known (and therefore *NTU* and C_R can be directly computed) and this information is used to determine the effectiveness (which can be used to provide the outlet temperatures). The second form assumes that the outlet temperatures and the capacitance rates of the two streams are known (and therefore ε and C_R can be directly computed) and this information is used to determine the *NTU* (which can be used to provide the *UA* that is required).

The most important results presented in this section are summarized below:

1. The *LMTD* and effectiveness-*NTU* solution methods are algebraically equivalent and they will provide exactly the same results.
2. The effectiveness (ε) and capacitance ratio (C_R) defined in the effectiveness-*NTU* method differ from the P and R factors that are used to calculate the F factor in the *LMTD* method. The P and R factors for the *LMTD* solution are always defined by Eqs. (8-49) and (8-50), and these definitions do not depend on which stream has the smaller capacitance rate. In the effectiveness-*NTU* method, C_R is defined such that it will always be less than one and the definition of ε changes as indicated in Eq. (8-57), depending on which stream has the minimum capacitance rate.
3. The effectiveness-*NTU* method is easier to use than the *LMTD* method in almost any situation and particularly when the *UA* and the capacitance rates of the two streams are known. This difference will be demonstrated in EXAMPLE 8.3-1.

EXAMPLE 8.3-1: PERFORMANCE OF A CROSS-FLOW HEAT EXCHANGER (REVISITED)

EXAMPLE 8.3-1: PERFORMANCE OF A CROSS-FLOW HEAT EXCHANGER (REVISITED)

The cross-flow heat exchanger investigated in EXAMPLE 8.1-1 and EXAMPLE 8.1-2 is used to heat air with hot water. Water enters the heat exchanger tubing with a mass flow rate, $\dot{m}_H = 0.03$ kg/s and temperature, $T_{H,in} = 60°C$. Air at $T_{C,in} = 20°C$ and atmospheric pressure is blown across the heat exchanger with a volumetric flowrate of $\dot{V}_C = 0.06$ m^3/s. The conductance of this heat exchanger has been calculated using several different techniques in Section 8.1; the best estimate of the conductance is $UA = 58.4$ W/K, based on the compact heat exchanger correlations.

a) **Determine the outlet temperatures of the water and air and the heat transfer rate using the ε-*NTU* method.**

This is the same problem that was solved in EXAMPLE 8.2-1 using the log-mean temperature difference method. Recall that the use of the *LMTD* method was not convenient and required a series of steps related to setting proper guess values and limits. Solving this problem using the ε-*NTU* method will provide a clear comparison of these heat exchanger solution methods.

The known information is entered in EES:

```
"EXAMPLE 8.3-1: Performance of a Cross-Flow Heat Exchanger (revisited)"

$UnitSystem SI MASS RAD PA K J
$Tabstops 0.2 0.4 0.6 3.5 in

"Inputs"
V_dot_C=0.06 [m^3/s]                         "volumetric flow rate of air"
p=1 [atm]*convert(atm,Pa)                    "atmospheric pressure"
T_C_in=convertTemp(C,K,20 [C])               "inlet air temperature"
T_H_in=convertTemp(C,K,60 [C])               "inlet water temperature"
m_dot_H=0.03 [kg/s]                          "water flow rate"
UA=58.4 [W/K]                                "conductance (from EXAMPLE 8.1-2)"
```

The density of air (ρ_C) is calculated at the inlet condition and used to compute the air-side mass flow rate:

$$\dot{m}_C = \rho_C \, \dot{V}_C$$

```
rho_C=density(Air,T=T_C_in,P=p)          "density of air"
m_dot_C=rho_C*V_dot_C                     "air mass flow rate"
```

The specific heat capacities of the air and the water should be evaluated at the average of the inlet and outlet temperatures for each fluid stream. However, these temperatures are not yet known. Reasonable values for $T_{C,out}$ and $T_{H,out}$ are assumed so that the problem can be solved sequentially; these values will be adjusted based on the solution.

```
T_C_out=convertTemp(C,K,25 [C])          "guess for the cold stream exit temp."
T_H_out=convertTemp(C,K,50 [C])          "guess for the hot stream exit temp."
```

The specific heat capacities are evaluated at the average of the inlet and outlet temperatures:

```
c_C=cP(Air,T=(T_C_in+T_C_out)/2)         "specific heat capacity of air"
c_H=cP(Water,T=(T_H_in+T_H_out)/2,P=p)   "specific heat capacity of water"
```

The capacitance rates of the fluids are calculated:

$$\dot{C}_C = \dot{m}_C \, c_C$$

$$\dot{C}_H = \dot{m}_H \, c_H$$

```
C_dot_C=m_dot_C*c_C                       "capacitance rate of the air"
C_dot_H=m_dot_H*c_H                       "capacitance rate of the water"
```

The minimum and maximum capacitance rates ($\dot{C}_{min}$ and $\dot{C}_{max}$) are evaluated using the **MIN** and **MAX** commands in EES:

```
C_dot_min=MIN(C_dot_C,C_dot_H)           "minimum capacitance rate"
C_dot_max=MAX(C_dot_C,C_dot_H)           "maximum capacitance rate"
```

The number of transfer units is calculated according to:

$$NTU = \frac{UA}{\dot{C}_{min}}$$

```
NTU=UA/C_dot_min                          "number of transfer units"
```

The effectiveness (ε) for a cross-flow heat exchanger with both fluids unmixed is obtained by selecting Heat Exchangers from the Function Information window and

EXAMPLE 8.3-1: PERFORMANCE OF A CROSS-FLOW HEAT EXCHANGER (REVISITED)

EXAMPLE 8.3-1: PERFORMANCE OF A CROSS-FLOW HEAT EXCHANGER (REVISITED)

then selecting NTU -> Effectiveness and scrolling to the correct configuration. Paste the function into the Equations Window and then change C_dot_1 to C_dot_H and C_dot_2 to C_dot_C in order to correspond to the variables used in this problem solution. (Note that the order in which you enter the capacity rates does not matter for this heat exchanger, but it would matter if one of the fluids were mixed.)

```
epsilon=HX('crossflow_both_unmixed', NTU, C_dot_C, C_dot_H, 'epsilon')
   "access effectiveness-NTU solution"
```

The maximum possible heat transfer rate ($\dot{q}_{max}$) is computed according to:

$$\dot{q}_{max} = \dot{C}_{min} (T_{H,in} - T_{C,in})$$

and the actual heat transfer rate ($\dot{q}$) is computed according to:

$$\dot{q} = \varepsilon \, \dot{q}_{max}$$

```
q_dot_max=C_dot_min*(T_H_in-T_C_in)          "maximum possible heat transfer rate"
q_dot=q_dot_max*epsilon                        "actual heat transfer rate"
```

The guess values are updated and the initial, guessed values for the outlet temperatures are commented out:

```
{T_C_out=convertTemp(C,K,25 [C])             "guess for the cold stream exit temp."
T_H_out=convertTemp(C,K,50 [C])              "guess for the hot stream exit temp."}
```

The outlet temperatures are computed using energy balances:

$$T_{C,out} = T_{C,in} + \frac{\dot{q}}{\dot{C}_C}$$

$$T_{H,out} = T_{H,in} - \frac{\dot{q}}{\dot{C}_H}$$

```
T_H_out=T_H_in-q_dot/C_dot_H                  "hot-fluid exit temperature"
T_H_out_C=converttemp(K,C,T_H_out)           "in C"
T_C_out=T_C_in+q_dot/C_dot_C                 "cold-fluid exit temperature"
T_C_out_C=converttemp(K,C,T_C_out)           "in C"
```

which leads to $\dot{q} = 1371$ W with $T_{C,out} = 38.9°C$ and $T_{H,out} = 49.1°C$. These results are identical to those obtained in EXAMPLE 8.2-1.

Notice that the problem could be solved directly without requiring any attention to the guess values and/or limits. This is the major advantage of the effectiveness-NTU method. Although it is algebraically identical to the LMTD method, it is formulated in a manner that allows direct determination of the heat transfer rate and outlet temperatures when the UA and capacitance rates are known.

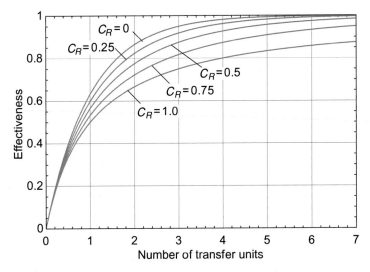

Figure 8-14: Effectiveness as a function of the number of transfer units for various values of capacity ratio for a counter-flow heat exchanger.

8.3.4 Further Discussion of Heat Exchanger Effectiveness

There are some additional, useful concepts that can be illustrated using the ε-*NTU* method. The different flow configurations listed in Table 8-1 and Table 8-2 require different relationships between ε, NTU, and C_R because the flow configuration influences how heat transfer affects the local temperature difference. For example, Figure 8-14, Figure 8-15, and Figure 8-16 illustrate ε as a function of NTU for various values of C_R for the counter-flow, parallel-flow, and cross-flow (with both fluids unmixed) configurations, respectively.

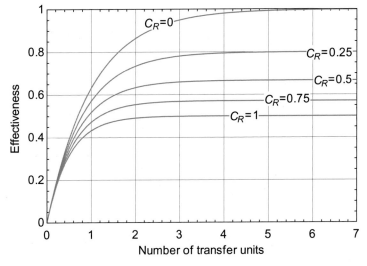

Figure 8-15: Effectiveness as a function of the number of transfer units for various values of capacity ratio for a parallel-flow heat exchanger.

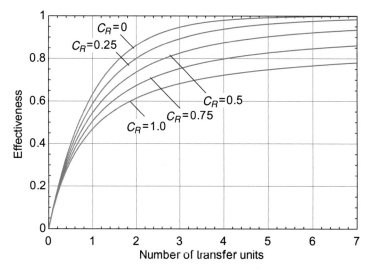

Figure 8-16: Effectiveness as a function of the number of transfer units for various values of capacity ratio for a cross-flow heat exchanger with both fluids unmixed.

Behavior as C_R Approaches Zero

The effect of the heat exchanger configuration on its performance is related to the interaction between the temperature changes of the two fluid streams. In the limit that the capacity ratio approaches zero (i.e., one fluid stream has a capacity rate that is much greater than the other), then there is no such interaction because one fluid stream does not change its temperature significantly. Figure 8-17(a) and (b) shows the temperature distributions that result within a counter-flow and parallel-flow heat exchanger, respectively, when C_R goes to zero because $\dot{C}_C \gg \dot{C}_H$.

Notice that the temperature distributions in Figure 8-17(a) and (b) are nearly identical and therefore configuration is not important. The temperature of the cold-fluid does not change significantly regardless of configuration; it does not matter if the cold-fluid temperature decreases slightly from left-to-right (as it does for the counter-flow case in Figure 8-17(a)) or increases slightly from left-to-right (as for the parallel-flow case in Figure 8-17(b)). In the limit that $C_R \rightarrow 0$, the effectiveness-*NTU* relationship for any

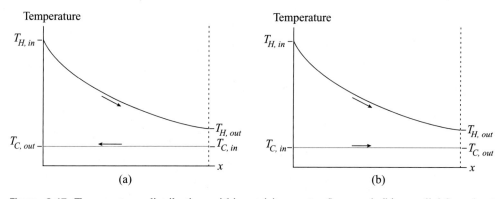

Figure 8-17: Temperature distribution within a (a) counter-flow and (b) parallel-flow heat exchanger as the capacity ratio approaches zero because $\dot{C}_c \gg \dot{C}_H$.

configuration reduces to:

$$\lim_{C_R \to 0} \varepsilon = 1 - \exp(-NTU) \tag{8-75}$$

Substituting the definition of effectiveness and *NTU* into Eq. (8-75) leads to:

$$\frac{T_{H,in} - T_{H,out}}{T_{H,in} - T_{C,in}} = 1 - \exp\left(-\frac{UA}{\dot{C}_H}\right) \tag{8-76}$$

Equation (8-76) is identical to the result obtained in Section 5.3.5 for an internal flow problem with a prescribed external temperature. Substituting T_∞ for $T_{C,in}$ and T_{in} and T_{out} for $T_{H,in}$ and $T_{H,out}$ and rearranging slightly leads to the same solution because it is the same situation – a single fluid stream interacting with an unchanging temperature. In Figure 8-14 through Figure 8-16, notice that the $C_R = 0$ curves (the upper-most curve in each figure) are identical even though the configuration changes.

The limit of $C_R \to 0$ is approached in many practical heat exchangers; e.g., the interaction of a flowing fluid with a constant temperature solid (as in a cold-plate) or a well-mixed tank of fluid or the situation that occurs when one of the two streams is undergoing constant pressure evaporation or condensation.

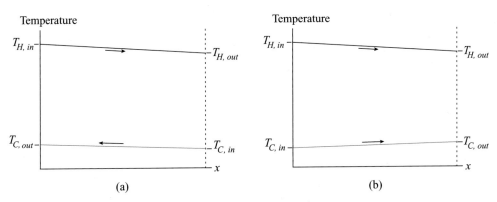

Figure 8-18: Temperature distribution within a (a) counter-flow and (b) parallel-flow heat exchanger as *NTU* approaches zero.

Behavior as NTU Approaches Zero

In the limit that *NTU* approaches zero, the configuration of the heat exchanger again does not matter. Figure 8-18(a) and (b) illustrate the temperature distribution within a counter-flow and parallel-flow heat exchanger, respectively, for a small *NTU*.

If *NTU* is small, then the heat exchanger is under-sized and the rate of heat transfer is not sufficient to change the temperature of either fluid substantially. As a result, the temperature difference in the heat exchanger is approximately constant and equal to $T_{H,in} - T_{C,in}$ throughout the entire heat exchanger. The configuration (e.g., counter-flow vs parallel-flow) is unimportant and does not impact the local temperature difference. In the limit that $NTU \to 0$, the heat transfer rate can be written as:

$$\dot{q} = UA\,(T_{H,in} - T_{C,in}) \tag{8-77}$$

because the temperature difference is constant. The definition of effectiveness is:

$$\varepsilon = \frac{\dot{q}}{C_{min}\,(T_{H,in} - T_{C,in})} \tag{8-78}$$

Substituting Eq. (8-77) into (8-78) leads to:

$$\varepsilon = \frac{UA \, (T_{H,in} - T_{C,in})}{\dot{C}_{min} \, (T_{H,in} - T_{C,in})} \tag{8-79}$$

or

$$\lim_{NTU \to 0} \varepsilon = NTU \tag{8-80}$$

In Figure 8-14 through Figure 8-16, notice that the curves for all values of C_R collapse near $NTU = 0$ and the slope of this single line is 1.0, regardless of configuration.

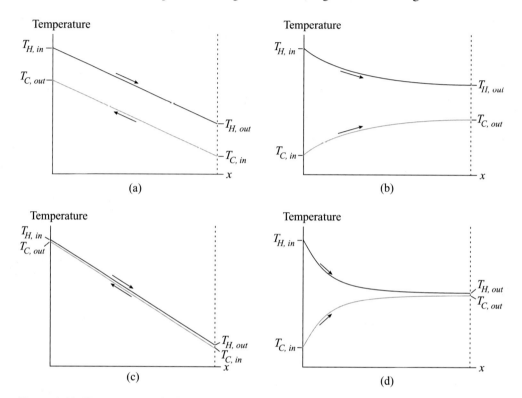

Figure 8-19: Temperature distribution within a balanced (i.e., $C_R = 1$) (a) counter-flow and (b) parallel-flow heat exchanger for a finite NTU. The temperature distribution within a balanced (c) counter-flow and (d) parallel-flow heat exchanger as $NTU \to \infty$.

Behavior as NTU Becomes Infinite

Section 8.3.4 shows that the configuration is not important when either the capacity ratio or the number of transfer units is small. For a finite capacity ratio, the configuration dictates the behavior of the heat exchanger as the number of transfer units becomes large. To see this clearly, consider a "balanced" heat exchanger; that is, a heat exchanger with $C_R = 1$ so that $\dot{C}_C = \dot{C}_H$. Figure 8-19 (a) and (b) illustrate the temperature distributions within counter-flow and parallel-flow heat exchangers, respectively, with a finite number of transfer units and $C_R = 1$. Notice that the temperature change experienced by both of the fluids is the same (consistent with a balanced heat exchanger), but that the temperature distributions are quite different.

The temperature difference somewhere within the heat exchanger will approach zero as the number of transfer units becomes large; this limit is shown in Figure 8-19 (c) and (d) for the counter-flow and parallel-flow heat exchangers, respectively. The location where the temperature difference approaches zero is referred to as the pinch point; the concept of a pinch point is explored more completely in Section 8.4.

Because the heat exchanger is balanced, the temperature change of the hot and cold streams must be equal regardless of configuration or *NTU*; energy balances constrain the effectiveness that can be achieved. For the counter-flow configuration shown in Figure 8-19(a) and (c), the effectiveness can approach 1.0 as the $NTU \to \infty$; this is evident because the cold-fluid outlet temperature can approach the hot-fluid inlet temperature (and, for the balanced case, the hot-fluid outlet temperature can approach the cold-fluid inlet temperature). Notice in Figure 8-14 that the $C_R = 1$ curve for a counter-flow heat exchanger asymptotically approaches 1.0 as $NTU \to \infty$. However, it approaches 1.0 very slowly because the temperature difference everywhere in the heat exchanger is being driven toward zero.

The same behavior does not occur for any other configuration. For example, notice in the parallel-flow configuration shown in Figure 8-19(b) and (d) that the effectiveness does not approach 1.0 even as $NTU \to \infty$; regardless of the size of the heat exchanger, the cold-fluid outlet temperature cannot approach the hot-fluid inlet temperature. For the balanced case, the hot-fluid outlet temperature can only get half-way there, corresponding to a maximum possible effectiveness of 0.5. Notice in Figure 8-15 that the $C_R = 1$ curve for a parallel-flow heat exchanger asymptotically approaches 0.5 as $NTU \to \infty$. However, it approaches 0.5 very quickly because the temperature difference in the heat exchanger remains quite large. The limit on the effectiveness of a parallel flow heat exchanger as $NTU \to \infty$ depends only on the capacity ratio. For the balanced case shown in Figure 8-19, the two streams experience the same temperature change. For smaller capacity ratios, one stream will change temperature by more than the other and therefore the effectiveness that can be achieved will increase. This result is evident by examining the behavior of Figure 8-19 as $NTU \to \infty$. In general, the limit to the performance of a parallel flow heat exchanger is:

$$\lim_{NTU \to \infty} \varepsilon_{pf} = \frac{1}{C_R + 1} \tag{8-81}$$

Notice that if $C_R = 0$ then Eq. (8-81) limits to 1.0, which is consistent with Eq. (8-75).

Other heat exchanger configurations fall somewhere between parallel-flow and counter-flow in terms of the effect of the capacity ratio on their performance. For example, examine the $C_R = 0.5$ curves for counter-flow, parallel-flow and cross-flow configurations shown in Figure 8-14, Figure 8-15, and Figure 8-16, respectively. At any value of *NTU*, the counter-flow configuration provides the highest effectiveness and the parallel-flow configuration the lowest; other configurations such as cross-flow fall between these limits. This result is illustrated in Figure 8-20, which shows the effectiveness as a function of *NTU* for various configurations at a constant capacity ratio of (a) $C_R = 1$, (b) $C_R = 0.5$, and (c) $C_R = 0.25$.

The difference between heat exchanger configurations is largest for the $C_R = 1$ condition (Figure 8-20(a)). As C_R is reduced, the effect of configuration diminishes and eventually all configurations collapse onto the same curve, given by Eq. (8-75), as $C_R \to 0$.

Heat Exchanger Design

The optimal design of a heat exchanger is a complex process that depends on the intended purpose of the heat exchanger and its cost. The design process will inevitably rely on an economic analysis in which the first cost of the heat exchanger is balanced against the operating cost. The first cost of the heat exchanger is directly related to its size and therefore its conductance, *UA*; for a given operating condition, the

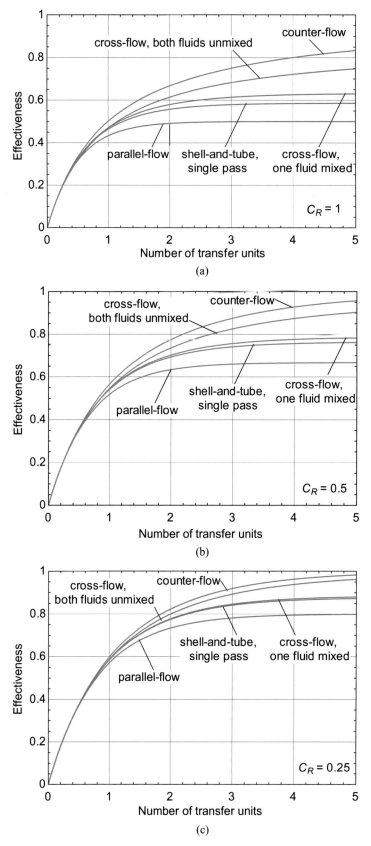

Figure 8-20: Effectiveness as a function of *NTU* for various configurations with a constant capacity ratio of (a) $C_R = 1$, (b) $C_R = 0.5$, and (c) $C_R = 0.25$.

conductance dictates the *NTU*. The operating cost will be inversely related to the effectiveness; a higher effectiveness will reduce the operating cost. Examination of any of the *ε-NTU* curves that are shown in Figure 8-14 through Figure 8-16 makes it clear that an optimally designed heat exchanger will be characterized by an *NTU* that is usually in the range of 1 to 2. At low values of *NTU*, the effectiveness can be increased (and therefore the operating cost reduced) substantially by making the heat exchanger larger; a little additional first cost will greatly reduce the operating cost. However, at high values of *NTU*, the effectiveness becomes insensitive to *NTU* so that it will take a large increase in the first cost to change the operating cost.

An introduction to economic analysis tools is provided in Appendix A.5, which can be found on the website associated with this text (www.cambridge.org/nellisandklein). These tools can be used to provide a true, optimal design that minimizes life cycle cost. However, a quick check on a heat exchanger design will involve calculating the *NTU*. If the *NTU* is large (say, greater than 10) then it is at least worth asking: what in the economic analysis is pushing you to this extreme? Is the first cost of the heat exchanger somehow very low (perhaps unrealistically) or is the operating cost very large.

8.4 Pinch Point Analysis

8.4.1 Introduction

Heat exchanger performance calculations using the log-mean temperature difference and effectiveness-*NTU* methods are presented in Sections 8.2 and 8.3, respectively. Both of these methods assume that the capacitance rates of the fluid streams are constant throughout the heat exchanger. In practice, this situation rarely occurs. Although the mass flow rates will not change with position for steady-state operation with no leakage, the specific heats of most fluids are temperature-dependent and therefore vary as the fluids progress through the heat exchanger. The variation in the specific heat capacities is approximately handled by evaluating the specific heat capacities at the average fluid temperatures within an *LMTD* or *ε-NTU* analysis. In many cases, this approximation provides results that are suitable for engineering purposes. This is particularly true in light of the uncertainty that is often associated with the other parameters, such as the heat transfer coefficients that are used to calculate the conductance.

However, there are situations where the capacitance rate of one or both fluids varies significantly within the heat exchanger and therefore the use of an *ε-NTU* analysis leads to significant error. One particularly important example occurs when the heat exchanger involves a phase change for one or both fluids. The discontinuity in the specific heat that occurs during a phase change can lead to internal "pinch points" that limit the rate of heat transfer to a much greater extent than would be expected based only on the inlet fluid temperatures. The identification of pinch-points and the associated analysis becomes important even if the capacitance rates of the streams are approximately constant but there are many different streams involved. For example, pinch point analysis is often used in the preliminary design of chemical process plants. In this section, the concept of a pinch point is explored and a very simple pinch point analysis is demonstrated.

8.4.2 Pinch Point Analysis for a Single Heat Exchanger

The concept of a pinch point is most clearly illustrated by example. The counter-flow heat exchanger shown in Figure 8-21 functions as a boiler. Water enters with mass flow rate $\dot{m}_w = 0.009$ kg/s and temperature $T_{w,in} = 30°C$. The water flows through the heat

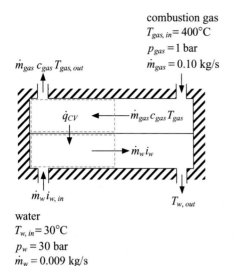

combustion gas
$T_{gas,\,in} = 400°C$
$p_{gas} = 1$ bar
$\dot{m}_{gas} = 0.10$ kg/s

Figure 8-21: Counter-flow heat exchanger using combustion gas to boil water.

water
$T_{w,\,in} = 30°C$
$p_w = 30$ bar
$\dot{m}_w = 0.009$ kg/s

exchanger at constant pressure $p_w = 30$ bar. The combustion gas used to heat the water enters at $T_{gas,in} = 400°C$ and flows at constant pressure of $p_{gas} = 1$ bar with mass flow rate $\dot{m}_{gas} = 0.10$ kg/s. The specific heat capacity of the combustion gas is approximately constant and equal to $c_{gas} = 1075$ J/kg-K. Before carrying out a detailed analysis of the heat exchanger, it is useful to answer the basic question: what is the maximum possible rate of heat transfer to the water and the corresponding combustion gas and water outlet temperatures? What is the limit of the performance of the boiler?

The known information is entered into EES:

```
$UnitSystem SI MASS RAD PA K J
$Tabstops 0.2 0.4 0.6 3.5 in

"Inputs"
T_w_in=converttemp(C,K,30 [C])              "Inlet water temperature"
m_dot_w=0.009 [kg/s]                        "Mass flow rate of water"
p_w=30 [bar] *convert(bar,Pa)               "Water pressure"
T_gas_in=converttemp(C,K,400 [C])           "Combustion gas inlet temperature"
m_dot_gas=0.10 [kg/s]                       "Combustion gas mass flow rate"
p_gas=1 [bar]*convert(bar,Pa)              "Gas pressure"
c_gas=1075 [J/kg-K]                         "Combustion gas specific heat capacity"
```

This problem appears to be simple; the maximum heat transfer rate corresponds to a heat exchanger with infinite conductance (UA or $NTU \rightarrow \infty$). It would seem, based on the discussion in Section 8.3.4, that in this limit the water and gas temperatures will approach each other at one end of the heat exchanger. The location where the temperature difference is smallest is referred to as the pinch point. Initially, we will assume that the temperature of the gas leaving the heat exchanger approaches the inlet temperature of the water (i.e., the pinch point is at the cold end):

$$T_{gas,out} = T_{w,in} \qquad (8\text{-}82)$$

```
T_gas_out=T_w_in                            "Assume that pinch point is at cold end"
```

The total heat transfer rate in the heat exchanger can be obtained using an energy balance on the gas-side of the heat exchanger:

$$\dot{q} = \dot{m}_{gas} c_{gas} \left(T_{gas,in} - T_{gas,out} \right) \quad (8\text{-}83)$$

```
q_dot=m_dot_gas*c_gas*(T_gas_in-T_gas_out)          "Energy balance on heat exchanger"
```

which leads to $\dot{q} = 39.8$ kW. Problem solved; unfortunately, this solution violates the second law of thermodynamics. The second law violation can be seen by plotting the temperatures of the water and combustion gas as a function of the rate of heat transfer from the gas to the water. The control volumes shown in Figure 8-21 extend from the cold end of the heat exchanger to an arbitrary location within the heat exchanger. The rate of heat transfer between the streams within either control volume, $\dot{q}_{CV}$ in Figure 8-21, is sometimes referred to as the duty. The duty increases from 0 to $\dot{q}$, the total rate of heat transfer, as the right edges of the control volumes are moved from the cold end to the warm end. The duty is defined in terms of the variable fd, which is fraction of the total heat exchanger duty:

$$\dot{q}_{CV} = fd\,\dot{q} \quad (8\text{-}84)$$

```
q_dot_CV=fd*q_dot                                   "duty"
```

The energy balance on the combustion gas control volume relates the gas temperature entering the control volume (T_{gas} in Figure 8-21) to the duty:

$$\dot{m}_{gas}\, c_{gas} \left(T_{gas} - T_{gas,out} \right) = \dot{q}_{CV} \quad (8\text{-}85)$$

```
q_dot_CV=m_dot_gas*c_gas*(T_gas-T_gas_out)          "energy balance on gas-side CV"
T_gas_C=converttemp(K,C,T_gas)                      "gas temp. in C"
```

Equation (8-85) can be rearranged:

$$T_{gas} = T_{gas,out} + \frac{\dot{q}_{CV}}{\dot{m}_{gas}\, c_{gas}} \quad (8\text{-}86)$$

Equation (8-86) shows that the temperature of a stream with a constant capacitance rate is a linear function of the duty, $\dot{q}_{CV}$; this fact is used to accomplish a pinch point analysis involving multiple, constant capacitance rate streams in Section 8.4.3.

An energy balance on the water control volume relates the specific enthalpy of the water leaving the control volume (i_w in Figure 8-21) to the duty:

$$\dot{m}_w \left(i_{w,in} - i_w \right) + \dot{q}_{CV} = 0 \quad (8\text{-}87)$$

where $i_{w,in}$ is the inlet specific enthalpy of the water, evaluated using EES' property routines:

```
i_w_in=enthalpy(Water,T=T_w_in,P=p_w)              "enthalpy of water entering the heat exchanger"
q_dot_CV=m_dot_w*(i_w-i_w_in)                       "energy balance on water-side CV"
```

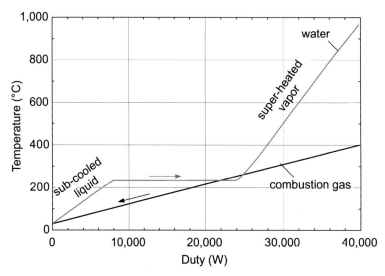

Figure 8-22: Temperature of the combustion gas and water as a function of the duty, assuming that the combustion gas exits at the water inlet temperature.

The water temperature is related to its specific enthalpy and pressure according to the thermodynamic properties of water, which are built into EES.

```
T_w=temperature(Water,h=i_w,P=p_w)          "water temperature"
T_w_C=converttemp(K,C,T_w)                   "water temp. in C"
```

A parametric table is generated that includes fd, $\dot{q}_{CV}$, T_{gas}, and T_w. The fractional duty, fd, is varied from 0 to 1 (corresponding to varying the duty, $\dot{q}_{CV}$, from 0 to the total rate of heat transfer, $\dot{q}$). The temperatures of the two streams as a function of the duty are shown in Figure 8-22.

Water enters at 30°C and 30 bar in a sub-cooled liquid state. Assuming negligible pressure losses, the water temperature increases steadily until it reaches its saturation temperature at 30 bar, 234.4°C. Then, boiling proceeds at constant temperature until the water is in a saturated vapor state. Further heating causes the water to exit the heat exchanger as superheated vapor. Because of the assumption of constant specific heat for the combustion gas, the combustion gas temperature varies linearly with duty. The violation of the second law is evident because the plots of temperature versus heat transfer rate cross; at both the cold and hot ends of the heat exchanger, the combustion gas is "heating" the water even though it is at a lower temperature than the water.

A pinch point analysis adjusts the temperature versus duty curves of the two streams until the second law violation disappears. This adjustment can be made using the EES code by increasing the assumed combustion gas outlet temperature:

```
{T_gas_out=T_w_in                            "Assume that pinch point is at cold end"}
T_gas_out_C=75 [C]                           "Slide gas outlet temperature up"
T_gas_out=converttemp(C,K,T_gas_out_C)
```

Figure 8-23(a) through (d) illustrate the temperature of the two streams as a function of duty for various values of the gas outlet temperature.

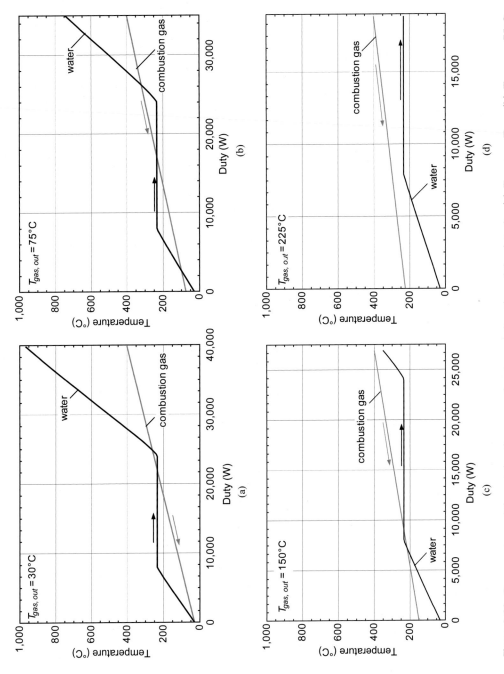

Figure 8-23: Temperature of the combustion gas and water as a function of the duty for (a) $T_{gas,out} = T_{w,in}$, (b) $T_{gas,out} = 75°C$, (c) $T_{gas,out} = 150°C$, and (d) $T_{gas,out} = 225°C$.

Notice that the second law violation is reduced and eventually disappears as the combustion gas temperature increases (i.e., as the rate of heat transfer decreases). In Figure 8-23(c), the water and combustion gas temperature nearly intersect at the location where the water begins to evaporate. The intersection occurs at the pinch point; for these conditions, the pinch point does not occur at either the cold or the hot end but rather at a location that is internal to the heat exchanger. The increase of the gas outlet temperature required to eliminate the second law violation coincides with a reduction in the total heat transferred in the heat exchanger; the total duty (the extent of the x-axis) in Figure 8-23(a) through (d) decreases.

The pinch point is defined as the location where the temperature difference between the fluid streams is a minimum. The pinch point temperature difference (ΔT_{pp}) is the temperature difference at the pinch point. Clearly, the pinch point temperature difference cannot be negative or the second law will be violated.

The EES code can be used to identify the pinch point temperature difference for a given value of $T_{gas,out}$. Calculate the temperature difference between the streams (ΔT):

```
DELTAT=T_gas-T_w                          "stream-to-stream temperature difference"
```

Update the guess values (select Update Guess Values from the Calculate menu) and then select Min/Max from the Calculate menu. Minimize the value of ΔT by varying fd, the fractional duty and set appropriate bounds and a guess value for fd. EES will determine that the minimum value of ΔT (i.e., ΔT_{pp}) is 64.78 K for $T_{gas,out} = 225°C$; this occurs at a duty of 7919 W, which is consistent with Figure 8-23(d). This analysis can be carried out over a range of outlet gas temperatures by using the Min/Max Table feature in EES. Set up a parametric table that includes fd, $T_{gas,out}$, $\dot{q}$, and ΔT. Comment out the specified value of $T_{gas,out}$

```
{T_gas_out_C=225 [C]}                      "Slide gas outlet temperature up"
```

and vary $T_{gas,out}$ from 50°C to 300°C in the table. Select Min/Max Table from the Calculate menu and again specify that ΔT should be minimized by adjusting fd. Figure 8-24 illustrates the pinch point temperature difference and heat transfer rate as a function of $T_{gas,out}$. The region of Figure 8-24 where $\Delta T_{pp} < 0$ corresponds to impossible operating conditions; it is clear that the performance is limited to $\dot{q} = 26000$ W according to the pinch point analysis. The $\Delta T_{pp} = 0$ limit corresponds to an infinitely large heat exchanger. As the pinch point temperature difference increases, the physical size of the heat exchanger that is required is reduced but the performance also drops.

8.4.3 Pinch Point Analysis for a Heat Exchanger Network

The pinch point analysis presented in Section 8.4.2 provides the basis for the design of heat exchanger networks in processing plants, as discussed by Linnhoff (1983). The pinch point design of heat exchanger networks is a complex and well-established method. Here, we will only discuss the underlying idea of a pinch point design.

It is often the case that the designer is confronted with a specific task and has several resources available to accomplish this task; in a large plant, there will be many such tasks and many resources. The initial design of the plant will consist of allocating these resources in order to most optimally accomplish the task. This preliminary design process can be accomplished most conveniently using an analysis that is based on the concept of a pinch point.

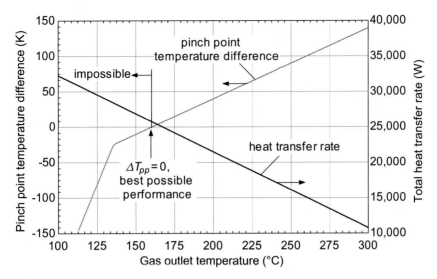

Figure 8-24: Pinch point temperature difference and total heat transfer rate as a function of $T_{gas,out}$.

For example, suppose that a stream of pressurized water with a mass flow rate of $\dot{m}_w = 1.2$ kg/s must be heated from $T_{w,in} = 25°C$ to $T_{w,out} = 200°C$. The water is pressurized sufficiently so that it does not change phase during this process and it has a nearly constant heat capacity, $c_w = 4200$ J/kg-K. The resources that are available to accomplish this heating process include a stream of exhaust air, a stream of waste water, and a combustor. The exhaust air stream is at $T_{a,in} = 300°C$ and has mass flow rate $\dot{m}_a = 0.75$ kg/s. The heat capacity of the exhaust air is constant, $c_a = 1007$ J/kg-K. The waste water stream is at $T_{ww,in} = 90°C$ and has mass flow rate $\dot{m}_{ww} = 2.5$ kg/s. The heat capacity of the waste water is constant, $c_{ww} = 4200$ J/kg-K.

Figure 8-25 illustrates the duty line associated with the water stream that must be heated. The duty line is simply the temperature of the water stream as a function of the heat transfer rate (the duty) that must be provided. The duty line for the water can be

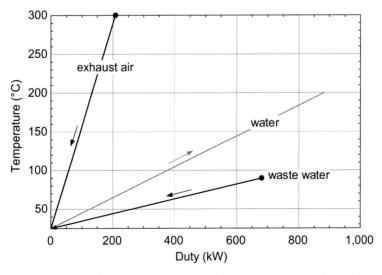

Figure 8-25: Duty lines of the water to be heated, the exhaust air, and the waste water.

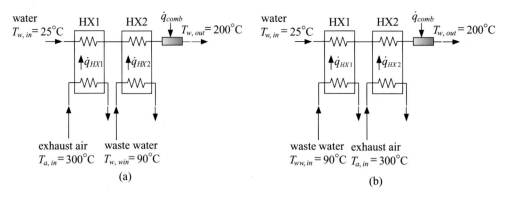

Figure 8-26: (a) Equipment configuration A and (b) equipment configuration B.

obtained by an energy balance on the water:

$$\dot{q}_{CV} = \dot{m}_w \, c_w \, (T_w - T_{w,in}) \tag{8-88}$$

where $\dot{q}_{CV}$ is the duty (the amount of heat transfer to the water). Equation (8-88) can be rearranged:

$$T_w = T_{w,in} + \frac{\dot{q}_{CV}}{\dot{m}_w \, c_w} \tag{8-89}$$

According to Eq. (8-89), the duty line for the water starts at $T_{w,in} = 25°C$ and increases linearly with duty; the slope of the line is the inverse of the capacitance rate of the water, $\dot{m}_w \, c_w = 0.2$ K/kW. The water must be heated to $T_{w,out} = 200°C$ which will require 882 kW (the extent of the duty line along the x-axis in Figure 8-25).

The objective of the system designer should be to minimize the amount of this heat transfer that must be provided by the combustor (which requires fuel and therefore adds to operating cost) by utilizing the exhaust air and waste water streams as completely as possible. There are several equipment configurations that can be used to accomplish this objective; two are shown in Figure 8-26. Both of the equipment configurations shown in Figure 8-26 include two counter-flow heat exchangers that are arranged in series. In configuration A, shown in Figure 8-26(a), the exhaust air passes through the first heat exchanger (HX1), providing a heat transfer rate of $\dot{q}_{HX1}$ to the water, and the waste water passes through the second heat exchanger (HX2), providing $\dot{q}_{HX2}$. In configuration B, shown in Figure 8-26(b), the waste water passes through the first heat exchanger and the exhaust air through the second heat exchanger. In either case, whatever additional energy that is required to bring the water to $T_{w,out} = 200°C$ is provided by the combustor.

It is not immediately obvious whether configuration A or B is optimal. A pinch point analysis can answer this question and therefore provide valuable system-level design guidance. The duty lines for the exhaust air and waste water streams are also shown in Figure 8-25. The lines terminate at the inlet temperature of the stream and have a slope that is the inverse of the capacitance rate of the stream. According to Figure 8-25, the exhaust air could provide 208 kW of heat transfer if it were cooled to 25°C while the waste water stream could provide 683 kW of heat transfer when cooled to 25°C. Therefore the total heat transfer available from the exhaust air and waste water is 891 kW, which is in excess of the 882 kW that is required to heat the water stream. Based strictly on the energy required for the task, the combustor should not be required; however, we will see that this is not the case.

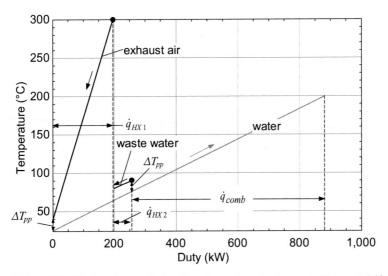

A pinch point analysis is graphical and intuitive. You simply slide the duty lines for the waste water and the air horizontally on Figure 8-25 so that they provide different portions of the duty that is required by the water stream. The temperature of the stream providing the energy must always be higher than the temperature of the water that it is heating. Designers will often constrain this process so that the minimum temperature difference within any heat exchanger is equal to a specified pinch point temperature difference that is based on experience. The pinch point temperature difference is related to the size of the heat exchanger that will eventually be required.

For example, in order to analyze configuration A, shown in Figure 8-26(a), the duty line for the exhaust air should be "slid" horizontally until a satisfactory pinch point temperature difference is maintained in HX1. This is shown in Figure 8-27, with a pinch point temperature difference of approximately 15 K. Figure 8-27 shows that the energy provided in HX1 is approximately $\dot{q}_{HX1} = 196\,\text{kW}$ and the temperature of the water leaving HX1 will be approximately 65°C. Next, the duty line for the waste water is moved horizontally until it meets as much of the remaining duty (i.e., above 196 kW) as is possible while maintaining a satisfactory pinch point temperature difference in HX2. According to Figure 8-27, HX2 will provide approximately $\dot{q}_{HX2} = 56\,\text{kW}$ and the temperature of the water leaving HX2 is approximately 75°C. The remaining duty required by the water is about 620 kW and this must be provided by the combustor ($\dot{q}_{comb}$).

A pinch analysis of configuration B, shown in Figure 8-26(b), is shown in Figure 8-28. The process used to generate Figure 8-28 is the same as was used to generate Figure 8-27 except that the duty line for the waste water stream is moved horizontally first followed by the duty line for the exhaust air. Figure 8-28 indicates that the waste water stream can provide $\dot{q}_{HX1} = 260\,\text{kW}$ and the exhaust air can provide $\dot{q}_{HX2} = 160\,\text{kW}$ if configuration B is used. The remaining duty is about 460 kW, which must be provided by the combustor. Based on this simple pinch analysis, it is clear that configuration B is capable of providing more of the required energy than configuration A. This is a very simple application of pinch point analysis, but it demonstrates the underlying idea as well as its value as an initial design tool.

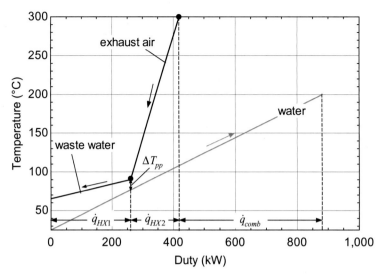

Figure 8-28: Pinch point analysis of configuration B, shown in Figure 8-26(b).

8.5 Heat Exchangers with Phase Change

8.5.1 Introduction

Section 8.4 discusses the concept of a pinch point and defines the pinch point temperature difference. Specifying the pinch point temperature difference indirectly fixes the heat exchanger conductance (UA) and therefore its physical size. It is much easier to specify the pinch point temperature difference and study the performance using a pinch analysis, as shown in Sections 8.4.2 and 8.4.3, than it is to specify the heat exchanger conductance and study the performance. Therefore, a pinch point analysis is recommended for early design studies. However, eventually a more detailed analysis of the heat exchanger will be required.

The log-mean temperature difference and effective-NTU methods presented in Sections 8.2 and 8.3 are not directly applicable for a heat exchanger operating under conditions where the capacitance rates are not constant. In this case, a more detailed analysis is required. One option is to develop a detailed numerical model, as discussed in Section 8.6. A simpler analysis method is possible for a heat exchanger in which a pure fluid is undergoing phase change; for example, the boiler that is shown in Figure 8-21. In this case, the heat exchanger can be analyzed by dividing it into several discrete sub-heat exchangers that correspond to the sub-cooled liquid, saturated, and superheated vapor regimes. Because the capacitance rates of the fluids are often nearly constant within each of these regimes, the log-mean temperature difference or, preferably, the effectiveness-NTU method can be applied using to each sub-heat exchanger.

8.5.2 Sub-Heat Exchanger Model for Phase-Change

In this section, the approach of separating the heat exchanger into sub-heat exchangers is illustrated in the context of a refrigeration condenser. In EXAMPLE 8.1-1 and EXAMPLE 8.1-2, the conductance of a cross-flow heat exchanger is evaluated for the operating condition where a flow of air is used to cool a single-phase flow of liquid water. In EXAMPLE 8.2-1 and EXAMPLE 8.3-1, the performance of the heat exchanger

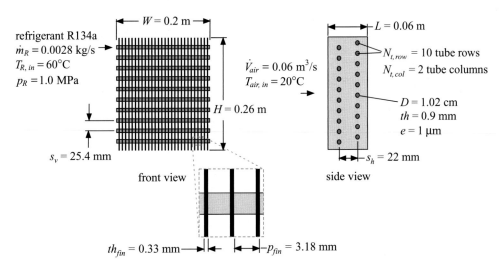

refrigerant R134a
$\dot{m}_R = 0.0028$ kg/s
$T_{R,\,in} = 60°C$
$p_R = 1.0$ MPa

$W = 0.2$ m

$\dot{V}_{air} = 0.06$ m³/s
$T_{air,\,in} = 20°C$

$H = 0.26$ m

$s_v = 25.4$ mm

front view

$L = 0.06$ m

$N_{t,\,row} = 10$ tube rows
$N_{t,\,col} = 2$ tube columns

$D = 1.02$ cm
$th = 0.9$ mm
$e = 1$ μm

$s_h = 22$ mm

side view

$th_{fin} = 0.33$ mm

$p_{fin} = 3.18$ mm

Figure 8-29: Schematic of a plate fin heat exchanger used as a condenser in a refrigeration cycle.

was computed using the *LMTD* method and effectiveness-*NTU* methods, respectively. These methods are appropriate because both streams had a constant capacity rate.

The same cross-flow heat exchanger core can be used as the condenser within a refrigeration cycle. Rather than water, refrigerant R134a vapor enters the cross-flow heat exchanger tubes with an inlet temperature $T_{R,in} = 60°C$, an inlet pressure $p_R = 1.0$ MPa and a mass flow rate of $\dot{m}_R = 0.0028$ kg/s.

The condenser geometry was previously presented in EXAMPLE 8.1-1, and is shown in Figure 8-29. The width and height of the front face of the heat exchanger are $W = 0.2$ m and $H = 0.26$ m, respectively. The fins are made of copper and the core geometry corresponds to finned circular tube surface 8.0-3/8T in the compact heat exchanger library. The fin thickness is $th_{fin} = 0.33$ mm and fin pitch is $p_{fin} = 3.18$ mm. Ten rows of tubes ($N_{t,row} = 10$) are arranged two columns ($N_{t,col} = 2$) and connected in series. The vertical and horizontal spacing between adjacent tubes is $s_v = 25.4$ mm and $s_h = 22$ mm, respectively. The length of the heat exchanger in the direction of the air flow is $L = 0.06$ m. The tubes are made of copper with an outer diameter $D_{out} = 1.02$ cm and a wall thickness $th = 0.9$ mm. The roughness of the inner surface of the tube is $e = 1.0$ μm.

Clean dry air is forced to flow through the heat exchanger perpendicular to the tubes (i.e., in cross-flow) with a volumetric flow rate $\dot{V}_{air} = 0.06$ m³/s. The inlet temperature of the air is $T_{air,in} = 20°C$ and the air is at atmospheric pressure.

A pressure-enthalpy diagram for R134a is shown in Figure 8-30; this plot can be generated quickly in EES by selecting Property Plot from the Plots menu and then scrolling to R134a. The inlet condition of the R134a at 60°C (333.2 K), 1 MPa is superheated. The saturation temperature at 1 MPa is 39.4°C. Assuming that the pressure drop associated with the flow of the refrigerant through the tubing is negligible, the outlet state of the R134a lies on the dotted line in Figure 8-30 corresponding to a constant pressure of 1 MPa. We do not know exactly where along this line the exit state will be (although, given that the heat exchanger is a "condenser," it seems likely that the state will be towards the liquid side of the vapor dome if the heat exchanger is well-designed).

The problem is solved by dividing the heat exchanger into a superheat section, a condensing section, and possibly a subcooling section if the outlet state in Figure 8-30 lies in the subcooled region. Each of the sections is investigated as if it were a cross-flow heat exchanger with both fluids unmixed.

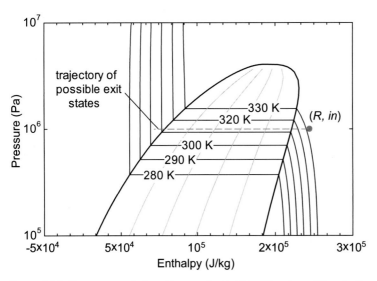

Figure 8-30: Pressure-enthalpy diagram for R134a. The dotted line indicates the trajectory of possible outlet states for the condenser.

The known information for the problem is entered into EES.

```
$UnitSystem SI MASS RAD PA K J
$Tabstops 0.2 0.4 0.6 3.5 in

"Inputs"
D_out=1.02 [cm]*convert(cm,m)              "outer diameter of tube"
th=0.9 [mm]*convert(mm,m)                  "tube wall thickness"
N_t_row=10 [-]                             "number of tube rows"
N_t_col=2 [-]                              "number of tube columns"
H=0.26 [m]                                 "height of heat exchanger face"
W=0.2 [m]                                  "width of heat exchanger face"
L=0.06 [m]                                 "length of heat exchanger in air flow direction"
V_dot_air=0.06 [m^3/s]                     "volumetric flow rate of air"
P=1 [atm]*convert(atm,Pa)                  "atmospheric pressure"
T_air_in=convertTemp(C,K,20 [C])           "inlet air temperature"
T_R_in=convertTemp(C,K,60 [C])            "inlet refrigerant temperature"
m_dot_R=0.0028 [kg/s]                      "refrigerant flow rate"
P_R=1.0 [MPa]*convert(MPa,Pa)              "refrigerant pressure"
s_v=25.4 [mm]*convert(mm,m)                "vertical separation distance between tubes"
s_h=22 [mm]*convert(mm,m)                  "horizontal separation distance between tubes"
th_fin=0.33 [mm]*convert(mm,m)             "fin thickness"
p_fin=3.18 [mm]*convert(mm,m)              "fin pitch"
e=1.0 [micron]*convert(micron,m)           "roughness of tube internal surface"
R$='R134a'                                 "refrigerant"
```

Note that the refrigerant name is contained in the string variable R$; any subsequent calls to the property functions are carried out with the string variable R$ as the first argument. This approach makes it is easy to run the model under conditions where a different refrigerant is flowing through the tubes.

The total air-side thermal resistance is determined first; this resistance will be allocated to the various sub-heat exchangers based on their size. The air-side resistance was previously evaluated for similar operating conditions in EXAMPLEs 8.1-1 and 8.1-2, and the same approach is used here.

The total tube length is:

$$L_{tube} = N_{t,row} \, N_{t.col} \, W \tag{8-90}$$

The total finned area, unfinned area, and air-side surface area are:

$$A_{s,fin,tot} = 2 \, \frac{W}{p_{fin}} \left(H \, L - N_{t,row} \, N_{t,col} \, \frac{\pi \, D_{out}^2}{4} \right) \tag{8-91}$$

$$A_{s,unfin} = \pi \, D_{out} \, L_{tube} \left(1 - \frac{th_{fin}}{p_{fin}} \right) \tag{8-92}$$

$$A_{tot} = A_{s,fin,tot} + A_{s,unfin} \tag{8-93}$$

```
"Air-side resistance"
L_tube=N_t_row*N_t_col*W                             "total tube length"
A_s_fin_tot=2*(W/p_fin)*(H*L-N_t_row*N_t_col*pi*D_out^2/4)   "total fin area"
A_s_unfin=pi*D_out*L_tube*(1-th_fin/p_fin)          "total un-finned area"
A_tot=A_s_fin_tot+A_s_unfin                         "total air-side surface area"
```

The average temperature:

$$\overline{T} = \frac{(T_{R,in} + T_{air,in})}{2} \tag{8-94}$$

is used to compute the air density (ρ_{air}) using EES' internal property routine. The mass flow rate of the air is:

$$\dot{m}_{air} = \rho_{air} \, \dot{V}_{air} \tag{8-95}$$

```
T_avg=(T_R_in+T_air_in)/2            "average temperature"
rho_air=density(Air,T=T_avg,P=P)     "density of air"
m_dot_air=rho_air*V_dot_air          "mass flow rate of air"
```

The compact heat exchanger library is used to compute the air-side heat transfer coefficient ($\overline{h}_{out}$); the heat exchanger geometry corresponds to finned circular tube surface 8.0-3/8T.

```
TypeHX$='fc_tubes_s80-38T'           "heat exchanger identifier name"
Call CHX_h_finned_tube(TypeHX$, m_dot_air, W*H, 'Air',T_avg, P:h_bar_out)
                                     "access compact heat exchanger procedure"
```

The plate fins are modeled as individual annular fins with effective radius, $r_{fin,eff}$, calculated according to:

$$A_{s,fin,tot} = 2 \frac{L_{tube}}{p_{fin}} \pi \left[r_{fin,eff}^2 - \left(\frac{D_{out}}{2} \right)^2 \right]$$ (8-96)

A_s_fin_tot=2*(L_tube/p_fin)*pi*(r_fin_eff^2-(D_out/2)^2)   "effective fin radius"

The conductivity of the copper fin (k_m) is obtained using EES' internal property routine. The fin efficiency (η_{fin}) is obtained using the function eta_fin_annular_rect:

k_m=k_('Copper',T_avg)   "tube conductivity"
eta_fin=eta_fin_annular_rect(th_fin, D_out/2, r_fin_eff, h_bar_out, k_m)
                                                                 "fin efficiency"

The overall surface efficiency is:

$$\eta_o = 1 - \frac{A_{s,fin,tot}}{A_{tot}} (1 - \eta_{fin})$$ (8-97)

The total thermal resistance on the air-side is:

$$R_{out} = \frac{1}{\eta_o \, \bar{h}_{out} \, A_{tot}}$$ (8-98)

eta_o=1-A_s_fin_tot*(1-eta_fin)/A_tot   "overall surface efficiency"
R_out=1/(eta_o*h_bar_out*A_tot)   "resistance on air-side"

We will start by determining the heat exchanger area (or length of tube) that is needed in the superheat region of the heat exchanger in order to change the state of the R134a from its superheated inlet condition to saturated vapor at 1 MPa. The variable F_{sh} is defined as the fraction of the total length of tube that is required in the superheat section (L_{sh}):

$$F_{sh} = \frac{L_{sh}}{L_{tube}}$$ (8-99)

The value of F_{sh} is initially assumed; this value will be adjusted in order to complete the first part of the problem:

"Superheat Section"
F_sh=0.2 [-]   "guess for fraction of heat exchanger req'd for superheat"
L_sh=L_tube*F_sh   "length req'd for superheat"

The thermal resistance on the air-side of the superheat section is:

$$R_{out,sh} = \frac{R_{out}}{F_{sh}}$$ (8-100)

```
R_out_sh=R_out/F_sh                    "air-side resistance in superheat section"
```

The thermal resistance on the refrigerant side of the superheat section can be obtained by evaluating the heat transfer coefficient associated with the single-phase internal flow of refrigerant vapor. The average refrigerant temperature in the superheat region is:

$$\overline{T}_{R,sh} = \frac{T_{R,sat} + T_{R,in}}{2} \tag{8-101}$$

where $T_{R,sat}$ is the temperature of saturated refrigerant vapor at p_R, evaluated using EES' internal property routine.

```
T_R_sat=temperature(R$,x=1,P=P_R)    "saturation temperature of refrigerant"
T_avg_R_sh=(T_R_sat+T_R_in)/2        "average temperature of refrigerant in superheat section"
```

The inner diameter of the tube is computed:

$$D_{in} = D_{out} - 2\,th \tag{8-102}$$

and the PipeFlow procedure is used to determine $\overline{h}_{R,sh}$, the average heat transfer coefficient in the superheat section:

```
D_in=D_out-2*th                      "tube inner diameter"
call PipeFlow(R$,T_avg_R_sh,P_R,m_dot_R,D_in,L_sh,e/D_in:&
   h_bar_R_sh,h_bar_R_sh_H,DELTAP_R_sh,Nusselt_T_R_sh,f_R_sh,Re_R_sh)
                   "access correlation for single-phase internal convection"
```

The thermal resistance on the refrigerant-side of the superheat region is:

$$R_{R,sh} = \frac{1}{\overline{h}_{R,sh}\,\pi\,D_{in}\,L_{sh}} \tag{8-103}$$

```
R_R_sh=1/(L_sh*pi*D_in*h_bar_R_sh)    "refrigerant-side resistance in superheat section"
```

The total resistance is the sum of the air- and refrigerant-side resistances (conduction through the tube is neglected)

$$R_{sh} = R_{R,sh} + R_{out,sh} \tag{8-104}$$

The conductance that is present in the superheated section (for the assumed value of F_{sh}) is:

$$UA_{sh} = \frac{1}{R_{sh}} \tag{8-105}$$

```
R_sh=R_out_sh+R_R_sh                  "total resistance in superheat section"
UA_sh=1/R_sh                          "conductance in superheat section"
```

The calculated conductance in the superheat section must match the conductance that is required to accomplish the de-superheating process; this condition is used to determine the actual value of F_{sh}. The specific heat capacity of vapor refrigerant ($c_{R,sh}$) is obtained at $\overline{T}_{R,sh}$ and used to evaluate the capacitance rate of the superheated refrigerant vapor:

$$\dot{C}_{R,sh} = \dot{m}_R \, c_{R,sh} \qquad (8\text{-}106)$$

```
c_R_sh=cP(R$,T=T_avg_R_sh,P=P_R)
                                    "specific heat capacity of refrigerant in superheat section"
C_dot_R_sh=m_dot_R*c_R_sh           "capacitance rate in superheat section"
```

The specific heat of air (c_{air}) is obtained at $\overline{T}$ and used to compute the capacitance rate of the air; note that the amount of the total air flow that passes across the superheat section is proportional to the fraction of the tube that is required for de-superheating:

$$\dot{C}_{air,sh} = \dot{m}_{air} \, c_{air} \, F_{sh} \qquad (8\text{-}107)$$

```
c_air=cP('Air',T=T_avg)             "specific heat capacity of air"
C_dot_air_sh=F_sh*m_dot_air*c_air   "capacitance rate of air in superheat section"
```

The minimum capacitance rate in the superheat section ($\dot{C}_{min,sh}$) is obtained:

```
C_dot_min_sh=MIN(C_dot_R_sh,C_dot_air_sh)            "minimum capacitance rate"
```

The actual rate of heat transfer required in the superheat section is given by an energy balance on the refrigerant:

$$\dot{q}_{sh} = \dot{m}_R(i_{R,in} - i_{R,v,sat}) \qquad (8\text{-}108)$$

where $i_{R,in}$ and $i_{R,v,sat}$ are the specific enthalpies of the inlet refrigerant and saturated refrigerant vapor, respectively:

```
i_R_in=enthalpy(R$,T=T_R_in,P=P_R)          "inlet enthalpy of refrigerant"
i_R_v_sat=enthalpy(R$,P=P_R,x=1)            "enthalpy of saturated refrigerant vapor"
q_dot_sh=m_dot_R*(i_R_in-i_R_v_sat)         "heat transfer rate in superheat section"
```

The superheat region of the heat exchanger is, approximately, a simple cross-flow heat exchanger that can be treated using the ε-NTU relations discussed in Section 8.3. The effectiveness of the superheat section is:

$$\varepsilon_{sh} = \frac{\dot{q}_{sh}}{\dot{q}_{max,sh}} \qquad (8\text{-}109)$$

where $\dot{q}_{max,sh}$ is the maximum possible heat transfer rate:

$$\dot{q}_{max,sh} = \dot{C}_{min,sh}(T_{R,in} - T_{air,in}) \qquad (8\text{-}110)$$

```
q_dot_max_sh=C_dot_min_sh*(T_R_in-T_air_in)
                                    "maximum possible q_dot in superheat section"
eff_sh=q_dot_sh/q_dot_max_sh        "effectiveness of the superheat section"
```

The number of transfer units required in the superheat section (NTU_{sh}) is obtained using the ε-NTU relations presented in Table 8-1 and Table 8-2 and implemented in the HX function in EES. Note that there is no explicit formula in Table 8-2 or EES function for NTU given ε for the cross-flow configuration with both fluids unmixed. Therefore, the function for ε given the NTU is implicit:

```
eff_sh=HX('crossflow_both_unmixed', NTU_sh, C_dot_air_sh, C_dot_R_sh, 'epsilon')
                                    "effectiveness-NTU relationship"
```

At this point, the conductance required by the superheat section can be computed from NTU_{sh}:

$$UA_{sh} = \dot{C}_{min,sh} \, NTU_{sh} \qquad (8\text{-}111)$$

The guess values are updated and the assumed value of F_{sh} is commented out before adding Eq. (8-111) to the EES code:

```
{F_sh=0.2 [-]}                   "guess for fraction of heat exchanger req'd for superheat"
UA_sh=NTU_sh*C_dot_min_sh        "UA req'd in superheat section"
```

The solution indicates that $F_{sh} = 0.164$; therefore, 16.4% of the tube length is required for the de-superheating process. Because $F_{sh} < 1$, it is evident that additional tube length is available to accomplish the condensing process and, perhaps, the subcooling process. Therefore, the calculation process is repeated for the condensing section.

The fraction of the heat exchanger tubing required for the condensing section:

$$F_{sat} = \frac{L_{sat}}{L_{tube}} \qquad (8\text{-}112)$$

is assumed. As with the superheat section, the value of F_{sat} will be adjusted so that the conductance in the condensing section matches the required conductance.

```
"Condensing Section"
F_sat=0.7 [-]           "guess for fraction of heat exchanger req'd for condensing"
L_sat=L_tube*F_sat      "length required for condensing"
```

The thermal resistance on the air-side in the condensing section is:

$$R_{out,sat} = \frac{R_{out}}{F_{sat}} \qquad (8\text{-}113)$$

```
R_out_sat=R_out/F_sat                        "air-side resistance in condensing section"
```

The average heat transfer coefficient for condensing refrigerant ($\overline{h}_{R,sat}$) is obtained using the procedure Cond_HorizontalTube_avg with an inlet quality of 1.0 and an exit quality of 0.0 (i.e., assuming complete condensation). If the analysis shows that complete condensation is not possible, then the exit quality should be adjusted. The wall temperature required by the procedure is estimated to be the air temperature. The refrigerant-side resistance in the condensing section is:

$$R_{R,sat} = \frac{1}{\pi \, D_{in} \, L_{sat} \, \overline{h}_{R,sat}} \qquad (8\text{-}114)$$

> Call Cond_HorizontalTube_avg(R$, m_dot_R, T_R_sat, T_air_in, D_in, 1.0, 0.0 : h_bar_R_sat)
>   "refrigerant-side coefficient in the condensing section"
> R_R_sat=1/(L_sat*pi*D_in*h_bar_R_sat)   "refrigerant-side resistance in condensing section"

The total resistance in the condensing section is:

$$R_{sat} = R_{R,sat} + R_{out,sat} \tag{8-115}$$

The conductance in the condensing region is calculated:

$$UA_{sat} = \frac{1}{R_{sat}} \tag{8-116}$$

> R_sat=R_out_sat+R_R_sat   "total resistance in condensing section"
> UA_sat=1/R_sat   "conductance in condensing section"

The actual heat transfer rate in the condensing section is calculated using an energy balance on the refrigerant:

$$\dot{q}_{sat} = \dot{m}_R(i_{R,v,sat} - i_{R,l,sat}) \tag{8-117}$$

where $i_{R,l,sat}$ is the specific enthalpy of saturated liquid.

> i_R_l_sat=enthalpy(R$,P=P_R,x=0)   "enthalpy of saturated refrigerant liquid"
> q_dot_sat=m_dot_R*(i_R_v_sat-i_R_l_sat)   "heat transfer rate in condensing section"

The capacitance rate of the air in the condensing section is:

$$\dot{C}_{air,sat} = \dot{m}_{air}\, c_{air}\, F_{sat} \tag{8-118}$$

The air-side capacitance rate is the minimum capacitance rate in the condensing section; the capacitance rate of the condensing refrigerant is effectively infinite. Therefore, the maximum possible heat transfer rate in the condensing section is:

$$\dot{q}_{max,sat} = \dot{C}_{air,sat}(T_{R,sat} - T_{air,in}) \tag{8-119}$$

> C_dot_air_sat=F_sat*m_dot_air*c_air   "capacitance rate of air in condensing section"
> q_dot_max_sat=C_dot_air_sat*(T_R_sat-T_air_in)
>   "maximum possible q_dot in condensing section"

The effectiveness of the condensing section is:

$$\varepsilon_{sat} = \frac{\dot{q}_{sat}}{\dot{q}_{max,sat}} \tag{8-120}$$

The number of transfer units required for any flow configuration as the capacitance ratio approaches zero is given in Table 8-2:

$$NTU_{sat} = -\ln(1 - \varepsilon_{sat}) \tag{8-121}$$

> eff_sat=q_dot_sat/q_dot_max_sat   "effectiveness of condensing section"
> NTU_sat=-ln(1-eff_sat)   "number of transfer units in condensing section"

The guess values are updated and the assumed value of F_{sat} is commented out. The conductance in the condensing section is computed using the required NTU_{sat}:

```
{F_sat=0.7 [-]}              "guess for fraction of heat exchanger req'd for condensing"
UA_sat=NTU_sat*C_dot_air_sat     "conductance required in the condensing section"
```

The solution indicates the $F_{sat} = 0.665$; therefore, 66.5% of the heat exchanger is required to accomplish the condensing process. Because F_{sh} and F_{sat} together to do not exceed 1.0, there is additional tubing available to accomplish some subcooling (i.e., the outlet state lies in the subcooled liquid region of Figure 8-30). The fraction of the heat exchanger that remains for the subcooling section is:

$$F_{sc} = 1 - F_{sat} - F_{sh} \qquad (8\text{-}122)$$

The length of tubing available for subcooling is:

$$L_{sc} = F_{sc} \, L_{tube} \qquad (8\text{-}123)$$

```
"Subcooling Section"
F_sc=1-F_sat-F_sh           "fraction of heat exchanger req'd for subcooling"
L_sc=F_sc*L_tube            "length of tube in subcooling region"
```

The air-side resistance in the subcooling section is:

$$R_{out,sc} = \frac{R_{out}}{F_{sc}} \qquad (8\text{-}124)$$

```
R_out_sc=R_out/F_sc         "air-side resistance in subcooling section"
```

The PipeFlow procedure is called in order to obtain the average heat transfer coefficient for single-phase, subcooled liquid $(\overline{h}_{R,sc})$ using the average temperature of the subcooled liquid:

$$\overline{T}_{R,sc} = \frac{T_{R,sat} + T_{air,in}}{2} \qquad (8\text{-}125)$$

```
T_avg_R_sc=(T_R_sat+T_air_in)/2           "average temp. in sub-cooled region"
call PipeFlow(R$,T_avg_R_sc,P_R,m_dot_R,D_in,L_sc,e/D_in:&
    h_bar_R_sc,h_bar_R_sc_H,DELTAP_R_sc,Nusselt_T_R_sc,f_R_sc,Re_R_sc)
        "access correlation for single-phase internal convection"
```

The refrigerant-side resistance in the subcooling section is:

$$R_{R,sc} = \frac{1}{\overline{h}_{R,sc} \, \pi \, D_{in} \, L_{sc}} \qquad (8\text{-}126)$$

The total resistance in the subcooling section is:

$$R_{sc} = R_{R,sc} + R_{air,sc} \qquad (8\text{-}127)$$

and the conductance in the subcooling section is:

$$UA_{sc} = \frac{1}{R_{sc}} \qquad (8\text{-}128)$$

```
R_R_sc=1/(L_sc*pi*D_in*h_bar_R_sc)    "refrigerant-side resistance in subcooling section"
R_sc=R_out_sc+R_R_sc                  "total resistance in subcooling section"
UA_sc=1/R_sc                          "conductance in superheat section"
```

The specific heat capacity of the subcooled liquid refrigerant ($c_{R,sc}$) is obtained at $\overline{T}_{R,sc}$ and used to compute the capacitance rate of the refrigerant in the subcooling section:

$$\dot{C}_{R,sc} = \dot{m}_R \, c_{R,sc} \qquad (8\text{-}129)$$

```
c_R_sc=cP(R$,T=T_avg_R_sc,P=P_R)
                  "specific heat capacity of refrigerant in subcooling section"
C_dot_R_sc=m_dot_R*c_R_sc             "capacitance rate in subcooling section"
```

The capacitance rate of the air in the subcooling section is:

$$\dot{C}_{air,sc} = \dot{m}_{air} \, c_{air} \, F_{sc} \qquad (8\text{-}130)$$

```
C_dot_air_sc=F_sc*m_dot_air*c_air     "capacitance rate of air in subcooling section"
```

The minimum capacitance rate in the subcooled section ($\dot{C}_{min,sc}$) is calculated. The number of transfer units in the subcooled section is:

$$NTU_{sc} = \frac{UA_{sc}}{\dot{C}_{min,sc}} \qquad (8\text{-}131)$$

```
C_dot_min_sc=MIN(C_dot_R_sc,C_dot_air_sc)       "minimum capacitance rate"
NTU_sc=UA_sc/C_dot_min_sc      "number of transfer units in the subcooling section"
```

The effectiveness of the subcooled section (ε_{sc}) is obtained using the HX function in EES. The maximum possible heat transfer rate in the subcooled section is:

$$\dot{q}_{sc,max} = \dot{C}_{min,sc}(T_{R,sat} - T_{air,in}) \qquad (8\text{-}132)$$

The actual heat transfer rate in the subcooled section is:

$$\dot{q}_{sc} = \varepsilon_{sc} \, \dot{q}_{sc,max} \qquad (8\text{-}133)$$

```
eff_sc=HX('crossflow_both_unmixed', NTU_sc, C_dot_air_sc, C_dot_R_sc, 'epsilon')
                  "effectiveness of the subcooling section"
q_dot_max_sc=C_dot_min_sc*(T_R_sat-T_air_in)
                  "maximum possible q_dot in the subcooled section"
q_dot_sc=q_dot_max_sc*eff_sc   "heat transfer rate in the subcooling section"
```

The specific enthalpy of the refrigerant leaving the subcooling section is:

$$i_{R,out} = i_{R,l,sat} - \frac{\dot{q}_{R,sc}}{\dot{m}_R} \qquad (8\text{-}134)$$

and the temperature of the refrigerant leaving the condenser is obtained using EES' internal property routine with the enthalpy and pressure known. The total capacity of the condenser is:

$$\dot{q} = \dot{q}_{sh} + \dot{q}_{sat} + \dot{q}_{sc} \qquad (8\text{-}135)$$

```
i_R_out=i_R_l_sat-q_dot_sc/m_dot_R
                    "enthalpy of refrigerant leaving the subcooling section"
T_R_out=temperature(R$,h=i_R_out,P=P_R)
                    "temperature of refrigerant leaving the condenser"
T_R_out_C=converttemp(K,C,T_R_out)    "in C"
q_dot=q_dot_sh+q_dot_sat+q_dot_sc    "total capacity of the condenser"
```

which leads to $\dot{q} = 539.6$ W and $T_{R,out} = 34.9°C$.

Determining the outlet conditions in the manner presented above is somewhat tedious as it requires that each of the sections of the heat exchanger be sequentially considered. An approximation that is sometimes used to simplify the analysis of a condenser is to neglect the superheat and subcooling sections. The rate of heat transfer is calculated assuming that the entire heat exchange process occurs in the condensing section with the refrigerant at the saturation temperature. The conductance in this case is provided by:

$$UA_{app} = \frac{1}{R_{out} + \dfrac{1}{L_{tube}\, \pi\, D_i\, \overline{h}_{R,sat}}} \tag{8-136}$$

The capacitance rate of the condensing refrigerant is infinite and therefore the number of transfer units is:

$$NTU_{app} = \frac{UA_{app}}{\dot{m}_{air}\, c_{air}} \tag{8-137}$$

```
"Approximate model, neglecting the subcool and superheat sections"
UA_app=1/(R_out+1/(L_tube*h_bar_R_sat*pi*D_in))    "total conductance"
NTU_app=UA_app/(m_dot_air*c_air)    "number of transfer units"
```

The effectiveness is computed assuming a capacitance ratio of zero using the appropriate formula in Table 8-1:

$$\varepsilon_{app} = 1 - \exp(-NTU_{app}) \tag{8-138}$$

The maximum possible heat transfer rate is:

$$\dot{q}_{max,app} = \dot{m}_{air}\, c_{air}(T_{R,sat} - T_{air,in}) \tag{8-139}$$

The actual heat transfer rate is:

$$\dot{q}_{app} = \varepsilon_{app}\, \dot{q}_{max,app} \tag{8-140}$$

```
eff_app=1-exp(-NTU_app)    "approximate effectiveness of entire coil"
q_dot_max_app=m_dot_air*c_air*(T_R_sat-T_air_in)
                    "maximum possible heat transfer rate"
q_dot_app=q_dot_max_app*eff_app    "heat transfer rate"
```

which leads to $\dot{q}_{app} = 689$ W. The rate of heat transfer is overestimated by about 28% if the subcooling and superheating processes are ignored in this case.

High-level modeling tools have been developed to analyze the heat transfer and pressure drop in phase-change heat exchangers. One such tool is the EVAP-COND program developed by NIST (Domanski, 2006), which carries out a tube-by-tube analysis of the heat exchanger. The performance predicted by EVAP-COND for this heat exchanger (using the program's default settings for calculating the heat transfer coefficients) is 520 W; this is within 4% of the value predicted using the model that is presented in this section.

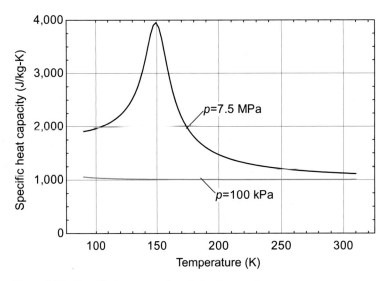

Figure 8-31: Specific heat capacity of air at p_H and p_C.

8.6 Numerical Model of Parallel- and Counter-Flow Heat Exchangers

8.6.1 Introduction

The heat exchanger modeling techniques presented in Sections 8.2 and 8.3 rely on analytical solutions to the governing equations that are made possible by simplifying assumptions such as constant specific heat capacity. This section discusses numerical techniques that are appropriate for considering parallel- and counter-flow heat exchangers under conditions where these assumptions are not valid. The numerical solutions are illustrated in the context of a plate heat exchanger operating between two streams of air. Air at $p_H = 7.5$ MPa and $T_{H,in} = 300$ K enters the hot-side and air at $p_C = 100$ kPa and $T_{C,in} = 90$ K enters the cold-side. The specific heat capacity of air at the two pressures over the temperature range of interest is shown in Figure 8-31; notice that the specific heat capacity of the high pressure stream varies substantially with temperature and therefore the use of either the *LMTD* or *ε-NTU* technique is not appropriate. Figure 8-31 was obtained from EES for the substance Air_ha which implements a real gas property routine for air.

8.6.2 Numerical Integration of Governing Equations

The most straight-forward numerical method is a direct extension of the numerical integration techniques discussed in Section 3.2 and elsewhere in this text. The state

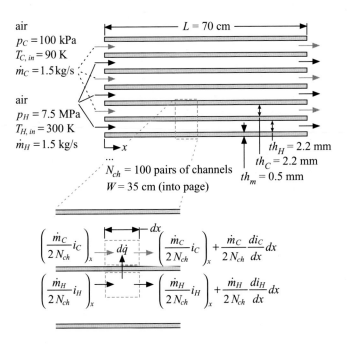

Figure 8-32: Plate heat exchanger in a parallel-flow configuration.

equations are derived and integrated for the heat exchanger operating in both a parallel-flow and counter-flow configuration.

Parallel-Flow Configuration

The plate heat exchanger operating in a parallel-flow configuration is shown schematically in Figure 8-32. In the hot-side of the heat exchanger, air at $p_H = 7.5$ MPa and $T_{H,in} = 300$ K enters one set of channels (i.e., every other channel). The total mass flow rate of air passing through all of the hot-side channels is $\dot{m}_H = 1.5$ kg/s. Air at $p_C = 100$ kPa and $T_{C,in} = 90$ K enters the other set of channels (the cold-side of the heat exchanger) with a total mass flow rate $\dot{m}_C = 1.5$ kg/s. Each plate is $th_m = 0.5$ mm thick and is composed of aluminum. The plates are $L = 70$ cm long in the flow direction and $W = 35$ cm wide (into the page). The plate separation distance (i.e., the channel height) is $th_H = th_C = 2.2$ mm. There are $N_{ch} = 100$ pairs of channels.

The first step in obtaining the numerical solution is to derive the state equations that must be integrated numerically through the heat exchanger. These are obtained from the differential energy balances on the hot stream and cold stream, shown in Figure 8-32. The energy balance on the hot stream is:

$$\left(\frac{\dot{m}_H}{2\,N_{ch}} i_H\right)_x = d\dot{q} + \left(\frac{\dot{m}_H}{2\,N_{ch}} i_H\right)_x + \frac{\dot{m}_H}{2\,N_{ch}} \frac{di_H}{dx} dx \qquad (8\text{-}141)$$

Note that the control volume encompasses one half of each channel and therefore the mass flow rate passing through the control volume is $\dot{m}_H / (2\,N_{ch})$. Equation (8-141) can be simplified to:

$$0 = d\dot{q} + \frac{\dot{m}_H}{2\,N_{ch}} \frac{di_H}{dx} dx \qquad (8\text{-}142)$$

The enthalpy derivative in Eq. (8-142) is expanded:

$$0 = d\dot{q} + \frac{\dot{m}_H}{2\,N_{ch}} \left[\underbrace{\left(\frac{\partial i_H}{\partial T}\right)_p}_{c_H} \frac{dT_H}{dx} + \left(\frac{\partial i_H}{\partial p}\right)_T \frac{dp_H}{dx} \right] dx \qquad (8\text{-}143)$$

The pressure driven enthalpy variation could be retained if the enthalpy of the fluid is a strong function of pressure or if large pressure gradients are expected. Here, the pressure driven change in the enthalpy is neglected:

$$0 = d\dot{q} + \frac{\dot{m}_H}{2\,N_{ch}} c_H \frac{dT_H}{dx} dx \qquad (8\text{-}144)$$

The heat transfer rate from the hot stream to the cold stream is:

$$d\dot{q} = \underbrace{\frac{(T_H - T_C)}{\left(\dfrac{1}{h_H\,W\,dx} + \dfrac{th_m}{k_m\,W\,dx} + \dfrac{1}{h_C\,W\,dx}\right)}}_{\text{thermal resistance in control volume}} \qquad (8\text{-}145)$$

where k_m is the conductivity of the plate and h_H and h_C are the local heat transfer coefficients on the hot-side and cold-side, respectively. Subsituting Eq. (8-145) into Eq. (8-144) leads to:

$$0 = \frac{(T_H - T_C)}{\left(\dfrac{1}{h_H\,W\,dx} + \dfrac{th_m}{k_m\,W\,dx} + \dfrac{1}{h_C\,W\,dx}\right)} + \frac{\dot{m}_H}{2\,N_{ch}} c_H \frac{dT_H}{dx} dx \qquad (8\text{-}146)$$

or

$$\boxed{\frac{dT_H}{dx} = -\frac{2N_{ch}(T_H - T_C)}{\dot{m}_H c_H \left(\dfrac{1}{h_H W} + \dfrac{th_m}{k_m W} + \dfrac{1}{h_c W}\right)}} \qquad (8\text{-}147)$$

A similar process carried out for the cold-fluid leads to:

$$\boxed{\frac{dT_C}{dx} = \frac{2N_{ch}(T_H - T_C)}{\dot{m}_C c_C \left(\dfrac{1}{h_H W} + \dfrac{th_m}{k_m W} + \dfrac{1}{h_c W}\right)}} \qquad (8\text{-}148)$$

Equations (8-147) and (8-148) are the state equations that must be integrated using one of the numerical techniques discussed previously in Section 3.2. Here, the Integral command in EES is applied to the problem.

The input parameters are entered in EES:

```
$UnitSystem SI MASS RAD PA K J
$Tabstops 0.2 0.4 0.6 3.5 in

"Inputs"
W=35 [cm]*convert(cm,m)                    "width of heat exchanger"
L=70 [cm]*convert(cm,m)                    "length of heat exchanger in flow direction"
N_ch=100 [-]                               "number of channel pairs"
th_H=2.2 [mm]*convert(mm,m)               "channel width on hot-side"
th_C=2.2 [mm]*convert(mm,m)               "channel width on cold-side"
th_m=0.5 [mm]*convert(mm,m)               "thickness of plate"
p_H=7.5 [MPa]*convert(MPa,Pa)             "hot-side pressure"
p_C=100 [kPa]*convert(kPa,Pa)             "cold-side pressure"
m_dot_H=1.5 [kg/s]                         "hot-side mass flow rate"
m_dot_C=1.5 [kg/s]                         "cold-side mass flow rate"
H$='Air_ha'                                "hot-side fluid"
C$='Air_ha'                                "cold-side fluid"
T_H_in=300 [K]                             "hot-side inlet temperature"
T_C_in=90 [K]                              "cold-side inlet temperature"
```

Note that the hot-side and cold-side fluid is specified to be Air_ha rather than Air.

In order to use the Integral command, it is first necessary to set up the EES code so that it can evaluate the integrands, Eqs. (8-147) and (8-148), at arbitrary values of the state and integration variables:

```
"arbitrary state variables to establish evaluation of the state equations"
x=0.1 [m]                                  "position in heat exchanger"
T_H=300 [K]                                "hot-side temperature"
T_C=100 [K]                                "cold-side temperature"
```

The specific heat capacity of the hot-side and cold-side streams, c_H and c_C, are evaluated using EES' internal property routine.

```
c_H=cP(H$,p=p_H,T=T_H)                     "hot-side specific heat capacity"
c_C=cP(C$,p=p_C,T=T_C)                     "cold-side specific heat capacity"
```

The local heat-transfer coefficients on the hot-side and cold-side (h_H and h_C) are obtained using the procedure DuctFlow_local, which returns the local heat transfer coefficient for flow through a rectangular duct. Note that the constant temperature boundary condition value of the heat transfer coefficient is used and that the pressure gradient in each channel ($\frac{dp_H}{dx}$ and $\frac{dp_C}{dx}$) is returned by the procedure.

```
call DuctFlow_local(H$,T_H,p_H,m_dot_H/N_ch,th_H,W,x,0 [-]:h_H, h_H_H , dpdx_H)
                                           "hot-side local heat transfer coefficient"
call DuctFlow_local(C$,T_C,p_C,m_dot_C/N_ch,th_C,W,x,0 [-]:h_C, h_H_C , dpdx_C)
                                           "cold-side local heat transfer coefficient"
```

The conductivity of aluminum (k_m) is evaluated at the average of the hot- and cold-side temperatures using EES' property function:

```
k_m=k_('Aluminum', (T_H+T_C)/2)          "metal conductivity at local average temperature"
```

The rate of change of the hot-side and cold-side temperatures are evaluated according to Eqs. (8-147) and (8-148):

```
dTHdx=-2*N_ch*(T_H-T_C)/(m_dot_H*c_H*(1/(h_H*W)+th_m/(k_m*W)+1/(h_C*W)))
                                         "state equation for T_H"
dTCdx=2*N_ch*(T_H-T_C)/(m_dot_C*c_C*(1/(h_H*W)+th_m/(k_m*W)+1/(h_C*W)))
                                         "state equation for T_C"
```

which leads to $\frac{dT_H}{dx} = -375$ K/m and $\frac{dT_C}{dx} = 405.6$ K/m; these results make sense, the hot-side temperature should fall and the cold-side rise in this parallel-flow configuration.

The arbitrary integration variables are commented out and the Integral command is used to integrate the state equations from $x = 0$ to $x = L$. The integration steps are specified according to:

$$\Delta x = \frac{L}{N} \tag{8-149}$$

where N is the number of integration steps.

```
{"arbitrary state variable to establish evaluation of the state equations"
x=0.1 [m]                                "position in heat exchanger"
T_H=300 [K]                              "hot-side temperature"
T_C=100 [K]                              "cold-side temperature"}
N=10 [-]                                 "number of integration steps"
DELTAx=L/N                               "integration step size"
T_H=T_H_in+Integral(dTHdx,x,0,L,DELTAx)  "integral state equation for T_H"
T_C=T_C_in+Integral(dTCdx,x,0,L,DELTAx)  "integral state equation for T_C"
```

For the parallel flow configuration, the fluid temperatures are known at $x = 0$ and therefore the initial conditions are known for the integration. When the heat exchanger is operated in a counter-flow configuration, only one of the two initial conditions at $x = 0$ are known and so iteration is required.

An Integral Table is created in order to record the temperatures as a function of position within the heat exchanger:

```
$IntegralTable x:DELTAx,T_H,T_C
```

Figure 8-33 illustrates the temperature distribution within the parallel-flow heat exchanger.

With any numerical solution it is necessary to ensure that a sufficient number of integration steps are used. In a heat exchanger problem, numerical error will show up as an energy unbalance; that is, the change in the enthalpy of the hot stream will not be exactly equal to the change in the enthalpy of the cold stream. This energy unbalance is an excellent figure of merit to use when evaluating numerical convergence.

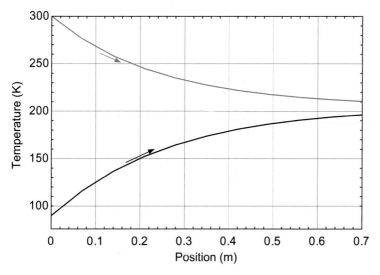

Figure 8-33: Temperature distribution within the parallel-flow heat exchanger.

The integral table is removed:

```
{$IntegralTable x:DELTAx,T_H,T_C}
```

The heat transfer rate from the hot-side fluid is calculated using an energy balance on the hot-fluid:

$$\dot{q}_H = \dot{m}_H \left(i_{H,in} - i_{H,out} \right) \tag{8-150}$$

where $i_{H,in}$ and $i_{H,out}$ are the specific enthalpies of the hot-fluid at the inlet and the exit states, respectively.

```
i_H_in=enthalpy(H$,T=T_H_in,P=p_H)        "enthalpy of hot-stream at inlet"
i_H_out=enthalpy(H$,T=T_H,P=p_H)          "enthalpy of hot-stream at exit"
q_dot_H=m_dot_H*(i_H_in-i_H_out)          "heat transfer rate to hot-stream"
```

The heat transfer rate to the cold-side fluid is calculated using an energy balance on the cold-fluid:

$$\dot{q}_C = \dot{m}_C \left(i_{C,in} - i_{C,out} \right) \tag{8-151}$$

where $i_{C,in}$ and $i_{C,out}$ are the specific enthalpies of the cold-fluid at the inlet and the exit states, respectively.

```
i_C_in=enthalpy(C$,T=T_C_in,P=p_C)        "enthalpy of cold-stream at inlet"
i_C_out=enthalpy(C$,T=T_C,P=p_C)          "enthalpy of cold-stream at exit"
q_dot_C=m_dot_C*(i_C_out-i_C_in)          "heat transfer rate to cold-stream"
```

The unbalance in the heat transfer rate, normalized by the rate of heat transfer, is evaluated:

$$UB = \frac{|\dot{q}_H - \dot{q}_C|}{\dot{q}_H} \tag{8-152}$$

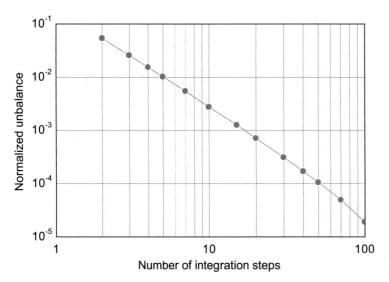

Figure 8-34: Normalized unbalance in the heat transfer rate as a function of the number of integration steps.

```
UB=abs(q_dot_H-q_dot_C)/q_dot_H                    "normalized unbalance"
```

Figure 8-34 illustrates the normalized unbalance in heat transfer as a function of the number of integration steps. The appropriate number of steps can be selected based on the desired level of accuracy. For example, to achieve 0.1% accuracy, approximately 20 integration steps are required. The predicted rate of heat transfer in the heat exchanger is $\dot{q} = 162.52\,\text{kW}$.

The numerical result should be checked against an analytical solution in an appropriate limit. In the limit that the capacitance rates, heat transfer coefficients, and metal conductivity are spatially uniform within the heat exchanger, it is possible to evaluate the total conductance and the performance using the ε-NTU technique discussed in Section 8.3.

In order to carry out this verification, the numerical model is modified so that it uses constant properties and the solution is also predicted using the ε-NTU solution. The specific heat capacities of the hot and cold streams are assumed to be constant and are evaluated at the average of the hot- and cold-side inlet temperatures (rather than at the local temperatures).

```
{c_H=cP(H$,p=P_H,T=T_H)          "hot-side heat capacity rate"
c_C=cP(C$,p=p_C,T=T_C)           "cold-side heat capacity rate"}
T_avg=(T_H_in+T_C_in)/2          "average temperature to evaluate properties"
c_H=cP(H$,p=P_H,T=T_avg)         "hot-side specific heat capacity at average temperature"
c_C=cP(C$,p=p_C,T=T_avg)         "cold-side specific heat capacity at average temperature"
```

The calculation of the local heat transfer coefficients on the hot- and cold-sides are commented out. Instead, the average heat transfer coefficient is calculated using the DuctFlow procedure. The local heat transfer coefficients are assumed to be constant and equal to the average heat transfer coefficients.

```
{call DuctFlow_local(H$,T_H,p_H,m_dot_H/N_ch,th_H,W,x,0 [-]:h_H, h_H_H , dpdx_H)
                                     "hot-side local heat transfer coefficient"
call DuctFlow_local(C$,T_C,p_C,m_dot_C/N_ch,th_C,W,x,0 [-]:h_C, h_H_C , dpdx_C)
                                     "cold-side local heat transfer coefficient"}
call DuctFlow(H$,T_avg,p_H,m_dot_H/N_ch,th_H,W,L, 0 [-]:&
    h_H, h_H_H ,DELTAp_H, Nusselt_T_H, f_H, Re_H)
                                     "hot-side average heat transfer coefficient"
call DuctFlow(C$,T_avg,p_C,m_dot_C/N_ch,th_C,W,L, 0 [-]:&
    h_C, h_C_H ,DELTAp_C, Nusselt_T_C, f_C, Re_C)
                                     "cold-side average heat transfer coefficient"
```

The metal conductivity is assumed to be constant and equal to the conductivity at the average temperature:

```
{k_m=k_('Aluminum', (T_H+T_C)/2)      "metal conductivity at local average temperature"}
k_m=k_('Aluminum', T_avg)             "metal conductivity at average temperature"
```

The total resistance is computed according to:

$$R_{tot} = \frac{1}{h_H \, 2 \, N_{ch} \, W \, L} + \frac{th_m}{k_m \, 2 \, N_{ch} \, W \, L} + \frac{1}{h_C \, 2 \, N_{ch} \, W \, L} \tag{8-153}$$

and the conductance is:

$$UA = \frac{1}{R_{tot}} \tag{8-154}$$

```
"effectiveness-NTU, constant property solution"
R_tot=1/(h_H*2*N_ch*W*L)+th_m/(k_m*2*N_ch*W*L)+1/(h_C*2*N_ch*W*L)
                                     "total resistance"
UA=1/R_tot                           "conductance"
```

The capacitance rates on the hot- and cold-sides are computed:

$$\dot{C}_C = c_C \, \dot{m}_C \tag{8-155}$$

$$\dot{C}_H = c_H \, \dot{m}_H \tag{8-156}$$

The minimum capacitance rate ($\dot{C}_{min}$) is identified and used to compute the number of transfer units:

$$NTU = \frac{UA}{\dot{C}_{min}} \tag{8-157}$$

```
C_dot_C=c_C*m_dot_C                  "capacitance rate of cold-fluid"
C_dot_H=c_H*m_dot_H                  "capacitance rate of cold-fluid"
C_dot_min=MIN(C_dot_C,C_dot_H)       "minimum capacitance rate"
NTU=UA/C_dot_min                     "number of transfer units"
```

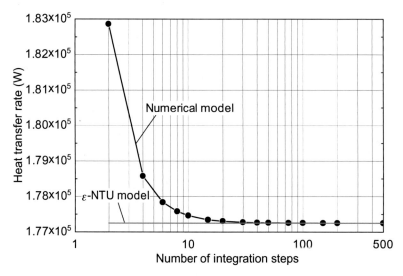

Figure 8-35: Heat transfer rate predicted by the numerical model in the constant property limit and by the effectiveness-*NTU* method.

The effectiveness (ε) is predicted using the HX function in EES. The rate of heat transfer predicted by the constant property model is:

$$\dot{q}_{cp} = \varepsilon\,\dot{C}_{min}\,(T_{h,in} - T_{c,in}) \tag{8-158}$$

```
eff=HX('parallelflow', NTU, C_dot_C, C_dot_H, 'epsilon')        "effectiveness"
q_dot_cp=eff*C_dot_min*(T_h_in-T_c_in)                          "heat transfer rate"
```

The heat transfer rate predicted by the numerical model in this limit is:

$$\dot{q} = \dot{m}_H\,c_H\,(T_{h,in} - T_{h,out}) \tag{8-159}$$

```
q_dot=m_dot_H*c_H*(T_H_in-T_H)        "heat transfer rate predicted by numerical solution"
```

Figure 8-35 illustrates the rate of heat transfer predicted by the numerical model (in the constant property limit) and the ε-*NTU* method as a function of the number of integration steps.

Note that the heat transfer rate predicted by the constant property model using the average properties is $\dot{q} = 170.21\,\text{kW}$; this is approximately 5% in error relative to $\dot{q} = 162.52\,\text{kW}$, predicted by the numerical model that explicitly considers the variation in specific heat capacity, heat transfer coefficient, and metal conductivity. The agreement is relatively good provided that the constant property model is implemented with suitably-averaged fluid properties.

Counter-Flow Configuration

This extended section can be found on the website www.cambridge.org/nellisandklein that accompanies this book. The plate heat exchanger shown in Figure 8-32 operating in a counter-flow configuration is analyzed by numerically integrating the governing differential equations. The steps are analogous to the analysis of the parallel-flow configuration.

8.6.3 Discretization into Sub-Heat Exchangers

An alternative method for obtaining a numerical solution to a heat exchanger problem is to divide the heat exchanger into sub-heat exchangers. It may be true that the assumptions underlying the ε-NTU solutions are not valid when applied to the entire heat exchanger (due, for example, to large changes in properties). However, the ε-NTU solutions become more valid for small sub-heat exchangers. Therefore, a computationally efficient model of the heat exchanger can be obtained by applying the ε-NTU solutions to each sub-heat exchanger and requiring that the boundary conditions associated with each sub-heat exchanger are consistent with the adjacent sub-heat exchangers.

The general procedure is as follows:

1. Assume an exit temperature on one side; using this exit temperature, compute the total heat transfer rate in the heat exchanger.
2. Divide the heat exchanger into sub-heat exchangers; each sub-heat exchanger is assigned an equal rate of heat transfer.
3. Carry out energy balances in order to determine the temperatures entering and leaving each sub-heat exchanger.
4. Apply ε-NTU solutions in order to determine the conductance that must be assigned to each sub-heat exchanger.
5. Based on the conductance and local operating conditions, determine the physical size of each sub-heat exchanger.
6. Vary the assumed exit temperature (from step 1) until the total physical size (or conductance) of the heat exchanger matches the actual physical size (or conductance) of the heat exchanger.

This process is illustrated for both a parallel-flow and counter-flow configuration.

Parallel-Flow Configuration

The plate heat exchanger operating in a parallel-flow configuration is shown schematically in Figure 8-32 and is used to demonstrate the sub-heat exchanger modeling methodology. The inputs are entered in EES:

```
$UnitSystem SI MASS RAD PA K J
$Tabstops 0.2 0.4 0.6 3.5 in

"Inputs"
W=35 [cm]*convert(cm,m)                    "width of heat exchanger"
L=70 [cm]*convert(cm,m)                    "length of heat exchanger in flow direction"
N_ch=100 [-]                               "number of channel pairs"
th_H=2.2 [mm]*convert(mm,m)                "channel width on hot-side"
th_C=2.2 [mm]*convert(mm,m)                "channel width on cold-side"
th_m=0.5 [mm]*convert(mm,m)                "thickness of plate"
p_H=7.5 [MPa]*convert(MPa,Pa)              "hot-side pressure"
p_C=100 [kPa]*convert(kPa,Pa)             "cold-side pressure"
m_dot_H=1.5 [kg/s]                         "hot-side mass flow rate"
m_dot_C=1.5 [kg/s]                         "cold-side mass flow rate"
H$='Air_ha'                                "hot-side fluid"
C$='Air_ha'                                "cold-side fluid"
T_H_in=300 [K]                             "hot-side inlet temperature"
T_C_in=90 [K]                              "cold-side inlet temperature"
```

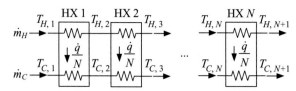

Figure 8-38: Sub-heat exchangers.

The hot-side exit temperature is assumed:

```
T_H_out=250 [K]                                 "assumed outlet temperature"
```

The total heat transfer rate is computed based on the assumed hot-side exit temperature by carrying out an energy balance on the hot-side fluid:

$$\dot{q} = \dot{m}_H \left(i_{H,in} - i_{H,out} \right) \tag{8-162}$$

where $i_{H,in}$ and $i_{H,out}$ are the inlet and outlet specific enthalpy of the hot-side fluid, respectively, calculated using EES' internal property routine:

```
i_H_in=enthalpy(H$,T=T_H_in,P=p_H)              "enthalpy of hot inlet fluid"
i_H_out=enthalpy(H$,T=T_H_out,P=p_H)            "enthalpy of hot outlet fluid"
q_dot=m_dot_H*(i_H_in-i_H_out)                  "total heat transfer rate"
```

The heat exchanger is divided into N sub-heat exchangers; the temperature of the hot- and cold-side fluids entering and leaving each sub-heat exchanger (see Figure 8-38) are obtained by an energy balance. The total rate of heat transfer in the heat exchanger increases as you move from left-to-right in Figure 8-38, including more sub-heat exchangers.

$$\dot{q}_i = \frac{\dot{q}}{N} (i-1) \quad \text{for } i = 1.. (N+1) \tag{8-163}$$

```
N=10 [-]                                        "number of sub-heat exchangers"
duplicate i=1,N
    q_dot[i]=i*q_dot/N                          "total heat transfer rate"
end
```

The temperatures $T_{H,1}$ and $T_{C,1}$ are the hot- and cold-side fluid inlet temperatures. The associated specific enthalpies of these states, $i_{H,1}$ and $i_{C,1}$, are computed:

```
"Obtain temperature distribution"
T_H[1]=T_H_in                                   "hot-side inlet temperature"
T_C[1]=T_C_in                                   "cold-side inlet temperature"
i_H[1]=i_H_in                                   "hot_side inlet enthalpy"
i_C[1]=enthalpy(C$,T=T_C[1],P=p_C)              "cold-side inlet enthalpy"
```

Recognizing that the rate of heat transfer in each sub-heat exchanger is $\dot{q}/N$; an energy balance on the hot-side of each of the sub-heat exchangers provides the enthalpy leaving the hot-side of each of the sub-heat exchangers:

$$i_{H,i+1} = i_{H,i} - \frac{\dot{q}}{N \dot{m}_H} \quad \text{for } i = 1..N \tag{8-164}$$

The temperature of the hot-side fluid leaving each sub-heat exchanger ($T_{H,i}$) is obtained from the enthalpy and pressure using EES' internal property routine:

```
duplicate i=2,(N+1)
  i_H[i]=i_H[i-1]-q_dot/(N*m_dot_H)
        "energy balance on hot-side of each sub-heat exchanger"
  T_H[i]=temperature(H$,h=i_H[i],P=p_H)
        "temperature leaving hot-side of each sub-heat exchanger"
end
```

An energy balance on the cold-side provides the enthalpy leaving the cold-side of each of the sub-heat exchangers:

$$i_{C,i+1} = i_{C,i} + \frac{\dot{q}}{N \dot{m}_C} \quad \text{for } i = 1..N \tag{8-165}$$

The temperature of the cold-side fluid leaving each sub-heat exchanger ($T_{C,i}$) is obtained from the enthalpy using EES' internal property routine:

```
duplicate i=2,(N+1)
  i_C[i]=i_C[i-1]+q_dot/(N*m_dot_C)
        "energy balance on cold-side of each sub-heat exchanger"
  T_C[i]=temperature(C$,h=i_C[i],P=p_C)
        "temperature leaving cold-side of each sub-heat exchanger"
end
```

The ε-NTU solution can be individually applied to each of the sub-heat exchangers. The capacity rates on the hot- and cold-side within each sub-heat exchanger are defined as:

$$\dot{C}_{H,i} = \dot{m}_H c_{H,i} \quad \text{for } i = 1..N \tag{8-166}$$

$$\dot{C}_{C,i} = \dot{m}_C c_{C,i} \quad \text{for } i = 1..N \tag{8-167}$$

where $c_{H,i}$ and $c_{C,i}$ are the specific heat capacities of the hot and cold-fluids, respectively. According to Figure 8-31, the specific heat capacity of the high pressure fluid is not constant throughout the heat exchanger. However, the specific heat capacity can be taken as being approximately constant within each sub-heat exchanger because the temperature range spanned by each sub-heat exchanger is small. The value of c_H and c_C used in Eqs. (8-166) and (8-167) should be an average value for the sub-heat exchanger. The most appropriate average value of the specific heat capacity is the ratio of the change in enthalpy to the change in temperature experienced by the fluid. (If you have a sub-heat exchanger where one fluid is pure and changing phase, then you will need to guard against dividing by zero because the specific heat capacity will become infinity in this case.)

$$\dot{C}_{H,i} = \dot{m}_H \underbrace{\frac{(i_{H,i} - i_{H,i+1})}{(T_{H,i} - T_{H,i+1})}}_{\substack{\text{average specific} \\ \text{heat capacity}}} \quad \text{for } i = 1..N \tag{8-168}$$

$$\dot{C}_{C,i} = \dot{m}_C \frac{(i_{C,i+1} - i_{C,i})}{(T_{C,i+1} - T_{C,i})} \quad \text{for } i = 1..N \qquad (8\text{-}169)$$

```
"Apply effectiveness-NTU solution"
duplicate i=1,N
  C_dot_H[i]=m_dot_H*(i_H[i]-i_H[i+1])/(T_H[i]-T_H[i+1])          "hot-side capacitance rate"
  C_dot_C[i]=m_dot_C*(i_C[i+1]-i_C[i])/(T_C[i+1]-T_C[i])          "cold-side capacitance rate"
end
```

The effectiveness of each sub-heat exchanger is computed. The actual rate of heat transfer in each sub-heat exchanger is known ($\dot{q}/N$) and the maximum possible heat transfer rate is the product of the minimum of $\dot{C}_{H,i}$ and $\dot{C}_{C,i}$ and the maximum temperature difference:

$$\varepsilon_i = \frac{\dot{q}/N}{\text{MIN}(\dot{C}_{C,i}, \dot{C}_{H,i})(T_{H,i} - T_{C,i})} \quad \text{for } i = 1..N \qquad (8\text{-}170)$$

The number of transfer units required by each sub-heat exchanger (NTU_i) is obtained using the ε-NTU solution for a parallel-flow heat exchanger, implemented by the function HX. The conductance required in each sub-heat exchanger is:

$$UA_i = NTU_i \, \text{MIN}(\dot{C}_{C,i}, \dot{C}_{H,i}) \qquad (8\text{-}171)$$

```
duplicate i=1,N
  eff[i]=q_dot/(N*MIN(C_dot_H[i],C_dot_C[i])*(T_H[i]-T_C[i]))
                                              "effectiveness of sub-heat exchanger"
  NTU[i]=HX('parallelflow', eff[i], C_dot_H[i], C_dot_C[i], 'NTU')
                                              "NTU required by sub-heat exchanger"
  UA[i]=NTU[i]*MIN(C_dot_H[i],C_dot_C[i])     "conductance in sub-heat exchanger"
end
```

It is possible to assume a hot-side exit temperature ($T_{H,out}$ in step 1) that is non-physical, in which case the temperature of the two streams will cross. For example, if the hot-side outlet temperature is assumed to be less than about 205 K for this problem then you will receive an error because the effectiveness of some of the sub-heat exchangers is greater than one and therefore the solution for the associated number of transfer units is undefined. Figure 8-39 illustrates the temperature of the hot- and cold-side fluids as a function of the heat transfer rate for the case where $T_{H,out} = 205$ K and shows that the temperatures nearly intersect at the outlet of the heat exchanger. Therefore, $T_{H,out} = 205$ K represents the limit of the performance of the heat exchanger.

The conductance for each of the sub-heat exchangers may be translated into its physical size based on the geometry and operating conditions. Each sub-heat exchanger corresponds to a certain small length of the overall plate heat exchanger in Figure 8-32. The conductance of an individual sub-heat exchanger is given by:

$$UA_i = \left(\frac{1}{h_{C,i} \, 2 \, N_{ch} \, W \, \Delta x_i} + \frac{th_m}{k_{m,i} \, 2 \, N_{ch} \, W \, \Delta x_i} + \frac{1}{h_{H,i} \, 2 \, N_{ch} \, W \, \Delta x_i} \right)^{-1} \quad \text{for } i = 1..N \qquad (8\text{-}172)$$

where $h_{C,i}$ and $h_{H,i}$ are the local cold- and hot-side heat transfer coefficients within the sub-heat exchanger, $k_{m,i}$ is the metal conductivity within the sub-heat exchanger, and

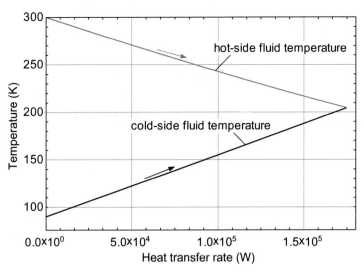

Figure 8-39: Temperature as a function of heat transfer rate for an assumed hot-side exit temperature $T_{H,out} = 205$ K.

Δx_i is the differential length of the sub-heat exchanger. Equation (8-172) must be solved in order to determine the physical distance, Δx_i, that each sub-heat exchanger occupies.

$$\Delta x_i = \frac{UA_i}{2\,N_{ch}\,W}\left(\frac{1}{h_{C,i}} + \frac{th_m}{k_{m,i}} + \frac{1}{h_{H,i}}\right) \quad \text{for } i = 1..N \tag{8-173}$$

The first sub-heat exchanger begins at $x_1 = 0$:

```
"determine length of each sub-heat exchanger"
x[1]=0 [m]                        "starting position of 1st sub-heat exchanger"
```

The local heat transfer coefficients are obtained using the DuctFlow_local function. Notice that average values of temperature are used and that the local heat transfer coefficient is evaluated at x_i plus some small value (0.001 m) in order to avoid potential problems with the local heat transfer coefficient calculation when $x_1 = 0$. The conductivity of aluminum is evaluated using EES' internal property routine. The differential length of each sub-heat exchanger is computed using Eq. (8-173) and used to evaluate the ending position of each sub-heat exchanger:

$$x_{i+1} = x_i + \Delta x_i \quad \text{for } i = 1..N \tag{8-174}$$

```
duplicate i=1,N
  call DuctFlow_local(H$,(T_H[i]+T_C[i])/2,p_H,m_dot_H/N_ch,th_H,W,x[i]+0.001[m],0 [-]:&
    h_H[i], h_H_H[i], dPHdx[i])            "hot-side local heat transfer coefficient"
  call DuctFlow_local(C$,(T_H[i]+T_C[i])/2,p_C,m_dot_C/N_ch,th_C,W,x[i]+0.001 [m],0 [-]: &
    h_C[i], h_C_H[i], dPCdx[i])            "cold-side local heat transfer coefficient"
  k_m[i]=k_('Aluminum', (T_H[i]+T_C[i])/2)  "metal conductivity at local average temperature"
  DELTAx[i]=UA[i]*(1/h_H[i]+th_m/k_m[i]+1/h_C[i])/(2*N_ch*W)
                                          "length of sub-heat exchanger"

  x[i+1]=x[i]+DELTAx[i]
end
```

Solving the EES code with $T_{H,out} = 250\,\text{K}$ leads to x_{N+1} (i.e., the length of heat exchanger required) of 0.1607 m; this is less than the specified length of the heat exchanger ($L = 0.70$ m). If the assumed value of $T_{H,out}$ is reduced (i.e., the performance of the heat exchanger is assumed to get better), then the predicted value of x_{N+1} will increase (the heat exchanger size must increase). The final step in obtaining a solution is to adjust $T_{H,out}$ until the predicted and specified physical size of the heat exchanger agree. Update guess values, comment out the assumed value of $T_{H,out}$ and specify that $x_{N+1} = L$:

```
{T_H_out=250 [K]}                        "assumed outlet temperature"
x[N+1]=L
```

It is likely that an error message will be displayed because EES has set a value of $T_{H,out}$ that is too low (i.e., below 205 K) and therefore obtained a non-physical temperature distribution. This problem can be avoided by setting up the solution as an optimization. Remove the specification that $x_{N+1} = L$ and instead define an objective function to be minimized:

$$err = \frac{|x_{N+1} - L|}{L} \tag{8-175}$$

```
{x[N+1]=L}
err=abs(x[N+1]-L)/L                      "objective function"
```

Select Min/Max from the Calculate menu and specify that the value of the variable err should be minimized by adjusting the variable T_H_out. Notice that the selection Stop if error occurs should not be checked because we do not want the optimization algorithm to stop working if a non-physical solution is encountered. If an error does occur, EES will set the value of the optimization target, in this case the variable err, to a large value as a penalty in order to ensure that a solution with an error is never identified as the optimum.

Select OK and EES should identify the correct solution: $T_{H,out} = 210.1$ K and $\dot{q} = 163.8\,\text{kW}$. Note that this solution is consistent with the solution obtained in Section 8.8.2 by numerically integrating the governing differential equations. Figure 8-40 illustrates the heat transfer rate predicted by the sub-heat exchanger model as a function of the number of sub-heat exchangers. Figure 8-40 shows that only about 5 sub-heat exchangers are required to obtain accurate results with the sub-heat exchanger modeling approach as compared to 20 integration steps required by direct integration of the state equations (see Figure 8-34).

Counter-Flow Configuration
This extended section can be found on the website www.cambridge.org/nellisandklein. The plate heat exchanger shown in Figure 8-32 operating in a counter-flow configuration is analyzed by dividing it into sub-heat exchangers. The steps are analogous to the analysis of the parallel-flow configuration.

8.6.4 Solution with Axial Conduction

This extended section can be found on the website www.cambridge.org/nellisandklein. Axial conduction is not commonly considered in heat exchanger analyses. However, the effect of axial conduction may be important for some heat exchangers; in particular,

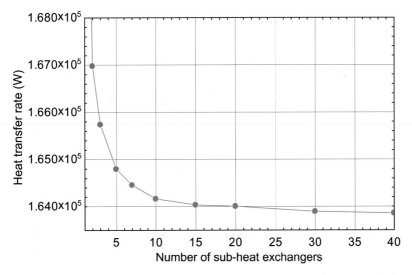

Figure 8-40: Heat transfer rate predicted by the sub-heat exchanger model as a function of the number of sub-heat exchangers.

axial conduction plays a critical role in the performance of very high effectiveness heat exchangers. Section 8.7 discusses the impact of axial conduction in a heat exchanger and provides several approximate models that can be used to estimate the associated performance degradation. The discussion in Section 8.7 is facilitated by the numerical model developed in this section that explicitly accounts for axial conduction. The numerical model is implemented in MATLAB. The numerical model can be accessed from EES using the AxialConductionHX procedure.

8.7 Axial Conduction in Heat Exchangers

8.7.1 Introduction

The heat exchanger modeling techniques that are presented in Sections 8.1 through 8.5 do not consider axial conduction through the fluid or, more importantly, through the heat exchanger structure. This approximation is appropriate since axial conduction has a negligible effect in many applications. In Section 8.6.4, a numerical model is developed that accounts for axial conduction in a counterflow heat exchanger; the results of this model show that axial conduction tends to reduce performance.

It is interesting to consider one of the challenges of heat exchanger design. In order to transfer energy from one stream to the other through the heat exchanger wall, you would like your heat exchanger structure to be very conductive. However, in order to avoid axial conduction penalties, you would like your heat exchanger structure to be very resistive. Practically, these conflicting demands are met by using long thin sheets of material (e.g., tubes or plates); the thermal resistance to conduction across a thin plate or tube will be much less than the thermal resistance to conduction along its length.

The numerical model developed in Section 8.6.4 can be used to explore the qualitative impact of axial conduction on the heat exchanger performance. Figure 8-47 illustrates the temperature distributions that are predicted for a balanced, counter-flow heat exchanger with varying amounts of axial conduction. Figure 8-47(a) illustrates the temperature distribution predicted if axial conduction is very small (accomplished by

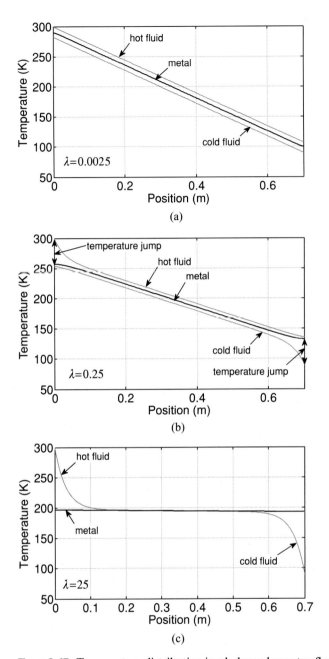

Figure 8-47: Temperature distribution in a balanced counter-flow heat exchanger with (a) low axial conduction ($\lambda = 0.0025$), (b) moderate axial conduction ($\lambda = 0.25$), and (c) high axial conduction ($\lambda = 25$).

making the conductivity of the heat exchanger wall very small). Figure 8-47(b) illustrates the temperature distribution predicted at the same conditions used to generate Figure 8-47(a), except with a larger amount of axial conduction (accomplished by increasing the conductivity of the wall). Finally, Figure 8-47(c) illustrates the temperature distribution predicted at the same conditions, but with a large amount of axial conduction (accomplished by setting the conductivity of the wall to a very high value). Recall that the resistance to conduction through the heat exchanger structure between the two fluid streams is ignored in the numerical model developed in Section 8.6.4; therefore, Figure 8-47(a) through (c) all correspond to the same values of UA and NTU.

The presence of axial conduction has a large impact on the behavior and performance of the heat exchanger. The effectiveness of the heat exchanger shown in Figure 8-47(c) is much less than the heat exchanger shown in Figure 8-47(a), as evidenced by the increase in the hot outlet temperature and the corresponding decrease in the cold outlet temperature. When the impact of axial conduction is significant but not dominant (Figure 8-47(b)), the temperature distribution begins to exhibit the "temperature jumps" (i.e., the rapid changes in temperature that occur very near the fluid inlets) that are characteristic of a heat exchanger suffering from axial conduction losses. When the impact of axial conduction becomes dominant (Figure 8-47(c)), the heat exchanger structure is so conductive that it cannot support a temperature gradient along its length and therefore it becomes essentially isothermal. In this limit, the two fluids separately experience a thermal equilibration with an isothermal plate.

The impact of axial conduction on a heat exchanger's performance is typically quantified by the dimensionless axial conduction parameter, λ. The axial conduction parameter is defined according to:

$$\lambda = \frac{1}{R_{ac}\,\dot{C}_{min}} \tag{8-219}$$

where R_{ac} is the thermal resistance to axial conduction and $\dot{C}_{min}$ is the minimum capacitance rate.

The axial conduction parameter is approximately related to the ratio of the rate of heat transfer along the length of the heat exchanger due to axial conduction to the rate of heat transfer between the streams. The rate of heat transfer due to axial conduction is calculated approximately according to:

$$\dot{q}_{ac} \approx \frac{(T_{H,in} - T_{C,in})}{R_{ac}} \tag{8-220}$$

where $T_{H,in}$ and $T_{C,in}$ are the hot- and cold-fluid inlet temperatures. The rate of heat transfer between the streams is approximately:

$$\dot{q}_{sts} \approx \dot{C}_{min}\,(T_{H,in} - T_{C,in}) \tag{8-221}$$

The ratio of these heat transfer rates is therefore:

$$\frac{\dot{q}_{ac}}{\dot{q}_{sts}} \approx \frac{1}{R_{ac}\,\dot{C}_{min}} = \lambda \tag{8-222}$$

Figure 8-47(a) through (c) were generated by setting the metal conductivity equal to values that provided $\lambda = 0.0025$, $\lambda = 0.25$, and $\lambda = 25.0$, respectively.

Axial conduction tends to be important in heat exchangers that have relatively large number of transfer units and high effectiveness. The approximate methods for quantifying the effects of axial conduction that are presented in Section 8.7.2 have been developed for heat exchangers with these characteristics.

8.7.2 Approximate Models for Axial Conduction

The numerical model developed in Section 8.6.4 explicitly and precisely includes the effect of axial conduction on the performance of a counterflow heat exchanger. However, it not convenient or necessary to generate a detailed numerical model that considers axial conduction in order to understand what the impact of this effect will be on performance. In this section, a set of models are discussed that can be used to approximately predict the performance of a heat exchanger that is subject to axial conduction

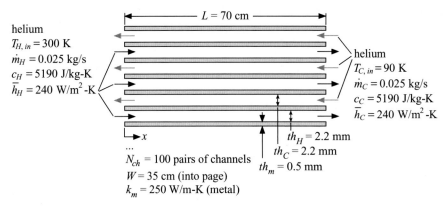

Figure 8-48: Plate heat exchanger operating in a counter-flow configuration.

losses. These approximate models are presented and investigated in the context of the plate heat exchanger operating in counter-flow that was previously considered in Section 8.6.4 and is shown again in Figure 8-48.

Figure 8-49 illustrates the effectiveness of the heat exchanger shown in Figure 8-48 predicted by the numerical model developed in Section 8.6.4 (ε) as a function of the mass flow rate (on both sides of the heat exchanger, $\dot{m}_H = \dot{m}_C$). Also shown in Figure 8-49 is the effectiveness predicted by the ε-NTU solution (ε_{nac}), that ignores axial conduction.

Notice that the discrepancy between the actual effectiveness (ε) and the ε-NTU solution (ε_{nac}) increases as the mass flow rate decreases. This behavior is consistent with the discussion in Section 8.7.1; the value of the dimensionless axial conduction parameter, λ, increases with decreasing mass flow rate. The resistance to axial conduction in the

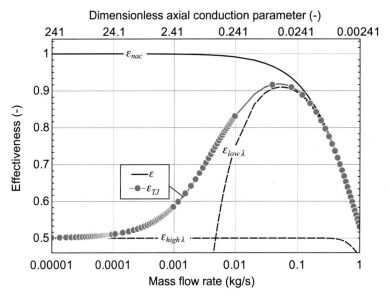

Figure 8-49: Effectiveness of the heat exchanger shown in Figure 8-48 in a balanced operating condition predicted by the numerical model derived in Section 8.6.4 (ε) as a function of mass flow rate. Also shown is the effectiveness without axial conduction predicted by the ε-NTU model (ε_{nac}), the effectiveness predicted by the low λ model ($\varepsilon_{low\,\lambda}$), the effectiveness predicted by the high λ model ($\varepsilon_{high\,\lambda}$) and the effectiveness predicted by the temperature jump model (ε_{TJ}).

plate heat exchanger shown in Figure 8-48 (R_{ac}) is computed according to:

$$R_{ac} = \frac{L}{N_{ch} \, 2 \, W \, th_m \, k_m} \tag{8-223}$$

The axial conduction parameter, λ, can be computed using Eq. (8-219). The code below is added to the MATLAB code developed in Section 8.6.4:

```
R_ac=L/(N_ch*2*th_m*W*k_m);        % resistance to axial conduction (K/W)
lambda=1/(R_ac*C_dot_min)          % dimensionless axial conduction parameter (-)
```

The upper x-axis in Figure 8-49 illustrates the value of λ corresponding to the mass flow rate on the lower x-axis.

Approximate Model at Low λ

The discrepancy between the ε-NTU solution and the performance predicted by the numerical model that considers axial conduction is small when λ is much less than one. In this region, the effect of axial conduction can be treated approximately by subtracting the rate of heat transfer due to axial conduction ($\dot{q}_{ac}$) from the rate of heat transfer that is expected without considering axial conduction ($\dot{q}_{nac}$).

$$\dot{q} \approx \dot{q}_{nac} - \dot{q}_{ac} \tag{8-224}$$

The rate of heat transfer without axial conduction is:

$$\dot{q}_{nac} = \varepsilon_{nac} \, \dot{C}_{min} \, (T_{H,in} - T_{C,in}) \tag{8-225}$$

Substituting Eqs. (8-220) and (8-225) into Eq. (8-224) leads to:

$$\dot{q} \approx \varepsilon_{nac} \, \dot{C}_{min} \, (T_{H,in} - T_{C,in}) - \frac{(T_{H,in} - T_{C,in})}{R_{ac}} \tag{8-226}$$

Equation (8-226) is divided by the maximum possible heat transfer rate:

$$\dot{q}_{max} = \dot{C}_{min} \, (T_{H,in} - T_{C,in}) \tag{8-227}$$

in order to obtain:

$$\varepsilon_{low \, \lambda} \approx \varepsilon_{nac} - \lambda \tag{8-228}$$

Figure 8-49 illustrates the effectiveness predicted by Eq. (8-228), $\varepsilon_{low \, \lambda}$, and shows that the model is appropriate at $\lambda < 0.1$.

Approximate Model at High λ

Figure 8-49 shows that the effectiveness of the heat exchanger at very high values of λ (or low $\dot{m}$) approaches a constant value of 0.5; this limit is consistent with the temperature distribution shown in Figure 8-47(c). At high λ, the heat exchanger structure is so conductive that it assumes a nearly uniform temperature, limiting the temperature change that can be experienced by either of the fluids. Therefore, at best the two fluid streams can exit at the same temperature. The performance of the heat exchanger is limited at high λ in much the same way that the performance of a heat exchanger operating in a parallel-flow configuration is limited, as discussed in Section 8.3.4.

The approximate model for a heat exchanger at very high λ treats the two streams separately by assuming that the metal is completely isothermal, as shown in Figure 8-50. The hot-stream is treated as a fluid transferring heat to a uniform temperature heat sink; that is, as a fluid flowing in a heat exchanger with a capacity ratio of zero. The

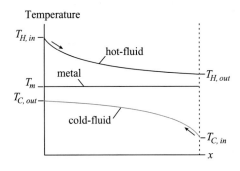

Figure 8-50: High λ model of a heat exchanger.

effectiveness of any heat exchanger where $C_R \to 0$ was discussed in Section 8.3.4 and is given in Table 8-2:

$$\lim_{C_R \to 0} \varepsilon = 1 - \exp(-NTU) \qquad (8\text{-}229)$$

Therefore, the effectiveness associated with the interaction between the hot-fluid and the metal is predicted by:

$$\varepsilon_H = 1 - \exp(-NTU_H) \qquad (8\text{-}230)$$

where NTU_H is the number of transfer units between the metal and the hot-fluid. For the plate heat exchanger shown in Figure 8-48, NTU_H is:

$$NTU_H = \frac{h_H \, 2 \, N_{ch} \, W \, L}{\dot{C}_H} \qquad (8\text{-}231)$$

The effectiveness associated with the interaction between the hot-fluid and the wall is the ratio of the actual to the maximum possible heat transfer rate:

$$\varepsilon_H = \frac{\dot{q}_H}{\dot{q}_{max,H}} = \frac{(T_{H,in} - T_{H,out})}{(T_{H,in} - T_m)} \qquad (8\text{-}232)$$

where T_m is the wall temperature. Similarly, the effectiveness associated with the interaction between the cold-fluid and the metal is predicted according to:

$$\varepsilon_C = 1 - \exp(-NTU_C) \qquad (8\text{-}233)$$

where NTU_C is the number of transfer units between the wall and the cold-fluid. For the plate heat exchanger shown in Figure 8-48, NTU_C is:

$$NTU_C = \frac{h_C \, 2 \, N_{ch} \, W \, L}{\dot{C}_C} \qquad (8\text{-}234)$$

The effectiveness associated with the interaction between the cold-fluid and the metal is:

$$\varepsilon_C = \frac{\dot{q}_C}{\dot{q}_{max,C}} = \frac{(T_{C,out} - T_{C,in})}{(T_m - T_{C,in})} \qquad (8\text{-}235)$$

The effectiveness of the heat exchanger (taken as a whole) in the high λ limit is:

$$\varepsilon_{high\,\lambda} = \frac{\dot{q}}{\dot{C}_{min}(T_{H,in} - T_{C,in})} \qquad (8\text{-}236)$$

Energy balances on the hot-side fluid and the cold-side fluid lead to:

$$\dot{q} = \dot{C}_H (T_{H,in} - T_{H,out}) = \dot{C}_C (T_{C,out} - T_{C,in}) \qquad (8\text{-}237)$$

Substituting Eq. (8-237) into Eq. (8-236) leads to:

$$\varepsilon_{high\,\lambda} = \frac{\dot{C}_H\,(T_{H,in} - T_{H,out})}{\dot{C}_{min}\,(T_{H,in} - T_{C,in})} \tag{8-238}$$

The denominator of Eq. (8-238) can be rearranged:

$$\varepsilon_{high\,\lambda} = \frac{\dot{C}_H\,(T_{H,in} - T_{H,out})}{\dot{C}_{min}\,[(T_{H,in} - T_m) - (T_{C,in} - T_m)]} \tag{8-239}$$

Substituting Eqs. (8-232) and (8-235) into Eq. (8-239) leads to:

$$\varepsilon_{high\,\lambda} = \frac{\dot{C}_H\,(T_{H,in} - T_{H,out})}{\dot{C}_{min}\left[\dfrac{(T_{H,in} - T_{H,out})}{\varepsilon_H} + \dfrac{(T_{C,out} - T_{C,in})}{\varepsilon_C}\right]} \tag{8-240}$$

Substituting Eq. (8-237) into Eq. (8-240) leads to:

$$\varepsilon_{high\,\lambda} = \frac{\dot{C}_H\,(T_{H,in} - T_{H,out})}{\dot{C}_{min}\left[\dfrac{(T_{H,in} - T_{H,out})}{\varepsilon_H} + \dfrac{\dot{C}_H\,(T_{H,in} - T_{H,out})}{\dot{C}_C\,\varepsilon_C}\right]} \tag{8-241}$$

or

$$\varepsilon_{high\,\lambda} = \frac{1}{\dot{C}_{min}\left[\dfrac{1}{\dot{C}_H\,\varepsilon_H} + \dfrac{1}{\dot{C}_C\,\varepsilon_C}\right]} \tag{8-242}$$

The high λ model is added to the MATLAB code developed in Section 8.6.4 according to:

```
% high lambda approximate model
NTU_H=h_H*2*N_ch*W*L/C_dot_H;                  % hot-side number of transfer units
NTU_C=h_C*2*N_ch*W*L/C_dot_C;                  % cold-side number of transfer units
eff_H=1-exp(-NTU_H);                           % hot-side effectiveness
eff_C=1-exp(-NTU_C);                           % cold-side effectiveness
eff_highlambda=1/(C_dot_min*(1/(C_dot_H*eff_H)+1/(C_dot_C*eff_C)))
                                               % effectiveness in the high lambda limit
```

The value of $\varepsilon_{high\,\lambda}$ is also shown in Figure 8-49 as a function of mass flow rate and matches the value predicted by the numerical model when λ is large, greater than approximately 50.

Temperature Jump Model

Figure 8-49 shows that there is a large range of λ where neither the low λ model nor the high λ model apply. An approximate model with a wider range of applicability is obtained using the concept of a temperature jump; the characteristic temperature jump associated with axial conduction is observed in Figure 8-47(b). Figure 8-51 illustrates, qualitatively, the temperature distribution that provides the basis of the temperature jump model. This behavior persists provided that the number of transfer units is relatively high; if the number of transfer units is not large then the temperature jumps disappear because the fluid is not well-connected to the wall and therefore the temperature jump model is not valid.

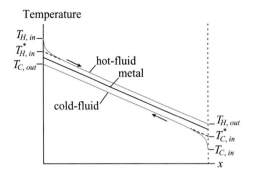

Figure 8-51: Temperature jump model.

The hot-fluid entering the heat exchanger transfers energy to the wall at a rate that is approximately equal to the rate of heat transfer associated with axial conduction; this energy transfer occurs almost immediately after the fluid enters the heat exchanger and causes the hot-fluid temperature to drop from $T_{H,in}$ to $T^*_{H,in}$, as shown in Figure 8-51.

$$\dot{q}_{TJ} = \dot{C}_H \left(T_{H,in} - T^*_{H,in}\right) = \frac{\left(T^*_{H,in} - T^*_{C,in}\right)}{R_{ac}} \tag{8-243}$$

where $\dot{q}_{TJ}$ is the heat transfer rate related to the temperature jump. A similar process occurs at the cold end. The cold-fluid entering the heat exchanger almost immediately receives energy from the metal at a rate that is approximately equal to the rate of heat transfer associated with axial conduction. This causes the cold-fluid temperature to increase from $T_{C,in}$ to $T^*_{C,in}$.

$$\dot{q}_{TJ} = \dot{C}_C \left(T^*_{C,in} - T_{C,in}\right) = \frac{\left(T^*_{H,in} - T^*_{C,in}\right)}{R_{ac}} \tag{8-244}$$

These spatially concentrated heat transfer rates result in the observed temperature jumps that occur near the heat exchanger inlets.

Solving Eq. (8-243) for $T^*_{C,in}$ leads to:

$$T^*_{C,in} = T^*_{H,in} - R_{ac}\,\dot{C}_H \left(T_{H,in} - T^*_{H,in}\right) \tag{8-245}$$

Substituting Eq. (8-245) into Eq. (8-244) leads to:

$$\dot{C}_C \left[T^*_{H,in} - R_{ac}\,\dot{C}_H \left(T_{H,in} - T^*_{H,in}\right) - T_{C,in}\right] = \frac{\left[T^*_{H,in} - T^*_{H,in} + R_{ac}\,\dot{C}_H \left(T_{H,in} - T^*_{H,in}\right)\right]}{R_{ac}} \tag{8-246}$$

which can be rearranged:

$$\dot{C}_C\,T^*_{H,in} - R_{ac}\,\dot{C}_C\,\dot{C}_H \left(T_{H,in} - T^*_{H,in}\right) - \dot{C}_C\,T_{C,in} = \dot{C}_H \left(T_{H,in} - T^*_{H,in}\right) \tag{8-247}$$

and solved for $T^*_{H,in}$:

$$T^*_{H,in} = \frac{R_{ac}\,\dot{C}_C\,\dot{C}_H\,T_{H,in} + \dot{C}_C\,T_{C,in} + \dot{C}_H\,T_{H,in}}{R_{ac}\,\dot{C}_C\,\dot{C}_H + \dot{C}_C + \dot{C}_H} \tag{8-248}$$

The modified cold-inlet temperature, $T^*_{C,in}$, can be predicted by substituting Eq. (8-248) into Eq. (8-245). The temperature jump model assumes that the heat exchanger behavior can be predicted by the ε-NTU solution using the modified inlet temperatures $T^*_{H,in}$

and $T^*_{C,in}$. The heat transfer rate that is not related to the temperature jump ($\dot{q}_{HX}$) is determined according to:

$$\dot{q}_{HX} = \varepsilon_{nac}\, \dot{C}_{min}\, \left(T^*_{H,in} - T^*_{C,in}\right) \tag{8-249}$$

where ε_{nac} is the effectiveness predicted using an ε-NTU solution that neglects axial conduction. The total rate of heat transfer in the heat exchanger is the sum of $\dot{q}_{HX}$ and $\dot{q}_{TJ}$. Therefore, the effectiveness predicted by the temperature jump model is given by:

$$\varepsilon_{TJ} = \frac{\dot{q}_{HX} + \dot{q}_{TJ}}{\dot{q}_{max}} = \frac{\varepsilon_{nac}\, \dot{C}_{min}\, \left(T^*_{H,in} - T^*_{C,in}\right) + \dfrac{\left(T^*_{H,in} - T^*_{C,in}\right)}{R_{ac}}}{\dot{C}_{min}\, \left(T_{H,in} - T_{C,in}\right)} \tag{8-250}$$

or

$$\varepsilon_{TJ} = (\varepsilon_{nac} + \lambda)\, \frac{\left(T^*_{H,in} - T^*_{C,in}\right)}{\left(T_{H,in} - T_{C,in}\right)} \tag{8-251}$$

The following code adds the temperature jump model to the MATLAB code from Section 8.6.4:

```
% temperature jump approximate model
T_H_in_star=(R_ac*C_dot_C*C_dot_H*T_H_in+C_dot_C*T_C_in+C_dot_H*T_H_in)...
    /(R_ac*C_dot_C*C_dot_H+C_dot_C+C_dot_H);
T_C_in_star=T_H_in_star-R_ac*C_dot_H*(T_H_in-T_H_in_star);
eff_TJ=(eff_nac+lambda)*(T_H_in_star-T_C_in_star)/(T_H_in-T_C_in);
```

The effectiveness predicted by the temperature jump model for the plate heat exchanger illustrated in Figure 8-48 is also shown in Figure 8-49. Note that the agreement with the numerical model is very good over the entire range of λ. The temperature jump model limits to the low conductivity model when λ is small and the high conductivity model when λ is high.

The agreement between the temperature jump model and the actual solution is not as good when the heat exchanger becomes unbalanced. For example, Figure 8-52 illustrates the effectiveness as a function of the cold-side mass flow rate with a capacity ratio $C_R = 0.75$ (i.e, $\dot{m}_H = 0.75\,\dot{m}_C$). The numerical solution, ε-NTU solution, and approximate models are all shown in Figure 8-52.

8.8 Perforated Plate Heat Exchangers

8.8.1 Introduction

In Section 8.7, we found that a heat exchanger structure with a low resistance to conduction in the axial (flow) direction will lead to substantial degradation in the performance of a heat exchanger. However, if the heat exchanger structure has a high resistance to conduction in the stream-to-stream direction then the performance will also be reduced. Therefore, the heat exchanger structure should be neither too conductive nor too resistive; in the limit that either $k_m \to 0$ or $k_m \to \infty$, the heat exchanger performance will be poor (where k_m is the conductivity of the structure material). In fact, an ideal heat exchanger structure is anisotropic with low conductivity in the axial direction but high conductivity in the stream-to-stream direction. It therefore seems natural to consider composite materials for heat exchanger structures. In Section 2.9, we found that the

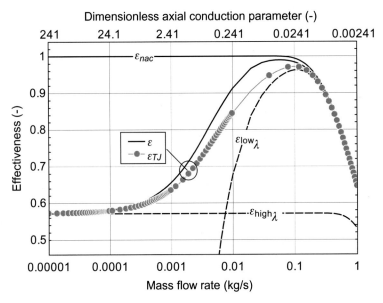

Figure 8-52: Effectiveness as a function of mass flow rate for the counter-flow heat exchanger shown in Figure 8-48 operating in an unbalanced condition ($\dot{m}_H = 0.75\,\dot{m}_C$) predicted by the numerical model developed in Section 8.6.4 (ε). Also shown is the effectiveness without axial conduction predicted by the ε-NTU model (ε_{nac}), the effectiveness predicted by the low λ model ($\varepsilon_{low\,\lambda}$), the effectiveness predicted by the high λ model ($\varepsilon_{high\,\lambda}$), and the effectiveness predicted by the temperature jump model (ε_{TJ}).

effective conductivity of a laminated structure composed of laminations that are alternately made of high and low conductivity material will be very different in the direction perpendicular to the laminations than it is in the direction parallel to the laminations. Heat exchangers that are designed based on this concept are referred to as perforated plate heat exchangers. A perforated plate heat exchanger is constructed of many plates that are oriented perpendicular to the flow, as shown in Figure 8-53.

The perforated plate heat exchanger is a composite structure. The plates are alternatively low conductivity spacers that limit axial conduction and high conductivity heat transfer plates that provide good thermal communication between the hot- and cold-fluids. The plates are joined together hermetically and the pattern of the spacers contains and directs the hot and cold-fluids through passages in the heat transfer plates.

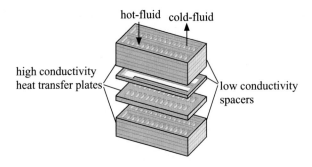

Figure 8-53: Perforated plate heat exchanger.

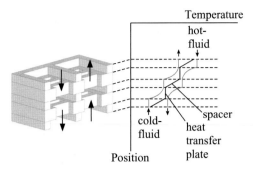

Figure 8-54: Qualitative temperature distribution in a spacer/heat transfer plate unit.

Figure 8-54 illustrates qualitatively the temperature distribution expected in the fluids and the spacer/heat transfer plate material.

Each heat transfer plate is an individual, small heat exchanger. The plates are composed of high conductivity material and the cross-sectional area for axial conduction in each plate is large. Therefore, the dimensionless axial conduction parameter associated with any individual plate will be large and each heat transfer plate is likely to be nearly isothermal. The hot- and cold-fluid entering the heat transfer plate will approach the plate temperature; this behavior is consistent with the temperature distribution shown earlier in Figure 8-50 for a high λ heat exchanger. The spacer is made of low conductivity material and the cross-sectional area available for axial conduction in each spacer is small. Therefore, the temperature gradient across each spacer is approximately linear. The fluid passing through the spacer does not change temperature significantly because there is very little surface area for heat transfer in a spacer and therefore the stream-to-stream thermal communication is poor.

8.8.2 Modeling Perforated Plate Heat Exchangers

An approximate model of a perforated plate heat exchanger treats each individual plate using the high λ model discussed in Section 8.7.2. The effectiveness of the series of heat exchangers represented by the stack of perforated plates is evaluated, ignoring the effect of axial conduction through the spacers. Finally, the resistance to axial conduction through the laminated stack is determined. Typically, the spacers limit axial conduction sufficiently that the perforated plate heat exchanger, considered as a whole, is in the low λ regime. Therefore, the low λ model discussed in Section 8.7.2 can be applied. This modeling methodology is illustrated in the context of a perforated plate heat exchanger that is used as the recuperative heat exchanger in a cryosurgical probe, shown in Figure 8-55.

Argon enters the hot end of the heat exchanger at $T_{H,in} = 300\,\text{K}$ with mass flow rate $\dot{m} = 0.0002\,\text{kg/s}$. Argon enters the cold end of the heat exchanger at $T_{C,in} = 190\,\text{K}$ with the same mass flow rate, $\dot{m} = 0.0002\,\text{kg/s}$. The pressure of the hot- and cold-sides are approximately constant at $p_H = 2.0\,\text{MPa}$ and $p_C = 100\,\text{kPa}$, respectively. There are $N_p = 20$ heat transfer plates composed of very high conductivity material; assume that the conductivity of the heat transfer plates is infinitely large. Each plate is $th_{HTP} = 0.125$ inch thick and has $N_H = 200$ holes of diameter $D_H = 0.006$ inch installed for the hot-fluid flow and $N_C = 150$ holes of diameter $D_C = 0.010$ inch installed for the cold-fluid flow. The heat transfer plates are separated by spacer plates that are $th_{sp} = 0.125$ inch thick. The spacer plates are composed of 304 stainless steel and the webs

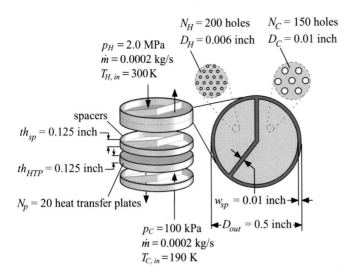

Figure 8-55: Perforated plate heat exchanger for a cryosurgical probe.

used to contain the fluids are $w_{sp} = 0.01$ inch thick. Both the spacer and heat transfer plates have outer diameter $D_{out} = 0.50$ inch.

The input parameters are entered in EES:

```
$UnitSystem SI MASS RAD PA K J
$Tabstops 0.2 0.4 0.6 3.5 in

"Inputs"
p_H=2.0 [MPa]*convert(MPa,Pa)                    "high pressure"
p_C=100 [kPa]*convert(kPa,Pa)                    "low pressure"
m_dot=0.0002 [kg/s]                              "mass flow rate"
T_H_in=300 [K]                                   "hot inlet temperature"
T_C_in=190 [K]                                   "cold inlet temperature"
N_H=200 [-]                                      "number of holes on the hot-side"
D_H=0.006 [inch]*convert(inch,m)                 "diameter of holes on the hot-side"
th_HTP=0.125 [inch]*convert(inch,m)              "thickness of heat transfer plate"
th_sp=0.125 [inch]*convert(inch,m)               "thickness of spacer plate"
N_C=150 [-]                                      "number of holes on the cold-side"
D_C=0.01 [inch]*convert(inch,m)                  "diameter of holes on the cold-side"
N_p=20 [-]                                       "number of plates"
D_out=0.5 [inch]*convert(inch,m)                 "outer diameter"
w_sp=0.010 [inch]*convert(inch,m)                "spacer width"
```

The model will not consider each plate individually. Rather, we will assume that the fluid and material properties can be considered constant and evaluated at the average temperature in the heat exchanger, $\overline{T}$:

$$\overline{T} = \frac{(T_{H,in} + T_{C,in})}{2} \qquad (8\text{-}252)$$

The fluid properties at the high and low pressures are evaluated using EES' internal property routine for argon; these include the conductivity (k_H and k_C), viscosity (μ_H and μ_C), density (ρ_H and ρ_C) and specific heat capacity (c_H and c_C):

```
T_avg=(T_H_in+T_C_in)/2                        "average temperature"
k_H=conductivity(Argon,T=T_avg,P=p_H)          "hot-side conductivity"
k_C=conductivity(Argon,T=T_avg,P=p_C)          "cold-side conductivity"
mu_H=viscosity(Argon,T=T_avg,P=p_H)            "hot-side viscosity"
mu_C=viscosity(Argon,T=T_avg,P=p_C)            "cold-side viscosity"
rho_H=density(Argon,T=T_avg,P=p_H)             "hot-side density"
rho_C=density(Argon,T=T_avg,P=p_C)             "cold-side density"
c_H=cP(Argon,T=T_avg,P=p_H)                    "hot-side specific heat capacity"
c_C=cP(Argon,T=T_avg,P=p_C)                    "cold-side specific heat capacity"
```

The conductivity of the spacer plates (k_{sp}) is determined:

```
k_sp=k_('SS304_cryogenic', T_avg)              "conductivity of spacer"
```

The performance of a single plate is determined using the model discussed in Section 8.7.2 for a heat exchanger with a large axial conduction parameter; note that the assumption that the conductivity of the heat transfer plate is effectively infinite corresponds to an infinite value of λ for any particular plate. The hot-fluid and cold-fluid capacitance rates are computed:

$$\dot{C}_C = \dot{m}\, c_C \qquad (8\text{-}253)$$

$$\dot{C}_H = \dot{m}\, c_H \qquad (8\text{-}254)$$

and the minimum of these capacitance rates ($\dot{C}_{min}$) is obtained:

```
C_dot_H=m_dot*c_H                              "hot-side capacitance rate"
C_dot_C=m_dot*c_C                              "cold-side capacitance rate"
C_dot_min=MIN(C_dot_H,C_dot_C)                 "minimum capacitance rate"
```

The average heat transfer coefficient on the hot- and cold-sides ($\bar{h}_H$ and $\bar{h}_C$) are computed using the **PipeFlow** procedure in EES; note that it is assumed that the fluid completely mixes in the spacer region between each plate.

```
call PipeFlow('Argon',T_avg,p_H,m_dot/N_H,D_H,th_HTP,0 [-]:&
  h_bar_H, h_H_bar_H ,DELTAP_H_plate, Nusselt_T_H, f_H, Re_H)   "hot-side h_bar"
call PipeFlow('Argon',T_avg,p_C,m_dot/N_C,D_C,th_HTP,0 [-]:&
  h_bar_C, h_H_bar_C ,DELTAP_C_plate, Nusselt_T_C, f_C, Re_C)   "cold-side h_bar"
```

The conductances that are associated with the interaction between the hot-side fluid and the plate material and the cold-side fluid and the plate material, UA_H and UA_C, are computed according to:

$$UA_H = \bar{h}_H\, \pi\, D_H\, th_{HTP}\, N_H \qquad (8\text{-}255)$$

$$UA_C = \bar{h}_C\, \pi\, D_C\, th_{HTP}\, N_C \qquad (8\text{-}256)$$

```
UA_H=h_bar_H*pi*D_H*th_HTP*N_H                 "conductance on hot-side"
UA_C=h_bar_C*pi*D_C*th_HTP*N_C                 "conductance on cold-side"
```

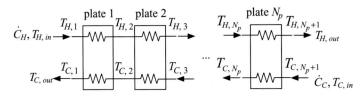

Figure 8-56: A stack of N_p heat exchangers (corresponding to the heat transfer plates) in series.

The number of transfer units associated with the hot-fluid to plate (NTU_H) and cold-fluid to plate (NTU_C) heat transfer are computed:

$$NTU_H = \frac{UA_H}{\dot{C}_H} \tag{8-257}$$

$$NTU_C = \frac{UA_C}{\dot{C}_C} \tag{8-258}$$

NTU_H=UA_H/C_dot_H   "number of transfer units on hot-side"
NTU_C=UA_C/C_dot_C   "number of transfer units on cold-side"

The effectiveness of the hot-fluid to plate (ε_H) and cold-fluid to plate (ε_C) interactions are computed:

$$\varepsilon_H = 1 - \exp\left(-NTU_H\right) \tag{8-259}$$

$$\varepsilon_C = 1 - \exp\left(-NTU_C\right) \tag{8-260}$$

eff_H=1-exp(-NTU_H)   "effectiveness of hot side"
eff_C=1-exp(-NTU_C)   "effectiveness of cold side"

The effectiveness of a single plate is computed using Eq. (8-242):

$$\varepsilon_{plate} = \frac{1}{\dot{C}_{min}\left[\dfrac{1}{\dot{C}_H\,\varepsilon_H} + \dfrac{1}{\dot{C}_C\,\varepsilon_C}\right]} \tag{8-261}$$

eff_plate=1/(C_dot_min*(1/(C_dot_H*eff_H)+1/(C_dot_C*eff_C)))
                                               "effectiveness of a single plate"

The performance of the stack of N_p plates (neglecting axial conduction along its length) is obtained by considering the plates to be heat exchangers in series, as shown in Figure 8-56.

The temperature of the hot-fluid entering the first plate is the hot-side inlet temperature:

$$T_{H,1} = T_{H,in} \tag{8-262}$$

and the temperature of the cold-fluid entering the last plate is the cold-side inlet temperature:

$$T_{C,N_p+1} = T_{C,in} \qquad (8\text{-}263)$$

An energy balance on each plate provides one set of equations:

$$\dot{C}_H \left(T_{H,i} - T_{H,i+1}\right) = \dot{C}_C \left(T_{C,i} - T_{C,i+1}\right) \quad \text{for } i = 1..N_p \qquad (8\text{-}264)$$

and the effectiveness of each plate provides another set of equations:

$$\dot{C}_H \left(T_{H,i} - T_{H,i+1}\right) = \varepsilon_{plate}\, \dot{C}_{min} \left(T_{H,i} - T_{C,i+1}\right) \quad \text{for } i = 1..N_p \qquad (8\text{-}265)$$

Equations (8-262) through (8-265) provides $2(N_p + 1)$ equations in the same number of unknown temperatures.

```
T_H[1]=T_H_in                                    "hot-side inlet temperature"
T_C[N_p+1]=T_C_in                                "cold-side inlet temperature"
duplicate i=1,N_p
  C_dot_H*(T_H[i]-T_H[i+1])=C_dot_C*(T_C[i]-T_C[i+1])    "energy balance on each plate"
  C_dot_H*(T_H[i]-T_H[i+1])=eff_plate*C_dot_min*(T_H[i]-T_C[i+1])
                                                 "performance of each plate"
end
```

The effectiveness of the perforated plate heat exchanger, neglecting axial conduction, is given by the ratio of the total rate of heat transfer to the hot-fluid to the maximum possible rate of heat transfer:

$$\varepsilon_{pp} = \frac{\dot{C}_H \left(T_{H,in} - T_{H,N_p+1}\right)}{\dot{C}_{min} \left(T_{H,in} - T_{C,in}\right)} \qquad (8\text{-}266)$$

```
eff_pp=C_dot_H*(T_H_in-T_H[N_p+1])/(C_dot_min*(T_H_in-T_C_in))
    "effectiveness of perforated plate heat exchanger (neglecting axial conduction"
```

which leads to $\varepsilon_{pp} = 0.9434$. It is instructive to compare ε_{pp} to the effectiveness that would be predicted if the heat exchanger were treated as a conventional, continuous heat exchanger using the $\varepsilon\text{-}NTU$ solution. The total stream-to-stream conductance in the heat exchanger is:

$$UA = \left[\frac{1}{\bar{h}_H\, \pi\, D_H\, th_{HTP}\, N_H\, N_p} + \frac{1}{\bar{h}_C\, \pi\, D_C\, th_{HTP}\, N_C\, N_p}\right]^{-1} \qquad (8\text{-}267)$$

and the total number of transfer units is:

$$NTU = \frac{UA}{\dot{C}_{min}} \qquad (8\text{-}268)$$

The $\varepsilon\text{-}NTU$ solution for a counter-flow heat exchanger is obtained using the function HX and provides an upper limit to the heat exchanger performance (ε_{limit}).

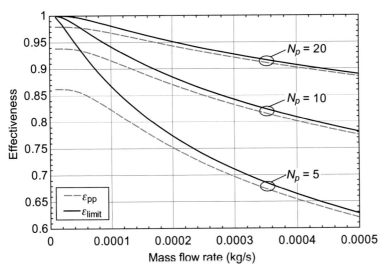

Figure 8-57: Effectiveness predicted by the perforated plate model (ε_{pp}) and by the ε-NTU solution (ε_{limit}) as a function of mass flow rate for various values of the number of plates.

UA=(1/(h_bar_H*pi*D_H*th_HTP*N_H*N_p)+1/(h_bar_C*pi*D_C*th_HTP*N_C*N_p))^(-1)
                                                                            "total conductance of heat exchanger"

NTU=UA/C_dot_min   "total number of transfer units"
eff_limit=HX('counterflow', NTU, C_dot_H, C_dot_C, 'epsilon')
                                                                            "limiting effectiveness"

which leads to $\varepsilon_{limit} = 0.9513$. Notice that ε_{pp} is very close to ε_{limit} which indicates that the discrete nature of the perforated plate heat exchanger has not affected its performance substantially relative to a heat exchanger that continuously distributes the same amount of conductance. This conclusion will generally be true provided that there are a relatively large number of plates and that the heat exchanger effectiveness is not exceptionally high. Examination of Figure 8-54 reveals that there is an unavoidable temperature difference between the hot- and cold-side fluids leaving each plate in a perforated plate heat exchanger. This temperature difference is related to the fact that the outlet fluid temperatures can, at best, approach the metal temperature in the last plate, but cannot approach the inlet temperature of the other fluid stream. Provided that this unavoidable temperature difference related to the perforated plate construction is small relative to the pinch point temperature difference discussed in Section 8.4, the performance of a perforated plate heat exchanger will not be substantially different from a continuous heat exchanger with the same total conductance. Figure 8-57 illustrates the effectiveness predicted by accounting for the perforated plate construction (ε_{pp}) and by using the ε-NTU solution (ε_{limit}) as a function of mass flow rate for various values of the number of plates. Notice that the discrepancy between ε_{pp} and ε_{limit} becomes larger as either the number of plates are reduced (which tends to increase the unavoidable, plate related temperature difference) or the mass flow rate is reduced (which tends to decrease the pinch point temperature difference).

The impact of axial conduction through the perforated plate heat exchanger is accounted for by computing the dimensionless axial conduction parameter associated with the entire stack of plates, λ. The cross-sectional area for axial conduction through the spacers is:

$$A_{c,sp} = \frac{\pi}{4}\left[D_{out}^2 - (D_{out} - 2w_{sp})^2\right] + (D_{out} - 2w_{sp})w_{sp} \qquad (8\text{-}269)$$

The resistance to axial conduction through the heat exchanger is only related to the resistance of the spacers (recall that the plates are assumed to be inifinitely conductive):

$$R_{ac} = \frac{(N_p + 2)th_{sp}}{k_{sp}A_{c,sp}} \qquad (8\text{-}270)$$

The axial conduction parameter is:

$$\lambda = \frac{1}{R_{ac}\dot{C}_{min}} \qquad (8\text{-}271)$$

> A_c_sp=pi*(D_out^2-(D_out-2*w_sp)^2)/4+(D_out-2*w_sp)*w_sp
>   "cross-sectional area of spacer"
>
> R_ac=(N_p+2)*{t\kern.5pt h}_sp/(k_sp*A_c_sp)   "resistance to axial conduction"
> lambda=1/(R_ac*C_dot_min)   "axial conduction parameter"

which leads to $\lambda = 0.025$. Because λ is much less than unity, it is appropriate to use the low λ model that is discussed in Section 8.7.2 and given by Eq. (8-228):

$$\varepsilon = \varepsilon_{pp} - \lambda \qquad (8\text{-}272)$$

> eff=eff_pp-lambda   "effectiveness"

which leads to $\varepsilon = 0.9184$.

8.9 Numerical Modeling of Cross-Flow Heat Exchangers

8.9.1 Introduction

Cross-flow heat exchangers are discussed in Section 8.1 and classified according to how well the fluid can mix laterally as it passes through the heat exchanger. The analytical solutions for the performance of cross-flow heat exchangers in the limit of constant properties and uniformly distributed conductance are presented as ε-NTU solutions in Table 8-1 and Table 8-2. In this section, the behavior of the cross-flow heat exchanger configuration is predicted using numerical models. These numerical models allow non-uniform properties and other effects that are not included in the ε-NTU solutions to be considered. In Section 8.9.2, a finite difference approach is utilized to predict the temperature distribution within a cross-flow heat exchanger when both fluids are unmixed with constant and with temperature-dependent properties. The website associated with this text (www.cambridge.org/nellisandklein) includes the derivation of a numerical model for the cases where one fluid is mixed and the other unmixed and where both fluids are mixed.

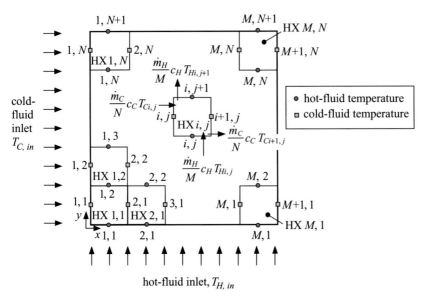

Figure 8-58: Cross-flow heat exchanger. The heat exchanger control volumes are labeled HX i,j and the location of the hot and cold-fluid temperature nodes are shown.

8.9.2 Finite Difference Solution

Both Fluids Unmixed with Uniform Properties

Figure 8-58 illustrates a cross-flow heat exchanger where the cold-fluid flows from left to right (in the x-direction) and the hot-fluid flows from bottom to top (in the y-direction). The solution presented in this section assumes that both fluids have constant properties. The hot-fluid enters with temperature $T_{H,in} = 400$ K with mass flow rate $\dot{m}_H = 0.05$ kg/s and specific heat capacity $c_H = 1005$ J/kg-K. The cold-fluid enters with temperature $T_{C,in} = 300$ K with mass flow rate $\dot{m}_C = 0.05$ kg/s and specific heat capacity $c_C = 1005$ J/kg-K. The total conductance of the heat exchanger is $UA = 200$ W/K. The solution is implemented in MATLAB and these inputs are entered:

```
clear all;
m_dot_H=0.05;                           % hot-side mass flow rate (kg/s)
m_dot_C=0.05;                           % cold-side mass flow rate (kg/s)
c_H=1005;                               % hot-side specific heat capacity (J/kg-K)
c_C=1005;                               % cold-side specific heat capacity (J/kg-K)
T_H_in=400;                             % hot-side inlet temperature (K)
T_C_in=300;                             % cold-side inlet temperature (K)
UA=200;                                 % conductance (W/K)
```

The solution where both fluids are unmixed corresponds to a heat exchanger geometry that includes barriers to prevent mixing in the direction lateral to the fluid flow. In Figure 8-58, the cold-fluid is not allowed to mix in the y-direction and the hot-fluid is not allowed to mix in the x-direction. In an unmixed situation, the fluid can support a temperature gradient in the direction that is perpendicular to its flow direction. For example, in Figure 8-58 the cold-fluid temperature may vary in the y-direction at any x-location. Therefore, the temperature distribution for both fluids will be 2-D, varying in both the flow direction and perpendicular to the flow direction.

The cross-flow heat exchanger with both fluids unmixed must be discretized in both directions for both fluids, as shown in Figure 8-58. The heat exchanger is divided into M divisions in the x-direction and N divisions in the y-direction. The locations of the nodes (the positions where the fluid temperatures will be predicted) are also shown in Figure 8-58. Note that the hot- and cold-fluid nodes are not positioned at coincident locations; instead, the nodes are positioned at the center of the face where the fluid enters and exits for convenience when writing the energy balances.

The dimensionless positions of the hot-side fluid nodes are:

$$\tilde{x}_{H\,i,j} = \frac{1}{M}\left(i - \frac{1}{2}\right) \quad \text{for } i = 1..M \quad \text{for } j = 1..(N+1) \tag{8-273}$$

$$\tilde{y}_{H\,i,j} = \frac{1}{N}(j - 1) \quad \text{for } i = 1..M \quad \text{for } j = 1..(N+1) \tag{8-274}$$

and the dimensionless positions of the cold-side fluid nodes are:

$$\tilde{x}_{C\,i,j} = \frac{1}{M}(i - 1) \quad \text{for } i = 1..(M+1) \quad \text{for } j = 1..N \tag{8-275}$$

$$\tilde{y}_{C\,i,j} = \frac{1}{N}\left(j - \frac{1}{2}\right) \quad \text{for } i = 1..(M+1) \quad \text{for } j = 1..N \tag{8-276}$$

```
M=20;                          % number of grids in the cold-flow direction, x (-)
N=20;                          % number of grids in the hot-flow direction, y (-)

% grid locations for hot-side temperatures
for i=1:M
  for j=1:(N+1)
    x_bar_H(i,j)=(i-1/2)/M;
    y_bar_H(i,j)=(j-1)/N;
  end
end

% grid locations for cold-side temperatures
for i=1:(M+1)
  for j=1:N
    x_bar_C(i,j)=(i-1)/M;
    y_bar_C(i,j)=(j-1/2)/N;
  end
end
```

The cross-flow heat exchanger is modeled using energy balances on the hot and cold-fluids passing through each of the heat exchanger control volumes. The hot-fluid energy balance for an arbitrary heat exchanger control volume, HX i, j shown in Figure 8-58, can be expressed as:

$$\frac{\dot{m}_H\, c_H}{M}\left(T_{H i,j} - T_{H i,j+1}\right) = \frac{UA}{MN}\left[\frac{(T_{H i,j} + T_{H i,j+1})}{2} - \frac{(T_{C i,j} + T_{C i+1,j})}{2}\right]$$

$$\text{for } i = 1..M, j = 1..N \tag{8-277}$$

The cold-fluid energy balance for HX i, j can be expressed as:

$$\frac{\dot{m}_C c_C}{N} (T_{Ci+1,j} - T_{Ci,j}) = \frac{UA}{MN} \left[\frac{(T_{Hi,j} + T_{Hi,j+1})}{2} - \frac{(T_{Ci,j} + T_{Ci+1,j})}{2} \right]$$

$$\text{for } i = 1..M, j = 1..N \qquad (8\text{-}278)$$

The boundary condition associated with the entering hot-fluid temperature is:

$$T_{Hi,1} = T_{H,in} \quad \text{for } i = 1..M \qquad (8\text{-}279)$$

The boundary condition associated with the entering cold-fluid temperature is:

$$T_{C1,j} = T_{C,in} \quad \text{for } j = 1..N \qquad (8\text{-}280)$$

Equations (8-277) through (8-280) represent $M(N+1) + N(M+1)$ equations in an equal number of unknown temperatures. In order to solve these equation in MATLAB, they must be placed in matrix format:

$$\underline{\underline{A}} \underline{X} = \underline{b} \qquad (8\text{-}281)$$

where $\underline{X}$ is a vector containing the unknown temperatures. One strategy for organizing the unknown temperatures in $\underline{X}$ is:

$$\underline{X} = \begin{bmatrix} X_1 = T_{H1,1} \\ X_2 = T_{H2,1} \\ \cdots \\ X_M = T_{HM,1} \\ X_{M+1} = T_{H1,2} \\ \cdots \\ X_{M(N+1)} = T_{HN+1,M} \\ X_{M(N+1)+1} = T_{C1,1} \\ \cdots \\ X_{M(N+1)+(M+1)N} = T_{CM+1,N} \end{bmatrix} \qquad (8\text{-}282)$$

According to Eq. (8-282), the unknown hot-fluid temperature $T_{Hi,j}$ corresponds to entry $(j-1)M + i$ of $\underline{X}$ and the unknown cold-fluid temperature $T_{Ci,j}$ corresponds to entry $(N+1)M + (j-1)(M+1) + i$ of $\underline{X}$. The governing equations must be placed into the rows of the matrix $\underline{\underline{A}}$; one strategy for organizing these equations is:

$$\underline{\underline{A}} = \begin{bmatrix} \text{row } 1 = \text{hot-fluid energy balance for HX } 1,1 \\ \text{row } 2 = \text{hot-fluid energy balance for HX } 2,1 \\ \cdots \\ \text{row } M = \text{hot-fluid energy balance for HX } M, 1 \\ \text{row } M+1 = \text{hot-fluid energy balance for HX } 1, 2 \\ \cdots \\ \text{row } MN = \text{hot-fluid energy balance for HX } M, N \\ \text{row } MN+1 = \text{cold-fluid energy balance for HX } 1,1 \\ \cdots \\ \text{row } 2\, MN = \text{cold-fluid energy balance for HX } M, N \\ \text{row } 2\, MN+1 = \text{hot-fluid boundary condition for } T_{H1,1} \\ \cdots \\ \text{row } 2\, MN+M = \text{hot-fluid boundary condition for } T_{HM,1} \\ \text{row } 2\, MN+M+1 = \text{cold-fluid boundary condition for } T_{C1,1} \\ \cdots \\ \text{row } 2\, MN+M+N = \text{cold-fluid boundary condition for } T_{C1,N} \end{bmatrix} \qquad (8\text{-}283)$$

According to Eq. (8-283), the hot-side energy balance for HX i, j corresponds to row $(j-1)M+i$ of $\underline{\underline{A}}$ and the cold-side energy balance for HX i, j corresponds to row $MN+(j-1)M+i$ of $\underline{\underline{A}}$. The hot-fluid boundary condition for node $T_{Hi,1}$ corresponds to row $2MN+i$ of $\underline{\underline{A}}$ and the cold-fluid boundary condition for node $T_{C1,j}$ corresponds to row $2MN+M+j$ of $\underline{\underline{A}}$.

Matrix $\underline{\underline{A}}$ and vector $\underline{b}$ are initialized. Note that matrix $\underline{\underline{A}}$ is defined as a sparse matrix and the maximum number of non-zero coefficients (the third argument of the declaration spalloc) is estimated by inspection of Eqs. (8-277) and (8-278).

```
% initialize matrices
A=spalloc(2*M*N+M+N,2*M*N+M+N,4*(2*M*N+M+N));
b=zeros(2*M*N+M+N,1);
```

Equations (8-277) through (8-280) are rearranged so that the coefficients multiplying the unknown temperatures are clear. The hot-fluid energy balances, Eq. (8-277), become:

$$T_{Hi,j} \underbrace{\left[\frac{\dot{m}_H c_H}{M} - \frac{UA}{2MN}\right]}_{A_{(j-1)M+i,(j-1)M+i}} + T_{Hi,j+1} \underbrace{\left[-\frac{\dot{m}_H c_H}{M} - \frac{UA}{2MN}\right]}_{A_{(j-1)M+i,(j+1-1)M+i}}$$

$$+ T_{Ci,j} \underbrace{\left[\frac{UA}{2MN}\right]}_{A_{(j-1)M+i,(N+1)M+(j-1)(M+1)+i}} + T_{Ci+1,j} \underbrace{\left[\frac{UA}{2MN}\right]}_{A_{(j-1)M+i,(N+1)M+(j-1)(M+1)+i+1}} \tag{8-284}$$

$$= 0 \quad \text{for } i=1..M, j=1..N$$

```
% hot-side energy balances
for i=1:M
  for j=1:N
    A((j-1)*M+i,(j-1)*M+i)=m_dot_H*c_H/M-UA/(M*N*2);
    A((j-1)*M+i,(j+1-1)*M+i)=-m_dot_H*c_H/M-UA/(M*N*2);
    A((j-1)*M+i,(N+1)*M+(j-1)*(M+1)+i)=UA/(M*N*2);
    A((j-1)*M+i,(N+1)*M+(j-1)*(M+1)+i+1)=UA/(M*N*2);
  end
end
```

The cold-fluid energy balances, Eq. (8-278), can be expressed as:

$$T_{Ci+1,j} \underbrace{\left[\frac{\dot{m}_C c_C}{N} + \frac{UA}{2MN}\right]}_{A_{MN+(j-1)M+i,(N+1)M+(j-1)(M+1)+i+1}} + T_{Ci,j} \underbrace{\left[-\frac{\dot{m}_C c_C}{N} + \frac{UA}{2MN}\right]}_{A_{MN+(j-1)M+i,(N+1)M+(j-1)(M+1)+i}}$$

$$+ T_{Hi,j} \underbrace{\left[-\frac{UA}{2MN}\right]}_{A_{MN+(j-1)M+i,(j-1)M+i}} + T_{Hi,j+1} \underbrace{\left[-\frac{UA}{2MN}\right]}_{A_{MN+(j-1)M+i,(j+1-1)M+i}} \tag{8-285}$$

$$= 0 \quad \text{for } i=1..M, j=1..N$$

```
% cold-side energy balances
for i=1:M
  for j=1:N
    A(M*N+(j-1)*M+i,(N+1)*M+(j-1)*(M+1)+i+1)=m_dot_C*c_C/N+UA/(2*M*N);
    A(M*N+(j-1)*M+i,(N+1)*M+(j-1)*(M+1)+i)=-m_dot_C*c_C/N+UA/(2*M*N);
    A(M*N+(j-1)*M+i,(j-1)*M+i)=-UA/(2*M*N);
    A(M*N+(j-1)*M+i,(j+1-1)*M+i)=-UA/(2*M*N);
  end
end
```

The hot-side boundary condition, Eq. (8-279), becomes:

$$T_{Hi,1} \underbrace{[1]}_{A_{2MN+i,i}} = \underbrace{T_{H,in}}_{b_{2MN+i}} \quad \text{for } i = 1..M \tag{8-286}$$

```
% hot-side inlet fluid temperature boundary condition
for i=1:M
  A(2*M*N+i,i)=1;
  b(2*M*N+i,1)=T_H_in;
end
```

The cold-side boundary condition, Eq. (8-280), becomes:

$$T_{C1,j} \underbrace{[1]}_{A_{2MN+M+j,(N+1)M+(j-1)(M+1)+1}} = \underbrace{T_{C,in}}_{b_{2MN+M+j}} \quad \text{for } j = 1..N \tag{8-287}$$

```
% cold-side inlet fluid temperature boundary condition
for j=1:N
  A(2*M*N+M+j,(N+1)*M+(j-1)*(M+1)+1)=1;
  b(2*M*N+M+j,1)=T_C_in;
end
```

The solution is obtained and the temperatures are placed into matrices $\underline{\underline{T_H}}$ and $\underline{\underline{T_C}}$ that hold the hot- and cold-fluid temperatures, respectively:

```
X=A\b;
for i=1:M
  for j=1:(N+1)
    T_H(i,j)=X((j-1)*M+i);
  end
end
for i=1:(M+1)
  for j=1:N
    T_C(i,j)=X((N+1)*M+(j-1)*(M+1)+i);
  end
end
```

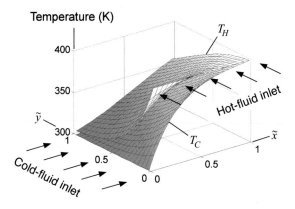

Figure 8-59: Temperature distribution associated with the hot- and cold-fluids for a cross-flow heat exchanger with both fluids unmixed.

It is possible to overlay two surface plots showing the temperature distributions of the hot- and cold-fluids. In the command window, enter:

```
>> hold off;
>> surf(x_bar_H,y_bar_H,T_H);
>> hold on;
>> surf(x_bar_C,y_bar_C,T_C);
```

which, with some formatting, leads to Figure 8-59. Notice that the temperature of both fluids varies in both the x- and y-directions because of the unmixed nature of the heat exchanger.

The numerical solution is implemented in the limit of a uniformly distributed UA and constant fluid properties. Therefore, the predicted performance can be compared directly with the ε-NTU solution discussed in Section 8.3. The total rate of heat transfer from the hot-fluid is computed according to:

$$\dot{q}_H = \frac{\dot{m}_H c_H}{M} \sum_{i=1}^{M} (T_{H,in} - T_{Hi,N+1}) \tag{8-288}$$

```
% total hot-side heat transfer rate
q_dot_H=m_dot_H*c_H*sum(T_H_in-T_H(:,N+1))/M;
```

The total rate of heat transfer to the cold-fluid is computed according to:

$$\dot{q}_C = \frac{\dot{m}_C c_C}{N} \sum_{j=1}^{N} (T_{CM+1,j} - T_{C,in}) \tag{8-289}$$

```
% total cold-side heat transfer rate
q_dot_C=m_dot_C*c_C*sum(T_C(M+1,:)-T_C_in)/N;
```

which leads to $\dot{q}_H = 3.628$ kW and $\dot{q}_C = 3.628$ kW; the agreement of $\dot{q}_H$ and $\dot{q}_C$ provides some verification of the numerical model. The minimum and maximum capacitance rates, $\dot{C}_{min}$ and $\dot{C}_{max}$, are computed:

```
C_dot_min=min([m_dot_H*c_H,m_dot_C*c_C]);        % minimum capacitance rate
C_dot_max=max([m_dot_H*c_H,m_dot_C*c_C]);        % maximum capacitance rate
```

The maximum possible rate of heat transfer is:

$$\dot{q}_{max} = \dot{C}_{min} \ (T_{H,in} - T_{C,in}) \qquad (8\text{-}290)$$

and the effectiveness of the heat exchanger, predicted by the numerical model, is:

$$\varepsilon = \frac{\dot{q}_H}{\dot{q}_{max}} \qquad (8\text{-}291)$$

```
q_dot_max=C_dot_min*(T_H_in-T_C_in);             % maximum possible capacitance rate
eff=q_dot_H/q_dot_max;                           % effectiveness
```

which leads to $\varepsilon = 0.7219$.

The number of transfer units is computed according to:

$$NTU = \frac{UA}{\dot{C}_{min}} \qquad (8\text{-}292)$$

The capacity ratio is:

$$C_R = \frac{\dot{C}_{min}}{\dot{C}_{max}} \qquad (8\text{-}293)$$

```
% eff-NTU solution
NTU=UA/C_dot_min;                                % number of transfer units
C_R=C_dot_min/C_dot_max;                         % capacitance ratio
```

The effectiveness predicted by the $\varepsilon\text{-}NTU$ solution for a cross-flow heat exchanger with both fluids unmixed ($\varepsilon_{crossflow\text{-}unmixed/unmixed}$) is obtained using the formula listed in Table 8-1:

$$\varepsilon_{crossflow\text{-}unmixed/unmixed} = 1 - \exp\left[\frac{NTU^{0.22}}{C_R}\{\exp(-C_R \, NTU^{0.78}) - 1\}\right] \qquad (8\text{-}294)$$

```
eff_crossflowmixedmixed=1-exp(NTU^0.22*(exp(-C_R*NTU^0.78)-1)/C_R);
% solution
```

which leads to $\varepsilon_{crossflow\text{-}unmixed/unmixed} = 0.7229$; this is within 0.2% of the numerical solution.

A numerical solution should be checked for convergence relative to the grid size. The script is made into a function in order to facilitate a study of the grid sensitivity. The header below is added to the top of the script; note that the input to the function is M, the number of heat exchanger control volumes in the x-direction. The number of heat exchanger control volumes in the y-direction (N) is also set equal to M. The single output is the effectiveness predicted by the numerical model

```
function[eff]=S8p9p2A(M)

% Input:
% M - number of heat exchanger control volumes (in both directions) (-)

% Output:
% eff - predicted effectiveness (-)

    % clear all;
    m_dot_H=0.05;        % hot-side mass flow rate (kg/s)
    m_dot_C=0.05;        % cold-side mass flow rate (kg/s)
    c_H=1005;            % hot-side specific heat capacity (J/kg-K)
    c_C=1005;            % cold-side specific heat capacity (J/kg-K)
    T_H_in=400;          % hot-side inlet temperature (K)
    T_C_in=300;          % cold-side inlet temperature (K)
    UA=200;              % conductance (W/K)

    %M=20;               % number of grids in the cold-flow direction, x (-)
    %N=20;               % number of grids in the hot-flow direction, y (-)
    N=M;
```

The script below parametrically varies the grid size and records the effectiveness predicted by the model.

```
clear all;
M=[1,2,4,7,10,15,20,30,40,50,70,100];
for i=1:12
    [eff(i)]=S8p9p2A(M(i))
end
```

Figure 8-60 illustrates the effectiveness of the heat exchanger predicted by the numerical model as a function of M (with $N = M$). Notice that the solution has converged for M greater than about 10.

Both Fluids Unmixed with Temperature-Dependent Properties

The solution presented in the previous section assumed that both fluids have constant properties. In this section, the solution will be modified in order to model a heat exchanger operating under conditions where the fluids cannot be assumed to have constant heat capacity. The hot-fluid is methane that enters with temperature $T_{H,in} = 350$ K, pressure $p_H = 500$ kPa, and mass flow rate $\dot{m}_H = 1.0$ kg/s. The cold-fluid is isobutane that enters with temperature $T_{C,in} = 150$ K, pressure $p_C = 5.0$ MPa, and mass flow rate $\dot{m}_C = 0.5$ kg/s. The heat exchanger has a total conductance $UA = 5000$ W/K.

Figure 8-61 shows the specific heat capacity of methane at 500 kPa and isobutane at 5.0 MPa, determined from EES, as a function of temperature (note that pressure drop in the heat exchanger is neglected).

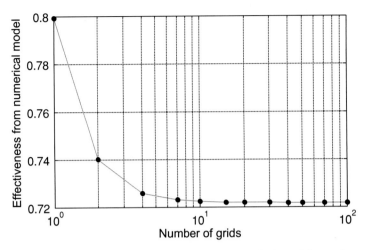

Figure 8-60: Effectiveness predicted by the numerical model as a function of the number of heat exchanger control volumes in the x- and y-directions, $M = N$.

The specific heat capacity of methane is represented in the numerical model by the curve fit (also shown in Figure 8-61):

$$c_H = 4782.92 \left[\frac{J}{kg\text{-}K}\right] - 32.8888 \left[\frac{J}{kg\text{-}K^2}\right] T + 0.148012 \left[\frac{J}{kg\text{-}K^3}\right] T^2$$
$$-0.000287288 \left[\frac{J}{kg\text{-}K^4}\right] T^3 + 2.19732 \times 10^{-7} \left[\frac{J}{kg\text{-}K^5}\right] T^4 \qquad (8\text{-}295)$$

where T is the temperature (in K). A sub-function (c_Hf) is declared at the bottom of the MATLAB function used to implement the numerical model in order to provide the specific heat capacity of the hot-fluid:

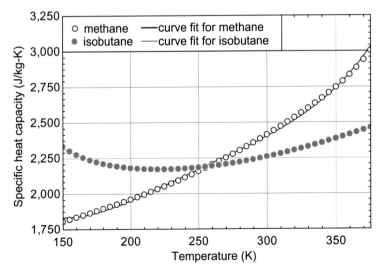

Figure 8-61: Specific heat capacity of methane at 500 kPa and isobutane at 5.0 MPa as a function of temperature. Also shown are the curve fits used in the numerical model.

```
function[c]=c_Hf(T)
% Input:
% T - temperature (K)
%
% Output:
% c - specific heat capacity (J/kg-K)

    c=4782.92-32.8888*T+0.148012*T^2-0.000287288*T^3+2.19732e-7*T^4;
end
```

Figure 8-61 also shows the specific heat capacity of isobutane at 5.0 MPa as a function of temperature; the specific heat capacity is represented in the numerical model by the curve fit:

$$
\begin{aligned}
c_C = 5823.01 \left[\frac{\text{J}}{\text{kg-K}}\right] - 72.4453 \left[\frac{\text{J}}{\text{kg-K}^2}\right] T + 0.463746 \left[\frac{\text{J}}{\text{kg-K}^3}\right] T^2 \\
-0.001240 \left[\frac{\text{J}}{\text{kg-K}^4}\right] T^3 + 1.24158 \times 10^{-6} \left[\frac{\text{J}}{\text{kg-K}^5}\right] T^4
\end{aligned}
\tag{8-296}
$$

where T is the temperature (in K). A sub-function (c_Cf) is declared in order to provide the specific heat capacity of the cold-fluid:

```
function[c]=c_Cf(T)
% Input:
% T - temperature (K)
%
% Output:
% c - specific heat capacity (J/kg-K)

    c=5823.01-72.4453*T+0.463746*T^2-0.00124*T^3+0.00000124158*T^4;
end
```

The inputs are entered at the top of the function:

```
function[x_bar_H,y_bar_H,T_H,x_bar_C,y_bar_C,T_C]=S8p9p2B(M)

% Input:
% M - number of heat exchanger volumes (N=M) (-)
%
% Outputs:
% x_bar_H, y_bar_H - matrices containing dimensionless positions of hot-fluid nodes(-)
% T_H - matrix containing hot-fluid temperatures
% x_bar_C, y_bar_C - matrices containing dimensionless positions of cold-fluid nodes(-)
% T_C - matrix containing cold-fluid temperatures
```

```
m_dot_H=1.0;                        % hot-side mass flow rate (kg/s)
m_dot_C=0.5;                        % cold-side mass flow rate (kg/s)
T_H_in=350;                         % hot-side inlet temperature (K)
T_C_in=150;                         % cold-side inlet temperature (K)
UA=5000;                            % conductance (W/K)
N=M;
```

The positions of each hot- and cold-side fluid node are setup as shown in Figure 8-58 and provided by Eqs. (8-273) through (8-276):

```
% grid locations for hot-side temperatures
for i=1:M
  for j=1:(N+1)
    x_bar_H(i,j)=(i-1/2)/M;
    y_bar_H(i,j)=(j-1)/N;
  end
end
% grid locations for cold-side temperatures
for i=1:(M+1)
  for j=1:N
    x_bar_C(i,j)=(i-1)/M;
    y_bar_C(i,j)=(j-1/2)/N;
  end
end
```

The matrix $\underline{A}$ and vector $\underline{b}$ are initialized as before:

```
% initialize matrices
A=spalloc(2*M*N+M+N,2*M*N+M+N,4*(2*M*N+M+N));
b=zeros(2*M*N+M+N,1);
```

The effect of the non-uniform specific heat capacity of the fluids will be accounted for using a successive substitution solution method; this technique is also discussed in Section 1.5.6 in the context of conduction problems where the conductivity of the material is a strong function of temperature. The hot-side energy balances previously provided by Eq. (8-277) are written below, accounting for the temperature dependent specific heat capacity of the hot-fluid:

$$\frac{\dot{m}_H\, c_{H,\,T=(T_{Hi,j}+T_{Hi,j+1})/2}}{M}\left(T_{Hi,j}-T_{Hi,j+1}\right) = \frac{UA}{MN}\left[\frac{(T_{Hi,j}+T_{Hi,j+1})}{2}-\frac{(T_{Ci,j}+T_{Ci+1,j})}{2}\right]$$

$$\text{for } i=1..M, \quad j=1..N$$

$$(8\text{-}297)$$

Equation (8-297) is rearranged:

$$T_{Hi,j}\left[\frac{\dot{m}_H\, c_{H,\,T=(T_{Hi,j}+T_{Hi,j+1})/2}}{M}-\frac{UA}{2MN}\right]+T_{Hi,j+1}\left[-\frac{\dot{m}_H\, c_{H,,\,T=(T_{Hi,j}+T_{Hi,j+1})/2}}{M}-\frac{UA}{2MN}\right]$$

$$+T_{Ci,j}\left[\frac{UA}{2MN}\right]+T_{Ci+1,j}\left[\frac{UA}{2MN}\right]=0 \quad \text{for } i=1..M, \quad j=1..N \tag{8-298}$$

The equations represented by Eq. (8-298) cannot be solved directly through matrix manipulation because they are non-linear. In order to apply successive substitution, the specific heat capacities in Eq. (8-298) are evaluated using a set of guess values for each of the temperatures ($\hat{T}_{Hi,j}$ and $\hat{T}_{Ci,j}$). Here, we will start by assigning the guess value for each of these temperatures as the average of the hot inlet and cold inlet temperature:

$$\hat{T}_{Hi,j}=\frac{(T_{H,in}+T_{C,in})}{2} \quad \text{for } i=1..M \quad \text{for } j=1..(N+1) \tag{8-299}$$

$$\hat{T}_{Ci,j}=\frac{(T_{H,in}+T_{C,in})}{2} \quad \text{for } i=1..(M+1) \quad \text{for } j=1..N \tag{8-300}$$

```
% initial guess values for the temperature distributions
T_H_g=((T_H_in+T_C_in)/2)*ones(M,N+1);
T_C_g=((T_H_in+T_C_in)/2)*ones(M+1,N);
```

Equation (8-298) is rewritten, using the guess values of the temperatures to evaluate the specific heat capacities:

$$T_{Hi,j}\underbrace{\left[\frac{\dot{m}_H\, c_{H,\,T=(\hat{T}_{Hi,j}+\hat{T}_{Hi,j+1})/2}}{M}-\frac{UA}{2MN}\right]}_{A_{(j-1)M+i,(j-1)M+i}}+T_{Hi,j+1}\underbrace{\left[-\frac{\dot{m}_H\, c_{H,\,T=(\hat{T}_{Hi,j}+\hat{T}_{Hi,j+1})/2}}{M}-\frac{UA}{2MN}\right]}_{A_{(j-1)M+i,(j+1-1)M+i}}$$

$$+T_{Ci,j}\underbrace{\left[\frac{UA}{2MN}\right]}_{A_{(j-1)M+i,(N+1)M+(j-1)(M+1)+i}}+T_{Ci+1,j}\underbrace{\left[\frac{UA}{2MN}\right]}_{A_{(j-1)M+i,(N+1)M+(j-1)(M+1)+i+1}}=0 \tag{8-301}$$

$$\text{for } i=1..M, \quad j=1..N$$

```
% hot-side energy balances
for i=1:M
  for j=1:N
    A((j-1)*M+i,(j-1)*M+i)=m_dot_H*c_Hf((T_H_g(i,j)+T_H_g(i,j+1))/2)/M-UA/(M*N*2);
    A((j-1)*M+i,(j+1-1)*M+i)=-m_dot_H*c_Hf((T_H_g(i,j)+T_H_g(i,j+1))/2)/M-UA/(M*N*2);
    A((j-1)*M+i,(N+1)*M+(j-1)*(M+1)+i)=UA/(M*N*2);
    A((j-1)*M+i,(N+1)*M+(j-1)*(M+1)+i+1)=UA/(M*N*2);
  end
end
```

Note that the MATLAB code above is the same as the code discussed previously for the constant property model, except that the specific heat capacity is evaluated using the sub-function c_Hf at the guess temperatures (i.e., the code indicated in bold has changed).

A similar process is carried out for the cold-fluid energy balances in Eq. (8-278), leading to:

$$
T_{Ci+1,j} \underbrace{\left[\frac{\dot{m}_C c_{C,\,T=(\hat{T}_{Ci+1,j}+\hat{T}_{Ci,j})}}{N} + \frac{UA}{2MN} \right]}_{A_{MN+(j-1)M+i,(N+1)M+(j-1)(M+1)+i+1}} + T_{Ci,j} \underbrace{\left[-\frac{\dot{m}_C c_{C,\,T=(\hat{T}_{Ci+1,j}+\hat{T}_{Ci,j})}}{N} + \frac{UA}{2MN} \right]}_{A_{MN+(j-1)M+i,(N+1)M+(j-1)(M+1)+i}}
$$

$$
+ T_{Hi,j} \underbrace{\left[-\frac{UA}{2MN} \right]}_{A_{MN+(j-1)M+i,(j-1)M+i}} + T_{Hi,j+1} \underbrace{\left[-\frac{UA}{2MN} \right]}_{A_{MN+(j-1)M+i,(j+1-1)M+i}} = 0
$$

$$ \text{(8-302)} $$

$$ \text{for } i = 1..M, \quad j = 1..N $$

```
% cold-side energy balances
for i=1:M
    for j=1:N
        A(M*N+(j-1)*M+i,(N+1)*M+(j-1)*(M+1)+i+1)=m_dot_C*c_Cf((T_C_g(i+1,j)+...
            T_C_g(i,j))/2)/N+UA/(2*M*N);
        A(M*N+(j-1)*M+i,(N+1)*M+(j-1)*(M+1)+i)=-m_dot_C*c_Cf((T_C_g(i+1,j)+...
            T_C_g(i,j))/2)/N+UA/(2*M*N);
        A(M*N+(j-1)*M+i,(j-1)*M+i)=-UA/(2*M*N);
        A(M*N+(j-1)*M+i,(j+1-1)*M+i)=-UA/(2*M*N);
    end
end
```

The hot and cold-fluid boundary conditions provided by Eqs. (8-286) and (8-287) are unchanged:

```
% hot-side inlet fluid temperature boundary condition
for i=1:M
  A(2*M*N+i,i)=1;
  b(2*M*N+i,1)=T_H_in;
end
```

```
% cold-side inlet fluid temperature boundary condition
for j=1:N
  A(2*M*N+M+j,(N+1)*M+(j-1)*(M+1)+1)=1;
  b(2*M*N+M+j,1)=T_C_in;
end
```

A solution is obtained and placed into the temperature matrices:

```
X=A\b;
for i=1:M
  for j=1:(N+1)
    T_H(i,j)=X((j-1)*M+i);
  end
end
for i=1:(M+1)
  for j=1:N
    T_C(i,j)=X((N+1)*M+(j-1)*(M+1)+i);
  end
end
```

The rms error between the solution and the most recent guess values (*err*) is computed:

$$err = \sqrt{ \frac{1}{M\,(N+1)} \sum_{i=1}^{M} \sum_{j=1}^{N+1} (T_{H\,i,j} - \hat{T}_{H\,i,j})^2 + \frac{1}{(M+1)\,N} \sum_{i=1}^{M+1} \sum_{j=1}^{N} (T_{C\,i,j} - \hat{T}_{C\,i,j})^2 }$$

(8-303)

```
err=0;
for i=1:M
  for j=1:(N+1)
    err=err+(T_H(i,j)-T_H_g(i,j))^2/(M*(N+1));
  end
end
for i=1:(M+1)
  for j=1:N
    err=err+(T_C(i,j)-T_C_g(i,j))^2/((M+1)*N);
  end
end
  err=sqrt(err)
end
```

To start the iteration process, the value of the variable err is set to a value that is larger than the variable tol, the tolerance criterion used to terminate the successive substitution process, in order to ensure that the while loop executes at least once. After the solution has been obtained, the rms error is computed and the guess temperatures are reset according to the most recent solution. The resulting code is shown below with the new lines highlighted in bold.

```
function[x_bar_H,y_bar_H,T_H,x_bar_C,y_bar_C,T_C]=S8p9p2B(M)

% Input:
% M - number of heat exchanger volumes (N=M) (-)
%
% Outputs:
% x_bar_H, y_bar_H - matrices containing dimensionless positions of hot-fluid nodes (-)
% T_H - matrix containing hot-fluid temperatures
% x_bar_C, y_bar_C - matrices containing dimensionless positions of cold-fluid nodes (-)
% T_C - matrix containing cold-fluid temperatures

    m_dot_H=1.0;           % hot-side mass flow rate (kg/s)
    m_dot_C=0.5;           % cold-side mass flow rate (kg/s)
    T_H_in=350;            % hot-side inlet temperature (K)
    T_C_in=150;            % cold-side inlet temperature (K)
    UA=5000;               % conductance (W/K)
    N=M;

    % grid locations for hot-side temperatures
    for i=1:M
        for j=1:(N+1)
            x_bar_H(i,j)=(i-1/2)/M;
            y_bar_H(i,j)=(j-1)/N;
        end
    end
```

```
% grid locations for cold-side temperatures
for i=1:(M+1)
  for j=1:N
    x_bar_C(i,j)=(i-1)/M;
    y_bar_C(i,j)=(j-1/2)/N;
  end
end

% initialize matrices
A=spalloc(2*M*N+M+N,2*M*N+M+N,4*(2*M*N+M+N));
b=zeros(2*M*N+M+N,1);

%initial guess values for the temperature distributions
T_H_g=((T_H_in+T_C_in)/2)*ones(M,N+1);
T_C_g=((T_H_in+T_C_in)/2)*ones(M+1,N);

tol=0.01;  %tolerance used to terminate the solution
err=tol+1; %rms error - initially set >tol so that while loop executes
while(err>tol)
  % hot-side energy balances
  for i=1:M
    for j=1:N
      A((j-1)*M+i,(j-1)*M+i)=m_dot_H*c_Hf((T_H_g(i,j)+T_H_g(i,j+1))/2)/M-UA/(M*N*2);
      A((j-1)*M+i,(j+1-1)*M+i)=-m_dot_H*c_Hf((T_H_g(i,j)+T_H_g(i,j+1))/2)/M-UA/(M*N*2);
      A((j-1)*M+i,(N+1)*M+(j-1)*(M+1)+i)=UA/(M*N*2);
      A((j-1)*M+i,(N+1)*M+(j-1)*(M+1)+i+1)=UA/(M*N*2);
    end
  end

  % cold-side energy balances
  for i=1:M
    for j=1:N
      A(M*N+(j-1)*M+i,(N+1)*M+(j-1)*(M+1)+i+1)=m_dot_C*...
        c_Cf((T_C_g(i+1,j)+T_C_g(i,j))/2)/N+UA/(2*M*N);
      A(M*N+(j-1)*M+i,(N+1)*M+(j-1)*(M+1)+i)=-m_dot_C*...
        c_Cf((T_C_g(i+1,j)+T_C_g(i,j))/2)/N+UA/(2*M*N);
      A(M*N+(j-1)*M+i,(j-1)*M+i)=-UA/(2*M*N);
      A(M*N+(j-1)*M+i,(j+1-1)*M+i)=-UA/(2*M*N);
    end
  end

  % hot-side inlet fluid temperature boundary condition
  for i=1:M
    A(2*M*N+i,i)=1;
    b(2*M*N+i,1)=T_H_in;
  end

  % cold-side inlet fluid temperature boundary condition
  for j=1:N
    A(2*M*N+M+j,(N+1)*M+(j-1)*(M+1)+1)=1;
    b(2*M*N+M+j,1)=T_C_in;
  end
```

```
X=A\b;
for i=1:M
  for j=1:(N+1)
    T_H(i,j)=X((j-1)*M+i);
  end
end
for i=1:(M+1)
  for j=1:N
    T_C(i,j)=X((N+1)*M+(j-1)*(M+1)+i);
  end
end

% compute rms error
err=0;
for i=1:M
  for j=1:(N+1)
    err=err+(T_H(i,j)-T_H_g(i,j))^2/(M*(N+1));
  end
end
for i=1:(M+1)
  for j=1:N
    err=err+(T_C(i,j)-T_C_g(i,j))^2/((M+1)*N);
  end
end
err=sqrt(err)
% replace guess values with solution
T_H_g=T_H;
T_C_g=T_C;
  end
end
```

Executing the function at the command line provides the rms error each time the function iterates (notice that the calculation of the variable **err** was not terminated in a semicolon):

```
>> [x_bar_H,y_bar_H,T_H,x_bar_C,y_bar_C,T_C]=S8p9p2B(20);
err =
   91.6980
err =
   7.2351
err =
   0.8409
err =
   0.0889
err =
   0.0072
```

The total rate of heat transfer from the hot-fluid is computed according to:

$$\dot{q}_H = \sum_{i=1}^{M}\sum_{j=1}^{N} \frac{\dot{m}_H c_{H,T=(T_{H\,i,j+1}+T_{H\,i,j})/2}\,(T_{H\,i,j}-T_{H\,i,j+1})}{M} \tag{8-304}$$

```
% total hot-side heat transfer rate
q_dot_H=0;
for i=1:M
  for j=1:N
    q_dot_H=q_dot_H+m_dot_H*c_Hf((T_H(i,j+1)+T_H(i,j))/2)*(T_H(i,j)-T_H(i,j+1))/M;
  end
end
```

which leads to $\dot{q}_H = 1.895 \times 10^5$ W. The total heat transfer rate to the cold-fluid is computed according to:

$$\dot{q}_C = \sum_{j=1}^{N}\sum_{i=1}^{M} \frac{\dot{m}_C c_{C,T=(T_{C\,i+1,j}+T_{C\,i,j})/2}\,(T_{C\,i+1,j}-T_{C\,i,j})}{N} \tag{8-305}$$

```
% total cold-side heat transfer rate
q_dot_C=0;
for j=1:N
  for i=1:M
    q_dot_C=q_dot_C+m_dot_C*c_Cf((T_C(i+1,j)+T_C(i,j))/2)*(T_C(i+1,j)-T_C(i,j))/N;
  end
end
```

which leads to $\dot{q}_C = 1.895 \times 10^5$ W. The agreement of these two values provides some verification of the model.

One Fluid Mixed, One Fluid Unmixed
This extended section can be found on the website www.cambridge.org/nellisandklein. This section presents the numerical model for a cross-flow heat exchanger in which the cold-fluid is mixed and the hot-fluid remains unmixed. In this situation, the temperature distribution associated with the hot-fluid remains 2-D but the temperature distribution associated with the cold-fluid becomes 1-D because mixing prevents any temperature gradient in the y-direction (i.e., the direction perpendicular to the flow). Therefore, a single cold-fluid temperature node ($T_{C\,i}$) represents the temperature for each x-location. This leads to differences in the spacing of the nodes and the algebraic equations associated with the energy balances.

Both Fluids Mixed
This extended section can be found on the website www.cambridge.org/nellisandklein. This section presents the numerical model for a cross-flow heat exchanger in which the both fluids are mixed. In this situation, the temperature distribution associated with the hot and cold-fluids are both 1-D.

8.10 Regenerators

8.10.1 Introduction

A regenerator is a type of heat exchanger in which the hot and cold-fluids occupy the same physical space, but at different times. The physical space is the called the heat exchanger "core," "bed," or "matrix"; it consists of a packed bed of discrete particles or channels within a solid material that provide high surface area and therefore allow good thermal contact between the flowing fluid and the solid material. The heat transfer from the hot-fluid to the cold-fluid in a regenerator does not occur directly through some separating wall, as in the heat exchangers studied in the earlier sections of this chapter. Rather, the heat transfer occurs indirectly. The hot-fluid transfers energy to the solid particles or matrix that makes up the regenerator bed for some period of time. The regenerator matrix material stores this energy until it is exposed to the cold-fluid, at which time the energy is transferred to the cold-fluid. Regenerators are therefore transient devices from the point of the view of the matrix material that is periodically exposed to hot and cold-fluid.

The two basic regenerator designs can be categorized as stationary (or fixed-bed) and rotary. A stationary regenerator requires valves that direct pressurized hot and cold-fluids to alternately flow through the matrix, as shown in Figure 8-65(a). The matrix is warmed as the hot-fluid passes through it during the hot-to-cold blow process that is shown on the left side of Figure 8-65(a). After a pre-determined time period, the valves controlling fluid flow are switched in order to terminate the flow of hot-fluid and initiate the flow of cold-fluid during the cold-to-hot blow process that is shown on the right side of Figure 8-65(a). After some time, the flow of cold-fluid is terminated and the process is repeated. Once the initial transients have decayed, the system achieves a periodic steady-state condition where the temperature distribution at the beginning of the hot-to-cold blow process exactly matches the temperature distribution at the conclusion of the cold-to-hot blow process. A continuous flow of hot and cold-fluid cannot be obtained from a stationary regenerator having a single bed, shown in Figure 8-65(a). However, a stationary regenerator system consisting of two beds, shown in Figure 8-65(b), allows continuous flow of the hot and cold-fluids while maintaining periodic flow in each of the beds.

In contrast to a stationary regenerator, the fluids in a rotary regenerator system flow continuously in one direction while the matrix material rotates, as shown in Figure 8-66. Systems of this type are common for building ventilation systems. The analysis methods that are used for rotary regenerators are similar to those used for stationary regenerators.

Regenerators offer some advantages relative to more conventional counter flow heat exchangers. One benefit is cost; the heat transfer matrix used in a regenerator can be as simple as a bed of particles. This design is a less expensive alternative than, for example, the plate heat exchanger shown in Figure 8-2 that requires elaborate plate geometries, gaskets and headers. The heat transfer passages within a regenerator can be made extremely small and therefore the ratio of the surface area within a bed to the volume of the bed (sometimes referred to as the compactness of the heat exchanger) can be very high. As a result, it is possible to obtain high effectiveness in a cheap and compact device. This is particularly important when both fluids are gases. A regenerator can often provide a specified heat transfer rate with a smaller volume and lower cost than a conventional counter- or cross-flow heat exchanger. Examples of applications that use regenerators for this reason are air-preheating systems in power-plants, building ventilation systems, and gas turbine energy recovery units. Regenerators are also

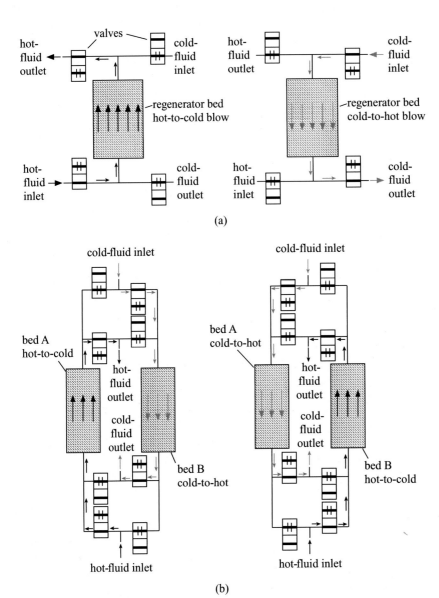

(a)

(b)

Figure 8-65: Schematic of a (a) single bed and (b) dual bed, stationary regenerator system.

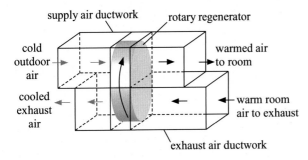

Figure 8-66: Schematic of a rotary regenerator system for a building ventilation application.

used in regenerative refrigeration cycles that operate at cryogenic temperatures where the working fluid is necessarily a gas (to avoid freezing) and where very high effectiveness is required in order for the cycle to function at all. Applications of regenerators in cryogenic refrigeration systems include pulse-tube, Gifford-McMahon, Stirling, and magnetic cycles. Regenerators are also central to the operation of Stirling heat engines.

Section 8.10.2 derives the partial differential equations that govern a regenerator. In Section 8.10.3, the specific case of a balanced and symmetric regenerator is considered; this is the most common case that is encountered in practice. The governing differential equations for the balanced, symmetric regenerator are made dimensionless in order to ascertain the dimensionless groups that govern the solution (number of transfer units and utilization). The concept of the regenerator effectiveness is introduced and the solution for the balanced, symmetric regenerator is presented graphically and made accessible as an EES function. In Section 8.10.4, the correlations that are required in order to estimate the thermal-fluid characteristics of typical regenerator matrix configurations are presented. These correlations are made accessible through charts and functions in EES. A numerical model of a regenerator is presented in Section 8.10.5 and used to examine regenerator behavior under different operating conditions; this numerical solution is also accessible as a procedure in EES.

8.10.2 Governing Equations

Regenerators necessarily operate in a transient mode; the temperatures throughout the system depend upon time as well as on position. Thus, the analysis of regenerators is more complicated than the steady state analysis of the heat exchangers discussed in Sections 8.1 through 8.9. Periodic steady-state refers to the operating condition where the regenerator material anywhere goes through a cycle; that is, the material at any location returns to the same temperature at the end of the cold-to-hot blow process that it started with at the beginning of the hot-to-cold blow process. Given sufficient time, a regenerator will reach a periodic steady-state operating condition and its performance at this condition is the quantity of interest in most regenerator analyses. Partial differential equations are needed to describe the time and spatial dependence of the temperature of the regenerator and the fluid. These equations can be identified using energy balances on the fluid and matrix material in a differential section of a regenerator, as shown in Figure 8-67.

The derivation that follows assumes constant matrix and fluid properties, negligible viscous dissipation, and no axial conduction in the fluid or matrix. Further, the temperature distribution is assumed to be only a function of x and t, where x is the distance in the flow direction measured from the hot inlet of the bed and t is time. The regenerator operation consists of two processes. During the hot-to-cold flow process (Figure 8-67(a)), a constant mass flow rate of hot-fluid ($\dot{m}_{HTCB}$) enters the matrix from the hot end (at $x = 0$) with temperature $T_{H,in}$. The duration of the hot-to-cold blow process is t_{HTCB}. During the cold-to-hot flow process (Figure 8-67(b)), a constant mass flow rate of cold-fluid ($\dot{m}_{CTHB}$) enters the matrix from the cold end (at $x = L$) with temperature $T_{C,in}$. The duration of the cold-to-hot blow process is t_{CTHB}.

An energy balance on the hot flow in the differential regenerator segment that is shown in Figure 8-67(a) leads to:

$$\dot{m}_{HTCB}\, c_f\, T_{f,x} = \dot{m}_{HTCB}\, c_f\, T_{f,x+dx} + \rho_f\, c_f\, V_f\, \frac{dx}{L}\, \frac{\partial T_f}{\partial t} + d\dot{q}_{HTCB} \quad \text{for } 0 < t < t_{HTCB}$$

$$(8\text{-}328)$$

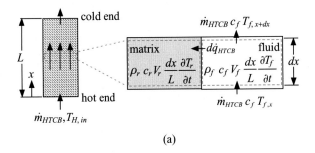

(a)

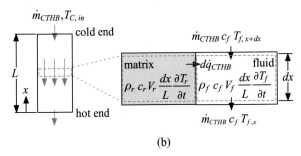

(b)

Figure 8-67: Differential section of a regenerator during the (a) hot-to-cold blow process and (b) the cold-to-hot blow process.

where c_f and ρ_f are the specific heat capacity and density of the fluid, respectively, T_f is the fluid temperature, V_f is the total volume of the regenerator bed that is occupied by fluid (i.e., the volume that is not occupied by the regenerator matrix, V_f is sometimes referred to as the pore volume or the dead volume), L is the length of the regenerator, and $d\dot{q}_{HTCB}$ is the rate of heat transfer from the fluid to the matrix within the control volume during the hot-to-cold blow process. The $x + dx$ term in Eq. (8-328) is expanded and the equation is simplified:

$$0 = \dot{m}_{HTCB}\, c_f\, \frac{\partial T_f}{\partial x}\, dx + \rho_f\, c_f\, V_f\, \frac{dx}{L}\, \frac{\partial T_f}{\partial t} + d\dot{q}_{HTCB} \quad \text{for } 0 < t < t_{HTCB} \qquad (8\text{-}329)$$

The rate of heat transfer between the fluid and the matrix during the hot-to-cold blow process is:

$$d\dot{q}_{HTCB} = h_{HTCB}\, A_s\, \frac{dx}{L}\, (T_f - T_r) \qquad (8\text{-}330)$$

where h_{HTCB} is the heat transfer coefficient during the hot-to-cold blow process, A_s is the total surface area of the regenerator matrix exposed to the fluid, and T_r is the regenerator matrix temperature. Substituting Eq. (8-330) into Eq. (8-329) and simplifying leads to:

$$0 = \underbrace{\dot{m}_{HTCB}\, c_f\, \frac{\partial T_f}{\partial x}}_{\text{enthalpy change of fluid}} + \underbrace{\frac{\rho_f\, c_f\, V_f}{L}\, \frac{\partial T_f}{\partial t}}_{\text{energy stored in fluid}} + \underbrace{\frac{h_{HTCB}\, A_s}{L}\, (T_f - T_r)}_{\text{energy transferred to matrix}} \quad \text{for } 0 < t < t_{HTCB}$$

$$(8\text{-}331)$$

The second term on the right side of Eq. (8-331) corresponds to the energy stored by the fluid in the pore volume (i.e., the entrained fluid) and is usually negligible, particularly when a gaseous working fluid is used. However, this term can be significant when the working fluid is a liquid.

An energy balance on the matrix material in the differential control volume during the hot-to-cold blow process leads to:

$$d\dot{q}_{HTCB} = \rho_r c_r V_r \frac{dx}{L} \frac{\partial T_r}{\partial t} \quad \text{for } 0 < t < t_{HTCB} \tag{8-332}$$

where ρ_r and c_r are the density and specific heat capacity of the regenerator material, respectively, and V_r is the total volume of regenerator material that is present in the regenerator bed (i.e., the actual volume of the solid material that makes up the regenerator). The matrix material is assumed to be lumped; that is, temperature gradients within individual particles that will occur due to conduction into and out of the matrix are neglected. In practice, this behavior is approached by using small particles or thin-walled flow channels. The performance of the regenerator is reduced if the material experiences substantial, local temperature gradients. Substituting Eq. (8-330) into Eq. (8-332) leads to:

$$\underbrace{h_{HTCB} A_s (T_f - T_r)}_{\text{energy transferred from fluid}} = \underbrace{\rho_r c_r V_r \frac{\partial T_r}{\partial t}}_{\text{energy stored in matrix}} \quad \text{for } 0 < t < t_{HTCB} \tag{8-333}$$

A differential energy balance on the fluid and regenerator during the cold-to-hot blow process is shown in Figure 8-67(b). An energy balance on the fluid leads to:

$$\dot{m}_{CTHB} c_f T_{f,x+dx} + d\dot{q}_{CTHB} = \dot{m}_{CTHB} c_f T_{f,x} + \rho_f c_f V_f \frac{dx}{L} \frac{\partial T_f}{\partial t}$$
$$\text{for } t_{HTCB} < t < (t_{HTCB} + t_{CTHB}) \tag{8-334}$$

The heat transfer rate within the control volume is given by:

$$d\dot{q}_{CTHB} = h_{CTHB} A_s \frac{dx}{L} (T_r - T_f) \tag{8-335}$$

where h_{CTHB} is the heat transfer coefficient during the cold-to-hot blow process. Substituting Eq. (8-335) into Eq. (8-334) and simplifying leads to:

$$\dot{m}_{CTHB} c_f \frac{\partial T_f}{\partial x} + \frac{h_{CTHB} A_s}{L} (T_r - T_f) = \frac{\rho_f c_f V_f}{L} \frac{\partial T_f}{\partial t} \quad \text{for } t_{HTCB} < t < (t_{HTCB} + t_{CTHB}) \tag{8-336}$$

An energy balance on the regenerator material leads to:

$$0 = \rho_r c_r V_r \frac{\partial T_r}{\partial t} + h_{CTHB} A_s (T_f - T_r) \quad \text{for } t_{HTCB} < t < (t_{HTCB} + t_{CTHB}) \tag{8-337}$$

Equations (8-331), (8-333), (8-336), and (8-337) are a coupled set of partial differential equations that are first order in time and position for T_f and first order in time for T_r. The boundary conditions required to completely specify the problem include the specified fluid inlet temperatures:

$$T_{f,x=0} = T_{H,in} \quad \text{for } 0 < t < t_{HTCB} \tag{8-338}$$

$$T_{f,x=L} = T_{C,in} \quad \text{for } t_{HTCB} < t < (t_{HTCB} + t_{CTHB}) \tag{8-339}$$

and the requirement of a periodic steady-state operating condition:

$$T_{f,t=0} = T_{f,t=(t_{HTCB}+t_{CTHB})} \tag{8-340}$$

$$T_{r,t=0} = T_{r,t=(t_{HTCB}+t_{CTHB})} \tag{8-341}$$

8.10.3 Balanced, Symmetric Flow with No Entrained Fluid Heat Capacity

The most common operating conditions encountered in practice approximately correspond to a balanced, symmetric flow condition with negligible entrained fluid heat capacity. Balanced flow indicates that the mass flow rates experienced during the hot-to-cold
and cold-to-hot blow processes are the same ($\dot{m}_{HTCB} = \dot{m}_{CTHB} = \dot{m}$) and the duration of
these flow periods are the same ($t_{HTCB} = t_{CTHB} = t_B$). Symmetric flow implies that the
heat transfer coefficient experienced in the hot-to-cold blow period is the same as in
the cold-to-hot blow period ($h_{HTCB} = h_{CTHB} = h$). Finally, if the fluid is a gas then the
thermal capacitance of the entrained fluid ($\rho_f c_f V_f$) is much smaller than the thermal
capacity of the matrix material ($\rho_r c_r V_r$) and therefore the term in the fluid energy balance that corresponds to energy storage in the entrained fluid can be neglected. With
these assumptions, the governing equations derived in Section 8.10.2 become:

$$\dot{m}\,c_f\,\frac{\partial T_f}{\partial x} + \frac{h\,A_s}{L}\,(T_f - T_r) = 0 \quad \text{for } 0 < t < t_B \tag{8-342}$$

$$\dot{m}\,c_f\,\frac{\partial T_f}{\partial x} + \frac{h\,A_s}{L}\,(T_r - T_f) = 0 \quad \text{for } t_B < t < 2\,t_B \tag{8-343}$$

$$0 = \rho_r\,c_r\,V_r\frac{\partial T_r}{\partial t} + h\,A_s\,(T_f - T_r) \quad \text{for } 0 < t < 2\,t_B \tag{8-344}$$

and the boundary conditions become:

$$T_{f,x=0} = T_{H,in} \quad \text{for } 0 < t < t_B \tag{8-345}$$

$$T_{f,x=L} = T_{C,in} \quad \text{for } t_B < t < 2\,t_B \tag{8-346}$$

$$T_{r,t=0} = T_{r,t=2\,t_B} \tag{8-347}$$

Utilization and Number of Transfer Units
The governing equations for a balanced, symmetric regenerator with no entrained fluid
heat capacity are made dimensionless by introducing a dimensionless position:

$$\tilde{x} = \frac{x}{L} \tag{8-348}$$

a dimensionless time:

$$\tilde{t} = \frac{t}{t_B} \tag{8-349}$$

and a dimensionless temperature difference:

$$\tilde{\theta} = \frac{(T - T_{C,in})}{(T_{H,in} - T_{C,in})} \tag{8-350}$$

Substituting Eqs. (8-348) through (8-350) into Eq. (8-342) leads to:

$$0 = \frac{\dot{m}\,c_f\,(T_{H,in} - T_{C,in})}{L}\,\frac{\partial \tilde{\theta}_f}{\partial \tilde{x}} + \frac{h\,A_s\,(T_{H,in} - T_{C,in})}{L}(\tilde{\theta}_f - \tilde{\theta}_r) \quad \text{for } 0 < \tilde{t} < 1 \tag{8-351}$$

Equation (8-351) is divided through by the quantity $\dot{m} c_f (T_{H,in} - T_{C,in})/L$ in order to make it dimensionless:

$$0 = \frac{\partial \tilde{\theta}_f}{\partial \tilde{x}} + \underbrace{\frac{h A_s}{\dot{m} c_f}}_{NTU} (\tilde{\theta}_f - \tilde{\theta}_r) \quad \text{for } 0 < \tilde{t} < 1 \tag{8-352}$$

The number of transfer units is the dimensionless term multiplying the second term in Eq. (8-352):

$$NTU = \frac{h A_s}{\dot{m} c_f} \tag{8-353}$$

which leads to:

$$0 = \frac{\partial \tilde{\theta}_f}{\partial \tilde{x}} + NTU (\tilde{\theta}_f - \tilde{\theta}_r) \quad \text{for } 0 < \tilde{t} < 1 \tag{8-354}$$

A similar process, applied to Eq. (8-343), leads to:

$$\frac{\partial \tilde{\theta}_f}{\partial x} + NTU (\tilde{\theta}_r - \tilde{\theta}_f) = 0 \quad \text{for } 1 < \tilde{t} < 2 \tag{8-355}$$

Substituting Eqs. (8-349) and (8-350) into Eq. (8-344) leads to:

$$0 = \frac{\rho_r c_r V_r (T_{H,in} - T_{C,in})}{t_B} \frac{\partial \tilde{\theta}_r}{\partial \tilde{t}} + h A_s (T_{H,in} - T_{C,in})(\tilde{\theta}_f - \tilde{\theta}_r) \quad \text{for } 0 < \tilde{t} < 2 \tag{8-356}$$

Equation (8-356) is divided through by the quantity $(T_{H,in} - T_{C,in}) \dot{m} c_f$ in order to make it dimensionless:

$$0 = \underbrace{\frac{\rho_r c_r V_r}{\dot{m} c_f t_B}}_{1/U} \frac{\partial \tilde{\theta}_r}{\partial \tilde{t}} + \underbrace{\frac{h A_s}{\dot{m} c_f}}_{NTU} (\tilde{\theta}_f - \tilde{\theta}_r) \quad \text{for } 0 < \tilde{t} < 2 \tag{8-357}$$

The inverse of the dimensionless group multiplying the first term in Eq. (8-357) is referred to as the utilization of the regenerator:

$$U = \frac{\dot{m} c_f t_B}{\rho_r c_r V_r} \tag{8-358}$$

The utilization is the ratio of the total heat capacity of the fluid that flows through the regenerator during one blow process to the total heat capacity of the regenerator material. The regenerator relies on having a large matrix heat capacity so that it can store energy from the fluid without changing temperature substantially. Therefore, a good regenerator design will be characterized by a small utilization as well as a large *NTU*.

Substituting Eqs. (8-353) and (8-358) into Eq. (8-357) leads to:

$$0 = \frac{1}{U} \frac{\partial \tilde{\theta}_r}{\partial \tilde{t}} + NTU (\tilde{\theta}_f - \tilde{\theta}_r) \quad \text{for } 0 < \tilde{t} < 2 \tag{8-359}$$

The boundary conditions, Eqs. (8-345) through (8-347), are expressed in terms of dimensionless variables:

$$\tilde{\theta}_{f,\tilde{x}=0} = 1 \quad \text{for } 0 < \tilde{t} < 1 \tag{8-360}$$

$$\tilde{\theta}_{f,\tilde{x}=1} = 0 \quad \text{for } 1 < \tilde{t} < 2 \tag{8-361}$$

$$\tilde{\theta}_{r,\tilde{t}=0} = \tilde{\theta}_{r,\tilde{t}=2} \tag{8-362}$$

The dimensionless temperature difference in the regenerator matrix and fluid must depend on dimensionless time and dimensionless position as well as on the utilization and number of transfer units:

$$\tilde{\theta}_r = \tilde{\theta}_r \left(\tilde{x}, \tilde{t}, NTU, U \right) \qquad (8\text{-}363)$$

$$\tilde{\theta}_f = \tilde{\theta}_f \left(\tilde{x}, \tilde{t}, NTU, U \right) \qquad (8\text{-}364)$$

Regenerator Effectiveness

The definition of the effectiveness for a regenerator is similar to the definition of the effectiveness for a steady flow heat exchanger, discussed in Section 8.3.3. The effectiveness is the ratio of the actual amount of energy transferred from the hot-fluid and to the cold-fluid (Q) to the maximum possible amount of energy that could be transferred in a perfect regenerator (Q_{max}) during one cycle.

$$\varepsilon = \frac{Q}{Q_{max}} \qquad (8\text{-}365)$$

The actual amount of energy transferred in the regenerator is obtained by carrying out an energy balance on the hot-fluid flowing during the hot-to-cold blow process.

$$Q_H = \int_{t=0}^{t_{HTCB}} \dot{m}_{HTCB} \, c_f \left(T_{H,in} - T_{f,x=L} \right) dt \qquad (8\text{-}366)$$

or on the cold-fluid during the cold-to-hot blow process:

$$Q_C = \int_{t=t_{HTCB}}^{t_{HTCB}+t_{CTHB}} \dot{m}_{CTHB} \, c_f \left(T_{f,x=0} - T_{C,in} \right) dt \qquad (8\text{-}367)$$

Equations (8-366) and (8-367) must be equal if the regenerator is operating in a periodic steady-state condition. The maximum possible amount of energy that can be transferred in the regenerator occurs when the fluid flow during the hot-to-cold blow process exits at the cold inlet temperature and/or the fluid flow during the cold-to-hot blow process exits at the hot inlet temperature. The fluid flow process that is characterized by the smallest total heat capacity ($\dot{m}_{HTCB} \, t_{HTCB} \, c_f$ or $\dot{m}_{CTHB} \, t_{CTHB} \, c_f$) will approach the inlet temperature in the same way that the fluid stream with the smallest capacitance rate establishes the maximum possible heat transfer rate in a steady-flow heat exchanger, as discussed in Section 8.3.2.

$$Q_{max} = \begin{cases} \dot{m}_{HTCB} \, c_f \, t_{HTCB} \left(T_{H,in} - T_{C,in} \right) & \text{if } \dot{m}_{HTCB} \, c_f \, t_{HTCB} < \dot{m}_{CTHB} \, c_f \, t_{CTHB} \\ \dot{m}_{CTHB} \, c_f \, t_{CTHB} \left(T_{H,in} - T_{C,in} \right) & \text{if } \dot{m}_{HTCB} \, c_f \, t_{HTCB} > \dot{m}_{CTHB} \, c_f \, t_{CTHB} \end{cases} \qquad (8\text{-}368)$$

For a balanced symmetric regenerator, the energy transfer from the hot-fluid, Eq. (8-366), can be written as:

$$Q_H = \int_{t=0}^{t_B} \dot{m} \, c_f \left(T_{H,in} - T_{f,x=L} \right) dt \qquad (8\text{-}369)$$

The energy transfer to the cold-fluid, Eq. (8-367), can be written as:

$$Q_C = \int_{t=t_B}^{2\,t_B} \dot{m}\, c_f \left(T_{f,x=0} - T_{C,in}\right) dt \tag{8-370}$$

The maximum possible amount of energy transfer, Eq. (8-368), can be written as:

$$Q_{max} = \dot{m}\, c_f\, t_B \left(T_{H,in} - T_{C,in}\right) \tag{8-371}$$

Substituting Eqs. (8-369) and (8-371) into Eq. (8-365) leads to:

$$\varepsilon = \frac{\displaystyle\int_{t=0}^{t_B} \dot{m}\, c_f \left(T_{H,in} - T_{f,x=L}\right) dt}{\dot{m}\, c_f\, t_B \left(T_{H,in} - T_{C,in}\right)} \tag{8-372}$$

or

$$\varepsilon = \frac{\displaystyle\int_{t=0}^{t_B} \left(T_{H,in} - T_{f,x=L}\right) dt}{t_B \left(T_{H,in} - T_{C,in}\right)} \tag{8-373}$$

Equation (8-373) is expressed in terms of the dimensionless variables $\tilde{\theta}_f$, $\tilde{x}$, and $\tilde{t}$:

$$\varepsilon = \int_{\tilde{t}=0}^{1} \left(1 - \tilde{\theta}_{f,\tilde{x}=1}\right) d\tilde{t} \tag{8-374}$$

According to Eq. (8-364), $\tilde{\theta}_f = \tilde{\theta}_f\left(\tilde{x}, \tilde{t}, NTU, U\right)$. Therefore, the effectiveness of a balanced, symmetric regenerator must be a function of the number of transfer units and utilization; only the fluid temperature at $\tilde{x} = 1$ is required in Eq. (8-374) and the integration removes the $\tilde{t}$ dependence.

$$\varepsilon = \varepsilon\left(NTU, U\right) \tag{8-375}$$

The partial differential equations presented by Eqs. (8-354), (8-355), and (8-359) have been solved exactly and numerically by Dragutinovic and Baclic (1998). A method for implementing a numerical solution to the regenerator equations is presented in Section 8.10.5. The effectiveness of a balanced, symmetric regenerator is shown in Figure 8-68 for various values of the utilization. The solution for a balanced, symmetric regenerator is also available in EES within the heat transfer function library. Select Function Info and Heat Exchangers from the drop-down menu; select NTU-> Effectiveness and scroll to the Regenerator option. The EES function HX was used to generate Figure 8-68. Figure 8-68 shows that a good regenerator (i.e., one with a high effectiveness) must have a large NTU (i.e., the fluid-to-bed thermal resistance must be low) and also have a low utilization (i.e., the heat capacity of the bed must be much larger than the heat capacity of the fluid passing through the bed). Figure 8-69 illustrates $\tilde{\theta}_f$ and $\tilde{\theta}_r$ as a function of $\tilde{t}$ for various values of $\tilde{x}$ for Case A in Figure 8-68 (where $NTU = 25$ and $U = 0.01$); note that Figure 8-69 through Figure 8-71 were generated using the numerical model presented in Section 8.10.5.

Figure 8-69 shows that the fluid-to-regenerator temperature difference is small for Case A because the NTU is high. The regenerator temperature does not change substantially with time during the cycle because the utilization is low. In the limit of $U \to 0$, the regenerator material does not change temperature with time at all and therefore

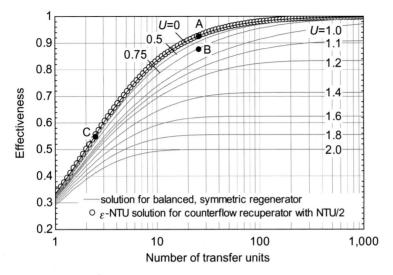

Figure 8-68: Effectiveness of a balanced, symmetric regenerator as a function of NTU for various values of the utilization, U.

the regenerator behavior becomes exactly analogous to that of a steady flow heat exchanger.

In the zero utilization limit, the hot-fluid transfers its energy via convection to the regenerator material at fixed temperature and then the regenerator transfers the energy via convection to the cold-fluid. This is no different from the steady flow heat exchanger where the hot-fluid transfers its energy via convection to the wall separating the hot-fluid from the cold-fluid. The separating wall is at a fixed temperature (with time) and transfers this energy on to the cold-fluid. The thermal behavior of the regenerator in the limit that $U \to 0$ can be predicted using the ε-NTU solution for a counter-flow heat exchanger. The total thermal resistance between the hot-fluid and the cold-fluid in the

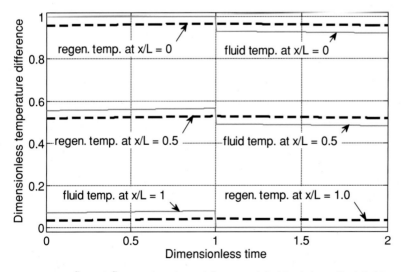

Figure 8-69: $\tilde{\theta}_f$ and $\tilde{\theta}_r$ as a function of $\tilde{t}$ at $\tilde{x} = 0.0$ (the hot end), 0.5 (the mid-point), and 1.0 (the cold end) for a balanced, symmetric regenerator with $NTU = 25$ and $U = 0.01$ (Case A in Figure 8-68).

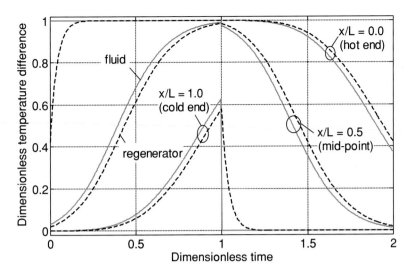

Figure 8-70: $\tilde{\theta}_f$ and $\tilde{\theta}_r$ as a function of $\tilde{t}$ at $\tilde{x} = 0.0$ (the hot end), 0.5 (the mid-point), and 1.0 (the cold end) for a balanced, symmetric regenerator with $NTU = 25$ and $U = 1.0$ (Case B in Figure 8-68).

regenerator is comprised of two convection resistances, one for each of the flow processes:

$$R_{tot} = \underbrace{\frac{1}{h\,A_s}}_{\substack{\text{resistance from hot}\\\text{fluid to regenerator}\\\text{in hot-to-cold blow}}} + \underbrace{\frac{1}{h\,A_s}}_{\substack{\text{resistance from}\\\text{regenerator to cold fluid}\\\text{in cold-to-hot blow}}} \qquad (8\text{-}376)$$

Therefore, the total conductance of an equivalent steady flow heat exchanger is:

$$UA = \frac{1}{R_{tot}} = \frac{h\,A_s}{2} \qquad (8\text{-}377)$$

The number of transfer units associated with the equivalent counter-flow heat exchanger is half of the number of transfer units for the regenerator, calculated using Eq. (8-353):

$$NTU_{cf} = \frac{NTU}{2} \qquad (8\text{-}378)$$

The capacitance ratio for the equivalent steady flow heat exchanger is $C_R = 1.0$ for a balanced regenerator. The ε-NTU solution for a balanced, counterflow heat exchanger (calculated with $NTU/2$, according to Eq. (8-378)) is overlaid onto Figure 8-68 and agrees exactly with the $U = 0$ solution for a balanced regenerator.

The performance of a regenerator is degraded if the heat capacity of the matrix becomes comparable to the heat capacity of the fluid; this situation corresponds to U approaching 1.0. Maintaining a low value of utilization is a particular problem for regenerators operating at cryogenic temperatures because the specific heat capacity of most solid materials decreases substantially at very low temperature. Figure 8-70 illustrates $\tilde{\theta}_f$ and $\tilde{\theta}_r$ as a function of $\tilde{t}$ for various values of $\tilde{x}$ for Case B in Figure 8-68 (where $NTU = 25$ and $U = 1.0$).

Notice that the fluid-to-regenerator temperature difference remains small in Figure 8-70 because the NTU is still high. However, the regenerator heat capacity is not

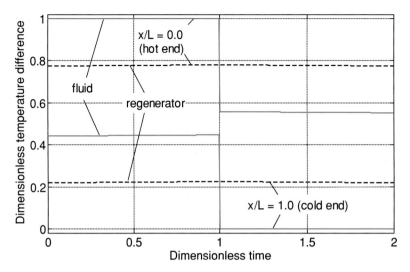

Figure 8-71: $\tilde{\theta}_f$ and $\tilde{\theta}_r$ as a function of $\tilde{t}$ at $\tilde{x} = 0.0$ (the hot end) and 1.0 (the cold end) for a balanced, symmetric regenerator with $NTU = 2.5$ and $U = 0.01$ (Case C in Figure 8-68).

large relative to the fluid heat capacity (the utilization is not low), and therefore the temperature of the regenerator material changes substantially with time during the cycle.

The performance of a regenerator is degraded as the number of transfer units is reduced in the same way that the performance of a conventional heat exchanger suffers at low NTU; a larger temperature difference is required to drive the heat transfer process. Figure 8-71 illustrates $\tilde{\theta}_f$ and $\tilde{\theta}_r$ as a function of $\tilde{t}$ for various values of $\tilde{x}$ for Case C in Figure 8-68 (where $NTU = 2.5$ and $U = 0.01$). Qualitatively, Figure 8-71 looks like the high NTU, low U case A shown in Figure 8-69. However, the fluid-to-regenerator temperature difference is much higher which reduces the performance of the device.

8.10.4 Correlations for Regenerator Matrices

The geometric configuration of the solid material used as the regenerator matrix must be selected in order to provide a high heat transfer coefficient and large amount of surface area between the fluid and the solid while minimizing the pressure drop of the fluid as it passes through the matrix. Temperature gradients that exist locally within the solid material will tend to reduce the effective energy storage capacity of the regenerator and thereby result in a degradation of performance. The requirement for uniform temperatures within the solid normally indicates a need for thin pieces of material with high thermal conductivity. However, regenerator performance is also degraded by axial conduction, in the same way that the performance of a steady flow heat exchanger suffers when axial conduction is high, as discussed in Section 8.7. Ideally, the regenerator matrix would have zero thermal conductivity in the fluid flow direction but infinite thermal conductivity in the direction normal to the flow. These conditions are assumed in the regenerator model discussed in Section 8.10.3 and in the numerical model developed in Section 8.10.5. Of course, no real regenerator packing material can exactly provide this behavior. An additional important consideration for a regenerator matrix is that the packing material should be uniform so that flow maldistribution does not occur.

The thermal performance of the matrix depends on the number of transfer units, NTU, defined as:

$$NTU = \frac{\bar{h} A_s}{\dot{m} c_f} \tag{8-379}$$

where $\bar{h}$ is the average heat transfer coefficient and A_s is the total surface area of the matrix that is exposed to the fluid. The total surface area is the product of the total volume of the regenerator bed and the specific surface area, α; the specific surface area is defined as the ratio of the surface area to the bed volume:

$$A_s = \alpha A_{fr} L \tag{8-380}$$

where A_{fr} is the frontal area of the regenerator bed and L is the length of the bed in the flow direction. Substituting Eq. (8-380) into Eq. (8-379) leads to:

$$NTU = \frac{\bar{h} \alpha A_{fr} L}{\dot{m} c_f} \tag{8-381}$$

The pressure loss of the fluid as it passes through the matrix is primarily a result of flow acceleration and core friction and can be estimated from the relation recommended by Kays and London (1984).

$$\Delta p \approx \frac{G^2 v_{in}}{2} \left[(1 + \phi^2) \left(\frac{v_{out}}{v_{in}} - 1 \right) + \frac{f L}{r_{char}} \left(\frac{v_{in} + v_{out}}{2 v_{in}} \right) \right] \tag{8-382}$$

where v_{in} and v_{out} are the specific volumes of the fluid at the inlet and outlet of the matrix, respectively, f is the friction factor associated with the matrix, ϕ is the matrix porosity (defined as the ratio of the regenerator bed volume that is occupied by the fluid to the total regenerator bed volume), r_{char} is the characteristic radius associated with the matrix (discussed subsequently), and G is the mass flux evaluated on the basis of the cross-sectional area for flow, A_c:

$$G = \frac{\dot{m}}{A_c} \tag{8-383}$$

The cross-sectional area for flow is the product of the frontal area and the porosity (ϕ):

$$A_c = A_{fr} \phi \tag{8-384}$$

Substituting Eq. (8-384) into Eq. (8-383) leads to:

$$G = \frac{\dot{m}}{\phi A_{fr}} \tag{8-385}$$

If the inlet and outlet specific volumes are not very different (as would be the case for regenerators that operate across a modest temperature range or with incompressible fluids) then Eq. (8-382) can be simplified to:

$$\Delta p \approx \frac{G^2 f L}{2 \rho_f r_{char}} \tag{8-386}$$

where ρ_f is the average density of the fluid.

The characteristic radius of the flow passage, r_{char}, is defined as four times the ratio of the volume in a flow passage to its surface area. (In Kays and London, r_{char} is referred to as the hydraulic radius, but it is defined differently from the hydraulic diameter that is used throughout this book and therefore we will refer to it as a characteristic radius in order to avoid confusion.)

$$r_{char} = \frac{L A_c}{A_s} \qquad (8\text{-}387)$$

In order to characterize a particular regenerator packing, it is necessary to know the details of the geometry (e.g., the porosity, specific surface area, etc.), the properties of the regenerator material, the properties of the fluid, and the thermal-fluid characteristics of the flow through the packing; specifically, the heat transfer coefficient in Eq. (8-381) and the friction factor in Eq. (8-382). The heat transfer coefficient for a particular packing is typically correlated in terms of the Colburn j_H factor, originally encountered in Section 8.1.6 in the context of compact heat exchanger correlations:

$$j_H = \frac{\bar{h}}{G c_f} Pr_f^{2/3} \qquad (8\text{-}388)$$

where c_f is the specific heat capacity of the fluid and Pr_f is the Prandtl number of the fluid. The Colburn j_H and friction factor for various packing geometries are correlated against the Reynolds number (e.g., see Kays and London (1984)), defined as:

$$Re = \frac{4 G r_{char}}{\mu_f} \qquad (8\text{-}389)$$

where μ_f is the viscosity of the fluid.

The matrix packing used in most regenerators can be classified as packed spheres, screens, or arrays of flow channels. The thermal-fluid characteristics of these packing geometries have been experimentally measured and correlated in order to facilitate their use to solve engineering problems. The correlations for several commonly encountered regenerator packing configurations are included in the EES convection heat transfer library.

Packed Bed of Spheres

Figure 8-72 illustrates the Colburn j_H factor and the friction factor for randomly packed spheres as a function of the Reynolds number. Figure 8-72 was generated using the EES procedure PackedSpheres_ND, which is based on interpolation of data provided in Kays and London (1984).

The characteristic radius for a packed bed of spheres can be estimated according to (Ackermann (1997)):

$$r_{char} = \frac{\phi d_p}{4 (1 - \phi)} \qquad (8\text{-}390)$$

where d_p is the diameter of the spherical particles. The porosity of a randomly packed bed of spheres of uniform diameter is usually in the range of $\phi = 0.32$ to $\phi = 0.37$. The surface area available for heat transfer per unit volume is:

$$\alpha = \frac{4 (1 - \phi)}{d_p} \qquad (8\text{-}391)$$

The dimensionless characteristics of a packed bed of spheres (i.e., j_H and f) can be obtained using the PackedSpheres_ND procedure in EES and the dimensional characteristics (e.g., the heat transfer coefficient $\bar{h}$) can be obtained using the PackedSpheres procedure in EES.

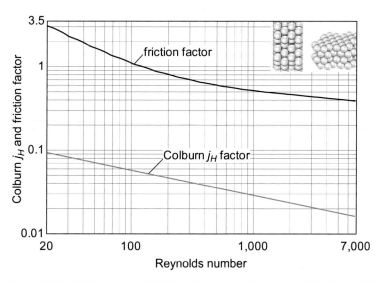

Figure 8-72: Colburn j_H and friction factor as a function of the Reynolds number for a matrix of packed spheres (generated using the PackedSpheres_ND procedure based on data from Kays and London (1984)).

Screens

Figure 8-73 illustrates the Colburn j_H factor and the friction factor for a perfectly stacked array of woven screens as a function of the Reynolds number for various values of porosity. Figure 8-73 was generated using the procedure Screens_ND which is based on interpolation of data provided in Kays and London (1984).

The geometric parameters that characterize a stacked bed of screens (e.g., the characteristic radius of the flow passage, porosity, etc.) can be estimated by considering a small segment of the screen, as shown in Figure 8-74 (Ackermann (1997)). The screen stack is characterized by the screen wire diameter (d_s) and mesh (m_s, the number of wires per unit length). The porosity of an ideal stack of woven screens (i.e., a stacking

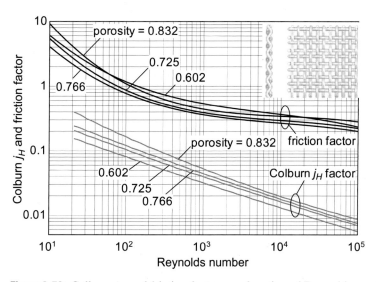

Figure 8-73: Colburn j_H and friction factor as a function of Reynolds number for screens of specified porosity (generated using the Screens_ND procedure based on data from Kays and London (1984)).

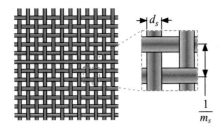

Figure 8-74: Small segment of a woven screen stack.

where the weaving causes no inclination of the wires and the screen layers are not separated) is:

$$\phi = 1 - \frac{\pi}{4} m_s d_s \tag{8-392}$$

The characteristic radius of the screens is:

$$r_{char} = \frac{\phi d_s}{4(1-\phi)} \tag{8-393}$$

The specific surface area of the screens is:

$$\alpha = \pi m_s \tag{8-394}$$

The dimensionless characteristics of a perfectly stacked screen pack can be obtained using the Screens_ND procedure and the dimensional characteristics can be obtained using the Screens procedure.

Triangular Passages

Figure 8-73 illustrates the Colburn j_H factor and the friction factor for a regenerator packing composed of an array of triangular passages as a function of the Reynolds number. Figure 8-75 was generated using the EES procedure Triangular_Channels_ND which is based on interpolation of data provided in Kays and London (1984).

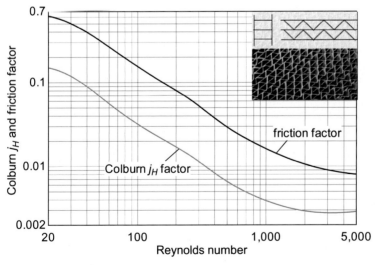

Figure 8-75: Colburn j_H and friction factor as a function of Reynolds number for triangular passages (generated using the Triangular_Channel_ND procedure based on data from Kays and London (1984)).

The geometric parameters that characterize a bed of triangular channels (e.g., the porosity, characteristic radius, and specific surface area) must be determined by examining the details of the structure. The dimensionless characteristics of a bed composed of triangular channels can be obtained using the Triangular_Channels_ND procedure in EES and the dimensional characteristics can be obtained using the Triangular_Channels procedure in EES.

EXAMPLE 8.10-1: AN ENERGY RECOVERY WHEEL

A building at the zoo houses primates, large cats, visitors and staff in four separate zones. The focus of this problem is on the zone that houses the primates. The total volume of the zone is $V_{zone} = 2500\,\text{m}^3$. In order to maintain the health of the animals, as well as to control odors so that the zoo is a pleasant place for visitors, it is necessary to ventilate the zone at a minimum rate of $ac = 2.5$ air changes per hour all of the time (i.e., 24 hours per day, 7 days a week). The outdoor air that replaces the ventilated air must be conditioned to $T_b = 20°C$. (Internal generation from lights and equipment provides the remaining heating needs.) The system is shown in Figure 1(a).

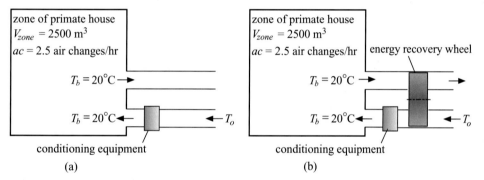

Figure 1: Ventilation and conditioning system (a) without energy recovery wheel, and (b) with energy recovery wheel.

Zoo personnel have found that the costs of heating (in the winter heating season) and cooling (in the summer cooling season) the outdoor air are substantial. Therefore, you have been asked to look at alternatives for cost savings. One possibility is the use of a rotary regenerator for recovering energy from the exhaust air and transferring it to the outside, ventilation air (an energy recovery wheel, see Figure 8-66); such a system is shown schematically in Figure 1(b).

During the heating system, the energy recovery wheel accepts heat from the warm building air leaving the zone and transfers it to the outside air, pre-heating the air in order to reduce the heating that must be provided by the building conditioning equipment. During the cooling season the opposite happens; the energy recovery wheel rejects heat to the (relatively) cool air leaving the building and accepts heat from the warm outdoor air, reducing the cooling that must be provided by the conditioning equipment. Therefore, the energy recovery wheel provides year-round savings and is particularly attractive in applications where large ventilation rates are required.

The energy recovery wheel being considered for this application is made of aluminum with density $\rho_r = 2700\,\text{kg/m}^3$ and $c_r = 900\,\text{J/kg-K}$. The packing is made up of triangular channels, as shown in Figure 2. The thickness of the aluminum

EXAMPLE 8.10-1: AN ENERGY RECOVERY WHEEL

EXAMPLE 8.10-1: AN ENERGY RECOVERY WHEEL

separating adjacent rows of channels is $th_b = 0.3$ mm and the thickness of the aluminum struts that separate adjacent passages is $th_s = 0.1$ mm. The channels themselves are $H_p = 2.5$ mm high and have a half-width of $W_p = 1.5$ mm. The diameter of the wheel is $D_r = 0.828$ m and the length of the wheel in the flow direction is $L = 0.203$ m. The wheel rotates at $N = 30$ rev/min and the matrix spends half of its time exposed to outside air and the other half exposed to building exhaust air.

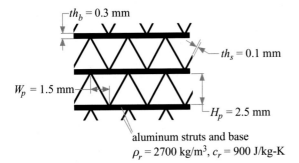

Figure 2: Structure of the energy recovery wheel.

The average outdoor temperature (T_o) for each month is provided in Table 1.

Table 1: Monthly-average ambient temperatures.

Month	Temperature (°C)
Jan	−8
Feb	−7
Mar	−1
Apr	7
May	13
Jun	19
Jul	21
Aug	20
Sep	15
Oct	10
Nov	2
Dec	−6

Notice that the cooling load is not substantial because the monthly average ambient temperature barely exceeds the building temperature in the warmest month. Therefore, the energy recovery wheel in this climate will primarily result in a savings in the heating energy required. The cost of providing heating is $hc = \$10/GJ$; neglect any cost savings associated with cooling. There is an operating cost associated with the additional fan power required to force the air through the rotary regenerator that must be considered. Assume that the fan efficiency is $\eta_{fan} = 0.5$ and that the cost of electricity is $ec = \$0.105/kW\text{-}hr$.

a) Estimate the annual cost savings that would result from installation of the rotary regenerator wheel.

The input information is entered in EES:

EXAMPLE 8.10-1: AN ENERGY RECOVERY WHEEL

"EXAMPLE 8.10-1: An Energy Recovery Wheel"

$UnitSystem SI MASS RAD PA K J
$Tabstops 0.2 0.4 0.6 3.5 in

"Inputs"

V_zone=2500 [m^3]	"volume of the zone that houses the primates"
ac=2.5 [1/hr]*convert(1/hr,1/s)	"air changes required for ventilation"
T_b=converttemp(C,K,20[C])	"conditioned air temperature"
rho_r=2700 [kg/m^3]	"density of regenerator material"
c_r=900 [J/kg-K]	"specific heat capacity of regenerator material"
th_b=0.3 [mm]*convert(mm,m)	"thickness of base plate between rows of channels"
th_s=0.1 [mm]*convert(mm,m)	"thickness of struts between adjacent channels"
H_p=2.5 [mm]*convert(mm,m)	"height of a channel"
W_p=1.5 [mm]*convert(mm,m)	"half-width of a channel"
D_r=0.828 [m]	"wheel diameter"
L=0.203 [m]	"wheel length"
N=30 [1/min]*convert(1/min,1/s)	"rotation rate"
hc=10 [$/GJ]*convert($/GJ,$/J)	"cost of heating"
ec=0.105 [$/kW-hr]*convert($/kW-hr,$/J)	"cost of electricity"
eta_fan=0.5 [-]	"fan efficiency"

Initially, the calculation will be carried out for the month of January; the result will subsequently be extended for an entire year. The number of days (N_{day}) and average outdoor temperature (T_o) for January are entered:

"Monthly conditions"

Month$='Jan'	"month"
N_days=31 [day]	"number of days in the month"
T_o_C=-8 [C]	"monthly average outdoor air temperature in C"
T_o=converttemp(C,K,T_o_C)	"monthly average outdoor air temperature"

The number of days in the month is used to determine the total operating time during the month (t_{month}):

t_month=N_days*convert(day,s)   "time associated with the month"

The matrix geometry shown in Figure 2 is analyzed in order to determine the characteristics of the packing. The cross-sectional area of a single passage is:

$$A_{c,p} = W_p H_p$$

and the wetted perimeter of a passage is:

$$per_p = 2 W_p + 2 \sqrt{H_p^2 + W_p^2}$$

EXAMPLE 8.10-1: AN ENERGY RECOVERY WHEEL

The surface area of a passage is:

$$A_{s,p} = L \, per_p$$

The characteristic radius of the passage is defined according to Eq. (8-387):

$$r_{char} = \frac{L \, A_{c,p}}{A_{s,p}}$$

"matrix geometry"
A_c_p=H_p*W_p   "cross-sectional area of a single passage"
p_p=2*W_p+2*sqrt(W_p^2+H_p^2)   "perimeter of a single passage"
A_s_p=p_p*L   "surface area for a single passage"
r_char=L*A_c_p/A_s_p   "characteristic radius of a single passage"

The frontal area of the regenerator wheel is:

$$A_{fr} = \pi \frac{D_r^2}{4}$$

The number of passages is the ratio of the frontal area to the approximate cross-sectional area of each passage, including the surrounding aluminum:

$$N_p = \frac{A_p}{A_{c,p} + th_s \sqrt{W_p^2 + H_p^2} + W_p \, th_b}$$

A_fr=pi*D_r^2/4   "frontal area of the regenerator"
N_p=A_fr/(A_c_p+sqrt(W_p^2+H_p^2)*th_s+W_p*th_b) "number of passages"

The total surface area associated with the matrix is the product of the number of passages and the surface area per passage:

$$A_s = N_p A_{s,p}$$

and the total void volume (i.e., the volume of the fluid entrained in the regenerator) is the product of the number of passages and the volume of each passage:

$$V_f = N_p A_{c,p} L$$

A_s=N_p*A_s_p   "total surface area"
V_f=L*A_c_p*N_p   "total void volume"

The total volume of the aluminum in the regenerator is:

$$V_r = A_{fr} L - V_f$$

EXAMPLE 8.10-1: AN ENERGY RECOVERY WHEEL

and the porosity of the bed is:

$$\phi = \frac{V_f}{A_{fr} L}$$

```
p=V_f/(A_fr*L)                          "porosity of regenerator wheel"
V_r=A_fr*L-V_f                          "regenerator material volume"
```

The air properties (ρ_a, μ_a, c_a, and Pr_a) are evaluated at the average temperature in the regenerator:

$$\overline{T}_a = \frac{T_o + T_b}{2}$$

```
"air properties"
T_a_avg=(T_o+T_b)/2                      "average air temperature"
rho_a=density(Air,T=T_a_avg,P=1 [atm]*convert(atm,Pa))   "density of air"
mu_a=viscosity(Air,T=T_a_avg)           "viscosity of air"
c_a=cP(Air,T=T_a_avg)                   "specific heat capacity of air"
Pr_a=Prandtl(Air,T=T_a_avg)             "conductivity of air"
```

The mass flow rate of air passing through the regenerator is evaluated according to the specified ventilation rate:

$$\dot{m}_a = V_{zone}\, ac\, \rho_a$$

The mass flux is evaluated based on the mass flow rate and the area available for the flow; note that only half of the wheel is exposed to each of the two flows and therefore the factor of 2 is required:

$$G = \frac{2\,\dot{m}_a}{A_{fr}\, p}$$

The Reynolds number is evaluated:

$$Re = \frac{4\,G\,r_{char}}{\mu_a}$$

```
m_dot_a=V_zone*ac*rho_a                 "mass flow rate of air"
G=2*m_dot_a/(A_fr*p)                    "mass flux based on open area"
Re=4*G*r_char/mu_a                      "Reynolds number"
```

The Colburn j_H and friction factors are obtained using the Triangular_channels_ND procedure in EES.

```
call Triangular_channels_ND(Re:f,j_H)   "access convection correlations for triangular channels"
```

EXAMPLE 8.10-1: AN ENERGY RECOVERY WHEEL

The heat transfer coefficient is obtained using the Colburn j_H factor in Eq. (8-388):

$$\bar{h} = \frac{j_H \, G \, c_a}{Pr_a^{2/3}}$$

and used to compute the number of transfer units:

$$NTU = \frac{\bar{h} \, A_s}{\dot{m}_a \, c_a}$$

```
h_bar=j_H*G*c_a/Pr_a^(2/3)                    "heat transfer coefficient"
NTU=h_bar*A_s/(m_dot_a*c_a)                    "number of transfer units"
```

The solution for the effectiveness of a balanced, symmetric regenerator (ε) is accessed using the HX procedure in EES. The solution requires the capacitance rate of the air and the equivalent capacitance rate of the matrix, defined as the total heat capacity of the matrix divided by the time that the matrix is in contact with each of the air streams (the blow time, t_B). The matrix material is exposed to each stream for half of a rotation, therefore the blow time is computed according to:

$$t_B = \frac{1}{2 \, N}$$

```
time_B=1/(2*N)                        "blow time period"
eff=HX('Regenerator', NTU, m_dot_a*c_a, V_r*c_r*rho_r/time_B, 'epsilon')
                                      "access solution for balanced, symmetric regenerator"
```

The effectiveness is defined as the ratio of the actual to the maximum possible amount of energy transfer to the outdoor air per rotation. On a rate basis, the effectiveness is therefore:

$$\varepsilon = \frac{\dot{q}}{\dot{m}_a \, c_a \, (T_b - T_o)}$$

where $\dot{q}$ is the average rate that heat is transferred to the outdoor air.

```
q_dot=eff*m_dot_a*c_a*(T_b-T_o)                    "rate of heat transfer"
```

The total amount of money saved in a month related to avoided heating costs is therefore:

$$\text{heating \$ saved} = t_{month} \, \dot{q} \, hc$$

```
heatingsavings=t_month*q_dot*hc                    "total avoided heating cost in a month"
```

which leads to a savings of $1161 in January.

The savings is mitigated by the cost of the electricity required to operate the fan. The pressure drop across the regenerator bed is estimated according to Eq. (8-386):

$$\Delta p = \frac{G^2 f L}{2 \, \rho_a \, r_{char}}$$

DELTAp=G^2*f*L/(2*rho_a*r_char)   "pressure drop across bed"

The fan power is therefore:

$$\dot{w}_{fan} = \frac{2 \, \Delta p \, \dot{m}_a}{\rho_a \, \eta_{fan}} \tag{1}$$

where the factor of two in Eq. (1) is related to the fact that fan power is required for both the building air and the outdoor air.

w_fan=2*DELTAp*m_dot_a/(rho_a*eta_fan)   "fan power consumption"

which leads to $\dot{w}_{fan} = 2.1\,\text{kW}$. The electrical cost associated with running the fans for a month is:

$$\text{fan \$} = \dot{w}_{fan} \, t_{month} \, ec$$

fancost=w_fan*t_month*ec   "fan operating cost"

The net savings is the heating cost avoided less the fan cost incurred:

$$\text{net \$ saved} = \text{heating \$ saved} - \text{fan \$}$$

savings=heatingsavings-fancost   "net monthly savings"

which leads to a net monthly savings of \$995 in January. The significance of this savings can be put into context by estimating the heating cost in January in the absence of the energy recovery wheel:

$$\text{heating \$ no energy recovery} = t_{month} \, \dot{m}_a \, c_a \, (T_b - T_o)$$

ventilationcost=m_dot_a*c_a*(T_b-T_o)*t_month*hc   "ventilation cost without energy recovery"

which leads to \$1653. Therefore, the energy recovery wheel can save the zoo about 60% of its heating costs in January.

The analysis can be extended over an entire year. A parametric table is generated that includes the parameters that characterize each month of the year as well as the interesting thermal and economic results of the analysis. (Note that those

EXAMPLE 8.10-1: AN ENERGY RECOVERY WHEEL

months where no heating is required are omitted.) The parameters that specify the characteristics of the month are commented out:

```
{Month$='Jan'                          "month"
N_days=31 [day]                        "number of days in the month"
T_o_C=-8 [C]                           "monthly average outdoor air temperature in C"}
```

and entered (from Table 1) into the parametric table. The parametric table is solved (Figure 3):

	Month$	N_{days} [day]	$T_{o,C}$ [C]	ventilationcost [$]	eff [-]	heatingsavings [$]	fancost [$]	savings [$]
Run 1	Jan	31	-8	1653	0.7023	1161	165.7	994.9
Run 2	Feb	28	-7	1437	0.7029	1010	149.9	860.2
Run 3	Mar	31	-1	1224	0.7067	865.2	167	698.3
Run 4	Apr	30	7	723.4	0.7115	614.7	102.8	351.9
Run 5	May	31	13	398.4	0.7149	284.8	169.1	115.7
Run 6	Jun	30	19	54.52	0.7179	39.14	164.3	-125.2
Run 7	Sep	30	15	274.4	0.7159	196.5	163.9	32.62
Run 8	Oct	31	10	572	0.7133	408.1	168.7	239.4
Run 9	Nov	30	2	1010	0.7085	715.8	162.1	553.8
Run 10	Dec	31	-6	1529	0.7035	1076	166.1	909.8

Figure 3: Parametric table containing the results of a month-by-month analysis of the energy recovery system.

It is possible to access statistics related to each column of the table by right-clicking the column header and selecting properties. One of the statistics is the sum of each of the entries in the column; the net savings over a year is $4631. It is more convenient to have EES automatically sum each column and place the results of this operation in a final row of the table; this is accomplished using the $SUMROW directive.

```
$SUMROW ON                             "create a sum row in the table"
```

When the parametric table is solved again, the sum row is placed at the bottom of the table. The energy recovery system will save $4631 annually or 52% of the $8877 heating cost that is incurred without the system.

The predicted cost savings are optimistic in that the effect of ice build-up during freezing conditions is not considered. Regenerators used for ventilation often include a desiccant to provide mass transfer as well as heat transfer. The desiccant transfers water vapor from the stream of higher relative humidity to the stream of lower relativity, which reduces the freeze-up problem. Regenerators that transfer both heat and mass are called enthalpy exchangers.

One method of defrosting a regenerator wheel is to reduce its rotation speed (and therefore increase the blow time) to the point where the effectiveness of the regenerator is significantly reduced; recall from Section 8.10.3 that the effectiveness of a regenerator is reduced if the utilization is increased. The reduction in effectiveness results in a lower rate of heat transfer from the exhaust air to the incoming ventilation air; therefore, the average temperature of the building air that is exhausted to outdoors increases. If the exhaust air temperature is above the freezing point then it will melt the ice that would otherwise form on the matrix.

b) Determine the rotation speed that will result in melting the ice in January.

EXAMPLE 8.10-1: AN ENERGY RECOVERY WHEEL

The characteristics of January are uncommented in the Equation Window:

Month$='Jan'   "month"
N_days=31 [day]   "number of days in the month"
T_o_C=-8 [C]   "monthly average outdoor air temperature in C"

The exhaust air temperature is calculated using an energy balance:

$$T_{exhaust} = T_b - \frac{\dot{q}}{\dot{m}_a\, c_a}$$

T_exhaust=T_b-q/(m_dot_a*c_a)   "exhaust air temperature"
T_exhaust_C=converttemp(K,C,T_exhaust)   "in C"

which leads to $T_{exhaust} = 0.34°C$; while this is above freezing, it is probably not sufficiently warm to melt the ice. The coldest temperature of the regenerator matrix surface will be approximately equal to the average of $T_{exhaust}$ and T_o:

$$T_{r,cold} \approx \frac{T_o + T_{exhaust}}{2}$$

T_r_cold=(T_exhaust+T_o)/2
                "approximate value of the coldest regenerator surface temperature"
T_r_cold_C=converttemp(K,C,T_r_cold)   "in C"

which leads to $T_{r,cold} = -3.8°C$. Figure 4 illustrates $T_{r,cold}$ as a function of the rotational speed.

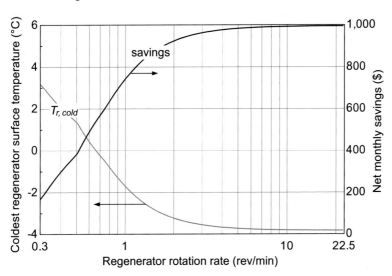

Figure 4: Approximate cold side regenerator surface temperature and net monthly savings for January as a function of the regenerator rotation rate.

Note that the regenerator surface temperature is greater than the freezing point of water when the rotation rate is reduced to $N = 0.55$ rev/min. However, the monthly savings that results when the regenerator is "turned down" to avoid frosting is only about $500, half of the amount that was predicted when frosting is ignored.

8.10.5 Numerical Model of a Regenerator with No Entrained Heat Capacity

This extended section can be found on the website www.cambridge.org/nellisandklein. This section presents the development of a flexible, numerical model of a regenerator in which the entrained heat capacity of the fluid is neglected. The numerical model is developed in MATLAB and can also be accessed in EES using the RegeneratorHX procedure.

Chapter 8: Heat Exchangers

The website associated with this book www.cambridge.org/nellisandklein provides many more problems.

Introduction to Heat Exchangers

8–1 Dry air at $T_{a,in} = 30°C$, and atmospheric pressure is blown at $\dot{V}_a = 1.0$ m³/s through a cross-flow heat exchanger in which refrigerant R134a is evaporating at a constant pressure of $p_R = 345$ kPa. The air exits the heat exchanger at $T_{a,out} = 13°C$. The tubes and fins of the heat exchanger are both made of copper. The tubes have an outer diamcter of $D_{out,t} = 1.64$ cm and $th_t = 1.5$ mm tube wall thickness. The fins are circular with a spacing that leads to 275 fins per meter, an outer diameter of $D_{out,f} = 3.1$ cm and a thickness of $th_f = 0.25$ mm. The heat transfer coefficient between the R134a and the inner tube wall is estimated to be $\bar{h}_R = 2{,}500$ W/m²-K. The heat transfer coefficient between the air and the surface of the tubes and the fins is estimated to be $\bar{h}_a = 70$ W/m²-K. The total length of finned tubes is $L = 110$ m.
   a.) Determine the rate of heat transfer from the air.
   b.) Determine the value of the heat exchanger conductance for this heat exchanger.
8–2 The cross-flow heat exchanger described in Problem 8-1 has geometry that is similar to compact heat exchanger core 'fc_tubes_sCF-70-58J.' The frontal area of the heat exchanger is $A_f = 0.5$ m² and the length of the heat exchanger in the flow direction is $W = 0.25$ m.
   a.) Use the compact heat exchanger library in EES to estimate the air-side conductance and the overall heat exchanger conductance assuming that the heat transfer coefficient between the R134a and the inner tube wall is $\bar{h}_R = 2{,}500$ W/m²-K.
   b.) Compare the result to the value determined in Problem 8-1.
8–3 A decision has been made to use chilled water, rather than R134a in the heat exchanger described in Problems 8-1 and 8-2. The mass flow rate of chilled water has been chosen so that the temperature rise of the water is $\Delta T_w = 4°C$ as it passes through the heat exchanger. The water side is configured so that the chilled water flows through $N_c = 10$ parallel circuits.
   a.) Estimate the overall heat transfer conductance and compare the result to your answers from Problems 8-1 and 8-2.
   b.) Estimate how much the overall heat transfer coefficient can be expected to drop over time due to fouling of the closed chilled water loop.

The Log-Mean Temperature Difference Method

8–4 In Problem 8-1, the inlet volumetric flowrate and the inlet and outlet temperatures of the air are known and therefore it is possible to determine the heat transfer rate without a heat exchanger analysis. However, you have just learned that the outlet air temperature was measured with a thermocouple in only one location in the duct and it is therefore not an accurate measurement of the mixed average outlet air temperature. Use the log-mean temperature difference method to estimate the average air outlet temperature.

8–5 Table P8-5 provides heat transfer data from a manufacturer's catalog for a counterflow oil cooler. The table provides the heat transfer rate for three different oil flow rates (expressed in gpm, gallons per minute). The values in the table are the heat transfer rate between the oil and water in units of Btu/min-ETD where ETD is the entering temperature difference in (°F). The entering temperature difference is the difference between the hot- and cold-fluid inlet temperatures. The density and specific heat of the oil are $\rho_o = 830\,\text{kg/m}^3$ and $c_o = 2.3$ kJ/kg-K, respectively. The water enters at $\dot{V}_w = 35$ gallons per minute at $T_{w,in} = 180°\text{F}$ and atmospheric pressure. The oil enters at $T_{o,in} = 240°\text{F}$.

Table P8-5: Heat transfer data (heat transfer rate/ETD) for different models and oil flow rates

	Oil flow rate		
	1 gpm	3 gpm	5 gpm
Model 1	2.5 Btu/min-deg. F	4.9 Btu/min-deg. F	
Model 2	2.9 Btu/min-deg. F	6.1 Btu/min-deg. F	8.1 Btu/min-deg. F
Model 3	3.1 Btu/min-deg. F	6.8 Btu/min-deg. F	9.7 Btu/min-deg. F

a.) Determine the outlet oil temperature, the log mean temperature difference, and the overall conductance for Model 2 at oil flow rates 1, 3 and 5 gallons/min:

b.) Plot the conductance of Model 2 as a function of the oil flow rate and provide an explanation for the observed variation.

c.) Using the results from part (b), estimate the oil outlet temperature if the oil enters the heat exchanger at $T_{o,in} = 225°\text{F}$ with a flow rate of $\dot{V}_o = 4$ gpm and the water enters with temperature $T_{w,in} = 180°\text{F}$ at a flow rate of $\dot{V}_w = 35$ gpm.

The Effectiveness-NTU Method

8–6 The plant where you work includes a process that results in a stream of hot combustion products at moderate temperature $T_{hg,in} = 150°\text{C}$ with mass flow rate $\dot{m}_{hg} = 0.25$ kg/s. The properties of the combustion products can be assumed to be the same as those for air. You would like to recover the energy associated with this flow in order to reduce the cost of heating the plant and therefore you are evaluating the use of the air-to-air heat exchanger shown in Figure P8-6.

The air-to-air heat exchanger is a cross-flow configuration. The length of the heat exchanger parallel to the two flow directions is $L = 10$ cm. The width of the heat exchanger in the direction perpendicular to the two flow directions is $W = 20$ cm. Cold air enters the heat exchanger from outdoors at $T_{cg,in} = -5°\text{C}$ with mass flow rate $\dot{m}_{cg} = 0.50$ kg/s and is heated by the combustion products to $T_{cg,out}$. The hot and cold air flows through channels that are rectangular (both sides of the heat exchanger have the same geometry). The width of the channels is $h_c = 1.0$ mm. There are fins placed in the channel in order to provide structural support and also increase the surface area for heat transfer. The fins can be assumed to run the complete length of the heat exchanger and are 100% efficient. The fins are spaced with pitch, $p_f = 0.5$ mm and the fins are $th_f = 0.10$ mm thick. The thickness of the metal separating the cold channels from the hot channels is $th_w = 0.20$ mm and

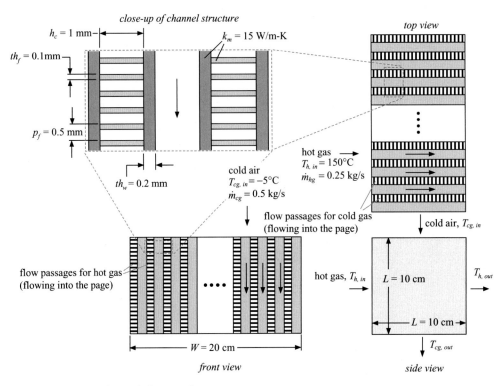

Figure P8-6: Air-to-air heat exchanger.

the conductivity of this metal is $k_m = 15$ W/m-K. Both the hot and cold flows are at nominally atmospheric pressure. The fouling factor associated with the flow of combustion gas through the heat exchanger is $R''_f = 0.0018$ K-m²/W. There is no fouling associated with the flow of outdoor air through the heat exchanger.

a.) Compute the heat transfer coefficient between the hot air and the channel walls and the cold air and the channel walls. Use the inlet temperatures of the air flows to compute the properties.

b.) Compute the total conductance of the heat exchanger.

c.) Determine the heat transferred in the heat exchanger and the temperature of the cold gas leaving the heat exchanger.

d.) Blowers are required to force the hot and cold flows through the heat exchanger. Assume that you have blowers that are $\eta_{blower} = 0.65$ efficient. Estimate the total blower power required to operate the energy recovery unit.

e.) If you pay $ec = 0.08$/kW-hr for electricity (to run the blowers) and 1.50$/therm for gas (to heat the plant) then estimate the net savings associated with the energy recovery system (neglect capital investment cost) for $time = 1$ year; this is the savings associated with the heat transferred in the heat exchanger less the cost to run the blower for a year. Assume that the plant runs continuously.

f.) Plot the net savings per year as a function of the mass flow rate of the cold air that is being heated. Your plot should show a maximum; explain why this maximum exists.

g.) Determine the optimal values of the mass flow rate of combustion gas and cold gas ($\dot{m}_{hg}$ and $\dot{m}_{cg}$) that maximize the net savings per year. You should use the Min/Max capability in EES to accomplish this. What is the maximum savings/

year? This is the most you could afford to pay for the blowers and heat exchanger if you wanted a 1 year pay-back.

8–7 A gas turbine engine is used onboard a small ship to drive the propulsion system. The engine consists of a compressor, turbine, and combustor as shown in Figure P8-7(a). Ambient air is pulled through the gas turbine engine with a mass flow rate of $\dot{m} = 0.1\,\text{kg/s}$ and enters the compressor at $T_1 = 20°\text{C}$ and $P_1 = 1\,\text{atm}$. The exit pressure of the compressor is $P_2 = 3.5\,\text{atm}$ and $T_2 = 167°\text{C}$. The air enters a combustor where it is heated to $T_3 = 810°\text{C}$ with very little loss of pressure so that $P_3 = 3.5\,\text{atm}$. The hot air leaving the combustor enters a turbine where it is expanded to $P_4 = 1\,\text{atm}$. The temperature of the air leaving the turbine is $T_4 = 522°\text{C}$. You may assume that the turbine and compressor are well-insulated and that the specific heat capacity of air is constant and equal to $c = 1000\,\text{J/kg-K}$. The difference between the power produced by the turbine and required by the compressor is used to drive the ship.

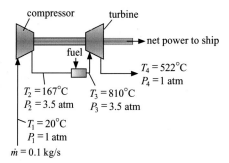

Figure P8-7(a): Unrecuperated gas turbine engine.

a.) Determine the efficiency of the gas turbine engine (the ratio of the net power to the ship to the heat transferred in the combustor).

b.) The combustor runs on fuel with a heating value of $HV = 44 \times 10^6\,\text{J/kg}$ and a mission lasts $t = 2$ days. What is the mass of fuel that the ship must carry?

In order to reduce the amount of fuel required, you have been asked to look at the option of adding a recuperative heat exchanger to the gas turbine cycle, as shown in Figure P8-7(b). You are considering the air-to-air heat exchanger that was evaluated in Problem 8-6 and is shown in Figure P8-6. The air-to-air heat exchanger is a cross-flow configuration. The length of the heat exchanger parallel to the two flow directions is $L = 10\,\text{cm}$. The width of the heat exchanger in the direction perpendicular to the two flow directions is $W = 20\,\text{cm}$, but this can easily be adjusted by adding additional plates. Air enters the heat exchanger from the compressor and is heated by the air leaving the turbine. The hot and cold air flows through channels that are rectangular (both sides of the heat exchanger have the same geometry). The width of the channels is $h_c = 1.0\,\text{mm}$. There are fins placed in the channel in order to provide structural support and also increase the surface area for heat transfer. The fins can be assumed to run the complete length of the heat exchanger and are 100% efficient. The fins are spaced with pitch, $p_f = 0.5\,\text{mm}$ and the fins are $th_f = 0.10\,\text{mm}$ thick. The thickness of the metal separating the cold channels from the hot channels is $th_w = 0.20\,\text{mm}$ and the conductivity of this metal is $k_m = 15\,\text{W/m-K}$. The hot and cold flows are at atmospheric pressure and the compressor discharge pressure, respectively. The fouling factor associated with the flow of the combustion products leaving the turbine is $R_f'' = 0.0018\ \text{K-m}^2/\text{W}$. There is no fouling associated with the flow of the air leaving the compressor.

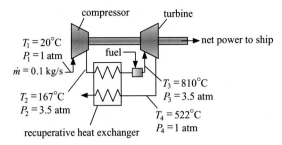

Figure P8-7(b): Recuperated gas turbine engine.

c.) Compute the heat transfer coefficient between the hot air from the turbine and the channel walls and between the colder air from the compressor and the channel walls. You may use the inlet temperatures of the air flows to compute the properties.
d.) Compute the total conductance of the heat exchanger.
e.) Determine the heat transferred in the heat exchanger and the temperature of the air entering the combustor.
f.) What is the efficiency of the recuperated gas turbine engine?
g.) What is the mass of fuel that must be carried by the ship for the 2 day mission if it uses a recuperated gas turbine engine?
h.) The density of the metal separating the cold channels from the hot channels is $\rho_m = 8000\,\mathrm{kg/m^3}$ and the density of the fins is $\rho_f = 7500\,\mathrm{kg/m^3}$. What is the mass of the heat exchanger?
i.) What is the net savings in mass associated with using the air-to-air recuperated gas turbine engine for the 2 day mission?
j.) Plot the net savings in mass as a function of the width of the heat exchanger (W). Your plot should show a maximum; explain why.

8–8 Buildings that have high ventilation rates can significantly reduce their heating load by recovering energy from the exhaust air stream. One way that this can be done is with a run-around loop, shown in Figure P8-8. As shown in the figure, a run-around loop consists of two conventional liquid to air cross-flow heat exchangers. An ethylene glycol solution with 35% mass percent glycol is pumped at a rate $\dot{m}_g = 1\,\mathrm{kg/s}$ through both heat exchangers. The specific heat of this glycol solution is $c_g = 3.58\,\mathrm{kJ/kg\text{-}K}$. (Note that the properties of glycol solutions can be determined using the brineprop2 function in EES.) During winter operation, the glycol solution is heated by the warm air exiting in the exhaust duct. The warm glycol solution is then used to preheat cold air entering from outdoors through the ventilation duct.

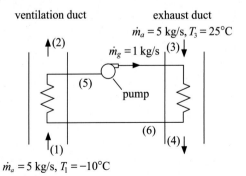

Figure P8-8: Run-around loop for energy recovery.

Outdoor air is blown into the building at a rate of $\dot{m}_a = 5\,\text{kg/s}$. The outdoor temperature is $T_1 = -10°\text{C}$. The building is tightly constructed so the exhaust air flow rate may be assumed to be equal to the ventilation air flow rate ($\dot{m}_a = 5\,\text{kg/s}$). The air leaving the building through the exhaust duct is at $T_3 = 25°\text{C}$. The cross-flow heat exchangers in the exhaust and ventilation streams are identical, each having a finned coil configuration and an estimated conductance $UA = 10\,\text{kW/K}$.

a.) Determine the effectiveness of the ventilation and exhaust heat exchangers.

b.) Determine the temperatures of the glycol solution at states (5) and (6).

c.) Determine the overall effectiveness of the run-around loop.

d.) It has been suggested that the performance of the run-around loop can be improved by optimizing the glycol flow rate. Plot the run-around loop overall effectiveness as a function of the glycol solution flow rate for $0.1\,\text{kg/s} < \dot{m}_g < 4\,\text{kg/s}$. Assume that the conductance of the heat exchangers vary with glycol solution flow rate to the 0.4 power based on a value of $UA = 10\,\text{kW/K}$ at $\dot{m}_g = 1\,\text{kg/s}$. What flow rate do you recommend?

8–9 A concentric tube heat exchanger is built and operated as shown in Figure P8-9. The hot stream is a heat transfer fluid with specific heat capacity $c_H = 2.5\,\text{kJ/kg-K}$. The hot stream enters at the axial center of the annular space at $T_{H,in} = 110°\text{C}$ with mass flow rate $\dot{m}_H = 0.64\,\text{kg/s}$ and then splits; an equal amount flows in both directions. The cold stream has specific heat capacity $c_C = 4.0\,\text{kJ/kg-K}$. The cold-fluid enters the center pipe at $T_{C,in} = 10°\text{C}$ with mass flow rate $\dot{m}_C = 0.2\,\text{kg/s}$. The outlet temperature of the hot-fluid that flows to the left is $T_{H,out,x=0} = 45°\text{C}$. The two sections of the heat exchanger have the same conductance.

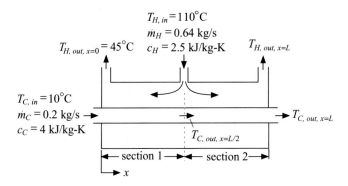

Figure P8-9: Concentric tube heat exchanger.

a.) Determine the temperature of the cold stream at the midpoint of the center tube ($T_{C,out,x=L/2}$) and the temperature of the cold-fluid leaving the heat exchanger ($T_{C,out,x=L}$).

b.) Calculate the overall conductance of this heat exchanger.

c.) How will the overall effectiveness be affected if the inlet temperature is increased to $400°\text{C}$. (Assume that the properties of the heat transfer fluid are independent of temperature.) Justify your answer.

d.) Is the overall effectiveness of this heat exchanger higher, lower, or the same as a counter-flow heat exchanger having the same inlet conditions? Justify your answer.

8–10 The power delivered to the wheels of a vehicle ($\dot{w}$) as a function of vehicle speed (V) is given by: $\dot{w} = -0.3937\,[\text{hp}] + 0.6300\,[\frac{\text{hp}}{\text{mph}}]\,V + 0.01417\,[\frac{\text{hp}}{\text{mph}^2}]\,V^2$

where power is in horsepower and velocity is in mph. The amount of heat rejected from the engine block ($\dot{q}_b$) is approximately equal to the amount of power delivered to the wheel (the rest of the energy from the fuel leaves with the exhaust gas). The heat is removed from the engine by pumping water through the engine block with a mass flow rate of $\dot{m} = 0.80$ kg/s. The thermal communication between the engine block and the cooling water is very good, therefore you may assume that the water will exit the engine block at the engine block temperature (T_b). For the purpose of this problem, you may model the water as having constant properties that are consistent with liquid water at 70°C. The heat is rejected from the water to the surrounding air using a radiator, as shown in Figure P8-10. When the car is moving, air is forced through the radiator due to the dynamic pressure associated with the relative motion of the car with respect to the air. That is, the air is forced through the radiator by a pressure difference that is equal to $\rho_a V^2 / 2$, where ρ_a is the density of air. Assume that the temperature of the ambient air is $T_\infty = 35$°C and model the air in the radiator assuming that it has constant properties consistent with this temperature.

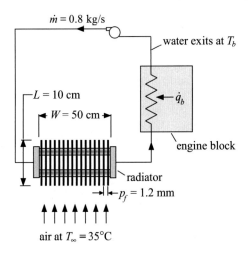

Figure P8-10: Engine block and radiator.

The radiator has a plate-fin geometry. There are a series of tubes installed in closely spaced metal plates that serve as fins. The fin pitch is $p_f = 1.2$ mm and therefore there are W/p_f plates available for heat transfer. The heat exchanger core has overall width $W = 50$ cm, height $H = 30$ cm (into the page), and length (in the flow direction) of $L = 10$ cm. For the purpose of modeling the air-side of the core, you may assume that the air flow is consistent with an internal flow through rectangular ducts with dimension $H \times p_f$. Assume that the fins are 100% efficient and neglect the thermal resistance between the fluid and the internal surface of the tubes. Also neglect convection from the external surfaces of the tubes as well as the reduction in the area of the plates associated with the presence of the tubes.

a.) Using the information above, develop an EES model that will allow you to predict the engine block temperature as a function of vehicle velocity. Prepare a plot showing T_b vs V and explain the shape of the plot. If necessary, produce additional plots to help with your explanation. If the maximum allowable temperature for the engine block is 100°C (in order to prevent vaporization of the water) then what range of vehicle speeds are allowed? You should see both a minimum and maximum limit.

It is not easy to overcome the maximum speed limit identified (a); however, to overcome the minimum speed limit (so that you can pull up to a stop sign without your car overheating) you decide to add a fan. The fan can provide at most 500 cfm ($\dot{V}_o$ – the open circuit flow) and can produce at most 2.0 inch H_2O (Δp_{dh} – the dead-head pressure). The transition from open circuit to dead-head is linear. The fan curve is given by:

$$\Delta p_{fan} = \Delta p_{dh}\left(1 - \frac{\dot{V}}{\dot{V}_o}\right)$$

b.) Modify your code to simulate the situation where the air is provided by the fan rather than the vehicle motion. Overlay a plot showing T_b vs V for this configuration on the one from (a); have you successfully overcome the lower speed limitation?

8–11 A parallel-flow heat exchanger has a total conductance $UA = 10$ W/K. The hot-fluid enters at $T_{h,in} = 400$ K and has a capacity rate $\dot{C}_h = 10$ W/K. The cold-fluid enters at $T_{c,in} = 300$ K and has a capacity rate $\dot{C}_c = 5$ W/K.

a.) Determine the number of transfer units (NTU), effectiveness (ε), heat transfer rate ($\dot{q}$), and exit temperatures ($T_{h,out}$ and $T_{c,out}$) for the heat exchanger.

b.) Sketch the temperature distribution within the heat exchanger.

c.) Sketch the temperature distribution within the heat exchanger if the conductance of the heat exchanger is very large; that is, what is the temperature distribution in the limit that $UA \rightarrow \infty$.

d.) Sketch how the hot exit temperature will change as the total conductance (UA) is varied, with all other quantities held constant at the values listed in the problem statement. Be sure to indicate how your plot behaves as UA approaches zero and as UA approaches infinity.

8–12 A heat exchanger has a core geometry that corresponds to finned circular tube core 'fc_tubes_s80_38T' in the compact heat exchanger library. The frontal area of the core has dimensions $W = 7.75$ inch and $H = 7.75$ inch. The length of the core is $L = 1.5$ inch. The core is integrated with a fan that has a head-flow curve given by: $\Delta p = a - b\dot{V}_a$ where Δp is the pressure rise across the fan, $\dot{V}_a$ is the volumetric flow rate of air, $a = 0.3927$ in H_2O and $b = 0.0021$ in H_2O/cfm are the coefficients of the fan curve. The manufacturer has tested the heat exchanger with atmospheric air at $T_{a,in} = 20°C$ and water at $T_{w,in} = 75°C$, $p_w = 65$ psia flowing through the tubes. The test data are shown in Table P8-12. The tubes are plumbed in series (i.e., all of the water flows through each tube) and the tube thickness is $th_t = 0.035$ inch.

Table P8-12: Manufacturer's data for heat exchanger.

Water flow rate	Water outlet temperature
0.13 gpm	44.3°C
0.25 gpm	51.1°C
0.5 gpm	60.1°C
1 gpm	66.9°C
2 gpm	70.8°C
4 gpm	72.9°C

a.) Develop a model using the effectiveness-NTU technique that can predict the outlet temperature of the water for a water flow rate of the water $\dot{V}_w$.

b.) Plot the outlet temperature of the water as a function of the water flow rate and overlay the manufacturer's data onto your plot.

Numerical Modeling of Parallel-Flow and Counter-Flow Heat Exchangers

8–13 A Joule-Thomson refrigeration cycle is illustrated in Figure P8-13.

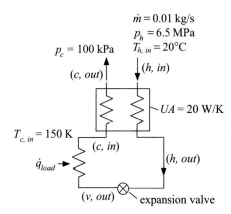

$\dot{m} = 0.01$ kg/s
$p_h = 6.5$ MPa
$p_c = 100$ kPa $T_{h,\,in} = 20°C$
(c, out)   (h, in)

$UA = 20$ W/K

$T_{c,\,in} = 150$ K
(c, in)
$\dot{q}_{load}$ (h, out)

(v, out)   expansion valve

Figure P8-13: Joule-Thomson refrigeration cycle.

The system uses pure argon as the working fluid. High pressure argon enters a counterflow heat exchanger with mass flow rate $\dot{m} = 0.01$ kg/s at $T_{h,in} = 20°C$ and $p_h = 6.5$ MPa. The argon flows through the heat exchanger where it is pre-cooled by the low pressure argon returning from the cold end of the cycle. The high pressure argon leaving the heat exchanger enters an expansion valve where it is expanded to $p_c = 100$ kPa. The argon passes through a load heat exchanger where it accepts a refrigeration load, $\dot{q}_{load}$, and is heated to $T_{c,in} = 150$ K. The conductance of the heat exchanger is $UA = 20$ W/K. Neglect pressure loss in the heat exchanger on both the hot and cold sides of the heat exchanger.

a.) Use the effectiveness-*NTU* method to estimate the effectiveness of the heat exchanger and the rate of heat transfer from the hot to the cold stream in the heat exchanger. Calculate the specific heat capacity of the high- and low-pressure argon using the average of the hot and cold inlet temperatures.

b.) Determine the refrigeration load provided by the cycle.

c.) Prepare a plot of refrigeration load as a function of cold inlet temperature for 85 K $< T_{c,in} < 290$ K and various values of the conductance. A negative refrigeration load is not physically possible (without some external cooling); therefore, terminate your plots at $\dot{q}_{load} = 0$ W.

d.) Instead of using the effectiveness-*NTU* method, divide the heat exchanger into sub-heat exchangers as discussed in Section 8.6.3. What is the heat transferred in the heat exchanger for the conditions listed in the problem statement?

e.) Determine the refrigeration load associated with your prediction from (d).

f.) Overlay on your plot from (c) the refrigeration load as a function of cold inlet temperature for the same values of the conductance.

8–14 A counter-flow heat exchanger has a total conductance of $UA = 130$ W/K. Air flows on the hot side. The air enters at $T_{h,in} = 500$ K with pressure $p_h = 1$ atm and mass flow rate $\dot{m}_h = 0.08$ kg/s. Carbon dioxide flows on the cold side. The

CO_2 enters at $T_{c,in} = 300$ K with pressure $p_c = 80$ atm and mass flow rate $\dot{m}_c = 0.02$ kg/s.

a.) Plot the specific heat capacity of air at 1 atm and carbon dioxide at 80 atm and comment on whether the ε-NTU solution can be applied to this heat exchanger.

b.) Prepare a solution to this problem by numerically integrating the governing equations using the Euler technique, as discussed in Section 8.6.2.

c.) Using your solution from (b), plot the temperature of the carbon dioxide and air as a function of the dimensionless axial position (x/L).

d.) Plot the rate of heat transfer predicted by the model as a function of the number of integration steps.

e.) Prepare a solution to this problem by sub-dividing the heat exchanger into sub-heat exchangers, as discussed in Section 8.6.3.

f.) Overlay on your plot from (c) the temperature distribution predicted by your model from (e).

g.) Overlay on your plot from (d) the rate of heat transfer predicted by your model from (e) as a function of the number of sub-heat exchangers.

Regenerators

8–15 A solar heating system is shown in Figure P8-15.

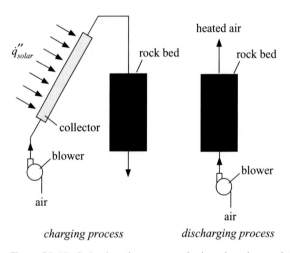

Figure P8-15: Solar heating system during charging and during discharging.

During the day, the solar heat is not required and air is blown through a series of solar collectors where it is heated as shown in Figure P8-15. The thermal energy is stored in a large rock bed regenerator. The rock bed is $L = 20$ ft long in the flow direction and 10 ft × 10 ft in cross-sectional area. The bed is filled with $D_p = 0.5$ inch diameter rocks with density $\rho_r = 100\,lb_m/ft^3$ and specific heat capacity $c_r = 0.2\,Btu/lb_m$-°F. The charging process goes on for $t_{charge} = 12$ hr. There are $N_{col} = 40$ solar collectors, each with area 8 ft × 4 ft. During the charging process, atmospheric air at $T_{indoor} = 70$°F enters the collectors where it is heated by the solar irradiation $\dot{q}''_{solar} = 750$ W/m². The efficiency of the collector is given by: $\eta_{collector} = 0.75 - 0.0015[K^{-1}](T_{r,in} - T_{outdoor})$ where $T_{outdoor} = 10$°F and $T_{r,in}$ is the temperature of the air leaving the collector and entering the regenerator. The collector efficiency is the ratio of the energy transferred to the air to the energy incident on the collector. During the night, the energy that was stored in the rock bed is used to provide heating, as shown in Figure P8-15. Air at $T_{indoor} = 70°F$ enters the rock bed where it is heated. The hot air is provided to the building. The blower used during both the charging and discharging process has an efficiency of $\eta_b = 0.6$ and a pressure/flow curve that goes linearly from $\Delta p_{dh} = 0.5$ inch of water at zero flow to $\dot{V}_{open} = 1800$ cfm at zero pressure rise. Neglect the pressure drop across the collectors and assume that the pressure drop that must be overcome by the blower is related to the flow through the rock bed. The porosity of the rock bed is $\phi = 0.35$. Assume that the rock bed is well-insulated.

a.) What is the temperature of the air entering the rock bed during the charging process and the mass flow rate of air during the charging and discharging process?

b.) What is the amount of heat transfer from the rock bed to the air during the discharge process?

c.) There are 100 heating days per year in this location. What is the total amount of heating energy saved over a 10 year period?

d.) If the cost of natural gas is $gc = 3.5\$/therm$ then what is the total heating cost saved over a 10 year period? (Neglect the time value of money for this analysis.)

e.) The cost of the solar collectors is $cc = 45\$/ft^2$ and the cost of the rock bed is $rc = 40\$/ton$. The cost of the electrical energy required to run the blowers is $ec = 0.12\$/kW\text{-hr}$. Determine the net savings associated with owning the equipment over a 10 year period.

f.) Plot the net savings as a function of the number of solar collectors. You should see that an optimal number of collectors exists. Provide an explanation for this observation.

g.) Plot the net savings as a function of the length of the rock bed (with $N_{col} = 40$). You should see that an optimal length of the rock bed. Explain this fact.

h.) Determine the optimal number of collectors and the optimal rock bed length.

8–16 A Stirling engine is shown in Figure P8-16.

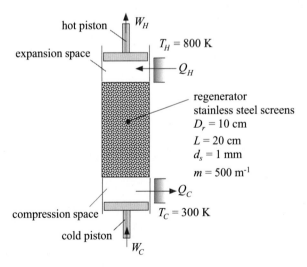

Figure P8-16: Stirling engine.

The mass of gas in the Stirling engine is $M_{gas} = 0.01$ kg. The gas is air and can be modeled as being an ideal gas with gas constant $R_a = 287.1$ J/kg-K and specific heat capacity ratio $\gamma = 1.4$. You may neglect the air entrained in the regenerator void volume and assume that all of the air is either in the compression or expansion space. The Stirling engine's performance will be estimated using a very simple model of the Stirling cycle. During the compression process, all of the gas is contained in the compression space and the cold piston is moved up until the pressure of the air goes from $P_{low} = 1.0$ MPa to $P_{high} = 1.5$ MPa. This process occurs isothermally at $T_C = 300$ K and will be modeled as being reversible. During the cold-to-hot blow process, the two pistons move together so that the gas moves from the compression space to the expansion space. To the extent that the regenerator is not 100% effective, the gas leaves the hot end of the regenerator at a temperature that is below $T_H = 800$ K and therefore a heat transfer occurs from the hot reservoir in order to heat this gas to T_H; this heat transfer is the manifestation of the regenerator loss. During the expansion process, all of the gas is contained in the expansion space and the hot piston is moved up until the pressure of the air goes from P_{high} to P_{low}. This process occurs isothermally at T_H and will also be modeled as being reversible. During the hot-to-cold blow process, the two pistons move together so that the gas moves from the expansion space to the compression space. The cycle occurs with a frequency of $f = 10$ Hz and each of the four processes take an equal amount of time.

a.) What is the efficiency of the cycle and the average power produced in the absence of any regenerator loss?

The regenerator is a cylinder filled with stainless steel screens. The regenerator diameter is $D_r = 10$ cm and the length is $L = 20$ cm. The screens have wire diameter $d_s = 1$ mm and mesh $m = 500$ m^{-1}.

b.) Estimate the regenerator loss per cycle and the efficiency including this regenerator loss.

c.) Plot the efficiency and average power as a function of the frequency of the Stirling engine. Explain the shape of your plot.

REFERENCES

Ackermann, R. A., *Cryogenic Regenerative Heat Exchangers*, Plenum Press, New York, (1997).

Domanski, P., EVAP-COND V2.1, National Institute of Standards and Technology, Gaithersburg, MD 20899-8631, (2006).

Dragutinovic, G. D. and B. S. Baclic, *Operation of Counterflow Regenerators*, Computational Mechanics Publications, Billerica, pp. 100–101, (1998).

Kakaç, S. and H. Liu, *Heat Exchangers – Selection, Rating and Thermal Design*, CRC Press, (1998).

Kays, W. M and A. L. London, *Compact Heat Exchangers, 3rd edition*, McGraw-Hill, (1984).

Linnhoff, B., "The Pinch Design for Heat Exchanger Networks," *Chemical Engineering Science*, Vol. 38, No. 5, pp. 745–763, (1983).

Rohsenow, W. M., J. P. Hartnett, and Y. I. Cho, Y. I., *Handbook of Heat Transfer, 3rd edition*, McGraw-Hill, (1998).

Turns, S. R., Thermal-Fluid Sciences, An Integrated Approach, Cambridge University Press, New York, (2006).

9 Mass Transfer

This extended chapter can be found on the website www.cambridge.org/nellisandklein. Mass transfer occurs whenever fluid flows; that is, some mass is transferred from one place to another. However, the focus in this chapter is on the transport of one chemical species (or component) within a mixture of chemical species that occurs as a direct result of a concentration gradient, independent of a pressure gradient. This type of mass transfer is called diffusion. Mass transfer, like momentum transfer, plays an important role in many important heat exchange processes and devices. For example, mass transfer is critical to the operation of cooling coils, cooling towers, and evaporative coolers and condensers that are commonly used in refrigeration and power systems. The energy transfer that occurs as a result of mass transfer can significantly improve the performance of these heat transfer devices. The processes of heat and mass transfer are analogous. The governing equations for heat and mass transfer are similar and therefore many of the relations and solution techniques that have been developed for heat transfer can be directly applied to mass transfer processes.

Chapter 9: Mass Transfer

The website associated with this book www.cambridge.org/nellisandklein provides many more problems.

Mass Transfer Concepts

9–1 A mixture is formed mixing $M_m = 0.25$ kg of methane (with molar mass $MW_m = 16$ kg/kgmol), $M_e = 0.15$ kg of ethane ($MW_e = 30$ kg/kgmol) and $M_n = 0.1$ kg of nitrogen ($MW_n = 28$ kg/kgmol). The mixture is placed in a container that is maintained at $T = 25°C$ and $p = 5$ bar. At these conditions, the mixture behaves in accordance with the ideal gas law. Determine:
   a.) the volume of the mixture
   b.) the equivalent molecular weight of the mixture
   c.) the density of the mixture on a mass basis
   d.) the density of the mixture on a molar basis
   e.) the mass fractions of each species
   f.) the mole fractions of each species
   g.) the mass concentration of each species
   h.) the molar concentration of each species
9–2 The composition of mixtures of air and water vapor are often reported in terms of the humidity ratio. The humidity ratio, ω, is defined as the mass of water vapor per mass of dry air. The humidity ratio is related to, but not exactly the same as the mass fraction. In a particular case, the humidity ratio is $\omega = 0.0078$ at temperature $T = 30°C$ and pressure $p = 101.3$ kPa. Determine:
   a.) the mass fraction of the water vapor

b.) the mole fraction of the water vapor

c.) the mass concentration of the water vapor

d.) the molar concentration of the water vapor

e.) the maximum possible value for the mole fraction of the water vapor at equilibrium.

Mass Diffusion and Fick's Law

9–3 The air-conditioning load for a building can be broken into latent and sensible contributions. The latent load represents the energy that must be expended in order to remove the water vapor from the building. Water vapor enters by infiltration as air from outdoor leaks inside and also by diffusion through the walls and ceiling. The building in question is rectangular with outer dimensions of 40 ft by 60 ft with 8 ft ceilings. The infiltration rate is estimated at 0.65 air changes per hour. The diffusion coefficient for water through 3/8 inch gypsum board (without a vapor barrier) is approximately 0.000045 ft^2/s at atmospheric pressure. Estimate and compare the rates of moisture transfer by infiltration and diffusion on a day in which the outdoor conditions are 95°F and 45% relative humidity and indoor conditions are 75°F, 40% relative humidity. Is the contribution by diffusion significant? If not, then why are people concerned with water vapor diffusion in a building?

9–4 Natural gas (methane) is transported at 25°C and 100 bar over long distances through 1.2 m diameter pipelines at a velocity of 10 m/s. The pipeline is made of steel with a wall thickness of 2.0 cm. It has been suggested that hydrogen gas could be transported in these same pipelines. However, hydrogen is a small molecule that diffuses through most materials. The diffusion coefficient for hydrogen in steel is about 7.9×10^{-9} m^2/s at 25°C.

a.) Calculate the power transported by methane (assuming it will be combusted) through the pipeline. The lower heating value of methane is 5.002×10^7 J/kg.

b.) Estimate the velocity required to provide the same power if hydrogen rather than methane is transported through the pipeline at the same temperature and pressure. The lower heating value of hydrogen is 1.200×10^8 J/kg.

c.) Compare the pumping power required to transport the natural gas and hydrogen a distance of 100 km.

d.) Estimate the rate of hydrogen loss due to diffusion from a 100 km pipeline. Do you believe this loss is significant?

9–5 A balloon made of a synthetic rubber is inflated with helium to a pressure of $P_{ini} = 130$ kPa at which point its diameter is $D_{ini} = 0.12$ m. The mass of the balloon material is $M_{bal} = 0.53$ gram and its thickness is $\delta = 0.085$ mm. The balloon is released in a room that is maintained at $T = 25°C$ filled with air (79% nitrogen, 21% oxygen by volume) at 100 kPa. Over a period of time, helium diffuses out of the balloon and oxygen and nitrogen diffuse in. The pressure in the balloon above atmospheric pressure is linearly proportional to the balloon volume. The diffusion coefficients for helium, oxygen and nitrogen through this synthetic rubber are 60×10^{-8}, 16×10^{-8}, and 15×10^{-8} cm^2/s, respectively.

a.) Prepare a numerical model of the balloon deflation process. Plot the volume and pressure within the balloon as a function of time. Plot the mass fraction of helium, oxygen, and nitrogen in the balloon as a function of time.

b.) At what time does the balloon lose its buoyancy?

Transient Diffusion through a Stationary Medium

9–6 A janitor is about to clean a large window at one end of a corridor with an ammonia-water solution. The corridor is 2.5 m high, 2 m wide and 3 m in length. The conditions in the corridor are 25°C, 101 kPa. The concentration of the ammonia that evaporates from the window is estimated to be 100 ppm. Many humans can detect ammonia by smell at levels of 1 ppm. Estimate the time required for a person standing at the other end of the corridor to detect the ammonia after the janitor starts to wash the window.

Mass Convection

9–7 In order to detect chemical threats that are being smuggled into the country within a shipping container, the government is working on a system that samples the air inside the container on the dock as it is being unloaded. The chance of detecting the chemical threat is strongly dependent upon its concentration distribution at the time that the container is sampled. Therefore, you have been asked to prepare a simple model of the migration of the threat species from its release point within a passage formed by the space between two adjacent boxes. The problem is not a simple diffusion problem because the threat chemical is adsorbed onto the walls of the passage. The situation is simplified as 1-D diffusion through a duct. One end of the duct is exposed to a constant concentration of the threat chemical that is equal to its saturation concentration, $c_{sat} = 0.026$ kg/m^3. The duct is filled with clean air and the walls of the duct are clean (i.e., at time $t = 0$ there is no threat chemicals either in the air in the duct or on the walls of the duct). The hydraulic diameter of the duct is $D_h = 10$ cm. The length of the duct is infinite. The diffusion coefficient for the threat chemical in air is $D = 2.2 \times 10^{-5}$ m^2/s. The mass of threat chemical per unit area adsorbed on the wall of the container (M_w'') is related to the concentration of the chemical in the air (c) according to:

$$\frac{M_w''}{M_{w,m}''} = \frac{A \dfrac{c}{c_{sat}}}{\left(1 - \dfrac{c}{c_{sat}}\right)\left[1 + (A - 1)\dfrac{c}{c_{sat}}\right]}$$

where $M_{w,m}'' = 4 \times 10^{-4}$ kg/m^2 is the mass per unit area associated with a single monolayer and $A = 20$ is a dimensionless constant. The total time available for diffusion between loading the container and unloading is $t_{transit} = 14$ days. Because the length of the duct is so much larger than the hydraulic diameter of the duct, it is reasonable to assume that the concentration distribution is 1-D. Further, because the concentration of the threat chemical is so small, it is reasonable to neglect any bulk velocity induced by the diffusion process; that is, only mass transfer by diffusion is considered.

a.) Prepare a 1-D transient model of the diffusion process using the ode45 solver in MATLAB.

b.) Plot the concentration distribution within the passage at various times.

c.) Plot the concentration distribution within the passage at $t = t_{transit}$ and overlay on this plot the zero-adsorption solution to show how adsorption has retarded the migration of the threat chemicals within the container.

9–8 Naphthalene is an aromatic hydrocarbon with a molecular weight of $MW = 128.2$ kg/kmol that sublimes at a relatively high rate at room temperature. Naphthalene was commonly used for moth balls, but is now considered to be a carcinogen.

At $T = 25°C$, solid naphthalene has a density of $\rho = 1.16\,\mathrm{g/cm^3}$ and its vapor pressure at this temperature is $p_v = 0.082$ mm Hg. An engineer has recognized that heat and mass transfer are analogous processes and he plans to estimate the heat transfer coefficient for an unusual geometry by measuring how much mass of naphthalene is sublimed over a fixed time period. A review of the literature indicates that the Schmidt number for naphthalene is $Sc = 2.5$. To test the accuracy of the heat/mass transfer analogy, the engineer first measures the mass of naphthalene that sublimes from a sphere of $D = 2.5$ cm diameter when exposed to a stream of pure air at temperature $T = 25°C$, pressure $p = 101.3$ kPa, and velocity $u_\infty = 10$ m/s. The test is run for $t_{test} = 2$ hr and during this time the mass of the naphthalene sphere is reduced by $\Delta m = 250$ mg.

a.) Determine the error relative to accepted correlations for this geometry.

9–9 Data for naphthalene at 25°C are provided in problem 9-8. Determine the time required for 90% of the mass in a 1.0 cm sphere of naphthalene to sublime into an air stream at 25°C and 100 kPa that is flowing at 5 m/s.

Simultaneous Heat and Mass Transfer

9–10 You have been asked to join the team of engineers responsible for the design of an air-washer. Your part of this project is to prepare an analysis that will determine the diameter, velocity, and temperature of droplets as they fall in an upward flowing air stream. You are considering a single water droplet with an initial diameter of 1.5 mm and an initial temperature of 45°C that is released into a 25°C, 35% relative humidity, 100 kPa air stream that is flowing upward at 30 m/s. Plot the diameter, velocity and temperature of the droplet as a function of time. Assume that the droplet remains spherical and that it can be considered to have a uniform temperature at any time.

9–11 One type of household humidifier operates by expelling water droplets into the air. The droplets have an average diameter of 10 μm. After leaving the dehumidifier, the droplets "float" around the room until they evaporate. In a particular case, the room is maintained at 25°C, 100 kPa and 25% relative humidity. You may assume that the droplet is at the temperature where energy transfer by evaporation and convection are balanced.

a.) Determine the temperature of a droplet.

b.) Plot the mass of the droplet as a function of time and determine the time required for the droplets to completely evaporate.

c.) The humidifier requires a work input to form the droplets. The work input is related to the change in surface area of the water as it is transformed from one large "drop" to many smaller droplets. Calculate the energy required to distribute 1 kg of droplets with this vaporizer and compare it to the energy needed to vaporize 1 kg of water at 25°C. Comment on whether you believe that this humidifier saves energy compared to traditional vaporization process based on boiling water.

Cooling Coil Analysis

9–12 Air enters a cooling coil with volumetric flow rate 20,000 cfm, temperature 90°F and 50% relative humidity where it is cooled and dehumidified by heat exchange with chilled water that enters the cooling coil with a mass flow rate of 80,000 lbm/hr and a temperature of 45°F. The total thermal resistance on the water-side of the heat exchanger is 4.44×10^{-6} hr °F/Btu. The air-side is finned and the total thermal resistance on the air side, including the effect of the fins, ranges from 1×10^{-5} hr-°F/Btu when the coil is completely dry to 3.33×10^{-6} hr-°F/Btu when the coil is completely wetted. The coil is large and employs many rows of tubes so that a counterflow heat transfer analysis is appropriate. Use the Dry Coil/Wet Coil analysis described in Section 9.6.2 to analyze the cooling coil.

a.) Estimate the fraction of the coil that is wetted.

b.) Determine the heat transfer rate between the chilled water and the air.

c.) Determine the outlet air temperature.

d.) Determine the rate of condensate.

e.) Determine the outlet temperature of the water.

9–13 Repeat Problem 9-12 using the enthalpy-effectiveness method described in Section 9.6.3.

9–14 Cooling towers are direct-contact heat and mass exchangers. The performance of a cooling tower can be analyzed using the enthalpy-based effectiveness method described in Section 9.6.3. In this case, the maximum rate of heat transfer between the air and water is based on the difference between enthalpy of the inlet air and enthalpy of saturated air exiting at the inlet water temperature. The saturation specific heat should be evaluated using the enthalpies of saturated air at the water inlet and air inlet temperatures, respectively. A steady flow of water enters an induced draft cooling tower with a mass flow rate of 15 kg/s and a temperature of 35°C. The fans provide 4.72 m³/s of ambient air at a dry-bulb temperature of 23°C and a relative humidity of 50%. Makeup water is supplied at 25°C. Use the enthalpy-based effectiveness technique to analyze this cooling tower.

a.) Prepare a plot of the outlet water temperature and the rate of water loss as a function of the number of transfer units associated with the cooling tower for *NTU* values between 0.5 and 5.

b.) Plot the range and approach as a function of *NTU*. The range is the difference between the inlet and outlet water temperatures. The approach is the difference between the outlet water temperature and the wet bulb temperature.

c.) Compare the rate of heat transfer associated with the cooling tower to the heat transfer rate that would be achieved by an air-cooled dry heat exchanger with the same air flow rate and *NTU*.

10 Radiation

10.1 Introduction to Radiation

10.1.1 Radiation

From a thermodynamic perspective, thermal energy can be transferred across a boundary (i.e., heat transfer can occur) by only two mechanisms: conduction and radiation. Conduction is the process in which energy exchange occurs due to the interactions of molecular (or smaller) scale energy carriers within a material. The conduction process is intuitive; it is easy to imagine energy carriers having a higher level of energy (represented by their temperature) colliding with neighboring particles and thereby transferring some of their energy to them. Convection is the process in which the surface of a solid material exchanges thermal energy with a fluid. Although convection is commonly treated as a separate heat transfer mechanism, it is more properly viewed as conduction in a substance that is also undergoing motion. The energy transfer by conduction and fluid motion are coupled, making convection problems more difficult to solve than conduction problems. However, convection is still an intuitive process since it can be explained by interactions between neighboring molecules with different energy levels. Radiation is a very different heat transfer process because energy is transferred without the benefit of any molecular interactions. Indeed, radiation energy exchange can occur over long distances through a complete vacuum. For example, the energy that our planet receives from the sun is a result of radiation exchange. The process of radiation heat transfer is not intuitive to most engineers.

The existence of radiation heat transfer can be confirmed by a relatively simple experiment. A system that is initially at a higher temperature than its surroundings will, in time, cool to the temperature of its surroundings. The process occurs even when conduction and convection mechanisms are eliminated by enclosing the system in a vacuum. All substances emit energy in the form of electromagnetic radiation as a result of molecular and atomic activity; molecular electronic, vibrational or rotational transitions result in the emission of energy in the form of radiation. The characteristics and amount of radiation emitted by a substance are dependent on its temperature as well as its surface properties. Energy is exchanged between a system and its surroundings by radiation even when they are at the same temperature. In this case, however, the net energy exchange is zero. The rate at which the system is emitting radiation is equal to the rate at which it is absorbing the incident radiation that was emitted from its surroundings. If this were not true, then it would be possible for the system energy (and therefore its temperature) to increase while the energy in the surroundings (and therefore the surrounding temperature) decreased. The result would be a heat transfer from a system at a colder temperature to a system at a warmer temperature, which would violate the second law of thermodynamics. This process has never been observed.

Several theories have been proposed to explain how energy can be transported by radiation. One theory, initiated by physicists in the early 1900s, views the transport process as occurring through waves, analogous to the energy dissipation and transfer associated with the waves that are induced when an object is dropped into water. The electromagnetic waves propagate through a vacuum at the speed of light. Evidence supporting this theory arises from experiments that show that electromagnetic radiation exhibits well-known, wave-like behaviors such as diffraction and constructive-destructive interference. Another theory assumes that energy is emitted by a substance in discrete quantities called photons that have particle-like behavior. Evidence for this theory has been provided by experiments that show that radiation can exert a pressure and thereby transport linear momentum. The solar sails that have been proposed for spacecraft propulsion are based on this behavior. The discrete nature of radiation emission is the basis for quantum theory. Thus thermal radiation must be viewed as having both wave-like and particle-like behaviors.

In this text, we will examine radiation from an engineering perspective with an eye toward solving radiation heat transfer problems. This chapter progresses by examining the behavior of a blackbody, a perfect absorber and emitter of radiation. A true blackbody (like a reversible process, in thermodynamics) does not actually exist. However, the concept of a blackbody provides a useful limiting case for radiation heat transfer. Section 10.2 examines how a blackbody emits radiation and Section 10.3 examines radiation exchange between systems of black bodies. Real surfaces are examined in Section 10.4. The behavior of a real surface is more complex, both in terms of how it emits radiation as well as what happens to radiation that is incident on it. In Section 10.5, methods for calculating the heat transfer associated with systems of real surfaces are presented. Section 10.6 examines the common situation where radiation is occuring together with heat transfer due to conduction or convection. Finally, Section 10.7 provides an introduction to Monte Carlo techniques applied to radiation heat transfer. The Monte Carlo modeling technique is a powerful tool as well as a rich and complex field that could, by itself, form the basis of another text book. In Section 10.7, the basic Monte Carlo modeling approach is discussed and applied to a few simple situations.

10.1.2 The Electromagnetic Spectrum

Electromagnetic radiation propagates through a vacuum at the speed of light, $c = 299,792$ km/s. The propagation speed is the product of the wavelength of the radiation (λ) and its frequency (ν):

$$c = \lambda \nu \tag{10-1}$$

When classified according to its wavelength, thermal radiation (i.e., the radiation that is emitted by an object by virtue of its temperature) is only a subset of electromagnetic radiation. The electromagnetic radiation spectrum is shown in Figure 10-1 and extends from the extremely energetic, high frequency (short wavelength) gamma rays to low energy (long wavelength) radio waves. Visible light, the portion of the electromagnetic spectrum that our eyes can detect, lies between 0.38 μm and 0.78 μm (representing colors ranging from violet to red). Our eyes are most sensitive to the green radiation that occurs at about the center of this range, 0.55 μm.

Approximately 10% of the energy emitted by the sun is ultraviolet (UV) radiation, which lies between the wavelengths 0.20 μm and 0.38 μm. Most of this UV radiation is absorbed by our atmosphere. UV radiation is sub-classified into 3 bands: UV-C (0.20 μm to 0.28 μm), UV-B (0.28 μm to 0.32 μm), and UV-A (0.32 μm to 0.38 μm). UV-A radiation has the lowest energy of the three bands and a large percentage of this

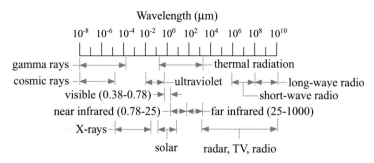

Figure 10-1: Electromagnetic spectrum (reproduced from Duffie and Beckman, 2006).

radiation is transmitted through our atmosphere. This is the type of radiation that is responsible for a suntan. Overexposure to UV-A without UV blocking sun creams can result in skin damage. However, UV-B radiation is of greater concern since radiation in this wavelength band can damage the DNA of plants and animals and cause skin cancer in humans. UV-B is strongly absorbed by the ozone (O_3) that resides in the stratosphere, approximately 25 miles above ground level. For this reason, the thinning of the ozone layer that has been caused by catalytic reactions with chlorofluorocarbon (CFC) refrigerants has raised concern and resulted in an international ban on CFC refrigerant production, an initiative that began in 1996. UV-C radiation is more energetic and potentially more dangerous than UV-B radiation. However, UV-C radiation is strongly absorbed by gases in our atmosphere and therefore essentially none of this radiation reaches ground level. Exposure to UV-C radiation is a concern for astronauts carrying out extra-vehicular activities.

Thermal radiation (i.e., radiation generated by a surface due to its temperature) is the portion of the electromagnetic spectrum between the wavelengths of approximately 0.2 μm and 1000 μm. The other portions of the spectrum are largely generated by non-thermal processes. For example, gamma rays are produced by radioactive disintegration and radio waves are artificially produced by electrical oscillations. The radiation problems considered in this chapter will involve only thermal radiation.

10.2 Emission of Radiation by a Blackbody

10.2.1 Introduction

Radiation heat transfer to a surface is a consequence of the difference between the amount of radiation that is emitted by the surface and the amount of radiation that is absorbed by the surface. Therefore, in order to understand radiation heat transfer, it is necessary to understand how the surface emits radiation as well as how it receives radiation, which is a complicated function of the orientation of the surface with respect to other surfaces as well as the temperatures of all of the surfaces involved.

We will start by considering the emission of thermal radiation by a surface. Thermal radiation is emitted by a surface due to its temperature. The magnitude of the radiation that is emitted by a surface at a given temperature may be a complicated function of wavelength (i.e., the radiation is distributed spectrally) and direction (i.e., the radiation is distributed directionally). A diffuse surface emits radiation uniformly in all directions. The spectral distribution of the radiation that is emitted by a real diffuse surface is shown in Figure 10-2. The information in Figure 10-2 is typically referred to as the spectral emissive power, E_λ, and has units power/(area-wavelength).

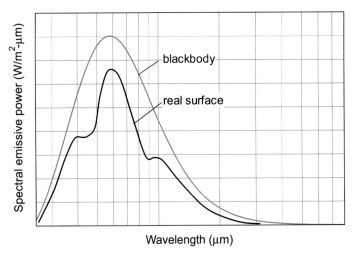

Figure 10-2: Spectral distribution of the radiation emitted by an actual surface and a blackbody at the same temperature.

Fortunately, the blackbody provides an ideal limit to the emissive behavior of any surface. The spectral emissive power of a blackbody is shown qualitatively in Figure 10-2 and is the absolute upper bound on the spectral emissive power that can be achieved by any real surface. Section 10.2.2 discusses the behavior of a blackbody. The behavior of a real surface, discussed in Section 10.4, is defined by comparison with this limiting case.

10.2.2 Blackbody Emission

The blackbody has several limiting characteristics; (1) it absorbs all of the radiation that is incident upon it (regardless of the direction or wavelength of the incident radiation), (2) it emits radiation uniformly in all directions (i.e., it is a diffuse emitter), and (3) it emits the maximum possible amount of radiation at a given temperature and wavelength (e.g., see Figure 10-2).

Planck's Law

The radiation that is emitted by a blackbody is a function of temperature and wavelength, as shown in Figure 10-3. Accurate measurements of the blackbody spectral emissive power ($E_{b,\lambda}$) were obtained using a cavity and published by Lummer and Pringsheim in 1900; these measurements were presented before any theory was available to explain the observed behavior.

A blackbody at a specific temperature will emit thermal radiation over a large range of wavelengths. The distribution of the blackbody spectral emissive power increases and shifts toward lower wavelengths as the temperature of the blackbody increases, as shown in Figure 10-3. The amount of thermal radiation emitted by a blackbody is an extremely strong function of temperature. (Note that the emissive power axis in Figure 10-3 is logarithmic.) Therefore, we can often neglect radiation when dealing with problems that occur near room temperature but radiation may dominate problems at high temperatures. (The relative importance of radiation in comparison to other heat transfer mechanisms is discussed in Section 10.6.3.) Even at room temperature, radiation can be significant in comparison with the heat transfer rate resulting from free convection.

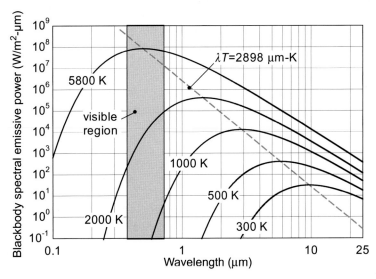

Figure 10-3: Blackbody spectral emissive power, $E_{b,\lambda}$, as a function of wavelength for various values of temperature.

The surface temperature of the sun is approximately 5800 K. Figure 10-3 shows that much of the radiation emitted by the sun is in the visible region (0.38–0.78 μm); it is not a coincidence that our eyes have evolved to make use of the radiation emitted at this temperature. Also notice that, as the temperature of an object is increased (starting at room temperature), the peak emission moves through invisible, infrared bands and then enters the visible band at wavelengths that we perceive as red. This behavior explains why most objects that are heated first appear to be red and eventually become "white" hot when there is emission at all of the wavelengths in the visible region.

There is a peak emissive power in the blackbody spectrum at any temperature that occurs at wavelength λ_{max}. The wavelength at maximum emissive power is related to temperature according to:

$$\lambda_{max} T = 2897.8 \ \mu\text{m-K} \tag{10-2}$$

Equation (10-2) is referred to as Wien's law, named for Wilhelm Wien who reported this observation in 1893 based on experimental data.

The area under any constant temperature curve in Figure 10-3 is the total blackbody emissive power, E_b, which has units of power/area and is the rate that radiation is emitted per unit area. The blackbody emissive power was shown experimentally by Stefan and theoretically by Boltzmann in 1879 to be proportional to the fourth power of temperature; note that the temperature referred to in any radiation problem must be the absolute temperature (K). The proportionality constant, σ, between blackbody emissive power and absolute temperature to the fourth power is called the Stefan-Boltzmann constant.

$$E_b = \int_0^\infty E_{b,\lambda} \, d\lambda = \sigma T^4 \quad \text{where} \quad \sigma = 5.67 \times 10^{-8} \frac{\text{W}}{\text{m}^2\text{-K}^4} \tag{10-3}$$

The development of a theory that describes the behavior of the spectral emissive power was of great interest among prominent physicists during the 1890–1900 era. The observations that electromagnetic radiation exhibited wave-like behavior prompted physicists in the early 1900s to attempt to derive the blackbody emission spectrum using classical statistical mechanics. Their solution agreed with the measured spectrum at long wavelengths, but failed to predict the observed peak and the subsequent reduction in the blackbody spectral emissive power at low wavelengths. This disagreement between the experimental data and the wave-based theory was called the "ultraviolet catastrophe" because the disagreement became most obvious in the UV wavelength region.

In 1901, Max Planck published an empirical equation that fit the experimental data at both long and short wavelengths. His equation, called Planck's law, is:

$$E_{b,\lambda} = \frac{C_1}{\lambda^5 \left[\exp\left(\dfrac{C_2}{\lambda T} \right) - 1 \right]} \qquad (10\text{-}4)$$

where $C_1 = 3.742 \times 10^8\,\text{W-}\mu\text{m}^4/\text{m}^2$ and $C_2 = 14,388\,\mu\text{m-K}$. Setting the derivative of $E_{b,\lambda}$, Eq. (10-4), with respect to wavelength equal to zero yields Wien's Law; this result is demonstrated by the following Maple program.

```
> restart;
> C_1:=3.742e8; C_2:=14388;
```
$$C\_1 := 0.3742\,10^9$$
$$C\_2 := 14388$$
```
> E_b_lambda:=C_1/(lambda^5*(exp(C_2/(lambda*T))-1));
```
$$E\_b\_lambda := \frac{0.3742\,10^9}{\lambda^5 \left(e^{\left(\frac{14388}{\lambda T}\right)} - 1 \right)}$$
```
> dE_b_lambdadlambda:=diff(E_b_lambda,lambda);
```
$$E\_b\_lambdadlambda := -\frac{0.18710\,10^{10}}{\lambda^6 \left(e^{\left(\frac{14388}{\lambda T}\right)} - 1 \right)} + \frac{0.53839896\,10^{13}\, e^{\left(\frac{14388}{\lambda T}\right)}}{\lambda^7 \left(e^{\left(\frac{14388}{\lambda T}\right)} - 1 \right)^2 T}$$
```
> lambda_max:=solve(dE_b_lambdadlambda=0,lambda);
```
$$lambda\_max := \frac{2897.818525}{T}$$

The integral of Eq. (10-4) over all wavelengths is in agreement with the Stefan-Boltzmann equation. The EES code below evaluates the blackbody spectral emissive power at a particular temperature and wavelength (1000 K and 1 μm) according to Eq. (10-4):

```
$UnitSystem SI MASS RAD PA K J
$Tabstops 0.2 0.4 0.6 3.5 in

T=1000 [K]                                          "temperature"
lambda=1 [micron]                                   "wavelength"
E_b_lambda=C1#/(lambda^5*(exp(C2#/(lambda*T))-1))   "blackbody emissive power"
```

which leads to $E_{b,\lambda} = 211.1\,\text{W/m}^2\text{-}\mu\text{m}$; this result is consistent with Figure 10-3. Note that C1# and C2# are built-in constants in EES and therefore these values are automatically provided. (See Constants under the Options menu.) The blackbody emissive power can be obtained by integrating the blackbody spectral emissive power from $\lambda = 0$ to ∞, according to Eq. (10-3). Numerical integration cannot be accomplished over this range since evaluating the EES code at $\lambda = 0$ will lead to an error and there is no way to evaluate the code at $\lambda = \infty$. Therefore, the integration is accomplished over a finite range of wavelengths that is sufficiently large so that it captures essentially all of the emitted radiation. Inspection of Figure 10-3 indicates that integration from $\lambda = 0.01\,\mu\text{m}$ to $1000\,\mu\text{m}$ should be sufficient. Therefore, the specified value of the wavelength is commented out and the Integral command is used to accomplish the integration (the final argument in the Integral function specifies the step size to use, 0.05 μm):

```
{lambda=1 [micron]}                                    "wavelength"
E_b=Integral(E_b_lambda,lambda,0.01 [micron],1000 [micron],0.05 [micron])
        "blackbody emissive power calculated by integrating the spectral emissive power"
```

which leads to $E_b = 56704\,\text{W/m}^2$. The blackbody emissive power predicted using Eq. (10-3) is also computed (note that the Stefan-Boltzmann constant, sigma#, is also a built-in constant):

```
E_b_SB=sigma#*T^4
        "blackbody emissive power calculated using the Stefan-Boltzmann constant"
```

which leads to $E_b = 56696\,\text{W/m}^2$. The small difference between the two answers arises from round-off errors in the values of the constants C1#, C2# and sigma# as well as numerical errors in the integration.

Following the publication of his empirical equation for the spectral blackbody emissive power, Planck spent years trying to establish a theory of electromagnetic radiation from classical physics; he was not successful in this endeavor. He then attempted a statistical approach using Boltzmann's statistics which was successful; this effort initiated the study of quantum theory.

Blackbody Emission in Specified Wavelength Bands

The blackbody spectral emissive power, Eq. (10-4), can be integrated over all possible wavelengths in order to arrive at the blackbody emissive power, σT^4:

$$E_b = \int_{0}^{\infty} \frac{C_1}{\lambda^5 \left[\exp\left(\dfrac{C_2}{\lambda T} \right) - 1 \right]} \, d\lambda = \sigma T^4 \tag{10-5}.$$

However, Eq. (10-4) cannot be integrated in closed form between arbitrary wavelength limits. For example, there is no analytical solution to the integral:

$$E_{b,0-\lambda_1} = \int_{0}^{\lambda_1} \frac{C_1}{\lambda^5 \left[\exp\left(\dfrac{C_2}{\lambda T} \right) - 1 \right]} \, d\lambda \tag{10-6}$$

where $E_{b,0-\lambda_1}$ is the amount of radiation per unit area emitted by a blackbody at wavelengths less than λ_1. It is necessary to use a graphical or tabular method or, if a computer is available, numerical integration in order to determine the blackbody emissive power corresponding to a specified wavelength band.

The graphical method is convenient for quick calculations. The fraction of the total emissive power that is emitted in the wavelength band from 0 to λ_1 ($F_{0-\lambda_1}$, sometimes

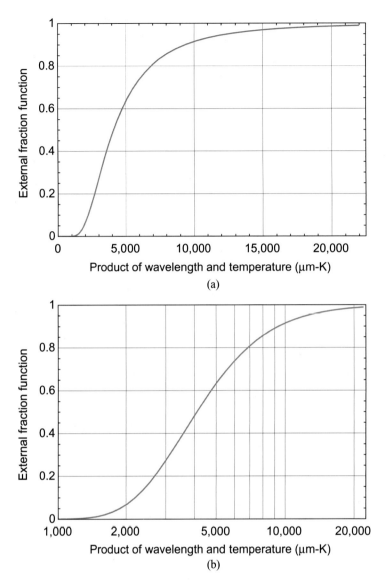

Figure 10-4: Fraction of the total blackbody emission that is emitted between wavelengths 0 and λ_1 at temperature T as a function of $\lambda_1 T$, illustrated using (a) a linear scale and (b) a logarithmic scale.

referred to as the external fractional function) is defined as the ratio of the integral given by Eq. (10-6) to the integral over all wavelengths, Eq. (10-5):

$$F_{0-\lambda_1} = \frac{E_{b,0-\lambda_1}}{E_b} = \int_0^{\lambda_1} \frac{C_1}{\sigma T^4 \lambda^5 \left[\exp\left(\frac{C_2}{\lambda T}\right) - 1 \right]} \, d\lambda \qquad (10\text{-}7)$$

The integral in Eq. (10-7) appears to be a function of both λ and T. However, if the integration variable is transformed from λ to the product of λ and T, then the integral can be written as:

$$F_{0-\lambda_1} = \int_0^{\lambda_1 T} \frac{C_1}{\sigma (\lambda T)^5 \left[\exp\left(\frac{C_2}{\lambda T}\right) - 1 \right]} \, d(\lambda T) \qquad (10\text{-}8)$$

Equation (10-8) has only one integration variable, the product λT, and therefore the quantity $F_{0-\lambda_1}$ can be calculated by numerical integration and tabulated numerically or graphically (as shown in Figure 10-4) as a function of $\lambda_1 T$, where λ_1 is the upper limit of the wavelength band. The value of the external fraction function, $F_{0-\lambda_1}$, can also be obtained using the **Blackbody** function that is provided in EES, as discussed in EXAMPLE 10.2-1.

The blackbody emissive power that is within a range of wavelengths from λ_1 and λ_2 can be found using the difference between two values of the external fraction function, as indicated by Eq. (10-9).

$$F_{\lambda_1 - \lambda_2} = \int_{\lambda_1}^{\lambda_2} \frac{C_1}{\sigma T^4 \lambda^5 \left[\exp\left(\dfrac{C_2}{\lambda T} \right) - 1 \right]} \, d\lambda$$

$$= \int_{0}^{\lambda_2} \frac{C_1}{\sigma T^4 \lambda^5 \left[\exp\left(\dfrac{C_2}{\lambda T} \right) - 1 \right]} \, d\lambda - \int_{0}^{\lambda_1} \frac{C_1}{\sigma T^4 \lambda^5 \left[\exp\left(\dfrac{C_2}{\lambda T} \right) - 1 \right]} \, d\lambda \quad (10\text{-}9)$$

$$= F_{0-\lambda_2} - F_{0-\lambda_1}$$

The emissive power in any specified wavelength region can be directly and more accurately determined using numerical integration, as demonstrated in the following example.

EXAMPLE 10.2-1: UV RADIATION FROM THE SUN

For the purposes of calculating the spectral distribution of the radiation received by Earth, the sun can be approximated as a blackbody that is at 5780 K.

a) Determine the fractions of the total radiation received from the sun that are in the UV-A (0.32 μm to 0.38 μm), UV-B (0.28 μm to 0.32 μm) and UV-C (0.20 μm to 0.28 μm) wavelength bands.

The fraction of the radiation occurring in any wavelength band (F) can be determined directly by numerically integrating Planck's law, Eq. (10-4), over the wavelength band and dividing the result by the total emissive power, σT^4. The calculations are accomplished with an EES program. The EES **Integral** command is employed to integrate between the limits lambda_1 and lambda_2 and the integration step size is specified to be 0.001 μm.

```
T=5780 [K]                                                    "temp. of the sun"
E_b_lambda=C1#/(lambda^5*(exp(C2#/(lambda*T))-1))             "Planck's law"
F=integral(E_b_lambda,lambda,lambda_1,lambda_2, 0.001 [micron])/(sigma#*T^4)
                                                              "fraction in band"
```

Since the same calculations must be accomplished for three separate ranges (corresponding to UV-A, UV-B, and UV-C), it is convenient to set up a parametric table with columns for the variables lambda_1, lambda_2, and F. The calculated fractions are therefore determined by solving the equations for the three lines in the parametric table, as shown in Figure 1.

EXAMPLE 10.2-1: UV RADIATION FROM THE SUN

EXAMPLE 10.2-1: UV RADIATION FROM THE SUN

	1 λ_1 [micron]	2 λ_2 [micron]	3 F [-]	4 F2 [-]
Run 1	0.2	0.28	0.0197	0.01969
Run 2	0.28	0.32	0.02424	0.02424
Run 3	0.32	0.38	0.05479	0.05479

Figure 1: Parametric table used to compute the UV radiation fractions.

Note that the EES library function, **BlackBody**, accomplishes the same numerical integration (albeit using a different integration technique). Information concerning the **Blackbody** function can be accessed by selecting Function Information from the Options menu and then selecting EES library routines; one of the library routines listed will be Blackbody.LIB. Selecting the Blackbody.LIB folder allows access to the online information for the **Blackbody** and **Eb** functions.

The Blackbody function is used to check the numerical integration.

F2=Blackbody(T,lambda_1,lambda_2)   "fraction determined using the Blackbody function"

The results associated with using the **Blackbody** function are also included in the parametric table shown in Figure 1 and agree with those obtained by numerical integration.

This problem could also be solved graphically using Figure 10-4. This process will be illustrated by calculating the fraction of the solar radiation that is contained in the entire UV wavelength range (i.e., the sum of the UV-A, UV-B, and UV-C radiation) that extends from $\lambda_1 = 0.20\,\mu m$ to $\lambda_2 = 0.38\,\mu m$. The product, $\lambda_1 T$ is:

$$\lambda_1 T = \frac{0.20\,\mu m}{} \left| \frac{5780\ K}{} \right. = 1156\,\mu m\text{-}K$$

and therefore the fraction of the total radiation emitted by the sun that is in the wavelength band between 0 and 0.2 μm is nearly zero, by inspection of Figure 10-4. The product $\lambda_2 T$ is:

$$\lambda_2 T = \frac{0.38\,\mu m}{} \left| \frac{5780\ K}{} \right. = 2196\,\mu m\text{-}K$$

and therefore the fraction of the total radiation emitted by the sun that is in the wavelength band between 0 and 0.38 μm is about 0.10, by careful inspection of Figure 10-4(b). The fraction in the wavelength band between 0.20 and 0.38 μm is the difference between these two results, which in this case is about 0.10. This result can be compared to 0.09853, which is the sum of the three values obtained by numerical integration for UV-A, B, and C in the parametric table. (This sum can be obtained by right clicking on the column containing the variable F and selecting Properties, one of the properties reported for the column is the sum of the entries.) The results obtained graphically using Figure 10-4 cannot be read with high precision, but high precision may not be required in many problems.

10.3 Radiation Exchange between Black Surfaces

10.3.1 Introduction

The net rate of radiation heat transfer to a surface is the difference between the rate of radiation that is emitted by the surface and the rate at which the radiation that is incident on the surface is absorbed. The emission of radiation from a blackbody was discussed in Section 10.2.

In this section, we will examine radiation heat transfer between black surfaces. Black surfaces provide a particularly simple place to start, because they absorb all incident radiation; none of the radiation is reflected or transmitted. As a result, it is only necessary to understand how much radiation is incident on the surface (by considering surfaces that are nearby) in order to complete the radiation problem. The net rate of radiation heat transfer from any black surface is the difference between the rate of radiation that it emits (calculated using Planck's law, as discussed in Section 10.2) and the amount of radiation that is striking the surface. The amount of incident radiation is determined by considering the radiation emitted by other surfaces and their geometric orientation with respect to the surface of interest. This section focuses on the latter part of the problem and is mainly concerned with calculating view factors. View factors are dimensionless ratios that characterize the degree to which two surfaces "see" one another and therefore how efficiently they exchange radiation.

10.3.2 View Factors

Diffuse surfaces emit radiation uniformly in all directions. This behavior reduces the complexity associated with determining the net radiation exchange between black surfaces. In the limit that the surfaces involved in a radiation problem are all diffuse emitters, the fraction of the radiation emitted by one surface that hits another depends only on the relative geometric orientation of the two surfaces, and not on the characteristics of the surfaces themselves. The geometric orientation is captured by the view factor[1]. The view factor, $F_{i,j}$, is defined as the fraction of the total radiation that leaves surface i and goes directly to surface j:

$$F_{i,j} = \frac{\text{radiation leaving surface } i \text{ that goes directly surface } j}{\text{total radiation leaving surface } i} \qquad (10\text{-}10)$$

The words "goes directly" in the definition of the view factor excludes the possibility of radiation emitted by surface i reflecting off of a third surface before finally reaching surface j; black surfaces do not reflect radiation in any case, but this possibility does exist for the non-black surfaces that are considered in Section 10.4. Also, the word "leaving" in the definition of the view factor recognizes that radiation will be reflected and emitted by non-black surfaces; for the black surfaces considered in this section, only radiation emitted by the surface will be "leaving" the surface.

Based on its definition, the view factor between any two surfaces must lie between 0 (for two surfaces that do not see each other at all, e.g., $F_{3,4} = 0$ in Figure 10-5) to unity (for two surfaces that are facing one another across a very small gap, e.g., $F_{1,2} = 1$ in Figure 10-5). In some cases, the view factor between two surfaces can be identified by inspection (for example, any two of the surfaces shown in Figure 10-5). However, in most cases the view factor is not obvious. The general formula that may be used to

---

[1] View factors are also sometimes referred to as shape factors, angle factors, and configuration factors.

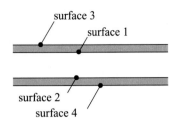

surface 3
surface 1
surface 2
surface 4

Figure 10-5: Four surfaces, numbered 1 through 4.

determine the view factor between two surfaces (i and j, as shown in Figure 10-6) is:

$$A_i F_{i,j} = A_j F_{j,i} = \int\limits_{A_j} \int\limits_{A_i} \frac{\cos(\theta_i)\cos(\theta_j)}{\pi r^2} \, dA_i \, dA_j \qquad (10\text{-}11)$$

where, as illustrated in Figure 10-6, r is the distance between differential areas dA_i and dA_j on the two surfaces, θ_i is the angle between the normal to dA_i and a vector connecting areas dA_i and dA_j, and θ_j is the angle between the normal to dA_j and a vector connecting areas dA_i and dA_j.

The actual setup and evaluation of Eq. (10-11) is usually difficult and often requires numerical integration. The integral provided by Eq. (10-11) is never explicitly set up and evaluated in this text. View factors for many geometries have already been determined are accessible from the view factor library in EES. The use of these view factor solutions is facilitated by the view factor rules and relationships that are presented in this section. The crossed-and-uncrossed strings method presented in this section can be used to compute the view factor between two arbitrary 2-D surfaces and Section 10.7 illustrates how the Monte Carlo technique can be used to determine view factors.

The Enclosure Rule
All of the radiation that is emitted by a surface that is part of an enclosure must strike some surface within the enclosure. For example, consider the enclosure shown in Figure 10-7. All of the radiation that is emitted by surface 1 must hit one of the surfaces in the enclosure (i.e., surface 1, 2, or 3); therefore:

$$F_{1,1} + F_{1,2} + F_{1,3} = 1 \qquad (10\text{-}12)$$

Similarly, all of the radiation emitted by surface 2 must hit surfaces 1, 2, or 3:

$$F_{2,1} + F_{2,2} + F_{2,3} = 1 \qquad (10\text{-}13)$$

In general, if surface i is part of an N surface enclosure then:

$$\boxed{\sum_{j=1}^{N} F_{i,j} = 1} \qquad (10\text{-}14)$$

Equation (10-14) can be written for each of the N surfaces in the enclosure.

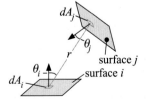

dA_j
θ_j
r
surface j
surface i
θ_i
dA_i

Figure 10-6: Two surfaces exchanging radiation.

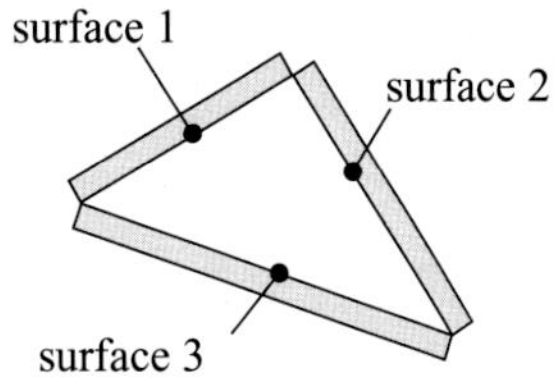

Figure 10-7: Three surfaces in an enclosure.

It is important to note that the view factor between a surface and itself ($F_{i,i}$) is not necessarily zero. For the flat surfaces shown in Figure 10-7, it is clear that $F_{1,1}$, $F_{2,2}$, and $F_{3,3}$ will all be zero because no part of these surfaces can see another part of the same surface. However, many surfaces are not flat. For example, radiation leaving surface 1 in Figure 10-8 will also hit surface 1 and therefore $F_{1,1}$ will not be zero and must be included in the enclosure rule.

It is often useful to define an imaginary surface where none exists in order to invoke the enclosure rule for a set of surfaces that would not otherwise form an enclosure. The imaginary surface characterizes radiation that passes through a region of space. For example, in Figure 10-8 surface 3 is defined in order to include all of the radiation that passes through the opening between the edges of surfaces 1 and 2.

Reciprocity

Figure 10-9 shows two black surfaces, i and j. Surface i has area A_i and is at temperature T_i while surface j has area A_j and is at temperature T_j. The total amount of radiation that is emitted by surface i is the product of its area and the blackbody emissive power of surface i:

$$\text{radiation emitted by surface } i = A_i \sigma T_i^4 \qquad (10\text{-}15)$$

According to the definition of the view factor, the radiation that is emitted by surface i and directly hits surface j is:

$$\text{radiation emitted by surface } i \text{ that hits surface } j = F_{i,j} A_i \sigma T_i^4 \qquad (10\text{-}16)$$

Similarly, the radiation that is emitted by surface j and hits surface i is:

$$\text{radiation emitted by surface } j \text{ that hits surface } i = F_{j,i} A_j \sigma T_j^4 \qquad (10\text{-}17)$$

The net radiation exchange from surface i to surface j, $\dot{q}_{i \text{ to } j}$, is the difference between Eq. (10-16) and Eq. (10-17).

$$\dot{q}_{i \text{ to } j} = F_{i,j} A_i \sigma T_i^4 - F_{j,i} A_j \sigma T_j^4 \qquad (10\text{-}18)$$

If surfaces i and j are otherwise adiabatic, then eventually their temperatures must equilibrate (i.e., eventually, $T_i = T_j = T$). At this point $\dot{q}_{i \text{ to } j} = 0$ and therefore:

$$F_{i,j} A_i \sigma T^4 - F_{j,i} A_j \sigma T^4 = 0 \qquad (10\text{-}19)$$

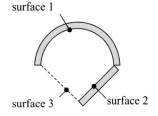

Figure 10-8: A three surface enclosure including a non-flat surface, surface 1, and an imaginary surface, surface 3.

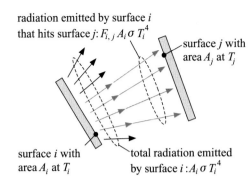

radiation emitted by surface i that hits surface j: $F_{i,j} A_i \sigma T_i^4$

surface j with area A_j at T_j

surface i with area A_i at T_i

total radiation emitted by surface i: $A_i \sigma T_i^4$

Figure 10-9: Two surfaces, i and j, exchanging radiation.

In order for Eq. (10-19) to be true, it is necessary that:

$$F_{i,j} A_i = F_{j,i} A_j \qquad (10\text{-}20)$$

Equation (10-20) was derived in the limit that the temperatures of surfaces 1 and 2 are equal. However, the areas and view factors in Eq. (10-20) are geometric quantities that do not depend on temperature; therefore, the relationship must be true whether or not the temperatures are equal. Equation (10-20) is called the reciprocity theorem and is a useful relationship between view factors.

Other View Factor Relationships

There are some additional helpful view factor relationships. Any two surfaces (j and k) can be combined; then, according to the definition of the view factor:

$$F_{i,jk} = F_{i,j} + F_{i,k} \qquad (10\text{-}21)$$

Using reciprocity together with Eq. (10-21) results in

$$(A_i + A_j) F_{ij,k} = A_i F_{i,k} + A_j F_{j,k} \qquad (10\text{-}22)$$

The property of symmetry can also sometimes be used to help determine view factors. For example, consider a sphere, surface 1, placed between two infinite parallel plates, surfaces 2 and 3, as shown in Figure 10-10. Since the sphere cannot see itself, Eq. (10-14) requires that $F_{1,2} + F_{1,3} = 1$. However, by symmetry (i.e., the fact that surface 1 "sees" surfaces 2 and 3 equally), $F_{1,2} = F_{1,3}$. Therefore, $F_{1,2} = F_{1,3} = 0.50$.

The Crossed and Uncrossed Strings Method

A useful method for calculating view factors in 2-D geometries is the crossed and uncrossed strings method, described in Hottel and Sarofim (1967) and McAdams (1954). The process of using the method is illustrated in Figure 10-11(a); notice the 4 "strings"

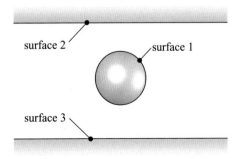

surface 2

surface 1

surface 3

Figure 10-10: Spherical surface between two infinite parallel plates.

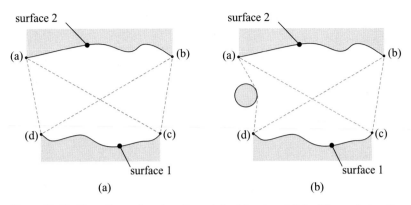

Figure 10-11: Two-dimensional surfaces (a) without and (b) with an obstruction.

in the figure corresponding to the four dotted lines (two "crossed strings" L_{ac} and L_{bd} and two "uncrossed" strings L_{ad} and L_{bc}). The crossed and uncrossed strings method provides the view factors between surfaces 1 and 2, according to:

$$A_1 F_{1,2} = A_2 F_{2,1} = W \frac{\sum L_{crossed} - \sum L_{uncrossed}}{2} = W \frac{(L_{ac} + L_{bd}) - (L_{ad} + L_{bc})}{2}$$

(10-23)

where L refers to the length of strings connecting the end points of the surfaces (L_{ac} and L_{bd} are the lengths of the crossed strings while L_{ad} and L_{bc} are the lengths of the uncrossed strings) and W is the width of the surfaces into the page. The method is applicable even if there is an obstruction between the surfaces, as indicated in Figure 10-11(b). In this case, the string must wrap around the obstruction, which affects its length. The method is only valid for 2-D geometries.

EXAMPLE 10.3-1: CROSSED AND UNCROSSED STRING METHOD

A long beam has a triangular cavity installed along its bottom edge that is $H_1 = 1.0$ m deep and $H_2 = 1.0$ m wide at its base, as shown in Figure 1. A plate that is $H_3 = 0.75$ m wide is positioned parallel to the beam at a distance $H_4 = 0.5$ m below its lower surface. The right edge of the plate is directly under the center of the triangular cavity and the left edge of the beam is directly above the left edge of the plate. The triangular cavity in the beam is referred to as surface 1, the top surface of the plate is referred to as surface 2, and the surroundings are referred to as surface 3.

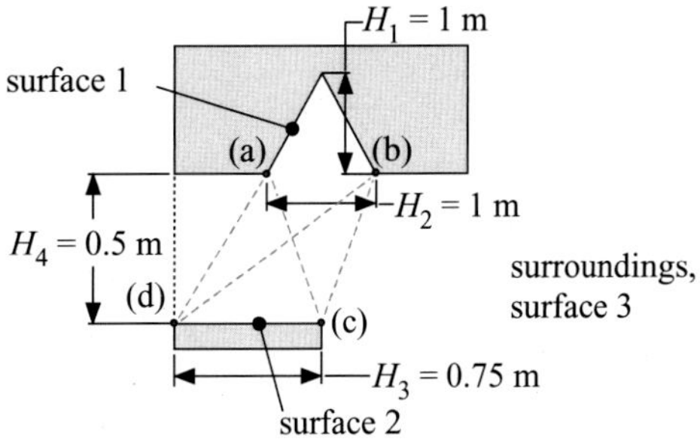

Figure 1: Long beam with triangular cavity.

EXAMPLE 10.3-1: CROSSED AND UNCROSSED STRING METHOD

EXAMPLE 10.3-1: CROSSED AND UNCROSSED STRING METHOD

a) Determine the view factor from the cavity to the plate ($F_{1,2}$).

The inputs are entered in EES:

```
"EXAMPLE 10.3-1: View Factor using the Crossed and Uncrossed Strings Method"

$UnitSystem SI MASS RAD PA K J
$Tabstops 0.2 0.4 0.6 3.5 in

"Inputs"
H_1=1.0 [m]                          "depth of the cavity"
H_2=1.0 [m]                          "width of the cavity"
H_3=0.75 [m]                         "width of the plate"
H_4=0.5 [m]                          "distance from the plate to the cavity opening"
```

Since the beam and plate extend for an indeterminate distance (W) into the paper, the geometry is two-dimensional and the problem is done on a per unit depth basis.

```
W=1 [m]                              "per unit length"
```

The simplest way to determine the view factors for this 2-D situation is to use the crossed and uncrossed string method. The area of the triangular cavity, A_1, is:

$$A_1 = 2 \, W \, \sqrt{H_1^2 + (H_2/2)^2} \tag{1}$$

The area of the plate, A_2, is

$$A_2 = W \, H_3 \tag{2}$$

```
A[1]=2*W*sqrt(H_1^2+(H_2/2)^2)       "area of the cavity"
A[2]=W*H_3                           "area of the plate"
```

The view factors between surfaces 1 and 2 are found using Eq. (10-23):

$$A_1 \, F_{1,2} = A_2 \, F_{2,1} = W \frac{\sum L_{\text{crossed}} - \sum L_{\text{uncrossed}}}{2} \tag{3}$$

or, for this problem:

$$A_1 \, F_{1,2} = A_2 \, F_{2,1} = W \frac{(L_{ac} + L_{bd}) - (L_{ad} + L_{bc})}{2} \tag{4}$$

The lengths of the "strings" L_{ac}, L_{bd}, L_{ad}, and L_{bc} in Eq. (4) are found from trigonometry:

$$L_{ac} = \sqrt{(H_2/2)^2 + H_4^2} \tag{5}$$

$$L_{bd} = \sqrt{(H_2/2 + H_3)^2 + H_4^2} \tag{6}$$

$$L_{ad} = \sqrt{(H_3 - H_2/2)^2 + H_4^2} \tag{7}$$

$$L_{bc} = \sqrt{(H_2/2)^2 + H_4^2} \tag{8}$$

EXAMPLE 10.3-1: CROSSED AND UNCROSSED STRING METHOD

```
L_ac=sqrt((H_2/2)^2+H_4^2)                    "crossed strings"
L_bd=sqrt((H_2/2+H_3)^2+H_4^2)
L_ad=sqrt((H_3-H_2/2)^2+H_4^2)                "uncrossed strings"
L_bc=sqrt((H_2/2)^2+H_4^2)
A[2]*F[2,1]=W*((L_ac+L_bc)-(L_ad+L_bc))/2     "crossed and uncrossed strings method"
A[1]*F[1,2]=A[2]*F[2,1]                        "reciprocity"
```

which leads to $F_{1,2} = 0.033$ and $F_{2,1} = 0.099$. Examination of Figure 1 suggests that these are reasonable values.

b) Determine the view factor between surface 1 and itself (i.e., find $F_{1,1}$).

The view factor $F_{1,1}$ can be found in several ways. Perhaps the easiest way to determine $F_{1,1}$ is to place an imaginary surface, call it surface 4, across the opening of the cavity (i.e., between points (a) and (b)) in order to form an enclosure that includes only surfaces 1 and 4, as shown in Figure 2.

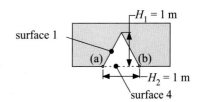

Figure 2: Surface 4 used to form a two-surface enclosure.

The area of surface 4 is:

$$A_4 = W\, H_2 \tag{9}$$

The enclosure rule written for surface 4 is:

$$F_{4,1} + F_{4,4} = 1.0 \tag{10}$$

Surface 4 is flat and cannot "see" itself; therefore, $F_{4,4}$ must be zero and:

$$F_{4,1} = 1.0 \tag{11}$$

By reciprocity:

$$A_4\, F_{4,1} = A_1\, F_{1,4} \tag{12}$$

so:

$$F_{1,4} = \frac{A_4}{A_1}\, F_{4,1} \tag{13}$$

The enclosure rule written for surface 1 is:

$$F_{1,1} + F_{1,4} = 1.0 \tag{14}$$

EXAMPLE 10.3-1: CROSSED AND UNCROSSED STRING METHOD

Substituting Eqs. (9) and (11) into Eq. (14) leads to:

$$F_{1,1} = 1 - \frac{W\, H_2}{A_1} \tag{15}$$

F[1,1]=1-W*H_2/A[1]   "view factor of surface 1 to itself"

which leads to $F_{1,1} = 0.553$; this is a reasonable value given that the two sides of the triangular enclosure are oriented towards one another.

c) Determine the view factor from the triangular enclosure to the surroundings (i.e., determine $F_{1,3}$).

All of the radiation that leaves surface 1 must either strike surface 2 (the plate) or surface 3 (the surroundings); therefore:

$$F_{1,3} = 1 - F_{1,1} - F_{1,2} \tag{16}$$

F[1,3]=1-F[1,1]-F[1,2]   "view factor between surface 1 to 3"

which leads to $F_{1,3} = 0.414$.

View Factor Library

The double integral that provides the view factor between two arbitrary surfaces, Eq. (10-11), is difficult to evaluate even for relatively simple geometries. Fortunately, view factors have already been determined for many common situations. For example, Siegel and Howell (2002) provide a summary of view factor formulae and a larger collection of view factors in either graphical or analytical form is provided on a web site compiled by Howell (http://www.me.utexas.edu/~howell/). A few of these view factor formulae are summarized in Table 10-1. A much more comprehensive set of view factor relations have been implemented as EES library functions; these are accessed from the Function Information Window by selecting Radiation View Factors from the drop down menu of heat exchanger functions. The view factors are sub-classified into 2-D, 3-D, or differential view factors. The 2-D view factors are appropriate when one dimension is much longer than the other dimensions in the problem so that this dimension can be assumed to be infinite. The view factors for these situations can be derived from the Hottel crossed and uncrossed string method presented in Section 10.3.2, but even so, the algebra involved is often cumbersome and therefore it is convenient to have a library of common 2-D view factors.

The 3-D view factors relate surfaces of finite size in all dimensions. Analytical solutions exist for simple geometries, such as for parallel or perpendicular plates of specified size. View factors for more complicated situations can often be determined from the simpler view factor relations using reciprocity and the other view factor relations discussed in Section 10.3.2.

The differential area view factors also involve 3-D geometries; however, one of the surfaces is of differential size and therefore numerical integration is required in order to determine the net radiation exchange. The advantage of the differential area view factors is that the surface represented by the differential element can be of arbitrary description; for example, the temperature may vary spatially.

Table 10-1: View factors.

Parallel plates $(L \gg W)$	Plates joined at an angle $(L \gg W)$	Plate to cylinder $(L \gg r, b_1)$

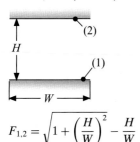

Parallel plates $(L \gg W)$

$$F_{1,2} = \sqrt{1 + \left(\frac{H}{W}\right)^2} - \frac{H}{W}$$

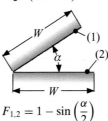

Plates joined at an angle $(L \gg W)$

$$F_{1,2} = 1 - \sin\left(\frac{\alpha}{2}\right)$$

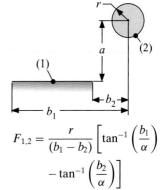

Plate to cylinder $(L \gg r, b_1)$

$$F_{1,2} = \frac{r}{(b_1 - b_2)}\left[\tan^{-1}\left(\frac{b_1}{\alpha}\right) - \tan^{-1}\left(\frac{b_2}{\alpha}\right)\right]$$

Semicircle to itself w/concentric cylinder $(L \gg r_1)$

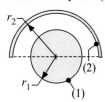

$$F_{2,2} = 1 - \frac{2}{\pi}\left[\sqrt{1 - \left(\frac{r_1}{r_2}\right)^2} + \frac{r_1}{r_2}\sin^{-1}\left(\frac{r_1}{r_2}\right)\right]$$

Sphere to coaxial disk

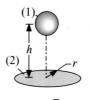

$$F_{1,2} = \frac{1}{2}\left[-\frac{1}{\sqrt{1 + \left(\frac{r}{h}\right)^2}}\right]$$

Cylinder to cylinder $(L \gg r)$

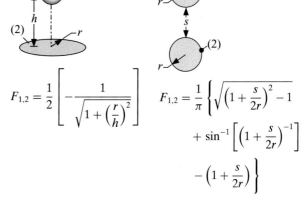

$$F_{1,2} = \frac{1}{\pi}\left\{\sqrt{\left(1 + \frac{s}{2r}\right)^2 - 1} + \sin^{-1}\left[\left(1 + \frac{s}{2r}\right)^{-1}\right] - \left(1 + \frac{s}{2r}\right)\right\}$$

Coaxial disks

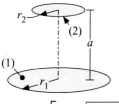

$$F_{1,2} = \frac{1}{2}\left[S - \sqrt{S^2 - 4\left(\frac{r_2}{r_1}\right)^2}\right]$$

$$S = 1 + \frac{1 + \left(\frac{r_2}{\alpha}\right)^2}{\left(\frac{r_1}{\alpha}\right)^2}$$

Sphere to a cylinder

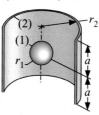

$$F_{1,2} = \frac{1}{\sqrt{1 + \left(\frac{r_2}{\alpha}\right)^2}}$$

Base of cylinder to internal sides

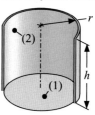

$$F_{1,2} = \frac{h}{r}\left(\sqrt{1 + \left(\frac{h}{2r}\right)^2} - \frac{h}{2r}\right)$$

EXAMPLE 10.3-2: THE VIEW FACTOR LIBRARY

An enclosure (Figure 1) is $H = 3$ m high with a base that is $L = 3$ m by $W = 1.5$ m.

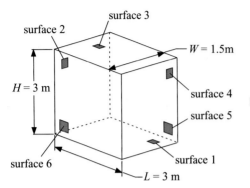

surface 3
surface 2
$W = 1.5$m
$H = 3$ m
surface 4
surface 5
Figure 1: Enclosure.
surface 6
surface 1
$L = 3$ m

a) Determine the view factors between all of the surfaces.

The inputs are entered in EES:

```
"EXAMPLE 10.3-2: The View Factor Library"

$UnitSystem SI MASS RAD PA K J
$Tabstops 0.2 0.4 0.6 3.5 in

L=3 [m]                                    "length of base"
W=1.5[m]                                   "width of base"
H=3 [m]                                    "height of enclosure"
```

The first step is to determine the area of each surface. It will be convenient to use array variables in this problem (and most radiation problems) since the calculations that are common for each surface can then be accomplished concisely using a duplicate loop in EES.

```
"area of each of the six surfaces"
A[1]=L*W                                   "area of floor"
A[2]=W*H                                   "area of left side"
A[3]=A[1]                                  "area of ceiling"
A[4]=A[2]                                  "area of right side"
A[5]=L*H                                   "area of rear"
A[6]=A[5]                                  "area of front"
```

It is necessary to determine each of the view factors, $F_{i,j}$. Since there are 6 surfaces, there are 36 view factors. All of the surfaces are flat, therefore $F_{i,i} = 0$ for $i = 1$ to 6.

```
duplicate i=1,6
        F[i,i]=0          "view factor for each surface to itself"
end
```

EXAMPLE 10.3-2: THE VIEW FACTOR LIBRARY

We start by specifying the other view factors for surface 1 to all other surfaces. Surfaces 1 and 2 share a common edge and are perpendicular to each other. The view factor is known for this geometry and a function that returns this view factor is provided in an EES library function. To see the existing library of view factors, select Function Info from the Options menu. Six radio buttons are provided at the top of this dialog. Click the bottom right radio button and then select Radiation View Factors from the control to the right of the button. View factors are classified as 2-Dimensional, 3-Dimensional and Differential. Select the 3-Dimensional subclass using the sub-class control. Next use the scroll bar below the picture to scroll to the desired geometry; the view factor for this geometry is provided by the function **F3D_2**. The geometry in the function **F3D_2** is represented in terms of dimensions a, b, and c. In our case, $a = L$, $b = H$, and $c = W$.

"view factors for surface 1"
F[1,2]=F3D_2(L,H,W)   "view factor from surface 1 to 4"

The view factor between surface 1 and surfaces 4, 5, and 6 can be obtained using the same view factor function:

F[1,2]=F3D_2(L,H,W)   "view factor from surface 1 to 2"
F[1,4]=F3D_2(L,H,W)   "view factor from surface 1 to 4"
F[1,5]=F3D_2(W,H,L)   "view factor from surface 1 to 5"
F[1,6]=F3D_2(W,H,L)   "view factor from surface 1 to 6"

Surfaces 1 and 3 are aligned and parallel and therefore the view factor is provided by the function **F3D_1**. The value of $F_{1,3}$ is provided by **F3D_1**, with $a = L$, $b = W$ and $c = H$.

F[1,3]=F3D_1(L,W,H)

The process is repeated for the other five surfaces. However, it is only necessary to specify the view factor values $F_{i,j}$ for which $i \leq j$ because the view factors for which $j < i$ can be determined subsequently using reciprocity. For surface 2,

"view factors for surface 2"
F[2,3]=F3D_2(H,L,W)
F[2,4]=F3D_1(H,W,L)
F[2,5]=F3D_2(W,L,H)
F[2,6]=F3D_2(W,L,H)

for surface 3,

"view factors for surface 3"
F[3,4]=F3D_2(L,H,W)
F[3,5]=F3D_2(W,H,L)
F[3,6]=F3D_2(W,H,L)

EXAMPLE 10.3-2: THE VIEW FACTOR LIBRARY

for surface 4,

```
"view factors for surface 4"
F[4,5]=F3D_2(W,L,H)
F[4,6]=F3D_2(W,L,H)
```

and for surface 5:

```
"view factor for surface 5"
F[5,6]=F3D_1(H,L,W)
```

Notice that the view factors in the Arrays Table is, at this point, upper triangular. Reciprocity relates $F_{i,j}$ to $F_{j,i}$ and can therefore be used to fill in the lower half of the view factor array matrix:

$$A_i\, F_{i,j} = A_j\, F_{j,i} \quad \text{for } i = 2\ldots 6 \quad \text{and} \quad j = 1\ldots(i-1)$$

```
"use reciprocity to get remaining view factors"
duplicate i=2,6
   duplicate j=1,(i-1)
      A[i]*F[i,j]=A[j]*F[j,i]
   end
end
```

Finally, it is good practice to check that the view factors for any surface to all of the other surfaces in the problem sum to 1 (i.e., to verify that the enclosure rule is satisfied):

$$\sum_{j=1}^{6} F_{i,j} = 1 \quad \text{for } i = 1\ldots 6$$

```
"check that enclosure rule is satisfied"
duplicate i=1,6
   sumF[i]=sum(F[i,1..6])
end
```

All of the 36 view factor values are displayed in the Arrays window (Figure 2) which shows also that the enclosure rule is satisfied.

Sort	A_i [m²]	$F_{i,1}$ [-]	$F_{i,2}$ [-]	$F_{i,3}$ [-]	$F_{i,4}$ [-]	$F_{i,5}$ [-]	$F_{i,6}$ [-]	sumF$_i$
[1]	4.5	0	0.1493	0.1167	0.1493	0.2924	0.2924	1
[2]	4.5	0.1493	0	0.1493	0.1167	0.2924	0.2924	1
[3]	4.5	0.1167	0.1493	0	0.1493	0.2924	0.2924	1
[4]	4.5	0.1493	0.1167	0.1493	0	0.2924	0.2924	1
[5]	9	0.1462	0.1462	0.1462	0.1462	0	0.4153	1
[6]	9	0.1462	0.1462	0.1462	0.1462	0.4153	0	1

Figure 2: Arrays table with view factor solution.

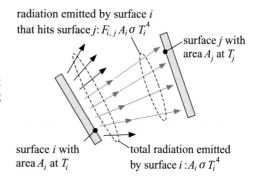

Figure 10-12: Two black surfaces, i and j, at different temperatures, T_i and T_j, exchanging radiation.

10.3.3 Blackbody Radiation Calculations

Figure 10-12 shows two black surfaces, i and j, that are exchanging radiation. Surface i has area A_i and is at temperature T_i while surface j has area A_j and is at temperature T_j. The total rate of radiation emitted by surface i is the product of its area and blackbody emissive power:

$$\text{radiation emitted by surface } i = A_i \sigma T_i^4 \qquad (10\text{-}24)$$

According to the definition of the view factor, the radiation that is emitted by surface i and incident on surface j is:

$$\text{radiation emitted by surface } i \text{ that hits surface } j = F_{i,j} A_i \sigma T_i^4 \qquad (10\text{-}25)$$

Similarly, the radiation that is emitted by surface j and hits surface i is:

$$\text{radiation emitted by surface } j \text{ that hits surface } i = F_{j,i} A_j \sigma T_j^4 \qquad (10\text{-}26)$$

The net rate of radiation exchange from surface i to surface j, $\dot{q}_{i\,\text{to}\,j}$, is the difference between Eq. (10-25) and Eq. (10-26).

$$\dot{q}_{i\,\text{to}\,j} = F_{i,j} A_i \sigma T_i^4 - F_{j,i} A_j \sigma T_j^4 \qquad (10\text{-}27)$$

Reciprocity, Eq. (10-20), requires that:

$$F_{i,j} A_i = F_{j,i} A_j \qquad (10\text{-}28)$$

so that Eq. (10-27) can be simplified:

$$\dot{q}_{i\,\text{to}\,j} = A_i F_{i,j} \sigma (T_i^4 - T_j^4) = A_i F_{i,j}(E_{b,i} - E_{b,j}) \qquad (10\text{-}29)$$

The Space Resistance

Equation (10-29) has the same form as a resistance equation. The net rate of radiation heat transfer between two black surfaces i and j ($\dot{q}_{i\,\text{to}\,j}$) is driven by a difference in their blackbody emissive powers, $E_{b,i} - E_{b,j}$, and the resistance to the radiation heat transfer is the inverse of the product of the area and view factor:

$$\dot{q}_{i\,\text{to}\,j} = \frac{(E_{b,i} - E_{b,j})}{R_{i,j}} \qquad (10\text{-}30)$$

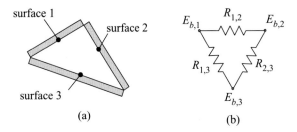

Figure 10-13: (a) Three surface enclosure and (b) corresponding resistance network.

The denominator of Eq. (10-30), $R_{i,j}$, is sometimes referred to as the surface-to-surface, geometrical, or space resistance:

$$R_{i,j} = \frac{1}{A_i F_{i,j}} = \frac{1}{A_j F_{j,i}} \tag{10-31}$$

The space geometric resistance tends to increase as either the area of the surface or the view factor between the surfaces is reduced; this makes sense as reducing the area or view factor will reduce the ease with which two surfaces can interact radiatively.

For radiation problems involving a few surfaces (for example, the three surface enclosure shown in Figure 10-13(a)) it is convenient to draw a resistance network where each node represents the emissive power of an isothermal surface and these nodes are connected by space resistances that are calculated using Eq. (10-31), as shown in Figure 10-13(b). The solution to black surface radiation exchange problems using a resistance network is illustrated in EXAMPLE 10.3-3.

EXAMPLE 10.3-3: APPROXIMATE TEMPERATURE OF THE EARTH

The average temperature of the earth can be estimated, very approximately, using the concepts discussed in this section. The earth interacts primarily with two surfaces, the sun and outer space (Figure 1). The effective temperature of the surface of the sun (surface 1) is approximately $T_1 = 5800$ K and the diameter of the sun, D_1, is 1.390×10^9 m. The effective temperature of space (surface 3) is approximately $T_3 = 2.7$ K. The diameter of the earth is $D_2 = 1.276 \times 10^7$ m. The distance between the earth and the sun varies throughout the year but is, on average, $R = 1.497 \times 10^{11}$ m.

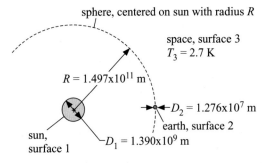

Figure 1: The earth and its relationship with the sun (not to scale).

a) Estimate the surface temperature of earth based on radiation exchange assuming that both the sun and the earth are black.

The inputs are entered in EES:

```
"EXAMPLE 10.3-3: Temperature of the Earth"

$UnitSystem SI MASS RAD PA K J
$Tabstops 0.2 0.4 0.6 3.5 in

"Inputs"
D_1=1.390e9 [m]                          "diameter of the sun"
T_1=5800 [K]                             "temperature of the sun"
T_3=2.7 [K]                              "temperature of space"
R=1.497e11 [m]                           "distance between the earth and the sun"
D_2=1.276e7 [m]                          "diameter of earth"
```

There are three surfaces involved in the problem; therefore, the resistance network that represents the situation must include three nodes, as shown in Figure 2. The nodes represent the blackbody emissive powers of each surface and the resistances between the nodes represent the degree to which the surfaces interact. The boundary conditions associated with the resistance network include the specified temperature of the sun (surface 1) and space (surface 3); their blackbody emissive powers are:

$$E_{b,1} = \sigma\, T_1^4$$

$$E_{b,3} = \sigma\, T_3^4$$

```
E_b_1=sigma#*T_1^4                       "emissive power of sun"
E_b_3=sigma#*T_3^4                       "emissive power of space"
```

We will assume that the earth does not store energy; that is, all of the energy received from the sun is radiated to space and therefore the earth is adiabatic.

Figure 2: Resistance network.

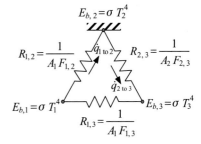

The sun-to-space space resistance is:

$$R_{1,3} = \frac{1}{A_1\, F_{1,3}}$$

where the surface area of the sun is:

$$A_1 = \pi\, D_1^2$$

EXAMPLE 10.3-3: APPROXIMATE TEMPERATURE OF THE EARTH

EXAMPLE 10.3-3: APPROXIMATE TEMPERATURE OF THE EARTH

The view factor between the sun and space is essentially, unity ($F_{1,3} \approx 1$; an observer on the surface of the sun would see much more of space than anything else).

```
A_1=pi*D_1^2                    "surface area of sun"
F_13=1                          "view factor between the sun and space"
R_13=1/(A_1*F_13)               "sun-to-space resistance"
```

Similarly, the earth-to-space space resistance ($R_{2,3}$) is:

$$R_{2,3} = \frac{1}{A_2 \, F_{2,3}}$$

where

$$A_2 = \pi \, D_2^2$$

The view factor between the earth and space is also nearly unity ($F_{2,3} \approx 1$) for the same reason.

```
A_2=pi*D_2^2                    "surface area of earth"
F_23=1                          "view factor between the earth and the space"
R_23=1/(A_2*F_23)               "earth-to-space resistance"
```

The view factor between the sun and the earth ($F_{1,2}$) is nearly zero. However, clearly $F_{1,2}$ is not exactly zero or the earth would be a very cold place. The value of $F_{1,2}$ can be determined by imagining a sphere of radius R centered at the sun (see Figure 1); the view factor is the ratio of the projected area of the earth on the sphere to the surface area of the sphere:

$$F_{1,2} = \frac{\pi \, \dfrac{D_2^2}{4}}{4 \, \pi \, R^2}$$

The sun-to-earth resistance ($R_{1,2}$) is:

$$R_{1,2} = \frac{1}{A_1 \, F_{1,2}}$$

```
F_12=(pi*D_2^2/4)/(4*pi*R^2)    "view factor between the sun and the earth"
R_12=1/(A_1*F_12)               "sun-to-earth resistance"
```

which leads to $F_{1,2} = 4.5 \times 10^{-10}$. This result indicates that about 5 out of every 10 billion photons emitted by the sun ultimately strike the surface of the earth. Because the earth-sun system is assumed to be at steady state, the rate at which heat is transferred from the sun to the earth and then re-radiated to space is:

$$\dot{q}_{1 \text{ to } 2} = \dot{q}_{2 \text{ to } 3} = \frac{(E_{b,1} - E_{b,3})}{R_{1,2} + R_{2,3}}$$

EXAMPLE 10.3-3: APPROXIMATE TEMPERATURE OF THE EARTH

```
q_dot_1to2=(E_b_1-E_b_3)/(R_12+R_23)          "sun-to-earth heat transfer rate"
q_dot_2to3=q_dot_1to2                          "earth-to-space heat transfer rate"
```

which leads to a sun-to-earth heat transfer rate of 1.74×10^{17} W. The blackbody emissive power of the earth ($E_{b,2}$) is:

$$E_{b,2} = E_{b,1} - \dot{q}_{1 \text{ to } 2} R_{1,2}$$

The temperature of the earth (T_2) is related to its blackbody emissive power:

$$T_2 = \left(\frac{E_{b,2}}{\sigma} \right)^{1/4}$$

```
E_b_2=E_b_1-q_dot_1to2*R_12          "blackbody emissive power of earth"
T_2=(E_b_2/sigma#)^(1/4)             "temperature of the earth"
T_2F=converttemp(K,F,T_2)            "temperature of the earth in deg. F"
T_2C=converttemp(K,C,T_2)            "temperature of the earth in deg. C"
```

These calculations lead to a temperature of the earth of 279.4 K (6.3°C or 43.3°F). This result leads into a discussion of several important issues, most notably the greenhouse effect and global warming. The calculated earth temperature is reasonable, but low relative to the observed average temperature of the earth. The average temperature of the earth varies somewhat year-to-year but is approximately 14°C or about 8°C higher than our prediction (6.3°C). There are several important effects that we've neglected in our analysis; however, one of the major reasons for the discrepancy between the observed and calculated temperature of the earth is the greenhouse effect that is caused by absorption of the radiation emitted by the earth in the atmosphere. The sun-to-earth radiation heat transfer that passes through the atmosphere ($\dot{q}_{1 \text{ to } 2}$) and the earth-to-space radiation heat transfer ($\dot{q}_{2 \text{ to } 3}$) are, according to our analysis, identical as we have assumed that the earth is adiabatic. However, as we learned in Section 10.2, the radiation associated with these two heat transfers is spectrally distributed very differently. The energy hitting the earth is emitted by an object at high temperature (the sun) and therefore, according to Planck's law, will be concentrated at lower wavelengths. The energy leaving the earth is emitted by an object at much lower temperature (the earth) and therefore is spread out over a wider range of much higher wavelengths. Figure 3 shows the spectral distribution of the total power emitted by the earth (assuming that it is at 280 K) and the sun (assuming that it is at 5800 K). Note that the integral of these spectral distributions over all wavelengths is the same (1.74×10^{17} W), although the logarithmic x-scale makes the areas look very different.

EXAMPLE 10.3-3: APPROXIMATE TEMPERATURE OF THE EARTH

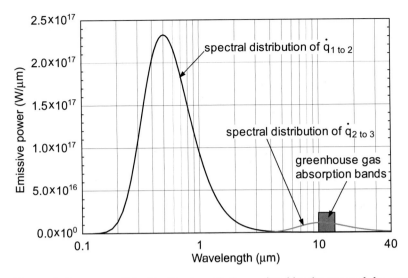

Figure 3: Spectrum of the blackbody radiation emitted by the sun and the earth.

There are strong absorption bands associated with the carbon dioxide and water vapor in the atmosphere. These bands lie at long wavelengths and therefore tend to preferentially "trap" the energy emitted by the earth so that some of it does not make it through the atmosphere. This selective absorption of energy in specific wavelength bands is referred to as the greenhouse effect and it is one reason that the temperature of the earth is higher than the value that we calculated. Global warming is primarily related to the gradual buildup of carbon dioxide in the atmosphere due to the combustion of fossil fuels. The carbon dioxide released by combustion tends to increase the greenhouse gas absorption bands shown in Figure 3. Therefore, the average temperature of the earth must rise in order to re-establish the thermal equilibrium, $\dot{q}_{1\,to\,2} = \dot{q}_{2\,to\,3}$.

N-Surface Solutions

The representation of a blackbody radiation problem with a resistance network, as discussed in the previous example, can be a useful method of visualizing the problem. However, for even a relatively modest number of surfaces this technique is not very practical. Typically, each surface interacts with all of the other surfaces involved in the problem and therefore a network involving more than 3 or 4 surfaces becomes more confusing than it is helpful. However, it is possible to systematically solve a blackbody radiation problem that involves an arbitrary number of surfaces provided that the view factors and areas of each surface are known and a boundary condition can be defined for each surface.

The net rate of radiation exchange from surface i to all of the other N surfaces in a problem is obtained by summing Eq. (10-29) over all of the surfaces involved:

$$\dot{q}_i = A_i \sigma \sum_{j=1}^{N} F_{i,j}\left(T_i^4 - T_j^4\right) = A_i \sum_{j=1}^{N} F_{i,j}\left(E_{b,i} - E_{b,j}\right) \quad \text{for } i = 1\ldots N \qquad (10\text{-}32)$$

Equation (10-32), written for each of the N surfaces, provides N equations in $2\,N$ unknowns (the temperature and the net heat transfer associated with each of the N surfaces). A complete set of boundary conditions will include a specification of either the temperature or net heat transfer rate (or a relationship between these quantities) for each of the surfaces, providing N additional equations and therefore a completely specified problem. The equations are non-linear with respect to temperature but they

can be solved directly using EES. The solution is facilitated by using arrays and duplicate loops, as demonstrated in the following example. The equations represented by Eq.
(10-32) together with a set of boundary conditions are linear in $\dot{q}_i$ and $E_{b,i}$, but not in
temperature, T_i. Therefore, they can be placed in matrix form and solved in MATLAB
provided that the blackbody emissive power (i.e., the temperature) or the heat flow is
specified for each surface. However, any boundary condition that involves both $\dot{q}_i$ and T_i
(e.g., a convectively cooled surface) will be nonlinear and therefore require a successive
substitution approach.

EXAMPLE 10.3-4: HEAT TRANSFER IN A RECTANGULAR ENCLOSURE

The bottom surface of the rectangular enclosure previously considered in EXAM-
PLE 10.3-2 is maintained at $T_1 = 250°C$ and the top surface is maintained at $T_3 = 20°C$, as shown in Figure 1. One of 1.5 m wide sidewalls, surface 2, is maintained
at $T_2 = 70°C$. The remaining three surfaces (surfaces 4, 5, and 6) are adiabatic. All
of the surfaces are black.

a) Determine the net rate of energy transfer required to maintain surfaces 1, 2, and
3 at their specified temperatures and the steady-state temperatures of surfaces
4, 5, and 6.

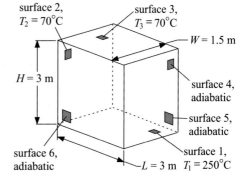

Figure 1: Black enclosure from EXAMPLE 10.3-2 with boundary conditions specified for each surface.

The net rate of heat transfer from any surface is given by:

$$\dot{q}_i = A_i \sigma \sum_{j=1}^{6} F_{i,j}\left(T_i^4 - T_j^4\right) = A_i \sum_{j=1}^{6} F_{i,j}(E_{b,i} - E_{b,j}) \tag{1}$$

Equation (1) can be written for all 6 surfaces involved in the problem:

$$\dot{q}_i = A_i \sigma \sum_{j=1}^{6} F_{i,j}\left(T_i^4 - T_j^4\right) = A_i \sum_{j=1}^{6} F_{i,j}(E_{b,i} - E_{b,j}) \quad \text{for } i = 1..6 \tag{2}$$

The areas and view factors appearing in Eq. (2) were determined previously in
EXAMPLE 10.3-2. The following EES code is appended to EXAMPLE 10.3-2 in
order to implement Eq. (2) within a duplicate loop:

```
"net heat transfer from each surface"
duplicate i=1,6
    q_dot[i]=A[i]*sigma#*sum(F[i,j] *(T[i]^4-T[j]^4),j=1,6)
end
```

Solving the problem will prompt EES to indicate that there are 6 more unknowns than equations; a complete set of boundary conditions must be specified. Three of the temperatures are known and three of the surfaces are adiabatic, i.e., they have a net heat transfer rate of zero. These boundary condition provide the additional 6 equations that are required to complete the solution:

```
"boundary conditions"
"specified temperatures"
T[1]=convertTemp(C,K,250 [C])
T[2]=convertTemp(C,K,70 [C])
T[3]=convertTemp(C,K,20 [C])

"specified heat transfers"
q_dot[4]=0
q_dot[5]=0
q_dot[6]=0
```

The results appear in the last two columns of the Arrays Table (Figure 2) after the calculations are completed.

Sort	A_i [m²]	$F_{i,1}$ [-]	$F_{i,2}$ [-]	$F_{i,3}$ [-]	$F_{i,4}$ [-]	$F_{i,5}$ [-]	$F_{i,6}$ [-]	sumF$_i$ [-]	T_i [K]	$\dot{q}_i$ [W]
[1]	4.5	0	0.1493	0.1167	0.1493	0.2924	0.2924	1	523.2	12307
[2]	4.5	0.1493	0	0.1493	0.1167	0.2924	0.2924	1	343.2	-5379
[3]	4.5	0.1167	0.1493	0	0.1493	0.2924	0.2924	1	293.2	-6928
[4]	4.5	0.1493	0.1167	0.1493	0	0.2924	0.2924	1	425.4	0
[5]	9	0.1462	0.1462	0.1462	0.1462	0	0.4153	1	423.7	0
[6]	9	0.1462	0.1462	0.1462	0.1462	0.4153	0	1	423.7	0

Figure 2: Arrays table with heat transfer and temperature solution.

The net rate of heat transfer to surface 1 is $\dot{q}_1 = 12.3$ kW and surfaces 2 and 3 must be cooled at a rate of $\dot{q}_2 = 5.4$ kW and $\dot{q}_3 = 6.9$ kW. The temperature of surfaces 4 through 6 are $T_4 = 425.4$ K and $T_5 = T_6 = 423.7$ K; the temperatures of surfaces 5 and 6 are identical, as expected by symmetry considerations.

Notice that it is easy to change the boundary conditions and obtain a new solution. For example, if the temperature of surface 4 were set to 75°C (rather than being adiabatic) then it is only necessary to change one line:

```
{q_dot[4]=0}
T[4]=convertTemp(C,K,75 [C])
```

It is not necessary that the surfaces be either adiabatic or have a specified temperature; any relationship between these quantities provides a suitable boundary condition. For example, suppose that surface 4 is convectively coupled to a flow of air at $T_\infty = 20$°C with average heat transfer coefficient $\bar{h} = 100$ W/m²-K. In this case, the net radiation heat transfer from surface 4 must be balanced by heat transfer from the fluid:

$$\dot{q}_4 = \bar{h} A_4 (T_\infty - T_4) \tag{3}$$

EXAMPLE 10.3-4

Equation (3) provides a relationship between the radiation heat transfer and temperature for surface 4 and therefore is an appropriate boundary condition. Update the guess values, remove the original boundary condition for surface 4, and replace it with Eq. (3):

```
{q_dot[4]=0}
h_bar=100 [W/m^2-K]                    "average heat transfer coefficient"
T_infinity=converttemp(C,K,20 [C])     "ambient temperature"
q_dot[4]=h_bar*A[4]*(T_infinity-T[4])  "convective boundary condition"
```

which leads to $T_4 = 304.8$ K.

10.3.4 Radiation Exchange between Non-Isothermal Surfaces

Non-isothermal surfaces can be considered by breaking the surface into small segments, either differential or finite, depending on whether an analytical or numerical technique will be used. In either case, the differential view factor library in EES facilitates the study of non-isothermal surfaces. The following example illustrates the use of the differential view factor library for this purpose.

EXAMPLE 10.3-5: DIFFERENTIAL VIEW FACTORS: RADIATION EXCHANGE BETWEEN PARALLEL PLATES

Two aligned parallel plates, each having width $W = 2$ m and length $L = 1$ m, are spaced a distance $H = 1.5$ m apart, as shown in Figure 1. The top plate (surface 2) is isothermal with temperature $T_2 = 500$ K. The temperature of the bottom plate (surface 1) varies linearly from $T_{1,LHS} = 500$ K at the left side (at $x = 0$) to $T_{1,RHS} = 1000$ K at the right side (at $x = W$) according to:

$$T_1 = T_{1,LHS} + (T_{1,RHS} - T_{1,LHS})\frac{x}{W} \tag{1}$$

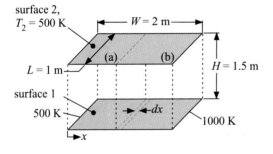

Figure 1: Radiation between two plates; the bottom plate is not isothermal.

a) Calculate the net rate of radiation heat transfer between plates 1 and 2.

The inputs are entered in EES:

"EXAMPLE 10.3-5: Radiation Exchange between Parallel Plates"

$UnitSystem SI MASS RAD PA K J
$Tabstops 0.2 0.4 0.6 3.5 in

"Inputs"
L=1 [m]   "plate length"
H=1.5 [m]   "plate-to-plate difference"
W=2 [m]   "width of plates"
T_1_LHS=500 [K]   "temperature of the left-hand side of plate 1"
T_1_RHS=1000 [K]   "temperature of the right-hand side of plate 1"
T_2=500 [K]   "temperature of plate 2"

The non-uniform temperature of the bottom plate (surface 1) is addressed by dividing surface 1 into differentially small segments of area dA_1. The rate of heat transfer from each differential segment is given by:

$$d\dot{q}_{1 \text{ to } 2} = F_{dA_{1,2}}\, \sigma (T_1^4 - T_2^4) dA_1$$

where $F_{dA_{1,2}}$ is the view factor between the differential area dA_1 and surface 2. For this problem, the differentially small segments have width dx and length L (see Figure 1).

$$d\dot{q}_{1 \text{ to } 2} = F_{dA_{1,2}}\, \sigma (T_1^4 - T_2^4) L\, dx$$

Differential view factors can be accessed from the View Factor Library in EES. The view factor between a differentially small strip that is positioned parallel to the edge of a plate is obtained using the FDiff_2 function. The view factor between the differential strip dA_1 and surface 2 must be obtained as the sum of the view factor between the strip and the portion of surface 2 that lies to the left of the strip ($F_{dA_1,a}$) and the strip and the portion of surface 2 that lies to the right of the strip ($F_{dA_1,b}$):

$$d\dot{q}_{1 \text{ to } 2} = (F_{dA_1,a} + F_{dA_1,b}) \sigma (T_1^4 - T_2^4) L\, dx \tag{2}$$

The total heat transfer from surface 1 to surface 2 is obtained by integrating Eq. (2) from $x = 0$ to $x = W$.

$$\dot{q}_{1 \text{ to } 2} = \int_{x=0}^{W} \left(F_{dA_1,a} + F_{dA_1,b} \right) \sigma (T_1^4 - T_2^4) L\, dx \tag{3}$$

In order to numerically evaluate the integral in Eq. (3), we first evaluate the integrand at a particular, arbitrary location ($x = 0.1$ m, for example).

x=0.1 [m]   "arbitrary position; this equation will subsequently be removed"

The temperature of surface 1 at this position is obtained using Eq. (1).

T_1=T_1_LHS+(T_1_RHS T_1_LHS)*x/W   "temperature of plate 1 surface"

The differential view factors, $F_{dA_1,a}$ and $F_{dA_1,b}$, are obtained using the FDiff_2 function:

```
F_dA1_a=FDiff_2(x,L,H)          "view factor from differential strip to plate on left"
F_dA1_b=FDiff_2((W-x),L,H)      "view factor from differential strip to plate on right"
```

The integrand of Eq. (3) is computed:

```
dq_dot_1to2dx=(F_dA1_a+F_dA1_b)*sigma#*(T_1^4-T_2^4)*L          "integrand"
```

which leads to $d\dot{q}_{1-2} = 110.8$ W/m at $x = 0.1$ m. Having managed to evaluate the integrand at a particular location, it is relatively easy to carry out the integration using the Integral command. Comment out the specified value of x and use the Integral command in order to allow EES to vary x and numerically evaluate the integral in Eq. (3).

```
{x=0.1 [m]}                         "arbitrary position"
q_dot=Integral(dq_dot_1to2dx,x,0,W)   "integral"
```

which leads to $\dot{q}_{1 \text{ to } 2} = 6342$ W.

If surface 1 is isothermal at 1000 K, then integration is not needed and the problem can be solved using the methodology discussed in Section 10.3.3. The rate of radiation heat transfer between plates 1 and 2 can be calculated directly and the result in this limiting case can be used to provide verification of the numerical calculation. Set $T_1 = 1000$ K and redo the integration:

```
{T_1=T_1_LHS+(T_1_RHS-T_1_LHS)*x/W      "temperature of plate 1 surface"}
T_1=1000 [K]                            "doublecheck with isothermal plate 1"
```

which leads to $\dot{q}_{1 \text{ to } 2} = 18702$ W.

The view factor between the two parallel plates is computed using the function F3D_1 that considers two aligned rectangular plates.

```
F_12=F3D_1(W,L,H)          "view factor between plates 1 and 2"
```

The heat transfer from surface 1 to 2 is calculated using Eq. (10-29):

$$\dot{q}_{1 \text{ to } 2} = A_1 F_{1,2} \sigma (T_1^4 - T_2^4)$$

where A_1 is the area of surface 1:

$$A_1 = L W$$

```
A_1=L*W                              "area of plate 1"
q_dot_check=A_1*F_12*sigma#*(T_1^4-T_2^4)   "doublecheck of result"
```

which also leads to $\dot{q}_{1 \text{ to } 2} = 18703$ W.

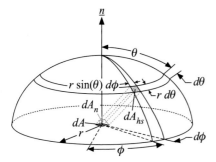

Figure 10-14: Radiation emitted from area dA that is intercepted by a hemisphere.

10.4 Radiation Characteristics of Real Surfaces

10.4.1 Introduction

Radiation heat transfer between two surfaces is a consequence of the difference between the rate of radiation that is emitted by a surface and the rate at which radiation is absorbed. The emission of radiation by a blackbody is investigated in Section 10.2 and in Section 10.3 radiation exchange between black surfaces is discussed. The radiation exchange calculations between blackbodies are simplified by the fact that blackbodies emit the maximum possible amount of radiation, given by Planck's law, and absorb all incident radiation. In this section, real surfaces are considered. Real surfaces emit less radiation than a blackbody and do not absorb all of the incident radiation; instead, some radiation is reflected or transmitted. The surface properties that characterize the emission, absorption, reflection, and transmission of radiation from a real surface are emissivity (or emittance), absorptivity (or absorptance), reflectivity (or reflectance), and transmittivity (or transmittance)[2]. For a given surface, these properties are generally complex functions of temperature, wavelength and direction.

10.4.2 Emission of Real Materials

Intensity
The radiation emitted by a real surface is a function of its temperature, wavelength, and direction (relative to the surface normal). Direction is most conveniently defined based on the location that the radiation intercepts a hemisphere that is placed above and centered on the differential surface area of interest (dA), as shown in Figure 10-14.

The simplest description of the situation is provided using spherical coordinates, θ and ϕ as defined in Figure 10-14. The differential area on the surface, dA, is emitting radiation in all directions. A second differential area, dA_{hs}, is positioned on the hemisphere (which has arbitrary radius r) and intercepts the radiation emitted that is from dA in direction θ and ϕ. In terms of these coordinates, the area on the hemisphere is:

$$dA_{hs} = r^2 \sin(\theta)\, d\theta\, d\phi \tag{10-33}$$

---

[2] Confusion exists about the difference between radiation terms that end in "-ance" versus "-ivity", e.g., absorptance versus absorptivity, transmittance versus transmissivity, reflectance versus reflectivity. Although there have been some efforts to standardize the nomenclature, no universal agreement has been reached. It has been proposed to use the terms that end with "-ivity" to refer to the property of a pure material. However, from a practical standpoint, radiation properties are never really properties of the material as they depend on surface characteristics such as oxidation, contamination (e.g., by fingerprints) and other factors that can strongly affect these parameters. In this text, these alternative terms are used interchangeably.

or

$$dA_{hs} = r^2 d\omega \qquad (10\text{-}34)$$

where $d\omega$ is referred to as the differential solid angle, defined as:

$$d\omega = \sin(\theta)\, d\theta\, d\phi \qquad (10\text{-}35)$$

Note that a typical angle is defined on a plane (e.g., the coordinate θ in Figure 10-14) and has units of radian (rad). However, a solid angle is defined in three-dimensional space and has units of steradian (sr). The integration of the solid angle over the entire hemisphere corresponds to integrating θ from 0 to $\pi/2$ rad and ϕ from 0 to 2π rad:

$$\int_0^{\omega_{hs}} d\omega = \int_0^{2\pi} \int_0^{\frac{\pi}{2}} \sin(\theta)\, d\theta\, d\phi \qquad (10\text{-}36)$$

Carrying out the integral in Eq. (10-36) leads to:

$$\omega_{hs} = \int_0^{2\pi} [-\cos(\theta)]_0^{\pi/2} d\phi = \int_0^{2\pi} d\phi = 2\pi \text{ [steradian]} \qquad (10\text{-}37)$$

Therefore, a solid angle of 2π steradians corresponds to a hemisphere.

The intensity of the emitted radiation, Ie, is formally defined as the rate of radiation emitted at wavelength λ per unit solid angle, ω, per unit surface area that is normal to the direction of emission defined by angles θ and ϕ (i.e., dA_n in Figure 10-14).

$$Ie_{\lambda,\theta,\phi} = \frac{\text{differential rate of radiation emitted}}{dA_n\, d\lambda\, d\omega} \qquad (10\text{-}38)$$

Note that the intensity is defined on the basis of the area that is perpendicular to the direction fixed by angles θ and ϕ. (Intensity is defined according to dA_n, which is the size of dA as viewed from the particular location on the hemisphere defined by the spherical coordinates θ and ϕ.) A differential area positioned at the top of the hemisphere (at $\theta = 0$) would see dA directly and therefore $dA_n = dA$, whereas a differential area that is located at the base of the hemisphere (at $\theta = \pi/2$) would not see dA at all and therefore $dA_n = 0$. In general, dA_n is related to dA according to:

$$dA_n = dA \cos(\theta) \qquad (10\text{-}39)$$

Substituting Eq. (10-39) into Eq. (10-38) leads to:

$$Ie_{\lambda,\theta,\phi} = \frac{\text{differential rate of radiation emitted}}{\cos(\theta)\, dA\, d\lambda\, d\omega} \qquad (10\text{-}40)$$

Note that the intensity of the radiation emitted by a surface at a particular temperature is, in general, a function of wavelength (λ) as well as direction (θ and ϕ); thus, intensity is written as $Ie_{\lambda,\theta,\phi}$.

The total rate of radiation emitted per unit area of the surface (dA) at a particular wavelength is the spectral emissive power of the surface (E_λ) and can be computed from the integral of the intensity over all angles intercepted by the hemisphere above the surface.

$$E_\lambda = \frac{\text{rate of radiation emitted}}{dA\, d\lambda} = \int_0^{2\pi} Ie_{\lambda,\theta,\phi} \cos(\theta)\, d\omega = \int_0^{2\pi} \int_0^{\pi/2} Ie_{\lambda,\theta,\phi} \cos(\theta) \sin(\theta)\, d\theta\, d\phi$$

$$(10\text{-}41)$$

Spectral, Directional Emissivity

A blackbody emits radiation uniformly in all directions; that is, the intensity of the radiation emitted by a blackbody is not a function of direction. Therefore, for a blackbody:

$$Ie_{\lambda,\theta,\phi} = Ie_{b,\lambda} \tag{10-42}$$

This characteristic of diffuse emission was assumed in Section 10.3 in order to calculate radiation exchange between black surfaces using view factors. Equation (10-41) is used to relate the blackbody emissive power ($E_{b,\lambda}$) to the blackbody intensity ($Ie_{b,\lambda}$):

$$E_{b,\lambda} = \int_0^{2\pi} \int_0^{\pi/2} Ie_{b,\lambda} \cos(\theta)\sin(\theta)\, d\theta\, d\phi \tag{10-43}$$

The intensity can be removed from the integrand since, for a blackbody, it is not a function of direction:

$$E_{b,\lambda} = Ie_{b,\lambda} \int_0^{2\pi} \int_0^{\pi/2} \cos(\theta)\sin(\theta)\, d\theta\, d\phi \tag{10-44}$$

The integration in Eq. (10-44) is carried out using Maple:

```
> int(int(cos(theta)*sin(theta),theta=0..Pi/2),phi=0..2*Pi);
                              π
```

so that:

$$E_{b,\lambda} = Ie_{b,\lambda}\,\pi \tag{10-45}$$

The spectral, directional emissivity (or emittance) of a surface is defined as the ratio of the intensity of the radiation emitted by a real surface ($Ie_{\lambda,\theta,\phi}$) to the intensity that would be emitted by a blackbody ($Ie_{b,\lambda}$); the spectral, directional emissivity ($\varepsilon_{\lambda,\theta,\phi}$) is a function of wavelength and direction at a specific temperature:

$$\varepsilon_{\lambda,\theta,\phi} = \frac{Ie_{\lambda,\theta,\phi}}{Ie_{b,\lambda}} \tag{10-46}$$

Typically, the emissivity of a surface will not depend strongly on the azimuthal angle, ϕ, but may depend substantially on θ. You may have observed the dependence of emissivity on θ if you have noticed that the reflectance of a surface (such as a computer screen) is quite high when viewed at a glancing angle (at θ near $\pi/2$) but substantially lower when viewed straight on (at θ near 0).

Hemispherical Emissivity

The hemispherical emissivity is the ratio of the rate of radiation emitted in all directions by a surface at a particular wavelength and temperature (E_λ, calculated according to Eq. (10-41)) to the rate of radiation emitted in all directions by a blackbody at the same wavelength and temperature, $E_{b,\lambda}$ The hemispherical emissivity of a surface is only a function of wavelength (ε_λ) at a specified temperature.

$$\varepsilon_\lambda = \frac{E_\lambda}{E_{b,\lambda}} \tag{10-47}$$

Substituting Eq. (10-41) into Eq. (10-47) leads to:

$$\varepsilon_\lambda = \frac{\int\limits_0^{2\pi} \int\limits_0^{\pi/2} I e_{\lambda,\theta,\phi} \cos(\theta) \sin(\theta) \, d\theta \, d\phi}{E_{b,\lambda}} \qquad (10\text{-}48)$$

Substituting Eq. (10-46) into Eq. (10-47) leads to:

$$\varepsilon_\lambda = \frac{\int\limits_0^{2\pi} \int\limits_0^{\pi/2} \varepsilon_{\lambda,\theta,\phi} I e_{b,\lambda} \cos(\theta) \sin(\theta) \, d\theta \, d\phi}{E_{b,\lambda}} \qquad (10\text{-}49)$$

Substituting Eq. (10-45) into Eq. (10-49) leads to:

$$\varepsilon_\lambda = \frac{\int\limits_0^{2\pi} \int\limits_0^{\pi/2} \varepsilon_{\lambda,\theta,\phi} E_{b,\lambda} \cos(\theta) \sin(\theta) \, d\theta \, d\phi}{\pi E_{b,\lambda}} \qquad (10\text{-}50)$$

Recognizing that $E_{b,\lambda}$ is not a function of direction allows Eq. (10-50) to be written as:

$$\varepsilon_\lambda = \frac{1}{\pi} \int\limits_0^{2\pi} \int\limits_0^{\pi/2} \varepsilon_{\lambda,\theta,\phi} \cos(\theta) \sin(\theta) \, d\theta \, d\phi \qquad (10\text{-}51)$$

Equation (10-51) shows that the hemispherical emissivity is the spectral, directional emissivity averaged over all directions.

Total Hemispherical Emissivity
The hemispherical emissivity is the value of the emissivity averaged over all directions. The directional dependence of the emissivity has been removed through the integration over θ and ϕ shown in Eq. (10-51). However, the hemispherical emissivity remains a function of wavelength. The total hemispherical emissivity is spectrally averaged (i.e., averaged over all wavelengths) as well. The total hemispherical emissivity of a surface is defined as the ratio of the total emissive power (E) to the total emissive power of a blackbody (E_b):

$$\varepsilon = \frac{E}{E_b} \qquad (10\text{-}52)$$

The total emissive power is obtained by integrating the spectral radiation of the surface over the entire wavelength band. The blackbody emissive power is given by Eq. (10-5). Therefore, Eq. (10-52) can be written as:

$$\varepsilon = \frac{\int\limits_0^\infty E_\lambda \, d\lambda}{\sigma T^4} \qquad (10\text{-}53)$$

Substituting Eq. (10-47) into Eq. (10-53) leads to:

$$\varepsilon = \frac{\displaystyle\int_0^\infty \varepsilon_\lambda\, E_{b,\lambda}\, d\lambda}{\sigma T^4} \tag{10-54}$$

Given information about the hemispherical emissivity of a surface as a function of wavelength, it is possible to use Eq. (10-54) to determine the total hemispherical emissivity of that surface through numerical integration. Total hemispherical emissivity data for a few surfaces are included in EES; to access this information, select Function Information from the Options menu and then select Solid/liquid properties. The epsilon_ function returns the total hemispherical emissivity. These values should be used with some caution as the emissivity of a surface may vary substantially depending on how it is handled.

The Diffuse Surface Approximation
A diffuse surface is defined as a surface that has an emissivity that is independent of direction (θ and ϕ). Therefore, the emissivity for a diffuse surface at a particular temperature is only a function of wavelength. Although the emissivity of a diffuse surface is independent of direction, it is important to understand that the diffuse surface emissivity is not the same as the hemispherical emissivity. The hemispherical emissivity is also independent of direction, but only because the directional dependence of the spectral, directional emissivity is removed through an averaging process over all directions.

The Diffuse Gray Surface Approximation
A diffuse gray surface is defined as a surface that has an emissivity that is independent of direction (θ and ϕ) and wavelength (λ). Again, it is important to distinguish between the emissivity of a diffuse gray surface and the total hemispherical emissivity. The total hemispherical emissivity is also independent of direction and wavelength, but this is because the spectral, directional emissivity is averaged over all directions and wavelengths. In contrast, the emissivity of a diffuse gray surface is assumed to be constant, independent of direction and wavelength.

The diffuse gray surface approximation is often made in order to simplify radiation exchange calculations. Methods for dealing with radiation exchange between diffuse gray surfaces are discussed in Sections 10.5 and 10.6. In order to rigorously deal with the directional and spectral characteristics of real surfaces, an advanced technique such as a Monte Carlo simulation is required. The most appropriate value of the emissivity to use for a diffuse gray surface model is the total hemispherical emissivity.

The Semi-Gray Surface
Figure 10-15 shows a plot of the spectral emittance for an idealized surface that has an emissivity that is constant within specific wavelength ranges. Surfaces that have different radiation properties in different wavelength bands are called radiation-selective; these surfaces can be useful for various applications, for example solar collectors. The term gray surface describes a surface that has an emissivity that is constant with respect to wavelength. A surface that can be represented by the idealized, step-change behavior of the emissivity shown in Figure 10-15 is referred to as a semi-gray surface. A semi-gray surface may have more than one step change in emissivity. The semi-gray assumption, like the gray surface assumption, is a modeling convenience; however, the

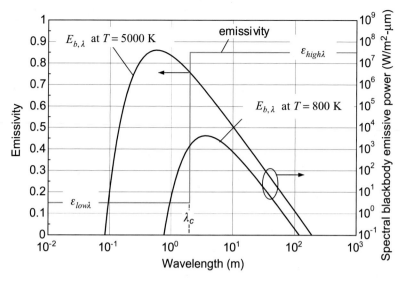

Figure 10-15: Spectral emittance as a function of wavelength for a semi-gray selective surface. Also shown are the spectral blackbody emissive powers for an object at 800 K and an object at 5000 K.

semi-gray model is useful to represent the selective surfaces that are encountered in many applications.

The surface shown in Figure 10-15 has an emissivity of $\varepsilon_{low\lambda} = 0.15$ for wavelengths that are less than $\lambda_c = 2\,\mu m$ and $\varepsilon_{high\lambda} = 0.85$ for wavelengths greater than $2\,\mu m$. The total hemispherical emissivity of a semi-gray surface can be evaluated relatively easily using the external fractional function that is introduced in Section 10.2.2 and is accessible from the **Blackbody** function in EES. The inputs for the surface shown in Figure 10-15, assumed to be at a uniform temperature $T = 800$ K, are entered in EES:

```
$UnitSystem SI MASS RAD PA K J
$Tabstops 0.2 0.4 0.6 3.5 in

"Inputs"
epsilon_lowlambda=0.15 [-]          "emissivity at low wavelengths"
epsilon_highlambda=0.85 [-]         "emissivty at high wavelengths"
lambda_c=2.0 [micron]               "cut-off wavelength"
T=800 [K]                           "temperature"
```

The total hemispherical emissivity is evaluated according to Eq. (10-54) by breaking the integral up according to the different wavelength bands:

$$\varepsilon = \frac{\displaystyle\int_0^{\lambda_c} \varepsilon_{low\lambda}\, E_{b,\lambda}\, d\lambda + \int_{\lambda_c}^{\infty} \varepsilon_{high\lambda}\, E_{b,\lambda}\, d\lambda}{\sigma\, T^4} \tag{10-55}$$

Within each wavelength band, the emissivity is constant and can be removed from the integral:

$$\varepsilon = \varepsilon_{low\lambda} \underbrace{\frac{\displaystyle\int_0^{\lambda_c} E_{b,\lambda}\,d\lambda}{\sigma T^4}}_{F_{0-\lambda_c}} + \varepsilon_{high\lambda} \underbrace{\frac{\displaystyle\int_{\lambda_c}^{\infty} E_{b,\lambda}\,d\lambda}{\sigma T^4}}_{1-F_{0-\lambda_c}} \qquad (10\text{-}56)$$

Equation (10-56) can be written in terms of the external fractional function:

$$\varepsilon = \varepsilon_{low\lambda}\, F_{0-\lambda_c} + \varepsilon_{high\lambda}\,(1 - F_{0-\lambda_c}) \qquad (10\text{-}57)$$

The external fractional function is evaluated using the **Blackbody** function in EES:

```
F_0lambdac=Blackbody(T,0 [micron],lambda_c)               "external fractional function"
epsilon=epsilon_lowlambda*F_0lambdac+epsilon_highlambda*(1-F_0lambdac)
        "total hemispherical emissivity"
```

which leads to $\varepsilon = 0.8362$. Notice that the surface at 800 K has a total hemispherical emissivity that is weighted towards $\varepsilon_{high\lambda}$ because most of radiation emitted by the surface is at higher wavelengths (see Figure 10-15). If the surface temperature is changed to $T = 5000$ K, then the total hemispherical emissivity is reduced to $\varepsilon = 0.2101$. This value is much closer to $\varepsilon_{low\lambda}$ because radiation from a surface at 5000 K is emitted at low wavelengths (see Figure 10-15).

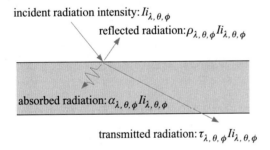

incident radiation intensity: $Ii_{\lambda,\theta,\phi}$

reflected radiation: $\rho_{\lambda,\theta,\phi} Ii_{\lambda,\theta,\phi}$

absorbed radiation: $\alpha_{\lambda,\theta,\phi} Ii_{\lambda,\theta,\phi}$

transmitted radiation: $\tau_{\lambda,\theta,\phi} Ii_{\lambda,\theta,\phi}$

Figure 10-16: Intensity of radiation striking a surface may be reflected, absorbed, or transmitted.

10.4.3 Reflectivity, Absorptivity, and Transmittivity

Section 10.4.2 discussed how a real surface emits radiation as a function of direction and wavelength and also considered a few simple models for such surfaces. In order to complete a radiation heat transfer problem, we must also consider how a real surface will deal with incident radiation. The intensity of the radiation that is incident on a surface at a particular wavelength (λ) from a particular direction (θ and ϕ, in spherical coordinates) is referred to as $Ii_{\lambda,\theta,\phi}$. In general, radiation that strikes a surface can be reflected, absorbed, or transmitted, as shown in Figure 10-16.

The ratio of the reflected intensity to the incident intensity at a given wavelength and direction is called the spectral, directional reflectivity, $\rho_{\lambda,\theta,\phi}$; the value of reflectivity must be between 0 and 1. Similar definitions are used for the spectral, directional absorptivity, $\alpha_{\lambda,\theta,\phi}$, and transmittivity, $\tau_{\lambda,\theta,\phi}$. At any wavelength, λ, and direction θ, ϕ, an

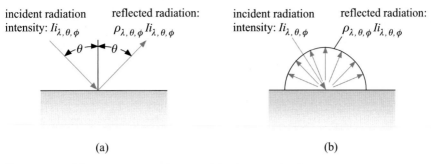

Figure 10-17: (a) Specular reflection and (b) Diffuse reflection.

energy balance requires that:

$$\rho_{\lambda,\theta,\phi} + \alpha_{\lambda,\theta,\phi} + \tau_{\lambda,\theta,\phi} = 1 \tag{10-58}$$

Opaque surfaces are defined as those that transmit no radiation. Therefore, $\tau_{\lambda,\theta,\phi} = 0$ in the wavelength range of interest. Black surfaces are defined as opaque surfaces that absorb all radiation regardless or wavelength and direction. Therefore $\alpha_{\lambda,\theta,\phi} = 1.0$ for a black surface. Equation (10-58) indicates that no radiation is reflected or transmitted from an opaque black surface ($\rho_{\lambda,\theta,\phi} = \tau_{\lambda,\theta,\phi} = 0$ for a black surface).

Most objects that we come into contact with in our day-to-day lives are not sufficiently hot to emit any substantial amount of radiation in the visible spectrum. Therefore, what our eyes perceive is usually the visible radiation from other sources (e.g., the sun or a light bulb) that is being reflected by these objects. The term blackbody arises because a surface that does not reflect any visible radiation will appear black to our eyes (unless it is so hot that it emits visible radiation). This terminology can be misleading, however, since the visible range is a small section of the electromagnetic spectrum. Snow, for example, reflects less than 2% of the infrared radiation that it receives; therefore, snow closely approximates a blackbody in the infrared wavelength band even though it clearly does not look black to our eyes. Many white paints behave in a similar manner. A true blackbody does not exist in nature since some radiation is always reflected from a surface; however, a few substances (e.g., carbon black) absorb nearly all incident radiation within the visible and near infrared range. Perhaps the closest approximation to a blackbody is a cavity with a pinhole opening. Radiation that enters the cavity through the pinhole may be reflected from the cavity walls many times before being absorbed; however, it has a very small chance of exiting through the pinhole. As a result, essentially all of the radiation entering the pinhole will ultimately be absorbed and none reflected.

Diffuse and Specular Surfaces

The directional dependence of the surface characteristics $\rho_{\lambda,\theta,\phi}$, $\tau_{\lambda,\theta,\phi}$ and $\alpha_{\lambda,\theta,\phi}$ is complicated; however, the complication is reduced by classifying surfaces as either diffuse or specular. The reflected radiation from a diffuse surface is assumed to be angularly uniform and completely independent of the direction of the incident radiation, as illustrated in Figure 10-17(b). At the other extreme, radiation that strikes a specular surface at a particular angle is reflected at the same angle, as indicated in Figure 10-17(a). The behavior of real surfaces is somewhere between these two extremes, being partially diffuse and partially specular. However, many materials tend to be nearly diffuse or nearly specular. For example, a highly polished metal surface or a mirror tends to exhibit specular behavior, at least within the narrow visible wavelength band that is detectable by our eyes; therefore, it is possible to see a clear reflection in a mirror. Roughened surfaces or surfaces with an oxidized coating tend to reflect radiation diffusely. Only radiation exchange involving diffuse surfaces are considered in this text. A detailed discussion of radiation exchange between surfaces that are specular or partially specular and partially diffuse is provided by Siegel and Howell (2002).

Hemispherical Reflectivity, Absorptivity, and Transmittivity

The hemispherical values of the reflectivity, absorptivity, and transmittivity are the ratios of the total rate of incident radiation at a particular wavelength that is reflected, absorbed, and transmitted, respectively, to the total rate of radiation that is incident on the surface. These hemispherical values are integrated over all directions and are therefore only a function of wavelength at a specified temperature (ρ_λ, α_λ, and τ_λ).

The total rate of radiation that is incident on a surface at a particular wavelength is referred to as the spectral irradiation, G_λ. The spectral irradiation is computed by integrating the incident radiation intensity over all directions according to:

$$G_\lambda = \int\limits_{0}^{2\pi} \int\limits_{0}^{\pi/2} Ii_{\lambda,\theta,\phi} \cos(\theta) \sin(\theta) \, d\theta \, d\phi \tag{10-59}$$

Note the similarity between the spectral irradiation, Eq. (10-59), and the spectral emissive power, Eq. (10-41). The intensity of the incident radiation, $Ii_{\lambda,\theta,\phi}$ in Eq. (10-59), depends on the orientation, temperature, and emissivity of the other surfaces that are in the vicinity of the surface in question and therefore it is not trivial to carry out the integration in Eq. (10-59).

The hemispherical reflectivity is calculated according to:

$$\rho_\lambda = \frac{\text{incident radiation at } \lambda \text{ that is reflected}}{\text{irradiation at } \lambda} = \frac{\displaystyle\int\limits_{0}^{2\pi} \int\limits_{0}^{\pi/2} \rho_{\lambda,\theta,\phi} \, Ii_{\lambda,\theta,\phi} \cos(\theta) \sin(\theta) \, d\theta \, d\phi}{G_\lambda} \tag{10-60}$$

The hemispherical absorptivity is calculated according to:

$$\alpha_\lambda = \frac{\text{incident radiation at } \lambda \text{ that is absorbed}}{\text{irradiation at } \lambda} = \frac{\displaystyle\int\limits_{0}^{2\pi} \int\limits_{0}^{\pi/2} \alpha_{\lambda,\theta,\phi} \, Ii_{\lambda,\theta,\phi} \cos(\theta) \sin(\theta) \, d\theta \, d\phi}{G_\lambda} \tag{10-61}$$

The hemispherical transmittivity is calculated according to:

$$\tau_\lambda = \frac{\text{incident radiation at } \lambda \text{ that is transmitted}}{\text{irradiation at } \lambda} = \frac{\displaystyle\int\limits_{0}^{2\pi} \int\limits_{0}^{\pi/2} \tau_{\lambda,\theta,\phi} \, Ii_{\lambda,\theta,\phi} \cos(\theta) \sin(\theta) \, d\theta \, d\phi}{G_\lambda} \tag{10-62}$$

Kirchoff's Law

Consider a black object that is placed within a black enclosure, as shown in Figure 10-18. The inner surface of the enclosure is maintained at a uniform temperature, T_{enc}, and therefore emits radiation according to Planck's law. The amount of radiation per unit

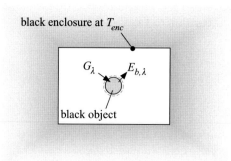

Figure 10-18: Black object within a black enclosure. The spectral irradiation incident on the object is G_λ and the spectral emissive power of the object is $E_{b,\lambda}$.

area at a specified wavelength, λ, that is incident on the blackbody is the spectral irradiation G_λ. According to Eq. (10-61), the total rate at which irradiation at λ is absorbed by the object ($\dot{q}_{abs,\lambda}$) is:

$$\dot{q}_{abs,\lambda} = \alpha_\lambda\, G_\lambda\, A \qquad (10\text{-}63)$$

where A is the surface area of the object. A blackbody has an absorptivity of unity, regardless of wavelength or direction ($\alpha_\lambda = 1$); therefore:

$$\dot{q}_{abs,\lambda} = G_\lambda\, A \qquad (10\text{-}64)$$

The radiation emitted by the object at wavelength λ per unit area is the blackbody spectral emissive power, $E_{b,\lambda}$. The total rate at which radiation is emitted by the black object at wavelength λ ($\dot{q}_{emit,\lambda}$) is:

$$\dot{q}_{emit,\lambda} = E_{b,\lambda}\, A \qquad (10\text{-}65)$$

If the enclosure and the object are in thermal equilibrium, then the temperature of the enclosure will be the same as the temperature of the object and it is necessary that the net heat transfer to the object be zero. Therefore, the amount of energy emitted by the object must be equal to the amount of energy absorbed by the object:

$$\dot{q}_{emit,\lambda} = \dot{q}_{abs,\lambda} \qquad (10\text{-}66)$$

Equation (10-66) must apply for all wavelengths. If this were not true, then the second law of thermodynamics could be violated by filtering out some part of the spectrum. Combining Eqs. (10-64) through (10-66) leads to:

$$E_{b,\lambda} = G_\lambda \qquad (10\text{-}67)$$

which implies that the body must be subjected to a spectral irradiation that is equal to the blackbody emissive power at the enclosure temperature.

Let's return to the object in thermal equilibrium with the enclosure (shown in Figure 10-18), but relax the assumption that the object is black. Instead, the surface of the object is characterized by a hemispherical emissivity, ε_λ, and hemispherical absorptivity, α_λ. The total rate at which radiation at wavelength λ is absorbed by the object ($\dot{q}_{abs,\lambda}$) is:

$$\dot{q}_{abs,\lambda} = \alpha_\lambda\, G_\lambda\, A \qquad (10\text{-}68)$$

The surroundings of the object have not changed and therefore the spectral irradiation is still equal to the blackbody emissive power, according to Eq. (10-67).

$$\dot{q}_{abs,\lambda} = \alpha_\lambda \, E_{b,\lambda} \, A \qquad (10\text{-}69)$$

The total rate at which radiation is emitted by the object at wavelength λ ($\dot{q}_{emit,\lambda}$) is now:

$$\dot{q}_{emit,\lambda} = \varepsilon_\lambda \, E_{b,\lambda} \, A \qquad (10\text{-}70)$$

The enclosure and the object are still at the same temperature, and therefore the net heat transferred to the object must remain zero; the emitted and absorbed energy must balance as before:

$$\dot{q}_{emit,\lambda} = \dot{q}_{abs,\lambda} \qquad (10\text{-}71)$$

Substituting Eqs. (10-69) and (10-70) into Eq. (10-71) leads to:

$$\boxed{\varepsilon_\lambda = \alpha_\lambda} \qquad (10\text{-}72)$$

which is known as Kirchoff's Law. For a blackbody, the absorptivity is 1 for all wavelengths and thus the emissivity is also 1 for all wavelengths. Thus, a blackbody is both a perfect absorber and a perfect emitter of radiation.

Note that Kirchoff's Law was derived here with reference to hemispherical emissivity and hemispherical absorptivity and therefore is strictly applicable only for the situation where either the irradiation or the surface is diffuse. However, Kirchoff's Law more generally relates the spectral, directional absorptivity to the spectral, directional emissivity:

$$\boxed{\varepsilon_{\lambda,\theta,\phi} = \alpha_{\lambda,\theta,\phi}} \qquad (10\text{-}73)$$

Equation (10-73) must be true because Eq. (10-71) involves integration over all directions of the emitted radiation (on the left side) and incident radiation (on the right side). The incident radiation is arbitrary and the only way that it is possible to guarantee that Eq. (10-71) is satisfied is if Eq. (10-73) is true.

Total Hemispherical Values

The hemispherical reflectivity, absorptivity, and transmittivity are the spectral, directional quantities averaged over all directions; therefore, the directional dependence has been removed. However, the hemispherical values are still functions of wavelength. The total hemispherical reflectivity, absorptivity, and transmittivity are spectrally averaged as well.

The total irradiation is the spectral irradiation averaged over all wavelengths:

$$G = \int_0^\infty G_\lambda \, d\lambda \qquad (10\text{-}74)$$

The total hemispherical reflectivity of a surface is defined as the fraction of the total irradiation that is reflected:

$$\rho = \frac{\text{reflected irradiation}}{\text{irradiation}} = \frac{\displaystyle\int_0^\infty \rho_\lambda \, G_\lambda \, d\lambda}{G} \qquad (10\text{-}75)$$

The total hemispherical absorptivity and transmittivity are defined similarly:

$$\alpha = \frac{\text{absorbed irradiation}}{\text{irradiation}} = \frac{\int_0^\infty \alpha_\lambda \, G_\lambda \, d\lambda}{G} \tag{10-76}$$

$$\tau = \frac{\text{transmitted irradiation}}{\text{irradiation}} = \frac{\int_0^\infty \tau_\lambda \, G_\lambda \, d\lambda}{G} \tag{10-77}$$

The Diffuse Surface Approximation

A diffuse surface is defined as a surface with an emissivity that is independent of direction (θ and ϕ). According to Kirchoff's Law, Eq. (10-73), the absorptivity is equal to the emissivity; therefore, the absorptivity of a diffuse surface must also be independent of direction. Further, if the diffuse surface is opaque then Eq. (10-58) indicates that the reflectivity is given by:

$$\rho_{\lambda,\theta,\phi} = 1 - \alpha_{\lambda,\theta,\phi} \tag{10-78}$$

Therefore, the reflectivity of an opaque, diffuse surface must also be independent of direction.

The Diffuse Gray Surface Approximation

A diffuse gray surface is defined as a surface that has an emissivity that is independent of direction (θ and ϕ) and wavelength (λ). According to Kirchoff's Law, the absorptivity of a diffuse gray surface must also be independent of direction and wavelength. An opaque, diffuse gray surface will have a reflectivity that is given by:

$$\rho = 1 - \alpha \tag{10-79}$$

The Semi-Gray Surface

Materials that have emissivity that is constant within different wavelength bands are referred to as semi-gray surfaces. By Kirchoff's Law, the absorptivity of the semi-gray surface must also be constant within different wavelength bands and equal to the emissivity. The reflectance of an opaque, semi-gray surface must be constant in different wavelength bands and equal to $1 - \alpha$.

EXAMPLE 10.4-1: ABSORPTIVITY AND EMISSIVITY OF A SOLAR SELECTIVE SURFACE

EXAMPLE 10.4-1: ABSORPTIVITY AND EMISSIVITY OF A SOLAR SELECTIVE SURFACE

A solar collector is a device that is designed to maximize the absorption of the irradiation that is received from the sun so that this energy can be used, for example, for domestic water heating. A typical solar collector for water heating consists of a metallic collector plate that is bonded to tubes through which the water to be heated flows, as shown in Figure 1. The collector plate is heated when it is exposed to solar radiation and this energy transfer is conducted through the plate to the tubes where it heats the water by convection. The plate is contained in an insulated enclosure in order to reduce the rate at which energy is lost from the sides and back of the plate. Some solar collectors use one or more transparent glazings that are placed above the plate. However, the solar collector in Figure 1 is designed for a low temperature swimming pool heating application and therefore does not employ a cover. Solar radiation is incident on the collector surface with irradiation $G = 800\,\text{W/m}^2$, assume that the irradiation is spectrally distributed as if it were emitted by a blackbody at $T_{sun} = 5780\,\text{K}$. The ambient temperature is $T_\infty = 25°\text{C}$ and the average convection heat transfer coefficient between the plate and the ambient air is $\bar{h} = 10\,\text{W/m}^2\text{-K}$. The collector plate has an average temperature of $T_{plate} = 45°\text{C}$. The plate exchanges radiation with surroundings at T_∞.

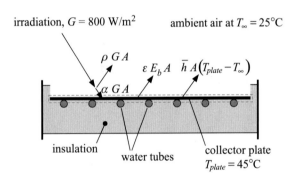

irradiation, $G = 800\,\text{W/m}^2$ ambient air at $T_\infty = 25°\text{C}$

Figure 1: Solar collector used for swimming pool heating.

The objective of the design of a solar collector is to maximize the absorption of solar radiation but minimize thermal loss from the plate to the surroundings due to radiation. Solar radiation is concentrated at relatively low wavelengths because it is emitted by a high temperature source, the sun. However, the collector the plate emits radiation at relatively high wavelengths because it is (by comparison to the sun) cold. Therefore, an ideal solar collector surface has an absorptivity (which, according to Kirchoff's Law, is equal to the emissivity) that is high at low wavelengths in order to capture the solar irradiation and an emissivity that is low at high wavelengths in order to minimize radiation heat loss. These types of selective surfaces are important for achieving high collector efficiency.

Figure 2 illustrates a semi-gray model of the selective surface that is used for the solar collector. The emissivity below $\lambda_c = 5.0\,\mu\text{m}$ is $\varepsilon_{low} = 0.95$ and the emissivity above λ_c is $\varepsilon_{high} = 0.05$; no real surface exhibits such a step-change behavior in its emissivity but this semi-gray model is useful to simulate some surfaces, such as black-chrome.

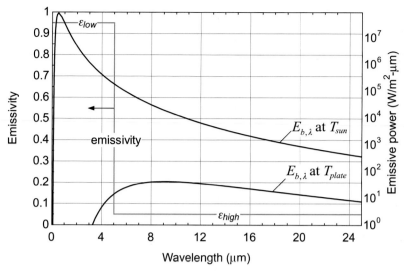

EXAMPLE 10.4-1: ABSORPTIVITY AND EMISSIVITY OF A SOLAR SELECTIVE SURFACE

Figure 2: Emissivity as a function of wavelength for the selective surface used for the collector plate. Also shown is the blackbody emissive power at T_{plate} and T_{sun}.

a) Calculate the steady-state rate of energy transfer to the water per unit area of collector surface.

The inputs are entered in EES:

```
"EXAMPLE 10.4-1"

$UnitSystem SI MASS RAD PA K J
$Tabstops 0.2 0.4 0.6 3.5 in

"Inputs"
epsilon_low=0.95 [-]                    "emissivity at low wavelengths"
epsilon_high=0.05 [-]                   "emissivty at high wavelengths"
lambda_c=5.0 [micron]                   "cut-off wavelength"
G=800 [W/m^2]                           "solar irradiation"
T_sun=5780 [K]                          "temperature of the sun"
A=1 [m^2]                               "unit area"
T_plate=converttemp(C,K,45 [C])         "plate temperature"
T_infinity=converttemp(C,K,25 [C])      "temperature of surroundings"
h_bar=10 [W/m^2-K]                      "average heat transfer coefficient"
```

A steady state energy balance on the plate is shown in Figure 1:

$$\dot{q}_{water} = \alpha\, G\, A - \varepsilon\, \sigma\, A\left(T_{plate}^4 - T_\infty^4\right) - \bar{h}\, A\left(T_{plate} - T_\infty\right) \tag{1}$$

where ε is the total hemispherical emissivity of the plate (evaluated relative to the emissive power of the plate) and α is the total hemispherical absorptivity of the plate (evaluated relative to irradiation from the sun). These quantities can be evaluated using the techniques discussed in Section 10.2.3 for semi-gray surfaces.

EXAMPLE 10.4-1: ABSORPTIVITY AND EMISSIVITY OF A SOLAR SELECTIVE SURFACE

The total hemispherical emissivity is obtained according to:

$$\varepsilon = \frac{1}{\sigma T_{plate}^4} \int_0^{\infty} \varepsilon_\lambda\, E_{b,\lambda}\, d\lambda$$

where $E_{b,\lambda}$ is evaluated using Planck's law, Eq. (10-4) at T_{plate}. For the semi-gray surface shown in Figure 2, the integral is broken up according to the different wavelength bands:

$$\varepsilon = \frac{\displaystyle\int_0^{\lambda_c} \varepsilon_{low}\, E_{b,\lambda}\, d\lambda + \int_{\lambda_c}^{\infty} \varepsilon_{high}\, E_{b,\lambda}\, d\lambda}{\sigma T_{plate}^4}$$

Within each wavelength band, the emissivity is constant and can be removed from the integrand:

$$\varepsilon = \varepsilon_{low} \underbrace{\frac{\displaystyle\int_0^{\lambda_c} E_{b,\lambda}\, d\lambda}{\sigma T_{plate}^4}}_{F_{0-\lambda_c,T_{plate}}} + \varepsilon_{high} \underbrace{\frac{\displaystyle\int_{\lambda_c}^{\infty} E_{b,\lambda}\, d\lambda}{\sigma T_{plate}^4}}_{1-F_{0-\lambda_c,T_{plate}}}$$

which can be written in terms of the external fractional function:

$$\varepsilon = \varepsilon_{low\lambda}\, F_{0-\lambda_c,T_{plate}} + \varepsilon_{high\lambda}\left(1 - F_{0-\lambda_c,T_{plate}}\right)$$

where $F_{0-\lambda_c,T_{plate}}$ is evaluated at T_{plate}. The external fractional function is obtained using the **Blackbody** function in EES.

```
epsilon=epsilon_low*Blackbody(T_plate,0 [micron],lambda_c)+&
    epsilon_high*(1-Blackbody(T_plate,0 [micron],lambda_c))
    "total hemispherical emissivity relative to the spectral blackbody emissive power of the plate"
```

which leads to $\varepsilon = 0.067$. Note that the total hemispherical emissivity of the collector plate is weighted towards ε_{high} because the plate emits primarily at high wavelengths (see Figure 2). The total hemispherical absorptivity of the plate can be evaluated in a similar manner according to:

$$\alpha = \alpha_{low} \underbrace{\frac{\displaystyle\int_0^{\lambda_c} E_{b,\lambda}\, d\lambda}{\sigma T_{sun}^4}}_{F_{0-\lambda_c,T_{sun}}} + \alpha_{high} \underbrace{\frac{\displaystyle\int_{\lambda_c}^{\infty} E_{b,\lambda}\, d\lambda}{\sigma T_{sun}^4}}_{1-F_{0-\lambda_c,T_{sun}}}$$

where $E_{b,\lambda}$ is evaluated using Planck's law, Eq. (10-4) at T_{sun}. Note that $\alpha_{low} = \varepsilon_{low}$ and $\alpha_{high} = \varepsilon_{high}$ according to Kirchoff's law. Therefore:

$$\alpha = \varepsilon_{low}\, F_{0-\lambda_c,T_{sun}} + \varepsilon_{high}\left(1 - F_{0-\lambda_c,T_{sun}}\right)$$

where $F_{0-\lambda_c,T_{sun}}$ is evaluated at T_{sun}.

EXAMPLE 10.4-1

```
alpha=epsilon_low*Blackbody(T_sun,0 [micron],lambda_c)+&
    epsilon_high*(1-Blackbody(T_sun,0 [micron],lambda_c))
    "total hemispherical absorptivity relative to the irradiation from the sun"
```

which leads to $\alpha = 0.945$. Note that the absorptivity is weighted toward ε_{low} because the irradiation from the sun is concentrated at low wavelengths. The energy balance, Eq. (1), is:

```
q_dot_water=G*A*alpha-epsilon*sigma#*A*(T_plate^4-T_infinity^4)-h_bar*A*(T_plate-T_infinity)
    "energy balance on plate"
```

which leads to $\dot{q}_{water} = 547$ W for the assumed $A = 1$ m^2 collector area. This result corresponds to a collector efficiency of 68.4% relative to capturing the solar irradiation, $G = 800$ W/m^2.

b) Compare your answer from (a) to the result that would be obtained if the plate had a constant spectral emissivity of 0.95 (i.e., if a selective surface were not used as the absorber plate but rather a plate with a uniform, high absorptivity were used instead).

The EES code is run again with $\varepsilon_{high} = 0.95$ which corresponds to a surface with a constant emissivity of 0.95 regardless of wavelength. The result is $\dot{q}_{water} = 433$ W/m^2; this is a 20% reduction in performance and illustrates the importance of using selective surfaces for solar collectors.

10.5 Diffuse Gray Surface Radiation Exchange

10.5.1 Introduction

A blackbody surface absorbs all of the radiation that hits it, regardless of wavelength or angle. None of the incident radiation is reflected or transmitted. A surface that exhibits this behavior has an absorptivity of 1 and therefore a blackbody is a perfect absorber of radiation. A blackbody is also a perfect emitter of radiation because, according to Kirchoff's law, a blackbody must have an emissivity of 1 regardless of wavelength. Radiation emitted by a blackbody surface is diffuse, i.e., the intensity is independent of angle.

No real surface exhibits the properties of a blackbody, although some materials approach this behavior within particular wavelength bands. Real surfaces have emissivity values that are lower than 1.0 and emissivity may vary with wavelength and direction, as described in Section 10.4.2. However, an average value of emissivity, the total hemispherical emissivity, can be defined by convolution of the emissivity and the wavelength dependence of emitted radiation, Eq. (10-54). The concept of a diffuse, gray surface was introduced in Section 10.4.3. A diffuse gray surface is another idealized surface, one that has a constant emissivity at all wavelengths and emits radiation uniformly in all directions. A semi-gray surface is a surface that has constant emissivity within specified radiation bands. No real material exhibits true gray or even semi-gray surface behavior. However, the results of diffuse gray surface radiation exchange calculations are more accurate than the blackbody radiation calculations presented in Section 10.3 and are often sufficient for engineering calculations. This section presents methods for calculating radiation exchange between diffuse gray surfaces that are opaque.

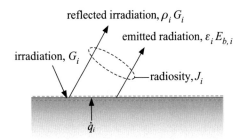

reflected irradiation, $\rho_i \, G_i$

emitted radiation, $\varepsilon_i \, E_{b,i}$

irradiation, G_i

radiosity, J_i

$\dot{q}_i$

Figure 10-19: Definition of radiosity for surface i.

10.5.2 Radiosity

Consider diffuse gray surface i that is receiving irradiation G_i, as shown in Figure 10-19. The source of the irradiation is not important; the radiation is likely a combination of radiation emitted and reflected from other surfaces in the vicinity of surface i (or even from surface i itself). The radiation leaving surface i includes the portion of the irradiation that is reflected from surface i ($\rho_i \, G_i$) as well as the radiation emitted by surface i ($\varepsilon_i \, E_{b,i}$). The sum of the reflected and emitted radiation per unit area is called radiosity, J_i; the radiosity is the rate of radiation that is leaving surface i per unit area. Note that the radiosity associated with a surface is a particularly complex function of wavelength since the spectral distribution of the radiation emitted from the surface may differ substantially from the spectral distribution of radiation reflected from the surface; this complexity can be ignored for gray surface calculations, since it is assumed that none of the surface characteristics depend on wavelength.

The radiosity is the sum of the reflected and emitted radiation in Figure 10-19:

$$J_i = \rho_i \, G_i + \varepsilon_i \, E_{b,i} \qquad (10\text{-}80)$$

where ε_i is the emissivity of the surface and ρ_i is the reflectivity of the surface. For an opaque, gray surface, the reflectivity and absorptivity are related by:

$$\rho_i = 1 - \alpha_i \qquad (10\text{-}81)$$

According to Kirchoff's Law, the absorptivity and emissivity for a gray surface must be equal and therefore:

$$\rho_i = 1 - \varepsilon_i \qquad (10\text{-}82)$$

Substituting Eq. (10-82) into Eq. (10-80) leads to:

$$J_i = \underbrace{(1 - \varepsilon_i) \, G_i}_{\text{reflected}} + \underbrace{\varepsilon_i \, E_{b,i}}_{\text{emitted}} \qquad (10\text{-}83)$$

The first term in Eq. (10-83) is the reflected irradiation and the second term is the emitted radiation. Note that if ε_i approaches 1 (i.e., the surface is black) then the radiosity will simply be the radiation that is emitted by the surface (the blackbody emissive power) because all incident radiation is absorbed. At the other extreme, if ε_i approaches 0 (i.e., the surface is a perfect reflector) then the radiosity will equal the irradiation, perfectly reflected from the surface with no additional emitted power.

The net rate of radiation heat transfer from surface i (which is the rate at which external energy is provided to surface i, $\dot{q}_i$) can be calculated using an energy balance on the surface (see Figure 10-19), expressed in terms of the radiosity from surface i.

$$\underbrace{\dot{q}_i}_{\substack{\text{energy transfer} \\ \text{to the surface}}} = \underbrace{A_i J_i}_{\text{radiosity}} - \underbrace{A_i G_i}_{\text{irradiation}} \tag{10-84}$$

Note that if the surface is adiabatic (i.e., no energy is provided to the surface, $\dot{q}_i = 0$), then Eq. (10-84) requires that the radiosity be equal to the irradiation. Rearranging Eq. (10-83) in order to solve for the irradiation, G_i, leads to:

$$G_i = \frac{J_i - \varepsilon_i E_{b,i}}{1 - \varepsilon_i} \tag{10-85}$$

Substituting Eq. (10-85) into Eq. (10-84) results in:

$$\dot{q}_i = A_i \left(J_i - \frac{J_i - \varepsilon_i E_{b,i}}{1 - \varepsilon_i} \right) = A_i \left(J_i \frac{1 - \varepsilon_i}{1 - \varepsilon_i} - \frac{J_i - \varepsilon_i E_{b,i}}{1 - \varepsilon_i} \right) = A_i \left(\frac{-\varepsilon_i J_i + \varepsilon_i E_{b,i}}{1 - \varepsilon_i} \right) \tag{10-86}$$

which can be rearranged:

$$\boxed{\dot{q}_i = \left(\frac{\varepsilon_i A_i}{1 - \varepsilon_i} \right) (E_{b,i} - J_i)} \tag{10-87}$$

Equation (10-87) is important because it relates the blackbody emissive power of the surface to the radiosity leaving the surface. The equivalent to Eq. (10-87) was not required for blackbody radiation exchange problems because the radiosity leaving a blackbody is equal to the blackbody emissive power.

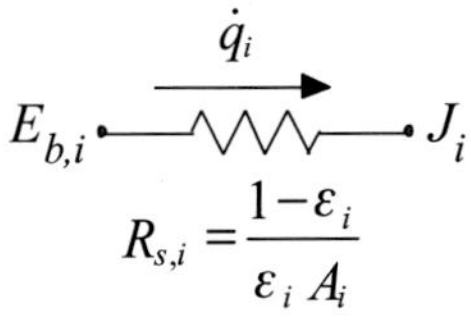

Figure 10-20: Definition of a surface resistance.

$$E_{b,i} \xrightarrow{\dot{q}_i} \!\!\!\!\! \mathord{-}\!\!\!\bigwedge\!\!\bigvee\!\!\bigwedge\!\!\mathord{-} \!\!\bullet J_i$$

$$R_{s,i} = \frac{1 - \varepsilon_i}{\varepsilon_i A_i}$$

10.5.3 Gray Surface Radiation Calculations

Equation (10-87) can be applied to any gray surface that is involved in radiation exchange. Notice that Eq. (10-87) has the form of a resistance equation. The driving force for heat transfer is the difference between the surface's blackbody emissive power and its radiosity ($E_{b,i} - J_i$) and the resistance to heat transfer is the quantity $(1 - \varepsilon_i)/(\varepsilon_i A_i)$, as illustrated in Figure 10-20. The resistance between the surface's blackbody emissive power and its radiosity is called the surface resistance ($R_{s,i}$) and should not be confused with the space resistance between surface i and another surface j ($R_{i,j}$), discussed in Section 10.3.3.

$$R_{s,i} = \frac{1 - \varepsilon_i}{A_i \varepsilon_i} \tag{10-88}$$

The surface resistance and the space resistance both have units m^{-2} (in the SI system). Note that if the surface is black (i.e., if $\varepsilon_i = 1$), then the surface resistance limits to 0 and the resistor "disappears"; this explains why it was not necessary to consider surface resistances when doing the blackbody radiation exchange problems in Section 10.3. Also, if the surface is a perfect reflector (i.e., if $\varepsilon_i = 0$), then the surface resistance becomes infinitely large. In this limit, the surface does not communicate radiatively with its environment; all incident radiation is reflected and the surface emits no radiation. It is often desirable to isolate a surface from radiation; for example, in a cryogenic experiment placed in a vacuum vessel or an apparatus that is launched in space. The surfaces of

Figure 10-21: The Hubble Space Telescope (http://www.utahskies.org/image_library/shallowsky/telescopes/hstorbitx.jpg).

such devices are often made as "shiny" as possible in order to closely emulate a perfect reflector and therefore maximize the value of the surface resistance in Eq. (10-88). Figure 10-21 shows the Hubble Space Telescope covered with a shiny layer of material. EXAMPLE 10.5-1 shows that layers of reflective material (called radiation shields or multi-layer insulation, MLI) can provide very high levels of thermal isolation.

The radiosity plays the same role that blackbody emissive power played in blackbody radiation exchange problems; the radiosity is the amount of radiation leaving each surface and interacting with other surfaces in the vicinity. The radiation exchange between surfaces can therefore be represented using the space resistances that were introduced in Section 10.3.3; however, these space resistances extend between the radiosity of each surface rather than the blackbody emissive power. Figure 10-22 shows two diffuse gray surfaces, i and j that are exchanging radiation. Surface i has area A_i, emissivity ε_i and is at temperature T_i while surface j has area A_j, emissivity ε_j and is at temperature T_j.

The total rate of radiation leaving surface i is the product of its area and the radiosity for surface i, J_i. According to the definition of the view factor, the radiosity leaving surface i that is incident on surface j is:

$$\text{radiation leaving surface } i \text{ that hits surface } j = F_{i,j} A_i J_i \qquad (10\text{-}89)$$

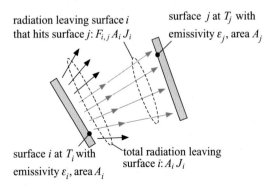

radiation leaving surface i
that hits surface j: $F_{i,j} A_i J_i$

surface j at T_j with
emissivity ε_j, area A_j

surface i at T_i with
emissivity ε_i, area A_i

total radiation leaving
surface i: $A_i J_i$

Figure 10-22: Two diffuse gray surfaces, i and j, at different temperatures, T_i and T_j, exchanging radiation.

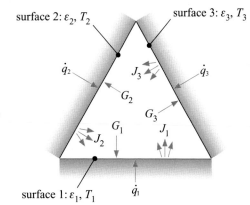

Figure 10-23: Three surface enclosure.

Similarly, the radiosity that leaves surface j and hits surface i is:

$$\text{radiation leaving surface } j \text{ that hits surface } i = F_{j,i} A_j J_j \qquad (10\text{-}90)$$

The net rate of radiation exchange between surface i and surface j, $\dot{q}_{i \text{ to } j}$, is the difference between Eq. (10-89) and Eq. (10-90).

$$\dot{q}_{i \text{ to } j} = F_{i,j} A_i J_i - F_{j,i} A_j J_j \qquad (10\text{-}91)$$

Reciprocity between view factors, Eq. (10-20), requires that:

$$F_{i,j} A_i = F_{j,i} A_j \qquad (10\text{-}92)$$

which allows Eq. (10-91) to be simplified:

$$\dot{q}_{i \text{ to } j} = A_i F_{i,j} (J_i - J_j) \qquad (10\text{-}93)$$

Note that Eq. (10-93) has the same form as Eq. (10-29). The net radiation heat transfer between two diffuse gray surfaces i and j is driven by the difference in their radiosities, $J_i - J_j$, and resisted by a space resistance that is the inverse of the product of the area and view factor:

$$\dot{q}_{i \text{ to } j} = \frac{(J_i - J_j)}{R_{i,j}} \qquad (10\text{-}94)$$

where $R_{i,j}$, remains the same space resistance that was used in Section 10.3 to model radiation exchange between black bodies:

$$R_{i,j} = \frac{1}{A_i F_{i,j}} = \frac{1}{A_j F_{j,i}} \qquad (10\text{-}95)$$

Based on this discussion, radiation exchange between diffuse gray surfaces is one step more complex than radiation exchange between blackbodies because the radiation exchange between surfaces is driven by radiosity and radiosity is only indirectly related to the blackbody emissive power of the surface itself. An extra resistance must be added to the resistance network that is used to model the problem; this extra resistance is the surface resistance that relates a surface's radiosity to its blackbody emissive power.

Consider the enclosure shown in Figure 10-23, which is composed of three surfaces. The temperature of each surface is uniform (at T_1, T_2, and T_3) and the rate at which energy must be provided to each of these surfaces in order to maintain these temperatures is $\dot{q}_1$, $\dot{q}_2$, and $\dot{q}_3$. If each of the surfaces is black ($\varepsilon_1 = \varepsilon_2 = \varepsilon_3 = 1.0$), then the resistance network that represents this enclosure is shown in Figure 10-24(a); the

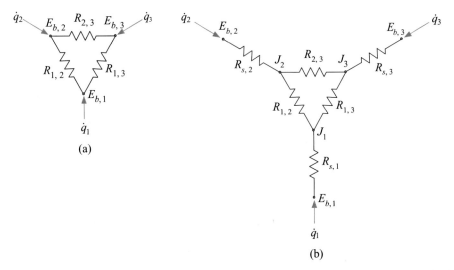

Figure 10-24: Resistance network that represents the enclosure shown in Figure 10-23 if (a) the surfaces are black and (b) the surfaces are diffuse and gray.

blackbody emissive power of each surface interacts directly via space resistances. If each surface is gray and diffuse, then the resistance network that represents the enclosure is shown in Figure 10-24(b); the blackbody emissive power of each surface is related to its radiosity by a surface resistance and the radiosities of each surface interact via space resistances.

The methodology for carrying out a diffuse gray surface problem using a resistance network is demonstrated in the following example.

EXAMPLE 10.5-1: RADIATION SHIELD

Radiation shields are used to reduce the rate of radiation heat transfer to or from an object that must be thermally isolated. Radiation shields are commonly used on spacecraft (for example, the Hubble Space Telescope in Figure 10-21), where radiation is the sole mechanism for heat transfer.

Consider a spherical object that is placed in a cubical enclosure, as shown in Figure 1. The diameter of the object is $D_p = 0.3$ m. The surface of the object is unpolished stainless steel; the surface can be considered diffuse and gray with an emittance of $\varepsilon_p = 0.30$. The object receives radiation from the cubical enclosure which is at a temperature $T_{enc} = 300$ K. The enclosure surfaces are black.

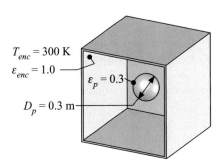

Figure 1: A spherical object in a cubical enclosure.

EXAMPLE 10.5-1: RADIATION SHIELD

a) Determine the rate of cooling that would be required to maintain the temperature of the object at $T_p = 10\,\text{K}$.

The inputs are entered in EES:

```
"EXAMPLE 10.5-1(a)"

$UnitSystem SI MASS RAD PA K J
$TABSTOPS 0.2 0.4 0.6 0.8 3.5 in


"Inputs"
T_enc=300 [K]            "temperature of the enclosure"
D_p=0.3 [m]             "diameter of object"
e_p=0.3 [-]             "emissivity of object"
T_p = 10 [K]           "temperature of object"
```

A network representation of the radiation heat transfer for this problem is shown in Figure 2. Note that the surface resistance for the enclosure is 0 because it is black and therefore $R_{s,enc}$ is not included in Figure 2.

Figure 2: Network representation for the radiation between the electronic package and the surroundings.

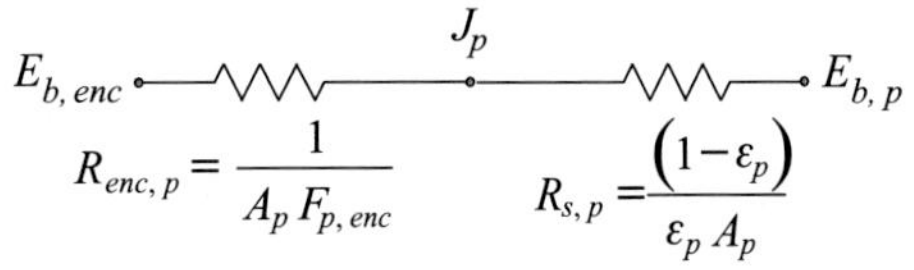

$$E_{b,\,enc} \bullet\!\!-\!\!\sqrt{\sqrt{\sqrt{}}}\!\!-\!\!\overset{J_p}{\bullet}\!\!-\!\!\sqrt{\sqrt{\sqrt{}}}\!\!-\!\!\bullet E_{b,\,p}$$

$$R_{enc,\,p} = \frac{1}{A_p\,F_{p,\,enc}} \qquad R_{s,\,p} = \frac{(1-\varepsilon_p)}{\varepsilon_p\,A_p}$$

The view factor between the object and the enclosure is one (all of the radiosity that leaves the object must hit the enclosure, $F_{p,enc} = 1.0$) and the surface area of the object is:

$$A_p = \pi\,D_p^2$$

The space resistance between the object and the enclosure is:

$$R_{enc,p} = \frac{1}{A_p\,F_{p,enc}}$$

```
F_p_enc=1.0 [-]            "view factor between the object and the enclosure"
A_p=pi*D_p^2              "surface area of object"
R_enc_p=1/(A_p*F_p_enc)   "space resistance between enclosure and object"
```

The surface resistance of the object is:

$$R_{s,p} = \frac{(1-\varepsilon_p)}{\varepsilon_p\,A_p} \qquad\qquad\qquad (1)$$

```
R_s_p=(1-e_p)/(e_p*A_p)           "surface resistance of the object"
```

The heat transfer rate from the enclosure to the object is given by:

$$\dot{q}_p = \frac{E_{b,enc} - E_{b,p}}{R_{s,p} + R_{enc,c}}$$

EXAMPLE 10.5-1: RADIATION SHIELD

where the blackbody emissive powers of the enclosure and object can be computed from their temperatures:

$$E_{b,p} = \sigma\, T_p^4 \qquad\qquad (2)$$

and

$$E_{b,enc} = \sigma\, T_{enc}^4 \qquad\qquad (3)$$

```
E_b_p=sigma#*T_p^4                          "blackbody emissive power of object"
E_b_enc=sigma#*T_enc^4                       "blackbody emissive power of enclosure"
q_dot_p=(E_b_enc-E_b_p)/(R_s_p+R_enc_p)      "heat transfer to object"
```

which leads to $\dot{q}_p = 39.0\,\mathrm{W}$.

b) The object is placed within a spherical, polished stainless steel radiation shield that has diameter $D_s = 0.5$ m, as shown in Figure 3. The inside and outside surfaces of this shield can be considered to be diffuse and gray with emissivity $\varepsilon_s = 0.17$. The shield is thin and conductive; therefore, both the inner and outer surfaces can be considered to be at the same temperature and diameter. Determine the heat transfer rate that would be required to maintain the object at $T_p = 10$ K with the radiation shield in place.

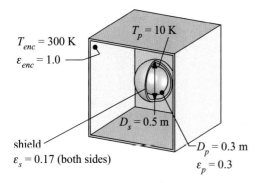

Figure 3: Object with a radiation shield.

The inputs are entered in EES:

```
"EXAMPLE 10.5-1(b)"

$UnitSystem SI MASS RAD PA K J
$TABSTOPS 0.2 0.4 0.6 0.8 3.5 in

"Inputs"
T_enc=300 [K]                    "temperature of the enclosure"
D_p=0.3 [m]                      "diameter of object"
e_p=0.3 [-]                      "emissivity of object"
T_p=10 [K]                       "temperature of object"
D_s=0.5 [m]                      "shield diameter"
e_s=0.17 [-]                     "shield emissivity"
```

EXAMPLE 10.5-1: RADIATION SHIELD

A network representation of the problem including the shield is shown in Figure 4; note that the internal surface of the shield is designated as surface (*si*) and the external surface as (*so*).

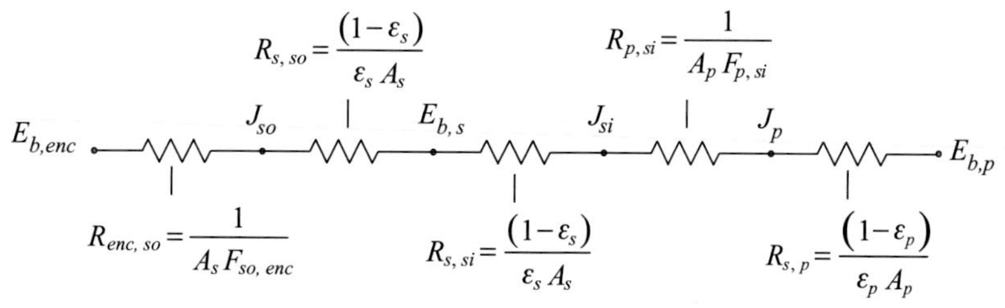

$$R_{s,so} = \frac{(1-\varepsilon_s)}{\varepsilon_s A_s} \qquad R_{p,si} = \frac{1}{A_p F_{p,si}}$$

$$E_{b,enc} \quad \text{—} \text{W} \text{—} \quad J_{so} \quad | \quad E_{b,s} \quad \text{—} \text{W} \text{—} \quad J_{si} \quad | \quad J_p \quad \text{—} \text{W} \text{—} \quad E_{b,p}$$

$$R_{enc,so} = \frac{1}{A_s F_{so,enc}} \qquad R_{s,si} = \frac{(1-\varepsilon_s)}{\varepsilon_s A_s} \qquad R_{s,p} = \frac{(1-\varepsilon_p)}{\varepsilon_p A_p}$$

Figure 4: Network representation with a radiation shield in place.

The inner and outer surfaces of the shield (*si* and *so*) are assumed to be at the same temperature because the shield is thin and conductive; therefore, $E_{b,si} = E_{b,so} = E_{b,s}$. The surface resistance associated with the object is given by Eq. (1). The surface resistances associated with the inner and outer surfaces of the shield are the same:

$$R_{s,si} = R_{s,so} = \frac{(1-\varepsilon_s)}{\varepsilon_s A_s}$$

where

$$A_s = \pi D_s^2$$

```
"Part b"
A_p=pi*D_p^2              "surface area of object"
R_s_p=(1-e_p)/(e_p*A_p)   "surface resistance of the object"
A_s=pi*D_s^2             "surface area of the shield"
R_s_si=(1-e_s)/(e_s*A_s)  "surface resistance of the shield – inner surface"
R_s_si=R_s_so            "surface resistance of the shield – outer surface"
```

The space resistance between the object and the inner surface of the shield is:

$$R_{p,si} = \frac{1}{A_p F_{p,si}}$$

where $F_{p,si}$ is 1.0. The space resistance between the outer surface of the shield and the enclosure is:

$$R_{enc,so} = \frac{1}{A_s F_{so,enc}}$$

where $F_{so,enc}$ is also 1.0.

```
F_p_si=1.0 [-]         "view factor between the object and the inner surface of the shield"
R_p_si=1/(A_p*F_p_si)  "space resistance between the object and the inner surface of the shield"
F_so_enc=1.0 [-]       "view factor between the outer surface of the shield and the enclosure"
R_enc_so=1/(A_s*F_so_enc)
                       "space resistance between the outer surface of the shield and the enclosure"
```

EXAMPLE 10.5-1: RADIATION SHIELD

The blackbody emissive powers that drive the heat transfer through the resistance network in Figure 4 are computed using Eqs. (2) and (3). The rate of heat transfer to the object is:

$$\dot{q}_p = \frac{E_{b,enc} - E_{b,p}}{R_{enc,so} + R_{s,si} + R_{s,so} + R_{p,si} + R_{s,p}}$$

E_b_p=sigma#*T_p^4   "blackbody emissive power of object"
E_b_enc=sigma#*T_enc^4   "blackbody emissive power of enclosure"
q_dot_p=(E_b_enc-E_b_p)/(R_enc_so+R_s_si+R_s_so+R_p_si+R_s_p)   "heat transfer to ojbect"

which leads to $\dot{q}_p = 18.0$ W. Note that the addition of the radiation shield has reduced the heat transfer rate by approximately 50% due to the additional resistances associated with the radiation shield (specifically, the two surface resistances $R_{s,so}$ and $R_{s,si}$ and the space resistance $R_{enc,so}$ in Figure 4). The surface resistances can be made large, and therefore the thermal isolation improved, by reducing the emissivity of the radiation shield surfaces. Figure 5 illustrates the heat transfer rate to the package as a function of the emissivity of the radiation shield.

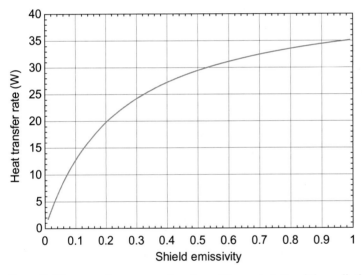

Figure 5: Heat transfer rate as a function of the emissivity of the radiation shield.

The representation of a diffuse, gray surface radiation problem using a resistance network can be a useful method of visualizing the problem. However, just as with the blackbody problems, if even a relatively modest number of surfaces are involved then the resistance diagram becomes hopelessly complicated. Each surface interacts with all (or most) of the other surfaces and therefore a network involving more than 3 or 4 surfaces becomes more confusing than useful.

It is possible to systematically solve a diffuse, gray surface radiation problem that involves an arbitrary number of surfaces, provided that the view factors and areas of each surface are known and a boundary condition can be defined for each surface. The system of equations that is required can be understood by examining Figure 10-24(b). The net rate of radiation exchange from any surface i to all of the other N surfaces is obtained from:

$$\dot{q}_i = \frac{\varepsilon_i A_i (E_{b,i} - J_i)}{(1 - \varepsilon_i)} \quad \text{for } i = 1...N \qquad (10\text{-}96)$$

Also, an energy balance written for each of the radiosity nodes leads to:

$$\dot{q}_i = A_i \sum_{j=1}^{N} F_{i,j} (J_i - J_j) \quad \text{for } i = 1...N \qquad (10\text{-}97)$$

When written for each of the N surfaces, Eqs. (10-96) and (10-97) provide 2N equations in 3N unknowns (the blackbody emissive power, net heat transfer rate, and radiosity for each of the N surfaces). A complete set of boundary conditions will include a specification of either the temperature or net heat transfer rate (or a relationship between these quantities) for each of the surfaces, providing N additional equations and therefore a completely specified problem.

EXAMPLE 10.5-2: EFFECT OF OVEN SURFACE PROPERTIES

A friend is planning to purchase a new oven and he has come to you (the engineer) for advice. He is considering two different ovens. In both ovens, the cooking space is cubical with each side of the cube measuring $L = 0.5$ m, as shown in Figure 1. Thermostatically controlled electric resistance heaters are located above the ceiling of both ovens and maintain the top surface (surface 2 in Figure 1) at $T_c = 200°C$. The other 5 oven walls (the floor and sides) are well insulated. In one oven model (oven A), all 6 of the oven walls have a dark coating with a high value of emissivity ($\varepsilon_w = 0.95$). In the other oven model (oven B), all 6 of the oven walls are reflective with a low value of emissivity ($\varepsilon_w = 0.15$). The manufacturer of oven B claims that the reflective walls cause foods to cook more quickly than they would if the walls are black.

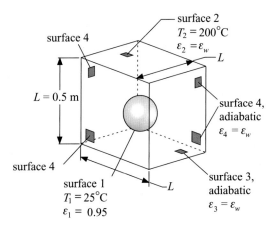

Figure 1: Roast placed in an oven.

To check the manufacturer's claim, you want to calculate the rate of radiation heat transfer to a roast with mass $M_r = 2.5$ kg when it is first placed in the oven at $T_r = 25°C$. Assume that the roast is located at the center of the oven and that the roast is approximately spherical with density $\rho_r = 1000$ kg/m³. The surface of the

EXAMPLE 10.5-2: EFFECT OF OVEN SURFACE PROPERTIES

roast has emissivity $\varepsilon_r = 0.95$. Ignore the effects of oven racks and assume that all surfaces are diffuse and gray.

a) Carry out a parametric study in order to evaluate the effect of the oven wall emissivity, ε_w, on the net rate of radiation heat transfer to the roast.

The spherical roast is considered to be surface 1. The oven ceiling, floor and vertical side walls are referred to as surfaces 2, 3, and 4, respectively. A resistance network for this four surface problem appears in Figure 2, with the surface and space resistances shown. Notice that the network is quite complicated, even for this relatively simple 4 surface problem.

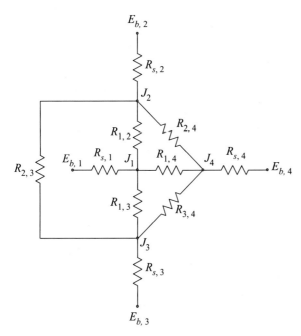

Figure 2: Resistance Network for oven and roast.

In general, the resistance network method is only really useful when there are three or fewer surfaces. Therefore, we will solve this problem using the more general method based on Eqs. (10-96) and (10-97).

The inputs are entered in EES:

```
"EXAMPLE 10.5-2"

$UnitSystem SI MASS RAD PA K J
$TABSTOPS 0.2 0.4 0.6 0.8 3.5 in

"Inputs"
L=0.5 [m]                          "size of the cube"
T_c=converttemp(C,K,200 [C])       "temperature of ceiling"
T_r=converttemp(C,K,25 [C])        "temperature of roast"
M_r=2.5 [kg]                       "mass of roast"
rho_r=1000 [kg/m^3]                "density of roast"
e_r=0.95 [-]                       "emissivity of roast surface"
e_w=0.95 [-]                       "emissivity of wall surface"
```

EXAMPLE 10.5-2: EFFECT OF OVEN SURFACE PROPERTIES

The first step is to determine the surface areas, emissivities, and view factors associated with the surfaces involved in the problem. In order to facilitate programming the system of equations given by Eqs. (10-96) and (10-97) using duplicate loops, it is useful to store this information in array format. The volume of the roast is:

$$V_r = \frac{M_r}{\rho_r}$$

The volume of the roast is related to its radius according to:

$$V_r = \frac{4}{3} \pi r_r^3$$

The surface area of the roast (surface 1) is:

$$A_{s,1} = 4 \pi r_r^2$$

The surface area of the oven ceiling and floor (surfaces 2 and 3) are:

$$A_{s,2} = A_{s,3} = L^2$$

The surface area of the 4 oven side walls (surface 4) is:

$$A_{s,4} = 4 L^2$$

```
V_r=M_r/rho_r                          "volume of roast"
V_r=4*pi*r_r^3/3                        "radius of roast"
"area of surfaces"
A_s[1]=4*pi*r_r^2                       "surface area of roast"
A_s[2]=L^2                             "surface area of ceiling"
A_s[3]=L^2                             "surface area of floor"
A_s[4]=4*L^2                           "surface area of walls"
```

The radius of the roast (r_r) is found to be 0.084 m; note that the roast is small relative to the size of the oven. The emissivities for all of the surfaces are entered:

```
"emissivity of surfaces"
e[1]=e_r                               "emissivity of roast"
e[2]=e_w                               "emissivity of ceiling"
e[3]=e_w                               "emissivity of floor"
e[4]=e_w                               "emissivity of walls"
```

The view factors for each of the four surfaces with respect to all of the others must be computed. Each surface is considered sequentially. The spherical roast cannot see itself, therefore:

$$F_{1,1} = 0$$

The roast sees each of the 6 faces of the cube equally, therefore:

$$F_{1,2} = F_{1,3} = \frac{1}{6}$$

$$F_{1,4} = \frac{2}{3}$$

EXAMPLE 10.5-2: EFFECT OF OVEN SURFACE PROPERTIES

"view factors"
F[1,1]=0   "sphere to itself"
F[1,2]=1/6   "sphere to ceiling"
F[1,3]=1/6   "sphere to floor"
F[1,4]=2/3   "sphere to walls"

The view factor between the ceiling and the roast is found by reciprocity:

$$F_{2,1} = \frac{A_{s,1}}{A_{s,2}} F_{1,2}$$

The ceiling is flat and therefore cannot see itself, therefore:

$$F_{2,2} = 0$$

F[2,1]=A_s[1]*F[1,2]/A_s[2]   "ceiling to sphere"
F[2,2]=0   "ceiling to ceiling"

We encounter a difficulty when computing the view factor between the ceiling (surface 2) and the remaining surfaces. The path between the ceiling and the floor (surface 2 to surface 3) is obscured in part by the roast (surface 1). If the oven were empty, then the view factor between surfaces 2 and 3 ($F_{2,3}$) could be computed exactly using the view factor function F3D_1 for two parallel plates (one of the built-in functions in EES' view factor library). It is tempting to assume that $F_{2,3}$ is the difference between the value obtained for two parallel plates and $F_{2,1}$, the fraction of radiation that is intercepted by the sphere. However, this assumption is not exactly correct, since not all of the radiation from surface 2 that is intercepted by the sphere would necessarily have hit surface 3; some may have hit surface 4, the walls. The error associated with this approximation becomes larger as the size of the sphere increases relative the oven dimension (i.e., if you were cooking a very large roast, then this would not be a good approximation). However, this assumption is employed here because the radius of the roast is much smaller than the oven dimension and the roast would not have been a perfect sphere anyway. In any case, the problem is primarily concerned with determining the relative merit of high as opposed to low emissivity oven walls and the conclusion with respect to this question will not be affected by a small error in this view factor.

F[2,3]=F3D_1(L,L,L)-F[2,1]   "ceiling to floor (approximate)"

The view factor between the ceiling and the walls is found using the enclosure rule:

$$F_{2,4} = 1 - F_{2,1} - F_{2,2} - F_{2,3}$$

F[2,4]=1-F[2,1]-F[2,2]-F[2,3]   "ceiling to walls"

EXAMPLE 10.5-2: EFFECT OF OVEN SURFACE PROPERTIES

By symmetry, the view factor between the floor and the sphere must be the same as the view factor between the ceiling and the sphere:

$$F_{3,1} = F_{2,1}$$

Also by symmetry, the view factor between the floor and the ceiling must be equal to the view factor between the ceiling and floor:

$$F_{3,2} = F_{2,3}$$

The view factor between the floor and itself is zero since the floor is flat:

$$F_{3,3} = 0$$

The view factor between the floor and the walls is found using the enclosure rule:

$$F_{3,4} = 1 - F_{3,1} - F_{3,2} - F_{3,3}$$

F[3,1]=F[2,1]	"floor to sphere"
F[3,2]=F[2,3]	"floor to ceiling"
F[3,3]=0	"floor to floor"
F[3,4]=1-F[3,1]-F[3,2]-F[3,3]	"floor to walls"

The view factor between the walls and the other surfaces can be found by reciprocity:

$$F_{4,1} = \frac{A_{s,1}}{A_{s,4}} F_{1,4}$$

$$F_{4,2} = \frac{A_{s,2}}{A_{s,4}} F_{2,4}$$

$$F_{4,3} = \frac{A_{s,3}}{A_{s,4}} F_{3,4}$$

The view factor from the walls to themselves is found by the enclosure rule:

$$F_{4,4} = 1 - F_{4,1} - F_{4,2} - F_{4,3}$$

F[4,1]=A_s[1]*F[1,4]/A_s[4]	"walls to sphere"
F[4,2]=A_s[2]*F[2,4]/A_s[4]	"walls to ceiling"
F[4,3]=A_s[3]*F[3,4]/A_s[4]	"walls to floor"
F[4,4]=1-F[4,1]-F[4,2]-F[4,3]	"walls to walls"

Solving the problem at this point will show that the areas, emissivities, and view factors are completely specified in the Arrays Table.

Equation (10-96) is written for each surface:

$$\dot{q}_i = \frac{\varepsilon_i A_{s,i}(E_{b,i} - J_i)}{(1 - \varepsilon_i)} \quad \text{for } i = 1...4$$

EXAMPLE 10.5-2: EFFECT OF OVEN SURFACE PROPERTIES

```
"surface heat transfer rates"
duplicate i=1,4
    q_dot[i]=e[i]*A_s[i]*(E_b[i]-J[i])/(1-e[i])
end
```

Equation (10-97) is also written for each surface:

$$\dot{q}_i = A_{s,i} \sum_{j=1}^{4} F_{i,j}(J_i - J_j) \quad \text{for } i = 1...4$$

```
"radiosity energy balances"
duplicate i=1,4
    q_dot[i]= A_s[i]*sum(F[i,j]*(J[i]-J[j]),j=1,4)
end
```

Finally, a complete set of boundary conditions are specified for each surface. The temperatures (and therefore blackbody emissive powers) of surfaces 1 (the roast) and 2 (the oven ceiling) are specified:

$$E_{b,1} = \sigma T_r^4$$
$$E_{b,2} = \sigma T_c^4$$

Surfaces 3 and 4 are adiabatic:

$$\dot{q}_3 = 0$$
$$\dot{q}_4 = 0$$

```
"boundary conditions"
E_b[1]=sigma#*T_r^4
E_b[2]=sigma#*T_c^4
q_dot[3]=0
q_dot[4]=0
```

The quantity of interest is the net rate of radiation heat transfer to the roast; this is the negative of $\dot{q}_1$, which is defined in Eqs. (10-96) and (10-97) as being the net rate of heat transfer provided to the surface:

$$\dot{q}_r = -\dot{q}_1$$

```
q_dot_r =-q_dot[1]                              "heat transferred to roast"
```

For the wall emissivity value used to set up the model, $\varepsilon_w = 0.95$ (consistent with oven A), the heat transfer to the roast is $\dot{q}_r = 161.3$ W. If all oven walls are reflective with an emissivity of $\varepsilon_w = 0.15$ (consistent with oven B), then $\dot{q}_r = 64.2$ W. Thus, radiation heat transfer is provided to the roast at a higher rate when the surfaces are black. This result may seem non-intuitive until the effect of the oven ceiling

EXAMPLE 10.5-2

is considered. Energy is radiated from the ceiling to the roast. When the ceiling emissivity is reduced, the surface resistance of the ceiling increases and therefore the rate that thermal energy can be transferred from the ceiling to the roast is also reduced. If the ceiling emissivity were maintained constant at 0.95 then the radiative heat transfer rate to the roast would be 161.3 W independent of the emissivity values for the other oven surfaces. This is because the oven floor and walls (surfaces 3 and 4) are re-radiating surfaces; that is, they are externally adiabatic. The value of the emissivity of a re-radiating surface does not affect the solution. To see this clearly, consider Figure 2; no net heat transfer flows into or out of the nodes labeled $E_{b,3}$ or $E_{b,4}$ because these surfaces are adiabatic. Therefore, no heat flows through the surface resistances associated with these surfaces ($R_{s,3}$ and $R_{s,4}$) and the radiosity and blackbody emissive power for these surfaces must be the same, regardless of the values of $R_{s,3}$ and $R_{s,4}$ (and therefore, regardless of the values of ε_3 and ε_4). The emissivity of the heater surface in the oven should be made as high as possible. However, the emissivity of the adiabatic walls does not affect the oven performance.

10.5.4 The $\hat{F}$ Parameter

The technique discussed in Section 10.5.3 is sufficient to solve gray surface problems. The result provides the temperature and net rate of heat transfer from each surface, which are the engineering quantities that are most likely to be of interest. However, the net rate of heat transfer from one surface to another surface cannot, in general, be ascertained from these solutions. The $\hat{F}$ (pronounced 'F-hat') parameter (Beckman (1968)) provides a useful method for examining the results of a gray body radiation exchange problem in order to more completely understand the radiation exchange process.

The $\hat{F}$ parameter incorporates the information that is contained in both the surface and space resistances. The factor $\hat{F}_{i,j}$ is the ratio of the energy leaving surface i that impinges on surface j, either directly or as the result of reflections from other surfaces, to the total energy leaving surface i. The definition of $\hat{F}_{i,j}$ is intentionally similar to the definition of the view factor $F_{i,j}$. Recall that $F_{i,j}$ is the ratio of the energy that is leaving from surface i that directly impinges on surface j to the total energy leaving surface i. The difference between $F_{i,j}$ and $\hat{F}_{i,j}$ is that $\hat{F}_{i,j}$ accounts for radiation that arrives at surface j indirectly. $\hat{F}_{i,j}$ and $F_{i,j}$ are identical in a radiation exchange problem in which all of the surfaces are black. However, they differ when one or more of the surfaces are gray. The similarity between $\hat{F}_{i,j}$ and $F_{i,j}$ is exploited in the $\hat{F}$ method to make gray surface radiation exchange problems look similar to the relatively simpler blackbody radiation exchange problems. The definition of the $\hat{F}$ parameter also allows the net rate of heat transfer between any two surfaces to be computed.

The difference between $\hat{F}_{i,j}$ and $F_{i,j}$ may become more clear by considering Figure 10-25, which shows four diffuse gray surfaces labeled i, j, a, and b. All of these surfaces are emitting thermal radiation to an extent that depends on their temperatures and surface properties. Consider surface i, which is emitting and reflecting radiation diffusely. Some of the radiation that is leaving surface i directly strikes surface j, as indicated by the solid arrow in Figure 10-25. The ratio of the radiation that directly strikes surface j to the radiation leaving surface i is the view factor, $F_{i,j}$. Provided that all surfaces are diffuse emitters, the view factor is only a function of geometry. Methods for calculating $F_{i,j}$ were discussed in Section 10.3.2. However, notice that some of the radiation that is leaving surface i also strikes surfaces a and b. Since these surfaces are assumed to be gray and diffuse (with an absorptivity that is less than one), a fraction of the radiation that strikes surfaces a and b is reflected and some of that reflected radiation may strike surface j, as shown by the dashed arrows in Figure 10-25.

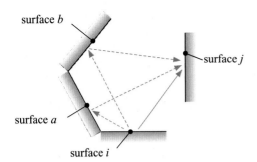

Figure 10-25: System of gray surfaces used to help understand $\hat{F}_{i,j}$.

The actual radiation exchange situation is much more complicated than is shown by Figure 10-25 as there are many other possible paths for the reflected radiation. For example, some of the radiation that is leaving surface i and directly strikes surface j may be reflected from surface j to surface b and then reflected back to surface j. In fact, there are an infinite number of paths by which radiation leaving surface i can take and ultimately impinge on surface j. The factor $\hat{F}_{i,j}$ is the fraction of the energy that leaves surface i and eventually impinges on surface j by all of these possible paths.

With so many paths available for the radiation emitted by surface i to reach surface j, it may seem difficult, if not impossible, to determine $\hat{F}_{i,j}$. Fortunately, it is actually quite easy to determine the $\hat{F}$ parameters as these depend only on the view factors and the reflectivities of the surfaces that are involved. For the system shown in Figure 10-25, $\hat{F}_{i,j}$ is given by:

$$\hat{F}_{i,j} = \underbrace{F_{i,j}}_{\text{directly strikes } j} + \underbrace{F_{i,i}\,\rho_i\,\hat{F}_{i,j} + F_{i,j}\,\rho_j\,\hat{F}_{j,j} + F_{i,a}\,\rho_a\,\hat{F}_{a,j} + F_{i,b}\,\rho_b\,\hat{F}_{b,j}}_{\text{indirectly strikes } j \text{ via reflections from } i,j,a, \text{ or } b} \qquad (10\text{-}98)$$

The first term on the right hand side of Eq. (10-98), $F_{i,j}$, is the fraction of the radiation leaving surface i that directly strikes surface j. The second and subsequent terms all have the same form. For example, consider the last term in Eq. (10-98). $F_{i,b}$ is the fraction of the radiation from surface i that directly hits surface b. The product $F_{i,b}\,\rho_b$ is therefore the fraction of energy that leaves surface i, hits surface b, and is then reflected from that surface. $\hat{F}_{b,j}$ is the fraction of the radiation that leaves surface b and subsequently hits surface j by any route. Therefore, the group $\rho_b\,F_{i,b}\,\hat{F}_{b,j}$ is the fraction of the radiation that was emitted by surface i and intercepted and reflected from surface b and then finally arrives at surface j by all possible paths. Note that $\hat{F}_{j,j}$ will generally be greater than zero, even if the surface cannot "see" itself (i.e., even if $F_{j,j} = 0$, $\hat{F}_{j,j}$ may not be zero).

The general formulation for $\hat{F}_{i,j}$ in an N-surface system is

$$\hat{F}_{i,j} = F_{i,j} + \sum_{k=1}^{N} \rho_k\, F_{i,k}\,\hat{F}_{k,j} \quad \text{for } i = 1..N \quad \text{and} \quad j = 1..N \qquad (10\text{-}99)$$

Recognizing that $\rho_i = 1\text{-}\varepsilon_i$ leads to:

$$\boxed{\hat{F}_{i,j} = F_{i,j} + \sum_{k=1}^{N} (1 - \varepsilon_k)\, F_{i,k}\,\hat{F}_{k,j} \quad \text{for } i = 1..N \quad \text{and} \quad j = 1..N} \qquad (10\text{-}100)$$

Assuming that all of the view factors and emissivity values are known, Eq. (10-100) forms a system of N^2 equations with N^2 unknowns. Actually Eq. (10-100) provides N sets of N equations and N unknowns, which is computationally a much simpler problem to solve than a set of N^2 equations. All of the $\hat{F}$ values must be determined in order to solve a gray body radiation exchange problem; however, the calculation of the $\hat{F}$ values is straightforward using an equation-solving program, like EES.

Equation (10-100) is sufficient to determine each of the $\hat{F}$ parameters, provided that each of the view factors and emissivities is known. However, there are some useful relations between $\hat{F}$ parameters that can be used to check your solution. These relations are analogous to the enclosure rule and reciprocity relationship for view factors that are presented in Section 10.3.2.

An energy balance requires that the radiation that leaves surface i must be eventually absorbed by one of the surfaces in the system. Thus, a statement of conservation of energy for surface i can be expressed as:

$$\alpha_1 \hat{F}_{i,1} + \alpha_2 \hat{F}_{i,2} + \cdots + \alpha_N \hat{F}_{i,N} = 1 \qquad (10\text{-}101)$$

or, recognizing that $\alpha_j = \varepsilon_j$ for gray surfaces:

$$\sum_{j=1}^{N} \varepsilon_j \hat{F}_{i,j} = 1 \quad \text{for any surface } i \qquad (10\text{-}102)$$

which is analogous to the enclosure rule for view factors, discussed in Section 10.3.2. There are N equations of the form of Eq. (10-102), one for each surface in the problem.

Reciprocity relations exist between $\hat{F}$ values just as they do for view factors. To see this, consider two gray surfaces, i and j. The energy emitted by surface i is $\varepsilon_i A_i \sigma T_i^4$. The fraction of this radiation that impinges on surface j, by any conceivable path, is $\hat{F}_{i,j}$. Thus the total radiation that is emitted by surface i and eventually strikes surface j is $\varepsilon_i A_i \sigma T_i^4 \hat{F}_{i,j}$. Since surface j is gray, only a fraction of this radiation, equal to the absorptivity of surface j, is absorbed by surface j. Thus, the radiation that is emitted by surface i and ultimately absorbed by surface j is $\alpha_j(\varepsilon_i A_i \sigma T_i^4 \hat{F}_{i,j})$. Using similar reasoning, the amount radiation that is emitted by surface j and ultimately absorbed by surface i is $\alpha_i(\varepsilon_j A_j \sigma T_j^4 \hat{F}_{j,i})$. The net energy exchange between surfaces i and j, $\dot{q}_{i \text{ to } j}$, is therefore:

$$\dot{q}_{i \text{ to } j} = \alpha_j \left(\varepsilon_i A_i \sigma T_i^4 \hat{F}_{i,j} \right) - \alpha_i \left(\varepsilon_j A_j \sigma T_j^4 \hat{F}_{j,i} \right) \qquad (10\text{-}103)$$

According to Kirchoff's law, the absorptivity of each gray surface must be equal to its emissivity; therefore, Eq. (10-103) can be written as:

$$\dot{q}_{i \text{ to } j} = \varepsilon_j \left(\varepsilon_i A_i \sigma T_i^4 \hat{F}_{i,j} \right) - \varepsilon_i \left(\varepsilon_j A_j \sigma T_j^4 \hat{F}_{j,i} \right) = \varepsilon_j \varepsilon_i \sigma \left(A_i \hat{F}_{i,j} T_i^4 - A_j \hat{F}_{j,i} T_j^4 \right)$$

$$(10\text{-}104)$$

Now, consider the situation where the two surfaces have the same temperature, $T_i = T_j$, the net energy exchange between these surfaces, $\dot{q}_{i \text{ to } j}$, must be zero or the second law of thermodynamics will be violated. Examination of Eq. (10-104) shows that the only way that $\dot{q}_{i \text{ to } j}$ can be zero is if:

$$A_i \hat{F}_{i,j} = A_j \hat{F}_{j,i} \qquad (10\text{-}105)$$

Substituting Eq. (10-105) into Eq. (10-104) allows the net heat transfer between any two surfaces i and j to be concisely calculated using the $\hat{F}$ parameters:

$$\boxed{\dot{q}_{i \text{ to } j} = \sigma \varepsilon_j \varepsilon_i A_i \hat{F}_{i,j} \left(T_i^4 - T_j^4 \right) = \sigma \varepsilon_j \varepsilon_i A_j \hat{F}_{j,i} \left(T_i^4 - T_j^4 \right)} \qquad (10\text{-}106)$$

It should be noted that a solution to a diffuse gray surface radiation problem can be obtained entirely using the $\hat{F}$ parameters, without ever invoking Eqs. (10-96) and (10-97). The net heat transfer from any surface i is the sum of Eq. (10-106) over all of the surfaces involved in the problem:

$$\dot{q}_i = \varepsilon_i A_i \sigma \sum_{j=1}^{N} \varepsilon_j \hat{F}_{i,j} \left(T_i^4 - T_j^4\right) \quad \text{for } i = 1..N \tag{10-107}$$

Equation (10-107) provides N equations in $2N$ unknowns (the temperatures, T_i, and heat transfer rates, $\dot{q}_i$). If a complete set of boundary conditions (one for each surface) is provided then there are $2N$ equations in an equal number of unknowns that can be solved. The solution obtained using Eq. (10-107) is algebraically equivalent to the solution obtained using Eqs. (10-96) and (10-97). The beauty of the $\hat{F}$ method is the simplicity that results because of the similarity of the gray surface relations, when expressed in terms of $\hat{F}$ factors, to the familiar black surface relations that were developed in Section 10.3. The additional advantage of using the $\hat{F}$ parameters is that Eq. (10-106) can be used to compute the net heat transfer between any two surfaces in the problem. In practice, a gray surface radiation problem is solved using the $\hat{F}$ method by first determining all of the $\hat{F}$ values using Eq. (10-100). Then Eq. (10-107) is solved together with the boundary conditions in order to predict the temperature and net rate of heat transfer from each surface.

EXAMPLE 10.5-3 illustrates the use of $\hat{F}$ parameters to calculate the net heat transfer between two parallel plates. EXAMPLE 10.5-3 is revisited in Section 10.7.3 where the problem is solved using a Monte Carlo simulation, which naturally leads to the net heat transfer between surfaces.

EXAMPLE 10.5-3: RADIATION HEAT TRANSFER BETWEEN PARALLEL PLATES

Figure 1 illustrates two aligned parallel plates, each with dimension $a = 1\,\text{m} \times b = 1\,\text{m}$, that are separated by $c = 1\,\text{m}$. Plate 1 is at a uniform temperature $T_1 = 600\,\text{K}$ with emissivity $\varepsilon_1 = 0.4$. Plate 2 is at a uniform temperature $T_2 = 350\,\text{K}$ with emissivity $\varepsilon_2 = 0.6$. The surroundings are black with temperature $T_3 = 300\,\text{K}$. Assume that the back side of each plate (i.e., the sides facing away from one another) are insulated so that radiation only occurs from the sides of the plates that are facing one another.

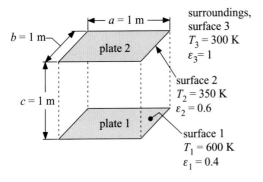

Figure 1: Parallel plates.

EXAMPLE 10.5-3: RADIATION HEAT TRANSFER BETWEEN PARALLEL PLATES

EXAMPLE 10.5-3: RADIATION HEAT TRANSFER BETWEEN PARALLEL PLATES

a) Determine the net rate of radiation heat transfer from plates 1 and 2.

This problem can be solved either using Eqs. (10-96) and (10-97) or, equivalently, using Eq. (10-100) to compute each of the $\hat{F}$ parameters and then using Eq. (10-107) to obtain the solution. Here, we will use Eqs. (10-96) and (10-97) to obtain a solution and then use the $\hat{F}$ parameter to interpret the solution in a way that would not otherwise be possible. Finally, the same solution will be obtained using Eq. (10-107).

The inputs are entered in EES; note that ε_3 is entered as a number close to but not equal to 1 in order to avoid dividing by zero in Eq. (10-96):

```
"Example 10.5-3"

$UnitSystem SI MASS RAD PA K J
$TABSTOPS 0.2 0.4 0.6 0.8 3.5 in

a=1 [m]                              "width of plate in y-direction"
b=1 [m]                              "width of plate in x-direction"
c=1 [m]                              "plate separation distance"
T[1]=600 [K]                         "temperature of plate 1"
T[2]=350 [K]                         "temperature of plate 2"
T[3]=300 [K]                         "temperature of surroundings"

"emissivities"
e[1]=0.4 [-]                         "emissivity of plate 1"
e[2]=0.6 [-]                         "emissivity of plate 2"
e[3]=0.9999 [-]                      "emissivity of surroundings"
```

The areas of each surface are computed. The areas of the two plates are:

$$A_1 = A_2 = a\,b$$

The area of the surroundings (A_3) is arbitrary and is set to a large value:

```
"areas"
A[1]=a*b                             "area of plate 1"
A[2]=a*b                             "area of plate 2"
A[3]=1e10 [m^2]                      "area of surroundings, ~infinite"
```

The view factors between the surfaces must be determined. The view factor from surface 1 to itself ($F_{1,1}$) is zero since plate 1 is flat. The view factor from surface 1 to surface 2 is obtained using the function **F3D_1** in EES view factor library. The view factor from surface 1 to the surroundings is obtained using the enclosure rule written for surface 1:

$$F_{1,3} = 1 - F_{1,1} - F_{1,2}$$

```
"view factors"
"surface 1"
F[1,1]=0                             "plate 1 cannot see itself"
F[1,2]=F3D_1(a,b,c)                  "view factor from surface 1 to 2"
F[1,3]=1-F[1,1]-F[1,2]              "enclosure rule written for plate 1"
```

EXAMPLE 10.5-3: RADIATION HEAT TRANSFER BETWEEN PARALLEL PLATES

The view factor from surface 2 to surface 1 is determined by reciprocity:

$$F_{2,1} = \frac{A_1}{A_2} F_{1,2}$$

The view factor from surface 2 to itself ($F_{2,2}$) must be zero since plate 2 is also flat. The view factor from surface 2 to surface 3 is obtained using the enclosure rule written for surface 2:

$$F_{2,3} = 1 - F_{2,1} - F_{2,2}$$

```
"surface 2"
F[2,1]=A[1]*F[1,2]/A[2]                    "reciprocity between surfaces 2 and 1"
F[2,2]=0                                    "plate 2 cannot see itself"
F[2,3]=1-F[2,1]-F[2,2]                     "enclosure rule written for plate 2"
```

The view factors between surface 3 and surfaces 1 and 2 are determined from reciprocity:

$$F_{3,1} = \frac{A_1}{A_3} F_{1,3}$$

$$F_{3,2} = \frac{A_2}{A_3} F_{2,3}$$

The view factor from the surroundings to itself is obtained using the enclosure rule written for surface 3:

$$F_{3,3} = 1 - F_{3,1} - F_{3,2}$$

```
"surface 3"
F[3,1]=A[1]*F[1,3]/A[3]                    "reciprocity between surfaces 1 and 3"
F[3,2]=A[2]*F[2,3]/A[3]                    "reciprocity between surfaces 2 and 3"
F[3,3]=1-F[3,1]-F[3,2]                     "enclosure rule written for surroundings"
```

Equations (10-96) and (10-97) are written for each surface:

$$\dot{q}_i = \frac{\varepsilon_i A_i (E_{b,i} - J_i)}{(1 - \varepsilon_i)} \quad \text{for } i = 1...3$$

$$\dot{q}_i = A_i \sum_{j=1}^{3} F_{i,j}(J_i - J_j) \quad \text{for } i = 1...3$$

```
"Radiosity method"
duplicate i=1,3
   q_dot[i]=e[i]*A[i]*(E_b[i]-J[i])/(1-e[i])
   q_dot[i]=A[i]*sum(F[i,j]*(J[i]-J[j]),j=1,3)
end
```

The boundary conditions for each surface correspond to the specified temperature:

$$E_{b,i} = \sigma T_i^4 \quad \text{for } i = 1...3$$

EXAMPLE 10.5-3: RADIATION HEAT TRANSFER BETWEEN PARALLEL PLATES

```
"boundary conditions"
E_b[1]=sigma#*T[1]^4
E_b[2]=sigma#*T[2]^4
E_b[3]=sigma#*T[3]^4
```

Solving the problem leads to the net heat transfer from each surface. The net heat transfer from plate 1 is $\dot{q}_1 = 2719$ W (i.e., 2719 W must be provided to surface 1 in order to keep it at $T_1 = 600$ K) and the net heat transfer from plate 2 is $\dot{q}_2 = -102$ W (i.e., 102 W must be removed from surface 2 in order to keep it at $T_2 = 350$ K).

b) Determine the rate of heat transfer from plate 1 to plate 2.

The heat transfer from surface 1 to surface 2 cannot be computed directly using the solution obtained in part (a); in order to determine $\dot{q}_{1\,to\,2}$, it is necessary to determine the $\hat{F}$ parameters for the system using Eq. (10-100):

$$\hat{F}_{i,j} = F_{i,j} + \sum_{k=1}^{3}(1 - \varepsilon_k)\,F_{i,k}\,\hat{F}_{k,j} \quad \text{for } i = 1..3 \quad \text{and } j = 1..3$$

```
"determine F-hat parameters"
duplicate i=1,3
  duplicate j=1,3
    F_hat[i,j]=F[i,j]+sum((1-e[k])*F[i,k]*F_hat[k,j],k=1,3)
  end
end
```

Once the view factors for the system have been determined, it is possible to use Eq. (10-106) to determine the net rate of heat transfer from plate 1 to plate 2:

$$\dot{q}_{1\,to\,2} = \sigma\,\varepsilon_2\,\varepsilon_1\,A_1\,\hat{F}_{1,2}\left(T_1^4 - T_2^4\right)$$

```
q_dot_1to2=sigma#*e[2]*e[1]*A[1]*F_hat[1,2]*(T[1]^4-T[2]^4)
```
"heat transfer between surfaces 1 and 2"

which leads to $\dot{q}_{1\,to\,2} = 314.6$ W.

It is interesting to note that the problem could be solved using the $\hat{F}$ parameters rather than using Eqs. (10-96) and (10-97). The corresponding equations are commented out:

```
{"Radiosity method"
duplicate i=1,3
  q_dot[i]=e[i]*A[i]*(E_b[i]-J[i])/(1-e[i])
  q_dot[i]=A[i]*sum(F[i,j]*(J[i]-J[j]),j=1,3)
end}
```

and instead Eq. (10-107) is written for each surface:

$$\dot{q}_i = \varepsilon_i\,A_i\,\sigma\sum_{j=1}^{3}\varepsilon_j\,\hat{F}_{i,j}\left(T_i^4 - T_j^4\right) \quad \text{for } i = 1..3$$

EXAMPLE 10.5-3

```
"solution using the F-hat equations"
duplicate i=1,3
    q_dot[i]=e[i]*A[i]*sigma#*sum(e[j]*F_hat[i,j]*(T[i]^4-T[j]^4),j=1,3)
end
```

which leads to $\dot{q}_1 = 2719\,\text{W}$ and $\dot{q}_2 = -102\,\text{W}$; the same answer that was found in part (a).

10.5.5 Radiation Exchange for Semi-Gray Surfaces

As discussed in Section 10.4.2, the emissivity of most surfaces is a complex function of temperature and wavelength. The gray surface approximation is useful for surfaces where the emissivity is either not strongly dependent upon wavelength or problems where the temperatures of the surfaces are not significantly different from each other. In many cases these conditions are not satisfied. For example, in the solar collector considered in EXAMPLE 10.4-1, the solar irradiation on the collector has a very different wavelength distribution than the thermal energy emitted from the relatively low temperature collector surface. The gray surface approximation is not appropriate for this situation because the characteristics of the collector surface at the shorter wavelengths associated with solar radiation differ from those at the longer wavelengths associated with the emitted radiation.

For situations where the gray surface approximation is not adequate, a semi-gray approximation can be employed in which the emissivity is assumed to have constant values in specified wavelength bands. The radiation heat transfer within each band is treated separately and then the heat transfer within each band is summed in order to obtain the total heat transfer from each surface.

For a problem with N surfaces and N_b wavelength bands, Eqs. (10-96) and (10-97) can be written for each surface and for each wavelength band.

$$\dot{q}_{i,w} = \frac{\varepsilon_{i,w}\, A_i\, (E_{b,i,w} - J_{i,w})}{(1 - \varepsilon_{i,w})} \quad \text{for } i = 1...N \quad \text{and} \quad w = 1..N_b \tag{10-108}$$

$$\dot{q}_{i,w} = A_i \sum_{j=1}^{N} F_{i,j}\, (J_{i,w} - J_{j,w}) \quad \text{for } i = 1...N \quad \text{and} \quad w = 1..N_b \tag{10-109}$$

where $\dot{q}_{i,w}$ is the net rate of heat transfer from surface i in wavelength band w, $\varepsilon_{i,w}$ is the emissivity of surface i in wavelength band w, $J_{i,w}$ is the radiosity from surface i in wavelength band w, and $E_{b,i,w}$ is the spectral blackbody emissive power associated with surface i integrated over all wavelengths that lie within wavelength band w:

$$E_{b,i,w} = \int_{\lambda_{low,w}}^{\lambda_{high,w}} E_{b,\lambda}\, d\lambda \text{ evaluated at } T_i \tag{10-110}$$

where $\lambda_{low,w}$ and $\lambda_{high,w}$ are the lower and upper bounds for wavelength band w. The total rate of heat transfer from each surface can then be found by summing the heat transfer within each wavelength band:

$$\dot{q}_i = \sum_{w=1}^{N_b} \dot{q}_{i,w} \quad \text{for } i = 1..N \tag{10-111}$$

The method is illustrated in EXAMPLE 10.5-4.

EXAMPLE 10.5-4: RADIATION EXCHANGE IN A DUCT WITH SEMI-GRAY SURFACES

The long triangular enclosure, shown in Figure 1, has one wall (surface 1) that is maintained at $T_1 = 300\,\mathrm{K}$ and another (surface 2) at $T_2 = 1000\,\mathrm{K}$. The third wall (surface 3) is adiabatic. The duct is very long (into the page) and therefore this is a 2-D problem. The width of each wall is $W = 0.1$ m.

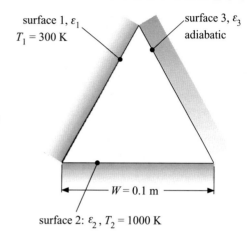

surface 1, ε_1
$T_1 = 300$ K

surface 3, ε_3
adiabatic

$W = 0.1$ m

surface 2: ε_2, $T_2 = 1000$ K

Figure 1: Long triangular duct with semi-gray surfaces.

The duct surfaces are not gray but can be modeled as being semi-gray. The emissivity of each of the 3 surfaces is illustrated in Figure 2.

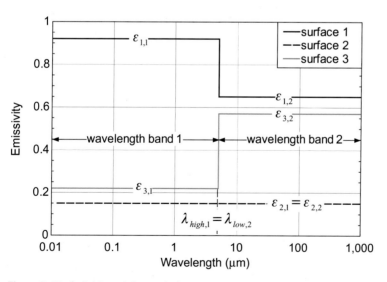

Figure 2: Emissivities of the surfaces.

There are two wavelength bands for this problem. Wavelength band 1 extends from $\lambda_{low,1} = 0\,\mu\mathrm{m}$ to $\lambda_{high,1} = 5.0\,\mu\mathrm{m}$ and wavelength band 2 extends from $\lambda_{low,2} = 5.0\,\mu\mathrm{m}$ to $\lambda_{high,2} = \infty$. In wavelength band 1, the emissivity of surfaces 1, 2, and 3 are $\varepsilon_{1,1} = 0.92$, $\varepsilon_{2,1} = 0.15$, and $\varepsilon_{3,1} = 0.22$, respectively. In wavelength band 2,

EXAMPLE 10.5-4: RADIATION EXCHANGE IN A DUCT WITH SEMI-GRAY SURFACES

EXAMPLE 10.5-4: RADIATION EXCHANGE IN A DUCT WITH SEMI-GRAY SURFACES

the emissivity of surfaces 1, 2, and 3 are $\varepsilon_{1,2} = 0.65$, $\varepsilon_{2,2} = 0.15$, and $\varepsilon_{3,2} = 0.57$, respectively.

a) **Determine the temperature of the adiabatic wall and the rate of heat transfer per unit length required to maintain the other two walls at their specified temperatures. Assume that only radiation heat transfer is occurring in the enclosure.**

The known information is entered into EES; note that $\lambda_{high,2}$ is set to a value that is large enough to include essentially all of the radiation emitted by each of the surfaces (see Figure 10-3).

```
"EXAMPLE 10.5-4"

$UnitSystem SI MASS RAD PA K J
$TABSTOPS 0.2 0.4 0.6 0.8 3.5 in

"Inputs"
W=0.1 [m]                        "width of the duct"
L=1 [m]                          "per unit length"
T[1]=300 [K]                     "temperature of surface 1"
T[2]=1000 [K]                    "temperature of surface 2"

"wavelength band 1"
lambda_low[1]=0.0 [micron]       "lower limit of wavelength band 1"
lambda_high[1]=5 [micron]        "upper limit of wavelength band 1"
epsilon[1,1]=0.92 [-]            "emissivity of surface 1 in wavelength band 1"
epsilon[2,1]=0.15 [-]            "emissivity of surface 2 in wavelength band 1"
epsilon[3,1]=0.22 [-]            "emissivity of surface 3 in wavelength band 1"

"wavelength band 2"
lambda_low[2]=5 [micron]         "lower limit of wavelength band 2"
lambda_high[2]=1000 [micron]     "upper limit of wavelength band 2"
epsilon[1,2]=0.65 [-]            "emissivity of surface 1 in wavelength band 2"
epsilon[2,2]=0.15 [-]            "emissivity of surface 2 in wavelength band 2"
epsilon[3,2]=0.57 [-]            "emissivity of surface 3 in wavelength band 2"
```

The area of each surface is the same:

$$A_1 = A_2 = A_3 = W\,L$$

```
"areas"
A[1]=W*L                                 "area of surface 1"
A[2]=W*L                                 "area of surface 2"
A[3]=W*L                                 "area of surface 3"
```

EXAMPLE 10.5-4: RADIATION EXCHANGE IN A DUCT WITH SEMI-GRAY SURFACES

Determination of the view factors is easy for this symmetric geometry and can be done by inspection; the view factor from any surface to itself must be zero (they are all flat) and the view factor from each surface to each of the other surfaces must be 0.5 by symmetry.

```
"view factors"
"surface 1"
F[1,1]=0 [-]
F[1,2]=0.5 [-]
F[1,3]=0.5 [-]

"surface 2"
F[2,1]=0.5 [-]
F[2,2]=0 [-]
F[2,3]=0.5 [-]

"surface 3"
F[3,1]=0.5 [-]
F[3,2]=0.5 [-]
F[3,3]=0 [-]
```

It is easiest to solve gray body problems by assuming a temperature for each surface (even those without specified temperatures) in order to allow the calculation of the emissive power of each surface within each wavelength band. The assumed surface temperatures must later be relaxed (i.e., commented out) in order to enforce the boundary conditions for the problem. For this problem, the temperature of surface 3 (the adiabatic surface) is initially assumed:

```
T[3]=500 [K]              "guess for temperature of surface 3 (this will be removed)"
```

The emissive power for each surface within each wavelength band is determined according to Eq. (10-110); the integral is obtained using the **Blackbody** function in EES:

$$E_{b,i,w} = \int_{\lambda_{low,w}}^{\lambda_{high,w}} E_{b,\lambda} \, d\lambda \text{ evaluated at } T_i \quad \text{for } i = 1..3 \quad \text{and} \quad w = 1..2$$

```
"blackbody emissive power for each surface in each wavelength band"
duplicate i=1,3
  duplicate w=1,2
    E_b[i,w]=Blackbody(T[i],lambda_low[w],lambda_high[w])*sigma#*T[i]^4
  end
end
```

The rates of heat transfer for each surface within each wavelength band are computed using the technique discussed in Section 10.5.5; Eqs. (10-108) and (10-109) are written for each surface and wavelength band:

$$\dot{q}_{i,w} = \frac{\varepsilon_{i,w}\, A_i (E_{b,i,w} - J_{i,w})}{(1 - \varepsilon_{i,w})} \quad \text{for } i = 1...3 \quad \text{and} \quad w = 1..2$$

$$\dot{q}_{i,w} = A_i \sum_{j=1}^{3} F_{i,j}(J_{i,w} - J_{j,w}) \quad \text{for } i = 1...3 \quad \text{and} \quad w = 1..2$$

```
"heat transfer rates for each surface in each wavelength band"
duplicate i=1,3
  duplicate w=1,2
    q_dot[i,w]=epsilon[i,w]*A[i]*(E_b[i,w]-J[i,w])/(1-epsilon[i,w])
    q_dot[i,w]=A[i]*sum(F[i,j]*(J[i,w]-J[j,w]),j=1,3)
  end
end
```

The net rate of radiation heat transfer for each surface is obtained by summing the radiation heat transfer within each wavelength band, according to Eq. (10-111):

$$\dot{q}_i = \sum_{w=1}^{2} \dot{q}_{i,w} \quad \text{for } i = 1..3$$

```
"total heat transfer to each surface"
duplicate i=1,3
  q_dot_total[i]=sum(q_dot[i,w],w=1,2)
end
```

Solving the problem will provide the heat transfer rate within each wavelength band as well as the net heat transfer for each surface, given the assumed temperature for surface 3 (Figure 3).

$\dot{q}_{i,1}$ [W]	$\dot{q}_{i,2}$ [W]	$\dot{q}_{totali}$ [W]	T_i [K]
-447.4	-244.4	-691.8	300
514	274.4	788.4	1,000
-66.61	-30.05	-96.6	500

Figure 3: Arrays table showing the heat transfer rate within each wavelength band as well as the net heat transfer rate from each surface.

Note that surface 3 is not adiabatic for the temperature that was assumed to solve the problem ($\dot{q}_3$ is −96.3 W according to Figure 3). The guess values are updated,

EXAMPLE 10.5-4

the assumed value of T_3 is commented out and the adiabatic boundary condition for surface 3 is enforced:

```
{T[3]=500 [K]}          "guess for temperature of surface 3 (this will be removed)"
q_dot_total[3]=0        "enforce that surface 3 is adiabatic"
```

which leads to $T_3 = 580\,\text{K}$, $\dot{q}_1 = -774.1\,\text{W}$ (i.e., 774.1 W must be removed from surface 1), and $\dot{q}_2 = 774.1\,\text{W}$ (i.e., 774.1 W must be provided to surface 2). The arrays table containing the details of the solution is shown in Figure 4.

Sort	7 $\lambda_{high,j}$ [micron]	8 $\lambda_{low,j}$ [micron]	9 $E_{b,j,2}$ [W/m²]	10 $E_{b,j,1}$ [W/m²]	11 $J_{j,2}$ [W/m²]	12 $J_{j,1}$ [W/m²]	13 $\dot{q}_{j,1}$ [W]	14 $\dot{q}_{j,2}$ [W]	15 $\dot{q}_{total,i}$ [W]	16 T_i [K]
[1]	5	0	453.3	5.901	2,110	411.4	-466.3	-307.8	-774.1	300
[2]	1,000	5	20,777	35,931	5,911	6,932	511.8	262.3	774.1	1,000
[3]			4,808	1,607	4,465	3,217	-45.43	45.43	0	579.8

Figure 4: Arrays table showing the solution within each wavelength band for each surface.

10.6 Radiation with other Heat Transfer Mechanisms

10.6.1 Introduction

Radiation is the only heat transfer mechanism that is considered in the problems discussed in Sections 10.1 through 10.5. However, radiation heat transfer will typically occur simultaneously with convection and conduction. These problems are called multi-mode problems. Section 10.6.2 discusses the relative importance of radiation and convection by revisiting the concept of a "radiation heat transfer coefficient" (which is also discussed in Section 1.2.6). Section 10.6.3 discusses the formal solution to multimode problems. The consideration of multi-mode problems is a straightforward extension of the techniques discussed in Sections 10.3 and 10.5 for radiation. The boundary conditions are complicated by the additional heat transfer mechanisms.

10.6.2 When Is Radiation Important?

A radiation heat transfer coefficient, $\bar{h}_{rad}$, that is analogous to the average convection heat transfer coefficient defined by Newton's Law of Cooling $(\bar{h})$ can be defined and used to approximate the effect of radiation and estimate its relative importance. Consider a surface at T_s that is experiencing both radiative and convective heat exchange. The rate of convective heat transfer is

$$\dot{q}_{conv} = \bar{h}\,A_s\,(T_s - T_\infty) \tag{10-112}$$

where A_s is the surface area and T_∞ is the temperature of the ambient air. Provided that the view factor between the surface and the surroundings is 1 (i.e., the surface sees only the surroundings at T_{sur}) and the surroundings are effectively black (i.e., no radiation is reflected back to the surface) then the rate of radiation heat transfer is:

$$\dot{q}_{rad} = \varepsilon\,A_s\,\sigma\,(T_s^4 - T_{sur}^4) \tag{10-113}$$

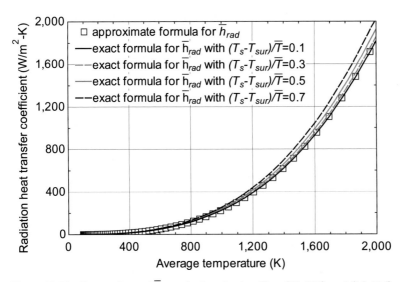

Figure 10-26: Comparison of $\overline{h}_{rad}$ calculated using Eqs. (10-115) and (10-116).

where ε is the emissivity of the surface. The radiation heat transfer coefficient, $\overline{h}_{rad}$, is defined so that Eq. (10-113) resembles Eq. (10-112):

$$\dot{q}_{rad} = \overline{h}_{rad} A_s \left(T_s - T_{sur}\right)$$ (10-114)

which requires that:

$$\overline{h}_{rad} = \varepsilon \sigma \left(T_s^2 + T_{sur}^2\right) \left(T_s + T_{sur}\right)$$ (10-115)

From Eq. (10-115), it may appear that $\overline{h}_{rad}$ is a strong function of both T_s and T_{sur}. However, these temperatures must be expressed on an absolute temperature scale, and therefore the dependence of $\overline{h}_{rad}$ on the individual temperatures is actually relatively small for most engineering applications. A reasonable engineering approximation of Eq. (10-115) is,

$$\overline{h}_{rad} \approx 4 \varepsilon \sigma \overline{T}^3$$ (10-116)

where

$$\overline{T} = \frac{T_s + T_{sur}}{2}$$ (10-117)

Figure 10-26 illustrates the exact and approximate formulae for the heat transfer coefficient, Eq. (10-115) and Eq. (10-116), respectively, for $\varepsilon_w = 1$ and different values of the normalized temperature range, $(T_s - T_{sur})/\overline{T}$. The agreement is nearly perfect when the temperature range spanned by the problem is less than 30% of the average absolute temperature involved in the problem and quite good even up to 50% or 70%.

Since convection and radiation occur in parallel, the net rate of heat transfer from the surface is:

$$\dot{q} = \dot{q}_{conv} + \dot{q}_{rad} = \overline{h} A_s \left(T_s - T_\infty\right) + \overline{h}_{rad} A_s \left(T_s - T_{sur}\right)$$ (10-118)

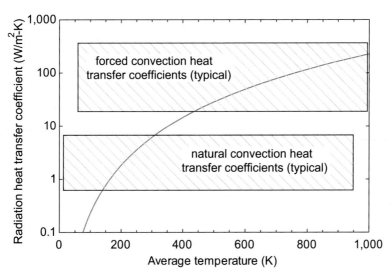

Figure 10-27: Radiation heat transfer coefficient (with $\varepsilon = 1$); also shown is the typical range of natural and forced convection heat transfer coefficients.

If the surroundings and the ambient air are at the same temperature (as is often the case) then Eq. (10-118) can be written as:

$$\dot{q} = \dot{q}_{conv} + \dot{q}_{rad} = \underbrace{(\overline{h} + \overline{h}_{rad})}_{\overline{h}_{eff}} A_s \, (T_s - T_\infty) \qquad (10\text{-}119)$$

An effective heat transfer coefficient is defined that approximately accounts for the combined effect of radiation and convection:

$$\overline{h}_{eff} = \overline{h} + \overline{h}_{rad} \qquad (10\text{-}120)$$

The concept of a radiation heat transfer coefficient provides an easy way to estimate when radiation is an important heat transfer mechanism. Figure 10-27 shows how the radiation heat transfer coefficient varies with $\overline{T}$ according to Eq. (10-116) assuming that $\varepsilon = 1.0$. Also shown in Figure 10-27 are the typical ranges for free and forced convection heat transfer coefficients.

Figure 10-27 shows that radiation is an important effect in nearly all natural convection problems. Radiation is not likely to be an important factor in most forced convection problems at moderate temperatures, but it is likely to become important at higher temperatures.

10.6.3 Multi-Mode Problems

Multi-mode problems must be solved using the techniques presented for blackbody and gray body (or even semi-gray body) radiation exchange, presented in Sections 10.3 and 10.5, respectively. The radiation portion of the problem will require boundary conditions for the blackbody emissive power and/or rate of heat transfer for each surface. In a multi-mode problem, these boundary conditions will typically be related to the solution of a separate but coupled convection or conduction problem.

No new concepts are required to solve multi-mode problems; it is only necessary that the radiation and conduction/convection problems be carefully solved and these solutions integrated through the boundary conditions. Because of the iterative nature of these solutions, it is often advisable to assume a set of surface temperatures and use these to separately solve the radiation and convection/conduction problems. The final step in the solution technique should be to adjust these surface temperatures in order to satisfy energy balances on the surfaces.

10.7 The Monte Carlo Method

10.7.1 Introduction

The Monte Carlo method is a stochastic method as opposed to the deterministic methods that have been considered elsewhere in this book. The Monte Carlo technique is a powerful simulation approach that can be used for a wide range of engineering problems; the method is most applicable when the physical system involves many variables and the relations between these variables are not easily expressed with analytical or numerical methods. At the heart of the Monte Carlo method is a random number generator that is used to statistically sample variables from a specified distribution. The statistical sampling is analogous to the activities that occur in games of chance such as poker or roulette. The Monte Carlo method is, in fact, named after a casino in Monaco.

Applied to radiation exchange calculations, the Monte Carlo method is very general and powerful; it can be used to solve problems that are otherwise intractable using deterministic methods. For example, the Monte Carlo method can determine the net rate of radiation heat transfer between surfaces of non-uniform temperature with wavelength, temperature and spatially-dependent properties. The disadvantage of the Monte Carlo method is that it requires substantial computational effort.

The application of Monte Carlo methods is a broad topic that is only covered briefly in this section. The technique is used in Section 10.7.2 to calculate view factors and in Section 10.7.3 to calculate the rate of radiation heat transfer between gray surfaces. A more thorough discussion of the Monte Carlo method for radiation problems can be found in Siegel and Howell (2002).

10.7.2 Determination of View Factors with the Monte Carlo Method

The direct determination of a view factor between two surfaces using the formal definition in Eq. (10-11) requires the evaluation of a complicated double integral. For many geometries, the integration cannot be done analytically and therefore multiple numerical integrations over multiple dimensions is required. The Monte Carlo method is an attractive alternative to Eq. (10-11).

In this section, the Monte Carlo method is used to determine the view factor for the relatively simple situation shown in Figure 10-28. Surfaces 1 and 2 are parallel rectangles and both are assumed to be diffuse emitters. This relatively simple geometry is selected because the view factor for this situation can also be determined by a deterministic method and is available in the EES view factor library as the function F3D_1. However, the Monte Carlo solution can be extended easily in order to determine the view factor for a more complex situation where no analytical solution is available.

In order to determine the view factor from surface 1 to surface 2, $F_{1,2}$, using the Monte Carlo method, it is necessary to track the fate of many 'rays' of radiation emitted from surface 1. The direction of the emission is randomly chosen from a known directional distribution. Each ray is considered to be a bundle with a fixed amount of energy, represented as a vector that is propagated from a known position on surface 1 in a randomly chosen direction. Some of the rays intercept surface 2. The ratio of the number of rays that strike surface 2 to the total number of rays emitted from surface 1 is an

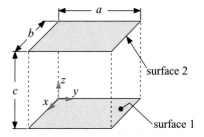

Figure 10-28: Parallel aligned surfaces. The view factor $F_{1,2}$ is determined using the Monte Carlo technique in this section.

estimate of the view factor; this estimate improves as more and more rays are produced and tracked.

The Monte Carlo method for determining view factor from surface 1 to surface 2 can be described as repetitions of the following steps:

1. Select a location on surface 1.
2. Select a direction of the ray.
3. Determine if the ray from surface 1 hits surface 2.

A formal programming language such as MATLAB is much better suited for the Monte Carlo technique than EES and therefore the solution to the problem shown in Figure 10-28 is implemented in MATLAB. A function is defined in MATLAB that takes as inputs the dimensions of the problem and the number of rays to generate:

```
function[F]=F_12(N,a,b,c)

% Inputs
% N              number of rays to generate
% a, b           width and height of the rectangles
% c              distance between the rectangles

% Outputs
% F              view factor
```

Two counters are initialized in order to count the number of rays that are tracked (*ict*) and the number of these rays that hit surface 2 (*hits*):

```
ict=0;          % counter for the number of rays
hits=0;         % counter for the number of impingements
```

The steps for the Monte Carlo process are placed within a while loop that terminates when *ict* reaches *N*, the user specified number of rays; the value of *ict* is incremented each time that the while loop executes:

```
while (ict<N)          % terminate loop when N rays have been generated
    ict=ict+1;         % increment the ray counter
```

The steps are discussed and implemented in the following sections.

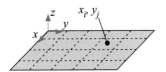

Figure 10-29: Surface 1 broken into a fixed number of cells.

Select a Location on Surface 1

In order for the Monte Carlo method to provide an accurate estimate of the view factor, it is necessary to generate rays from a statistically representative number of locations on surface 1. The locations can be selected deterministically or statistically. In the deterministic approach, surface 1 is broken into a specified number of cells, as shown in Figure 10-29. The center point of each cell on the surface is chosen to be the emission point or origin for the rays. The same number of rays is emitted from each cell.

In the statistical approach, a location on surface 1 (x, y) is chosen stochastically; the location x is set to a uniformly distributed, random number between 0 (the upper edge of surface 1 in Figure 10-28) and b (the bottom edge of surface 1). A second uniformly distributed random number is used to set the location of y between 0 and a.

```
x=rand*b;                    % randomly select a ray origin
y=rand*a;
```

Note that the function rand in MATLAB generates a random number between 0 and 1 with a uniform probability distribution. The next two steps involve setting the direction of the ray emitted from position (x, y) and then determining whether it strikes surface 2 and, if so, where. Because these calculations must be repeated from arbitrary locations, they are placed in a subfunction Ray that takes as inputs the origin of the ray as well as the geometry of the surfaces (a, b, and c for the problem in Figure 10-28). The subfunction Ray provides as outputs the binary indicator *hit* that is 1 if the ray strikes surface 2 and 0 if it misses as well as the location on surface 2 that the ray hits, (x_i, y_i).

```
function[hit, x_i, y_i]=Ray(x, y, a, b, c)
% Inputs
% x, y     origin of the ray (m)
% a, b     width and height of the rectangles (m)
% c        distance between the rectangles (m)
% Outputs
% hit      flag indicating whether the ray hits surface 2 (0 or 1)
% x_i, y_i the intersection point (m)
```

Select the Direction of the Ray

View factors are determined assuming that the surfaces are diffuse emitters. A diffuse emitter is defined as a surface that emits radiation uniformly in all directions. As discussed in Section 10.4.2, the direction of the ray emitted from a selected point is most conveniently described in spherical coordinates by its polar and azimuthal angles, θ and ϕ, respectively (see Figure 10-30). The ray is represented as the unit vector $\underline{r}$:

$$\underline{r} = [\cos(\phi) \sin(\theta)]\underline{i} + [\sin(\phi) \sin(\theta)]\underline{j} + [\cos(\theta)]\underline{k} \qquad (10\text{-}121)$$

where $\underline{i}$, $\underline{j}$, and $\underline{k}$ are unit vectors in the x, y, and z directions, respectively.

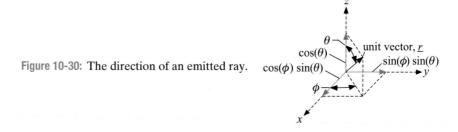

Figure 10-30: The direction of an emitted ray.

The polar and azimuthal angles that determine the direction of the unit vector are chosen randomly from a probability distribution that is representative of a diffuse emitter. The cumulative probability of emission from a diffusely emitting surface at polar angles between 0 and θ for all azimuthal angles, P_θ, is (Brewster (1992), Siegel and Howell (2002)):

$$P_\theta = \sin^2 (\theta) \tag{10-122}$$

The cumulative probability P_ϕ of emission from a diffusely emitting surface at azimuthal angles between 0 and ϕ for all polar angles is:

$$P_\phi = \frac{\phi}{2\pi} \tag{10-123}$$

The cumulative probability functions P_θ and P_ϕ have values that range from 0 to 1 as θ increases from 0 to $\pi/2$ and ϕ increases from 0 to 2π, respectively. A randomly selected polar angle can be chosen from the probability distribution by generating a random number between 0 and 1 and setting it equal to P_θ. The corresponding polar angle is found by solving Eq. (10-122).

$$\theta = \sin^{-1} \left(\sqrt{P_\theta} \right) \tag{10-124}$$

A randomly chosen azimuthal angle is obtained in the same manner using Eq. (10-123).

$$\phi = 2\pi P_\phi \tag{10-125}$$

```
Ptheta=rand;              % uniformly distributed random number between 0 and 1
theta=asin(sqrt(Ptheta)); % determine the polar angle
Pphi=rand;                % uniformly distributed random number between 0 and 1
phi=Pphi*2*pi;            % determine the azimuthal angle
```

The unit vector representing the randomly selected ray is then determined by Eq. (10-121).

Determine whether the Ray from Surface 1 Strikes Surface 2
The procedure for estimating a view factor $F_{1,2}$ with the Monte Carlo method is to determine the fraction of the rays emitted from surface 1 that strike surface 2. It is therefore necessary to determine whether the ray leaving surface 1 at the randomly selected location (x, y) from Step 1 in the randomly selected direction θ, ϕ from Step 2 intercepts surface 2. The location of intersection is also of interest; in Section 10.7.3, the rate of radiation heat transfer between the plates is calculated using the Monte Carlo method

and this requires knowledge of the intersection point in order to continue tracking the path of reflected rays.

The problem is reduced to finding the point where an extended vector representing the ray intercepts the plane that is defined by surface 2. The intersection point on the plane is then tested to see if it lies within the boundaries defining surface 2. This calculation is very simple if the planes containing surfaces 1 and 2 are parallel (as they are for the geometry indicated in Figure 10-28), but in general it can be geometrically complex. For this problem, the coordinate z can be set to 0 for the plane containing surface 1 and z is set to c for the plane containing surface 2 (where c is the distance between the surfaces, see Figure 10-28). The ray position ($\underline{R}$) for an arbitrary length L emanating from a particular (x, y) location on surface 1 $(z = 0)$ is obtained by multiplying the unit vector from Eq. (10-121) by L and adding the initial location:

$$
\begin{aligned}
\underline{R} &= x\underline{i} + y\underline{j} + 0\underline{k} + \underline{r}\,L \\
&= \underbrace{[x + L\cos(\phi)\sin(\theta)]}_{x_i}\underline{i} + \underbrace{[y + L\sin(\phi)\sin(\theta)]}_{y_i}\underline{j} + \underbrace{[L\cos(\theta)]}_{z_i}\underline{k} \quad (10\text{-}126)
\end{aligned}
$$

The length of the ray when it intersects surface 2 is obtained by setting the z-coordinate of Eq. (10-126) to c:

$$
L\cos(\theta) = c \qquad (10\text{-}127)
$$

or

$$
L = \frac{c}{\cos(\theta)} \qquad (10\text{-}128)
$$

The corresponding x and y coordinates at the intersection point (x_i, y_i) are obtained by substituting Eq. (10-128) into Eq. (10-126):

$$
x_i = x + c\cos(\phi)\tan(\theta) \qquad (10\text{-}129)
$$

$$
y_i = y + c\sin(\phi)\tan(\theta) \qquad (10\text{-}130)
$$

```
x_i=x+c*tan(theta)*cos(phi);              % determine the intersection point
y_i=y+c*tan(theta)*sin(phi);
```

The intersection point is next tested to determine if it is within the boundaries of surface 2. If so, the ray is counted as a hit.

$$
hit = \begin{cases} 0 & \text{if } x_i < 0 \quad \text{or} \quad x_i > b \quad \text{or} \quad y_i < 0 \quad \text{or} \quad y_i > a \\ 1 & \text{otherwise} \end{cases} \qquad (10\text{-}131)
$$

```
if (x_i<0)                    % check to ensure that the ray the rectangle
    hit=0;                    % 0 indicates that the ray does not hit the surface
elseif (x_i>b)
    hit=0;
elseif (y_i<0)
    hit=0;
```

```
    elseif (y_i>a)
        hit=0;
    else
        hit=1;
    end;
end
```

Steps 1 through 3 are repeated many times until the standard deviation of the estimated view factor reaches a sufficiently small value. The number of repetitions required to achieve a specified accuracy can be determined by examining the predicted result for convergence.

Returning to the body of the function F_12; the subfunction Ray is called using the location on surface 1 that was determined in Step 1:

```
[hit,x_i,y_i]=Ray(x,y,a,b,c);          % see if ray hits surface 2
```

if *hit* is 1, then the counter for the number of rays that hit surface 2, *hits*, is incremented by 1:

```
if (hit==1)
    hits=hits+1;          % if ray hits then increment hit counter
end
```

When the while loop is terminated (because the specified number of rays have been emitted and tracked), the view factor is estimated according to the ratio of the two counters *hits* (the number of rays that hit surface 2) and *ict* (the number of rays that were emitted).

$$F_{1,2} = \frac{hits}{ict} \tag{10-132}$$

```
    end
    F=hits/ict;
end
```

The function F_12 with its subfunction Ray is listed below:

```
function[F]=F_12(N,a,b,c)
% Inputs
% N      number of rays to generate (-)
% a, b   width and height of the rectangles (m)
% c      distance between the rectangles (m)
% Outputs
```

```
% F      view factor (-)

   ict=0;                              % counter for the number of rays
   hits=0;                             % counter for the number of impingements
   while (ict<N)                       % terminate loop when N rays have been generated
     ict=ict+1;                        % increment the ray counter
     x=rand*b;                         % randomly select a ray origin
     y=rand*a;
     [hit,x_i,y_i]=Ray(x,y,a,b,c);     % see if ray hits surface 2
     if (hit==1)
        hits=hits+1;                   % if ray hits then increment hit counter
     end
   end
   F=hits/ict;
end

function[hit, x_i, y_i]=Ray(x, y, a, b, c)
% Inputs
% x, y    origin of the ray (m)
% a, b    width and height of the rectangles (m)
% c       distance between the rectangles (m)
% Outputs
% hit     flag indicating whether the ray hits surface 2 (0 or 1)
% x_i, y_i the intersection point (m)

   Ptheta=rand;                        % uniformly distributed random number between 0 and 1
   theta=asin(sqrt(Ptheta));           % determine the polar angle
   Pphi=rand;                          % uniformly distributed random number
   phi=Pphi*2*pi;                      % determine the azimuthal angle
   x_i=c*tan(theta)*cos(phi)+x;        % determine the intersection point
   y_i=c*tan(theta)*sin(phi)+y;

   if (x_i<0)                          % check to ensure that the ray the rectangle
      hit=0;                           % 0 indicates that the ray does not hit the surface
   elseif (x_i>b)
      hit=0;
   elseif (y_i<0)
      hit=0;
   elseif (y_i>a)
      hit=0;
   else
      hit=1;
   end;
end
```

The function F_12 is tested using the inputs $a = b = c = 1$ m and $N = 1000$ rays:

```
>> [F]=F_12(1000,1,1,1)
F =
   0.1990
```

The analytically determined value of the view factor is obtained using EES with the function F3D_1.

```
$UnitSystem SI MASS RAD PA K J
$TABSTOPS 0.2 0.4 0.6 0.8 3.5 in

F=F3D_1(1 [m],1 [m], 1 [m])                    "view factor, determined analytically"
```

which leads to $F_{1,2} = 0.20$.

The script listed below runs the function F_12 many times (25) for various values of N; because the Monte Carlo technique is stochastic rather than deterministic, the predicted value of $F_{1,2}$ will not be the same each time the program is called even if the input parameters are identical. The script records the average result and the standard deviation of the results (which is an estimate of the error associated with the Monte Carlo technique) in order to assess the number of rays that is required to provide a useful answer.

```
a=1;                                    % plate dimension a (m)
b=1;                                    % plate dimension b (m)
c=1;                                    % plate separation distance (m)
Nv=[10,20,30,40,50,80,100,150,200,300,400,500,800,1000,1500,2000,...
    2500,3000,4000,5000,8000,10000];    % number of rays
for i=1:22
  for j=1:25
    [F(i,j)]=F_12(Nv(i),a,b,c);
  end
end
F_bar=mean(F');                         % compute average value of view factors
F_std=std(F');                          % compute standard deviation of view factors
```

Figure 10-31 illustrates the average value of the predicted view factor for the 25 repeats as a function of the number of rays. The error bars in Figure 10-31 indicate the standard deviation of the 25 repeats. The analytical value of the view factor is 0.20; notice that the view factor is reliably determined using the Monte Carlo technique to within 5% of its correct value for $N > 1,000$ and to within less than 2% for $N > 10,000$.

A major advantage of the Monte Carlo method is that it can be adapted to geometries that would be difficult to accommodate through either analytical or numerical integration of Eq. (10-11). For example, a view factor relation is not available if surface 2 is a disk of radius r centered above the rectangular surface 1, as shown in Figure 10-32.

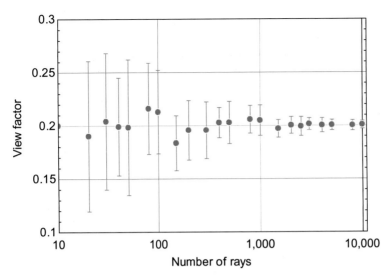

Figure 10-31: Average view factor for 25 repeats as a function of the number of rays, N. The error bars show the standard deviation of the 25 repeats.

However, this geometry can easily be considered by modifying the function F_12. The function is renamed F_12_disk and the header is modified to accept the disk radius as an input:

```
function[F]=F_12_disk(N,a,b,c,r)
% Inputs
% N       number of rays to generate (-)
% a, b    width and height of the rectangle (m)
% c       distance between the rectangle and the disk (m)
% r       radius of the disk (m)
% Outputs
% F       view factor (-)
```

The process of stochastically selecting the position to emit a ray from surface 1 does not change. The subfunction Ray must be altered in order to include the disk radius as an input:

```
function[hit, x_i, y_i]=Ray(x, y, a, b, c, r)
% Inputs
% x, y      origin of the ray (m)
```

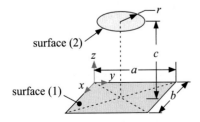

Figure 10-32: Disk centered above a parallel rectangle.

```
% a, b      width and height of the rectangles (m)
% c         distance between the rectangles (m)
% r         radius of disk (m)
% Outputs
% hit       flag indicating whether the ray hits surface 2 (0 or 1)
% x_i, y_i  the intersection point (m)
```

The subfunction Ray is changed so that it checks whether the intersection point in the plane containing surface 2 is within the disk (rather than within the rectangle):

$$hit = \begin{cases} 1 & \text{if } \sqrt{\left(x_i - \dfrac{b}{2}\right)^2 + \left(y_i - \dfrac{a}{2}\right)^2} \leq r \\ 0 & \text{if } \sqrt{\left(x_i - \dfrac{b}{2}\right)^2 + \left(y_i - \dfrac{a}{2}\right)^2} > r \end{cases} \qquad (10\text{-}133)$$

```
if(sqrt((x_i-b/2)^2+(y_i-a/2)^2)<=r)
    hit=1;
else
    hit=0;
end
```

The MATLAB function F_12_disk with its subfunction Ray is shown below:

```
function[F]=F_12_disk(N,a,b,c,r)
% Inputs
% N       number of rays to generate (-)
% a, b    width and height of the rectangle (m)
% c       distance between the rectangle and the disk (m)
% r       radius of the disk (m)
% Outputs
% F       view factor (-)

    ict=0;                          % counter for the number of rays
    hits=0;                         % counter for the number of impingements
    while (ict<N)                   % terminate loop when N rays have been generated
        ict=ict+1;                  % increment the ray counter
        x=rand*b;                   % randomly select a ray origin
        y=rand*a;
        [hit,x_i,y_i]=Ray(x,y,a,b,c,r);   % see if ray hits surface 2
        if (hit==1)
            hits=hits+1;            % if ray hits then increment hit counter
        end
    end
    F=hits/ict;
end

function[hit, x_i, y_i]=Ray(x, y, a, b, c, r)
% Inputs
% x, y      origin of the ray (m)
```

```
% a, b      width and height of the rectangles (m)
% c         distance between the rectangles (m)
% r         radius of disk (m)
% Outputs
% hit       flag indicating whether the ray hits surface 2 (0 or 1)
% x_i, y_i  the intersection point (m)

    Ptheta=rand;                    % uniformly distributed random number between 0 and 1
    theta=asin(sqrt(Ptheta));       %determine the polar angle
    Pphi=rand;                      % uniformly distributed random number
    phi=Pphi*2*pi;                  % determine the azimuthal angle
    x_i=c*tan(theta)*cos(phi)+x;    % determine the intersection point
    y_i=c*tan(theta)*sin(phi)+y;


    if(sqrt((x_i-b/2)^2+(y_i-a/2)^2)>=r)
        hit=1;
    else
        hit=0;
    end
end
```

The same computer program (implemented in a formal programming language) is used in the radiation view factor library in EES in order to provide the view factor between a co-planar rectangle and disk in function F3D_21.

10.7.3 Radiation Heat Transfer Determined by the Monte Carlo Method

The Monte Carlo method can be extended to handle essentially all of the complexities that arise in radiation heat transfer problems. For example, the method can be applied to calculate the net rate of heat transfer between surfaces that have wavelength, directional, or spatially-dependent surface properties and/or spatially varying temperatures. The Monte Carlo method is often the only computational tool that can be applied in these situations, but the computational effort involved can be significant. A detailed discussion of the Monte Carlo method is not presented in this text and the interested reader is referred to Brewster (1992) or Siegel and Howell (2002) for additional information on Monte Carlo methods.

In this section, the Monte Carlo method for computing radiation exchange is introduced by considering heat transfer between the two parallel plates previously considered in EXAMPLE 10.5-3 and Section 10.7.2. The problem is shown in Figure 10-33.

Two aligned parallel plates, each with dimension $a = 1\,\text{m} \times b = 1\,\text{m}$, are separated by $c = 1\,\text{m}$. Plate 1 is at a uniform temperature $T_1 = 600\,\text{K}$ with emissivity $\varepsilon_1 = 0.4$. Plate 2 is at a uniform temperature $T_2 = 350\,\text{K}$ with emissivity $\varepsilon_2 = 0.6$. The surroundings are black with temperature $T_3 = 300\,\text{K}$. In this section, the net rate of radiation heat transfer between the plates is computed using the Monte-Carlo method. In EXAMPLE 10.5-3, the net rate of radiation heat transfer from plate 1 to plate 2 is computed using the deterministic solution techniques for gray body radiation exchange, discussed in Section 10.5. Therefore, it is possible to check the result of the Monte Carlo calculation against the result obtained in EXAMPLE 10.5-3. However, the gray surface radiation calculation methods presented in Section 10.5 are not applicable if surfaces 1 and 2 are not isothermal or if the emissivities are functions of position or wavelength, whereas the Monte Carlo method can handle these complexities with little additional effort.

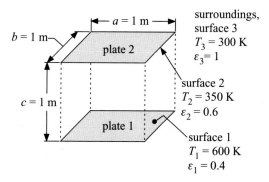

Figure 10-33: Two parallel plates exchanging radiation.

The Monte Carlo calculation proceeds using the ray-tracing approach that is discussed in Section 10.7.2. The steps are outlined in this section and laid out in the flow chart shown in Figure 10-34.

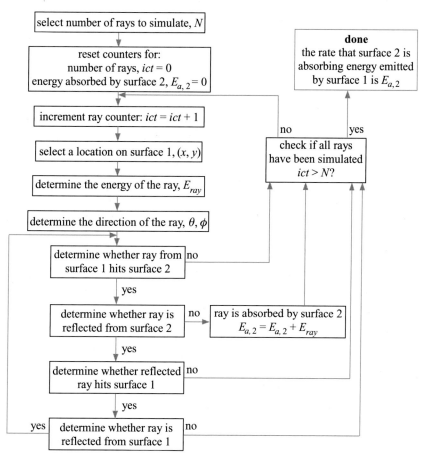

Figure 10-34: Flow chart showing the process of using the Monte Carlo method to compute the rate at which radiation emitted by plate 1 is absorbed by plate 2 for the problem shown in Figure 10-33. A similar calculation provides the rate at which radiation emitted by plate 2 is absorbed by plate 1; the difference in these quantities is the net radiation transfer between plate 1 and plate 2.

The amount of energy emitted by plate 1 and absorbed by plate 2 is determined by repeating the sequence of calculations shown in Figure 10-34 many times. The steps are:

1. Select a location on surface 1.
2. Determine the energy of the ray.
3. Select the direction of the ray.
4. Determine whether the ray from surface 1 strikes surface 2. If the ray does strike surface 2, then go to step 5. If the ray does not strike surface 2, then it is "lost" (i.e., it is absorbed by the surroundings which are black) and the next ray can be simulated (i.e., return to step 1).
5. Determine whether the ray that strikes surface 2 is reflected or absorbed. If the ray is absorbed then add the energy associated with the ray to the running tally of energy absorbed by surface 2; the next ray can be simulated (i.e., return to step 1). If the ray is reflected then go to step 6.
6. Determine whether the ray that is reflected from surface 2 strikes surface 1. If the ray does strike surface 1, then go to step 7. If the ray does not strike surface 1, then it is "lost" and the next ray can be simulated.
7. Determine whether the ray that strikes surface 1 is reflected or absorbed. If the ray is absorbed then it is "lost" and the next ray can be simulated (i.e., return to step 1). If the ray is reflected then go to step 3 in order to continue to follow the path of the ray until it is either absorbed by surface 2 or "lost."

Clearly the process for determining the radiation exchange is more involved than the process for finding a view factor, discussed in Section 10.7.2. However, conceptually, the simulation process is the same. You are keeping track of one ray at a time and simulating its path in order to understand the energy flow. By simulating "enough" rays, the solution converges on the actual behavior of the plate-to-plate radiation exchange problem. The steps discussed above and shown in Figure 10-34 are repeated in order to determine the rate that energy emitted by surface 2 is absorbed by surface 1. The difference between these two quantities is the net rate at which plates 1 and 2 are exchanging radiation.

The Monte Carlo simulation is programmed using MATLAB and placed in a function, q_net, that takes as input the number of rays to use in the simulation and provides as output the net rate of heat transfer between surfaces 1 and 2.

```
function[q_dot_1to2]=q_net(N)

% Inputs
% N        number of rays to generate (-)

% Outputs
% q_dot_1to2    net rate heat transfer from surface 1 to surface 2 (W)
```

The specified parameters are assigned:

```
a=1;              % plate size in y-direction (m)
b=1;              % plate size in x-direction (m)
c=1;              % plate separation (m)
T1=600;           % plate 1 temperature (K)
```

```
T2=350;              % plate 2 temperature (K)
e1=0.4;              % emissivity of plate 1 (-)
e2=0.6;              % emissivity of plate 2 (-)
sigma=5.67e-8;       % Stefan-Boltzmann constant (W/m^2-K^4)
```

The simulation begins by calculating the rate at which radiation emitted by surface 1 is absorbed by surface 2. The counters *ict* (the number of rays simulated) and $E_{a,2}$ (the rate at which energy is absorbed by surface 2) are reset:

```
% transfer of energy from surface 2 to surface 1
Ea2=0;       % counter for the amount of energy absorbed by 1 that was emitted by 2
ict=0;       % counter to track number of rays emitted by 2
```

The simulation of each ray is placed in a while loop that terminates when *ict* reaches *N*, the number of rays; the counter *ict* is incremented each time the while loop executes.

```
while(ict<N)
    ict=ict+1;                   % emit a ray
```

The location on surface 1 from which the ray emanates is selected stochastically, as discussed in Section 10.7.2:

```
x=rand*b;                % x-location of emitted ray (m)
y=rand*a;                % y-location of emitted ray (m)
```

The energy of the ray is determined based on the temperature at the location that it is emitted, the emissivity of the plate and the differential area of the plate associated with the ray; the differential area is the total area divided by the number of rays:

$$E_{ray} = \varepsilon_1 \, \sigma \, T_1^4 \frac{a\,b}{N} \qquad (10\text{-}134)$$

Note that if the plate temperature vary spatially, then the temperature in Eq. (10-134) would be calculated according to the temperature distribution. Further, if the plate had spectrally dependent properties, then the wavelength of the ray would have to be stochastically determined from a specified distribution and the energy of the ray would depend on its wavelength.

```
Eray=e1*sigma*T1^4*a*b/N;                        % determine energy of ray (W)
```

The ray is traced until it is "lost"; that is, it is absorbed by one of the surfaces in the problem. The binary variable *lost* is set to 0 (indicating that the ray is not yet lost) and the ray tracing steps are placed in a while loop that terminates when *lost* is equal to 1 (indicating that the ray has been absorbed):

```
lost=0;                              % indicator to see if ray is "lost"
while(lost==0)                       % loop is terminated once the ray is "lost"
```

The intersection of the ray with the plane that is defined by surface 2 is calculated by the subfunction **Ray** that was created in Section 10.7.2. The subfunction **Ray** returns the binary variable *hit* (which is 0 if the ray misses surface 2 and 1 if the ray hits surface 2) and the location on surface 2 where the ray hits (x_i, y_i):

```
[hit,x_i,y_i]=Ray(x,y,a,b,c);        % see if ray leaving 1 hits surface 2
```

An if statement is used to check if the ray hits surface 2 or misses. If the ray misses surface 2 then it is absorbed by the black surroundings and the variable *lost* is set to 1:

```
if(hit==0)        % ray misses surface 2 and is "lost"
   lost=1;
else              % ray hits surface 2
```

If the ray hits surface 2, then it is necessary to see if it is reflected or absorbed. The absorptivity of surface 2 (which is equal to the emissivity of surface 2, according to Kirchoff's law for this gray surface) represents the likelihood that the ray will be absorbed. A random number between 0 and 1 is generated and compared to the emissivity. If the random number is less than the emissivity, then the ray is considered to have been absorbed and the counter $E_{a,2}$ is increased by the energy of the ray. In this case, the absorbed ray is lost and the variable *lost* is set to 1.

```
if(rand<e2)          % ray is absorbed by surface 2
   Ea2=Ea2+Eray;     % add ray's energy to that absorbed by surface 2
   lost=1;           % ray is absorbed and "lost"
else                 % ray is reflected from surface 2
```

If the ray is reflected from surface 2, then it is necessary to see if it subsequently hits surface 1. The subfunction **Ray** is used again to determine the intersection of the reflected ray with the plane defined by surface 1:

```
[hit,x_i,y_i]=Ray(x_i,y_i,a,b,c);       % see if ray reflected from surface 2 hits surface 1
```

If the ray misses surface 1 then it is absorbed by the surroundings and lost; therefore, the variable *lost* is set to 1:

```
if(hit==0)        % reflected ray misses surface 1 and is "lost"
   lost=1;
else
```

If the ray hits surface 1, then it is necessary to determine whether it is absorbed or reflected. A random number is generated and compared to the emissivity of surface 1. If the random number is less than ε_1 then the ray is absorbed and therefore lost (the variable *lost* is set to 1):

```
if(rand<e1) % reflected ray is absorbed by surface 1
   lost=1;
else        % ray is reflected from surface 1
```

If the ray is reflected from surface 1 then the process of tracking the ray begins again, starting with the new ray location:

```
         x=x_i;   % start process over, with new emission point
         y=y_i;
       end
     end
   end
 end
end
end
```

The result of the calculation, when completed N times, is a prediction of the rate at which energy emitted by surface 1 is absorbed by surface 2 ($E_{a,2}$). The entire section of the code is shown below:

```
% transfer of radiation from surface 1 to 2
Ea2=0;     % counter to track amount of energy absorbed by 2 that was emitted by 1
ict=0;     % counter to track number of rays emitted by 1
while(ict<N)
  ict=ict+1;                              % emit a ray
  x=rand*b;                               % x-location of emitted ray (m)
  y=rand*a;                               % y-location of emitted ray (m)
  Eray=e1*sigma*T1^4*a*b/N;               % determine energy of ray (W)
  lost=0;                                 % indicator to see if ray is "lost"
  while(lost==0)                          % loop is terminated once the ray is "lost"
    [hit,x_i,y_i]=Ray(x,y,a,b,c);         % see if ray leaving 1 hits surface 2
    if(hit==0)                            % ray misses surface 2 and is "lost"
      lost=1;
    else                                  % ray hits surface 2
      if(rand<e2)                         % ray is absorbed by surface 2
        Ea2=Ea2+Eray;                     % add ray's energy to that absorbed by surface 2
        lost=1;                           % ray is absorbed and "lost"
      else                                % ray is reflected from surface 2
        [hit,x_i,y_i]=Ray(x_i,y_i,a,b,c);
        %see if ray reflected from surface 2 hits surface 1
        if(hit==0)                        % reflected ray misses surface 1 and is "lost"
          lost=1;
        else
          if(rand<e1)                     % reflected ray is absorbed by surface 1
            lost=1;
```

```
              else                        % ray is reflected from surface 1
                x=x_i;                    % start process over, with new emission point
                y=y_i;
              end
            end
          end
        end
      end
    end
```

Another N rays are simulated; these rays are emitted from surface 2 and traced until they are absorbed by surface 1 or lost. The process of writing this section of code follows naturally from the previous discussion:

```
% transfer of radiation from surface 2 to 1
Ea1=0;    % counter to track amount of energy absorbed by 1 that was emitted by 2
ict=0;    % counter to track number of rays emitted by 2
while(ict<N)
  ict=ict+1;                              % emit a ray
  x=rand*b;                               % x-location of emitted ray (m)
  y=rand*a;                               % y-location of emitted ray (m)
  Eray=e2*sigma*T2^4*a*b/N;               % determine energy of ray (W)
  lost=0;                                 % indicator to see if ray is "lost"
  while(lost==0)                          % loop is terminated once the ray is "lost"
    [hit,x_i,y_i]=Ray(x,y,a,b,c);         % see if ray leaving 2 hits surface 1
    if(hit==0)                            % ray misses surface 1 and is "lost"
      lost=1;
    else                                  % ray hits surface 1
      if(rand<e1)                         % ray is absorbed by surface 1
        Ea1=Ea1+Eray;                     % add ray's energy to that absorbed by surface 1
        lost=1;                           % ray is absorbed and "lost"
      else                                % ray is reflected from surface 1
      [hit,x_i,y_i]=Ray(x_i,y_i,a,b,c);
      % see if ray reflected from surface 1 hits surface 2
        if(hit==0)                        % reflected ray misses surface 2 and is "lost"
          lost=1;
        else
          if(rand<e2)                     % reflected ray is absorbed by surface 2
            lost=1;
          else                            % ray is reflected from surface 2
            x=x_i;                        % start process over, with new emission point
            y=y_i;
          end
        end
      end
    end
  end
end
```

The result of this second calculation is $E_{a,1}$, the rate at which energy emitted by surface 2 is absorbed by surface 1. The net rate of radiation heat transfer between the plates is the difference between $E_{a,2}$ and $E_{a,1}$:

```
    q_dot_1to2=Ea2-Ea1;          % net rate of radiation exchange from 1 to 2
end
```

Running the program from the command line with $N = 50,000$ rays leads to:

```
>> [q]=q_net(50000)
q =
    313.3539
```

Note that the Monte Carlo technique is stochastic and therefore a different answer will be obtained each time the simulation is run. The solution obtained in EXAMPLE 10.5-3 was $\dot{q}_{1\ to\ 2} = 314.6\,\text{W}$. The script below calls the function q_net 25 times using the same value of N and computes the average and standard deviation of the predicted net heat transfer. The process is repeated with 22 different values of N in order to study the convergence of the solution.

```
Nv=[100,150,200,300,400,500,800,1000,1500,2000,2500,3000,4000,5000,8000,...
    10000,20000,30000,50000,100000,200000,500000]'; % number of rays
for i=1:22
    Nv(i)
    for j=1:25
        [q_dot_1to2(i,j)]=q_net(Nv(i));
    end
end
q_dot_1to2_bar=mean(q_dot_1to2');     % average value of heat transfer rate
q_dot_1to2_std=std(q_dot_1to2');      % standard deviation of heat transfer rate
```

The solution obtained in EXAMPLE 10.5-3 using the techniques for solving gray surface problems is also shown in Figure 10-35; notice that the Monte Carlo technique converges to the deterministic solution as N increases.

The advantage of the Monte Carlo technique is that it is comparatively easy to deal with complications such as non-uniform temperature or emissivity. For example, suppose that the temperature of plate 1 varies according to:

$$T_1 = 300\,[\text{K}] + 300\,[\text{K}]\,\exp[(x - 0.5\,[\text{m}])^2 + (y - 0.5\,[\text{m}])^2] \qquad (10\text{-}135)$$

The spatial variation of the temperature of plate 1 is shown in Figure 10-36 and provided by a subfunction T_1p which is placed at the end of the previous MATLAB code:

```
function[T]=T_1p(x,y)
%Inputs
%x,y - location on plate 1 (m)

%Outputs
%T - temperature (K)
    T=300+300*exp((x-0.5)^2+(y-0.5)^2);
end
```

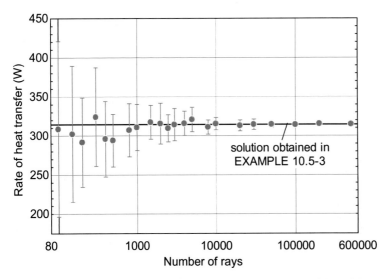

Figure 10-35: Average heat transfer rate between surface 1 and 2 for 25 repeats of the Monte Carlo solution as a function of the number of rays, N. The error bars show the standard deviation of the 25 repeats. The solution obtained in EXAMPLE 10.5-3 is indicated by the dark line.

In order to simulate the heat transfer process with this temperature distribution, it is necessary to compute the energy of the ray emitted by surface 1 according to the local temperature. The line:

```
% Eray=e1*sigma*T1^4*a*b/N;    %determine energy of ray (W)
```

is replaced with:

```
Eray=e1*sigma*T_1p(x,y)^4*a*b/N;    % determine energy of ray (W)
```

Running the simulation leads to $\dot{q}_{1 \text{ to } 2} = 463.2\,\text{W}$ (depending on the number of rays used). This problem would be difficult or impossible to solve in any way other than a Monte Carlo model.

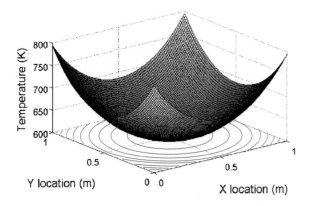

Figure 10-36: Spatial distribution of the temperature of plate 1.

Chapter 10: Radiation

The website associated with this book www.cambridge.org/nellisandklein provides many more problems.

Emission of Radiation by a Blackbody

10–1 Radiation that passes through the atmosphere surrounding our planet is absorbed to an extent that depends on its wavelength due to the presence of gases such as water vapor, oxygen, carbon dioxide and methane. However, there is a large range of wavelengths between 8 and 13 microns for which there is relatively little absorption in the atmosphere and thus the transmittance of atmosphere is high. This wavelength band is called the atmospheric window. Infrared detectors on satellites measure the relative amount of infrared radiation emitted from the ground in this wavelength band in order provide an indication of the ground temperature.

a.) What fraction of the radiation from the sun is in the atmospheric window? The sun can be approximated as a blackbody source at 5780 K.

b.) Prepare a plot of the fraction of the thermal radiation emitted by the ground between 8 and 13 microns to the total radiation emitted by the ground for ground temperatures between $-10°C$ to $30°C$.

c.) Based on your answers to a) and b), indicate whether radiation measurements in the atmospheric window can provide a clear indication of surface temperature to satellite detectors.

10–2 Photovoltaic cells convert a portion of the radiation that is incident on their surface into electrical power. The efficiency of the cells is defined as the ratio of the electrical power produced to the incident radiation. The efficiency of solar cells is dependent upon the wavelength distribution of the incident radiation. An explanation for this behavior was originally provided by Einstein and initiated the discovery of quantum theory. Radiation can be considered to consist of a flux of photons. The energy per photon (e) is: $e = hc/\lambda$ where h is Planck's constant, c is the speed of light, and λ is the wavelength of the radiation. The number of photons per unit area and time is the ratio of the spectral emissive power of the emitting surface, $E_{b,\lambda}$, to the energy of a single photon, e. When radiation strikes a material, it may dislodge electrons. However, the electrons are held in place by forces that must be overcome. Only those photons that have energy above a material-specific limit, called the band-gap energy limit (i.e., photons with wavelengths lower than $\lambda_{bandgap}$) are able to dislodge an electron. In addition, photons having energy above the band-gap limit are still only able to dislodge one electron per photon; therefore, only a fraction of their energy, equal to $\lambda/\lambda_{bandgap}$, is useful for providing electrical current. Assuming that there are no imperfections in the material that would prevent dislodging of an electron and that none of the dislodged electrons recombine (i.e, a quantum efficiency of 1), the efficiency of a photovoltaic cell can be expressed as:

$$\eta = \frac{\displaystyle\int_0^{\lambda_{bandgap}} \frac{\lambda}{\lambda_{bandgap}} E_{b,\lambda} d\lambda}{\displaystyle\int_0^{\infty} E_{b,\lambda} d\lambda}$$

a.) Calculate the maximum efficiency of a silicon solar cell that has a band-gap wavelength of $\lambda_{bandgap} = 1.12\,\mu m$ and is irradiated by solar energy having an equivalent blackbody temperature of 5780 K.

b.) Calculate the maximum efficiency of a silicon solar cell that has a band-gap wavelength of $\lambda_{bandgap} = 1.12\,\mu m$ and is irradiated by incandescent light produced by a black tungsten filament at 2700 K.

c.) Repeat part (a) for a gallium arsenide cell that has a band-gap wavelength of $\lambda_{bandgap} = 0.73\,\mu m$, corresponding to a band gap energy of 1.7 ev.

d.) Plot the efficiency versus bandgap wavelength for solar irradiation. What bandgap wavelength provides the highest efficiency?

10–3 A novel hybrid solar lighting system has been proposed in which concentrated solar radiation is collected and then filtered so that only radiation in the visible range (0.38 – 0.78 µm) is transferred to luminaires in the building by a fiber optic bundle. The unwanted heating of the building caused by lighting can be reduced in this manner. The non-visible energy at wavelengths greater than 0.78 µm can be used to produce electricity with thermal photovoltaic cells. Solar radiation can be approximated as radiation from a blackbody at 5780 K. See Problem 10-2 for a discussion of a model of the efficiency of a photovoltaic cell.

a.) Determine the maximum theoretical efficiency of silicon photovoltaic cells ($\lambda_{bandgap} = 1.12\,\mu m$) if they are illuminated with solar radiation that has been filtered so that only wavelengths greater than 0.78 µm are available.

b.) Determine the band-gap wavelength ($\lambda_{bandgap}$) that maximizes the efficiency of the photovoltaic cell for this application.

10–4 Light is "visually evaluated radiant energy," i.e., radiant energy that your eyes are sensitive to (just like sound is pressure waves that your ears are sensitive to). Because light is both radiation and an observer-derived quantity, two different systems of terms and units are used to describe it: radiometric (related to its fundamental electromagnetic character) and photometric (related to the visual sensation of light). The radiant power ($\dot{q}$) is the total amount of radiation emitted from a source and is a radiometric quantity (with units W). The radiant energy emitted by a blackbody at a certain temperature is the product of the blackbody emissive power (E_b, which is the integration of blackbody spectral emissive power over all wavelengths) and the surface area of the object (A).

$$\dot{q} = A\,E_b = A \int_{\lambda=0}^{\infty} E_{b,\lambda}\,d\lambda = A\,\sigma\,T^4$$

On the other hand, luminous power (F) is the amount of "light" emitted from a source and is a photometric quantity (with units of lumen, which are abbreviated lm). The radiant and luminous powers are related by:

$$F = A\,K_m \int_{0}^{\infty} E_{b,\lambda}\,(\lambda)\,V\,(\lambda)\,d\lambda$$

where K_m is a constant (683 lm/W photopic) and $V(\lambda)$ is the relative spectral luminous efficiency curve. Notice that without the constant K_m, the luminous power is just the radiant power filtered by the function $V(\lambda)$ and has units of W; the constant K_m can be interpreted as converting W to lumen, the photometric unit of light. The filtering function $V(\lambda)$ is derived based on the sensitivity of the human eye to different wavelengths (in much the same way that sound meters use a scale based on the sensitivity of your ear in order to define the acoustic unit, decibel or dB). The function $V(\lambda)$ is defined as the ratio of the sensitivity of the human eye to radiation at a particular wavelength to the sensitivity of your eye to radiation at 0.555 μm; 0.555 μm is selected because your eye is most sensitive to this wavelength (which corresponds to green). An approximate equation for $V(\lambda)$ is: $V(\lambda) = \exp[-285.4 (\lambda - 0.555)^2]$ where λ is the wavelength in micron. The luminous efficiency of a light source (η_l) is defined as the number of lumens produced per watt of radiant power:

$$\eta_l = \frac{F}{\dot{q}} = \frac{K_m \displaystyle\int_0^\infty E_{b,\lambda}(\lambda)\, V(\lambda)\, d\lambda}{\sigma\, T^4}$$

The conversion factor from W to lumen, K_m, is defined so that the luminous efficiency of sunlight is 100 lm/W; most other, artificial light sources will be less than this value. The most commonly used filament in an incandescent light bulb is tungsten; tungsten will melt around 3650 K. An incandescent light bulb with a tungsten filament is typically operated at 2770 K in order to extend the life of the bulb. Determine the luminous efficiency of an incandescent light bulb with a tungsten filament.

Radiation Exchange between Black Surfaces

10–5 Find the view factor $F_{1,2}$ for the geometry shown in Figure P10-5 in the following two ways and compare the results.
   a.) Use the view factor function F3D-2 in EES (you will need to call the function more than once).
   b.) Use the differential view factor relation FDiff_4 and do the necessary integration.

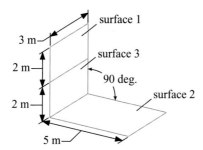

Figure P10-5: Determine the view factor $F_{1,2}$.

10–6 A rectangular building warehouse has dimensions of 50 m by 30 m with a ceiling height of 10 m. The floor of this building is heated. On a cold day, the inside surface temperature of the walls are found to be 16°C, the ceiling surface is 12°C, and the heated floor is at a temperature of 32°C. Estimate the radiant heat transfer from the floor to walls and the ceiling assuming that all surfaces are black. What fraction of the heat transfer is radiated to the ceiling?

10–7 A furnace wall has a 4 cm hole in the insulated wall for visual access. The wall is 8 cm wide. The temperature inside the furnace is 1900°C and it is 25°C on the outside of the furnace. Assuming that the insulation acts as a black surface at a uniform temperature, estimate the radiative heat transfer through the hole.

10–8 A homeowner has installed a skylight in a room that measures 6 m × 4 m with a 2.5 m ceiling height, as shown in Figure P10-8. The skylight is located in the center of the ceiling and it is square, 2 m on each side. A desk is to be located in a corner of the room. The surface of the desk is 0.9 m high and the desk surface is 0.5 by 1 m in area. The skylight has a diffusing glass so that the visible light that enters the skylight should be uniformly distributed.

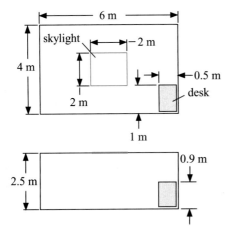

Figure P10-8: Desk and skylight in a room.

Determine the fraction of the light emanating from skylight that will directly illuminate the desktop. Does it matter which wall the desk is positioned against?

10–9 The bottom surface of the cylindrical cavity shown in Figure P10-9 is heated to $T_{bottom} = 750°C$ while the top surface is maintained at $T_{top} = 100°C$. The side of the cavity is insulated externally and isothermal (i.e., the side is made of a conductive material and therefore comes to a single temperature).

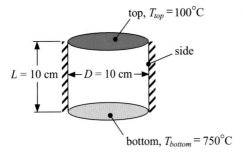

Figure P10-9: Cylindrical cavity heated from the bottom and cooled on top.

The diameter of the cylinder is $D = 10$ cm and its length is $L = 10$ cm. Assume that the cylinder is evacuated so that the only mechanism for heat transfer is radiation. All surfaces are black ($\varepsilon = 1.0$).

a.) Calculate the net rate of heat transfer from the bottom to the top surface. How much of this energy is radiated directly from the bottom surface to the top and how much is transferred indirectly (from the bottom to the side to the top)?

b.) What is the temperature of the side of the container?

c.) If the side was not insulated but rather also cooled to $T_{side} = 100°C$ then what would be the total heat transfer from the bottom surface?

10–10 A homeowner has inadvertently left a spray can near the barbeque grill, as shown in Figure P10-10. The spray can is $H = 8$ inch high with a diameter of

$D = 2.25$ inch. The side of the barbeque grill is $H = 8$ inch high and $W = 18$ inch wide. The spray can is located with its center aligned with the center of the grill wall and it is $a = 6$ inch from the wall, as shown in Figure P10-10. Assume the can to be insulated on its top and bottom.

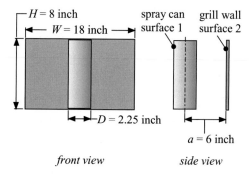

front view   *side view*

Figure P10-10: Spray can near a grill.

a.) What is the view factor between the spray can, surface 1, and the grill wall, surface 2?

b.) Assuming both surfaces to be black, what is the heat transfer rate to the spray can when the grill wall is at $T_2 = 350°F$ and the spray can exterior is $T_1 = 75°F$?

c.) The surroundings, surface 3 are at $75°F$. What is the equilibrium temperature of the spray can if it can be assumed to be isothermal and radiation is the only heat transfer mechanism?

10–11 You are working on an advanced detector for biological agents; the first step in the process is to ablate (i.e., vaporize) individual particles in an air stream so that their constituent molecules can be identified through mass spectrometry. There are various methods available for providing the energy to the particle that is required for ablation; for example, using multiple pulses of a high power laser. You are analyzing a less expensive technique for vaporization that utilizes radiation energy. A very high temperature element is located at the bottom of a cylinder, as shown in Figure P10-11.

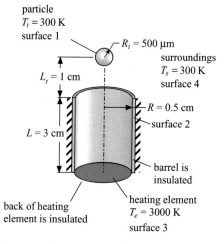

Figure P10-11: Radiation vaporization technique.

The length of the cylinder which is the "barrel" of the heat source is $L = 3.0$ cm and the radius of the cylinder and the heating element is $R = 0.5$ cm. The heating element is maintained at a very high temperature, $T_e = 3000$ K. The back side of the heating element and the external surfaces of the barrel of the heat source are insulated. The particle that is being ablated may be modeled as a sphere with radius $R_s = 500$ μm and is located $L_t = 1.0$ cm from the mouth of the barrel. The particle is located on the centerline of the barrel. The particle is at $T_t = 300$ K and the surroundings are at $T_s = 300$ K. All surfaces are black. For this problem, the particle is surface 1, the cylindrical barrel is surface 2, the disk shaped heating element is surface 3, and the surroundings is surface 4.

a.) Determine the areas of all surfaces and the view factors between each surface.

b.) Determine the net radiation heat transfer to the target.

c.) What is the efficiency of the ablation system? (i.e., what is the ratio of the energy delivered to the particle to the energy required by the element?)

d.) The particle has density $\rho_t = 7000$ kg/m^3 and specific heat capacity $c_t = 300$ J/kg-K. Use your radiation model as the basis of a transient, lumped capacitance numerical model of the particle that can predict the temperature of the particle as a function of time. Assume that the particle is initially at $T_{t,in} = 300$ K. Use the Integral function in EES and prepare a plot showing the particle temperature as a function of time.

Radiation Characteristics of Real Surfaces

10–12 A 10,000 sq. ft. office building requires approximately $\dot{q}''_v = 1.0$ W/ft^2 of visible radiant energy for lighting; this is energy emitted between the wavelengths $\lambda_{v,low} = 0.38$ μm and $\lambda_{v,high} = 0.78$ μm. The efficiency of a lighting system (η_v) can be calculated as the ratio of the visible radiant energy that is emitted to the total amount of energy emitted.

a.) Compute the efficiency of a light source that consists of a blackbody at $T = 2800$ K.

b.) Plot the efficiency of a black body lighting system as a function of the temperature of the light source.

There are two costs associated with providing the lighting that is required by the office. The electricity required to heat the blackbody to its temperature and the electricity that is required to run the cooling system that must remove the energy provided by the light source (note that both the visible and the invisible radiation is deposited as thermal energy in the building). Assume that the building cooling system has an average coefficient of performance of $COP = 3.0$ and the building is occupied for 5 days per week and 12 hours per day. Assume that the cost of electricity is \$0.12/kW-hr.

c.) What is the total cost associated with providing lighting to the office building for one year? How much of this cost is direct (that is, associated with buying electricity to run the light bulbs) versus indirect (that is, associated with running air conditioning equipment in order to remove the energy dumped into the building by the light bulbs). Assume that you are using a light bulb that is a blackbody with a temperature of 2800 K.

An advanced light bulb has been developed that is not a blackbody but rather has an emissivity that is a function of wavelength. The temperature of the advanced light bulb remains 2800 K, but the filament can be modeled as being semi-gray; the emissivity is equal to $\varepsilon_{low} = 0.80$ for wavelengths from 0 to $\lambda_c = 1.0$ μm and $\varepsilon_{high} = 0.25$ for wavelengths above 1.0 μm.

d.) What is the efficiency of the new light bulb?

e.) What is the yearly savings in electricity that can be realized by replacing your old light bulbs (the blackbody at 2800 K) with the advanced light bulbs?

10–13 The intensity of a surface has been measured as a function of the elevation angle and correlated with the following relation:

$$I = I_{b,\lambda}(1 - \exp(-0.0225 - 6.683\cos(\theta) + 5.947\cos^2(\theta) - 2.484\cos^3(\theta)))$$

where $I_{b,\lambda}$ is the intensity of a blackbody at wavelength λ.

a.) Determine the spectral emissive power for this surface at the wavelength where it is maximum if it is maintained at 1200 K.

b.) What is the hemispherical emissivity of this surface?

10–14 A surface has wavelength-dependent properties as listed in Table P10-14. The surface is maintained at 500 K.

Table P10-14: Wavelength-dependent absorption.

Wavelength Range [μm]	α^λ
0–0.6	0.8
0.6–2.6	0.25
2.6–100	0.10

a.) Determine the total hemispherical absorptivity of this surface for solar radiation.

b.) Determine the total hemispherical emissivity of this surface.

10–15 Calculate and plot the total reflectivity of polished aluminum at 697 K for radiation emitted from sources between 300 K and 6000 K. The hemispherical emissivity of polished aluminum is provided in the EES Radiation Properties folder as the table Aluminum-Spectral.lkt.

Diffuse Gray Surface Radiation Exchange

10–16 Three metal plates, each $W = 40$ cm by $L = 60$ cm, are parallel and centered as shown in Figure P10-16. Each of the plates have an emissivity of $\varepsilon = 0.15$. The top and bottom plates (surfaces 1 and 3) are separated by a vertical distance of $H = 50$ cm. The bottom and middle plates (surfaces 1 and 2) are separated by a vertical distance a. The temperature of the bottom plate is maintained at $T_1 = 584°C$. The plates radiatively interact with the surroundings at $T_4 = 25°C$. The underside of the bottom plate is insulated.

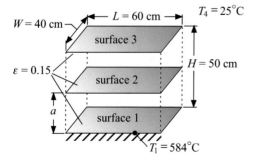

Figure P10-16: Three metal plates.

Calculate and plot the temperature of the upper plate and the net rate of radiative heat transfer from the lower plate as a function of a for 1 cm $< a <$ 49 cm.

10–17 Consider two parallel plates that are separated by a distance of $H = 0.5$ m. The
plates are each $L = 1$ m by $W = 2$ m. The lower plate (surface 1) is maintained
at $T_1 = 400$ K and has emissivity $\varepsilon_1 = 0.4$. The surroundings (surface 2) are at
$T_2 = 4$ K. The upper plate has a temperature profile that varies linearly in the x-
direction from $T_C = 500$ K at one edge ($x = 0$) to $T_H = 1000$ K at the other edge
($x = W$). The temperature is uniform in the y-direction. The emissivity of the
upper plate is $\varepsilon_3 = 0.6$. This problem can be solved numerically by discretizing
the upper plate into N equal area segments, each at a constant temperature that
is equal to the temperature of upper plate at the center of the segment. Assume
that the upper surface of the upper plate and the lower surface of the lower plate
are both insulated.

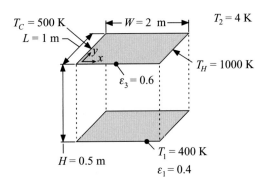

Figure P10-17: Two plates.

a.) Calculate the total rate at which energy must be provided to the upper plate.
b.) Plot the total rate of energy provided to the upper plate as a function of N
for $N = 1$ to 10. From your results, how many segments do you believe are
needed to represent the effect of the temperature distribution in the upper
plate?

10–18 A cylindrical heating element is used to heat a flow of water to an appliance. Typically, the element is exposed to water and therefore it is well-cooled. However,
you have been asked to assess the fire hazard associated with a scenario in which
the appliance is suddenly drained (i.e., the water is removed) but the heating
element is not deactivated. You want to determine the maximum temperature
that the element will reach under this condition. The heating element and passage wall are shown in Figure P10-18. The length of the element is $L = 9.0$ cm
and its diameter is $D_1 = 0.5$ cm. The element is concentric to a passage wall with
diameter $D_2 = 2.0$ cm. The emissivity of the element is $\varepsilon_1 = 0.5$ and the emissivity of the passage wall is $\varepsilon_2 = 0.9$. The surroundings are at $T_3 = 25°C$. The worst
case situation occurs if the outer passage wall is assumed to be insulated externally (i.e., there is no conduction or convection from the passage). The heating
element dissipates $\dot{q}_e = 60$ W.

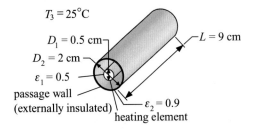

Figure P10-18: Heating element.

a.) What is the temperature of the element? Assume that radiation is the only important heat transfer mechanism for this problem. Note that your problem should include three surfaces (the element, the passage, and the surroundings); that is, you should not neglect the radiation exchange between the element and passage and the surroundings. However, you may assume that the flat edges of the element are adiabatic.

b.) What is the temperature of the passage wall?

c.) Other calculations have shown that the passage wall will not reach temperatures greater than $80°C$ because it is thermally communicating with surroundings. If the passage wall is maintained at $T_2 = 80°C$ then what is the maximum temperature that the heating element will reach?

10–19 This problem considers a (fictitious) power generation system for a spacecraft orbiting the planet Mercury. The surface of Mercury can reach 700 K and therefore you are considering the possibility of collecting radiation emitted from Mercury in order to operate a heat engine. The details of the collector are shown schematically in Figure P10-19(a).

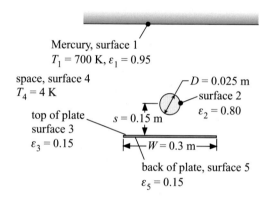

Figure P10-19(a): Energy collection system.

The collector geometry consists of a pipe and a backing plate; this geometry is 2-D, so the problem will be solved on a per unit length basis, $L = 1\,\text{m}$, into the page. The diameter of the pipe is $D = 0.025\,\text{m}$. The pipe surface (surface 2) is maintained at a constant temperature (T_2) and has emissivity $\varepsilon_2 = 0.8$. Energy that is transferred to the pipe is provided to the power generation system. The pipe is oriented so that it is parallel to the surface of the planet (surface 1) which is at $T_1 = 700\,\text{K}$ and has an emissivity of $\varepsilon_1 = 0.95$. You may assume that the surface of the planet extends infinitely in all directions. There is a back plate positioned $s = 0.15\,\text{m}$ away from the centerline of the collector pipe. The back plate is $W = 0.30\,\text{m}$ wide and is centered with respect to the pipe. The top surface of the back plate (the surface oriented towards the collector pipe, surface 3) has emissivity $\varepsilon_3 = 0.15$. The bottom surface of the back plate (the surface oriented towards space, surface 5) also has emissivity $\varepsilon_5 = 0.15$. The collector and back plate are surrounded by outer space, which has an effective temperature $T_4 = 4\,\text{K}$; assume that the collector is shielded from the sun. Assume that the back plate is isothermal.

a.) Prepare a plot showing the net rate of radiation heat transfer to the collector from Mercury as a function of the collector temperature, T_2.

The energy transferred to the collector pipe is provided to the hot end of a heat engine that operates between T_2 and $T_{radiator}$, where T_2 is the collector pipe temperature and $T_{radiator}$ is the temperature of a radiator panel that is used to reject heat, as shown in Figure P10-19(b). The heat engine has a second law efficiency $\eta_2 = 0.30$; that is, the heat engine produces 30% of the power that a reversible heat engine would produce if it were operating between the same temperature limits (T_2 and $T_{radiator}$). The heat engine radiator rejects heat to space. Assume that the radiator panel has an emissivity $\varepsilon_{radiator} = 0.90$ and a surface area $A_{radiator} = 10\,m^2$. Also, assume that the radiator only sees space at $T_4 = 4\,K$.

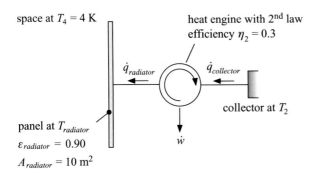

Figure P10-19(b): Schematic of the power generation system.

b.) Prepare a plot showing the amount of power generated by the heat engine ($\dot{w}$) and the radiator temperature (T_6) as a function of the collector temperature, T_2.

Radiation with other Heat Transfer Mechanisms

10–20 A photovoltaic panel having dimensions of 1 m by 2 m is oriented directly towards the sun (i.e., south) at a 45° angle. The panel is exposed to solar radiation at 720 W/m². The efficiency of the panel, defined as the electrical power produced divided by the incident solar radiation, is 11.2%. The back side of the photovoltaic panel is well-insulated. The emissivity of the photovoltaic material is estimated to be 0.90. The ambient and ground temperature during the test is 22°C and there is no measurable wind. The sky is clear and the equivalent temperature of the sky for radiation is 7°C. Estimate the steady-state surface temperature of the photovoltaic panel assuming that all of the radiation that strikes the panel is absorbed. What fraction of the thermal energy transfer to the air is due to radiation?

10–21 A thermocouple has a diameter $D_{tc} = 0.02\,m$. The thermocouple is made of a material with density $\rho = 8000\,kg/m^3$ and specific heat capacity $c = 450\,J/kg\text{-}K$. The temperature of the thermocouple (you may assume that the thermocouple is at a uniform temperature) is $T_{tc} = 320\,K$ and the emissivity of the thermocouple's surface is $\varepsilon_{tc} = 0.50$. The thermocouple is placed between two very large (assume infinite in all directions) black plates. One plate is at $T_1 = 300\,K$ and the other is at $T_2 = 500\,K$. The thermocouple is also exposed to a flow of air at $T_a = 300\,K$. The het transfer coefficient between the air and thermocouple is $\bar{h} = 50\,W/m^2\text{-}K$. The situation is shown in Figure P10-21.

a.) What is the rate of convective heat transfer from the thermocouple?

b.) What is the net rate of radiative heat transfer to the thermocouple?

c.) What is the rate of temperature change of the thermocouple?

d.) If you want the thermocouple to accurately measure the temperature of the air (and therefore be unaffected by radiation), would you try to increase or decrease its emissivity? Justify your answer.

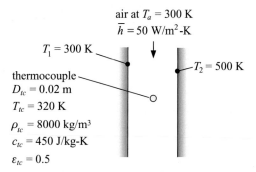

Figure P10-21: Thermocouple placed between two plates.

10–22 Figure P10-22 illustrates a set of three reactor beds that are heated radiantly by three heating elements.

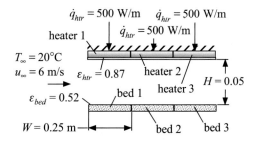

Figure P10-22: Reactor beds with heaters.

The reactants are provided as a flow over the beds. The temperature of the reactant flow is $T_\infty = 20°C$ and the free stream velocity is $u_\infty = 6$ m/s; you may assume that the properties of the reactant flow are consistent with those of air at atmospheric pressure. All of the heaters and beds are each $W = 0.25$ m wide and very long (the problem is two-dimensional). The heaters and beds are separated by $H = 0.05$ m. The beds are insulated on their back-sides but transfer heat to the free stream by convection. The surface of the beds has emissivity, $\varepsilon_{bed} = 0.52$. The heaters are each provided with $\dot{q}_{htr} = 500$ W/m; there is a piece of glass that protects the heaters from the reactants and prevents convective heat loss from the heaters. The upper surfaces of the heaters are insulated. You may assume that the 3 heaters and 3 beds are all isothermal (i.e., they are each at a unique but uniform temperature). The surface of the heaters has emissivity, $\varepsilon_{htr} = 0.87$. The surroundings are at $T_{sur} = 20°C$.
a.) Determine the temperature of each of the beds.
b.) What is the efficiency of the heating system?
c.) Determine the heater power that should be applied to each of the 3 heaters in order to keep each of the 3 beds at $T_{bed} = 65°C$.

10–23 The earth radiates to space, which has an effective temperature of about 4 K. However, the earth is surrounded by an atmosphere consisting of gases that absorb radiation in specific wavelength bands. For this reason, the equivalent blackbody temperature of the sky is greater than 4 K but generally lower than the ambient temperature by 5 to 30°C, depending on the extent of cloud cover and amount of moisture in the air. The largest difference between the ambient and equivalent blackbody sky temperature occurs during nights in which there is no cloud cover and low humidity. An important multimode heat transfer problem is related to determining the nighttime temperature at which there is a danger that citrus fruit will freeze. Consider the following situation. During a clear calm night, an orange with diameter $D = 6.5$ cm experiences radiation heat transfer with the sky and the ground as well as convection to the ambient air. The ground temperature is approximately $T_{ground} = 10°C$, regardless of the ambient temperature, and is constant during the night. The equivalent blackbody temperature of the sky, T_{sky}, is $\Delta T_{sky} = 15°C$ lower than the ambient temperature, T_∞. The emissivity of the ground is $\varepsilon_{ground} = 0.8$ and the sky can be considered to be black. The emissivity of the orange is $\varepsilon_{orange} = 0.5$. Estimate the ambient temperature, T_∞, at which the orange will freeze; assume that the orange achieves a steady-state condition. Oranges consist of mostly water and therefore they freeze at about $0°C$.

The Monte Carlo Method

10–24 Two parallel rectangular surfaces, each 2 m by 3 m, are aligned with one another and separated by a distance of 1 m. Surface 2 has a 1 m diameter hole in it. The center of the hole is located at the center of the rectangle.
   a.) Determine the view factor from surface 1 to surface 2 using the Monte Carlo method.
   b.) Compare the value obtained in (a) to the value obtained from the view factor library.
   c.) Determine the view factor between the surfaces if both surfaces have a 1 m diameter hole at their centers.

10–25 Two parallel rectangular surfaces, each 2 m by 3 m are separated by a distance of 1 m and aligned with one another. Each surface has a hole with a diameter of 1 m located at their center. The emissivity of surfaces 1 and 2 are 0.8 and 0.6, respectively. Surface 1 is at 700 K and surface 2 is at 300 K.
   a.) Determine the net rate of heat transfer from surface 1 to surface 2 using the Monte Carlo method.
   b.) Using the view factor determined for this geometry in Problem 10-24, calculate the net rate of heat transfer between surfaces 1 and 2 using the $\hat{F}$ method and compare your answer to the result from part (a).

REFERENCES

Beckman, W. A., "Temperature Uncertainties in Systems with Combined Radiation and Conduction Heat Transfer," ASME, 1968 Aviation & Aerospace Conference, June, (1968).

Brewster, M. Q., *Thermal Radiative Transfer and Properties*, Wiley, New York, (1992).

Duffie, J. A. and Beckman, W. A., *Solar Engineering of Thermal Processes, 3rd Ed*, Wiley Interscience, Second edition, (2006).

Hottel, H. C. and A. F. Sarofim, *Radiative Transfer*, McGraw-Hill, New York, (1967).

Lummer, O. and E. Pringsheim, *Transactions of the German Physical Society* 2 (1900), p. 163.

McAdams, W. H., *Heat Transmission*, 3rd edition, McGraw-Hill, New York, (1954).

Siegel, R. and Howell, J. R., *Thermal Radiation Heat Transfer, 4th edition*, Taylor and Francis, (2002).

Appendices

A.1: Introduction to EES

This extended section of the book can be found on the website www.cambridge.org/ nellisandklein. EES (pronounced 'ease') is an acronym for Engineering Equation Solver. The basic function provided by EES is the numerical solution of non-linear algebraic and differential equations. EES is an equation-solver, rather than a programming language, since it does not require the user to enter instructions for iteratively solving non-linear equations. EES provides capability for unit checking of equations, parametric studies, optimization, uncertainty analyses, and high-quality plots. It provides array variables that can be used in finite-difference calculations. In addition, EES provides high-accuracy thermodynamic and transport property functions for many fluids and solid materials that can be integrated with the equations. The combination of these capabilities together with an extensive library of heat transfer functions, discussed throughout this text, makes EES a very powerful tool for solving heat transfer problems. This appendix provides a tutorial that will allow you to become familiar with EES.

A.2: Introduction to Maple

This extended section of the book can be found on the website www.cambridge.org/ nellisandklein. Maple is an application that can be used to solve algebraic and differential equations. Maple has the ability to do mathematics in symbolic form and therefore it can determine the analytical solution to algebraic and differential equations. Maple provides a very convenient mathematical reference; if, for example, you've forgotten that the derivative of sine is cosine, it is easy to use Maple to quickly provide this information. Maple can replace numerous mathematical reference books that might otherwise be required to carry out all of the integration, differentiation, simplification, etc. required to solve many engineering problems. Maple and EES can be used effectively together; Maple can determine the analytical solution to a problem and these symbolic expressions can subsequently be copied (almost directly) into EES for convenient numerical evaluation and manipulation in the context of a specific application. This appendix summarizes the commands that are the most useful and are used throughout this text.

A.3: Introduction to MATLAB

This extended section of the book can be found on the website www.cambridge.org/ nellisandklein. MATLAB is a sophisticated software package and we are only going to touch upon a few of its capabilities within this book. MATLAB is essentially a powerful programming language; one of the reasons that it is so powerful is that its basic data structure is the array. Therefore, you can solve problems that involve large vectors and matrices intuitively and manipulate the results easily. MATLAB is used in this book almost exclusively for carrying out numerical simulations. This tutorial allows you to familiarize yourself with some of the basic features of MATLAB.

A.4: Introduction to FEHT

This extended section of the book can be found on the website www.cambridge.org/ nellisandklein. FEHT (pronounced 'feet') is an acronym for Finite Element Heat Transfer. The basic function provided by FEHT is the numerical solution of 2-D steady state and transient heat transfer problems. FEHT is intuitive to use and therefore can be learned very quickly by students and researchers. This tutorial allows you to quickly familiarize yourself with some of the basic features of FEHT.

A.5: Introduction to Economics

This extended section of the book can be found on the website www.cambridge.org/ nellisandklein. Thermal systems are generally capital intensive. That is, the equipment needed for a specific application, such as a heat exchanger or a furnace, is relatively expensive. Proper design of thermal systems is almost always based on an economic analysis that balances these capital costs with potential energy savings that are realized over time. Such an analysis is complicated by the fact that the value of money is not constant but rather a function of time. This appendix provides an introduction to economic analysis that is sufficient for most thermal system design problems.

INDEX

Note that page numbers starting with EX- indicate that these pages can be found on the extended section EX, which is available at www.cambridge.org/nellisandklein. For example, E23–31 is pg. 31 of extended section E23.

Printed in the United States
By Bookmasters